AF615380

# PROBABILITY, MARKOV CHAINS, QUEUES, AND SIMULATION

# PROBABILITY, MARKOV CHAINS, QUEUES, AND SIMULATION

## The Mathematical Basis of Performance Modeling

William J. Stewart

PRINCETON UNIVERSITY PRESS

PRINCETON AND OXFORD

Published by Princeton University Press, 41 William street, Princeton, New Jersey 08540
In the United Kingdom: Princeton University Press, 6 Oxford Street,
Woodstock, Oxfordshire OX20 1TW



Library of Congress Cataloging-in-Publication Data
Stewart, William J., 1946–
Probability, Markov chains, queues and simulation: the mathematical basis of performance modeling/William J. Stewart. – 1st ed.
p. cm.
ISBN 978-0-691-14062-9 (cloth : alk. paper) 1. Probability–Computer simulation.
2. Markov processes. 3. Queueing theory. I. Title.
QA273.S7532 2009
519.201′13–dc22        2008041122

British Library Cataloging-in-Publication Data is available

This book has been composed in Times
Printed on acid-free paper. ∞
press.princeton.edu



Typeset by S R Nova Pvt Ltd, Bangalore, India
Printed in the United States of America

10 9 8 7 6 5 4 3 2 1

**This book is dedicated to all those whom I love, especially**

*My dear wife, Kathie,*
*and my wonderful children*
*Nicola, Stephanie, Kathryn, and William*

*My father, William J. Stewart and*
*the memory of my mother, Mary (Marshall) Stewart*

# Contents

# Preface and Acknowledgments

This book has been written to provide a complete, yet elementary and pedagogic, treatment of the mathematical basis of systems performance modeling. Performance modeling is of fundamental importance to many branches of the mathematical sciences and engineering as well as to the social and economic sciences. Advances in methodology and technology have now provided the wherewithal to build and solve sophisticated models. The purpose of this book is to provide the student and teacher with a modern approach for building and solving probability based models with confidence.

The book is divided into four major parts, namely, "Probability," "Markov Chains," "Queueing Models," and "Simulation." The eight chapters of Part I provide the student with a comprehensive and thorough knowledge of probability theory. Part I is self-contained and complete and should be accessible to anyone with a basic knowledge of calculus. Newcomers to probability theory as well as those whose knowledge of probability is rusty should be equally at ease in their progress through Part I. The first chapter provides the fundamental concepts of set-based probability and the probability axioms. Conditional probability and independence are stressed as are the laws of total probability and Bayes' rule. Chapter 2 introduces combinatorics—the art of counting—which is so important for the correct evaluation of probabilities. Chapter 3 introduces the concepts of random variables and distribution functions including functions of a random variable and conditioned random variables. This chapter prepares the ground work for Chapters 4 and 5: Chapter 4 introduces joint and conditional distributions and Chapter 5 treats expectations and higher moments. Discrete distribution functions are the subject of Chapter 6 while their continuous counterparts, continuous distribution functions, are the subject of Chapter 7. Particular attention is paid to phase-type distributions due to the important role they play in modeling scenarios and the chapter also includes a section on fitting phase-type distributions to given means and variances. The final chapter in Part I is devoted to bounds and limit theorems, including the laws of large numbers and the central limit theorem.

Part II contains two rather long chapters on the subject of Markov chains, the first on theoretical aspects of Markov chains, and the second on their numerical solution. In Chapter 9, the basic concepts of discrete and continuous-time Markov chains and their underlying equations and properties are discussed. Special attention is paid to irreducible Markov chains and to the potential, fundamental, and reachability matrices in reducible Markov chains. This chapter also contains sections on random walk problems and their applications, the property of reversibility in Markov chains, and renewal processes. Chapter 10 deals with numerical solutions, from Gaussian elimination and basic iterative-type methods for stationary solutions to ordinary differential equation solvers for transient solutions. Block methods and iterative aggregation-disaggregation methods for nearly completely decomposable Markov chains are considered. A section is devoted to matrix geometric and matrix analytic methods for structured Markov chains. Algorithms and computational considerations are stressed throughout this chapter.

Queueing models are presented in the five chapters that constitute Part III. Elementary queueing theory is presented in Chapter 11. Here an introduction to the basic terminology and definitions is followed by an analysis of the simplest of all queueing models, the $M/M/1$ queue. This is then generalized to birth-death processes, which are queueing systems in which the underlying Markov chain matrix is tridiagonal. Chapter 12 deals with queues in which the arrival process need no longer be Poisson and the service time need not be exponentially distributed. Instead, interarrival times and service times can be represented by phase-type distributions and the underlying Markov chain is now block tridiagonal. The following chapter, Chapter 13, explores the $z$-transform approach for solving similar types of queues. The $M/G/1$ and $G/M/1$ queues are the subject of Chapter 14. The approach used is that of the embedded Markov chain. The Pollaczek-Khintchine mean value and transform equations are derived and a detailed discussion of residual time and busy period follows. A thorough discussion of nonpreemptive and preempt-resume scheduling policies as well as shortest-processing-time-first scheduling is presented. An analysis is also provided for the case in which only a limited number of customers can be accommodated in both the $M/G/1$ and $G/M/1$ queues. The final chapter of Part III, Chapter 15, treats queueing networks. Open networks are introduced via Burke's theorem and Jackson's extensions to this theorem. Closed queueing networks are treated using both the convolution algorithm and the mean value approach. The "flow-equivalent server" approach is also treated and its potential as an approximate solution procedure for more complex networks is explored. The chapter terminates with a discussion of product form in queueing networks and the BCMP theorem for open, closed, and mixed networks.

The final part of the text, Part IV, deals with simulation. Chapter 16 explores how uniformly distributed random numbers can be applied to obtain solutions to probabilistic models and other time-independent problems—the "Monte Carlo" aspect of simulation. Chapter 17 describes the modern approaches for generating uniformly distributed random numbers and how to test them to ensure that they are indeed uniformly distributed and independent of each other. The topic of generating random numbers that are not uniformly distributed, but satisfy some other distribution such as Erlang or normal, is dealt with in Chapter 18. A large number of possibilities exist and not all are appropriate for every distribution. The next chapter, Chapter 19, provides guidelines for writing simulation programs and a number of examples are described in detail. Chapter 20 is the final chapter in the book. It concerns simulation measurement and accuracy and is based on sampling theory. Special attention is paid to the generation of confidence intervals and to variance reduction techniques, an important means of keeping the computational costs of simulation to a manageable level.

The text also includes two appendixes; the first is just a simple list of the letters of the Greek alphabet and their spellings; the second is a succinct, yet complete, overview of the linear algebra used throughout the book.

## Genesis and Intent

This book saw its origins in two first-year graduate level courses that I teach, and have taught for quite some time now, at North Carolina State University. The first is entitled "An Introduction to Performance Evaluation;" it is offered by the Computer Science Department and the Department of Electrical and Computer Engineering. This course is required for our networking degrees. The second is a course entitled "Queues and Stochastic Service Systems" and is offered by the Operations Research Program and the Industrial and Systems Engineering Department. It follows then that this book has been designed for students from a variety of academic disciplines in which stochastic processes constitute a fundamental concept, disciplines that include not only computer science and engineering, industrial engineering, and operations research, but also mathematics, statistics, economics, and business, the social sciences—in fact all disciplines in which stochastic performance modeling plays a primary role. A calculus-based probability course is a prerequisite for both these courses so it is expected that students taking these classes are already familiar with probability theory. However, many of the students who sign up for these courses are returning students, and it is often the case that it has been several years and in some cases a decade or more, since they last studied probability. A quick review of probability is hardly sufficient to bring them up to the required level. Part I of the book has been designed with them in mind. It provides the prerequisite probability background needed to fully understand and appreciate the material in the remainder of the text. The presentation, with its numerous examples and exercises, is such that it facilitates an independent review so the returning student in a relatively short period of time, preferably prior to the beginning of class, will once again have mastered probability theory. Part I can then be used as a reference source as and when needed.

The entire text has been written at a level that is suitable for upper-level undergraduate students or first-year graduate students and is completely self-contained. The entirety of the text can be covered in a two-semester sequence, such as the stochastic processes sequence offered by the Industrial Engineering (IE) Department and the Operations Research (OR) Program at North Carolina State University. A two-semester sequence is appropriate for classes in which students have limited (or no) exposure to probability theory. In such cases it is recommended that the first semester be devoted to the Chapters 1–8 on probability theory, the first five sections of Chapter 9, which introduce the fundamental concepts of discrete-time Markov chains, and the first three sections of Chapter 11, which concern elementary queueing theory. With this background clearly understood, the student should have no difficulty in covering the remaining topics of the text in the second semester.

The complete content of Parts II–IV might prove to be a little too much for some one-semester classes. In this case, an instructor might wish to omit the later sections of Chapter 10 on the numerical solution of Markov chains, perhaps covering only the basic direct and iterative methods. In this case the material of Chapter 12 should also be omitted since it depends on a knowledge of the matrix geometric method of Chapter 10. Because of the importance of computing numerical solutions, it would be a mistake to omit Chapter 10 in its entirety. Some of the material in Chapter 18 could also be eliminated: for example, an instructor might include only the first three sections of this chapter. In my own case, when teaching the OR/IE course, I concentrate on covering all of the Markov chain and queueing theory chapters. These students often take simulation as an individual course later on. When teaching the computer science and engineering course, I omit some of the material on the numerical solution of Markov chains so as to leave enough time to cover simulation.

Numerous examples with detailed explanations are provided throughout the text. These examples are designed to help the student more clearly understand the theoretical and computational aspects of the material and to be in a position to apply the acquired knowledge to his/her own areas of interest. A solution manual is available for teachers who adopt this text for their courses. This manual contains detailed explanations of the solution of all the exercises.

Where appropriate, the text contains program modules written in Matlab or in the Java programming language. These programs are not meant to be robust production code, but are presented so that the student may experiment with the mathematical concepts that are discussed. To free the student from the hassle of copying these code segments from the book, a listing of all of the code used can be freely downloaded from the web page:
`http://press.princeton.edu/titles/8844.html`

## Acknowledgments

As mentioned just a moment ago, this book arose out of two courses that I teach at North Carolina State University. It is, therefore, ineluctable that the students who took these courses contributed immeasurably to its content and form. I would like to express my gratitude to them for their patience and input. I would like to cite, in particular, Nishit Gandhi, Scott Gerard, Rong Huang, Kathryn Peding, Amirhosein Norouzi, Robert Shih, Hui Wang, Song Yang, and Shengfan Zhang for their helpful comments. My own doctoral students, Shep Barge, Tugrul Dayar, Amy Langville, Ning Liu, and Bin Peng, were subjected to different versions of the text and I owe them a particular expression of thanks. One person who deserves special recognition is my daughter Kathryn, who allowed herself to be badgered by her father into reading over selected probability chapters.

I would like to thank Vickie Kearn and the editorial and production staff at Princeton University Press for their help and guidance in producing this book. It would be irresponsible of me not to mention the influence that my teachers, colleagues, and friends have had on me. I owe them a considerable debt of gratitude for helping me understand the vital role that mathematics plays, not only in performance modeling, but in all aspects of life.

Finally, and most of all, I would like to thank my wife Kathie and our four children, Nicola, Stephanie, Kathryn, and William, for all the love they have shown me over the years.

# Part I

# PROBABILITY

# Chapter 1

# *Probability*

## 1.1 Trials, Sample Spaces, and Events

The notions of trial, sample space, and event are fundamental to the study of probability theory. Tossing a coin, rolling a die, and choosing a card from a deck of cards are examples that are frequently used to explain basic concepts of probability. Each toss of the coin, roll of the die, or choice of a card is called a *trial* or *experiment*. We shall use the words trial and experiment interchangeably. Each execution of a trial is called a *realization* of the probability experiment.

At the end of any trial involving the examples given above, we are left with a head or a tail, an integer from one through six, or a particular card, perhaps the queen of hearts. The result of a trial is called an *outcome*. The set of all possible outcomes of a probability experiment is called the *sample space* and is denoted by $\Omega$. The outcomes that constitute a sample space are also referred to as *sample points* or *elements*. We shall use $\omega$ to denote an element of the sample space.

**Example 1.1** The sample space for coin tossing has two sample points, a head $(H)$ and a tail $(T)$. This gives $\Omega = \{H, T\}$, as shown in Figure 1.1.

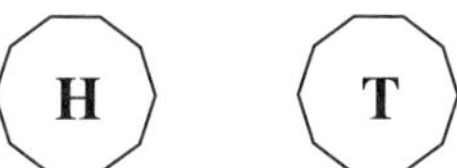


Figure 1.1. Sample space for tossing a coin has two elements $\{H, T\}$.

**Example 1.2** For throwing a die, the sample space is $\Omega = \{1, 2, 3, 4, 5, 6\}$, Figure 1.2, which represents the number of spots on the six faces of the die.

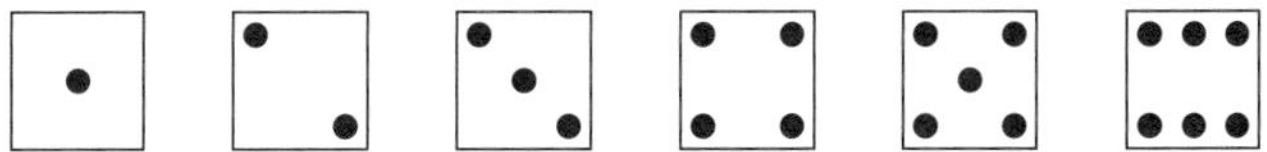

Figure 1.2. Sample space for throwing a die has six elements $\{1, 2, 3, 4, 5, 6\}$.

**Example 1.3** For choosing a card, the sample space is a set consisting of 52 elements, one for each of the 52 cards in the deck, from the ace of spades through the king of hearts.

**Example 1.4** If an experiment consists of three tosses of a coin, then the sample space is given by

$$\{HHH,\ HHT,\ HTH,\ THH,\ HTT,\ THT,\ TTH,\ TTT\}.$$

Notice that the element *HHT* is considered to be different from the elements *HTH* and *THH*, even though all three tosses give two heads and one tail. The position in which the tail occurs is important.

A sample space may be finite, denumerable (i.e., infinite but countable), or infinite. Its elements depend on the experiment and how the outcome of the experiment is defined. The four illustrative examples given above all have a finite number of elements.

**Example 1.5** The sample space derived from an experiment that consists of observing the number of email messages received at a government office in one day may be taken to be denumerable. The sample space is denumerable since we may tag each arriving email message with a unique integer $n$ that denotes the number of emails received prior to its arrival. Thus, $\Omega = \{n | n \in \mathcal{N}\}$, where $\mathcal{N}$ is the set of nonnegative integers.

**Example 1.6** The sample space that arises from an experiment consisting of measuring the time one waits at a bus stop is infinite. Each outcome is a nonnegative real number $x$ and the sample space is given by $\Omega = \{x | x \geq 0\}$.

If a finite number of trials is performed, then, no matter how large this number may be, there is no guarantee that every element of its sample space will be realized, even if the sample space itself is finite. This is a direct result of the essential probabilistic nature of the experiment. For example, it is possible, though perhaps not very likely (i.e., not very probable) that after a very large number of throws of the die, the number 6 has yet to appear.

Notice with emphasis that the sample space is a *set*, the set consisting of all the elements of the sample space, i.e., all the possible outcomes, associated with a given experiment. Since the sample space is a set, all permissible set operations can be performed on it. For example, the notion of subset is well defined and, for the coin tossing example, four subsets can be defined: the null subset $\phi$; the subsets $\{H\}$, $\{T\}$; and the subset that contains all the elements, $\Omega = \{H, T\}$. The set of subsets of $\Omega$ is

$$\phi, \ \{H\}, \ \{T\}, \ \{H, \ T\}.$$

### *Events*

The word *event* by itself conjures up the image of something having happened, and this is no different in probability theory. We toss a coin and get a head, we throw a die and get a five, we choose a card and get the ten of diamonds. Each experiment has an outcome, and in these examples, the outcome is an element of the sample space. These, the elements of the sample space, are called the *elementary events* of the experiment. However, we would like to give a broader meaning to the term event.

**Example 1.7** Consider the event of tossing three coins and getting exactly two heads. There are three outcomes that allow for this event, namely, $\{HHT, HTH, THH\}$. The single tail appears on the third, second, or first toss, respectively.

**Example 1.8** Consider the event of throwing a die and getting a prime number. Three outcomes allow for this event to occur, namely, $\{2, \ 3, \ 5\}$. This event comes to pass so long as the throw gives neither one, four, nor six spots.

In these last two examples, we have composed an event as a subset of the sample space, the subset $\{HHT, HTH, THH\}$ in the first case and the subset $\{2, \ 3, \ 5\}$ in the second. This is how we define an *event* in general. Rather than restricting our concept of an event to just another name for the elements of the sample space, we think of events as subsets of the sample space. In this case, the *elementary* events are the singleton subsets of the sample space, the subsets $\{H\}$, $\{5\}$, and $\{10$ of diamonds$\}$, for example. More complex events consist of subsets with more than one outcome. Defining an event as a subset of the sample space and not just as a subset that contains a single element provides us with much more flexibility and allows us to define much more general events.

The event is said "to occur" if and only if, the outcome of the experiment is any one of the elements of the subset that constitute the event. They are assigned names to help identify and manipulate them.

**Example 1.9** Let $\mathcal{A}$ be the event that a throw of the die gives a number greater than 3. This event consists of the subset $\{4, 5, 6\}$ and we write $\mathcal{A} = \{4, 5, 6\}$. Event $\mathcal{A}$ occurs if the outcome of the trial, the number of spots obtained when the die is thrown, is any one of the numbers 4, 5, and 6. This is illustrated in Figure 1.3.

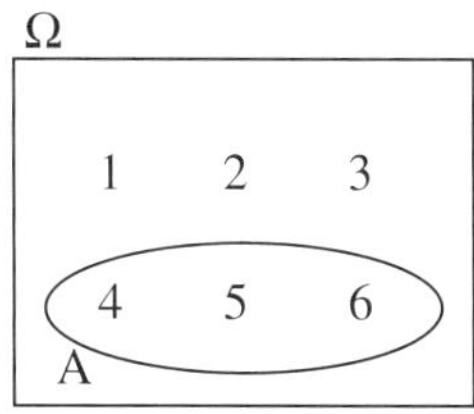


Figure 1.3. Event $\mathcal{A}$: "Throw a number greater than 3."

**Example 1.10** Let $\mathcal{B}$ be the event that the chosen card is a 9. Event $\mathcal{B}$ is the subset containing four elements of the sample space: the 9 of spades, the 9 of clubs, the 9 of diamonds, and the 9 of hearts. Event $\mathcal{B}$ occurs if the card chosen is one of these four.

**Example 1.11** The waiting time in minutes at a bus stop can be any nonnegative real number. The sample space is $\Omega = \{t \in \Re \mid t \geq 0\}$, and $\mathcal{A} = \{2 \leq t \leq 10\}$ is the event that the waiting time is between 2 and 10 minutes. Event $\mathcal{A}$ occurs if the wait is 2.1 minutes or 3.5 minutes or 9.99 minutes, etc.

To summarize, the standard definition of an event is a subset of the sample space. It consists of a set of outcomes. The null (or empty) subset, which contains none of the sample points, and the subset containing the entire sample space are legitimate events—the first is called the "null" or impossible event (it can never occur); the second is called the "universal" or certain event and is sure to happen no matter what the outcome of the experiments gives. The execution of a trial, or observation of an experiment, *must* yield one and only one of the outcomes in the sample space. If a subset contains *none* of these outcomes, the event it represents cannot happen; if a subset contains *all* of the outcomes, then the event it represents must happen. In general, for each outcome in the sample space, either the event occurs (if that particular outcome is in the defining subset of the event) or it does not occur.

Two events $\mathcal{A}$ and $\mathcal{B}$ defined on the same sample space are said to be *equivalent* or *identical* if $\mathcal{A}$ occurs if and only if $\mathcal{B}$ occurs. Events $\mathcal{A}$ and $\mathcal{B}$ may be specified differently, but the elements in their defining subsets are identical. In set terminology, two sets $\mathcal{A}$ and $\mathcal{B}$ are equal (written $\mathcal{A} = \mathcal{B}$) if and only if $\mathcal{A} \subset \mathcal{B}$ and $\mathcal{B} \subset \mathcal{A}$.

**Example 1.12** Consider an experiment that consists of simultaneously throwing two dice. The sample space consists of all pairs of the form $(i, j)$ for $i = 1, 2, \ldots, 6$ and $j = 1, 2, \ldots, 6$. Let $\mathcal{A}$ be the event that the sum of the number of spots obtained on the two dice is even, i.e., $i + j$ is an even number, and let $\mathcal{B}$ be the event that both dice show an even number of spots or both dice show an odd number of spots, i.e., $i$ and $j$ are even *or* $i$ and $j$ are odd. Although event $\mathcal{A}$ has been stated differently from $\mathcal{B}$, a moment's reflection should convince the reader that the sample points in both defining subsets must be exactly the same, and hence $\mathcal{A} = \mathcal{B}$.

Viewing events as subsets allows us to apply typical set operations to them, operations such as set union, set intersection, set complementation, and so on.

1. If $\mathcal{A}$ is an event, then the *complement* of $\mathcal{A}$, denoted $\mathcal{A}^c$, is also an event. $\mathcal{A}^c$ is the subset of all sample points of $\Omega$ that are *not* in $\mathcal{A}$. Event $\mathcal{A}^c$ occurs only if $\mathcal{A}$ does *not* occur.
2. The *union* of two events $\mathcal{A}$ and $\mathcal{B}$, denoted $\mathcal{A} \cup \mathcal{B}$, is the event consisting of all the sample points in $\mathcal{A}$ and in $\mathcal{B}$. It occurs if *either* $\mathcal{A}$ or $\mathcal{B}$ occurs.

3. The *intersection* of two events $\mathcal{A}$ and $\mathcal{B}$, denoted $\mathcal{A} \cap \mathcal{B}$, is also an event. It consists of the sample points that are in both $\mathcal{A}$ and $\mathcal{B}$ and occurs if *both* $\mathcal{A}$ and $\mathcal{B}$ occur.
4. The *difference* of two events $\mathcal{A}$ and $\mathcal{B}$, denoted by $\mathcal{A} - \mathcal{B}$, is the event that $\mathcal{A}$ occurs and $\mathcal{B}$ does not occur. It consists of the sample points that are in $\mathcal{A}$ but not in $\mathcal{B}$. This means that

$$\mathcal{A} - \mathcal{B} = \mathcal{A} \cap \mathcal{B}^c.$$

   It follows that $\Omega - \mathcal{B} = \Omega \cap \mathcal{B}^c = \mathcal{B}^c$.
5. Finally, notice that if $\mathcal{B}$ is a subset of $\mathcal{A}$, i.e., $\mathcal{B} \subset \mathcal{A}$, then the event $\mathcal{B}$ implies the event $\mathcal{A}$. In other words, if $\mathcal{B}$ occurs, it must follow that $\mathcal{A}$ has also occurred.

**Example 1.13** Let event $\mathcal{A}$ be "throw a number greater than 3" and let event $\mathcal{B}$ be "throw an odd number." Event $\mathcal{A}$ occurs if a 4, 5, or 6 is thrown, and event $\mathcal{B}$ occurs if a 1, 3, or 5 is thrown. Thus both events occur if a 5 is thrown (this is the event that is the *intersection* of events $\mathcal{A}$ and $\mathcal{B}$) and neither event occurs if a 2 is thrown (this is the event that is the *complement* of the union of $\mathcal{A}$ and $\mathcal{B}$). These are represented graphically in Figure 1.4. We have

$$\mathcal{A}^c = \{1, 2, 3\}; \quad \mathcal{A} \cup \mathcal{B} = \{1, 3, 4, 5, 6\}; \quad \mathcal{A} \cap \mathcal{B} = \{5\}; \quad \mathcal{A} - \mathcal{B} = \{4, 6\}.$$

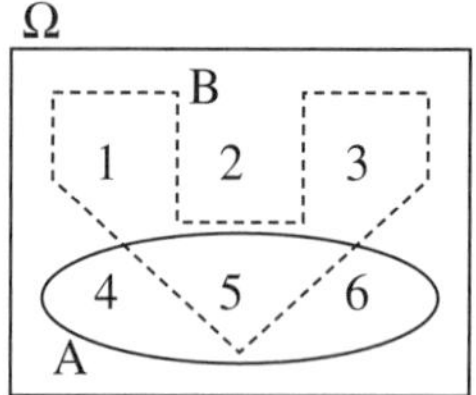


Figure 1.4. Two events on the die-throwing sample space.

**Example 1.14** Or again, consider the card-choosing scenario. The sample space for the deck of cards contains 52 elements, each of which constitutes an elementary event. Now consider two events. Let event $\mathcal{A}$ be the subset containing the 13 elements corresponding to the diamond cards in the deck. Event $\mathcal{A}$ occurs if any one of these 13 cards is chosen. Let event $\mathcal{B}$ be the subset that contains the elements representing the four queens. This event occurs if one of the four queens is chosen. The event $\mathcal{A} \cup \mathcal{B}$ contains 16 elements, the 13 corresponding to the 13 diamonds plus the queens of spades, clubs, and hearts. The event $\mathcal{A} \cup \mathcal{B}$ occurs if any one of these 16 cards is chosen: i.e., if one of the 13 diamond cards is chosen *or* if one of the four queens is chosen (logical OR). On the other hand, the event $\mathcal{A} \cap \mathcal{B}$ has a single element, the element corresponding to the queen of diamonds. The event $\mathcal{A} \cap \mathcal{B}$ occurs only if a diamond card is chosen *and* that card is a queen (logical AND). Finally, the event $\mathcal{A} - \mathcal{B}$ occurs if any diamond card, *other than the queen of diamonds*, occurs.

Thus, as these examples show, the union of two events is also an event. It is the event that consists of all of the sample points in the two events. Likewise, the intersection of two events is the event that consists of the sample points that are simultaneously in both events. It follows that the union of an event and its complement is the universal event $\Omega$, while the intersection of an event and its complement is the null event $\phi$.

The definitions of union and intersection may be extended to more than two events. For $n$ events $\mathcal{A}_1, \mathcal{A}_2, \ldots, \mathcal{A}_n$, they are denoted, respectively, by

$$\bigcup_{i=1}^{n} \mathcal{A}_i \quad \text{and} \quad \bigcap_{i=1}^{n} \mathcal{A}_i.$$

In the first case, the event $\bigcup_{i=1}^{n} \mathcal{A}_i$ occurs if any one of the events $\mathcal{A}_i$ occurs, while the second event, $\bigcap_{i=1}^{n} \mathcal{A}_i$ occurs only if all the events $\mathcal{A}_i$ occur. The entire logical algebra is available for use with events, to give an "algebra of events." Commutative, associative, and distributive laws, the laws of DeMorgan and so on, may be used to manipulate events. Some of the most important of these are as follows (where $\mathcal{A}$, $\mathcal{B}$, and $\mathcal{C}$ are subsets of the universal set $\Omega$):

| Intersection | Union |
|---|---|
| $\mathcal{A} \cap \Omega = \mathcal{A}$ | $\mathcal{A} \cup \Omega = \Omega$ |
| $\mathcal{A} \cap \mathcal{A} = \mathcal{A}$ | $\mathcal{A} \cup \mathcal{A} = \mathcal{A}$ |
| $\mathcal{A} \cap \phi = \phi$ | $\mathcal{A} \cup \phi = \mathcal{A}$ |
| $\mathcal{A} \cap \mathcal{A}^c = \phi$ | $\mathcal{A} \cup \mathcal{A}^c = \Omega$ |
| $\mathcal{A} \cap (\mathcal{B} \cap \mathcal{C}) = (\mathcal{A} \cap \mathcal{B}) \cap \mathcal{C}$ | $\mathcal{A} \cup (\mathcal{B} \cup \mathcal{C}) = (\mathcal{A} \cup \mathcal{B}) \cup \mathcal{C}$ |
| $\mathcal{A} \cap (\mathcal{B} \cup \mathcal{C}) = (\mathcal{A} \cap \mathcal{B}) \cup (\mathcal{A} \cap \mathcal{C})$ | $\mathcal{A} \cup (\mathcal{B} \cap \mathcal{C}) = (\mathcal{A} \cup \mathcal{B}) \cap (\mathcal{A} \cup \mathcal{C})$ |
| $(\mathcal{A} \cap \mathcal{B})^c = \mathcal{A}^c \cup \mathcal{B}^c$ | $(\mathcal{A} \cup \mathcal{B})^c = \mathcal{A}^c \cap \mathcal{B}^c$ |

Venn diagrams can be used to illustrate these results and can be helpful in establishing proofs. For example, an illustration of DeMorgan's laws is presented in Figure 1.5.

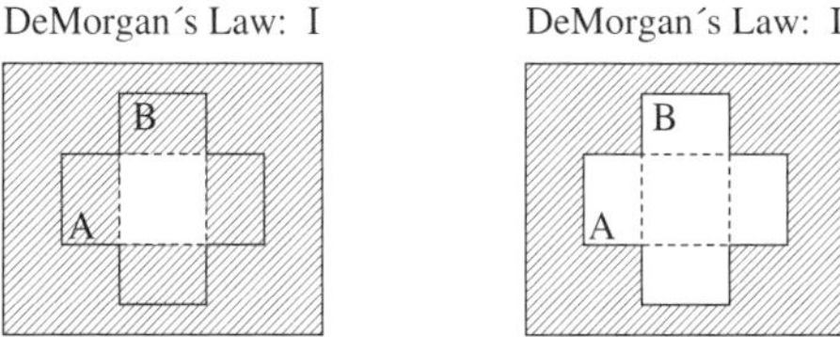


Figure 1.5. DeMorgan's laws: $(\mathcal{A} \cap \mathcal{B})^c = \mathcal{A}^c \cup \mathcal{B}^c$ and $(\mathcal{A} \cup \mathcal{B})^c = \mathcal{A}^c \cap \mathcal{B}^c$.

### *Mutually Exclusive and Collectively Exhaustive Events*

When two events $\mathcal{A}$ and $\mathcal{B}$ contain no element of the sample space in common (i.e., $\mathcal{A} \cap \mathcal{B}$ is the null set), the events are said to be *mutually exclusive* or *incompatible*. The occurrence of one of them precludes the occurrence of the other. In the case of multiple events, $\mathcal{A}_1, \mathcal{A}_2, \ldots, \mathcal{A}_n$ are mutually exclusive if and only if $\mathcal{A}_i \cap \mathcal{A}_j = \phi$ for all $i \neq j$.

**Example 1.15** Consider the four events, $\mathcal{A}_1$, $\mathcal{A}_2$, $\mathcal{A}_3$, $\mathcal{A}_4$ corresponding to the four suits in a deck of cards, i.e., $\mathcal{A}_1$ contains the 13 elements corresponding to the 13 diamonds, $\mathcal{A}_2$ contains the 13 elements corresponding to the 13 hearts, etc. Then none of the sets $\mathcal{A}_1$ through $\mathcal{A}_4$ has any element in common. The four sets are mutually exclusive.

**Example 1.16** Similarly, in the die-throwing experiment, if we choose $\mathcal{B}_1$ to be the event "throw a number greater than 5," $\mathcal{B}_2$ to be the event "throw an odd number," and $\mathcal{B}_3$ to be the event "throw a 2," then the events $\mathcal{B}_1$ through $\mathcal{B}_3$ are mutually exclusive.

In all cases, an event $\mathcal{A}$ and its complement $\mathcal{A}^c$ are mutually exclusive. In general, a *list of events* is said to be mutually exclusive if no element in their sample space is in more than one event. This is illustrated in Figure 1.6.

When all the elements in a sample space can be found in at least one event in a list of events, then the list of events is said to be *collectively exhaustive*. In this case, no element of the sample space is omitted and a single element may be in more than one event. This is illustrated in Figure 1.7.

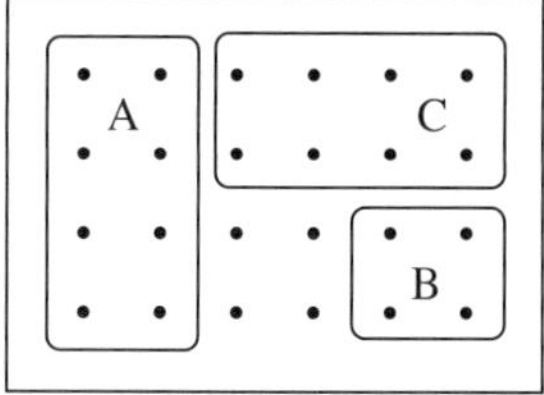


Figure 1.6. Events $\mathcal{A}$, $\mathcal{B}$, and $\mathcal{C}$ are mutually exclusive.

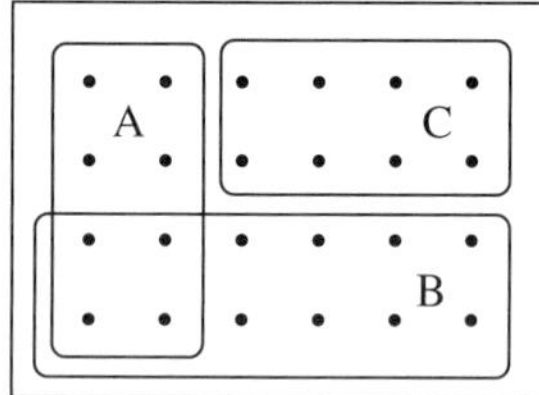


Figure 1.7. Events $\mathcal{A}$, $\mathcal{B}$, and $\mathcal{C}$ are collectively exhaustive.

Events that are both mutually exclusive and collectively exhaustive, such as those illustrated in Figure 1.8, are said to form a *partition* of the sample space. Additionally, the previously defined four events on the deck of cards, $\mathcal{A}_1$, $\mathcal{A}_2$, $\mathcal{A}_3$, $\mathcal{A}_4$, are both mutually exclusive and collectively exhaustive and constitute a partition of the sample space. Furthermore, since the elementary events (or outcomes) of a sample space are mutually exclusive and collectively exhaustive they too constitute a partition of the sample space. Any set of mutually exclusive and collectively exhaustive events is called an *event space*.

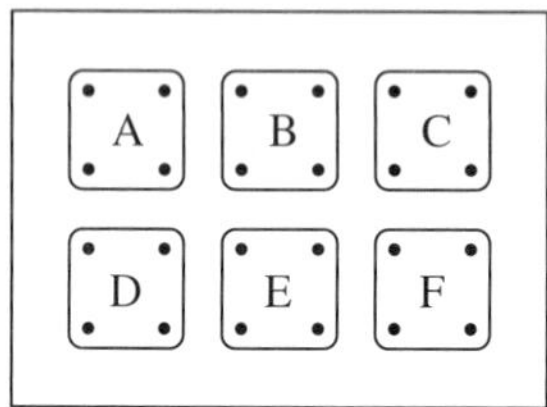


Figure 1.8. Events $\mathcal{A}$–$\mathcal{F}$ constitute a partition.

**Example 1.17** Bit sequences are transmitted over a communication channel in groups of five. Each bit may be received correctly or else be modified in transit, which occasions an error. Consider an experiment that consists in observing the bit values as they arrive and identifying them with the letter *c* if the bit is correct and with the letter *e* if the bit is in error.

The sample space consists of 32 outcomes from *ccccc* through *eeeee*, from zero bits transmitted incorrectly to all five bits being in error. Let the event $\mathcal{A}_i$, $i = 0, 1, \ldots, 5$, consist of all outcomes in which *i* bits are in error. Thus $\mathcal{A}_0 = \{ccccc\}$, $\mathcal{A}_1 = \{ecccc, ceccc, ccecc, cccec, cccce\}$, and so on up to $\mathcal{A}_5 = \{eeeee\}$. The events $\mathcal{A}_i$, $i = 0, 1, \ldots, 5$, partition the sample space and therefore constitute an event space. It may be much easier to work in this small *event space* rather than in the larger *sample space*, especially if our only interest is in knowing the number of bits transmitted in error. Furthermore, when the bits are transmitted in larger groups, the difference becomes even more important. With 16 bits per group instead of five, the event space now contains 17 events, whereas the sample space contains $2^{16}$ outcomes.

## 1.2 Probability Axioms and Probability Space

### *Probability Axioms*

So far our discussion has been about trials, sample spaces, and events. We now tackle the topic of probabilities. Our concern will be with assigning probabilities to events, i.e., providing some measure of the relative likelihood of the occurrence of the event. We realize that when we toss a fair coin, we have a 50–50 chance that it will give a head. When we throw a fair die, the chance of getting a 1 is the same as that of getting a 2, or indeed any of the other four possibilities. If a deck of cards is well shuffled and we pick a single card, there is a one in 52 chance that it will be the queen of hearts. What we have done in these examples is to associate probabilities with the elements of the sample space; more correctly, we have assigned probabilities to the elementary events, the events consisting of the singleton subsets of the sample space.

Probabilities are real numbers in the closed interval $[0, 1]$. The greater the value of the probability, the more likely the event is to happen. If an event has probability zero, that event cannot occur; if it has probability one, then it is certain to occur.

**Example 1.18** In the coin-tossing example, the probability of getting a head in a single toss is 0.5, since we are equally likely to get a head as we are to get a tail. This is written as

$$\text{Prob}\{H\} = 0.5 \quad \text{or} \quad \text{Prob}\{\mathcal{A}_1\} = 0.5,$$

where $\mathcal{A}_1$ is the event $\{H\}$.

Similarly, the probability of throwing a 6 with a die is 1/6 and the probability of choosing the queen of hearts is 1/52. In these cases, the elementary events of each sample space all have equal probability, or equal likelihood, of being the outcome on any given trial. They are said to be *equiprobable* events and the outcome of the experiment is said to be *random*, since each event has the same chance of occurring. In a sample space containing $n$ equally likely outcomes, the probability of any particular outcome occurring is $1/n$. Naturally, we can assign probabilities to events other than elementary events.

**Example 1.19** Find the probability that should be associated with the event $\mathcal{A}_2 = \{1, 2, 3\}$, i.e., throwing a number smaller than 4 using a fair die. This event occurs if any of the numbers 1, 2, or 3 is the outcome of the throw. Since each has a probability of 1/6 and there are three of them, the probability of event $\mathcal{A}_2$ is the sum of the probabilities of these three elementary events and is therefore equal to 0.5.

This holds in general: the probability of any event is simply the sum of the probabilities associated with the (elementary) elements of the sample space that constitute that event.

**Example 1.20** Consider Figure 1.6 once again (reproduced here as Figure 1.9), and assume that each of the 24 points or elements of the sample space is equiprobable.

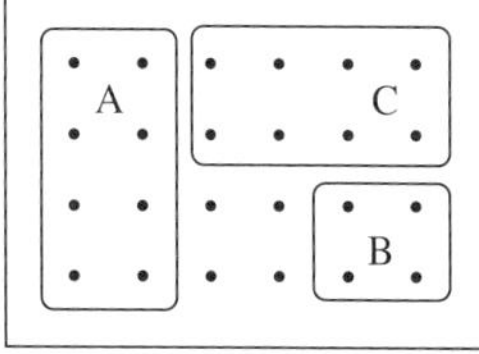


Figure 1.9. Sample space with 24 equiprobable elements.

Then event $\mathcal{A}$ contains eight elements, and so the probability of this event is

$$\text{Prob}\{\mathcal{A}\} = \frac{1}{24} + \frac{1}{24} + \frac{1}{24} + \frac{1}{24} + \frac{1}{24} + \frac{1}{24} + \frac{1}{24} + \frac{1}{24} = 8 \times \frac{1}{24} = \frac{1}{3}.$$

Similarly, $\text{Prob}\{\mathcal{B}\} = 4/24 = 1/6$ and $\text{Prob}\{\mathcal{C}\} = 8/24 = 1/3$.

Assigning probabilities to events is an extremely important part of developing probability models. In some cases, we know in advance the probabilities to associate with elementary events, while in other cases they must be estimated. If we assume that the coin and the die are fair and the deck of cards completely shuffled, then it is easy to associate probabilities with the elements of the sample space and subsequently to the events described on these sample spaces. In other cases, the probabilities must be guessed at or estimated.

Two approaches have been developed for defining probabilities: the *relative frequency* approach and the *axiomatic* approach. The first, as its name implies, consists in performing the probability experiment a great many times, say $N$, and counting the number of times a certain event occurs, say $n$. An estimate of the probability of the event may then be obtained as the relative frequency $n/N$ with which the event occurs, since we would hope that, in the limit (limit in a probabilistic sense) as $N \rightarrow \infty$, the ratio $n/N$ tends to the correct probability of the event. In mathematical terms, this is stated as follows: Given that the probability of an event is $p$, then

$$\lim_{N\rightarrow\infty} \text{Prob}\left\{\left|\frac{n}{N} - p\right| > \epsilon\right\} = 0$$

for any small $\epsilon > 0$. In other words, no matter how small we choose $\epsilon$ to be, the probability that the difference between $n/N$ and $p$ is greater than $\epsilon$ tends to zero as $N \rightarrow \infty$. Use of relative frequencies as estimates of probability can be justified mathematically, as we shall see later.

The axiomatic approach sets up a small number of laws or axioms on which the entire theory of probability is based. Fundamental to this concept is the fact that it is possible to manipulate probabilities using the same logic algebra with which the events themselves are manipulated. The three basic axioms are as follows.

**Axiom 1:** For any event $\mathcal{A}$, $0 \leq \text{Prob}\{\mathcal{A}\} \leq 1$; i.e., probabilities are real numbers in the interval $[0, 1]$.

**Axiom 2:** $\text{Prob}\{\Omega\} = 1$; The universal or certain event is assigned probability 1.

**Axiom 3:** For any *countable collection* of events $\mathcal{A}_1, \mathcal{A}_2, \ldots$ that are mutually exclusive,

$$\text{Prob}\left\{\bigcup_{i=1}^{\infty} \mathcal{A}_i\right\} \equiv \text{Prob}\{\mathcal{A}_1 \cup \mathcal{A}_2 \cup \cdots \cup \mathcal{A}_n \cup \cdots\} = \sum_{i=1}^{\infty} \text{Prob}\{\mathcal{A}_i\}.$$

In some elementary texts, the third axiom is replaced with the simpler

**Axiom 3$^\dagger$:** For two mutually exclusive events $\mathcal{A}$ and $\mathcal{B}$, $\text{Prob}\{\mathcal{A} \cup \mathcal{B}\} = \text{Prob}\{\mathcal{A}\} + \text{Prob}\{\mathcal{B}\}$

and a comment included stating that this extends in a natural sense to any finite or denumerable number of mutually exclusive events.

These three axioms are very natural; the first two are almost trivial, which essentially means that all of probability is based on unions of mutually exclusive events. To gain some insight, consider the following examples.

**Example 1.21** If $\text{Prob}\{\mathcal{A}\} = p_1$ and $\text{Prob}\{\mathcal{B}\} = p_2$ where $\mathcal{A}$ and $\mathcal{B}$ are two mutually exclusive events, then the probability of the events $\mathcal{A} \cup \mathcal{B}$ and $\mathcal{A} \cap \mathcal{B}$ are given by

$$\text{Prob}\{\mathcal{A} \cup \mathcal{B}\} = p_1 + p_2 \quad \text{and} \quad \text{Prob}\{\mathcal{A} \cap \mathcal{B}\} = 0.$$

**Example 1.22** If the sets $\mathcal{A}$ and $\mathcal{B}$ are not mutually exclusive, then the probability of the event $\mathcal{A} \cup \mathcal{B}$ will be less than $p_1 + p_2$ since some of the elementary events will be present in both $\mathcal{A}$ and $\mathcal{B}$, but can only be counted once. The probability of the event $\mathcal{A} \cap \mathcal{B}$ will be greater than zero; it will be the sum of the probabilities of the elementary events found in the intersection of the two subsets. It follows then that

$$\text{Prob}\{\mathcal{A} \cup \mathcal{B}\} = \text{Prob}\{\mathcal{A}\} + \text{Prob}\{\mathcal{B}\} - \text{Prob}\{\mathcal{A} \cap \mathcal{B}\}.$$

Observe that the probability of an event $\mathcal{A}$, formed from the union of a set of mutually exclusive events, is equal to the sum of the probabilities of those mutually exclusive events, i.e.,

$$\mathcal{A} = \text{Prob}\left\{\bigcup_{i=1}^{n} A_i\right\} = \text{Prob}\{\mathcal{A}_1 \cup \mathcal{A}_2 \cup \cdots \cup \mathcal{A}_n\} = \sum_{i=1}^{n} \text{Prob}\{\mathcal{A}_i\}.$$

In particular, the probability of *any* event is equal to the sum of the probabilities of the outcomes in the sample space that constitute the event since outcomes are elementary events which are mutually exclusive.

A number of the most important results that follow from these definitions are presented below. The reader should make an effort to prove these independently.

- For any event $\mathcal{A}$, $\text{Prob}\{\mathcal{A}^c\} = 1 - \text{Prob}\{\mathcal{A}\}$. Alternatively, $\text{Prob}\{\mathcal{A}\} + \text{Prob}\{\mathcal{A}^c\} = 1$.
- For the impossible event $\phi$, $\text{Prob}\{\phi\} = 0$ (since $\text{Prob}\{\Omega\} = 1$).
- If $\mathcal{A}$ and $\mathcal{B}$ are any events, not necessarily mutually exclusive,

$$\text{Prob}\{\mathcal{A} \cup \mathcal{B}\} = \text{Prob}\{\mathcal{A}\} + \text{Prob}\{\mathcal{B}\} - \text{Prob}\{\mathcal{A} \cap \mathcal{B}\}.$$

  Thus $\text{Prob}\{\mathcal{A} \cup \mathcal{B}\} \le \text{Prob}\{\mathcal{A}\} + \text{Prob}\{\mathcal{B}\}$.
- For arbitrary events $\mathcal{A}$ and $\mathcal{B}$,

$$\text{Prob}\{\mathcal{A} - \mathcal{B}\} = \text{Prob}\{\mathcal{A}\} - \text{Prob}\{\mathcal{A} \cap \mathcal{B}\}.$$

- For arbitrary events $\mathcal{A}$ and $\mathcal{B}$ with $\mathcal{B} \subset \mathcal{A}$,

$$\text{Prob}\{\mathcal{B}\} \le \text{Prob}\{\mathcal{A}\}.$$

It is interesting to observe that an event having probability zero does not necessarily mean that this event cannot occur. The probability of no heads appearing in an infinite number of throws of a fair coin is zero, but this event can occur.

### *Probability Space*

The set of subsets of a given set, which includes the empty subset and the complete set itself, is sometimes referred to as the *superset* or *power set* of the given set. The superset of a set of elements in a sample space is therefore the set of all possible events that may be defined on that space. When the sample space is finite, or even when it is countably infinite (denumerable), it is possible to assign probabilities to each event in such a way that all three axioms are satisfied. However, when the sample space is not denumerable, such as the set of points on a segment of the real line, such an assignment of probabilities may not be possible. To avoid difficulties of this nature, we restrict the set of events to those to which probabilities satisfying all three axioms can be assigned. This is the basis of "measure theory:" for a given application, there is a particular family of events (a class of subsets of $\Omega$), to which probabilities can be assigned, i.e., given a "measure." We shall call this family of subsets $\mathcal{F}$. Since we will wish to apply set operations, we need to insist that $\mathcal{F}$ be closed under countable unions, intersections, and complementation. A collection of subsets of a given set $\Omega$ that is closed under countable unions and complementation is called a $\sigma$*-field* of subsets of $\Omega$.

The term *σ-algebra* is also used. Using DeMorgan's law, it may be shown that countable intersections of subsets of a σ-field $\mathcal{F}$ also lie in $\mathcal{F}$.

**Example 1.23** The set $\{\Omega,\ \phi\}$ is the smallest σ-field defined on a sample space. It is sometimes called the trivial σ-field over Ω and is a subset of every other σ-field over Ω. The superset of Ω is the largest σ-field over Ω.

**Example 1.24** If $\mathcal{A}$ and $\mathcal{B}$ are two events, then the set containing the events $\Omega,\ \phi,\ \mathcal{A},\ \mathcal{A}^c,\ \mathcal{B}$, and $\mathcal{B}^c$ is a σ-field.

**Example 1.25** In a die-rolling experiment having sample space $\{1, 2, 3, 4, 5, 6\}$, the following are all σ-fields:

$$\mathcal{F} = \{\Omega,\ \phi\},$$
$$\mathcal{F} = \{\Omega,\ \phi,\ \{2, 4, 6\},\ \{1, 3, 5\}\},$$
$$\mathcal{F} = \{\Omega,\ \phi,\ \{1, 2, 4, 6\},\ \{3, 5\}\},$$

but the sets $\{\Omega,\ \phi,\ \{1, 2\},\ \{3, 4\},\ \{5, 6\}\}$ and $\{\Omega,\ \phi,\ \{1, 2, 4, 6\},\ \{3, 4, 5\}\}$ are not.

We may now define a *probability space* or *probability system*. This is defined as the triplet $\{\Omega, \mathcal{F}, \text{Prob}\}$, where Ω is a set, $\mathcal{F}$ is a σ-field of subsets of Ω that includes Ω, and Prob is a probability measure on $\mathcal{F}$ that satisfies the three axioms given above. Thus, $\text{Prob}\{\cdot\}$ is a function with domain $\mathcal{F}$ and range [0, 1] which satisfies axioms 1–3. It assigns a number in [0, 1] to events in $\mathcal{F}$.

## 1.3 Conditional Probability

Before performing a probability experiment, we cannot know precisely the particular outcome that will occur, nor whether an event $\mathcal{A}$, composed of some subset of the outcomes, will actually happen. We may know that the event is likely to take place, if $\text{Prob}\{\mathcal{A}\}$ is close to one, or unlikely to take place, if $\text{Prob}\{\mathcal{A}\}$ is close to zero, but we cannot be sure until after the experiment has been conducted. $\text{Prob}\{\mathcal{A}\}$ is the *prior probability* of $\mathcal{A}$. We now ask how this prior probability of an event $\mathcal{A}$ changes if we are informed that some other event, $\mathcal{B}$, has occurred. In other words, a probability experiment has taken place and one of the outcomes that constitutes an event $\mathcal{B}$ observed to have been the result. We are not told which particular outcome in $\mathcal{B}$ occurred, just that this event was observed to occur. We wish to know, given this additional information, how our knowledge of the probability of $\mathcal{A}$ occurring must be altered.

**Example 1.26** Let us return to the example in which we consider the probabilities obtained on three throws of a fair coin. The elements of the sample space are

$$\{HHH,\ HHT,\ HTH,\ THH,\ HTT,\ THT,\ TTH,\ TTT\}$$

and the probability of each of these events is 1/8. Suppose we are interested in the probability of getting three heads, $\mathcal{A} = \{HHH\}$. The prior probability of this event is $\text{Prob}\{\mathcal{A}\} = 1/8$. Now, how do the probabilities change if we know the result of the first throw?

If the first throw gives tails, the event $\mathcal{B}$ is constituted as $\mathcal{B} = \{THH,\ THT,\ TTH,\ TTT\}$ and we know that we are not going to get our three heads! Once we know that the result of the first throw is tails, the event of interest becomes impossible, i.e., has probability zero.

If the first throw gives a head, i.e., $\mathcal{B} = \{HHH,\ HHT,\ HTH,\ HTT\}$, then the event $\mathcal{A} = \{HHH\}$ is still possible. The question we are now faced with is to determine the probability of getting $\{HHH\}$ given that we know that the first throw gives a head. Obviously the probability must now be greater than 1/8. All we need to do is to get heads on the second and third throws, each of which is obtained with probability 1/2. Thus, given that the first throw yields heads, the probability of getting the event *HHH* is 1/4. Of the original eight elementary events, only four of them can now be assigned positive probabilities. From a different vantage point, the event $\mathcal{B}$ contains four equiprobable outcomes and is known to have occurred. It follows that the probability of any one of these four equiprobable outcomes, and in particular that of *HHH*, is 1/4.

The effect of knowing that a certain event has occurred changes the original probabilities of other events defined on the sample space. Some of these may become zero; for some others, their associated probability is increased. For yet others, there may be no change.

**Example 1.27** Consider Figure 1.10 which represents a sample space with 24 elements all with probability 1/24. Suppose that we are told that event $\mathcal{B}$ has occurred. As a result the prior probabilities associated with elementary events outside $\mathcal{B}$ must be reset to zero and the sum of the probabilities of the elementary events inside $\mathcal{B}$ must sum to 1. In other words, the probabilities of the elementary events must be renormalized so that only those that can possibly occur have strictly positive probability and these probabilities must be coherent, i.e., they must sum to 1. Since the elementary events in $\mathcal{B}$ are equiprobable, after renormalization, they must each have probability $1/12$.

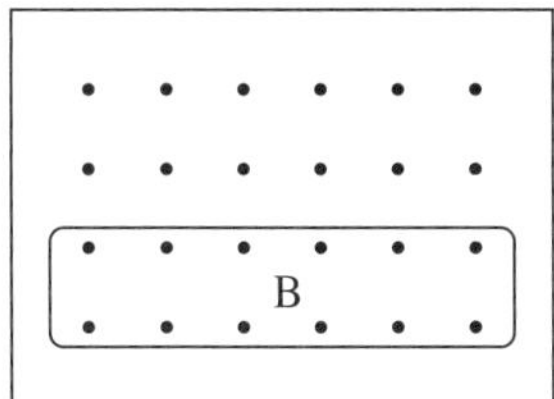


Figure 1.10. Sample space with 24 equiprobable elements.

We let Prob$\{\mathcal{A}|\mathcal{B}\}$ denote the probability of $\mathcal{A}$ given that event $\mathcal{B}$ has occurred. Because of the need to renormalize the probabilities so that they continue to sum to 1 after this given event has taken place, we must have

$$\text{Prob}\{\mathcal{A}|\mathcal{B}\} = \frac{\text{Prob}\{\mathcal{A} \cap \mathcal{B}\}}{\text{Prob}\{\mathcal{B}\}}. \tag{1.1}$$

Since it is known that event $\mathcal{B}$ occurred, it must have positive probability, i.e., Prob$\{\mathcal{B}\} > 0$, and hence the quotient in Equation (1.1) is well defined. The quantity Prob$\{\mathcal{A}|\mathcal{B}\}$ is called the *conditional probability of event $\mathcal{A}$ given the hypothesis $\mathcal{B}$*. It is defined only when Prob$\{\mathcal{B}\} \neq 0$. Notice that a rearrangement of Equation (1.1) gives

$$\text{Prob}\{\mathcal{A} \cap \mathcal{B}\} = \text{Prob}\{\mathcal{A}|\mathcal{B}\}\text{Prob}\{\mathcal{B}\}. \tag{1.2}$$

Similarly,

$$\text{Prob}\{\mathcal{A} \cap \mathcal{B}\} = \text{Prob}\{\mathcal{B}|\mathcal{A}\}\text{Prob}\{\mathcal{A}\}$$

provided that Prob$\{\mathcal{A}\} > 0$.

Since conditional probabilities are probabilities in the strictest sense of the term, they satisfy all the properties that we have seen so far concerning ordinary probabilities. In addition, the following hold:

- Let $\mathcal{A}$ and $\mathcal{B}$ be two mutually exclusive events. Then $\mathcal{A} \cap \mathcal{B} = \phi$ and hence Prob$\{\mathcal{A}|\mathcal{B}\} = 0$.
- If event $\mathcal{B}$ implies event $\mathcal{A}$, (i.e., $\mathcal{B} \subset \mathcal{A}$), then Prob$\{\mathcal{A}|\mathcal{B}\} = 1$.

**Example 1.28** Let $\mathcal{A}$ be the event that a red queen is pulled from a deck of cards and let $\mathcal{B}$ be the event that a red card is pulled. Then Prob$\{\mathcal{A}|\mathcal{B}\}$, the probability that a red queen is pulled *given* that a red card is chosen, is

$$\text{Prob}\{\mathcal{A}|\mathcal{B}\} = \frac{\text{Prob}\{\mathcal{A} \cap \mathcal{B}\}}{\text{Prob}\{\mathcal{B}\}} = \frac{2/52}{1/2} = 1/13.$$

Notice in this example that Prob$\{\mathcal{A} \cap \mathcal{B}\}$ and Prob$\{\mathcal{B}\}$ are *prior probabilities*. Thus the event $\mathcal{A} \cap \mathcal{B}$ contains two of the 52 possible outcomes and the event $\mathcal{B}$ contains 26 of the 52 possible outcomes.

**Example 1.29** If we observe Figure 1.11 we see that Prob$\{\mathcal{A} \cap \mathcal{B}\} = 1/6$, that Prob$\{\mathcal{B}\} = 1/2$, and that Prob$\{\mathcal{A}|\mathcal{B}\} = (1/6)/(1/2) = 1/3$ as expected. We know that $\mathcal{B}$ has occurred and that event $\mathcal{A}$ will occur if one of the four outcomes in $\mathcal{A} \cap \mathcal{B}$ is chosen from among the 12 equally probable outcomes in $\mathcal{B}$.

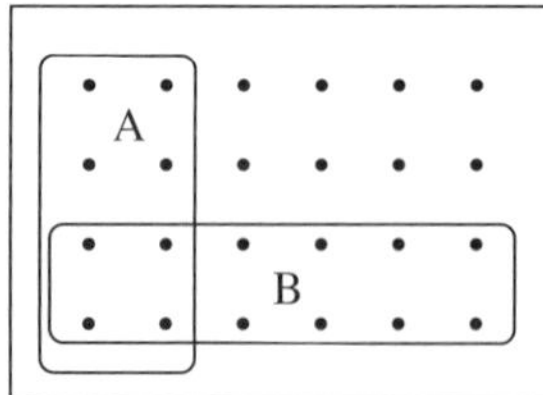


Figure 1.11. Prob$\{\mathcal{A}|\mathcal{B}\} = 1/3$.

Equation (1.2) can be generalized to multiple events. Let $\mathcal{A}_i$, $i = 1, 2, \ldots, k$, be $k$ events for which Prob$\{\mathcal{A}_1 \cap \mathcal{A}_2 \cap \cdots \cap \mathcal{A}_k\} > 0$. Then

$$\begin{aligned}\text{Prob}\{\mathcal{A}_1 \cap \mathcal{A}_2 \cap \cdots \cap \mathcal{A}_k\} &= \text{Prob}\{\mathcal{A}_1\}\text{Prob}\{\mathcal{A}_2|\mathcal{A}_1\}\text{Prob}\{\mathcal{A}_3|\mathcal{A}_1 \cap \mathcal{A}_2\} \cdots \\ &\times \text{Prob}\{\mathcal{A}_k|\mathcal{A}_1 \cap \mathcal{A}_2 \cap \cdots \cap \mathcal{A}_{k-1}\}.\end{aligned}$$

The proof is by induction. The base clause ($k = 2$) follows from Equation (1.2):

$$\text{Prob}\{\mathcal{A}_1 \cap \mathcal{A}_2\} = \text{Prob}\{\mathcal{A}_1\}\text{Prob}\{\mathcal{A}_2|\mathcal{A}_1\}.$$

Now let $\mathcal{A} = \mathcal{A}_1 \cap \mathcal{A}_2 \cap \cdots \cap \mathcal{A}_k$ and assume the relation is true for $k$, i.e., that

$$\text{Prob}\{\mathcal{A}\} = \text{Prob}\{\mathcal{A}_1\}\text{Prob}\{\mathcal{A}_2|\mathcal{A}_1\}\text{Prob}\{\mathcal{A}_3|\mathcal{A}_1 \cap \mathcal{A}_2\} \cdots \text{Prob}\{\mathcal{A}_k|\mathcal{A}_1 \cap \mathcal{A}_2 \cap \cdots \cap \mathcal{A}_{k-1}\}.$$

That the relation is true for $k + 1$ follows immediately, since

$$\text{Prob}\{\mathcal{A}_1 \cap \mathcal{A}_2 \cap \cdots \cap \mathcal{A}_k \cap \mathcal{A}_{k+1}\} = \text{Prob}\{\mathcal{A} \cap \mathcal{A}_{k+1}\} = \text{Prob}\{\mathcal{A}\}\text{Prob}\{\mathcal{A}_{k+1}|\mathcal{A}\}.$$

**Example 1.30** In a first-year graduate level class of 60 students, ten students are undergraduates. Let us compute the probability that three randomly chosen students are all undergraduates. We shall let $\mathcal{A}_1$ be the event that the first student chosen is an undergraduate student, $\mathcal{A}_2$ be the event that the second one chosen is an undergraduate, and so on. Recalling that the intersection of two events $\mathcal{A}$ and $\mathcal{B}$ is the event that occurs when both $\mathcal{A}$ and $\mathcal{B}$ occur, and using the relationship

$$\text{Prob}\{\mathcal{A}_1 \cap \mathcal{A}_2 \cap \mathcal{A}_3\} = \text{Prob}\{\mathcal{A}_1\}\text{Prob}\{\mathcal{A}_2|\mathcal{A}_1\}\text{Prob}\{\mathcal{A}_3|\mathcal{A}_1 \cap \mathcal{A}_2\},$$

we obtain

$$\text{Prob}\{\mathcal{A}_1 \cap \mathcal{A}_2 \cap \mathcal{A}_3\} = \frac{10}{60} \times \frac{9}{59} \times \frac{8}{58} = 0.003507.$$

## 1.4 Independent Events

We saw previously that two events are *mutually exclusive* if and only if the probability of the *union* of these two events is equal to the *sum* of the probabilities of the events, i.e., if and only if

$$\text{Prob}\{\mathcal{A} \cup \mathcal{B}\} = \text{Prob}\{\mathcal{A}\} + \text{Prob}\{\mathcal{B}\}.$$

Now we investigate the probability associated with the *intersection* of two events. We shall see that the probability of the intersection of two events is equal to the *product* of the probabilities of the events if and only if the outcome of one event does not influence the outcome of the other, i.e., if and only if the two events are *independent* of each other.

Let $\mathcal{B}$ be an event with positive probability, i.e., $\text{Prob}\{\mathcal{B}\} > 0$. Then event $\mathcal{A}$ is said to be *independent* of event $\mathcal{B}$ if

$$\text{Prob}\{\mathcal{A}|\mathcal{B}\} = \text{Prob}\{\mathcal{A}\}. \tag{1.3}$$

Thus the fact that event $\mathcal{B}$ occurs with positive probability has no effect on event $\mathcal{A}$. Equation (1.3) essentially says that the probability of event $\mathcal{A}$ occurring, given that $\mathcal{B}$ has already occurred, is just the same as the unconditional probability of event $\mathcal{A}$ occurring. It makes no difference at all that event $\mathcal{B}$ has occurred.

**Example 1.31** Consider an experiment that consists in rolling two colored (and hence distinguishable) dice, one red and one green. Let $\mathcal{A}$ be the event that the sum of spots obtained is 7, and let $\mathcal{B}$ be the event that the red die shows 3. There are a total of 36 outcomes, each represented as a pair $(i, j)$, where $i$ denotes the number of spots on the red die, and $j$ the number of spots on the green die. Of these 36 outcomes, six, namely, $(1, 6)$, $(2, 5)$, $(3, 4)$, $(4, 3)$, $(5, 2)$, $(6, 1)$, result in event $\mathcal{A}$ and hence $\text{Prob}\{\mathcal{A}\} = 6/36$. Also six outcomes result in the occurrence of event $\mathcal{B}$, namely, $(3, 1)$, $(3, 2)$, $(3, 3)$, $(3, 4)$, $(3, 5)$, $(3, 6)$, but only one of these gives event $\mathcal{A}$. Therefore $\text{Prob}\{\mathcal{A}|\mathcal{B}\} = 1/6$. Events $\mathcal{A}$ and $\mathcal{B}$ must therefore be independent since

$$\text{Prob}\{\mathcal{A}|\mathcal{B}\} = 1/6 = 6/36 = \text{Prob}\{\mathcal{A}\}.$$

If event $\mathcal{A}$ is independent of event $\mathcal{B}$, then event $\mathcal{B}$ must be independent of event $\mathcal{A}$; i.e., independence is a *symmetric* relationship. Substituting $\text{Prob}\{\mathcal{A}\cap\mathcal{B}\}/\text{Prob}\{\mathcal{B}\}$ for $\text{Prob}\{\mathcal{A}|\mathcal{B}\}$ it must follow that, for independent events

$$\text{Prob}\{\mathcal{A}|\mathcal{B}\} = \frac{\text{Prob}\{\mathcal{A} \cap \mathcal{B}\}}{\text{Prob}\{\mathcal{B}\}} = \text{Prob}\{\mathcal{A}\}$$

or, rearranging terms, that

$$\text{Prob}\{\mathcal{A} \cap \mathcal{B}\} = \text{Prob}\{\mathcal{A}\}\text{Prob}\{\mathcal{B}\}.$$

Indeed, this is frequently taken as a definition of independence. Two events $\mathcal{A}$ and $\mathcal{B}$ are said to be *independent* if and only if

$$\text{Prob}\{\mathcal{A} \cap \mathcal{B}\} = \text{Prob}\{\mathcal{A}\}\text{Prob}\{\mathcal{B}\}.$$

Pursuing this direction, it then follows that, for two independent events

$$\text{Prob}\{\mathcal{A}|\mathcal{B}\} = \frac{\text{Prob}\{\mathcal{A} \cap \mathcal{B}\}}{\text{Prob}\{\mathcal{B}\}} = \frac{\text{Prob}\{\mathcal{A}\}\text{Prob}\{\mathcal{B}\}}{\text{Prob}\{\mathcal{B}\}} = \text{Prob}\{\mathcal{A}\},$$

which conveniently brings us back to the starting point.

**Example 1.32** Suppose a fair coin is thrown twice. Let $\mathcal{A}$ be the event that a head occurs on the first throw, and $\mathcal{B}$ the event that a head occurs on the second throw. Are $\mathcal{A}$ and $\mathcal{B}$ independent events?

Obviously $\text{Prob}\{\mathcal{A}\} = 1/2 = \text{Prob}\{\mathcal{B}\}$. The event $\mathcal{A} \cap \mathcal{B}$ is the event of a head occurring on the first throw *and* a head occurring on the second throw. Thus, $\text{Prob}\{\mathcal{A} \cap \mathcal{B}\} = 1/2 \times 1/2 = 1/4$ and since $\text{Prob}\{\mathcal{A}\} \times \text{Prob}\{\mathcal{B}\} = 1/4$, the events $\mathcal{A}$ and $\mathcal{B}$ must be independent, since

$$\text{Prob}\{\mathcal{A} \cap \mathcal{B}\} = 1/4 = \text{Prob}\{\mathcal{A}\}\text{Prob}\{\mathcal{B}\}.$$

**Example 1.33** Let $\mathcal{A}$ be the event that a card pulled randomly from a deck of 52 cards is red, and let $\mathcal{B}$ be the event that this card is a queen. Are $\mathcal{A}$ and $\mathcal{B}$ independent events? What happens if event $\mathcal{B}$ is the event that the card pulled is the queen of hearts?

The probability of pulling a red card is $1/2$ and the probability of pulling a queen is $1/13$. Thus $\text{Prob}\{\mathcal{A}\} = 1/2$ and $\text{Prob}\{\mathcal{B}\} = 1/13$. Now let us find the probability of the event $\mathcal{A} \cap \mathcal{B}$ (the probability of pulling a red queen) and see if it equals the product of these two. Since there are two red queens, the probability of choosing a red queen is $2/52$, which is indeed equal to the product of $\text{Prob}\{\mathcal{A}\}$ and $\text{Prob}\{\mathcal{B}\}$ and so the events are independent.

If event $\mathcal{B}$ is now the event that the card pulled is the queen of hearts, then $\text{Prob}\{\mathcal{B}\} = 1/52$. But now the event $\mathcal{A} \cap \mathcal{B}$ consists of a single outcome: there is only one red card that is the queen of hearts, and so $\text{Prob}\{\mathcal{A} \cap \mathcal{B}\} = 1/52$. Therefore the two events are *not* independent since

$$1/52 = \text{Prob}\{\mathcal{A} \cap \mathcal{B}\} \neq \text{Prob}\{\mathcal{A}\}\text{Prob}\{\mathcal{B}\} = 1/2 \times 1/52.$$

We may show that, if $\mathcal{A}$ and $\mathcal{B}$ are independent events, then the pairs $(\mathcal{A}, \mathcal{B}^c)$, $(\mathcal{A}^c, \mathcal{B})$, and $(\mathcal{A}^c, \mathcal{B}^c)$ are also independent. For example, to show that $\mathcal{A}$ and $\mathcal{B}^c$ are independent, we proceed as follows. Using the result, $\text{Prob}\{\mathcal{A}\} = \text{Prob}\{\mathcal{A} \cap \mathcal{B}\} + \text{Prob}\{\mathcal{A} \cap \mathcal{B}^c\}$ we obtain

$$\begin{aligned}\text{Prob}\{\mathcal{A} \cap \mathcal{B}^c\} &= \text{Prob}\{\mathcal{A}\} - \text{Prob}\{\mathcal{A} \cap \mathcal{B}\} \\ &= \text{Prob}\{\mathcal{A}\} - \text{Prob}\{\mathcal{A}\}\text{Prob}\{\mathcal{B}\} \\ &= \text{Prob}\{\mathcal{A}\}(1 - \text{Prob}\{\mathcal{B}\}) = \text{Prob}\{\mathcal{A}\}\text{Prob}\{\mathcal{B}^c\}.\end{aligned}$$

The fact that, given two independent events $\mathcal{A}$ and $\mathcal{B}$, the four events $\mathcal{A}$, $\mathcal{B}$, $\mathcal{A}^c$, and $\mathcal{B}^c$ are *pairwise independent*, has a number of useful applications.

**Example 1.34** Before being loaded onto a distribution truck, packages are subject to two independent tests, to ensure that the truck driver can safely handle them. The weight of the package must not exceed 80 lbs and the sum of the three dimensions must be less than 8 feet. It has been observed that 5% of packages exceed the weight limit and 2% exceed the dimension limit. What is the probability that a package that meets the weight requirement fails the dimension requirement?

The sample space contains four possible outcomes: $(ws, ds)$, $(wu, ds)$, $(ws, du)$, and $(wu, du)$, where $w$ and $d$ represent weight and dimension, respectively, and $s$ and $u$ represent satisfactory and unsatisfactory, respectively. Let $\mathcal{A}$ be the event that a package satisfies the weight requirement, and $\mathcal{B}$ the event that it satisfies the dimension requirement. Then $\text{Prob}\{\mathcal{A}\} = 0.95$ and $\text{Prob}\{\mathcal{B}\} = 0.98$. We also have $\text{Prob}\{\mathcal{A}^c\} = 0.05$ and $\text{Prob}\{\mathcal{B}^c\} = 0.02$.

The event of interest is the single outcome $\{(ws, du)\}$, which is given by $\text{Prob}\{\mathcal{A} \cap \mathcal{B}^c\}$. Since $\mathcal{A}$ and $\mathcal{B}$ are independent, it follows that $\mathcal{A}$ and $\mathcal{B}^c$ are independent and hence

$$\text{Prob}\{(ws, du)\} = \text{Prob}\{\mathcal{A} \cap \mathcal{B}^c\} = \text{Prob}\{\mathcal{A}\}\text{Prob}\{\mathcal{B}^c\} = 0.95 \times 0.02 = 0.0019.$$

### *Multiple Independent Events*

Consider now multiple events. Let $\mathcal{Z}$ be an arbitrary class of events, i.e.,

$$\mathcal{Z} = \mathcal{A}_1, \ \mathcal{A}_2, \ldots, \ \mathcal{A}_n, \ldots.$$

These events are said to be *mutually independent* (or simply independent), if, for every finite subclass $\mathcal{A}_1, \ \mathcal{A}_2, \ldots, \ \mathcal{A}_k$ of $\mathcal{Z}$,

$$\text{Prob}\{\mathcal{A}_1 \cap \mathcal{A}_2 \cap \cdots \cap \mathcal{A}_k\} = \text{Prob}\{\mathcal{A}_1\}\text{Prob}\{\mathcal{A}_2\} \cdots \text{Prob}\{\mathcal{A}_k\}.$$

In other words, any pair of events $(\mathcal{A}_i, \ \mathcal{A}_j)$ must satisfy

$$\text{Prob}\{\mathcal{A}_i \cap \mathcal{A}_j\} = \text{Prob}\{\mathcal{A}_i\}\text{Prob}\{\mathcal{A}_j\};$$

any triplet of events $(\mathcal{A}_i, \ \mathcal{A}_j, \ \mathcal{A}_k)$ must satisfy

$$\text{Prob}\{\mathcal{A}_i \cap \mathcal{A}_j \cap \mathcal{A}_k\} = \text{Prob}\{\mathcal{A}_i\}\text{Prob}\{\mathcal{A}_j\}\text{Prob}\{\mathcal{A}_k\};$$

and so on, for quadruples of events, for quintuples of events, etc.

**Example 1.35** The following example shows the need for this definition. Figure 1.12 shows a sample space with 16 equiprobable elements and on which three events $\mathcal{A}$, $\mathcal{B}$, and $\mathcal{C}$, each with probability 1/2, are defined. Also, observe that

$$\begin{aligned}\text{Prob}\{\mathcal{A} \cap \mathcal{B}\} &= \text{Prob}\{\mathcal{A} \cap \mathcal{C}\} = \text{Prob}\{\mathcal{B} \cap \mathcal{C}\} \\ &= \text{Prob}\{\mathcal{A}\}\text{Prob}\{\mathcal{B}\} = \text{Prob}\{\mathcal{A}\}\text{Prob}\{\mathcal{C}\} = \text{Prob}\{\mathcal{B}\}\text{Prob}\{\mathcal{C}\} = 1/4\end{aligned}$$

and hence $\mathcal{A}$, $\mathcal{B}$, and $\mathcal{C}$ are *pairwise* independent.

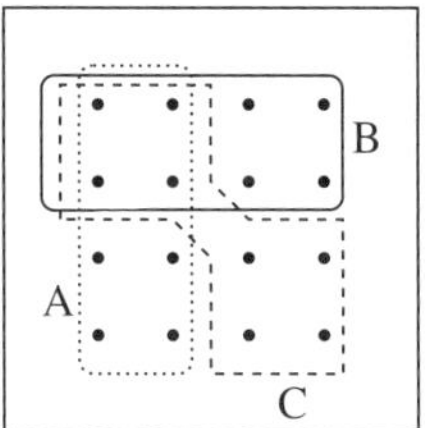


Figure 1.12. Sample space with 16 equiprobable elements.

However, they are not *mutually* independent since

$$\text{Prob}\{\mathcal{C}|\mathcal{A} \cap \mathcal{B}\} = \frac{\text{Prob}\{\mathcal{A} \cap \mathcal{B} \cap \mathcal{C}\}}{\text{Prob}\{\mathcal{A} \cap \mathcal{B}\}} = \frac{1/4}{1/4} = 1 \neq \text{Prob}\{\mathcal{C}\}.$$

Alternatively,

$$1/4 = \text{Prob}\{\mathcal{A} \cap \mathcal{B} \cap \mathcal{C}\} \neq \text{Prob}\{\mathcal{A}\}\text{Prob}\{\mathcal{B}\}\text{Prob}\{\mathcal{C}\} = 1/8.$$

In conclusion, we say that the three events $\mathcal{A}$, $\mathcal{B}$, and $\mathcal{C}$ defined above, are *not* independent; they are simply pairwise independent. Events $\mathcal{A}$, $\mathcal{B}$, and $\mathcal{C}$ are *mutually* independent only if *all* the following conditions hold:

$$\begin{aligned}\text{Prob}\{\mathcal{A} \cap \mathcal{B}\} &= \text{Prob}\{\mathcal{A}\}\text{Prob}\{\mathcal{B}\}, \\ \text{Prob}\{\mathcal{A} \cap \mathcal{C}\} &= \text{Prob}\{\mathcal{A}\}\text{Prob}\{\mathcal{C}\}, \\ \text{Prob}\{\mathcal{B} \cap \mathcal{C}\} &= \text{Prob}\{\mathcal{B}\}\text{Prob}\{\mathcal{C}\}, \\ \text{Prob}\{\mathcal{A} \cap \mathcal{B} \cap \mathcal{C}\} &= \text{Prob}\{\mathcal{A}\}\text{Prob}\{\mathcal{B}\}\text{Prob}\{\mathcal{C}\}.\end{aligned}$$

**Example 1.36** Consider a sample space that contains four equiprobable outcomes denoted $a, b, c,$ and $d$. Define three events on this sample space as follows: $\mathcal{A} = \{a, b\}$, $\mathcal{B} = \{a, b, c\}$, and $\mathcal{C} = \phi$. This time

$$\text{Prob}\{\mathcal{A} \cap \mathcal{B} \cap \mathcal{C}\} = 0 \quad \text{and} \quad \text{Prob}\{\mathcal{A}\}\text{Prob}\{\mathcal{B}\}\text{Prob}\{\mathcal{C}\} = 1/2 \times 3/4 \times 0 = 0$$

but

$$1/2 = \text{Prob}\{\mathcal{A} \cap \mathcal{B}\} \neq \text{Prob}\{\mathcal{A}\}\text{Prob}\{\mathcal{B}\} = 1/2 \times 3/4.$$

The events $\mathcal{A}$, $\mathcal{B}$, and $\mathcal{C}$ are not independent, nor even pairwise independent.

## 1.5 Law of Total Probability

If $\mathcal{A}$ is any event, then it is known that the intersection of $\mathcal{A}$ and the universal event $\Omega$ is $\mathcal{A}$. It is also known that an event $\mathcal{B}$ and its complement $\mathcal{B}^c$ constitute a partition. Thus

$$\mathcal{A} = \mathcal{A} \cap \Omega \quad \text{and} \quad \mathcal{B} \cup \mathcal{B}^c = \Omega.$$

Substituting the second of these into the first and then applying DeMorgan's law, we find

$$\mathcal{A} = \mathcal{A} \cap (\mathcal{B} \cup \mathcal{B}^c) = (\mathcal{A} \cap \mathcal{B}) \cup (\mathcal{A} \cap \mathcal{B}^c). \tag{1.4}$$

Notice that the events $(\mathcal{A} \cap \mathcal{B})$ and $(\mathcal{A} \cap \mathcal{B}^c)$ are mutually exclusive. This is illustrated in Figure 1.13 which shows that, since $\mathcal{B}$ and $\mathcal{B}^c$ cannot have any outcomes in common, the *intersection* of $\mathcal{A}$ and $\mathcal{B}$ cannot have any outcomes in common with the *intersection* of $\mathcal{A}$ and $\mathcal{B}^c$.

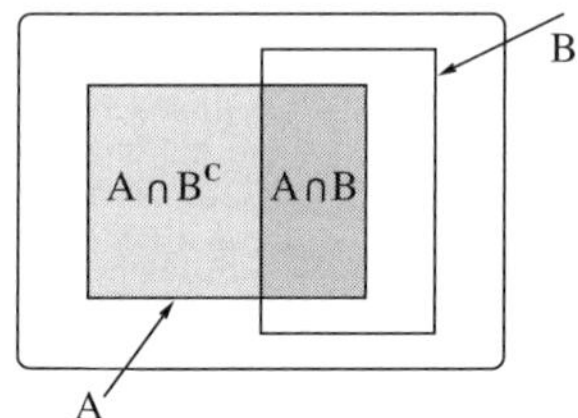


Figure 1.13. Events $(\mathcal{A} \cap \mathcal{B})$ and $(\mathcal{A} \cap \mathcal{B}^c)$ are mutually exclusive.

Returning to Equation (1.4), using the fact that $(\mathcal{A} \cap \mathcal{B})$ and $(\mathcal{A} \cap \mathcal{B}^c)$ are mutually exclusive, and applying Axiom 3, we obtain

$$\text{Prob}\{\mathcal{A}\} = \text{Prob}\{\mathcal{A} \cap \mathcal{B}\} + \text{Prob}\{\mathcal{A} \cap \mathcal{B}^c\}.$$

This means that to evaluate the probability of the event $\mathcal{A}$, it is sufficient to find the probabilities of the intersection of $\mathcal{A}$ with $\mathcal{B}$ and $\mathcal{A}$ with $\mathcal{B}^c$ and to add them together. This is frequently easier than trying to find the probability of $\mathcal{A}$ by some other method.

The same rule applies for any partition of the sample space and not just a partition defined by an event and its complement. Recall that a partition is a set of events that are mutually exclusive and collectively exhaustive. Let the $n$ events $\mathcal{B}_i,\ i = 1, 2, \ldots, n$, be a partition of the sample space $\Omega$. Then, for any event $\mathcal{A}$, we can write

$$\text{Prob}\{\mathcal{A}\} = \sum_{i=1}^{n} \text{Prob}\{\mathcal{A} \cap \mathcal{B}_i\}, \quad n \geq 1.$$

This is the *law of total probability*. To show that this law must hold, observe that the sets $\mathcal{A} \cap \mathcal{B}_i,\ i = 1, 2, \ldots, n$, are mutually exclusive (since the $\mathcal{B}_i$ are) and the fact that $\mathcal{B}_i,\ i = 1, 2, \ldots, n$, is a partition of $\Omega$ implies that

$$\mathcal{A} = \bigcup_{i=1}^{n} \mathcal{A} \cap \mathcal{B}_i, \quad n \geq 1.$$

Hence, using Axiom 3,

$$\text{Prob}\{\mathcal{A}\} = \text{Prob}\left\{ \bigcup_{i=1}^{n} \mathcal{A} \cap \mathcal{B}_i \right\} = \sum_{i=1}^{n} \text{Prob}\{\mathcal{A} \cap \mathcal{B}_i\}.$$

**Example 1.37** As an illustration, consider Figure 1.14, which shows a partition of a sample space containing 24 equiprobable outcomes into six events, $\mathcal{B}_1$ through $\mathcal{B}_6$.

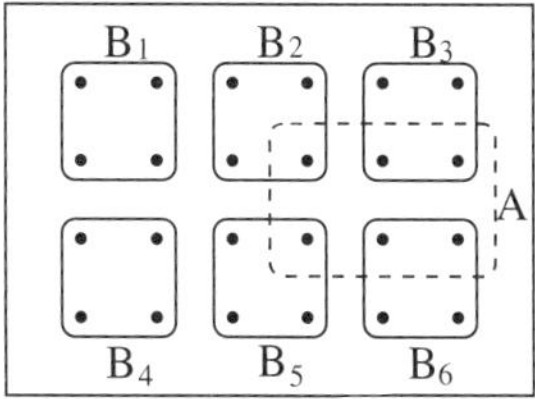


Figure 1.14. Law of total probability.

It follows then that the probability of the event $\mathcal{A}$ is equal to 1/4, since it contains six of the sample points. Because the events $\mathcal{B}_i$ constitute a partition, each point of $\mathcal{A}$ is in one and only one of the events $\mathcal{B}_i$ and the probability of event $\mathcal{A}$ can be found by adding the probabilities of the events $\mathcal{A} \cap \mathcal{B}_i$ for $i = 1, 2, \ldots, 6$. For this particular example it can be seen that these six probabilities are given by 0, 1/24, 1/12, 0, 1/24, and 1/12 which when added together gives 1/4.

The law of total probability is frequently presented in a different context, one that explicitly involves conditional probabilities. We have

$$\text{Prob}\{\mathcal{A}\} = \sum_{i=1}^{n} \text{Prob}\{\mathcal{A} \cap \mathcal{B}_i\} = \sum_{i=1}^{n} \text{Prob}\{\mathcal{A}|\mathcal{B}_i\}\text{Prob}\{\mathcal{B}_i\}, \tag{1.5}$$

which means that we can find Prob$\{\mathcal{A}\}$ by first finding the probability of $\mathcal{A}$ given $\mathcal{B}_i$, for all $i$, and then computing their weighted average. This often turns out to be a much more convenient way of computing the probability of the event $\mathcal{A}$, since in many instances we are provided with information concerning *conditional probabilities* of an event and we need to use Equation (1.5) to *remove* these conditions to find the unconditional probability, Prob$\{\mathcal{A}\}$.

To show that Equation (1.5) is true, observe that, since the events $\mathcal{B}_i$ form a partition, we must have $\bigcup_{i=1}^{n} \mathcal{B}_i = \Omega$ and hence

$$\mathcal{A} = \bigcup_{i=1}^{n} \mathcal{A} \cap \mathcal{B}_i.$$

Thus

$$\text{Prob}\{\mathcal{A}\} = \text{Prob}\left\{\bigcup_{i=1}^{n} \mathcal{A} \cap \mathcal{B}_i\right\} = \sum_{i=1}^{n} \text{Prob}\,\{\mathcal{A} \cap \mathcal{B}_i\} = \sum_{i=1}^{n} \frac{\text{Prob}\{\mathcal{A} \cap \mathcal{B}_i\}}{\text{Prob}\{\mathcal{B}_i\}}\text{Prob}\{\mathcal{B}_i\}$$

and the desired result follows.

**Example 1.38** Suppose three boxes contain a mixture of white and black balls. The first box contains 12 white and three black balls; the second contains four white and 16 black balls and the third contains six white and four black balls. A box is selected and a single ball is chosen from it. The choice of box is made according to a throw of a fair die. If the number of spots on the die is 1, the first box is selected. If the number of spots is 2 or 3, the second box is chosen; otherwise (the number of spots is equal to 4, 5, or 6) the third box is chosen. Suppose we wish to find Prob$\{\mathcal{A}\}$ where $\mathcal{A}$ is the event that a white ball is drawn.

In this case we shall base the partition on the three boxes. Specificially, let $\mathcal{B}_i$, $i = 1, 2, 3$, be the event that box $i$ is chosen. Then Prob$\{\mathcal{B}_1\} = 1/6$, Prob$\{\mathcal{B}_2\} = 2/6$, and Prob$\{\mathcal{B}_3\} = 3/6$. Applying the law of total probability, we have

$$\text{Prob}\{\mathcal{A}\} = \text{Prob}\{\mathcal{A}|\mathcal{B}_1\}\text{Prob}\{\mathcal{B}_1\} + \text{Prob}\{\mathcal{A}|\mathcal{B}_2\}\text{Prob}\{\mathcal{B}_2\} + \text{Prob}\{\mathcal{A}|\mathcal{B}_3\}\text{Prob}\{\mathcal{B}_3\},$$

which is easily computed using

$$\text{Prob}\{\mathcal{A}|\mathcal{B}_1\} = 12/15, \quad \text{Prob}\{\mathcal{A}|\mathcal{B}_2\} = 4/20, \quad \text{Prob}\{\mathcal{A}|\mathcal{B}_3\} = 6/10.$$

We have

$$\text{Prob}\{\mathcal{A}\} = 12/15 \times 1/6 + 4/20 \times 2/6 + 6/10 \times 3/6 = 1/2.$$

## 1.6 Bayes' Rule

It frequently happens that we are told that a certain event $\mathcal{A}$ has occurred and we would like to know which of the mutually exclusive and collectively exhaustive events $\mathcal{B}_j$ has occurred, at least probabilistically. In other words, we would like to know $\text{Prob}\{B_j|A\}$ for any $j$. Consider some oft-discussed examples. In one scenario, we may be told that among a certain population there are those who carry a specific disease and those who are disease-free. This provides us with the partition of the sample space (the population) into two disjoint sets. A certain, not entirely reliable, test may be performed on patients with the object of detecting the presence of this disease. If we know the ratio of diseased to disease-free patients and the reliability of the testing procedure, then given that a patient is declared to be disease-free by the testing procedure, we may wish to know the probability that the patient in fact actually has the disease (the probability that the patient falls into the first (or second) of the two disjoint sets). The same scenario may be obtained by substituting integrated circuit chips for the population, and partitioning it into defective and good chips, along with a tester which may sometimes declare a defective chip to be good and vice versa. Given that a chip is declared to be defective, we wish to know the probability that it is in fact defective. The transmission of data over a communication channel subject to noise is yet a third example. In this case the partition is the information that is sent (usually 0's and 1s) and the noise on the channel may or may not alter the data. Scenarios such as these are best answered using Bayes' rule.

We obtain Bayes' rule from our previous results on conditional probability and the theorem of total probability. We have

$$\text{Prob}\{B_j|A\} = \frac{\text{Prob}\{A \cap B_j\}}{\text{Prob}\{A\}} = \frac{\text{Prob}\{A|B_j\}\text{Prob}\{B_j\}}{\sum_i \text{Prob}\{A|B_i\}\text{Prob}\{B_i\}}.$$

Although it may seem that this complicates matters, what we are in fact doing is dividing the problem into simpler pieces. This becomes obvious in the following example where we choose a sample space partitioned into three events, rather than into two as is the case with the examples outlined above.

**Example 1.39** Consider a university professor who observes the students who come into his office with questions. This professor determines that 60% of the students are BSc students whereas 30% are MSc and only 10% are PhD students. The professor further notes that he can handle the questions of 80% of the BSc students in less than five minutes, whereas only 50% of the MSc students and 40% of the PhD students can be handled in five minutes or less. The next student to enter the professor's office needed only two minutes of the professor time. What is the probability that student was a PhD student?

To answer this question, we will let $\mathcal{B}_i$, $i = 1, 2, 3$, be the event "the student is a BSc, MSc, PhD" student, respectively, and we will let event $\mathcal{A}$ be the event "student requires five minutes or less." From the theorem of total probability, we have

$$\begin{aligned}\text{Prob}\{A\} &= \text{Prob}\{A|B_1\}\text{Prob}\{B_1\} + \text{Prob}\{A|B_2\}\text{Prob}\{B_2\} + \text{Prob}\{A|B_3\}\text{Prob}\{B_3\}\\ &= 0.8 \times 0.6 + 0.5 \times 0.3 + 0.4 \times 0.1 = 0.6700.\end{aligned}$$

This computation gives us the denominator for insertion into Bayes' rule. It tells us that approximately two-thirds of all students' questions can be handled in five minutes or less. What we would now like to compute is Prob$\{B_3|A\}$, which from Bayes' rule, is

$$\text{Prob}\{B_3|A\} = \frac{\text{Prob}\{A|B_3\}\text{Prob}\{B_3\}}{\text{Prob}\{A\}} = \frac{0.4 \times 0.1}{0.6700} = 0.0597,$$

or about 6% and is the answer that we seek.

The critical point in answering questions such as these is in determining which set of events constitutes the partition of the sample space. Valuable clues are usually found in the question posed. If we remember that we are asked to compute Prob$\{\mathcal{B}_j|\mathcal{A}\}$ and relate this to the words in the question, then it becomes apparent that the words "student was a PhD student" suggest a partition based on the status of the student, and the event $\mathcal{A}$, the information we are given, relates to the time taken by the student. The key is in understanding that we are given Prob$\{\mathcal{A}|\mathcal{B}_j\}$ and we are asked to find Prob$\{\mathcal{B}_j|\mathcal{A}\}$. In its simplest form, Bayes' law is written as

$$\text{Prob}\{B|A\} = \frac{\text{Prob}\{A|B\}\text{Prob}\{B\}}{\text{Prob}\{A\}}.$$

## 1.7 Exercises

**Exercise 1.1.1** A multiprocessing system contains six processors, each of which may be up and running, or down and in need of repair. Describe an element of the sample space and find the number of elements in the sample space. List the elements in the event $\mathcal{A}$ ="at least five processors are working."

**Exercise 1.1.2** A gum ball dispenser contains a large number of gum balls in three different colors, red, green, and yellow. Assuming that the gum balls are dispensed one at a time, describe an appropriate sample space for this scenario and list all possible events.

A determined child continues to buy gum balls until he gets a yellow one. Describe an appropriate sample space in this case.

**Exercise 1.1.3** A brother and a sister arrive at the gum ball dispenser of the previous question, and each of them buys a single gum ball. The boy always allows his sister to go first. Let $\mathcal{A}$ be the event that the girl gets a yellow gum ball and let $\mathcal{B}$ be the event that at least one of them gets a yellow gum ball.

(a) Describe an appropriate sample space in this case.
(b) What outcomes constitute event $\mathcal{A}$?
(c) What outcomes constitute event $\mathcal{B}$?
(d) What outcomes constitute event $\mathcal{A} \cap \mathcal{B}$?
(e) What outcomes constitute event $\mathcal{A} \cap \mathcal{B}^c$?
(f) What outcomes constitute event $\mathcal{B} - \mathcal{A}$?

**Exercise 1.1.4** The mail that arrives at our house is for father, mother, or children and may be categorized into junk mail, bills, or personal letters. The family scrutinizes each piece of incoming mail and observes that it is one of nine types, from *jf* (junk mail for father) through *pc* (personal letter for children). Thus, in terms of trials and outcomes, each trial is an examination of a letter and each outcome is a two-letter word.

(a) What is the sample space of this experiment?
(b) Let $\mathcal{A}_1$ be the event "junk mail." What outcomes constitute event $\mathcal{A}_1$?
(c) Let $\mathcal{A}_2$ be the event "mail for children." What outcomes constitute event $\mathcal{A}_2$?
(d) Let $\mathcal{A}_3$ be the event "not personal." What outcomes constitute event $\mathcal{A}_3$?
(e) Let $\mathcal{A}_4$ be the event "mail for parents." What outcomes constitute event $\mathcal{A}_4$?
(f) Are events $\mathcal{A}_2$ and $\mathcal{A}_4$ mutually exclusive?
(g) Are events $\mathcal{A}_1$, $\mathcal{A}_2$, and $\mathcal{A}_3$ collectively exhaustive?
(h) Which events imply another?

**Exercise 1.1.5** Consider an experiment in which three different coins (say a penny, a nickel, and a dime in that order) are tossed and the sequence of heads and tails observed. For each of the following pairs of events,

$\mathcal{A}$ and $\mathcal{B}$, give the subset of outcomes that defines the events and state whether the pair of events are mutually exclusive, collectively exhaustive, neither or both.

(a) $\mathcal{A}$: The penny comes up heads. $\mathcal{B}$: The penny comes up tails.
(b) $\mathcal{A}$: The penny comes up heads. $\mathcal{B}$: The dime comes up tails.
(c) $\mathcal{A}$: At least one of the coins shows heads. $\mathcal{B}$: At least one of the coins shows tails.
(d) $\mathcal{A}$: There is exactly one head showing. $\mathcal{B}$: There is exactly one tail showing.
(e) $\mathcal{A}$: Two or more heads occur. $\mathcal{B}$: Two or more tails occur.

**Exercise 1.1.6** A brand new light bulb is placed in a socket and the time it takes until it burns out is measured. Describe an appropriate sample space for this experiment. Use mathematical set notation to describe the following events:

(a) $\mathcal{A} =$ the light bulb lasts at least 100 hours.
(b) $\mathcal{B} =$ the light bulb lasts between 120 and 160 hours.
(c) $\mathcal{C} =$ the light bulb lasts less than 200 hours.
(d) $\mathcal{A} \cap \mathcal{C}^c$.

**Exercise 1.2.1** An unbiased die is thrown once. Compute the probability of the following events.

(a) $\mathcal{A}_1$ : The number of spots shown is odd.
(b) $\mathcal{A}_2$ : The number of spots shown is less than 3.
(c) $\mathcal{A}_3$ : The number of spots shown is a prime number.

**Exercise 1.2.2** Two unbiased dice are thrown simultaneously. Describe an appropriate sample space and specify the probability that should be assigned to each. Also, find the probability of the following events:

(a) $\mathcal{A}_1$ : The number on each die is equal to 1.
(b) $\mathcal{A}_2$ : The sum of the spots on the two dice is equal to 3.
(c) $\mathcal{A}_3$ : The sum of the spots on the two dice is greater than 10.

**Exercise 1.2.3** A card is drawn from a standard pack of 52 well-shuffled cards. What is the probability that it is a king? Without replacing this first king card, a second card is drawn. What is the probability that the second card pulled is a king? What is the probability that the first four cards drawn from a standard deck of 52 well-shuffled cards are all kings. Once drawn, a card is not replaced in the deck.

**Exercise 1.2.4** Prove the following relationships.

(a) $\text{Prob}\{\mathcal{A} \cup \mathcal{B}\} = \text{Prob}\{\mathcal{A}\} + \text{Prob}\{\mathcal{B}\} - \text{Prob}\{\mathcal{A} \cap \mathcal{B}\}$.
(b) $\text{Prob}\{\mathcal{A} \cap \mathcal{B}^c\} = \text{Prob}\{\mathcal{A} \cup \mathcal{B}\} - \text{Prob}\{\mathcal{B}\}$.

**Exercise 1.2.5** A card is drawn at random from a standard deck of 52 well-shuffled cards. Let $\mathcal{A}$ be the event that the card drawn is a queen and let $\mathcal{B}$ be the event that the card pulled is red. Find the probabilities of the following events and state in words what they represent.

(a) $\mathcal{A} \cap \mathcal{B}$.
(b) $\mathcal{A} \cup \mathcal{B}$.
(c) $\mathcal{B} - \mathcal{A}$.

**Exercise 1.2.6** A university professor drives from his home in Cary to his university office in Raleigh each day. His car, which is rather old, fails to start one out of every eight times and he ends up taking his wife's car. Furthermore, the rate of growth of Cary is so high that traffic problems are common. The professor finds that 70% of the time, traffic is so bad that he is forced to drive fast his preferred exit off the beltline, Western Boulevard, and take the next exit, Hillsborough street. What is the probability of seeing this professor driving to his office along Hillsborough street, in his wife's car?

**Exercise 1.2.7** A prisoner in a Kafkaesque prison is put in the following situation. A regular deck of 52 cards is placed in front of him. He must choose cards one at a time to determine their color. Once chosen, the card is replaced in the deck and the deck is shuffled. If the prisoner happens to select three consecutive red cards, he is executed. If he happens to selects six cards before three consecutive red cards appear, he is granted freedom. What is the probability that the prisoner is executed.

**Exercise 1.2.8** Three marksmen fire simultaneously and independently at a target. What is the probability of the target being hit at least once, given that marksman one hits a target nine times out of ten, marksman two hits a target eight times out of ten while marksman three only hits a target one out of every two times.

**Exercise 1.2.9** Fifty teams compete in a student programming competition. It has been observed that 60% of the teams use the programming language C while the others use C++, and experience has shown that teams who program in C are twice as likely to win as those who use C++. Furthermore, ten teams who use C++ include a graduate student, while only four of those who use C include a graduate student.

(a) What is the probability that the winning team programs in C?
(b) What is the probability that the winning team programs in C and includes a graduate student?
(c) What is the probability that the winning team includes a graduate student?
(d) Given that the winning team includes a graduate student, what is the probability that team programmed in C?

**Exercise 1.3.1** Let $\mathcal{A}$ be the event that an odd number of spots comes up when a fair die is thrown, and let $\mathcal{B}$ be the event that the number of spots is a prime number. What is Prob$\{\mathcal{A}|\mathcal{B}\}$ and Prob$\{\mathcal{B}|\mathcal{A}\}$?

**Exercise 1.3.2** A card is drawn from a well-shuffled standard deck of 52 cards. Let $\mathcal{A}$ be the event that the chosen card, is a heart, let $\mathcal{B}$ be the event that it is a black card, and let $\mathcal{C}$ be the event that the chosen card is a red queen. Find Prob$\{\mathcal{A}|\mathcal{C}\}$ and Prob$\{\mathcal{B}|\mathcal{C}\}$. Which of the events $\mathcal{A}$, $\mathcal{B}$, and $\mathcal{C}$ are mutually exclusive?

**Exercise 1.3.3** A family has three children. What is the probability that all three children are boys? What is the probability that there are two girls and one boy? Given that at least one of the three is a boy, what is the probability that all three children are boys. You should assume that Prob$\{boy\}$ = Prob$\{girl\} = 1/2$.

**Exercise 1.3.4** Three cards are placed in a box; one is white on both sides, one is black on both sides, and the third is white on one side and black on the other. One card is chosen at random from the box and placed on a table. The (uppermost) face that shows is white. Explain why the probability that the hidden face is black is equal to 1/3 and not 1/2.

**Exercise 1.4.1** If Prob$\{\mathcal{A} \mid \mathcal{B}\}$ = Prob$\{\mathcal{B}\}$ = Prob$\{\mathcal{A} \cup \mathcal{B}\} = 1/2$, are $\mathcal{A}$ and $\mathcal{B}$ independent?

**Exercise 1.4.2** A flashlight contains two batteries that sit one on top of the other. These batteries come from different batches and may be assumed to be independent of one another. Both batteries must work in order for the flashlight to work. If the probability that the first battery is defective is 0.05 and the probability that the second is defective is 0.15, what is the probability that the flashlight works properly?

**Exercise 1.4.3** A spelunker enters a cave with two flashlights, one that contains three batteries in series (one on top of the other) and another that contains two batteries in series. Assume that all batteries are independent and that each will work with probability 0.9. Find the probability that the spelunker will have some means of illumination during his expedition.

**Exercise 1.5.1** Six boxes contain white and black balls. Specifically, each box contains exactly one white ball; also box $i$ contains $i$ black balls, for $i = 1, 2, \ldots, 6$. A fair die is tossed and a ball is selected from the box whose number is given by the die. What is the probability that a white ball is selected?

**Exercise 1.5.2** A card is chosen at random from a deck of 52 cards and inserted into a second deck of 52 well-shuffled cards. A card is now selected at random from this augmented deck of 53 cards. Show that the probability of this card being a queen is exactly the same as the probability of drawing a queen from the first deck of 52 cards.

**Exercise 1.5.3** A factory has three machines that manufacture widgets. The percentages of a total day's production manufactured by the machines are 10%, 35%, and 55%, respectively. Furthermore, it is known that 5%, 3%, and 1% of the outputs of the respective three machines are defective. What is the probability that a randomly selected widget at the end of the day's production runs will be defective?

**Exercise 1.5.4** A computer game requires a player to find safe haven in a secure location where her enemies cannot penetrate. Four doorways appear before the player, from which she must choose to enter one and only one. The player must then make a second choice from among two, four, one, or five potholes to descend, respectively depending on which door she walks through. In each case one pothole leads to the safe haven.

The player is rushed into making a decision and in her haste makes choices randomly. What is the probability of her safely reaching the haven?

**Exercise 1.5.5** The first of two boxes contains $b_1$ blue balls and $r_1$ red balls; the second contains $b_2$ blue balls and $r_2$ red balls. One ball is randomly chosen from the first box and put into the second. When this has been accomplished, a ball is chosen at random from the second box and put into the first. A ball is now chosen from the first box. What is the probability that it is blue?

**Exercise 1.6.1** Returning to Exercise 1.5.1, given that the selected ball is white, what is the probability that it came from box 1?

**Exercise 1.6.2** In the scenario of Exercise 1.5.3, what is the probability that a defective, randomly selected widget was produced by the first machine? What is the probability that it was produced by the second machine. And the third?

**Exercise 1.6.3** A bag contains two fair coins and one two-headed coin. One coin is randomly selected, tossed three times, and three heads are obtained. What is the probability that the chosen coin is the two-headed coin?

**Exercise 1.6.4** A most unusual Irish pub serves only Guinness and Harp. The owner of this pub observes that 85% of his male customers drink Guinness as opposed to 35% of his female customers. On any given evening, this pub owner notes that there are three times as many males as females. What is the probability that the person sitting beside the fireplace drinking Guinness is female?

**Exercise 1.6.5** Historically on St. Patrick's day (March 17), the probability that it rains on the Dublin parade is 0.75. Two television stations are noted for their weather forecasting abilities. The first, which is correct nine times out of ten, says that it will rain on the upcoming parade; the second, which is correct eleven times out of twelve, says that it will not rain. What is the probability that it will rain on the upcoming St. Patrick's day parade?

**Exercise 1.6.6** 80% of the murders committed in a certain town are committed by men. A dead body with a single gunshot wound in the head has just been found. Two detectives examine the evidence. The first detective, who is right seven times out of ten, announces that the murderer was a male but the second detective, who is right three times out of four, says that the murder was committed by a woman. What is the probability that the author of the crime was a woman?

# Chapter 2

# *Combinatorics—The Art of Counting*

In many of the examples we have seen so far, it has been necessary to determine the number of ways in which a given event can occur, and hence, knowing the probability of each of these occurrences and Axiom 3, we can work out the probability of the event. In this section, we shall consider techniques, based on the law of total probability, for counting all possibilities. It is usual to describe this problem in terms of how *distinguishable* balls may be chosen from a box and it is also usual to assume that the choice is random, i.e., each ball in the box is equally likely to be chosen. Alternatively, the problem may be phrased in terms of how *indistinguishable* balls may be inserted into *distinguishable* boxes. The former is called the *selection problem*; the latter is called the *allocation problem*. We shall formulate our discussion in terms of the selection problem.

In the context of the selection problem, once one of the distinguishable balls has been chosen from the box, and prior to the next ball being selected, a decision must be taken as to what should be done with the first ball chosen. If it is put back into the box, it is said that the selection is made *with replacement*. In this case, the same ball may be chosen a second time, then a third time, and a fourth time, and so on. The second possibility is that the ball is set aside and never returned to the box. In this case, the selection is made *without replacement*.

A final point concerns the *order* in which the balls are selected. In some cases, this order is important. For example, it may be necessary to know whether the black ball was chosen before or after the white ball. When the order is important, the term used is *permutation*. In other cases, all that is needed is to know that both a black ball and a white ball were chosen, and not the order in which they were chosen. This is known as a *combination*. Combinations do not distinguish selections by their order: permutations do.

## 2.1 Permutations

An arrangement of items, also called an *ordered sequence* of items, is said to be a *permutation* of the items. An ordered sequence of $n$ items is called an $n$*-permutation*. With $n$ *distinct* items, there are $n!$ permutations possible. This is the number of different ways in which the $n$ distinct items can be arranged.

**Example 2.1** Given three different letters $A$, $B$, and $C$ there are $3! = 6$ permutations (or arrangements), namely,

$$ABC,\ ACB,\ BAC,\ BCA,\ CAB,\ CBA.$$

Notice that, not only there are three distinct items $A$, $B$, and $C$, but there are also three different places into which they may be placed, i.e., the first, second, and third positions of a three-letter string. For this reason, $ABC$ is considered to be a different permutation from $ACB$.

Now consider the number of permutations that may be generated from items that are not all distinct.

**Example 2.2** Consider the word *OXO*. This time not all letters are distinct. There are two $O$'s and one $X$, and only three different arrangements can be found, namely,

$$OXO, \ XOO, \ OOX.$$

Since the character $O$ appears twice, the number $n! = 3!$ obtained with $n = 3$ distinct characters must be divided by 2!. In this case, we have $3!/2! = 3$ arrangements.

**Example 2.3** Consider the word *PUPPY*. If all five characters were distinct, the number of permutations that would be obtained is 5!. However, since $P$ appears three times in *PUPPY*, the 5! must be divided by 3! to yield $5!/3! = 5 \times 4 = 20$. The number of different ways in which the characters of the word *PUPPY* can be arranged is therefore 20.

**Example 2.4** As a final example, consider the word *NEEDED*. It contains three $E$'s, two $D$'s, and one $N$. The number of different ways in which the characters in this word may be arranged is

$$\frac{6!}{3! \times 2!} = 60.$$

Thus, the number of permutations that may be obtained from $n$ items, not all of which are distinct, may be obtained by first assuming them to be distinct (which gives $n!$), and then, for each multiple item, dividing by the factorial of its multiplicity. We are now ready to examine the various versions of the selection problem. We shall first consider permutations with and without replacement and then consider combinations without and with replacements.

## 2.2 Permutations with Replacements

In the case of permutations with replacement, our goal is to count the number of ways in which $k$ balls may be selected from among $n$ distinguishable balls. After each ball is chosen and its characteristics recorded, it is replaced in the box and the next ball is chosen. An alternative way to state this is to say that after each ball is selected, that ball is set aside and a ball that is identical to it takes its place in the box. In this way it is possible for the same ball to be chosen many times. If $k = 1$, i.e., only one ball is to be chosen, then the number of possible permutations is $n$, since any of the $n$ balls may be chosen. If $k = 2$, then any of the $n$ balls may be chosen as the first ball and then replaced in the box. For each of these $n$ choices for first ball, the next ball chosen may also be any of the $n$ distinguishable balls, and hence the number of permutations when $k = 2$ is $n \times n = n^2$. It is evident that this reasoning may be continued to show that the number of permutations obtained for any $k$ is $n^k$.

**Example 2.5** The number of four-digit codes that can be obtained using the decimal number system (with ten digits from 0 to 9 inclusive) is $10^4 = 10{,}000$. These are the codes ranging from 0000 through 9999.

Suppose now that there are $n_1$ ways of choosing a first item, and $n_2$ ways of choosing a second item. Then the number of distinct ordered pairs is equal to $n_1 n_2$. In terms of distinguishable balls in boxes, this may be viewed as the number of ways in which one ball may be chosen from a first box containing $n_1$ distinguishable balls and a second ball chosen from a second box containing $n_2$ distinguishable balls. The extension to more than two different items is immediate. If there are $n_i$ ways of choosing an $i^{\text{th}}$ item, for $i = 1, 2, \ldots, k$, then the number of distinct ordered $k$-tuples is equal to $\prod_{i=1}^{k} n_i = n_1 n_2 \ldots n_k$.

**Example 2.6** Suppose a shirt may be chosen from among $n_1 = 12$ different shirts and a tie from among $n_2 = 20$ different ties. Then the number of shirt and tie combinations is $n_1 n_2 = 240$.

## 2.3 Permutations without Replacement

We now move on to the problem of counting the number of different permutations obtained in selecting $k$ balls from among $n$ distinguishable balls in the case that once a particular ball is chosen, it is *not* returned to the box. This is equal to the number of *ordered sequences* of $k$ distinguishable objects, i.e., the number of *k-permutations*, that can be obtained from $n$ distinguishable objects. We denote this number $P(n, k)$ and assign the values $P(n, 0) = 1,\ n = 0, 1, \ldots,$ by convention. We say that $P(n, k)$ is the number of permutations of $n$ objects taken $k$ at a time.

If $k = 1$, then any ball may be chosen and the total number of permutations is just $n$. If $k = 2$, then any of the $n$ balls may be chosen as the first. For each of these $n$ possible choices, there remain $n - 1$ balls in the box, any one of which may be chosen as the second. Thus the total number of permutations for $k = 2$ is equal to $n(n - 1)$. With $k = 3$, there are $n$ different possibilities for the first ball, but only $n-1$ for the second and $n-2$ for the third, which gives $P(n, 3) = n(n-1)(n-2)$. We may now generalize this to any arbitrary $k \le n$ to obtain

$$P(n, k) = n(n-1)(n-2)\cdots(n-k+1) = \frac{n!}{(n-k)!}.$$

A number of relationships exist among the values of $P(n, k)$ for different values of the parameters $n$ and $k$. For example,

$$P(n, k) = nP(n-1, k-1), \quad k = 1, 2, \ldots, n, \quad n = 1, 2, \ldots,$$

for all permissible $n$ and $k$. This formula may be obtained by observing that

$$P(n, k) = \frac{n!}{(n-k)!} = n\frac{(n-1)!}{[(n-1)-(k-1)]!} = nP(n-1, k-1).$$

It may also be arrived by simply reasoning as follows. The first ball chosen can be any of the $n$ distinguishable balls. This leaves $n - 1$ balls from which $k - 1$ must still be chosen: the number of ways in which $k - 1$ balls may be chosen from $n - 1$ is equal to $P(n - 1, k - 1)$.

**Example 2.7** Let $n = 4$, $k = 3$ and distinguish the four ball by means of the four letters $A$, $B$, $C$, and $D$. We have

$$P(4, 3) = 4 \times 3 \times 2 = 24.$$

These 24 different possibilities are given as

$$\begin{array}{cccccc} ABC, & ABD, & ACB, & ACD, & ADB, & ADC, \\ BAC, & BAD, & BCA, & BCD, & BDA, & BDC, \\ CAB, & CAD, & CBA, & CBD, & CDA, & CDB, \\ DAB, & DAC, & DBA, & DBC, & DCA, & DCB. \end{array}$$

Notice that we distinguish between, for example, $ABC$ and $ACB$, since the order is important, but we do not include the same letter more than once (e.g., $AAB$ is not present), since letters once used cannot be used again.

**Example 2.8** Suppose we wish to find the number of ways that a four-digit code, in which all the digits are different, can be selected. Given that we work with ten digits (0 through 9), the first digit can be any of the ten, the second any of the remaining nine, and so on. Continuing in this fashion, we see that the total number of permutations is given as $10 \times 9 \times 8 \times 7$, which naturally is the same value when computed directly from the formula.

$$P(10, 4) = \frac{10!}{(10-4)!} = 5040.$$

Problems such as these are frequently formulated as probability questions, requesting one, for example, to find the probability that a randomly generated four-digit code has all different digits. As we have just seen, the total number of four-digit codes having all digits different is 5,040, while the total number of four-digit numbers is 10,000. If it is assumed that all choices of four-digit codes are equally likely, then the probability of getting one in which all the digits are different is 5,040/10,000 or about 0.5. Such questions require us to find the size of the sample space of permutations without replacement and to compare it to the size of the sample space with replacement. Obviously, this only works correctly when all the choices are equally likely (random selection). Two further example will help clarify this.

**Example 2.9** Suppose $r$ pigeons returning from a race are equally likely to enter any of $n$ homing nests, where $n > r$. We would like to know the probability of all pigeons ending up in different nests?

Let $k_1$ be the nest entered by the first pigeon, $k_2$ that entered by the second pigeon, and so on. We have $1 \leq k_i \leq n$, for $i = 1, 2, \ldots, r$. The total number of possibilities available to the pigeons is equal to $N = n^r$, since pigeon number 1 can fly into any of the $n$ nests, and the same for pigeon number 2, and so on. This is the situation of permutations *with* replacement. We are told that each of these $N$ possible choices is equally likely. There are thus $n^r$ distinct and equiprobable arrangements of the $r$ pigeons in the $n$ nests. These $n^r$ events are the elementary events that constitute the sample space of the experiment.

Let $\mathcal{A}$ denote the event that all pigeons end up in different nests. This event occurs if all the $k_i, i = 1, 2, \ldots, k_r$, are distinct: each pigeon flies into a different nest. This is the situation of permutation without replacement. The first pigeon may enter any one of the $n$ nests, the second may enter any of the remaining $n - 1$ nests, and so on. The number of possible ways in which the $r$ pigeons can be arranged in the $n$ nests so that none share a nest is therefore given by $n(n-1)\cdots(n-r+1)$. This is the number of outcomes that result in the event $\mathcal{A}$, and since there are a total of $n^r$ equiprobable outcomes in the sample space, the probability of $\mathcal{A}$ must be

$$\text{Prob}\{\mathcal{A}\} = \frac{n(n-1)\cdots(n-r+1)}{n^r} = \frac{n!/(n-r)!}{n^r} = \frac{\text{w/o replacement}}{\text{w/ replacement}}.$$

**Example 2.10** Consider the well-known birthday problem, that of determining the likelihood that among a group of $k$ persons at least two have the same birthday (day and month only). To find the probability that at least two people have the same birthday, it is easier to compute the complement of this event, the probability that no two individuals have the same birthday. If we assume that a year has 365 days, then the number of permutations with replacement is $365^k$. This is the size of the sample space from which all possible birthday permutations are drawn. The number of permutations without replacement is $365!/(365-k)!$. This is the number of outcomes in the sample space in which no two birthdays in the permutation are the same. The probability of finding among the $k$ individuals, no two with the same birthday is the ratio of these two, i.e.,

$$\begin{aligned}\frac{365!/(365-k)!}{365^k} &= \frac{365}{365} \times \frac{364}{365} \times \frac{363}{365} \times \cdots \times \frac{365-k+1}{365} \\ &= \left(1 - \frac{1}{365}\right)\left(1 - \frac{2}{365}\right)\cdots\left(1 - \frac{k-1}{365}\right).\end{aligned}$$

It may be observed that this probability is less than 0.5 for what may appear to be a relatively small value of $k$, namely, $k = 23$. In other words, in a class with 23 students, the likelihood that no two students have the same birthday is less than 50%.

## 2.4 Combinations without Replacement

We now turn to the case in which balls are selected without replacement, but the order in which they are selected is unimportant. All that matters is that certain balls have been chosen and the point at which any one was selected is not of interest. For example, we may have chosen a green ball, a red ball, and a black ball, but we no longer know or care which of these three was chosen first, second, or third. We shall let $C(n, k)$ denote the number of ways in which $k$ balls may be selected without replacement and irrespective of their order, from a box containing $n$ balls. This is called the number of combinations of $n$ items taken $k$ at a time and it is understood to be without regard for order.

As we saw previously, any collection of $k$ distinct items may be placed into $k!$ different permutations. For example, with $k = 3$ and using the letters $ABC$, we have the $3! = 6$ permutations

$$ABC,\ ACB,\ BAC,\ BCA,\ CAB,\ CBA.$$

These are all different permutations because the order is important. If the order is not important, then the only information we need is that there is an $A$, a $B$, and a $C$. All $3!$ permutations yield only a single combination. This provides us with a means to relate permutations without replacement to combinations without replacement. Given that any sequence of $k$ items may be arranged into $k!$ permutations, it follows that there must be $k!$ times as many permutations (without replacement) of $k$ distinct items as there are combinations (without replacement), since the actual order matters in permutations but not in combinations. In other words, we must have

$$k!C(n, k) = P(n, k) \quad \text{for } n = 0, 1, 2, \ldots \text{ and } k = 0, 1, \ldots, n.$$

This leads to

$$C(n, k) = \frac{P(n, k)}{k!} = \frac{n!}{k!(n-k)!} \quad \text{for } n = 0, 1, 2, \ldots \text{ and } k = 0, 1, \ldots, n.$$

**Example 2.11** Let $A$, $B$, $C$, $D$, and $E$ be five distinguishable items from which two are to be chosen (which implicitly implies, *without replacement*, and it matters not which one comes first: i.e., combination). The number of ways that these two can be chosen is given by $C(5, 2) = 5!/(2!3!) = 10$. These are

$$\begin{array}{cccc} AB & AC & AD & AE \\ & BC & BD & BE \\ & & CD & CE \\ & & & DE. \end{array}$$

If the choices are made randomly, then the probability of getting any particular one of these ten combinations is 1/10. Let us compute the probability of choosing exactly one of $(A, B)$ and one of $(C, D, E)$. This is the event that contains the outcomes $\{AC, AD, AE, BC, BD, BE\}$. The number of ways in which such a combination can appear is the product of the number of ways in which we can choose one from two (i.e., $C(2, 1)$) times the number of ways we can choose one from the remaining three (i.e., $C(3, 1)$). The probability of such a choice is then

$$\frac{C(2, 1) \times C(3, 1)}{C(5, 2)} = \frac{2!/(1!1!) \times 3!/(1!2!)}{5!/(2!3!)} = \frac{2 \times 3}{10} = 0.6.$$

$C(n, k)$ is called a *binomial coefficient* since, as we shall see in the next section, it is the coefficient of the term $p^k q^{n-k}$ in the expansion of the binomial $(p+q)^n$. Alternative ways for writing this binomial coefficient are

$$C(n, k) = C_k^n = \binom{n}{k},$$

and we shall use all three, depending on the context. It is read as "$n$ choose $k$."

To compute binomial coefficients, notice that

$$\binom{n}{k} = \frac{n!}{k!(n-k)!} = \frac{n}{k}\left(\frac{(n-1)!}{(k-1)!(n-k)!}\right)$$
$$= \frac{n}{k}\left(\frac{(n-1)!}{(k-1)!([n-1]-[k-1])!}\right) = \frac{n}{k}\binom{n-1}{k-1},$$

which eventually leads to

$$\binom{n}{k} = \frac{n}{k} \times \frac{n-1}{k-1} \times \cdots \times \frac{n-(k-1)}{k-(k-1)} = \prod_{j=1}^{k} \frac{n-j+1}{j},$$

and is computationally more robust than forming factorials and taking a ratio.

The binomial coefficients possess a large number of interesting properties, only four of which we list below. Their proof and interpretation is left as an exercise. For $0 \leq j \leq k \leq n$:

(a)
$$\binom{n}{k} = \binom{n}{n-k}.$$

(b)
$$\binom{n}{k} = \frac{n}{k}\binom{n-1}{k-1}.$$

(c)
$$\binom{n}{k} = \binom{n-1}{k} + \binom{n-1}{k-1}.$$

(d)
$$\binom{n}{k}\binom{k}{j} = \binom{n}{j}\binom{n-j}{k-j}.$$

Consider now the problem of placing $k$ distinguishable balls into $n$ different boxes in such a way that the number of balls in box $i$ is $k_i$, for $i = 1, 2, \ldots, n$. We assume that $\sum_{i=1}^{n} k_i = k$, so that each ball is put into one of the boxes and none are left over. The number of combinations obtained in selecting the first $k_1$ balls for box 1 is given as $C(k, k_1)$. This leaves $k - k_1$ balls, from which $k_2$ are chosen and put into box 2. The number of combinations obtained in selecting these $k_2$ from $k - k_1$ balls is $C(k - k_1, k_2)$, so that the total obtained with the first two boxes is $C(k, k_1) \times C(k - k_1, k_2)$. This now leaves $k - k_1 - k_2$ from which $k_3$ must be chosen and put into box 3. Continuing in this fashion, we see that the total number of combinations is given by

$$C(k, k_1) \times C(k - k_1, k_2) \times C(k - k_1 - k_2, k_3) \times \cdots \times C\left(k - \sum_{i=1}^{n-1} k_i, k_n\right),$$

where $C(k - \sum_{i=1}^{n-1} k_i, k_n) = C(k_n, k_n)$. Substituting in the formula for the number of combinations, we have

$$\frac{k!}{k_1!(k-k_1)!} \times \frac{(k-k_1)!}{k_2!(k-k_1-k_2)!} \times \frac{(k-k_1-k_2)!}{k_3!(k-k_1-k_2-k_3)!} \times \cdots$$
$$= \frac{k!}{k_1!k_2!\cdots k_n!} \equiv \binom{k}{k_1, k_2, \ldots, k_n}. \tag{2.1}$$

These are called the *multinomial coefficients*. Observe that, when $k = k_1 + k_2$, we have

$$\binom{k}{k_1, k_2} = \binom{k}{k_1} = \binom{k}{k_2}.$$

**Example 2.12** Let us compute the number of ways in which five cards can be dealt from a regular deck of 52 cards. Since it does not matter in which order the cards are drawn, our concern is with combinations rather than permutations so the answer is $C(52, 5) = 2,598,960$. Let $\mathcal{K}$ be the event that there are exactly three kings among the selected cards. The number of outcomes that result in event $\mathcal{K}$ is the product of $C(4, 3)$ and $C(48, 2)$ since three kings must be drawn from four cards and two cards from the remaining 48. The probability of obtaining exactly three kings is

$$\frac{C(4, 3) \times C(48, 2)}{C(52, 5)} = 0.001736.$$

**Example 2.13** During quality control of a batch of 144 widgets, 12 are chosen at random and inspected. If any of the 12 is defective, the batch is rejected; otherwise it is accepted. We wish to compute the probability that a batch containing 10 defective widgets is accepted.

The number of ways in which the inspector can choose the 12 widgets for inspection is the number of combinations of 144 items taken 12 at a time, and is thus equal to

$$N = C(144, 12) = \frac{144!}{12!\,132!}.$$

We are told that these are all equiprobable, since the inspector chooses the 12 widgets at random.

Let $\mathcal{A}$ be the event that the batch is accepted, i.e., none of the 10 defective widgets appears in the sample of 12 chosen by the inspector. This means that all 12 selected widgets belong to the 134 good ones. The number of ways in which this can happen, denoted by $N(\mathcal{A})$, is given as

$$N(\mathcal{A}) = C(134, 12) = \frac{134!}{12!\,122!}.$$

It now follows that

$$\text{Prob}\{\mathcal{A}\} = \frac{N(\mathcal{A})}{N} = \frac{134!\,12!\,132!}{12!\,122!\,144!} = 0.4066.$$

## 2.5 Combinations with Replacements

It only remains to determine the number of combinations possible when $k$ balls are selected *with replacement* from a box containing $n$ distinguishable balls. It may be shown that this is identical to the problem of counting the number of combinations when $k$ balls are selected *without* replacement from a box containing a total of $n + k - 1$ distinguishable balls, i.e.,

$$\frac{(n + k - 1)!}{(n - 1)!\,k!}.$$

It is useful to describe different scenarios that result in the same number of combinations. Consider, for example, the one we have been using so far, that of counting the number of different combinations possible in selecting $k$ balls from a box containing $n$ distinguishable balls. As an example, we shall let $n = 4$ and differentiate among the balls by calling them $A$, $B$, $C$, and $D$. If we choose $k = 3$ then the formula tells us that there are $(4 + 3 - 1)!/(3! \times 3!) = 20$ different combinations. These are

$$\begin{array}{cccccccccc} AAA & AAB & AAC & AAD & ABB & ABC & ABD & ACC & ACD & ADD \\ & & & & BBB & BBC & BBD & BCC & BCD & BDD \\ & & & & & & & CCC & CCD & CDD \\ & & & & & & & & & DDD. \end{array}$$

Notice that, although the four balls are distinguishable, the same ball may occur more than once in a given combination. This is because after a ball is chosen, it is replaced and may be chosen once again. Notice also that we do not include combinations such as *BAA* or *CBA* because with combinations, the order is unimportant and these are equivalent to *AAB* and *ABC*, respectively.

A second equivalent scenario changes things around. It may be shown that counting the number of combinations in selecting $k$ balls from a box of $n$ distinguishable balls is equivalent to counting the number of combinations obtained in distributing $k$ *indistinguishable* balls among $n$ *distinguishable* boxes. This is the assignment problem mentioned previously. Consider the example with $n = 4$ boxes and $k = 3$ balls and let the four distinguishable boxes be represented by a vector of four integer components, distinguished according to their position in the vector. The 20 different combinations are then given as

$$\begin{array}{llll} (0\,0\,0\,3) & (1\,0\,0\,2) & (2\,0\,0\,1) & (3\,0\,0\,0) \\ (0\,0\,1\,2) & (1\,0\,1\,1) & (2\,0\,1\,0) & \\ (0\,0\,2\,1) & (1\,0\,2\,0) & (2\,1\,0\,0) & \\ (0\,0\,3\,0) & (1\,1\,0\,1) & & \\ (0\,1\,0\,2) & (1\,1\,1\,0) & & \\ (0\,1\,1\,1) & (1\,2\,0\,0) & & \\ (0\,1\,2\,0) & & & \\ (0\,2\,0\,1) & & & \\ (0\,2\,1\,0) & & & \\ (0\,3\,0\,0) & & & \end{array}$$

where, for example, (1 0 0 2) indicates that one ball is in box 1 and two balls are in box 4 and we are not able to distinguish among the balls. In this case, we need to distinguish between, for example, (1 0 0 2) and (2 0 0 1) since, in the first case, the first box contains one ball and the fourth two balls, while in the second case, the first box contains two balls and the fourth only one ball.

This latter scenario is useful in determining the number of states in certain queuing network models and Markov chains. The problem is to determine the number of ways in which $k$ identical customers may be distributed among $n$ different queuing centers. This is the same as the number of integer vectors of length $n$ that satisfy the constraints

$$(k_1, k_2, \ldots, k_n) \text{ with } 0 \le k_i \le k \text{ and } \sum_{i=1}^{n} k_i = k$$

and is given by

$$\frac{(n + k - 1)!}{(n - 1)!\,k!}.$$

We summarize all four possibilities below, using an example of four distinguishable items, called $A$, $B$, $C$, and $D$, taken two at a time under the various possibilities.

$$\left.\begin{array}{cccc} AA & AB & AC & AD \\ BA & BB & BC & BD \\ CA & CB & CC & CD \\ DA & DB & DC & DD \end{array}\right\} \text{Permutations with replacement}$$

$$\left.\begin{array}{cccc} & AB & AC & AD \\ BA & & BC & BD \\ CA & CB & & CD \\ DA & DB & DC & \end{array}\right\} \text{Permutations without replacement}$$

$$\left.\begin{array}{cccc} AA & AB & AC & AD \\ & BB & BC & BD \\ & & CC & CD \\ & & & DD \end{array}\right\} \text{Combinations with replacement}$$

$$\left.\begin{array}{ccc} AB & AC & AD \\ & BC & BD \\ & & CD \end{array}\right\} \text{Combinations without replacement}$$

The formulae for the four different possibilities are collected together and shown below.

| | With replacement | Without replacement |
|---|---|---|
| Permutations | $n^k$ | $P(n, k) = \dfrac{n!}{(n-k)!}$ |
| Combinations | $\dbinom{n+k-1}{k} = \dfrac{(n+k-1)!}{k!\,(n-1)!}$ | $C(n, k) = \dbinom{n}{k} = \dfrac{n!}{k!(n-k)!}$ |

## 2.6 Bernoulli (Independent) Trials

We saw previously that in a single toss of a fair coin the probability of getting heads is one-half. Now let us consider what happens when this coin is tossed $n$ times. An element of the sample space may be written as $(s_1, s_2, \ldots, s_n)$, i.e., a string of length $n$ in which each letter $s_i$ is either $s_i = H$ or $s_i = T$, $i = 1, 2, \ldots, n$. It follows that the size of the sample space is $2^n$ and all the elementary events in this space are equiprobable since each toss is assumed to be *independent* of all previous tosses. Thus, for example, the probability of getting all heads is $2^{-n}$ and is the same as that of getting all tails, or exactly $k$ heads followed by $n - k$ tails, for $0 \leq k \leq n$.

Let $\mathcal{A}$ be the event that there are exactly $k$ heads among the $n$ tosses. The number of outcomes that result in event $\mathcal{A}$ is $C(n, k)$, since this is the number of ways in which the $k$ positions containing $H$ can be selected in a string of length $n$. The probability of obtaining exactly $k$ heads on $n$ tosses of the coin is therefore given as

$$\binom{n}{k} 2^{-n}.$$

We now modify the results for the case in which the coin is not a fair coin. Suppose the probability of obtaining heads on one toss of the coin is given as $p$ and the probability of getting tails is $q$, where $p + q = 1$. The sample space is the same as before; only the probabilities of the elementary events change. The probability of tossing $n$ heads is $p \times p \times \cdots \times p = p^n$, that of tossing $n$ tails is $q \times q \times \cdots \times q = q^n$, and the probability of tossing $k$ heads followed by $n - k$ tails is now $p \times \cdots \times p \times q \times \cdots \times q = p^k q^{n-k}$. The probability of any single outcome having $k$ heads and $n - k$ tails in any order is also given by $p^k q^{n-k}$, because the individual $p$'s and $q$'s may be interchanged. The probability of obtaining exactly $k$ heads in $n$ tosses is now given as

$$\binom{n}{k} p^k q^{n-k}.$$

Substituting $p = q = 1/2$ gives the same result as before. Notice that if we sum over all possible values of $k$ we obtain, from the binomial theorem,

$$1 = (p + q)^n = \sum_{k=0}^{n} \binom{n}{k} p^k q^{n-k},$$

which must be the case for a proper probability assignment.

Sequences of $n$ independent repetitions of a probability experiment such as this are referred to as *Bernoulli sequences*. Often, the outcomes of each trial, rather than being termed heads and tails, are called success and failure, or good and defective, and so on, depending on the particular application at hand.

**Example 2.14** An experiment has probability of success equal to 0.7. Let us find the probability of three successes and two failures in a sequence of five independent trials of the experiment.

Let the probability of success be denoted by $p$ and the probability of failure be $q = 1 - p$. The probability of obtaining three successes and two failures is equal to $p^3 q^2 = 0.7^3 \times 0.3^2 = 0.03087$. The number of ways in which this can happen, i.e., the number of outcomes with exactly three successes, is $C(5, 3) = 10$. Thus, the answer we seek is equal to 0.3087.

**Example 2.15** Consider a binary communication channel that sends words coded into $n$ bits. The probability that any single bit is received correctly is $p$. The probability that it is received in error is therefore $q = 1 - p$. Using code correcting capabilities, it is possible to correct as many as $e$ errors per word sent. Let us find the probability that the transmission of any word is successful, assuming that the transmission of individual bits is independent.

To solve this problem, notice that, as long as the number of bits in error is $e$ or less, the word will be received correctly. The probability that the word is received having exactly $k$ bits in error is given as

$$\binom{n}{k} (1 - p)^k p^{n-k}.$$

Therefore, the probability that the word is correctly received is

$$\sum_{k=0}^{e} \binom{n}{k} (1 - p)^k p^{n-k}.$$

Notice that, in a sequence of $n$ independent trials having probability of success equal to $p$, the probability of $n_1$ successes and $n_2$ failures, with $n = n_1 + n_2$, is given by

$$\binom{n}{n_1} p^{n_1}(1-p)^{n-n_1} = \binom{n}{n_2}(1-p)^{n_2}p^{n-n_2}.$$

Bernoulli trials often appear in the context of *reliability* modeling. A system is represented by a set of independent components each of which has its own probability of success. Components may be organized into *series* in which the output from one becomes the input for the next. Success is achieved if a successful outcome is obtained with *each* of the components in the series. If any of the components fail, then the entire system fails. Alternatively, components may be arranged in *parallel* and a successful outcome is achieved if one or more of the components succeeds in completing its task correctly. The purpose of arranging components in parallel is often to increase the overall reliability of the system: in the event of one component failing, others continue to function and a successful outcome for the system as a whole remains possible. Combinations of components in series and in parallel are also possible.

Consider the case of $n$ components in series, in which the probability of success of each component is $p$. The probability that the complete operation is successful is equal to the probability that all $n$ components function correctly and is equal to $p^n$. The probability that the overall operation is unsuccessful is $1 - p^n$.

Now consider the case when the $n$ components are placed in parallel and success is achieved if *at least one* of the components is successful. In this case, it is easier to first compute the probability of the event that the overall operation is *not* a success and then to compute the probability of the complement of this event. In general, in probability related questions, it is often easier to compute the probability of the *intersection* of events than it is to compute the *union* of a set of events. Use of the techniques of complementation and DeMorgan's law frequently allows us to choose the easier option. In the present situation, a system failure occurs if *all* $n$ components fail (*all* implies intersection, whereas *at least one* implies union). The probability of failure in a system of $n$ components arranged in parallel, in which the probability of failure of each component is $1 - p$, is equal to $(1 - p)^n$. From this we can compute the probability of success of the overall system to be $1 - (1 - p)^n$.

**Example 2.16** In life-critical environments, it is not uncommon to have multiple computers perform exactly the same sequence of tasks so that, if one computer fails, the mission is not compromised. The mission will continue, and will be deemed to be successful, as long as at least one of the computers executes its tasks correctly. Consider a system in which three computers perform the same sequence of six tasks one after the other. Consistency tests are built into each task. A computer is assumed to be functioning correctly if its state satisfies the consistency tests. Let the probability that a computer fails to satisfy the consistency test during any task be equal to 0.04. We wish to compute the probability that the mission is successful.

This situation may be represented as a reliability model which consists of three parts in parallel (the three computers) and in which each part contains six components in series. Let $p = 0.96$ be the probability that a computer satisfies the consistency test on any given task. Then the probability that a single computer successfully completes all six tasks is equal to $p^6 = 0.7828$; on the other hand, $p' = 1 - p^6$ is the probability that it will not. Since there are three computers in parallel, the probability that all three will fail is given by $(1 - p')^3 = 0.01025$. Thus the probability of a successful mission is $1 - (1 - p')^3 = 0.9897$.

The concept of a Bernoulli trial may be generalized from two possible outcomes to many possible outcomes. Suppose that instead of two possibilities (heads and tails, or 0 and 1, or good and bad) there are $m$ possibilities $h_1, h_2, \ldots, h_m$ with probabilities $p_1, p_2, \ldots, p_m$, and $\sum_{j=1}^{m} p_j = 1$. On each trial, one and only one of the $m$ possibilities can occur. This time the sample space consists of

$n$-tuples in which each element is one of the $m$ choices $h_i, i = 1, 2, \ldots, m$. Suppose, for example, that the number of trials is given by $n = 6$ and that the number of possible outcomes on each trial is $m = 4$. Let us denote the different outcomes by $a,\ b,\ c$, and $d$. Some possible elements of the sample space are

$$(a,\ a,\ a,\ a,\ a,\ a),\quad (c,\ a,\ d,\ d,\ a,\ b),\quad (a,\ b,\ b,\ c,\ c,\ c),\quad \ldots.$$

Now let us consider those particular elements of the sample space for which $h_i$ occurs $n_i$ times, for $i = 1, 2, \ldots, m$. We must have $\sum_{j=1}^{m} n_j = n$. For example, choosing $n_1 = 1, n_2 = 2, n_3 = 0$ and $n_4 = 3$, some possibilities are

$$(a,\ b,\ b,\ d,\ d,\ d),\quad (b,\ a,\ d,\ d,\ d,\ b),\quad (d,\ a,\ d,\ d,\ b,\ b),\quad \ldots.$$

The total number of elementary events that have the property that $h_i$ occurs $n_i$ times, subject to the conditions above, is given by the *multinomial coefficient*

$$\binom{n}{n_1, n_2, \ldots, n_m} = \frac{n!}{n_1!n_2!\ldots n_m!}.$$

We leave it as an exercise to show that indeed the number of ways of placing $h_1$ in $n_1$ slots, $h_2$ in $n_2$ slots, and so on, with $\sum_{j=1}^{m} n_j = n$, is given by the multinomial coefficient. The probability of occurrence of any elementary event is given by $p_1^{n_1} p_2^{n_2} \ldots p_m^{n_m}$ and hence the probability that $h_1$ will occur $n_1$ times, $h_2$ will occur $n_2$ times, and so on is given by

$$p(n_1, n_2, \ldots, n_m) = \frac{n!}{n_1!n_2!\ldots n_m!} p_1^{n_1} p_2^{n_2} \ldots p_m^{n_m}.$$

**Example 2.17** An introductory graduate level class on applied probability has 10 undergraduate students, 30 masters students and 20 doctoral students. During the professor's office hours, students arrive independently to seek answers. What is the probability that, of the last eight students to visit the professor, two were undergraduate, five were masters, and one was a doctoral student?

We assume that the probability of an undergraduate student seeking the professor during office hours is 10/60, that of a master's student 30/60, and that of a doctoral student 20/60. In this case, the required answer is given by

$$\binom{8}{2,\ 5,\ 1}\left(\frac{10}{60}\right)^2\left(\frac{30}{60}\right)^5\left(\frac{20}{60}\right)^1 = \frac{7}{144}.$$

## 2.7 Exercises

**Exercise 2.1.1** For each of the following words, how many different arrangements of the letters can be found: *RECIPE, BOOK, COOKBOOK*?

**Exercise 2.1.2** Find a word for which the number of different arrangements of its letters is equal to 30.

**Exercise 2.2.1** A pizza parlor offers ten different toppings on its individually sized pizzas. In addition, it offers thin crust and thick crust. William brings his friends Jimmy and Jon with him and each purchases a single one-topping pizza. How many different possible commands can be sent to the kitchen?

**Exercise 2.2.2** Billy and Kathie bring two of their children to a restaurant in France. The menu is composed of an entree, a main dish, and either dessert or cheese. The entree is chosen from among four different possibilities, one of which is snails. Five different possibilities, including two seafood dishes, are offered as the main dish. To the dismay of the children, who always choose dessert over cheese, only two possibilities are offered for dessert. The cheese plate is a restaurant-specific selection and no other choice of cheese is possible. Although

Billy will eat anything, neither Kathie nor the children like snails and one of the children refuses to eat seafood. How many different possible commands can be sent to the kitchen?

**Exercise 2.2.3** Consider a probability experiment in which four fair dice are thrown simultaneously. How many different outcomes (permutations) can be found? What is the probability of getting four 6's?

**Exercise 2.2.4** Is it more probable to obtain at least one 6 in four throws of a fair die than it is to obtain at least one pair of 6's when two fair dice are thrown simultaneously 24 times?

**Exercise 2.3.1** Show that

$$P(n,k) = P(n-1,k) + kP(n-1,k-1), \quad k = 1,2,\ldots,n, \quad n = 1,2,\ldots.$$

Interpret this result in terms of permutations of $n$ items taken $k$ at a time.

**Exercise 2.3.2** William Stewart's four children come down for breakfast at random times in the morning. What is the probability that they appear in order from the oldest to the youngest.

Note: William Stewart does not have twins (nor triplets, nor quadruplets!).

**Exercise 2.3.3** The Computer Science Department at North Carolina State University wishes to market a new software product. The name of this package must consist of three letters only.

(a) How many possibilities does the department have to choose from?
(b) How many possibilities are there if exactly one vowel is included?
(c) How many possibilities are there if all three characters must be different and exactly one must be a vowel?

**Exercise 2.4.1** For each of the following properties of the binomial coefficient, prove their correctness, first using an algebraic argument and then using an intuitive "probabilistic" argument.

(a)
$$\binom{n}{k} = \binom{n}{n-k}.$$

(b)
$$\binom{n}{k} = \frac{n}{k}\binom{n-1}{k-1}.$$

(c)
$$\binom{n}{k} = \binom{n-1}{k} + \binom{n-1}{k-1}.$$

(d)
$$\binom{n}{k}\binom{k}{j} = \binom{n}{j}\binom{n-j}{k-j}.$$

**Exercise 2.4.2** Three cards are chosen at random from a full deck. What is the probability that all three are kings?

**Exercise 2.4.3** A box contains two white balls and four black ones.

(a) Two balls are chosen at random. What is the probability that they are of the same color?
(b) Three balls are chosen at random. What is the probability that all three are black?

**Exercise 2.4.4** Yves Bouchet, guide de haute montagne, leads three clients on "la traversé de la Meije." All four spend the night in the refuge called "La Promotoire" and must set out on the ascension at 3:00 a.m., i.e., in the dark. Yves is the first to arise and, since it is pitch dark, randomly picks out two boots from the four pairs left by him and his clients.

(a) What is the probability that Yves actually gets his own boots?
(b) If Yves picks two boots, what is the probability that he chooses two left boots?
(c) What is the probability of choosing one left and one right boot?
(d) If, instead of picking just two boots, Yves picks three, what is the probability that he finds his own boots among these three?

**Exercise 2.4.5** A water polo team consists of seven players, one of whom is the goalkeeper, and six reserves who may be used as substitutes. The Irish water polo selection committee has chosen 13 players from which to form a team. These 13 include two goalkeepers, three strikers, four "heavyies," i.e., defensive players, and four generalists, i.e., players who function anywhere on the team.

(a) How many teams can the coach form, given that he wishes to field a team containing a goalkeeper, one striker, two heavyies, and three generalists?
(b) Toward the end of the last period, Ireland is leading its main rival, England, and, with defense in mind, the coach would like to field a team with all four heavyies (keeping, of course, one goalkeeper, one striker, and one generalist). From how many teams can he now choose?

**Exercise 2.4.6** While in France, William goes to buy baguettes each morning and each morning he takes the exact change, 2.30 euros, from his mother's purse. One morning he finds a 2 euro coin, two 1 euro coins, five 0.20 euro coins, and four 0.10 euro coins in her purse. William has been collecting coins and observes that no coin of the same value comes from the same European country (i.e., all the coins are distinct). How many different possibilities has William from which to choose the exact amount?

**Exercise 2.4.7** An apple grower finds that 5% of his produce goes bad during transportation. A supermarket receives a batch of 160 apples per day. The manager of the supermarket randomly selects six apples.

(a) What is the probability that none of the six are bad?
(b) What is the probability that the supermarket manager chooses at least two bad apples?

**Exercise 2.4.8** A box contains two white balls, two red balls, and a black ball. Balls are chosen without replacement from the box.

(a) What is the probability that a red ball is chosen before the black ball?
(b) What is the probability of choosing the two white balls before choosing any other ball?

**Exercise 2.4.9** Three boxes, one large, one medium, and one small, contain white and red balls. The large box, which is chosen half the time, contains fifteen white balls and eight red ones; the medium box, which is chosen three times out of ten, contains nine white balls and three red ones. The small box contains four white balls and five red ones.

(a) On choosing two balls at random from the large box, what is the probability of
  (i) getting two white balls?
  (ii) getting one white and one red ball?
(b) After selecting one of the boxes according to the prescribed probabilities, what is the probability of getting one white and one red ball from this box?
(c) Given that a box is chosen according to the prescribed probabilities, and that a white ball is randomly picked from that box, what is the probability that the ball was actually chosen from the large box?

**Exercise 2.4.10** A deck of 52 cards is randomly dealt to four players so that each receives 13 cards. What is the probability that each player holds an ace?

**Exercise 2.4.11** Consider a system that consists of $p$ processes that have access to $r$ identical units of a resource. Suppose each process alternates between using a single unit of the resource and not using any resource.

(a) Describe a sample space that illustrates the situation of the $p$ processes.
(b) What is the size of this sample space when (a) $p = 6,\ r = 8$ and (b) $p = 6,\ r = 4$?
(c) In this first case ($p = 6,\ r = 8$), define the following events.
  - $\mathcal{A}$: Either two or three processes are using the resource.
  - $\mathcal{B}$: At least three but not more than five processes are using the resource.
  - $\mathcal{C}$: Either all the processes are using the resource or none are.

  Assuming that each process is equally likely as not to be using a unit of resource, compute the probabilities of each of these three events.
(d) Under the same assumption, what do the following events represent, and what are their probabilities?
  - $\mathcal{A} \cap \mathcal{B}$.
  - $\mathcal{A} \cup \mathcal{B}$.
  - $\mathcal{A} \cup \mathcal{B} \cup \mathcal{C}$.

**Exercise 2.5.1** Show that, as the population size $n$ grows large, the difference between the number of combinations *without* replacement for a fixed number of items $k$ becomes equal to the number of combinations *with* replacement.

**Exercise 2.6.1** The probability that a message is successfully transmitted over a communication channel is known to be $p$. A message that is not received correctly is retransmitted until such time as it is received correctly. Assuming that successive transmissions are independent, what is the probability that no retransmissions are needed? What is the probability that exactly two retransmissions are needed?

**Exercise 2.6.2** The outcome of a Bernoulli trial may be any one of $m$ possibilities which we denote $h_1, h_2, \ldots, h_m$. Given $n$ positions, show that the number of ways of placing $h_1$ in $n_1$ positions, $h_2$ in $n_2$ positions and so on, with $\sum_{j=1}^{m} n_j = n$, is given by the binomial coefficient,

$$\binom{n}{n_1, n_2, \ldots, n_m} = \frac{n!}{n_1!n_2!\cdots n_m!}.$$

# Chapter 3

# *Random Variables and Distribution Functions*

## 3.1 Discrete and Continuous Random Variables

Random variables, frequently abbreviated as RV, define functions, with domain and range, rather than variables. Simply put, a random variable is a function whose domain is a sample space $\Omega$ and whose range is a subset of the real numbers, $\Re$. Each elementary event $\omega \in \Omega$ is mapped by the random variable into $\Re$. Random variables, usually denoted by uppercase latin letters, assign real numbers to the outcomes in their sample space.

**Example 3.1** Consider a probability experiment that consists of throwing two dice. There are 36 possible outcomes which may be represented by the pairs $(i, j); i = 1, 2, \ldots, 6, j = 1, 2, \ldots, 6$. These are the 36 elements of the sample space. It is possible to associate many random variables having this domain. One possibility is the random variable $X$, defined explicitly by enumeration as

$$
\begin{aligned}
&X(1,1) = 2;\\
&X(1,2) = X(2,1) = 3;\\
&X(1,3) = X(2,2) = X(3,1) = 4;\\
&X(1,4) = X(2,3) = X(3,2) = X(4,1) = 5;\\
&X(1,5) = X(2,4) = X(3,3) = X(4,2) = X(5,1) = 6;\\
&X(1,6) = X(2,5) = X(3,4) = X(4,3) = X(5,2) = X(6,1) = 7;\\
&X(2,6) = X(3,5) = X(4,4) = X(5,3) = X(6,2) = 8;\\
&X(3,6) = X(4,5) = X(5,4) = X(6,3) = 9;\\
&X(4,6) = X(5,5) = X(6,4) = 10;\\
&X(5,6) = X(6,5) = 11;\\
&X(6,6) = 12.
\end{aligned}
$$

Thus $X$ maps the outcome $(1, 1)$ into the real number 2, the outcomes $(1, 2)$ and $(2, 1)$ into the real number 3, and so on. It is apparent that the random variable $X$ represents the sum of the spots obtained when two dice are thrown simultaneously. As indicated in Figure 3.1, its domain is the 36 elements of the sample space and its range is the set of integers from 2 through 12.

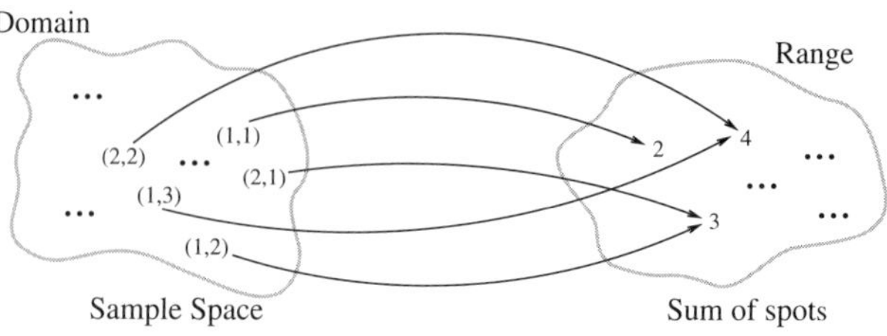


Figure 3.1. Domain and range of $X$ (not all points are shown).

Observe that the random variable $X$ of Example 3.1 partitions the sample space into 11 subsets; i.e., into 11 events that are mutually exclusive and collectively exhaustive: in other words, these 11 events form an *event space*. The subsets are distinguished according to the particular value taken by the function. Thus $\{(1, 3), (2, 2), (3, 1)\}$ is the subset consisting of elements of the sample space that are mapped onto the real number 4; $\{(5, 6), (6, 5)\}$ is the subset consisting of the elements that are mapped onto the real number 11; and so on. Each and every outcome in the sample space is associated with a real number; several may be associated with the same real number: for example, $(1, 3)$, $(2, 2)$, and $(3, 1)$ are all associated with the integer 4, but none is associated with more than one real number. This is shown graphically in Figure 3.1 and follows because $X$ has been defined as a function. Different random variables may be defined on the same sample space. They have the same domain, but their range can be different.

**Example 3.2** Consider a different random variable $Y$ whose domain is the same sample space as that of $X$ but which is defined as the *difference* in the number of spots on the two dice. In this case we have

$$Y(1,1) = Y(2,2) = Y(3,3) = Y(4,4) = Y(5,5) = Y(6,6) = 0;$$

$$Y(2,1) = Y(3,2) = Y(4,3) = Y(5,4) = Y(6,5) = 1;$$

$$Y(1,2) = Y(2,3) = Y(3,4) = Y(4,5) = Y(5,6) = -1;$$

$$Y(3,1) = Y(4,2) = Y(5,3) = Y(6,4) = 2;$$

$$Y(1,3) = Y(2,4) = Y(3,5) = Y(4,6) = -2;$$

$$Y(4,1) = Y(5,2) = Y(6,3) = 3;$$

$$Y(1,4) = Y(2,5) = Y(3,6) = -3;$$

$$Y(5,1) = Y(6,2) = 4;$$

$$Y(1,5) = Y(2,6) = -4;$$

$$Y(6,1) = 5;$$

$$Y(1,6) = -5.$$

Observe that the random variable $Y$ has the same domain as $X$ but the range of $Y$ is the set of integers between $-5$ and $+5$ inclusive. As illustrated in Figure 3.2, $Y$ also partitions the sample space into 11 subsets, but not the same 11 as those obtained with $X$.

Yet a third random variable, $Z$, may be defined as the *absolute* value of the difference between the spots obtained on the two dice. Again, $Z$ has the same domain as $X$ and $Y$, but now its range is the set of integers between 0 and 5 inclusive. This random variable partitions the sample space into six subsets: the first subset containing elements of the sample space in which the number of spots differ by 5, irrespective of which is first and which is second; the second subset of the partition contains elements in which the number of spots differ by 4, and so on. This partition is shown in Figure 3.3.

This is an important property of random variables, that they partition the sample space into more meaningful representations of the probability experiment. If our concern is only with the sum obtained in throwing two dice, it makes more sense to work with the 11 events obtained by the partition generated by the random variable $X$, than the 36 elementary events of the sample space. In this way, random variables are used as a means of simplification.

**Example 3.3** Let $R$ be the random variable that counts the number of heads obtained in tossing three coins. Then $R$ assigns the value 3 to the outcome $HHH$; the value 2 to each of the outcomes

(1,1)
(1,2)
(2,1)
(1,3)
(2,2)
(3,1)
(1,4)
(2,3)
(3,2)
(4,1)
(1,5)
(2,4)
(3,3)
(4,2)
(5,1)
(1,6)
(2,5)
(3,4)
(4,3)
(5,2)
(6,1)
(2,6)
(3,5)
(4,4)
(5,3)
(6,2)
(3,6)
(4,5)
(5,4)
(6,3)
(4,6)
(5,5)
(6,4)
(5,6)
(6,5)
(6,6)
X=2
X=3
X=4
X=12
Y=-5 Y=-4 ······
······ Y=3 Y=4 Y=5

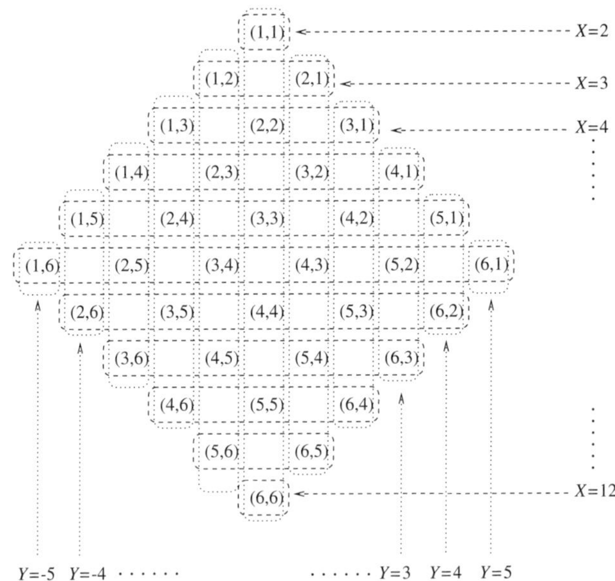

Figure 3.2. Partitions according to $X$ (horizontal and dashed) and $Y$ (vertical and dotted).

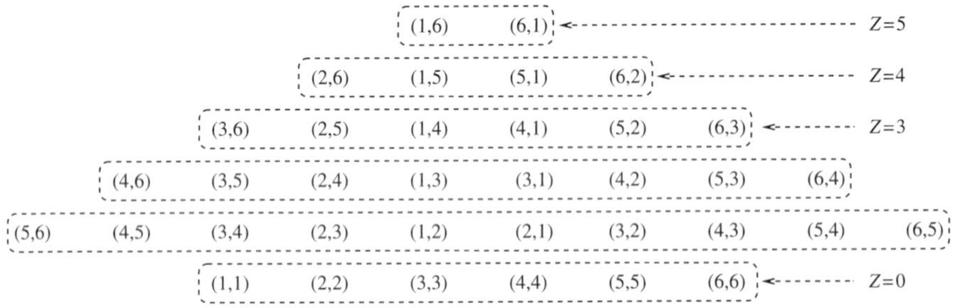

(1,6) (6,1) Z=5
(2,6) (1,5) (5,1) (6,2) Z=4
(3,6) (2,5) (1,4) (4,1) (5,2) (6,3) Z=3
(4,6) (3,5) (2,4) (1,3) (3,1) (4,2) (5,3) (6,4)
(5,6) (4,5) (3,4) (2,3) (1,2) (2,1) (3,2) (4,3) (5,4) (6,5)
(1,1) (2,2) (3,3) (4,4) (5,5) (6,6) Z=0

Figure 3.3. Partition according to $Z$.

*HHT*, *HTH* and *THH*; the value 1 to each of the outcomes *HTT*, *THT*, and *TTH*. Finally, $R$ assigns the value 0 to the outcome *TTT*. If our concern is uniquely with the number of heads obtained, then it is easier to work with $R$ than with the eight outcomes of the probability experiment.

Since a random variable is a function, it is not possible for two or more values in the range to be assigned to the same outcome in the sample space. For a random variable $X$, the set of elements of the sample space that are assigned the value $x$ is denoted by $A_x$ or $X^{-1}(x)$. Notice that this describes an event, since it consists of a subset of the sample space. We have

$$A_x \equiv [X = x] \equiv \{\omega \in \Omega | X(\omega) = x\}$$

which is the set of all outcomes in the sample space (domain) for which the random variable $X$ has the value $x$ (an element of the range). In the example of two dice thrown simultaneously and $X$ the random variable describing the sum of spots obtained, we find, for example, that $A_3$ is the set $\{(1, 2),\ (2, 1)\}$, and $A_5$ is the set $\{(1, 4),\ (2, 3)\ (3, 2)\ (4, 1)\}$. As we previously observed, the set of all such events is called an *event space* and it is frequently more convenient to work in this event space than in the original sample space. The event space of our random variable $X$ defines 11 events, each corresponding to a value of $x = 2, 3, \ldots, 12$. For all other values of $x$, $A_x$ is the null set.

Random variables which take on values from a discrete set of numbers, (i.e., whose range consists of isolated points on the real line) are called *discrete random variables*. The set may be infinite but countable. The examples considered so far have all been discrete and finite. Discrete random variables are used to represent distinct indivisible items, like people, cars, trees, and so on. *Continuous random variables* take on values from a continuous interval. For example, a random variable that represents the time between two successive arrivals to a queueing system, or that represents the temperature in a nuclear reactor, is an example of a continuous random variable. Time, in the first case, or temperature in the second, are reals belonging to a continuous subset of the real number system. Of course, random variables that represent quantities like time, temperature, distance, area, volume, etc. may be discretized into minutes, hours, yards, miles, and so on, in which case it is appropriate to use *discrete* random variables to represent them. All random variables defined on a discrete sample space must be discrete. However, random variables defined on a continuous sample space may be discrete or continuous.

## 3.2 The Probability Mass Function for a Discrete Random Variable

The probability mass function (frequently abbreviated to *pmf*) for a *discrete* random variable $X$, gives the *probability* that the value obtained by $X$ on the outcome of a probability experiment is equal to $x$. It is denoted by $p_X(x)$. Sometimes the term *discrete density function* is used in place of probability mass function. We have

$$p_X(x) \equiv \text{Prob}\{[X = x]\} = \text{Prob}\{\omega | X(\omega) = x\} = \sum_{X(\omega)=x} \text{Prob}\{\omega\}.$$

The notation $[X = x]$ defines an *event*: the event that contains all the outcomes that map into the real number $x$. If none of the outcomes map into $x$, then $[X = x]$ is the null event and has probability zero. Most often we simply write $\text{Prob}\{[X = x]\}$ as $\text{Prob}\{X = x\}$. Since $p_X(x)$ is a probability, we must have

$$0 \leq p_X(x) \leq 1 \text{ for all } x \in \Re. \tag{3.1}$$

The probability mass function $p_X(x)$ is often written more simply as $p(x)$; i.e., the subscript $X$ is omitted when there can be no confusion.

**Example 3.4** In the case of two dice being thrown simultaneously, and $X$ the random variable representing the sum of spots obtained, we have $p_X(2) = 1/36$, since only one of the 36 equiprobable outcomes in the sample space gives the sum 2 (when each die tossed yields a single spot). Similarly, $p_X(3) = 2/36$ since there are two outcomes that give a total number of spots equal to 3, namely, $(1, 2)$ and $(2, 1)$. We may continue in this fashion. So long as $x$ is equal to one of the integers in the interval $[2, 12]$, then $p_X(x)$ is strictly positive. For all other values of $x$, $p_X(x) = 0$.

Discrete random variables are completely characterized by their probability mass functions. These functions, and hence their associated discrete random variables, are frequently illustrated either in tabular form or in the form of bar graphs. In bar graph form, the probability mass function of the random variable $X$ (sum of spots obtained in throwing two dice) is shown in Figure 3.4. The corresponding tabular form of this bar graph is shown in Table 3.1.

Table 3.1. Probability mass function of $X$ in tabular form.

| $x_i$ | 2 | 3 | 4 | 5 | 6 | 7 | 8 | 9 | 10 | 11 | 12 |
|---|---|---|---|---|---|---|---|---|---|---|---|
| $p_X(x_i)$ | 1/36 | 2/36 | 3/36 | 4/36 | 5/36 | 6/36 | 5/36 | 4/36 | 3/36 | 2/36 | 1/36 |

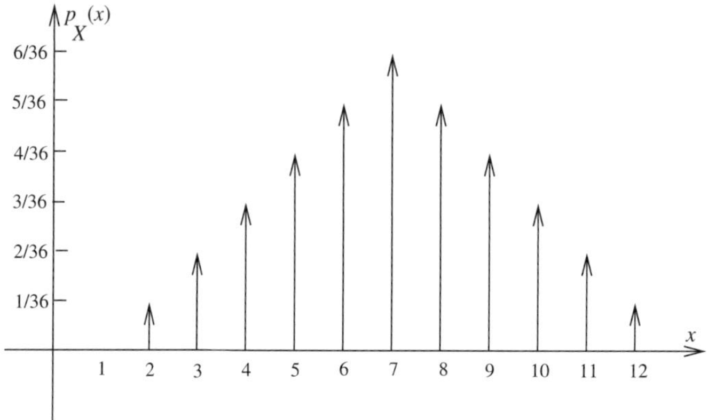


Figure 3.4. Bar graph representation of $X$.

Notice that

$$\sum_{x \in \Re} p_X(x) = 1$$

since the random variable $X$ assigns some value $x \in \Re$ to each and every sample point $\omega \in \Omega$, so that when we sum over every possible value of $x$, we sum over the probabilities of all the outcomes in the sample space, and this must be equal to 1.

When (as is the case here) the random variable is discrete, then the set of values assumed by $x$ is denumerable, and constitutes a subset of the reals. We shall denote successive elements of this set by $\{x_1, x_2, \ldots\}$. The set itself is called the *image of* $X$ and it is usual to denote $\text{Prob}\{X = x_i\}$ by $p_i$. It follows that

$$\sum_k p_X(x_k) = \sum_k p_k = 1. \tag{3.2}$$

It may be shown that any real-valued function defined on $\Re$ and satisfying Equations (3.1) and (3.2) is the probability mass function for some random variable, $X$.

**Example 3.5** Let us return to the experiment that consists of tossing a fair coin three times. The sample space is given by

$$\{HHH,\ HHT,\ HTH,\ HTT,\ THH,\ THT,\ TTH,\ TTT\}$$

and each of these has probability $1/8 = 0.125$. Let $R$ be the discrete random variable defined as the number of heads obtained during this sequence of tosses. Since $R$ counts the number of heads obtained it can only have the values 0, 1, 2, or 3. Only one of the eight elementary events gives zero heads, so the probability that $R = 0$ is $1/8$, i.e., $p_R(0) = \text{Prob}\{R = 0\} = 0.125$. Three of the elementary events yield one head (*HTT*, *THT*, and *TTH*) and hence $p_R(1) = 0.375$, and so on. The values that $R$ can assume and their corresponding probabilities are shown in Table 3.2.

Table 3.2. Probability mass function of $R$.

| $x_i$ | 0 | 1 | 2 | 3 |
|---|---|---|---|---|
| $p_R(x_i)$ | 0.125 | 0.375 | 0.375 | 0.125 |

In probability theory, the same basic formulation of a probability mass function can often be used in different scenarios. Such probability mass functions are defined with a small number of parameters and various scenarios captured by simply changing the values of the parameters.

Random variables whose probability mass functions are identical up to the values assigned to these parameters are called *families of random variables*. These include Bernoulli random variables, binomial random variables, and Poisson random variables among the class of discrete random variables, and the exponential, Erlang, and standard normal random variables among the class of continuous random variables. These and others are considered in detail in later chapters. At this point, and for illustration purposes only, we present two families of random variables, *Bernoulli* random variables and *Poisson* random variables.

A random variable $X$ is said to be a Bernoulli random variable if its probability mass function has the form

$$p_X(x) = \begin{cases} 1-p, & x=0, \\ p, & x=1, \\ 0 & \text{otherwise.} \end{cases}$$

In this case, the parameter is $p$ and must satisfy the condition $0 < p < 1$. Taking $p = 1/2$ satisfies the requirements for probability experiments dealing with tossing a fair coin; other values of $p$ are appropriate for experiments using a biased coin.

The probability mass function for a *Poisson* random variable, $X$, is given by

$$p_X(n) = \begin{cases} \alpha^n e^{-\alpha}/n!, & n = 0, 1, 2, \dots, \\ 0 & \text{otherwise,} \end{cases}$$

where the parameter $\alpha$ is strictly positive. Poisson random variables are frequently used to model arrival processes as follows. Let $\lambda$ be the rate of arrivals to a facility and $t$ an interval of observation. Then, during the time interval $t$, the number of arrivals could be taken to be a Poisson random variable with parameter $\alpha = \lambda t$. Indeed, we shall frequently write the probability mass function of a Poisson random variable as

$$p_X(n) = \frac{(\lambda t)^n}{n!} e^{-\lambda t} \text{ for } n = 0, 1, 2, \dots; \quad \text{otherwise } p_X(n) = 0.$$

**Example 3.6** The number of students who visit a professor during office hours is a Poisson random variable with an average of $\lambda = 12$ students per hour.

(a) Compute the probability that no students arrive in an interval of 15 minutes.
Using $\lambda = 12$ and $t = 0.25$, $\lambda t = 3$ and the probability of $n = 0, 1, 2, \dots$ students arriving is

$$p_X(n) = \frac{3^n}{n!} e^{-3}.$$

Therefore the probability that no students ($n = 0$) arrive in a 15 minute period is

$$p_X(0) = e^{-3} = 0.04979.$$

(b) Compute the probability that not more than three students arrive in an interval of 30 minutes.
In this case, the time interval is 1/2 hour, so $\lambda t = 6$. To compute the probability that not more than three students arrive in 30 minutes we need to add the probabilities of zero, one, two, and three students arriving. These four probabilities are given, respectively, by

$$p_X(0) = e^{-6} = 0.002479, \quad p_X(1) = 6e^{-6} = 0.014873,$$

$$p_X(2) = 6^2 e^{-6}/2 = 0.044618, \quad p_X(3) = 6^3 e^{-6}/6 = 0.089235,$$

and their sum is given by $p_X(0) + p_X(1) + p_X(2) + p_X(3) = 0.1512$.

For the most part, in this section on the probability mass function of a discrete random variable, our concern has been with determining the probability that the random variable $X$ has a certain value

$x$: $p_X(x) = \text{Prob}\{X = x\}$. In the next section, we consider the probability that the random variable has any one of the values in some specified set of values. The last part of the previous example has already set the stage.

## 3.3 The Cumulative Distribution Function

For any discrete random variable $X$, there is a denumerable number of reals that are assigned to $X$. These reals are denoted by the set $\{x_1, x_2, \ldots\}$ and it is natural to arrange these reals in increasing order of magnitude. For example, these were arranged from 2 through 12 for the random variable $X$ and from 0 through 3 for the random variable $R$ as has been shown in Tables 3.1 and 3.2, respectively. Given this situation, we may now wish to find the probability that a random variable $X$ assumes a value that is less than or equal to a given $x_i$, i.e., the probability $\text{Prob}\{X \le x_i\}$. Notice that this corresponds to the probability of an event, the event that consists of all elements $\omega$ of the sample space for which $X(\omega) = x_k$ with $x_k \le x_i$.

**Example 3.7** Consider once again, the random variable $X$, defined as the sum of spots obtained in simultaneously throwing two fair dice. We previously examined the probability mass function for $X$. Perhaps now we are interested in knowing the probability that the random variable $X$ has a value less than 6. Let us denote this event as $\mathcal{A}$ and observe that its probability is just the sum of the probabilities

$$\text{Prob}\{\mathcal{A}\} = \text{Prob}\{X = 2\} + \text{Prob}\{X = 3\} + \text{Prob}\{X = 4\} + \text{Prob}\{X = 5\},$$

which can be denoted more simply by $\text{Prob}\{X < 6\}$. Hence we have

$$\text{Prob}\{X < 6\} = \frac{1}{36} + \frac{2}{36} + \frac{3}{36} + \frac{4}{36} = \frac{10}{36}.$$

Observe that the event $\mathcal{A}$ is defined as the subset consisting of the ten outcomes:

$$\mathcal{A} = \{(1,1),\ (1,2),\ (2,1),\ (1,3),\ (2,2),\ (3,1),\ (1,4),\ (2,3),\ (3,2),\ (4,1)\}.$$

We can go further. Suppose we would like to know the probability that the random variable $X$ has a value that lies between 5 and 7 inclusive. In this case, we may write

$$\text{Prob}\{\mathcal{A}\} = \text{Prob}\{X = 5\} + \text{Prob}\{X = 6\} + \text{Prob}\{X = 7\},$$

which again may be written more succinctly as $\text{Prob}\{5 \le X \le 7\}$, and we have

$$\text{Prob}\{5 \le X \le 7\} = \frac{4}{36} + \frac{5}{36} + \frac{6}{36} = \frac{15}{36}.$$

Now consider a subset $S$ that contains an arbitrary collection of the values that the random variable $X$ can assume. Then, the set $\{\omega | X(\omega) \in S\}$ contains all the outcomes for which the random variable $X$ has a value in $S$. By definition, this constitutes an event which is denoted by $[X \in S]$ and we have

$$[X \in S] = \{\omega | X(\omega) \in S\} = \bigcup_{x_k \in S} \{\omega | X(\omega) = x_k\}.$$

If $p_X(x)$ is the probability mass function for the discrete random variable $X$, then

$$\text{Prob}\{X \in S\} = \sum_{x_k \in S} p_X(x_k).$$

**Example 3.8** Let us return to the example of two dice and the random variable $X$ that describes the sum of spots obtained and let $S$ be the set that contains only the two values $x_1 = 2$ and $x_{10} = 11$, i.e., $S = \{2, 11\}$. Then $\{\omega | X(\omega) = x_1\} = (1, 1)$ and $\{\omega | X(\omega) = x_{10}\} = \{(5, 6), (6, 5)\}$. Hence $[X \in S] = \{(1, 1), (5, 6), (6, 5)\}$ and

$$\text{Prob}\{X \in S\} = p_X(2) + p_X(11) = 1/36 + 2/36 = 1/12.$$

When the set $S$ contains all possible values within a range of values that $X$ can assume, say $(a, b]$, then it is more usual to write $\text{Prob}\{X \in S\}$ as $\text{Prob}\{a < X \leq b\}$ as illustrated previously. In the particular case when $S$ contains all values that are less than or equal to some specified value $x$, then $\text{Prob}\{X \in S\}$ is called the *cumulative distribution function (CDF)* of the random variable $X$. This function is denoted by $F_X(x)$ and is also referred to as the *probability distribution function (PDF)*. It is defined for real values of $x$ in the range $(-\infty < x < \infty)$. We have

$$F_X(x) \equiv \text{Prob}\{-\infty < X \leq x\} = \text{Prob}\{X \leq x\}.$$

Observe that, if $X$ is a *discrete* random variable, then $x$ does not have to be one of the values that $X$ can assume; i.e., $x$ need not be one of the $x_i$. For example, when the $x_i$ are integers,

$$F_X(x) = \sum_{-\infty < x_i \leq \lfloor x \rfloor} p_X(x_i)$$

where $\lfloor x \rfloor$ is the floor of $x$ and denotes the largest integer less than or equal to $x$.

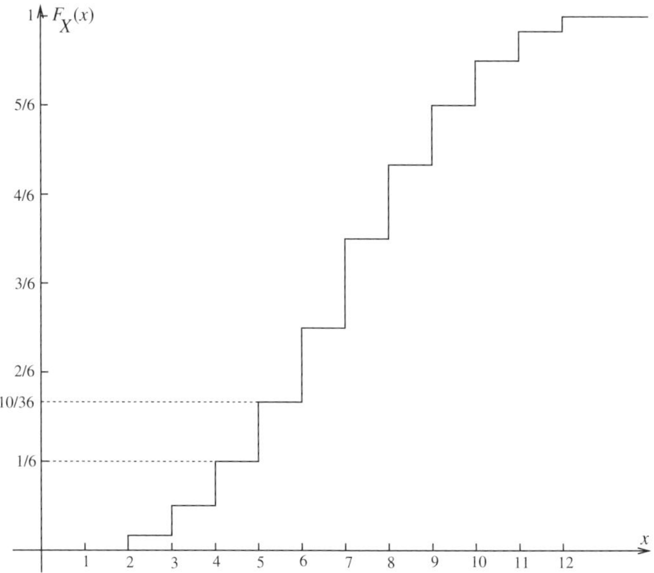


Figure 3.5. Cumulative distribution function of $X$.

To illustrate this, consider Figure 3.5, which shows the cumulative distribution function of the random variable $X$, defined as the sum of spots found on two simultaneously thrown dice. It may be observed from this figure that for any value of $x$ in the interval $[4, 5)$, for example, $F_X(x) = 1/6$. This is the probability that the number of spots obtained is 4 or less. If $x$ lies in the interval $[5, 6)$, then $F_X(x) = 10/36$, the probability that the number of spots is 5 or less. For the random variable $R$, which denotes the number of heads obtained in three tosses of a fair coin, the cumulative distribution function is shown in Figure 3.6. If $x$ lies somewhere in the interval $[1, 2)$, then $F_R(x) = 1/2$ which

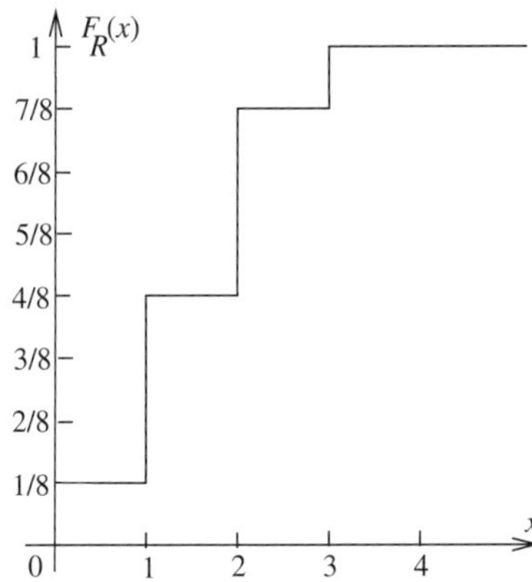


Figure 3.6. Cumulative distribution function of $R$.

is the probability of throwing one or zero heads, i.e., the event $\mathcal{A} = \{TTT, HTT, THT, TTH\}$. In both of these figures, the step "risers" have been included simply to show the steplike feature of the cumulative distribution function of a discrete random variable; the value assumed by the cumulative distribution function at $x_i$ is the value of the top of the riser.

Observe that, in both figures, the value of the cumulative distribution function is zero for sufficiently small values of $x$. The function is nondecreasing with increasing values of $x$ and is equal to 1 for sufficiently large values of $x$. This is due to the fact that the probability that the random variable will assume a value smaller than the smallest $x_k$ must be zero while the probability that it will have a value less than or equal to the largest $x_k$ must be 1. These properties are inherent in all cumulative distribution functions.

The disjointed steplike structure of the two functions shown is a distinctive feature of discrete random variables. These examples illustrate the fact that the distribution function of a discrete random variable is *not* continuous. They have at least one point at which the approach of the function from the left and from the right do not meet: the approach from the right is strictly greater than the approach from the left. In the examples, these are the points $x_i$ that may be assumed by the random variable. For continuous random variables, the monotonic nondecreasing property of the cumulative distribution function is maintained, but the increase is continuous and not steplike. The cumulative distribution function of a continuous random variable is a continuous function of $x$ for all $-\infty < x < \infty$.

Whereas the probability mass function is defined only for discrete random variables, the cumulative distribution function applies to both discrete and continuous random variables. The following definition covers both cases:

**Definition 3.3.1 (Cumulative distribution function)** *The cumulative distribution function $F_X$ of a random variable $X$ is defined to be the function*

$$F_X(x) = \text{Prob}\{X \leq x\} \quad \text{for} \quad -\infty < x < \infty.$$

*Simply put, it is just the probability that the random variable $X$ does not exceed a given real number $x$.*

**Example 3.9** The random variable $X$ whose distribution function has the value zero when $x < 0$, the value one when $x \geq 1$, and which increases *uniformly* from the value zero at $x = 0$ to the value

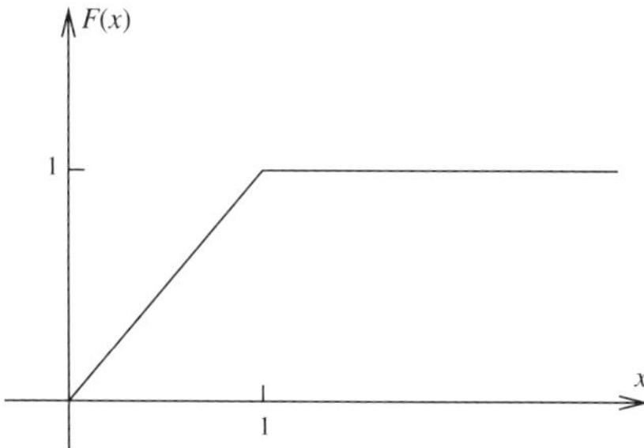


Figure 3.7. Cumulative distribution function of the uniform continuous random variable.

one at $x = 1$ is called the uniform continuous random variable on the interval $[0, 1]$. Its distribution function is usually written as

$$F(x) = \begin{cases} 0, & x < 0, \\ x, & 0 \le x < 1, \\ 1, & x \ge 1, \end{cases}$$

and is illustrated in Figure 3.7.

Observe that the cumulative distribution function illustrated in Figure 3.7 has a derivative everywhere except at $x = 0$ and $x = 1$. The distribution functions of continuous random variables that we shall consider will be *absolutely* continuous, meaning that they are continuous *and* their derivatives exist everywhere, except possibly at a finite number of points.

**Example 3.10** Consider a second continuous random variable, $M$, defined as

$$F_M(x) = \begin{cases} 1 - e^{-x} & \text{if } x \ge 0, \\ 0 & \text{otherwise.} \end{cases}$$

A random variable having this distribution function is said to be an (negative) *exponential* random variable with parameter value equal to one. A graph of this function is shown in Figure 3.8.

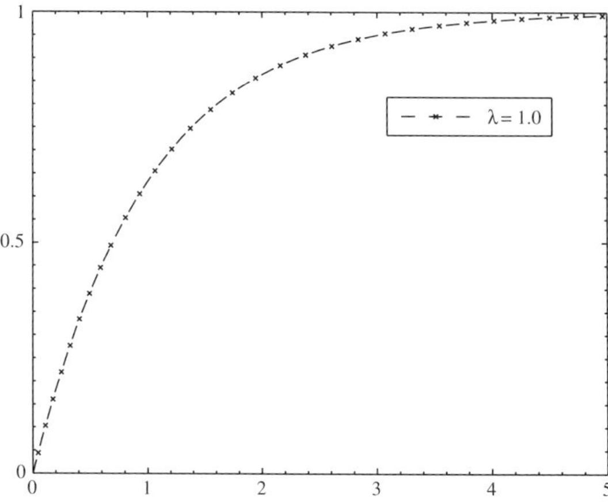


Figure 3.8. CDF of an exponential random variable: $F_M(x) = 1 - e^{-x}$ for $x \ge 0$.

Distribution functions have a number of mathematical properties, the most important of which we denote below. The reader may wish to verify that these properties are indeed evident in Figures 3.5, 3.6, 3.7, and 3.8.

- $0 \leq F(x) \leq 1$ for $-\infty < x < \infty$, since $F(x)$ is a probability.
- $\lim_{x \to -\infty} F(x) = 0$ and $\lim_{x \to \infty} F(x) = 1$.
- If the random variable $X$ has a finite image, then
$F(x) = 0$ for all $x$ sufficiently small, and
$F(x) = 1$ for all $x$ sufficiently large.
- $F(x)$ is a nondecreasing function of $x$. Since $(-\infty, x_1] \in (-\infty, x_2]$ for $x_1 \leq x_2$, it follows that $F(x_1) \leq F(x_2)$ for $x_1 \leq x_2$.
- $F(x)$ is *right continuous*. This means that for any $x$ and any *decreasing* sequence $x_k$ with $k \geq 1$ that converges to $x$, we must have $\lim_{k \to \infty} F(x_k) = F(x)$.

Notice that for any random variable $X$, discrete or continuous, we have

$$\text{Prob}\{a < X \leq b\} = \text{Prob}\{X \leq b\} - \text{Prob}\{X \leq a\} = F(b) - F(a).$$

Now let $a \to b$ and observe what happens. We obtain

$$\text{Prob}\{X = b\} = F(b) - F(b^-)$$

where $F(b^-)$ is the limit of $F(x)$ *from the left*, at the point $b$. There are now two possibilities:

- $F(b^-) = F(b)$. In this case $F(x)$ is continuous at the point $b$ and thus $\text{Prob}\{X = b\} = 0$. The event $X = b$ has zero probability *even though that event may occur.*
- $F(b^-) \neq F(b)$. Here $F(x)$ has a discontinuity (a jump) at the point $b$ and $\text{Prob}\{X = b\} > 0$.

As we have seen, the cumulative distribution function of a discrete random variable grows only by jumps, while the cumulative distribution function of a continuous random variable has no jumps, but grows continuously. When $X$ is a continuous random variable, the probability that $X$ has any given value must be zero. All we can do is assign a positive probability that $X$ falls into some finite interval, such as $[a, b]$, on the real axis.

For a discrete random variable, $F(x)$ has a staircaselike appearance. At each of the points $x_i$, it has a positive jump equal to $p_X(x_i)$. Between these points, i.e., in the interval $[x_i, x_{i+1})$, it has a constant value. In other words, as shown in Figure 3.9, we must have

$$F(x) = F(x_i) \quad \text{for } x_i \leq x < x_{i+1},$$
$$F(x_{i+1}) = F(x_i) + p_X(x_{i+1}).$$

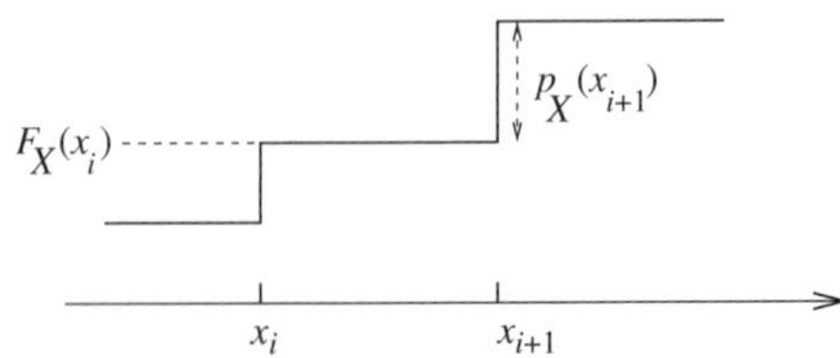


Figure 3.9. Cumulative distribution function of a discrete random variable.

Finally, we point out that any continuous, monotonically nondecreasing function $\Phi_X(x)$ for which

$$\lim_{x \to -\infty} \Phi_X(x) = 0 \quad \text{and} \quad \lim_{x \to \infty} \Phi_X(x) = 1$$

can serve as the distribution function of a continuous random variable $X$. We previously noted that the probability mass function provides a complete specification of a discrete random variable. The same can be said of a cumulative distribution function; that it completely specifies a random variable. The two distributions are closely related. Given one, the other can be easily determined.

**Example 3.11** Let $X$ be a discrete random variable with probability distribution function given by

$$F_X(x) = \begin{cases} 0, & x \in (-\infty, -2), \\ 1/10, & x \in [-2, -1), \\ 3/10, & x \in [-1, 1), \\ 6/10, & x \in [1, 2), \\ 1, & x \in [2, \infty). \end{cases}$$

Let us compute the probability mass function of this random variable. Observe first that this random variable has discontinuities at the points $x = -2, -1, 1$, and 2. These are the values that $X$ can assume and are the only points at which $p_X(x)$ can have a nonzero value. By subtracting the value of the function just prior to one of these points from the value of the function at the point itself, we obtain the following probability mass function of $X$:

$$P_X(x) = \begin{cases} 1/10, & x = -2, \\ 2/10, & x = -1, \\ 3/10, & x = 1, \\ 4/10, & x = 2, \\ 0 & \text{otherwise.} \end{cases}$$

## 3.4 The Probability Density Function for a Continuous Random Variable

For a continuous random variable $X$, the function defined as

$$f(x) = dF(x)/dx$$

is called the *probability density function* of $X$. This function fulfills much the same role for a continuous random variable as the role played by the probability mass function of a discrete random variable. In this case, the summation used earlier is replaced by integration. Observe that

$$f(x) = \lim_{\Delta x \to 0} \frac{F(x + \Delta x) - F(x)}{\Delta x} = \lim_{\Delta x \to 0} \frac{\text{Prob}\{x < X \leq x + \Delta x\}}{\Delta x}.$$

Thus, when $\Delta x$ is small, we have

$$f(x)\Delta x \approx \text{Prob}\{x < X \leq x + \Delta x\},$$

i.e., $f(x)\Delta x$ is approximately equal to the probability that $X$ lies in a small interval, $(x, x + \Delta x]$. As $\Delta x$ tends to the infinitesimally small $dx$, we may write

$$\text{Prob}\{x < X \leq x + dx\} = f(x)dx.$$

It must follow that $f(x) \geq 0$ for all $x$.

If we know the density function of a continuous random variable, we can obtain its cumulative distribution function by integration. We have

$$F_X(x) = \text{Prob}\{X \leq x\} = \int_{-\infty}^{x} f_X(t)dt \quad \text{for } -\infty < x < \infty.$$

Since $F(\infty) = 1$, it follows that

$$\int_{-\infty}^{\infty} f_X(x)dx = 1.$$

From these considerations, we may state that a function $f(x)$ is the probability density function for some continuous random variable if and only if it satisfies the two properties:

$$f(x) \geq 0 \text{ for all } x \in \Re$$

and

$$\int_{-\infty}^{\infty} f(x)dx = 1.$$

The probability that $X$ lies in the interval $(a, b]$ is obtained from

$$\text{Prob}\{X \in (a, b]\} = \text{Prob}\{a < X \leq b\} = \text{Prob}\{X \leq b\} - \text{Prob}\{X \leq a\}$$

$$= \int_{-\infty}^{b} f_X(t)dt - \int_{-\infty}^{a} f_X(t)dt = \int_{a}^{b} f_X(t)dt.$$

Thus, the probability that the random variable lies in a given interval on the real axis is equal to the area under the probability density curve on this interval.

**Example 3.12** The density function of the continuous random variable $M$, whose cumulative distribution function is given by

$$F_M(x) = \begin{cases} 1 - e^{-x} & \text{if } x \geq 0, \\ 0 & \text{otherwise,} \end{cases}$$

is

$$f_M(x) = \begin{cases} e^{-x} & \text{if } x \geq 0, \\ 0 & \text{otherwise,} \end{cases}$$

and is shown in Figure 3.10. We have

$$F_M(x) = \int_{-\infty}^{x} f_M(t)dt = \int_{0}^{x} e^{-t}dt = -e^{-t}\Big|_0^x = -e^{-x} - (-1) = 1 - e^{-x}$$

and $\int_{-\infty}^{\infty} f_M(x)dx = \int_0^{\infty} e^{-x}dx = 1$.

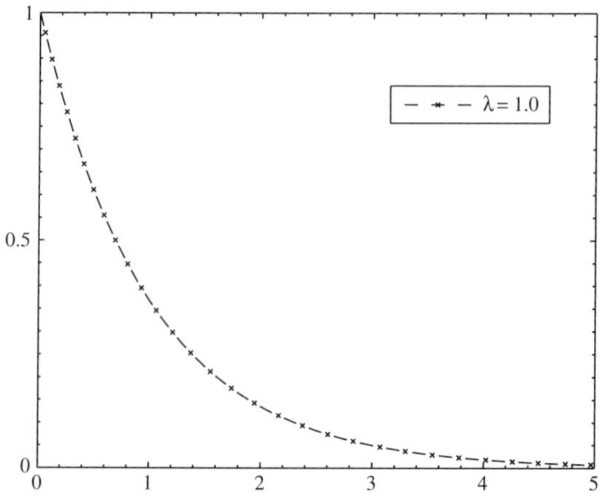


Figure 3.10. Density function for an exponentially distributed random variable.

Notice that the values of the density function are *not* probabilities and may have values greater than 1 at some point. This is illustrated in the following example.

**Example 3.13** The density function of the continuous random variable with cumulative distribution function

$$F(x) = \begin{cases} 0, & x < 0, \\ 4x, & 0 \le x < 0.25, \\ 1, & x \ge 0.25, \end{cases}$$

is given by

$$f(x) = \begin{cases} 0, & x < 0, \\ 4, & 0 \le x \le 0.25, \\ 0, & x > 0.25. \end{cases}$$

We mentioned earlier that the probability of a continuous random variable having a given value, say $c$, must be zero. This now becomes evident since the probability of the event $[X = c] = \{\omega | X(\omega) = c\}$ is given by

$$\text{Prob}\{X = c\} = \text{Prob}\{c \le X \le c\} = \int_c^c f_X(t)dt = 0.$$

The set $\{\omega | X(\omega) = c\}$ is not necessarily empty; it is just that the probability assigned to the event $\omega$ is zero. As a final remark, notice that, since the probability assigned to individual points must be zero, it follows that it makes little difference whether end points are included or not into intervals. We have

$$\begin{aligned} \text{Prob}\{a \le x \le b\} &= \text{Prob}\{a < x \le b\} = \text{Prob}\{a \le x < b\} \\ &= \text{Prob}\{a < x < b\} = \int_a^b f_X(x)dx = F_X(b) - F_X(a). \end{aligned}$$

## 3.5 Functions of a Random Variable

We have seen that a random variable $X$ is a function that assigns values in $\Re$ to each outcome in a sample space. Given a random variable $X$ we may define other random variables that are functions of $X$. For example, the random variable $Y$ defined as being equal to $2X$ takes each outcome in the sample space and assigns it a real that is equal to twice the value that $X$ assigns to it. The random variable $Y = X^2$ assigns an outcome that is the square of the value that $X$ assigns to it, and so on. In general, if a random variable $X$ assigns the value $x$ to an outcome and if $g(X)$ is a function of $X$, then $Y = g(X)$ is a random variable and assigns the value $g(x)$ to that outcome. The random variable $Y$ is said to be a *derived* random variable.

Consider now the case of a discrete random variable $X$ whose probability mass function is $p_X(x)$. Given some function $g(X)$ of $X$ we would like to create the probability mass function of the random variable $Y = g(X)$, i.e., we seek to find $p_Y(y)$. We shall proceed by means of some examples.

**Example 3.14** Let $X$ be a discrete random variable with probability mass function given by

$$p_X(x) = \begin{cases} 1/10, & x = 1, \\ 2/10, & x = 2, \\ 3/10, & x = 3, \\ 4/10, & x = 4, \\ 0 & \text{otherwise.} \end{cases}$$

With this random variable, each outcome in the sample space is mapped into one of the four real numbers 1, 2, 3, or 4. Consider now the random variable $Y = 2X$. In this case each outcome is mapped into one of the integers 2, 4, 6, or 8. But what about the probability mass function of $Y$? It is evident that, if the probability that $X$ maps an outcome into 1 is $1/10$, then the probability that $Y$ maps an outcome into 2 must also be 1/10; if $X$ maps an outcome into 2 with probability 2/10, then this must also be the probability that $Y$ maps an outcome into 4, and so on. In this specific example, we have

$$p_Y(y) = \text{Prob}\{Y = g(x)\} = \text{Prob}\{X = x\} = p_X(x).$$

However, as the next example shows, this is not always the case.

**Example 3.15** Let $X$ be a discrete random variable with probability mass function given by

$$p_X(x) = \begin{cases} 1/10, & x = -2, \\ 2/10, & x = -1, \\ 3/10, & x = 1, \\ 4/10, & x = 2, \\ 0 & \text{otherwise.} \end{cases}$$

With this random variable $X$, each outcome in the sample space is mapped into one of the four real numbers $-2$, $-1$, 1, or 2. Now consider the derived random variable $Y = X^2$. In this case, each outcome is mapped into 1 or 4. If $X$ maps an outcome into either $-1$ or 1, then $Y$ will map that outcome into 1; while if $X$ maps an outcome into $-2$ or 2, then $Y$ will map that outcome into 4. The random variable $Y$ has the following probability mass function, which is not the same as that of $X$:

$$p_Y(y) = \begin{cases} 2/10 + 3/10, & y = 1, \\ 1/10 + 4/10, & y = 4, \\ 0 & \text{otherwise.} \end{cases}$$

These examples illustrate the rule that the probability mass function of a random variable $Y$, which is a function of a random variable $X$, is equal to the probability mass function of $X$, if $g(x_1) \neq g(x_2)$ when $x_1 \neq x_2$. Otherwise the probability mass function of $Y$ is obtained from the general relation (general in the sense that it also covers the above case):

$$p_Y(y) = \sum_{x:\ g(x)=y} p_X(x).$$

Here the summation is over all $x$ for which $g(x) = y$. One final example of the applicability of this rule is now given.

**Example 3.16** Let $X$ be a discrete random variable that is defined on the integers in the interval $[-3, 4]$. Let the probability mass function of $X$ be as follows.

$$p_X(x) = \begin{cases} 0.05, & x \in \{-3, 4\}, \\ 0.10, & x \in \{-2, 3\}, \\ 0.15, & x \in \{-1, 2\}, \\ 0.20, & x \in \{0, 1\}, \\ 0 & \text{otherwise.} \end{cases}$$

We wish to find the probability mass function of the random variable $Y$ defined as $Y = X^2 - |X|$.

Since the possible values of $X$ are given by $[-3, -2, -1, 0, 1, 2, 3, 4]$, it follows that the corresponding values of $Y = X^2 - |X|$ are $[6, 2, 0, 0, 0, 2, 6, 12]$ and hence the range of $Y$ is

[0, 2, 6, 12]. This allows us to compute the probability mass function of $Y$ as

$$p_Y(y) = \begin{cases} 0.15 + 0.20 + 0.20 = 0.55 & \text{if } y = 0 \quad (g(x) = y \text{ for } x = -1, 0, 1), \\ 0.10 + 0.15 = 0.25 & \text{if } y = 2 \quad (g(x) = y \text{ for } x = -2, 2), \\ 0.05 + 0.10 = 0.15 & \text{if } y = 6 \quad (g(x) = y \text{ for } x = -3, 3), \\ 0.05 = 0.05 & \text{if } y = 12 \quad (g(x) = y \text{ for } x = 4), \\ 0 & \text{otherwise.} \end{cases}$$

We now consider functions of a *continuous* random variable $X$, and again we shall first provide some illustrative examples.

**Example 3.17** Let $X$ be a continuous random variable that is uniformly distributed in the interval $[-1, +1]$. Let $Y = \alpha X + \beta$, with $\alpha > 0$, be a derived random variable. Observe that the probability density function of $X$ is given by

$$f_X(x) = \begin{cases} 1/2, & -1 \le x \le 1, \\ 0 & \text{otherwise,} \end{cases}$$

since $X$ is uniformly distributed and the area of the density function between $x = -1$ and $x = 1$ must be equal to 1. This density function has the shape of a rectangle with base $[-1, +1]$ and height 1/2. The cumulative distribution function of $X$ is given by

$$F_X(x) = \begin{cases} 0, & x \le -1, \\ (x+1)/2, & -1 \le x \le 1, \\ 1, & x \ge 1. \end{cases}$$

Now consider the random variable $Y$. The effect of multiplying $X$ by $\alpha$ is to multiply the interval $[-1, +1]$ by $\alpha$. This becomes the interval over which $Y$ has nonzero values. Since $Y$ is also uniformly distributed, its density function will be rectangular in shape, but its height must be $1/\alpha$ times the height of $f_X(x)$ so that the area beneath $f_Y(y)$ between $-\alpha$ and $+\alpha$ remains equal to 1. The effect of adding $\beta$ is to shift the interval over which nonzero values of $Y$ occur. Taking these together, we obtain

$$f_Y(y) = \begin{cases} 1/2\alpha, & \beta - \alpha \le y \le \beta + \alpha, \\ 0 & \text{otherwise.} \end{cases}$$

The cumulative distribution of $Y$ is given as

$$F_Y(y) = \begin{cases} 0, & y \le \beta - \alpha, \\ (y + \alpha - \beta)/2\alpha, & \beta - \alpha \le y \le \beta + \alpha, \\ 1, & y \ge \beta + \alpha. \end{cases}$$

Observing that the interval $\beta - \alpha \le y \le \beta + \alpha$ may be written as

$$-1 \le \frac{y - \beta}{\alpha} \le 1,$$

the probability density function and the cumulative distribution of $Y$ may be written in terms of those same functions of $X$. We have

$$f_Y(y) = \frac{1}{\alpha} f_X\left(\frac{y-\beta}{\alpha}\right) \quad \text{and} \quad F_Y(y) = F_X\left(\frac{y-\beta}{\alpha}\right).$$

These relations hold in general for any continuous random variable $X$ and a random variable $Y$ derived from $X$ by multiplication with a positive constant $\alpha$ and shifted by an amount $\beta$. In the absence of a shift $\beta$, these relations simplify to

$$f_Y(y) = \frac{1}{\alpha} f_X(y/\alpha) \quad \text{and} \quad F_Y(y) = F_X(y/\alpha)$$

as the following example shows.

**Example 3.18** Let $X$ be a continuous random variable having a wedge-shaped probability density function on the interval [0, 1]. Specifically, let

$$f_X(x) = \begin{cases} 2x, & 0 \le x \le 1, \\ 0 & \text{otherwise,} \end{cases} \quad \text{and} \quad F_X(x) = \begin{cases} 0, & x \le 0, \\ x^2, & 0 \le x \le 1, \\ 1, & x \ge 1, \end{cases}$$

where $F_X(x)$ has been derived from $f_X(x)$ by integration. The probability density function and cumulative probability function of the derived random variable $Y = 5X$ are given by

$$f_Y(y) = \begin{cases} 2y/25, & 0 \le y \le 5, \\ 0 & \text{otherwise,} \end{cases} \quad \text{and} \quad F_Y(y) = \begin{cases} 0, & y \le 0, \\ y^2/25, & 0 \le y \le 5, \\ 1, & y \ge 5. \end{cases}$$

For $Y = \alpha X$, these functions are given by

$$f_Y(y) = \begin{cases} 2y/\alpha^2, & 0 \le y \le \alpha, \\ 0 & \text{otherwise,} \end{cases} \quad \text{and} \quad F_Y(y) = \begin{cases} 0, & y \le 0, \\ y^2/\alpha^2, & 0 \le y \le \alpha, \\ 1, & y \ge \alpha, \end{cases}$$

and again

$$f_Y(y) = \frac{1}{\alpha} f_X(y/\alpha) \quad \text{and} \quad F_Y(y) = F_X(y/\alpha).$$

To conclude this section, we consider two examples of derived random variables that are more complex than those obtained by scalar multiplication and shift.

**Example 3.19** The cumulative distribution function of an exponential random variable $X$ with parameter $\lambda$ is given by

$$\text{Prob}\{X \le x\} = \begin{cases} 1 - e^{-\lambda x}, & x \ge 0, \\ 0 & \text{otherwise.} \end{cases}$$

Let us compute the cumulative distribution function and the density function of a random variable $Y$ defined as $Y = X^2$. We have

$$\text{Prob}\{Y \le y\} = \text{Prob}\{X^2 \le y\} = \text{Prob}\{X \le \sqrt{y}\} = 1 - e^{-\lambda\sqrt{y}} \quad \text{for } y \ge 0.$$

We may now compute the density function of $Y$ by differentiation. We obtain

$$f_Y(y) = \frac{d}{dy}\text{Prob}\{Y \le y\} = \lambda e^{-\lambda y^{1/2}} \times \frac{1}{2} y^{-1/2} = \frac{\lambda}{2\sqrt{y}} e^{-\lambda\sqrt{y}} \quad \text{for } y > 0.$$

In simulation experiments it is frequently the case that one has access to sequences of random numbers that are uniformly distributed over the interval (0, 1), but what is really required are random numbers from a nonuniform distribution. In particular, in queuing scenarios, random numbers that are distributed according to a (negative) exponential distribution are often needed. This last example shows how such numbers may be obtained.

**Example 3.20** Let $X$ be a random variable that is uniformly distributed on (0, 1) and define a derived random variable $Y$ as $Y = -\ln(1 - X)/\lambda$. The cumulative distribution function of $Y$ is given by

$$\begin{aligned} F_Y(y) &= \text{Prob}\{Y \le y\} = \text{Prob}\{-\ln(1 - X)/\lambda \le y\} \\ &= \text{Prob}\{\ln(1 - X) \ge -\lambda y\} = \text{Prob}\{1 - X \ge e^{-\lambda y}\} \\ &= \text{Prob}\{X \le 1 - e^{-\lambda y}\} = F_X(1 - e^{-\lambda y}). \end{aligned}$$

Since $X$ is uniformly distributed on the interval (0, 1), we must have $F_X(x) = x$ on this interval. In particular $F_X(1 - e^{-\lambda y}) = 1 - e^{-\lambda y}$ and therefore $F_Y(y) = 1 - e^{-\lambda y}$ for $0 \le 1 - e^{-\lambda y} \le 1$.

Since all nonnegative values of $y$ satisfy these bounds, we have

$$F_Y(y) = \begin{cases} 0, & y \le 0, \\ 1 - e^{-\lambda y}, & y \ge 0. \end{cases}$$

This defines the exponential distribution with parameter $\lambda$.

A function $g(x)$ of $x$ is said to be an *increasing* function of $x$ if for $x_1 \le x_2$, $g(x_1) \le g(x_2)$. Similarly, it is said to be a *decreasing* function of $x$ if for $x_1 \le x_2$, $g(x_1) \ge g(x_2)$. The next example shows that, when the random variable $Y$ is an increasing or decreasing function of $X$, and the density function of $X$ is provided, then it is *not* necessary to first find the cumulative distributions of $X$ and then $Y$ to compute the density function of $Y$.

**Example 3.21** Let the probability density function of a random variable $X$ be given by

$$f_X(x) = \begin{cases} \alpha/x^5, & 1 \le x < \infty, \\ 0 & \text{otherwise.} \end{cases}$$

Let a new random variable be defined as $Y = 1/X$. We seek the probability density function of $Y$.

We shall first answer the question using the standard approach. Our first task is to compute the value of $\alpha$. Since we must have

$$1 = \int_{-\infty}^{\infty} f_X(x) = \int_1^{\infty} \frac{\alpha}{x^5} = -\alpha \frac{x^{-4}}{4}\bigg|_1^{\infty} = \frac{\alpha}{4},$$

we must have $\alpha = 4$. The cumulative distribution function of $X$ may now be computed as

$$F_X(x) = \int_1^x f_X(t)dt = \int_1^x \frac{4}{t^5}dt = -t^{-4}\big|_1^x = 1 - \frac{1}{x^4}, \quad x \ge 1.$$

Now using probabilities, and observing that the range of $Y$ is $(0, 1]$, we can compute the cumulative distribution function of $Y$ as

$$\text{Prob}\{Y \le y\} = \text{Prob}\{1/X \le y\} = \text{Prob}\{X \ge 1/y\} = 1 - \text{Prob}\{X < 1/y\} = 1 - F_X(1/y) = y^4$$

for $0 < Y \le 1$. Finally, we may now compute the probability density function of $Y$ by taking derivatives. We obtain

$$f_Y(y) = \frac{d}{dy}F_Y(y) = \frac{d}{dy}y^4 = 4y^3, \quad 0 < y \le 1.$$

In examples like this last one, in which $Y$ is a decreasing function of $X$ (as $x$ gets larger, $y = 1/x$ gets smaller), an easier approach may be taken. Indeed, it may be shown that if $Y = g(X)$ is an increasing or a decreasing function of $X$ then

$$f_Y(y) = f_X(x)\left|\frac{dx}{dy}\right|, \tag{3.3}$$

where, in this expression, $x$ is the value obtained when $y = g(x)$ is solved for $x$.

To show that Equation (3.3) is true for increasing functions, consider the following. Let $Y = g(X)$ be an increasing function of $X$. Then, because $g(X)$ is an increasing function

$$F_Y(y) = \text{Prob}\{Y \le y\} = \text{Prob}\{g(X) \le g(x)\} = \text{Prob}\{X \le x\} = F_X(x).$$

It now follows that

$$f_Y(y) = \frac{d}{dy}F_Y(y) = \frac{d}{dy}F_X(x) = F'_X(x)\frac{dx}{dy} = f_X(x)\frac{dx}{dy}.$$

In the example, $y = 1/x$, so $x = 1/y$ and substituting into Equation (3.3), we have

$$f_Y(y) = \frac{4}{(1/y)^5}\left|\frac{-1}{y^2}\right| = 4y^3$$

as before. We leave the proof of the case when $Y$ is a *decreasing* function of $X$ as an exercise.

## 3.6 Conditioned Random Variables

We saw earlier that an event $\mathcal{A}$ may be conditioned by a different event $\mathcal{B}$ and that the probability assigned to $\mathcal{A}$ may change, stay the same, or even become zero as a result of knowing that the event $\mathcal{B}$ occurs. We wrote this as Prob$\{\mathcal{A}|\mathcal{B}\}$, the probability of event $\mathcal{A}$ given the event $\mathcal{B}$. Since a random variable $X$ defines events on a sample space, it follows that the probabilities assigned to random variables may also change upon knowing that a certain event $\mathcal{B}$ has occurred. The event $[X = x]$ contains all the outcomes which are mapped by the random variable $X$ onto the real number $x$ and its probability, Prob$\{X = x\}$, is equal to the sum of the probabilities of these outcomes. Knowing that an event $\mathcal{B}$ has occurred may alter Prob$\{X = x\}$ for all possible values of $x$. The conditional probabilities Prob$\{X = x|\mathcal{B}\}$ are defined whenever Prob$\{\mathcal{B}\} > 0$. When the random variable $X$ is discrete, Prob$\{X = x|\mathcal{B}\}$ is called the *conditional probability mass function* of $X$ and is denoted by $p_{X|\mathcal{B}}(x)$. From our prior definition of conditional probability for two events, we may write

$$p_{X|\mathcal{B}}(x) = \frac{\text{Prob}\{[X = x] \cap \mathcal{B}\}}{\text{Prob}\{\mathcal{B}\}}.$$

In many practical examples, the event $[X = x]$ is contained entirely within the event $\mathcal{B}$, or else the intersect of $[X = x]$ and $\mathcal{B}$ is the null event. In the first case we have

$$p_{X|\mathcal{B}}(x) = \frac{p_X(x)}{\text{Prob}\{\mathcal{B}\}} \quad \text{if} \quad [X = x] \subset \mathcal{B} \tag{3.4}$$

while in the second $[X = x] \cap \mathcal{B} = \phi$ and hence $p_{X|\mathcal{B}}(x) = 0$. We now give three examples.

**Example 3.22** Let $X$ be the random variable that counts the number of spots obtained when two fair dice are thrown. The probability mass function for $X$ is

| $x_i$ | 2 | 3 | 4 | 5 | 6 | 7 | 8 | 9 | 10 | 11 | 12 |
|---|---|---|---|---|---|---|---|---|---|---|---|
| $p_X(x_i)$ | 1/36 | 2/36 | 3/36 | 4/36 | 5/36 | 6/36 | 5/36 | 4/36 | 3/36 | 2/36 | 1/36 |

Let $\mathcal{B}$ be the event that the throw gives an even number of spots on one of the dice and an odd number of spots on the other. It then follows that the event $\mathcal{B}$ contains 18 outcomes. Each outcome is a pair with the property that the first can be any of six numbers, but the second must be one of the three even numbers (if the first is odd) or one of three odd numbers (if the first is even). Since all outcomes are equally probable, we have Prob$\{\mathcal{B}\} = 1/2$. Given this event, we now compute the conditional probability mass function, $p_{X|\mathcal{B}}$. Since the intersection of $\mathcal{B}$ and $[X = x]$ when $x$ is an even number is empty, it immediately follows that the probability of the sum being an even number is zero. For permissible odd values of $x$, the event $[X = x]$ is entirely contained within the event $\mathcal{B}$ and we may use Equation (3.4). In summary, the conditional probability mass function of $X$ is obtained as

$$p_{X|\mathcal{B}}(x) = \begin{cases} p_X(x)/\text{Prob}\{\mathcal{B}\}, & x = 3, 5, 7, 9, 11, \\ 0 & \text{otherwise}, \end{cases}$$

and is presented in tabular form as follows.

| $x_i$ | 2 | 3 | 4 | 5 | 6 | 7 | 8 | 9 | 10 | 11 | 12 |
|---|---|---|---|---|---|---|---|---|---|---|---|
| $p_{X\|\mathcal{B}}(x_i)$ | 0 | 2/18 | 0 | 4/18 | 0 | 6/18 | 0 | 4/18 | 0 | 2/18 | 0 |

**Example 3.23** Let $X$ be the same random variable as before, but now let $\mathcal{B}$ be the event that the sum obtained is strictly greater than 8, i.e., $\mathcal{B} = \{X > 8\}$. There are a total of ten outcomes that give a sum greater than 8, which means that $\text{Prob}\{X > 8\} = 10/36$. The conditional probability mass function, now written as $p_{X|X>8}(x)$, may be obtained from Equation (3.4) and we find

| $x_i$ | 2 | 3 | 4 | 5 | 6 | 7 | 8 | 9 | 10 | 11 | 12 |
|---|---|---|---|---|---|---|---|---|---|---|---|
| $p_{X\|X>8}(x_i)$ | 0 | 0 | 0 | 0 | 0 | 0 | 0 | 4/10 | 3/10 | 2/10 | 1/10 |

If $\mathcal{B}_i$, $i = 1, 2, \ldots, n$, is a set of mutually exclusive and collectively exhaustive events, i.e., an event space, then

$$p_X(x) = \sum_{i=1}^{n} p_{X|\mathcal{B}_i}(x)\text{Prob}\{\mathcal{B}_i\}.$$

This result follows immediately from applying the law of total probability to the event $[X = x]$.

**Example 3.24** Let $X$ be a random variable that denotes the duration in months that it takes a newly graduated computer science student to find suitable employment. Statistics indicate that the number of months, $n$ is distributed according to the formula $\alpha(1-\alpha)^{n-1}$, with $\alpha = 0.75$ for students whose GPA exceeds 3.5, $\alpha = 0.6$ for students whose GPA is higher that 2.8 but does not exceed 3.5, and $\alpha = 0.2$ for all other students. This leads to the following conditional probability mass functions, in which we indicate student performance by the letters $H$, $M$, or $L$ depending on whether their GPA is high, medium, or low, respectively:

$$\begin{aligned} p_{X|H}(n) &= 0.75(0.25)^{n-1} \quad \text{for } n = 1, 2, \ldots, \\ p_{X|M}(n) &= 0.60(0.40)^{n-1} \quad \text{for } n = 1, 2, \ldots, \\ p_{X|L}(n) &= 0.20(0.80)^{n-1} \quad \text{for } n = 1, 2, \ldots. \end{aligned}$$

If we know that 20% of the students have GPAs in excess of 3.5 and 35% have GPAs less than 2.8, then the probability mass function of $X$ is given by

$$p_X(n) = 0.2 \times 0.75(0.25)^{n-1} + 0.45 \times 0.60(0.40)^{n-1} + 0.35 \times 0.20(0.80)^{n-1} \quad \text{for } n = 1, 2, \ldots.$$

These concepts carry over in a natural fashion to random variables that are continuous. For a continuous random variable $X$, and a conditioning event $\mathcal{B}$, we may write

$$f_{X|\mathcal{B}}(x)dx = \text{Prob}\{x < X \leq x + dx|\mathcal{B}\} = \frac{\text{Prob}\{[x < X \leq x + dx] \cap \mathcal{B}\}}{\text{Prob}\{\mathcal{B}\}}.$$

Thus

$$f_{X|\mathcal{B}}(x) = \begin{cases} f_X(x)/\text{Prob}\{\mathcal{B}\} & \text{if } [X = x] \subset \mathcal{B}, \\ 0 & \text{if } [X = x] \cap \mathcal{B} = \phi. \end{cases}$$

**Example 3.25** Let $X$ be a continuous random variable with probability density function given by

$$f_X(x) = \begin{cases} (1/2)e^{-x/2} & \text{if } x \geq 0, \\ 0 & \text{otherwise.} \end{cases}$$

Let $\mathcal{B}$ be the event that $X < 1$ and we wish to find $f_{X|X<1}(x)$. We first compute $\text{Prob}\{X < 1\}$ as

$$\text{Prob}\{X < 1\} = \int_0^1 (1/2)e^{-x/2}dx = -e^{-x/2}\Big|_0^1 = 1 - e^{-1/2}.$$

The conditional probability density function of $X$ is then given by

$$f_{X|X<1}(x) = \begin{cases} f_X(x)/\text{Prob}\{X < 1\} & \text{for } 0 \le x < 1, \\ 0 & \text{otherwise,} \end{cases}$$

$$= \begin{cases} (1/2)e^{-x/2}/(1 - e^{-1/2}) & \text{for } 0 \le x < 1, \\ 0 & \text{otherwise.} \end{cases}$$

The conditional cumulative distribution function, $F_{X|\mathcal{B}}(x|\mathcal{B})$, of a random variable $X$, given that the event $\mathcal{B}$ occurs, is defined as

$$F_{X|\mathcal{B}}(x|\mathcal{B}) = \frac{\text{Prob}\{X \le x, \mathcal{B}\}}{\text{Prob}\{\mathcal{B}\}},$$

where $(X \le x, \mathcal{B})$ is the intersection of the events $[X \le x]$ and $\mathcal{B}$. Note that we must have

$$F_{X|\mathcal{B}}(-\infty|\mathcal{B}) = 0 \text{ and } F_{X|\mathcal{B}}(\infty|\mathcal{B}) = 1.$$

Also,

$$\text{Prob}\{x_1 < X \le x_2|\mathcal{B}\} = F_{X|\mathcal{B}}(x_2|\mathcal{B}) - F_{X|\mathcal{B}}(x_1|\mathcal{B}) = \frac{\text{Prob}\{[x_1 < X \le x_2], \mathcal{B}\}}{\text{Prob}\{\mathcal{B}\}}.$$

The conditional density function may be obtained as the derivative:

$$f_{X|\mathcal{B}}(x|\mathcal{B}) = \frac{d}{dx}F_{X|\mathcal{B}}(x|\mathcal{B}),$$

which must be nonnegative and have area equal to 1.

## 3.7 Exercises

**Exercise 3.1.1** A fair die is thrown once. Alice wins \$2.00 if one, two, or three spots appear and \$4.00 if four or five spots appear, but pays Bob \$6.00 if six spots appear. Let $W$ be the random variable that represents the amount Alice wins or loses. What is the domain and range of $W$? Give the partition of the state space induced by this random variable. What is the set $A_4$?

**Exercise 3.1.2** A four-sided die (a tetrahedron having the numbers 1, 2, 3, and 4 printed on it, one per side) is thrown twice. Let $M$ be the random variable that denotes the maximum number obtained on the two throws. What is the domain and range of $M$? Give the partition of the state space induced by this random variable. What is the set $A_3$?

**Exercise 3.1.3** A survey reveals that 20% of students at Tobacco Road High School smoke while 15% drink beer, both considered to be bad habits. Let $X$ be the random variable that denotes the number of these bad habits indulged by a randomly chosen student. What is the domain and range of $X$? Describe the partition of the state space induced by this random variable.

**Exercise 3.2.1** Let $X$ be a random variable whose probability mass function is given as

$$p_X(x) = \begin{cases} \alpha x^2, & x = 1, 2, 3, 4, \\ 0 & \text{otherwise.} \end{cases}$$

(a) What is the value of $\alpha$?
(b) Compute $\text{Prob}\{X = 4\}$.
(b) Compute $\text{Prob}\{X \le 2\}$.

**Exercise 3.2.2** Let $X$ be a random variable whose probability mass function is given as

$$p_X(x) = \begin{cases} \alpha/x, & x = 1, 2, 3, 4, \\ 0 & \text{otherwise.} \end{cases}$$

(a) What is the value of $\alpha$?
(b) Compute Prob$\{X$ is odd$\}$.
(c) Compute Prob$\{X > 2\}$.

**Exercise 3.2.3** Balls are drawn from an urn containing two white balls and five black balls until a white ball appears. Let $X$ be the random variable that denotes the number of black balls drawn *before* a white ball appears. What is the domain and range of $X$ and what is the partition induced on the sample space by $X$? Give a table of the probability mass function of $X$.

**Exercise 3.2.4** One of the enclosures in a wildlife park contains a herd of sixteen elephants, six rhinos, and ten hippos. Off in an isolated corner of the enclosure, four of these animals (maybe young males?) have become agitated and are menacing each other. Construct a table of the probability mass function of the random variable that counts the number of tusks in this group of four. For the purposes of answering this question, you should assume that the four in question are equally likely to come from any of the 32 animals in the enclosure. *Assume that elephants have two tusks, rhinos one, and hippos none.*

**Exercise 3.2.5** An urn contains seven white balls and five black ones. Suppose $n$ balls are chosen at random. Let the random variable $X$ denote the number of white balls in the sample. What is the probability mass function of $X$ if the $n$ balls are chosen (a) without replacement? (b) with replacement?

**Exercise 3.3.1** A discrete random variable $X$ takes the value 1 if the number 6 appears on a single throw of a fair die and the value 0 otherwise. Sketch the probability mass function and the probability distribution function of $X$.

**Exercise 3.3.2** The cumulative distribution function of a discrete random variable $X$ is given as

$$F_X(x) = \begin{cases} 0, & x < 0, \\ 1/4, & 0 \le x < 1, \\ 1/2, & 1 \le x < 2, \\ 1, & x \ge 2. \end{cases}$$

Find the probability mass function of $X$.

**Exercise 3.3.3** The cumulative distribution function of a discrete random variable $N$ is given as

$$F_N(x) = 1 - \frac{1}{2^n} \quad \text{if } x \in [n, n+1) \ \text{ for } n = 1, 2, \ldots$$

and has the value 0 if $x < 1$.

(a) What is the probability mass function of $N$?
(b) Compute Prob$\{4 \le N \le 10\}$.

**Exercise 3.3.4** A bag contains two fair coins and a third coin which is biased: the probability of tossing a head on this third coin is 3/4. A coin is pulled at random and tossed three times. Let $X$ be the random variable that counts the number of heads obtained in these three tosses. Give the probability mass function and the cumulative density function of $X$.

**Exercise 3.3.5** Let $X$ be a continuous random variable whose cumulative distribution function is

$$F_X(x) = \begin{cases} 0, & x < 0, \\ x/5, & 0 \le x \le 5, \\ 1, & x > 5. \end{cases}$$

Compute the following probabilities:

(a) Prob$\{X \le 1\}$.
(b) Prob$\{X > 3\}$.
(c) Prob$\{2 < X \le 4\}$.

**Exercise 3.4.1** The cumulative distribution function of a continuous random variable $X$ is given by

$$F_X(x) = \begin{cases} 0, & x < 0, \\ x^2/4, & 0 \le x \le 2, \\ 1, & x > 2. \end{cases}$$

Find the probability density function of $X$.

**Exercise 3.4.2** Let $X$ be a continuous random variable whose probability density function is given by

$$f_X(x) = \begin{cases} \alpha(1+x)^{-3}, & x > 0, \\ 0 & \text{otherwise.} \end{cases}$$

Find $\alpha$ and Prob$\{0.25 \le x \le 0.5\}$.

**Exercise 3.4.3** The cumulative distribution function of a continuous random variable $E$ is given by

$$F_E(x) = \begin{cases} 0, & x < 0, \\ 1 - e^{-\mu x} - \mu x e^{-\mu x}, & 0 \le x < \infty, \end{cases}$$

and $\mu > 0$. Find the probability density function of $E$.

**Exercise 3.4.4** Let the probability density function of a continuous random variable $X$ be given by

$$f_X(x) = \begin{cases} 1/4, & -2 \le x \le 2, \\ 0 & \text{otherwise.} \end{cases}$$

Find the cumulative distribution function of $X$.

**Exercise 3.4.5** The probability density function for a continuous "Rayleigh" random variable $X$ is given by

$$f_X(x) = \begin{cases} \alpha^2 x e^{-\alpha^2 x^2/2}, & x > 0, \\ 0 & \text{otherwise.} \end{cases}$$

Find the cumulative distribution of $X$.

**Exercise 3.4.6** Let $f_1(x),\ f_2(x),\ \ldots,\ f_n(x)$ be a set of $n$ probability density functions and let $p_1, p_2, \ldots, p_n$ be a set of probabilities for which $\sum_{i=1}^n p_i = 1$. Prove that $\sum_{i=1}^n p_i f_i(x)$ is a probability density function.

**Exercise 3.5.1** The cumulative distribution function of a discrete random variable $X$ is given as

$$F_X(x) = \begin{cases} 0, & x < 0, \\ 1/4, & 0 \le x < 1, \\ 1/2, & 1 \le x < 2, \\ 1, & x \ge 2. \end{cases}$$

Find the probability mass function and the cumulative distribution function of $Y = X^2$.

**Exercise 3.5.2** A discrete random variable $X$ assumes each of the values of the set $\{-10, -9, \ldots, 9, 10\}$ with equal probability. In other words, $X$ is a discrete integer-valued random variable that is uniformly distributed on the interval $[-10, 10]$. Compute the following probabilities:

$$\text{Prob}\{4X \le 2\}, \qquad \text{Prob}\{4X + 4 \le 2\}, \qquad \text{Prob}\{X^2 - X \le 3\}, \qquad \text{Prob}\{|X - 2| \le 2\}.$$

**Exercise 3.5.3** Homeowners in whose homes wildlife creatures take refuge call "Critter Control" to rid them of these "pests." Critter Control charges \$25 for each animal trapped and disposed of as well as a flat fee of \$45 per visit. Experience has shown that the distribution of trapped animals found per visit is as follows.

| $x_i$ | 0 | 1 | 2 | 3 | 4 |
|---|---|---|---|---|---|
| $p_X(x_i)$ | 0.5 | 0.25 | 0.15 | 0.05 | 0.05 |

Find the probability mass function of the amount of money Critter Control collects per visit.

**Exercise 3.5.4** Let the probability distribution function of a continuous random variable $X$ be given by

$$F_X(x) = \begin{cases} 1 - e^{-2x}, & 0 < x < \infty, \\ 0 & \text{otherwise.} \end{cases}$$

Find the cumulative distribution function of $Y = e^X$.

**Exercise 3.5.5** Show that if the derived random variable $Y = g(X)$ is a *decreasing* function of a random variable $X$, then

$$f_Y(y) = f_X(x) \left| \frac{dx}{dy} \right|,$$

where, in this expression, $x$ is the value obtained when $y = g(x)$ is solved for $x$.

**Exercise 3.6.1** The probability mass function of a discrete integer-valued random variable is

$$p_X(x) = \begin{cases} 1/4, & x = -2, \\ 1/8, & x = -1, \\ 1/8, & x = 0, \\ 1/2, & x = 1, \\ 0 & \text{otherwise.} \end{cases}$$

Find $p_{X|\mathcal{B}}(x)$ where $\mathcal{B} = [X < 0]$.

**Exercise 3.6.2** The probability mass function of a discrete random variable is given by

$$p_X(x) = \begin{cases} \alpha/x, & x = 1, 2, 3, 4, \\ 0 & \text{otherwise.} \end{cases}$$

Find $p_{X|\mathcal{B}}(x)$ where $\mathcal{B} = [X \text{ is odd}]$.

**Exercise 3.6.3** The density function of a continuous random variable is given by

$$f_X(x) = \begin{cases} 1/20, & -10 \le x \le 10, \\ 0 & \text{otherwise.} \end{cases}$$

Find the conditional density functions $f_{X|X<0}(x)$ and $f_{X|X>5}(x)$.

**Exercise 3.6.4** Let $X$ be a continuous random variable whose probability density function is given by

$$f_X(x) = \begin{cases} \alpha(1+x)^{-3}, & x > 0, \\ 0 & \text{otherwise.} \end{cases}$$

Find the conditional density function $f_{X|0.25 \le X \le 0.5}(x)$. Verify that your answer is correct by integrating $f_{X|0.25 \le X \le 0.5}(x)$ with respect to $x$ between the limits 0.25 and 0.5 and showing that this gives the value 1.

# Chapter 4

# *Joint and Conditional Distributions*

## 4.1 Joint Distributions

In the previous chapter we discussed individual random variables that were either discrete or continuous. In this chapter we shall examine the relationships that may exist between two or more random variables. This is often a more common modeling problem than one that involves a single random variable. For example, we may be interested in relating the random variable $X$ that measures the strength of a cell phone signal at some location and the random variable $Y$ that measure the distance of that location from the nearest coverage tower. Or we may be interested in relating a random variable $X$ that defines the arrival process of patients to a doctor's office and the random variable $Y$ that measures the length of time those customers spend in the waiting room. As a final example, we may be interested in examining a situation in which jobs arrive at a machine shop in which the machines are subject to failure. A first random variable may be used to indicate the number of jobs present and a second may be used to represent the number of working machines.

Sums, differences, and products of random variables are often of importance, as are relations such as the maximum and minimum. Consider the execution of a computer program which consists of different sections whose time to complete depends on the amount and type of data provided. It follows that the time required to execute the entire program is also data dependent. The execution time of the $i^{\text{th}}$ section may be represented by a random variable $X_i$. Let the random variable $X$ represent the time needed to execute the entire program. If the sections are executed one after the other, then $X$ is given by $X = \sum_i X_i$. If the sections are executed in parallel, then $X = \max\{X_i;\ i = 1, 2, \ldots\}$.

We begin by defining the joint cumulative distribution function for two random variables and then consider joint probability mass functions for two discrete random variables and the joint probability density function for two continuous random variables. These latter two are generally more useful in actual applications that the joint cumulative distribution function.

## 4.2 Joint Cumulative Distribution Functions

The joint cumulative distribution function of two random variables $X$ and $Y$ is given by

$$F_{X,Y}(x, y) = \text{Prob}\{X \le x, Y \le y\};\quad -\infty < x < \infty,\quad -\infty < y < \infty.$$

To be a bona fide joint CDF, a function must possess the following properties, similar to those found in the CDF of a single random variable:

- $0 \le F_{X,Y}(x, y) \le 1$ for $-\infty < x < \infty; -\infty < y < \infty$, since $F_{X,Y}(x, y)$ is a probability.
- $F_{X,Y}(x, -\infty) = F_{X,Y}(-\infty, y) = 0$,
  $F_{X,Y}(-\infty, -\infty) = 0$, $F_{X,Y}(+\infty, +\infty) = 1$.
- $F_{X,Y}$ must be a nondecreasing function of both $x$ and $y$, i.e.,

$$F_{X,Y}(x_1, y_1) \le F_{X,Y}(x_2, y_2) \;\text{ if }\; x_1 \le x_2 \;\text{ and }\; y_1 \le y_2.$$

**Example 4.1** The cumulative distribution function of a single discrete random variable $X$, given in tabular form, is as follows.

| | $-2 < x$ | $-2 \le x < -1$ | $-1 \le x < 0$ | $0 \le x < 1$ | $1 \le x$ |
|---|---|---|---|---|---|
| $F_X(x)$ | 0 | 1/8 | 1/4 | 1/2 | 1 |

This tells us, for example, that Prob$\{X < 1.0\} = 1/2$. Also, Prob$\{X \le 0.5\} = 1/2$, Prob$\{X \le 0.0\} = 1/2$, and so on. Similarly, Prob$\{X < -1.0\} = 1/8$, Prob$\{X \le -1.5\} = 1/8$, and so on. Using the same approach, we may display the joint cumulative distribution function of two discrete random variables $X$ and $Y$ in tabular form. An example is the following:

| | | | | | |
|---|---|---|---|---|---|
| $y \ge 2$ | 0 | 1/8 | 1/4 | 1/2 | 1 |
| $0 \le y < 2$ | 0 | 3/32 | 3/16 | 3/8 | 3/4 |
| $-2 \le y < 0$ | 0 | 1/32 | 1/16 | 1/8 | 1/4 |
| $y < -2$ | 0 | 0 | 0 | 0 | 0 |
| $F_{X,Y}(x, y)$ | $-2 < x$ | $-2 \le x < -1$ | $-1 \le x < 0$ | $0 \le x < 1$ | $x \ge 1$ |

From this table, it may be seen, for example, that Prob$\{X < 1, Y < 0\} = 1/8$, that Prob$\{X \le 0.5, Y < 0\} = 1/8 =$ Prob$\{X \le 0.0, Y < 0\}$, that Prob$\{X \le -1.3, Y \le 1.3\} = 3/32$, that Prob$\{X \ge 1, Y < 1\} = 3/4$, and so on. Observed that the conditions detailed above for a function to be a bona fide CDF all hold in this example.

Whereas a single random variable maps outcomes into points on the real line, joint random variables map outcomes into points in a higher-dimensional space. For example, two random variables map outcomes into the $(x, y)$ plane, three random variables map outcomes into three-dimensional space $(x, y, z)$, and so on. For a single random variable, the value of the cumulative distribution function at a point $x$ is equal to the probability of the event that contains all outcomes that are mapped into points less than or equal to $x$. The value of the cumulative distribution of two random variables $X$ and $Y$ at a point $(x, y)$ is the probability of the event that contains all outcomes that are mapped into the infinite area that lies to the left of a vertical line through $X = x$ and below a horizontal line through $Y = y$.

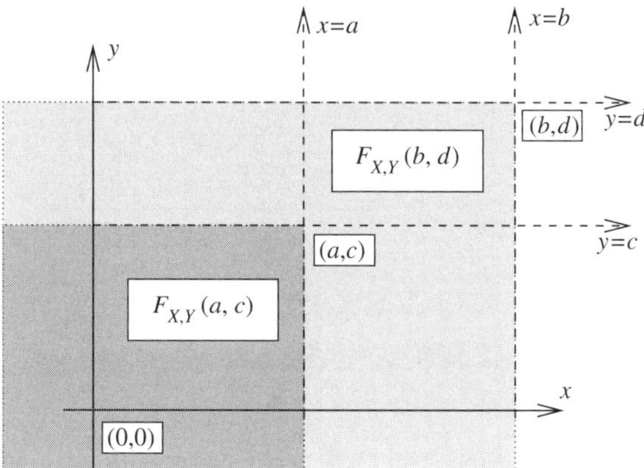


Figure 4.1. Events $\{X \le b, Y \le d\}$ and $\{X \le a, Y \le c\}$ represented as shaded areas.

It may be observed from Figure 4.1 that $F_{X,Y}(b, d) =$ Prob$\{X \le b, Y \le d\}$ is equal to the probability of the event that contains all the outcomes within the shaded areas (both light and heavy shading), while $F_{X,Y}(a, c) =$ Prob$\{X \le a, Y \le c\}$ is equal to the probability of the event that contains all the outcomes within the more heavily shaded area. Other results follow. For example, we have

$$\text{Prob}\{a < X \le b,\ Y \le c\} = F_{X,Y}(b, c) - F_{X,Y}(a, c)$$

and

$$\text{Prob}\{X \le a,\ c < Y \le d\} = F_{X,Y}(a,d) - F_{X,Y}(a,c).$$

Use of the joint cumulative distribution function of two random variables to compute the probability that $(X, Y)$ lies in a *rectangle* in the plane is not unduly complicated. We may use the result

$$\text{Prob}\{a < X \le b,\ c < Y \le d\} = F_{X,Y}(b,d) - F_{X,Y}(a,d) - F_{X,Y}(b,c) + F_{X,Y}(a,c), \quad (4.1)$$

which requires that the joint cumulative distribution function be evaluated at each of the four points of the rectangle. Unfortunately, when the required probability is found in a nonrectangular region, the joint CDF becomes much more difficult to use. This is one of the reasons why joint CDFs are infrequently used. Nevertheless, Equation (4.1) is a useful special case whose proof constitutes an enlightening exercise.

We proceed as follows: let $A$ represent the rectangle with corners $(a, c)$, $(b, c)$, $(b, d)$, and $(a, d)$ (counterclockwise order), i.e., the rectangle within which we seek the probability that $X, Y$ can be found. Then $\text{Prob}\{A\} = \text{Prob}\{a < X \le b,\ c < Y \le d\}$. Let $B$ represent the semi-infinite rectangle that lies to the left of $A$, and let $C$ represent the semi-infinite rectangle that lies below $A$, as illustrated in Figure 4.2. Observe that $A$, $B$, and $C$ are events that have no points in common, i.e., they are mutually exclusive, and so

$$\text{Prob}\{A \cup B \cup C\} = \text{Prob}\{A\} + \text{Prob}\{B\} + \text{Prob}\{C\},$$

and hence

$$\text{Prob}\{A\} = \text{Prob}\{A \cup B \cup C\} - \text{Prob}\{B\} - \text{Prob}\{C\}.$$

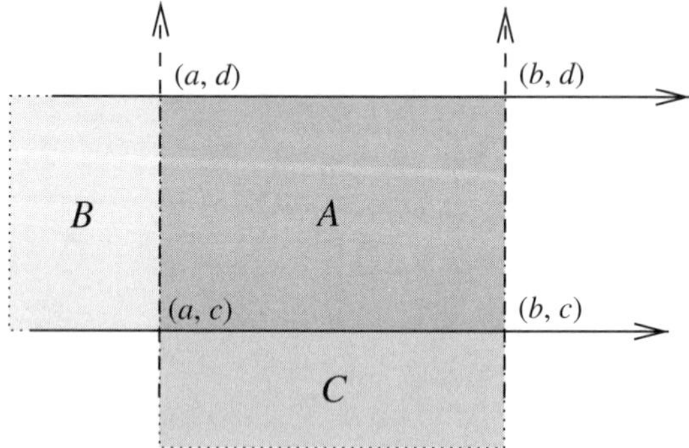


Figure 4.2. Disjoint areas $A$, $B$, and $C$ representing mutually exclusive events.

Furthermore, from our definition of the joint CDF, we have the following relationships:

$$\begin{aligned}\text{Prob}\{B\} &= F_{X,Y}(a,d) - F_{X,Y}(a,c),\\ \text{Prob}\{C\} &= F_{X,Y}(b,c) - F_{X,Y}(a,c),\\ \text{Prob}\{A \cup B \cup C\} &= F_{X,Y}(b,d) - F_{X,Y}(a,c),\end{aligned}$$

and we may conclude that

$$\begin{aligned}\text{Prob}\{A\} &= F_{X,Y}(b,d) - F_{X,Y}(a,c) - [F_{X,Y}(a,d) - F_{X,Y}(a,c) + F_{X,Y}(b,c) - F_{X,Y}(a,c)]\\ &= F_{X,Y}(b,d) - F_{X,Y}(a,d) - F_{X,Y}(b,c) + F_{X,Y}(a,c).\end{aligned}$$

**Example 4.2** The joint CDF of two continuous random variables is given by

$$F_{X,Y}(x, y) = (1 - e^{-\lambda x})(1 - e^{-\mu y}), \quad 0 \le x < \infty,\ 0 \le y < \infty,$$

and is zero otherwise, where both $\lambda$ and $\mu$ are strictly positive. Let us compute the probability Prob$\{0 \le x \le 1,\ 0 \le y \le 1\}$. From the above relationship, we have

$$\begin{aligned}\text{Prob}\{0 \le x \le 1,\ 0 \le y \le 1\} &= F_{X,Y}(1,1) - F_{X,Y}(0,1) - F_{X,Y}(1,0) + F_{X,Y}(0,0)\\ &= (1-e^{-\lambda})(1-e^{-\mu}) - (1-e^{-\lambda})(1-e^{0}) - (1-e^{0})(1-e^{-\mu})\\ &\quad + (1-e^{0})(1-e^{0})\\ &= (1-e^{-\lambda})(1-e^{-\mu}).\end{aligned}$$

We now turn our attention to *marginal* cumulative distribution functions derived from joint CDFs by considering what happens as one, but not both, of $X$ and $Y$ goes to infinity. In the limit as $y \to \infty$, the event $\{X \le x, Y \le y\}$ tends to the event $\{X \le x, Y < \infty\} = \{X \le x\}$, where we have used $Y < \infty$ to signify that $Y$ may take on *any* value. This implies that the only restriction is that $X \le x$, i.e., the event $\{X \le x, Y < \infty\}$ reduces to the event $\{X \le x\}$. But now, the probability of the event $\{X \le x\}$ is simply the cumulative distribution of $X$ and hence

$$\lim_{y\to\infty} F_{X,Y}(x, y) = F_X(x).$$

Similarly,

$$\lim_{x\to\infty} F_{X,Y}(x, y) = F_Y(y).$$

These are called the *marginal* distributions, and may be computed from the joint distribution.

**Example 4.3** Returning to Example 4.1, we may compute the marginal distributions as

| | $-2 < x$ | $-2 \le x < -1$ | $-1 \le x < 0$ | $0 \le x < 1$ | $x \ge 1$ |
|---|---|---|---|---|---|
| $F_X(x)$ | 0 | 1/8 | 1/4 | 1/2 | 1 |

for $X$ as before, and for $Y$:

| | $-2 < y$ | $-2 \le y < 0$ | $0 \le y < 2$ | $y \ge 2$ |
|---|---|---|---|---|
| $F_Y(y)$ | 0 | 1/4 | 3/4 | 1 |

These are just the top row and the rightmost column of the CDF table provided earlier.

Similarly, the marginal CDFs of $X$ and $Y$, derived from the joint CDF of Example 4.2, are easily computed, and are respectively given by

$$\begin{aligned}F_X(x) &= (1-e^{-\lambda x})(1-0) = (1-e^{-\lambda x}),\\ F_Y(y) &= (1-0)(1-e^{-\mu y}) = (1-e^{-\mu y}).\end{aligned}$$

The random variables $X$ and $Y$ are said to be *independent* if

$$F_{X,Y}(x, y) = F_X(x)F_Y(y), \quad -\infty < x < \infty, \quad -\infty < y < \infty.$$

In other words, if $X$ and $Y$ are independent random variables, their joint cumulative distribution function factors into the product of their marginal cumulative distribution functions. The reader may wish to verify that the random variables $X$ and $Y$ of Examples 4.1 and 4.2 are independent. If, in addition, $X$ and $Y$ have the same distribution function (for example, when $\lambda = \mu$ in Example 4.2), then the term *independent and identically distributed* is applied to them. This is usually abbreviated to *iid*.

The generalization of the definition of the joint cumulative distribution function to joint distribution functions of more than two random variables, referred to as *random vectors*, is straightforward. Given $n$ random variables $X_1, X_2, \ldots X_n$, their joint cumulative distribution function is defined for all real $x_i$, $-\infty < x_i < \infty$, $i = 1, 2, \ldots, n$, as

$$F(x_1, x_2, \ldots, x_n) = \text{Prob}\{X_1 \le x_1,\ X_2 \le x_2,\ \ldots,\ X_n \le x_n\}.$$

In this case, $F$ must be a nondecreasing function of all the $x_i$, $F(-\infty, -\infty, \ldots, -\infty) = 0$ and $F(\infty, \infty, \ldots, \infty) = 1$. The marginal distribution of $X_i$ is obtained from

$$\text{Prob}\{X_i \le x_i\} = F_{X_i}(x_i) = F(\infty, \ldots, \infty, x_i, \infty \ldots \infty).$$

The random variables $X_1, X_2, \ldots, X_n$ are said to be *independent* if, for all real $x_1, x_2, \ldots, x_n$,

$$F(x_1, x_2, \ldots, x_n) = \prod_{i=1}^{n} F_{X_i}(x_i).$$

If in addition, all $n$ random variables, $X_i$, $i = 1, 2, \ldots, n$, have the same distribution function, they are said to be independent and identically distributed.

## 4.3 Joint Probability Mass Functions

The probability mass function for a single discrete random variable $X$ is denoted $p_X(x)$ and gives the probability that $X$ has the value $x$, i.e., $p_X(x)$ is the probability of the event that consists of all outcomes that are mapped into the value $x$. If $X$ and $Y$ are two discrete random variables, an outcome is a point in the $(x, y)$ plane and an event is a subset of these two-dimensional points: $[X = x, Y = y]$ is the event that consists of all outcomes for which $X$ is mapped into $x$ *and* $Y$ is mapped into $y$. Their joint probability mass function is denoted $p_{X,Y}(x, y)$ and is equal to the probability of the event $[X = x, Y = y]$. We write

$$p_{X,Y}(x, y) = \text{Prob}\{X = x, Y = y\}.$$

The joint probability mass function of two discrete random variables may be conveniently represented in tabular form. For example, we may place realizable values of $X$ across the top of the table and those of $Y$ down the left side. The element of the table corresponding to column $X = x$ and row $Y = y$ denotes the probability of the event $[X = x, Y = y]$.

**Example 4.4** Let $X$ be a random variable that has values in $\{1, 2, 3, 4, 5\}$ and $Y$ a random variable with values in $\{-1, 0, 1, 2\}$ and let their joint probability mass function be given by $p_{X,Y}(x, y) = \alpha$ for all possible values of $x$ and $y$. To compute the value of $\alpha$, we observe that only 20 points in the $(x, y)$ plane have positive probability, and furthermore the probability of each is the same. This gives $\alpha = 1/20 = 0.05$. The following table displays the joint probability mass function of $X$ and $Y$.

| | $X=1$ | $X=2$ | $X=3$ | $X=4$ | $X=5$ |
|---|---|---|---|---|---|
| $Y=-1$ | 1/20 | 1/20 | 1/20 | 1/20 | 1/20 |
| $Y=0$ | 1/20 | 1/20 | 1/20 | 1/20 | 1/20 |
| $Y=1$ | 1/20 | 1/20 | 1/20 | 1/20 | 1/20 |
| $Y=2$ | 1/20 | 1/20 | 1/20 | 1/20 | 1/20 |

Every event $\mathcal{A}$ is a subset of points in the $(x, y)$ plane and we may write

$$\text{Prob}\{\mathcal{A}\} = \sum_{(x,y)\in\mathcal{A}} p_{X,Y}(x, y),$$

where the summation is over all points $(x, y)$ which are in the set $\mathcal{A}$. Figure 4.3 illustrates two events, $\mathcal{A} = [X + Y < 4]$ and $\mathcal{B} = [|X| + |Y| \le 3]$, where all possible values of $X$ and $Y$ are marked by dots.

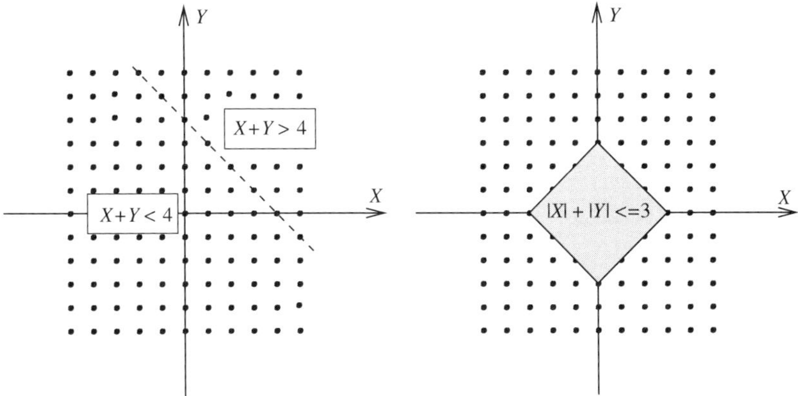


Figure 4.3. The events $\mathcal{A} = \{X + Y < 4\}$ and $\mathcal{B} = \{|X| + |Y| \le 3\}$.

**Example 4.5** Let $X$ and $Y$ be discrete random variables whose joint probability mass function is given by the previous example. Let $\mathcal{A}$ be the event that $X$ has a value no greater than $Y$, written as $[X \le Y]$. The only points in the plane having positive probability and for which the value of $X$ is less than or equal to the value of $Y$ are $(1, 1)$, $(1, 2)$, and $(2, 2)$. Therefore, the probability of this event is given by $\text{Prob}\{\mathcal{A}\} = \text{Prob}\{X \le Y\} = 3/20$.

Now let $\mathcal{B}$ be the event $[XY = 0]$. This event contains five outcomes: $(1, 0)$, $(2, 0)$, $(3, 0)$, $(4, 0)$, and $(5, 0)$, and hence $\text{Prob}\{\mathcal{B}\} = \text{Prob}\{XY = 0\} = 5/20$.

Even though a joint probability mass function specifies the probability of joint events, denoted $[X = x, Y = y]$, it is possible to consider the probability of events that concern only one of the two random variables, events such as $[X = x]$. Our interest lies only in the value of $X$; the random variable $Y$ may assume *any* value since it does not interest us. For example, returning to Example 4.4, the event $[X = 2]$ consists of the four outcomes $(2, -1)$, $(2, 0)$, $(2, 1)$, and $(2, 2)$ and has probability equal to 4/20. In this way, we may use the joint probability mass function of the two random variables to construct the probability mass function for either of them. These are the *marginal* probability mass functions. The marginal probability mass function of $X$ is obtained as

$$p_X(x) = \text{Prob}\{X = x\} = \sum_k \text{Prob}\{X = x, Y = y_k\} = \sum_k p_{X,Y}(x, y_k)$$

and is a direct consequence of the fact that, for any event $\mathcal{A}$, $\text{Prob}\{\mathcal{A}\} = \sum_{(x,y)\in\mathcal{A}} p_{X,Y}(x, y)$. The marginal distribution of $Y$ is similarly obtained.

Random variables $X$ and $Y$ are said to be *independent* if

$$\text{Prob}\{X = x_i,\ Y = y_i\} = \text{Prob}\{X = x_i\}\text{Prob}\{Y = y_i\}$$

for all $x_i$ and $y_i$. In other words, $X$ and $Y$ are independent if their joint probability mass function factors into the product of their individual probability mass functions.

**Example 4.6** Let $X$ and $Y$ be two discrete random variables with joint probability mass function:

| | $X = -1$ | $X = 0$ | $X = 1$ |
|---|---|---|---|
| $Y = -1$ | 1/12 | 3/12 | 1/12 |
| $Y = 0$ | 1/12 | 0/12 | 1/12 |
| $Y = 1$ | 1/12 | 3/12 | 1/12 |

Thus, both discrete random variables $X$ and $Y$ take values from the set $\{-1, 0, 1\}$. The table indicates that $\text{Prob}\{X = 0; Y = 0\} = 0$, that $\text{Prob}\{X = 0; Y = 1\} = 1/4$ and so on. The *marginal* probability mass function of $X$, $p_X(x)$, is obtained by adding the entries in each column of the table; that of $Y$, $p_Y(y)$, is found by adding the entries in each row of the table. We have

| | $X = -1$ | $X = 0$ | $X = 1$ | $p_Y(y)$ |
|---|---|---|---|---|
| $Y = -1$ | 1/12 | 3/12 | 1/12 | 5/12 |
| $Y = 0$ | 1/12 | 0/12 | 1/12 | 1/6 |
| $Y = 1$ | 1/12 | 3/12 | 1/12 | 5/12 |
| $p_X(x)$ | 1/4 | 1/2 | 1/4 | 1 |

In this example, $X$ and $Y$ are *not* independent since

$$1/12 = \text{Prob}\{X = 1, Y = 1\} \neq \text{Prob}\{X = 1\}\text{Prob}\{Y = 1\} = 3/12 \times 5/12.$$

When arranged in tabular form like this, with the marginal probability mass functions along the bottom and right-hand side, it is not difficult to see where the name *marginals* comes from.

**Example 4.7** The Town of Cary has a men's soccer team and a women's soccer team. The probability mass functions of the number of goals scored by each team in a typical game are

| | 0 | 1 | 2 | 3 | $> 3$ |
|---|---|---|---|---|---|
| Men | 0.5 | 0.2 | 0.1 | 0.1 | 0.1 |
| Women | 0.3 | 0.1 | 0.3 | 0.2 | 0.1 |

We wish to compute the probability that the women's team scores more goals than the men's team during a randomly chosen match, and the probability that both score the same number of goals.

To proceed, we make the assumption that the number of goals scores by one team is independent of the number scored by the other. This being the case, we form the joint probability mass function from $p_{X,Y}(x, y) = p_X(x)p_Y(y)$, where $X$ is the number of goals scored by the men's team and $Y$ the number scored by the women's team, and obtain

| $p_{X,Y}(x, y)$ | $X = 0$ | $X = 1$ | $X = 2$ | $X = 3$ | $X > 3$ |
|---|---|---|---|---|---|
| $Y = 0$ | 0.15 | 0.06 | 0.03 | 0.03 | 0.03 |
| $Y = 1$ | 0.05 | 0.02 | 0.01 | 0.01 | 0.01 |
| $Y = 2$ | 0.15 | 0.06 | 0.03 | 0.03 | 0.03 |
| $Y = 3$ | 0.10 | 0.04 | 0.02 | 0.02 | 0.02 |
| $Y > 3$ | 0.05 | 0.02 | 0.01 | 0.01 | 0.01 |

The probability that the women's team scores more goals that the men's team is given by

$$\begin{aligned}\text{Prob}\{X < Y\} &= 0.05 + (0.15 + 0.06) + (0.10 + 0.04 + 0.02) + (0.05 + 0.02 + 0.01 + 0.01)\\ &= 0.51.\end{aligned}$$

The probability that both score the same number of goals is

$$\text{Prob}\{X = Y\} = 0.15 + 0.02 + 0.03 + 0.02 + 0.01 = 0.23.$$

## 4.4 Joint Probability Density Functions

Recall that the probability density function $f_X(x)$ of a single continuous random variable $X$ is defined by means of the relation

$$F_X(x) = \int_{-\infty}^{x} f_X(t)dt \quad \text{for} - \infty < x < \infty.$$

We also saw that for an infinitesimally small $dx$, Prob$\{x < X \leq x + dx\} = f_X(x)dx$, $\int_{-\infty}^{\infty} f_X(x)dx = 1$, and that Prob$\{a < X \leq b\} = \int_a^b f_X(x)dx$. These same ideas carry over to two dimensions and higher. If $X$ and $Y$ are continuous random variables, we can usually find a *joint probability density function* $f_{X,Y}(x, y)$ such that

$$F_{X,Y}(x, y) = \int_{-\infty}^{y} \int_{-\infty}^{x} f_{X,Y}(u, v)dudv, \tag{4.2}$$

and which satisfies the conditions $f_{X,Y}(x, y) \geq 0$ for all $(x, y)$, and

$$\int_{-\infty}^{\infty} \int_{-\infty}^{\infty} f_{X,Y}(u, v)dudv = 1.$$

If the partial derivatives of $F_{X,Y}(x, y)$ with respect to $x$ and $y$ exist, then

$$f_{X,Y}(x, y) = \frac{\partial^2 F_{X,Y}(x, y)}{\partial x \partial y}.$$

These relations imply that

$$\text{Prob}\{a < X \leq b, c < Y \leq d) = \int_a^b \int_c^d f_{X,Y}(x, y)dydx.$$

This means that, whereas the probability density function of a single random variable is measured per unit length, the joint probability density function of two random variables is measured per unit area. Furthermore, for infinitesimal $dx$ and $dy$,

$$\text{Prob}\{x < X \leq x + dx,\ y < Y \leq y + dy\} \approx f_{X,Y}(x, y)dxdy.$$

This property leads us to take $f_{X,Y}(x, y)$ as the continuous analog of Prob$\{X = x, Y = y\}$.

With two joint random variables that are continuous, an event corresponds to an area in the $(x, y)$ plane. The event occurs if the random variable $X$ assigns a value $x_1$ to the outcome, $Y$ assigns a value $y_1$ to the outcome and the point $(x_1, y_1)$ lies within the area that defines the event. The probability of the occurrence of an event $\mathcal{A}$ is given by

$$\text{Prob}\{\mathcal{A}\} = \iint_{\{x,y\,|\,(x,y)\in\mathcal{A}\}} f_{X,Y}(u, v)dudv = \iint_{\mathcal{A}} f_{X,Y}(u, v)dudv.$$

**Example 4.8** Let $X$ and $Y$ be two continuous random variables whose joint probability density function has the form

$$f_{X,Y}(x, y) = \alpha(x + y), \quad 0 \leq x \leq y \leq 1,$$

and is equal to zero otherwise. Find the value of the constant $\alpha$ and sketch the region of positive probability, i.e., the joint domain of $f_{X,Y}(x, y)$ for which $f_{X,Y}(x, y) > 0$. Also, sketch the region for which $X + Y \leq 1$ and integrate over this region to find Prob$\{X + Y \leq 1\}$.

The region of the plane which contains positive probability is shown in Figure 4.4.

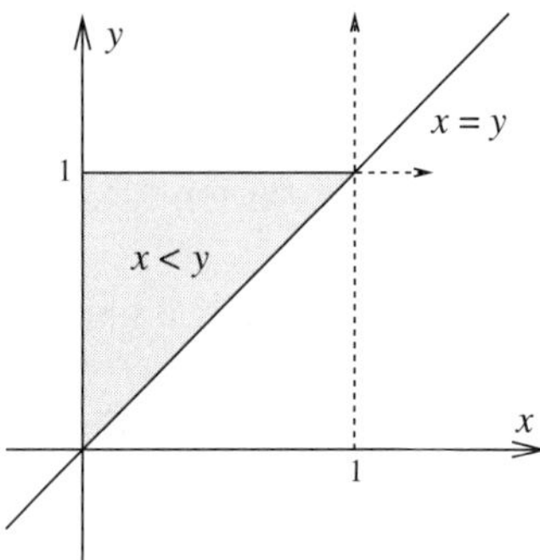


Figure 4.4. The region for which $0 \leq x \leq y \leq 1$.

To satisfy the requirements of a probability density function, $f_{X,Y}(x, y)$ must be non-negative everywhere and $\int_{-\infty}^{\infty}\int_{-\infty}^{\infty} f_{X,Y}(x, y)dxdy$ must be equal to 1. Observe that $f_{X,Y}(x, y) \geq 0$ for all $x$ and $y$ and that

$$1 = \int_{-\infty}^{\infty}\int_{-\infty}^{\infty} f_{X,Y}(x, y)dxdy = \int_0^1\int_0^y \alpha(x + y)dxdy = \int_0^1 \left(\alpha x^2/2 + \alpha xy|_0^y\right) dy$$

$$= \int_0^1 \frac{3\alpha y^2}{2} dy = \alpha/2.$$

It follows that we must have $\alpha = 2$ for this to be a probability density function. The region for which $X + Y \leq 1$ (within the region of positive probability) is shown in Figure 4.5. It is over this region that we must integrate.

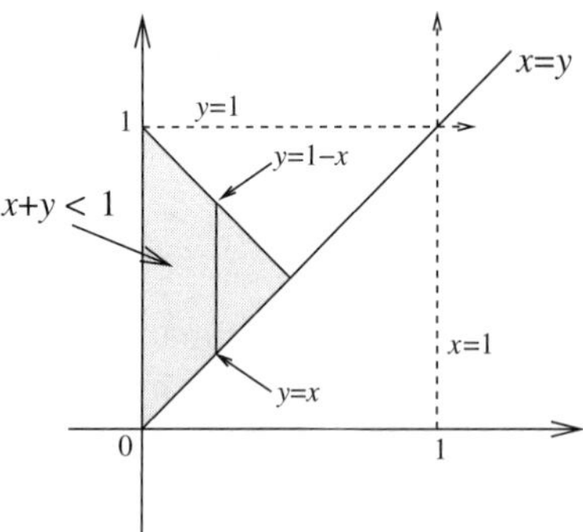


Figure 4.5. Limits of integration for the event $\{X + Y \leq 1\}$.

We have

$$\text{Prob}\{X + Y \leq 1\} = \int_{x=0}^{1/2}\int_{y=x}^{1-x} (2x + 2y)\,dy\,dx = \int_{x=0}^{1/2} (2xy + y^2)\,|_x^{1-x}\,dx$$

$$= \int_{x=0}^{1/2} \left[2x(1-x) + (1-x)^2 - 2x^2 - x^2\right]\,dx$$

$$= \int_{x=0}^{1/2} (1 - 4x^2)\,dx = (x - 4x^3/3)\,|_0^{1/2} = 1/2 - 1/6 = 1/3.$$

**Example 4.9** Two continuous random variables have joint probability density function given by

$$f_{X,Y}(x,y) = \begin{cases} \alpha x y^2, & 0 \le x \le 1,\ 0 \le y \le 1, \\ 0 & \text{otherwise.} \end{cases}$$

Let us find the value of $\alpha$ and the following probabilities:

$$\text{(a) Prob}\{Y < X\}, \quad \text{(b) Prob}\{Y < X^2\}, \quad \text{and (c) Prob}\{\max(X,Y) \ge 1/2\}.$$

The value of $\alpha$ is found by integrating the joint density function over all $(x, y)$ and setting the result to 1. We have

$$1 = \int_{-\infty}^{\infty}\int_{-\infty}^{\infty} f_{X,Y}(x,y)\,dy\,dx = \int_0^1\int_0^1 \alpha x y^2\,dy\,dx$$

$$= \int_0^1 \left(\alpha x \frac{y^3}{3}\right)\bigg|_0^1 dx = \int_0^1 \frac{\alpha x}{3}\,dx = \frac{\alpha x^2}{6}\bigg|_0^1 = \frac{\alpha}{6},$$

and hence $\alpha = 6$. We now compute the required probabilities.

$$\text{(a) Prob}\{Y < X\} = \int_0^1\int_0^x 6xy^2\,dy\,dx = \int_0^1 6x\frac{y^3}{3}\bigg|_0^x dx = \int_0^1 2x^4 dx = \frac{2x^5}{5}\bigg|_0^1 = 2/5.$$

$$\text{(b) Prob}\{Y < X^2\} = \int_0^1\int_0^{x^2} 6xy^2\,dy\,dx = \int_0^1 6x\frac{y^3}{3}\bigg|_0^{x^2} dx = \int_0^1 2x^7 dx = \frac{2x^8}{8}\bigg|_0^1 = 1/4.$$

$$\text{(c) Prob}\{\max(X,Y) \le 1/2\} = \int_0^{1/2}\int_0^{1/2} 6xy^2\,dy\,dx = \int_0^{1/2} 2x\,dx \int_0^{1/2} 3y^2\,dy$$

$$= \left(x^2\big|_0^{1/2}\right)\left(y^3\big|_0^{1/2}\right) = \left(\frac{1}{2}\right)^5 = \frac{1}{32}.$$

### *Uniformly Distributed Joint Random Variables*

When the random variables $X$ and $Y$ are *uniformly* distributed over a region of the plane, it becomes possible to use geometric arguments to compute certain probabilities. For example, if $X$ and $Y$ are continuous random variables, uniformly distributed over the unit square, and we seek the probability $X \le Y$, we note that the region $0 \le x \le 1$, $0 \le y \le 1$, and $x \le y$ constitutes one-half of the unit square, i.e., the triangle bounded by the $y$-axis, the lines $y = 1$ and $x = y$, and so $\text{Prob}\{X \le Y\} = 0.5$. More generally, if $R$ is a finite region of the plane into which $(X, Y)$ is *certain* to fall, the probability that $(X, Y)$ actually lies in a subregion $A$ of $R$ is proportional to the size of $A$, relative to the size of $R$. In other words,

$$\text{Prob}\{(X,Y) \in A\} = \frac{\text{Area of } A}{\text{Area of } R}.$$

**Example 4.10** The joint probability density function of two continuous random variables, $X$ and $Y$, is given by

$$f_{X,Y}(x,y) = \begin{cases} \alpha, & x \ge 0,\ y \ge 0 \text{ and } x + y \le 1, \\ 0 & \text{otherwise.} \end{cases}$$

Thus $X$ and $Y$ are uniformly distributed over the region of positive probability. Describe this region and find the value of the constant $\alpha$. Also, compute the probability $\text{Prob}\{X + Y \le 0.5\}$.

The region of positive probability is the right-angled triangle whose base is the unit interval along the $x$-axis, having unit height and whose hypotenuse is a segment of the line $x + y = 1$, i.e., the right angle of the triangle is at the origin. Such a shape is sometimes referred to as a *wedge*. The usual way to find the constant $\alpha$ is by integrating over this region and setting the resulting value to 1. With this approach, we obtain

$$1 = \int_0^1 \int_0^{1-x} \alpha \, dydx = \int_0^1 \alpha \, y|_0^{1-x} \, dx = \int_0^1 (\alpha - \alpha x)dx = \alpha x - \alpha x^2/2\big|_0^1 = \alpha/2,$$

and hence $\alpha = 2$. An alternative way to obtain this result is to observe that $X$ and $Y$ are *uniformly* distributed over the region, and since the region occupies one-half of the unit square, the value of $\alpha$ must be 2. Also, although the required probability may be found as

$$\text{Prob}\{X + Y \le 0.5\} = \int_0^{0.5} \int_0^{0.5-x} 2 \, dydx = \int_0^{0.5} 2y\big|_0^{0.5-x} dx$$

$$= \int_0^{0.5} (1 - 2x)dx = (x - x^2)\big|_0^{0.5} = 0.25,$$

a geometric argument shows that the area of the right-angle triangle with both base and height equal to $1/2$ is $1/2 \times 1/2 \times 1/2 = 1/8$ and hence

$$\text{Prob}\{(X, Y) \in A\} = \frac{\text{Area of } A}{\text{Area of } R} = \frac{1/8}{1/2} = 1/4,$$

as before.

We consider one further example of this type.

**Example 4.11** Let $X$ and $Y$ be two continuous random variables, uniformly distributed on the unit square. We wish to find $\text{Prob}\{X^2 + Y^2 \le r^2\}$ for $r \le 1$. The joint probability density function of $X$ and $Y$ is given by

$$f_{X,Y}(x, y) = \begin{cases} 1, & 0 \le x \le 1,\ 0 \le y \le 1, \\ 0 & \text{otherwise.} \end{cases}$$

The area of the region of positive probability, i.e., the area of $R$, is equal to 1, Observe that $x^2 + y^2 = r^2$ defines a circle centered on the origin and with radius $r$. The area of this circle is known to be $\pi r^2$. The portion of it that lies in the positive quadrant, where the joint density function is positive, is $\pi r^2/4$. Thus, in our terminology, the area of $A$ is $\pi r^2/4$ and hence

$$\text{Prob}\left\{X^2 + Y^2 \le r^2\right\} = \frac{\pi r^2/4}{1} = \frac{\pi r^2}{4}.$$

### *Derivation of Joint CDFs from Joint Density Functions*

Equation (4.2) relates the joint density function and the joint cumulative distribution function of two random variables, which leads us consider using one to obtain the other. However, as we now show in an example, deriving the joint CDF from a joint probability density function is frequently not straightforward. The difficulty arises from the fact that, whereas the density function may be nonzero over a relatively simple region and zero elsewhere, the joint CDF will often be nonzero over a much larger region.

**Example 4.12** Random variables $X$ and $Y$ have joint probability density function given by

$$f_{X,Y}(x, y) = \begin{cases} 2(x + y), & 0 \le x \le y \le 1, \\ 0 & \text{otherwise.} \end{cases}$$

We wish to construct the joint CDF of $X$ and $Y$. We have

$$F_{X,Y}(x, y) = \int_{-\infty}^{y} \int_{-\infty}^{x} f_{X,Y}(u, v)dudv,$$

but we need to pay special attention to the limits of integration. There are five cases that need to be considered. Each depends on the region into which the point $(x, y)$ falls and is illustrated in Figure 4.6.

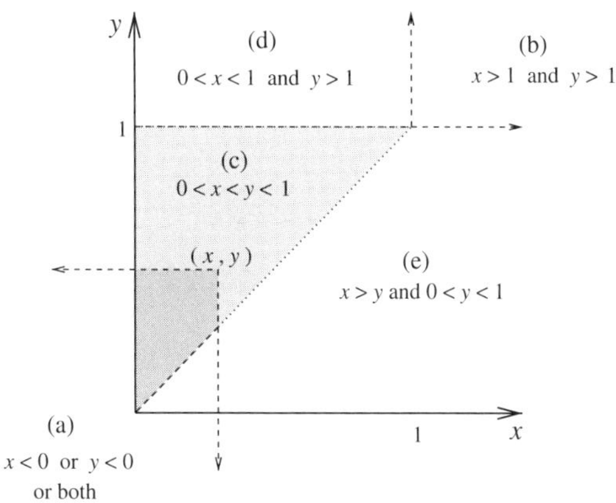


Figure 4.6. Limits of integration for the event $\{X + Y \leq 1\}$.

(a) $x < 0$ or $y < 0$ or both. In this case, none of the area of positive probability falls into the area of integration, which means that, in this case, $F_{X,Y}(x, y) = 0$.

(b) $x > 1$ and $y > 1$. In this case, all of the region of positive probability falls into the area of integration, which means that, in this case, $F_{X,Y}(x, y) = 1$.

(c) $0 \leq x \leq y \leq 1$.

$$F_{X,Y}(x, y) = \int_{-\infty}^{x} \int_{-\infty}^{y} F_{X,Y}(u, v)\, dv\, du = \int_{0}^{x} \int_{u}^{y} (2u + 2v)\, dv\, du = \int_{0}^{x} (2uv + v^2) \mid_{u}^{y} du$$

$$= \int_{0}^{x} (2uy + y^2 - 2u^2 - u^2)\, du = (u^2 y + uy^2 - u^3) \mid_{0}^{x} = x^2 y + xy^2 - x^3.$$

(d) $0 < x < 1$ and $y > 1$.

$$F_{X,Y}(x, y) = \int_{-\infty}^{x} \int_{-\infty}^{y} F_{X,Y}(u, v)\, dv\, du = \int_{0}^{x} \int_{u}^{1} (2u + 2v)\, dv\, du = \int_{0}^{x} (2uv + v^2) \mid_{u}^{1} du$$

$$= \int_{0}^{1} (2u + 1 - 2u^2 - u^2)\, du = (u + u^2 - u^3) \mid_{0}^{x} = x + x^2 - x^3.$$

(e) $x > y$ and $0 < y < 1$.

$$F_{X,Y}(x, y) = \int_{-\infty}^{x} \int_{-\infty}^{y} F_{X,Y}(u, v)\, dv\, du = \int_{0}^{y} \int_{0}^{v} (2u + 2v)\, du\, dv = \int_{0}^{y} (u^2 + 2uv) \mid_{0}^{v} dv$$

$$= \int_{0}^{y} (v^2 + 2v^2)\, dv = v^3 \mid_{0}^{y} = y^3.$$

A complete description of the CDF is therefore given by

$$F_{X,Y}(x, y) = \begin{cases} 0, & x < 0 \text{ or } y < 0 \text{ or both}, \\ 1, & x > 1 \text{ and } y > 1, \\ x^2y + xy^2 - x^3, & 0 \leq x \leq y \leq 1, \\ x + x^2 - x^3, & 0 < x < 1 \text{ and } y > 1, \\ y^3, & x > y \text{ and } 0 < y < 1. \end{cases}$$

### *Marginal Probability Density Functions and Independence*

As in the discrete case, we may also form marginal distributions. Since

$$F_X(x) = \int_{-\infty}^{x} \int_{-\infty}^{\infty} f_{X,Y}(u, y)dydu,$$

the marginal density functions $f_X$ and $f_Y$ may be formed as

$$f_X(x) = \int_{-\infty}^{\infty} f_{X,Y}(x, y)dy \quad \text{and} \quad f_Y(y) = \int_{-\infty}^{\infty} f_{X,Y}(x, y)dx. \tag{4.3}$$

Joint continuous random variables $X$ and $Y$ are *independent* if $f_{X,Y}(x, y) = f_X(x)f_Y(y)$. When $X$ is a discrete random variable and $Y$ is a continuous random variable (or vice versa), we may still form their joint distribution. In this case, the condition for independence is

$$\text{Prob}\{X = x, Y \leq y\} = p_X(x)F_Y(y), \quad \text{for all } x, y.$$

**Example 4.13** Let the joint probability density function of two continuous random variables, $X$ and $Y$ be given by

$$f_{X,Y}(x, y) = \begin{cases} (2x + 5y)/100, & 0 < x \leq 5, \ 0 < y \leq 2, \\ 0 & \text{otherwise}. \end{cases}$$

Let us form the marginal probability density function of each random variable and decide whether they are independent or not. The marginal probability density function of $X$ is given by

$$f_X(x) = \int_{-\infty}^{\infty} f_{X,Y}(x, y)dy = \int_0^2 \frac{2x + 5y}{100} dy = \frac{x}{50} \int_0^2 dy + \frac{1}{20} \int_0^2 ydy = \frac{x}{25} + \frac{y^2}{40}\bigg|_0^2 = \frac{2x + 5}{50}.$$

Similarly, for $Y$, we find

$$f_Y(y) = \int_{-\infty}^{\infty} f_{X,Y}(x, y)dx = \int_0^5 \frac{2x + 5y}{100} dx = \frac{1}{50} \int_0^5 xdx + \frac{y}{20} \int_0^5 dx = \frac{x^2}{100}\bigg|_0^5 + \frac{5y}{20} = \frac{1 + y}{4}.$$

In this case, $X$ and $Y$ are *not independent* since $f_{X,Y}(x, y) \neq f_X(x)f_Y(y)$.

**Example 4.14** Consider two continuous random variables $X$ and $Y$ with joint density function given by

$$f_{X,Y}(x, y) = \alpha\beta e^{-(\alpha x + \beta y)}, \quad 0 < x < \infty, \quad 0 < y < \infty.$$

We seek the marginal densities of the two random variables and their joint cumulative distribution function. We would also like to compute the probability that $X$ is greater than $Y$.

The marginal density function of $X$ is computed as

$$f_X(x) = \int_{-\infty}^{\infty} f_{X,Y}(x, y)dy = \int_0^{\infty} \alpha\beta e^{-(\alpha x + \beta y)}dy = \alpha e^{-\alpha x}, \quad 0 < x < \infty.$$

In the same way, we find the marginal density of $Y$ to be

$$f_Y(y) = \beta e^{-\beta y} \quad 0 < y < \infty.$$

Observe that $X$ and $Y$ are independent random variables, since $f_{X,Y}(x, y) = f_X(x) f_Y(y)$.

The joint cumulative distribution function of $X$ and $Y$ is computed from

$$F_{X,Y}(x, y) = \int_{-\infty}^{x} \int_{-\infty}^{y} f_{X,Y}(u, v) du dv = \int_0^x \int_0^y \alpha\beta e^{-(\alpha u + \beta v)} du dv$$
$$= \left(1 - e^{-\alpha x}\right)\left(1 - e^{-\beta y}\right), \quad 0 < x < \infty, \quad 0 < y < \infty,$$

and is equal to zero otherwise. Because of the simplicity of the region of positive probability (the entire positive quadrant), the usual difficulties associated with computing joint CDFs from joint probability density functions are avoided in this example. Also, observe that since $X$ and $Y$ are independent, we could have derived this result by first finding the CDF of each of the marginals and multiplying them together. Finally, we have

$$\text{Prob}\{X \ge Y\} = \int_{x=0}^{\infty} \int_{y=0}^{x} f_{X,Y}(x, y) dy dx = \int_{x=0}^{\infty} \left( \int_{y=0}^{x} \beta e^{-\beta y} dy \right) \alpha e^{-\alpha x} dx$$
$$= \int_{x=0}^{\infty} \left(1 - e^{-\beta x}\right) \alpha e^{-\alpha x} dx = \int_{x=0}^{\infty} \left[\alpha e^{-\alpha x} - \alpha e^{-(\alpha+\beta)x}\right] dx$$
$$= -e^{-\alpha x}\Big|_0^{\infty} + \frac{\alpha}{\alpha+\beta} e^{-(\alpha+\beta)x}\Big|_0^{\infty} = 1 - \frac{\alpha}{\alpha+\beta} = \frac{\beta}{\alpha+\beta}.$$

## 4.5 Conditional Distributions

In the previous chapter we considered the effect of knowing that a certain event $\mathcal{B}$ had occurred on the probability distribution of a random variable $X$. We now extend this to the case of two random variables $X$ and $Y$ and our concern is with the probability distribution of one of these when the conditioning event is defined by the value taken by the second.

### *Discrete Random Variables*

Consider first the case when $X$ and $Y$ are two *discrete* random variables with joint probability mass function $p_{XY}(x, y)$. Then the conditional probability mass function of $X$ given the event $\{Y = y\}$ is

$$p_{X|Y}(x|y) = \text{Prob}\{X = x | Y = y\} = \frac{\text{Prob}\{X = x,\ Y = y\}}{\text{Prob}\{Y = y\}} = \frac{p_{XY}(x, y)}{p_Y(y)} \tag{4.4}$$

so long as $p_Y(y) > 0$. In other words, we choose a specific value for $y$ and find $\text{Prob}\{X = x | Y = y\}$ as a function of $x$.

**Example 4.15** Let $X$ and $Y$ be two discrete random variables with joint probability mass function:

| | $X = -1$ | $X = 0$ | $X = 1$ | $p_Y(y)$ |
|---|---|---|---|---|
| $Y = -1$ | 1/12 | 3/12 | 1/12 | 5/12 |
| $Y = 0$ | 1/12 | 0/12 | 1/12 | 2/12 |
| $Y = 1$ | 1/12 | 3/12 | 1/12 | 5/12 |
| $p_X(x)$ | 3/12 | 6/12 | 3/12 | |

Let us compute $\text{Prob}\{X = x|Y = 1\}$. We have

$$\text{Prob}\{X = -1|Y = 1\} = \text{Prob}\{X = -1, Y = 1\}/\text{Prob}\{Y = 1\} = (1/12)/(5/12) = 1/5,$$
$$\text{Prob}\{X = 0|Y = 1\} = \text{Prob}\{X = 0, Y = 1\}/\text{Prob}\{Y = 1\} = (3/12)/(5/12) = 3/5,$$
$$\text{Prob}\{X = 1|Y = 1\} = \text{Prob}\{X = 1, Y = 1\}/\text{Prob}\{Y = 1\} = (1/12)/(5/12) = 1/5.$$

Observe that for each different value $Y$ can assume, we obtain a different probability mass function $p_{XY}(x|y)$. In Example 4.15, three different probability density function corresponding to $y = -1, 0, 1$ may be obtained, only the last ($Y = 1$) of which was computed.

**Example 4.16** In addition to its men's and women's soccer teams, the town of Cary also has a youth team, some of whose players also play on the men's team. The number of goals scored by the men's team in a typical match is not independent of the number of goals scored by the youth team. The joint probability mass function of the men's and youth team is given by

| $p_{X,Y}(x, y)$ | $X = 0$ | $X = 1$ | $X = 2$ | $X = 3$ | $X > 3$ | $p_Y(y)$ |
|---|---|---|---|---|---|---|
| $Y = 0$ | 0.04 | 0.02 | 0.05 | 0.03 | 0.01 | 0.15 |
| $Y = 1$ | 0.03 | 0.05 | 0.10 | 0.05 | 0.02 | 0.25 |
| $Y = 2$ | 0.04 | 0.03 | 0.08 | 0.07 | 0.02 | 0.24 |
| $Y = 3$ | 0.04 | 0.04 | 0.04 | 0.04 | 0.04 | 0.20 |
| $Y > 3$ | 0.01 | 0.02 | 0.06 | 0.04 | 0.03 | 0.16 |
| $p_X(x)$ | 0.16 | 0.16 | 0.33 | 0.23 | 0.12 | 1.00 |

where $Y$ now indicates the number of goals scored by the youth team. For convenience, we have also included the marginals along the right-hand side and bottom. Let us compute the probability that the men's team scores two goals, given that the youth team scores one goal, and the probability that the youth team scores two goals, given that the men's team scores one goal. These probabilities are respectively given by

$$\text{Prob}\{X = 2|Y = 1\} = 0.10/0.25 = 0.40$$

and

$$\text{Prob}\{Y = 2|X = 1\} = 0.03/0.16 = 0.19.$$

From Equation (4.4) we see that

$$p_{XY}(x, y) = p_Y(y)p_{X|Y}(x|y) \quad \text{for } p_Y(y) > 0.$$

Similarly, we may show that

$$p_{XY}(x, y) = p_X(x)p_{Y|X}(y|x) \quad \text{for } p_X(x) > 0.$$

In this manner, we may compute the joint probability mass function of $X$ and $Y$, from their marginal distributions and their conditional probability mass functions. The procedure applies whether they are independent or not. If they are independent, then

$$p_{X|Y}(x|y) = p_X(x) \quad \text{and} \quad p_{Y|X}(y|x) = p_Y(y)$$

and $p_{XY}(x, y) = p_X(x)p_Y(y)$ as previously seen. Also, observe that the marginal probability may be formed as

$$p_X(x) = \sum_{\text{all } y} p_{XY}(x, y) = \sum_{\text{all } y} p_{X|Y}(x|y)p_Y(y),$$

which is another form of the theorem of total probability. Continuing with the case of two discrete random variables, we may form the conditional cumulative distribution function as

$$F_{X|Y}(x|y) = \text{Prob}\{X \le x \mid Y = y\} = \frac{\text{Prob}\{X \le x,\ Y = y\}}{\text{Prob}\{Y = y\}}$$

for all possible values of $x$ and values of $y$ for which $\text{Prob}\{Y = y\} > 0$. This can be obtained from the conditional probability mass function as

$$F_{X|Y}(x|y) = \frac{\sum_{u \le x} p_{XY}(u, y)}{p_Y(y)} = \sum_{u \le x} p_{X|Y}(u|y).$$

**Example 4.17** Returning to the joint random variables $X$ and $Y$ of Example 4.15, let us compute the conditional cumulative distribution function of $X$ given $Y = 1$. We have

$$F_{X|Y}(x|1) = \begin{cases} 0, & x < -1, \\ 12/5 \times 1/12 = 1/5, & -1 \le x < 0, \\ 12/5 \times (1/12 + 3/12) = 4/5, & 0 \le x < 1, \\ 12/5 \times (1/12 + 3/12 + 1/12) = 1, & x \ge 1. \end{cases}$$

### *Continuous Random Variables*

When $X$ and $Y$ are *not* discrete random variables problems arise since, for a continuous random variable, the denominator, $\text{Prob}\{Y = y\}$, is zero. However, we may define conditional density functions as

$$f_{X|Y}(x|y) = \frac{f_{X,Y}(x, y)}{f_Y(y)} \quad \text{for} \quad 0 < f_Y(y) < \infty. \tag{4.5}$$

This mimics Equation (4.4) where we use $f_{X,Y}(x, y)$ in place of $\text{Prob}\{X = x, Y = y\}$ and $f_Y(y)$ in place of $\text{Prob}\{Y = y\}$.

**Example 4.18** Compute the conditional marginal probability density function $f_{X|Y}(x|Y = y)$ from the following joint density function of two continuous random variables $X$ and $Y$:

$$f_{X,Y}(x, y) = \begin{cases} e^{-y}, & 0 < x < y < \infty, \\ 0 & \text{otherwise.} \end{cases}$$

We first compute the marginal probability density functions. We have

$$f_X(x) = \int_x^\infty e^{-y} dy = -e^{-y}\Big|_x^\infty = e^{-x}$$

since $f_X(x) > 0$ only when $y > x$. The other marginal is more straightforward. We have

$$f_Y(y) = \int_0^y e^{-y} dx = ye^{-y}.$$

Let us now form the conditional marginal distribution $f_{X|Y}(x|Y = y)$. From Equation (4.5),

$$f_{X|Y}(x|Y = y) = \frac{e^{-y}}{ye^{-y}} = \frac{1}{y} \quad \text{for } 0 < x < y.$$

This conditional density function is *uniformly* distributed on $(0, y)$.

Corresponding to Equation (4.4), we may rewrite Equation (4.5) as

$$f_{X,Y}(x, y) = f_Y(y) f_{X|Y}(x|y) = f_X(x) f_{Y|X}(y|x) \ \text{for } 0 < f_Y(y) < \infty,\ \ 0 < f_X(x) < \infty. \tag{4.6}$$

When $X$ and $Y$ are independent random variables, then

$$f_{X,Y}(x, y) = f_X(x) f_Y(y)$$

and it must follow that

$$f_{Y|X}(y|x) = f_Y(y). \tag{4.7}$$

Indeed, Equation (4.7) is a necessary and sufficient condition for two jointly distributed random variables $X$ and $Y$ to be independent. Notice also that Equation (4.5) may be used to compute the joint probability density function for two continuous random variables, as the next example shows.

**Example 4.19** Let $X$ be uniformly distributed on $[0, 1]$ and let $Y$ be uniformly distributed on $[0, X]$. We would like to find the joint probability density function of $X$ and $Y$. The statement of the problem allows us to write

$$f_X(x) = 1, \;\; 0 \le x \le 1,$$
$$f_{Y|X}(y|x) = 1/x, \;\; 0 \le y \le x.$$

Therefore, using Equation (4.6), we obtain the required joint probability density function as

$$f_{X,Y}(x, y) = f_X(x) f_{Y|X}(y|x) = 1 \times 1/x = 1/x, \quad 0 \le y \le x \le 1.$$

There are a number of other results that may be obtained from the definition of the joint density function. For example, we can compute the marginal density of $X$ in terms of the conditional density as

$$f_X(x) = \int_{-\infty}^{\infty} f(x, y)dy = \int_{-\infty}^{\infty} f_Y(y) f_{X|Y}(x|y)dy,$$

which is the continuous analog of the theorem of total probability. We may further obtain the continuous analog of Bayes' rule as

$$f_{Y|X}(y|x) = \frac{f_Y(y) f_{X|Y}(x|y)}{\int_{-\infty}^{\infty} f_Y(y) f_{X|Y}(x|y)dy}.$$

Finally, from Equation (4.5) we may define conditional cumulative distribution functions as

$$F_{X|Y}(x|y) = \text{Prob}\{X \le x | Y = y\} = \frac{\int_{-\infty}^{x} f(u, y)du}{f_Y(y)} = \int_{-\infty}^{x} f_{X|Y}(u|y)du.$$

**Example 4.20** Let us return to Example 4.19 to compute the marginal distribution of $Y$ and the conditional probability density function of $X$ given $Y$. We have

$$f_Y(y) = \int_{-\infty}^{\infty} f_{X,Y}(x, y)dx = \int_y^1 \frac{1}{x} dx = -\ln y$$

and

$$f_{X|Y}(x|y) = \frac{f_{X,Y}(x, y)}{f_Y(y)} = \frac{1/x}{-\ln y}, \quad y \le x \le 1.$$

## 4.6 Convolutions and the Sum of Two Random Variables

Suppose we are given the probability mass functions of two independent discrete random variables $X$ and $Y$ which take their values on the nonnegative integers, and we wish to find the probability mass function of their sum, $S = X + Y$. Let

$$p_i = \text{Prob}\{X = i\} \text{ and } q_i = \text{Prob}\{Y = i\}.$$

When $S$ has the value $k$, neither $X$ nor $Y$ can be greater than $k$ but both must lie in the interval $[0, k]$ with sum equal to $k$. For example, if $X = i$ and $0 \leq i \leq k$, then we must have $Y = k - i$. Therefore, for all $k = 0, 1, \ldots,$

$$\text{Prob}\{S = k\} = \text{Prob}\{X + Y = k\} = \sum_{i=0}^{k} \text{Prob}\{X = i, Y = k - i\}.$$

If $X$ and $Y$ are *independent* random variables, then

$$\text{Prob}\{S = k\} = \sum_{i=0}^{k} \text{Prob}\{X = i\}\text{Prob}\{Y = k - i\} = \sum_{i=0}^{k} p_i q_{k-i}.$$

This is called the *convolution* of the series $p_i$ and $q_i$.

**Example 4.21** Let $X$ and $Y$ be two independent, discrete random variables with probability mass functions given by

$$p_X(i) = \text{Prob}\{X = i\} = \frac{\lambda^i}{i!}e^{-\lambda},$$

$$p_Y(i) = \text{Prob}\{Y = i\} = \frac{\mu^i}{i!}e^{-\mu}, \quad \text{for} \quad i = 0, 1, 2, \ldots \text{ and } \lambda, \mu > 0,$$

respectively. In a later chapter, we shall call such random variables *Poisson* random variables. Let us compute the probability mass function of $X + Y$. Using the convolution result, we obtain

$$\text{Prob}\{X + Y = k\} = \sum_{i=0}^{k} \frac{\lambda^i}{i!}e^{-\lambda}\frac{\mu^{k-i}}{(k-i)!}e^{-\mu} = e^{-(\lambda+\mu)}\frac{1}{k!}\sum_{i=0}^{k}\frac{k!}{i!(k-i)!}\lambda^i\mu^{k-i}$$

$$= e^{-(\lambda+\mu)}\frac{1}{k!}\sum_{i=0}^{k}\binom{k}{i}\lambda^i\mu^{k-i} = \frac{(\lambda+\mu)^k}{k!}e^{-(\lambda+\mu)},$$

which is also a Poisson distributed random variable with parameter $\lambda + \mu$.

Similar results hold for continuous random variables. Let $X$ and $Y$ be continuous random variables with probability density functions given by $f_X(x)$ and $f_Y(y)$, respectively, and let $S = X + Y$. Then

$$F_S(s) = \text{Prob}\{X + Y \leq s\} = \int_{-\infty}^{\infty}\int_{-\infty}^{s-x} f_{X,Y}(x, y)dydx$$

and

$$f_S(s) = \frac{d}{ds}F_S(s) = \int_{-\infty}^{\infty}\frac{d}{ds}\left(\int_{-\infty}^{s-x} f_{X,Y}(x, y)dy\right)dx = \int_{-\infty}^{\infty} f_{X,Y}(x, s-x)dx,$$

for values $-\infty < s < \infty$. Similarly, or more simply by substituting $y = s - x$, we may also obtain

$$f_S(s) = \int_{-\infty}^{\infty} f_{X,Y}(s-y, y)dy, \quad -\infty < s < \infty.$$

When $X$ and $Y$ are *independent* random variables, then

$$f_S(s) = \int_{-\infty}^{\infty} f_X(x)f_Y(s-x)dx = \int_{-\infty}^{\infty} f_X(s-y)f_Y(y)dy, \quad -\infty < s < \infty.$$

This is the convolution formula for *continuous* random variables and can be viewed as a special case of the density function of $X + Y$ when, as a result of their independence, $f_{X,Y}(x, y) = f_X(x)f_Y(y)$.

If, in addition, $X$ and $Y$ are *nonnegative* random variables, then

$$f_S(s) = \int_0^s f_X(x)f_Y(s-x)dx = \int_0^s f_X(s-y)f_Y(y)dy, \quad 0 < s < \infty,$$

i.e., the lower limit changes from $-\infty$ to 0 because $f_X(x) = 0$ for all $x < 0$ and the upper limit changes from $\infty$ to $s$ because $f_Y(s-x) = 0$ for $x > s$.

**Example 4.22** Let $X$ and $Y$ be two independent and nonnegative, continuous random variables with probability density functions

$$f_X(x) = \begin{cases} e^{-x}, & x \geq 0, \\ 0 & \text{otherwise,} \end{cases} \qquad f_Y(y) = \begin{cases} 2e^{-2y}, & y \geq 0, \\ 0 & \text{otherwise.} \end{cases}$$

We would like to find the probability density function of $S = X + Y$. Since both $X$ and $Y$ are nonnegative, $S$ must also be nonnegative. We have

$$f_S(s) = \int_{-\infty}^{\infty} f_X(s-y)f_Y(y)dy = \int_0^s e^{-(s-y)}2e^{-2y}dy = 2e^{-s}\int_0^s e^{-y}dy = 2e^{-s}(1-e^{-s}), \quad s > 0.$$

For $s \leq 0$, $f_S(s) = 0$.

## 4.7 Exercises

**Exercise 4.1.1** Give an example (possibly from your own experience) of an instance in which two discrete random variables are jointly needed to represent a given scenario.

**Exercise 4.1.2** Give an example (possibly from your own experience) of an instance in which two continuous random variables are jointly needed to represent a given scenario.

**Exercise 4.1.3** Give an example (possibly from your own experience) of an instance in which two random variables, one discrete and one continuous, are jointly needed to represent a given scenario.

**Exercise 4.1.4** Give an example (possibly from your own experience) of an instance in which *three* random variables are jointly needed to represent a given scenario.

**Exercise 4.2.1** Construct the joint cumulative probability distribution function of two discrete random variables $X$ and $Y$ from their individual cumulative distributions given below, under the assumption that $X$ and $Y$ are independent.

| | $x < 0$ | $0 \leq x < 1$ | $1 \leq x < 2$ | $2 \leq x < 3$ | $3 \leq x < 4$ | $x \geq 4$ |
|---|---|---|---|---|---|---|
| $F_X(x)$ | 0.0 | 0.2 | 0.5 | 0.5 | 0.6 | 1.0 |

| | $y < 0$ | $0 \leq y < 1$ | $1 \leq y < 2$ | $2 \leq y < 3$ | $3 \leq y < 4$ | $y \geq 4$ |
|---|---|---|---|---|---|---|
| $F_Y(y)$ | 0.0 | 0.3 | 0.5 | 0.6 | 0.7 | 1.0 |

**Exercise 4.2.2** The joint cumulative distribution function of two discrete random variables is given below. Compute the following probabilities: (a) Prob$\{X < 1, Y \leq 1\}$, (b) Prob$\{X \leq 1, Y < 1\}$, (c) Prob$\{-2 < X \leq 2, 1 < Y \leq 2\}$, (d) Prob$\{-1 < X \leq 1, 0 < Y \leq 2\}$. Are $X$ and $Y$ independent?

| | | | | | | |
|---|---|---|---|---|---|---|
| $y \geq 2$ | 0.000 | 0.200 | 0.440 | 0.720 | 0.940 | 1.000 |
| $1 \leq y < 2$ | 0.00 | 0.120 | 0.264 | 0.432 | 0.564 | 0.600 |
| $0 \leq y < 1$ | 0.00 | 0.040 | 0.088 | 0.144 | 0.188 | 0.200 |
| $y < 0.000$ | 0.000 | 0.000 | 0.000 | 0.000 | 0.000 | 0.000 |
| $F_{X,Y}(x, y)$ | $x < -3$ | $-3 \leq x < -1$ | $-1 \leq x < 0$ | $0 \leq x < 1$ | $1 \leq x < 3$ | $x \geq 3$ |

**Exercise 4.2.3** The joint cumulative distribution function of two continuous random variables $X$ and $Y$ is

$$F_{X,Y}(x, y) = \begin{cases} 1 - e^{-2x} - e^{-3y} + e^{-(2x+3y)}, & 0 \leq x < \infty,\ 0 \leq y < \infty, \\ 0 & \text{otherwise.} \end{cases}$$

(a) Find Prob$\{X \leq 2,\ Y \leq 1\}$.
(b) Use Equation (4.1) to find Prob$\{0 < X \leq 1,\ 1 < Y \leq 2\}$.
(c) Show that $X$ and $Y$ are independent.
(d) How could the knowledge that $X$ and $Y$ are independent facilitate the computation of the probability in part (b)?

**Exercise 4.3.1** Consider two discrete random variables $X$ and $Y$ whose joint probability mass function is given in the table below.

| | $x = 0$ | $x = 1$ | $x = 2$ | $x = 3$ |
|---|---|---|---|---|
| $y = 0$ | $a$ | $2a$ | $2a$ | $a$ |
| $y = 1$ | $b$ | $2b$ | $2b$ | $b$ |

What restrictions must be placed on the values of $a$ and $b$ so that this table represents a valid joint probability density function?

**Exercise 4.3.2** Two discrete random variables $X$ and $Y$ have joint probability mass function as follows:

| | $X = 1$ | $X = 2$ | $X = 3$ | $X = 4$ |
|---|---|---|---|---|
| $Y = 0$ | 1/6 | 1/12 | 0 | 1/3 |
| $Y = 1$ | 1/6 | 0 | 1/4 | 0 |

(a) Compute the marginal probability mass functions of $X$ and $Y$ and check for independence.
(b) Compute the probability of each of the following four events:
(i) $X < 3$; (ii) $X$ is odd; (iii) $XY$ is odd; (iv) $Y$ is odd given that $X$ is odd.

**Exercise 4.3.3** Let $X$ and $Y$ be two discrete random variables with joint probability mass function

| | $X = -2$ | $X = 0$ | $X = 2$ |
|---|---|---|---|
| $Y = 1$ | 0 | $2a$ | $a$ |
| $Y = 2$ | $2a$ | 0 | $2a$ |
| $Y = 4$ | $a$ | $2a$ | 0 |

Let $S$ and $Z$ be two additional discrete random variables such that $S = X + Y$ and $Z = X - Y$.

(a) Find $a$ and the marginal probability mass function of $X$.
(b) Are $X$ and $Y$ independent?
(c) Construct the table of the joint probability mass function of $S$ and $Z$.
(d) Are $S$ and $Z$ independent?

**Exercise 4.3.4** Let $X$, $Y$, and $Z$ be three discrete random variables for which

$$\begin{aligned}\text{Prob}\{X = 0,\ Y = 0,\ Z = 0\} &= 6/24,\\ \text{Prob}\{X = 0,\ Y = 1,\ Z = 0\} &= 8/24,\\ \text{Prob}\{X = 0,\ Y = 1,\ Z = 1\} &= 6/24,\\ \text{Prob}\{X = 1,\ Y = 0,\ Z = -1\} &= 1/24,\\ \text{Prob}\{X = 1,\ Y = 0,\ Z = 1\} &= 1/24,\\ \text{Prob}\{X = 1,\ Y = 1,\ Z = 0\} &= 2/24.\end{aligned}$$

Let $S$ be a new random variable for which $S = X + Y + Z$.

(a) Find the marginal probability mass function of $X$.
(b) Are $X$ and $Y$ independent?
(c) Are $X$ and $Z$ independent?
(d) Find the marginal probability mass function of $S$.

**Exercise 4.3.5** Let $X$, $Y$, and $Z$ be three discrete random variables for which

$$\begin{aligned}\text{Prob}\{X = 1,\ Y = 1,\ Z = 0\} &= p,\\ \text{Prob}\{X = 1,\ Y = 0,\ Z = 1\} &= (1-p)/2,\\ \text{Prob}\{X = 0,\ Y = 1,\ Z = 1\} &= (1-p)/2,\end{aligned}$$

where $0 < p < 1$. What is the joint probability mass function of $X$ and $Y$?

**Exercise 4.3.6** Let $X$ be a random variable that has the value $\{-1,\ 0,\ 1\}$ depending upon whether a child in school is performing below, at, or above grade level, respectively. Let $Y$ be a random variable that is equal to zero if a child comes from an impoverished family and equal to one otherwise. In a particular class, it is observed that 20% of the children come from impoverished families and are performing below grade level, that 20% are from impoverished families and are performing at grade level, and that 6% are from impoverished families and are performing above grade level. Of the remaining children, half are performing at grade level and one-third are performing above grade level. Construct a table of the joint probability mass function of $X$ and $Y$ and compute the marginal probability density function of both random variables. Are $X$ and $Y$ independent?

**Exercise 4.4.1** Two continuous random variables $X$ and $Y$ have joint probability density function given by

$$f_{X,Y}(x, y) = \begin{cases} 2xy + x, & 0 \le x \le 1,\ 0 \le y \le 1,\\ 0 & \text{otherwise.}\end{cases}$$

Show that $f_{X,Y}(x, y)$ is a bona fide joint density function and find (a) $\text{Prob}\{X \ge Y\}$ and (b) $\text{Prob}\{\min(X, Y) \le 1/2\}$.

**Exercise 4.4.2** $X$ and $Y$ are two independent and identically distributed continuous random variables. The probability density function of $X$ is

$$f_X(x) = \begin{cases} 3x^2, & 0 \le x \le 1,\\ 0 & \text{otherwise.}\end{cases}$$

That of $Y$ is similarly defined. Write down the joint probability density function of $X$ and $Y$ and find the probability that $\text{Prob}\{Y - X \ge 1/2\}$.

**Exercise 4.4.3** The roots of a quadratic polynomial $ax^2 + bx + c = 0$ are real when $b^2 - 4ac \ge 0$. Use this result to find the probability that the roots of $r^2 + 2Xr + Y = 0$ are real, where $X$ and $Y$ are independent, continuous random variables with $X$ uniformly distributed over $(-1, 1)$ and $Y$ uniformly distributed over $(0, 1)$.

**Exercise 4.4.4** Two friends agree to meet between 6:00 p.m. and 7:00 p.m. for an afterwork drink at the local pub. Each arrives randomly within this hour, but both being impatient will only wait for at most ten minutes for the other to arrive. Use geometric arguments to find the probability that the two friends actually meet.

**Exercise 4.4.5** Let $X$ and $Y$ be two continuous random variables whose joint density function is given by

$$f_{X,Y}(x, y) = \begin{cases} \alpha e^{-y}, & 0 < x < y < \infty, \\ 0 & \text{otherwise.} \end{cases}$$

Compute the value of $\alpha$ and then find $F_{X,Y}(2, y)$.

**Exercise 4.4.6** Let $X$ and $Y$ be two continuous random variables whose joint probability density function is given by

$$f_{X,Y}(x, y) = \begin{cases} 1 - \alpha(x + y), & 0 \le x \le 1,\ 0 \le y \le 2, \\ 0 & \text{otherwise.} \end{cases}$$

Find the value of the constant $\alpha$ and the probabilities Prob$\{X \ge 1/2, Y \le 1\}$ and Prob$\{X < Y\}$.

**Exercise 4.4.7** Let the joint probability density function of two continuous random variables $X$ and $Y$ be given by

$$f_{X,Y}(x, y) = \begin{cases} \alpha, & a < x \le b,\ 0 < y \le c, \\ 0 & \text{otherwise.} \end{cases}$$

Find the value of $\alpha$ and the marginal density functions of $X$ and $Y$. Are $X$ and $Y$, independent random variables?

**Exercise 4.4.8** The joint probability density function of two continuous random variables $X$ and $Y$ is given by

$$f_{X,Y}(x, y) = \begin{cases} \alpha, & x^2 + y^2 \le r^2, \\ 0 & \text{otherwise.} \end{cases}$$

Compute the value of the constant $\alpha$ and the marginal density functions of $X$ and $Y$. Are $X$ and $Y$ independent?

**Exercise 4.5.1** Let $X$ and $Y$ be two discrete random variables whose joint probability mass function is as follows.

| $p_{X,Y}(x, y)$ | $X = -3$ | $X = -1$ | $X = 0$ | $X = 1$ | $X = 3$ |
|---|---|---|---|---|---|
| $Y = 0$ | 0.10 | 0.04 | 0.03 | 0.02 | 0.01 |
| $Y = 1$ | 0.10 | 0.10 | 0.10 | 0.10 | 0.00 |
| $Y = 2$ | 0.00 | 0.10 | 0.15 | 0.10 | 0.05 |

Find Prob$\{X = 0 \mid Y = 0\}$ and Prob$\{Y = 2 \mid X = 1\}$.

**Exercise 4.5.2** Two discrete random variables $X$ and $Y$, which are independent, have probability mass functions given by

| | 0 | 1 | 2 | 3 | 4 |
|---|---|---|---|---|---|
| $X$ | 0.2 | 0.3 | 0.0 | 0.1 | 0.4 |
| $Y$ | 0.3 | 0.2 | 0.1 | 0.1 | 0.3 |

Find Prob$\{X = 1 \mid Y > 2\}$ and Prob$\{X = 1 \mid Y \ne 2\}$.

**Exercise 4.5.3** The joint probability density function of two random variables $X$ and $Y$ is given by

$$f_{X,Y}(x, y) = \begin{cases} 2, & 0 \le y \le x \le 1, \\ 0 & \text{otherwise.} \end{cases}$$

Derive the conditional probability density functions $f_{X|Y}(x|y)$ and $f_{Y|X}(y|x)$.

**Exercise 4.5.4** Random variables $X$ and $Y$ have joint probability density function given by

$$f_{X,Y}(x, y) = \begin{cases} 2(x + y), & 0 \le x \le y \le 1, \\ 0 & \text{otherwise.} \end{cases}$$

Find the conditional probability density functions $f_{X|Y}(x|y)$ and $f_{Y|X}(y|x)$.

**Exercise 4.5.5** The joint probability density function of two continuous random variables $X$ and $Y$ is given by

$$f_{X,Y}(x, y) = \begin{cases} \alpha x^2 y, & 0 \le x \le 1,\ 0 \le y \le 1, \\ 0 & \text{otherwise.} \end{cases}$$

Find the value of $\alpha$ and $\text{Prob}\{X \le 0.5 | Y = 0.2\}$.

**Exercise 4.5.6** Let $X$ and $Y$ be two continuous random variables whose joint probability density function is given by

$$f_{X,Y}(x, y) = \begin{cases} 1 - (x + y)/3, & 0 \le x \le 1,\ 0 \le y \le 2, \\ 0 & \text{otherwise.} \end{cases}$$

What are the marginal density functions of $X$ and $Y$? Compute the probability density function of $Y$ given $X$ and hence the following probabilities: (a) $\text{Prob}\{Y \le 1 \mid X = 0.5\}$ and (b) $\text{Prob}\{Y \le 1 \mid X = 0.25\}$.

**Exercise 4.6.1** The probability mass function of a discrete random variable $X$, in tabular form, is given below. Random variable $Y$ is identically distributed to, and independent of, $X$.

| | $X = 1$ | $X = 2$ | $X = 3$ | $X = 4$ |
|---|---|---|---|---|
| $p_X(x)$ | 0.1 | 0.2 | 0.3 | 0.4 |

(a) Express the probability mass function of $X$ as a formula.
(b) Use the convolution equation to compute the distribution of $X + Y$.
(c) Construct a table of the joint probability mass function $p_{(X,Y)}(x, y)$ and from this table find $\text{Prob}\{X + Y = 4\}$.
(d) Verify your answer to (c) against that computed by inserting the appropriate values into your answer to part (b).

**Exercise 4.6.2** Let $X$ and $Y$ be two discrete random variables that are independent and identically distributed with probability mass function given by

$$p_X(k) = \text{Prob}\{X = k\} = \begin{cases} p(1 - p)^k, & k = 1, 2, \ldots,\ 0 \le p \le 1, \\ 0 & \text{otherwise.} \end{cases}$$

Such random variables are called *geometric* random variables, and are discussed in Chapter 6. Find the distribution of $X + Y$.

**Exercise 4.6.3** Let $X$ and $Y$ be two independent random variables that are identically and uniformly distributed on the unit interval. Use the convolution formula to compute the probability density function of the random variable $S = X + Y$.

# Chapter 5

# *Expectations and More*

## 5.1 Definitions

A random variable is completely characterized by its cumulative distribution function or equivalently, by its probability mass/density function. However, it is not always possible, and in many cases not necessary, to have this complete information. In some instances a single number that captures some average characteristic is sufficient. Such numbers include the *mean value, median*, and *mode* of the random variable. The mean value of a random variable is computed in a similar manner to that used to compute the average value of a set of numbers. The average value of a set of $n$ numbers is formed by multiplying each number by $1/n$ and adding the results together. The mean value of a discrete random variable which assumes the values $\{x_1, x_2, \ldots, x_n\}$ is obtained by multiplying each by the proportion of time it occurs, and then adding, i.e., by forming $\sum_{i=1}^{n} x_i p_i$. The mean value of a random variable $X$ is more commonly called the *expectation* of $X$ and we shall have much more to say about this in just a moment. The *median* of a set of numbers is that number (or numbers) which lies in the middle of the set once the numbers are arranged into ascending or descending order. The median of a random variable $X$ is any number $m$ that satisfies

$$\text{Prob}\{X < m\} \leq 1/2 \quad \text{and} \quad \text{Prob}\{X > m\} \leq 1/2,$$

or equivalently,

$$\text{Prob}\{X < m\} = \text{Prob}\{X > m\}.$$

The *mode* is the most likely possible value. For a random value, it is the value for which the probability mass function or probability density function attains its maximum. If $m$ is a mode of a random variable $X$, then $\text{Prob}\{X = m\} \geq \text{Prob}\{X = x\}$ for all values of $x$. Like the median, and unlike the mean value, a random variable may have multiple modes.

**Example 5.1** If $X$ is the random variable that denotes the number of spots obtained in one throw of a fair die, then there are six modes (since each number of spots is equally likely and all outcomes attain the same maximum value of 1/6); the median is *any* number in the interval (3, 4); there is only one mean value, which is equal to $(1 + 2 + 3 + 4 + 5 + 6)/6 = 3.5$.

The definition of the mean and the fact that it is a unique number make it, from a mathematical point of view, much more attractive to work with than either the median or the mode. As we shall now see, it may be readily extended to provide a complete characterization of a random variable. Indeed, a random variable may be *completely* characterized by a set of values, called *moments*. These moments are defined in terms of the distribution function, but generally do not need the distribution function to be computed. It may be shown that, if two random variables have the same moments of all orders, then they also have the same distribution.

The first moment is the *mean* or *expectation* of the random variable $X$ and is given by the Riemann-Stieltjes integral,

$$E[X] = \int_{-\infty}^{\infty} x dF(x).$$

For a continuous random variable $X$, this becomes

$$E[X] = \int_{-\infty}^{\infty} x f(x) dx,$$

where $f(x)$ is the probability density function of $X$. If $X$ is a discrete random variable whose values $\{x_1, x_2, \ldots\}$ have corresponding probabilities $\{p_1, p_2, \ldots\}$, then the integral becomes a sum and we have, as previously indicated,

$$E[X] = \sum_{i=1}^{\infty} x_i p_i.$$

When a discrete random variable takes its values on the set of nonnegative integers, we may write its expectation as

$$\begin{aligned} E[X] &= p_1 + 2p_2 + 3p_3 + 4p_4 + \cdots \\ &= (p_1 + p_2 + p_3 + p_4 + \cdots) + (p_2 + p_3 + p_4 + \cdots) + (p_3 + p_4 + \cdots) + \cdots, \end{aligned}$$

which gives the useful formula

$$E[X] = \sum_{i=1}^{\infty} \text{Prob}\{x \geq i\}.$$

The $n^{\text{th}}$ moment of a random variable $X$ is defined similarly. For a continuous random variable, it is given by

$$E[X^n] = \int_{-\infty}^{\infty} x^n f(x) dx,$$

while for a discrete random variable $X$ it is

$$E[X^n] = \sum_i x_i^n p_i.$$

The first two moments are the most important. The first, as we have seen, is called the expectation or mean value of $X$; the second moment is called the *mean-square* value of $X$. It is given by

$$E[X^2] = \begin{cases} \displaystyle\int_{-\infty}^{\infty} x^2 f(x) dx & \text{if } X \text{ is continuous,} \\ \displaystyle\sum_i x_i^2 p_i & \text{if } X \text{ is discrete.} \end{cases}$$

Central moments are also important. They are the moments of the difference between a random variable $X$ and its mean value $E[X]$. The $n^{\text{th}}$ central moment is given by

$$E[(X - E[X])^n] = \begin{cases} \displaystyle\int_{-\infty}^{\infty} (x - E[X])^n f(x) dx & \text{if } X \text{ is continuous,} \\ \displaystyle\sum_i (x_i - E[X])^n p_i & \text{if } X \text{ is discrete.} \end{cases}$$

The second central moment is more commonly called the *variance* of the random variable $X$ and is written as

$$\sigma_X^2 \equiv \text{Var}[X] \equiv E[(X - E[X])^2] = \begin{cases} \int_{-\infty}^{\infty} (x - E[X])^2 f(x)dx & \text{if } X \text{ is continuous,} \\ \sum_i (x_i - E[X])^2 p_i & \text{if } X \text{ is discrete.} \end{cases}$$

A frequently used formula for the variance, and one we shall derive shortly (see Equation 5.5), is

$$\text{Var}[X] = E[X^2] - (E[X])^2. \tag{5.1}$$

Since $E[X^2]$ is the mean-square value of $X$, the variance is equal to the mean-square value of $X$ minus the square of the mean value of $X$. The variance, which cannot be negative, characterizes the dispersion of the random variable $X$ about its mean value. Its square root, denoted by $\sigma_X$, is called the *standard deviation* of the random variable $X$. The standard deviation yields a number whose units are the same as those of the random variable and for this reason provides a clearer picture of the dispersion of the random variable about its mean. One final characterization of a random variable that we shall need, is its *coefficient of variation*, written as $C_X$ and defined as

$$C_X = \frac{\sigma_X}{E[X]},$$

i.e., the standard deviation of $X$ divided by its expected value. Its usefulness lies in the fact that it is a dimensionless measure of the variability of $X$. Often, in place of the coefficient of variation, the *squared coefficient of variation* is used instead. This is defined as

$$C_X^2 = \frac{\text{Var}[X]}{(E[X])^2}.$$

**Example 5.2** Consider the random variable $X$ which denotes the total number of spots obtained when two fair dice are thrown simultaneously. Previously, we saw that the probability mass function of $X$ is given by

| $x_i$ | 2 | 3 | 4 | 5 | 6 | 7 | 8 | 9 | 10 | 11 | 12 |
|---|---|---|---|---|---|---|---|---|---|---|---|
| $p_X(x_i)$ | 1/36 | 2/36 | 3/36 | 4/36 | 5/36 | 6/36 | 5/36 | 4/36 | 3/36 | 2/36 | 1/36 |

In this case the expectation of $X$ is given by

$$\begin{aligned} E[X] &= \sum_{i=1}^{\infty} x_i p_i = \sum_{i=2}^{12} x_i p_i \\ &= 2\frac{1}{36} + 3\frac{2}{36} + 4\frac{3}{36} + 5\frac{4}{36} + 6\frac{5}{36} + 7\frac{6}{36} + 8\frac{5}{36} + 9\frac{4}{36} + 10\frac{3}{36} + 11\frac{2}{36} + 12\frac{1}{36} \\ &= \frac{252}{36} = 7.0. \end{aligned}$$

This is also the value of the median and the mode of $X$. To compute the variance of $X$, we proceed as follows:

$$\begin{aligned} \text{Var}[X] &= \sum_{i=2}^{12} (x_i - 7)^2 p_i \\ &= 5^2\frac{1}{36} + 4^2\frac{2}{36} + 3^2\frac{3}{36} + 2^2\frac{4}{36} + 1^2\frac{5}{36} + 0^2\frac{6}{36} + 1^2\frac{5}{36} + 2^2\frac{4}{36} + 3^2\frac{3}{36} + 4^2\frac{2}{36} \\ &\quad + 5^2\frac{1}{36} = \frac{210}{36} = 5.8333. \end{aligned}$$

It follows that the standard deviation of $X$ is $\sqrt{5.8333} = 2.4152$.

**Example 5.3** Let $X$ be a continuous random variable whose probability density function is

$$f_X(x) = \begin{cases} \alpha x^2, & -1 \le x \le 1, \\ 0 & \text{otherwise,} \end{cases}$$

for some constant $\alpha$. We wish to compute the value of $\alpha$ and then find the expectation and standard deviation of $X$. Since

$$1 = \int_{-\infty}^{\infty} f_X(x)dx = \int_{-1}^{1} \alpha x^2 \, dx = \alpha x^3/3\big|_{-1}^{1} = 2\alpha/3,$$

we must have $\alpha = 3/2$. We may now compute $E[X]$ as

$$E[X] = \int_{-\infty}^{\infty} x f_X(x)dx = \int_{-1}^{1} 3x^3/2 \, dx = 3x^4/8\big|_{-1}^{1} = 0.$$

The variance is computed from

$$\text{Var}[X] = \int_{-\infty}^{\infty} (x - E[X])^2 f_X(x)dx = \int_{-\infty}^{\infty} x^2 f_X(x)dx,$$

which is just equal to the second moment! In this case,

$$\text{Var}[X] = E[X^2] = \int_{-1}^{1} 3x^4/2 \, dx = 3x^5/10\big|_{-1}^{1} = 3/5$$

and the standard deviation is given by $\sigma_X = \sqrt{3/5}$.

**Example 5.4** Consider the continuous random variable $M$ whose probability density function is given by

$$f_M(x) = \begin{cases} e^{-x} & \text{if } x \ge 0, \\ 0 & \text{otherwise.} \end{cases}$$

In this case, the expectation is given by

$$E[M] = \int_{-\infty}^{\infty} x f(x)dx = \int_{0}^{\infty} x e^{-x} dx = -e^{-x}\big|_0^{\infty} = 1,$$

and its variance is computed as

$$\text{Var}[M] = \int_{0}^{\infty} (x - E[M])^2 e^{-x} dx = \int_{0}^{\infty} (x-1)^2 e^{-x} dx$$

$$= \int_{0}^{\infty} x^2 e^{-x} dx - 2\int_{0}^{\infty} x e^{-x} dx + \int_{0}^{\infty} e^{-x} dx = 2 - 2 + 1 = 1,$$

which is just the same as its mean value!

We stated previously that the standard deviation of a random variable is a number whose units are the same as those of the random variable and is an appropriate measure of the dispersion of the random variable about its mean. If a particular value for $X$, say $x$, lies in an interval of $\sigma_X$ about $E[X]$, i.e., $x \in [E[X] - \sigma_X, E[X] + \sigma_X]$, then it is considered to be close to the expected value of $X$. Usual (or typical) values of $X$ lie inside this range. If $x$ lies outside $2\sigma_X$ of $E[X]$, i.e., $x < E[x] - 2\sigma_x$ or $x > E[x] + 2\sigma_x$, then $x$ is considered to be far from the mean.

Chebychev's inequality, which is proven in a later section, makes these concepts more precise. Let $X$ be a random variable whose expectation is $E[X] = \mu_X$ and whose standard deviation is given by $\sigma_X$. Chebychev's inequality states that

$$\text{Prob}\{|X - \mu_X| \leq k\sigma_X\} \geq 1 - \frac{1}{k^2}.$$

In words, this says that the probability a random variable lies within $\pm k$ standard deviations of its mean value is at least $1 - 1/k^2$. Alternatively, the probability that a random variable lies outside $\pm k$ standard deviations of its mean is at most $1/k^2$. This means that 75% $(= 1 - 1/2^2)$ of the time, a random variable is within two standard deviations of its mean, and 89% $(= 1 - 1/3^2)$ of the time, it falls within three standard deviations. In fact it turns out that in most cases, these numbers are underestimations of what generally occurs. Indeed, for many distributions, the probability that a random variable lies within two standard deviations can be as high as 95% or greater.

**Example 5.5** In Ms. Wright's class of 32 eighth graders, the height of her students, rounded to the nearest inch, is given in the following table.

| Height | $\leq 62$ | 63 | 64 | 65 | 66 | 67 | 68 | 69 | 70 | 71 | $\geq 72$ |
|---|---|---|---|---|---|---|---|---|---|---|---|
| Students | 1 | 2 | 1 | 3 | 4 | 3 | 6 | 5 | 4 | 2 | 1 |

Are Peter, whose height is 5′4″ and Paul, whose height is 5′11″ outside the norm for this class?

We first compute the mean and standard deviation of student height. We find the expectation $E[H]$ to be

$$(62+2\times63+64+3\times65+4\times66+3\times67+6\times68+5\times69+4\times70+2\times71+1\times72)/32 = 67.47$$

and the variance to be

$$\begin{aligned}\text{Var}[H] = \big(&(62-67.47)^2 + 2(63-67.47)^2 + (64-67.47)^2 + 3(65-67.47)^2 \\ &+ 4(66-67.47)^2 + 3(67-67.47)^2 + 6(68-67.47)^2 + 5(69-67.47)^2 \\ &+ 4(70-67.47)^2 + 2(71-67.47)^2 + (72-67.47)^2\big)/32 = 6.06,\end{aligned}$$

which means that the standard deviation is $\sqrt{6.06} = 2.46$. It follows that students whose heights lie between $67.47 - 2.46$ and $67.47 + 2.46$, i.e., in the interval $[65.01, 69.93]$, or since we have rounded to the nearest inch, students whose height lies between 65 and 70 inches, should be considered normal for this class. Since Peter's and Paul's heights both lie outside this, neither is "average" for this class. However, neither student's height lies outside the 2 standard deviation range of $67.47 \pm 2 \times 2.46 = [62.55, 72.39]$.

The random variables that we shall consider all have finite expectation. The expectation is said not to exist unless the corresponding integral or summation is *absolutely convergent*, i.e., the expectations exist only if

$$\int_{-\infty}^{\infty} |x| f(x) dx < \infty \quad \text{and} \quad \sum_{i=1}^{\infty} |x_i| p_i < \infty.$$

**Example 5.6** The function

$$f_X(x) = \frac{1}{\pi(1 + x^2)}, \qquad -\infty < x < \infty,$$

is a probability density function but does *not* have an expectation. To see this, observe that

$$\int_{-\infty}^{\infty} \frac{dx}{\pi(1+x^2)} = \frac{1}{\pi} \arctan(x)\Big|_{-\infty}^{\infty} = \frac{1}{\pi}\left(\frac{\pi}{2} - \frac{-\pi}{2}\right) = 1,$$

and so $f_X(x)$ is a density function. However, this does not converge *absolutely* since

$$\int_{-\infty}^{\infty} \left|\frac{x}{\pi(1+x^2)}\right| dx = \frac{1}{\pi}\int_0^{\infty} \frac{2x}{1+x^2} dx = \frac{1}{\pi} \ln(1+x^2)\Big|_0^{\infty} = \infty.$$

## 5.2 Expectation of Functions and Joint Random Variables

### *Expectation of Functions of a Random Variable*

Given a discrete random variable $X$, the expectation of a new random variable $Y$ which is a function of $X$, i.e., $Y = h(X)$, may be written as

$$E[Y] = E[h(X)] = \sum_{-\infty}^{\infty} h(X) p_X(x). \tag{5.2}$$

Similarly, for a continuous random variable $X$, the expectation of some function $h(X)$ of $X$ is given as

$$E[h(X)] = \int_{-\infty}^{\infty} h(X) f_X(x) dx. \tag{5.3}$$

**Example 5.7** For the discrete random variable $R$ which denotes the number of heads obtained in three tosses of a fair coin, we saw that its probability mass function is given by

| $x_k$ | 0 | 1 | 2 | 3 |
|---|---|---|---|---|
| $p_R(x_k)$ | 0.125 | 0.375 | 0.375 | 0.125 |

Consider now a gambling situation in which a player loses \$3 if no heads appear, loses \$2 if one head appears, wins nothing and gains nothing if two heads appear but gains \$7 dollars if three heads appears. Let us compute the mean number of dollars won or lost in one game. To do so, we define the derived random variable $Y$ as follows:

$$Y = h(X) = \begin{cases} -3, & X = 0, \\ -2, & X = 1, \\ 0, & X = 2, \\ 7, & X = 3, \\ 0 & \text{otherwise.} \end{cases}$$

We now compute $E[Y]$ and discover that the player should expect to lose 25 cents each time he plays, since

$$E[Y] = \sum_{x=0,1,2,3} h(X) p_X(x) = -3 \times 1/8 - 2 \times 3/8 + 7 \times 1/8 = -1/4.$$

**Example 5.8** Let $X$ be a continuous random variable with probability density function

$$f_X(x) = \begin{cases} 2x, & 0 \le x \le 1, \\ 0 & \text{otherwise,} \end{cases}$$

and let $Y$ be the derived random variable $Y = 4X + 2$. The expectation of $X$ is given by

$$E[X] = \int_0^1 2x^2\,dx = \left.\frac{2x^3}{3}\right|_0^1 = 2/3.$$

Let us now compute the expectation of $Y$ in two ways: first integrating its density function over the appropriate interval, and second using the easier approach offered by Equation (5.3).

Using the approach of Section 3.5, we find the probability density function of $Y$ by first finding its cumulative distribution function. However, we first need to find the cumulative distribution function of $X$. This is

$$F_X(x) = \int_0^x 2u\,du = x^2 \text{ for } 0 \le x \le 1; \qquad (5.4)$$

is equal to zero for $x < 0$, and is equal to 1 for $x > 1$. We may now compute $F_Y(y)$ as

$$F_Y(y) = \text{Prob}\{Y \le y\} = \text{Prob}\{4X + 2 \le y\} = \text{Prob}\left\{X \le \frac{y-2}{4}\right\} = F_X\left(\frac{y-2}{4}\right).$$

To find the correct limits, we observe that $F_X(x) = 0$ for $x < 0$, which allows us to assert that $F_Y(y) = 0$ for $(y-2)/4 < 0$, i.e., for $y < 2$. Also, $F_X(x) = 1$ for $x > 1$, so that $F_Y(y) = 1$ for $(y-2)/4 > 1$, i.e., for $y > 6$. It follows from Equation (5.4) that

$$F_Y(y) = \left(\frac{y-2}{4}\right)^2 \quad \text{for } 2 \le y \le 6.$$

Finally, by differentiating $F_Y(y)$ with respect to $y$, we may now form the density function of $Y$ as

$$f_Y(y) = \begin{cases} 2(y-2)/16, & 2 \le y \le 6, \\ 0 & \text{otherwise,} \end{cases}$$

and compute its expectation as

$$E[Y] = \int_2^6 2y(y-2)/16\,dy = (2y^3/3 - 2y^2)/16|_2^6 = 4.6667.$$

Using Equation (5.3), we obtain

$$E[Y] = E[h(X)] = \int_{-\infty}^{\infty} h(X) f_X(x)\,dx = \int_0^1 (4x+2)2x\,dx = \int_0^1 (8x^2 + 4x)\,dx$$
$$= (8x^3/3 + 2x^2)\Big|_0^1 = 14/3,$$

the same as before, but with considerably less computation.

The extension to functions of $n$ random variables is immediate. If a random variable $Y$ is defined as a function $h$ of the $n$ random variables $X_1, X_2, \ldots, X_n$, its expectation may be found using the formula

$$E[Y] = E[h(X_1, X_2, \ldots, X_n)] = \begin{cases} \int_{-\infty}^{\infty}\int_{-\infty}^{\infty}\cdots\int_{-\infty}^{\infty} h(x_1, x_2, \ldots, x_n) f(x_1, x_2, \ldots, x_n) dx_1 dx_2 \cdots dx_n \\ \sum_{x_1}\sum_{x_2}\cdots\sum_{x_n} h(x_1, x_2, \ldots, x_n) p(x_1, x_2, \ldots, x_n) \end{cases}$$

for the continuous and discrete cases, respectively. If the function $h$ is a linear function, i.e.,

$$h(x_1, x_2, \ldots, x_n) = \alpha_1 x_1 + \alpha_2 x_2 + \cdots + \alpha_n x_n,$$

where $\alpha_i,\ i = 1, 2, \ldots, n$, are constants, then

$$E[Y] = E[h(X_1, X_2, \ldots, X_n)] = h(E[X_1], E[X_2], \ldots, E[X_n]).$$

Notice that the random variables $X_1, X_2, \ldots, X_n$ need not be independent. However, this result does not generally apply to functions $h$ that are not linear.

**Example 5.9** Returning to Example 5.8, and observing that $h(x) = 4x + 2$ is a linear function, we may compute $E[Y]$ as

$$E[Y] = E[h(X)] = h(E[X]) = h(2/3) = 4(2/3) + 2 = 14/3,$$

as before.

### *Expectation of Jointly Distributed Random Variables*

When $X$ and $Y$ are jointly distributed random variables and $Z$ is a random variable that is a function of $X$ and $Y$, i.e., $Z = h(X, Y)$, then the expectation of $Z$ may be found directly from the joint distribution. When $X$ and $Y$ are discrete random variables, then

$$E[Z] = \sum_{\text{all } x} \sum_{\text{all } y} h(x, y) p_{X,Y}(x, y),$$

and when they are continuous random variables,

$$E[Z] = \int_{-\infty}^{\infty} \int_{-\infty}^{\infty} h(x, y) f_{X,Y}(x, y)\, dx\, dy.$$

**Example 5.10** Let $X$ and $Y$ be two continuous random variables whose joint probability density function is

$$f_{X,Y}(x, y) = \begin{cases} 1 - (x + y)/3, & 0 \le x \le 1,\ 0 \le y \le 2, \\ 0 & \text{otherwise.} \end{cases}$$

We wish to find $E[XY]$. We have

$$E[XY] = \int_{y=0}^{2} \int_{x=0}^{1} xy[1 - (x + y)/3]\, dx\, dy = \int_{y=0}^{2} \int_{x=0}^{1} \left( xy - \frac{x^2 y}{3} - \frac{xy^2}{3} \right) dx\, dy$$

$$= \int_{y=0}^{2} \left( \frac{x^2 y}{2} - \frac{x^3 y}{9} - \frac{x^2 y^2}{6} \right) \Bigg|_{x=0}^{1} dy = \int_{y=0}^{2} \left( \frac{y}{2} - \frac{y}{9} - \frac{y^2}{6} \right) dy$$

$$= \left( \frac{7y^2}{36} - \frac{y^3}{18} \right) \Bigg|_{0}^{2} = \frac{12}{36} = 1/3.$$

Observe that, when $X$ and $Y$ are independent random variables, then $E[XY] = E[X]E[Y]$ since we have

$$E[XY] = \int_{-\infty}^{\infty} \int_{-\infty}^{\infty} xy f_{XY}(x, y) dx dy = \int_{-\infty}^{\infty} \int_{-\infty}^{\infty} xy f_X(x) f_Y(y) dx dy$$

$$= \int_{-\infty}^{\infty} \int_{-\infty}^{\infty} x f_X(x)\, y f_Y(y)\, dx dy = \int_{-\infty}^{\infty} E[X]\, y f_Y(y) dy = E[X]E[Y].$$

The case in which a random variable is defined as the sum of two other random variables, warrants special attention. We have

$$\begin{aligned}
E[Z] = E[X+Y] &= \int_{-\infty}^{\infty}\int_{-\infty}^{\infty}(x+y)f_{XY}(x,y)dxdy \\
&= \int_{-\infty}^{\infty}\int_{-\infty}^{\infty}xf_{XY}(x,y)dxdy + \int_{-\infty}^{\infty}\int_{-\infty}^{\infty}yf_{XY}(x,y)dxdy \\
&= \int_{-\infty}^{\infty}x\left[\int_{-\infty}^{\infty}f_{XY}(x,y)dy\right]dx + \int_{-\infty}^{\infty}y\left[\int_{-\infty}^{\infty}f_{XY}(x,y)dx\right]dy \\
&= \int_{-\infty}^{\infty}xf_X(x)dx + \int_{-\infty}^{\infty}yf_Y(y)dy = E[X]+E[Y].
\end{aligned}$$

In developing this result, we have used Equation (4.3), which defines the marginal probabilities $f_X(x)$ and $f_Y(y)$. Notice that this property, called the *linearity property* of the expectation, holds whether the random variables are independent or not. More generally, the expectation of a sum of random variables is the sum of the expectations of those random variables. That is, let $X_1$, $X_2, \ldots, X_n$ be a set of $n$ random variables. Then

$$E[X_1 + X_2 + \cdots + X_n] = E[X_1] + E[X_2] + \cdots + E[X_n].$$

The same result does *not* apply to variances: the variance of a sum of two random variables is equal to the sum of the variances only if the random variables are *independent*. To show this we let $\mu_X = E[X]$ and $\mu_Y = E[Y]$. Then

$$\begin{aligned}
\text{Var}\,[X+Y] &= E[(X+Y-\mu_X-\mu_Y)^2] \\
&= E[(X-\mu_X)^2] + 2E[(X-\mu_X)(Y-\mu_Y)] + E[(Y-\mu_Y)^2] \\
&= E[(X-\mu_X)^2] + E[(Y-\mu_Y)^2] = \text{Var}\,(X) + \text{Var}\,(Y),
\end{aligned}$$

where we have used the fact that, when $X$ and $Y$ are independent,

$$E[(X-\mu_X)(Y-\mu_Y)] = E[X-\mu_X]E[Y-\mu_Y] = 0,$$

since, for example, $E[X-\mu_X] = E[X] - E[\mu_X] = \mu_X - \mu_X = 0$.

To compute the variance of the sum of two random variables $X$ and $Y$ that are not necessarily independent, we use the relation $\text{Var}\,[X] = E[X^2] - E[X]^2$, and proceed as follows:

$$\begin{aligned}
\text{Var}\,[X+Y] &= E[(X+Y)^2] - (E[X]+E[Y])^2 \\
&= E[X^2] + 2E[XY] + E[Y^2] - (E[X])^2 - 2E[X]E[Y] - (E[Y])^2
\end{aligned}$$

and hence

$$\text{Var}\,[X+Y] = \text{Var}\,[X] + \text{Var}\,[Y] + 2(E[XY] - E[X]E[Y]).$$

Thus the variance of the sum of two random variables is equal to the sum of the variances only if $E[XY] - E[X]E[Y]$ is zero. This is the case, for example, when $X$ and $Y$ are independent. The quantity $E[XY] - E[X]E[Y]$ is called the *covariance* of $X$ and $Y$ and is denoted

$$\text{Cov}\,[X,Y] = E[XY] - E[X]E[Y].$$

The variance of a sum of random variables, $X_1, X_2, \ldots, X_n$ is given by

$$\text{Var}\,[X_1 + X_2 + \cdots + X_n] = \sum_{k=1}^{n}\text{Var}\,[X_k] + 2\sum_{j<k}\text{Cov}\,[X_j, X_k]$$

Some properties of the covariance of two random variables are

- $\text{Cov}\,(\alpha X, \beta Y) = \alpha\beta\,\text{Cov}(X, Y)$,
- $\text{Cov}\,(X + \alpha, Y + \beta) = \text{Cov}(X, Y)$,
- $\text{Cov}\,(X, \alpha X + \beta) = \alpha \text{Var}\,(X)$,

where $\alpha$ and $\beta$ are scalars. As a general rule, the covariance is positive if above-average values of $X$ are associated with above-average values of $Y$ or if below-average values of $X$ are associated with below-average values of $Y$. If above-average values of one are associated with below-average values of the other, then the covariance will be negative. Sometime difficulties arise in interpreting the magnitude of a covariance, and for this reason, a new quantity, called the *correlation* , is used. Just as the standard deviation of a random variable often gives a better feel than the variance for the deviation of a random variable from its expectation, the correlation of two random variables normalizes the covariance to make it unitless. The correlation of two random variables is defined as

$$\text{Corr}\,[X, Y] = \frac{\text{Cov}\,[X, Y]}{\sigma_X \sigma_Y}$$

where $\sigma_X$ and $\sigma_Y$ are the standard deviations of $X$ and $Y$, respectively. For any two random variables, we always have $-1 \leq \text{Corr}\,[X, Y] \leq 1$.

**Example 5.11** Let us compute the covariance and correlation of the two random variables $X$ and $Y$ whose joint probability density function is

$$f_{X,Y}(x, y) = \begin{cases} 6x, & 0 < x < y < 1, \\ 0 & \text{otherwise.} \end{cases}$$

Let us find $E[XY]$, $\text{Cov}\,[X, Y]$, $\text{Corr}\,[X, Y]$, $E[X + Y]$, and $\text{Var}\,[X + Y]$.

We begin by computing the marginal density function of both $X$ and $Y$, and from these we shall compute the expectation and variance of $X$ and of $Y$. With this information we can then compute the requested items.

$$f_X(x) = \int_{-\infty}^{\infty} f_{XY}(x, y)\,dy = \int_x^1 6x\,dy = 6xy|_x^1 = 6x(1 - x), \quad 0 \leq x \leq 1,$$

and is zero otherwise.

$$f_Y(y) = \int_{-\infty}^{\infty} f_{XY}(x, y)\,dx = \int_0^y 6x\,dx = 3x^2|_0^y = 3y^2, \quad 0 \leq y \leq 1,$$

and is zero otherwise.

$$E[X] = \int_{-\infty}^{\infty} x f_X(x) dx = \int_0^1 6x^2(1 - x)\,dx = \left(2x^3 - \frac{3x^4}{2}\right)\Big|_0^1 = 1/2,$$

$$E[Y] = \int_{-\infty}^{\infty} y f_Y(y) dy = \int_0^1 3y^3\,dy = 3/4,$$

$$E[X^2] = \int_{-\infty}^{\infty} x^2 f_X(x) dx = \int_0^1 6x^3(1 - x)\,dx = \left(\frac{3x^4}{2} - \frac{6x^5}{5}\right)\Big|_0^1 = 3/10,$$

$$E[Y^2] = \int_{-\infty}^{\infty} y^2 f_Y(y) dy = \int_0^1 3y^4\,dy = 3/5,$$

$$\text{Var}\,[X] = E[X^2] - E[X]^2 = 3/10 - 1/4 = 1/20,$$

$$\text{Var}\,[Y] = E[Y^2] - E[Y]^2 = 3/5 - 9/16 = 3/80.$$

We are now in a position to compute the required quantities. We have

$$E[XY] = \int_{x=0}^{1}\int_{y=x}^{1} 6x^2y\,dy\,dx = \int_0^1 3x^2y^2\Big|_{y=x}^{1}\,dx = \int_0^1 (3x^2-3x^4)\,dx = \left(x^3 - \frac{3x^5}{5}\right)\Big|_0^1 = 2/5,$$

$$\mathrm{Cov}\,[X,Y] = E[XY] - E[X]E[Y] = 2/5 - 1/2 \times 3/4 = 1/40,$$

$$\mathrm{Corr}\,[X,Y] = \frac{\mathrm{Cov}[X,Y]}{\sigma_X\sigma_Y} = \frac{1/40}{\sqrt{1/20}\sqrt{3/80}} = \frac{1}{\sqrt{3}} = \sqrt{3}/3,$$

$$E[X+Y] = E[X] + E[Y] = 1/2 + 3/4 = 5/4,$$

$$\mathrm{Var}\,[X+Y] = \mathrm{Var}\,[X] + \mathrm{Var}\,[Y] + 2\,\mathrm{Cov}\,[X,Y] = 1/20 + 3/80 + 1/20 = 11/80.$$

The covariance of two random variables $X$ and $Y$ captures the degree to which they are correlated. $X$ and $Y$ are uncorrelated if $E[XY] = E[X]E[Y]$. If, for example, $X$ and $Y$ are independent, then $E[XY] = E[X]E[Y]$ and $X$ and $Y$ are also uncorrelated. However, if $X$ and $Y$ are uncorrelated, it does not necessarily follow that they are also independent.

**Example 5.12** Let $X$ and $Y$ be two discrete random variables with joint probability mass function given by

| | $X = -1$ | $X = 0$ | $X = 1$ | $p_Y(y)$ |
|---|---|---|---|---|
| $Y = -1$ | 1/12 | 3/12 | 1/12 | 5/12 |
| $Y = 0$ | 1/12 | 0/12 | 1/12 | 2/12 |
| $Y = 1$ | 1/12 | 3/12 | 1/12 | 5/12 |
| $p_X(x)$ | 3/12 | 6/12 | 3/12 | 1 |

From this table we have

$$E[X] = -1 \times 3/12 + 1 \times 3/12 = 0, \quad E[Y] = -1 \times 5/12 + 1 \times 5/12 = 0,$$

and

$$E[XY] = (-1)(-1)1/12 + (-1)(1)1/12 + (1)(-1)1/12 + (1)(1)1/12 = 0.$$

It follows then that $X$ and $Y$ are uncorrelated, since $E[XY] = E[X]E[Y] = 0$, but they are not independent since

$$p_{X,Y}(0,0) = 0 \neq p_X(0)p_Y(0) = 6/12 \times 2/12.$$

In summary, $X$ and $Y$ are uncorrelated if $\mathrm{Corr}\,[X,Y] = 0$, or if $\mathrm{Cov}\,[X,Y] = 0$ or if $E[XY] = E[X]E[Y]$; independent random variables are uncorrelated, but uncorrelated random variables are not necessarily independent.

Consider now the variance of the *difference* of two random variables $X$ and $Y$:

$$\mathrm{Var}\,[X-Y] = \mathrm{Var}\,[X] + \mathrm{Var}\,[Y] - 2(E[XY] - E[X]E[Y])$$

which is obtained by replacing $X+Y$ with $X-Y$ in the formula for the variance of a sum. Thus the variance of the sum of two random variables is also equal to the variance of their difference, if they are uncorrelated.

### *Conditional Expectation and Variance*

Given a conditional density function $f_{X|Y}(x|y)$ of joint random variables $X$ and $Y$, we may form the *conditional expectation* of $X$ given $Y$ as

$$E[X|Y=y] \equiv E[X|y] = \int_{-\infty}^{\infty} x f_{X|Y}(x|y)dx = \frac{\int_{-\infty}^{\infty} x f_{X,Y}(x,y)dx}{f_Y(y)}, \quad 0 < f_Y(y) < \infty.$$

More generally, the conditional expectation of a function $h$ of $X$ and $Y$ is given by

$$E[h(X,Y)|Y=y] = \int_{-\infty}^{\infty} h(x,y) f_{X|Y}(x|y)dx.$$

When $X$ and $Y$ are discrete random variables, we have

$$E[X|Y=y] = \sum_{all\ x} x \text{ Prob}\{X=x|Y=y) = \sum_{all\ x} x p_{X|Y}(x|y),$$

and if $h$ is a function of $X$ and $Y$, then

$$E[h(X,Y)|Y] = \sum_{\text{all } x} h(x,y) p_{X|Y}(x|y).$$

These results lead to the law of total expectation:

$$E[X] = E[\,E[X|Y]\,],$$

i.e., the expectation of the conditional expectation of $X$ given $Y$ is equal to the expectation of $X$, which is quite a mouthful. For continuous random variables, we have

$$E[X] = E[\,E[X|Y]\,] = \int_{-\infty}^{\infty} E[X|Y=y] f_Y(y)dy,$$

while for discrete random variables

$$E[X] = E[\,E[X|Y]\,] = \sum_y E[X|Y=y]\text{Prob}\{Y=y\}.$$

We prove this in the discrete case only, the continuous case being analogous. We have

$$\begin{aligned} E[X] = E[\,E[X|Y]\,] &= \sum_y E[X|Y=y]\text{ Prob}\{Y=y\} \\ &= \sum_y \left( \sum_x x \text{ Prob}\{X=x|Y=y) \right) \text{Prob}\{Y=y\} \\ &= \sum_x x \left( \sum_y \text{Prob}\{X=x|Y=y)\text{Prob}\{Y=y\} \right) \\ &= \sum_x x \text{ Prob}\{X=x\} = E[X], \end{aligned}$$

since, by the law of total probability, $\sum_y \text{Prob}\{X=x|Y=y)\text{Prob}\{Y=y\} = \text{Prob}\{X=x\}$.

For two random variables $X$ and $Y$, the *conditional variance* of $Y$ given $X=x$ is defined as

$$\text{Var}\,[Y|X] = E[(Y - E[Y|X])^2\,|\,X],$$

or by the following equivalent formulation which is frequently more useful,

$$\text{Var}\,[Y|X] = E[Y^2|X] - (\,E[Y|X]\,)^2.$$

We also have

$$\text{Var}\,[Y] = E\,[\text{Var}\,[Y|X]] + \text{Var}\,[E[Y|X]]\,.$$

**Example 5.13** Let $X$ and $Y$ be two continuous random variables whose joint probability density function is given by

$$f_{X,Y}(x,y) = \begin{cases} 1-(x+y)/3, & 0 \le x \le 1,\ 0 \le y \le 2, \\ 0 & \text{otherwise.} \end{cases}$$

We wish to find $E[X|Y = 0.5]$ and $E[Y|X = 0.5]$. This requires a knowledge of the marginal distributions of $X$ and $Y$ as well as the conditional marginal distributions so we begin by computing these quantities. We have

$$f_X(x) = \int_{-\infty}^{\infty} f_{X,Y}(x,y)dy = \int_{y=0}^{2} [1-(x+y)/3]\,dy = (y - xy/3 - y^2/6)\Big|_0^2$$
$$= 4/3 - 2x/3, \quad 0 \le x \le 1,$$

and

$$f_Y(y) = \int_{-\infty}^{\infty} f_{X,Y}(x,y)dx = \int_{x=0}^{1} [1-(x+y)/3]\,dx = (x - xy/3 - x^2/6)\Big|_0^1$$
$$= 5/6 - y/3, \quad 0 \le y \le 2.$$

Also

$$f_{X|Y}(x|0.5) = \frac{f_{X,Y}(x,0.5)}{f_Y(0.5)} = \frac{1 - x/3 - 0.5/3}{5/6 - 0.5/3} = \frac{5}{4} - \frac{x}{2},$$

$$f_{Y|X}(y|0.5) = \frac{f_{X,Y}(0.5,y)}{f_X(0.5)} = \frac{1 - 0.5/3 - y/3}{4/3 - 1/3} = \frac{5}{6} - \frac{y}{3}.$$

The required results may now be formed. We have

$$E[X|Y=0.5] = \int_{-\infty}^{\infty} x f_{X|Y}(x|0.5)dx = \int_0^1 x\left(\frac{5}{4} - \frac{x}{2}\right)dx = \left(\frac{5x^2}{8} - \frac{x^3}{6}\right)\Big|_0^1 = \frac{11}{24},$$

$$E[Y|X=0.5] = \int_{-\infty}^{\infty} y f_{Y|X}(y|0.5)dy = \int_0^2 y\left(\frac{5}{6} - \frac{y}{3}\right)dy = \left(\frac{5y^2}{12} - \frac{y^3}{9}\right)\Big|_0^2 = \frac{7}{9}.$$

**Example 5.14** Let us return to Example 5.11 and find $E[Y|x]$. For convenience, we rewrite the joint density function and the previously computed result for the marginal density of $X$. These were

$$f_{X,Y}(x,y) = \begin{cases} 6x, & 0 < x < y < 1, \\ 0 & \text{otherwise,} \end{cases}$$

and

$$f_X(x) = 6x(1-x), \quad 0 \le x \le 1,$$

respectively. We may now compute $E[Y|x]$ as

$$E[Y|x] = \int_{-\infty}^{\infty} y f_{Y|X}(y|x)\,dy = \int_{y=x}^{1} \frac{6xy}{6x(1-x)}\,dy = \frac{1}{1-x}\int_{y=x}^{1} y\,dy = \frac{1}{1-x}\left(\frac{1-x^2}{2}\right)$$
$$= \frac{1+x}{2}.$$

### *Properties of the Expectation and Variance*

Some of the most important properties of the expectation of random variables $X$ and $Y$ are collected together and given below.

1. $E[1] = 1$. Indeed, for any constant $c$, $E[c] = c$.
2. If $c$ is a constant, then $E[cX] = cE[X]$.
3. $|E[X]| \le E[|X|]$.
4. If $X \ge 0$, then $E[X] \ge 0$. Furthermore, if $X \le Y$, then $E[X] \le E[Y]$.
5. Linearity: If $X$ and $Y$ are two random variables, not necessarily independent, then $E[X + Y] = E[X] + E[Y]$.
6. If $X$ and $Y$ are *independent* random variables, then $E[XY] = E[X]E[Y]$.
7. If $X_1, X_2, \ldots$ is an increasing or decreasing sequence of random variables converging to $X$, then $E[X] = \lim_{i\to\infty} E[X_i]$.

A number of other important properties can be deduced from these. For example, from properties 2 and 5, we have

$$E\left[\sum_{i=1}^{n} c_i X_i\right] = \sum_{i=1}^{n} c_i E[X_i].$$

In particular, for given constants $a$ and $b$,

$$E[aX + b] = aE[X] + b.$$

Taking $a = 0$ and $b = 1$ yields property 1 again. Taking $a = 1$ and $b = -E[X]$ gives

$$E[X - E[X]] = 0,$$

and thus the first central moment must always be zero. Finally, the frequently used formula for the variance of a random variable, namely,

$$\text{Var}[X] = E[X^2] - (E[X])^2, \tag{5.5}$$

may be derived from these properties. We have

$$\begin{aligned}\text{Var}[X] &= E[(X - E[X])^2] = E[X^2 - 2XE[X] + (E[X])^2] \\ &= E[X^2] - 2E[X]E[X] + (E[X])^2 = E[X^2] - (E[X])^2.\end{aligned}$$

Some properties of the variance are given below.

1. $\text{Var}[1] = 0$.
2. $\text{Var}[X] = E[X^2] - (E[X])^2$.
3. $\text{Var}[cX] = c^2\text{Var}[X]$.
4. If $X$ and $Y$ are *independent* random variables, then
$$\text{Var}[X + Y] = \text{Var}[X] + \text{Var}[Y].$$
5. If $X$ and $Y$ are *uncorrelated* random variables, then
$$\text{Var}[X - Y] = \text{Var}[X + Y].$$

## 5.3 Probability Generating Functions for Discrete Random Variables

Given a discrete random variable $X$ that assumes only nonnegative integer values $\{0, 1, 2, \ldots\}$ with $p_k = \text{Prob}\{X = k\}$, we may define the probability generating function of $X$ by

$$G_X(z) = \sum_{i=0}^{\infty} p_i z^i = p_0 + p_1 z + p_2 z^2 + \cdots + p_k z^k + \cdots.$$

$G_X(z)$ is also called the $z$-transform of $X$ and may be shown to converge for any complex number $z$ for which $|z| < 1$. Since $\sum_{i=0}^{\infty} p_i = 1$, it follows that $G_X(1) = 1$.

**Example 5.15** The random variable $R$ denoting the number of heads obtained in three tosses of a fair coin has generating function given by

$$G_R(z) = \sum_{i=0}^{3} p_i z_i = z^0\frac{1}{8} + z^1\frac{3}{8} + z^2\frac{3}{8} + z^3\frac{1}{8} = \frac{1}{8}(1 + 3z + 3z^2 + z^3).$$

When $z = 1$ we find

$$G_R(1) = \frac{1}{8}(1 + 3 + 3 + 1) = 1.$$

The probability distribution of a discrete random variable $X$ is *uniquely* determined by its generating function. It follows then that if two discrete random variables $X$ and $Y$ have the same probability generating function, then they must also have the same probability distribution and probability mass function. Observe that the $k^{\text{th}}$ derivative of a probability generating function evaluated at $z = 0$, can be used to obtain the probability $p_k$. We have

$$\frac{1}{k!}G_X^{(k)}(0) = p_X(k), \quad k = 0, 1, 2, \ldots .$$

**Example 5.16** (continued from Example 5.15). From the second derivative of $G_R(z)$ evaluated at $z = 0$ we obtain

$$p_2 = \frac{1}{2!}\frac{d^2}{dz^2}\left(\frac{1}{8}(1 + 3z + 3z^2 + z^3)\right)\Bigg|_{z=0} = \frac{1}{16}(6 + 6z)\Bigg|_{z=0} = \frac{6}{16} = \frac{3}{8},$$

the probability of getting a single head in three tosses of a fair coin.

In addition to selected probabilities, expectations and higher moments of a random variable may be obtained directly from the probability generating function by differentiation and evaluation at $z = 1$. We have

$$E[X] = \sum_{k=0}^{\infty} kp_k = \lim_{z\to 1}\left(\sum_{k=0}^{\infty} kp_k z^{k-1}\right) = \lim_{z\to 1} G'_X(z) = G'_X(1).$$

**Example 5.17** (continued from Example 5.15). For the random variable $R$, we have

$$E[R] = \frac{d}{dz}\left(\frac{1}{8}(1 + 3z + 3z^2 + z^3)\right)\Bigg|_{z=1} = \left(\frac{1}{8}(3 + 6z + 3z^2)\right)\Bigg|_{z=1} = \frac{1}{8}(3{+}6{+}3) = \frac{12}{8} = 1.5.$$

Higher moments are obtained from higher derivatives. Observe that

$$E[X] = \sum_{k=0}^{\infty} kp_k,$$

$$E[X(X-1)] = E[X^2] - E[X] = \sum_{k=0}^{\infty} k(k-1)p_k,$$

$$E[X(X-1)(X-2)] = E[X^3] - 3E[X^2] + 2E[X] = \sum_{k=0}^{\infty} k(k-1)(k-2)p_k,$$

$$\vdots$$

$$E[X(X-1)\cdots(X-n+1)] = \sum_{k=0}^{\infty} k(k-1)\cdots(k-n+1)p_k.$$

These are called the *factorial moments* of $X$. The $n^{\text{th}}$ derivative of the probability generating function of $X$ evaluated at $z = 1$ gives the $n^{\text{th}}$ factorial moment. For example,

$$G''_X(z) = \sum_{k=0}^{\infty} k(k-1)z^{k-2}p_k,$$

and hence

$$G''_X(1) = \lim_{z\to 1} G''_X(z) = \lim_{z\to 1} \sum_{k=0}^{\infty} k(k-1)z^{k-2}p_k = \sum_{k=0}^{\infty} k(k-1)p_k = E[X(X-1)].$$

In general, the $k^{\text{th}}$ derivative of $G_X(z)$ evaluated at $z = 1$ provides the $k^{\text{th}}$ factorial moment, i.e.,

$$G_X^{(k)}(1) = E[X(X-1)\cdots(X-k+1)].$$

This provides a rather straightforward procedure for computing regular moments once factorial moments have been computed. For example, since $E[X(X-1)] = E[X^2] - E[X]$, then given the second factorial moment and the expectation, the second regular moment may be computed as

$$E[X^2] = E[X] + E[X(X-1)].$$

**Example 5.18** Let $X$ be a discrete random variable with probability mass function

$$p_k = p_X(k) = \begin{cases} \binom{n}{k} p^k q^{n-k}, & 0 \le k \le n, \\ 0 & \text{otherwise,} \end{cases}$$

where $p + q = 1$. The corresponding probability generating function is

$$G_X(z) = \sum_{k=0}^{n} z^k \binom{n}{k} p^k q^{n-k} = \sum_{k=0}^{n} \binom{n}{k} (zp)^k q^{n-k} = (zp+q)^n.$$

Taking the first and second derivatives with respect to $z$, we obtain

$$G'_X(z) = n(zp+q)^{n-1}p \quad \text{and} \quad G''_X(z) = (n-1)n(zp+q)^{n-2}p^2.$$

Evaluating these at $z = 1$ and relating them to the first two factorial moments, we obtain

$$E[X] = G'_X(1) = np \quad \text{and} \quad E[X(X-1)] = G''_X(1) = (n-1)np^2.$$

It follows that

$$E[X^2] = (n-1)np^2 + np = (np)^2 - np^2 + np.$$

The variance of $X$ may now be computed as

$$\text{Var}\,[X] = E[X^2] - E[X]^2 = (np)^2 - np^2 + np - (np)^2 = np(1-p) = npq.$$

Observe that for fixed $z$, the probability generating function $G_X(z)$ is just the mathematical expectation of the random variable $h(X) = z^X$, i.e.,

$$G_X(z) = E[z^X], \quad |z| \le 1,$$

since from the previous properties of the expectation we have

$$E[h(X)] = \sum_{-\infty}^{\infty} h(X)p_X(x) = \sum_{k=0}^{\infty} z^X \text{Prob}\{X = k\} = \sum_{k=0}^{\infty} z^k p_k.$$

Thus, if $X_1, X_2, \ldots, X_n$ are $n$ independent random variables taking the values $0, 1, 2, \ldots$, then, for fixed $z$, the random variables $z^{X_i}$ for $i = 1, 2, \ldots, n$ are also independent, and it follows that

$$E[z^{X_1+X_2+\cdots+X_n}] = E[z^{X_1}z^{X_2}\cdots z^{X_n}] = E[z^{X_1}]E[z^{X_2}]\cdots E[z^{X_n}],$$

and hence

$$G_X(z) = G_{X_1}(z)G_{X_2}(z)\cdots G_{X_n}(z). \tag{5.6}$$

This expresses the generating function of a sum of random variables ($X = X_1 + X_2 + \cdots + X_n$) in terms of the generating functions of the individual terms of the summand. To compute the probability mass function of a sum of discrete random variables that are independent, it suffices therefore to form the probability generating function for each of them, to compute the product of these generating functions and then to determine the required probabilities by continuous differentiation of the product of the generating functions.

**Example 5.19** Let $X$ be the random variable that denotes the number of heads obtained when two coins are simultaneously tossed and let $Y$ be the random variable that denotes the number of spots obtained when a fair die is thrown. Let us find the probability generating function of $X + Y$. First we compute the probability generating function of both $X$ and $Y$. Since $p_X(0) = p_X(2) = 1/4$; $p_X(1) = 1/2$ and $p_Y(k) = 1/6$ for $k = 1, 2, \ldots, 6$, we have

$$G_X(z) = \frac{1}{4} + \frac{z}{2} + \frac{z^2}{4} \quad \text{and} \quad G_Y(z) = \frac{z + z^2 + z^3 + \cdots + z^6}{6}.$$

Therefore

$$G_{X+Y}(z) = G_X(z)G_Y(z) = \frac{z}{24} + \frac{z^2}{8} + \frac{z^3 + z^4 + z^5 + z^6}{6} + \frac{z^7}{8} + \frac{z^8}{24}.$$

Finally, when $X_1$ and $X_2$ are two independent discrete-valued random variables and the derived random variable $X$ is equal to $X_1$ with probability $r$ and equal to $X_2$ with probability $(1 - r)$, then

$$G_X(z) = r\,G_{X_1}(z) + (1 - r)\,G_{X_2}(z).$$

## 5.4 Moment Generating Functions

The moment generating function is defined for both discrete and continuous random variables. For a given random variable $X$, its moment generating function, written as $\mathcal{M}_X(\theta)$, is defined for real values of $\theta$, as

$$\mathcal{M}_X(\theta) = E[e^{\theta X}].$$

It exists if this expectation is finite for all $\theta$ in an open interval around zero, i.e., if the expectation exists for all $-c < \theta < c$, for some positive constant $c$. The range of values of $\theta$ for which $\mathcal{M}_X(\theta)$ exists is called the *region of convergence*.

If $X$ is a discrete random variable, then from our previous definition of expectation,

$$\mathcal{M}_X(\theta) = \sum_{\text{all } x} e^{\theta x} p_X(x).$$

A similar result holds for continuous random variables. Let $f_X(x)$ be the density function of a continuous random variable, $X$. Then the moment generating function of $X$ is

$$\mathcal{M}_X(\theta) = \int_{-\infty}^{\infty} e^{\theta x} f_X(x)dx, \tag{5.7}$$

providing the integral exists.

**Example 5.20** Let $X$ be a discrete random variable whose probability mass function is

$$p_X(x) = \begin{cases} 1/n, & x = 1, 2, \ldots, n, \\ 0 & \text{otherwise.} \end{cases}$$

The moment generating function of $X$ is given by

$$\mathcal{M}_X(\theta) = \sum_{\text{all } x} e^{\theta x} p_X(x) = \sum_{x=1}^{n} \frac{e^{\theta x}}{n} = \frac{1}{n}(e^{\theta} + e^{2\theta} + \cdots + e^{n\theta}) = \frac{1}{n}\left(\frac{e^{\theta} - e^{(n+1)\theta}}{1 - e^{\theta}}\right).$$

**Example 5.21** Let $X$ be a continuous random variable whose probability density function is

$$f_X(x) = \begin{cases} \lambda e^{-\lambda x}, & x \geq 0, \\ 0 & \text{otherwise,} \end{cases}$$

where $\lambda > 0$. The moment generating function of $X$ is

$$\mathcal{M}_X(\theta) = \int_0^{\infty} e^{\theta x} \lambda e^{-\lambda x} dx = \lambda \int_0^{\infty} e^{(\theta - \lambda)x} dx.$$

This integral exists only if $\theta < \lambda$ and this defines the range of convergence of this moment generating function. For $\theta < \lambda$ we find

$$\mathcal{M}_X(\theta) = \frac{\lambda}{\theta - \lambda} e^{(\theta - \lambda)x} \Big|_0^{\infty} = \frac{\lambda}{\lambda - \theta}.$$

It should be apparent that there is a connection between the probability generating function, $G_X(z)$, of a discrete random variable $X$ defined in the previous section, and its moment generating function, $\mathcal{M}_X(\theta)$. Specifically, we have

$$\mathcal{M}_X(\theta) = G_X(e^{\theta}).$$

Indeed, given the probability generating function of $X$, we may immediately write its moment generating function, using the above equation.

**Example 5.22** We saw in Example 5.18 that the probability generating function for the random variable $X$ with probability mass function

$$p_k = p_X(k) = \begin{cases} \binom{n}{k} p^k q^{n-k}, & 0 \leq k \leq n, \\ 0 & \text{otherwise,} \end{cases}$$

where $p + q = 1$ is given by

$$G_X(z) = (pz + q)^n.$$

It must follow that the moment generating function of $X$ is given by

$$\mathcal{M}_X(\theta) = (pe^{\theta} + q)^n.$$

Observe that, if we expand the exponential in Equation (5.7) and take expectations, we find

$$\begin{aligned} \mathcal{M}_X(\theta) &= \int_{-\infty}^{\infty} \left(1 + \theta x + \frac{(\theta x)^2}{2!} + \frac{(\theta x)^3}{3!} + \cdots\right) f_X(x)dx \\ &= \int_{-\infty}^{\infty} f_X(x)dx + \theta \int_{-\infty}^{\infty} x f_X(x)dx + \frac{\theta^2}{2!}\int_{-\infty}^{\infty} x^2 f_X(x)dx + \cdots \\ &= 1 + \theta E[X] + \frac{\theta^2 E[X^2]}{2!} + \frac{\theta^3 E[X^3]}{3!} + \cdots. \end{aligned}$$

Hence

$$\mathcal{M}_X(0) = 1$$

and the $k^{\text{th}}$ moment of $X$ is obtained from the $k^{\text{th}}$ derivative evaluated at $\theta = 0$,

$$\left.\frac{d^k}{d\theta^k}\mathcal{M}_X(\theta)\right|_{\theta=0} = E[X^k], \quad k = 1, 2, \ldots,$$

which justifies the name of moment generating function. From a practical point of view, when computing moments of a random variable $X$, it is frequently easier to first find $\mathcal{M}_X(\theta)$ and then to compute the moments by differentiation as described above, than the alternative "first principles" approach of integrating $x^n f_X(x)$.

**Example 5.23** Let us return to Example 5.21 and compute the first, second, and higher moments of $X$. Previously we computed the moment generating function of $X$ to be

$$\mathcal{M}_X(\theta) = \frac{\lambda}{\lambda - \theta}.$$

Hence

$$E[X] = \left.\frac{d\mathcal{M}_X(\theta)}{d\theta}\right|_{\theta=0} = \left.\frac{\lambda}{(\lambda-\theta)^2}\right|_{\theta=0} = \frac{1}{\lambda}$$

and

$$E[X^2] = \left.\frac{d^2\mathcal{M}_X(\theta)}{d\theta^2}\right|_{\theta=0} = \left.\frac{2\lambda}{(\lambda-\theta)^3}\right|_{\theta=0} = \frac{2}{\lambda^2}.$$

By continuous differentiation with respect to $\theta$,

$$\frac{d^k}{d\theta^k}\mathcal{M}_X(\theta) = \frac{\lambda k!}{(\lambda-\theta)^{k+1}} \qquad \text{for } k = 1, 2, \ldots.$$

Hence, setting $\theta = 0$,

$$E[X^k] = \frac{k!}{\lambda^k}.$$

The same approach applies to discrete random variables as illustrated in the next example.

**Example 5.24** The probability mass function of a discrete random variable $X$ is given by

$$p_X(x) = \begin{cases} 1-p, & x = 0, \\ p, & x = 1, \\ 0 & \text{otherwise.} \end{cases}$$

Then

$$\mathcal{M}_X(\theta) = \sum_{\text{all } x} e^{\theta x} p_X(x) = e^0(1-p) + e^\theta p = 1 - p + pe^\theta.$$

Hence

$$E[X] = \left.\frac{d\mathcal{M}_X(\theta)}{d\theta}\right|_{\theta=0} = pe^\theta\big|_{\theta=0} = p$$

and

$$E[X^2] = \left.\frac{d^2\mathcal{M}_X(\theta)}{d\theta^2}\right|_{\theta=0} = pe^\theta\big|_{\theta=0} = p,$$

and the same for all higher moments.

Moment generating functions turn out to be especially useful in analyzing sums of *independent* random variables. If $X_1$ and $X_2$ are two independent random variables and $Y = X_1 + X_2$, then

$$\mathcal{M}_Y(\theta) = E[e^{\theta(X_1+X_2)}] = E[e^{\theta X_1}e^{\theta X_2}] = E[e^{\theta X_1}]E[e^{\theta X_2}] = \mathcal{M}_{X_1}(\theta)\mathcal{M}_{X_2}(\theta).$$

In words, the moment generating function of a *sum* of two independent random variables is equal to the product of their generating functions. The proof follows directly as a consequence of the result $E[X_1X_2] = E[X_1]E[X_2]$ for independent random variables $X_1$ and $X_2$ and the additional fact that if $X_1$ and $X_2$ are independent, then the derived random variables $e^{\theta X_1}$ and $e^{\theta X_2}$ are also independent. This result generalizes directly to the sum of $n$ independent random variables. If $X_1, X_2, \ldots, X_n$ are $n$ independent random variables and $Y = X_1 + X_2 + \cdots + X_n$, then

$$\mathcal{M}_Y(\theta) = E[e^{\theta X_1}e^{\theta X_2}\cdots e^{\theta X_n}] = \mathcal{M}_{X_1}(\theta)\mathcal{M}_{X_2}(\theta)\cdots\mathcal{M}_{X_n}(\theta). \tag{5.8}$$

If, in addition to being independent, the $n$ random variables are also identically distributed, then $\mathcal{M}_{X_i}(\theta) = \mathcal{M}_X(\theta)$ for all $i = 1, 2, \ldots, n$ and

$$\mathcal{M}_X(\theta) = [\mathcal{M}_X(\theta)]^n.$$

**Example 5.25** Let $X_1$ and $X_2$ defined below be two discrete random variables that are independent.

$$p_{X_1}(j) = \begin{cases} 0.4, & j = 1, \\ 0.6, & j = 2, \\ 0 & \text{otherwise,} \end{cases} \qquad p_{X_2}(k) = \begin{cases} 0.2, & k = -1, \\ 0.8, & k = 4, \\ 0 & \text{otherwise.} \end{cases}$$

Let $Y = X_1 + X_2$. We wish to find the expectation of $Y$ and its probability mass function. From the definition, $\mathcal{M}_X(\theta) = \sum_{\text{all } x} e^{\theta x} p_X(x)$, we have

$$\mathcal{M}_{X_1}(\theta) = 0.4e^{\theta} + 0.6e^{2\theta} \quad \text{and} \quad \mathcal{M}_{X_2}(\theta) = 0.2e^{-\theta} + 0.8e^{4\theta},$$

and hence

$$\mathcal{M}_Y(\theta) = \left(0.4e^{\theta} + 0.6e^{2\theta}\right) \times \left(0.2e^{-\theta} + 0.8e^{4\theta}\right) = 0.08 + 0.12e^{\theta} + 0.32e^{5\theta} + 0.48e^{6\theta},$$

The expectation of $Y$ is given by

$$E[Y] = \left.\frac{d}{d\theta}\mathcal{M}_Y(\theta)\right|_{\theta=0} = \left(0.12e^{\theta} + 5(0.32)e^{5\theta} + 6(0.48)e^{6\theta}\right)\big|_{\theta=0} = 4.6.$$

The probability mass function can be found directly from the previously computed value of $\mathcal{M}_Y(\theta)$ and matching coefficients.

$$\begin{aligned}\mathcal{M}_Y(\theta) &= \sum_{\text{all } y} e^{\theta y} p_Y(y) = p_Y(0) + p_Y(1)e^{\theta} + p_Y(2)e^{2\theta} + \cdots \\ &= 0.08 + 0.12e^{\theta} + 0.32e^{5\theta} + 0.48e^{6\theta}.\end{aligned}$$

We find

$$p_Y(y) = \begin{cases} 0.08, & y = 0, \\ 0.12, & y = 1, \\ 0.32, & y = 5, \\ 0.48, & y = 6, \\ 0 & \text{otherwise.} \end{cases}$$

**Example 5.26** Consider a random variable $Y$ with probability density function

$$f_Y(y) = \begin{cases} \lambda^2 y e^{-\lambda y}, & y \geq 0, \\ 0 & \text{otherwise}, \end{cases}$$

and whose parameter $\lambda$ is strictly greater than zero. Then

$$\mathcal{M}_Y(\theta) = \int_0^\infty e^{\theta y} \lambda^2 y e^{-\lambda y} dy = \lambda^2 \int_0^\infty y e^{-(\lambda-\theta)y} dy = \frac{\lambda^2}{(\lambda-\theta)} \int_0^\infty y(\lambda-\theta)e^{-(\lambda-\theta)y} dy.$$

Using integration by parts with $u = y$ and $dv = (\lambda - \theta)e^{-(\lambda-\theta)y}dy$, and observing the constraint $\lambda > \theta$, we obtain

$$\int_0^\infty y(\lambda-\theta)e^{-(\lambda-\theta)y} dy = uv|_0^\infty - \int_0^\infty v du = ye^{-(\lambda-\theta)y}\Big|_0^\infty - \int_0^\infty e^{-(\lambda-\theta)y} dy = \frac{1}{\lambda-\theta}.$$

It follows then that

$$\mathcal{M}_Y(\theta) = \frac{\lambda^2}{(\lambda-\theta)^2} = \left(\frac{\lambda}{\lambda-\theta}\right)^2.$$

However, previously we saw that a random variable having probability density function

$$f_X(x) = \begin{cases} \lambda e^{-\lambda x}, & x \geq 0, \\ 0 & \text{otherwise}, \end{cases}$$

had moment generating function given by

$$\mathcal{M}_X(\theta) = \frac{\lambda}{\lambda-\theta}.$$

It must follow that the random variable $Y$ is equal to the sum of two independent and identically distributed random variables $X_1$ and $X_2$ with

$$f_{X_1}(x) = f_{X_2}(x) = \begin{cases} \lambda e^{-\lambda x}, & x \geq 0, \\ 0 & \text{otherwise}. \end{cases}$$

Moment generating functions satisfy the linear translation property . If $\alpha$ and $\beta$ are constants and $Y = \alpha X + \beta$, then

$$\mathcal{M}_Y(\theta) = e^{\beta\theta}\mathcal{M}_X(\alpha\theta).$$

To see this, observe that

$$\mathcal{M}_Y(\theta) = E[e^{\theta Y}] = E[e^{\theta(\alpha X+\beta)}] = E[e^{\beta\theta}e^{\alpha\theta X}] = e^{\beta\theta}E[e^{\alpha\theta X}] = e^{\beta\theta}\mathcal{M}_X(\alpha\theta).$$

The following is a list of the properties of moment generating functions that were presented in this section:

- $\mathcal{M}_X(\theta) = G_X(e^\theta)$, when $X$ is a discrete random variable.
- $\mathcal{M}_X^{(k)}(\theta)\Big|_{\theta=0} = E[X^k]$, for $k = 1, 2, \ldots$.
- $\mathcal{M}_{X+Y}(\theta) = \mathcal{M}_X(\theta)\mathcal{M}_Y(\theta)$, for independent random variables $X$ and $Y$.
- $\mathcal{M}_Y(\theta) = e^{\beta\theta}\mathcal{M}_X(\alpha\theta)$, for $Y = \alpha X + \beta$.

The moment generating function is closely related to the Laplace transform . The Laplace transform of a real-valued function $h(t)$, $t \geq 0$, is generally denoted by $h^*(s)$ and is defined as

$$h^*(s) = \int_0^\infty e^{-sx}h(x)dx,$$

where $s$, unlike $\theta$ in the moment generating function, may be a complex number. This definition applies to functions $h(x)$ that need not be density functions, and indeed the Laplace transform is found to be useful in many domains other than statistics and probability. In the case that $h(t),\ t > 0$, does in fact correspond to a density function, then the Laplace transform is equal to the moment generating function of $X$ evaluated at $-s$. In our current context of random variables, it is usual to denote the Laplace transform as $\mathcal{L}_X(s)$ and we have

$$\mathcal{L}_X(s) \equiv h^*(s) = \mathcal{M}_X(-s).$$

It follows then that

$$\mathcal{L}_X(s) = E[e^{-sX}]$$

and

$$E[X^k] = (-1)^k \left. \frac{d^k}{ds^k} \mathcal{L}_X(s) \right|_{s=0} \quad \text{for } k = 1, 2, \ldots.$$

**Example 5.27** Making the substitution $s = -\theta$ in Example 5.21, it immediately follows that the Laplace transform of the continuous random variable $X$ with probability density function $f_X(x) = \lambda e^{-\lambda x}$, $x \geq 0$ and zero otherwise, is given by

$$\mathcal{L}_X(s) = \frac{\lambda}{\lambda + s}.$$

There is one other function of a random variable $X$ that warrants our attention, namely, the *characteristic function*, defined as

$$\mathcal{C}_X(\theta) = E[e^{i\theta X}],$$

where $i = \sqrt{-1}$. It is related to the moment generating function since $\mathcal{C}_X(\theta) = \mathcal{M}_X(i\theta)$ and in contexts outside probability and statistics, is called the *Fourier transform*. Although the characteristic function has the disadvantage of involving complex numbers, it has two important advantages over moment generating functions. First, the characteristic function is finite for all random variables and all real values of $\theta$ and second, an inversion formula exists which may facilitate the construction of the probability mass function for a discrete random variable, or the probability density of a continuous random variable, if it indeed exists. The characteristic function possesses properties that are similar to those that are listed above for the moment generating function.

## 5.5 Maxima and Minima of Independent Random Variables

Let $X_1, X_2, \ldots, X_n$ be $n$ independent random variables whose cumulative distribution functions are given by $F_{X_1}(x), F_{X_2}(x), \ldots, F_{X_n}(x)$, respectively. We wish to compute the distribution functions that are defined as the *maximum* and *minimum* of these $n$ random variables. In other words, we seek the distribution function of the random variables $X_{\max}$ and $X_{\min}$ defined as

$$X_{\max} = \max(X_1, X_2, \ldots, X_n) \quad \text{and} \quad X_{\min} = \min(X_1, X_2, \ldots, X_n).$$

Observe first that $X_{\max} \leq x$ if and only if $X_i \leq x$ for all $i = 1, 2, \ldots, n$ and therefore

$$F_{\max}(x) = \text{Prob}\{X_1 \leq x, X_2 \leq x, \ldots, X_n \leq x\}.$$

Hence, from our assumption that the $n$ random variables are independent, it follows that

$$F_{\max}(x) = \prod_{i=1}^{n} \text{Prob}\{X_i \leq x\} = \prod_{i=1}^{n} F_{X_i}(x).$$

**Example 5.28** Let us compare the expected time to perform three data-dependent tasks on a multiprocessor computer system to the expected time to perform these tasks on single processor. Let $X_1$, $X_2$, and $X_3$ be the random variables that describe the times needed to perform the three tasks, and, due to data dependence, let us further assume that each has an identical (exponential) cumulative distribution given by

$$F_{X_i} = 1 - e^{-t/4}, \quad i = 1, 2, 3.$$

Thus, the mean processing time for each of the three tasks is given by $E[X_i] = 4$. Let $X$ be the random variable that describes the time needed to complete all three tasks. Since all three tasks on the multiprocessor will finish as soon as the last of them has been completed, we have $X = \max\{X_1, X_2, X_3\}$. Thus the cumulative distribution function of $X$ is

$$F_X(t) = (1 - e^{-t/4})^3$$

and its density function, obtained by differentiating $F_X(t)$, is given by

$$f_X(t) = 3(1 - e^{-t/4})^2\,(1/4)e^{-t/4} = \frac{3}{4}e^{-t/4}(1 - e^{-t/4})^2.$$

As a side note, observe that the distribution of the maximum of a set of exponentially distributed random variables is *not* exponential. The expectation of $X$ may now be computed. Keeping in mind the previously computed result,

$$\int_0^\infty \lambda t e^{-\lambda t} = \frac{1}{\lambda}, \quad t \geq 0,$$

we obtain

$$\begin{aligned}
E[X] &= \frac{3}{4}\int_0^\infty te^{-t/4}(1 - e^{-t/4})^2 dt \\
&= 3\int_0^\infty \frac{t}{4}e^{-t/4}(1 - 2e^{-t/4} + e^{-t/2})dt \\
&= 3\left(\int_0^\infty \frac{t}{4}e^{-t/4}dt - \int_0^\infty \frac{t}{2}e^{-t/2}dt + \frac{1}{3}\int_0^\infty \frac{3t}{4}e^{-3t/4}dt\right) \\
&= 3\left(4 - 2 + \frac{4}{9}\right) = 7.3333.
\end{aligned}$$

Notice that this computed expected value for the last of the three tasks to complete, namely, 7.3333, is quite a bit longer than the expected time to complete any one of the tasks, namely, 4.

On a single processor, the random variable that describes the total time required is now given by $X = X_1 + X_2 + X_3$ and we compute its expectation simply as

$$E[X] = \sum_{i=1}^{3} E[X_i] = 12.$$

Now consider the random variable $X_{\min} = \min(X_1, X_2, \ldots, X_n)$. In this case, it is easier to first compute $\text{Prob}\{X_{\min} > x\}$ and to form $F_{\min}(x)$ from the relationship

$$F_{\min}(x) = 1 - \text{Prob}\{X_{\min} > x\}.$$

Notice that $X_{\min} > x$ if and only if $X_i > x$ for all $i = 1, 2, \ldots, n$ and therefore

$$\text{Prob}\{X_{\min} > x\} = \text{Prob}\{X_1 > x, X_2 > x, \ldots, X_n > x\}.$$

Once again, reverting to the independence of the $n$ random variables $X_1, X_2, \ldots, X_n$, we have

$$\text{Prob}\{X_{\min} > x\} = \prod_{i=1}^{n} \text{Prob}\{X_i > x\} = \prod_{i=1}^{n} (1 - F_i(x))$$

and so

$$F_{\min}(x) = 1 - \prod_{i=1}^{n} (1 - F_i(x)).$$

**Example 5.29** Let us find the distribution of the time until the first of the three tasks on the multiprocessor computer system of the previous example finishes. The random variable $X$ that describes this situation is given by $X = \min\{X_1, X_2, X_3\}$ and its cumulative distribution function is given by

$$F_X(t) = 1 - e^{-t/4}e^{-t/4}e^{-t/4} = 1 - e^{-3t/4}, \quad t \geq 0.$$

Thus the mean time for the first task to finish is equal to 4/3 which is considerably less than the mean of any of the individual tasks!

Notice, in the above example, that the distribution obtained as the minimum of the set of three exponential distributions is itself an exponential distribution. This holds true in general: unlike the maximum of a set of exponential distributions, the minimum of a set of exponential distributions is itself an exponential distribution.

## 5.6 Exercises

**Exercise 5.1.1** Find the median, mode, and expected value of the discrete random variable whose cumulative distribution function is as follows:

$$F_X(x) = \begin{cases} 0, & x < -1, \\ 1/8, & -1 \leq x < 0, \\ 1/4, & 0 \leq x < 1, \\ 1/2, & 1 \leq x < 2, \\ 1, & x \geq 2. \end{cases}$$

**Exercise 5.1.2** The probability mass function of a discrete random variable $X$ is as follows:

$$p_X(x) = \begin{cases} 0.1, & x = -2, \\ 0.2, & x = 0, \\ 0.3, & x = 2, \\ 0.4, & x = 5, \\ 0 & \text{otherwise.} \end{cases}$$

Find the expectation, second moment, variance, and standard deviation of $X$.

**Exercise 5.1.3** A random variable $X$ takes values $0, 1, 2, \ldots, n, \ldots$ with probabilities

$$\frac{1}{2}, \frac{1}{3}, \frac{1}{3^2}, \frac{1}{3^3}, \ldots, \frac{1}{3^n}, \ldots.$$

Show that this represents a genuine probability mass function. Find $E[X]$, the expected value of $X$.

**Exercise 5.1.4** Two fair dice are thrown. Let $X$ be the random variable that denotes the number of spots shown on the first die and $Y$ the number of spots that show on the second die. It follows that $X$ and $Y$ are independent and identically distributed. Compute $E[X^2]$ and $E[XY]$ and observe that they are not the same.

**Exercise 5.1.5** Balls are drawn from an urn containing $w$ white balls and $b$ black balls until a white ball appears. Find the mean value and the variance of the number of balls drawn, assuming that each ball is replaced after being drawn.

**Exercise 5.1.6** Find the expectation, second moment, variance, and standard deviation of the continuous random variable with probability density function given by

$$f_X(x) = \begin{cases} 3x^2/16, & -2 \le x \le 2, \\ 0 & \text{otherwise.} \end{cases}$$

**Exercise 5.1.7** Let $X$ be a continuous random variable with probability density function

$$f_X(x) = \begin{cases} 1/10, & 2 \le x \le 12, \\ 0 & \text{otherwise.} \end{cases}$$

Find the expectation and variance of $X$.

**Exercise 5.1.8** The probability density function of a random variable $X$ is given as

$$f_X(x) = \begin{cases} 8/x^3, & x \ge 2, \\ 0 & \text{otherwise.} \end{cases}$$

Find the expectation, second moment, and variance of $X$.

**Exercise 5.1.9** The joint probability density function of two random variables $X$ and $Y$ is given by

$$f_{X,Y}(x, y) = \begin{cases} \sin x \sin y, & 0 \le x \le \pi/2,\ 0 \le y \le \pi/2, \\ 0 & \text{otherwise.} \end{cases}$$

Find the mean and variance of the random variable $X$.

**Exercise 5.1.10** A stockbroker is interested in studying the manner in which his clients purchase and sell stock. He has observed that these clients may be divided into two equal groups according to their trading habits. Clients belonging to the first group buy and sell stock very quickly (day traders and their ilk); those in the second group buy stocks and hold on to them for a long period. Let $X$ be the random variable that denotes the length of time that a client holds a given stock. The stockbroker observes that the holding time for clients of the first type has mean $m_1$ and variance $\lambda$; for those of the second type, the corresponding numbers are $m_2$ and $\mu$, respectively. Find the variance of $X$ in terms of $m_1$, $m_2$, $\lambda$, and $\mu$.

**Exercise 5.2.1** The probability mass function of a discrete random variable $X$ is

$$p_X(x) = \begin{cases} 0.2, & x = -2, \\ 0.3, & x = -1, \\ 0.4, & x = 1, \\ 0.1, & x = 2, \\ 0 & \text{otherwise.} \end{cases}$$

Find the expectation of the following functions of $X$:

$$\text{(a) } Y = 3X - 1, \quad \text{(b) } Z = -X, \quad \text{and (c) } W = |X|.$$

**Exercise 5.2.2** The cumulative distribution function of a continuous random variable $X$ is given as

$$F_X(x) = \begin{cases} 1 - e^{-2x}, & 0 < x < \infty, \\ 0 & \text{otherwise.} \end{cases}$$

Find the probability density function and expectation of the derived random variable $Y = e^X$.

**Exercise 5.2.3** Let $X$ and $Y$ be two continuous random variables with probability density functions

$$f_X(x) = \begin{cases} 1/5, & -3 \le x \le 2, \\ 0 & \text{otherwise,} \end{cases} \quad \text{and} \quad f_Y(y) = \begin{cases} 8/y^3, & x \ge 2, \\ 0 & \text{otherwise.} \end{cases}$$

Find the expectation of the random variable $Z = 4X - 3Y$.

**Exercise 5.2.4** Let $X$ and $Y$ be two discrete random variables with joint probability mass function

| | $X = -1$ | $X = 0$ | $X = 1$ |
|---|---|---|---|
| $Y = -1$ | 0.15 | 0.05 | 0.20 |
| $Y = 0$ | 0.05 | 0.00 | 0.05 |
| $Y = 1$ | 0.10 | 0.05 | 0.35 |

Find the expectation of $Z = XY$ and $W = X + Y$.

**Exercise 5.2.5** Let $X$ and $Y$ be two random variables whose joint probability distribution function is given below.

| | $X = 1$ | $X = 2$ | $X = 3$ | $X = 4$ |
|---|---|---|---|---|
| $Y = -1$ | $c$ | 0 | 0 | 0 |
| $Y = 0$ | $a$ | $2a$ | $2a$ | $a$ |
| $Y = 1$ | $b$ | $2b$ | $2b$ | $b$ |

(a) Under what conditions does this table represent a proper joint distribution function for the random variables $X$ and $Y$?
(b) Compute the expectation of the random variables $X$ and $Y$.
(c) Compute $E[Y^2]$ and $E[(Y - E[Y])^2]$.

**Exercise 5.2.6** The joint probability density function of two random variables $X$ and $Y$ is given by

$$f_{X,Y}(x, y) = \begin{cases} 2, & 0 \le x \le y \le 1, \\ 0 & \text{otherwise.} \end{cases}$$

Find the following quantities:
(a) $E[X]$ and $E[Y]$, (b) $\text{Var}[X]$ and $\text{Var}[Y]$, (c) $E[XY]$ and $\text{Cov}[X, Y]$, and (d) $E[X + Y]$ and $\text{Var}[X + Y]$.

**Exercise 5.2.7** Let $X$ and $Y$ be two random variables whose joint probability density function is

$$f_{X,Y}(x, y) = \begin{cases} x + y, & 0 < x < 1,\ 0 < y < 1, \\ 0 & \text{otherwise.} \end{cases}$$

Find $E[XY]$, $\text{Cov}[X, Y]$, $\text{Corr}[X, Y]$, $E[X + Y]$, and $\text{Var}[X + Y]$.

**Exercise 5.2.8** Let $X$, $Y$, and $Z$ be three discrete random variables for which

$$\begin{aligned} \text{Prob}\{X = 1,\ Y = 1,\ Z = 0\} &= p, \\ \text{Prob}\{X = 1,\ Y = 0,\ Z = 1\} &= (1 - p)/2, \\ \text{Prob}\{X = 0,\ Y = 1,\ Z = 1\} &= (1 - p)/2, \end{aligned}$$

where $0 < p < 1$. Determine the covariance matrix for $X$ and $Y$.

**Exercise 5.2.9** The joint density function of two continuous random variables $X$ and $Y$ is given below.

$$f_{X,Y}(x, y) = \begin{cases} 9x^2y^2, & 0 \le x \le 1,\ 0 \le y \le 1, \\ 0 & \text{otherwise.} \end{cases}$$

Find $E[Y|x]$.

**Exercise 5.2.10** The joint probability density function of $X$ and $Y$ is given by

$$f_{X,Y}(x, y) = \begin{cases} 1/2, & -1 \le x \le y \le 1, \\ 0 & \text{otherwise.} \end{cases}$$

Find $f_{X|Y}(x|y)$, $f_{Y|X}(x|y)$, $E[X|y]$, and $E[Y|x]$.

**Exercise 5.2.11** Let $X$ and $Y$ be two random variables. Prove the following:

(a) $E[XY] = E[X]E[Y]$, when $X$ and $Y$ are independent.
(b) $E[\,E[Y|X]\,] = E[Y]$ for jointly distributed random variables.
(c) $\text{Var}\,(\alpha X + \beta) = \alpha^2 X$ for constants $\alpha$ and $\beta$.
(d) For jointly distributed random variables $X$ and $Y$, show that the variance of $Y$ is equal to the sum of the expectation of the conditional variance of $Y$ given $X$ and the variance of the conditional expectation of $Y$ given $X$, i.e., that

$$\text{Var}\,[Y] = E\,[\text{Var}\,[Y|X]] + \text{Var}\,[E[Y|X]]\,.$$

**Exercise 5.3.1** A discrete random variable $X$ takes the value 1 if the number 6 appears on a single throw of a fair die and takes the value 0 otherwise. Find its probability generating function.

**Exercise 5.3.2** Find the probability generating function of a discrete random variable with probability mass function given by

$$p_X(k) = q^{k-1}p, \quad k = 1, 2, \ldots,$$

where $p$ and $q$ are probabilities such that $p + q = 1$. We shall see later that this is called the geometric distribution function.

**Exercise 5.3.3** Let $X$ be a discrete random variable with probability mass function given by

$$p_X(x) = \begin{cases} 1/10, & x = 1, \\ 2/10, & x = 2, \\ 3/10, & x = 3, \\ 4/10, & x = 4, \\ 0 & \text{otherwise.} \end{cases}$$

Find the probability generating function of $X$ and of $2X$. Now find $\text{Prob}\{2X = 3\}$ and $E[2X]$.

**Exercise 5.3.4** Consider a discrete random variable $X$ whose probability generating function is given by

$$G_X(z) = e^z - e + 2 - z.$$

Observe that when $z = 1$, $G_X(z) = 1$. Find the probabilities $p_X(0)$, $p_X(1)$, $p_X(2)$, $p_X(3)$, and $p_X(4)$. What do you suppose is the probability of $p_X(k)$? Find $E[X]$ and $\text{Var}\,[X]$.

**Exercise 5.3.5** The probability mass function of a random variable $X$ is given by

$$p_k = \text{Prob}\{X = k) = p_X(k) = \begin{cases} \binom{n}{k} p^k q^{n-k}, & 0 \le k \le n, \\ 0 & \text{otherwise,} \end{cases}$$

where $p$ and $q$ are nonzero probabilities such that $p + q = 1$. This distribution is referred to as the *binomial distribution* with parameters $n$ and $p$. Find the probability generating function of $X$. Also find the probability generating function of the sum of $m$ such random variables, $X_i$, $i = 1, 2, \ldots, m$, with parameters $n_i$ and $p$, respectively.

**Exercise 5.3.6** Consider a discrete random variable $X$ whose probability generating function is given by

$$p_k = p_X(k) = \begin{cases} \alpha^k e^{-\alpha}/k!, & k = 0, 1, 2, \ldots, \quad \alpha > 0, \\ 0 & \text{otherwise.} \end{cases}$$

This distribution is referred to as the *Poisson distribution* with parameter $\alpha$. Find the probability generating function of $X$. Also find the probability generating function of the sum of $m$ such random variables, $X_i$, $i = 1, 2, \ldots, m$, with parameters $\alpha_i$, respectively.

**Exercise 5.4.1** Find the moment generating function of the discrete random variable $X$ with probability mass function given in Exercise 5.3.3. Derive the value of $E[X]$ from this moment generating function.

**Exercise 5.4.2** Find the moment generating function of the discrete random variable $X$ with probability mass function given in Exercise 5.3.2. Also find $E[X]$ and $\text{Var}\,[X]$.

**Exercise 5.4.3** Consider a discrete random variable $X$ whose probability mass function is given by

$$p_X(k) = \begin{cases} e, & k = 0, \\ 0, & k = 1, \\ 1/k!, & k = 2, 3, \ldots, \\ 0 & \text{otherwise.} \end{cases}$$

Find the moment generating function of $X$ and from it compute $E[X]$.

**Exercise 5.4.4** The moment generating function of a discrete random variable $X$ is

$$\mathcal{M}_X(\theta) = \frac{e^{\theta}}{12} + \frac{e^{3\theta}}{3} + \frac{e^{6\theta}}{6} + \frac{e^{9\theta}}{3} + \frac{e^{12\theta}}{12}.$$

Find the probability mass function of $X$.

**Exercise 5.4.5** The probability density function of a continuous random variable $X$ is given by

$$f_X(x) = \begin{cases} 1/(b-a), & 0 \leq x \leq b, \\ 0 & \text{otherwise.} \end{cases}$$

Find the moment generating function of $X$ and from it the expectation $E[X]$.

**Exercise 5.4.6** The probability density function of a continuous random variable $X$ is given by

$$f_X(x) = \frac{\alpha}{2} e^{-\alpha|x|}, \quad -\infty < x < \infty,$$

for $\alpha > 0$. Prove that $f_X(x)$ is indeed a density function. Find the moment generating function of $X$ and $E[X]$. (A random variable having this density function is called a *Laplace* random variable.)

**Exercise 5.5.1** Three individuals of equal ability set out to run a one mile race. Assume that the number of minutes each needs to run a mile is a random variable whose cumulative distribution function is given by

$$F_X(x) = \begin{cases} 0, & x \leq 5, \\ (x-5)/4, & 5 \leq x \leq 9, \\ 1, & x \geq 9. \end{cases}$$

Find the cumulative distribution of the number of minutes it takes for all three to have passed the finish line, and compute the expectation of this distribution. What is the probability that all three runners finish in less than six minutes?

**Exercise 5.5.2** Using the scenario of Exercise 5.5.1, find the distribution of the time it takes for the first runner to arrive, and its expected value. What is the probability that this time exceeds seven minutes?

**Exercise 5.5.3** Alice and Bob, working with an architect, have drawn up plans for their new home, and are ready to send them out to the best builders in town to get estimates on the cost of building this home. At present they have identified five different builders. Their architect has told them that the estimates should be uniformly distributed between \$500,000 and \$600,000. What is the cumulative distribution and the expectation of the lowest estimate, assuming that all builders, and hence their estimates, are independent? Compute the probability that the lowest estimate does not exceed \$525,000. What would this distribution, expectation, and probability be, if Alice and Bob were able to identify 20 builders? What conclusions could be drawn concerning the size of the lowest estimate and the number of builders?

**Exercise 5.5.4** The shelf lives of three different products all have the same type of cumulative distribution function, namely,

$$F_X(x) = \begin{cases} 1 - e^{-\alpha x}, & x \geq 0, \\ 0 & \text{otherwise,} \end{cases}$$

where $\alpha > 0$ is different for each of the three products: more specifically, the different values of $\alpha$ are given by 1/2, 1/3, and 1/6. Find the cumulative distribution function and the expectation of the random variable that characterizes the time until the shelf life of one of these three products expires and that product needs to be replaced. You should assume that the shelf life of each product is independent of the others.

# Chapter 6

# *Discrete Distribution Functions*

On a number of occasions in previous chapters, we alluded to random variables of one sort or another and to their distribution functions by specific names, such as *exponential*, *geometric*, and so on. It frequently happens in practice, that the distribution functions of random variables fall into different *families* and that one random variable differs from another only in the value assigned to a single parameter, or in the value of a small number of parameters. For example, the probability density function of one continuous random variable may be given by $f(x) = 2e^{-2x}$ for $x \geq 0$ and zero otherwise while for another random variable it may be $f(x) = e^{-x/2}/2$ for $x \geq 0$ and zero otherwise. Both random variables belong to the family of (exponential) random variables whose probability density function may be written $f(x) = \lambda e^{-\lambda x}$ for $x \geq 0$, zero otherwise, and having $\lambda > 0$: they differ only in the value assigned to $\lambda$. Being able to associate a given random variable as a member of a family of random variables gives us a greater understanding of the situation represented by the given random variable. Furthermore, known results and theorems concerning that family of random variables can be directly applied to the given random variable.

In this chapter we consider some of the most common distribution functions of discrete random variables; in the next, we turn our attention to those of continuous random variables.

## 6.1 The Discrete Uniform Distribution

Let $X$ be a discrete random variable with finite image $\{x_1, x_2, \ldots, x_n\}$. The uniform distribution is obtained when each value in the image has equal probability and, since there are $n$ of them, each must have probability equal to $1/n$. The uniform probability mass function is given as

$$p_X(x_i) = \text{Prob}\{X = x_i\} = \begin{cases} 1/n, & i = 1, 2, \ldots, n, \\ 0 & \text{otherwise.} \end{cases}$$

The corresponding cumulative distribution function is given by

$$F_X(t) = \sum_{i=1}^{\lfloor t \rfloor} p_X(x_i) = \frac{\lfloor t \rfloor}{n} \quad \text{for} \quad 1 \leq t \leq n,$$

while $F_x(t) = 0$ for $t < 1$ and $F_X(t) = 1$ for $t > n$.

With the discrete uniform distribution, it often happens that the $x_i$ are equally spaced. In this case, a bar graph of the probability mass function of this distribution will show $n$ equally spaced bars all of height $1/n$, while a plot of the cumulative distribution function will display a perfect staircase function of $n$ identical steps, each of height $1/n$, beginning at zero just prior to the first step at $x_1$ and rising to the value 1 at the last step at $x_n$.

In the specific case in which $x_i = i$ for $i = 1, 2, \ldots, n$, the moments of the discrete uniform distribution are given by

$$E[X^k] = \sum_{i=1}^{n} i^k/n.$$

In particular, we have

$$E[X] = \sum_{i=1}^{n} i/n = \frac{1}{n}\sum_{i=1}^{n} i = \frac{1}{n}\frac{n(n+1)}{2} = \frac{n+1}{2},$$

and, using the well-known formula for the sum of the squares of the first $n$ integers,

$$E[X^2] = \sum_{i=1}^{n} i^2/n = \frac{1}{n}\sum_{i=1}^{n} i^2 = \frac{1}{n}\left(\frac{n(n+1)(2n+1)}{6}\right) = \frac{(n+1)(2n+1)}{6}.$$

The variance is then computed as

$$\mathrm{Var}\,[X] = E[X^2] - (E[X])^2 = \frac{(n+1)(2n+1)}{6} - \left(\frac{n+1}{2}\right)^2 = \frac{n^2-1}{12}.$$

When the $x_i$ are $n$ equally spaced points beginning at $x_1 = a$ and ending at $x_n = b = a + n - 1$, we find the corresponding formulae for the mean and variance to be respectively given by

$$E[X] = \frac{a+b}{2}, \qquad \mathrm{Var}\,[X] = \frac{(b-a+2)(b-a)}{12}.$$

The probability generating function for a discrete uniformly distributed random variable $X$ is

$$G_X(z) = \sum_{i=1}^{n} z^i/n = \frac{1}{n}\sum_{i=1}^{n} z^i = \frac{z(1-z^n)}{n(1-z)}.$$

Its moment generating function $\mathcal{M}_X(\theta)$ is obtained by setting $z = e^\theta$, since $\mathcal{M}_X(\theta) = G_X(e^\theta)$.

**Example 6.1** An example of a discrete uniform random variable is the random variable that denotes the number of spots that appear when a fair die is thrown once. Its probability mass function is given by $p(x_i) = 1/6$ if $x_i \in \{1, 2, \ldots, 6\}$ and is equal to zero otherwise.

## 6.2 The Bernoulli Distribution

A discrete random variable $X$ that can assume only two values, zero and one (for example, representing success or failure, heads or tails, etc.) gives rise to a Bernoulli distribution. It arises from the realization of a single Bernoulli trial. Its probability mass function is given by

$$p_X(0) = \mathrm{Prob}\{X = 0\} = 1 - p,$$
$$p_X(1) = \mathrm{Prob}\{X = 1\} = p,$$

where $0 < p < 1$. In our discussions on Bernoulli random variables, we shall often set $q = 1 - p$. The cumulative distribution function of a Bernoulli random variable is given as

$$F_X(x) = \begin{cases} 0, & x < 0, \\ q, & 0 \le x < 1, \\ 1, & x \ge 1. \end{cases}$$

The probability mass and cumulative distribution functions are shown in Figure 6.1. The moments are computed from

$$E[X^j] = 0^j q + 1^j p = p \quad \text{for all } j = 1, 2, \ldots.$$

Also,

$$\mathrm{Var}\,[X] = E[X^2] - (E[X])^2 = p - p^2 = p(1-p) = pq$$

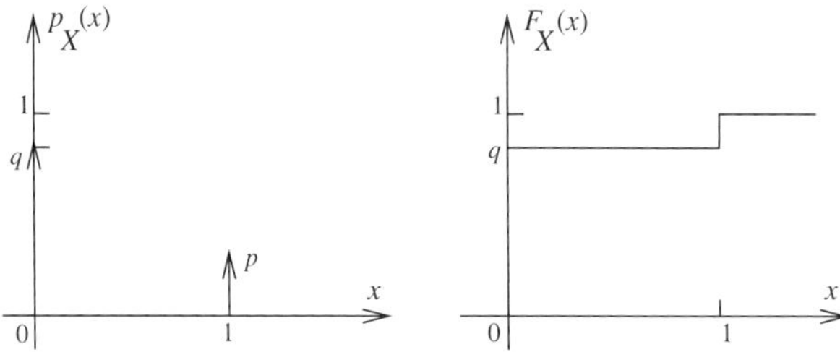


Figure 6.1. Probability mass function and CDF of the Bernoulli distribution.

and

$$C_X^2 = \frac{pq}{p^2} = \frac{q}{p}.$$

The probability generating function for a Bernoulli random variable is given by

$$G_X(z) = qz^0 + pz^1 = q + pz.$$

Its moment generating function is

$$\mathcal{M}_X(e^\theta) = q + pe^\theta.$$

**Example 6.2** A survey reveals that 20% of students at Tobacco Road High School smoke. A student is chosen at random. Let $X = 0$ if that student smokes and $X = 1$ if the student does not smoke. Then $X$ is a Bernoulli distributed random variable with probability mass function given by

$$p_X(x) = \begin{cases} 1/5, & x = 0, \\ 4/5, & x = 1, \\ 0 & \text{otherwise.} \end{cases}$$

## 6.3 The Binomial Distribution

The binomial distribution arises from a sequence of independent Bernoulli trials, with probability of success equal to $p$ on each trial. It is used whenever a series of trials is made for which

- Each trial has two mutually exclusive outcomes: called success and failure.
- The outcomes of successive trials are mutually independent.
- The probability of success, denoted by $p$, is the same on each trial.

An elementary event $\omega$ in a probability experiment consisting of $n$ Bernoulli trials may be written as a string of zeros and ones such as

$$\underbrace{011010100\ldots010100101111}_{n \text{ times}}$$

where a 0 in position $j$ denotes failure at the $j^{\text{th}}$ trial and a 1 denotes success. Since the Bernoulli trials are independent, the probability that an elementary event $\omega$ has exactly $k$ successes (and $n-k$ failures) is given by $p^k q^{n-k}$.

Now let $X$ be the random variable that describes the number of successes in $n$ trials. The domain of $X$, the set of all possible outcomes, is the set of all strings of length $n$ that are composed of zeros and ones, and its image is the set $\{0, 1, \ldots, n\}$. For example, with $n = 5$, an element of the domain is $\omega = 00110$ and the value assumed by $X$ is given as $X(\omega) = 2$. In other words, the value assigned by $X$ to any outcome is the number of 1's (i.e., the number of successes) it contains.

Thus, $X(\omega) = k$ if there are exactly $k$ ones in $\omega$. It follows then that the number of distinct outcomes with $k$ successes is equal to the number of distinct sequences that contain $k$ ones and $n-k$ zeros. This number is given by the binomial coefficient

$$C(n,k) = \binom{n}{k} = \frac{n!}{k!(n-k)!}.$$

These $C(n,k)$ elementary events all have probability equal to $p^k q^{n-k}$ and hence

$$\text{Prob}\{X = k\} = C(n,k)p^k q^{n-k}.$$

Thus the probability mass function of a binomial random variable $X$ is given by

$$b(k;n,p) = \text{Prob}\{X = k) = p_X(k) = \begin{cases} \binom{n}{k} p^k q^{n-k}, & 0 \le k \le n, \\ 0 & \text{otherwise.} \end{cases}$$

A random variable $X$ having this probability mass function is called *binomial random variable.* Its probability mass function is denoted by $b(k;n,p)$ and gives the probability of $k$ successes in $n$ independent Bernoulli trials with probability of success at each trial being equal to $p$. The binomial theorem may be used to show that this defines a proper probability mass function, (and thereby lends its name to the distribution). We have

$$\sum_{k=0}^{n} p_k = \sum_{k=0}^{n} \binom{n}{k} p^k(1-p)^{n-k} = [p + (1-p)]^n = 1.$$

The interested reader may wish to draw some plots of the probability mass function of a binomial random variable. When $p$ is close to 0.5, the distribution tends to be symmetric about the midpoint $(n/2)$; as the value of $p$ moves further and further away from 1/2, the distribution becomes more and more skewed. The size of $n$ also has an effect on the shape of the distribution: larger values of $n$ result in more symmetry. As a rule of thumb, approximately 95% of the distribution of a binomial random variables falls in an interval of $\pm 2\sigma_X$ (two standard deviations) of its expectation.

The cumulative distribution function of a binomial random variable is denoted by $B(t;n,p)$. It has the value 0 for $t < 0$; the value 1 for $t > n$, while for $0 \le t \le n$,

$$B(t;n,p) = \sum_{k=0}^{\lfloor t \rfloor} \binom{n}{k} p^k(1-p)^{n-k}.$$

Values of the binomial cumulative distribution function have been tabulated for some values of $n$ and $p$ and are available in many texts. In using these tables, the following identities may prove useful:

$$B(t;n,p) = 1 - B(n-t-1;n,1-p),$$
$$b(t;n,p) = B(t;n,p) - B(t-1;n,p).$$

Since a binomial random variable $X$ is the sum of $n$ mutually independent Bernoulli random variables, $X_i,\ i = 1, 2, \ldots, n$, i.e.,

$$X = \sum_{i=1}^{n} X_i,$$

we have, using the linear property of expectation and variance,

$$E[X] = \sum_{i=1}^{n} E[X_i] = np,$$

$$\text{Var}\,[X] = \sum_{i=1}^{n} \text{Var}\,[X_i] = npq,$$

$$C_X^2 = \frac{npq}{n^2p^2} = \frac{q}{np}.$$

Finally, from Equation (5.6), we may compute the probability generating function of the binomial distribution as

$$G_X(z) = \prod_{i=1}^{n} G_{X_i}(z) = (q + pz)^n = [1 - p(1 - z)]^n,$$

since the generating function for each Bernoulli random variable $X_i$ is, as we have just seen, $q + pz$. Its moment generating function is found from $\mathcal{M}_X(\theta) = G_X(e^\theta)$.

**Example 6.3** A grocery store receives a very large number of apples each day. Since past experience has shown that 5% of them will be bad, the manager chooses six at random for quality control purposes. Let $X$ be the random variable that denotes the number of bad apples in the sample. What values can $X$ assume? For each possible value of $k$, what is the value of $p_X(k)$? In particular, what is the probability that the manager finds no bad apples?

The number of bad apples found by the manager (and hence the value assumed by the random variable $X$) can be any number from 0 through 6. The probability associated with any value $k$ is the same as that of obtaining $k$ "successes" in six trials with a probability of success equal to 0.05, i.e., from the binomial probability mass function as

$$b(k; 6, 0.05) = C(6, k) \times 0.05^k \times 0.95^{6-k}.$$

The following is a table of the values of $p_X(k)$.

| $k$ | 0 | 1 | 2 | 3 | 4 | 5 | 6 |
|---|---|---|---|---|---|---|---|
| $p_X(k)$ | 0.7351 | 0.2321 | 0.03054 | 0.002143 | $8.4609 \times 10^{-5}$ | $1.78125 \times 10^{-6}$ | $1.5625 \times 10^{-8}$ |

Thus the probability that the manager finds no bad apples is 0.7351.

This question may be posed in many different forms, as, for example, in terms of the transmission of binary digits through a communication channel that is subject to error, i.e., a bit is transmitted in error with probability $p$ (equal to 0.05 in our bad apple example) and is independent of previous errors. The probability of an error-free transmission is computed in just the same way as that used to compute the probability that the manager finds no bad apples.

The binomial distribution may be used in polling situations to gauge the sentiments of voters prior to an election. A subset of size $n$ of the population is chosen and their view of a particular candidate running for office, or their opinion of a bond issue, etc., is solicited. Each person providing an opinion may be viewed as a Bernoulli random variable and since we shall assume that the opinions of the respondents are independent, the polling experiment may be viewed as a binomial random variable, $X$. However, in this situation, the actual number of respondents in favor of the candidate is less important that the *percentage* of them who favor the candidate and so it is the induced random variable $Y = X/n$ that is most useful. Since $X$ is a binomial random variable with expectation

$E[X] = np$ and $\text{Var}[X] = np(1-p)$, the expectation and variance of $Y$ are given by

$$E[Y] = E[X/n] = E[X]/n = p,$$
$$\text{Var}[Y] = \text{Var}[X/n] = \text{Var}[X]/n^2 = \frac{p(1-p)}{n}.$$

This now begs the question of the accuracy of the polling results. We have formed the random variable $Y$ and wish to know how well it performs in computing the correct percentage of the population in favor of the candidate, $p$. A good indication may be obtained from the $2\sigma$ bounds. Observe that the variance, and hence the standard deviation, is maximized when $p = 1/2$ (a simple calculus exercise shows that the maximum of $p(1-p)$ occurs when $p = 1/2$). The maximum value of $\sigma_X$ is $\sqrt{1/4n}$ and the $2\sigma$ bound cannot exceed $1/\sqrt{n}$. Therefore, if the sample size is 100, it is likely that the value obtained from the polling results differ from the true value by no more than $\pm 1/\sqrt{100} = \pm 0.1$, while if the sample size is 900, they differ from the true result by no more than $\pm 0.033$. Putting this in terms of our political candidate, if the poll of 900 voters shows that 45% of the population supports this candidate, then that candidate should expect to receive between 42% and 48% of the vote. We shall return to this concept of polling in a later chapter, when we shall examine it in much more detail.

## 6.4 Geometric and Negative Binomial Distributions

Like a binomial random variable, a geometric random variable can be associated with a sequence of Bernoulli trials. However, instead of counting the number of successes in a fixed number of trials, a *geometric* random variable counts the number of trials up to, and including, the first success; a *modified geometric* random variable counts the number of trials *before* the first success. A discrete random variable that has a *negative binomial* distribution, counts the number of trials up to and including the $k^{\text{th}}$ success. We begin with the geometric distribution.

### *The Geometric Distribution*

Let $X$ be a geometric random variable with parameter $p$, $0 < p < 1$, the probability of success in a single Bernoulli trial. If 0 denotes failure and 1 success, then the (infinite) sample space of a geometric probability experiment consists of all sequences of a string of 0's followed by a single 1, i.e., $\{1, 01, 001, 0001, \ldots\}$. The geometric random variable $X$ assumes the value of the number of digits in any string. For example, $X(0001) = 4$ and $X(000001) = 6$. Its image is the set of integers greater than or equal to 1, and its probability mass function is given by

$$p_X(n) = \begin{cases} p(1-p)^{n-1}, & n = 1, 2, \ldots, \\ 0 & \text{otherwise.} \end{cases}$$

(see Figure 6.2). The probability $p_X(n) = p(1-p)^{n-1} = q^{n-1}p$ is that of obtaining a sequence of $n-1$ failures (each with probability $q = 1-p$) followed by a single success (with probability $p$). Thus the random variable $X$ denotes the index of the first success in a random experiment consisting of a number of independent trials each with probability of success equal to $p$ and probability of failure equal to $q = 1-p$. Notice that, using the formula for the sum of a geometric series (hence the name), we have

$$\sum_{n=1}^{\infty} p_X(n) = \sum_{n=1}^{\infty} pq^{n-1} = \frac{p}{1-q} = 1.$$

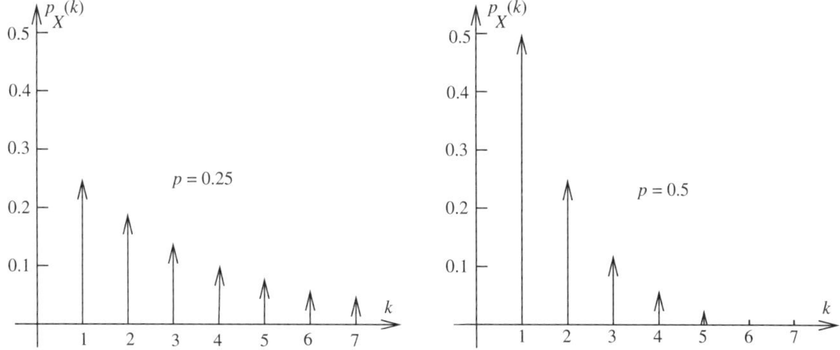


Figure 6.2. The geometric probability mass function for two values of $p$.

Since

$$\text{Prob}\{X \le t\} = \sum_{n=1}^{\lfloor t \rfloor} p(1-p)^{n-1} = p\frac{1-(1-p)^{\lfloor t \rfloor}}{1-(1-p)} = 1-(1-p)^{\lfloor t \rfloor} = 1 - q^{\lfloor t \rfloor} \quad \text{for } t \ge 0,$$

the cumulative distribution function for a geometric random variable is

$$F_X(t) = \begin{cases} 0, & t \le 0, \\ 1 - q^{\lfloor t \rfloor}, & t \ge 0. \end{cases}$$

The expectation and variance of a geometric random variable are easily found by differentiating its probability generating function to obtain factorial moments. The first (factorial) moment is the expectation, while the second (factorial) moment and the expectation can be used to form the variance. However, we shall leave this as an exercise and instead we shall compute the expectation and variance directly from their definitions so that we may introduce an effective way to evaluate series of the form $\sum_{n=1}^{\infty} n^j q^n$ with $0 < q < 1$ and $j \ge 1$. We do so by associating $n^j q^n$ with its $j^{\text{th}}$ derivative with respect to $q$ and interchanging the summation and the derivative. This procedure will become apparent in the computation of the expectation (with $j = 1$) and variance (with $j = 2$) of a geometric random variable.

In computing the expectation, we use the fact that $nq^{n-1} = d(q^n)/dq$ and proceed as follows:

$$E[X] = \sum_{n=1}^{\infty} npq^{n-1} = p\sum_{n=1}^{\infty} nq^{n-1} = p\sum_{n=0}^{\infty}\frac{d}{dq}q^n = p\frac{d}{dq}\sum_{n=0}^{\infty} q^n = p\frac{d}{dq}\frac{1}{1-q} = \frac{p}{(1-q)^2} = \frac{1}{p}.$$

The variance is computed in a similar manner. This time using $n(n-1)q^{n-2} = d^2(q^n)/dq^2$, we find

$$\begin{aligned} E[X^2] &= \sum_{n=1}^{\infty} n^2 pq^{n-1} = \sum_{n=1}^{\infty} n(n-1)pq^{n-1} + \sum_{n=1}^{\infty} npq^{n-1} = q\sum_{n=1}^{\infty} n(n-1)pq^{n-2} + \sum_{n=1}^{\infty} npq^{n-1} \\ &= pq\frac{d^2}{dq^2}\sum_{n=1}^{\infty} q^n + p\frac{d}{dq}\sum_{n=1}^{\infty} q^n = pq\frac{d^2}{dq^2}\frac{q}{1-q} + p\frac{d}{dq}\frac{q}{1-q} \\ &= pq\frac{2}{(1-q)^3} + p\frac{1}{(1-q)^2} = \frac{2pq}{p^3} + \frac{p}{p^2} = \frac{2q+p}{p^2} = \frac{1+q}{p^2}. \end{aligned}$$

Hence

$$\text{Var}\,[X] = E[X^2] - E[X]^2 = \frac{1+q}{p^2} - \frac{1}{p^2} = \frac{q}{p^2} = \frac{1-p}{p^2}.$$

We may now form the squared coefficient of variation for a geometric random variable. It is given as

$$C_X^2 = 1 - p.$$

To compute the probability generating function, we have

$$G_X(z) = \sum_{n=1}^{\infty} pq^{n-1}z^n = \frac{p}{q}\sum_{n=1}^{\infty}(qz)^n = \frac{p}{q} \times \frac{qz}{1-qz} = \frac{pz}{1-(1-p)z}.$$

Its moment generating function is

$$\mathcal{M}_X(e^\theta) = \frac{pe^\theta}{1-(1-p)e^\theta}.$$

The geometric distribution is the only discrete probability distribution that has the *Markov* or *memoryless* property, something that we shall address (and prove) in a later chapter. It implies that a sequence of $n-1$ unsuccessful trials has no effect on the probability of getting a success on the $n^{\text{th}}$ attempt.

**Example 6.4** Locating a bug in a computer program involves running the program with randomly selected data sets until the bug manifests itself. The probability that any particular data set isolates the bug is known to be 0.2. It follows that the expectation and standard deviation of the number of data sets needed to find the bug are given by $1/0.2 = 5$ and $\sqrt{(1-0.2)/0.2^2} = 4.47$, respectively.

**Example 6.5** Consider a machine that at the start of each day either is in working condition or has broken down and is waiting to be repaired. If at the start of any day it is working, then with probability $p_1$ it will break down. On the other hand, if at the start of the day, it is broken, then with probability $p_2$ it will be repaired. Let $X$ denote the number of consecutive days that the machine is found to be in working condition. Then $X$ is geometrically distributed since the probability that the machine is in working order at the beginning of exactly $n$ consecutive days and breaks down during the $n^{\text{th}}$ day is equal to $(1-p_1)^{n-1}p_1$. The mean and variance of the number of consecutive days that the machine begins the day in working order is given by

$$E[X] = \frac{1}{p_1}, \quad \text{Var}\,[X] = \frac{1-p_1}{p_1^2}.$$

Similar results can be found concerning the number of days that the machine spends undergoing repairs. In this case, the parameter used is $p_2$. The term *sojourn time* is used to designate the time spent in a given state and later during our analysis of discrete-time Markov chains we shall see that the sojourn time in any state of a discrete-time Markov chain is geometrically distributed.

### *The Modified Geometric Distribution*

Whereas a *geometric* random variable counts the number of trials up to and including the first success, a *modified geometric* random variable counts the number of trials *before* the first success. Its probability mass function is given by

$$p_X(n) = \begin{cases} p(1-p)^n, & n = 0, 1, 2, \ldots, \\ 0 & \text{otherwise.} \end{cases}$$

Observe that the modified geometric distribution is just a regular geometric distribution, but defined on the set $\{0, 1, 2, \ldots\}$ instead of the set $\{1, 2, 3, \ldots\}$. The cumulative distribution function of a modified geometric random variable is

$$F_X(t) = \sum_{n=0}^{\lfloor t \rfloor} p(1-p)^n = p\left(\frac{1-(1-p)^{\lfloor t \rfloor + 1}}{1-(1-p)}\right) = 1-(1-p)^{\lfloor t+1 \rfloor} \quad \text{for } t \geq 0,$$

and zero otherwise. Various characteristics are given by

$$E[X] = \frac{1-p}{p}, \qquad \text{Var}\,[X] = \frac{1-p}{p^2}, \qquad \text{and} \qquad C_X^2 = \frac{1}{1-p}.$$

Its probability generating function and moment generating functions are, respectively,

$$G_X(z) = \frac{p}{1-(1-p)z} \qquad \text{and} \qquad \mathcal{M}_X(\theta) = \frac{p}{1-(1-p)e^{\theta}}.$$

**Example 6.6** Professor Don Bitzer keeps a large jar full of jellybeans in his office and each day he eats a certain number of them. It is well known that once you start eating jellybeans, it is difficult to stop (and Professor Bitzer is no exception to this rule) so some days he does not even start eating them. So, before eating each jellybean (including the first), Professor Bitzer considers the situation and with probability $p$ decides to eat no more. If $X$ is the number of jellybeans he eats on a typical day, then $X$ is a modified geometric random variable whose probability mass function is

$$p_X(n) = \begin{cases} p(1-p)^n, & n = 0, 1, 2, \ldots, \\ 0 & \text{otherwise.} \end{cases}$$

If we take $p = 0.04$, since jellybeans are hard to resist, then the average number he eats on a typical day is $E[X] = (1-p)/p = 0.96/0.04 = 24$, and the probability he eats more than ten on a typical day is given by

$$\begin{aligned} \text{Prob}\{X > 10\} &= 1 - \text{Prob}\{X \le 10\} = 1 - \left[1 - (1-p)^{10+1}\right] = (1-p)^{11} = (0.96)^{11} \\ &= 0.6382. \end{aligned}$$

### *The Negative Binomial Distribution*

Let us now turn our attention to random variables that have a negative binomial distribution. Such random variables count the number of Bernoulli trials, with parameter $p$, up to and including the occurrence of the $k^{\text{th}}$ success. Thus, a geometric random variable is a negative binomial random variable with $k = 1$. The probability mass function of a negative binomial random variable $X$ with parameters $p$ and $k$ is given by

$$p_X(n) = \binom{n-1}{k-1} p^k (1-p)^{n-k}, \qquad k \ge 1, \quad n = k, k+1, \ldots. \tag{6.1}$$

If the $k^{\text{th}}$ success is to occur on exactly trial $n$, then there must have been exactly $k-1$ successes scattered among the first $n-1$ trials. The probability of obtaining a success on the $k^{\text{th}}$ trial is $p$ and the probability of obtaining $k-1$ successes in the first $n-1$ trials, obtained from the binomial formula, is

$$\binom{n-1}{k-1} p^{k-1} (1-p)^{(n-1)-(k-1)}.$$

The product of these two probabilities gives Equation (6.1).

To find the expectation and variance of a negative binomial random variable, $X$, it is convenient to consider $X$ as the sum of $k$ independent *geometric* random variables, $X_1, X_2, \ldots, X_k$, each with expectation $1/p$ and variance $(1-p)/p^2$. In this context $X_1$ represents the number of trials up to and including the first success, $X_2$ represents the number of trials from the first success up to and including the second and so on, up to $X_k$ which represents the number of trial from success $k-1$

up to and including success number $k$. Then

$$E[X] = \sum_{i=1}^{k} E[X_i] = \frac{k}{p},$$

$$\text{Var}\,[X] = \sum_{i=1}^{k} \text{Var}\,[X_i] = \frac{k(1-p)}{p^2}.$$

**Example 6.7** The chance of winning a prize at a certain booth at the state fair is 0.15 and each attempt costs \$1. What is the probability that I will win a prize for each of my four children with my last \$10 bill?

Let $X$ denote the (negative binomial) random variable which gives the number of attempts up to and including the $k^{\text{th}}$ success. The \$10 bill gives me $n = 10$ attempts to win $k = 4$ prizes. The probability of winning the fourth prize on my tenth try can be computed from the negative binomial distribution as

$$\text{Prob}\{X = 10\} = \binom{10-1}{4-1} (0.15)^4\,(0.85)^6 = 0.0160.$$

The probability that I will need at least \$7 may be found by subtracting the probability of taking four, five, or six attempts to win four prizes from 1, i.e.,

$$\begin{aligned}\text{Prob}\{X \geq 7\} &= 1 - \sum_{i=4}^{6} \text{Prob}\{X = i\} \\ &= 1 - \binom{3}{3} (0.15)^4\,(0.85)^0 - \binom{4}{3} (0.15)^4\,(0.85)^1 - \binom{5}{3} (0.15)^4\,(0.85)^2 \\ &= 1 - 0.0005 - 0.0017 - 0.0073 = 0.9905,\end{aligned}$$

rather large odds, indeed. In fact, the average number of dollars I had need to spend to win four prizes is given by

$$E[X] = \frac{k}{p} = \frac{4}{0.15} = 26.67.$$

## 6.5 The Poisson Distribution

The Poisson probability mass function, denoted as $f(k, \alpha)$, is given by

$$f(k,\alpha) = p_X(k) = \begin{cases} \alpha^k e^{-\alpha}/k!, & k = 0, 1, 2, \ldots, \quad \alpha > 0, \\ 0 & \text{otherwise.} \end{cases}$$

Notice that this defines a genuine probability mass function since

$$\sum_{k=0}^{\infty} f(k,\alpha) = e^{-\alpha} \sum_{k=0}^{\infty} \frac{\alpha^k}{k!} = e^{-\alpha} e^{\alpha} = 1.$$

The cumulative distribution function is given by

$$F(k) = \text{Prob}\{X \leq k\} = e^{-\alpha} \sum_{j=0}^{k} \frac{\alpha^j}{j!}.$$

A *Poisson process*, extensively used in queueing theory and discussed in that context in a later chapter, is a counting process in which the number of events that occur within a given time period has a *Poisson distribution*. If $\lambda$ is the rate at which these events occur, and $t$ is the time period over which we observe these events, then the parameter of interest is $\lambda t$ which, to all intents and purposes, is the number of events that occurred in time $t$. In this case we set $\alpha = \lambda t$.

**Example 6.8** Customers arrive at a queueing system according to a Poisson distribution at a rate of 12 customers per hour.

- What is the probability that exactly six customers will arrive in the next 30 minutes?
- What is the probability that three or more customers will arrive in the next 15 minutes?
- What is the probability that two, three, or four customers will arrive in the next 5 minutes?

In problems of this nature, we begin by setting $\alpha = \lambda t$. For all three parts of the question we have $\lambda = 12$ customers per hour and hence the parameter for the Poisson distribution is $12t$. Therefore

$$\text{Prob}\{X_t = k\} = \frac{(12t)^k}{k!}e^{-12t}.$$

(a) Since the question is set in units of an hour, we take $t = 0.5$ for 30 minutes and get

$$\text{Prob}\{\text{Exactly 6 customers in next 30 minutes}\} = \text{Prob}\{X_{0.5} = 6\} = \frac{6^6}{6!}e^{-6} = 0.1606.$$

(b) The probability of getting three or more arrivals is equal to 1 minus the probability of having zero, one, or two arrivals in the next 15 minutes. So, using $\lambda t = 12 \times (1/4) = 3$, we have

$$1 - \text{Prob}\{0, 1, 2 \text{ arrivals }\} = 1 - e^{-3}\left(1 + \frac{3}{1} + \frac{3^2}{2!}\right) = 0.5768.$$

(c) Since five minutes is equal to 1/12 hours, the parameter $\alpha$ is now equal to 1. We have

$$\text{Prob}\{2, 3, 4 \text{ customers}\} = \sum_{k=2}^{4} \frac{(1)^k}{k!}e^{-1} = e^{-1}\frac{17}{24} = 0.2606.$$

We now derive the expectation and variance of a Poisson random variable. Its expectation is given by

$$E[X] = \sum_{k=0}^{\infty} k p_X(k) = e^{-\alpha}\sum_{k=0}^{\infty} k\frac{\alpha^k}{k!}$$

$$= \alpha e^{-\alpha}\sum_{k=1}^{\infty}\frac{\alpha^{k-1}}{(k-1)!} = \alpha e^{-\alpha}\sum_{j=0}^{\infty}\frac{\alpha^j}{j!} = \alpha e^{-\alpha}e^{\alpha} = \alpha.$$

Therefore the expected number of events that occur in $(0, t]$ is equal to $\alpha = \lambda t$. For the variance, we first compute the (factorial) moment

$$E[X(X-1)] = \sum_{k=0}^{\infty} k(k-1)p_X(k) = e^{-\alpha}\sum_{k=0}^{\infty} k(k-1)\frac{\alpha^k}{k!}$$

$$= e^{-\alpha}\alpha^2\sum_{k=2}^{\infty}\frac{\alpha^{k-2}}{(k-2)!} = e^{-\alpha}\alpha^2\sum_{k=0}^{\infty}\frac{\alpha^k}{k!} = \alpha^2 = (\lambda t)^2.$$

Now we form the variance as follows:

$$\sigma_X^2 = E[X^2] - E^2[X] = (E[X(X-1)] + E[X]) - (E[X])^2 = \alpha^2 + \alpha - \alpha^2 = \alpha.$$

Thus, the mean and variance of a Poisson random variable are identical. The expectation and variance may also be conveniently computed from the probability generating function, which we now derive. We have

$$G_X(z) = E[z^X] = \sum_{k=0}^{\infty} z^k p_X(k) = \sum_{k=0}^{\infty} e^{-\alpha} \frac{(\alpha z)^k}{k!} = e^{-\alpha+\alpha z} = e^{\alpha(z-1)}.$$

The moment generating function of a Poisson random variable is given by

$$\mathcal{M}_X(e^\theta) = e^{\alpha(e^\theta - 1)}.$$

The probability mass function of a Poisson random variable may be viewed as the limiting case of the probability mass function of a binomial random variable $b(k; n, p)$, when $n$ is large and $np$ is moderately sized. Such is the case when the number of trials, $n$, is large, the probability of success, $p$, is small (so that each success is a rare event), but nevertheless the average number of successes, $np$, is moderate. This happens, for example, if as $n \to \infty$ then $p \to 0$ in such a way that their product $np = \alpha$ remains constant. We proceed as follows. Let $X$ be a discrete binomial random variable such that $np = \alpha$. Then

$$\begin{aligned}\text{Prob}\{X = k\} &= \frac{n!}{(n-k)!k!} p^k (1-p)^{n-k} = \frac{n!}{(n-k)!k!} \left(\frac{\alpha}{n}\right)^k \left(1 - \frac{\alpha}{n}\right)^{n-k} \\ &= \frac{n(n-1)\cdots(n-k+1)}{n^k} \left(\frac{\alpha^k}{k!}\right) \left(1 - \frac{\alpha}{n}\right)^n \left(1 - \frac{\alpha}{n}\right)^{-k}.\end{aligned}$$

Now letting $n$ tend to infinity, we find

$$\lim_{n \to \infty} \frac{n(n-1)\cdots(n-k+1)}{n^k} = 1,$$

$$\lim_{n \to \infty} \left(1 - \frac{\alpha}{n}\right)^n = e^{-\alpha} \quad \text{and} \quad \lim_{n \to \infty} \left(1 - \frac{\alpha}{n}\right)^{-k} = 1.$$

Hence, assuming that $p \to 0$ and $np = \alpha$ is constant as $n \to \infty$, we obtain

$$\lim_{n \to \infty} \text{Prob}\{X = k\} = e^{-\alpha} \frac{\alpha^k}{k!},$$

which is a Poisson probability mass function.

**Example 6.9** In a lottery, a total of $N = 1{,}000$ tickets are printed of which $M = 5$ are winning tickets. How many tickets should I buy to make the probability of winning equal to $p = 3/4$.

The probability that any given ticket is a winning ticket is given by $M/N = 1/200$. Thus, each ticket that I buy can be considered as a separate trial with a probability of success equal to 1/200. If I buy $n$ tickets, then this corresponds to a series of $n$ independent trials. Since the probability of winning is rather small (1/200) and the probability that I would like to have for winning is rather high (3/4), it is clear that a rather large number of tickets must be bought ($n$ must be large). It follows that the number of winning tickets among those purchased is a random variable with approximately Poisson distribution. The probability that there are exactly $k$ winning tickets among the $n$ purchased is

$$p_X(k) = \text{Prob}\{k \text{ winning tickets}\} = \frac{\alpha^k}{k!} e^{-\alpha}$$

where $\alpha = n \times (M/N) = n/200$. The probability that at least one of the tickets is a winning ticket is

$$1 - P(0) = 1 - e^{-\alpha},$$

and so the number of tickets that must be bought to have a probability of 3/4 to have a winning ticket is the smallest integer $n$ satisfying

$$e^{-n/200} \leq 1 - 3/4 = 0.25$$

Some representative values are provided in the table below.

| $n$ | 100 | 200 | 277 | 278 | 300 | 500 | 1,000 | 2,000 |
|---|---|---|---|---|---|---|---|---|
| $e^{-n/200}$ | 0.6065 | 0.3679 | 0.2503 | 0.2491 | 0.2231 | 0.0821 | 0.0067 | 0.00004540 |

The answer in this case is given by $n = 278$. Notice that the approximations we have made result in the erroneous situation that even if I buy all tickets, I am still not guaranteed to win!

## 6.6 The Hypergeometric Distribution

The hypergeometric distribution arises in situations in which objects fall into two distinct categories, such as black and white, good and bad, male and female, etc. Suppose a collection consists of $N$ such objects, $r$ of which are of type 1 and $N - r$ of type 2. Now let $n$ of the $N$ objects be chosen at random *and without replacement* from the set. The random variable $X$ that denotes the number of type 1 from the selected $n$, is a hypergeometric random variable and its probability mass function is given by

$$p_X(k) = \frac{\binom{r}{k}\binom{N-r}{n-k}}{\binom{N}{n}}, \qquad k = \max(0, n - N + r), \ldots, \min(n, r),$$

and is equal to zero otherwise.

The expectation and variance of a hypergeometric random variable $X$, with parameters $n$, $N$, and $r$, are given by

$$E[X] = n\left(\frac{r}{N}\right),$$

$$\text{Var}\,[X] = n\left(\frac{r}{N}\right)\left(1 - \frac{r}{N}\right)\left(\frac{N-n}{N-1}\right).$$

**Example 6.10** A box contains twelve white balls and eight black balls. Seven are chosen at random. Let $X$ denote the number of white balls in this chosen set of seven. The probability that only six of the chosen seven are white is found by taking the ratio of the number of ways that six white balls and one black ball can be chosen from the box to the total number of ways of selecting seven balls from the box. This gives

$$\text{Prob}\{X = 6\} = \frac{\binom{12}{6}\binom{8}{1}}{\binom{20}{7}} = 0.09536.$$

This same result is obtained by substituting the values $N = 20$, $n = 7$, $r = 12$, and $k = 6$ into the probability mass function given previously. The probability of getting $k$ white balls is

$$\text{Prob}\{X = k\} = \frac{\binom{12}{k}\binom{8}{7-k}}{\binom{20}{7}}, \qquad k = 0, 1, \ldots, 7.$$

The expected number of white balls chosen and the standard deviation are

$$E[X] = n\left(\frac{r}{N}\right) = 7 \times \frac{12}{20} = 4.2,$$

$$\sigma_X = \sqrt{n\left(\frac{r}{N}\right)\left(1-\frac{r}{N}\right)\left(\frac{N-n}{N-1}\right)} = \sqrt{4.2\left(\frac{8}{20}\right)\left(\frac{13}{19}\right)} = \sqrt{1.1495} = 1.0721.$$

If $n = 14$ balls are selected rather than seven, the permissible values of $k$ run from $k = \max(0, n-N+r) = 6$ to $k = \min(n, r) = 12$ and the distribution of white balls is

$$\text{Prob}\{X = k\} = \frac{\binom{12}{k}\binom{8}{14-k}}{\binom{20}{14}}, \qquad k = 6, 7, \ldots, 12.$$

The hypergeometric distribution allows us to be more precise in answering certain types of questions that we previously answered using the binomial distribution: questions that pertain to determining the probability of finding a certain number of defective parts from a pool of manufactured parts, for example, or the probability of finding bad apples in a consignment of apples. When faced with such questions, we assumed that the probability of choosing a defective part was always the same. However, this is not the case. Suppose ten apples in a batch of 1000 are bad. Then the probability of choosing one bad apple is 1/100; the probability of choosing two bad apples is $1/100 \times 9/999$ and not $(1/100)^2$, since the selection is made *without replacement*.

**Example 6.11** If a jar contains 1000 jellybeans, 200 of which are red, and if I choose eight at random, then the probability that I get exactly three red ones among the eight selected is

$$\binom{8}{3} \times \left[\left(\frac{200}{1000}\frac{199}{999}\frac{198}{998}\right)\left(\frac{800}{997}\frac{799}{996}\frac{798}{995}\frac{797}{994}\frac{796}{993}\right)\right].$$

The term in the square brackets may be *approximated* by $(0.2)^3\ (0.8)^5$ which is what is obtained from the binomial distribution under the assumption that the probability of choosing a red jellybean is always $p = 2/10$.

The key phrase is that, in situations involving a hypergeometric random variable, the selection is made *without replacement*, which indeed is what happens frequently in practice. The binomial distribution can be an extremely useful and easy to use *approximation*, when the size of the selection is small compared to the size of the population from which the selection is made. However, it remains an approximation.

## 6.7 The Multinomial Distribution

The multinomial distribution arises in probability experiments that involve $n$ independent and identical trials in which each outcome falls into one and only one of $k$ mutually exclusive classes. We seek the distribution of the $n$ trials among the $k$ classes. Let $p_i,\ i = 1, 2, \ldots, k$, be the *probability* that an outcome falls into class $i$ and let $X_i$ be the random variable that denotes the *number* of outcomes that fall into class $i,\ i = 1, 2, \ldots, k$. The multinomial probability mass function is given by

$$p_{X_1, X_2, \ldots, X_k}(n_1, n_2, \ldots, n_k) = \frac{n!}{n_1! n_2! \cdots n_k!} p_1^{n_1} p_2^{n_2} \cdots p_k^{n_k}$$

subject to $\sum_{i=1}^k n_i = n$ and $\sum_{i=1}^k p_i = 1$. Like the binomial distribution, and unlike the hypergeometric distribution, the multinomial distribution is associated with the concept of selection *with replacement*, since the probabilities $p_i$, $i = 1, 2, \ldots, k$, are constant and do not change with the trial number.

The distributions of the $k$ random variables $X_i$, $i = 1, 2, \ldots, k$, may be computed from this joint probability mass function and their expectations and variances subsequently found. In fact each $X_i$ is a binomial random variable with parameter $p_i$, since $X_i$ simply provides a count of the number of successes in (i.e., the number of times that an outcome falls into) class $i$. We have

$$E[X_i] = np_i \quad \text{and} \quad \text{Var}\,[X_i] = np_i(1 - p_i).$$

**Example 6.12** Assume that mountain climbing expeditions fail as a result of three mutually exclusive events. With probability 0.6 an expedition fails due to adverse weather conditions; with probability 0.25, it fails due to climber injuries, while with probability 0.15 the failure is due to insufficient or lost equipment. We shall label these failure causes 1, 2, and 3, respectively. On a recent summer, four expeditions failed in their attempt to climb *La Meije*. Let $N_1$, $N_2$, and $N_3$ be three random variables that denote the number of these four failures that were due to causes 1, 2, and 3, respectively.

The joint distribution of $N_1$, $N_2$, and $N_3$ is the multinomial distribution and is given by

$$p_{N_1,N_2,N_3}(n_1, n_2, n_3) = \frac{4!}{n_1!n_2!n_3!}p_1^{n_1}p_2^{n_2}p_3^{n_3} = \frac{4!}{n_1!n_2!n_3!}0.6^{n_1}0.25^{n_2}0.15^{n_3}$$

for $0 \leq n_i \leq 4$, $i = 1, 2, 3$ with $n_1 + n_2 + n_3 = 4$, and is equal to zero otherwise.

Thus, the probability that two failures were due to inclement weather and the other two were due to equipment failures (causes 1 and 3) is given by

$$p_{N_1,N_2,N_3}(2, 0, 2) = \frac{4!}{2!\,0!\,2!}0.6^2 0.25^0 0.15^2 = 0.0486,$$

while the probability that all four failures were due to bad weather is

$$p_{N_1,N_2,N_3}(4, 0, 0) = \frac{4!}{4!\,0!\,0!}0.6^4 0.25^0 0.15^0 = 0.1296.$$

Observe that the probability of exactly two failures being caused by adverse weather conditions (the distribution of the other two causes is now irrelevant) is given by the binomial distribution—since we are looking for two successes (where here a success actually means a failure to reach the summit) out of four tries. The probability of exactly $n_1$ failures due to adverse weather conditions is

$$p_{N_1}(n_1) = \frac{4!}{n_1!\,(4 - n_1)!}0.6^{n_1}(1 - 0.6)^{4-n_1}$$

and so

$$p_{N_1}(2) = \frac{4!}{2!\,2!}0.6^2(0.4)^2 = 0.3456.$$

Indeed, the complete marginal distribution of $N_1$ is

$$p_{N_1}(n_1) = \frac{4!}{n_1!\,(4 - n_1)!}0.6^{n_1}0.4^{4-n_1} = \begin{cases} 0.4^4 = 0.0256, & n_1 = 0, \\ 4(0.6)0.4^3 = 0.1536, & n_1 = 1, \\ 6(0.6)^2(0.4)^2 = 0.3456, & n_1 = 2, \\ 4(0.6)^3 0.4 = 0.3456, & n_1 = 3, \\ 0.6^4 = 0.1296, & n_1 = 4. \end{cases}$$

Its expectation and variance are

$$E[X_1] = np_1 = 2.4, \qquad \text{Var}\,[X_1] = np_1(1 - p_1) = 0.96.$$

## 6.8 Exercises

**Exercise 6.1.1** The number of letters, $X$, delivered to our home each day is uniformly distributed between 3 and 10. Find

(a) The probability mass function of $X$.
(b) $\text{Prob}\{X < 8\}$.
(c) $\text{Prob}\{X > 8\}$.
(d) $\text{Prob}\{2 \leq X \leq 5\}$.

**Exercise 6.2.1** A bag contains six white balls and twelve black balls. A probability experiment consists of choosing a ball at random from the bag and inspecting its color. Identify this situation with the Bernoulli random variable and give its probability mass function. What is the expected value of the random variable and what does this mean in terms of the probability experiment?

**Exercise 6.3.1** A school has 800 students. What is the probability that exactly four students were born on July 31? (Assume 365 days in a year.)

**Exercise 6.3.2** The probability of winning a lottery is 0.0002. What is the probability of winning at least twice in 1,000 tries?

**Exercise 6.3.3** What is the probability of getting a passing grade of 70% on a ten-question true-false test, if all answers are guessed? How does this probability change if the number of questions on the test is increased to twenty?

**Exercise 6.3.4** A manufacturer produces widgets which are sold in packets of 144. In the most recent batch of 21,600, it is estimated that 5% are defective. Let $X$ be the random variable that denotes the number of defective widgets in a packet. Compute the probability mass function of $X$ under the simplifying assumption that for every widget in the package the probability that it is defective is 5% (which essentially reduces to the case of selection *with* replacement). What is the probability that a package contains more than ten defective widgets?

**Exercise 6.3.5** A family has eight children. What is the probability of there being seven boys and one girl, assuming that a boy is as likely to be born as a girl? What is the probability of there being between three and five boys? How many children should be born to be 95% sure of having a girl?

**Exercise 6.3.6** A scientific experiment is carried out a number of times in the hope that a later analysis of the data finds at least one success. Let $n$ be the number of times that the experiment is conducted and suppose that the probability of success is $p = 0.2$. Assuming that the experiments are conducted independently from one another, what is the number of experiments that must be conducted to be 95% sure of having at least one success?

**Exercise 6.3.7** A coin is tossed 400 times and the number of heads that appear is equal to 225. What is the likelihood that this coin is biased?

**Exercise 6.3.8** A political candidate polls 100 likely voters from a large population and finds that only 44 have the intention of voting for him. Should he abandon his run for office? What should he do if he polls 2,500 likely voters and finds that 1,100 indicate a preference for him?

**Exercise 6.4.1** Write down the probability generating function of a geometric random variable with parameter $p$, $0 < p < 1$, and use this to find its expectation and variance.

**Exercise 6.4.2** A basketball player at the free-throw line has a 80% chance of making the basket, a statistic that does not change during the course of the game. What is the average number of free throws he makes before his first miss? What is the probability that his first miss comes on his fourth try? On his fifth try?

**Exercise 6.4.3** Nicola, Stephanie, Kathryn, and William, in that order, take turns at rolling a fair die until one of them throws a 6. What is the probability that William is the first to throw a 6?

**Exercise 6.4.4** Messages relayed over a communication channel have probability $p$ of being received correctly. A message that is not received correctly is retransmitted until it is. What value should $p$ have so that the probability of more than one retransmission is less than 0.05?

**Exercise 6.4.5** A young couple decides to have children until their first son is born. Assuming that each child born is equally likely to be a boy or a girl, what is the probability that this couple will have exactly four children? What is the most probable range for the number of children this couple will have?

**Exercise 6.4.6** A different couple from those of Exercise 6.4.5 intend to continue having children until they have two boys. What is the probability they will have exactly two children? exactly three children? exactly four children? What is the most probable range for the number of children this couple will have?

**Exercise 6.4.7** Let $X$ be an integer valued random variable whose probability mass function is given by

$$\text{Prob}\{X = n\} = \alpha t^n \quad \text{for all } n \geq 0,$$

where $0 < t < 1$. Find $G_X(z)$, the value of $\alpha$, the expectation of $X$, and its variance.

**Exercise 6.4.8** The boulevards leading into Paris appear to drivers as an endless string of traffic lights, one after the other, extending as far as the eye can see. Fortunately, these successive traffic lights are synchronized. Assume that the effect of the synchronization is that a driver has a 95% chance of not being stopped at any light, independent of the number of green lights she has already passed. Let $X$ be the random variable that counts the number of green lights she passes before being stopped by a red light.

(a) What is the probability distribution of $X$?
(b) How many green lights does she pass on average before having to stop at a red light?
(c) What is the probability that she will get through 20 lights before having to stop for the first time?
(d) What is the probability that she will get through 50 lights before stopping for the fourth time?

**Exercise 6.4.9** A popular morning radio show offers free entrance tickets to the local boat show to the sixth caller who rings the station with the correct answer to a question. Assume that all calls are independent and have probability $p = 0.7$ of being correct. Let $X$ be the random variable that counts the number of calls needed to find the winner.

(a) What is the probability mass function of $X$?
(b) What is the probability of finding a winner on the $12^{\text{th}}$ call?
(c) What is the probability that it will take more than ten calls to find a winner?

**Exercise 6.4.10** In a best of seven sports series, the first person to win four games is declared the overall winner. One of the players has a 55% chance of winning each game, independent of any other game. What is the probability that this player wins the series?

**Exercise 6.4.11** Returning to Exercise 6.4.10, find the probability that the series ends after game 5. Once again, assume that one of the players has a 55% change of winning each game, independent of any other game.

**Exercise 6.4.12** Given the expansion

$$(1-q)^{-k} = \sum_{i=0}^{\infty} \binom{i+k-1}{k-1} q^i,$$

show that Equation (6.1) defines a bona fide probability mass function. The name *negative* binomial comes from this expansion with its negative exponent, $-k$.

**Exercise 6.5.1** Write down the probability generating function for a Poisson random variable $X$, and use it to compute its expectation and variance.

**Exercise 6.5.2** Let $X$ be a Poisson random variable with mean value $\alpha = 3$ that denotes the number of calls per minute received by an airline reservation center. What is the probability of having no calls in a minute? What is the probability of having more than three calls in a minute?

**Exercise 6.5.3** During rush hours at a Paris metro station, trains arrive according to a Poisson distribution with an expected number of ten trains per 60 minute period.

(a) What is the probability mass function of train arrivals in a period of length $t$ minutes?
(b) What is the probability that two trains will arrive in a three minute period?
(c) What is the probability that no trains will arrive in a ten minute period?
(d) Find the probability that at least one train will arrive in a period of length $t$ minutes and use this to compute how long is needed to be 95% sure that a train will arrive?

**Exercise 6.5.4** Five percent of the blood samples taken at a doctor's office need to be sent off for additional testing. From the binomial probability mass function, what is the probability that among a sample of 160, ten have to be sent off? What answer is obtained when the Poisson approximation to the binomial is used instead?

**Exercise 6.5.5** On average, one widget in 100 manufactured at a certain plant is defective. Assuming that defects occur independently, what is the distribution of defective parts in a batch of 48? Use the Poisson approximation to the binomial to compute the probability that there is more than one defective part in a batch of 48? What is the probability that there are more than two defective parts in a batch of 48?

**Exercise 6.5.6** Consider a situation in which certain events occur randomly in time, such as arrivals to a queueing system. Let $X(t)$ be the number of these events that occur during a time interval of length $t$. We wish to compute the distribution of the random variable $X(t)$, under the following three assumptions:

(1) The events are independent of each other. This means that $X(\Delta t_1), X(\Delta t_2), \ldots$ are independent if the intervals $\Delta t_1, \Delta t_2, \ldots$ do not overlap.
(2) The system is *stationary*, i.e., the distribution of $X(\Delta t)$ depends only on the length of $\Delta t$ and not on the actual time of occurrence.
(3) We have the following probabilities:
- Prob{at least 1 event in $\Delta t$} $= \lambda\Delta t + o(\Delta t)$,
- Prob{more than 1 event in $\Delta t$} $= o(\Delta t)$.

where $o(\Delta t)$ is a quantity that goes to zero faster than $\Delta t$, i.e.,

$$\lim_{\Delta t \to 0} \frac{o(\Delta t)}{\Delta t} = 0,$$

and $\lambda$ is a positive real number that denotes the rate of occurrence of the events. Show that $X$ is a Poisson random variable.

**Exercise 6.5.7** Prove that the sum of $n$ independent Poisson random variables $X_i$, $i = 1, 2, \ldots, n$, with parameters $\lambda_1, \lambda_2, \ldots, \lambda_n$, respectively, is also Poisson distributed.

**Exercise 6.6.1** I have four pennies and three dimes in my pocket. Using the hypergeometric distribution, compute the probability that, on pulling two coins at random from my pocket, I have enough to purchase a 20-cent newspaper. Compute the answer again, this time using only probabilistic arguments.

**Exercise 6.6.2** Let $X$ be a hypergeometric random variable with parameters $N = 12$, $r = 8$, and $n = 6$.

(a) What are the possible values for $X$?
(b) What is the probability that $X$ is greater than 2?
(c) Compute the expectation and variance of $X$.

**Exercise 6.6.3** Last year, a wildlife monitoring program tagged 12 wolves from a population of 48 wolves. This year they returned and captured 16.

(a) How many tagged wolves should they expect to find among those captured?
(b) What is the probability of capturing exactly four tagged wolves?
(c) Suppose seven tagged wolves are captured. Should the wildlife team assume that the population of wolves has increased, decreased or stayed the same over the past year?

**Exercise 6.6.4** Cardinal Gibbons High School has 1,200 students, 280 of whom are seniors. In a random sample of ten students, compute the expectation and variance of the number of included seniors using both the hypergeometric distribution and its binomial approximation.

**Exercise 6.7.1** A large jar of jellybeans contains 200 black ones, 300 brown ones, 400 green ones, 500 red ones, and 600 yellow ones, all mixed randomly together. Use the multinomial distribution to estimate the probability that a fistful of 15 jellybeans contains 5 green, 5 red, and 5 yellow ones? Write an expression (without evaluating it) for the probability of obtaining no black, brown, or green jellybeans in a fistful of 15.

**Exercise 6.7.2** Each Friday evening, Kathie and Billy go to one of four local pubs. They come to a decision as to which one to go to on any particular Friday night by throwing a fair die. If the die shows 1, they go to RiRa's, if it shows 2, they go to Tir na n'Og, if it shows 3, they go to the Fox and Hound, while if it shows a number greater than 3, they go to their favorite, the Hibernian. What is the probability that during the 14 weeks of summer, they visit RiRa's twice, Tir na n'Og three times, the Fox and Hound three times, and the Hibernian six times? What is the probability that during the four consecutive Fridays in June, they visit each pub exactly once?

**Exercise 6.7.3** Revels Tractor Works observes that riding lawnmowers break down for one of four different (read mutually exclusive) reasons, which we simply label as type 1, 2, 3, and 4, with probabilities $p_1 = 0.1$, $p_2 = 0.2$, $p_3 = 0.3$, and $p_4 = 0.4$, respectively. On Monday, five of these mowers are brought in for repair.

(a) Give an expression for the probability mass function $p_{X_1,X_2,X_3,X_4}(n_1, n_2, n_3, n_4)$, where $X_i$ is the number requiring repairs of type $i$, $i = 1, 2, 3, 4$.
(b) Use your knowledge of the binomial distribution to compute the probability that three require repairs of type 4.
(c) What is the probability that three require repairs of type 4 and the other two require repairs of type 1?

# Chapter 7

# *Continuous Distribution Functions*

## 7.1 The Uniform Distribution

A continuous random variable $X$ that is equally likely to take any value in a range of values $(a, b)$, with $a < b$, gives rise to the uniform distribution. Such a distribution is *uniformly distributed* on its range. The probability density function of $X$ is given as

$$f_X(x) = \begin{cases} c, & a < x < b, \\ 0 & \text{otherwise.} \end{cases}$$

Thus the range is given by $(a, b)$ and the value assumed by $X$ on this range is $c$. This means that

$$c = \frac{1}{b-a}$$

since the sum of probabilities must be 1, i.e.,

$$1 = \int_{-\infty}^{\infty} f_X(x)dx = \int_a^b c\,dx = c(b-a).$$

The cumulative distribution function is given by

$$F_X(x) = \int_{-\infty}^{x} f_X(t)dt = \frac{1}{b-a}t\,\Big|_{-\infty}^{x} = \begin{cases} 0, & x \le a, \\ (x-a)/(b-a), & a \le x \le b, \\ 1, & x \ge b. \end{cases}$$

Figure 7.1 shows the cumulative distribution function and the probability density function of this random variable. Observe that, if $(r, s)$ is any subinterval of $(a, b)$, the probability that $X$ takes a value in this subinterval does not depend on the actual position of the subinterval in $(a, b)$ but only on the length of the subinterval. In other words, we have

$$\text{Prob}\{r < X < s\} = F_X(s) - F_X(r) = \frac{s-r}{b-a}.$$

The values of $f_X(x)$ at the endpoints $a$ and $b$ do not affect the probabilities defined by areas under the graph.

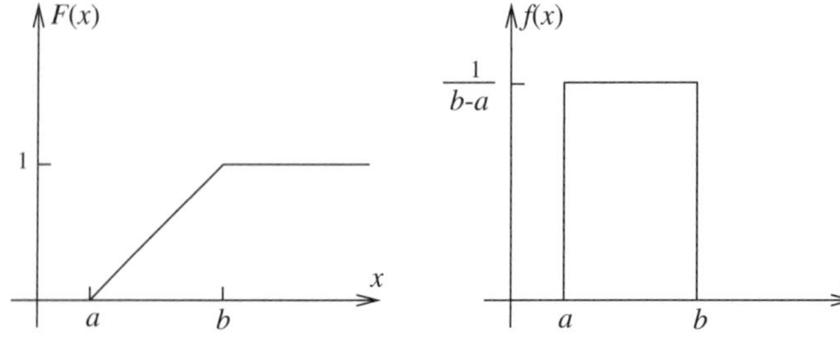


Figure 7.1. The CDF and the probability density function of the continuous uniform distribution.

The mean of the uniform distribution is obtained as

$$E[X] = \int_{-\infty}^{\infty} xf(x)dx = c\int_a^b x dx = \frac{a+b}{2}.$$

Its variance is computed from

$$\text{Var}\,[X] = E[X^2] - (E[X])^2 = \int_a^b \frac{x^2}{b-a}dx - \left(\frac{a+b}{2}\right)^2 = \frac{(b-a)^2}{12}.$$

We may now compute the squared coefficient of variation as

$$C_X^2 = \frac{(b-a)^2}{12} \times \frac{4}{(b+a)^2} = \frac{(b-a)^2}{3(b+a)^2}.$$

**Example 7.1** A continuous uniformly distributed random variable $X$ has two parameters which are the endpoints of its range $(a, b)$. If $E[X] = 5$ and $\text{Var}\,[X] = 3$, we can generate two equations with which to obtain $a$ and $b$. We have

$$E[X] = \frac{a+b}{2} = 5, \quad \text{Var}\,[X] = \frac{(b-a)^2}{12} = 3,$$

which gives $a+b = 10$ and $(b-a)^2 = 36$. The second of these gives $b - a = \pm 6$ and if we assume that $a < b$, we end up with two equations in two unknowns,

$$a + b = 10, \quad b - a = 6,$$

which we solve to obtain $a = 2$ and $b = 8$. The probability density function of $X$ is given by

$$f_X(x) = \begin{cases} 1/6, & 2 < x < 8, \\ 0 & \text{otherwise.} \end{cases}$$

The astute reader will have noticed that, although the expectation of a *continuous* random variable $X$ uniformly distributed on $(a, b)$ is equal to that of a *discrete* random variable $Y$ uniformly distributed over the same interval, the same is *not* true for the variance. The variances in the continuous and in the discrete case are respectively given by

$$\text{Var}\,[X] = \frac{(b-a)^2}{12} \text{ and } \text{Var}\,[Y] = \frac{(b-a+2)(b-a)}{12}.$$

The variance of $X$ is less than the variance of $Y$. Discretization is the process of deriving a discrete random variable from a continuous one. Let $a$ and $b$ be integers with $b > a$ and let $j = b - a$. Let $X$ be the continuous uniform distribution on $(a, b)$. The discrete uniform random variable $Y$, with probability mass function $p_Y(i) = 1/(b-a)$ for $i = a+1, a+2, \ldots, a+j = b$ and zero otherwise, is obtained by setting $Y = \lceil X \rceil$, i.e., the ceiling of $X$. This discretizes or lumps the event $\{i - 1 < x \le i\}$ into the event $\{Y = i\}$ with the result that $Y$ is a discrete uniform distribution on $[a+1, b]$, since

$$p_Y(i) = \int_{i-1}^{i} f_X(x)\,dx = \int_{i-1}^{i} \frac{1}{b-a}dx = \frac{1}{b-a}, \quad i = a+1, a+2, \ldots, b,$$

and zero otherwise. As we have just seen, this process of discretization has a tendency to increase variability, the variance of $Y$ being greater than the variance of $X$.

To compute the Laplace transform of the continuous uniform distribution, we assume, without loss of generality, that $0 \le a < b$. Then

$$\mathcal{L}_X(s) = \int_a^b e^{-sx}\frac{1}{b-a}dx = \frac{e^{-as} - e^{-bs}}{s(b-a)}.$$

In later sections, we shall discuss simulation as a modeling technique. At this point, it is worthwhile pointing out that the continuous uniform distribution on the unit interval ($0 \le x \le 1$) is widely used in simulation, since other distributions, both continuous *and* discrete, may be generated from it. Of all distribution functions, the uniform distribution is considered the most random in the sense that it offers the least help in predicting the value that will be taken by the random variable $X$.

**Example 7.2** let $X$ be a continuous random variable that is uniformly distributed over $[-2, 2]$. Then the probability density function of $X$ is

$$f_X(x) = \begin{cases} 1/4, & -2 \le x \le 2, \\ 0 & \text{otherwise.} \end{cases}$$

Its cumulative distribution function is $F_X(x) = 0$ for $x \le -2$ and $F_X(x) = 1$ for $x \ge 2$. For values of $x$ between $-2$ and $+2$, it is

$$F_X(x) = \int_{-2}^{x} f_X(t)dt = \frac{1}{4}\int_{-2}^{x} dt = \frac{x+2}{4}.$$

Let us compute $E[X^3]$, the third moment of $X$:

$$E[X^3] = \frac{1}{4}\int_{-2}^{2} x^3 dx = \left.\frac{x^4}{16}\right|_{-2}^{2} = 0.$$

It should be obvious that all odd moments, including the expectation, of this particular random variable are equal to zero. Continuing, let us compute $E[e^X]$ and the Laplace transform, $\mathcal{L}_X(s)$. We have

$$E[e^X] = \frac{1}{4}\int_{-2}^{2} e^x dx = \left.\frac{e^x}{4}\right|_{-2}^{2} = \frac{e^2 - e^{-2}}{4} = 1.8134$$

and, by substitution,

$$\mathcal{L}_X(s) = \frac{e^{2s} - e^{-2s}}{4s}.$$

## 7.2 The Exponential Distribution

The cumulative distribution function for an exponential random variable, $X$, with parameter $\lambda > 0$, is given by

$$F(x) = \begin{cases} 1 - e^{-\lambda x}, & x \ge 0, \\ 0 & \text{otherwise.} \end{cases}$$

Graphs of this distribution for various values of $\lambda$ are shown in Figure 7.2. The corresponding probability density function is obtained simply by taking the derivative of $F(x)$ with respect to $x$:

$$f(x) = \begin{cases} \lambda e^{-\lambda x}, & x \ge 0, \\ 0 & \text{otherwise.} \end{cases}$$

Again some plots for various values of $\lambda$ are given in Figure 7.3. Observe that the probability density function intercepts the $y = f(x)$ axis at the point $\lambda$, the parameter of the distribution.

One of the most important properties of the exponential distribution is that it possesses the memoryless property. This means that the past history of a random variable that is exponentially

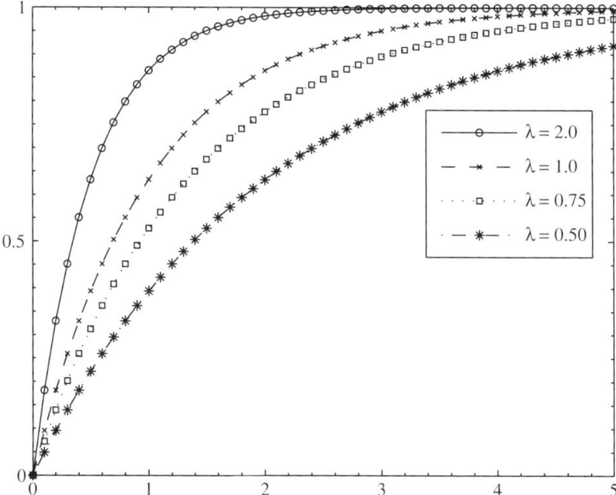


Figure 7.2. CDFs of some exponentially distributed random variables.

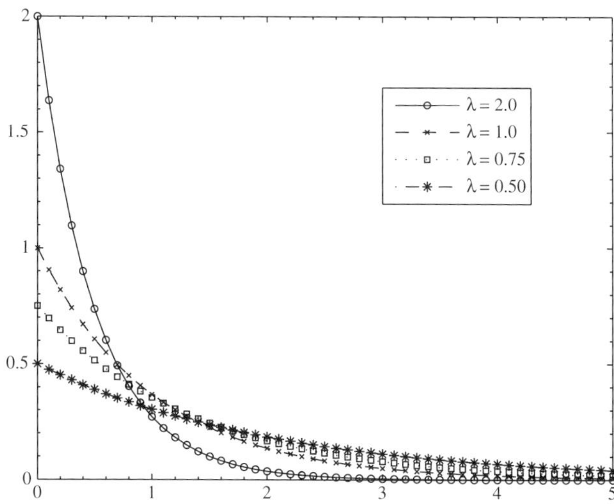


Figure 7.3. Some probability density functions for the exponential distribution.

distributed plays no role in predicting its future. For example, let $X$ be the random variable that denotes the length of time that a customer spends in service (called the *service time*) at some facility and let us assume that $X$ is exponentially distributed. Then the probability that the customer in service finishes at some future time $t$ is independent of how long that customer has already been in service. Similarly, if the time between arrivals (the *interarrival time*) of patients to a doctor's office is exponentially distributed, then the probability that an arrival occurs by time $t$ is independent of the length of time that has elapsed from the previous arrival.

To explore this further, let us assume that an arrival occurs at time 0. Let $t_0$ seconds pass during which no arrivals occur. Now, what is the probability that the next arrival occurs $t$ seconds from now? We assume that $X$ is the random variable defined as the time between successive arrivals, and that $X$ is exponentially distributed with parameter $\lambda$, i.e., $F(x) = 1 - e^{-\lambda x}$. Then

$$\begin{aligned}
\text{Prob}\{X \le t_0 + t | X > t_0\} &= \frac{\text{Prob}\{t_0 < X \le t_0 + t\}}{\text{Prob}\{X > t_0\}} \\
&= \frac{\text{Prob}\{X \le t_0 + t\} - \text{Prob}\{X \le t_0\}}{\text{Prob}\{X > t_0\}} \\
&= \frac{1 - e^{-\lambda(t+t_0)} - (1 - e^{-\lambda t_0})}{1 - (1 - e^{-\lambda t_0})} \\
&= \frac{e^{-\lambda t_0} - e^{-\lambda t_0} e^{-\lambda t}}{e^{-\lambda t_0}} = 1 - e^{-\lambda t} \\
&= \text{Prob}\{X \le t\},
\end{aligned}$$

which shows that the distribution of remaining time until the next arrival, given that $t_0$ seconds has already elapsed since the last arrival, is identically equal to the unconditional distribution of the interarrival time.

**Example 7.3** Assume that the amount of time a patient spends in a dentist's office is exponentially distributed with mean equal to 40 minutes ($\lambda = 1/40$). We first compute the probability that a patient spends more than 60 minutes in the dentist's office. Let $X$ be the exponentially distributed random variable that describes the time a patient spends in the office. Then the probability that a patient spends more than one hour there is given by

$$\text{Prob}\{X > 60\} = e^{-60\lambda} = e^{-1.5} = 0.2231.$$

Second, we find the probability that a patient will spend 60 minutes in the dentist's office given that she has already spent 40 minutes there. In this case, we seek the probability that the patient will spend a further 20 minutes in the office, but since the exponential distributon has the memoryless property, this is just equal to the unconditional probability that she will spend 20 minutes there, which is given by

$$\text{Prob}\{X > 20\} = e^{-20\lambda} = e^{-.5} = 0.6065.$$

The exponential distribution is the only *continuous* distribution that exhibits this memoryless property, which is also called the *Markov property*. Furthermore, it may be shown that, if $X$ is a nonnegative continuous random variable having this memoryless property, then the distribution of $X$ *must* be exponential. For a discrete random variable, the *geometric* distribution is the only distribution with this property. Indeed, if $X$ is an exponentially distributed random variable with parameter $\lambda$, then the *discrete* random variable $Y$ obtained from $X$ by discretization as $Y = \lceil X \rceil$ is a geometric random variable with probability of success given by $p = 1 - e^{-\lambda}$. We have

$$\begin{aligned}
p_Y(i) = \text{Prob}\{i - 1 < X \le i\} &= F_X(i) - F_X(i-1) \\
&= 1 - e^{-\lambda i} - 1 + e^{-\lambda(i-1)} \\
&= e^{-\lambda(i-1)} \left(1 - e^{-\lambda}\right) \\
&= (1-p)^{i-1} p.
\end{aligned}$$

The mean $E[X]$ of the exponential distribution may be found as

$$\begin{aligned}
E[X] &= \int_{-\infty}^{\infty} x f(x) dx = \int_0^{\infty} \lambda x e^{-\lambda x} dx = -\lambda \frac{\partial}{\partial \lambda} \int_0^{\infty} e^{-\lambda x} dx \\
&= -\lambda \frac{\partial}{\partial \lambda} \left[ -\frac{1}{\lambda} e^{-\lambda x} \Big|_0^{\infty} \right] = -\lambda \frac{\partial}{\partial \lambda} \left[ \frac{1}{\lambda} - \frac{1}{\lambda} e^{-\infty} \right] = -\lambda \frac{\partial}{\partial \lambda} \left[ \frac{1}{\lambda} \right] = -\lambda \left[ -\frac{1}{\lambda^2} \right] = \frac{1}{\lambda}.
\end{aligned}$$

This same result may also be obtained directly by integration by parts. If an exponentially distributed random variable represents the interarrival time of customers to a queue, then the average interval between arrivals is equal to $1/\lambda$, which is a nice result, since the probability of an arrival in an interval of length $\Delta t$ is $\lambda\Delta t$, which implies that $\lambda$ is equal to the average rate of arrivals.

To compute the variance, we write

$$E[X^2] = \int_0^\infty x^2\lambda e^{-\lambda x}dx = \lambda\frac{\partial^2}{\partial\lambda^2}\int_0^\infty e^{-\lambda x}dx = \lambda\frac{\partial^2}{\partial\lambda^2}\left(\frac{1}{\lambda}\right) = \frac{2}{\lambda^2}.$$

Thus the variance is computed as

$$\sigma_X^2 = E[X^2] - E^2[X] = \frac{2}{\lambda^2} - \left(\frac{1}{\lambda}\right)^2 = \frac{1}{\lambda^2}.$$

We use an induction proof to show that the $n^{\text{th}}$ moment is given by

$$E[X^n] = \int_0^\infty x^n\lambda e^{-\lambda x}dx = \frac{n!}{\lambda^n}.$$

We have already shown that the basis clause, $E[X] = 1/\lambda$, is true. Assuming that $E[X^{n-1}] = (n-1)!/\lambda^{n-1}$ we now show that $E[X^n] = n!/\lambda^n$. Using integration by parts, with $u = x^n$ and $dv = \lambda e^{-\lambda x}dx$, we obtain

$$\begin{aligned} E[X^n] &= \int_0^\infty x^n\lambda e^{-\lambda x}dx = -x^n e^{-\lambda x}\Big|_0^\infty + \int_0^\infty nx^{n-1}e^{-\lambda x}dx \\ &= 0 + \frac{n}{\lambda}\int_0^\infty x^{n-1}\lambda e^{-\lambda x}dx \\ &= \frac{n}{\lambda}E[X^{n-1}] = \frac{n!}{\lambda^n}. \end{aligned}$$

Alternatively, these same results can be found from the moment generating function of an exponentially distributed random variable, which is given by

$$\mathcal{M}_X(\theta) = E[e^{\theta X}] = \int_0^\infty e^{\theta x}\lambda e^{-\lambda x}dx = \lambda\int_0^\infty e^{-(\lambda-\theta)x}dx = \frac{\lambda}{\lambda-\theta} \quad \text{for } \theta < \lambda.$$

Now, for example, we find the second moment as

$$E[X^2] = \frac{d^2}{d\theta^2}\mathcal{M}_X(\theta)\Big|_{\theta=0} = \frac{d^2}{d\theta^2}\frac{\lambda}{\lambda-\theta}\Big|_{\theta=0} = \frac{2\lambda}{(\lambda-\theta)^3}\Big|_{\theta=0} = \frac{2}{\lambda^2}.$$

**Example 7.4** The time spent waiting for a bus on the North Carolina State Campus may be represented by an exponential random variable with a mean of three minutes. We would like to know the probability of having to wait more than five minutes and the probability that the time spent waiting is within $\pm 2$ standard deviations of the mean.

We have $E[X] = 1/\lambda = 3$, which gives $\lambda = 1/3$. The probability of having to wait more than five minutes is

$$\text{Prob}\{X > 5\} = 1 - \text{Prob}\{X \le 5\} = 1 - F_X(5) = 1 - (1 - e^{-1/3\times 5}) = e^{-5/3} = 0.1889.$$

For an exponential random variable, the standard deviation is equal to the mean. Thus we have $\sigma_X = 3$, and the probability that the time spent waiting is within two standard deviations of the mean is

$$\text{Prob}\{3 - 6 \le X \le 3 + 6\} = \text{Prob}\{X \le 9\} = 1 - e^{-1/3\times 9} = 0.9502.$$

The Laplace transform of the exponential probability density function is given by

$$F^*(s) = \int_0^\infty e^{-sx} f(x)dx = \int_0^\infty e^{-sx}\lambda e^{-\lambda x}dx = \lambda \int_0^\infty e^{-(s+\lambda)x}dx = \frac{\lambda}{s+\lambda}.$$

Let $X_1$ and $X_2$ be independent and identically distributed exponential random variables with mean $1/\lambda$. The probability distribution of their sum is given by

$$\begin{aligned} F_{X_1+X_2}(x) = \text{Prob}\{X_1 + X_2 \le x\} &= \int_0^x \text{Prob}\{X_1 \le x - s\}\lambda e^{-\lambda s}ds \\ &= \int_0^x \left(1 - e^{-\lambda(x-s)}\right)\lambda e^{-\lambda s}ds \\ &= \lambda \int_0^x \left(e^{-\lambda s} - e^{-\lambda x}\right) ds \\ &= \lambda \int_0^x e^{-\lambda s}ds - \lambda e^{-\lambda x}\int_0^x ds \\ &= 1 - e^{-\lambda x} - \lambda x e^{-\lambda x}. \end{aligned}$$

The corresponding density function is obtained by differentiation, which gives

$$f_{X_1+X_2}(x) = \lambda^2 x e^{-\lambda x}, \quad x \ge 0.$$

This result may also be obtained using the convolution approach and is left to the exercises.

Let $X_1, X_2, \ldots, X_n$ be $n$ independent exponential random variables having parameters $\lambda_1, \lambda_2, \ldots, \lambda_n$, respectively. Let $X_{\min} = \min(X_1, X_2, \ldots, X_n)$ and $X_{\max} = \max(X_1, X_2, \ldots, X_n)$. Then, from Section 5.5, the cumulative distribution functions of $X_{\min}$ and $X_{\max}$ are given, respectively, by

$$\text{Prob}\{X_{\min} \le x\} = 1 - \prod_{i=1}^n \left(1 - (1 - e^{-\lambda_i x})\right) = 1 - \prod_{i=1}^n e^{-\lambda_i x} = 1 - e^{-(\lambda_1+\lambda_2+\cdots+\lambda_n)x},$$

$$\text{Prob}\{X_{\max} \le x\} = \prod_{i=1}^n \left(1 - e^{-\lambda_i x}\right).$$

Observe that the first of these is exponentially distributed with parameter $\lambda_1 + \lambda_2 + \cdots + \lambda_n$; the second is not exponentially distributed.

Now let $X_1$ and $X_2$ be independent exponential random variables with means $1/\lambda_1$ and $1/\lambda_2$, respectively. The probability that one is smaller than the other is given by

$$\begin{aligned} \text{Prob}\{X_1 < X_2\} &= \int_0^\infty \text{Prob}\{X_1 < X_2 \mid X_2 = x\}\lambda_2 e^{-\lambda_2 x}dx \\ &= \int_0^\infty \text{Prob}\{X_1 < x\}\lambda_2 e^{-\lambda_2 x}dx \\ &= \int_0^\infty \left(1 - e^{-\lambda_1 x}\right)\lambda_2 e^{-\lambda_2 x}dx \\ &= \int_0^\infty \lambda_2 e^{-\lambda_2 x}dx - \lambda_2 \int_0^\infty e^{-(\lambda_1+\lambda_2)x}dx \\ &= 1 - \frac{\lambda_2}{\lambda_1+\lambda_2} = \frac{\lambda_1}{\lambda_1+\lambda_2}. \end{aligned}$$

More generally,

$$\begin{aligned}\text{Prob}\{X_i = \min(X_1, X_2, \ldots, X_n)\} &= \int_0^\infty \prod_{j \neq i} \text{Prob}\{X_j > x\}\lambda_i e^{-\lambda_i x} dx \\ &= \int_0^\infty \lambda_i e^{-(\lambda_1+\lambda_2+\cdots+\lambda_n)x} dx \\ &= \frac{\lambda_i}{(\lambda_1 + \lambda_2 + \cdots + \lambda_n)}.\end{aligned}$$

**Example 7.5** In order for a flashlight to function correctly, both the bulb and the battery must be in working order. If the average lifetime of a bulb is 800 hours and that of the battery is 300 hours, and if the lifetime of both is independent and exponentially distributed, then the probability that the flashlight fails due to an exhausted battery rather than a burnt out bulb can be obtained from the above formula as

$$\frac{\lambda_1}{\lambda_1 + \lambda_2} = \frac{1/300}{1/300 + 1/800} = 0.7273.$$

The exponential distribution is closely related to the Poisson distribution. Specifically, if a random variable describing an arrival process has a Poisson distribution, then the associated random variable defined as the time between successive arrivals (the interarrival time) has an exponential distribution. To see this, notice that since the arrival process is Poisson, we have

$$p_n(t) = e^{-\lambda t}\frac{(\lambda t)^n}{n!} \quad (\text{with } p_0(t) = e^{-\lambda t}),$$

which describes the number of arrivals that have occurred by time $t$. Let $X$ be the random variable that denotes the time between successive events (arrivals). Its probability distribution function $A(t)$ is given by

$$A(t) = \text{Prob}\{X \leq t\},$$

i.e.,

$$A(t) = 1 - \text{Prob}\{X > t\}.$$

But $\text{Prob}\{X > t\} = \text{Prob}\{0 \text{ arrivals in } (0, t]\} = p_0(t)$. Thus

$$A(t) = 1 - p_0(t) = 1 - e^{-\lambda t}, \quad t \geq 0.$$

Differentiating, we obtain the associated density function

$$a(t) = \lambda e^{-\lambda t}, \quad t \geq 0,$$

which is none other than the exponential density function. In other words, $X$ is exponentially distributed with mean $1/\lambda$.

## 7.3 The Normal or Gaussian Distribution

The *normal* or *Gaussian* density function, which has the familiar bell-shaped curve, is omnipresent throughout the world of statistics and probability. If $X$ is a Gaussian random variable then its probability density function $f_X(x)$, sometimes denoted by $N(\mu, \sigma^2)$, is

$$f_X(x) = N(\mu, \sigma^2) = \frac{1}{\sigma\sqrt{2\pi}}e^{-(x-\mu)^2/2\sigma^2} \text{ for } -\infty < x < \infty. \tag{7.1}$$

It may be shown that expectation is $E[X] = \mu$ and the variance $\text{Var}[X] = \sigma^2$. Thus the mean and variance are the actual parameters of the distribution. The normal density function for selected values of its parameters is shown in Figure 7.4.

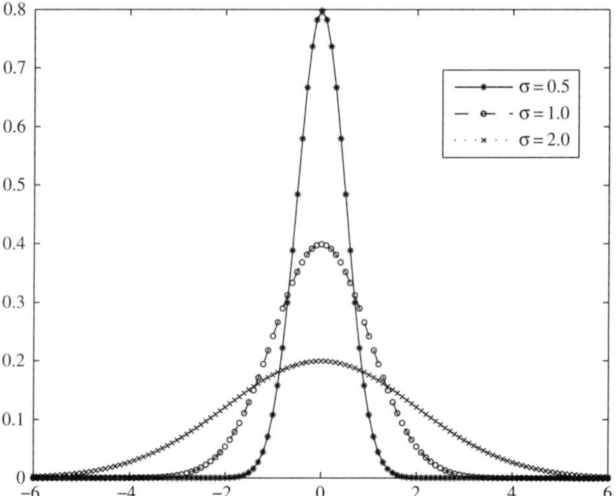


Figure 7.4. The normal distribution for different values of $\sigma$.

As is apparent from this figure, the normal distribution is symmetric about its mean value $\mu$. It is short and flat when $\sigma$ is large and tall and skinny when $\sigma$ is small. As we shall see momentarily, approximately 68% of the area under $f_X(x)$ lies within one standard deviation of its mean while 95% lies within two standard deviations.

The cumulative distribution function for a random variable that is normally distributed has no closed form, i.e., we cannot write $F_X(x)$ in the form of an equation as we have been able to do with other distributions. This means that we must have recourse to precomputed tables. Usually these tables provide data relating to the *standard* normal distribution $N(0, 1)$, having mean 0 and variance 1. The probability density function of a standard normal random variable is obtained by substituting the values $\mu = 0$ and $\sigma^2 = 1$ into Equation (7.1) . Because of its importance, and to avoid confusion, the probability density function of a standard normal random variable $X$ is often denoted $\phi_X(x)$ and its cumulative distribution function by $\Phi_X(x)$.

Mathematical tables generally provide the values of $\Phi(x)$ in increments of 0.01 from $x = 0$ (where $\Phi(0) = 0.50$) to $x = 3$ (where $\Phi(3) = 0.9987$). Some tables include values of $x$ up to $x = 5$ ($\Phi(5) = 0.9999997$). Since a standard normal random variable $X$ has variance and standard deviation equal to 1, the quantity $\Phi(1) - \Phi(-1)$ is the probability that $X$ lies within one standard deviation of its mean, $\Phi(2) - \Phi(-2)$ is the probability that it lies within two standard deviations of its mean, and so on. The symmetry of the normal distribution allows us to use the standard tables to find $\Phi(x)$ for negative arguments. For example, the symmetry of the standard normal distribution around zero implies that $\text{Prob}\{X \le -1\} = \text{Prob}\{X \ge 1\}$ and hence $\text{Prob}\{X \le -1\} = 1 - \text{Prob}\{X \le 1\}$. Obviously, this holds more generally: for any nonnegative value $\alpha$, we have

$$\text{Prob}\{X \le -\alpha\} = 1 - \text{Prob}\{X \le \alpha\}.$$

The values provided by tables of the standard normal distribution can be used to find the values of all normal distributions, $N(\mu, \sigma^2)$. Indeed, if a random variable $X$ has the distribution $N(\mu, \sigma^2)$, then the random variable $Z = (X - \mu)/\sigma$ has a standard normal distribution. To see this, observe that

$$\begin{aligned}\text{Prob}\{Z \le z\} &= \text{Prob}\{(X-\mu)/\sigma \le z\} = \text{Prob}\{X \le \sigma z + \mu\}\\ &= \int_{-\infty}^{\sigma z+\mu} \frac{1}{\sqrt{2\pi}\sigma} e^{-(x-\mu)^2/2\sigma^2} dx\\ &= \int_{-\infty}^{z} \frac{1}{\sqrt{2\pi}\sigma} e^{-t^2/2} dt\end{aligned}$$

for $t = (x-\mu)/\sigma$. Thus, if $X$ has a normal distribution with expectation $\mu$ and standard deviation $\sigma$, then, for any constant $\alpha$,

$$\text{Prob}\{X \le \alpha\} = \text{Prob}\{(X-\mu)/\sigma \le (\alpha-\mu)/\sigma\} = \text{Prob}\{Z \le (\alpha-\mu)/\sigma\}.$$

In words, the probability that $X$, a normally distributed random variable, is less than $\alpha$ is equal to the probability that the *standard* normal random variable is less than $(\alpha - \mu)/\sigma$. Observe also that if a random variable $X$ has distribution $N(\mu, \sigma^2)$, then the derived random variable $Y = aX + b$ has a normal distribution with mean $\mu_Y = a\mu + b$ and standard deviation $\sigma_Y = |a|\sigma$, i.e., $Y$ is $N(a\mu + b, (a\sigma)^2)$.

**Example 7.6** Assume that test scores in CSC579 are normally distributed with mean $\mu = 83$ and standard deviation $\sigma = 8$. Since we wish to use the normal distribution, we shall treat the test scores as real numbers rather than integer values. We wish to compute the probability of scores between 75 and 95.

$$\begin{aligned}\text{Prob}\{75 \le X \le 95\} &= \text{Prob}\{X \le 95\} - \text{Prob}\{X \le 75\}\\ &= \text{Prob}\{Z \le (95-83)/8\} - \text{Prob}\{Z \le (75-83)/8\}\\ &= \text{Prob}\{Z \le 1.5\} - \text{Prob}\{Z \le -1\}\\ &= \Phi(1.5) - \Phi(-1)\\ &= 0.9332 - 0.1587 = 0.7745.\end{aligned}$$

Observe that if we were to replace 95 with 91 in the above analysis, we would effectively compute the probability of scores that lies within one standard deviation of the mean (i.e., scores in the range $83 \pm 8$). This yields

$$\text{Prob}\{75 \le X \le 91\} = \text{Prob}\{Z \le 1\} - \text{Prob}\{Z \le -1\} = 0.8413 - 0.1587 = 0.6826.$$

This result is not restricted to the example given above. Approximately two-thirds of the outcomes associated with normally distributed random variable lie within one standard deviation of the mean. Furthermore, since $\Phi(2) - \Phi(-2) = 0.9773 - (1 - 0.9773) = 0.9546$, approximately 95% of the outcomes of a normally distributed random variable lie within two standard deviations of the mean.

We shall now compute the moment generating function for the normal distribution. We first consider the standard normal case. Let the random variable $Z$ be normally distributed with mean 0 and variance 1. Then

$$\begin{aligned}\mathcal{M}_Z(\theta) = E(e^{\theta Z}) &= \int_{-\infty}^{\infty} e^{\theta z} f_Z(z)dz = \int_{-\infty}^{\infty} \frac{1}{\sqrt{2\pi}} e^{\theta z - z^2/2} dz\\ &= \int_{-\infty}^{\infty} \frac{1}{\sqrt{2\pi}} e^{-(z^2-2\theta z+\theta^2-\theta^2)/2} dz = \int_{-\infty}^{\infty} \frac{1}{\sqrt{2\pi}} e^{-(z-\theta)^2/2+\theta^2/2} dz\\ &= e^{\theta^2/2} \int_{-\infty}^{\infty} \frac{1}{\sqrt{2\pi}} e^{-(z-\theta)^2/2} dz = e^{\theta^2/2} \int_{-\infty}^{\infty} f_Z(z)dz = e^{\theta^2/2}.\end{aligned}$$

Now let $X$ be a normally distributed random variable with mean $\mu$ and variance $\sigma^2$ and let $Z$ be standard normal. Then $X = \mu + \sigma Z$ and

$$\begin{aligned}\mathcal{M}_X(\theta) &= E[e^{\theta X}] = E[e^{\theta(\mu+\sigma Z)}] = e^{\mu\theta} E[e^{\sigma\theta Z}] \\ &= e^{\mu\theta}\mathcal{M}_Z(\sigma\theta) = e^{\mu\theta} e^{\sigma^2\theta^2/2} = e^{\mu\theta+\sigma^2\theta^2/2}.\end{aligned}$$

Let us now consider a random variable $X$ constructed as a sum of normally distributed random variables. Recall from Equation (5.8), that if $X_1, X_2, \ldots, X_n$ are $n$ independent random variables and $X = X_1 + X_2 + \cdots + X_n$, then

$$\mathcal{M}_X(\theta) = E[e^{\theta X_1} e^{\theta X_2} \cdots e^{\theta X_n}] = \mathcal{M}_{X_1}(\theta)\mathcal{M}_{X_2}(\theta)\cdots\mathcal{M}_{X_n}(\theta).$$

If these $n$ random variables are *normally distributed*, with expectations $\mu_i$ and variances $\sigma_i^2$ respectively, then

$$\begin{aligned}\mathcal{M}_X(\theta) &= e^{\mu_1\theta+\sigma_1^2\theta^2/2}\, e^{\mu_2\theta+\sigma_2^2\theta^2/2} \cdots e^{\mu_n\theta+\sigma_n^2\theta^2/2} \\ &= e^{(\mu_1+\mu_2+\cdots+\mu_n)\theta+(\sigma_1^2+\sigma_2^2+\cdots+\sigma_n^2)\theta^2/2}.\end{aligned}$$

Setting $\mu = \mu_1 + \mu_2 + \cdots + \mu_n$ and $\sigma^2 = \sigma_1^2 + \sigma_2^2 + \cdots + \sigma_n^2$, it becomes apparent that $X$ is normally distributed with mean $\mu$ and variance $\sigma^2$.

This result concerning sums of normally distributed random variables leads us to an important application of the normal distribution. It turns out that the normal distribution, in certain cases *may* be used as an approximation to the discrete binomial distribution. If the random variable $X$ has a binomial distribution for which both $n$ and the mean value $np$ are large, then the density function of $X$ is close to that of the normal distribution. In other words, the binomial distribution becomes more and more normal as $n$ becomes large. We provide the following theorem, stated without proof.

**Theorem 7.3.1 (De Moivre-Laplace)** *Let $X_1, X_2, \ldots, X_n$ be $n$ independent and identically distributed Bernoulli random variables each taking the value 1 with probability $p$ and the value 0 with probability $q = 1 - p$. Let*

$$X = \sum_{k=1}^{n} X_k \quad \textit{and} \quad X^* = \frac{X - E[X]}{\sqrt{\text{Var}\,[X]}}.$$

*Then*

$$\lim_{n\to\infty} \text{Prob}\{a \le X^* \le b\} = \frac{1}{\sqrt{2\pi}} \int_a^b e^{-x^2/2} dx.$$

Observe that $X$ is a binomial random variable and denotes the number of successes in $n$ independent Bernoulli trials with expectation $np$ and standard deviation $\sqrt{npq}$. The random variable $X^*$ is the normalized sum that takes the values

$$\xi = \frac{k - np}{\sqrt{npq}}, \quad k = 0, 1, \ldots, n,$$

with probabilities

$$\text{Prob}\{X^* = \xi\} = C_k^n p^k q^{n-k}, \quad k = 0, 1, \ldots, n.$$

**Example 7.7** As an example of the use of the normal distribution in approximating a binomial distribution, let us compare the results obtained for binomial distribution with $n = 10$ and $p = 0.5$ and with the normal distribution with mean $\mu = np = 5$ and $\sigma = \sqrt{npq} = \sqrt{2.5}$. We shall just check a single value, Prob$\{X \leq 8\}$. For the binomial distribution we have

$$\sum_{k=0}^{8} C(10,k)0.5^k 0.5^{10-k} = 0.9893.$$

For the normal distribution, we obtain

$$\text{Prob}\{X \leq 8\} = \text{Prob}\{Z \leq (8-5)/\sqrt{2.5}\} = \text{Prob}\{Z \leq 1.8974\} = 0.9713.$$

The approximation is not particularly accurate, but the reason is that $n$ is not sufficiently large. It is possible to add a *continuity correction* which attempts to adjust for the fact that we are approximating a discrete distribution with a continuous one. However, for small values of $n$ it is relatively easy to compute binomial coefficients. This task becomes much harder for large $n$, but it is precisely these cases that the approximation by the normal distribution becomes more accurate.

One final comment on the normal distribution: the central limit theorem, which is discussed in Section 8.5, states that the sum of $n$ independent random variables tends to the normal distribution in the limit as $n \to \infty$. This is an important result in many branches of statistics and probability for it asserts that no matter which distribution the $n$ random variables have, so long as they are independent, the distribution of their sum can be approximated by the normal distribution. Thus our previous comments concerning approximating a binomial distribution with a normal distribution should not be too surprising since a binomial random variable is a sum of $n$ independent Bernoulli random variables.

## 7.4 The Gamma Distribution

The *gamma distribution* gets its name from the fact that the *gamma function* appears in the denominator of its density function. The gamma function itself is defined as an integral over the right half axis as

$$\Gamma(\alpha) = \int_0^\infty y^{\alpha-1}e^{-y}dy$$

and has a number of interesting properties, including

$$\begin{aligned}
\Gamma(1) &= 1, \\
\Gamma(1/2) &= \sqrt{\pi}, \\
\Gamma(\alpha) &= (\alpha-1)\Gamma(\alpha-1) \quad \text{for } \alpha > 1, \\
\Gamma(n) &= (n-1)! \quad \text{for } n = 1, 2, \ldots.
\end{aligned}$$

The first of these can be found from simple integration, the second by appealing to the standard normal density function, the third by using integration by parts, and the fourth from the application of the third.

The probability density function of a random variable $X$ having a gamma distribution is defined with positive parameters $\alpha, \beta > 0$ as

$$f_X(x) \equiv f(x;\beta,\alpha) = \frac{1}{\beta^\alpha \Gamma(\alpha)} x^{\alpha-1} e^{-x/\beta}, \quad x > 0, \tag{7.2}$$

and is equal to zero for other values of $x$. The parameter $\alpha$ is called the *shape* parameter since the shape of the density curve assumes different forms for different values of $\alpha$. The parameter $\beta$ is called the *scale* parameter. It does not change the shape of the curve, but scales it horizontally and vertically. The corresponding cumulative distribution function is

$$F_X(x) \equiv F(x;\beta,\alpha) = \int_0^x \frac{1}{\beta^\alpha \Gamma(\alpha)} u^{\alpha-1} e^{-u/\beta}\, du.$$

When $\alpha = 1$, we obtain

$$f_X(x) = \frac{1}{\beta} e^{-x/\beta}, \quad x > 0,$$

which is the exponential distribution seen earlier in this chapter. Setting $\alpha = n/2$ and $\beta = 2$ gives the density function for a random variable having a *chi-square* distribution, namely,

$$f_X(x) = \frac{x^{(n/2)-1} e^{-x/2}}{2^{n/2}\Gamma(n/2)}, \quad n = 1, 2, \ldots, \quad x > 0.$$

In this case, the parameter $n$ is called the *degrees of freedom* of the distribution. The gamma distribution is also related to the Erlang-$r$ distribution which we shall consider in detail in the next section. As we shall see, the Erlang-$r$ distribution is defined as

$$f_X(x) = \frac{\mu(\mu x)^{r-1} e^{-\mu x}}{(r-1)!}, \quad x > 0, \ \lambda > 0,$$

where $r = 1, 2, \ldots$ and can be derived from Equation (7.2) by making the substitutions $\alpha = r$, $\beta = 1/\mu$, and invoking the property $\Gamma(r) = (r-1)!$. We shall also see later that the Erlang distribution may be equated to a sum of independent and identically distributed exponential distributions and if these exponentials are associated with arrivals subject to a Poisson distribution, then the gamma distribution may also be associated with the Poisson distribution. The gamma distribution describes the waiting time until the $k^{\text{th}}$ Poisson arrival. In this case, the reciprocal of the shape parameter $\beta$ is called the *rate* parameter.

When $\alpha$ is a positive integer, the cumulative distribution function of a gamma random variable $X$ may be written in terms of a summation instead of an integral. Letting $k = \alpha$ be a positive integer, repeated integration by parts of the cumulative distribution function yields

$$F_X(x) \equiv F(x;\beta,k) = 1 - \sum_{i=0}^{k-1} \frac{(x/\beta)^i}{i!} e^{-x/\beta},$$

which is a Poisson distribution with parameter $x/\beta$. It follows that precomputed tables of the Poisson cumulative distribution function may be used to evaluate the CDF of a gamma random variable whose $\alpha$ parameter is a positive integer.

**Example 7.8** A random variable $X$ has a gamma distribution with parameters $\alpha \equiv k = 10$ and $\beta = 0.5$. Let us find Prob$\{X > 4\}$.

$$\text{Prob}\{X > 4\} = 1 - \int_0^4 \frac{1}{(0.5)^{10}\,\Gamma(10)} x^{10-1} e^{-x/0.5}\, dx$$
$$= 1 - F(4; 0.5, 10) = \sum_{i=0}^{9} \frac{(4/0.5)^i}{i!} e^{-4/0.5} = \sum_{i=0}^{9} \frac{8^i}{i!} e^{-8} = 0.7166.$$

The expectation and variance of a random variable $X$ having a gamma distribution with parameters $\alpha$ and $\beta$ are given by

$$E[X] = \alpha\beta \quad \text{and} \quad \text{Var}\,[X] = \alpha\beta^2,$$

respectively. Conversely, given an expectation $E[X]$ and variance $\text{Var}[X]$, the parameters of the gamma distribution that fits these characteristics are given by

$$\alpha = \frac{E[X]^2}{\text{Var}[X]} \quad \text{and} \quad \beta = \frac{\text{Var}[X]}{E[X]}.$$

We now derive the formula for the expectation. To do so, we work to make the value of the integral equal to 1 by integrating the gamma density function with parameter $\alpha + 1$ between 0 and infinity (line 2) and then apply the third property of the gamma function given previously, namely, $\alpha\Gamma(\alpha) = \Gamma(\alpha + 1)$:

$$\begin{aligned} E[X] &= \int_0^\infty x \frac{1}{\beta^\alpha \Gamma(\alpha)} x^{\alpha-1} e^{-x/\beta}\, dx = \frac{1}{\beta^\alpha \Gamma(\alpha)} \int_0^\infty x^{(1+\alpha)-1} e^{-x/\beta}\, dx \\ &= \frac{\beta^{1+\alpha}\Gamma(1+\alpha)}{\beta^\alpha \Gamma(\alpha)} \int_0^\infty \frac{1}{\beta^{1+\alpha}\Gamma(1+\alpha)} x^{(1+\alpha)-1} e^{-x/\beta}\, dx \\ &= \frac{\beta^{1+\alpha}\Gamma(1+\alpha)}{\beta^\alpha \Gamma(\alpha)} = \beta \frac{\Gamma(1+\alpha)}{\Gamma(\alpha)} = \beta \frac{\alpha\Gamma(\alpha)}{\Gamma(\alpha)} = \alpha\beta. \end{aligned}$$

The second moment, and as a result the variance, of $X$ may be found using a similar analysis. We obtain

$$E[X^2] = \alpha(\alpha + 1)\beta^2$$

and

$$\text{Var}[X] = E[X^2] - E[X]^2 = \alpha(\alpha + 1)\beta^2 - \alpha^2\beta^2 = \alpha\beta^2.$$

When $\alpha$ is a positive integer, tables of the Poisson cumulative distribution function can be used to obtain values of the cumulative distribution function of a gamma random variable. For other values of $\alpha$, other tables are available. These are not tables for the gamma distribution itself, but more usually tables of the *incomplete gamma function*. This function is the cumulative distribution of a gamma distributed random variable for which the parameter $\beta$ has the value 1. In other words, they are tables of

$$\int_0^x \frac{u^{\alpha-1} e^{-u}}{\Gamma(\alpha)}\, du,$$

and provide the value of this function for selected values of $x$ and $\alpha$. The probabilities of gamma random variables for which the parameter $\beta = 1$ may be read off directly from such tables. For other values of $\beta$ we need to work with a derived random variable. We now show that, if $X$ is a random variable whose distribution is gamma with parameters $\alpha$ and $\beta \neq 1$, then the cumulative distribution of the derived random variable $Y = X/\beta$ is an incomplete gamma function. We have

$$\text{Prob}\{Y \leq y\} = \text{Prob}\{X/\beta \leq y\} = \text{Prob}\{X \leq y\beta\} = \int_0^{y\beta} \frac{x^{\alpha-1} e^{-x/\beta}}{\beta^\alpha \Gamma(\alpha)} dx.$$

Introducing the change of variables $u = x/\beta$ (i.e., $x = \beta u$), the integral between $x = 0$ and $x = y\beta$ becomes the integral between $u = 0$ and $u = y$ and we obtain

$$\text{Prob}\{Y \leq y\} = \int_0^y \frac{(u\beta)^{\alpha-1} e^{-u}}{\beta^\alpha \Gamma(\alpha)} \beta du, = \int_0^y \frac{u^{\alpha-1} e^{-u}}{\Gamma(\alpha)} du,$$

which is the incomplete gamma function. This allows us to compute the probability that a gamma distributed random variable $X$, with parameters $\alpha$ and $\beta$, is less than some value $x$, i.e., $\text{Prob}\{X \le x\}$, since

$$\text{Prob}\,\{X \le x\} = \text{Prob}\left\{\frac{X}{\beta} \le \frac{x}{\beta}\right\} = \text{Prob}\left\{Y \le \frac{x}{\beta}\right\}.$$

**Example 7.9** Let $X$ be a gamma random variable with expectation $E[X] = 12$ and variance $\text{Var}\,[X] = 48$. Let us compute $\text{Prob}\{X > 15\}$ and $\text{Prob}\{X \le 9\}$.

We first compute the parameters $\alpha$ and $\beta$ of the gamma distribution as

$$\alpha = \frac{E[X]^2}{\text{Var}\,[X]} = \frac{12 \times 12}{48} = 3 \quad \text{and} \quad \beta = \frac{\text{Var}\,[X]}{E[X]} = \frac{48}{12} = 4.$$

Then

$$\text{Prob}\{X > 15\} = 1 - \text{Prob}\{X \le 15\} = 1 - \text{Prob}\left\{\frac{X}{\beta} \le \frac{15}{4}\right\} = 1 - \text{Prob}\{Y \le 3.75\}.$$

However, a small problem now arises for most tables of the incomplete gamma function do not provide the value of the function at $y = 3.75$. To overcome this problem we need to interpolate between the entries of the table. If our tables provide for integer values of $y$ only, then we obtain the values $\text{Prob}\{Y \le 3\} = 0.577$ and $\text{Prob}\{Y \le 4\} = 0.762$, and since 3.75 lies three-quarters of the way between 3 and 4, we estimate

$$\text{Prob}\{Y \le 3.75\} \approx 0.577 + \frac{3}{4}(0.762 - 0.577) = 0.7158,$$

which gives $\text{Prob}\{X > 15\} \approx 1 - 0.7158 = 0.2842$.

Similarly,

$$\text{Prob}\{X \le 9\} = \text{Prob}\left\{\frac{X}{\beta} \le \frac{9}{4}\right\} = \text{Prob}\{Y \le 2.25\}.$$

From the tables, we find $\text{Prob}\{Y \le 2\} = 0.323$ and $\text{Prob}\{Y \le 3\} = 0.577$ and we may estimate

$$\text{Prob}\{Y \le 2.25\} \approx 0.323 + \frac{1}{4}(0.577 - 0.323) = 0.3865.$$

Since $\alpha = 3$ is a positive integer, we could have answered this question by referring to the associated Poisson distribution, but here again, we may need to interpolate between values in our tables. Let us compute $\text{Prob}\{X > 15\}$ in this manner. We have

$$\text{Prob}\{X > 15\} = 1 - F(15; 4, 3) = \sum_{i=0}^{2} \frac{(15/4)^i}{i!} e^{-15/4},$$

a Poisson distribution with parameter 3.75. If our tables are at integer values, we obtain

$$\sum_{i=0}^{2} \frac{3^i}{i!} e^{-3} = 0.4232 \quad \text{and} \quad \sum_{i=0}^{2} \frac{4^i}{i!} e^{-4} = 0.2381$$

and interpolating between them gives

$$\text{Prob}\{X > 15\} \approx 0.4232 + \frac{3}{4}(0.2381 - 0.4232) = 0.2844.$$

Other tables provide values at intervals of 0.2 in which case we can bound 3.75 between 3.6 and 3.8. Given that

$$\sum_{i=0}^{2} \frac{(3.6)^i}{i!} e^{-3.6} = 0.303 \quad \text{and} \quad \sum_{i=0}^{2} \frac{(3.8)^i}{i!} e^{-3.8} = 0.269,$$

interpolating now gives the more accurate result

$$\text{Prob}\{X > 15\} \approx 0.303 + \frac{3}{4}(0.269 - 0.303) = 0.2775.$$

## 7.5 Reliability Modeling and the Weibull Distribution

### *The Weibull Distribution*

The Weibull distribution has the remarkable property that it can be made to assume a wide variety of shapes, a property that makes it very attractive for reliability modeling. The Weibull probability density function is described, sometimes as having one, sometimes as having two, and sometimes as having three parameters. For the *three-parameter* version, these parameters are the *scale* parameter $\eta$, the shape parameter $\beta$, and the location parameter $\xi$. The *two-parameter* version is obtained by setting the location parameter to zero, $\xi = 0$. We note that the location parameter simply shifts the position of the density curve along the $x$-axis, but does not otherwise alter its shape or scale. Other density functions could also be annotated with a location parameter, but we have refrained from doing so. The *one-parameter* Weibull density function is obtained by setting the location parameter to zero and the shape parameter to a constant value, i.e., $\beta = c$, so that the only remaining variable is the scale parameter $\eta$. The three-parameter probability density function is defined as

$$f_X(x) = \frac{\beta}{\eta}\left(\frac{x-\xi}{\eta}\right)^{\beta-1} e^{-\left(\frac{x-\xi}{\eta}\right)^{\beta}}$$

with $x \geq 0$, $\beta > 0$, $\eta > 0$, and $-\infty < \xi < \infty$. For other values of $x$, the Weibull probability density function is zero. Notice that when $\beta = 1$ the Weibull distribution reduces to the two-parameter *exponential* distribution. As may readily be verified by differentiation, the corresponding Weibull cumulative distribution function is given by

$$F_X(x) = 1 - e^{-\left(\frac{x-\xi}{\eta}\right)^{\beta}}.$$

Obviously substitution of $\xi = 0$ will yield the two-parameter Weibull probability density and distribution functions, while substitution of $\xi = 0$ and $\beta = c$ will yield the one-parameter versions. In what follows, we shall set $\xi = 0$ and consider the three-parameter version no further. The expectation and variance of a two-parameter Weibull random variable are given by

$$E[X] = \eta\,\Gamma\left(1 + \frac{1}{\beta}\right)$$

and

$$\sigma_X^2 = \eta^2\left[\Gamma\left(1 + \frac{2}{\beta}\right) - \Gamma\left(1 + \frac{1}{\beta}\right)^2\right],$$

respectively, where $\Gamma(1 + 1/\beta)$ is the gamma function evaluated at the point $1 + 1/\beta$. We now derive the expectation of a Weibull random variable, but leave a similar analysis for the variance as an exercise. We have

$$E[X] = \int_0^\infty x\frac{\beta}{\eta^\beta}x^{\beta-1}e^{-(x/\eta)^\beta}\,dx = \frac{\beta}{\eta^\beta}\int_0^\infty e^{-(x/\eta)^\beta}x^\beta\,dx.$$

Making the substitution $u = (x/\eta)^\beta$ (or $x = \eta u^{1/\beta}$), we find $x^{\beta-1}dx = (\eta^\beta/\beta)du$ and so $x^\beta dx = (\eta^{\beta+1}/\beta)u^{1/\beta}du$. Hence

$$E[X] = \frac{\beta}{\eta^\beta}\int_0^\infty e^{-u}\frac{\eta^{\beta+1}}{\beta}u^{1/\beta}du = \eta\int_0^\infty u^{1/\beta}e^{-u}du = \eta\,\Gamma\left(1 + \frac{1}{\beta}\right).$$

**Example 7.10** A random variable $X$ has a Weibull distribution with $\eta = 1/4$ and $\beta = 1/2$. Then, its probability density function is given as

$$f_X(x) = \frac{4}{2}(4x)^{-1/2} e^{-(4x)^{1/2}}, \quad x \geq 0,$$

and is equal to zero otherwise. Its cumulative distribution function is

$$F_X(x) = 1 - e^{-(4x)^{1/2}}, \quad x \geq 0,$$

and is otherwise equal to zero. Also, using the result $\Gamma(k) = (k-1)!$ for integer $k$, we have

$$\begin{aligned} E[X] &= 0.25\,\Gamma(3) = 0.25 \times 2! = 0.5, \\ \text{Var}\,[X] &= 0.25^2\,(\Gamma(5) - \Gamma(3)^2) = 0.0625(4! - 2!^2) = 1.25, \\ \text{Prob}\{X > 1\} &= 1 - \text{Prob}\{X \leq 1\} = e^{-(4)^{1/2}} = 0.1353. \end{aligned}$$

### *Reliability: Components in Series and in Parallel*

As was mentioned previously, the versatility of the Weibull distribution has resulted in it being widely applied in reliability and life-data analysis and so we take this opportunity to consider reliability modeling in more detail. Let $X$ be a random variable that denotes the time to failure of some system (component, product, or process). Then the *reliability* at any time $t$, denoted by $R_X(t)$, of that system is defined as

$$R_X(t) = \text{Prob}\{X > t\} = 1 - \text{Prob}\{X \leq t\} = 1 - F_X(t). \tag{7.3}$$

In words, the reliability of the system is simply the probability that the time to failure will exceed $t$ or equivalently, it is the probability that the system will survive until time $t$. As can be inferred from Equation (7.3), $R_X(t)$ is a monotone nonincreasing function of $t$ and is equal to zero in the limit as $t \to \infty$: no system lasts forever. In biological applications, the term *survivor function* is used in the place of reliability function. It is generally denoted by $S_X(t)$ and is defined identically to the reliability function as $S_X(t) = \text{Prob}\{X > t\} = 1 - F_X(t)$.

**Example 7.11** If the distribution of the time to failure of a system is exponentially distributed with parameter $\mu$, then the reliability function is

$$R_X(t) = 1 - \left(1 - e^{-\mu t}\right) = e^{-\mu t}.$$

For example, if the mean time to failure is five years, then $\mu = 1/5$ and the reliability after four years is

$$R_X(4) = e^{-4/5} = 0.4493.$$

This is the probability that the system is still functioning after four years. If the distribution of the time to failure is normally distributed, then we need to consult tables of standard normal distribution. For example, if the expectation of the time to failure is five years and the standard deviation is one year, then the reliability after four years is

$$R_X(4) = \text{Prob}\{X > 4\} = \text{Prob}\{Z > (4-5)/1\} = \text{Prob}\{Z > -1\} = 0.8413.$$

In practical situations, a system whose reliability is to be gauged sometimes consists of components in series or in parallel, and these individual components have their own failure distributions. In the case of components in series, the failure of any one component results in a complete system failure. This is sometimes represented graphically as a string (or series) of components connected together in a linear fashion so that if one component fails, the line is broken and the overall system fails. Let $X$ be the random variable that represents the time to failure of the overall system and let $X_i$ be the random variable representing the time to failure of component $i$, for $i = 1, 2, \ldots, n$. We shall take $F_{X_i}(t)$ to be the cumulative distribution function of $X_i$ and $R_{X_i}(t) = 1 - F_{X_i}(t)$ to be its reliability function. The system will continue to work after time $t$ if and only if each of the $n$ components continues to function after time $t$, i.e., if and only if the minimum of the $X_i$ is greater than $t$. It follows then that the reliability of the system, denoted by $R_X(t)$ is given by

$$R_X(t) = \text{Prob}\{X > t\} = \text{Prob}\{\min(X_1, X_2, \ldots, X_n) > t\}.$$

From earlier results on the minimum of a set of independent random variables, independent because we shall now assume that the components fail independently of one another, this leads to

$$R_X(t) = \prod_{i=1}^{n} \left(1 - F_{X_i}(t)\right) = \prod_{i=1}^{k} R_{X_i}(t).$$

It is apparent that the more components are added to a system in series, the more unreliable that system becomes. If the probability that component $i$ operates correctly is $p_i$, then the probability that all $n$ components function correctly is just $\prod_{i=1}^{n} p_i$.

**Example 7.12** As an example of a system of four components in series (but without the typical illustration of four in a line), consider the reliability of installing four new, identical, tires on a car. Let $X$ be a Weibull random variable that denotes the number of miles a tire will last before it fails. We shall assume that the tires fail independently and that the parameters of the Weibull distribution are $\eta = 72$ and $\beta = 2$, given that the mileage is denoted in thousands of miles. Failure of the system (of four tires) occurs at the moment the first tire fails.

The cumulative distribution function and the reliability function of a Weibull random variable with parameters $\eta = 72$ and $\beta = 2$ are respectively given by

$$F_{X_i}(x) = 1 - e^{-(x/72)^2} \quad \text{and} \quad R_{X_i}(x) = e^{-(x/72)^2}.$$

The expectation and variance are

$$E[X_i] = 72\,\Gamma(1 + 1/2) = 36\Gamma(1/2) = 36\sqrt{\pi} = 63.81,$$
$$\text{Var}\,[X_i] = 72^2\,(\Gamma(2) - \pi/4) = 72^2(0.2146) = 1{,}112.5.$$

The reliability of four tires each having this reliability function is

$$R_X(x) = \prod_{i=1}^{4} e^{-(x/72)^2} = e^{-4(x/72)^2} = e^{-(x/36)^2}.$$

Observe that this itself is a Weibull reliability function with parameter $\eta = 36$ and $\beta = 2$ and that its expectation is $E[X] = 36\Gamma(1 + 1/2) = 18\sqrt{\pi} = 31.90$, considerably less than the expectation of a single tire. The probability of driving 30,000 miles before a tire failure is

$$\text{Prob}\{X > 30\} = R_X(30) = e^{-(30/36)^2} = e^{-0.6944} = 0.4994.$$

For the case of components in parallel, as long as one of the components is working, the system has not yet failed. This is the idea behind *redundancy*: additional components are incorporated into the system to increase its reliability. Usually such systems are drawn as a collection of components, one above the other, with separate lines through each one, so that if one line disappears, the others are still present and a system-wide failure is avoided. As before, let $X_i$ be the random variable that represents the time to failure of component $i$, but now let $X$ be the random variable that describes the time to failure for the system when its $i$ components are arranged in parallel. To derive an equation for $R_X(t)$ ($= \text{Prob}\{X > t\}$, the probability that the overall system will survive until time $t$), it is more convenient to first find $\text{Prob}\{X \le t\}$ and then to compute $R_X(t)$ from the relation $R_X(t) = 1 - \text{Prob}\{X \le t\}$. For a system of independent components in parallel, $\{X \le t\}$ if and only if $\max\{X_1, X_2, \ldots, X_n\} \le t$. Therefore

$$\text{Prob}\{X \le t\} = \text{Prob}\{\max(X_1, X_2, \ldots, X_n) \le t\} = \prod_{i=1}^{n} F_{X_i}(t) = \prod_{i=1}^{n}(1 - R_{X_i}(t)),$$

and it follows that the probability of the time to failure exceeding $t$ is given by

$$R_X(t) = 1 - \prod_{i=1}^{k}\left(1 - R_{X_i}(t)\right).$$

**Example 7.13** Consider a system consisting of $n$ components in parallel. Assume that the time to failure of the $i^{\text{th}}$ component $X_i$ is exponentially distributed with parameter $\mu_i$, i.e., $\text{Prob}\{X_i \le t\} = 1 - e^{-\mu_i t}$ for $i = 1, 2, \ldots, n$. Then

$$R_X(t) = 1 - \prod_{i=1}^{n}(1 - e^{-\mu_i t}).$$

Suppose $n = 3$ and the mean number of years to failure of the three components is 1.0, 2.0, and 2.5, respectively; then the reliability functions of the three components are

$$R_1(t) = e^{-t}, \quad R_2(t) = e^{-0.5t}, \quad \text{and} \quad R_3(t) = e^{-0.4t},$$

respectively. The reliability function of the three-component parallel system is

$$R_X(t) = 1 - (1 - e^{-1.0t})(1 - e^{-0.5t})(1 - e^{-0.4t}).$$

The probability that the system lasts more than two years is

$$R_X(2) = 1 - (1 - e^{-2.0})(1 - e^{-1.0})(1 - e^{-0.8}) = 1 - (0.8647)(0.6321)(0.5507) = 0.6990.$$

Just one final point concerning this example. When all $n$ components are identically, exponentially distributed, i.e., $\mu_i = \mu$ for all $i$, then it may be shown that

$$E[X] = \left(1 + \frac{1}{2} + \cdots + \frac{1}{n}\right)\mu,$$

which shows that the mean time to failure increases as additional components are added, but with diminishing returns.

It is possible, and indeed happens often, that some components in a system are arranged in parallel while elsewhere in the same system, other components are arranged in series. This gives rise to a mixed system, but provides no additional difficulty in its analysis.

**Example 7.14** Consider a system consisting of seven components labeled $A$ through $G$ and arranged into series and parallel sections as shown in Figure 7.5. For the system to function it is necessary that $A$ and $G$ have not failed and that one of ($B$ and $C$) or $D$ or ($E$ and $F$) continues to work.

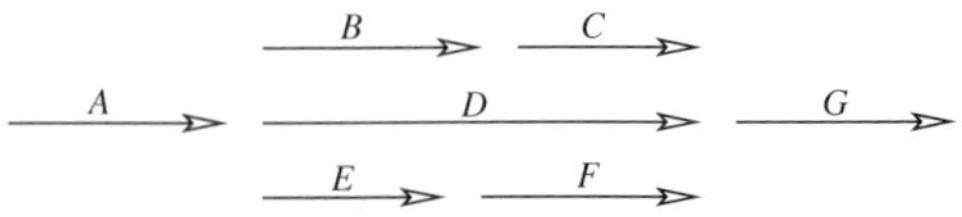

Figure 7.5. Components in series and in parallel.

Assume that the time to failure of the components are exponentially distributed and measured in thousands of hours with mean times to failure given by $1/\mu_i$, $i = 1, 2, \ldots, 7$, for components $A$ through $F$, respectively. Components $B$ and $C$ and components $E$ and $F$ both constitute a subsystem in series whose reliability functions, denoted $R_{BC}(t)$ and $R_{EF}(t)$ are given by

$$R_{BC}(t) = R_B(t)R_C(t) = e^{-\mu_2 t}e^{-\mu_3 t} = e^{-(\mu_2+\mu_3)t},$$

$$R_{EF}(t) = R_E(t)R_F(t) = e^{-\mu_5 t}e^{-\mu_6 t} = e^{-(\mu_5+\mu_6)t}.$$

Both these are arranged in parallel with component $D$ so that the reliability of the subsystem consisting of components $B$, $C$, $D$, $E$, and $F$ is given by

$$\begin{aligned} R_{BCDEF}(t) &= 1 - [1 - R_{BC}(t)]\,[1 - R_D(t)]\,[1 - R_{EF}(t)] \\ &= 1 - \left[1 - e^{-(\mu_2+\mu_3)t}\right]\left[1 - e^{-\mu_4 t}\right]\left[1 - e^{-(\mu_5+\mu_6)t}\right]. \end{aligned}$$

Finally, this subsystem is arranged in series with component $A$ before it and component $G$ after it. The reliability of the overall system is therefore given by

$$\begin{aligned} R_X(t) &= R_A(t)\,R_{BCDEF}(t)\,R_G(t) \\ &= e^{-\mu_1 t}\left(1 - \left[1 - e^{-(\mu_2+\mu_3)t}\right]\left[1 - e^{-\mu_4 t}\right]\left[1 - e^{-(\mu_5+\mu_6)t}\right]\right)e^{-\mu_7 t}. \end{aligned}$$

Given the following mean times to failure, in thousands of hours,

$$\frac{1}{\mu_1} = 4,\quad \frac{1}{\mu_2} = 1,\quad \frac{1}{\mu_3} = 0.5,\quad \frac{1}{\mu_4} = 2,\quad \frac{1}{\mu_5} = 0.25,\quad \frac{1}{\mu_6} = 0.75,\quad \frac{1}{\mu_7} = 5,$$

we find, for example, the probability that the time to system failure exceeds 1,000 hours is

$$R_X(1) = e^{-1/4}\left(1 - \left[1 - e^{-3}\right]\left[1 - e^{-1/2}\right]\left[1 - e^{-16/3}\right]\right)e^{-1/5} = 0.4004.$$

### *The Hazard or Instantaneous Failure Rate*

Related to the reliability function just discussed, the *hazard* rate is also used in reliability modeling. Alternate names for the hazard rate are *force of mortality, instantaneous failure rate, failure-rate function, intensity function*, and *conditional failure rate*. Let $X$ be the random variable that denotes the time to failure of a system and let $f_X(t)$ be its density function and $F_X(t)$, its cumulative distribution function. We have just seen that the reliability function is given by $R_X(t) = 1 - F_X(t)$. With these definitions in hand, we define the hazard rate to be

$$h_X(t) = \frac{f_X(t)}{R_X(t)} = \frac{f_X(t)}{1 - F_X(t)}. \tag{7.4}$$

The hazard rate may be viewed as the conditional probability that the system will fail in the next small interval of time, $\Delta t$, given that it has survived to time $t$. To see this, we have

$$\text{Prob}\{X \le t + \Delta t \mid X > t\} = \frac{\text{Prob}\{X \le t + \Delta t,\, X > t\}}{\text{Prob}\{X > t\}} = \frac{\int_t^{t+\Delta t} f_X(u)du}{R_X(t)} \approx h_X(t)\Delta t,$$

where we have approximated the integral over the small interval $\Delta t$ by $f_X(t)\Delta t$. The reader should clearly distinguish between $f_X(t)\Delta t$, which is (approximately) the *unconditional* probability that the system will fail in the interval $(t, t + \Delta t]$, and $h(t)\Delta t$, which is (approximately) the *conditional* probability that the system will fail in the interval $(t, t + \Delta t]$ *given* that it has survived to time $t$. It follows that $h_X(t)$ must always be greater than $f_X(t)$. This may also be seen from Equation (7.4) since the denominator in the right hand side is always less than 1.

**Example 7.15** Let $X$ be the random variable that denotes the time to failure of a system and let $f_X(t) = \lambda e^{-\lambda t}$, i.e., $X$ is exponentially distributed. Then the hazard rate is constant since

$$h_X(t) = \frac{\lambda e^{-\lambda t}}{1 - (1 - e^{-\lambda t})} = \frac{\lambda e^{-\lambda t}}{e^{-\lambda t}} = \lambda.$$

**Example 7.16** Consider a system consisting of two components in parallel and in which the lifetime of each component is independently and identically distributed with cumulative distribution functions $F_{X_1}(t) = F_{X_2}(t) = F_{X_c}(t)$, respectively. We wish to find the hazard rate of the two-component system, $h_X(t)$. The cumulative distribution function of the time to failure of the system is given by

$$F_X(t) = F_{\max}(t) = \prod_{i=1}^{n} F_{X_i}(t) = [F_{X_c}(t)]^2$$

with corresponding density function obtained by differentiation,

$$f_X(t) = 2f_{X_c}(t)F_{X_c}(t).$$

The hazard rate of the two component system is therefore given by

$$\begin{aligned} h_X(t) &= \frac{f_X(t)}{1 - F_X(t)} = \frac{2f_{X_c}(t)F_{X_c}(t)}{1 - [F_{X_c}(t)]^2} \\ &= \frac{2F_{X_c}(t)}{1 + F_{X_c}(t)} \times \frac{f_{X_c}(t)}{1 - F_{X_c}(t)} \\ &= \left(\frac{2F_{X_c}(t)}{1 + F_{X_c}(t)}\right) h_{X_c}(t). \end{aligned}$$

Since the term in parentheses can never exceed 1, the failure rate of the system is always less than that of the individual components, but since the term in parentheses tends to 1 as $t \to \infty$, the failure rate of the system tends to that of an individual component as $t \to \infty$.

Given a reliability function $R_X(t)$, the corresponding hazard rate may be found by forming the cumulative distribution of $X$ from $F_X(x) = 1 - R_X(t)$, which in turn may be used to obtain the density function $f_X(x)$, and the hazard rate determined as the ratio of $f_X(t)$ to $R_X(t)$. Similarly, given the hazard rate $h_X(t)$, the corresponding reliability function may be derived from it. We have

$$\frac{d}{dt}R_X(t) = \frac{d}{dt}(1 - F_X(t)) = -f_X(t)$$

and hence

$$-h_X(t) = \frac{-f_X(t)}{R_X(t)} = \frac{1}{R_X(t)}\frac{d}{dt}R_X(t) = \frac{d}{dt}\ln R_X(t),$$

by applying the result that if $u$ is any differentiable function of $t$, then $d/dt\,(\ln u) = (1/u)\,d/dt\,(u)$. Thus $\ln R_X(t) = -\int_0^t h_X(u)du$, and

$$R_X(t) = e^{-\int_0^t h_X(u)du}.$$

The integral in the above expression is called the *cumulative hazard* and is written as $H_X(t)$, i.e.,

$$H_X(t) = \int_0^t h_X(u)\,du \quad \text{and} \quad R_X(t) = e^{-H_X(t)}.$$

**Example 7.17** Let the hazard rate of a component be given as $h_X(t) = e^{\lambda t}$. The corresponding reliability function is obtained as

$$R_X(t) = e^{-\int_0^t e^{\lambda u}du} = e^{(1-e^{\lambda t})/\lambda}.$$

**Example 7.18** Assume the hazard rate of a component is given by $h_X(t) = 2t$. We would like to compute the probability that the component will still be functioning after two years. We first compute the reliability function as

$$R_X(t) = e^{-\int_0^t 2u\,du} = e^{-t^2}.$$

Substitution of $t = 2$ into this expression gives $R_X(2) = e^{-4} = 0.0183$, the probability that the component is still functioning after two years. Since the cumulative distribution is $F_X(t) = 1 - e^{-t^2}$, we can proceed to compute the density function as

$$f_X(t) = 2te^{-t^2}.$$

If the hazard rate is an *increasing* function of $t$, then the system wears out with time. This is an easily understood concept and applies to very many systems. On the other hand, if the hazard rate is a decreasing function of $t$, then the system becomes more reliable with time. This is a less common occurrence, but it can be observed in the early stages of some processes such as the ageing of some fine wines. Some electronic components also experience a "burn-in" period—a period over which the components experience increased reliability with time.

This brings us back to the Weibull distribution. It has the property that the hazard rate of a random variable with this distribution can be made increasing or decreasing, depending upon whether the parameter $\beta$ is greater than or less than 1. We have already seen that, when $\beta = 1$, the Weibull distribution reduces to the exponential distribution which has a constant hazard rate. Substituting for the density and distribution function of the two-parameter Weibull distribution into Equation (7.4) we obtain

$$h_X(t) = \frac{\frac{\beta}{\eta}\left(\frac{t}{\eta}\right)^{\beta-1} e^{-(t/\eta)^\beta}}{e^{-(t/\eta)^\beta}} = \frac{\beta}{\eta}\left(\frac{t}{\eta}\right)^{\beta-1}.$$

The Weibull distribution allows us to handle situations in which a system experiences an initial period during which reliability improves (the hazard rate decreases, $\beta < 1$), a generally long second period, the useful life of the system, during which the reliability is constant ($\beta = 1$), and then an end period during which the reliability decreases (the hazard rate increases, $\beta > 1$). The typical graph of the hazard rate of such a scenario exhibits a "bathtub" shape. One well-known example concerns the human lifespan: an initial period in which the small child is susceptible to childhood illnesses, a long period from youth to mature adult, and a final period as the person progresses into old age. The advantage of the Weibull distribution is that it allows us to cover all these by simply adjusting a parameter.

## 7.6 Phase-Type Distributions

The exponential distribution is very widely used in performance modeling. The reason, of course, is the exceptional mathematical tractability that flows from the memoryless property of this distribution. But sometimes mathematical tractability is not sufficient to overcome the need to model processes for which the exponential distribution is simply not adequate. This leads us to explore ways in which we can model more general distributions while maintaining some of the tractability of the exponential. This is precisely what phase-type distributions permit us to do. Additionally, as we shall see in later sections, phase-type distributions prove to be very useful when it is necessary to form a distribution having some given expectation and variance.

### *7.6.1 The Erlang-2 Distribution*

Phase-type distributions (the word "stage" is also used) get their name from the fact that they can be represented as the passage through a succession of exponential phases or stages. We begin by examining the exponential distribution which consists of a single exponential phase. To make our discussion more concrete, we shall consider a random variable $X$ which denotes the service time of a customer at a service center. This is the time that the customer spends receiving service and does not include the time, if any, that customer spends waiting for service. We assume that this service time is exponentially distributed with parameter $\mu > 0$. This is represented graphically in Figure 7.6 where the single exponential phase is represented by a circle that contains the parameter of the exponential distribution. Customers are served by entering the phase from the left, spending an amount of time that is exponentially distributed with parameter $\mu$ within the phase and then exiting to the right. We could equally well have chosen the random variable $X$ to represent the interarrival time of customers to a service center, or indeed any of the previous examples in which $X$ is exponentially distributed.

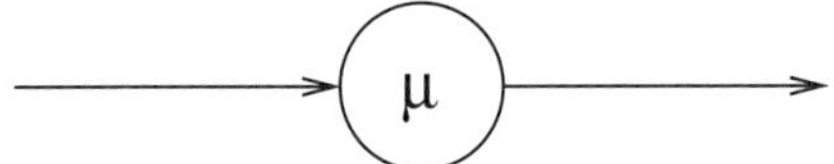


Figure 7.6. An exponential service phase.

Recall that the density function of the exponential distribution is given by

$$f_X(x) \equiv \frac{dF(x)}{dx} = \mu e^{-\mu x}, \quad x \geq 0,$$

and has expectation and variance $E[X] = 1/\mu, \ \ \sigma_X^2 = 1/\mu^2$.

Now consider what happens when the service provided to a customer can be expressed as one exponential phase followed by a second exponential phase. We represent this graphically in Figure 7.7.

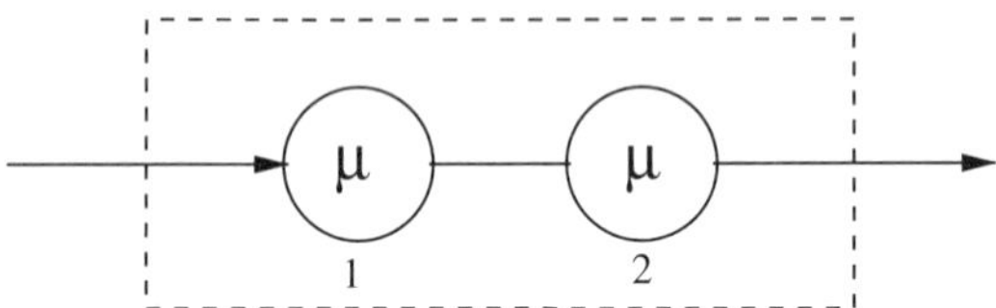


Figure 7.7. Two exponential service phases in tandem.

A customer enters the servicing process (drawn as a box) and immediately begins to receive service that is exponentially distributed with parameter $\mu$. Upon completion of this phase of its service, the customer enters the second phase and immediately begins to receive service that is once again exponentially distributed with parameter $\mu$. At the end of this second phase, the customer departs and the servicing process is now ready to begin the service of another customer. Notice that although both service phases are exponentially distributed with the same parameter, they are completely independent.

The servicing process does *not* contain two independent servers. Rather, think of it as consisting of a single service provider that operates in one phase or the other (but not both) at any given instant of time. With this interpretation it is easy to see that no more than one customer can be receiving service at any time, since a single server cannot serve two customers at the same time. A second customer cannot enter phase 1 of the servicing process until after the previous customer has exited from phase 2, nor can the customer receiving service choose to terminate its service after phase 1 and depart without receiving service at phase 2.

To analyze this situation, we shall assume that the probability density function of each of the phases is given by

$$f_Y(y) = \mu e^{-\mu y}, \quad y \geq 0,$$

with expectation and variance $E[Y] = 1/\mu$, $\sigma_Y^2 = 1/\mu^2$. A customer first spends an amount of time that is randomly chosen from $f_Y(y)$. Upon completion, the customer again spends an amount of time randomly (independently) chosen from $f_Y(y)$. After this second random interval expires, the customer departs and only then a new customer can begin to receive service. We now seek the distribution of the total time spent by a customer in the service facility. Obviously it is a random variable that is the sum of two independent and identically distributed exponential random variables. Let $Y$ be an exponentially distributed random variable with parameter $\mu$ and let $X = Y + Y$. From our previous results (see Section 4.6) concerning the convolution of two independent random variables, we have

$$\begin{aligned} f_X(x) &= \int_{-\infty}^{\infty} f_Y(y) f_Y(x-y) dy \\ &= \int_0^x \mu e^{-\mu y} \mu e^{-\mu(x-y)} dy \\ &= \mu^2 e^{-\mu x} \int_0^x dy = \mu^2 x e^{-\mu x}, \quad x \geq 0, \end{aligned}$$

and is equal to zero for $x \leq 0$. This is the probability density function for a random variable that has an Erlang-2 distribution. We shall use the notation $E_2$ to denote this distribution. The corresponding cumulative distribution function is given by

$$F_X(x) = 1 - e^{-\mu x} - \mu x e^{-\mu x} = 1 - e^{-\mu x}(1 + \mu x), \quad x \geq 0.$$

It is a useful exercise to compute the density function using Laplace transforms. We can form the Laplace transform of the overall service time probability density function as the product of the Laplace transform of the independent phases. Let the Laplace transform of the overall service time distribution be

$$\mathcal{L}_X(s) \equiv \int_0^{\infty} e^{-sx} f_X(x) dx$$

and let the Laplace transform of each of the exponential phases be

$$\mathcal{L}_Y(s) \equiv \int_0^{\infty} e^{-sy} f_Y(y) dy.$$

Then

$$\mathcal{L}_X(s) = E[e^{-sx}] = E[e^{-s(y_1+y_2)}] = E[e^{-sy_1}]E[e^{-sy_2}] = \mathcal{L}_Y(s)\mathcal{L}_Y(s) = \left(\frac{\mu}{s+\mu}\right)^2,$$

since the Laplace transform of the exponential distribution with mean $\mu$ is $\mu/(s+\mu)$. To obtain $f_X(x)$ we must now invert this transform, i.e., we need to find the function of $x$ whose transform is $\mu^2/(s+\mu)^2$. The easiest way to accomplish this is to look up tables of transform pairs and to pick off the appropriate answer. One well-known transform pair is the transform $1/(s+a)^{r+1}$ and its associated function $x^r e^{-ax}/r!$. It is usual to write such a pair as

$$\frac{1}{(s+a)^{r+1}} \iff \frac{x^r}{r!} e^{-ax}. \tag{7.5}$$

Use of this transform pair with $a = \mu$ and $r = 1$ allows us to invert $\mathcal{L}_X(s)$ to obtain

$$f_X(x) = \mu^2 x e^{-\mu x} = \mu(\mu x)e^{-\mu x}, \quad x \geq 0,$$

as before.

The expectation and higher moments may be found from the Laplace transform as

$$E[X^k] = (-1)^k \left. \frac{d^k}{ds^k} \mathcal{L}_X(s) \right|_{s=0} \quad \text{for } k = 1, 2, \ldots.$$

This allows us to find the mean as

$$E[X] = -\left. \frac{d}{ds} \mathcal{L}_X(s) \right|_{s=0} = -\mu^2 \left. \frac{d}{ds}(s+\mu)^{-2} \right|_{s=0} = \mu^2\, 2(s+\mu)^{-3}\big|_{s=0} = \frac{2}{\mu}.$$

This should hardly be surprising since the time spent in service is the sum of two independent and identically distributed random variables and hence the expected time in service is equal to the sum of the expectations of each, i.e., $E[X] = E[Y] + E[Y] = 1/\mu + 1/\mu = 2/\mu$. Similarly, we may show that the variance of an Erlang-2 random variable is given by

$$\sigma_X^2 = \sigma_Y^2 + \sigma_Y^2 = \left(\frac{1}{\mu}\right)^2 + \left(\frac{1}{\mu}\right)^2 = \frac{2}{\mu^2}.$$

**Example 7.19** Let us compare the expectation and variance of an exponential random variable with parameter $\mu$ and a two-phase Erlang-2 random variable, each of whose phases has parameter $2\mu$. We find

| | Mean | Variance |
|---|---|---|
| Exponential | $1/\mu$ | $1/\mu^2$ |
| Erlang-2 | $1/\mu$ | $1/2\mu^2$ |

Notice that the mean time in service in the single-phase and the two-phase system is the same (which is to be expected since we sped up each phase in the two-phase system by a factor of 2). However, notice the variance of the two-phase system is one-half the variance of the one-phase system. This shows that an Erlang-2 random variable has less variability than an exponentially distributed random variable when both have the same mean.

### 7.6.2 The Erlang-r Distribution

It is easy to generalize the Erlang-2 distribution to the Erlang-$r$ distribution, which we shall designate by $E_r$. We can visualize an $E_r$ distribution as a succession of $r$ identical, but independent, exponential phases with parameter $\mu$ as represented in Figure 7.8. A customer entering a service facility that consists of a single Erlang-$r$ server must spend $r$ consecutive intervals of time, each drawn from an exponential distribution with parameter $\mu$, before its service is completed. No other customer can receive service during this time, nor can the customer in service leave until all $r$ phases have been completed.

To analyze this situation, let the time a customer spends in phase $i$ be drawn from the density function $f_Y(y) = \mu e^{-\mu y}$, $y \geq 0$. Then the expectation and variance per phase are given by

$$E[Y] = 1/\mu \quad \text{and} \quad \sigma_Y^2 = 1/\mu^2, \quad \text{respectively.}$$

Since the total time spent in the service facility is the sum of $r$ independent and identically distributed random variables, the mean and variance of the overall service received by a customer

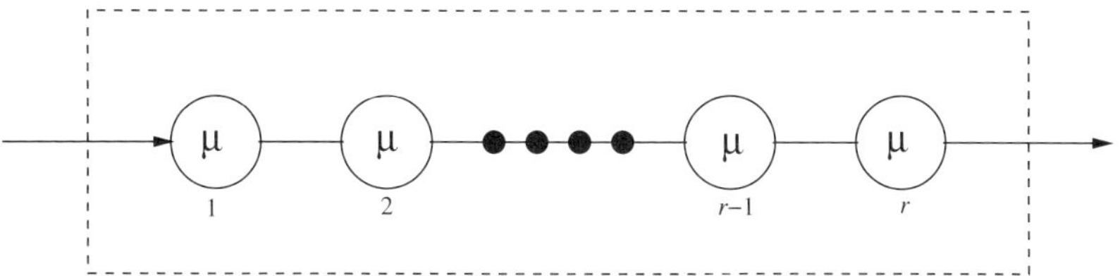


Figure 7.8. $r$ exponential service phases in tandem.

are given by

$$E[X] = r\left(\frac{1}{\mu}\right) = \frac{r}{\mu}, \quad \sigma_X^2 = r\left(\frac{1}{\mu}\right)^2 = \frac{r}{\mu^2}, \text{ respectively.}$$

The Laplace transform of the service time is

$$\mathcal{L}_X(s) = \left(\frac{\mu}{s+\mu}\right)^r,$$

which, using the same transform pair as before

$$\frac{1}{(s+a)^{r+1}} \iff \frac{x^r}{r!}e^{-ax},$$

with $a = \mu$, inverts to give

$$f_X(x) = \frac{\mu(\mu x)^{r-1}e^{-\mu x}}{(r-1)!}, \quad x \geq 0. \tag{7.6}$$

This is the Erlang-$r$ probability density function. The corresponding cumulative distribution function is given by

$$F_X(x) = 1 - e^{-\mu x}\sum_{i=0}^{r-1}\frac{(\mu x)^i}{i!}, \quad x \geq 0 \text{ and } r = 1, 2, \ldots. \tag{7.7}$$

As we mentioned in a previous chapter, an Erlang-$r$ random variable may be viewed as the time spent waiting until the $r^{\text{th}}$ arrival in a scenario in which customers arrive (alternatively events/successes occur) according to a Poisson distribution. To consider this further, consider a situation in which customers arrive according to a Poisson distribution whose rate is given by $\mu$. Let $N(t)$ be the random variable that denotes the number of arrivals during the time period $[0, t]$. Therefore $N(t)$ is a Poisson random variable with parameter $\mu t$ and so the probability of $r-1$ or fewer customer arrivals is given by the cumulative distribution of $N(t)$ as

$$\text{Prob}\{N(t) \leq r-1\} = \sum_{k=0}^{r-1}\frac{(\mu t)^k}{k!}e^{-\mu t}.$$

We wish to find the probability distribution of the random variable $W_r$, the time we need to wait until the first $r$ customers arrive. However, rather than trying to compute $\text{Prob}\{W_r \leq t\}$ directly, we shall first find $\text{Prob}\{W_r > t\}$. Notice that the arrival of the $r^{\text{th}}$ customer will be greater than $t$ if, and only if, $r-1$ or fewer customers arrive by time $t$. In other words,

$$\text{Prob}\{W_r > t\} = \text{Prob}\{N(t) \leq r-1\}.$$

It now follows that

$$\text{Prob}\{W_r \le t\} = 1 - \text{Prob}\{W_r > t\} = 1 - \sum_{k=0}^{r-1} \frac{(\mu t)^k}{k!} e^{-\mu t}, \tag{7.8}$$

which is just the Erlang-$r$ distribution.

**Example 7.20** Let $N(t)$ be a Poisson arrival process with rate $\mu = 0.5$ and let $W_4$ be the waiting time until the fourth arrival. Let us find the density function and the cumulative distribution function of $W_4$, as well as its expectation and standard deviation. We shall also find the probability that the wait is longer than 12 time units.

Replacing $\mu$ with 0.5 and $r$ with 4 in Equation (7.8), we immediately obtain the cumulative distribution as

$$F_{W_4}(t) = \text{Prob}\{W_4 \le t\} = 1 - e^{-t/2} \sum_{k=0}^{3} \frac{(t/2)^k}{k!}, \quad t \ge 0.$$

The density function may be found from this, or more simply, by substituting directly into the formula for the density of an Erlang-$r$ random variable. We find

$$f_{W_4}(t) = \frac{0.5(t/2)^3 e^{-t/2}}{3!} = \frac{1}{96} t^3 e^{-t/2}, \quad t \ge 0.$$

In this same way, the expectation and standard deviation may be obtained directly from the corresponding formulae for an Erlang-$r$ random variable. We have

$$E[W_4] = \frac{4}{0.5} = 8 \quad \text{and} \quad \sigma_{W_4} = \sqrt{\frac{4}{.25}} = 4,$$

$$\text{Prob}\{W_4 > 12\} = 1 - \text{Prob}\{W_4 \le 12\} = e^{-6} \sum_{k=0}^{3} \frac{6^k}{k!} = e^{-6}\left(1 + 6 + \frac{36}{2} + \frac{216}{6}\right) = 0.1512.$$

By differentiating $F_X(x)$ with respect to $x$ we may show that Equation (7.7) is indeed the distribution function with corresponding density function given by Equation (7.6). We have

$$\begin{aligned}
f_X(x) = \frac{d}{dx} F_X(x) &= \mu e^{-\mu x} \sum_{k=0}^{r-1} \frac{(\mu x)^k}{k!} - e^{-\mu x} \sum_{k=0}^{r-1} \frac{k(\mu x)^{k-1} \mu}{k!} \\
&= \mu e^{-\mu x} + \mu e^{-\mu x} \sum_{k=1}^{r-1} \frac{(\mu x)^k}{k!} - e^{-\mu x} \sum_{k=1}^{r-1} \frac{k(\mu x)^{k-1} \mu}{k!} \\
&= \mu e^{-\mu x} - \mu e^{-\mu x} \sum_{k=1}^{r-1} \left( \frac{k(\mu x)^{k-1}}{k!} - \frac{(\mu x)^k}{k!} \right) \\
&= \mu e^{-\mu x} \left\{ 1 - \sum_{k=1}^{r-1} \left( \frac{(\mu x)^{k-1}}{(k-1)!} - \frac{(\mu x)^k}{k!} \right) \right\} \\
&= \mu e^{-\mu x} \left\{ 1 - \left( 1 - \frac{(\mu x)^{r-1}}{(r-1)!} \right) \right\} = \frac{\mu (\mu x)^{r-1}}{(r-1)!} e^{-\mu x}.
\end{aligned}$$

To show that the area under this density curve is equal to 1, let

$$I_r = \int_0^\infty \frac{\mu^r x^{r-1} e^{-\mu x}}{(r-1)!} dx, \quad r = 1, 2, \ldots.$$

Then $I_1$ is the area under the exponential density curve and has previously been shown to be equal to 1. Now, using integration by parts $\left(\int u dv = uv - \int v du \text{ with } u = \mu^{r-1}x^{r-1}/(r-1)! \text{ and } dv = \mu e^{-\mu x}dx\right)$, we have

$$I_r = \int_0^\infty \frac{\mu^{r-1}x^{r-1}}{(r-1)!}\mu e^{-\mu x}dx = \left.\frac{-\mu^{r-1}x^{r-1}}{(r-1)!}e^{-\mu x}\right|_0^\infty + \int_0^\infty \frac{\mu^{r-1}x^{r-2}}{(r-2)!}e^{-\mu x}dx = 0 + I_{r-1},$$

and it must follow that $I_r = 1$ for all $r \geq 1$.

It is interesting to examine the squared coefficient of variation, $C_X^2$, for the family of Erlang-$r$ distributions. These are

$$C_X^2 = \frac{r/\mu^2}{(r/\mu)^2} = \frac{1}{r} < 1 \qquad \text{for } r \geq 2.$$

These coefficients of variation are less than that of the exponential distribution, which implies that Erlang random variables are "more regular" than exponential random variables. Different possible values for the squared coefficient of variation using the Erlang-$r$ distribution are

$$\frac{1}{2}, \frac{1}{3}, \frac{1}{4}, \ldots.$$

This suggests a means of approximating a *constant* distribution. By allowing $r$, the number of phases to increase and setting the parameter at each phase to be $r\mu$, we obtain distributions with increasingly smaller variance. In the limit as $r \to \infty$, the variance tends to zero (i.e., the constant distribution) while the expectation remains equal to $1/\mu$.

Mixing an Erlang-$(r-1)$ distribution with an Erlang-$r$ distribution gives a distribution whose squared coefficient of variation lies between $1/(r-1)$ and $1/r$. This is illustrated in Figure 7.9, where, with probability $\alpha$, the top series of $r-1$ exponential phases is taken and with probability $1-\alpha$, the bottom series of $r$ phases is taken. Such a distribution is denoted by $E_{r-1,r}$ and its probability density function is

$$f_X(x) = \alpha\frac{\mu(\mu x)^{r-2}e^{-\mu x}}{(r-2)!} + (1-\alpha)\frac{\mu(\mu x)^{r-1}e^{-\mu x}}{(r-1)!}, \quad x \geq 0.$$

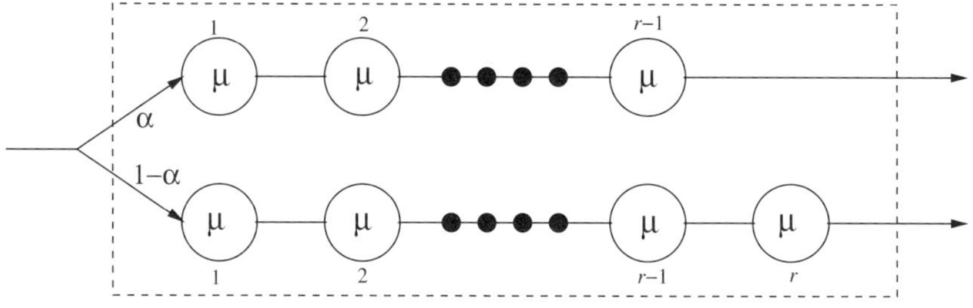


Figure 7.9. Mixed Erlang representation.

If $\alpha = 1$, then the squared coefficient of variation of the mixed distribution is $1/(r-1)$, while if $\alpha = 0$, it is equal to $1/r$. For intermediate values of $\alpha$, intermediate values of $C_X^2$ are obtained. For a given expectation, $E[X]$, and squared coefficient of variation $C_X^2$ lying in the interval $[1/r,\ 1/(r-1)]$, the values of $\alpha$ and $\mu$ to be used are given by the formulae

$$\alpha = \frac{1}{1+C_X^2}\left(rC_X^2 - \sqrt{r(1+C_X^2) - r^2C_X^2}\right) \quad \text{and} \quad \mu = \frac{r-\alpha}{E[X]}. \tag{7.9}$$

### 7.6.3 The Hypoexponential Distribution

As we have just seen, it is possible to use a sequence of exponential phases, an Erlang-$r$ distribution, to model random variables that have less variability than an exponentially distributed random variable while maintaining the desirable mathematical properties of the exponential distribution. Any given value $\mu$ for the mean can be obtained by an appropriate choice for the mean of each phase: with $r$ phases, each should have mean equal to $r\mu$. However, with a single series of phases, the choice of variance is limited. Only values that lead to squared coefficients of variation equal to $1/i$ for integer $i$, are possible. We saw one approach to overcome this difficulty lies in a mixture of Erlang distributions. An alternative approach is to modify the phase-type representation of the Erlang-$r$ distribution by permitting each phase to have a different parameter; i.e., we allow phase $i$ to provide service that is exponentially distributed with parameter $\mu_i$, which may be different for each phase. This leads to the *hypoexponential* distribution and is illustrated in Figure 7.10.

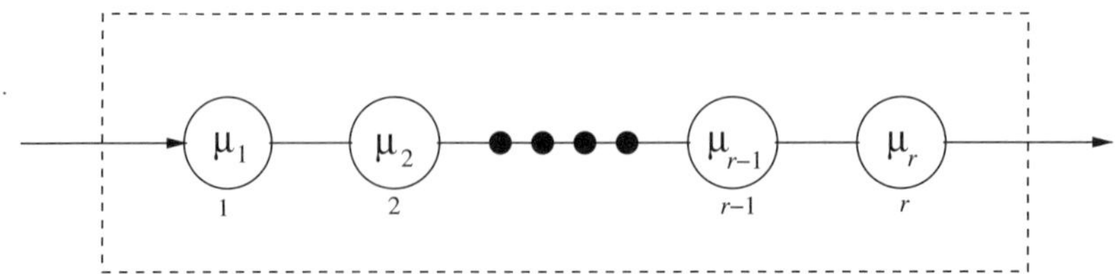


Figure 7.10. Hypoexponential distribution.

Consider first the case of two exponential phases. Let $Y_1$ and $Y_2$ be two independent random variables that are exponentially distributed with parameters $\mu_1$ and $\mu_2$, respectively, and let $X = Y_1 + Y_2$. Then the probability density function of $X$ is obtained from the convolution of the two exponential distributions. We have

$$\begin{aligned} f_X(x) &= \int_{-\infty}^{\infty} f_{Y_1}(y) f_{Y_2}(x-y)dy \\ &= \int_0^x \mu_1 e^{-\mu_1 y}\, \mu_2 e^{-\mu_2(x-y)}dy \\ &= \mu_1\mu_2 e^{-\mu_2 x}\int_0^x e^{-(\mu_1-\mu_2)y}dy \\ &= \frac{\mu_1\mu_2}{\mu_1-\mu_2}\left(e^{-\mu_2 x} - e^{-\mu_1 x}\right) \end{aligned}$$

for $x \geq 0$ and is 0 otherwise. Its corresponding cumulative distribution function is given by

$$F_X(x) = 1 - \frac{\mu_2}{\mu_2-\mu_1}e^{-\mu_1 x} + \frac{\mu_1}{\mu_2-\mu_1}e^{-\mu_2 x}, \quad x \geq 0.$$

Its expectation, variance, and squared coefficient of variation are respectively given by

$$E[X] = \frac{1}{\mu_1} + \frac{1}{\mu_2}, \quad \text{Var}\,[X] = \frac{1}{\mu_1^2} + \frac{1}{\mu_2^2}, \quad \text{and} \quad C_X^2 = \frac{\mu_1^2+\mu_2^2}{(\mu_1+\mu_2)^2} < 1,$$

and its Laplace transform by

$$\mathcal{L}_X(s) = \left(\frac{\mu_1}{s+\mu_1}\right)\left(\frac{\mu_2}{s+\mu_2}\right).$$

For more than two phases, the analysis is more tedious. The Laplace transform for the probability density function of an $r$-phase hypoexponential random variable $X$ is

$$\mathcal{L}_X(s) = \left(\frac{\mu_1}{s+\mu_1}\right)\left(\frac{\mu_2}{s+\mu_2}\right)\cdots\left(\frac{\mu_r}{s+\mu_r}\right).$$

The density function $f_X(x)$ is the convolution of $r$ exponential densities each with its own parameter $\mu_i$ and is given by

$$f_X(x) = \sum_{i=1}^{r} \alpha_i \mu_i e^{-\mu_i x}, \quad x > 0, \quad \text{where } \alpha_i = \prod_{j=1,\, j\neq i}^{r} \frac{\mu_j}{\mu_j - \mu_i},$$

and is equal to zero otherwise. The expectation, variance, and squared coefficient of variation are as follows:

$$E[X] = \sum_{i=1}^{r} \frac{1}{\mu_i}, \quad \text{Var}\,[X] = \sum_{i=1}^{r} \frac{1}{\mu_i^2}, \quad \text{and} \quad C_X^2 = \frac{\sum_i 1/\mu_i^2}{\left(\sum_i 1/\mu_i\right)^2} \le 1.$$

To show that $C_X^2$ cannot be greater than 1, we use the fact that for real $a_i \ge 0$, $\sum_i a_i^2 \le \left(\sum_i a_i\right)^2$, since the right-hand side contains the left-hand side plus the sum of all the nonnegative cross-terms. Now taking $a_i = 1/\mu_i$ implies that $C_X^2 \le 1$.

**Example 7.21** Consider a random variable modeled as three consecutive exponential phases with parameters $\mu_1 = 1$, $\mu_2 = 2$, and $\mu_3 = 3$. Let us find the expectation, variance, and squared coefficient of variation of $X$, as well as its probability density function.

The expectation of $X$ is just equal to the sum of the expectations of each phase, and since the three exponential phase are independent of each other, the variance is also equal to the sum of the variances of each phase. Hence

$$E[X] = \sum_{i=1}^{3} \frac{1}{\mu_i} = \frac{1}{1} + \frac{1}{2} + \frac{1}{3} = \frac{11}{6},$$

$$\text{Var}\,[X] = \sum_{i=1}^{3} \frac{1}{\mu_i^2} = \frac{1}{1} + \frac{1}{4} + \frac{1}{9} = \frac{49}{36}.$$

The squared coefficient of variation is

$$C_X^2 = \frac{49/36}{121/36} = \frac{49}{121} = 0.4050.$$

$$\alpha_1 = \prod_{j=1,\, j\neq i}^{r} \frac{\mu_j}{\mu_j - \mu_1} = \frac{\mu_2}{\mu_2 - \mu_1} \times \frac{\mu_3}{\mu_3 - \mu_1} = \frac{2}{1} \times \frac{3}{2} = 3,$$

$$\alpha_2 = \prod_{j=1,\, j\neq i}^{r} \frac{\mu_j}{\mu_j - \mu_2} = \frac{\mu_1}{\mu_1 - \mu_2} \times \frac{\mu_3}{\mu_3 - \mu_2} = \frac{1}{-1} \times \frac{3}{1} = -3,$$

$$\alpha_3 = \prod_{j=1,\, j\neq i}^{r} \frac{\mu_j}{\mu_j - \mu_3} = \frac{\mu_1}{\mu_1 - \mu_3} \times \frac{\mu_2}{\mu_2 - \mu_3} = \frac{1}{-2} \times \frac{2}{-1} = 1.$$

It follows then that

$$f_X(x) = \sum_{i=1}^{3} \alpha_i \mu_i e^{-\mu_i x} = 3e^{-x} - 6e^{-2x} + 3e^{-3x}, \quad x > 0.$$

Neither the hypoexponential distribution nor a mixture of Erlang distributions can be used if coefficients of variation greater than 1 are required. To proceed in this direction, we turn to phases in parallel rather than phases in series as tried to this point.

### 7.6.4 The Hyperexponential Distribution

Our goal now is to find a phase-type arrangement that gives larger coefficients of variation than the exponential. Consider the configuration presented in Figure 7.11, in which $\alpha_1$, is the probability that the upper phase is taken and $\alpha_2 = 1 - \alpha_1$ the probability that the lower phase is taken. If such a distribution is used to model a service facility, then a customer entering service will, with probability $\alpha_1$, receive service that is exponentially distributed with parameter $\mu_1$ and then exit the server, or else, with probability $\alpha_2$ receive service that is exponentially distributed with parameter $\mu_2$ and then exit the server. Once again, only one customer can be in the process of receiving service at any one time, i.e., both phases cannot be active at the same time.

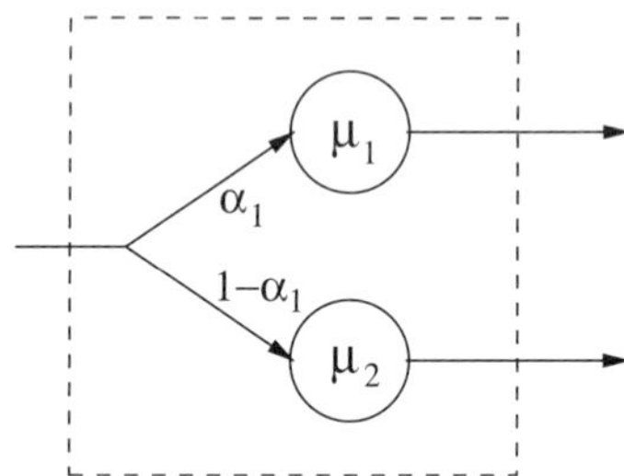


Figure 7.11. Two exponential phases in parallel.

The density function of the service time received by a customer is given by

$$f_X(x) = \alpha_1\mu_1 e^{-\mu_1 x} + \alpha_2\mu_2 e^{-\mu_2 x}, \quad x \geq 0,$$

and is otherwise equal to zero, while the cumulative distribution function is

$$F_X(x) = \alpha_1(1 - e^{-\mu_1 x}) + \alpha_2(1 - e^{-\mu_2 x}), \quad x \geq 0.$$

Its Laplace transform is

$$\mathcal{L}_X(s) = \alpha_1 \frac{\mu_1}{s + \mu_1} + \alpha_2 \frac{\mu_2}{s + \mu_2}.$$

The first and second moments are respectively given by

$$E[X] = \frac{\alpha_1}{\mu_1} + \frac{\alpha_2}{\mu_2} \quad \text{and} \quad E[X^2] = \frac{2\alpha_1}{\mu_1^2} + \frac{2\alpha_2}{\mu_2^2}.$$

This allows us to compute the variance as $E[X^2] - (E[X])^2$ and from it the squared coefficient of variation as

$$C_X^2 = \frac{E[X^2] - (E[X])^2}{(E[X])^2} = \frac{E[X^2]}{(E[X])^2} - 1 = \frac{2\alpha_1/\mu_1^2 + 2\alpha_2/\mu_2^2}{(\alpha_1/\mu_1 + \alpha_2/\mu_2)^2} - 1 \geq 1.$$

The proof that the squared coefficient of variation is not less than 1 is given in the more general context of multiple parallel phases, which is discussed next.

**Example 7.22** Let us find the expectation, standard deviation, and squared coefficient of variation of a two-phase hyperexponential random variable $X$ with parameters $\alpha_1 = 0.4$, $\mu_1 = 2$, and

$\mu_2 = 1/2$. We have, directly from the formulae,

$$E[X] = \frac{0.4}{2} + \frac{0.6}{0.5} = 1.40, \quad E[X^2] = \frac{0.8}{4} + \frac{1.2}{0.25} = 5,$$
$$\sigma_X = \sqrt{5 - 1.4^2} = \sqrt{3.04} = 1.7436,$$
$$C_X^2 = \frac{5}{1.4^2} - 1 = 2.5510 - 1.0 = 1.5510,$$

and we see that the squared coefficient of variation is greater than 1.

We now extend this phase-type service facility by increasing the number of phases that it contains. With $r$ parallel phases and branching probabilities $\sum_{i=1}^{r} \alpha_i = 1$, we have the situation illustrated in Figure 7.12. The density function and Laplace transform for this phase-type distribution are given by

$$f_X(x) = \sum_{i=1}^{r} \alpha_i \mu_i e^{-\mu_i x}, \quad x \geq 0, \quad \text{and} \quad \mathcal{L}_X(s) = \sum_{i=1}^{r} \alpha_i \frac{\mu_i}{s + \mu_i}, \quad \text{respectively.}$$

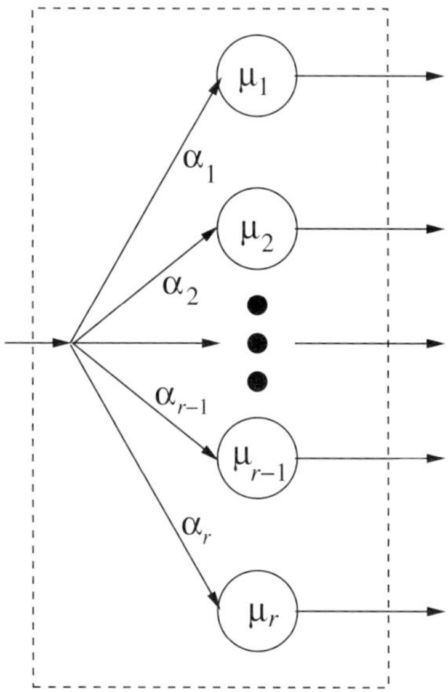


Figure 7.12. Multiple exponential phases in parallel.

This distribution is referred to as the *hyperexponential* distribution. Its first two moments are

$$E[X] = \sum_{i=1}^{r} \frac{\alpha_i}{\mu_i} \quad \text{and} \quad E[X^2] = 2\sum_{i=1}^{r} \frac{\alpha_i}{\mu_i^2},$$

which allows us to form its squared coefficient of variation as

$$C_X^2 = \frac{E[X^2]}{(E[X])^2} - 1 = \frac{2\sum_{i=1}^{r} \alpha_i/\mu_i^2}{\left(\sum_{i=1}^{r} \alpha_i/\mu_i\right)^2} - 1.$$

We now show that this squared coefficient of variation is greater than or equal to 1. Observe that it suffices to show that

$$\left(\sum_{i=1}^{r} \alpha_i/\mu_i\right)^2 \le \sum_{i=1}^{r} \alpha_i/\mu_i^2.$$

This requires the use of the Cauchy-Schwartz inequality, which for real $a_i$ and $b_i$, states that

$$\left(\sum_i a_i b_i\right)^2 \le \left(\sum_i a_i^2\right)\left(\sum_i b_i^2\right).$$

Substituting $a_i = \sqrt{\alpha_i}$ and $b_i = \sqrt{\alpha_i}/\mu_i$ implies that

$$\begin{aligned}
\left(\sum_i \frac{\alpha_i}{\mu_i}\right)^2 &= \left(\sum_i \sqrt{\alpha_i}\frac{\sqrt{\alpha_i}}{\mu_i}\right)^2 \\
&\le \sum_i \sqrt{\alpha_i}^2 \sum_i \left(\frac{\sqrt{\alpha_i}}{\mu_i}\right)^2, \qquad \text{using Cauchy-Schwartz} \\
&= \left(\sum_i \alpha_i\right)\left(\sum_i \frac{\alpha_i}{\mu_i^2}\right) = \sum_i \frac{\alpha_i}{\mu_i^2}, \qquad \text{since } \sum_i \alpha_i = 1.
\end{aligned}$$

Therefore $C_X^2 \ge 1$.

### 7.6.5 The Coxian Distribution

It is possible to construct phase-type distributions that are a mixture of hypoexponential and hyperexponential distributions to obtain increasingly complex representations. With the introduction of the mathematically expedient notion of "complex probabilities" and "complex rates," Cox in 1955 showed how any distribution having a rational Laplace transform could be represented by a sequence of exponential phases. The sequence of phases could be arranged one after the other in series formation, with the provision of permitting termination after the completion of any phase. Thus, after receiving service at phase $i$, the process, with probability $\alpha_i$, continues on to phase $i+1$ to continue service there, or with probability $1-\alpha_i$, the service process is terminated. This is represented graphically in Figure 7.13.

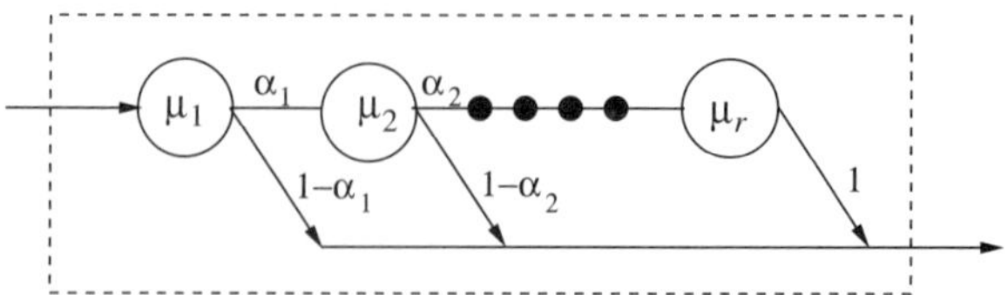


Figure 7.13. The Coxian distribution.

It is apparent from Figure 7.13 that with probability $p_1 = 1 - \alpha_1$ the service process terminates after the first phase of service; with probability $p_2 = \alpha_1(1 - \alpha_2)$ it completes the first two service phases and then terminates, and so on. Continuing with this line of reasoning, it becomes apparent that the probability of only the first $k$ service phases being completed before the process is terminated is given by $p_k = (1 - \alpha_k)\prod_{i=1}^{k-1} \alpha_i$. It follows then that a Coxian distribution may be represented as a probabilistic choice from among $r$ *hypoexponential* distributions as shown in Figure 7.14.

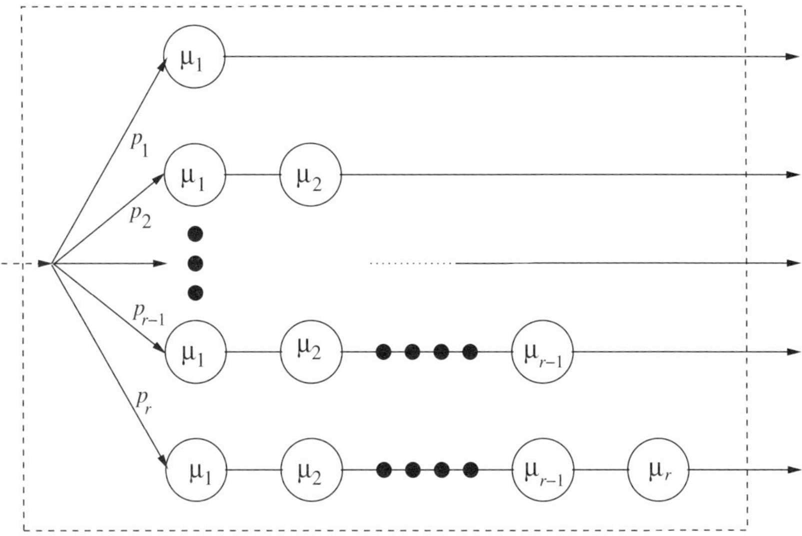


Figure 7.14. The extended Erlang distribution.

The expectation of a Coxian random variable may be found by referring to either Figure 7.13 or Figure 7.14. From Figure 7.13, we see that the first phase is always executed and has expectation $E[X_1] = 1/\mu_1$; the second phase has expectation $E[X_2] = 1/\mu_2$ but is only executed with probability $\alpha_1$; and so on. Phase $k > 1$, which has expectation $1/\mu_k$, is executed with probability $\prod_{j=1}^{k-1} \alpha_j$. Since the expectation of a sum is equal to the sum of the expectations, we find the expectation of a Coxian-$r$ random variable to be

$$E[X] = \frac{1}{\mu_1} + \frac{\alpha_1}{\mu_2} + \frac{\alpha_1\alpha_2}{\mu_3} + \cdots + \frac{\alpha_1\alpha_2\cdots\alpha_{r-1}}{\mu_r} = \sum_{k=1}^{r} \frac{A_k}{\mu_k},$$

where $A_1 = 1$ and, for $k > 1$, $A_k = \prod_{j=1}^{k-1} \alpha_j$. The case of a Cox-2 random variable is especially important. Its expectation is

$$E[X] = \frac{1}{\mu_1} + \alpha\frac{1}{\mu_2} = \frac{\mu_2 + \alpha\mu_1}{\mu_1\mu_2}, \tag{7.10}$$

We now derive a formula for the squared coefficient of variation for a Cox-2 random variable $X$, with parameters $\mu_1$, $\mu_2$, and $\alpha$. We begin with the Laplace transform from which we compute the second moment, and from this, the variance and the squared coefficient of variation. We have

$$\mathcal{L}_X(s) = (1-\alpha)\frac{\mu_1}{s+\mu_1} + \alpha\frac{\mu_1}{s+\mu_1}\frac{\mu_2}{s+\mu_2},$$

$$
\begin{aligned}
E[X^2] &= (-1)^2 \frac{d^2}{ds^2} \left( \frac{(1-\alpha)\mu_1}{s+\mu_1} + \frac{\alpha\mu_1\mu_2}{(s+\mu_1)(s+\mu_2)} \right)\Bigg|_{s=0} \\
&= \frac{d}{ds} \left( \frac{-(1-\alpha)\mu_1}{(s+\mu_1)^2} + \alpha\mu_1\mu_2 \left[ \frac{-1}{(s+\mu_1)(s+\mu_2)^2} + \frac{-1}{(s+\mu_1)^2(s+\mu_2)} \right] \right)\Bigg|_{s=0} \\
&= \left( \frac{2(1-\alpha)\mu_1}{(s+\mu_1)^3} + \frac{2\alpha\mu_1\mu_2}{(s+\mu_1)(s+\mu_2)^3} + \frac{\alpha\mu_1\mu_2}{(s+\mu_1)^2(s+\mu_2)^2} \right. \\
&\quad \left. + \frac{2\alpha\mu_1\mu_2}{(s+\mu_1)^3(s+\mu_2)} + \frac{\alpha\mu_1\mu_2}{(s+\mu_1)^2(s+\mu_2)^2} \right)\Bigg|_{s=0} \\
&= \frac{2(1-\alpha)}{\mu_1^2} + \frac{2\alpha}{\mu_2^2} + \frac{\alpha}{\mu_1\mu_2} + \frac{2\alpha}{\mu_1^2} + \frac{\alpha}{\mu_1\mu_2} \\
&= \frac{2}{\mu_1^2} + \frac{2\alpha}{\mu_2^2} + \frac{2\alpha}{\mu_1\mu_2}, \\
\text{Var}\,[X] &= \left( \frac{2}{\mu_1^2} + \frac{2\alpha}{\mu_2^2} + \frac{2\alpha}{\mu_1\mu_2} \right) - \left( \frac{1}{\mu_1} + \frac{\alpha}{\mu_2} \right)^2 \\
&= \frac{2\mu_2^2 + 2\alpha\mu_1^2 + 2\alpha\mu_1\mu_2}{\mu_1^2\mu_2^2} - \frac{(\mu_2+\alpha\mu_1)^2}{\mu_1^2\mu_2^2} \\
&= \frac{\mu_2^2 + 2\alpha\mu_1^2 - \alpha^2\mu_1^2}{\mu_1^2\mu_2^2} = \frac{\mu_2^2 + \alpha\mu_1^2(2-\alpha)}{\mu_1^2\mu_2^2}.
\end{aligned}
$$

Finally, we obtain the coefficient as $C_X^2 = \text{Var}\,[X]/E[X]^2$, although we also point out that we could have formed it without first computing $\text{Var}\,[X]$, by using the formulation $C_X^2 = E[X^2]/E[X]^2 - 1$,

$$
C_X^2 = \frac{\text{Var}\,[X]}{E[X]^2} = \frac{\mu_2^2 + \alpha\mu_1^2(2-\alpha)}{\mu_1^2\mu_2^2} \times \frac{\mu_1^2\mu_2^2}{(\mu_2+\alpha\mu_1)^2} = \frac{\mu_2^2 + \alpha\mu_1^2(2-\alpha)}{(\mu_2+\alpha\mu_1)^2}. \tag{7.11}
$$

**Example 7.23** The expectation, second moment, variance, and squared coefficient of variation of a Coxian-2 random variable with parameters $\mu_1 = 2$, $\mu_2 = 0.5$, and $\alpha = 0.25$, are

$$
\begin{aligned}
E[X] &= \frac{1}{\mu_1} + \frac{\alpha}{\mu_2} = \frac{1}{2} + \frac{1/4}{1/2} = 1, \\
E[X^2] &= \frac{2}{\mu_1^2} + \frac{2\alpha}{\mu_2^2} + \frac{2\alpha}{\mu_1\mu_2} = \frac{2}{4} + \frac{1/2}{1/4} + \frac{1/2}{1} = 3, \\
\text{Var}\,[X] &= E[X^2] - E[X]^2 = 3 - 1 = 2, \\
C_X^2 &= \text{Var}\,[X]/E[X]^2 = 2.
\end{aligned}
$$

### *7.6.6 General Phase-Type Distributions*

Phase-type distributions need not be restricted to linear arrangements, but can be very general indeed. Consider a collection of $k$ phases such that the random variable that describes the time spent in phase $i$, $i = 1, 2, \ldots, k$, is exponentially distributed with parameter $\mu_i$. It is possible to envisage a general phase type probability distribution defined on these phases as the total time spent moving in some probabilistic fashion among the $k$ different phases. It suffices to specify:

- The initial probabilities, i.e., the probabilities $\sigma_i$, $i = 1, 2, \ldots, k$ that a given phase $i$ is the first phase entered: $\sum_{i=1}^{k} \sigma_i = 1$.
- The *routing* probabilities $r_{ij}$, $i, j = 1, 2, \ldots, k$, which give the probabilities that, after spending an exponentially distributed amount of time with mean $1/\mu_i$ in phase $i$, the next phase entered is phase $j$. For all $i$, $r_{ii} < 1$ while for at least one value of $i$, $\sum_{j=1}^{k} r_{ij} < 1$.
- The terminal probabilities, i.e., the probabilities $\eta_i$, $i = 1, 2, \ldots, k$, that the process terminates on exiting from phase $i$. At least one must be strictly positive. For all $i = 1, 2, \ldots, k$, we must have $\eta_i + \sum_{j=1}^{k} r_{ij} = 1$: on exiting from phase $i$, either another phase $j$ is entered (with probability $r_{ij}$) or the process terminates (with probability $\eta_i$).

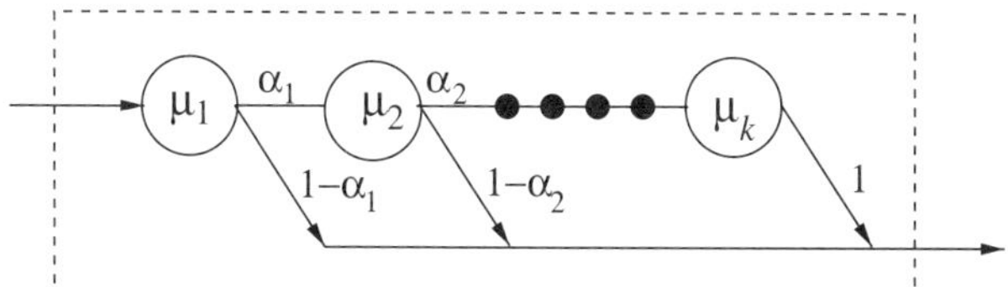


Figure 7.15. The Coxian distribution, again.

**Example 7.24** Consider the Coxian distribution described earlier and shown again in Figure 7.15. The initial probabilities are (in vector form) $\sigma = (1, 0, 0, \ldots, 0)$; the terminal probabilities are $\eta = (1 - \alpha_1, 1 - \alpha_2, \ \ldots, \ 1 - \alpha_{k-1}, 1)$ and the probabilities $r_{ij}$ are the elements of the matrix

$$R = \begin{pmatrix} 0 & \alpha_1 & 0 & \cdots & 0 \\ 0 & 0 & \alpha_2 & \cdots & 0 \\ \vdots & \vdots & & \ddots & \vdots \\ 0 & 0 & 0 & & \alpha_{k-1} \\ 0 & 0 & 0 & \cdots & 0 \end{pmatrix}.$$

The vectors $\sigma$ and $\eta$ and the matrix $R$ together with the parameters of the exponential distributions completely characterize a Coxian distribution.

**Example 7.25** A general phase-type distribution having four phases and exponential parameters $\mu_i, i = 1, 2, 3, 4$; vectors $\sigma = (0, .4, 0, .6)$ and $\eta = (0, 0, .1, 0)$ and routing probability matrix

$$R = \begin{pmatrix} 0 & .5 & 0 & .5 \\ 0 & 0 & 1 & 0 \\ .2 & 0 & 0 & .7 \\ 1 & 0 & 0 & 0 \end{pmatrix}$$

begins life in phase 2 with probability .4 or in phase 4 with probability .6, and moves among the phases until it eventually departs from phase 3. It has the graphical interpretation of Figure 7.16.

General phase-type distributions frequently have an extra phase appended to represent the exterior into which the process finally departs. Once in this phase, called a sink or an absorbing phase, the process remains there forever. The phase-type probability distribution is that of the random variable which represents the time to absorption, from some initial phase(s), into this sink. In this case, it is usual to combine the parameters of the exponential distributions of the phases and the routing probabilities into a single matrix $Q$ whose elements $q_{ij}$ give the rate of transition

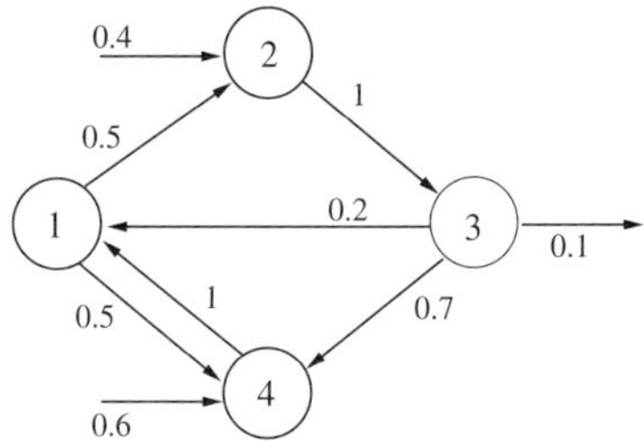


Figure 7.16. A general phase-type distribution.

from phase $i$ to some other phase $j$, i.e., $q_{ij} = \mu_i r_{ij}$. It is also usual to set the diagonal element in each row of this matrix equal to the negated sum of the off-diagonal elements in that row, i.e., $q_{ii} = -\sum_{j \neq i} q_{ij}$. Thus the sum of all the elements in any row is zero. As we shall see later, this matrix together with an initial starting vector describes a continuous-time Markov chain with a single absorbing state. The matrix is called the *transition-rate matrix* or *infinitesimal generator* of the Markov chain.

**Example 7.26** For the Coxian distribution of Example 7.24,

$$\text{Initial distribution} : (1, 0, 0, \ldots, 0 \mid 0) = (\sigma \mid 0),$$

$$Q = \left(\begin{array}{cccccc|c} -\mu_1 & \mu_1\alpha_1 & 0 & \cdots & 0 & & \mu_1(1-\alpha_1) \\ 0 & -\mu_2 & \mu_2\alpha_2 & \cdots & 0 & & \mu_2(1-\alpha_2) \\ \vdots & \vdots & & \ddots & \vdots & & \\ 0 & 0 & 0 & & \mu_{k-1}\alpha_{k-1} & & \mu_{k-1}(1-\alpha_{k-1}) \\ 0 & 0 & 0 & \cdots & -\mu_k & & \mu_k \\ \hline 0 & 0 & 0 & \cdots & 0 & & 0 \end{array}\right).$$

For the general phase-type distribution of Example 7.25,

$$\text{Initial distribution: } (0, .4, 0, .6 \mid 0) = (\sigma \mid 0),$$

$$Q = \left(\begin{array}{cccc|c} -\mu_1 & .5\mu_1 & 0 & .5\mu_1 & 0 \\ 0 & -\mu_2 & \mu_2 & 0 & 0 \\ .2\mu_3 & 0 & -\mu_3 & .7\mu_3 & .1\mu_3 \\ \mu_4 & 0 & 0 & -\mu_4 & 0 \\ \hline 0 & 0 & 0 & 0 & 0 \end{array}\right).$$

To proceed, let us consider an arbitrary phase-type distribution consisting of $k$ phases plus the absorbing phase. Such a distribution is defined by a transition rate matrix $Q$ and an initial probability distribution $\sigma'$. Let us partition $Q$ and $\sigma'$ as follows:

$$Q = \begin{pmatrix} S & S^0 \\ 0 & 0 \end{pmatrix}, \quad \sigma' = (\sigma \mid \sigma_{k+1}) = (\sigma_1, \sigma_2, \ldots, \sigma_k \mid \sigma_{k+1}).$$

Here $S$ is a $k \times k$ matrix that contains the transition rates among the nonabsorbing states and with diagonal elements equal to the negated sum across the rows of $Q$; $S^0$ is a column vector of length $k$ whose $i^{\text{th}}$ component gives the rate at which the process enters the absorbing state from state $i$; $\sigma'_i$, $i = 1, 2, \ldots, k+1$, is the probability that the starting phase is phase $i$. Notice that it is possible to have $\sigma_{k+1} > 0$, i.e., for the process to begin in the absorbing phase, phase $k+1$.

The probability distribution of the phase-type random variable defined by $Q$ and $\sigma'$ is identical to the probability distribution of the time to absorption into the sink state, state $k+1$. It is given by

$$F_X(x) = 1 - \sigma e^{Sx} e, \quad x \geq 0,$$

where $e$ is a column vector of length $k$ whose elements are all equal to 1. Its probability density function has a jump of magnitude $\sigma_{k+1}$ at the origin. Its density on $(0, \infty)$ is

$$f_X(x) = F'_X(x) = \sigma e^{Sx} S^0,$$

and its moments, which are all finite, are given by

$$E[X^j] = (-1)^j j! \sigma S^{-j} e, \quad j = 1, 2, \ldots .$$

In particular, the expectation of a phase-type distribution is just the expected time to absorption in the Markov chain. We have

$$E[X] = -\sigma S^{-1} e \equiv xe,$$

where $x = -\sigma S^{-1}$ is a row vector of length $k$. Since the effect of multiplying $x$ with $e$ is to sum the components of $x$, we may take $x_i$, $i = 1, 2, \ldots, k$, to be the mean time spent in phase $i$ prior to absorption. The proof of these results, obtained in the Markov chain context, can be found in Chapters 9 and 10.

**Example 7.27** Consider a Coxian distribution with four phases and rates $\mu_1 = 1$, $\mu_2 = 2$, $\mu_3 = 4$, $\mu_4 = 8$. On completion of phase $i = 1, 2, 3$, the process proceeds to phase $i + 1$ with probability .5 or enters the sink phase with probability .5. This allows us to write

$$\sigma' = (1,\ 0,\ 0,\ 0,\ |\ 0) = (\sigma \mid 0),$$

$$Q = \left(\begin{array}{cccc|c} -\mu_1 & .5\mu_1 & 0 & 0 & .5\mu_1 \\ 0 & -\mu_2 & .5\mu_2 & 0 & .5\mu_2 \\ 0 & 0 & -\mu_3 & .5\mu_3 & .5\mu_3 \\ 0 & 0 & 0 & -\mu_4 & \mu_4 \\ \hline 0 & 0 & 0 & 0 & 0 \end{array}\right) = \left(\begin{array}{cccc|c} -1 & .5 & 0 & 0 & .5 \\ 0 & -2 & 1 & 0 & 1 \\ 0 & 0 & -4 & 2 & 2 \\ 0 & 0 & 0 & -8 & 8 \\ \hline 0 & 0 & 0 & 0 & 0 \end{array}\right) = \begin{pmatrix} S & S^0 \\ 0 & 0 \end{pmatrix}.$$

Given that

$$S^{-1} = \begin{pmatrix} -1 & -.25 & -.0625 & -.015625 \\ 0 & -.5 & -.125 & -.03125 \\ 0 & 0 & -.25 & -.0625 \\ 0 & 0 & 0 & -.125 \end{pmatrix},$$

we find

$$-\sigma S^{-1} = (1,\ .25,\ .0625,\ .015625).$$

Therefore the expectation of this Coxian (and the mean time to absorption in the Markov chain) is

$$-\sigma S^{-1} e = 1.328125.$$

Also, the mean time spent in

$$\begin{aligned} &\text{phase1}: && 1/\mu_1 = 1 = x_1, \\ &\text{phase2}: && .5/\mu_2 = .25 = x_2, \\ &\text{phase3}: && (.5 \times .5)/\mu_3 = .0625 = x_3, \\ &\text{phase4}: && .5^3/\mu_4 = .015625 = x_4. \end{aligned}$$

### *7.6.7 Fitting Phase-Type Distributions to Means and Variances*

Phase-type distributions play an important role in modeling scenarios in which only the first and second moments of a distribution are known or indeed when the only information one possesses is a sequence of data points from which one can compute moments. Even in instances where the underlying distribution is known, or suspected, it may be advantageous to fit a phase-type distribution to the first and second moments, rather than using the original distribution. Such is frequently the case of the normal distribution whose mathematical formulation makes it difficult to incorporate into certain modeling techniques such as queuing networks. Phase-type distributions more readily lend themselves to analysis, a result of the mathematical tractability of the exponential distribution itself and its memoryless property.

In previous sections we saw that an Erlang-$r$ distribution provides the means to match any prespecified expectation and squared coefficient of variation, $C_X^2$, of the form $1/r$ for integer $r$, while mixing two Erlang distributions permits any $C_X^2 \leq 1$ to be obtained. To match expectations and coefficient of variations greater than 1, hyperexponential distributions could be used. Coxian distributions may be used to match any value of the squared coefficient of variations: it suffices to choose the number of phases, branching probabilities, and phase parameters, appropriately. One criterion in selecting these properties should be to use the smallest number of phases possible, since the number of phases can sometimes adversely affect computation time and memory needs during a modeling experiment.

It is best to differentiate between the two cases $C_X \leq 1$ and $C_X > 1$ when constructing Coxian distributions to match a given expectation $E[X]$ and a given $C_X^2$. For the case $C_X^2 \leq 1$, the following choices are recommended [4]:

$$\mu_i = \mu, \ \ i = 1, 2, \ldots, r,$$
$$\alpha_i = \alpha, \ \ i = 2, 3, \ldots, r - 1.$$

This is shown in Figure 7.17.

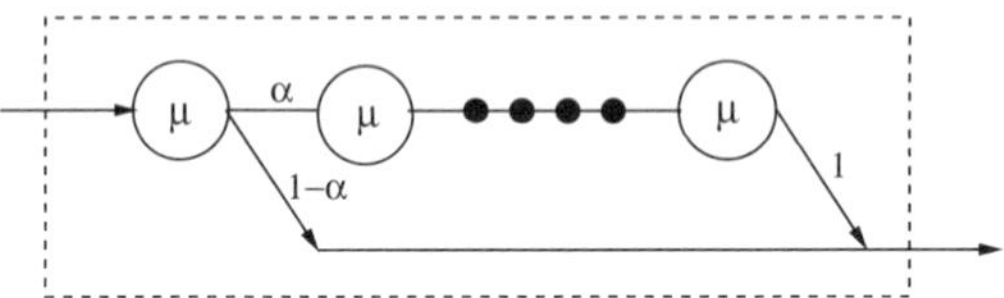


Figure 7.17. Suggested Coxian for $C_X^2 < 1$.

To obtain the values that we must assign to $\mu$ and to $\alpha$, we must first compute the mean and the squared coefficient of variation of the distribution represented by Figure 7.17. As usual we begin with the Laplace transform,

$$\mathcal{L}_X(s) = (1-\alpha)\frac{\mu}{s+\mu} + \alpha\prod_{i=1}^{r}\frac{\mu}{s+\mu} = (1-\alpha)\frac{\mu}{s+\mu} + \alpha\frac{\mu^r}{(s+\mu)^r}.$$

Then

$$\begin{aligned} E[X] &= -\frac{d}{ds}\left((1-\alpha)\frac{\mu}{s+\mu} + \alpha\frac{\mu^r}{(s+\mu)^r}\right)\bigg|_{s=0} \\ &= \left((1-\alpha)\frac{\mu}{(s+\mu)^2} + \alpha\frac{\mu^r r}{(s+\mu)^{r+1}}\right)\bigg|_{s=0} \\ &= (1-\alpha)\frac{1}{\mu} + \alpha\frac{r}{\mu}. \end{aligned} \tag{7.12}$$

For the second moment

$$\begin{aligned}
E[X^2] &= \frac{d^2}{ds^2}\left((1-\alpha)\frac{\mu}{s+\mu} + \alpha\frac{\mu^r}{(s+\mu)^r}\right)\bigg|_{s=0} \\
&= \frac{d}{ds}\left(-(1-\alpha)\frac{\mu}{(s+\mu)^2} - \alpha\frac{\mu^r r}{(s+\mu)^{r+1}}\right)\bigg|_{s=0} \\
&= \left((1-\alpha)\frac{2\mu}{(s+\mu)^3} + \alpha\frac{\mu^r r(r+1)}{(s+\mu)^{r+2}}\right)\bigg|_{s=0} \\
&= (1-\alpha)\frac{2}{\mu^2} + \alpha\frac{r(r+1)}{\mu^2}.
\end{aligned} \tag{7.13}$$

It follows that the variance is given by

$$\text{Var}\,[X] = E[X^2] - E[X]^2 = \frac{2(1-\alpha) + \alpha r(r+1) - (1-\alpha+\alpha r)^2}{\mu^2}$$

and the squared coefficient of variation by

$$C_X^2 = \frac{\text{Var}\,[X]}{E[X]^2} = \frac{2(1-\alpha) + \alpha r(r+1) - (1-\alpha+\alpha r)^2}{(1-\alpha+\alpha r)^2}. \tag{7.14}$$

We are required to select three parameters, $r$, $\alpha$, and $\mu$, that satisfy the two equations (7.12) and (7.14). Additionally, as the analysis of the Erlang-$r$ distribution suggests, we should choose $r$ to be greater than $1/C_X^2$. Since it is advantageous to choose $r$ as small as possible, we shall set

$$r = \left\lceil \frac{1}{C_X^2} \right\rceil.$$

Having thus chosen a value for $r$, and with our given value of $C_X^2$, we may use Equation (7.14), which involves only $r$, $C_X^2$, and $\alpha$, to find an appropriate value for $\alpha$. Marie [32] recommends the choice

$$\alpha = \frac{r - 2C_X^2 + \sqrt{r^2 + 4 - 4rC_X^2}}{2(C_X^2+1)(r-1)}.$$

Finally, we may now compute $\mu$ from Equation (7.12) and obtain

$$\mu = \frac{1+\alpha(r-1)}{E[X]}.$$

**Example 7.28** Let us construct a phase-type distribution having expectation $E[X] = 4$ and variance $\text{Var}\,[X] = 5$.

With these parameters, we have $C_X^2 = 5/16 = 0.3125$ which is less than 1 and we may use the analysis just developed. We choose parameters for a Coxian distribution as represented in Figure 7.17:

$$r = \left\lceil \frac{1}{C_X^2} \right\rceil = \left\lceil \frac{1}{0.3125} \right\rceil = \lceil 3.2 \rceil = 4,$$

$$\alpha = \frac{r - 2C_X^2 + \sqrt{r^2 + 4 - 4rC_X^2}}{2(C_X^2+1)(r-1)} = \frac{4 - 2(0.3125) + \sqrt{16 + 4 - 16(0.3125)}}{2(0.3125+1)(3)} = 0.9204,$$

$$\mu = \frac{1+\alpha(r-1)}{E[X]} = \frac{1 + 3(0.9204)}{4} = 0.9403.$$

Let us check our answers by computing the expectation and variance of this Coxian:

$$E[X] = (1-\alpha)\frac{1}{\mu} + \alpha\frac{r}{\mu} = (0.0796)\frac{1}{0.9403} + (0.9204)\frac{4}{0.9403} = 0.0847 + 3.9153 = 4.0,$$

$$\begin{aligned}\text{Var}[X] &= \frac{2(1-\alpha) + \alpha r(r+1) - (1-\alpha+\alpha r)^2}{\mu^2} \\ &= \frac{2(0.0796) + (0.9204)20 - [0.0796 + 4(0.9204)]^2}{(0.9403)^2} = \frac{4.4212}{0.8841} = 5.0.\end{aligned}$$

For coefficients of variation greater than 1, it suffices to use a two-phase Coxian. This is represented in Figure 7.18, where it is apparent that we need to find three parameters, namely, $\mu_1$, $\mu_2$, and $\alpha$.

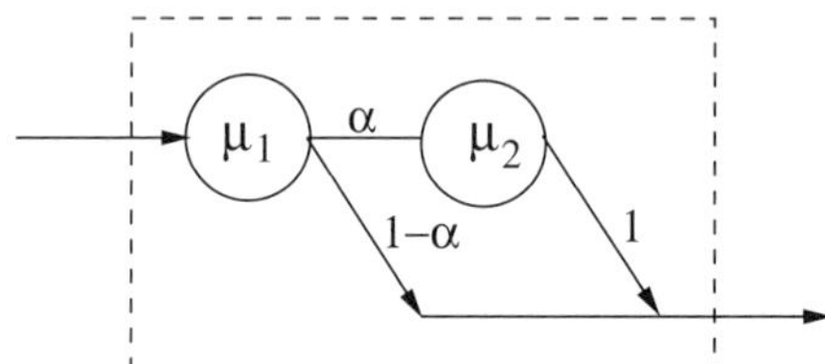


Figure 7.18. Suggested Coxian for $C_X^2 \geq 0.5$.

Previously, we found the expectation and squared coefficient of variation of this Coxian to be given by

$$E[X] = \frac{\mu_2 + \alpha\mu_1}{\mu_1\mu_2},$$

$$C_X^2 = \frac{\mu_2^2 + \alpha\mu_1^2(2-\alpha)}{(\mu_2+\alpha\mu_1)^2}.$$

Given $E[X]$ and $C_X^2$, our task is to use these equations to find the three parameters $\mu_1$, $\mu_2$, and $\alpha$. From the infinite number of solutions possible, the following is frequently recommended since it yields particularly simple forms [32]:

$$\mu_1 = \frac{2}{E[X]}, \quad \alpha = \frac{1}{2C_X^2}, \quad \text{and } \mu_2 = \frac{1}{E[X]\,C_X^2} = \alpha\mu_1.$$

This distribution is valid for values of $C_X^2$ that satisfy $C_X^2 \geq 0.5$, and not just for those values greater than or equal to 1.

**Example 7.29** A random variable $X$ having expectation $E[X] = 3$ and standard deviation equal to $\sigma_X = 4$ may be modeled as a two-phase Coxian. Given that $E[X] = 3$ and $C_X^2 = 16/9$, we may take the parameters of the Coxian distribution to be

$$\mu_1 = \frac{2}{E[X]} = \frac{2}{3}, \quad \alpha = \frac{1}{2C_X^2} = \frac{9}{32}, \quad \text{and } \mu_2 = \frac{1}{E[X]\,C_X^2} = \frac{3}{16}.$$

To check, we find the expectation and standard deviation of this Coxian

$$\frac{\mu_2 + \alpha\mu_1}{\mu_1\mu_2} = \frac{\frac{3}{16} + \frac{9}{32}\frac{2}{3}}{\frac{2}{3}\frac{3}{16}} = \frac{\frac{6}{16}}{\frac{1}{8}} = 3$$

$$\frac{\mu_2^2 + \alpha\mu_1^2(2-\alpha)}{(\mu_2 + \alpha\mu_1)^2} = \frac{\frac{9}{256} + \frac{9}{32}\frac{4}{9}\frac{55}{32}}{\left(\frac{3}{16} + \frac{9}{32}\frac{2}{3}\right)^2} = \frac{0.25}{0.1406} = 1.7778 = \frac{16}{9}.$$

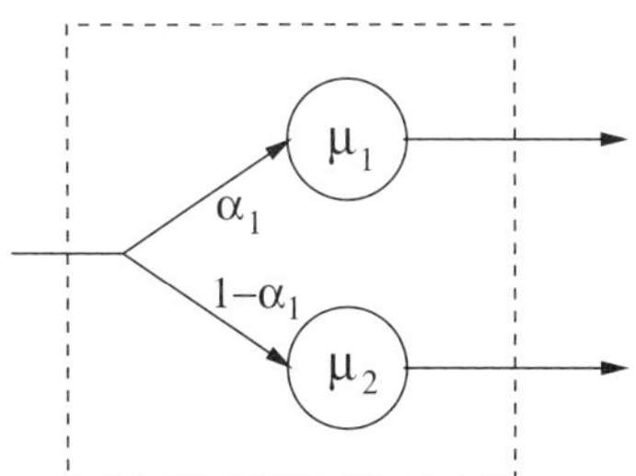


Figure 7.19. Two-phase hyperexponential distribution.

As an alternative to the Coxian-2, a two-phase hyperexponential distribution such as that shown in Figure 7.19 may also be used to model distributions for which $C_X^2 \geq 1$. As is the case for a Coxian-2 distribution, a two-phase hyperexponential distribution is not defined uniquely by its first two moments so that there is a considerable choice of variations possible. One that has attracted attention is to *balance* the means by imposing the additional condition

$$\frac{\alpha}{\mu_1} = \frac{1-\alpha}{\mu_2}.$$

This leads to the formulae

$$\alpha = \frac{1}{2}\left(1 + \sqrt{\frac{C_X^2 - 1}{C_X^2 + 1}}\right), \quad \mu_1 = \frac{2\alpha}{E[X]}, \quad \text{and} \quad \mu_2 = \frac{2(1-\alpha)}{E[X]}.$$

**Example 7.30** Let us return to the previous example of a random variable with expectation $E[X] = 3$ and squared coefficient of variation $C_X^2 = 16/9$ and see what parameters we obtain for the two-phase hyperexponential:

$$\alpha = \frac{1}{2}\left(1 + \sqrt{\frac{C-1}{C+1}}\right) = \frac{1}{2}\left(1 + \sqrt{\frac{7/9}{25/9}}\right) = 0.7646,$$

$$\mu_1 = \frac{2\alpha}{E[X]} = \frac{1.5292}{3} = 0.5097, \quad \text{and} \quad \mu_2 = \frac{2(1-\alpha)}{E[X]} = \frac{0.4709}{3} = 0.1570.$$

We now check these results. The expectation of the two-phase hyperexponential is given by

$$E[X] = \frac{\alpha}{\mu_1} + \frac{1-\alpha}{\mu_2} = \frac{0.7646}{0.5097} + \frac{0.2354}{0.1570} = 1.50 + 1.50 = 3.0.$$

Similarly, the squared coefficient of variation is given by

$$C_X^2 = \frac{2\alpha/\mu_1^2 + 2(1-\alpha)/\mu_2^2}{(\alpha/\mu_1 + (1-\alpha)/\mu_2)^2} - 1$$

$$= \frac{1.5292/0.2598 + 0.4708/0.0246}{(0.7646/0.5097 + 0.2354/0.1570)^2} - 1$$

$$= \frac{25}{9} - 1 = \frac{16}{9}.$$

Although we considered the possibility of matching only the first two moments of a distribution with phase-type representations, techniques exist for matching higher moments. However, this is beyond what we seek to accomplish in this text. The interested reader requiring further information may wish to consult the relevant literature, particularly [41] and the references therein.

## 7.7 Exercises

**Exercise 7.1.1** Let $X$ be a continuous, uniformly distributed random variable with both expectation and variance equal to 3. Find the probability density function of $X$ and the probability that $X > 2$.

**Exercise 7.1.2** Let $X$ be uniformly distributed on the interval $(0, 4)$. What is the probability that the roots of $z^2 + 2Xz - 2X + 15 = 0$ are real?

**Exercise 7.1.3** Let $X_1$ and $X_2$ be two independent random variables with probability density functions given by $p_{X_1}(x_1)$ and $p_{X_2}(x_2)$ respectively. Let $Y$ be the random variable $Y = X_1 + X_2$. Find the probability density function of $Y$ when both $X_1$ and $X_2$ are uniformly distributed on $[0, 1]$. Illustrate the probability density function of $Y$ by means of a figure.

**Exercise 7.2.1** Derive the expectation and variance of an exponential random variable from its Laplace transform.

**Exercise 7.2.2** The density function of an exponential random variable is given by

$$f_X(x) = \begin{cases} .5e^{-x/2}, & x > 0, \\ 0 & \text{otherwise.} \end{cases}$$

What is the expectation and standard deviation of $X$? Write down its cumulative distribution function and find the probability that $X$ is greater than 4.

**Exercise 7.2.3** What is the probability that the value of an exponential random variable with parameter $\lambda$ lies within $2\sigma$ of its expectation?

**Exercise 7.2.4** Let $X$ be an exponential random variable whose variance is equal to 16. Find $\text{Prob}\{1 \leq X \leq 2\}$.

**Exercise 7.2.5** The time spent waiting in the emergency room of a hospital before one is called to see a doctor often seems to be exponentially distributed with mean equal to 60 minutes. Let $X$ be the random variable that describes this situation. Find the following probabilities:
(a) $\text{Prob}\{X > 30\}$, (b) $\text{Prob}\{X > 90\}$, (c) $\text{Prob}\{X > 90 | X > 60\}$.

**Exercise 7.2.6** When Charles enters his local bank, he finds that the two tellers are busy, one serving Alice and the other serving Bob. Charles' service will begin as soon as the first of the two tellers becomes free. Find the probability that Charles will not be the last of Alice, Bob, and himself to leave the bank, assuming that the time taken to serve any customer is exponentially distributed with mean $1/\lambda$.

**Exercise 7.2.7** Let $X$ be an exponentially distributed random variable and let $\alpha$ and $\beta$ be two constants such that $\alpha \geq 0$ and $\beta \geq 0$. Prove that

$$\text{Prob}\{X > \alpha + \beta\} = \text{Prob}\{X > \alpha\}\text{Prob}\{X > \beta\}$$

and from it deduce that

$$\text{Prob}\{X > \alpha + \beta | X > \alpha\} = \text{Prob}\{X > \beta\}.$$

**Exercise 7.2.8** Let $X$ be an exponentially distributed random variable with parameter $\lambda$. Find the probability density function and the cumulative distribution function of (a) $Y = X^2$ and (b) $Z = \sqrt{X}$.

**Exercise 7.2.9** Let $X_1$ and $X_2$ be two independent and identically distributed exponential random variables with parameter $\lambda$. Find the density function of $Y = X_1 + X_2$ using the convolution approach.

**Exercise 7.2.10** Show that a geometric sum of exponentially distributed random variables is itself exponentially distributed.

*Hint: Let $N$ be a geometrically distributed random variable, let $X_1, X_2, \ldots$ be independent and identically distributed exponential random variables with parameter $\mu$, and let $S = \sum_{n=1}^{N} X_n$. Now evaluate the Laplace transform of $S$, $E[e^{-sS}]$, using a conditioning argument.*

**Exercise 7.3.1** Show, by taking derivatives of the moment generating function of the standard normal distribution, that this distribution has mean value zero and variance equal to 1.

**Exercise 7.3.2** Using tables of the cumulative distribution of a standard normal, random variable, compute the following probabilities of a normally distributed random variable with $\mu = 10$ and $\sigma = 4$. (a) Prob$\{X < 0\}$, (b) Prob$\{X > 12\}$, and (c) Prob$\{-2 \leq X \leq 8\}$.

**Exercise 7.3.3** The time spent waiting for a bus is normally distributed with mean equal to 10 minutes and standard deviation equal to 10 minutes. Find (a) the probability of waiting less than 12 minutes and (b) the probability of waiting more than 15 minutes.

**Exercise 7.3.4** A normally distributed random variable has mean $\mu = 4$. However, instead of being given the standard deviation, we are told that Prob$\{2 \leq X \leq 6\} = 0.4$. Use this to find the standard deviation of $X$.

**Exercise 7.3.5** The most recent census population estimates that there are 12.5 million octogenarians in a total population of 250 million. Assume that the age (in years) of the population is normally distributed with mean equal to 38 years. Find the standard deviation of this distribution and estimate the total number of individuals older than 55.

**Exercise 7.3.6** The age of a randomly selected person in a certain population is a normally distributed random variable $X$. Furthermore, it is known that Prob$\{X \leq 40\} = 0.5$ and Prob$\{X \leq 30\} = 0.25$. Find the mean $\mu$ and standard deviation $\sigma$ of $X$. Also find Prob$\{X > 65\}$ and the percentage of those over 65 who are older than 85.

**Exercise 7.3.7** It is estimated that 15% of traffic accidents are caused by impatient drivers. Of the 400 accidents that occurred last week, estimate the probability that no more than 50 were due to impatient drivers.

**Exercise 7.3.8** Suppose 2,880 fair dice are thrown all at once. Find the probability that the number of 1's that show lies between 450 and 470.

**Exercise 7.4.1** Prove that the variance of a random variable having a gamma distribution with parameters $\alpha$ and $\beta$ is given by Var$[X] = \alpha\beta^2$.

**Exercise 7.4.2** A random variable $X$ has a gamma distribution with parameters $\beta = 1$ and $\alpha = 6$. Find (a) Prob$\{X \leq 5\}$, (b) Prob$\{X > 4\}$, and (c) Prob$\{4 \leq X \leq 8\}$.

**Exercise 7.4.3** The length of time that a supermarket can keep milk on its shelves before it goes bad is a gamma random variable $X$ with parameters $\beta = 3$ and $\alpha = 4$. Find (a) Prob$\{X \leq 12\}$, (b) Prob$\{12 \leq X < 15\}$, and (c) the expected shelf life of a carton of milk in this supermarket.

**Exercise 7.4.4** The time it takes Kathie to cook dinner has a gamma distribution with mean 20 minutes and variance 40 minutes. Find the probability that it will take more than 30 minutes to cook dinner, (a) using tables of the Poisson cumulative distribution function, and (b) using tables of the incomplete gamma function.

**Exercise 7.4.5** The time it takes to eat breakfast has a gamma distribution with mean five minutes and standard deviation two minutes. Find the parameters $\alpha$ and $\beta$ of the distribution and the probability that it takes more than eight minutes to eat breakfast.

**Exercise 7.5.1** Derive the second moment of a Weibull random variable and from it show that

$$\text{Var}[X] = \eta^2 \left[ \Gamma\left(1 + \frac{2}{\beta}\right) - \Gamma\left(1 + \frac{1}{\beta}\right)^2 \right].$$

**Exercise 7.5.2** The time in minutes to execute a certain task is represented as a Weibull random variable $X$ with parameters $\eta = 8$ and $\beta = 2$. Give the probability density function and the cumulative distribution of $X$. Also find the expectation and standard deviation of $X$ and the probability that the task takes longer than 16 minutes.

**Exercise 7.5.3** Let the lifetime of component $i$, $i = 1, 2, \ldots, n$, in a series of independent components have a Weibull distribution with parameters $\eta_i$, $i = 1, 2, \ldots, n$, and $\beta$. Show that the lifetime of the complete system has a Weibull distribution and find its parameters.

**Exercise 7.5.4** A system contains two components arranged in series, so that when one of the components fails, the whole system fails. The time to failure of each component is a Weibull random variable with parameters $\eta_c = 8$ and $\beta_c = 2$. What is the expected lifetime of each component and the expected lifetime of the system. Find the probability that the system survives until time $t = 4$.

**Exercise 7.5.5** The time to failure of each of two embedded computers in a time-critical space mission is an exponential random variable with mean time to failure equal to 5,000 hours. At least one of the computers must function for a successful mission outcome. What is the longest time that the mission can last so that the probability of success is 0.999?

**Exercise 7.5.6** Consider a system that contains four components with reliability functions given respectively by

$$R_1(t) = e^{-\alpha t}, \quad R_2(t) = e^{-\beta t}, \quad R_1(t) = e^{-\gamma t}, \quad \text{and} \quad R_1(t) = e^{-\delta t}.$$

What is the reliability function of the system that has these four components arranged (a) in series and (b) in parallel? A different system arranges these components so that the first two are in series, the last two are in series, but the two groups of two are in parallel with each other. What is the reliability function in this case?

**Exercise 7.5.7** Find the hazard rate of a product whose lifetime is uniformly distributed over the interval $(a, b)$, with $a < b$.

**Exercise 7.5.8** The hazard rate of a certain component is given by

$$h(t) = \frac{e^{t/4}}{5}, \quad t > 0.$$

What are the cumulative hazard function and the reliability function of this component? What is the probability that it survives until $t = 2$.

**Exercise 7.5.9** The hazard rate of a component is given by

$$h_X(t) = \frac{\alpha}{2} t^{-1/2} + \frac{\beta}{4} t^{-3/4}.$$

Find the reliability function of the component.

**Exercise 7.6.1** The number of telephone calls to a central office is a Poisson process with rate $\mu = 4$ per five minute period. Find the probability that the waiting time for three or more calls to arrive is greater than two minutes. Give also the mean and standard deviation of the time to observe three calls arrive.

**Exercise 7.6.2** Let $X$ be an Erlang-$r$ random variable. Construct phase-type representations of $X$ when its expectation and standard deviation are as follows: (a) $E[X] = 4$, $\sigma_X = 2$, (b) $E[X] = 6$, $\sigma_X = 2$, and (c) $E[X] = 3$, $\sigma_X = 2$. In each case, specify the number of exponential phases, $r$, in the representation and the parameter of each phase $\mu$. Also in each case give the probability density function of X.

**Exercise 7.6.3** It has been observed that the time bank customers spend with a teller has an average value of four minutes and a variance of nine minutes. This sevice time is to be modeled as a mixed Erlang distribution whose probability density function is

$$f_X(t) = \alpha\,\mu e^{-\mu t} + (1-\alpha)\,\mu^2 t e^{-\mu t}, \quad t \geq 0.$$

What values should be assigned to the parameters $\alpha$ and $\mu$?

**Exercise 7.6.4** The manufacturing process for widgets involves three steps. Assume that the time spent in each of these three steps is an exponential random variable with parameter values equal to 2, 4, and 5 respectively. Find the expected duration, variance, and squared cefficient of variation of the manufacturing process. Also find the probability density function of the manufacturing time.

**Exercise 7.6.5** A person chooses to go to McDonald's for breakfast in the morning with probability 0.5. Otherwise, with probability 0.3 he goes to Hardee's and with probability 0.2 he goes to IHOP. If the mean time spent eating breakfast is 15 minutes in McDonald's, 12 minutes in Hardee's, and 20 minutes in IHOP and if the distribution is exponential, find the expectation and standard deviation of the time spent eating breakfast. Also compute the squared coefficient of variation and show that it is greater than 1.

**Exercise 7.6.6** Prove that the first and second moments of a hyperexponential-2 random variable $X$, with parameters $\mu_1$, $\mu_2$, and $\alpha_1$, are respectively given by

$$E[X] = \frac{\alpha_1}{\mu_1} + \frac{\alpha_2}{\mu_2} \quad \text{and} \quad E[X^2] = \frac{2\alpha_1}{\mu_1^2} + \frac{2\alpha_2}{\mu_2^2},$$

where $\alpha_2 = 1 - \alpha_1$.

**Exercise 7.6.7** Professor Nozit has arranged his lectures into modules. The time it takes to present the theoretical aspects of a module is exponentially distributed with a mean of 45 minutes. Eighty percent of the time, Professor Nozit also provides an example of the application of the theory, and the time to complete this is also exponentially distributed with a mean of 20 minutes. Find the mean and standard deviation of the length of Professor Nozit's lectures.

**Exercise 7.6.8** Derive the expectation of the Coxian-$r$ random variable that is represented graphically in Figure 7.14. Show that the answer you compute using the notation of Figure 7.14 is the same as that derived with reference to Figure 7.13.

**Exercise 7.6.9** Show that a Cox-2 random variable with parameters $\mu_1 = 2/E$, $\mu_2 = 1/(EC)$, and $\alpha = 1/2C$ has expectation equal to $E$ and squared coefficient of variation equal to $C$.

# Chapter 8

# *Bounds and Limit Theorems*

We shall begin this section with three theorems that allow us to bound probabilities. They are of interest when bounds are sought on the *tail* of the distribution, i.e., on the probability that a random variable $X$ exceeds some given quantity—a quantity that is usually greater than the mean. The first, the Markov inequality, uses only the expectation of a random variable and provides bounds that are often rather loose. The second, the Chebychev inequality, uses both the expectation and the variance of a random variable and usually generates tighter bounds. The third, the Chernoff bound, requires a knowledge of the moment generating function of a random variable. Since a knowledge of a moment generating function implicitly implies a knowledge of moments of all orders, we should expect the Chernoff bound to provide tighter bounds than those obtained from either the Markov or Chebychev inequalities.

## 8.1 The Markov Inequality

Let $X$ be a random variable and $h$ a nondecreasing, nonnegative function. The expectation of $h(X)$, assuming it exists, is given by

$$E[h(X)] = \int_{-\infty}^{\infty} h(u) f_X(u) du,$$

and we may write

$$\int_{-\infty}^{\infty} h(u) f_X(u) du \geq \int_{t}^{\infty} h(u) f_X(u) du \geq h(t) \int_{t}^{\infty} f_X(u) du = h(t)\text{Prob}\{X \geq t\}.$$

This leads directly to the so-called Markov inequality, given by

$$\text{Prob}\{X \geq t\} \leq \frac{E[h(X)]}{h(t)}. \tag{8.1}$$

It is frequently applied in the case when $X$ is nonnegative and $h(x) = x$, when it reduces to

$$\text{Prob}\{X \geq t\} \leq \frac{E[X]}{t}, \quad t > 0.$$

This inequality is exceedingly simple, requiring only the mean value of the distribution. It is used primarily when $t$ is large (in bounding the tail of a distribution), when $E[X]/t$ is small and a relatively tight bound can be obtained. Otherwise the bound can be very loose. Bear in mind also that this inequality is applicable only to random variables that are nonnegative.

**Example 8.1** Let $X$ be the random variable that denotes the age (in years) of a randomly selected child in Centennial Campus Middle school. If the average child's age at that school is 12.5 years, then, using the Markov inequality, the probability that a child is at least 20 years old satisfies the inequality

$$\text{Prob}\{X \geq 20\} \leq 12.5/20 = 0.6250,$$

which is a remarkably loose bound. While this bound is obviously correct, the probability that there is a child aged 20 in a middle school should be very close to zero.

## 8.2 The Chebychev Inequality

The Markov inequality is a first-order inequality in that it requires only knowledge of the mean value. The Chebychev inequality is second order: it requires both the mean value and the variance of the distribution. It may be derived from the Markov inequality as follows. Let the variance $\sigma_X^2$ be finite and define a new random variable $Y$ as

$$Y \equiv (X - E[X])^2.$$

As in the simple form of the Markov inequality, let $h(x) = x$. Then, from the Markov inequality,

$$\text{Prob}\{Y \geq t^2\} \leq \frac{E[Y]}{t^2}.$$

Observe that

$$\text{Prob}\{Y \geq t^2\} = \text{Prob}\{(X - E[X])^2 \geq t^2\} = \text{Prob}\{|X - E[X]| \geq t\}$$

and that

$$E[Y] = E[(X - E[X])^2] = \sigma_X^2,$$

from which we immediately have the Chebychev inequality

$$\text{Prob}\{|X - E[X]| \geq t|\} \leq \frac{\sigma_X^2}{t^2}. \tag{8.2}$$

From Equation (8.2), it is apparent that the random variable $X$ does not stray far from its mean value $E[X]$ when its variance $\sigma_X^2$ is small. If we set $t = c\,\sigma_X$, for some positive constant $c$, we obtain

$$\text{Prob}\{|X - E[X]| \geq c\,\sigma_X\} \leq \frac{1}{c^2}, \tag{8.3}$$

and thus the probability that a random variable is greater than $c$ standard deviations from its mean is less than $1/c^2$. Setting $c = 2$ for example, shows that the probability that a random variable (indeed any random variable) is more than two standard deviations from its mean is less that 1/4. An alternative form of Equation (8.3) is

$$\text{Prob}\{|X - E[X]| \leq c\,\sigma_X\} \geq 1 - \frac{1}{c^2}.$$

This form of the Chebychev inequality is often used to compute confidence intervals in simulation experiments as we shall see in later chapters.

As we have mentioned, the Chebychev inequality may be applied to any random variable, unlike the Markov inequality which is applicable only to random variables that are nonnegative. Furthermore, it generally provides a better bound, since it incorporates the variance as well as the expected value of the random variable into the computation of the bound.

**Example 8.2** Let us return to the same example of middle school children and recompute the bound using the Chebychev inequality. In this case we also need the variance of the ages of the children, which we take to be 3. We seek the probability that a child at the school could be as old as 20. We need to first put this into the form needed by the Chebychev equation:

$$\text{Prob}\{X \geq 20\} = \text{Prob}\{(X - E[X]) \geq (20 - E[X])\} = \text{Prob}\{(X - E[X]) \geq 7.5\}.$$

However,

$$\text{Prob}\{(X - 12.5) \geq 7.5\} \neq \text{Prob}\{|X - 12.5| \geq 7.5\} = \text{Prob}\{5 \leq X \geq 20\},$$

so that we cannot apply the Chebychev inequality to directly compute $\text{Prob}\{X \geq 20\}$. Instead, we can compute a bound for $\text{Prob}\{5 \leq X \geq 20\}$ and obtain

$$\text{Prob}\{|X - E[X]| \geq 7.5\} \leq \frac{3}{(7.5)^2} = 0.0533,$$

which is still a much tighter bound than that obtained previously.

## 8.3 The Chernoff Bound

As is the case the Chebychev inequality, the Chernoff bound may be derived from the Markov inequality. Setting $h(x) = e^{\theta x}$ for some $\theta \geq 0$ in Equation (8.1), and using the fact that the moment generating function of a random variable is given by

$$\mathcal{M}_X(\theta) = E[e^{\theta X}],$$

we obtain

$$\text{Prob}\{X \geq t\} \leq \frac{E[e^{\theta X}]}{e^{\theta t}} = e^{-\theta t}\mathcal{M}_X(\theta) \quad \text{for all } \theta \geq 0.$$

Since this holds for all $\theta \geq 0$ and since it is in our interest to have the smallest lowest bound possible, we may write

$$\text{Prob}\{X \geq t\} \leq \min_{\text{all } \theta \geq 0} e^{-\theta t}\mathcal{M}_X(\theta),$$

which is known as the Chernoff bound.

**Example 8.3** The moment generating function of a normally distributed random variable with mean value $\mu = 4$ and variance $\sigma^2 = 1$ is given by

$$\mathcal{M}_X = e^{\mu\theta + \sigma^2\theta^2/2} = e^{4\theta + \theta^2/2}.$$

The Chernoff bound yields

$$\text{Prob}\{X \geq 8\} \leq \min_{\theta \geq 0} e^{-8\theta} e^{4\theta + \theta^2/2} = \min_{\theta \geq 0} e^{(\theta^2 - 8\theta)/2}.$$

Observe that the upper bound is minimized when $\theta^2 - 8\theta$ is minimized. Taking the derivative of this function and setting it equal to zero gives $2\theta - 8 = 0$ so that the minimum is achieved when $\theta = 4$. The Chernoff bound then gives

$$\text{Prob}\{X \geq 8\} \leq e^{(\theta^2 - 8\theta)/2}\Big|_{\theta=4} = e^{-8} = 0.0003355.$$

The reader may wish to note that the bound computed from the Markov inequality is 0.5; that obtained from the Chebychev inequality is 0.0625, while the exact value of this probability is 0.0000317.

## 8.4 The Laws of Large Numbers

Let $X_1, X_2, \ldots, X_n$ be $n$ independent and identically distributed random variables. We may view these $n$ random variables as $n$ independent trials of the same probability experiment. In this light they are sometimes referred to as a *random sample* of size $n$ from the experimental distribution. We shall assume that the mean and variance are both finite and given by $E[X]$ and $\sigma_X^2$, respectively. We would expect that as more experiments are conducted, i.e., as $n$ becomes large, the average value obtained in the $n$ experiments should approach the expected value, $E[X]$. This is exactly what happens. Setting

$$S_n \equiv \sum_{i=1}^{n} X_i,$$

the average value obtained by these $n$ experiments is given by $S_n/n$. Other names for this *computed mean* value $S_n/n$ are the *arithmetic mean* and the *statistical average*. As $n$ becomes large, we expect $S_n/n$ to be close to $E[X]$: the *weak law of large numbers* provides justification for this assertion. Notice that

$$E\left[S_n/n\right] = E[X] \quad \text{and} \quad \text{Var}\left[S_n/n\right] = \frac{n\sigma_X^2}{n^2} = \frac{1}{n}\sigma_X^2, \tag{8.4}$$

which shows that the mean is independent of $n$ whereas the variance decreases as $1/n$. Thus, as $n$ becomes large, the variance approaches 0 and the distribution of $S_n/n$ becomes more concentrated about $E[X]$.

Let us now apply the Chebychev inequality to $S_n/n$. Substituting from (8.4), we obtain

$$\text{Prob}\left\{\left|\frac{S_n}{n} - E[X]\right| \geq \epsilon\right\} \leq \frac{\sigma_X^2/n}{\epsilon^2} = \frac{\sigma_X^2}{n\epsilon^2}. \tag{8.5}$$

This leads directly to the weak law of large numbers. Taking the limit of both sides of (8.5) as $n \to \infty$ and for any fixed value of $\epsilon$, no matter how small, we find

$$\lim_{n\to\infty} \text{Prob}\left\{\left|\frac{S_n}{n} - E[X]\right| \geq \epsilon\right\} = 0. \tag{8.6}$$

We may interpret this inequality as follows. Irrespective of how small $\epsilon$ may be, it is possible to select a large enough value of $n$ so that the probability of $S_n/n$ being separated by more than $\epsilon$ from $E[X]$ converges to zero. Alternatively, we may write

$$\lim_{n\to\infty} \text{Prob}\left\{\left|\frac{S_n}{n} - E[X]\right| < \epsilon\right\} = 1,$$

i.e., $\text{Prob}\{|S_n/n - E[X]| < \epsilon\}$ approaches 1 as $n \to \infty$. Thus, as the number of experiments increases, it becomes less likely that the statistical average differs from $E[X]$.

In the above derivation we assumed that both the expectation and variance of the random variables $X_1, X_2, \ldots$ were finite. However, the weak law of large numbers requires only that the expectation be finite. Both Equations (8.5) and (8.6) hold when the variance is finite, but only Equation (8.6) holds when the variance is infinite. Equation (8.5) is a more precise statement since it provides a bound in terms of $n$. We state the theorem as follows:

**Theorem 8.4.1 (Weak law of large numbers)** *Let $X_1, X_2, \ldots, X_n$ be $n$ independent and identically distributed random variables with finite mean $E[X]$. Then, setting $S_n = X_1 + X_2 + \cdots + X_n$ and given any $\epsilon > 0$, however small,*

$$\lim_{n\to\infty} \text{Prob}\left\{\left|\frac{S_n}{n} - E[X]\right| \geq \epsilon\right\} = 0.$$

Convergence in this sense is said to be convergence in a probabilistic sense: a sequence of random variables $X_n$ is said to *converge in probability* to a random variable $X$ if, for any $\epsilon > 0$,

$$\lim_{n\to\infty} \text{Prob}\{|X_n - X| \geq \epsilon\} = 0.$$

In making explicit reference to a *weak* law of large numbers, we have implicitly implied the existence of a *strong* law of large numbers, and indeed such a law exists. The *strong law of large numbers*, a proof of which may be found in Feller [15], states that *with probability* 1, the sequence of numbers $S_n/n$ converges to $E[X]$ as $n \to \infty$. We have

**Theorem 8.4.2 (Strong law of large numbers)** *Let $X_1, X_2, \ldots, X_n$ be $n$ independent and identically distributed random variables with finite mean $E[X]$. Then, setting $S_n = X_1 + X_2 + \cdots + X_n$ and given any $\epsilon > 0$, however small,*

$$\text{Prob}\left\{\lim_{n\to\infty}\left|\frac{S_n}{n} - E[X]\right| \geq \epsilon\right\} = 0.$$

The difference between the weak and strong laws of large numbers is subtle. Observe that, in contrast with the weak law of large numbers, the limit in the strong law is taken *inside* the probability braces. The weak law concerns the *average* behavior of many sample paths, some of which might not converge to $E[X]$, whereas the strong law is very precise and states that each sequence converges to $E[N]$ for sufficiently large $n$. It follows that the strong law of large numbers implies the weak law. In practical applications of probability theory, the difference is usually inconsequential.

## 8.5 The Central Limit Theorem

In our discussion of the normal distribution, we saw that the sum of $n$ normally distributed random variables was itself normally distributed. Furthermore, we stated that the normal distribution could be used to approximate the discrete binomial distribution. In this section, we take this one step further. The *central limit theorem*, which we state below without proof, asserts that the sum of $n$ independent random variables may be approximated by a normal distribution, irrespective of the particular distribution of the $n$ random variables. It is for this reason that the normal distribution plays such an important role in practical applications.

We consider first the case when the $n$ random variables are independent and identically distributed having mean value $E[X_i] = \mu$ and variance $\text{Var}[X_i] = \sigma^2$ for $i = 1, 2, \ldots, n$. Then the sum of these $n$ random variables, $X_1 + X_2 + \cdots + X_n$, for $n$ sufficiently large, has an approximate normal distribution with mean $n\mu$ and variance $n\sigma^2$. We write this as

$$\text{Prob}\left\{\frac{X - n\mu}{\sigma\sqrt{n}} \leq x\right\} \approx \text{Prob}\{Z \leq x\}. \tag{8.7}$$

**Example 8.4** The tube of toothpaste that sits on our bathroom sink has a mean lifetime of 30 days and a standard deviation of 5 days. When one tube is finished it is immediately replaced by a new tube. Kathie has been to the sales and has returned with 12 new tubes of toothpaste. We shall let $X_i, i = 1, 2, \ldots, n$, be the random variable that represents the lifetime of the $i^{\text{th}}$ tube and it seems reasonable to assume that these $n$ random variables are independent and identically distributed. Kathie would like to know the probability that these 12 tubes will last a full year of 365 days or more.

The random variable $X = X_1 + X_2 + \cdots + X_n$ represents the lifetime of these 12 tubes of toothpaste. Under the assumption that $n = 12$ is sufficiently large to apply the normal approximation, we obtain the mean value and standard deviation of $X$ to be

$$E[X] = 12 \times 30 = 360 \text{ days} \quad \text{and} \quad \sigma_X = 5 \times \sqrt{12} = 17.3205.$$

Taking these for the mean and standard deviation of the normal distribution, the probability that the 12 tubes will last 365 days or less is

$$\text{Prob}\{X \le 365\} = \text{Prob}\left\{Z \le \frac{365 - 360}{17.3205}\right\} = \text{Prob}\{Z \le 0.2887\} = 0.6141.$$

The probability that they will last more than one year is 0.3859.

Consider next the case of the central limit theorem applied to a sample mean $S_n = (X_1 + X_2 + \cdots + X_n)/n$. Once again, the $n$ random variables $X_i$, $i = 1, 2, \ldots, n$, are independent and identically distributed with mean values $E[X_i] = \mu$ and variances $\text{Var}[X_i] = \sigma^2$. In this case, for sufficiently large values of $n$, the sample mean $S_n$ is approximately normally distributed with mean $\mu$ and variance $\sigma^2/n$. We write this as

$$\text{Prob}\left\{\frac{S_n - \mu}{\sigma/\sqrt{n}} \le x\right\} \approx \text{Prob}\{Z \le x\}. \tag{8.8}$$

**Example 8.5** A survey is being conducted to compute the mean income of North Carolina families. Let us suppose that this is actually equal to \$32,000 and has a standard deviation of \$10,000. These figures are unknown to those conducting the survey. What is the probability that the mean value computed by the survey is within 5% of the true mean if (a) 100 families are surveyed, and (b) 225 families are surveyed?

We seek the probability that the survey will yield a mean value between \$30,400 and \$33,600 (since 5% of \$32,000 is \$1,600). Using $E[S_n] = 32{,}000$ and for the first case when only 100 families are surveyed, $\sigma_{S_n} = 10{,}000/10 = 1{,}000$, we have

$$\text{Prob}\{30{,}400 < S_n < 33{,}600\} = \text{Prob}\left\{\frac{30{,}400 - 32{,}000}{1{,}000} < Z < \frac{33{,}600 - 32{,}000}{1{,}000}\right\}$$

$$= \text{Prob}\{-1.6 < Z < 1.6\} = 0.9452 - 0.0548 = 0.8904.$$

When 225 families are surveyed, we have $E[S_n] = 32{,}000$ and $\sigma_{S_n} = 10,000/15 = 666.67$. This time we get

$$\text{Prob}\{30{,}400 < S_n < 33{,}600\} = \text{Prob}\left\{\frac{30{,}400 - 32{,}000}{666.67} < Z < \frac{33{,}600 - 32{,}000}{666.67}\right\}$$

$$= \text{Prob}\{-2.4 < Z < 2.4\} = 0.9918 - 0.0082 = 0.9836.$$

This example shows that it is not necessary to sample large numbers of families to get an accurate estimate of the mean. The difficulty with sampling of course lies elsewhere: it lies in ensuring that those families chosen are randomly selected and representative of the population as a whole.

In the general statement of the central limit theorem, the random variables do not need to be identically distributed.

**Theorem 8.5.1 (Central limit theorem)** *Let $X_1, X_2, \ldots, X_n$ be independent random variables whose expectations $E[X_i] = \mu_i$ and variances $\text{Var}[X_i] = \sigma_i^2$ are both finite. Let $Y_n$ be the normalized random variable*

$$Y_n = \frac{\sum_{i=1}^n X_i - \sum_{i=1}^n \mu_i}{\sqrt{\sum_{i=1}^n \sigma_i^2}}.$$

*Then* $E[Y_n] = 0$ *and* $\text{Var}[Y_n] = 1$, *and under certain rather broad conditions,*

$$\lim_{n\to\infty} F_{Y_n}(t) = \text{Prob}\{Y_n \le t\} = \int_{-\infty}^{t} \frac{1}{\sqrt{2\pi}} e^{-u^2/2} du.$$

*In other words, the limiting distribution of* $Y_n$ *is standard normal,* $N(0, 1)$.

When the random variables are all identically distributed, as well as being independent, then $Y_n$ simplifies to

$$Y_n = \frac{\sum_{i=1}^n X_i - n\mu}{\sigma\sqrt{n}},$$

where $\mu = \sum_{i=1}^n \mu_i/n$ and $\sigma^2 = \sum_{i=1}^n \sigma_i^2/n$.

**Example 8.6** A fair coin is tossed 400 times. We shall use the central limit theorem to compute an approximation to the probability of getting at least 205 heads. Let $X_i$ be the random variable that has the value 1 if a head appears on toss $i$ and the value 0 otherwise. The number of heads obtained in $n$ tosses is given by $Y_n = X_1 + X_2 + \cdots + X_n$. With $n = 400$, $\mu = 1/2$, and $\sigma^2 = 1/4$, we obtain

$$\text{Prob}\{Y_n \ge 205\} = \text{Prob}\left\{\frac{Y_n - n\mu}{\sigma\sqrt{n}} \ge \frac{205 - n\mu}{\sigma\sqrt{n}}\right\} = \text{Prob}\left\{\frac{Y_n - 200}{10} \ge 0.5\right\} = \text{Prob}\{Z \ge 0.5\}.$$

Since $\text{Prob}\{Z \ge 0.5\} = 1 - 0.6915$, the computed approximation is 0.3085.

The above example is instructive. Using the same approach, we may compute $\text{Prob}\{Y_n \le 204\}$ and we find that this works out to be equal to 0.6554 (see Exercise 8.5.3). We know that $\text{Prob}\{Y_n \le 204\} + \text{Prob}\{Y_n \ge 205\}$ must be equal to 1, but the sum of our two approximations gives only $0.6554 + 0.3085 = 0.9639$! The reason for this error is primarily because we are approximating a discrete distribution with the continuous normal distribution. This difficulty can be circumvented by replacing each of the integers 204 and 205 with 204.5, in essence dividing the range (204, 205) between the two of them. This correction is called the *continuity correction* or the *histogram correction* and should be used whenever the central limit theorem approximation is applied to integer-valued random variables. If we now compute $\text{Prob}\{Y_n \ge 204.5\}$ (this is all we need since we have now ensured that $\text{Prob}\{Y_n \le 204.5\} = 1 - \text{Prob}\{Y_n \ge 204.5\}$), we find

$$\text{Prob}\{Y_n \ge 204.5\} = \text{Prob}\left\{\frac{Y_n - 200}{10} \ge 0.45\right\} = \text{Prob}\{Z \ge 0.45\} = 1 - 0.6736 = 0.3264,$$

which turns out to be a much better approximation than that obtained previously.

The same approach may be used when we are required to find the probability that $X$ lies in some interval $[a, b]$. The formula to use in this case is

$$\text{Prob}\{a \le X \le b\} \approx \Phi\left(\frac{b + \frac{1}{2} - \mu}{\sigma}\right) - \Phi\left(\frac{a - \frac{1}{2} - \mu}{\sigma}\right).$$

**Example 8.7** Returning to the previous example, let us now compute $\text{Prob}\{201 \le X \le 204\}$. Inserting $a = 201$ and $b = 204$ into the approximation given above, we obtain

$$\text{Prob}\{201 \le X \le 204\} \approx \text{Prob}\left\{\frac{Y_n - 200}{10} \le 0.45\right\} - \text{Prob}\left\{\frac{Y_n - 200}{10} \le 0.05\right\}$$

$$= \text{Prob}\{Z \le 0.45\} - \text{Prob}\{Z \le 0.05\} = 0.6736 - 0.5199 = 0.1537.$$

Indeed, even when this interval shrinks down to a single point, the continuity approximation provides a meaningful approximation, even though for a continuous random variable the probability of it assuming a single point is zero.

**Example 8.8** To use the normal approximation to find the probability of getting exactly 204 heads in 400 tosses of a fair coin, we set both $a$ and $b$ equal to 204 and write

$$\text{Prob}\{X = 204\} \approx \text{Prob}\left\{\frac{Y_n - 200}{10} \leq 0.45\right\} - \text{Prob}\left\{\frac{Y_n - 200}{10} \leq 0.35\right\}$$

$$= \text{Prob}\{Z \leq 0.45\} - \text{Prob}\{Z \leq 0.35\} = 0.6736 - 0.6368 = 0.0368.$$

The use of $a - \frac{1}{2}$ and $b + \frac{1}{2}$ rather than $a$ and $b$ is essential when approximating a binomial distribution for which $\sqrt{np(1-p)}$ is small. It is also essential when $a$ and $b$ are close together, irrespective of the value of $\sqrt{np(1-p)}$. As we noted above, the continuity correction should be applied when the normal distribution is used to approximate an integer-valued random variable. In our examples, the range of the discrete random variable has always been the set of integers. However, the range of a discrete random variable can be the set $\{0, \pm\delta, \pm 2\delta, \pm 3\delta, \ldots\}$ for any real number $\delta$, and the continuity correction still conveniently applies.

A final word of warning. Not all distributions obey the central limit theorem. One example is the Cauchy random variable, which has the probability density function

$$f(x) = \frac{1}{\pi(1 + x^2)}.$$

The expectation and all higher moments of a Cauchy random variable are not defined. In particular, it does not have a finite variance and therefore does not satisfy the conditions of the central limit theorem.

## 8.6 Exercises

**Exercise 8.1.1** Let $X$ be the random variable that denotes the number of trials until the first success in a probability experiment consisting of a sequence of Bernoulli trials, each having probability of success $p = 1/4$. Use the Markov inequality to find an upper bound on the probability that $X > 5$ and compare this with the exact probability, computed from the geometric distribution.

**Exercise 8.1.2** Let ten seconds be the mean time between the arrival of electronic orders to purchase stocks. The orders are (almost) instantaneously routed by a clerk from his computer to a trader on the stock floor. The clerk would like to go next door to buy a sandwich for lunch, but expects that it will take about two minutes. What can be said about the probability of no orders arriving during the time the clerk is off buying lunch?

**Exercise 8.1.3** Construct an example for which the upper bound obtained by the Markov inequality gives a value in excess of 1.0 (and therefore serves no useful purpose, since all probabilities must be less than or equal to 1).

**Exercise 8.2.1** An *unfair* coin, which produces heads only once out of every five attempts, is tossed 100 times. Let $X$ be the random variable that denotes the number of heads obtained. Show that $E[X] = 20$ and that $\text{Var}[X] = 16$. Now find an upper bound on the probability that $X \geq 60$, first using the Markov inequality and second using the Chebychev inequality.

**Exercise 8.2.2** Complete Example 8.3 by finding the bounds obtained from the Markov and Chebychev inequalities as well as the exact value of the probability.

**Exercise 8.2.3** Consider an experiment in which two fair dice are repeatedly thrown. Let $X$ be the random variable that denotes the number of throws needed to observe the second occurrence of two 6's. Apply both the Markov inequality and the Chebychev inequality to bound the probability that at least 200 tosses are needed.

**Exercise 8.2.4** The probability of an event occurring in one trial is 0.5. Use Chebychev's inequality to show that the probability of this event occurring between 450 and 550 times in 1000 independent trials exceeds 0.90.

**Exercise 8.2.5** The local garden shop keeps an average of 60 rose bushes in stock with variance equal to 64. A customer arrives and wishes to purchase 36 rose bushes. Use Chebychev's inequality to determine the likelihood of this request being met.

**Exercise 8.2.6** A farmer agrees to sell batches of 12 dozen apples to grocery stores. He estimates that 2% of his apples are bad and wishes to provide the grocery stores with guarantees whereby, if more than $k$ apples per batch are bad, he will refund the entire cost of the batch to the grocery store. How large should $k$ be so that the farmer will not have to reimburse grocery store owners more than 5% of the time?

**Exercise 8.2.7** Let $X$ be an exponentially distributed random variable with parameter $\lambda = 1$. Compare the upper bound on the probability Prob$\{X > 4\}$ obtained from the Chebychev inequality and the exact value of this probability.

**Exercise 8.3.1** Let $X_i$, $i = 1, 2, 3, 4, 5$, be independent exponentially distributed random variables each with the same parameter $\mu = 1/4$ and let $X = \sum_{i=1}^{5} X_i$. It follows from earlier remarks that $X$ has an Erlang-5 distribution. Use the Chernoff bound to find an upper bound on Prob$\{X > 40\}$. Compare this to the exact probability.

**Exercise 8.4.1** Show that the weak law of large numbers provides justification for the frequency interpretation of probability.

**Exercise 8.4.2** Apply Equation (8.5) to the case of $n$ independent Bernoulli random variables which take the value 1 with probability $p$, to show that

$$\text{Prob}\left\{\left|\frac{S_n}{n} - p\right| \geq \epsilon\right\} \leq \frac{p(1-p)}{n\epsilon^2}.$$

Now show that $p(1-p) \leq 1/4$ for $0 < p < 1$ and hence that

$$\text{Prob}\left\{\left|\frac{S_n}{n} - p\right| \geq \epsilon\right\} \leq \frac{1}{4n\epsilon^2}.$$

Observe that the upper bound in this second case is independent of $p$.

Compute both these bounds for the two separate cases $p = 1/2$ and $p = 1/10$. Now let $\epsilon = 0.1$. What do these two inequalities tell us about the value of $n$ needed in this case ($\epsilon = 0.1$) to obtain an upper bound of 0.01.

**Exercise 8.5.1** A coded message of 5 million bits is transmitted over the internet. Each bit is independent and the probability that any bit is 1 is $p = 0.6$. Use the central limit theorem approximation to estimate the probability that the number of 1's transmitted lies between 2,999,000 and 3,001,000.

**Exercise 8.5.2** Apply the central limit theorem to compute an approximation to the probability sought in Exercise 8.3.1. How does this approximation compare to the exact answer and the Chernoff bound?

**Exercise 8.5.3** A fair coin is tossed 400 times. Use the central limit theorem to compute an approximation to the probability of getting at most 204 heads.

**Exercise 8.5.4** Consider a system in which parts are routed by conveyor belts from one workstation to the next. Assume the time spent at each workstation is uniformly distributed on the interval [0, 4] minutes and that the time to place an object on the conveyor belt, transport it to and then remove it from the next station, is a constant time of 15 seconds. An object begins the process by being placed on the first conveyor belt which directs it to the first station and terminates immediately after being served at the eighth workstation. Find

(a) the mean and variance of the time an object spends in the system,
(b) the probability that the total time is greater than 25 minutes, and
(c) the probability that the total time is less than 20 minutes.

**Exercise 8.5.5** It is known that a sum of independent Poisson random variables is itself Poisson. In particular, if $X_1$ is Poisson with parameter $\lambda_1$ and $X_2$ is Poisson with parameter $\lambda_2$, then $X = X_1 + X_2$ is Poisson with parameter $\lambda_1 + \lambda_2$. Use this result to apply the central limit theorem approximation with continuity correction to find the probability that fewer than 380 customers arrive during peak demand hours, when it is known that the random variable that describes the arrival process during this time is Poisson with mean value $\lambda = 400$.

# Part II

# MARKOV CHAINS

# Chapter 9

# *Discrete- and Continuous-Time Markov Chains*

Figure 9.1. Andrei Andreevich Markov (1856–1922). Photo courtesy of the Markov family.

## 9.1 Stochastic Processes and Markov Chains

It is often possible to represent the behavior of a system, physical or mathematical, by describing all the different states it may occupy (possibly an infinite number) and by indicating how it moves among these states. The system being modeled is assumed to occupy one and only one state at any moment in time and its evolution is represented by transitions from state to state. These transitions are assumed to occur instantaneously; in other words, the actual business of moving from one state to the next consumes zero time. If the future evolution of the system depends only on its current state and not on its past history, then the system may be represented by a *Markov process*. Even when the system does not possess this *Markov property* explicitly, it is often possible to construct a corresponding implicit representation. Examples of the use of Markov processes may be found extensively throughout the biological, physical, and social sciences as well as in business and engineering.

As an example, consider the behavior of a frog who leaps from lily pad to lily pad on a pond or lake. The lily pads constitute the states of the system. Where the frog jumps to depends only upon what information it can deduce from its current lily pad—it has no memory and thus recalls nothing about the states (lily pads) it visited prior to its current position, nor even the length of time it has been on the present lily pad. The assumption of instantaneous transitions is justified by the fact that the time the frog spends in the air between two lily pads is negligible compared to the time it spends sitting on a lily pad. The information we would like to obtain concerning a system is the probability of being in a given state or set of states at a certain time after the system becomes operational. Often this time is taken to be sufficiently long that all influence of the initial starting state has been erased. Other measures of interest include the time taken until a certain state is reached for the first time. For example, in the case of the frog and the lily pond, if one of the lily pads were not really a lily pad but the nose of an alligator waiting patiently for breakfast, we would be interested in knowing how long the frog survives before being devoured by the alligator. In Markov chain terminology, such a state is referred to as an *absorbing state*, and in such cases we seek to determine the mean time to absorption.

A *Markov process* is a special type of *stochastic process*. A stochastic process is defined as a family of random variables $\{X(t), t \in T\}$. In other words, each $X(t)$ is a random variable and as such is defined on some probability space. The parameter $t$ usually represents time, so $X(t)$ denotes the value assumed by the random variable at time $t$. $T$ is called the *index set* or *parameter space* and is a subset of $(-\infty, +\infty)$. If the index set is discrete, e.g., $T = \{0, 1, 2, \ldots\}$, then we have a *discrete-(time) parameter* stochastic process; otherwise, if $T$ is continuous, e.g., $T = \{t : 0 \le t < +\infty\}$, we call the process a *continuous-(time) parameter* stochastic process. The values assumed by the random variables $X(t)$ are called *states*. The set of all possible states forms the *state space* of the process and this may be discrete or continuous. If the state space is *discrete*, the process is referred to as a *chain* and the states are usually identified with the set of natural numbers $\{0, 1, 2, \ldots\}$ or a subset of it. An example of a discrete state space is the number of customers at a service facility. An example of a continuous state space is the length of time a customer has been waiting for service. Other examples of a continuous state space could include the level of water in a dam or the temperature inside a nuclear reactor where the water level/temperature can be any real number in some continuous interval of the real axis. Thus, a stochastic process can have a discrete state space or a continuous state space, and may evolve at a discrete set of time points or continuously in time.

In certain models, the statistical characteristics in which we are interested, may dependent on the time $t$ at which the system is started. The evolution of a certain process that begins in mid-morning may be very different from the evolution of that same process if it begins in the middle of the night. A process whose evolution depends on the time at which it is initiated is said to be *nonstationary*. On the other hand, a stochastic process is said to be *stationary* when it is invariant under an arbitrary shift of the time origin. Mathematically, we say that a stochastic process is stationary if its joint distribution is invariant to shifts in time, i.e., if for any constant $\alpha$,

$$\begin{aligned}&\text{Prob}\{X(t_1) \le x_1,\ X(t_2) \le x_2,\ \ldots,\ X(t_n) \le x_n\}\\&= \text{Prob}\{X(t_1+\alpha) \le x_1,\ X(t_2+\alpha) \le x_2,\ \ldots,\ X(t_n+\alpha) \le x_n\}\end{aligned}$$

for all $n$ and all $t_i$ and $x_i$ with $i = 1, 2, \ldots, n$. Stationarity does not imply that transitions are not allowed to depend on the current situation. Indeed, the transition probabilities can depend on the amount of time that has elapsed from the moment the process is initiated (which is equal to $t$ time units if the process begins at time zero). When the transitions in a stochastic process do in fact depend upon the amount of time that has elapsed, the stochastic process is said to be *nonhomogeneous*. When these transitions are independent of the elapsed time, it is said to be *homogeneous*. In either case (homogeneous or nonhomogeneous) the stochastic process may, or may not, be stationary. If it is stationary, its evolution may change over time (nonhomogeneous) but this evolution will be the same irrespective of when the process was initiated. Henceforth, in this text, our concern is with stationary processes only.

A *Markov process* is a stochastic process whose conditional probability distribution function satisfies the *Markov* or *memoryless* property. Consequently, we may define discrete-time Markov chains (DTMCs) and continuous-time Markov chains (CTMCs) as well as discrete-time Markov processes and continuous-time Markov processes. Our interest lies only in (discrete-state) Markov chains in both discrete and continuous time and we shall leave (continuous-state) Markov processes to more advanced texts.

## 9.2 Discrete-Time Markov Chains: Definitions

For a discrete-time Markov chain, we observe its state at a discrete, but infinite, set of times. Transitions from one state to another can only take place, or fail to take place, at these, somewhat abstract, time instants—time instants that are mostly taken to be one time unit apart. Therefore we may represent, without loss of generality, the discrete index set $T$ of the underlying stochastic process by the set of natural numbers $\{0, 1, 2, \ldots\}$. The successive observations define the random variables $X_0, X_1, \ldots, X_n, \ldots$ at time steps $0, 1, \ldots, n, \ldots$, respectively. Formally, a discrete-time Markov chain $\{X_n,\ n = 0, 1, 2, \ldots\}$ is a stochastic process that satisfies the following relationship, called the *Markov property*:

For all natural numbers $n$ and all states $x_n$,

$$\begin{aligned}\text{Prob}\{X_{n+1} &= x_{n+1}|X_n = x_n, X_{n-1} = x_{n-1}, \ldots, X_0 = x_0\}\\ &= \text{Prob}\{X_{n+1} = x_{n+1}|X_n = x_n\}.\end{aligned} \tag{9.1}$$

Thus, the fact that the system is in state $x_0$ at time step 0, in state $x_1$ at time step 1, and so on, up to the fact that it is in state $x_{n-1}$ at time step $n-1$ is completely irrelevant. The state in which the system finds itself at time step $n+1$ depends only on where it is at time step $n$. The fact that the Markov chain is in state $x_n$ at time step $n$ is the sum total of all the information concerning the history of the chain that is relevant to its future evolution.

To simplify the notation, rather than using $x_i$ to represent the states of a Markov chain, henceforth we shall use single letters, such as $i$, $j$, and $k$. The conditional probabilities $\text{Prob}\{X_{n+1} = x_{n+1}|X_n = x_n\}$, now written as $\text{Prob}\{X_{n+1} = j|X_n = i\}$, are called the *single-step transition probabilities*, or just the transition probabilities, of the Markov chain. They give the conditional probability of making a transition from state $x_n = i$ to state $x_{n+1} = j$ when the time parameter increases from $n$ to $n+1$. They are denoted by

$$p_{ij}(n) = \text{Prob}\{X_{n+1} = j|X_n = i\}. \tag{9.2}$$

The matrix $P(n)$, formed by placing $p_{ij}(n)$ in row $i$ and column $j$, for all $i$ and $j$, is called the *transition probability matrix* or *chain matrix*. We have

$$P(n) = \begin{pmatrix} p_{00}(n) & p_{01}(n) & p_{02}(n) & \cdots & p_{0j}(n) & \cdots \\ p_{10}(n) & p_{11}(n) & p_{12}(n) & \cdots & p_{1j}(n) & \cdots \\ p_{20}(n) & p_{21}(n) & p_{22}(n) & \cdots & p_{2j}(n) & \cdots \\ \vdots & \vdots & \vdots & \vdots & \vdots & \vdots \\ p_{i0}(n) & p_{i1}(n) & p_{i2}(n) & \cdots & p_{ij}(n) & \cdots \\ \vdots & \vdots & \vdots & \vdots & \vdots & \vdots \end{pmatrix}.$$

Notice that the elements of the matrix $P(n)$ satisfy the following two properties:

$$0 \le p_{ij}(n) \le 1,$$

and, for all $i$,

$$\sum_{\text{all } j} p_{ij}(n) = 1.$$

A matrix that satisfies these properties is called a *Markov matrix* or *stochastic matrix*.

A Markov chain is said to be *(time-)homogeneous* if for all states $i$ and $j$

$$\text{Prob}\{X_{n+1} = j|X_n = i\} = \text{Prob}\{X_{n+m+1} = j|X_{n+m} = i\}$$

for $n = 0, 1, 2, \ldots$ and $m \geq 0$. To elaborate on this a little further, we have, for a *homogeneous* Markov chain,

$$p_{ij} = \text{Prob}\{X_1 = j|X_0 = i\} = \text{Prob}\{X_2 = j|X_1 = i\} = \text{Prob}\{X_3 = j|X_2 = i\} = \cdots$$

and we have replaced $p_{ij}(n)$ with $p_{ij}$, since transitions no longer depend on $n$.

For a *nonhomogeneous* Markov chain,

$$p_{ij}(0) = \text{Prob}\{X_1 = j|X_0 = i\} \neq \text{Prob}\{X_2 = j|X_1 = i\} = p_{ij}(1).$$

As we have just seen, in a homogeneous discrete-time Markov chain the transition probabilities $p_{ij}(n) = \text{Prob}\{X_{n+1} = j|X_n = i\}$ are independent of $n$ and are consequently written as

$$p_{ij} = \text{Prob}\{X_{n+1} = j|X_n = i\}, \quad \text{for all } n = 0, 1, 2, \ldots.$$

It follows then that the matrix $P(n)$ should be replaced with the matrix $P$.

The evolution of the Markov chain in time is as follows. Assume that it begins at some initial time 0 and in some initial state $i$. Given some (infinite) set of time steps, the chain may change state at each of these steps but only at these time steps. For example, if at time step $n$ the Markov chain is in state $i$, then its next move will be to state $j$ with probability $p_{ij}(n)$. In particular, if $p_{ii}(n) > 0$, a situation referred to as a *self-loop*, the chain will, with probability $p_{ii}(n)$, remain in its current state.

The probability of being in state $j$ at time step $n+1$ *and* in state $k$ at time step $n+2$, given that the Markov chain is in state $i$ at time step $n$, is

$$\begin{aligned}
&\text{Prob}\{X_{n+2} = k, X_{n+1} = j \mid X_n = i\}\\
&= \text{Prob}\{X_{n+2} = k \mid X_{n+1} = j, X_n = i\}\text{Prob}\{X_{n+1} = j \mid X_n = i\}\\
&= \text{Prob}\{X_{n+2} = k \mid X_{n+1} = j\}\text{Prob}\{X_{n+1} = j \mid X_n = i\}\\
&= p_{jk}(n+1)p_{ij}(n),
\end{aligned}$$

where we have first used properties of conditional probability and second used the Markov property itself. A sequence of states visited by the chain is called a *sample path*: $p_{jk}(n+1)p_{ij}(n)$ is the probability of the sample path $i, j, k$ that begins in state $i$ at time step $n$. More generally,

$$\begin{aligned}
&\text{Prob}\{X_{n+m} = a, X_{n+m-1} = b, \ldots, X_{n+2} = k, X_{n+1} = j \mid X_n = i\} \qquad (9.3)\\
&= \text{Prob}\{X_{n+m} = a \mid X_{n+m-1} = b\}\text{Prob}\{X_{n+m-1} = b \mid X_{n+m-2} = c\}\cdots\\
&\quad \cdots\text{Prob}\{X_{n+2} = k|X_{n+1} = j\}\text{Prob}\{X_{n+1} = j \mid X_n = i\}\\
&= p_{ba}(n+m-1)p_{cb}(n+m-2)\cdots p_{jk}(n+1)p_{ij}(n).
\end{aligned}$$

When the Markov chain is homogeneous, we find

$$\text{Prob}\{X_{n+m} = a, X_{n+m-1} = b, \ldots, X_{n+2} = k, X_{n+1} = j|X_n = i\} = p_{ij}\, p_{jk} \cdots p_{cb}\, p_{ba}$$

for all possible values of $n$.

Markov chains are frequently illustrated graphically. Circles or ovals are used to represent states. Single-step transition probabilities are represented by directed arrows, which are frequently, but not always, labeled with the values of the transition probabilities. The absence of a directed arrow indicates that no single-step transition is possible.

**Example 9.1** A weather model. Consider a homogeneous, discrete-time Markov chain that describes the daily weather pattern in Belfast, Northern Ireland (well known for its prolonged periods of rainy days). We simplify the situation by considering only three types of weather pattern: rainy, cloudy, and sunny. These three weather conditions describe the three states of our Markov chain: state 1 ($R$) represents a (mostly) rainy day; state 2 ($C$), a (mostly) cloudy day; and state 3 ($S$), a (mostly) sunny day. The weather is observed daily. On any given rainy day, the probability that it will rain the next day is estimated at 0.8; the probability that the next day will be cloudy is 0.15, while the probability that tomorrow will be sunny is only 0.05. Similarly, probabilities may be assigned when a particular day is cloudy or sunny as shown in Figure 9.2.This is the probability of the sample

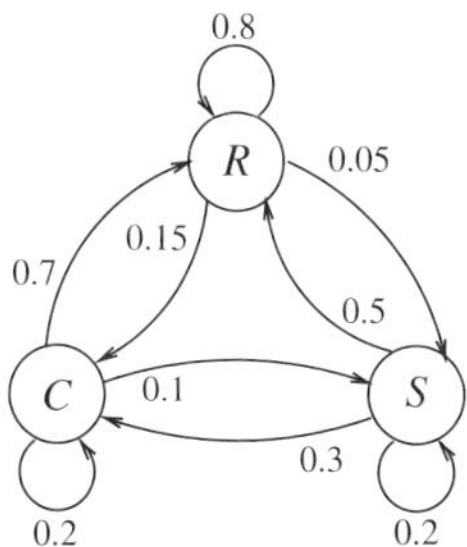


Figure 9.2. Transition diagram for weather at Belfast.

The transition probability matrix $P$ for this Markov chain is

$$P = \begin{pmatrix} 0.8 & 0.15 & 0.05 \\ 0.7 & 0.2 & 0.1 \\ 0.5 & 0.3 & 0.2 \end{pmatrix}. \tag{9.4}$$

Note that the elements in $P$ represent *conditional probabilities*. For example, the element $p_{32}$ tells us that the probability that tomorrow is cloudy, *given* that today is sunny, is 0.3. Using Equation (9.4), we may compute quantities such as the probability that tomorrow is cloudy and the day after is rainy, given that it is sunny today. In this case we have, for all $n = 0, 1, 2, \ldots$,

$$\text{Prob}\{X_{n+2} = R, X_{n+1} = C \mid X_n = S\} = p_{SC}\, p_{CR} = 0.3 \times 0.7 = 0.21.$$

This is the probability of the sample path $S \rightarrow C \rightarrow R$.

**Example 9.2** A social mobility model. Sociologists broadly categorize the population of a country into upper-, middle-, and lower-class brackets. One of their concerns is to monitor the movement of successive generations among these three classes. This may be modeled by a discrete-time Markov chain $\{X_n,\ n \geq 0\}$, where $X_n$ gives the class of the $n^{\text{th}}$ generation of a family. To satisfy the Markov property, it is assumed that the class of any generation depends only on the class of its parent generation and not on the class of its grandparent generation or other ancestors. A possible transition probability matrix for this Markov is given by

$$P = \begin{pmatrix} 0.45 & 0.50 & 0.05 \\ 0.15 & 0.65 & 0.20 \\ 0.00 & 0.50 & 0.50 \end{pmatrix},$$

where states $U = 1$, $M = 2$, and $L = 3$ represent the upper, middle, and lower class respectively. Thus, with probabilities 0.45, 0.5, and 0.05, a generation born into the upper class will itself remain in the upper class, move to the middle class, or even to the lower class, respectively. On the other hand, the generation born into the lower class is equally likely to remain in the lower class or move to the middle class. According to this modeling scenario, a generation born into the lower class cannot move, in one generation, into the upper class. To find the probability that a second generation will end up in the upper class we need to compute

$$\text{Prob}\{X_{n+2} = U, X_{n+1} = M | X_n = L\} = p_{LM}p_{MU} = 0.5 \times 0.15 = 0.075,$$

i.e., the probability of the sample path $L \to M \to U$.

**Example 9.3** The Ehrenfest model. This model arises from considering two volumes of gas that are connected by a small opening. This opening is so small that only one molecule can pass through it at any time. The model may also be viewed as two buckets containing balls. At each time instant, a ball is chosen at random from one of the buckets and moved into the other. A state of the system is given by the number of balls (or molecules) in the first bucket (or volume of gas). Let $X_n$ be the random variable that gives the number of balls in the first bucket after the $n^{\text{th}}$ selection (or time step). Then $\{X_n, n = 1, 2, \ldots\}$ is a Markov chain, since the number of balls in the first bucket at time step $n+1$ depends only on the number that is present in this bucket at time step $n$. Suppose that initially there is a total of $N$ balls partitioned between the two buckets. If at any time step $n$, there are $k < N$ balls in bucket 1, the probability that there are $k + 1$ after the next exchange is given by

$$\text{Prob}\{X_{n+1} = k + 1 \mid X_n = k\} = \frac{N - k}{N}.$$

If the number in the first bucket is to increase, the ball that is exchanged must be one of the $N - k$ that are in the second bucket and, since a ball is chosen at random, the probability of choosing one of these $N - k$ is equal to $(N - k)/N$. Similarly, if $k \geq 1$,

$$\text{Prob}\{X_{n+1} = k - 1 \mid X_n = k\} = \frac{k}{N}.$$

These are the only transitions that are possible. For example, if $N = 6$, the transition probability matrix has seven rows and columns corresponding to the states in which there are zero through six balls in the first bucket and we have

$$P = \begin{pmatrix} 0 & 6/6 & 0 & 0 & 0 & 0 & 0 \\ 1/6 & 0 & 5/6 & 0 & 0 & 0 & 0 \\ 0 & 2/6 & 0 & 4/6 & 0 & 0 & 0 \\ 0 & 0 & 3/6 & 0 & 3/6 & 0 & 0 \\ 0 & 0 & 0 & 4/6 & 0 & 2/6 & 0 \\ 0 & 0 & 0 & 0 & 5/6 & 0 & 1/6 \\ 0 & 0 & 0 & 0 & 0 & 6/6 & 0 \end{pmatrix}.$$

**Example 9.4** A nonhomogeneous Markov chain. Let us now take an example of a nonhomogeneous discrete-time Markov chain. Consider a two-state Markov chain $\{X_n, n = 1, 2, \ldots, \}$ in which we denote the two states $a$ and $b$, respectively. At time step $n$, the probability that the Markov chain remains in its current state is given by $p_{aa}(n) = p_{bb}(n) = 1/n$, while the probability that it changes state is given by $p_{ab}(n) = p_{ba}(n) = (n - 1)/n$—see Figure 9.3. The first four transition probability

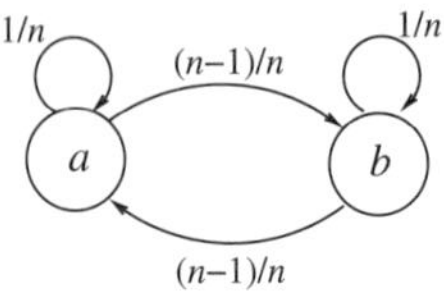


Figure 9.3. Two-state nonhomogeneous Markov chain.

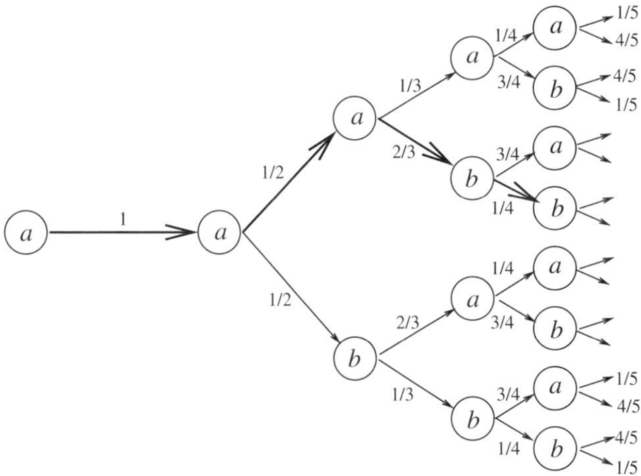


Figure 9.4. Sample paths in a nonhomogeneous Markov chain.

matrices are given by

$$P(1) = \begin{pmatrix} 1 & 0 \\ 0 & 1 \end{pmatrix}, \quad P(2) = \begin{pmatrix} 1/2 & 1/2 \\ 1/2 & 1/2 \end{pmatrix}, \quad P(3) = \begin{pmatrix} 1/3 & 2/3 \\ 2/3 & 1/3 \end{pmatrix}, \quad P(4) = \begin{pmatrix} 1/4 & 3/4 \\ 3/4 & 1/4 \end{pmatrix},$$

while the $n^{\text{th}}$ is

$$P(n) = \begin{pmatrix} 1/n & (n-1)/n \\ (n-1)/n & 1/n \end{pmatrix}.$$

As a consequence, the probability of changing state at each time step increases with time, while the probability of remaining in the same state at each time step decreases with time. The important point to note is that, although these probabilities change with time, the process is still Markovian and that at any point, the future evolution does not depend on what has happened in the past, but only on the current state in which the system finds itself. To see this more clearly, let us examine some sample paths beginning in state $a$. These are shown graphically in Figure 9.4.

Starting in state $a$, then, according to $P(1)$, after a single time step, the Markov chain remains in state $a$. In the next time step, according to $P(2)$, the chain either stays in state $a$ with probability $1/2$ or moves to state $b$ with the same probability. And so the process unfolds. Figure 9.4 highlights a particular sample path in bold type. This is the path that begins in state $a$, stays in state $a$ after the first and second time steps, moves to state $b$ on the third time step, and remains in state $b$ on the fourth time step. The probability that this path is taken is the product of the probabilities of taking each segment and, using Equation (9.3), is

$$\begin{aligned} &\text{Prob}\{X_5 = b, X_4 = b, X_3 = a, X_2 = a | X_1 = a\} \\ &= p_{aa}(1) p_{aa}(2) p_{ab}(3) p_{bb}(4) \\ &= 1 \times 1/2 \times 2/3 \times 1/4 = 1/12. \end{aligned}$$

Other paths lead to state $b$ after four transitions, and have different probabilities according to the route they follow. What is important is that, no matter which route is chosen, once the Markov chain arrives in state $b$ after four steps, the future evolution is specified by $P(5)$, and not any other $P(i)$, $i \leq 4$. On the figure, all transition probabilities leading out of any $b$ in the right most column, are the same. The future evolution o f the system depends only on the fact that the system is in state $b$ at time step 4.

### $k$-Dependent Markov Chains

A stochastic process is not a Markov chain if its evolution depends on more than its current state, for instance, if transitions at step $n+1$ depend not only on the state occupied at time step $n$, but also on the state occupied by the process at time step $n-1$. Suppose, in the Belfast weather example, we are told that, if there are two rainy days in a row, then the probabilities that the following day is rainy, cloudy, or sunny are given as $(0.6, 0.3, 0.1)$, respectively. On the other hand, if a sunny or cloudy day is followed by a rainy day, the probabilities that tomorrow is rainy, cloudy, or sunny are as before, i.e., $(0.8, 0.15, 0.05)$, respectively. Thus the probability of transitions from a rainy day depend not only on the fact that today it is raining, but also on the weather yesterday; hence the stochastic process with states $R$, $C$, and $S$ is not a Markov chain. However, at the cost of increasing the number of states, it can be made into a Markov chain. In this example, it suffices to add a single state, one corresponding to the situation in which it has rained for two consecutive days. Let us denote this new state $RR$. The previous rainy state, which we now denote just by $R$ describes the situation when the day prior to a rainy day was either cloudy or sunny. The transition probability diagram for this new Markov chain is shown in Figure 9.5 where we have implicitly assumed that the remaining probabilities (cloudy day followed by rainy, cloudy, or sunny day, i.e., the previous row 2; and sunny day followed by rainy, cloudy, or sunny day, i.e., the previous row 3) are unchanged.

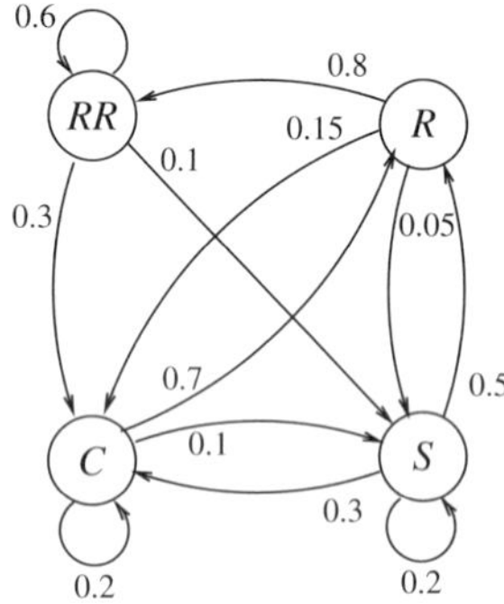


Figure 9.5. Modified transition diagram for Belfast weather.

This technique of converting a non-Markovian process to a Markov chain by incorporating additional states may be generalized. If a stochastic process has $s$ states and is such that transitions from each state depend on the history of the process during the two prior steps, then a new process consisting of $s^2$ states may be defined as a Markov chain. If the weather in Belfast on any particular day depends on the weather during the preceding two days, then we may define a nine-state Markov chain in which each state is a pair, $RR$, $RC$, $RS$, $CR$, $CC$, $CS$, $SR$, $SC$, $SS$, each of which characterizes the weather on the previous two days. Naturally, we may extend this even further. If a stochastic process that consists of $s$ states is such that transitions depend on the state of the system during the prior $k$ time steps, then we may construct a Markov chain with $s^k$ states. Referring back to the definition of the Markov property, Equation (9.1), we state this more formally as follows:

Let $\{X_n,\ n \geq 0\}$ be a stochastic process such that there exists an integer $k$ for which

$$\begin{aligned}\text{Prob}\{X_{n+1} &= x_{n+1} | X_n = x_n, \ldots, X_{n-k+1} = x_{n-k+1}, X_{n-k} = x_{n-k}, \ldots, X_0 = x_0\}\\ &= \text{Prob}\{X_{n+1} = x_{n+1} | X_n = x_n, \ldots, X_{n-k+1} = x_{n-k+1}\}\end{aligned}$$

for all $n \geq k$. Thus, the future of the process depends on the previous $k$ states occupied. Such a process is said to be a *$k$-dependent process*. If $k = 1$, then $X_n$ is a Markov chain. For $k > 1$, a new stochastic process $\{Y_n,\ n \geq 0\}$, with

$$Y_n = (X_n, X_{n+1}, \ldots, X_{n+k-1}),$$

is a Markov chain. If the set of states of $X_n$ is denoted by $\mathcal{S}$, then the states of the Markov chain $Y_n$ are the elements of the cross product

$$\underbrace{\mathcal{S} \times \mathcal{S} \times \cdots \times \mathcal{S}}_{k \text{ terms}}.$$

**The Sojourn Time**

In Examples 9.1, 9.2, and 9.4, the diagonal elements of the transition probability matrices are all nonzero and strictly less than 1. This means that at any time step the system may remain in its current state. The number of consecutive time periods that a Markov chain remains in a given state is called the *sojourn time* or *holding time* of that state. At each time step, the probability of leaving any state $i$ is independent of what occurred at previous time steps and, for a homogeneous Markov chain, is equal to

$$\sum_{i \neq j} p_{ij} = 1 - p_{ii}.$$

This process may be identified with a sequence of Bernoulli trials with probability of success (taken to mean an exit from state $i$) at each trial equal to $1 - p_{ii}$. The probability that the sojourn time is equal to $k$ time steps is that of having $k - 1$ consecutive Bernoulli failures followed by a single success. It follows immediately that the sojourn time at state $i$, which we denote by $R_i$, has probability mass function

$$\text{Prob}\{R_i = k\} = (1 - p_{ii})p_{ii}^{k-1}, \quad k = 1, 2, \ldots, \qquad 0 \text{ otherwise},$$

which is the geometric distribution with parameter $1 - p_{ii}$. In other words, the distribution of the sojourn times in any state of a homogeneous discrete-time Markov chain has a geometric distribution. Recall that the geometric distribution is the only discrete distribution that has the memoryless property—a sequence of $j - 1$ unsuccessful trials has no effect on the probability of success on the $j^{\text{th}}$ attempt. The mean and variance of the sojourn time are respectively given by

$$E[R_i] = \frac{1}{1 - p_{ii}} \quad \text{and} \quad \text{Var}\,[R_i] = \frac{p_{ii}}{(1 - p_{ii})^2}. \tag{9.5}$$

We stress that. this result holds only when the Markov chain is homogeneous. When the Markov chain is nonhomogeneous, and in state $i$ at time step $n$, the probability mass function of $R_i(n)$, the random variable that represents the remaining number of time steps in state $i$, is

$$\text{Prob}\{R_i(n) = k\} = p_{ii}(n) \times p_{ii}(n + 1) \times \cdots \times p_{ii}(n + k - 2) \times [1 - p_{ii}(n + k - 1)].$$

This is not a geometric distribution—it becomes geometric when $p_{ii}(n) = p_{ii}$ for all $n$.

**The Embedded Markov Chain**

In the last example, that of the extended state space for the weather at Belfast, the state $R$ must be exited after spending exactly one time unit in that state. In the diagram, there is no self-loop on the state $R$, unlike the other states. A rainy day is followed by another rainy day (the state $RR$), or a cloudy day $C$, or a sunny day $S$. These are the only possibilities. This leads to an alternative way of viewing a homogeneous Markov chain. Until now, we considered the evolution of the system at each time step, and this included the possibility of the state being able to transition back to itself (or simply remain where it is) at any time step. We now view the Markov chain only at those time steps at which an actual change of state takes place, neglecting (or more correctly, postponing) any consideration of the sojourn time in the different states. For all rows $i$ for which $0 < p_{ii} < 1$, we need to alter that row by removing the self-loop—by setting $p_{ii} = 0$—and replacing the elements $p_{ij}$

with probabilities *conditioned* on the fact that the system transitions out of state $i$—the probabilities $p_{ij}$, $i \neq j$, must be replaced with $p_{ij}/(1 - p_{ii})$. Notice that the new matrix remains a stochastic transition probability matrix since, for all $i$,

$$\sum_j \frac{p_{ij}}{1 - p_{ii}} = \frac{p_{ii}}{1 - p_{ii}} + \sum_{j \neq i} \frac{p_{ij}}{1 - p_{ii}} = \sum_{j \neq i} \frac{p_{ij}}{1 - p_{ii}} = \frac{1}{1 - p_{ii}} \sum_{j \neq i} p_{ij} = \frac{1}{1 - p_{ii}} \times (1 - p_{ii}) = 1.$$

It is the transition probability matrix of a different discrete-time Markov chain. This new Markov chain is called the *embedded* Markov chain, since it is a Markov chain embedded in the original Markov chain—the embedding points of time being precisely those at which an actual transition out of a state occurs. An embedded Markov chain has no self-loops and to identify it with the chain from which it originates, we must also specify the sojourn time in each state. Embedded Markov chains are more commonly associated with continuous-time processes that may or may not be Markovian, and we shall return to this in later sections, where we shall see that a study of the embedded chain together with information concerning the sojourn time in the states can often be used to facilitate the analysis of continuous-time stochastic processes. Finally, notice that the diagonal elements of Example 9.3 are all zero and that with this example, we only observe the Markov chain when a molecule (ball) actually moves.

**Example 9.5** The original transition probability matrix $P$ for the example of the weather at Belfast is given by

$$P = \begin{pmatrix} 0.8 & 0.15 & 0.05 \\ 0.7 & 0.2 & 0.1 \\ 0.5 & 0.3 & 0.2 \end{pmatrix}.$$

The transition diagram for the corresponding embedded Markov chain is given in Figure 9.6.

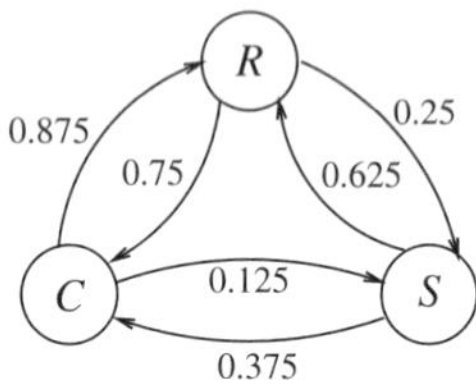


Figure 9.6. Embedded Markov chain for weather at Belfast.

Its transition probability matrix is given by

$$\begin{pmatrix} 0.0 & 0.75 & 0.25 \\ 0.875 & 0.0 & 0.125 \\ 0.625 & 0.375 & 0.0 \end{pmatrix},$$

and the parameters (probability of success) for the geometric distributions representing the sojourn in the three states are respectively given by 0.2, 0.8, and 0.8.

## 9.3 The Chapman-Kolmogorov Equations

In the previous section we saw how, in a discrete-time Markov chain and using Equation 9.3, we could compute the probability of following a sample path from a given starting state. For example, in the weather example, we saw that the probability of rain tomorrow, followed by clouds the day after, given that today is sunny, is $p_{SR}p_{RC} = 0.075$. We would now like to answer questions such as how to find the probability that it is cloudy two days from now, given that it is sunny today. To do so, we must examine the sample paths that includes both sunny and cloudy days as the intermediate day (tomorrow) as well as the one we have just computed (intermediate day is rainy)—the weather on the intermediate day can be any of these three possibilities. Thus, the probability that it is cloudy two days from now, given that today is sunny, is

$$\begin{aligned} &p_{SR}p_{RC} + p_{SC}p_{CC} + p_{SS}p_{SC} \\ &= \sum_{w=R,C,S} p_{Sw}p_{wC} \\ &= 0.5 \times 0.15 + 0.3 \times 0.2 + 0.2 \times 0.3 = 0.195, \end{aligned}$$

as shown graphically in Figure 9.7.

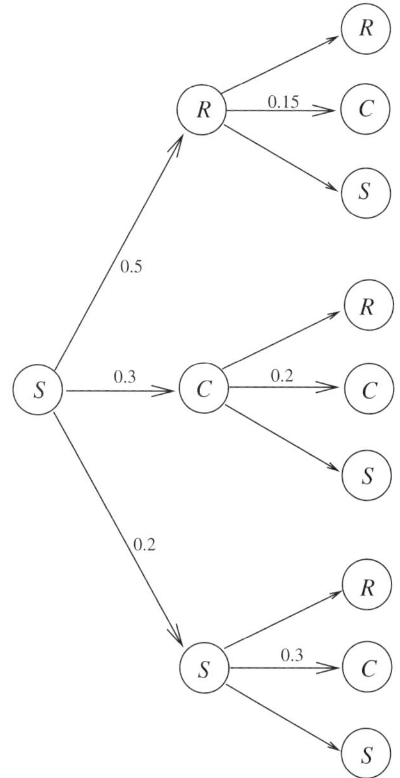


Figure 9.7. Sample paths of length 2 beginning on a sunny day.

This result may be obtained directly from the single-step transition probability matrix of the Markov chain. Recall that this is given by

$$P = \begin{pmatrix} 0.8 & 0.15 & 0.05 \\ 0.7 & 0.2 & 0.1 \\ 0.5 & 0.3 & 0.2 \end{pmatrix}.$$

The result we seek is obtained when we form the inner product of row 3 of $P$, i.e., $r_3 = (.5, .3, .2)$, which corresponds to transitions *from* a sunny day, with column 2 of $P$, i.e., $c_2 = (.15, .2, .3)^T$, which corresponds to transitions *into* a cloudy day. This gives $r_3c_2 = 0.5 \times 0.15 + 0.3 \times 0.2 + 0.2 \times 0.3 = 0.195$ as before. Now suppose that the weather today is given by $w_1$, where $w_1 \in \{R, C, S\}$.

The probability that, two days from now, the weather is $w_2$ where $w_2 \in \{R, C, S\}$, is obtained by forming the inner product of row $w_1$ with column $w_2$. All nine different possibilities constitute the elements of the matrix $P^2$. We have

$$P^2 = \begin{pmatrix} 0.8 & 0.15 & 0.05 \\ 0.7 & 0.2 & 0.1 \\ 0.5 & 0.3 & 0.2 \end{pmatrix} \times \begin{pmatrix} 0.8 & 0.15 & 0.05 \\ 0.7 & 0.2 & 0.1 \\ 0.5 & 0.3 & 0.2 \end{pmatrix} = \begin{pmatrix} .770 & .165 & .065 \\ .750 & .175 & .075 \\ .710 & .195 & .095 \end{pmatrix}.$$

Thus the (2, 3) element, which is equal to 0.075, gives the probability that it is sunny in two days time, given that it is cloudy today. In a more general context, the $ij$ element of the square of a transition probability matrix of a homogeneous discrete-time Markov chain is the probability of being in state $j$ two time steps from now, given that the current state is state $i$. In a similar fashion, the elements of $P^3$ give the conditional probabilities three steps from now, and so on. For this particular example, it turns out that if we continue to take higher and higher powers of the matrix $P$, then this sequence converges to a matrix in which all rows are identical. We have

$$\lim_{n\to\infty} P^n = \begin{pmatrix} .76250 & .16875 & .06875 \\ .76250 & .16875 & .06875 \\ .76250 & .16875 & .06875 \end{pmatrix}. \tag{9.6}$$

In obtaining these results we have essentially used the theorem of total probability. We are required to sum over all sample paths of length 2 that lead from state $i$ to state $k$. For a homogeneous discrete-time Markov chain, we have, for any $n = 0, 1, 2, \ldots,$

$$\text{Prob}\{X_{n+2} = k,\ X_{n+1} = j | X_n = i\} = p_{ij}\, p_{jk},$$

and from the theorem of total probability, we obtain

$$\text{Prob}\{X_{n+2} = k | X_n = i\} = \sum_{\text{all } j} p_{ij}\, p_{jk}.$$

Observe that the right-hand side defines the $ik^{\text{th}}$ element of the matrix obtained when $P$ is multiplied with itself, i.e., the element $(P^2)_{ik} \equiv p_{ik}^{(2)}$. Continuing in this fashion, and observing that

$$\text{Prob}\{X_{n+3} = l,\ X_{n+2} = k,\ X_{n+1} = j | X_n = i\} = p_{ij}\, p_{jk}\, p_{kl},$$

we obtain, once again with the aid of the theorem of total probability,

$$\text{Prob}\{X_{n+3} = l | X_n = i\} = \sum_{\text{all } j}\sum_{\text{all } k} p_{ij}\, p_{jk}\, p_{kl} = \sum_{\text{all } j} p_{ij} \sum_{\text{all } k} p_{jk}\, p_{kl} = \sum_{\text{all } j} p_{ij}\, p_{jl}^{(2)},$$

which is the $il^{\text{th}}$ element of $P^3$, $p_{jl}^{(2)}$ being the $jl$ element of $P^2$.

It follows that we may generalize the single-step transition probability matrix of a homogeneous Markov chain to an $m$-step transition probability matrix whose elements

$$p_{ij}^{(m)} = \text{Prob}\{X_{n+m} = j | X_n = i\}$$

can be obtained from the single-step transition probabilities. Observe that $p_{ij} = p_{ij}^{(1)}$. We now show that $p_{ij}^{(m)}$ can be computed from the following recursive formula, called the *Chapman-Kolmogorov* equation:

$$p_{ij}^{(m)} = \sum_{\text{all } k} p_{ik}^{(l)}\, p_{kj}^{(m-l)} \quad \text{for } 0 < l < m.$$

For a homogeneous discrete-time Markov chain, we have

$$\begin{aligned} p_{ij}^{(m)} &= \text{Prob}\{X_m = j | X_0 = i\} \\ &= \sum_{\text{all k}} \text{Prob}\{X_m = j, X_l = k | X_0 = i\} \quad \text{for } 0 < l < m \\ &= \sum_{\text{all } k} \text{Prob}\{X_m = j | X_l = k, X_0 = i\}\text{Prob}\{X_l = k | X_0 = i\}. \end{aligned}$$

Now applying the Markov property, we find

$$\begin{aligned} p_{ij}^{(m)} &= \sum_{\text{all } k} \text{Prob}\{X_m = j | X_l = k\}\text{Prob}\{X_l = k | X_0 = i\} \\ &= \sum_{\text{all } k} p_{kj}^{(m-l)} p_{ik}^{(l)} \quad \text{for } 0 < l < m. \end{aligned}$$

In matrix notation, the Chapman-Kolmogorov equations are written as

$$P^{(m)} = P^{(l)} P^{(m-l)},$$

where, by definition, $P^{(0)} = I$, the identity matrix. This relation states that it is possible to write any $m$-step homogeneous transition probability matrix as the sum of products of $l$-step and $(m-l)$-step transition probability matrices. To go from $i$ to $j$ in $m$ steps, it is necessary to go from $i$ to an intermediate state $k$ in $l$ steps, and then from $k$ to $j$ in the remaining $m-l$ steps. By summing over all possible intermediate states $k$, we consider all possible distinct paths leading from $i$ to $j$ in $m$ steps. Note in particular that

$$P^{(m)} = PP^{(m-1)} = P^{(m-1)} P.$$

Hence, the matrix of $m$-step transition probabilities is obtained by multiplying the matrix of one-step transition probabilities by itself $(m-1)$ times. In other words, $P^{(m)} = P^m$.

For a nonhomogeneous discrete-time Markov chain, the matrices $P(n)$ may depend on the particular time step $n$. In this case the product $P^2$ must be replaced by the product $P(n)P(n+1)$, $P^3$ with $P(n)P(n+1)P(n+2)$ and so on. It follows that

$$P^{(m)}(n, n+1, \ldots, n+m-1) = P(n)P(n+1) \cdots P(n+m-1)$$

is a matrix whose $ij$ element is $\text{Prob}\{X_{n+m} = j | X_n = i\}$.

Let $\pi_i^{(0)}$ be the probability that the Markov chain begins in state $i$, and let $\pi^{(0)}$ be the row vector whose $i^{\text{th}}$ element is $\pi_i^{(0)}$. Then the $j^{\text{th}}$ element of the vector that results from forming the product $\pi^{(0)} P(0)$ gives the probability of being in state $j$ after the first time step. We write this as

$$\pi^{(1)} = \pi^{(0)} P(0).$$

For a homogeneous Markov chain, this becomes

$$\pi^{(1)} = \pi^{(0)} P.$$

The elements of the vector $\pi^{(1)}$ provide the probability of being in the various states of the Markov chain (the probability distribution) after the first time step. For example, consider the Belfast weather example and assume that observations begin at time step 0 with the weather being cloudy. Thus, $\pi^{(0)} = (0,\ 1,\ 0)$ and

$$\pi^{(1)} = \pi^{(0)} P(0) = (0,\ 1,\ 0) \begin{pmatrix} 0.8 & 0.15 & 0.05 \\ 0.7 & 0.2 & 0.1 \\ 0.5 & 0.3 & 0.2 \end{pmatrix} = (0.7,\ 0.2,\ 0.1),$$

which just returns the second row of the matrix $P$. So, starting on day zero with a cloudy day, the probability that day one is also cloudy is 0.2, as previously computed. To compute the probability of being in any state (the probability distribution) after two time steps, we need to form

$$\pi^{(2)} = \pi^{(1)} P(1) = \pi^{(0)} P(0) P(1).$$

For a homogeneous Markov chain, this becomes

$$\pi^{(2)} = \pi^{(1)} P = \pi^{(0)} P^2,$$

and for the Belfast example, we find

$$\pi^{(2)} = \pi^{(1)} P = \pi^{(0)} P^2 = (0,\ 1,\ 0) \begin{pmatrix} .770 & .165 & .065 \\ .750 & .175 & .075 \\ .710 & .195 & .095 \end{pmatrix} = (0.750,\ 0.175,\ 0.075).$$

Thus the probability that it is cloudy two days after observations begin on a cloudy day is given by 0.175. In computing this quantity, we have summed over all sample paths of length 2 that begin and end in state 2.

In general, after $n$ steps, the probability distribution is given by

$$\pi^{(n)} = \pi^{(n-1)} P(n-1) = \pi^{(0)} P(0) P(1) \cdots P(n-1)$$

or, for a homogeneous Markov chain,

$$\pi^{(n)} = \pi^{(n-1)} P = \pi^{(0)} P^n.$$

In forming the $j^{\text{th}}$ component of this vector, we effectively compute the summation over all sample paths of length $n$ which start, with probability $\pi_i^{(0)}$, in state $i$ at time step 0 and end in state $j$. Letting $n \to \infty$, for the Belfast example, we find

$$\lim_{n\to\infty} \pi^{(n)} = \pi^{(0)} \lim_{n\to\infty} P^n = (0,\ 1,\ 0) \begin{pmatrix} .76250 & .16875 & .06875 \\ .76250 & .16875 & .06875 \\ .76250 & .16875 & .06875 \end{pmatrix} = (0.76250,\ 0.16875,\ 0.06875).$$

We hasten to point out that the limit, $\lim_{n\to\infty} \pi^{(n)}$, does not necessarily exist for all Markov chains, nor even for all finite-state Markov chains. We shall return to this topic in a later section. Over the course of the remainder of this chapter, we shall consider homogeneous Markov chains only, the sole exception being certain definitions in the section devoted to *continuous-time* Markov chains. From time to time, however, we shall include the word "homogeneous" just as a gentle reminder.

## 9.4 Classification of States

We now proceed to present a number of important definitions concerning the individual states of (homogeneous) discrete-time Markov chains. Later we shall address the classification of groups of states, and indeed, the classification of the Markov chain itself. We distinguish between states that are *recurrent*, meaning that the Markov chain is guaranteed to return to these states infinitely often, and states that are *transient*, meaning that there is a nonzero probability that the Markov chain will never return to such a state. This does not mean that a transient state cannot be visited many times, only that the probability of never returning to it is nonzero. A transient state can be visited only a finite number of times.

Figure 9.8 illustrates different possibilities. No probabilities are drawn on the paths in this figure: instead, it is implicitly assumed that the sum of the probabilities on paths exiting from any

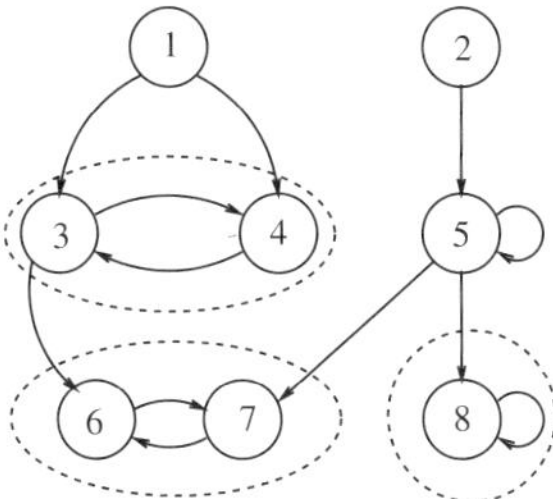


Figure 9.8. Recurrent and transient states.

state is 1. We may make the following observations:

- States 1 and 2 are transient states. Furthermore, the figure shows that the Markov chain can be in state 1 or 2 only at the very first time step. Transient states that cannot exist beyond the initial time step are said to be *ephemeral* states.
- States 3 and 4 are transient states. The Markov chain can enter either of these states and move from one to the other for a number of time steps, but eventually it will exit from state 3 to enter state 6.
- State 5 is also a transient state. The Markov chain will enter this state from state 2 at the first time step, if state 2 is the initial state occupied. Once in state 5, the Markov chain may, for some finite number of time steps, remain in state 5, but eventually it will move on to either state 7 or state 8.
- States 6 and 7 are recurrent states. If the Markov chain reaches one of these states, then all subsequent transitions will take it from one to the other. Notice that, when the Markov chain is in state 6, it returns to state 6 every second transition. The same is true of state 7. Returns to the states in this group must take place at time steps that are multiples of 2. Such states are said to be *periodic* of period 2. Furthermore, it is apparent that the mean recurrence time (which is equal to 2) is finite. Recurrent states whose mean recurrence time is finite are said to be *positive recurrent*. Recurrent states whose mean recurrence time are infinite are said to be *null recurrent*. Infinite mean recurrence times can only occur when the Markov chain has an infinite number of states.
- Finally, state 8 is also a recurrent state. If the Markov chain ever arrives in this state, it will remain there forever. A state with this property is said to be an *absorbing* state. State $i$ is an absorbing state if and only if $p_{ii} = 1$. A nonabsorbing state $i$ must have $p_{ii} < 1$ and may be either transient or recurrent.

We now proceed to a more formal description of these characteristics. In the previous section we saw that $p_{jj}^{(n)}$ is the probability that the Markov chain is once again in state $j$, $n$ time steps after leaving it. In these intervening steps, it is possible that the process visited many different states as well as state $j$ itself. Based on this definition, let us define a new conditional probability, the probability that on leaving state $j$ the *first return* to state $j$ occurs $n$ steps later. We shall denote this quantity $f_{jj}^{(n)}$ and we have

$$\begin{aligned} f_{jj}^{(n)} &= \text{Prob}\,\{\text{first return to state } j \text{ occurs exactly } n \text{ steps after leaving it}\} \\ &= \text{Prob}\,\{X_n = j, X_{n-1} \neq j, \ldots, X_1 \neq j | X_0 = j\} \quad \text{for } n = 1, 2, \ldots. \end{aligned}$$

This should not be confused with $p_{jj}^{(n)}$, which is the probability of returning to state $j$ in $n$ steps, without excluding the possibility that state $j$ was visited at one or more intermediate steps.

We shall now relate $p_{jj}^{(n)}$ and $f_{jj}^{(n)}$ and construct a recurrence relation that permits us to compute $f_{jj}^{(n)}$. We have already seen how to compute $p_{jj}^{(n)}$ by taking higher powers of the single-step transition probability matrix $P$. Notice that $f_{jj}^{(1)} = p_{jj}^{(1)} = p_{jj}$; i.e., the probability that the first return to state

$j$ occurs one step after leaving it, is just the single-step transition probability $p_{jj}$. Since $p_{jj}^{(0)} = 1$, we may write this as

$$p_{jj}^{(1)} = f_{jj}^{(1)} p_{jj}^{(0)}.$$

Now consider $p_{jj}^{(2)}$, the probability of being in state $j$ two time steps after leaving it. This can happen because the Markov chain simply does not move from state $j$ at either time step or else because it leaves state $j$ on the first time step and returns on the second. In order to fit our analysis to the recursive formulation, we interpret these two possibilities as follows.

1. The Markov chain "leaves" state $j$ and "returns" to it for the first time after one step, which has probability $f_{jj}^{(1)}$, and then "returns" again at the second step, which has probability $p_{jj}^{(1)}$. The verbs, "leaves" and "returns" are in quotation marks since in this case the Markov chain simply stays in state $j$ at each of these two time steps, rather than actually changing state.
2. The Markov chain leaves state $j$ and does *not* return for the first time until two steps later, which has probability $f_{jj}^{(2)}$.

Thus

$$p_{jj}^{(2)} = f_{jj}^{(1)} p_{jj}^{(1)} + f_{jj}^{(2)} = f_{jj}^{(1)} p_{jj}^{(1)} + f_{jj}^{(2)} p_{jj}^{(0)}$$

and hence $f_{jj}^{(2)}$ may be computed from

$$f_{jj}^{(2)} = p_{jj}^{(2)} - f_{jj}^{(1)} p_{jj}^{(1)}.$$

In a similar manner, we may write an expression for $p_{jj}^{(3)}$, the probability that the process is in state $j$ three steps after leaving it. This occurs if the first return to state $j$ is after one step and in the remaining two steps the process may have gone elsewhere, but has returned to state $j$ by the end of these two steps; or if the process returns to state $j$ for the first time after two steps and in the third step remains in state $j$; or finally, if the first return to state $j$ is three steps after leaving it. These are the only possibilities and combining them gives

$$p_{jj}^{(3)} = f_{jj}^{(1)} p_{jj}^{(2)} + f_{jj}^{(2)} p_{jj}^{(1)} + f_{jj}^{(3)} p_{jj}^{(0)},$$

from which $f_{jj}^{(3)}$ is easily computed as

$$f_{jj}^{(3)} = p_{jj}^{(3)} - f_{jj}^{(1)} p_{jj}^{(2)} - f_{jj}^{(2)} p_{jj}^{(1)}.$$

Continuing in this fashion, essentially by applying the theorem of total probability and using $p_{jj}^{(0)} = 1$, it may be shown that

$$p_{jj}^{(n)} = \sum_{l=1}^{n} f_{jj}^{(l)} p_{jj}^{(n-l)}, \qquad n \geq 1. \tag{9.7}$$

Hence $f_{jj}^{(n)}$, $n \geq 1$, may be calculated recursively as

$$f_{jj}^{(n)} = p_{jj}^{(n)} - \sum_{l=1}^{n-1} f_{jj}^{(l)} p_{jj}^{(n-l)}, \qquad n \geq 1.$$

The probability of ever returning to state $j$ is denoted by $f_{jj}$ and is given by

$$f_{jj} = \sum_{n=1}^{\infty} f_{jj}^{(n)}.$$

If $f_{jj} = 1$, then state $j$ is said to be *recurrent*. In other words, state $j$ is recurrent if and only if, beginning in state $j$, the probability of returning to $j$ is 1, i.e., the Markov chain is guaranteed to return to this state in the future. In this case, we have $p_{jj}^{(n)} > 0$ for some $n > 0$. In fact, since the Markov chain is guaranteed to return to this state, it must return to it infinitely often. Thus the expected number of visits that the Markov chain makes to a recurrent state $j$ given that it starts in state $j$ is infinite. We now show that the expected number of visits that the Markov chain makes to state $j$ given that it starts in state $j$ is equal to $\sum_{n=0}^{\infty} p_{jj}^{(n)}$ and hence it must follow that $\sum_{n=0}^{\infty} p_{jj}^{(n)} = \infty$ when state $j$ is a recurrent state. Let $I_n = 1$ if the Markov chain is in state $j$ at time step $n$ and $I_n = 0$ otherwise. Then $\sum_{n=0}^{\infty} I_n$ is the total number of time steps that state $j$ is occupied. Conditioning on the fact that the Markov chain starts in state $j$, the expected number of time steps it is in state $j$ is

$$E\left[\sum_{n=0}^{\infty} I_n | X_0 = j\right] = \sum_{n=0}^{\infty} E\left[I_n | X_0 = j\right] = \sum_{n=0}^{\infty} \text{Prob}\{X_n = j | X_0 = j\} = \sum_{n=0}^{\infty} p_{jj}^{(n)}. \tag{9.8}$$

Thus, when state $j$ is recurrent,

$$\sum_{n=0}^{\infty} p_{jj}^{(n)} = \infty.$$

If $f_{jj} < 1$, then state $j$ is said to be *transient*. There is a nonzero probability that the Markov chain will never return to this state. The expected number of times the Markov chain returns to state $j$ is therefore finite. On each occasion that the Markov chain is in state $j$, the probability that it will never return is $1 - f_{jj}$. The returns to state $j$ may be identified with a sequence of Bernoulli trials which counts the number of trials up to and including the first "success," where "success" is taken to mean that the Markov chain will not return to state $j$. The probability that the Markov chain, upon leaving state $j$, will return exactly $n - 1$ times ($n \geq 1$) and then never return again, is equal to $(1 - f_{jj})f_{jj}^{n-1}$ which is none other than the geometric probability mass function. Thus the mean number of times the Markov chain will return to state $j$ is equal to $1/(1 - f_{jj})$ which is finite. It follows that, when state $j$ is transient,

$$\sum_{n=0}^{\infty} p_{jj}^{(n)} < \infty.$$

When state $j$ is recurrent, i.e., when $f_{jj} = 1$, we define the *mean recurrence time* $M_{jj}$ of state $j$ as

$$M_{jj} = \sum_{n=1}^{\infty} n f_{jj}^{(n)}.$$

This is the average number of steps taken to return to state $j$ for the first time after leaving it. A recurrent state $j$ for which $M_{jj}$ is finite is called a *positive recurrent* state or a *recurrent nonnull* state. If $M_{jj} = \infty$, we say that state $j$ is a *null recurrent* state. Theorem 9.4.1 follows immediately from these definitions.

**Theorem 9.4.1** *In a finite Markov chain*

- *No state is null recurrent.*
- *At least one state must be positive recurrent, i.e., not all states can be transient.*

If all its states were transient, a Markov chain would spend a finite amount of time in each of them, after which it would have nowhere else to go. But this is impossible, and so there must be at least one positive-recurrent state.

**Example 9.6** Consider the discrete-time Markov chain whose transition diagram is shown in Figure 9.9.

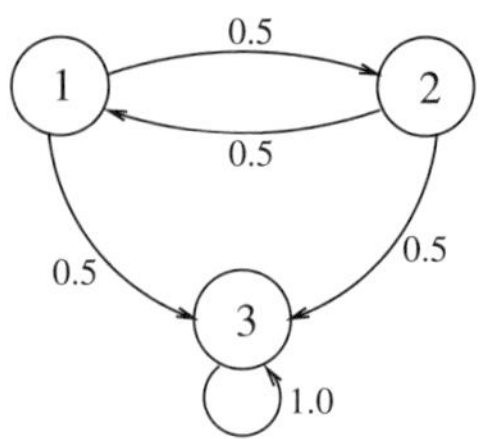


Figure 9.9. Transition diagram for Example 9.6.

Its transition probability matrix is given by

$$P = \begin{pmatrix} 0 & 1/2 & 1/2 \\ 1/2 & 0 & 1/2 \\ 0 & 0 & 1 \end{pmatrix}.$$

Observe that successive powers of $P$ are given by

$$P^k = \begin{cases} \begin{pmatrix} 0 & (1/2)^k & 1-(1/2)^k \\ (1/2)^k & 0 & 1-(1/2)^k \\ 0 & 0 & 1 \end{pmatrix} & \text{if } k = 1, 3, 5, \ldots, \\[2ex] \begin{pmatrix} (1/2)^k & 0 & 1-(1/2)^k \\ 0 & (1/2)^k & 1-(1/2)^k \\ 0 & 0 & 1 \end{pmatrix} & \text{if } k = 2, 4, 6, \ldots. \end{cases}$$

It follows then that

$$\begin{aligned} f_{11}^{(1)} &= p_{11}^{(1)} = 0, \\ f_{11}^{(2)} &= p_{11}^{(2)} - f_{11}^{(1)} p_{11}^{(1)} = (1/2)^2 - 0 = (1/2)^2, \\ f_{11}^{(3)} &= p_{11}^{(3)} - f_{11}^{(2)} p_{11}^{(1)} - f_{11}^{(1)} p_{11}^{(2)} = 0 - (1/2)^2 \times 0 - 0 \times (1/2)^2 = 0, \\ f_{11}^{(4)} &= p_{11}^{(4)} - f_{11}^{(3)} p_{11}^{(1)} - f_{11}^{(2)} p_{11}^{(2)} - f_{11}^{(1)} p_{11}^{(3)} = (1/2)^4 - 0 - (1/2)^2 \times (1/2)^2 - 0 = 0, \end{aligned}$$

and in general,

$$f_{11}^{(k)} = 0 \text{ for all } k \geq 3.$$

In fact, from the transition probability diagram, it is evident that the first return to state 1 *must* occur after two steps; the first return cannot be at any other time. Thus, $f_{11} = \sum_{n=1}^{\infty} f_{11}^{(n)} = 1/4 < 1$ and hence state 1 is *transient*. A similar result applies to state 2. Since this is a finite Markov chain consisting of three states, two of which are transient, it must follow that the third state is positive recurrent. We may show this explicitly as follows. We have

$$\begin{aligned} f_{33}^{(1)} &= p_{33}^{(1)} = 1, \\ f_{33}^{(2)} &= p_{33}^{(2)} - f_{33}^{(1)} p_{33}^{(1)} = 1 - 1 = 0, \\ f_{33}^{(3)} &= p_{33}^{(3)} - f_{33}^{(1)} p_{33}^{(2)} - f_{33}^{(2)} p_{33}^{(1)} = 1 - 1 - 0 = 0, \end{aligned}$$

and so on. We have

$$f_{33}^{(k)} = 0 \text{ for all } k \geq 2.$$

which, once again, must evidently be the case, as seen in Figure 9.9. Thus $f_{33} = \sum_{n=1}^{\infty} f_{33}^{(n)} = 1$ and state 3 is recurrent. Its mean recurrence time is

$$M_{33} = \sum_{n=1}^{\infty} n f_{33}^{(n)} = 1,$$

and since this is finite, state 3 is positive recurrent.

So far in this section, we have been concerned with transitions from any state back to that same state again. Now let us consider transitions between two different states. Corresponding to $f_{jj}^{(n)}$, let us define $f_{ij}^{(n)}$ for $i \neq j$ as the probability that, starting from state $i$, the first passage to state $j$ occurs in exactly $n$ steps. We then have $f_{ij}^{(1)} = p_{ij}$, and as was the case for Equation 9.7, we may derive

$$p_{ij}^{(n)} = \sum_{l=1}^{n} f_{ij}^{(l)} p_{jj}^{(n-l)}, \qquad n \geq 1.$$

This equation may be rearranged to obtain

$$f_{ij}^{(n)} = p_{ij}^{(n)} - \sum_{l=1}^{n-1} f_{ij}^{(l)} p_{jj}^{(n-l)},$$

which is more convenient for finding $f_{ij}^{(n)}$. The probability $f_{ij}$ that state $j$ is ever reached from state $i$ is given by

$$f_{ij} = \sum_{n=1}^{\infty} f_{ij}^{(n)}.$$

If $f_{ij} < 1$, the process starting from state $i$ may never reach state $j$. When $f_{ij} = 1$, the expected value of the sequence $f_{ij}^{(n)}$, $n = 1, 2, \ldots$, of first passage probabilities for a fixed pair $i$ and $j$ ($i \neq j$) is called the *mean first passage time* and is denoted by $M_{ij}$. We have

$$M_{ij} = \sum_{n=1}^{\infty} n f_{ij}^{(n)} \quad \text{for } i \neq j.$$

The $M_{ij}$ uniquely satisfy the equation

$$M_{ij} = p_{ij} + \sum_{k \neq j} p_{ik}(1 + M_{kj}) = 1 + \sum_{k \neq j} p_{ik} M_{kj}, \tag{9.9}$$

since the process in state $i$ either goes to state $j$ in one step (with probability $p_{ij}$), or else goes first to some intermediate state $k$ in one step (with probability $p_{ik}$) and then eventually on to $j$ in an additional $M_{kj}$ steps. If $i = j$, then $M_{ij}$ is the mean recurrence time of state $i$ and Equation 9.9 continues to hold.

We now write Equation 9.9 in matrix form. We shall use the letter $e$ to denote a (column) vector whose components are all equal to 1 and whose length is determined by its context. Likewise, we shall use $E$ to denote a square matrix whose elements are all equal to 1. Notice that $E = ee^T$. Letting diag$\{M\}$ be the diagonal matrix whose $i^{\text{th}}$ diagonal element is $M_{ii}$, it follows that

$$M_{ij} = 1 + \sum_{k \neq j} p_{ik} M_{kj} = 1 + \sum_{k} p_{ik} M_{kj} - p_{ij} M_{jj}.$$

Equation (9.9) may be written in matrix form as

$$M = E + P(M - \text{diag}\{M\}). \tag{9.10}$$

The diagonal elements of $M$ are the mean recurrence times, whereas the off-diagonal elements are the mean first passage times. The matrix $M$ may be obtained iteratively from the equation

$$M^{(k+1)} = E + P(M^{(k)} - \text{diag}\{M^{(k)}\}) \quad \text{with } M^{(0)} = E. \tag{9.11}$$

As Example 9.8 will show, not all of the elements of $M^{(k+1)}$ need converge to a finite limit. The mean recurrence times of certain states as well as certain mean first passage times may be infinite, and the elements of $M^{(k+1)}$ that correspond to these situations will diverge as $k \to \infty$. Divergence of some elements in this context is an "acceptable" behavior for this iterative approach.

The matrix $F$, whose $ij^{\text{th}}$ component is $f_{ij}$, is called the *reachability matrix*. This matrix is examined in more detail in Section 9.6, where we explore an alternative way to compute the quantities $f_{ij}$, the probability of ever visiting state $j$ on leaving state $i$. In that section, we shall also describe how the expected number of visits to any state $j$ after leaving some state $i$ may be found.

A final property of the states of a Markov chain that is of interest to us is that of *periodicity*. A state $j$ is said to be *periodic with period $p$*, or *cyclic of index $p$*, if on leaving state $j$ a return is possible only in a number of transitions that is a multiple of the integer $p > 1$. In other words, the period of a state $j$ is defined as the greatest common divisor of the set of integers $n$ for which $p_{jj}^{(n)} > 0$. A state whose period is $p = 1$ is said to be *aperiodic*. Observe, in Example 9.6, that states 1 and 2 are periodic. On leaving either of these states, a return is possible only in a number of steps that is an integer multiple of 2. On the other hand, state 3 is aperiodic. A state that is positive recurrent and aperiodic is said to be *ergodic*. If all the states of a Markov chain are ergodic, then the Markov chain itself is said to be ergodic.

To conclude this section, we present some initial results concerning the limiting behavior of $P^n$ as $n \to \infty$. In a later section, we shall delve much more deeply into $P^n$ and its limit as $n \to \infty$. Recalling that the $ij^{\text{th}}$ element of the $n^{\text{th}}$ power of $P$, i.e., $P^n$ is written as $p_{ij}^{(n)}$, we have the following result.

**Theorem 9.4.2** *Let $j$ be a state of a discrete-time Markov chain.*

- *If state $j$ is a null recurrent or transient state, and $i$ is any state of the chain, then*

$$\lim_{n\to\infty} p_{ij}^{(n)} = 0.$$

- *If $j$ is positive recurrent and aperiodic (i.e., ergodic), then*

$$\lim_{n\to\infty} p_{jj}^{(n)} > 0,$$

*and for any other state $i$, positive recurrent, transient, or otherwise,*

$$\lim_{n\to\infty} p_{ij}^{(n)} = f_{ij} \lim_{n\to\infty} p_{jj}^{(n)}.$$

**Example 9.7** The Markov chain with transition probability $P$ given below has two transient states 1 and 2, and two ergodic states 3 and 4. The matrix $P$ and $\lim_{n\to\infty} P^n$ are

$$P = \begin{pmatrix} .4 & .5 & .1 & 0 \\ .3 & .7 & 0 & 0 \\ 0 & 0 & 0 & 1 \\ 0 & 0 & .8 & .2 \end{pmatrix}, \qquad \lim_{n\to\infty} P^n = \begin{pmatrix} 0 & 0 & 4/9 & 5/9 \\ 0 & 0 & 4/9 & 5/9 \\ 0 & 0 & 4/9 & 5/9 \\ 0 & 0 & 4/9 & 5/9 \end{pmatrix}.$$

Since states 1 and 2 are transient, $\lim_{n\to\infty} p_{ij}^{(n)} = 0$ for $i = 1, 2, 3, 4$ and $j = 1, 2$. Since states 3 and 4 are ergodic, $\lim_{n\to\infty} p_{jj}^{(n)} > 0$ for $j = 3, 4$ while since $f_{ij} = 1$ for $i = 1, 2, 3, 4,\ j = 3, 4$ and $i \neq j$:

$$\lim_{n\to\infty} p_{ij}^{(n)} = f_{ij} \lim_{n\to\infty} p_{jj}^{(n)} = \lim_{n\to\infty} p_{jj}^{(n)} > 0, \quad i = 1, 2, 3, 4, \quad j = 3, 4, \quad i \neq j.$$

**Example 9.8** Consider a homogeneous discrete-time Markov chain whose transition probability matrix is

$$P = \begin{pmatrix} a & b & c \\ 0 & 0 & 1 \\ 0 & 1 & 0 \end{pmatrix} \tag{9.12}$$

with $0 < a < 1$. Note that, in this example,

$$\begin{aligned} p_{11}^{(n)} &= a^n && \text{for } n = 1, 2, \ldots, \\ p_{22}^{(n)} &= p_{33}^{(n)} = 0 && \text{for } n = 1, 3, 5, \ldots, \\ p_{22}^{(n)} &= p_{33}^{(n)} = 1 && \text{for } n = 0, 2, 4, \ldots, \end{aligned}$$

and that

$$p_{12}^{(n)} = ap_{12}^{(n-1)} + b \times 1_{\{n \text{ is odd}\}} + c \times 1_{\{n \text{ is even}\}},$$

where $1_{\{\cdot\}}$ is an *indicator function*, which has the value 1 when the condition inside the braces is true and the value 0 otherwise. Then

$$\begin{aligned} f_{11}^{(1)} &= p_{11}^{(1)} = a, \\ f_{11}^{(2)} &= p_{11}^{(2)} - f_{11}^{(1)} p_{11}^{(1)} = a^2 - a \times a = 0, \\ f_{11}^{(3)} &= p_{11}^{(3)} - f_{11}^{(1)} p_{11}^{(2)} - f_{11}^{(2)} p_{11}^{(1)} = p_{11}^{(3)} - f_{11}^{(1)} p_{11}^{(2)} = 0. \end{aligned}$$

It immediately follows that $f_{11}^{(n)} = 0$ for all $n \geq 2$, and thus the probability of ever returning to state 1 is given by $f_{11} = a < 1$. State 1 is therefore a transient state. Also,

$$\begin{aligned} f_{22}^{(1)} &= p_{22}^{(1)} = 0, \\ f_{22}^{(2)} &= p_{22}^{(2)} - f_{22}^{(1)} p_{22}^{(1)} = p_{22}^{(2)} = 1, \\ f_{22}^{(3)} &= p_{22}^{(3)} - f_{22}^{(1)} p_{22}^{(2)} - f_{22}^{(2)} p_{22}^{(1)} = p_{22}^{(3)} = 0, \end{aligned}$$

and again it immediately follows that $f_{22}^{(n)} = 0$ for all $n \geq 3$. We then have $f_{22} = \sum_{n=1}^{\infty} f_{22}^{(n)} = f_{22}^{(2)} = 1$, which means that state 2 is recurrent. Furthermore, it is positive recurrent, since $M_{22} = \sum_{n=1}^{\infty} n f_{22}^{(n)} = 2 < \infty$. In a similar fashion, it may be shown that state 3 is also positive recurrent.

Now consider $f_{12}^{(n)}$:

$$\begin{aligned} f_{12}^{(1)} &= b, \\ f_{12}^{(2)} &= p_{12}^{(2)} - f_{12}^{(1)} p_{22}^{(1)} = p_{12}^{(2)} = ap_{12}^{(1)} + c = ab + c, \\ f_{12}^{(3)} &= p_{12}^{(3)} - f_{12}^{(1)} p_{22}^{(2)} - f_{12}^{(2)} p_{22}^{(1)} = p_{12}^{(3)} - f_{12}^{(1)} \\ &= (a^2 b + ac + b) - b = a^2 b + ac. \end{aligned}$$

Continuing in this fashion, we find

$$\begin{aligned} f_{12}^{(4)} &= a^3 b + a^2 c, \\ f_{12}^{(5)} &= a^4 b + a^3 c, \\ &\text{etc.,} \end{aligned}$$

and it is easy to show that in general we have

$$f_{12}^{(n)} = a^{n-1} b + a^{n-2} c.$$

It follows that the probability that state 2 is ever reached from state 1 is

$$f_{12} = \sum_{n=1}^{\infty} f_{12}^{(n)} = \frac{b}{1-a} + \frac{c}{1-a} = 1.$$

Similarly, we may show that $f_{13} = 1$. Also, it is evident that

$$f_{23}^{(1)} = f_{32}^{(1)} = 1$$

and

$$f_{23}^{(n)} = f_{32}^{(n)} = 0 \quad \text{for } n \geq 2$$

so that $f_{23} = f_{32} = 1$. However, note that

$$f_{21}^{(n)} = f_{31}^{(n)} = 0 \quad \text{for } n \geq 1,$$

and so state 1 can never be reached from state 2 or from state 3.

To examine the matrix $M$ of mean first passage times (with diagonal elements equal to the mean recurrence times), we shall give specific values to the variables $a$, $b$, and $c$. Let

$$P = \begin{pmatrix} 0.7 & 0.2 & 0.1 \\ 0.0 & 0.0 & 1.0 \\ 0.0 & 1.0 & 0.0 \end{pmatrix} \quad \text{and take } M^{(0)} = \begin{pmatrix} 1.0 & 1.0 & 1.0 \\ 1.0 & 1.0 & 1.0 \\ 1.0 & 1.0 & 1.0 \end{pmatrix}.$$

Then, using the iterative formula (9.11) we find

$$M^{(1)} = \begin{pmatrix} 1.3 & 1.8 & 1.9 \\ 2.0 & 2.0 & 1.0 \\ 2.0 & 1.0 & 2.0 \end{pmatrix}, \qquad M^{(2)} = \begin{pmatrix} 1.6 & 2.36 & 2.53 \\ 3.0 & 2.0 & 1.0 \\ 3.0 & 1.0 & 2.0 \end{pmatrix}, \qquad \text{etc.}$$

The iterative process tends to the matrix

$$M^{(\infty)} = \begin{pmatrix} \infty & 11/3 & 12/3 \\ \infty & 2 & 1 \\ \infty & 1 & 2 \end{pmatrix},$$

and it may be readily verified that this matrix satisfies Equation (9.10). Thus, the mean recurrence time of state 1 is infinite, as are the mean first passage times from states 2 and 3 to state 1. The mean first passage time from state 2 to state 3 or vice versa is given as 1, which must obviously be true since, on leaving either of these states, the process immediately enters the other. The mean recurrence time of both state 2 and state 3 is 2. States 2 and 3 are each periodic with period 2, since, on leaving either one of these states, a return to that same state is only possible in a number of steps that is a multiple of 2. These states are not ergodic.

Part 1 of Theorem 9.4.2 allows us to assert that the first column of $\lim_{n\to\infty} P^n$ contains only zero elements, since state 1 is transient. Since states 2 and 3 are both periodic, we are not in a position to apply the second part of this theorem.

## 9.5 Irreducibility

Having thus discussed the classification of individual states, we now move to classifications concerning groups of states. Let $S$ be the set of all states in a Markov chain, and let $S_1$ and $S_2$ be two subsets of states that partition $S$. The subset of states $S_1$ is said to be *closed* if no one-step transition is possible from any state in $S_1$ to any state in $S_2$. This is illustrated in Figure 9.10, where the subset consisting of states $\{4, 5, 6\}$ is closed. The subset containing states 1 through 3 is *not* closed. Notice also, that the set that contains all six states is closed. More generally, any nonempty

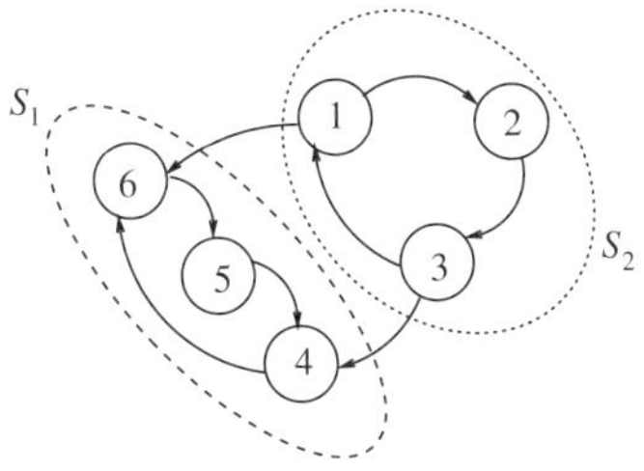


Figure 9.10. $S_1$ is a closed subset of states.

subset $S_1$ of $S$ is said to be closed if *no state* in $S_1$ leads to any state outside $S_1$ (in any number of steps), i.e.,

$$p_{ij}^{(n)} = 0 \quad \text{for } i \in S_1,\ j \notin S_1,\ n \geq 1.$$

If the closed subset $S_1$ consists of a single state, then that state is an absorbing state. Naturally enough, a set of states that is not closed is said to be *open*. It is apparent that any finite set of transient states must constitute an open set. Any individual state that is not an absorbing state constitutes by itself, an open set. If the set of all states $S$ is closed and does not contain any proper subset that is closed, then the Markov chain is said to be *irreducible*. On the other hand, if $S$ contains proper subsets that are closed, the chain is said to be *reducible*. The Markov chain drawn in Figure 9.10 is not an irreducible Markov chain. A closed subset of states is said to be an *irreducible subset* if it contains no proper subset that is closed. In Figure 9.10, the subset consisting of states 4, 5, and 6 is the unique irreducible subset of this Markov chain. In this example, no other subset of states constitutes an irreducible subset. Any proper subset of an irreducible subset constitutes a set of states that is open.

The matrix of transition probabilities of the Markov chain shown in Figure 9.10 has the following nonzero structure:

$$P = \left(\begin{array}{ccc|ccc} 0 & * & 0 & 0 & 0 & * \\ 0 & 0 & * & 0 & 0 & 0 \\ * & 0 & 0 & * & 0 & 0 \\ \hline 0 & 0 & 0 & 0 & 0 & * \\ 0 & 0 & 0 & * & 0 & 0 \\ 0 & 0 & 0 & 0 & * & 0 \end{array}\right) = \begin{pmatrix} D_{11} & U_{12} \\ L_{21} & D_{22} \end{pmatrix}.$$

In this matrix, the symbol $*$ represents a nonzero probability corresponding to a transition of the Markov chain. The matrix has been decomposed according to the partition $\{1, 2, 3\}$, $\{4, 5, 6\}$ into two diagonal blocks $D_{11}$ and $D_{22}$, and two off-diagonal blocks, $U_{12}$ and $L_{21}$, all of size $3 \times 3$. Observe that the lower off-diagonal block $L_{21}$ is identically equal to zero. This means that no transition is possible from any state represented by diagonal block $D_{22}$ to any state represented by diagonal block $D_{11}$. On the other hand, the upper off-diagonal block $U_{12}$ does contain nonzero elements signifying that transitions do occur from the states of $D_{11}$ to the states of $D_{22}$.

An alternative definition for a Markov chain to be irreducible can be enunciated in terms of the reachability of the states. State $j$ is said to be *reachable* or *accessible* from state $i$ if there exists a path from state $i$ to state $j$. We write this as $i \rightarrow j$. Thus in the above example, there exists a path from state 3 through states 4 and 6 to state 5 (with nonzero probability equal to $p_{34}p_{46}p_{65} > 0$) so state 5 is accessible from state 3. On the other hand, there is no path from state 5 to state 3 and hence state 3 is not reachable from state 5. A discrete-time Markov chain is *irreducible* if every state is reachable from every other state, i.e., if there exists an integer $n$ for which $p_{ij}^{(n)} > 0$ for every pair of states $i$ and $j$. In the example of Figure 9.10, it is not possible for any of the states 4, 5, or 6 to reach states 1, 2, or 3. The converse is possible, i.e., it is possible to reach states 4, 5, and 6 from any of states 1, 2, or 3. This Markov chain becomes irreducible if we insert a path from any of states 4, 5, or 6 to state 1, 2, or 3.

If state $j$ is reachable from state $i$ ($i \rightarrow j$) and state $i$ is reachable from state $j$ ($j \rightarrow i$) then states $i$ and $j$ are said to be *communicating* states and we write $i \leftrightarrow j$. By its very nature, this communication property is symmetric, transitive, and reflexive and thus constitutes an equivalence relationship. We have, for any states $i$, $j$, and $k$,

$$i \leftrightarrow j \Longrightarrow j \leftrightarrow i,$$
$$i \leftrightarrow j \text{ and } j \leftrightarrow k \Longrightarrow i \leftrightarrow k,$$
$$i \leftrightarrow j \text{ and } j \leftrightarrow i \Longrightarrow i \leftrightarrow i.$$

The first of these must obviously hold from the definitions. To show that the second holds, consider the following: $i \leftrightarrow j$ implies that $i \rightarrow j$ and thus there exists an $n_1 > 0$ for which $p_{ij}^{(n_1)} > 0$. Similarly, $j \leftrightarrow k$ implies that $j \rightarrow k$ and there exists an $n_2 > 0$ for which $p_{jk}^{(n_2)} > 0$. Set $n = n_1 + n_2$. Then, from the Chapman-Kolmogorov equation, we have

$$p_{ik}^{(n)} = \sum_{\text{all } l} p_{il}^{(n_1)} p_{lk}^{(n_2)} \geq p_{ij}^{(n_1)} p_{jk}^{(n_2)} > 0,$$

and so $i \rightarrow k$. It may similarly be shown that $k \rightarrow i$, and the required result follows. The third (reflexive) relationship follows directly from the second (transitive). A state that communicates with itself in this fashion is called a *return* state. A *nonreturn* state is one that does not communicate with itself. As its name implies, once a Markov chain leaves a nonreturn state, it will never return.

The set of all states that communicate with state $i$ forms a *class* and is denoted by $C(i)$. This may be the empty set since it is possible that a state communicates with no other state, not even itself. For example, this is the case of an ephemeral state, a transient state that can only be occupied initially and which departs to some other state at the very first time step. On the other hand, any state $i$ for which $p_{ii} > 0$ is a return state. It follows that the states of a Markov chain may be partitioned into communicating classes and nonreturn states. Furthermore, the communicating classes may, or may not, be closed. If state $i$ is recurrent, the communicating class to which it belongs is closed. Only transient states can belong to nonclosed communicating classes.

Notice that, if state $i$ is recurrent and $i \rightarrow j$, then state $j$ must communicate with state $i$, i.e., $i \leftrightarrow j$. There is a path from $i$ to $j$ and since $i$ is recurrent, after leaving $j$ we must return at some point to $i$, which shows that there must be a path from $j$ to $i$. Furthermore, in this case, state $j$ must be also be recurrent. Since we return to state $i$ infinitely often, and $j$ is reachable from $i$, we can also return to state $j$ infinitely often. Since $i$ and $j$ communicate, we have

$$p_{ij}^{(n_1)} > 0 \quad \text{and} \quad p_{ji}^{(n_2)} > 0 \quad \text{for some } n_1 > 0,\ n_2 > 0,$$

and since state $i$ is recurrent, there exists an integer $n > 0$ such that

$$p_{jj}^{(n_2+n+n_1)} \geq p_{ji}^{(n_2)} p_{ii}^{(n)} p_{ij}^{(n_1)} > 0.$$

Since $i$ is recurrent, $\sum_{n=1}^{\infty} p_{ii}^{(n)} = \infty$, and it follows that state $j$ is recurrent $\left(\sum_{m=1}^{\infty} p_{jj}^{(m)} = \infty\right)$, since

$$\sum_{n=1}^{\infty} p_{ji}^{(n_2)} p_{ii}^{(n)} p_{ij}^{(n_1)} = p_{ji}^{(n_2)} p_{ij}^{(n_1)} \sum_{n=1}^{\infty} p_{ii}^{(n)} = \infty.$$

Thus recurrent states can only reach other recurrent states: no transient state can be reached from a recurrent state and the set of recurrent states must be closed. If state $i$ is a recurrent state, then $C(i)$ is an irreducible closed set and contains only recurrent states. Furthermore, all these states must be positive recurrent or they all must be null recurrent. A Markov chain in which all the states belong to the same communicating class is *irreducible*. The following theorems concerning *irreducible* discrete-time Markov chains follow immediately from these definitions.

**Theorem 9.5.1** *An irreducible, discrete-time Markov chain is positive recurrent or null recurrent or transient; i.e.,*

- *all the states are positive recurrent, or*
- *all the states are null recurrent, or*
- *all the states are transient.*

*Furthermore, all states are periodic with the same period p, or else all states are aperiodic.*

Examples of irreducible Markov chains in which all states are null recurrent or all are transient, are given in Section 9.7.

**Theorem 9.5.2** *In a finite, irreducible Markov chain, all states are positive recurrent.*

We saw previously that in a finite Markov chain, no states are null recurrent, and at least one state must be positive recurrent. Adding the irreducible property means that all states must be positive recurrent.

**Theorem 9.5.3** *The states of an aperiodic, finite, irreducible Markov chain are ergodic.*

The conditions given in this last theorem are *sufficient* conditions only. The theorem must not be taken as a definition of ergodicity.

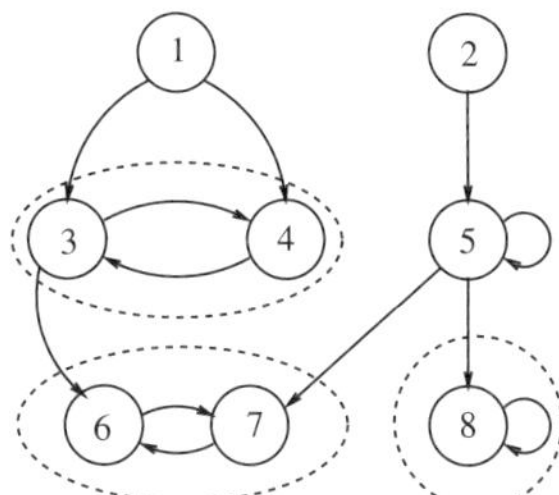


Figure 9.11. Communicating classes and nonreturn states.

We return to Figure 9.8, reproduced again as Figure 9.11, to illustrate these new definitions. In the Markov chain of this figure, state 1 and state 2 are each nonreturn states, states 3 and 4 constitute a communicating class that is not closed, state 5 is a return state and constitutes a nonclosed communicating class by itself (if state 5 did not have a self-loop, then it would be a nonreturn state), states 6 and 7 together form a closed communicating class, and state 8 is a return state that forms a one-state closed communicating class. State 8 is an absorbing state. Thus, the states are partitioned as follows:

$$\{1\},\ \{2\},\ \{3,4\},\ \{5\},\ \{6,7\},\ \{8\}.$$

The Markov chain is reducible, states 1 through 5 are transient states, and states 6 through 8 are positive-recurrent states. The set of recurrent states $\{6, 7, 8\}$ is closed and contains two closed proper subsets $\{6, 7\}$ and $\{8\}$.

The state space of a Markov chain may be partitioned into two subsets, the first containing only transient states and the second subset containing only recurrent states. This second group may be further partitioned into irreducible, closed communicating classes. After a possible renumbering of the states of a Markov chain, it is always possible to bring the transition probability matrix to the following "normal" form:

$$P = \begin{pmatrix} T_{11} & T_{12} & T_{13} & \cdots & T_{1N} \\ 0 & R_2 & 0 & \cdots & 0 \\ 0 & 0 & R_3 & \cdots & 0 \\ \vdots & \vdots & \vdots & \ddots & \vdots \\ 0 & 0 & 0 & \cdots & R_N \end{pmatrix},$$

in which there are $N - 1$ closed communicating classes $R_k$, $k = 2, \ldots, N$, and a set containing all transient states $T$. Once the process enters one of the closed classes, it remains there. The set of transient states may contain multiple communicating classes, but, at least for the moment, we need not be concerned about this possibility. If some $T_{1k}$ is not identically equal to zero, then transitions are possible from at least one of the transient states into the closed set $R_k$. Evidently, each of the $R_k$ may be considered as an irreducible Markov chain in its own right. This means that many of the properties of $R_k$ may be determined independently of other states.

The matrix corresponding to the Markov chain of Figure 9.11 is shown in partitioned form below, where the nonzero elements are shown with an asterisk.

$$P = \left(\begin{array}{ccccc|cc|c} 0 & 0 & * & * & 0 & 0 & 0 & 0 \\ 0 & 0 & 0 & 0 & * & 0 & 0 & 0 \\ 0 & 0 & 0 & * & 0 & * & 0 & 0 \\ 0 & 0 & * & 0 & 0 & 0 & 0 & 0 \\ 0 & 0 & 0 & 0 & * & 0 & * & * \\ \hline 0 & 0 & 0 & 0 & 0 & 0 & * & 0 \\ 0 & 0 & 0 & 0 & 0 & * & 0 & 0 \\ 0 & 0 & 0 & 0 & 0 & 0 & 0 & * \end{array}\right).$$

## 9.6 The Potential, Fundamental, and Reachability Matrices

In Section 9.4, we defined the *reachability matrix* $F$ as the matrix whose $ij$ element $f_{ij}$ is the probability of ever reaching state $j$ from state $i$. Our interest in this matrix $F$ lies chiefly in that it allows us to compute the probability of ending up in a particular recurrent state when the Markov chain begins in a given transient state. Now, let $r_{ij}$ be the expected number of times that state $j$ is visited, given that the Markov chain starts in state $i$, and let $R$ be the matrix whose $ij^{\text{th}}$ element is $r_{ij}$. The matrix $R$ is called the *potential matrix*. By means of a conditioning argument similar to that used in Equation (9.8), we obtain

$$R = \sum_{n=0}^{\infty} P^n. \tag{9.13}$$

When the states are arranged in such a way that transient states precede recurrent states, the upper left-hand corner of the potential matrix (the part that concerns transitions that leave and return to transient states) is called the *fundamental matrix*. The fundamental matrix is denoted[1] by $S$ and its $ij^{\text{th}}$ element gives the expected number of times the Markov chain is in transient state $j$, given

[1] The reader may have noticed a somewhat unfortunate, but generally accepted, choice of letters to designate these different matrices: the reachability matrix (which might have been called $R$) is denoted $F$, the fundamental matrix (which might have been $F$) is denoted $S$, and the potential matrix is called $R$.

that it started in transient state $i$. These matrices allow us to compute some properties of particular importance including

- The mean and variance of the random variable that describes the number of visits to a particular transient state starting from a (possibly) different transient state, before being absorbed into a recurrent state.
- Beginning in a given transient state, the mean and variance of the random variable that describes the total number of steps the Markov chain makes before being absorbed into a recurrent state. This mean is often designated MTTA, the *mean time to absorption.*

In general, it is easier to first compute the matrix $R$ and from it the matrix $F$, and this is what we now proceed to do. The next section, Section 9.6.1, is concerned with the potential and fundamental matrices and from them, the computation of the *mean number of steps* taken until absorption; Section 9.6.2 is concerned with the reachability matrix and the *probabilities* of absorption into different recurrent classes.

### 9.6.1 Potential and Fundamental Matrices and Mean Time to Absorption

Some elements of the potential matrix $R$ may be infinite, other may be zero, and yet others positive finite real numbers. We consider a state $i$ and the elements on row $i$ of $R$. A number of different possibilities occur according to the classification of the initial state and the final state. Recall that $r_{ij}$ is the expected number of times that state $j$ is visited, given that the Markov chain starts in state $i$.

***State $i$ is recurrent***

When state $i$ is recurrent, then we know that the Markov chain returns to state $i$ an infinite number of times and so the element in column position $i$ of row $i$ (the diagonal element) must be infinite. We write this as $r_{ii} = \infty$. Further, the elements in column positions $j$ corresponding to states that communicate with state $i$, must also be infinite ($r_{ij} = \infty$). This includes all states $j$ that are in the same closed communicating class, $C(i)$, as $i$. All other elements in row $i$ must be zero, since it is not possible to go from a recurrent state to a transient state, nor to a recurrent state that is in a different irreducible class. Thus, if $i$ is a recurrent state

$$r_{ij} = \begin{cases} \infty, & j \in C(i), \\ 0 & \text{otherwise.} \end{cases}$$

***State $i$ is transient and state $j$ is recurrent***

If a transient state $i$ can reach any state in recurrent class $C(j)$, we must have $r_{ik} = \infty$ for all $k \in C(j)$. In this case, after leaving $i$, the Markov chain can enter $C(j)$ and stay in it forever. If transient state $i$ cannot reach any state of recurrent class $C(j)$, then $r_{ik} = 0$ for all $k \in C(j)$. Thus, given that $i$ is transient and $j$ is recurrent:

If there is an $n$ for which $p_{ij}^{(n)} > 0$

$$r_{ik} = \infty \quad \text{for all } k \in C(j);$$

otherwise

$$r_{ik} = 0 \quad \text{for all } k \in C(j).$$

***Both states $i$ and $j$ are transient***

When both $i$ and $j$ are transient states, we use Equation (9.13). Let us assume that the states are numbered in such a way that all transient states come before all recurrent states. The transition probability matrix may then be written as

$$P = \begin{pmatrix} T & U \\ 0 & V \end{pmatrix},$$

in which the submatrix $T$ represents transitions among transient states only, $U$ represents transitions from transient states to recurrent states, and $V$ represents transitions among recurrent states. Then

$$P^n = \begin{pmatrix} T^n & \bar{U}(n) \\ 0 & V^n \end{pmatrix}$$

for some matrix $\bar{U}(n)$. Thus

$$R = \sum_{n=0}^{\infty} P^n = \begin{pmatrix} \sum_{n=0}^{\infty} T^n & \sum_{n=0}^{\infty} \bar{U}(n) \\ 0 & \sum_{n=0}^{\infty} V^n \end{pmatrix} \equiv \begin{pmatrix} S & \sum_{n=0}^{\infty} \bar{U}(n) \\ 0 & \sum_{n=0}^{\infty} V^n \end{pmatrix}.$$

The quantities we seek, the expected number of visits to transient state $j$ given that the Markov chain starts in transient state $i$, are the elements of $S = \sum_{n=0}^{\infty} T^n$. These are the only elements of the potential matrix $R$ that can be different from zero or infinity. As we mentioned previously, the matrix $S$ is called the *fundamental matrix* of the Markov chain.

The $ij$ element of the fundamental matrix, $s_{ij}$, gives the expected number of times the Markov chain is in transient state $j$ starting from transient state $i$. The matrix $S$ has a number of interesting formulations. Since

$$S = I + T + T^2 + \cdots$$

we have

$$S - I = T + T^2 + T^2 + \cdots = TS \tag{9.14}$$

and hence

$$S - TS = I \quad \text{or} \quad (I - T)S = I.$$

It is equally easy to show that $S$ also satisfies $S(I - T) = I$. Notice that

$$ST = TS = T + T^2 + \cdots,$$

which is the original series without the identity matrix. When the number of transient states is finite, then, since $T^n$ gives the probability of moving from one transient state to another transient state after $n$ steps, the matrices $T^n$ must eventually tend to zero. In the final analysis, the Markov chain cannot avoid leaving the set of transient states and entering a recurrent state. Thus, as $n \to \infty$, $T^n \to 0$ and the series $I + T + \cdots + T^n$ converges. From

$$I = (I + T + T^2 + \cdots)(I - T)$$

we have

$$S = \sum_{k=0}^{\infty} T^k = I + T + T^2 + \cdots = (I - T)^{-1}.$$

The proof of the nonsingularity of $(I - T)$ may also be justified on the basis that $T$ is strictly substochastic (i.e., it has at least one row whose sum is strictly less than 1). Thus, when the number of transient states is finite, the matrix $I - T$ is nonsingular and we may compute the elements of $S$ as the inverse matrix of $I - T$:

$$S = (I - T)^{-1}.$$

If the number of transient states is not finite, then it may be shown that $S$ is the minimal nonnegative solution of

$$(I - T)X = I, \quad X \geq 0.$$

**Example 9.9** Let us compute the potential matrix for the Markov chain whose transition probability matrix is given by

$$P = \left(\begin{array}{ccc|cc|c|cc} .4 & .2 & & .2 & & & & .2 \\ .3 & .3 & & & .1 & & .2 & .1 \\ & & .1 & .3 & .1 & & .5 & \\ \hline & & & .7 & .3 & & & \\ & & & .5 & .5 & & & \\ \hline & & & & & 1.0 & & \\ \hline & & & & & & .9 & .1 \\ & & & & & & .1 & .9 \end{array}\right).$$

In this example, states 1, 2, and 3 are transient, states 4 and 5 constitute an irreducible subset, state 6 is an absorbing state, and states 7 and 8 constitute a different irreducible subset. Rows 4 through 8 of the matrix $R$ are therefore easy to compute. In these five rows all elements are zero except for the elements in diagonal blocks, which represent transitions among recurrent states of the same closed class and which are all equal to $\infty$. In rows 1 through 3, the elements in column positions 4, 5, 7, and 8 must all be $\infty$: the Markov chain eventually will move from these states to the irreducible subset containing states 4 and 5, or to the irreducible subset containing states 7 and 8 and once there will remain there. The elements in column position 6 must be zero, since there is no path from any of the transient states to the absorbing state 6. This leaves only the fundamental matrix $S = (I - T)^{-1}$ to be computed:

$$T = \begin{pmatrix} .4 & .2 & 0 \\ .3 & .3 & 0 \\ 0 & 0 & .1 \end{pmatrix}, \quad (I - T) = \begin{pmatrix} .6 & -.2 & 0 \\ -.3 & .7 & 0 \\ 0 & 0 & .9 \end{pmatrix},$$

$$(I - T)^{-1} = \begin{pmatrix} 1.9444 & 0.5556 & 0.0 \\ 0.8333 & 1.6667 & 0.0 \\ 0.0 & 0.0 & 1.1111 \end{pmatrix}.$$

Therefore, the matrix $R$, whose $ij^{\text{th}}$ element $r_{ij}$ is the expected number of times that state $j$ is visited, given that the Markov chain begins in state $i$, is

$$R = \left(\begin{array}{ccc|cc|c|cc} 1.9444 & 0.5556 & 0.0 & \infty & \infty & 0 & \infty & \infty \\ 0.8333 & 1.6667 & 0.0 & \infty & \infty & 0 & \infty & \infty \\ 0.0 & 0.0 & 1.1111 & \infty & \infty & 0 & \infty & \infty \\ \hline 0 & 0 & 0 & \infty & \infty & 0 & 0 & 0 \\ 0 & 0 & 0 & \infty & \infty & 0 & 0 & 0 \\ \hline 0 & 0 & 0 & 0 & 0 & \infty & 0 & 0 \\ \hline 0 & 0 & 0 & 0 & 0 & 0 & \infty & \infty \\ 0 & 0 & 0 & 0 & 0 & 0 & \infty & \infty \end{array}\right).$$

Let $N_{ij}$ be a random variable, the total number of visits to transient state $j$ beginning from transient state $i$ and let $N$ be the matrix whose elements are $N_{ij}$. We have just seen that $E[N_{ij}] = s_{ij}$, i.e.,

$$E[N] = S.$$

For completeness, we give the second moments of the random variables, $N$, namely,

$$E[N^2] = S(2\,\text{diag}\{S\} - I).$$

The variance of $N$ may now be computed as

$$\text{Var}\,[N] = E[N^2] - E[N]^2 = S(2\,\text{diag}\{S\} - I) - sq\{S\}$$

where $sq\{S\}$ is the matrix whose $ij$ element is $s_{ij}^2$ ($= s_{ij} \times s_{ij}$, which is not the same as $s_{ij}^{(2)} = (S^2)_{ij}$).

Since the $ij$ element of the fundamental matrix gives the number of times the Markov chain is in transient state $j$ starting from transient state $i$, it follows that the sum of the elements in row $i$ of $S$ gives the mean number of steps that the Markov chain, beginning in state $i$, makes before being absorbed. This is the $i^{\text{th}}$ element of the vector $Se$. It is the mean time to absorption (MTTA) beginning from state $i$. The variance of the total time to absorption beginning from state $i$ is given by

$$(2S - I)Se - sq\{Se\}$$

where $sq\{Se\}$ is a vector whose $i^{\text{th}}$ element is the square of the $i^{\text{th}}$ element of $Se$. We summarize these results in the following theorem, under the assumption that $T$ is finite and thus $(I - T)^{-1}$ is well defined.

**Theorem 9.6.1** *Consider a discrete-time Markov chain having a finite number of transient states and whose transition probability matrix is written as*

$$P = \begin{pmatrix} T & U \\ 0 & V \end{pmatrix}. \tag{9.15}$$

*Assume this Markov chain starts in some transient state i. Then*

- *The mean number of times the Markov chain visits transient state $j$ is given by the $ij^{th}$ element of*

$$S = (I - T)^{-1}.$$

- *The variance of the number of times the Markov chain visits transient state $j$ is given by the $ij^{th}$ element of*

$$S(2\,\text{diag}\{S\} - I) - sq\{S\}.$$

- *The mean time to absorption (the mean number of steps among transient states before moving into a recurrent state) is given by the $i^{th}$ element of*

$$Se = (I - T)^{-1}e.$$

- *The variance of the time to absorption is given by the $i^{th}$ element of*

$$(2S - I)Se - sq\{Se\}.$$

If the Markov chain begins in transient state $i$ with probability $\alpha_i$, then the above results must be modified accordingly. Let $\alpha$ be the probability vector whose $i^{\text{th}}$ component is $\alpha_i$. Then

- $\alpha S$ is a vector whose $j^{\text{th}}$ component gives the mean number of visits to state $j$ before absorption,
- $\alpha Se$ is a real number that gives the expected total number of steps before absorption, and
- $\alpha S[2\,\text{diag}\{S\} - I] - sq\{\alpha S\}$ and $\alpha(2S - I)Se - (\alpha Se)^2$ are the corresponding variances.

In Theorem 9.6.1, the submatrix $T$ usually represents the set of all transient states. However, it also holds when $T$ represents *any* open subset of states. Let us use this theorem to investigate properties of an individual nonabsorbing state, $i$, by letting $T$ be the submatrix that consists of the single element $p_{ii}$. If the state of interest is the first, then the matrix $T$ in (9.15) reduces to the single element, $p_{11}$ which must be strictly less than 1, since this is a nonabsorbing state. In this case, the matrix $S$ consists of a single element, namely $1/(1 - p_{11})$ and from the first part of the theorem is equal to the mean number of steps that the Markov chain remains in this nonabsorbing state.

Additionally the variance may be computed from the second part of Theorem 9.6.1. It is given by

$$\frac{1}{(1-p_{11})}\left(\frac{2}{1-p_{11}}-1\right)-\left(\frac{1}{1-p_{11}}\right)^2=\frac{2}{(1-p_{11})^2}-\frac{1}{(1-p_{11})}-\frac{1}{(1-p_{11})^2}$$

$$=\frac{1}{(1-p_{11})^2}-\frac{1}{1-p_{11}}=\frac{p_{11}}{(1-p_{11})^2}.$$

If the nonabsorbing state is state $i$, then the mean number of time steps the Markov chain spends in this singleton subset of transient states before moving to a recurrent state is $1/(1-p_{ii})$ and the variance is given by $p_{ii}/(1-p_{ii})^2$. The reader may recognize that these results were derived earlier and in a different context, when we identified the sojourn time in a state of a Markov chain with a sequence of Bernoulli trials, Equation (9.5).

**Example 9.10** Let us return to Example 9.9, where we saw that the fundamental matrix is given by

$$S=(I-T)^{-1}=\begin{pmatrix}1.9444 & 0.5556 & 0.0\\ 0.8333 & 1.6667 & 0.0\\ 0.0 & 0.0 & 1.1111\end{pmatrix}.$$

The mean time to absorption when the Markov chain begins in state $i=1,2,3$ are the elements of

$$Se=(I-T)^{-1}e=\begin{pmatrix}1.9444 & 0.5556 & 0.0\\ 0.8333 & 1.6667 & 0.0\\ 0.0 & 0.0 & 1.1111\end{pmatrix}\begin{pmatrix}1\\1\\1\end{pmatrix}=\begin{pmatrix}2.5\\2.5\\1.1111\end{pmatrix}.$$

The variance of the number of times the Markov chain visits state $j$ beginning from state $i$ is given by the $ij^{\text{th}}$ element of

$$S(2\,\text{diag}\{S\}-I)-sq\{S\}$$

$$=\begin{pmatrix}1.9444 & 0.5556 & 0.0\\ 0.8333 & 1.6667 & 0.0\\ 0.0 & 0.0 & 1.1111\end{pmatrix}\begin{pmatrix}2.8889 & 0.0 & 0.0\\ 0.0 & 2.3333 & 0.0\\ 0.0 & 0.0 & 1.2222\end{pmatrix}-\begin{pmatrix}3.7809 & 0.3086 & 0.0\\ 0.6944 & 2.7778 & 0.0\\ 0.0 & 0.0 & 1.2346\end{pmatrix}$$

$$=\begin{pmatrix}1.8364 & 0.9877 & 0.0\\ 1.7130 & 1.1111 & 0.0\\ 0.0 & 0.0 & 0.1235\end{pmatrix},$$

while the variance of the total time to absorption beginning from state $i$ is given by the elements of

$$[2S-I](Se)-sq\{Se\}$$

$$=\begin{pmatrix}2.8889 & 1.1111 & 0.0\\ 1.6667 & 2.3333 & 0.0\\ 0.0 & 0.0 & 1.2222\end{pmatrix}\begin{pmatrix}2.5000\\2.5000\\1.1111\end{pmatrix}-\begin{pmatrix}6.2500\\6.2500\\1.2346\end{pmatrix}=\begin{pmatrix}3.7500\\3.7500\\0.1235\end{pmatrix}.$$

### *9.6.2 The Reachability Matrix and Absorption Probabilities*

We now turn to the computation of the elements of the reachability matrix $F$. We shall separate this into different categories depending on the classification of the initial and terminal states. Recall that the element $f_{ij}$ of the reachablility matrix $F$ gives the probability of ever reaching state $j$ upon leaving state $i$.

***Both states are recurrent and belong to the same closed communicating class***

In this case

$$f_{ij}=1.$$

***Both states are recurrent but belong to different closed communicating classes***
In this case

$$f_{ij} = 0.$$

***State $i$ is recurrent and state $j$ is transient***
Here

$$f_{ij} = 0.$$

***Both states are transient***
In this case, the probability of ever visiting state $j$ from state $i$ may be nonzero. Recall that $s_{ij}$ is the expected number of visits to transient state $j$ starting from transient state $i$ and $f_{ij}$ is the probability of ever visiting state $j$ from state $i$. Therefore, for a finite number of states:

$$s_{ij} = 1_{\{i=j\}} + f_{ij}s_{jj},$$

where $1_{\{i=j\}}$ is equal to 1 if $i = j$ and is equal to 0 otherwise. If we let $H$ denote the upper left corner of $F$ corresponding to transient states only, then, in matrix notation, the above equation becomes

$$S = I + H\left[\text{diag}\{S\}\right],$$

or alternatively

$$H = (S - I)[\text{diag}\{S\}]^{-1}. \tag{9.16}$$

The inverse must exist since $s_{ii} \geq 1$ for all $i$. Thus we may write the individual elements of $S$ in terms of those of $H$ as

$$s_{ii} = \frac{1}{1 - h_{ii}} \quad \text{and} \quad s_{ij} = h_{ij}s_{jj} \quad \text{for } i \neq j,$$

or alternatively, since at this point we require the elements of $H$,

$$h_{ii} = 1 - \frac{1}{s_{ii}} \quad \text{and} \quad h_{ij} = \frac{s_{ij}}{s_{jj}} \quad \text{for } i \neq j.$$

Two additional results concerning the case when states $i$ and $j$ are both transient are of interest. The first concerns the *probability of visiting a particular transient state $j$ a fixed number of times*, assuming the Markov chain starts in transient state $i$. To visit state $j$ exactly $k > 0$ times, the Markov chain must make a transition from $i$ to $j$ at least once (with probability $h_{ij}$), return from state $j$ to state $j$ a total of $k - 1$ times (with probability $h_{jj}^{k-1}$) and then never return from state $j$ to state $j$ again (probability $1 - h_{jj}$). Writing this in matrix terms, we obtain

$$H \times \text{diag}\{H\}^{k-1} \times [I - \text{diag}\{H\}].$$

Using the substitution $H = (S - I)[\text{diag}\{S\}]^{-1}$, observing that $\text{diag}\{H\} = I - [\text{diag}\{S\}]^{-1}$, and using the fact that diagonal matrices commute under matrix multiplication, this becomes

$$(S - I)[\text{diag}\{S\}]^{-1} \times [\text{diag}\{S\}]^{-1} \times (I - [\text{diag}\{S\}]^{-1})^{k-1}.$$

The probability that state $j$ is visited zero times, when the Markov chain begins in state $i$ must be zero if $i = j$ and equal to $1 - h_{ij}$ if $i \neq j$.

The second result concerns the *mean number of different transient states visited before absorption* into a recurrent class, assuming that the Markov chain starts in transient state $i$. This is equal to the sum of the probabilities of getting to the different transient states. Since the probability of ever visiting state $i$ from state $i$ is 1 and $h_{ij}$ is the probability of reaching state $j$ from state $i$, the mean number of transient states visited before absorption is $1 + \sum_{j \neq i} h_{ij}$. In matrix terminology, this is obtained by summing across row $i$ of $(H - \text{diag}\{H\} + I)$. In terms of the fundamental matrix $S$, this becomes

$$\begin{aligned}
[H - \text{diag}\{H\} + I]e &= [H + (I - \text{diag}\{H\})]e \\
&= ((S - I)[\text{diag}\{S\}]^{-1} + (I - (I - [\text{diag}\{S\}]^{-1}))e \\
&= ((S - I)[\text{diag}\{S\}]^{-1} + [\text{diag}\{S\}]^{-1})e \\
&= S[\text{diag}\{S\}]^{-1}e.
\end{aligned}$$

***State $i$ is transient and state $j$ is recurrent***

Observe that if a recurrent state $j$ can be reached from a transient state $i$, then all the states that are in $C(j)$, the recurrent class containing $j$, can be reached from $i$, and with the same probability, i.e.,

$$f_{ik} = f_{ij} \ \text{ for all } k \in C(j).$$

It suffices then to determine the probability of entering any state of a recurrent class. These probabilities are called *absorption probabilities*. To simplify matters, we combine all states in an irreducible recurrent set into a single state, an absorbing state, and compute the probability of entering this state from transient state $i$. We shall assume that the states are arranged in normal form, i.e.,

$$P = \begin{pmatrix} T_{11} & T_{12} & T_{13} & \cdots & T_{1N} \\ 0 & R_2 & 0 & \cdots & 0 \\ 0 & 0 & R_3 & \cdots & 0 \\ \vdots & \vdots & \vdots & \ddots & \vdots \\ 0 & 0 & 0 & \cdots & R_N \end{pmatrix}.$$

Each of the matrices $R_k$ is replaced with the value 1 (indicating a single absorbing state), the matrix $T_{11}$ is left unaltered, and the matrices $T_{1k}$ are replaced by vectors $t_k$, created by summing across each row of $T_{ik}$, i.e., $t_k = T_{1k}e$, This gives

$$\bar{P} = \begin{pmatrix} T_{11} & t_2 & t_3 & \cdots & t_N \\ 0 & 1 & 0 & \cdots & 0 \\ 0 & 0 & 1 & \cdots & 0 \\ \vdots & \vdots & \vdots & \ddots & \vdots \\ 0 & 0 & 0 & \cdots & 1 \end{pmatrix}.$$

A Markov chain whose transition matrix is in this form is referred to as an *absorbing chain.*

**Example 9.11** Consider the discrete-time Markov chain with transition probability matrix $P$:

$$P = \left(\begin{array}{ccc|cc|c|cc}
.4 & .2 & & .2 & & & & .2 \\
.3 & .3 & & & .1 & & .2 & .1 \\
 & & .1 & .3 & .1 & & .5 & \\
\hline
 & & & .7 & .3 & & & \\
 & & & .5 & .5 & & & \\
\hline
 & & & & & 1.0 & & \\
\hline
 & & & & & & .9 & .1 \\
 & & & & & & .1 & .9
\end{array}\right).$$

The absorbing chain is given by the matrix $\bar{P}$:

$$\bar{P} = \left(\begin{array}{ccc|c|c|c} .4 & .2 & & .2 & & .2 \\ .3 & .3 & & .1 & & .3 \\ & & .1 & .4 & & .5 \\ \hline & & & 1.0 & & \\ \hline & & & & 1.0 & \\ \hline & & & & & 1.0 \end{array}\right).$$

To simplify the notation, we write the absorbing chain as

$$\bar{P} = \left(\begin{array}{c|c} T & B \\ \hline 0 & I \end{array}\right).$$

Taking higher powers of $\bar{P}$ we obtain

$$\bar{P}^n = \left(\begin{array}{c|c} T^n & (I + T + \cdots + T^{n-1})B \\ \hline 0 & I \end{array}\right) = \left(\begin{array}{c|c} T^n & B_n \\ \hline 0 & I \end{array}\right),$$

where $B_n = (I+T+\cdots+T^{n-1})B$. We are now in a position to compute the absorption probabilities. The probability that, starting from transient state $i$, the Markov chain enters a state of the $j^{\text{th}}$ irreducible recurrent class by time step $n$ is given by the $ij$ element of $B_n$. Notice that the states of the first recurrent class are those whose transitions are associated with diagonal block $R_2$, those of the second by diagonal block $R_3$, and so on. The probability of ever being absorbed from state $i$ into the $j^{\text{th}}$ recurrent class is then given by the $ij^{\text{th}}$ element of $\lim_{n\to\infty} B_n$.

The matrix $A = \lim_{n\to\infty} B_n$ is called the absorption probability matrix. If the number of transient states is finite we may obtain $A$ directly from the fundamental matrix, $S$. We have

$$A = \lim_{n\to\infty} B_n = \lim_{n\to\infty} (I + T + \cdots + T^{n-1})B = \left(\sum_{k=0}^{\infty} T^k\right) B = (I - T)^{-1}B = SB.$$

The $ij^{\text{th}}$ element of $A$ gives the probability of ever reaching the $j^{\text{th}}$ recurrent class starting from transient state $i$. For every state $k$ in this class,

$$f_{ik} = a_{ij}.$$

If the probabilities of absorption into only a single absorbing state/recurrent class are required, it suffices to form the product of $S$ and the corresponding column of $B$, rather than perform the complete matrix–matrix product. The $i^{\text{th}}$ element of the resulting vector gives the absorption probability when the Markov chain starts in state $i$.

A final property concerning the matrix $A$ is of note. Let

$$\bar{A} = \left(\begin{array}{c|c} 0 & A \\ \hline 0 & I \end{array}\right).$$

Then

$$\bar{P}\bar{A} = \left(\begin{array}{c|c} T & B \\ \hline 0 & I \end{array}\right)\left(\begin{array}{c|c} 0 & A \\ \hline 0 & I \end{array}\right) = \left(\begin{array}{c|c} 0 & A \\ \hline 0 & I \end{array}\right) = \bar{A}$$

since

$$TA + B = TSB + B = (S - I)B + B = SB = A,$$

where we have first used the fact that $A = SB$ and second, from Equation (9.14), that $TS = S - I$. Hence, a column of $\bar{A}$ may be found as the vector $\bar{\alpha}$ which satisfies

$$\bar{P}\bar{\alpha} = \bar{\alpha}.$$

In other words $\bar{\alpha}$ is the right-hand eigenvector corresponding to a unit eigenvalue of the matrix $\bar{P}$ and may be computed using standard eigenvalue/eigenvector methods with resorting to matrix inversion.

**Example 9.12** Referring back to the running example,

$$P = \left(\begin{array}{ccc|cc|c|cc} .4 & .2 & & .2 & & & & .2 \\ .3 & .3 & & & .1 & & .2 & .1 \\ & & .1 & .3 & .1 & & .5 & \\ \hline & & & .7 & .3 & & & \\ & & & .5 & .5 & & & \\ \hline & & & & & 1.0 & & \\ \hline & & & & & & .9 & .1 \\ & & & & & & .1 & .9 \end{array}\right),$$

we can immediately fill in the elements on rows 4 through 8 of $F$. These elements are equal to 1 inside the diagonal blocks and zero elsewhere. To compute the first $3 \times 3$ block, we use Equation (9.16). We have

$$H = F_{3\times 3} = (S - I)[\mathrm{diag}\{S\}]^{-1}$$

$$= \begin{pmatrix} 0.9444 & 0.5556 & 0.0 \\ 0.8333 & 0.6667 & 0.0 \\ 0.0 & 0.0 & 0.1111 \end{pmatrix} \begin{pmatrix} 1.9444 & 0.0 & 0.0 \\ 0.0 & 1.6667 & 0.0 \\ 0.0 & 0.0 & 1.1111 \end{pmatrix}^{-1} = \begin{pmatrix} 0.4857 & 0.3333 & 0.0 \\ 0.4286 & 0.4000 & 0.0 \\ 0.0 & 0.0 & 0.1000 \end{pmatrix}.$$

Only the absorption probabilities still remain to be found. From

$$A = \lim_{n\to\infty} B_n = (I - T)^{-1} B = SB,$$

$$A = \begin{pmatrix} 0.6 & -0.2 & 0.0 \\ -0.3 & 0.7 & 0.0 \\ 0.0 & 0.0 & 0.9 \end{pmatrix}^{-1} \begin{pmatrix} 0.2 & 0.0 & 0.2 \\ 0.1 & 0.0 & 0.3 \\ 0.4 & 0.0 & 0.5 \end{pmatrix} = \begin{pmatrix} 0.4444 & 0.0 & 0.5556 \\ 0.3333 & 0.0 & 0.6667 \\ 0.4444 & 0.0 & 0.5556 \end{pmatrix}.$$

Observe that for each starting state, the sum of the absorption probabilities is 1. The $ij$ element of $A$ gives the probability of being absorbed into the $j^{\text{th}}$ recurrent class, starting from transient state $i$.
Thus the entire matrix $F$ is written as

$$F = \left(\begin{array}{ccc|cc|c|cc} 0.4857 & 0.3333 & 0.0 & 0.4444 & 0.4444 & 0.0 & 0.5556 & 0.5556 \\ 0.4286 & 0.4000 & 0.0 & 0.3333 & 0.3333 & 0.0 & 0.6667 & 0.6667 \\ 0.0 & 0.0 & 0.1000 & 0.4444 & 0.4444 & 0.0 & 0.5556 & 0.5556 \\ \hline 0.0 & 0.0 & 0.0 & 1.0 & 1.0 & 0.0 & 0.0 & 0.0 \\ 0.0 & 0.0 & 0.0 & 1.0 & 1.0 & 0.0 & 0.0 & 0.0 \\ \hline 0.0 & 0.0 & 0.0 & 0.0 & 0.0 & 1.0 & 0.0 & 0.0 \\ \hline 0.0 & 0.0 & 0.0 & 0.0 & 0.0 & 0.0 & 1.0 & 1.0 \\ 0.0 & 0.0 & 0.0 & 0.0 & 0.0 & 0.0 & 1.0 & 1.0 \end{array}\right). \tag{9.17}$$

The results we have given so far relate to Markov chains with both transient and recurrent states. We have been concerned with probabilities of moving from transient states to one or more of the closed communicating classes. It is interesting to observe that these results may also be used to extract useful information in the context of Markov chains that have *no* transient states. For example, given a finite irreducible Markov chain, we may wish to compute the probability of reaching some state $i$ before state $j$ when starting from any third state $k$ (with $i \neq j \neq k$). This can be achieved by separating out states $i$ and $j$ from the remainder of the states and forming the fundamental matrix corresponding to all other states—all states other than $i$ and $j$ become transient states. The set of all states other than states $i$ and $j$ forms an *open* subset of states and the results of Theorem 6.1.9 can be applied. In other words, the problem posed can be solved by turning states $i$ and $j$ into absorbing states and determining the probability that, starting in some nonabsorbing state $k$, the Markov chain enters absorbing state $i$ before it enters absorbing state $j$.

Assume that there are a total of $n$ states and that these are ordered as follows: the set of $(n-2)$ states which does not contain $i$ or $j$, but which does contain state $k$, appears first, followed by state $i$ and finally state $j$. The probability that state $i$ is entered before state $j$ is given by the elements of the vector formed as the product of $S$ and a vector, denoted $v_{n-1}$, whose components are the first $n-2$ elements of column $n-1$ of $P$, i.e., the elements that define the conditional probabilities of entering state $i$ given that the Markov chain is in state $k$, $k = 1, 2, \ldots, n-2$. Similarly, the probability that state $j$ is reached before state $i$ is computed from the elements of the vector obtained when $S$ is multiplied by a vector, denoted $v_n$, whose components are the first $n-2$ elements of the last column of $P$.

**Example 9.13** Consider the six-state Markov chain whose transition probability matrix is given by

$$P = \left(\begin{array}{cccc|cc} .25 & .25 & & & .50 & \\ .50 & & .50 & & & \\ & .50 & & .50 & & \\ & & .50 & & & .50 \\ \hline & & & .50 & & .50 \\ & & & & .50 & .50 \end{array}\right).$$

We have

$$S = \begin{pmatrix} .75 & -.25 & & \\ -.50 & 1.0 & -.50 & \\ & -.50 & 1.0 & -.50 \\ & & -.50 & 1.0 \end{pmatrix}^{-1} = \frac{2}{9}\begin{pmatrix} 8.0 & 3.0 & 2.0 & 1.0 \\ 6.0 & 9.0 & 6.0 & 3.0 \\ 4.0 & 6.0 & 10.0 & 5.0 \\ 2.0 & 3.0 & 5.0 & 7.0 \end{pmatrix}.$$

With

$$v_5 = (0.5\ 0.0\ 0.0\ 0.0)^T \quad \text{and} \quad v_6 = (0.0\ 0.0\ 0.0\ 0.5)^T,$$

we obtain

$$Sv_5 = (0.8889\ 0.6667\ 0.4444\ 0.2222)^T \quad \text{and} \quad Sv_6 = (0.1111\ 0.3333\ 0.5556\ 0.7778)^T.$$

Thus, for example, the probability that state $i = 5$ is reached before state $j = 6$ given state $k = 1$ as the starting state is 0.8889. The probability that state $j$ is reached before state $i$ is only 0.1111. Notice that combining $v_5$ and $v_6$ into a single $4 \times 2$ matrix gives what we previously referred to as the matrix $B$, so what we are actually computing is just $SB = A$, the matrix of absorption probabilities.

## 9.7 Random Walk Problems

Picture a drunken man trying to walk along an imagined straight line on his way home from the pub. Sometimes he steps to the right of his imagined straight line and sometimes he steps to the left of it. This colorful scenario leads to an important class of Markov chain problems, that of the random walk. The states of the Markov chain are the integers $0, \pm 1, \pm 2, \ldots$ (the drunkard's straight line) and the only transitions from any state $i$ are to neighboring states $i+1$ (a step to the right) with probability $p$ and $i-1$ (a step to the left) with probability $q = 1 - p$. The state transition diagram is shown in Figure 9.12 and the process begins in state 0. When $p = 1/2$, the process is said to be a *symmetric* random walk.

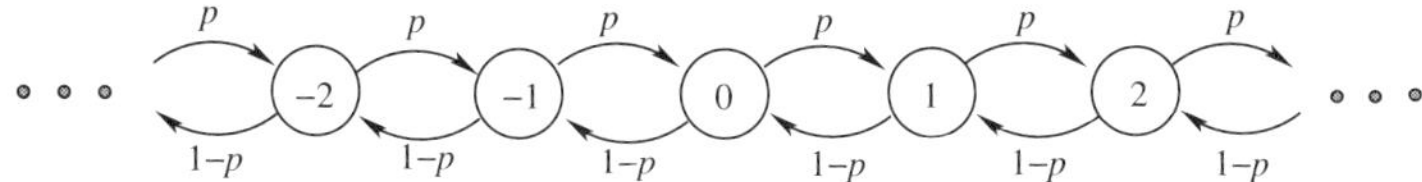


Figure 9.12. State transition diagram for a random walk on the integers.

There are many variations on this basic theme. The case in which the state space is finite is sometimes called the "gambler's ruin" problem. A gambler starts with a certain amount of money, say $i$ dollars (in terms of the Markov chain, the process begins in state $i > 0$) and on each play the gambler can either win a dollar with probability $p$ or lose a dollar with probability $1 - p$. The interest is in determining whether the gambler is ruined, i.e., loses all his money (the Markov chain moves to state 0, which is taken to be an absorbing state) or wins a fortune (the Markov chain moves into absorbing state $N > i$, where $N$ is large). Other variations are based on a random walk that is infinite on one side only, the state space is taken to be the set of nonnegative integers, and the Markov chain begins in state 0. Random walk problems can also be defined on more than one dimension. Two-dimensional random walks take place on a grid that can be infinite and cover the entire plane; infinite but span only the upper right quadrant; or finite. In these cases, each state communicates with at most its four neighbors to the north, south, east, and west. The process is symmetric when each transition probability is equal to 1/4. Random walks on higher dimensions are readily envisaged.

We begin by considering random walk problems in which the Markov chain is irreducible, when there is a path from any state to any other state. In such circumstances we know that all of the states are either positive recurrent or all are null recurrent or all are transient. Our interest is in finding out which applies. In the original one-dimensional random walk on the integers $0, \pm 1, \pm 2, \ldots$, the states are all transient except in the symmetric case, $p = 1/2$. The proof of this is outlined in the exercises. The same result holds in two dimensions, but in three dimensions or higher, all the states are transient, even in the symmetric case. We now analyze the case of a random walk on the nonnegative integers and leave some of the other cases to the exercises. The following two theorems (stated without proof) are useful in this respect.

**Theorem 9.7.1** *Let $P$ be the single-step transition probability matrix of an irreducible Markov chain. Then all the states of this Markov chain are positive recurrent if and only if the system of linear equations*

$$z = zP,$$

*in which $z$ is a row vector, has a solution with $\sum_{\text{all } j} z_j = 1$. If such a solution exists, then we must have $z_j > 0$ for all $j$ and there are no other solutions.*

We shall see momentarily that any vector $z \neq 0$ which satisfies $z = zP$, not necessarily one for which *all* the components of $z$ are strictly positive, is called a *stationary distribution* of the Markov chain. The second theorem that we shall need is the following.

**Theorem 9.7.2** *Let $P$ be the single-step transition probability matrix of an irreducible Markov chain. Let $P'$ be the matrix obtained from $P$ by deleting row $k$ and column $k$. Then all states are recurrent if and only if the solution $y$ of*

$$y = P'y, \quad 0 \leq y_i \leq 1,$$

*is the zero vector, i.e., a column vector whose components are all equal to zero.*

Consider the infinite (denumerable) Markov chain whose transition probability matrix is given by

$$P = \begin{pmatrix} 0 & 1 & 0 & \cdots & & \\ q & 0 & p & 0 & \cdots & \\ 0 & q & 0 & p & 0 & \cdots \\ 0 & 0 & q & 0 & p & \\ \vdots & \vdots & \vdots & \vdots & \vdots & \ddots \end{pmatrix},$$

where $p$ is a positive probability and $q = 1 - p$. Observe that every time the Markov chain reaches state 0, it must leave it again at the next time step. State 0 is said to constitute a *reflecting barrier*. In other applications it may be more appropriate to set $p_{11} = 1 - p$ and $p_{12} = p$, called a *Bernoulli barrier*, or $p_{11} = 1$ and $p_{12} = 0$, called an *absorbing barrier*. We pursue our analysis with the reflecting barrier option. As we pointed out previously, since every state can reach every other state, the Markov chain is irreducible and hence all the states are positive recurrent or all the states are null recurrent or all the states are transient. Notice also that a return to any state is possible only in a number of steps that is a multiple of 2. Thus the Markov chain is periodic with period equal to 2. We now wish to use Theorems 9.7.1 and 9.7.2 to classify the states of this chain. Consider the system of equations $z = zP$.

$$(z_0\ z_1\ z_2 \ldots) = (z_0\ z_1\ z_2 \ldots) \begin{pmatrix} 0 & 1 & 0 & \cdots & & \\ q & 0 & p & 0 & \cdots & \\ 0 & q & 0 & p & 0 & \cdots \\ 0 & 0 & q & 0 & p & \\ \vdots & \vdots & \vdots & \vdots & \vdots & \ddots \end{pmatrix}.$$

Taking these equations one at a time, we find

$$z_0 = z_1 q \implies z_1 = \frac{1}{q} z_0,$$

$$z_1 = z_0 + q z_2 \implies z_2 = \frac{1}{q}(z_1 - z_0) = \frac{1}{q}(1 - q) z_1 = \left(\frac{p}{q}\right) z_1 = \frac{1}{q}\left(\frac{p}{q}\right) z_0.$$

All subsequent equations are of the form

$$z_j = p z_{j-1} + q z_{j+1} \quad \text{for } j \geq 2$$

and have the solution

$$z_{j+1} = \left(\frac{p}{q}\right) z_j,$$

which may be proven by induction as follows. Base clause, $j = 2$:

$$z_2 = p z_1 + q z_3 \implies z_3 = \frac{1}{q}(z_2 - p z_1) = \frac{1}{q} z_2 - z_2 = \left(\frac{p}{q}\right) z_2.$$

We now assume this solution to hold for $j$ and prove it true for $j + 1$. From

$$z_j = p z_{j-1} + q z_{j+1}$$

we obtain

$$z_{j+1} = \frac{1}{q}(z_j - p z_{j-1}) = \left(\frac{1}{q}\right) z_j - \left(\frac{p}{q}\right) z_{j-1} = (1/q - 1) z_j = \left(\frac{p}{q}\right) z_j$$

which completes the proof. This solution allows us to write any component in terms of $z_0$. We have

$$z_j = \left(\frac{p}{q}\right) z_{j-1} = \left(\frac{p}{q}\right)^2 z_{j-2} = \cdots = \left(\frac{p}{q}\right)^{j-1} z_1 = \frac{1}{q}\left(\frac{p}{q}\right)^{j-1} z_0, \quad j \geq 1$$

and summing over all $z_j$, we find

$$\sum_{j=0}^{\infty} z_j = z_0 \left[ 1 + \frac{1}{q} \sum_{j=1}^{\infty} \left(\frac{p}{q}\right)^{j-1} \right].$$

Notice that the summation inside the square bracket is finite if and only if $p < q$ in which case

$$\sum_{j=1}^{\infty} \left(\frac{p}{q}\right)^{j-1} = \sum_{j=0}^{\infty} \left(\frac{p}{q}\right)^{j} = \frac{1}{1 - p/q} \quad \text{(iff } p < q\text{)}.$$

The sum of all $z_j$ being equal to 1 implies that

$$1 = z_0 \left[ 1 + \frac{1}{q} \left( \frac{1}{1 - p/q} \right) \right].$$

Simplifying the term inside the square brackets, we obtain

$$1 = z_0 \left[ 1 + \frac{1}{q - p} \right] = z_0 \left[ \frac{q - p + 1}{q - p} \right] = z_0 \left[ \frac{2q}{q - p} \right]$$

and hence

$$z_0 = \frac{q - p}{2q} = \frac{1}{2}\left(1 - \frac{p}{q}\right) \quad \text{for } p < q.$$

The remaining $z_j$ are obtained as

$$z_j = \frac{1}{q}\left(\frac{p}{q}\right)^{j-1} z_0 = \frac{1}{2q}\left(1 - \frac{p}{q}\right)\left(\frac{p}{q}\right)^{j-1}.$$

This solution satisfies the conditions of Theorem 9.7.1 (it exists and its components sum to 1) and hence all states are positive recurrent (when $p < q$). We also see that the second part of this theorem is true (all $z_j$ are strictly positive). Finally, this theorem tells us that there is no other solution.

We now examine the other possibilities, namely, $p = q$ and $p > q$. Under these conditions, either all states are null recurrent or else all states are transient (from Theorems 9.5.1 and 9.7.1). We shall now use Theorem 9.7.2 to determine which case holds. Let the matrix $P'$ be the matrix obtained from $P$ when the first row and column of $P$ is removed and let us consider the system of equations $y = P'y$. We have

$$\begin{pmatrix} y_1 \\ y_2 \\ y_3 \\ \vdots \end{pmatrix} = \begin{pmatrix} 0 & p & 0 & 0 & \cdots \\ q & 0 & p & 0 & \cdots \\ 0 & q & 0 & p & \cdots \\ \vdots & \vdots & \ddots & \ddots & \ddots \end{pmatrix} \begin{pmatrix} y_1 \\ y_2 \\ y_3 \\ \vdots \end{pmatrix}.$$

Again taking these equations one at a time, we have

$$y_1 = py_2,$$
$$y_2 = qy_1 + py_3,$$

and in general

$$y_j = qy_{j-1} + py_{j+1}.$$

Writing the left-hand side as $py_j + qy_j$, we have

$$py_j + qy_j = qy_{j-1} + py_{j+1} \quad \text{for } j \geq 2,$$

which yields

$$p(y_{j+1} - y_j) = q(y_j - y_{j-1}) \quad \text{for } j \geq 2. \tag{9.18}$$

When $j = 1$, a similar rearrangement gives

$$p(y_2 - y_1) = qy_1.$$

It follows from Equation (9.18) that

$$\begin{aligned} y_{j+1} - y_j &= \left(\frac{q}{p}\right)(y_j - y_{j-1}) = \cdots = \left(\frac{q}{p}\right)^{j-1}(y_2 - y_1) \\ &= \left(\frac{q}{p}\right)^{j-1}\left(\frac{q}{p}\right)y_1 = \left(\frac{q}{p}\right)^{j} y_1 \quad \text{for } j \geq 1. \end{aligned}$$

Thus

$$y_{j+1} - y_j = \left(\frac{q}{p}\right)^{j} y_1 \quad \text{for } j \geq 1$$

and

$$\begin{aligned} y_{j+1} &= (y_{j+1} - y_j) + (y_j - y_{j-1}) + \cdots + (y_2 - y_1) + y_1 \\ &= \left(\frac{q}{p}\right)^{j} y_1 + \left(\frac{q}{p}\right)^{j-1} y_1 + \cdots + \left(\frac{q}{p}\right)^{1} y_1 + y_1 \\ &= \left[1 + \left(\frac{q}{p}\right) + \left(\frac{q}{p}\right)^2 + \cdots + \left(\frac{q}{p}\right)^j\right] y_1. \end{aligned}$$

Consider now the case in which $p = q$. We obtain

$$y_j = jy_1 \quad \text{for } j \geq 1.$$

In order to satisfy the conditions of Theorem 9.7.2, we need to have $0 \leq y_j \leq 1$ for all $j$. Therefore the only possibility in this case ($p = q$) is that $y_1 = 0$, in which case $y_j = 0$ for all $j$. Since the only solution is $y = 0$, we may conclude from Theorem 9.7.2 that all the states are *recurrent* states and since we know that they are not *positive recurrent*, they must be null recurrent.

Finally, consider the case in which $p > q$. In this case, $q/p$ is a fraction and the summation converges. We have

$$y_j = \left[1 + \left(\frac{q}{p}\right) + \left(\frac{q}{p}\right)^2 + \cdots + \left(\frac{q}{p}\right)^{j-1}\right] y_1 = \sum_{k=0}^{j-1} \left(\frac{q}{p}\right)^k y_1 = \frac{1 - (q/p)^j}{1 - (q/p)} y_1.$$

It is apparent that if we now set

$$y_1 = 1 - \frac{q}{p}$$

we obtain a value of $y_1$ that satisfies $0 \leq y_1 \leq 1$ and, furthermore, we also obtain

$$y_j = 1 - \left(\frac{q}{p}\right)^j,$$

which also satisfies $0 \leq y_j \leq 1$ for all $j \geq 1$. From Theorem 9.7.2 we may now conclude that all the states are transient when $p > q$.

Let us now consider a different random walk, the gambler's ruin problem. The gambler begins with $i$ dollars (in state $i$) and on each play either wins a dollar with probability $p$ or loses a dollar with probability $q = 1 - p$. If $X_n$ is the amount of money he has after playing $n$ times, then $\{X_n, n = 0, 1, 2, \ldots\}$ is a Markov chain and the transition probability matrix is given by

$$P = \begin{pmatrix} 1 & 0 & 0 & 0 & & \cdots & & & 0 \\ q & 0 & p & 0 & & \cdots & & & 0 \\ 0 & q & 0 & p & & \cdots & & & 0 \\ & & & & & & & & \\ \vdots & & \ddots & \ddots & \ddots & & & & \vdots \\ & & & & & & & & \\ 0 & & \cdots & & q & 0 & p & & 0 \\ 0 & & \cdots & & 0 & q & 0 & & p \\ 0 & & \cdots & & 0 & 0 & 0 & & 1 \end{pmatrix}.$$

We would like to find the probability that the gambler will eventually make his fortune by arriving in state $N$ given that he starts in state $i$. Let $x_i$ be this probability. We immediately have that $x_0 = 0$ and $x_N = 1$, since the gambler cannot start with zero dollars and win $N$ dollars, while if he starts with $N$ dollars he has already made his fortune. Given that the gambler starts with $i$ dollars, after the first play he has $i + 1$ dollars with probability $p$, and $i - 1$ dollars with probability $q$. It follows that the probability of ever reaching state $N$ from state $i$ is the same as the probability of reaching $N$ beginning in $i + 1$ with probability $p$ plus the probability of reaching state $N$ beginning in state $i - 1$ with probability $q$. In other words,

$$x_i = px_{i+1} + qx_{i-1}.$$

This is the equation that will allow us to solve our problem. It holds for all $1 \le i \le N - 1$. Be aware that it does not result from the multiplication of a vector by the transition probability matrix, but rather by conditioning on the result of the first play. Writing this equation as

$$x_i = px_i + qx_i = px_{i+1} + qx_{i-1}$$

allows us to derive the recurrence relation

$$x_{i+1} - x_i = \frac{q}{p}(x_i - x_{i-1}), \quad i = 1, 2, \ldots, N - 1.$$

Now, using the fact that $x_0 = 0$, we have

$$\begin{aligned} x_2 - x_1 &= \frac{q}{p}(x_1 - x_0) = \frac{q}{p}x_1, \\ x_3 - x_2 &= \frac{q}{p}(x_2 - x_1) = \left(\frac{q}{p}\right)^2 x_1, \\ &\vdots \\ x_i - x_{i-1} &= \frac{q}{p}(x_{i-1} - x_{i-2}) = \left(\frac{q}{p}\right)^{i-1} x_1, \qquad i = 2, 3, \ldots, N. \end{aligned}$$

Adding these equations, we find

$$x_i - x_1 = \left[\left(\frac{q}{p}\right) + \left(\frac{q}{p}\right)^2 + \cdots + \left(\frac{q}{p}\right)^{i-1}\right] x_1,$$

i.e.,

$$x_i = \left[1 + \left(\frac{q}{p}\right) + \left(\frac{q}{p}\right)^2 + \cdots + \left(\frac{q}{p}\right)^{i-1}\right] x_1 = \sum_{k=0}^{i-1} \left(\frac{q}{p}\right)^k x_1.$$

We must consider the two possible cases for this summation, $p \neq q$ and $p = q = 1/2$. When $p \neq q$, then

$$x_i = \frac{1 - (q/p)^i}{1 - q/p} x_1 \quad \text{and in particular} \quad \frac{1 - (q/p)^N}{1 - q/p} x_1 = x_N = 1.$$

This allows us to compute $x_1$ as

$$x_1 = \frac{1 - q/p}{1 - (q/p)^N},$$

and from this, all the remaining values of $x_i$, $i = 2, 3, \ldots, N-1$, can be found. Collecting these results together, we have

$$x_i = \frac{1 - (q/p)^i}{1 - (q/p)} \times \frac{1 - (q/p)}{1 - (q/p)^N} = \frac{1 - (q/p)^i}{1 - (q/p)^N}, \quad i = 1, 2, \ldots N \text{ and } p \neq q.$$

Notice the limits as $N$ tends to infinity:

$$\lim_{N \to \infty} x_i = \begin{cases} 1 - (q/p)^i, & p > 1/2, \\ 0, & p < 1/2. \end{cases}$$

This means that, when the game favors the gambler ($p > 1/2$), there is a positive probability that the gambler will make his fortune. However, when the game favors the house, then the gambler is sure to lose all his money. This same sad result holds when the game favors neither the house nor the gambler ($p = 1/2$). We leave it as an exercise for the reader to show that, when $p = q = 1/2$, then $x_i = i/N$, for $i = 1, 2, \ldots, N$, and hence $x_i$ approaches zero as $N \to \infty$.

**Example 9.14** Billy and his brother Gerard play marbles. At each game Billy has a 60% chance of winning (perhaps he cheats) while Gerard has only a 40% chance of winning. If Billy starts with 16 marbles and Gerard with 20, what is the probability that Billy ends up with all the marbles? Suppose Billy starts with only 4 marbles (and Gerard with 20), what is the probability that Gerard ends up with all the marbles?

We shall use the following equation generated during the gambler's ruin problem:

$$x_i = \frac{1 - (q/p)^i}{1 - (q/p)^N}, \quad i = 1, 2, \ldots, N \text{ and } p \neq q.$$

Substituting in the values $p = 0.6$, $i = 16$, and $N = 36$, we have

$$x_{16} = \frac{1 - (.4/.6)^{16}}{1 - (.4/.6)^{36}} = 0.998478.$$

To answer the second part, the probability that Gerard, in a more advantageous initial setup, wins all the marbles, we first find $x_4$, the probability that Billy wins them all. We have

$$x_4 = \frac{1 - (.4/.6)^4}{1 - (.4/.6)^{24}} = 0.802517.$$

So the probability that Gerard takes all of Billy's marbles is only $1 - 0.802517 = 0.197483$.

## 9.8 Probability Distributions

We now turn our attention to probability distributions defined on the states of a (homogeneous) discrete-time Markov chain. Of particular interest is the probability that the chain is in a given state at a particular time step. We shall denote by $\pi_i(n)$ the probability that a Markov chain is in state $i$ at step $n$, i.e.,

$$\pi_i(n) = \text{Prob}\{X_n = i\}.$$

In vector notation, we let $\pi(n) = (\pi_1(n), \pi_2(n), \ldots, \pi_i(n), \ldots)$. Note that the vector $\pi$ is a row vector. We adopt the convention that all probability vectors are row vectors. All other vectors are taken to be column vectors unless specifically stated otherwise. The state probabilities at any time step $n$ may be obtained from a knowledge of the initial probability distribution (at time step 0) and the matrix of transition probabilities. We have, from the theorem of total probability,

$$\pi_i(n) = \sum_{\text{all } k} \text{Prob}\{X_n = i | X_0 = k\}\pi_k(0).$$

The probability that the Markov chain is in state $i$ at step $n$ is therefore given by

$$\pi_i(n) = \sum_{\text{all } k} p_{ki}^{(n)} \pi_k(0), \tag{9.19}$$

which in matrix notation becomes

$$\pi(n) = \pi(0)P^{(n)} = \pi(0)P^n,$$

where $\pi(0)$ denotes the initial state distribution and $P^{(n)} = P^n$ since we assume the chain to be homogeneous. The probability distribution $\pi(n)$ is called a *transient distribution*, since it gives the probability of being in the various states of the Markov chain at a particular instant in time, i.e., at step $n$. As the Markov chain evolves onto step $n+1$, the distribution at time step $n$ is discarded—hence it is only transient. Transient distributions should not be confused with transient states, which were discusssed in an earlier section. For a discrete-time Markov chain, transient distributions are computed directly from the defining equation, Equation (9.19). In the remainder of this section, we shall be concerned with stationary, limiting, and steady-state distributions, rather than with transient distributions.

**Definition 9.8.1 (Stationary distribution)** *Let $P$ be the transition probability matrix of a discrete-time Markov chain, and let the vector $z$, whose elements $z_j$ denote the probability of being in state $j$, be a probability distribution; i.e.,*

$$z_j \in \Re, \quad 0 \le z_j \le 1, \quad \textit{and} \ \sum_{\text{all } j} z_j = 1.$$

*Then $z$ is said to be a* stationary distribution *if and only if $zP = z$.*

Thus

$$z = zP = zP^2 = \cdots = zP^n = \cdots.$$

In other words, if $z$ is chosen as the initial probability distribution, i.e., $\pi_j(0) = z_j$ for all $j$, then for all $n$, we have $\pi_j(n) = z_j$. The reader might recognize the similarity of this distribution and the vector quantity introduced in Theorem 9.7.1 in the context of *irreducible* Markov chains. In that theorem, all the components of the vector $z$ must be strictly positive in which case its existence implies that the states are positive recurrent. In some texts, the term *invariant measure* or *invariant distribution* is used in place of stationary distribution. We prefer to reserve the term invariant measure for any nonzero vector quantity that satisfies $zP = z$ without necessarily forcing the elements of $z$ to be probabilities. Thus, an invariant measure is any left-hand eigenvector corresponding to a unit eigenvalue of $P$.

**Example 9.15** Let

$$P = \begin{pmatrix} 0.4 & 0.6 & 0 & 0 \\ 0.6 & 0.4 & 0 & 0 \\ 0 & 0 & 0.5 & 0.5 \\ 0 & 0 & 0.5 & 0.5 \end{pmatrix}.$$

It is easy to verify, by direct substitution, that the following probability distributions all satisfy $zP = z$ and hence are stationary distributions:

$$\begin{aligned} z &= (1/2,\ 1/2,\ 0,\ 0), \\ z &= (0,\ 0,\ 1/2,\ 1/2), \\ z &= (\alpha/2,\ \alpha/2,\ (1-\alpha)/2,\ (1-\alpha)/2), \quad 0 \le \alpha \le 1. \end{aligned}$$

The vector $(1,\ 1,\ -1,\ -1)$ is an invariant vector but it is not a stationary distribution.

In many cases of interest, a discrete-time Markov chain has a *unique* stationary distribution. An example is the following.

**Example 9.16** Let us compute the stationary distribution of the social mobility model discussed earlier. The transition probability matrix for this Markov is given by

$$P = \begin{pmatrix} 0.45 & 0.50 & 0.05 \\ 0.15 & 0.65 & 0.20 \\ 0.00 & 0.50 & 0.50 \end{pmatrix},$$

where states 1, 2, and 3 represent the upper, middle, and lower class respectively. To find the stationary distribution of this Markov chain, we observe that $zP = z$ gives rise to the homogeneous system of equations $z(P - I) = 0$ which must have a singular coefficient matrix $(P - I)$, for otherwise $z = 0$. This singular coefficient matrix gives rise to only two (not three) linearly independent equations. We choose the first two, namely,

$$\begin{aligned} z_1 &= 0.45z_1 + 0.15z_2, \\ z_2 &= 0.5z_1 + 0.65z_2 + 0.5z_3. \end{aligned}$$

Temporarily setting $z_1 = 1$, we have, from the first of these two equations, $z_2 = 0.55/0.15 = 3.666667$. Now substituting the values of $z_1$ and $z_2$ into the second equation, we find $z_3 = 2[-0.5 + 0.35(0.55/0.15)] = 1.566667$. Our computed solution at this point is $(1,\ 3.666667,\ 1.566667)$. We must now normalize this solution so that its components sum to 1. By adding $z_1 + z_2 + z_3$ and dividing each component by this sum, we obtain $z = (0.160428,\ 0.588235,\ 0.251337)$.

The effect of incorporating this normalizing equation has been to convert the system of equations with singular coefficient matrix, namely, $z(P - I) = 0$, into a system of equations with nonsingular coefficient matrix and nonzero right-hand side:

$$(z_1, z_2, z_3) \begin{pmatrix} -0.55 & 0.50 & 1.0 \\ 0.15 & -0.35 & 1.0 \\ 0.00 & 0.50 & 1.0 \end{pmatrix} = (0, 0, 1)$$

—the third equation has been replaced by the normalizing equation. The resulting system of equations has a unique, nonzero solution—the unique stationary distribution of the Markov chain.

It may be observed that

$$zP = (0.160428, 0.588235,\ 0.251337)\begin{pmatrix}0.45 & 0.50 & 0.05\\ 0.15 & 0.65 & 0.20\\ 0.00 & 0.50 & 0.50\end{pmatrix}$$
$$= (0.160428, 0.588235,\ 0.251337) = z.$$

What this result tell us is that, in a country where class mobility can be represented by this Markov chain, approximately 16% of the population is in the upper-class bracket, 59% is in the middle-class bracket, and 25% is in the lower-class bracket.

A unique stationary distribution exists when the Markov chain is finite and irreducible. In this case, when one of the equations in the system of linear equations $z(P - I) = 0$ is replaced by the normalizing equation, the resulting system has a nonsingular coefficient matrix and nonzero right-hand side, meaning that it has a unique solution.

We now turn our attention to the existence or nonexistence of $\lim_{n\to\infty} \pi(n) = \lim_{n\to\infty} \pi(0)P^{(n)}$ for some initial probability distribution $\pi(0)$.

**Definition 9.8.2 (Limiting distribution)** *Let $P$ be the transition probability matrix of a homogeneous discrete-time Markov chain and let $\pi(0)$ be an initial probability distribution. If the limit*

$$\lim_{n\to\infty} P^{(n)} = \lim_{n\to\infty} P^n$$

*exists, then the probability distribution*

$$\pi = \lim_{n\to\infty} \pi(n) = \lim_{n\to\infty} \pi(0)P^{(n)} = \pi(0) \lim_{n\to\infty} P^{(n)} = \pi(0) \lim_{n\to\infty} P^n$$

*exists and is called a limiting distribution of the Markov chain.*

When the states of the Markov chain are positive recurrent and aperiodic (i.e., ergodic) then the limiting distribution always exists and indeed is unique. This same result also holds when the Markov chain is finite, irreducible and aperiodic, since the first two of these properties implies positive recurrency.

**Example 9.17** We have previously seen that the transition probability matrix $P$ and $\lim_{n\to\infty} P^n$ for the Markov chain weather example are given by

$$P = \begin{pmatrix}0.8 & 0.15 & 0.05\\ 0.7 & 0.2 & 0.1\\ 0.5 & 0.3 & 0.2\end{pmatrix}, \qquad \lim_{n\to\infty} P^n = \begin{pmatrix}.76250 & .16875 & .06875\\ .76250 & .16875 & .06875\\ .76250 & .16875 & .06875\end{pmatrix}.$$

Since the elements in a probability vector lie in the interval $[0, 1]$ and sum to 1, it follows that multiplication of $\lim_{n\to\infty} P^n$ by any probability vector yields the limiting distribution $\pi = (.76250,\ .16875,\ .06875)$.

Limiting distributions can also exist when the transition probability matrix is reducible, so long as all the states are positive recurrent.

**Example 9.18** Let

$$P = \begin{pmatrix}0.4 & 0.6 & 0 & 0\\ 0.6 & 0.4 & 0 & 0\\ 0 & 0 & 0.5 & 0.5\\ 0 & 0 & 0.5 & 0.5\end{pmatrix}.$$

Observe that

$$\lim_{n\to\infty} P^{(n)} = \lim_{n\to\infty} P^n = \begin{pmatrix} 0.5 & 0.5 & 0 & 0 \\ 0.5 & 0.5 & 0 & 0 \\ 0 & 0 & 0.5 & 0.5 \\ 0 & 0 & 0.5 & 0.5 \end{pmatrix}.$$

It follows that

$$(1,0,0,0)\lim_{n\to\infty} P^{(n)} = (.5,.5,0,0),$$

$$(0,0,.5,.5)\lim_{n\to\infty} P^{(n)} = (0,0,.5,.5),$$

$$(\alpha,1-\alpha,0,0)\lim_{n\to\infty} P^{(n)} = (.5,.5,0,0) \quad \text{for } 0 \le \alpha \le 1,$$

$$(.375,.375,.125,.125)\lim_{n\to\infty} P^{(n)} = (.375,.375,.125,.125)$$

all satisfy the conditions necessary for a probability distribution to be a limiting distribution.

On the other hand, the aperiodicity property is necessary.

**Example 9.19** Given the following transition probability matrix:

$$P = \begin{pmatrix} 0 & 1 & 0 \\ 0 & 0 & 1 \\ 1 & 0 & 0 \end{pmatrix},$$

then $\lim_{n\to\infty} P^n$ does not exist and $P$ does not have a limiting distribution. In this case, successive powers of $P$ alternate as follows:

$$\begin{pmatrix} 0 & 1 & 0 \\ 0 & 0 & 1 \\ 1 & 0 & 0 \end{pmatrix} \longrightarrow \begin{pmatrix} 0 & 0 & 1 \\ 1 & 0 & 0 \\ 0 & 1 & 0 \end{pmatrix} \longrightarrow \begin{pmatrix} 1 & 0 & 0 \\ 0 & 1 & 0 \\ 0 & 0 & 1 \end{pmatrix} \longrightarrow \begin{pmatrix} 0 & 1 & 0 \\ 0 & 0 & 1 \\ 1 & 0 & 0 \end{pmatrix} \longrightarrow \cdots$$

i.e.,

$$P, \quad P^2, \quad P^3, \quad P^4 = P, \quad P^5 = P^2, \quad P^6 = P^3, \quad P^7 = P, \quad \ldots .$$

Hence this Markov chain has no limiting distribution.

**Definition 9.8.3 (Steady-state distribution)** *A limiting distribution $\pi$ is a steady-state distribution if it converges, independently of the initial starting distribution $\pi(0)$, to a vector whose components are strictly positive (i.e., $\pi_i > 0$ for all states $i$) and sum to 1. If a steady-state distribution exists, it is* unique.

**Example 9.20** We have just seen that the Markov chain of Example 9.18 has multiple limiting distributions, each determined by an initial starting distribution. Consequently, this Markov chain has no steady-state distribution. The Markov chain of Example 9.19 does not have a limiting distribution, and hence neither does it possess a steady-state distribution.

Component $i$ of a steady-state distribution is generally thought of as the long-run proportion of time the Markov chain spends in state $i$, or, equivalently, the probability that a random observer sees the Markov chain in state $i$ after the process has evolved over a long period of time. We stress that the existence of a steady-state distribution implies that both the vector $\pi(n)$ and the matrix $P^{(n)}$ converge *independently* of the initial starting distribution $\pi(0)$. Steady-state distributions are also called *equilibrium distributions* and *long-run distributions*, to highlight the sense that the effect of the initial state distribution $\pi(0)$ has disappeared. When a steady-state distribution exists for a Markov chain, then that distribution is also the unique *stationary distribution* of the chain. Observe that a steady-state distribution is the unique vector $\pi$ that satisfies

$$\pi = \pi(0) \lim_{n\to\infty} P^{(n)} = \pi(0) \lim_{n\to\infty} P^{(n+1)} = \left[\pi(0) \lim_{n\to\infty} P^{(n)}\right] P = \pi P,$$

i.e., $\pi = \pi P$, and consequently $\pi$ is the (unique) stationary probability vector. In the queueing theory literature, the equations $\pi = \pi P$ are called the *global balance equations* since they equate the flow into and out of states. To see this, notice that the $i^{\text{th}}$ equation, $\pi_i = \sum_{\text{all } j} \pi_j p_{ji}$, may be written as

$$\pi_i(1 - p_{ii}) = \sum_{j,\ j\neq i} \pi_j p_{ji}$$

or

$$\pi_i \sum_{j,\ j\neq i} p_{ij} = \sum_{j,\ j\neq i} \pi_j p_{ji}.$$

The left-hand side represents the total flow from out of state $i$ into states other than $i$, while the right-hand side represents the total flow out of all states $j \neq i$ into state $i$.

**Example 9.21** Steady-state distribution—a two-state Markov chain. Let $\{X_n, n \geq 0\}$ be a two-state (0 and 1) Markov chain, whose transition probability matrix is given by

$$P = \begin{pmatrix} 1-p & p \\ q & 1-q \end{pmatrix},$$

where $0 < p < 1$ and $0 < q < 1$. Let $\text{Prob}\{X_0 = 0\} = \pi_0(0)$ be the probability that the Markov chain begins in state 0. It follows that $\text{Prob}\{X_0 = 1\} = 1 - \pi_0(0) = \pi_1(0)$. Let us find the probability distribution at time step $n$. We shall use the relationship

$$\begin{aligned}\text{Prob}\{X_{n+1} = 0\} &= \text{Prob}\{X_{n+1} = 0 \mid X_n = 0\}\text{Prob}\{X_n = 0\} \\ &\quad + \text{Prob}\{X_{n+1} = 0 \mid X_n = 1\}\text{Prob}\{X_n = 1\}.\end{aligned}$$

Let $n = 0$;

$$\begin{aligned}\text{Prob}\{X_1 = 0\} &= \text{Prob}\{X_1 = 0 \mid X_0 = 0\}\text{Prob}\{X_0 = 0\} \\ &\quad + \text{Prob}\{X_1 = 0 \mid X_0 = 1\}\text{Prob}\{X_0 = 1\} \\ &= (1-p)\pi_0(0) + q[1 - \pi_0(0)] \\ &= (1-p-q)\pi_0(0) + q.\end{aligned}$$

Let $n = 1$;

$$\begin{aligned}\text{Prob}\{X_2 = 0\} &= \text{Prob}\{X_2 = 0 \mid X_1 = 0\}\text{Prob}\{X_1 = 0\} \\ &\quad + \text{Prob}\{X_2 = 0 \mid X_1 = 1\}\text{Prob}\{X_1 = 1\} \\ &= (1-p)[\,(1-p-q)\pi_0(0) + q\,] + q[\,1 - (1-p-q)\pi_0(0) - q\,] \\ &= (1-p-q)[\,(1-p-q)\pi_0(0) + q\,] + q \\ &= (1-p-q)^2\pi_0(0) + (1-p-q)q + q.\end{aligned}$$

This leads us to suspect (the proof, by induction, is left as Exercise 9.8.3) that for arbitrary $n$ we have

$$\begin{aligned}\text{Prob}\{X_n = 0\} &= (1-p-q)^n\pi_0(0) + q\sum_{j=0}^{n-1}(1-p-q)^j, \\ \text{Prob}\{X_n = 1\} &= 1 - \text{Prob}\{X_n = 0\}.\end{aligned}$$

Since $\sum_{j=0}^{n-1}(1-p-q)^j$ is the sum of a finite geometric progression, we have

$$\sum_{j=0}^{n-1}(1-p-q)^j = \frac{1-(1-p-q)^n}{p+q},$$

which gives

$$\text{Prob}\{X_n = 0\} = (1-p-q)^n \pi_0(0) + \frac{q}{p+q} - q\frac{(1-p-q)^n}{p+q}$$

$$= \frac{q}{p+q} + (1-p-q)^n \left[\pi_0(0) - \frac{q}{p+q}\right].$$

Also

$$\text{Prob}\{X_n = 1\} = \frac{p}{p+q} + (1-p-q)^n \left[\pi_1(0) - \frac{p}{p+q}\right].$$

We now find the limiting distribution by examining what happens as $n \to \infty$. Since $|1-p-q| < 1$, $\lim_{n\to\infty}(1-p-q)^n = 0$ and hence

$$\lim_{n\to\infty} \text{Prob}\{X_n = 0\} = \frac{q}{p+q} \quad \text{and} \quad \lim_{n\to\infty} \text{Prob}\{X_n = 1\} = \frac{p}{p+q}.$$

Also,

$$\lim_{n\to\infty} P^n = \lim_{n\to\infty} \begin{pmatrix} 1-p & p \\ q & 1-q \end{pmatrix}^n = \frac{1}{p+q}\begin{pmatrix} q & p \\ q & p \end{pmatrix}$$

since

$$\frac{1}{p+q}\begin{pmatrix} q & p \\ q & p \end{pmatrix}\begin{pmatrix} 1-p & p \\ q & 1-q \end{pmatrix} = \frac{1}{p+q}\begin{pmatrix} q & p \\ q & p \end{pmatrix}.$$

Finally, for $0 \leq \alpha \leq 1$, the product of $(\alpha, 1-\alpha)$ and $\lim_{n\to\infty} P^n$ gives $(q/(p+q),\ p/(p+q))$, i.e.,

$$(\alpha, 1-\alpha)\begin{pmatrix} q/(p+q) & p/(p+q) \\ q/(p+q) & p/(p+q) \end{pmatrix} = (q/(p+q),\ p/(p+q)).$$

Hence $(q/(p+q), p/(p+q))$ is the unique steady-state distribution of this Markov chain. That this unique steady-state distribution is also the unique stationary distribution of this Markov chain may be verified by checking that

$$[q/(p+q), p/(p+q)]\begin{pmatrix} 1-p & p \\ q & 1-q \end{pmatrix} = [q/(p+q), p/(p+q)].$$

As we have just seen, in some examples the unique stationary distribution is also the limiting distribution. In some other cases, a Markov chain may possess a stationary distribution but not a limiting distribution, and so on. To explore this further we consider various categories of Markov chains and investigate the existence or nonexistence of stationary, limiting, and steady-state distributions.

**Irreducible Markov Chains that are Null Recurrent or Transient**

An example of such a Markov chain is the reflective barrier, random walk problem examined in detail in the previous section where, when $p = q$, we obtained an irreducible, null-recurrent

Markov chain and when $p > q$, we obtained an irreducible, transient Markov chain. Recall that $p$ is the probability of moving from any state $i > 0$ to state $i + 1$, $q$ is the probability of moving from any state $i > 0$ to $i - 1$. In such Markov chains, there is no stationary probability vector. The only solution to the system of equations $z = zP$ is the vector whose components are all equal to zero. Furthermore, if a limiting distribution exists, its components must all be equal to zero.

### Irreducible Markov Chains that are Positive Recurrent

An example of an irreducible, positive-recurrent Markov chain is the same random walk problem but this time with $p < q$. In an irreducible, positive-recurrent Markov chain, the system of equations $z = zP$ has a unique and strictly positive solution. This solution is the stationary probability distribution $\pi$, and its elements are given by

$$\pi_j = 1/M_{jj}, \tag{9.20}$$

where $M_{jj}$ is the mean recurrence time of state $j$ (which, for a positive-recurrent state, is finite). Equation (9.20) is readily verified by multiplying both sides of Equation 9.10 by the vector $\pi$ and observing that $\pi P = \pi$. We get

$$\pi M = \pi E + \pi P(M - \text{diag}\{M\}) = e^T + \pi(M - \text{diag}\{M\}) = e^T + \pi M - \pi\text{diag}\{M\}$$

and thus $\pi\text{diag}\{M\} = e^T$. Conversely, the states of an irreducible Markov chain which has a unique stationary probability vector, are positive recurrent. An irreducible, positive-recurrent Markov chain does not necessarily have a limiting probability distribution. This is the case when the Markov chain is periodic as the following example shows.

**Example 9.22** Consider the four-state irreducible, positive-recurrent Markov chain whose transition probability matrix is

$$P = \begin{pmatrix} 0 & 1 & 0 & 0 \\ 0 & 0 & 1 & 0 \\ 0 & 0 & 0 & 1 \\ 1 & 0 & 0 & 0 \end{pmatrix}.$$

It may readily be verified that the vector (.25, .25, .25, .25) is the unique stationary distribution of this Markov chain, but that no matter which starting state is chosen, there is no limiting distribution. This Markov chain is periodic with period 4 which means that if it is in state 1 at time step $n$ it will move to state 2 at time step $n + 1$, to state 3 at time step $n + 2$, to state 4 at time step $n + 3$, and back to state 1 at time step $n + 4$. It will alternate forever in this fashion and will never settle into a limiting distribution. As illustrated in the second part of Example, 9.18, the limit $\lim_{n\to\infty} P^{(n)}$ does not exist.

The existence of a unique stationary distribution of a Markov chain does *not necessarily* mean that the Markov chain has a limiting distribution.

### Irreducible, Aperiodic Markov Chains

An example of an irreducible and aperiodic Markov chain is the semi-infinite random walk problem with a Bernoulli barrier. From state 0, rather than moving to state 1 with probability 1, the Markov chain remains in state 0 with probability $q$ or moves to state 1 with probability $p$. This introduces a self-loop on state 0 and destroys the periodicity property of the original chain. The three previous characteristics of the states, transient, null recurrent, and positive recurrent, remain in effect according to whether $p > q$, $p = q$, or $p < q$, respectively.

In an *irreducible and aperiodic* Markov chain, the *limiting distribution* always exists and is independent of the initial probability distribution. Moreover, exactly one of the following conditions must hold:

1. All states are transient or all states are null recurrent, in which case $\pi_j = 0$ for all $j$, and there exists *no stationary distribution* (even though the limiting distribution exists). The state space in this case must be infinite.
2. All states are positive recurrent (which, together with the aperiodicity property, makes them ergodic), in which case $\pi_j > 0$ for all $j$, and the probabilities $\pi_j$ constitute a stationary distribution. The $\pi_j$ are *uniquely* determined by means of

$$\pi_j = \sum_{\text{all } i} \pi_i p_{ij} \quad \text{and} \quad \sum_j \pi_j = 1.$$

In matrix terminology, this is written as

$$\pi = \pi P, \quad \pi e = 1.$$

When the Markov chain is irreducible and contains only a finite number of states, then these states are all positive recurrent and there exists a unique stationary distribution. If the Markov chain is also aperiodic, the aperiodicity property allows us to assert that this stationary distribution is also the unique steady-state distribution. The states of an irreducible, finite, and aperiodic Markov chain are ergodic as is the Markov chain itself.

**Irreducible, Ergodic Markov Chains**

Recall that in an ergodic discrete-time Markov chain all the states are positive recurrent and aperiodic. This, together with the irreducibility property, implies that in such a Markov chain the probability distribution $\pi(n)$, as a function of $n$, always converges to a limiting distribution $\pi$, which is independent of the initial state distribution. This limiting (steady-state) distribution is also the unique stationary distribution of the Markov chain. It follows from Equation (9.19) that

$$\pi_j(n+1) = \sum_{\text{all } i} p_{ij}\pi_i(n),$$

and taking the limit as $n \to \infty$ of both sides gives

$$\pi_j = \sum_{\text{all } i} p_{ij}\pi_i.$$

Thus, the equilibrium probabilities may be uniquely obtained by solving the matrix equation

$$\pi = \pi P \quad \text{with } \pi > 0 \quad \text{and} \quad \|\pi\|_1 = 1. \tag{9.21}$$

It may be shown that, as $n \to \infty$, the rows of the $n$-step transition matrix $P^{(n)} = P^n$ all become identical to the vector of stationary probabilities. Letting $p_{ij}^{(n)}$ denote the $ij^{\text{th}}$ element of $P^{(n)}$, we have

$$\pi_j = \lim_{n\to\infty} p_{ij}^{(n)} \quad \text{for all } i \text{ and } j,$$

i.e., the stationary distribution is replicated on each row of $P^n$ in the limit as $n \to \infty$. This property may be observed in the example of the Markov chain that describes the evolution of the weather in Belfast. The matrix given in Equation (9.6) consists of rows that are all identical and equal to the steady-state probability vector. We have $\pi = (.76250,\ .16875,\ .06875)$.

The following are some performance measurements often deduced from the steady-state probability vector of irreducible, ergodic Markov chains.

- $\nu_j(\tau)$, the average time spent by the chain in state $j$ in a fixed period of time $\tau$ at steady state, is equal to the product of the steady-state probability of state $j$ and the duration of the observation period:

$$\nu_j(\tau) = \pi_j \tau.$$

  The steady-state probability $\pi_j$ itself may be interpreted as the proportion of time that the process spends in state $j$, averaged over the long run. Returning to the weather example, the mean number of sunny days per week is only 0.48125, while the average number of rainy days is 5.33750.
- $1/\pi_j$ is the average number of steps between successive visits to state $j$. For example, the average number of transitions from one sunny day to the next sunny day in the weather example is $1/.06875 = 14.55$. Hence the average number of days *between* two sunny days is 13.55.
- $\nu_{ij}$ is the average time spent by the Markov chain in state $i$ at steady state between two successive visits to state $j$. It is equal to the ratio of the steady-state probabilities of states $i$ and $j$:

$$\nu_{ij} = \pi_i / \pi_j.$$

  The quantity $\nu_{ij}$ is called the *visit ratio*, since it indicates the average number of visits to state $i$ between two successive visits to state $j$. In our example, the mean number of rainy days between two sunny days is 11.09 while the mean number of cloudy days between two sunny days is 2.45.[2]

**Irreducible, Periodic Markov Chains**

We now investigate the effects that periodicity introduces when we seek limiting distributions and higher powers of the single-step transition matrix. In an irreducible discrete-time Markov chain, when the number of single-step transitions required on leaving any state to return to that same state (by any path) is a multiple of some integer $p > 1$, the Markov chain is said to be *periodic of period* $p$, or *cyclic of index* $p$. One of the fundamental properties of such a Markov chain, is that it is possible by a permutation of its rows and columns to transform it to the form, called the *normal form*,

$$P = \begin{pmatrix} 0 & P_{12} & 0 & \dots & 0 \\ 0 & 0 & P_{23} & \dots & 0 \\ \vdots & \vdots & \vdots & \ddots & \vdots \\ 0 & 0 & 0 & \dots & P_{p-1,p} \\ P_{p1} & 0 & 0 & \dots & 0 \end{pmatrix}, \tag{9.22}$$

in which the diagonal submatrices $P_{ii}$ are square and the only nonzero submatrices are $P_{12}, P_{23}, \dots, P_{p1}$. This corresponds to a partitioning of the states of the system into $p$ distinct subsets and an ordering imposed on the subsets. These subsets are referred to as the *cyclic classes* of the Markov chain. The imposed ordering is such that once the system is in a state of subset $i$, it must exit this subset in the next time step and enter a state of subset $(i \bmod p) + 1$.

[2] We would like to point out to our readers that the weather in Northern Ireland is not as bad as the numbers in this example, set up for illustrative purposes only, would have us believe.

**Example 9.23** Consider the Markov chain whose transition diagram is given in Figure 9.13, in which the states have been ordered according to their cyclic classes.

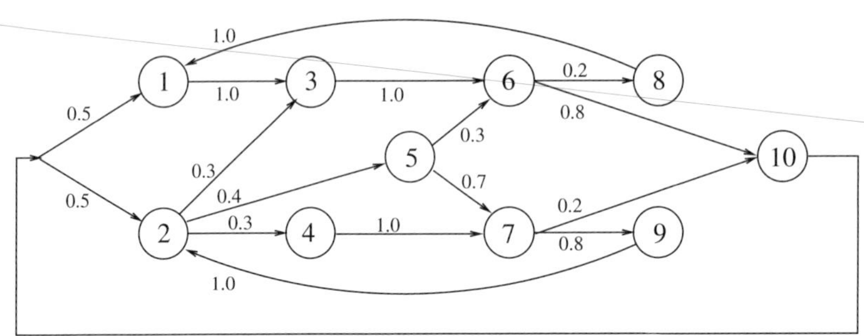


Figure 9.13. A cyclic Markov chain.

Its transition probability matrix is given by

$$P = \left(\begin{array}{cc|ccc|cc|ccc}
 & & 1.0 & & & & & & & \\
 & & 0.3 & 0.3 & 0.4 & & & & & \\
\hline
 & & & & & 1.0 & & & & \\
 & & & & & & 1.0 & & & \\
 & & & & & 0.3 & 0.7 & & & \\
\hline
 & & & & & & & 0.2 & & 0.8 \\
 & & & & & & & & 0.8 & 0.2 \\
\hline
1.0 & & & & & & & & & \\
 & 1.0 & & & & & & & & \\
0.5 & 0.5 & & & & & & & & 
\end{array}\right).$$

Evidently, this Markov chain has periodicity $p = 4$. There are four cyclic classes $C_1$ through $C_4$ given by

$$C_1 = \{1, 2\},\ C_2 = \{3, 4, 5\},\ C_3 = \{6, 7\},\ \text{and } C_4 = \{8, 9, 10\}.$$

On leaving any state of class $C_i$, $i = 1, 2, 3, 4$, the Markov chain can only go to states of class $C_{(i \bmod p)+1}$ in the next time step and therefore it can only return to a starting state after four, or some multiple of four, steps.

Our interest in periodic Markov chains such as this is in determining its behavior at time step $n$ in the limit as $n \to \infty$. Specifically, we wish to investigate the behavior of $P^n$ as $n \to \infty$, the existence or nonexistence of a limiting probability distribution and the existence or nonexistence of a stationary distribution. We begin by examining the case when the Markov chain possesses four cyclic classes, i.e., the case of Equation (9.22) with $p = 4$. We have

$$P = \begin{pmatrix} 0 & P_{12} & 0 & 0 \\ 0 & 0 & P_{23} & 0 \\ 0 & 0 & 0 & P_{34} \\ P_{41} & 0 & 0 & 0 \end{pmatrix},$$

and taking successive powers, we obtain

$$P^2 = \begin{pmatrix} 0 & 0 & P_{12}P_{23} & 0 \\ 0 & 0 & 0 & P_{23}P_{34} \\ P_{34}P_{41} & 0 & 0 & \\ 0 & P_{41}P_{12} & 0 & 0 \end{pmatrix},$$

$$P^3 = \begin{pmatrix} 0 & 0 & 0 & P_{12}P_{23}P_{34} \\ P_{23}P_{34}P_{41} & 0 & 0 & 0 \\ 0 & P_{34}P_{41}P_{12} & 0 & \\ 0 & 0 & P_{41}P_{12}P_{23} & 0 \end{pmatrix},$$

$$P^4 = \begin{pmatrix} P_{12}P_{23}P_{34}P_{41} & 0 & 0 & 0 \\ 0 & P_{23}P_{34}P_{41}P_{12} & 0 & 0 \\ 0 & 0 & P_{34}P_{41}P_{12}P_{23} & 0 \\ 0 & 0 & 0 & P_{41}P_{12}P_{23}P_{34} \end{pmatrix}.$$

These successive powers show that, beginning in one of the states of $C_1$, a state of $C_2$ may be reached in one step, at least two steps are needed to reach a state of $C_3$, and a minimum of three steps to reach a state of $C_4$. After four steps, the Markov chain is back in a state of $C_1$. Thus, after four steps, the transition matrix is block diagonal, and each block is a stochastic matrix. It follows then that at any time step $4n$, $n = 1, 2, \ldots$, the transition matrix has this block structure and each block represents the transition probability matrix of an irreducible, recurrent and *aperiodic* Markov chain; aperiodic since a single-step in the new chain corresponds to four steps in the original chain. This means that we may now apply the previously discussed theory on irreducible, recurrent and aperiodic Markov chains to each of the blocks. In particular, each has a limit as $n \to \infty$. It follows that $\lim_{n\to\infty} P^{4n}$ exists.

**Example 9.24** Consider the Markov chain of Example 9.23. We find that $P^4 =$

| | | | | | | | | | |
|---|---|---|---|---|---|---|---|---|---|
| 0.6000 | 0.4000 | | | | | | | | |
| 0.3100 | 0.6900 | | | | | | | | |
| | | 0.7200 | 0.1200 | 0.1600 | | | | | |
| | | 0.3700 | 0.2700 | 0.3600 | | | | | |
| | | 0.4750 | 0.2250 | 0.3000 | | | | | |
| | | | | | 0.7680 | 0.2320 | | | |
| | | | | | 0.4780 | 0.5220 | | | |
| | | | | | | | 0.2000 | 0.0000 | 0.8000 |
| | | | | | | | 0.0840 | 0.4640 | 0.4520 |
| | | | | | | | 0.1420 | 0.2320 | 0.6260 |

and $\lim_{n\to\infty} P^{4n} =$

| | | | | | | | | | |
|---|---|---|---|---|---|---|---|---|---|
| 0.4366 | 0.5634 | | | | | | | | |
| 0.4366 | 0.5634 | | | | | | | | |
| | | 0.6056 | 0.1690 | 0.2254 | | | | | |
| | | 0.6056 | 0.1690 | 0.2254 | | | | | |
| | | 0.6056 | 0.1690 | 0.2254 | | | | | |
| | | | | | 0.6732 | 0.3268 | | | |
| | | | | | 0.6732 | 0.3268 | | | |
| | | | | | | | 0.1347 | 0.2614 | 0.6039 |
| | | | | | | | 0.1347 | 0.2614 | 0.6039 |
| | | | | | | | 0.1347 | 0.2614 | 0.6039 |

Each diagonal block is treated as the transition probability matrix of a finite, aperiodic, irreducible Markov chain whose limiting distribution is then equal to its stationary distribution. This stationary distribution may be computed for each of the four blocks separately and, in the limit as $n \to \infty$, each block row must become equal to this distribution. Now observe that concatenating the four stationary distributions together yields a vector $z$ for which $z = zP$. We have

$$z = (0.4366, 0.5634, 0.6056, 0.1690, 0.2254, 0.6732, 0.3268, 0.1367, 0.2614, 0.6039),$$

which when normalized becomes the stationary distribution of the original Markov chain:

$$(0.1092, 0.1408, 0.1514, 0.0423, 0.0563, 0.1683, 0.0817, 0.0337, 0.0654, 0.1510).$$

Notice that this assumes (correctly) that the Markov chain spends equal amounts of time in all periodic classes. Since an irreducible, positive-recurrent Markov chain has a unique and strictly positive stationary distribution, it follows that this computed solution is the only possible solution.

**Reducible Markov Chains**

We now turn to Markov chains that are reducible in that they contain multiple transient and irreducible closed classes. In this case, a Markov chain will possess multiple (rather than a unique) stationary distributions. Also any linear combination of the different stationary distributions, once normalized to yield a probability vector, is also a stationary distribution. In such a case, it suffices to treat each irreducible closed class separately. To illustrate this, we return to a previous example.

**Example 9.25** Consider the Markov chain whose transition probability matrix is

$$P = \left(\begin{array}{ccc|cc|c|cc}
.4 & .2 & & .2 & & & & .2 \\
.3 & .3 & & & .1 & & .2 & .1 \\
& & .1 & .3 & .1 & & .5 & \\
\hline
& & & .7 & .3 & & & \\
& & & .5 & .5 & & & \\
\hline
& & & & & 1.0 & & \\
\hline
& & & & & & .9 & .1 \\
& & & & & & .1 & .9
\end{array}\right).$$

This Markov chain has three transient states, namely, states 1, 2, and 3, and three irreducible classes, consisting of states $\{4, 5\}$, $\{6\}$, and $\{7, 8\}$, respectively. The stationary distributions of the three irreducible classes are unique and given by

$$(0.6250,\ 0.3750) = (0.6250,\ 0.3750)\begin{pmatrix} 0.7 & 0.3 \\ 0.5 & 0.5 \end{pmatrix},$$

$$(1.0) = (1.0)(1),$$

$$\text{and}\quad (0.5,\ 0.5) = (0.5,\ 0.5)\begin{pmatrix} 0.9 & 0.1 \\ 0.1 & 0.9 \end{pmatrix}.$$

If any of these three distributions is padded with zeros to produce a vector of length 8, then the resulting vector is a stationary distribution of the complete Markov chain. For example,

$$(0, 0, 0, 0.6250,\ 0.3750, 0, 0, 0)\left(\begin{array}{ccc|cc|c|cc}
.4 & .2 & & .2 & & & & .2 \\
.3 & .3 & & & .1 & & .2 & .1 \\
& & .1 & .3 & .1 & & .5 & \\
\hline
& & & .7 & .3 & & & \\
& & & .5 & .5 & & & \\
\hline
& & & & & 1.0 & & \\
\hline
& & & & & & .9 & .1 \\
& & & & & & .1 & .9
\end{array}\right)$$

$$= (0, 0, 0, 0.6250,\ 0.3750, 0, 0, 0),$$

and similarly for the other two. Indeed, any linear combination of all three, once normalized, is a stationary distribution. For example, adding all three together gives $(0, 0, 0, 0.6250, 0.3750, 1, 0.5, 0.5)$ which when normalized is $(0, 0, 0, .2083, .1250, .3333, .1667, .1667)$, and it is easy to verify that

$$(0, 0, 0, .2083, .1250, .3333, .1667, .1667)\left(\begin{array}{ccc|cc|c|cc} .4 & .2 & & .2 & & & & .2 \\ .3 & .3 & & & .1 & & .2 & .1 \\ & & .1 & .3 & .1 & & .5 & \\ \hline & & & .7 & .3 & & & \\ & & & .5 & .5 & & & \\ \hline & & & & & 1.0 & & \\ \hline & & & & & & .9 & .1 \\ & & & & & & .1 & .9 \end{array}\right)$$

$$= (0, 0, 0, .2083, .1250, .3333, .1667, .1667).$$

Taking successively higher powers of the matrix $P$, we obtain

$$\lim_{n\to\infty} P^n = \left(\begin{array}{ccc|cc|c|cc} 0.0 & 0.0 & 0.0 & 0.2778 & 0.1667 & 0.0 & 0.2778 & 0.2778 \\ 0.0 & 0.0 & 0.0 & 0.2083 & 0.1250 & 0.0 & 0.3333 & 0.3333 \\ 0.0 & 0.0 & 0.0 & 0.2778 & 0.1667 & 0.0 & 0.2778 & 0.2778 \\ \hline & & & 0.6250 & 0.3750 & & & \\ & & & 0.6250 & 0.3750 & & & \\ \hline & & & & & 1.0 & & \\ \hline & & & & & & 0.5 & 0.5 \\ & & & & & & 0.5 & 0.5 \end{array}\right).$$

Although not all rows are equal, all rows are stationary distributions and are equal to some linear combination of the three individual solutions. Finally, the elements of $\lim_{n\to\infty} P^n$ are those predicted by Theorem 9.4.2. All transition probabilities into transient states are zero and transition probabilities from ergodic states back to the same ergodic state are strictly positive. As for the final part of this theorem, namely, that the transition probability from any state $i$ into an ergodic state $j$ is given by

$$\lim_{n\to\infty} p_{ij}^{(n)} = f_{ij} \lim_{n\to\infty} p_{jj}^{(n)},$$

we must first compute the elements $f_{ij}$ of the reachability matrix $F$. Fortunately, we have already computed these in a previous example, Equation (9.17), which allows us to compute results such as

$$\lim_{n\to\infty} p_{14}^{(n)} = 0.4444 \times 0.6250 = 0.2778,$$

$$\lim_{n\to\infty} p_{28}^{(n)} = 0.6667 \times 0.5000 = 0.3333,$$

$$\lim_{n\to\infty} p_{35}^{(n)} = 0.4444 \times 0.3750 = 0.1667,$$

and so on.

To conclude, the states of a Markov chain may be partitioned into disjoint subsets according to its irreducible closed classes and transient states. The states of each irreducible closed subset are recurrent and may be analyzed independently as separate Markov chains.

## 9.9 Reversibility

We defined a discrete-time Markov chain $\{X_n,\ n = 0, 1, 2, \ldots\}$, as a stochastic process that satisfies the Markov property: for all natural numbers $n$ and all states $x_n$,

$$\text{Prob}\{X_{n+1} = x_{n+1}|X_n = x_n, X_{n-1} = x_{n-1}, \ldots, X_0 = x_0\} = \text{Prob}\{X_{n+1} = x_{n+1}|X_n = x_n\}.$$

The state in which the system finds itself at time step $n + 1$ depends only on the state it occupied at time step $n$ and not on the states occupied at time steps prior to $n$. Our Markov chain evolves as time moves forward: $n = 0, 1, 2, \ldots$. Beginning the process at time step 0 is convenient, but not necessary. We may define the process so that it begins at $-\infty$ and continues on to $+\infty$. To accommodate this possibility the previous definition should be modified to allow $-\infty < n < +\infty$ with the understanding that, for all permissible values of $n$, $X_n$ is a state of the Markov chain. Suppose now, instead of watching the process move forward in time, we watch it moving *backward* in time, from the state it occupies at some time step $m$ to the state it occupies at time step $m - 1$ to its state at time step $m - 2$, and so on. In an analogy with watching a movie, the original Markov chain $\{X_n, n \geq 0\}$ represents the usual showing of the movie with one frame per value of $n$, while watching the process in reverse mode corresponds to showing the movie backward, frame by frame. When time is turned backward in this fashion, the process is called the *reversed process*. In most cases, the original process, which we shall now refer to as the *forward* process, is not the same as the reversed process. However, when both are stochastically identical, the Markov chain is said to be *reversible*. Using the accepted (and unfortunate) current terminology, all Markov chains can be "reversed", but not all Markov chains are "reversible." If a movie were reversible (which is rather difficult to imagine with our understanding of the term), one would not be able to tell if the operator were running it in the forward direction or in the backward direction. For a Markov chain to be reversible, it must be stationary and homogeneous, hence we remove the dependence on $n$ and write the transition probabilities as $p_{ij}$ and not $p_{ij}(n)$.

We consider an irreducible, discrete-time Markov chain that is stationary and homogeneous with transition probability matrix $P$ and a unique stationary distribution $\pi$. Recall that an irreducible Markov chain has a unique stationary distribution $\pi$ with $\pi = \pi P$ if and only if all states are positive recurrent. The probability of moving from state $i$ at time step $n$ to state $j$ at time step $n - 1$ is obtained as

$$\begin{aligned} \text{Prob}\{X_{n-1} = j \mid X_n = i\} &= \frac{\text{Prob}\{X_{n-1} = j,\ X_n = i\}}{\text{Prob}\{X_n = i\}} \\ &= \frac{\text{Prob}\{X_n = i \mid X_{n-1} = j\}\text{Prob}\{X_{n-1} = j\}}{\text{Prob}\{X_n = i\}} \\ &= \frac{p_{ji}\pi_j}{\pi_i}. \end{aligned} \tag{9.23}$$

Let $r_{ij} = p_{ji}\pi_j/\pi_i$. Then the matrix $R$, whose $ij$ element is $r_{ij}$, is just the single-step transition matrix for the reversed Markov chain. Observe that $R$ is stochastic since $r_{ij} \geq 0$ for all $i$ and $j$ and

$$\sum_{\text{all } j} r_{ij} = \frac{1}{\pi_i}\sum_{\text{all } j} \pi_j p_{ji} = \frac{1}{\pi_i}\pi_i = 1 \quad \text{for all } i.$$

In matrix terminology, we have

$$R = \text{diag}\{\pi\}^{-1}\ P^T\ \text{diag}\{\pi\},$$

where $\text{diag}\{\pi\}$ is a diagonal matrix whose $i^{\text{th}}$ diagonal element is $\pi_i$ and $P^T$ is the transpose of $P$.

**Example 9.26** Consider a Markov chain whose transition probability matrix is

$$P = \begin{pmatrix} .8 & .15 & .05 \\ .7 & .2 & .1 \\ .5 & .3 & .2 \end{pmatrix}.$$

Its unique stationary distribution is given by (.76250, .16875, .06875). In fact, this is the transition matrix for the Belfast weather example. The transition probability matrix of the reversed chain is

$$R = \begin{pmatrix} .76250 & 0 & 0 \\ 0 & .16875 & 0 \\ 0 & 0 & .06875 \end{pmatrix}^{-1} \begin{pmatrix} .8 & .7 & .5 \\ .15 & .2 & .3 \\ .05 & .1 & .2 \end{pmatrix} \begin{pmatrix} .76250 & 0 & 0 \\ 0 & .16875 & 0 \\ 0 & 0 & .06875 \end{pmatrix}$$

$$= \begin{pmatrix} .80000 & .15492 & .04508 \\ .67778 & .20000 & .12222 \\ .55455 & .24545 & .20000 \end{pmatrix}.$$

Observe that $R$ is a stochastic matrix: all elements lie between zero and one and, the row sums are all equal to 1.

A Markov chain is said to be (time) reversible if and only if $r_{ij} = p_{ij}$ for all $i$ and $j$; in other words, if and only if $R = P$. Thus the Belfast weather example is *not* reversible. From Equation (9.23), we see that this condition ($R = P$) is equivalent to

$$\pi_i p_{ij} = \pi_j p_{ji} \quad \text{for all } i, j. \tag{9.24}$$

Equations (9.24) are called the *detailed balance* equations. Detailed balance is a necessary and sufficient condition for reversibility. Importantly, detailed balance also implies *global balance*. By summing Equation (9.24) over all $i$, we obtain

$$\sum_{\text{all } i} \pi_i p_{ij} = \sum_{\text{all } i} \pi_j p_{ji} \quad \text{for all } j$$

and since $\sum_{\text{all } i} \pi_j p_{ji} = \pi_j \sum_{\text{all } i} p_{ji} = \pi_j$

$$\pi_j = \sum_{\text{all } i} \pi_i p_{ij} \quad \text{for all } j,$$

which in matrix form gives the global balance equations

$$\pi = \pi P.$$

Global balance, on the other hand, does not imply detailed balance. The detailed balance equations imply that $\pi$ exists and is the stationary distribution vector of both the forward and backward Markov chains.

**Example 9.27** Consider a random walk on the integers $0, 1, 2, \ldots$ with $p_{01} = 1$, $p_{i,i+1} = p$ and $p_{i,i-1} = 1 - p = q$ for $i > 0$. The transition probability matrix is given by

$$P = \begin{pmatrix} 0 & 1 & 0 & \cdots & & & \\ q & 0 & p & 0 & \cdots & & \\ 0 & q & 0 & p & 0 & \cdots & \\ 0 & 0 & q & 0 & p & & \\ \vdots & \vdots & \vdots & \vdots & \vdots & \ddots & \end{pmatrix}.$$

We saw previously that all the states in this Markov chain are positive recurrent if and only if $p < q$ and we assume that this condition holds in this example. Since the Markov chain is irreducible, it now follows that it has a unique stationary probability vector $\pi$ and we may write $\pi = \pi P$. We wish to show that the detailed balance equations are satisfied, i.e., that

$$\pi_i p_{ij} = \pi_j p_{ji} \quad \text{for all } i, j.$$

First observe that this holds when $|i-j| \geq 2$, since in this case $p_{ij} = p_{ji} = 0$. It also holds trivially when $i = j$. To show that the detailed balanced equations hold in the remaining cases, we expand the equations $\pi = \pi P$.

$$\begin{aligned} \pi_0 = \pi_1 q \quad &\Rightarrow \quad \pi_0 p_{01} = \pi_1 p_{10}, \\ \pi_1 = \pi_0 + q\pi_2 \quad \Rightarrow \quad \pi_1(1-q) = \pi_2 q \quad &\Rightarrow \quad \pi_1 p_{12} = \pi_2 p_{21}, \\ \pi_2 = (1-q)\pi_1 + q\pi_3 \quad \Rightarrow \quad \pi_2(1-q) = \pi_3 q \quad &\Rightarrow \quad \pi_2 p_{23} = \pi_3 p_{32}, \end{aligned}$$

and, as can be shown by induction,

$$\pi_i = (1-q)\pi_{i-1} + q\pi_{i+1} \quad \Rightarrow \quad \pi_i(1-q) = \pi_{i+1} q \quad \Rightarrow \quad \pi_i p_{i,i+1} = \pi_{i+1} p_{i+1,i}$$

for all remaining values of $i$. Hence this Markov chain is reversible.

This example can be generalized in a straightforward manner to cover the case when the transition probability matrix has the form

$$P = \begin{pmatrix} 0 & 1 & 0 & \cdots & & \\ q_1 & 0 & p_1 & 0 & \cdots & \\ 0 & q_2 & 0 & p_2 & 0 & \cdots \\ 0 & 0 & q_3 & 0 & p_3 & \\ \vdots & \vdots & \vdots & \vdots & \vdots & \ddots \end{pmatrix},$$

where $p_i + q_i = 1$ and where the values of $p_i$ are $q_i$ are such that a unique stationary distribution exists. Later we shall see that all *birth–death processes*, an important class of queueing models that we discuss in detail in Chapter 12, fall into the continuous-time version of this category, and so we may conclude that all birth–death processes are reversible. More generally, a positive-recurrent, tree-structured Markov chain is reversible. A discrete-time Markov chain is said to be tree structured if between any two distinct states $i$ and $j$ there is one, and only one, path consisting of distinct states $k_1, k_2, \ldots, k_m$ for which

$$p_{ik_1} p_{k_1k_2} \cdots p_{k_{m-1}k_m} p_{k_m j} > 0.$$

Figure 9.14 illustrates a tree-structured Markov chain.

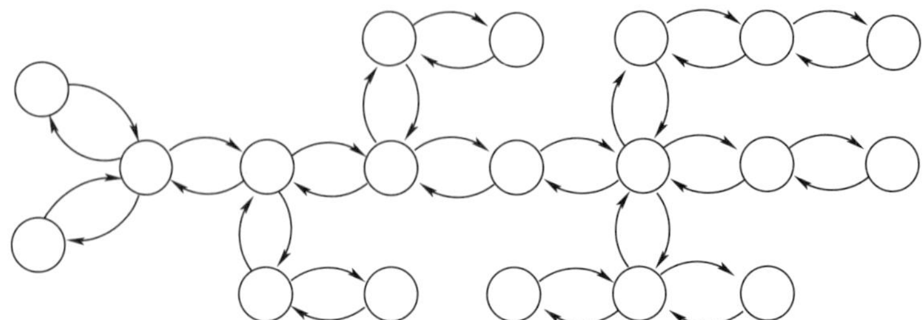

Figure 9.14. Transition diagram of a tree-structured Markov chain.

The curious reader may wonder why we have not as yet tried to pigeonhole *ergodic* Markov chains into our discussion of reversibility, instead confining ourselves to *irreducible, positive-recurrent* Markov chains. The reason lies in the aperiodicity property of an ergodic Markov chain. An ergodic Markov chain is both positive recurrent and aperiodic, but as the previous example shows, a periodic Markov chain can be reversible. The Markov chain of Example 9.27 has period 2. However, a Markov chain having a period greater than 2 *cannot* be reversible. Consider any two distinct states $i$ and $j$ for which $p_{ij} > 0$ in a Markov chain with period strictly greater than 2. Then, in a single step, this Markov chain can move from state $i$ to state $j$ and we have $\pi_i p_{ij} > 0$. But there is no possibility of the chain being able to move from $j$ to $i$ in a single step, for otherwise the chain could not have period greater than 2. Hence $\pi_j p_{ji} = 0$ and the detailed balance equations are not satisfied: a Markov chain with period strictly greater than 2 cannot be reversible. Ergodicity is not a necessary condition for a Markov chain to be reversible whereas irreducibility and positive recurrency are necessary. Reversible Markov chains having period equal to 2, as illustrated by the birth–death examples, form an important class of reversible Markov chains.

An irreducible, positive-recurrent Markov chain whose transition probability matrix is symmetric, i.e., $p_{ij} = p_{ji}$ for all $i, j$, or in matrix terms, $P = P^T$, is reversible. Such a matrix is doubly stochastic: both its row sums and its column sums are equal to 1. It follows that all the elements in the stationary distribution are equal and hence the detailed balance equations are satisfied. However, a matrix that is doubly stochastic is reversible only if it is symmetric. Since all the components in the stationary distribution of a doubly stochastic matrix are equal, the two diagonal matrices in $R = \text{diag}\{\pi\}^{-1} P^T \text{diag}\{\pi\}$ cancel each other and we are left with $R = P^T$.

A matrix $A$ is said to have a *symmetric structure* if for all $a_{ij} \neq 0$ we have $a_{ji} \neq 0$. A necessary condition for an irreducible, positive-recurrent Markov chain to be reversible is that its transition probability matrix $P$ have a symmetric structure. It follows that if $p_{ij} > 0$ and $p_{ji} = 0$ the detailed balance equations cannot be satisfied.

Except in some special cases such as tree-structured Markov chains, it can be sometimes difficult to determine if a Markov chain is reversible. In the absence of any knowledge that might be obtained from the system concerning its reversibility, the usual approach is to first compute the stationary distribution and verify that the detailed balance equations are satisfied. There is fortunately one other characteristic of reversible Markov chains that can sometimes be valuable. This is encapsulated in Kolmogorov's criterion.

**Kolmogorov's criterion** *An irreducible, positive-recurrent Markov chain is reversible if and only if the probability of traversing any closed path (cycle) is the same in both the forward and backward directions.*

To see why this is true, consider an irreducible, positive-recurrent Markov chain with transition probability matrix $P$ and the closed path

$$p_{i_0 i_1} p_{i_1 i_2} p_{i_2 i_3} \cdots p_{i_{m-1} i_m} p_{i_m i_0} > 0.$$

If this Markov chain is reversible, and $R$ is the transition probability matrix of the reversed chain, then $R = P$ and

$$\begin{aligned} p_{i_0 i_1} p_{i_1 i_2} p_{i_2 i_3} \cdots p_{i_{m-1} i_m} p_{i_m i_0} &= r_{i_0 i_1} r_{i_1 i_2} r_{i_2 i_3} \cdots r_{i_{m-1} i_m} r_{i_m i_0} \\ &= \left(p_{i_1 i_0} \pi_{i_1} / \pi_{i_0}\right) \left(p_{i_2 i_1} \pi_{i_2} / \pi_{i_1}\right) \left(p_{i_3 i_2} \pi_{i_3} / \pi_{i_2}\right) \cdots \left(p_{i_0 i_m} \pi_{i_0} / \pi_{i_m}\right) \\ &= p_{i_1 i_0} p_{i_2 i_1} p_{i_3 i_2} \cdots p_{i_0 i_m} \\ &= p_{i_0 i_m} \cdots p_{i_3 i_2} p_{i_2 i_1} p_{i_1 i_0}, \end{aligned}$$

and the probabilities in the forward and backward directions are the same. The converse is equally easy to show. It follows that the reversibility property of a Markov chain can be verified by checking that the Kolmogorov criterion is satisfied along all closed paths.

An important case which gives rise to a reversible Markov chain occurs in the following context. Consider an arbitrarily connected undirected graph with positive weights on each edge. Let the weight on the edge connecting node $i$ and $j$ be $w_{ij}$, for all $i$ and $j$. If the nodes correspond to cities, for example, then the weights might represent the distance in miles between cities. The weighted adjacency matrix of such a graph is symmetric. We now transform this adjacency matrix to the transition probability matrix of a Markov chain by setting $p_{ij} = w_{ij}/w_i$ with $w_i = \sum_j w_{ij} < \infty$. The matrix $P$ obtained in this fashion, although not necessarily symmetric, does have a symmetric structure. If $w = \sum_i w_i < \infty$, then the $i^{\text{th}}$ component of the stationary distribution is given by

$$\pi_i = \frac{\sum_j w_{ij}}{\sum_k \sum_j w_{kj}} = \frac{w_i}{w} \quad \text{for all } i.$$

Observe that $\pi_i > 0$ for all $i$, that

$$\sum_i \pi_i = \sum_i \frac{w_i}{w} = 1,$$

and

$$\pi_i = \frac{w_i}{w} = \sum_k \frac{w_{ik}}{w} = \sum_k \frac{w_{ki}}{w} = \sum_k \frac{w_k}{w}\frac{w_{ki}}{w_k} = \sum_k \pi_k p_{ki},$$

i.e., that $\pi = \pi P$ and hence $\pi$ is indeed the stationary distribution. Furthermore, since

$$\pi_i p_{ij} = \frac{w_i}{w}\frac{w_{ij}}{w_i} = \frac{w_{ij}}{w} = \frac{w_{ji}}{w} = \frac{w_{ji}}{w}\frac{w_j}{w_j} = \frac{w_j}{w}\frac{w_{ji}}{w_j} = \pi_j p_{ji},$$

the detailed balance equations are satisfied and this Markov chain is reversible.

**Example 9.28** Consider the undirected graph whose adjacency matrix is

$$\begin{pmatrix} 0 & 5 & 4 \\ 5 & 0 & 1 \\ 4 & 1 & 0 \end{pmatrix}.$$

Converting this to the transition probability matrix of a Markov chain gives

$$P = \begin{pmatrix} 0 & 5/9 & 4/9 \\ 5/6 & 0 & 1/6 \\ 4/5 & 1/5 & 0 \end{pmatrix}$$

and

$$\pi = (9/20,\ 6/20,\ 5/20).$$

The transition probability matrix of the reversed chain is

$$R = \begin{pmatrix} 20/9 & 0 & 0 \\ 0 & 20/6 & 0 \\ 0 & 0 & 20/5 \end{pmatrix} \begin{pmatrix} 0 & 5/6 & 4/5 \\ 5/9 & 0 & 1/5 \\ 4/9 & 1/6 & 0 \end{pmatrix} \begin{pmatrix} 9/20 & 0 & 0 \\ 0 & 6/20 & 0 \\ 0 & 0 & 5/20 \end{pmatrix}$$

$$= \begin{pmatrix} 0 & 5/9 & 4/9 \\ 5/6 & 0 & 1/6 \\ 4/5 & 1/5 & 0 \end{pmatrix} = P.$$

To close this section on reversible Markov chains, we consider the case of *truncated* chains. Let $P$ be the transition probability matrix of a reversible Markov chain and consider what happens when some of the states are removed leaving a truncated process. In the state transition diagram not only are the states removed but all edges leading to, or emanating from, the eliminated states are also removed. The states that remain must be able to communicate with each other so that the irreducibility property is conserved. For example, if the original Markov chain is tree structured, and hence reversible, truncated Markov chains can be formed by removing some of the branches but not by sawing the trunk in two. Without loss of generality, we may assume that the state space is ordered in such a way that states remaining after the truncation precede those that are removed. This means that after the truncation, only the upper left-hand portion of $P$ contains nonzero probabilities and we illustrate this as follows

$$P = \left( \begin{array}{c|c} P_{11} & P_{12} \\ \hline P_{21} & P_{22} \end{array} \right), \qquad \hat{P} = \left( \begin{array}{c|c} P_{11} & 0 \\ \hline 0 & 0 \end{array} \right),$$

where $\hat{P}$ represents the truncated process. Notice that $\hat{P}$, or more precisely $P_{11}$, is not stochastic. Since the original Markov chain is irreducible, $P_{12}$ cannot be identically zero and therefore the sum of the probabilities across at least one row of $P_{11}$ must be strictly less than 1. However, if the rows of $P_{11}$ are normalized so that they sum to 1 (replace $p_{ij}$ with $p_{ij}/\sum_j p_{ij}$) then $P_{11}$ becomes stochastic and, furthermore, if the Markov chain represented by $P_{11}$ is irreducible, then it is also reversible. Let $\pi$ be the stationary distribution of the original Markov chain and let $\mathcal{E}$ be the set of states in the truncated chain. Since $\pi_i p_{ij} = \pi_j p_{ji}$ for all states prior to truncation, it must also hold for the states in $\mathcal{E}$ after truncation. The restriction of $\pi$ to the states of $\mathcal{E}$ is not a proper distribution since $\sigma = \sum_{i\in\mathcal{E}} \pi_i < 1$. However, when divided by $\sigma$ it become a proper distribution and is in fact the stationary distribution of the truncated chain. To summarize, an irreducible, truncated Markov chain, obtained from a reversible Markov chain with stationary probability vector $\pi$, is itself reversible and its stationary distribution is given by $\pi/\sigma$. Truncated processes arise frequently in queueing models with finite waiting rooms.

## 9.10 Continuous-Time Markov Chains

In a discrete-time Markov chain, we define an infinite (denumerable) sequence of time steps at which the chain may either change state or remain in its current state. The transition probability matrix in this case can have nonzero probabilities on the diagonal, called self-loops, which allows for the possibility of the Markov chain remaining in its current state at any time step. With a continuous-time Markov chain, a change of state may occur at *any* point in time. Analogous to the definition of a discrete-time Markov chain, we say that a stochastic process $\{X(t),\ t \geq 0\}$ is a continuous-time Markov chain if for all integers (states) $n$, and for any sequence $t_0, t_1, \ldots, t_n, t_{n+1}$ such that $t_0 < t_1 < \ldots < t_n < t_{n+1}$,

$$\begin{aligned}\text{Prob}\{X(t_{n+1}) = x_{n+1} | X(t_n) = x_n, X(t_{n-1}) = x_{n-1}, \ldots, X(t_0) = x_0\} \\ = \text{Prob}\{X(t_{n+1}) = x_{n+1} | X(t_n) = x_n\}.\end{aligned}$$

Notice from this definition that not only is the sequence of previously visited states irrevelant to the future evolution of the chain, but so too is the amount of time already spent in the current state. Equivalently, we say that the stochastic process $\{X(t),\ t \geq 0\}$ is a continuous-time Markov chain if for integers (states) $i, j, k$ and for all time instants $s, t, u$ with $t \geq 0,\ s \geq 0$, and $0 \leq u \leq s$, we have

$$\text{Prob}\{X(s+t) = k | X(s) = j, X(u) = i\} = \text{Prob}\{X(s+t) = k | X(s) = j\}.$$

If a continuous-time Markov chain is nonhomogeneous, we write

$$p_{ij}(s,t) = \text{Prob}\{X(t) = j | X(s) = i\},$$

where $X(t)$ denotes the state of the Markov chain at time $t \geq s$. When the continuous-time Markov chain is homogeneous, these transition probabilities depend on the difference $\tau = t - s$ rather than on the actual values of $s$ and $t$. In this case, we simplify the notation by writing

$$p_{ij}(\tau) = \text{Prob}\{X(s+\tau) = j | X(s) = i\} \text{ for all } s \geq 0.$$

This denotes the probability of being in state $j$ after an interval of length $\tau$, given that the current state is state $i$. It depends on the length $\tau$ but not on $s$, the specific moment at which this time interval begins. It follows that

$$\sum_{\text{all } j} p_{ij}(\tau) = 1 \quad \text{for all values of } \tau.$$

## *9.10.1 Transition Probabilities and Transition Rates*

Whereas in a discrete-time Markov chain the interactions among the states are given in terms of transition probabilities, the interactions in a continuous-time Markov chain are usually specified in terms of the rates at which transitions occur. This simplifies the notation and the analysis. A continuous-time Markov chain in some state $i$ at time $t$ will move to some other state $j$ at rate $q_{ij}(t)$ per unit time. In this manner, while a discrete-time Markov chain is represented by its matrix of *transition probabilities*, $P(n)$, at time step $n$, a continuous-time Markov chain is represented by its matrix of *transition rates*, $Q(t)$, at time $t$. It therefore behooves us to examine the relationship between these two quantities (probabilities and rates) in the context of Markov chains.

The *probability* that a transition occurs from a given source state depends not only on the source state itself but also on the length of the interval of observation. In what follows, we shall consider a period of observation $\tau = \Delta t$. Let $p_{ij}(t, t+\Delta t)$ be the probability that a transition occurs from state $i$ to state $j$ in the interval $[t, t+\Delta t)$. As the duration of this interval becomes very small, the probability that we observe a transition also becomes very small. In other words, as $\Delta t \to 0$, $p_{ij}(t, t+\Delta t) \to 0$ for $i \neq j$. It then follows from conservation of probability that $p_{ii}(t, t+\Delta t) \to 1$ as $\Delta t \to 0$. On the other hand, as $\Delta t$ becomes large, the probability that we observe a transition increases. As $\Delta t$ becomes even larger, the probability that we observe multiple transitions becomes nonnegligible. Observation periods are chosen sufficiently small that the probability of observing multiple events in any observation period is of order $o(\Delta t)$. Recall that $o(\Delta t)$, "little oh," is a quantity for which

$$\lim_{\Delta t \to 0} \frac{o(\Delta t)}{\Delta t} = 0.$$

It is something small compared to $\Delta t$; in particular, $o(\Delta t)$ tends to zero faster than $\Delta t$.

A *rate* of transition does not depend on the length of an observation period; it is an *instantaneously* defined quantity that denotes the number of transitions that occur per unit time. Let $q_{ij}(t)$ be the rate at which transitions occur from state $i$ to state $j$ at time $t$. Note that the rate of transition in a nonhomogeneous Markov chain, like the probability of transition, may depend on the time $t$. However, unlike the transition probability, it does not depend on a time *interval* $\Delta t$. To help understand these concepts, the reader may wish to associate probability with distance and rate with speed and recall the relationship between distance and speed, where speed may be thought of as the instantaneous rate at which distance is covered. We have

$$q_{ij}(t) = \lim_{\Delta t \to 0} \left\{ \frac{p_{ij}(t, t+\Delta t)}{\Delta t} \right\} \quad \text{for } i \neq j. \tag{9.25}$$

It then follows that

$$p_{ij}(t, t+\Delta t) = q_{ij}(t)\Delta t + o(\Delta t) \quad \text{for } i \neq j, \tag{9.26}$$

which in words simply states that, correct to terms of order $o(\Delta t)$, the probability that a transition occurs from state $i$ at time $t$ to state $j$ in the next $\Delta t$ time units is equal to the rate of transition at time $t$ multiplied by the length of the time period $\Delta t$—exactly as it should be. In terms of the distance/speed analogy, this simply says that the distance covered in a certain time ($= p_{ij}(t, t+\Delta t)$) is equal to the speed ($= q_{ij}(t)$) multiplied by the time interval ($= \Delta t$).

From conservation of probability and Equation (9.26), we find

$$1 - p_{ii}(t, t + \Delta t) = \sum_{j \neq i} p_{ij}(t, t + \Delta t) \tag{9.27}$$

$$= \sum_{j \neq i} q_{ij}(t)\Delta t + o(\Delta t). \tag{9.28}$$

Dividing by $\Delta t$ and taking the limit as $\Delta t \to 0$, we get

$$\lim_{\Delta t \to 0} \left\{ \frac{1 - p_{ii}(t, t + \Delta t)}{\Delta t} \right\} = \lim_{\Delta t \to 0} \left\{ \frac{\sum_{j \neq i} q_{ij}(t)\Delta t + o(\Delta t)}{\Delta t} \right\} = \sum_{j \neq i} q_{ij}(t).$$

In continuous-time Markov chains, it is usual to let this be equal to $-q_{ii}(t)$, or, alternatively, the transition rate corresponding to the system remaining in place is defined by the equation

$$q_{ii}(t) = -\sum_{j \neq i} q_{ij}(t). \tag{9.29}$$

When state $i$ is an absorbing state, $q_{ii}(t) = 0$. The fact that $q_{ii}(t)$ is negative should not be surprising. This quantity denotes a transition *rate* and as such is defined as a derivative. Given that the system is in state $i$ at time $t$, the probability that it will transfer to a different state $j$ increases with time, whereas the probability that it remains in state $i$ must decrease with time. It is appropriate in the first case that the derivative at time $t$ be positive, and in the second that it be negative. Substitution of Equation (9.29) into Equation 9.28 provides the analog of (9.26). For convenience we write them both together,

$$p_{ij}(t, t + \Delta t) = q_{ij}(t)\Delta t + o(\Delta t) \quad \text{for } i \neq j,$$
$$p_{ii}(t, t + \Delta t) = 1 + q_{ii}(t)\Delta t + o(\Delta t).$$

When the continuous-time Markov chain is homogeneous, we have

$$q_{ij} = \lim_{\Delta t \to 0} \left( \frac{p_{ij}(\Delta t)}{\Delta t} \right), \quad i \neq j; \qquad q_{jj} = \lim_{\Delta t \to 0} \left( \frac{p_{jj}(\Delta t) - 1}{\Delta t} \right). \tag{9.30}$$

In a continuous-time Markov chain, unlike its discrete-time counterpart, we do not associate self-loops with states—continuously leaving and simultaneously returning to a state makes little sense.

The matrix $Q(t)$ whose $ij^{\text{th}}$ element is $q_{ij}(t)$ is called the *infinitesimal generator* or *transition-rate matrix* for the continuous-time Markov chain. In matrix form, we have

$$Q(t) = \lim_{\Delta t \to 0} \left\{ \frac{P(t, t + \Delta t) - I}{\Delta t} \right\},$$

where $P(t, t + \Delta t)$ is the transition probability matrix, its $ij^{\text{th}}$ element is $p_{ij}(t, t + \Delta t)$, and $I$ is the identity matrix. It is apparent from Equation (9.29) that the sum of all elements in any row of $Q(t)$ must be zero. When the continuous-time Markov chain is homogeneous, the transition rates $q_{ij}$ are independent of time, and the matrix of transition rates is written simply as $Q$.

Let $\{X(t), t \geq 0\}$ be a homogeneous continuous-time Markov chain and suppose that at time $t = 0$ the Markov chain is in nonabsorbing state $i$. Let $T_i$ be the random variable that describes the time until a transition out of state $i$ occurs. Then, with $s > 0,\ t > 0$,

$$\begin{aligned}\text{Prob}\{T_i > s+t \mid X(0)=i\} &= \text{Prob}\{T_i > s+t \mid X(0)=i,\ T_i > s\}\,\text{Prob}\{T_i > s \mid X(0)=i\}\\ &= \text{Prob}\{T_i > s+t \mid X(s)=i\}\,\text{Prob}\{T_i > s \mid X(0)=i\}\\ &= \text{Prob}\{T_i > t \mid X(0)=i\}\,\text{Prob}\{T_i > s \mid X(0)=i\},\end{aligned} \tag{9.31}$$

and hence

$$\hat{F}(s+t) = \hat{F}(t)\,\hat{F}(s),$$

where $\hat{F}(t) = \text{Prob}\{T_i > t \mid X(0) = i\},\ t > 0$. This equation is satisfied if and only if $\hat{F}(t) = e^{-\mu_i t}$ for some positive parameter $\mu_i > 0$ and $t > 0$. Thus the sojourn time in state $i$ must be exponentially distributed. Equation (9.31) affords a different perspective on this. If it is known that the Markov chain started at time $t = 0$ in state $i$ and has not moved from state $i$ by time $s$, i.e., $\text{Prob}\{T_i > s \mid X(0) = i\} = 1$, then

$$\text{Prob}\{T_i > s+t \mid T_i > s\} = \text{Prob}\{T_i > t\},$$

and the continuous random variable $T_i$ is memoryless. Since the only continuous distribution that has the memoryless property—the distribution of residual time being equal to the distribution itself—is the exponential distribution, it follows that the duration of time until a transition occurs from state $i$ is exponentially distributed.

Given a homogeneous continuous-time Markov chain in which state $i$ is a nonabsorbing state, there is one or more states $j \neq i$ to which a transition from state $i$ can occur. As we have just seen, the memoryless property of the Markov chain forces the duration of time until such a transition occurs to be exponentially distributed. Furthermore, we have also seen that the rate of transition from state $i$ to state $j \neq i$ is $q_{ij}$. Putting these together, it follows that the distribution of the time to reach some state $j \neq i$ is exponentially distributed with rate $q_{ij}$. Furthermore, on exiting state $i$, it is possible that many different states can be reached. A *race condition* is thereby established and the actual transition occurs to the state that wins, i.e., the state which minimizes the sojourn time in state $i$. Since the minimum value of a number of exponentially distributed random variables is also an exponentially distributed random variable with rate equal to the *sum* of the original rates, it follows that the time spent in state $i$ of a homogeneous, continuous-time Markov chain is exponentially distributed and that the parameter of the exponential distribution of this sojourn time is $\mu_i = \sum_{j\neq i} q_{ij}$. Thus the probability distribution of the sojourn time in state $i$ is given by

$$F_i(x) = 1 - e^{-\mu_i x}, \quad x \geq 0,$$

where

$$\mu_i = \sum_{j\neq i} q_{ij} = -q_{ii}. \tag{9.32}$$

The arguments outlined above show that the sojourn time in any state of a *homogeneous*, continuous-time Markov chain must be exponentially distributed. This does not hold if the Markov chain is nonhomogeneous. The sojourn time in a nonhomogeneous continuous-time Markov chain is *not* exponentially distributed, just as we saw previously that the sojourn time in a nonhomogeneous discrete-time Markov chain is not geometrically distributed.

**Example 9.29** Cars arrive at a QuikLube service center at an average rate of five per hour and it takes on average ten minutes to service each car. To represent this situation as a homogeneous continuous-time Markov chain, we must first specify the state space of the model. We shall assume that our only concern is with the number of cars in the QuickLube center at any time, which means that we can use the nonnegative integers $0, 1, 2, \ldots$ to represent the situation in which there are $0, 1, 2, \ldots$ cars in the center. This is our state space. Our next concern is the interaction among the states. If we assume that no more than one car can arrive at any moment, that no more than one car can exit from service at any moment and that cars do not arrive and depart simultaneously,[3] then the transitions among states can only be to nearest neighbors. Possible transitions among the states are shown in Figure 9.15.

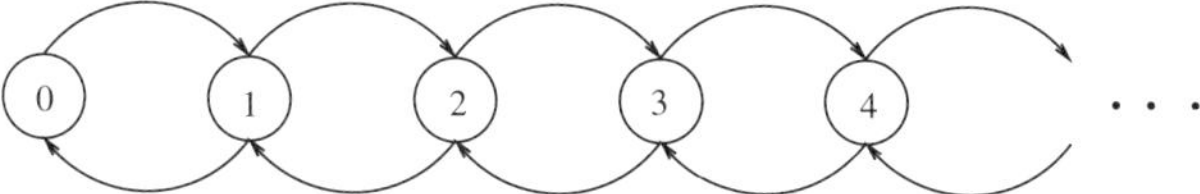


Figure 9.15. The only permissible transitions are between neighboring states.

Transitions from any state $i$ to its next highest neighbor $i+1$ occur as the result of an arrival and we are told that the mean time between arrivals is 1/5 hours. To satisfy the Markov property, this interarrival time must be exponentially distributed with mean $1/5$. In other words, the rate of going from any state $i$ to state $i+1$ is $q_{i,i+1} = $ 5/hour for $i \geq 0$. Transitions from state $i+1$ to state $i$ occur as the result of a car departing from the service center. Again, to satisfy the Markov property, the distribution of time spent servicing a car must be exponentially distributed, with mean service time equal to ten minutes. Thus the rate of going from any state $i+1$ to state $i$ is $q_{i+1,i} = $ 6/hour for $i \geq 0$.

Since the only transitions possible are to nearest neighbors, the infinitesimal generator (or transition-rate matrix) $Q$ must be tridiagonal with superdiagonal elements all equal to 5 and subdiagonal elements all equal to 6. The diagonal elements, as specified by Equation (9.29), are the negated sums of the off-diagonal elements. It follows then that

$$Q = \begin{pmatrix} -5 & 5 & 0 & 0 & 0 & \cdots \\ 6 & -11 & 5 & 0 & 0 & \cdots \\ 0 & 6 & -11 & 5 & 0 & \cdots \\ 0 & 0 & 6 & -11 & 5 & \\ \vdots & \vdots & & \ddots & \ddots & \ddots \end{pmatrix}.$$

### 9.10.2 The Chapman-Kolmogorov Equations

The Chapman-Kolmogorov equations for a nonhomogeneous continuous-time Markov chain may be obtained directly from the Markov property. They are specified by

$$p_{ij}(s,t) = \sum_{\text{all } k} p_{ik}(s,u)p_{kj}(u,t) \quad \text{for } i, j = 0, 1, \ldots \text{ and } s \leq u \leq t.$$

In passing from state $i$ at time $s$ to state $j$ at time $t$ $(s < t)$, we must pass through some intermediate state $k$ at some intermediate time $u$. When the continuous-time Markov chain is homogeneous, the Chapman-Kolmogorov equation may be written as

$$\begin{aligned} p_{ij}(t+\Delta t) &= \sum_{\text{all } k} p_{ik}(t)p_{kj}(\Delta t) \quad \text{for } t,\ \Delta t \geq 0 \\ &= \sum_{k \neq j} p_{ik}(t)p_{kj}(\Delta t) + p_{ij}(t)p_{jj}(\Delta t). \end{aligned} \tag{9.33}$$

[3] We shall see later in our study of queueing systems that the exponential nature of interarrival and service times implies that the probability of multiple arrivals, multiple departures, and simultaneous arrivals and departures are all equal to $o(\Delta t)$ and disappear when certain limits are taken.

Thus

$$\begin{aligned}\frac{p_{ij}(t+\Delta t)-p_{ij}(t)}{\Delta t} &= \sum_{k\neq j} p_{ik}(t)\frac{p_{kj}(\Delta t)}{\Delta t} + p_{ij}(t)\frac{p_{jj}(\Delta t)}{\Delta t} - \frac{p_{ij}(t)}{\Delta t}\\ &= \sum_{k\neq j} p_{ik}(t)\frac{p_{kj}(\Delta t)}{\Delta t} + p_{ij}(t)\left(\frac{p_{jj}(\Delta t)-1}{\Delta t}\right).\end{aligned}$$

Taking the limit as $\Delta t \to 0$ and recalling Equation (9.30),

$$\frac{dp_{ij}(t)}{dt} = \sum_{k\neq j} p_{ik}(t)q_{kj} + p_{ij}(t)q_{jj}.$$

In other words,

$$\frac{dp_{ij}(t)}{dt} = \sum_{\text{all } k} p_{ik}(t)q_{kj} \quad \text{for } i, j = 0, 1, \ldots .$$

These are called the *Kolmogorov forward equations*. In matrix form they are written as

$$\frac{dP(t)}{dt} = P(t)Q. \tag{9.34}$$

In a similar manner, by writing Equation (9.33) in the form

$$p_{ij}(t+\Delta t) = \sum_{\text{all } k} p_{ik}(\Delta t)p_{kj}(t) \quad \text{for } t,\ \Delta t \geq 0,$$

we may derive the *Kolmogorov backward equations*, which are

$$\frac{dp_{ij}(t)}{dt} = \sum_{\text{all } k} q_{ik}p_{kj}(t) \quad \text{for } i, j = 0, 1, \ldots ,$$

or, in matrix form,

$$\frac{dP(t)}{dt} = QP(t).$$

The solution of the Kolmogorov forward equations (9.34) is given by the matrix exponential

$$P(t) = ce^{Qt} = e^{Qt} = \left(I + \sum_{n=1}^{\infty}\frac{Q^n t^n}{n!}\right). \tag{9.35}$$

for some constant of integration $c = P(0) = I$. Unfortunately, the computation of the matrix exponential can be rather difficult and unstable to compute. We shall take up this topic again in a later section. For a nonhomogeneous Markov chain, the forward and backward Kolmogorov equations are given by

$$\frac{dP(t)}{dt} = P(t)Q(t) \quad \text{and} \quad \frac{dP(t)}{dt} = Q(t)P(t)$$

respectively. In the remainder of this chapter, unless otherwise stated, we consider only homogeneous continuous-time Markov chains.

**Example 9.30** The Matlab function `expm` $(Q)$ may be used to find the matrix exponential of a matrix $Q$. Let us use it to compute $P(t)$ for $t = 0.0001, 0.01, 1, 2, 5, 10,$ and $100$ when the infinitesimal generator matrix is given by

$$Q = \begin{pmatrix} -4 & 4 & 0 & 0 \\ 3 & -6 & 3 & 0 \\ 0 & 2 & -4 & 2 \\ 0 & 0 & 1 & -1 \end{pmatrix}.$$

We find

$$P(0.0001) = \begin{pmatrix} 0.9996 & 0.0004 & 0.0000 & 0.0000 \\ 0.0003 & 0.9994 & 0.0003 & 0.0000 \\ 0.0000 & 0.0002 & 0.9996 & 0.0002 \\ 0.0000 & 0.0000 & 0.0001 & 0.9999 \end{pmatrix},$$

$$P(0.01) = \begin{pmatrix} 0.9614 & 0.0380 & 0.0006 & 0.0000 \\ 0.0285 & 0.9426 & 0.0286 & 0.0003 \\ 0.0003 & 0.0190 & 0.9612 & 0.0195 \\ 0.0000 & 0.0001 & 0.0098 & 0.9901 \end{pmatrix},$$

$$P(1) = \begin{pmatrix} 0.2285 & 0.2572 & 0.2567 & 0.2576 \\ 0.1929 & 0.2282 & 0.2573 & 0.3216 \\ 0.1283 & 0.1715 & 0.2503 & 0.4499 \\ 0.0644 & 0.1072 & 0.2249 & 0.6035 \end{pmatrix}, \quad P(2) = \begin{pmatrix} 0.1514 & 0.1891 & 0.2470 & 0.4125 \\ 0.1418 & 0.1803 & 0.2450 & 0.4329 \\ 0.1235 & 0.1633 & 0.2409 & 0.4723 \\ 0.1031 & 0.1443 & 0.2362 & 0.5164 \end{pmatrix},$$

$$P(5) = \begin{pmatrix} 0.1208 & 0.1608 & 0.2402 & 0.4782 \\ 0.1206 & 0.1605 & 0.2401 & 0.4788 \\ 0.1201 & 0.1601 & 0.2400 & 0.4798 \\ 0.1195 & 0.1596 & 0.2399 & 0.4810 \end{pmatrix},$$

$$P(10) = P(100) = \begin{pmatrix} 0.1200 & 0.1600 & 0.2400 & 0.4800 \\ 0.1200 & 0.1600 & 0.2400 & 0.4800 \\ 0.1200 & 0.1600 & 0.2400 & 0.4800 \\ 0.1200 & 0.1600 & 0.2400 & 0.4800 \end{pmatrix}.$$

### *9.10.3 The Embedded Markov Chain and State Properties*

If we ignore the time actually spent in any state of a continuous-time Markov chain $\{X(t),\ t \geq 0\}$ and consider only the sequence of transitions that actually occur, we define a new, discrete-time, Markov chain —the *embedded Markov chain* (*EMC*), also called the *jump chain*. Let the $n^{\text{th}}$ state visited by the continuous-time Markov chain be denoted by $Y_n$. Then $\{Y_n,\ n = 0, 1, 2, \ldots\}$ is the embedded Markov chain derived from $\{X(t),\ t \geq 0\}$. For a continuous-time Markov chain with no absorbing states, the one-step transition probabilities of its embedded discrete-time Markov chain, denoted by $w_{ij}$, i.e., $w_{ij} \equiv \text{Prob}\,\{Y_{n+1} = j | Y_n = i\}$, are equal to zero if $i = j$ and

$$w_{ij} = \frac{q_{ij}}{\sum_{j \neq i} q_{ij}}, \quad j \neq i.$$

Since no state is absorbing, the denominator $\sum_{j \neq i} q_{ij}$ cannot be zero. If we let $W$ denote the transition probability matrix of the EMC, then in matrix terms we have

$$W = I - D_Q^{-1} Q,$$

where $D_Q = \text{diag}\{Q\}$ is a diagonal matrix whose nonzero diagonal elements are equal to the diagonal elements of $Q$. Note that all the elements of $W$ satisfy $0 \leq w_{ij} \leq 1$ and that $\sum_{j,\, j \neq i} w_{ij} = 1$

for all $i$. Thus $W$ possesses the characteristics of a transition probability matrix for a discrete-time Markov chain.

**Example 9.31** Consider a continuous-time Markov with infinitesimal generator matrix

$$Q = \begin{pmatrix} -4 & 4 & 0 & 0 \\ 3 & -6 & 3 & 0 \\ 0 & 2 & -4 & 2 \\ 0 & 0 & 1 & -1 \end{pmatrix};$$

we find its discrete-time embedded chain to be

$$W = \begin{pmatrix} 0 & 1 & 0 & 0 \\ .5 & 0 & .5 & 0 \\ 0 & .5 & 0 & .5 \\ 0 & 0 & 1 & 0 \end{pmatrix}.$$

Many of the properties of the states of a continuous-time Markov chain $\{X(t), t \geq 0\}$ can be deduced from those of its corresponding embedded chain. In a continuous-time Markov chain, $\{X(t), t \geq 0\}$, we say that state $i$ can reach state $j$, for any $i \neq j$, if there is a $t \geq 0$ such that $p_{ij}(t) > 0$. This allows us to define the concept of irreducibility in a continuous-time Markov chain. A continuous-time Markov chain is irreducible if, for any two states $i$ and $j$, there exists reals $t_1 \geq 0,\ t_2 \geq 0$ such that $p_{ij}(t_1) > 0$ and $p_{ji}(t_2) > 0$. It is then easy to show that a continuous-time Markov chain is irreducible if and only if its embedded chain is irreducible. Notice from the relationship $W = I - D_Q^{-1}Q$ that, for $i \neq j$, $w_{ij} = 0$ if and only if $q_{ij} = 0$, and thus, if $Q$ is irreducible, $W$ is irreducible, and vice versa.

The concepts of communicating class, closed class, absorbing state, recurrence, transience, and irreducibility are all inherited directly from the embedded Markov chain. For example, a state $i$ is transient in a continuous-time Markov chain if and only if it is transient in its embedded chain. Also, a state $j$ is recurrent in a continuous-time Markov chain if and only if it is recurrent in its embedded chain. However, a certain amount of care must be exercised, for it is possible for a state $j$ to be positive recurrent in a continuous-time Markov chain and null recurrent in its corresponding embedded chain, and vice versa!

The concept of periodicity has no meaning in a continuous-time Markov chain, since there is no concept of time steps at which transitions either occur or do not occur. Thus, a state of a continuous-time Markov chain is said to be ergodic if it is positive recurrent and the Markov chain itself is said to be ergodic if all its states are ergodic. In particular, a finite, irreducible continuous-time Markov chain is ergodic.

When a continuous-time Markov chain is reducible with closed communicating classes, we often wish to find the probabilities of absorption into closed classes as well as the mean time to absorption. Previous results, obtained in the context of discrete-time Markov chains and now written in terms of the elements of $Q$, can be applied to the embedded Markov chain to produce corresponding results for continuous-time Markov chains. Let $A$ be a closed subset of states of a continuous-time Markov chain. Then $h_i^A$, the probability of hitting a state of $A$ from any state $i$, is the $i^{\text{th}}$ component of the solution to the system of equations

$$h_i^A = 1 \quad \text{for } i \in A,$$
$$\sum_j q_{ij} h_j^A = 0 \quad \text{for } i \notin A.$$

**Example 9.32** Consider a reducible, continuous-time Markov chain whose infinitesimal generator is

$$Q = \begin{pmatrix} -4 & 2 & 1 & 1 \\ 3 & -6 & 2 & 1 \\ 0 & 0 & 0 & 0 \\ 0 & 0 & 0 & 0 \end{pmatrix},$$

and let us find the probability of absorption from states 1 and 2 into the closed class $A = \{3\}$. We have

$$\begin{aligned} h_3^A &= 1, \\ q_{11}h_1^A + q_{12}h_2^A + q_{13}h_3^A + q_{14}h_4^A &= 0, \\ q_{21}h_1^A + q_{22}h_2^A + q_{23}h_3^A + q_{24}h_4^A &= 0, \\ q_{41}h_1^A + q_{42}h_2^A + q_{43}h_3^A + q_{44}h_4^A &= 0. \end{aligned}$$

Since $h_3^A = 1$ and $h_4^A = 0$, we obtain

$$\begin{aligned} -4h_1^A + 2h_2^A + 1 &= 0, \\ 3h_1^A - 6h_2^A + 2 &= 0, \end{aligned}$$

which yields the result $h_1^A = 0.5556$ and $h_2^A = 0.6111$.

In a similar manner, it may be shown that $k_i^A$, the mean time to absorption into a closed subset $A$ from some state $i$ in a reducible, continuous-time Markov chain, is given as the $i^{\text{th}}$ component of the solution of the system of equations

$$\begin{aligned} k_i^A &= 0 \quad \text{for } i \in A, \\ -\sum_j q_{ij}k_j^A &= 1 \quad \text{for } i \notin A. \end{aligned}$$

**Example 9.33** Consider a reducible, continuous-time Markov chain whose infinitesimal generator is

$$Q = \begin{pmatrix} -4 & 4 & 0 & 0 \\ 3 & -6 & 2 & 1 \\ 0 & 2 & -4 & 2 \\ 0 & 0 & 0 & 0 \end{pmatrix},$$

and let us find the mean time to absorption from state 1 into the closed class (absorbing state) $A = \{4\}$. We have

$$\begin{aligned} k_4^A &= 0, \\ -q_{11}k_1^A - q_{12}k_2^A - q_{13}k_3^A - q_{14}k_4^A &= 1, \\ -q_{21}k_1^A - q_{22}k_2^A - q_{23}k_3^A - q_{24}k_4^A &= 1, \\ -q_{31}k_1^A - q_{32}k_2^A - q_{33}k_3^A - q_{34}k_4^A &= 1. \end{aligned}$$

Since $k_4^A = 0$, we obtain

$$\begin{aligned} 4k_1^A - 4k_2^A &= 1, \\ -3k_1^A + 6k_2^A - 2k_3^A &= 1, \\ -2k_2^A + 4k_3^A &= 1, \end{aligned}$$

which has the solution (1.3750, 1.1250, 0.8125). It follows that the mean time to absorption beginning from state 1 is 1.3750.

These equations have a simple, intuitive justification. Starting in state 1, we spend a time equal to $-1/q_{11} = 1/4$ in that state before moving to state 2. The mean time to absorption from state 1, which we have called $k_1$, is therefore equal to 1/4 plus the mean time to absorption from state 2, i.e., $k_1 = 1/4 + k_2$ or $4k_1 - 4k_2 = 1$ which gives us the first equation. Similarly, starting in state 2, we spend a time equal to $-1/q_{22} = 1/6$ there before going to state 1 with probability 3/6, or to state 3 with probability 2/6 or to state 4 with probability 1/6. Therefore the mean time to absorption from state 2 is given by $k_2 = 1/6 + k_1/2 + k_3/3$, i.e., the second equation $-3k_1 + 6k_2 - 2k_3 = 1$, and so on.

### *9.10.4 Probability Distributions*

**Transient Distributions**

Consider a system that is modeled by a continuous-time Markov chain. Let $\pi_i(t)$ be the probability that the system is in state $i$ at time $t$, i.e.,

$$\pi_i(t) = \text{Prob}\{X(t) = i\}.$$

The probability that the system is in state $i$ at time $t + \Delta t$, correct to terms of order $o(\Delta t)$, must be equal to the probability that it is in state $i$ at time $t$ and it does not change state in the period $[t, \Delta t)$, plus the probability that it is in some state $k \neq i$ at time $t$ and moves to state $i$ in the interval $\Delta t$, i.e.,

$$\pi_i(t + \Delta t) = \pi_i(t)\left(1 - \sum_{\text{all } j\neq i} q_{ij}(t)\Delta t\right) + \left(\sum_{\text{all } k\neq i} q_{ki}(t)\pi_k(t)\right)\Delta t + o(\Delta t).$$

Since $q_{ii}(t) = -\sum_{\text{all } j\neq i} q_{ij}(t)$, we have

$$\pi_i(t + \Delta t) = \pi_i(t) + \left(\sum_{\text{all } k} q_{ki}(t)\pi_k(t)\right)\Delta t + o(\Delta t)$$

and

$$\lim_{\Delta t\to 0}\left(\frac{\pi_i(t + \Delta t) - \pi_i(t)}{\Delta t}\right) = \lim_{\Delta t\to 0}\left(\sum_{\text{all } k} q_{ki}(t)\pi_k(t) + o(\Delta t)/\Delta t\right),$$

i.e.,

$$\frac{d\pi_i(t)}{dt} = \sum_{\text{all } k} q_{ki}(t)\pi_k(t).$$

In matrix notation, this gives

$$\frac{d\pi(t)}{dt} = \pi(t)Q(t).$$

When the Markov chain is homogeneous, we may drop the dependence on time and simply write

$$\frac{d\pi(t)}{dt} = \pi(t)Q.$$

It follows that the solution $\pi(t)$ is given by

$$\pi(t) = \pi(0)e^{Qt} = \pi(0)\left(I + \sum_{n=1}^{\infty}\frac{Q^n t^n}{n!}\right). \tag{9.36}$$

Observe that this could have been obtained directly from Equation (9.35) since, by definition

$$\pi(t) = \pi(0)P(t) = \pi(0)e^{Qt}.$$

**Example 9.34** Returning to Example 9.31, and assuming that the Markov chain begins in state 1, i.e., $\pi(0) = (1, 0, 0, 0)$, we find

$$\begin{aligned}
\pi(0.0001) &= (0.9996,\ 0.0004,\ 0.0000,\ 0.0000),\\
\pi(0.01) &= (0.9614,\ 0.0381,\ 0.0006,\ 0.0000),\\
\pi(1) &= (0.2285,\ 0.2572,\ 0.2567,\ 0.2576),\\
\pi(2) &= (0.1514,\ 0.1891,\ 0.2470,\ 0.4125),\\
\pi(5) &= (0.1208,\ 0.1608,\ 0.2402,\ 0.4782),\\
\pi(10) = \pi(100) &= (0.1200,\ 0.1600,\ 0.2400,\ 0.4800).
\end{aligned}$$

### Stationary, Limiting, and Steady-State Distributions

In a manner similar to that described for a discrete-time Markov chain, we define an invariant vector of a homogeneous continuous-time Markov chain with generator matrix $Q$ as any nonzero vector $z$ for which $zQ = 0$. If this system of equations has a solution $z$ which is a probability vector ($z_i \geq 0$ for all $i$ and $\|z\|_1 = 1$), then $z$ is a stationary distribution. If replacement of one of the equations by a normalizing equation causes the coefficient matrix to become nonsingular, then the stationary distribution is unique. This is the case when the Markov chain is irreducible and finite.

We now turn to limiting distributions. In a homogeneous, continuous-time Markov chain, the $i^{\text{th}}$ element of the vector $\pi(t)$ is the probability that the chain is in state $i$ at time $t$ and we have just seen that these state probabilities are governed by the system of differential equations

$$\frac{d\pi(t)}{dt} = \pi(t)Q.$$

If the evolution of the Markov chain is such that there arrives a point in time at which the rate of change of the probability distribution vector $\pi(t)$ is zero, then the left-hand side of this equation, $d\pi(t)/dt$, is identically equal to zero. In this case, the system has reached a limiting distribution, which is written simply as $\pi$ in order to show that it no longer depends on time $t$.

When the limiting distribution exists, when all its components are strictly positive, and when it is independent of the initial probability vector $\pi(0)$, then it is unique and it is called the steady-state distribution, also referred to as the equilibrium or long-run probability vector—its $i^{\text{th}}$ element $\pi_i$ is the probability of being in state $i$ at statistical equilibrium. For a finite, irreducible, continuous-time Markov chain, the limiting distribution always exists and is identical to the stationary distribution of the chain. The steady-state distribution may be obtained by solving the system of linear equations

$$\pi Q = 0, \tag{9.37}$$

subject to the condition that $\|\pi\|_1 = 1$. Just as in the discrete-time case, these equations are called the *global balance equations*. From $\sum_{\text{all } i} \pi_i q_{ij} = 0$, we have

$$\sum_{i,\, i\neq j} \pi_i q_{ij} = -\pi_j q_{jj} \quad = \pi_j \sum_{i,\, i\neq j} q_{ji}.$$

The left-hand side represents the total flow from all states $i$, different from $j$, into state $j$, while the right-hand side represents the total flow out of state $j$ into all other states $i \neq j$. Thus these equations equate the flow into and out of states.

**Example 9.35** Consider the homogeneous continuous-time Markov chain of Example 9.31:

$$Q = \begin{pmatrix} -4 & 4 & 0 & 0 \\ 3 & -6 & 3 & 0 \\ 0 & 2 & -4 & 2 \\ 0 & 0 & 1 & -1 \end{pmatrix}.$$

We derive its stationary/limiting distribution by solving the system of equations

$$\pi Q = (\pi_1, \pi_2, \pi_3, \pi_4) \begin{pmatrix} -4 & 4 & 0 & 0 \\ 3 & -6 & 3 & 0 \\ 0 & 2 & -4 & 2 \\ 0 & 0 & 1 & -1 \end{pmatrix} = (0, 0, 0, 0).$$

Arbitrarily setting $\pi_1 = 1$, we obtain from the first equation

$$-4\pi_1 + 3\pi_2 = 0 \implies \pi_2 = 4/3.$$

Now, using the second equation,

$$4\pi_1 - 6\pi_2 + 2\pi_3 \implies \pi_3 = 2,$$

and from the third equation,

$$3\pi_2 - 4\pi_3 + \pi_4 = 0 \implies \pi_4 = 4.$$

The vector $\pi = (1, 4/3, 2, 4)$ must be normalized by dividing each component by $25/3$ to obtain the same distribution as before, namely, $\pi = (.12, .16, .24, .48)$.

The stationary distribution of an irreducible, continuous-time Markov chain may be found from the stationary distribution of its embedded chain, assuming it exists. Let $\phi$ be the stationary distribution of the embedded chain, i.e.,

$$\phi(I - W) = 0 \quad \text{and} \quad \|\phi\|_1 = 1,$$

where

$$W = I - D_Q^{-1} Q \quad \text{or} \quad D_Q^{-1} Q = (I - W). \tag{9.38}$$

Premultiplying Equation (9.38) by $\phi$, we see that

$$\phi D_Q^{-1} Q = \phi(I - W) = 0$$

and hence $-\phi D_Q^{-1}$ (whose $i^{\text{th}}$ component is $-\phi_i/q_{ii}$) is an invariant vector for $Q$. This becomes a stationary distribution of $Q$ once we normalize it, so that the sum of its components is one. Specifically, the stationary distribution of $Q$ is given as

$$\pi = \frac{-\phi D_Q^{-1}}{\|\phi D_Q^{-1}\|_1}.$$

**Example 9.36** It is easy to check that the stationary probability vector of the embedded Markov chain of Example 9.31 is given by

$$\phi = (1/6, 1/3, 1/3, 1/6).$$

Since

$$D_Q = \begin{pmatrix} -4 & 0 & 0 & 0 \\ 0 & -6 & 0 & 0 \\ 0 & 0 & -4 & 0 \\ 0 & 0 & 0 & -1 \end{pmatrix},$$

we have

$$-\phi D_Q^{-1} = (1/6,\ 1/3,\ 1/3,\ 1/6)\begin{pmatrix} 1/4 & 0 & 0 & 0 \\ 0 & 1/6 & 0 & 0 \\ 0 & 0 & 1/4 & 0 \\ 0 & 0 & 0 & 1 \end{pmatrix} = (1/24,\ 1/18,\ 1/12,\ 1/6),$$

which when normalized yields the stationary probability vector of the continuous-time Markov chain as

$$\pi = (.12,\ .16,\ .24,\ .48)$$

—the same result as that obtained previously.

### *9.10.5 Reversibility*

The concept of reversibility as discussed previously in the context of discrete-time Markov chains carries over in a straightforward manner to continuous-time Markov chains. Given a homogeneous continuous-time Markov chain with infinitesimal generator $Q$ and stationary probability vector $\pi$, its reversed chain can be constructed. If $\tilde{Q}$ denotes the infinitesimal generator of the reversed chain, then

$$\tilde{q}_{ij} = q_{ji}\frac{\pi_j}{\pi_i},$$

and the Markov chain is reversible if and only if $\tilde{Q} = Q$. The detailed balance equation are given by

$$\pi_i q_{ij} = \pi_j q_{ji} \quad \text{for all } i,\ j.$$

Kolomogorov's criterion continues to apply: a positive-recurrent, continuous-time Markov chain is reversible if and only if the product of the transition rates along any circuit is the same in both the forward and backward directions. Finally, a continuous-time Markov chain is reversible if and only if its embedded Markov chain is reversible.

## 9.11 Semi-Markov Processes

*Semi-Markov Processes (SMPs)* are generalizations of Markov chains in that the future evolution of the process is independent of the sequence of states visited prior to the current state and independent of the time spent in each of the previously visited states, as is the case for discrete- and continuous-time Markov chains. Semi-Markov processes differ from Markov chains in that the probability distribution of the remaining time in any state can depend on the length of time the process has already spent in that state. Let $\{X_n,\ n = 0, 1, 2, \ldots\}$ be a discrete-time Markov chain where $X_n$ is the $n^{\text{th}}$ state visited by a certain stochastic process. Let $\tau_n$ be the time at which the $n^{\text{th}}$ transition of the stochastic process occurs, with $\tau_0 = 0$. Then the (homogeneous) continuous-time process $\{Z(t),\ t \geq 0\}$ defined by

$$Z(t) = X_n \ \text{ on } \ \tau_n \leq t < \tau_{n+1}$$

is a semi-Markov process and is represented by its associated semi-Markov kernel,

$$\begin{aligned} K_{ij}(t) &= \text{Prob}\{X_{n+1} = j,\ \tau_{n+1} - \tau_n \leq t \mid X_0, X_1, \ldots, X_n = i;\ \tau_0, \tau_1, \ldots \tau_n\} \\ &= \text{Prob}\{X_{n+1} = j,\ \tau_{n+1} - \tau_n \leq t \mid X_n = i\}, \quad t \geq 0. \end{aligned}$$

If the semi-Markov process is nonhomogeneous, its kernel is given by

$$K_{ij}(s, t) = \text{Prob}\{X_{n+1} = j,\ \tau_{n+1} \leq t \mid X_n = i,\ \tau_n = s\}, \quad 0 \leq s \leq t.$$

However, we consider only homogeneous semi-Markov processes in this text. Define

$$p_{ij} = \lim_{t\to\infty} K_{ij}(t) = \text{Prob}\{\, X_{n+1} = j \mid X_n = i\}, \quad n = 0, 1, 2, \ldots .$$

These are the elements of the transition probability matrix $P$ of the homogeneous discrete-time (embedded) Markov chain which determines the state to which the process will next go on leaving state $i$.

Thus, a semi-Markov process consists of two components: (i) a discrete-time Markov chain $\{X(n),\ n = 0, 1, \ldots\}$ with transition probability matrix $P$ which describes the sequence of states visited by the process, and (ii) $H_{ij}(t)$, the conditional distribution function of a random variable $T_{ij}$ which describes the time spent in state $i$ from the moment the process last entered that state

$$H_{ij}(t) = \text{Prob}\{T_{ij} \le t\} = \text{Prob}\{\tau_{n+1} - \tau_n \le t \mid X_{n+1} = j, X_n = i\}, \quad t \ge 0.$$

The random variable $T_{ij}$ is the sojourn time in state $i$ *per visit* to state $i$ prior to jumping to state $j$. Observe that $H_{ij}(t)$ can depend on both the source state $i$ and the destination state $j$ of the transition. The probability distributions $H_{ij}(t)$ and $K_{ij}(t)$ are related. Let us make the assumption that if $p_{ij} = 0$ for some $i$ and $j$; then $H_{ij}(t) = 1$ and $K_{ij}(t) = 0$ for all $t$. Then

$$p_{ij} H_{ij}(t) = K_{ij}(t),$$

since

$$\text{Prob}\{\tau_{n+1} - \tau_n \le t \mid X_{n+1} = j, X_n = i\}\, \text{Prob}\{\, X_{n+1} = j \mid X_n = i\}$$

$$= \text{Prob}\{X_{n+1} = j,\ \tau_{n+1} - \tau_n \le t \mid X_n = i\}.$$

The evolution of a semi-Markov process is as follows:

1. The moment the semi-Markov process enters any state $i$, it randomly selects the next state to visit $j$ according to $P$, its transition probability matrix.
2. If state $j$ is selected, then the time the process remains in state $i$ before moving to state $j$ is a random variable $T_{ij}$ with probability distribution $H_{ij}(t)$.

Thus the next state to visit is chosen first and the time to be spent in state $i$ chosen second, which allows the sojourn time per visit to depend on the destination state as well as the source state. A semi-Markov process defined in reverse order would force $H_{ij}(t) = H_i(t)$ for all $j$—the same distribution for all destination states. A semi-Markov process in which $H_{ij}(t)$ is an exponential distribution, e.g.,

$$H_{ij}(t) = 1 - e^{-\mu_{ij} t} \quad \text{or equivalently} \quad K_{ij}(t) = p_{ij}\left(1 - e^{-\mu_{ij} t}\right), \quad t \ge 0,$$

is a (homogeneous) continuous-time Markov chain and whereas the Markov (memoryless) property of a semi-Markov process holds only at the moments of transition $t = \tau_k,\ k = 0, 1, 2, \ldots,$ it holds at *all* times $t$ when $H_{ij}(t)$ has an exponential form and *only* when $H_{ij}(t)$ has this form. Thus although a semi-Markov.process provides a great deal of latitude in the choice of sojourn times $T_{ij}$, it does so at the cost of losing the memoryless property at all but a sequence of points in time. When a semi-Markov process evolves over a discrete set of time points and when the time between any two transitions is taken to be one unit, i.e., $\tau_{n+1} - \tau_n = 1$ for all $n$, then $H_{ij}(t) = 1$ for all $i, j$ and the semi-Markov process is a discrete-time Markov chain.

Since the random variable $T_{ij}$ denotes the *conditional* sojourn time in state $i$ given that the next state to visit is state $j$, the random variable $T_i$, defined as

$$T_i = \sum_j p_{ij} T_{ij},$$

is the *unconditional* sojourn time per visit to state $i$ and has probability distribution function

$$H_i(t) = \text{Prob}\{T_i \leq t\} = \sum_j p_{ij} H_{ij}(t) = \sum_j K_{ij}(t), \quad t \geq 0.$$

Sample paths in a semi-Markov process are specified by providing both the sequence of states visited and the times at which each state along the path is entered—a sequence of paired state and entrance times, such as

$$(x_0, \tau_0), \; (x_1, \tau_1), \; (x_2, \tau_2), \; \ldots, \; (x_k, \tau_k), \; (x_{k+1}, \tau_{k+1}), \; \ldots,$$

signifying that the process enters state $x_k$ at instant $\tau_k$ and remains there during the period $[\tau_k, \tau_{k+1})$.

To find the stationary distribution $\pi$ of a semi-Markov process,we need to find the stationary distribution of its embedded Markov chain and the mean sojourn time in each state. Let $\phi$ be the stationary distribution of the embedded Markov chain, i.e., $\phi = \phi P$ and $\|\phi\|_1 = 1$, and let $T_i$ be the random variable that describes the per visit sojourn time in state $i$ and $T_{ij}$ the random variable that describes the time in state $i$ prior to a jump to state $j$. The mean sojourn time in state $i$ is

$$E[T_i] = \sum_j p_{ij} E[T_{ij}]$$

and the stationary distribution is the vector whose $i^{\text{th}}$ component is

$$\pi_i = \frac{\phi_i E[T_i]}{\sum_{\text{all } j} \phi_j E[T_j]},$$

where the denominator forces the probabilities to sum to 1. In words, this says that the stationary probability of the semi-Markov process being in any state $i$ is just the (normalized) product of the stationary probability that its embedded Markov chain is in state $i$ and the average amount of time the semi-Markov process spends in state $i$.

## 9.12 Renewal Processes

A renewal process is a special type of *counting process*. A counting process, $\{N(t), \; t \geq 0\}$, is a stochastic process which counts the number of events that occur up to (and including) time $t$. Thus any counting process $N(t)$ is integer valued with the properties that $N(t) \geq 0$ and $N(t_1) \leq N(t_2)$ if $t_1 \leq t_2$. We shall let $X_n, n \geq 1$, be the random variable that denotes the time which elapses between events $(n - 1)$ and $n$. This allows us to define a renewal process as follows:

**Definition 9.11.1 (Renewal process)** *A counting process $\{N(t), \; t \geq 0\}$ is a renewal process if the sequence $\{X_1, X_2, \ldots\}$ of nonnegative random variables that represent the time between events are independent and identically distributed.*

Observe that this definition permits events to occur simultaneously ($X_n = 0$), but we shall restrict ourselves to renewal processes where this cannot happen. In our definition, $X_1$, the time until the first event, is independent of, and has the same distribution as, all other random variables $X_i, \; i > 1$. This may be interpreted in terms of a zeroth event which occurs at time $t = 0$, giving $X_1$ the same interevent time meaning as the other $X_i$. An alternative approach is to assume that the time until the occurrence of the first event has a different distribution from the random variables $X_i, \; i > 1$.

The word "renewal" is appropriate since on the occurrence of each event, the process essentially renews itself: at the exact moment of occurrence of an event, the distribution of time until the next event is independent of everything that has happened previously. The term "recurrent process" is sometimes used in place of "renewal process."

**Example 9.37**

1. Suppose the interarrival times of pieces of junk mail are independent and identically distributed. Then $\{N(t),\ t \geq 0\}$, the number of pieces of junk mail that have arrived by time $t$, is a renewal process.
2. Assume an infinite supply of standard flashlight batteries whose lifetimes are independent and identically distributed. As soon as one battery dies it is replaced by another. Then $\{N(t),\ t \geq 0\}$, the number of batteries that have failed by time $t$, is a renewal process.
3. A biased coin is tossed at times $t = 1, 2, \ldots$. The probability of a head appearing at any time is $\rho$, $0 < \rho < 1$. Then $\{N(t),\ t \geq 0\}$, with $N(0) = 0$, the number of heads obtained up to and including time $t$, is a renewal process. The time between renewals in this case all have the same geometric probability distribution function:

$$\text{Prob}\{X_n = i\} = \rho(1-\rho)^{i-1}, \quad i \geq 1.$$

The renewal process that results is called a *binomial process*.

Let $\{N(t), t \geq 0\}$ be a renewal process with interevent (or *interrenewal*) periods $X_1,\ X_2, \ldots$ and let $S_n$ be the time at which the $n^{\text{th}}$ event/renewal occurs, i.e.,

$$S_0 = 0, \quad S_n = X_1 + X_2 + \cdots + X_n, \quad n \geq 1.$$

In other words, the process renews itself for the $n^{\text{th}}$ time at time $S_n$. The sequence $\{S_n, n \geq 0\}$ is called a *renewal sequence*. The period between renewals is called a *cycle*; a cycle is completed the moment a renewal occurs.

**Example 9.38** Let $\{Y_n,\ n \geq 0\}$ be a homogeneous, discrete-time Markov chain whose state space is the nonnegative integers and assume that at time $t = 0$ the chain is in state $k$. Let $S_n$ denote the time at which the $n^{\text{th}}$ visit to state $k$ begins. Since $\{X_n = S_n - S_{n-1},\ n \geq 1\}$ is a sequence of independent and identically distributed random variables, it follows that $\{S_n,\ n \geq 0\}$ is a renewal sequence and $\{N(t),\ t \geq 0\}$, the number of visits to state $k$ in $(0, t]$, is a renewal process associated with state $k$. The initial visit to state $k$ (the process starting at time $t = 0$ in state $k$) must *not* be included in this count. A similar statement can be made with respect to a continuous-time Markov chain.

Let the distribution function of the random variables $\{X_n, n \geq 1\}$ be denoted by $F_X(t)$, i.e.,

$$\text{Prob}\{X_n \leq t\} = F_X(t),$$

and let us find the distribution function of the renewal process $\{N(t), t \geq 0\}$. First note that since the random variables $X_n$ are independent and identically distributed, the distribution of the renewal sequence $S_n = X_1 + X_2 + \cdots + X_n$ is given as $F_X^{(n)}(t)$, the $n$-fold convolution of $F_X(t)$ with itself.[4] The only way the number of renewals can exceed or equal $n$ at time $t$ ($N(t) \geq n$) is if the $n^{\text{th}}$ renewal occurs no later than time $t$ ($S_n \leq t$). The converse is also true: the only way that the $n^{\text{th}}$ renewal can occur no later than time $t$ is if the number of renewals prior to $t$ is at least equal to $n$. This means that $N(t) \geq n$ if and only if $S_n \leq t$ and we may write

$$\text{Prob}\{N(t) \geq n\} = \text{Prob}\{S_n \leq t\}. \tag{9.39}$$

[4] Recall that if $H(t)$ is the distribution function that results when two random variables having distribution functions $F(t)$ and $G(t)$ are added, then the convolution is defined as

$$H(t) = \int_0^t F(t-x)\,dG(x) = \int_0^t G(t-x)\,dF(x).$$

Less rigorously, we may write $H(t) = \int_0^t F(t-x)g(x)dx = \int_0^t G(t-x)f(x)dx$, where $f(x)$ and $g(x)$ are the corresponding probability density functions.

Therefore the probability of exactly $n$ renewals by time $t$ is given by

$$\text{Prob}\{N(t) = n\} = \text{Prob}\{N(t) \geq n\} - \text{Prob}\{N(t) \geq n + 1\}$$

and using Equation (9.39) we conclude

$$\text{Prob}\{N(t) = n\} = \text{Prob}\{S_n \leq t\} - \text{Prob}\{S_{n+1} \leq t\} = F_X^{(n)}(t) - F_X^{(n+1)}(t).$$

**Example 9.39** Assume that interrenewal times are exponentially distributed with parameter $\lambda$, i.e.,

$$F_X(t) = 1 - e^{-\lambda t}, \quad t \geq 0.$$

Then

$$F_X^{(n)}(t) = 1 - \sum_{k=0}^{n-1} e^{-\lambda t} \frac{(\lambda t)^k}{k!}, \quad t \geq 0,$$

an Erlang distribution, and

$$\text{Prob}\{N(t) = n\} = F_X^{(n)}(t) - F_X^{(n+1)}(t) = e^{-\lambda t} \frac{(\lambda t)^n}{n!}$$

—the renewal process $\{N(t), t \geq 0\}$ has a Poisson distribution and is called a *Poisson process*.

We now show that it is not possible for an infinite number of renewals to occur in a finite period of time. Let $\omega = E[X_n]$, $n \geq 1$, be the mean time between successive renewals. Since $\{X_n, n \geq 1\}$ are nonnegative random variables and we have chosen the simplifying assumption that $\text{Prob}\{X_n = 0\} = F_X(0) < 1$, it follows that $\omega$ must be strictly greater than zero, i.e., $0 < \omega \leq \infty$. In instances in which the mean interrenewal time is infinite, we shall interpret $1/\omega$ as zero. We now show that

$$\text{Prob}\{N(t) < \infty\} = 1 \quad \text{for all finite } t \geq 0.$$

From the strong law of large numbers

$$\text{Prob}\left\{\frac{S_n}{n} \to \omega\right\} = 1 \quad \text{as } n \to \infty,$$

and since $0 < \omega \leq \infty$, $S_n$ must tend to infinity as $n$ tends to infinity:

$$\text{Prob}\{\lim_{n\to\infty} S_n = \infty\} = 1.$$

Now using the fact that $N(t) = \max\{n : S_n \leq t\}$, we find, for finite $t$,

$$\text{Prob}\{N(t) = \infty\} = \text{Prob}\{\max\{n : S_n \leq t\} = \infty\} = \text{Prob}\{\lim_{n\to\infty} S_n \leq t < \infty\}$$

$$= 1 - \text{Prob}\{\lim_{n\to\infty} S_n = \infty\} = 0.$$

This result is for finite $t$ only and does not hold as $t \to \infty$. When $t \to \infty$, we have $\lim_{t\to\infty} N(t) = \infty$.

Finding the probability distribution of an arbitrary renewal function can be difficult so that frequently only $E[N(t)]$, the expected number of renewals by time $t$, is computed. The mean $E[N(t)]$ is called the *renewal function* and is denoted by $M(t)$. We have (see also Exercise 9.12.2)

$$M(t) = E[N(t)] = \sum_{n=1}^{\infty} \text{Prob}\{N(t) \geq n\} = \sum_{n=1}^{\infty} \text{Prob}\{S_n \leq t\}$$

$$= \sum_{n=1}^{\infty} F_X^{(n)}(t).$$

We now show that $M(t)$ uniquely determines the renewal process. Let

$$\tilde{M}(s) = \int_0^\infty e^{-st} dM(t), \quad \tilde{F}_X(s) = \int_0^\infty e^{-st} dF_X(t), \quad \text{and} \quad \tilde{F}_X^{(n)}(s) = \int_0^\infty e^{-st} dF_X^{(n)}(t)$$

be the Laplace-Stiltjes transform (LST) of $M(t)$, $F_X(t)$, and $F_X^{(n)}(t)$, respectively. Then

$$\tilde{M}(s) = \int_0^\infty e^{-st} d\left(\sum_{n=1}^\infty F_X^{(n)}(t)\right) = \sum_{n=1}^\infty \int_0^\infty e^{-st} dF_X^{(n)}(t) = \sum_{n=1}^\infty \tilde{F}_X^{(n)}(s)$$

$$= \sum_{n=1}^\infty \left(\tilde{F}_X(s)\right)^n = \frac{\tilde{F}_X(s)}{1 - \tilde{F}_X(s)}.$$

Thus

$$\tilde{M}(s) = \frac{\tilde{F}_X(s)}{1 - \tilde{F}_X(s)} \quad \text{or} \quad \tilde{F}_X(s) = \frac{\tilde{M}(s)}{1 + \tilde{M}(s)}$$

so that once $\tilde{M}(s)$ is known completely, so also is $\tilde{F}_X(s)$, and vice versa. Since a renewal process is completely characterized by $F_X(t)$, it follows that it is also completely characterized by the renewal function $M(t)$. Continuing with the renewal function, observe that

$$M(t) = \sum_{n=1}^\infty F_X^{(n)}(t) = F_X^{(1)}(t) + \sum_{n=1}^\infty F_X^{(n+1)}(t) = F_X(t) + \sum_{n=1}^\infty \int_0^t F_X^{(n)}(t-s) dF_X(s)$$

$$= F_X(t) + \int_0^t \sum_{n=1}^\infty F_X^{(n)}(t-s) dF_X(s) = F_X(t) + \int_0^t M(t-s) dF_X(s). \tag{9.40}$$

This is called the *fundamental renewal equation.* Taking the LST of both sides gives the previous result, namely,

$$\tilde{M}(s) = \tilde{F}_X(s) + \tilde{M}(s)\tilde{F}_X(s).$$

**Example 9.40** Suppose that interrenewal times are exponentially distributed with parameter $\lambda$, i.e., $F_X(t) = 1 - e^{-\lambda t}$. Then

$$\tilde{F}_X(s) = \frac{\lambda}{\lambda + s}$$

and

$$\tilde{M}(s) = \frac{\lambda/(\lambda + s)}{1 - \lambda/(\lambda + s)} = \frac{\lambda}{s}$$

which when inverted gives

$$M(t) = \lambda t.$$

We may conclude that, when the expected number of renewals increases linearly with time, the renewal process is a Poisson process. Furthermore, from the uniqueness property, the Poisson process is the only renewal process with a linear mean-value (renewal) function.

We now give two important results on the limiting behavior of renewal processes. We have previously seen that the limit as $t$ tends to infinity of $N(t)$ is infinite. Our first result concerns the rate at which $N(t) \to \infty$. Observe that $N(t)/t$ is the average number of renewals per unit time.

We now show that, with probability 1,

$$\frac{N(t)}{t} \to \frac{1}{\omega} \quad \text{as} \quad t \to \infty,$$

and point out why $1/\omega$ is called the *rate* of the renewal process. Recall that $S_n$ is the time at which the $n^{\text{th}}$ renewal occurs. Since $N(t)$ is the number of arrivals that occur prior to or at time $t$, it must follow that $S_{N(t)}$ is just the time at which the last renewal *prior to or at* time $t$ occurred. Likewise, $S_{N(t)+1}$ is the time at which the first renewal after time $t$ occurs. Therefore

$$S_{N(t)} \le t < S_{N(t)+1}$$

and

$$\frac{S_{N(t)}}{N(t)} \le \frac{t}{N(t)} < \frac{S_{N(t)+1}}{N(t)}.$$

Since $S_{N(t)}/N(t)$ is the average of $N(t)$ independent and identically distributed random variables:

$$\frac{S_{N(t)}}{N(t)} = \frac{1}{N(t)} \sum_{i=1}^{N(t)} X_i.$$

Given that $\omega = E[X_n]$ is the mean time between (successive) renewals, it follows from the strong law of large numbers that

$$\lim_{N(t)\to\infty} \frac{S_{N(t)}}{N(t)} = \omega,$$

which is the same as writing

$$\lim_{t\to\infty} \frac{S_{N(t)}}{N(t)} = \omega,$$

since $N(t) \to \infty$ when $t \to \infty$. Also, by means of a similar argument,

$$\lim_{N(t)\to\infty} \frac{S_{N(t)+1}}{N(t)} = \lim_{N(t)\to\infty} \frac{S_{N(t)+1}}{N(t)+1} \frac{N(t)+1}{N(t)} = \left( \lim_{N(t)\to\infty} \frac{S_{N(t)+1}}{N(t)+1} \right) \left( \lim_{N(t)\to\infty} \frac{N(t)+1}{N(t)} \right) = \omega.$$

Thus we see that $t/N(t)$ is sandwiched between two numbers each of which tends to $\omega$ as $t \to \infty$. Therefore $t/N(t)$ must also converge to $\omega$ as $t \to \infty$. Now using the fact that $1/x$ is a continuous function, we have, with probability 1,

$$\lim_{t\to\infty} \frac{N(t)}{t} = \frac{1}{\omega}.$$

Intuitively, this result makes sense. $N(t)/t$ is the average number of renewals per unit time over the interval $(0, t]$ and in the limit as $t$ tends to infinity, it is the average *rate* of renewals. In other words, the expected *rate of renewals* for all renewal processes is $1/\omega$, i.e., the reciprocal of the expected time between renewals. Given an average interrenewal time of six minutes ($\omega = 6$) we would expect to get renewals at the rate of ten per hour, or $1/6$ per minute.

The second result, which we state without proof, is similar. It can be shown that, with probability 1,

$$\lim_{t\to\infty} \frac{M(t)}{t} = \frac{1}{\omega}.$$

The derivative of the renewal function $M(t) = E[N(t)]$ is called the *renewal density* and is denoted by $m(t)$.

$$m(t) = \frac{dM(t)}{dt} = \sum_{n=1}^{\infty} f_X^{(n)}(t),$$

where $f_X^{(n)}(t)$ is the derivative of $F_X^{(n)}(t)$. This leads to the *renewal equation* (as opposed to the *fundamental* renewal equation, Equation (9.40)):

$$m(t) = f_X(t) + \int_0^t m(t-s) f_X(s) ds.$$

It may be shown that

$$\lim_{t\to\infty} m(t) = \frac{1}{\omega}.$$

For small values of $\Delta t$, $m(t)\Delta t$ gives the probability that a renewal will occur in the interval $(t, t + \Delta t]$. For a Poisson process, we saw that $M(t) = \lambda t$. Hence $m(t) = \lambda$ when a renewal process is Poisson, as we might have expected.

**Renewal Reward Processes**

Consider a renewal process $\{N(t),\ t \geq 0\}$ and let $X_n$ be the $n^{\text{th}}$ interrenewal time. Assume that on the completion of a cycle, a "reward" is received or alternatively a "cost" is paid. Let $R_n$ be the reward (positive or negative) obtained at the $n^{\text{th}}$ renewal. We assume that the rewards $R_n$ are independent and identically distributed. This does not prevent $R_n$ from depending on $X_n$, the length of the cycle in which the reward $R_n$ is earned. The total reward received by time $t$ is given by

$$R(t) = \sum_{n=1}^{N(t)} R_n.$$

Observe that, if $R_n = 1$ for all $n$, then $R(t) = N(t)$, the original renewal process. We now show that

$$\lim_{t\to\infty} \frac{R(t)}{t} = \frac{E[R]}{E[X]},$$

where $E[R]$ is the expected reward obtained in any cycle, and $E[X]$ is the expected duration of a cycle. We have

$$\frac{R(t)}{t} = \frac{\sum_{n=1}^{N(t)} R_n}{t} = \frac{\sum_{n=1}^{N(t)} R_n}{N(t)} \times \frac{N(t)}{t}.$$

Taking the limits of both sides as $t \to \infty$, and using the fact that the strong law of large numbers allows us to write

$$\lim_{t\to\infty} \frac{\sum_{n=1}^{N(t)} R_n}{N(t)} = E[R],$$

gives us the desired result:

$$\lim_{t\to\infty} \frac{R(t)}{t} = E[R] \times \lim_{t\to\infty} \frac{N(t)}{t} = \frac{E[R]}{E[X]}.$$

This tells us that the long run *rate of reward* is equal to the expected reward per cycle divided by the mean cycle length. It may also be shown that

$$\lim_{t\to\infty} \frac{E[R(t)]}{t} = \frac{E[R]}{E[X]},$$

i.e., the *expected reward* per unit time in the long run is also equal to the expected reward per cycle divided by the mean cycle length.

Renewal reward models arise in the context of an ongoing process in which a product, such as a car or piece of machinery, is used for a period of time (a cycle) and then replaced. In order to have a renewal process, the new car/machine is assumed to have identical characteristics (more precisely, an identical lifetime function) to the one that is replaced. A replacement policy specifies a recommended time $T$ at which to purchase the new product and the cost $c_1$ of doing so at this time. A cost $c_2$ over and above the replacement cost $c_1$ must be paid if for some reason (e.g., the car/machine breaks down) replacement must take place prior to the recommended time $T$. In some scenarios, a third factor, the resale value of the car/machine, is also included.

Let $Y_n,\ n \geq 1$, be the lifetime of the $n^{\text{th}}$ machine and assume that the $Y_n$ are independent and identically distributed with probability distribution function $F(t)$. If $S_n$ is the time of the $n^{\text{th}}$ replacement, then the sequence $\{S_n,\ n \geq 0\}$ is a renewal sequence. Let $X_n$ be the time between two replacements, i.e., $X_n$ is the duration of the $n^{\text{th}}$ cycle and we have

$$X_n = \min\{Y_n, T\}.$$

In the absence of a resale value, the reward (or more properly, cost) $R_n$ is given by

$$R_n = \begin{cases} c_1, & Y_n \geq T, \\ c_1 + c_2, & Y_n < T, \end{cases}$$

and $R(t)$, the total cost up to time $t$, is a renewal reward process:

$$R(t) = \sum_{n=1}^{N(t)} R_n, \qquad N(t) > 0.$$

Using $X$ to denote the duration of an arbitrary cycle and $Y$ the lifetime of an arbitrary machine, the expected cost per cycle is

$$E[R] = c_1 \text{Prob}\{Y \geq T\} + (c_1 + c_2)\text{Prob}\{Y < T\} = c_1 + c_2 F(T).$$

We now need to compute $E[X]$, the expected length of a cycle. Since the length of the $n^{\text{th}}$ cycle is $\min\{Y_n, T\}$ we have

$$E[X] = \int_0^T x f(x)dx + \int_T^\infty T f(x)dx = \int_0^T x f(x)dx + T(1 - F(T)),$$

where $f(t)$ is the lifetime density function of an arbitrary machine. Hence the long-run average cost is

$$\frac{E[R]}{E[X]} = \frac{c_1 + c_2 F(T)}{\int_0^T x f(x)dx + T(1 - F(T))}.$$

It makes sense to try to find the optimal value of $T$, the value which minimizes the expected cost. If $T$ is chosen to be small, the number of replacements will be high but the cost due to failure will be small. If $T$ is chosen to be large, then the opposite occurs.

**Example 9.41** A manufacturing company produces widgets from a machine called an "Inneall" whose lifetimes $X_n$ are independent and identically uniformly distributed between three and five years. In the context of the renewal reward model just described, we shall let $c_1 = 5$ and $c_2 = 1$ (hundred thousand dollars). Let us find the optimal replacement policy. The probability distribution and density functions of the lifetime of the Inneall machine are

$$F_X(x) = \begin{cases} 0, & x < 3, \\ (x-3)/2, & 3 \leq x \leq 5, \\ 1, & x > 5, \end{cases} \quad \text{and} \quad f_X(x) = \begin{cases} 0, & x < 3, \\ 1/2, & 3 \leq x \leq 5, \\ 0, & x > 5. \end{cases}$$

Then

$$\frac{E[R]}{E[X]} = \frac{c_1 + c_2F(T)}{\int_0^T xf_X(x)dx + T(1 - F(T))} = \frac{c_1 + c_2(T-3)/2}{T^2/4 + T(1 - [(T-3)/2])} = \frac{7/2 + T/2}{5T/2 - T^2/4}.$$

To find the minimum, we take the derivative of this function and set the numerator to zero:

$$\left(\frac{5T}{2} - \frac{T^2}{4}\right)\frac{1}{2} - \left(\frac{T}{2} + \frac{7}{2}\right)\left(\frac{5}{2} - \frac{T}{2}\right) = 0.$$

Simplifying, this reduces to

$$T^2 + 14T - 70 = 0,$$

which has roots equal to $-17.9087$ and $3.9087$, the latter of which gives the value of $T$ that minimizes the long-run average cost of the Inneall machine.

It is interesting to observe what happens when the lifetime of the product has an exponential distribution with parameter $\mu$, i.e., $F_Y(t) = 1 - e^{-\mu t}$, $f_Y(t) = \mu e^{-\mu t}$. In this case

$$\frac{E[R]}{E[X]} = \frac{c_1 + c_2\left(1 - e^{-\mu T}\right)}{\int_0^T \mu x e^{-\mu x}dx + Te^{-\mu T}} = \frac{c_1 + c_2\left(1 - e^{-\mu T}\right)}{-xe^{-\mu x}|_0^T + \int_0^T e^{-\mu x}dx + Te^{-\mu T}}$$

$$= \frac{c_1 + c_2\left(1 - e^{-\mu T}\right)}{\left(1 - e^{-\mu T}\right)/\mu} = \frac{\mu c_1}{1 - e^{-\mu T}} + \mu c_2,$$

—a monotonically decreasing function of $T$! The exponential property of the lifetime distribution holds that at any point in time, the remaining lifetime is identical to the lifetime of a brand new machine.

**Alternating Renewal Processes**

An *alternating renewal process* is a renewal process in which the random variables $X_n, n \geq 1$, representing the time between renewals, are constructed from the sum of two or more random variables. In the simplest case there are just two random variables which we denote by $Y_n$ and $Z_n$. Thus $X_n = Y_n + Z_n$. If $Y_n$ and $Z_n$ are independent and identically distributed, and are independent of each other, then the process with interrenewal periods $X_n = Y_n + Z_n$ is an alternating renewal process. The extension to more than two random variables is immediate. Our interest with alternating renewal processes is generally in determining the proportion of time that the process spends in one of the phases that constitute a part of a cycle. This is accomplished by associating a reward with the duration of the selected phase.

**Example 9.42** Consider a machine that works correctly for a time and then breaks down. As soon as it breaks down a repair process is begun and the machine brought back to the functioning state. In this case, a cycle consists of a working period followed by a repair period. If $Y_n$ is the random variable that describes the length of the working period in the $n^{\text{th}}$ cycle, and $Z_n$ the corresponding repair period and if $Y_n$ and $Z_n$ have the independence and distribution properties described above, then the process is an alternating renewal process.

To relate this to renewal reward processes, we let the reward be the length of time that the machine functions correctly. This allows us to compute the proportion of time that the machine is functioning as

$$\frac{E[Y]}{E[X]} = \frac{E[Y]}{E[Y] + E[Z]}.$$

It may be shown that this is also the long-run probability that the machine is working.

## 9.13 Exercises

**Exercise 9.1.1** Give a real (or imagined) world example of

(a) a stationary, homogeneous stochastic process,
(b) a stationary, nonhomogeneous stochastic process,
(c) a nonstationary, homogeneous stochastic process, and
(d) a nonstationary, nonhomogeneous stochastic process.

In each case, state whether your process is a discrete-state or a continuous-state process and whether it is a discrete-time or a continuous-time process.

**Exercise 9.1.2** In the four scenarios of your answer to the previous question, state whether each stochastic process is, or is not, a Markov process. Be sure to justify your answer.

**Exercise 9.2.1** Brand preference model.

Suppose that there are five types of breakfast cereal, which we call $A$, $B$, $C$, $D$, and $E$. Customers tend to stick to the same brand. Those who choose type $A$ choose it the next time around with probability 0.8; those who choose type $B$ choose it next time with probability 0.9. The probabilities for types $C$, $D$, and $E$ are given by 0.7, 0.8, and 0.6, respectively. When customers do change brand, they choose one of the other four equally probably. Explain how this may be modeled by a Markov chain and give the transition probability matrix.

**Exercise 9.2.2** Inventory chain model.

Consider a store that sells television sets. If at the end of the day there is one or zero sets left, then that evening, after the store has closed, the shopkeeper brings an enough new sets so that the number of sets in stock for the next day is equal to five. This means that each morning, at store opening time, there are between two and five television sets available for sale. Such a policy is said to be an $(s, S)$ inventory control policy. Here we have assigned the values $s = 1$, $S = 5$. The shopkeeper knows from experience that the probabilities of selling 0 through 5 sets on any given day are $0.4,\ 0.3,\ 0.15,\ 0.15,\ 0.0$, and $0.0$.

Explain how this scenario may be modeled by a Markov chain $\{X_n,\ n = 1, 2, \ldots\}$, where $X_n$ is the random variable that defines the number of television sets left at the end of the $n^{\text{th}}$ day. Write down and explain the structure of the transition probability matrix.

**Exercise 9.2.3** The transition probability matrix of a discrete-time Markov chain is given by

$$P = \begin{pmatrix} 0 & 0 & 0 & 0 & 1.0 \\ 0 & 0 & 1.0 & 0 & 0 \\ 0 & 0 & 0 & 1.0 & 0 \\ 0 & 0.8 & 0.2 & 0 & 0 \\ 0.4 & 0 & 0.6 & 0 & 0 \end{pmatrix}.$$

Draw all sample paths of length 4 that begin in state 1. What is the probability of being in each of the states 1 through 5 after four steps beginning in state 1?

**Exercise 9.2.4** Consider a discrete-time Markov chain consisting of four states $a,\ b,\ c$, and $d$ and whose transition probability matrix is given by

$$P = \begin{pmatrix} 0.0 & 0.0 & 1.0 & 0.0 \\ 0.0 & 0.4 & 0.6 & 0.0 \\ 0.8 & 0.0 & 0.2 & 0.0 \\ 0.2 & 0.3 & 0.0 & 0.5 \end{pmatrix}.$$

Compute the following probabilities:

(a) $\text{Prob}\{X_4 = c, X_3 = c, X_2 = c, X_1 = c \mid X_0 = a\}$.
(b) $\text{Prob}\{X_6 = d, X_5 = c, X_4 = b \mid X_3 = a\}$.
(c) $\text{Prob}\{X_5 = c, X_6 = a, X_7 = c, X_8 = c \mid X_4 = b, X_3 = d\}$.

**Exercise 9.2.5** A Markov chain with two states $a$ and $b$ has the following conditional probabilities: If it is in state $a$ at time step $n$, $n = 0, 1, 2, \ldots$, then it stays in state $a$ with probability $0.5(0.5)^n$. If it is in state $b$ at time step $n$, then it stays in state $b$ with probability $0.75(0.25)^n$. If the Markov chain begins in state $a$ at time step $n = 0$, compute the probabilities of the following sample paths:

$$a \longrightarrow b \longrightarrow a \longrightarrow b \quad \text{and} \quad a \longrightarrow a \longrightarrow b \longrightarrow b.$$

**Exercise 9.2.6** The transition probability matrix of an embedded Markov chain is

$$P^E = \begin{pmatrix} 0.0 & 0.3 & 0.4 & 0.3 \\ 0.1 & 0.0 & 0.2 & 0.7 \\ 0.3 & 0.2 & 0.0 & 0.5 \\ 0.4 & 0.4 & 0.2 & 0.0 \end{pmatrix}.$$

Given that the homogeneous, discrete-time Markov chain $P$ from which this embedded chain is extracted, spends on average, 1 time unit in state 1, 2 time units in state 2, 4 time units in state 3, and 2.5 time units in state 4, derive the original Markov chain, $P$.

**Exercise 9.3.1** The following matrix is the single-step transition probability matrix of a discrete-time Markov chain which describes the weather. State 1 represents a sunny day, state 2 a cloudy day, and state 3 a rainy day.

$$P = \begin{array}{c} \\ Sunny \\ Cloudy \\ Rainy \end{array} \begin{array}{c} \begin{array}{ccc} Sunny & Cloudy & Rainy \end{array} \\ \begin{pmatrix} 0.7 & 0.2 & 0.1 \\ 0.3 & 0.5 & 0.2 \\ 0.2 & 0.6 & 0.2 \end{pmatrix} \end{array}.$$

(a) What is the probability of a sunny day being followed by two cloudy days?
(b) Given that today is rainy, what is the probability that the sun will shine the day after tomorrow?
(c) What is the mean length of a rainy period?

**Exercise 9.3.2** Consider the four-state discrete-time Markov chain whose transition probability matrix at time step $n$, $n = 0, 1, \ldots,$ is

$$P(n) = \begin{pmatrix} 0 & 0.6 & 0.4 & 0 \\ 0.8 & 0 & 0 & 0.2 \\ 0 & 0 & 0.5(0.5)^n & 1 - 0.5(0.5)^n \\ 0 & 0 & 0.8(0.8)^n & 1 - 0.8(0.8)^n \end{pmatrix}.$$

What is the probability distribution after two steps if the Markov chain is initiated in (a) state 1; (b) state 4?

**Exercise 9.3.3** William, the collector, enjoys collecting the toys in McDonald's' Happy Meals. And now McDonald's has come out with a new collection containing five toy warriors. Each Happy Meal includes one randomly chosen warrior. Naturally William has to collect all five different types.

(a) Use a discrete-time Markov chain to represent the process that William will go through to collect all five warriors and draw the state transition diagram.
(b) Construct the stochastic transition probability matrix for this discrete-time Markov chain and compute the probability distribution after William has eaten three happy meals.
(c) Let $T$ denote the total number of Happy Meals that William will eat to enable him to get all five warriors. Compute $E[T]$ and Var $[T]$.

**Exercise 9.3.4** A prisoner in a Kafkaesque prison is put in the following situation. A regular deck of 52 cards is placed in front of him. He must choose cards one at a time to determine their color. Once chosen, the card is replaced in the deck and the deck is shuffled. If the prisoner happens to select three consecutive red cards, he is executed. If he happens to select six cards before three consecutive red cards appear he is granted freedom.

(a) Represent the prisoner's situation with a Markov chain and draw its transition diagram.
*Hint: let state i denote the number of successive red cards obtained.*
(b) Construct the transition probability matrix, and the initial state probability vector.
(c) Determine the probability that the prisoner will be set free.

**Exercise 9.4.1** Give example state transition diagrams for the following types of Markov chain states. Justify your answers. (a) Transient state, (b) positive-recurrent state, (c) periodic state, (d) absorbing state, (e) closed set of states, and (f) irreducible chain.

**Exercise 9.4.2** Give as precise a classification as possible to each of the states of the discrete-time Markov chain whose transition probability matrix is

$$P = \begin{pmatrix} 0 & 0 & 0 & 0 & .3 & 0 & .7 & 0 & 0 & 0 \\ 0 & 0 & 0 & 0 & .6 & 0 & 0 & .4 & 0 & 0 \\ 0 & 0 & 0 & 0 & 0 & .5 & 0 & 0 & .5 & 0 \\ 0 & 0 & 0 & 0 & 0 & .2 & 0 & 0 & 0 & .8 \\ 0 & 0 & 0 & 0 & 1 & 0 & 0 & 0 & 0 & 0 \\ 0 & 0 & 0 & 0 & 0 & 1 & 0 & 0 & 0 & 0 \\ 0 & 0 & 0 & 0 & 0 & 0 & 0 & .9 & .1 & 0 \\ 0 & 0 & 0 & 0 & 0 & 0 & .8 & 0 & 0 & .2 \\ 0 & 0 & 0 & 0 & 0 & 0 & 0 & 0 & 0 & 1 \\ 0 & 0 & 0 & 0 & 0 & 0 & 0 & 0 & 1 & 0 \end{pmatrix}.$$

**Exercise 9.4.3** Find the mean recurrence time of state 2 and the mean first passage time from state 1 to state 2 in the discrete-time Markov chain whose transition probability matrix is

$$P = \begin{pmatrix} 0.1 & 0.3 & 0.6 \\ 0.2 & 0.5 & 0.3 \\ 0.4 & 0.4 & 0.2 \end{pmatrix}.$$

*Hint: use Matlab.*

**Exercise 9.5.1** In the Markov chain whose transition probability matrix $P$ is given below, identify examples of the following (if they exist):

(a) a return state,
(b) a nonreturn state,
(c) an absorbing state,
(d) a closed communicating class,
(e) an open communicating class,
(f) a closed communicating class containing recurrent states,
(g) an open communicating class containing recurrent states,
(h) a closed communicating class containing transient states,
(i) an open communicating class containing transient states, and
(j) a communicating class with both transient and recurrent states.

$$P = \begin{pmatrix} 0 & 0.5 & 0 & 0.5 & 0 & 0 & 0 & 0 \\ 0.5 & 0 & 0.5 & 0 & 0 & 0 & 0 & 0 \\ 0 & 0 & 0 & 0.5 & 0 & 0.5 & 0 & 0 \\ 0 & 0 & 0 & 0 & 1.0 & 0 & 0 & 0 \\ 0 & 0 & 0 & 0 & 0 & 0 & 0 & 1.0 \\ 0 & 0 & 0 & 0 & 0 & 1.0 & 0 & 0 \\ 0 & 0 & 0 & 0 & 1.0 & 0 & 0 & 0 \\ 0 & 0 & 0 & 0 & 0 & 0 & 1.0 & 0 \end{pmatrix}.$$

**Exercise 9.5.2** In the Markov chain whose transition probability matrix $P$ is given below, what are the communicating classes to which the following states belong: (a) state 2, (b) state 3, (c) state 4, and (d) state 5.

$$P = \begin{pmatrix} 0.5 & 0.25 & 0.25 & 0 & 0 & 0 & 0 \\ 0.25 & 0.5 & 0 & 0.25 & 0 & 0 & 0 \\ 0 & 0 & 0.5 & 0.5 & 0 & 0 & 0 \\ 0 & 0 & 0 & 0 & 0.5 & 0.5 & 0 \\ 0 & 0 & 0 & 0 & 0.5 & 0.25 & 0.25 \\ 0 & 0 & 0 & 0 & 0.25 & 0.5 & 0.25 \\ 0 & 0 & 0 & 0 & 0.25 & 0.25 & 0.5 \end{pmatrix}.$$

**Exercise 9.5.3** Consider a discrete-time Markov chain with transition probabilities given by

$$p_{ij} = e^{-\lambda} \sum_{n=0}^{j} \binom{i}{n} p^n q^{i-n} \frac{\lambda^{j-n}}{(j-n)!},$$

where $p + q = 1, 0 \leq p \leq 1$ and $\lambda > 0$.

(a) Is this chain reducible? Explain.
(b) Is this chain periodic? Explain.

**Exercise 9.5.4** Prove that if one state in a communicating class $C$ is transient, then all states in $C$ are transient.

**Exercise 9.5.5** Let $C$ be a nonclosed communicating class. Show that no state in $C$ can be recurrent. (This means that every recurrent class is closed.)

**Exercise 9.6.1** Compute the potential matrix of the Markov chain whose transition probability matrix is

$$P = \begin{pmatrix} .5 & .5 & 0 & 0 \\ .5 & 0 & .5 & 0 \\ .5 & 0 & 0 & .5 \\ 0 & 0 & 0 & 1 \end{pmatrix}$$

and from it find (a) the expected number of visits to state 3 beginning from state 1, (b) the expected number of visits to state 1 beginning in state 3, and (c) the expected number of visits to state 2 beginning in state 4.

**Exercise 9.6.2** Consider the gambler's ruin problem in which a gambler begins with \$50, wins \$10 on each play with probability $p = 0.45$, or loses \$10 with probability $q = 0.55$. The gambler will quit once he doubles his money or has nothing left of his original \$50.

(a) What is the expected number of times he has \$90 before quitting?
(b) What is the expected number of times he has \$50 before quitting?
(c) What is the expected number of times he has \$10 before quitting?

**Exercise 9.6.3** Consider a discrete-time Markov chains whose transition probability matrix is

$$P = \begin{pmatrix} .2 & .8 & 0 & 0 & 0 \\ 0 & .4 & .6 & 0 & 0 \\ 0 & 0 & .6 & .4 & 0 \\ 0 & 0 & 0 & .8 & .2 \\ 0 & 0 & 0 & 0 & 1 \end{pmatrix}.$$

Compute the mean and variance of the time to absorption from all transient starting states.

**Exercise 9.6.4** The transition probability matrix of a discrete-time Markov chain is

$$P = \begin{pmatrix} .6 & .4 & 0 & 0 & 0 & 0 & 0 & 0 \\ 0 & .2 & .6 & 0 & .2 & 0 & 0 & 0 \\ .2 & 0 & .3 & 0 & 0 & 0 & 0 & .5 \\ 0 & 0 & 0 & 0 & 0 & .9 & 0 & .1 \\ 0 & 0 & 0 & 0 & 0 & 1.0 & 0 & 0 \\ 0 & 0 & 0 & 0 & .5 & 0 & .5 & 0 \\ 0 & 0 & 0 & 0 & 1.0 & 0 & 0 & 0 \\ 0 & 0 & 0 & 0 & 0 & 0 & 0 & 1.0 \end{pmatrix}.$$

Assume that the initial state is 2 with probability 0.4 and 4 with probability 0.6. Compute

(a) the mean and variance of the number of times the Markov chain visits state 1 before absorption;
(b) the mean and variance of the total number of steps prior to absorption.

**Exercise 9.6.5** Consider a discrete-time Markov chain whose transition probability matrix is

$$P = \begin{pmatrix} .2 & .8 & 0 & 0 & 0 \\ 0 & .4 & .6 & 0 & 0 \\ 0 & 0 & .6 & .4 & 0 \\ .2 & 0 & 0 & .6 & .2 \\ 0 & 0 & 0 & 0 & 1 \end{pmatrix}.$$

(a) Compute the part of the reachability matrix that corresponds to transitions among transient states (the matrix called $H$ in the text).
(b) Beginning in state 1, what is the probability of going to state 4 exactly four times?
(c) Beginning in state 1, how many different states are visited, on average, prior to absorption?

**Exercise 9.6.6** In the gambler's ruin scenario of Exercise 9.6.2:

(a) What is the mean number of plays he makes (starting with \$50) before stopping?
(b) How many different amounts of money does he have during the time he is playing?

**Exercise 9.6.7** Compute the *absorbing chain* and from it, the matrix of absorbing probabilities, of the Markov chain whose transition probability matrix is

$$P = \begin{pmatrix} .6 & .4 & 0 & 0 & 0 & 0 & 0 & 0 \\ 0 & .2 & .6 & 0 & .2 & 0 & 0 & 0 \\ .2 & 0 & .3 & 0 & 0 & 0 & 0 & .5 \\ 0 & 0 & 0 & 0 & 0 & .9 & 0 & .1 \\ 0 & 0 & 0 & 0 & 0 & 1.0 & 0 & 0 \\ 0 & 0 & 0 & 0 & .5 & 0 & .5 & 0 \\ 0 & 0 & 0 & 0 & 1.0 & 0 & 0 & 0 \\ 0 & 0 & 0 & 0 & 0 & 0 & 0 & 1.0 \end{pmatrix}.$$

What are the probabilities of being absorbed into state 8 from states 1, 2, and 3?

**Exercise 9.6.8** Consider a random walk problem on the set of integers, $0, 1, \ldots, 10$. Let the transition probabilities be given by $p_{i,i+1} = 0.45$; $p_{i+1,i} = 0.55$, $i = 0, 1, \ldots, 9$, $p_{00} = 0.55$, $p_{10,10} = 0.45$. Beginning in state 5, what is the probability of reaching state 0 before state 10?

**Exercise 9.7.1** Random walk on the integers $0, \pm 1, \pm 2, \ldots$.
Consider the Markov chain whose transition probability matrix is

$$P = \begin{pmatrix} \ddots & \vdots & \vdots & \vdots & \vdots & \vdots & \vdots & \cdots \\ \cdots & 0 & p & 0 & 0 & 0 & 0 & \cdots \\ \cdots & q & 0 & p & 0 & 0 & 0 & \cdots \\ \cdots & 0 & q & 0 & p & 0 & 0 & \cdots \\ \cdots & 0 & 0 & q & 0 & p & 0 & \cdots \\ \cdots & 0 & 0 & 0 & q & 0 & p & \cdots \\ \cdots & 0 & 0 & 0 & 0 & q & 0 & \cdots \\ \vdots & \vdots & \vdots & \vdots & \vdots & \vdots & \vdots & \ddots \end{pmatrix}.$$

Show that

$$p_{00}^{(2n+1)} = 0 \quad \text{and} \quad p_{00}^{(2n)} = \binom{2n}{n} p^n q^n, \quad \text{for } n = 1, 2, 3, \ldots.$$

Use Sterling's formula, $n! \approx n^{n+1/2} e^{-n} \sqrt{2\pi}$, to show that $(2n)!/(n!n!) = 4^n/\sqrt{n\pi}$ and hence that

$$p_{00}^{(2n)} \approx \frac{(4pq)^n}{\sqrt{n\pi}}.$$

Write an expression for $\sum_{n=1}^{\infty} p_{00}^{(n)}$ and show that this is infinite if and only if $p = 1/2$, and hence that the Markov chain is recurrent if and only if $p = 1/2$ and is transient for all other values of $p$.

**Exercise 9.7.2** Consider a Markov chain whose transition probability matrix is given by

$$P = \begin{pmatrix} q & p & 0 & \cdots & & \\ q & 0 & p & 0 & \cdots & \\ 0 & q & 0 & p & 0 & \cdots \\ \vdots & \vdots & \vdots & \vdots & \ddots & \vdots \end{pmatrix},$$

where $p > 0$, $q > 0$, and $p + q = 1$. This is a variant of the random walk problem with Bernoulli boundary. By considering the system of equations $z = zP$ with $ze = 1$ and Theorem 9.7.1, find conditions under which the states of the Markov chain are ergodic and, assuming that these conditions hold, find the values of $z_k$ for $k = 0, 1, \ldots$.

**Exercise 9.7.3** Complete the analysis for the gambler's ruin problem by considering the case when $p = q = 1/2$. In particular, show that the probability of winning $N$ dollars when starting with $i$ dollars is $x_i = i/N$, for $i = 1, 2, \ldots, N$.

**Exercise 9.7.4** One of the machines at a gambling casino pays a return that is slightly favorable to the player, rather than to the house. On each play, the player has a 51% chance of winning. What is the probability that a gambler with \$20 doubles his money? What is the probability of doubling his money if he starts with \$100?

**Exercise 9.7.5** On each play in roulette, a player wins \$10 with probability 18/38 and loses \$10 with probability 20/38. Kathie has \$100 in her purse but really needs \$250 to purchase the pair of shoes she has just seen in the shop next to the casino. What is the probability of her converting her \$100 into \$250 at the roulette table?

**Exercise 9.7.6** Harold Nunn is an entrepreneur. He wishes to establish a high-stakes gambling saloon based only on a single roulette table. Each bet costs a player \$500. Harold knows that the odds of 20/38 are in his favor. Suppose he starts with a capital of \$100,000, what is the probability that Harold makes an unlimited fortune?

**Exercise 9.8.1** A square matrix is said to be *stochastic* if each of its elements lies in the interval [0, 1] and the sum of the elements in each row is equal to 1. It is said to be *doubly stochastic* if, in addition, the sum of elements in each *column* is equal to 1. Show that each element of the stationary distribution of an irreducible, finite, $K$-state Markov chain whose transition probability matrix is doubly stochastic is equal to $1/K$. Is this stationary distribution unique? Does this same result hold when the Markov chain is reducible? Explain your answer. Can an irreducible Markov chain whose transition probability matrix is doubly stochastic have transient states? Explain your answer. Does the *limiting distribution* necessarily exist for an irreducible Markov chain whose transition probability matrix is doubly stochastic? Explain your answer.

**Exercise 9.8.2** Let $\{X_n, n \geq 0\}$ be a two-state Markov chain, whose transition probability matrix is given by

$$P = \begin{pmatrix} 1-p & p \\ q & 1-q \end{pmatrix}$$

with $0 < p, q < 1$. Use the fact that $P^{n+1} = P^n P$ to show

$$p_{11}^{(n+1)} = (1-p-q)p_{11}^{(n)} + q.$$

Now prove by induction that

$$p_{11}^{(n)} = \frac{q}{p+q} + \frac{p}{p+q}(1-p-q)^n$$

and hence find the limit: $\lim_{n\to\infty} P^n$.

**Exercise 9.8.3** Let $\{X_n, n \geq 0\}$ be a two-state Markov chain, whose transition probability matrix is given by

$$P = \begin{pmatrix} 1-p & p \\ q & 1-q \end{pmatrix},$$

where $0 < p < 1$ and $0 < q < 1$. Let $\text{Prob}\{X_0 = 0\} = \pi_0(0)$ be the probability that the Markov chain begins in state 0. Prove by induction that

$$\text{Prob}\{X_n = 0\} = (1-p-q)^n \pi_0(0) + q\sum_{j=0}^{n-1}(1-p-q)^j.$$

**Exercise 9.8.4** In the scenario of Exercise 9.3.1,

(a) What is the unconditional probability of having a sunny day?
(b) What is the mean number of rainy days in a month of 31 days?
(c) What is the mean recurrence time of sunny days?

**Exercise 9.8.5** For each of the following Markov chains ($A, B, C$, and $D$), represented by transition probability matrices given below, state whether it has (i) a stationary distribution, (ii) a limiting distribution, (iii) a steady-state distribution. If your answer is negative, explain why? If your answer is positive, give the distribution.

$$P_A = \begin{pmatrix} 0.5 & 0.5 & 0.0 \\ 0.0 & 0.5 & 0.5 \\ 0.5 & 0.0 & 0.5 \end{pmatrix}, \quad P_B = \begin{pmatrix} 0.0 & 1.0 & 0.0 \\ 0.0 & 0.0 & 1.0 \\ 1.0 & 0.0 & 0.0 \end{pmatrix}, \quad P_C = \begin{pmatrix} 0.5 & 0.5 & 0.0 & 0.0 \\ 0.5 & 0.5 & 0.0 & 0.0 \\ 0.0 & 0.0 & 0.5 & 0.5 \\ 0.0 & 0.0 & 0.5 & 0.5 \end{pmatrix}.$$

$$P_D(n) = \begin{pmatrix} 1/n & (n-1)/n \\ (n-1)/n & 1/n \end{pmatrix}, \quad n = 1, 2, \ldots.$$

**Exercise 9.8.6** Consider a Markov chain having state space $S = \{0, 1, 2\}$ and transition matrix

$$\begin{array}{c} \\ 0 \\ 1 \\ 2 \end{array}\begin{array}{c} \begin{array}{ccc} 0 & 1 & 2 \end{array} \\ \begin{pmatrix} 1/3 & 1/3 & 1/3 \\ 1/4 & 1/2 & 1/4 \\ 1/6 & 1/3 & 1/2 \end{pmatrix} \end{array}.$$

Show that this chain has a unique stationary distribution $\pi$ and find $\pi$.

**Exercise 9.8.7** Consider a machine that at the start of any particular day is either broken down or in operating condition. Assume that if the machine is broken down at the start of the $n^{\text{th}}$ day, the probability that it will be successfully repaired and in operating condition at the start of the $(n+1)^{\text{th}}$ day is $p$. Assume also that if the machine is in operating condition at the start of the $n^{\text{th}}$ day, the probability that it will fail and be broken down at the start of the $(n+1)^{\text{th}}$ day is $q$. Let $\pi_0(0)$ denote the probability that the machine is broken down initially.

(a) Find the following probabilities:

$$\text{Prob}\{\, X_{n+1} = 1 \mid X_n = 0 \,\}, \;\; \text{Prob}\{\, X_{n+1} = 0 \mid X_n = 1 \,\}, \;\; \text{Prob}\{\, X_{n+1} = 0 \mid X_n = 0 \,\}.$$

$$\text{Prob}\{\, X_{n+1} = 1 \mid X_n = 1 \,\} \text{ and } \text{Prob}\{\, X_0 = 1 \,\}.$$

(b) Compute $\text{Prob}\{X_n = 0\}$ and $\text{Prob}\{X_n = 1\}$ in terms of $p$, $q$, and $\pi_0(0)$.

(c) Find the steady-state distribution $\lim_{n\to\infty} \text{Prob}\{X_n = 0\}$ and $\lim_{n\to\infty} \text{Prob}\{X_n = 1\}$.

**Exercise 9.8.8** Consider a Markov chain defined on the nonnegative integers and having transition probabilities given by

$$p_{n,n+1} = p \text{ and } p_{n,0} = 1 - p$$

for all $n$, where $0 < p < 1$.

(a) Compute the mean recurrence time of state 0.
(b) Show that the Markov chain is positive recurrent.
(c) Prove that the limiting probabilities $\pi$ exist.
(d) Find $\pi$ (in terms of $p$).

**Exercise 9.8.9** Remaining lifetime.
Analyze the Markov chain whose transition probability matrix is given by

$$P = \begin{pmatrix} p_0 & p_1 & p_2 & p_3 & \cdots \\ 1 & 0 & 0 & 0 & \cdots \\ 0 & 1 & 0 & 0 & \cdots \\ 0 & 0 & 1 & 0 & \cdots \\ 0 & 0 & 0 & 1 & \cdots \\ \vdots & \vdots & \vdots & \ddots & \vdots \end{pmatrix}.$$

This Markov chain arises in models of a system component which fails and which is instantaneously replaced with an identical component. When this identical component also fails, it too is instantaneously replaced with another identical component, and so on. The Markov chain, $\{X_n;\, n \geq 0\}$, denotes the remaining lifetime of the component in use at time step $n$. The probability $p_k$ denotes the probability that a newly installed component will last for $k$ time steps.

(a) Under what condition(s) does a stationary distribution exist for this Markov chain?
(b) Compute this distribution assuming your condition(s) hold.

**Exercise 9.9.1** The transition probability matrices of a number of Markov chains are given below. For each different $P$, find the transition probability matrix of the reversed chain.

$$P_1 = \begin{pmatrix} 0 & 1 & 0 \\ 0 & 0 & 1 \\ 1 & 0 & 0 \end{pmatrix}, \quad P_2 = \begin{pmatrix} .4 & .2 & .4 \\ .1 & .3 & .6 \\ .5 & .5 & 0 \end{pmatrix}, \quad P_3 = \begin{pmatrix} 1/3 & 1/3 & 1/3 \\ 1/4 & 1/2 & 1/4 \\ 1/6 & 1/3 & 1/2 \end{pmatrix}, \quad P_4 = \begin{pmatrix} 0 & 1 & 0 & 0 \\ 0 & 0 & .6 & .4 \\ 0 & 0 & 0 & 1 \\ .6 & .4 & 0 & 0 \end{pmatrix}.$$

**Exercise 9.9.2** Which of the transition probability matrices given below belong to reversible Markov chains? Give reasons for your answer.

$$P_1 = \begin{pmatrix} .25 & .25 & .25 & .25 \\ .25 & .50 & .25 & 0.0 \\ .50 & 0.0 & .25 & .25 \\ .10 & .20 & .30 & .40 \end{pmatrix}, \quad P_2 = \begin{pmatrix} 0 & 3/4 & 1/4 & 0 \\ 1/3 & 0 & 4/9 & 2/9 \\ 1/10 & 4/10 & 0 & 5/10 \\ 0 & 2/7 & 5/7 & 0 \end{pmatrix},$$

$$P_3 = \begin{pmatrix} .1 & 0 & 0 & .9 & 0 & 0 \\ 0 & 0 & 0 & 0 & 1 & 0 \\ 0 & 0 & 0 & 0 & 1 & 0 \\ .4 & 0 & 0 & 0 & .2 & .4 \\ 0 & .5 & .3 & .2 & 0 & 0 \\ 0 & 0 & 0 & 1 & 0 & 0 \end{pmatrix}, \qquad P_4 = \begin{pmatrix} .5 & .1 & .4 \\ .2 & .4 & .4 \\ .3 & .2 & .5 \end{pmatrix}.$$

**Exercise 9.10.1** A triple redundancy system consists of three computing units and a"voter." The purpose of the voter is to compare the results produced by the three computing units and to select the answer given by at least two of them. The computing units as well as the voter are all susceptible to failure. The rate at which a computing unit fails is taken to be $\lambda$ failures per hour and the rate at which the voter fails is $\xi$ failures per hour. All items are repairable: computing units are repaired at a rate of $\mu$ per hour and the voter at a rate of $\nu$ per hour. During the repair of the voter, each computing unit is inspected and repaired if necessary. The time for repair of the computing units in this case is included in the $1/\nu$ hours needed to repair the voter. This system is to be represented as a continuous-time Markov chain.

(a) What conditions are imposed by representing this system as a continuous-time Markov chain?
(b) What is a suitable state descriptor for this Markov chain?
(c) Draw the state transition rate diagram.
(d) Write down the infinitesimal generator for this Markov chain.

**Exercise 9.10.2** The time between arrivals at a service center is exponentially distributed with a mean interarrival time of fifteen minutes. Each arrival brings one item to be serviced with probability $p$ $(0 < p < 1)$ or two items with probability $1 - p$. Items are serviced individually at a rate of five per hour. Represent this system as a continuous-time Markov chain and give its infinitesimal generator $Q$.

**Exercise 9.10.3** Construct the embedded chain of the continuous-time Markov chains whose infinitesimal generators are given below:

$$Q_1 = \begin{pmatrix} -5 & 5 \\ 2 & -2 \end{pmatrix}, \qquad Q_2 = \begin{pmatrix} -3 & 2 & 1 \\ 0 & -5 & 5 \\ 0 & 2 & -2 \end{pmatrix}.$$

**Exercise 9.10.4** Consider a six-state $(0, 1, \ldots, 5)$ continuous-time Markov chain with a tridiagonal infinitesimal generator given by

$$Q = \begin{pmatrix} 0 & 0 & & & & \\ 2 & -3 & 1 & & & \\ & 4 & -6 & 2 & & \\ & & 6 & -9 & 3 & \\ & & & 8 & -12 & 4 \\ & & & & 0 & 0 \end{pmatrix}.$$

(a) Find the probability of being absorbed into state 0 from state 3.
(b) Find the probability of being absorbed into state 5 from state 2.
(c) Let $B = \{0, 5\}$. Find the mean time to absorption from state 1 into the closed subset $B$.
(d) Find the mean time to absorption from state 4 into $B$.

**Exercise 9.10.5** Suppose a single repairman has been assigned the responsibility of maintaining three machines. For each machine, the probability distribution of up time (machine is functioning properly) before a breakdown is exponential with a mean of nine hours. The repair time is also exponentially distributed with a mean of two hours. Calculate the steady-state probability distribution and the expected number of machines that are not running.

**Exercise 9.10.6** Consider a continuous-time Markov chain with transition rate matrix

$$Q = \begin{pmatrix} 0 & 0 & & & & \\ 2 & -3 & 1 & & & \\ & 4 & -6 & 2 & & \\ & & 6 & -9 & 3 & \\ & & & 8 & -12 & 4 \\ & & & & 0 & 0 \end{pmatrix}.$$

Use Matlab and its matrix exponential function to compute the probability of being in state 1, state 4, and state 6 at times $t = 0.01, 0.1, 0.5, 1.0, 2.0,$ and $5.0$, given that the system begins in state 4.

**Exercise 9.10.7** Compute the stationary probability distribution of the continuous-time Markov chain whose infinitesimal generator $Q$ is shown below.

(a) From the stationary distribution of its embedded chain, and
(b) By solving the system of equation $\pi Q = 0$.

$$Q = \begin{pmatrix} -1 & 1 & & & & \\ & -2 & 2 & & & \\ & & -3 & 3 & & \\ & & & -4 & 4 & \\ & & & & -5 & 5 \\ 6 & & & & & -6 \end{pmatrix}$$

**Exercise 9.12.1** Suppose the interrenewal time in a renewal process has a Poisson distribution with mean $\lambda$. Find $\text{Prob}\{S_n \leq t\}$, the probability distribution of the renewal sequence $S_n$, and calculate $\text{Prob}\{N(t) = n\}$.

**Exercise 9.12.2** As it relates to renewal processes, use the definition

$$\text{Prob}\{N(t) = n\} = F_X^{(n)}(t) - F_X^{(n+1)}(t)$$

to show that

$$E[N(t)] = \sum_{n=1}^{\infty} F_X^{(n)}(t).$$

**Exercise 9.12.3** Brendan replaces his car every $T$ years, unless it has been involved in a major accident prior to $T$ years, in which case he replaces it immediately. If the lifetimes of the cars he purchases are independent and identically distributed with distribution function $F_X(x)$, show that, in the long run, the rate at which Brendan replaces cars is given by

$$\frac{1}{\left[\int_0^T x f_X(x)dx + T(1 - F_X(T))\right]},$$

where $f_X(x)$ is the density function of his cars' lifetimes.

**Exercise 9.12.4** The builder of the Inneall machine of Example 9.41 is prepared to offer the widget manufacturer a trade-in $W(T)$ on the purchase of a new machine. Write an expression for the long-run average cost given this trade-in function. Now, using the parameters of Example 9.41, find the value of $T$ that minimizes the long-run average cost if the trade-in function is given by $W(T) = 3 - T/2$.

**Exercise 9.12.5** Consider a queueing system that consists of a single server and no waiting positions. An arriving customer who finds the server busy is lost—i.e., departs immediately without receiving service. If the arrival process is Poisson with rate $\lambda$ and the distribution of service time has expectation $1/\mu$, show that the queueing process is an alternating renewal process and find the probability that the server is busy.

# Chapter 10

# *Numerical Solution of Markov Chains*

## 10.1 Introduction

We now turn our attention to the computation of stationary and transient distributions of Markov chains. For the computation of stationary distributions, we consider first the case when the number of states can be large but finite. In particular, we examine direct methods based on Gaussian elimination, point and block iterative methods such as point and block Gauss–Seidel and decompositional methods that are especially well suited to Markov chains that are almost reducible. Second, when the Markov chain has an infinite state space and the pattern of nonzero elements in the transition matrix is block repetitive, we develop and analyze the matrix geometric and matrix analytic methods. As for the computation of transient distributions, we describe methods that are appropriate for small-scale Markov chains, as well as the popular uniformization method and methods based on differential equation solvers for large-scale chains. Throughout this chapter, our concern is with stationary, homogeneous Markov chains only.

### *10.1.1 Setting the Stage*

We begin by considering finite, irreducible Markov chains. Such Markov chains have a unique stationary probability distribution $\pi$ whose elements are strictly greater than zero. When the Markov chain is also aperiodic, this unique stationary distribution is also the steady-state distribution. Let $P$ be the transition probability matrix of a finite, irreducible discrete-time Markov chain. Then $\pi$ is the left-hand eigenvector corresponding to the dominant (and simple) unit eigenvalue of $P$, normalized so that its elements sum to 1:

$$\pi P = \pi \quad \text{with } \pi e = 1. \tag{10.1}$$

If the Markov chain evolves in continuous time rather than in discrete time and its infinitesimal generator is denoted by $Q$, then its stationary distribution may be found from the system of linear equations

$$\pi Q = 0 \quad \text{with } \pi e = 1. \tag{10.2}$$

Observe that both these equations may be put into the same form. We may write the first, $\pi P = \pi$, as $\pi(P - I) = 0$ thereby putting it into the form of Equation (10.2). Observe that $(P - I)$ has all the properties of an infinitesimal generator, namely, all off-diagonal elements are nonnegative; row sums are equal to zero and diagonal elements are equal to the negated sum of off-diagonal row elements. On the other hand, we may discretize a continuous-time Markov chain. From

$$\pi Q = 0 \quad \text{with } \pi e = 1$$

we may write

$$\pi(Q\Delta t + I) = \pi \quad \text{with } \pi e = 1, \tag{10.3}$$

thereby posing it in the form of Equation (10.1). In the discretized Markov chain, transitions take place at intervals $\Delta t$, $\Delta t$ being chosen sufficiently small that the probability of two transitions taking place in time $\Delta t$ is negligible, i.e., of order $o(\Delta t)$. One possibility is to take

$$\Delta t \leq \frac{1}{\max_i |q_{ii}|}.$$

In this case, the matrix $(Q\Delta t + I)$ is stochastic and the stationary[1] probability vector $\pi$ of the continuous-time Markov chain, obtained from $\pi Q = 0$, is identical to that of the discretized chain, obtained from $\pi(Q\Delta t + I) = \pi$. It now follows that numerical methods designed to compute the stationary distribution of discrete-time Markov chains may be used to compute the stationary distributions of continuous-time Markov chains, and vice versa.

**Example 10.1** In the previous chapter we saw that one very simple method for computing the stationary distribution of a discrete-time Markov chain, is to successively compute the probability distribution at each time step until no further change in the distribution is observed. We shall now apply this method (which we shall later refer to as the *power method*) to the *continous-time* Markov chain whose infinitesimal generator is given by

$$Q = \begin{pmatrix} -4 & 4 & 0 & 0 \\ 3 & -6 & 3 & 0 \\ 0 & 2 & -4 & 2 \\ 0 & 0 & 1 & -1 \end{pmatrix}.$$

Setting $\Delta t = 1/6$, we obtain

$$Q\Delta t + I = \begin{pmatrix} 1-4/6 & 4/6 & 0 & 0 \\ 3/6 & 1-6/6 & 3/6 & 0 \\ 0 & 2/6 & 1-4/6 & 2/6 \\ 0 & 0 & 1/6 & 1-1/6 \end{pmatrix} = \begin{pmatrix} 1/3 & 2/3 & 0 & 0 \\ 1/2 & 0 & 1/2 & 0 \\ 0 & 1/3 & 1/3 & 1/3 \\ 0 & 0 & 1/6 & 5/6 \end{pmatrix}.$$

Beginning with $\pi^{(0)} = (1, 0, 0, 0)$, and successively computing $\pi^{(n)} = \pi^{(n-1)}(Q\Delta t + I)$ for $n = 1, 2, \ldots$, we obtain

$$\begin{aligned}
\pi^{(0)} &= (1.0000 \quad 0.0000 \quad 0.0000 \quad 0.0000),\\
\pi^{(1)} &= (0.3333 \quad 0.6667 \quad 0.0000 \quad 0.0000),\\
\pi^{(2)} &= (0.4444 \quad 0.2222 \quad 0.3333 \quad 0.0000),\\
\pi^{(3)} &= (0.2593 \quad 0.4074 \quad 0.2222 \quad 0.1111),\\
&\vdots \\
\pi^{(10)} &= (0.1579 \quad 0.1929 \quad 0.2493 \quad 0.3999),\\
&\vdots \\
\pi^{(25)} &= (0.1213 \quad 0.1612 \quad 0.2403 \quad 0.4773),\\
&\vdots \\
\pi^{(50)} &= (0.1200 \quad 0.1600 \quad 0.2400 \quad 0.4800),
\end{aligned}$$

and, correct to four decimal places, no further change is observed in computing the distribution at higher step values. Thus we may take (0.1200, 0.1600, 0.2400, 0.4800) to be the stationary distribution of the continuous-time Markov chain represented by the infinitesimal generator $Q$.

Thus, depending on how we formulate the problem, we may obtain the stationary distribution of a Markov chain (discrete *or* continuous time) by solving either $\pi P = \pi$ or $\pi Q = 0$.

[1] We caution the reader whose interest is in transient solutions (i.e., probability distributions at an arbitrary time $t$) that those of the discretized chain, represented by the transition probability matrix $Q\Delta t + I$, are not the same as those of the continuous-time chain, represented by the infinitesimal generator $Q$. However, both have the same stationary distribution.

From the perspective of a numerical analyst, these are two different problems. The first is an *eigenvalue/eigenvector* problem in which the stationary solution vector $\pi$ is the left-hand eigenvector corresponding to a unit eigenvalue of the transition probability matrix $P$. The second is a *linear equation* problem in which the desired vector $\pi$ is obtained by solving a homogeneous (right-hand side identically equal to zero) system of linear equations with singular coefficient matrix $Q$. In light of this, it behooves us to review some eigenvalue/eigenvector properties of stochastic matrices and infinitesimal generators.

### *10.1.2 Stochastic Matrices*

A matrix $P \in \Re^{n \times n}$ is said to be a *stochastic* matrix if it satisfies the following three conditions:

1. $p_{ij} \geq 0$ for all $i$ and $j$.
2. $\sum_{\text{all } j} p_{ij} = 1$ for all $i$.
3. At least one element in each column differs from zero.

Matrices that obey condition 1 are called nonnegative matrices, and stochastic matrices form a proper subset of them. Condition 2 implies that a transition is guaranteed to occur from state $i$ to at least one state in the next time period (which may be state $i$ again). Condition 3 specifies that, since each column has at least one nonzero element, there are no *ephemeral* states, i.e., states that could not possibly exist after the first time transition. In much of the literature on stochastic matrices this third condition is omitted (being considered trivial). In the remainder of this text we also shall omit this condition. We shall now list some important properties concerning the eigenvalues and eigenvectors of stochastic matrices.

**Property 10.1.1** *Every stochastic matrix has an eigenvalue equal to 1.*

**Proof:** Since the sum of the elements of each row of $P$ is 1, we must have

$$Pe = e,$$

where $e = (1, 1, \ldots, 1)^T$. It immediately follows that $P$ has a unit eigenvalue.

**Corollary 10.1.1** *Every infinitesimal generator $Q$ has a zero eigenvalue.*

**Proof:** This follows immediately from the fact that $Qe = 0$.

**Property 10.1.2** *The eigenvalues of a stochastic matrix must have modulus less than or equal to 1.*

**Proof:** To prove this result we shall use the fact that for any matrix $A$,

$$\rho(A) \leq \|A\|_\infty = \max_j \left( \sum_{\text{all } k} |a_{jk}| \right),$$

where $\rho(A)$ denotes the *spectral radius* (magnitude of the largest eigenvalue) of $A$. The *spectrum* of $A$ is the set of eigenvalues of $A$. For a stochastic matrix $P$, $\|P\|_\infty = 1$, and therefore we may conclude that

$$\rho(P) \leq 1.$$

Hence, no eigenvalue of a stochastic matrix $P$ can exceed 1 in modulus.

Notice that this property, together with Property 10.1.1, implies that the spectral radius of a stochastic matrix is 1, i.e.,

$$\rho(P) = 1.$$

**Property 10.1.3** *The right-hand eigenvector corresponding to a unit eigenvalue $\lambda_1 = 1$ of a stochastic matrix $P$ is given by $e$ where $e = (1, 1, \ldots)^T$.*

**Proof:** Since the sum of each row of $P$ is 1, we have $Pe = e = \lambda_1 e$.

**Property 10.1.4** *The vector $\pi$ is a stationary probability vector of a stochastic matrix $P$ iff it is a left-hand eigenvector corresponding to a unit eigenvalue.*

**Proof:** By definition, a stationary probability vector $\pi$, does not change when post-multiplied by a stochastic transition probability matrix $P$, which implies that

$$\pi = \pi P.$$

Therefore $\pi$ satisfies the eigenvalue equation

$$\pi P = \lambda_1 \pi \quad \text{for } \lambda_1 = 1.$$

The converse is equally obvious.

The following additional properties apply when $P$ is the stochastic matrix of an *irreducible* Markov chain.

**Property 10.1.5** *The stochastic matrix of an irreducible Markov chain possesses a simple unit eigenvalue.*

The theorem of Perron and Frobenius is a powerful theorem which has applicability to irreducible Markov chains [50]. Property 10.1.5 follows directly from the Perron–Frobenius theorem.

**Property 10.1.6** *For any irreducible Markov chain with stochastic transition probability matrix $P$, let*

$$P(\alpha) = I - \alpha(I - P), \tag{10.4}$$

*where $\alpha \in \Re' \equiv (-\infty, \infty)\backslash\{0\}$ (i.e., the real line with zero deleted). Then 1 is a simple eigenvalue of every $P(\alpha)$, and associated with this unit eigenvalue is a uniquely defined positive left-hand eigenvector of unit 1-norm, which is precisely the stationary probability vector $\pi$ of $P$.*

**Proof:** Notice that the spectrum of $P(\alpha)$ is given by $\lambda(\alpha) = 1 - \alpha(1 - \lambda)$ for $\lambda$ in the spectrum of $P$. Furthermore, as can be verified by substituting from Equation (10.4), the left-hand eigenvectors of $P$ and $P(\alpha)$ agree in the sense that

$$x^T P = \lambda x^T \quad \text{if and only if} \quad x^T P(\alpha) = \lambda(\alpha) x^T \quad \text{for all } \alpha.$$

In particular, this means that regardless of whether or not $P(\alpha)$ is a stochastic matrix, $\lambda(\alpha) = 1$ is a simple eigenvalue of $P(\alpha)$ for all $\alpha$, because $\lambda = 1$ is a simple eigenvalue of $P$. Consequently, the entire family $P(\alpha)$ has a unique positive left-hand eigenvector of unit 1-norm associated with $\lambda(\alpha) = 1$, and this eigenvector is precisely the stationary distribution $\pi$ of $P$.

**Example 10.2** Consider the $3 \times 3$ stochastic matrix given by

$$P = \begin{pmatrix} .99911 & .00079 & .00010 \\ .00061 & .99929 & .00010 \\ .00006 & .00004 & .99990 \end{pmatrix}. \tag{10.5}$$

Its eigenvalues are 1.0, .9998, and .9985. The maximum value of $\alpha$ that we can choose so that $P(\alpha)$ is stochastic is $\alpha_1 = 1/.00089$. The resulting stochastic matrix that we obtain is given by

$$P(\alpha_1) = \begin{pmatrix} 0 & .88764 & .11236 \\ .68539 & .20225 & .11236 \\ .06742 & .04494 & .88764 \end{pmatrix},$$

and its eigenvalues are 1.0, .77528, and −.68539. Yet another choice, $\alpha_2 = 10{,}000$, yields a $P(\alpha_2)$ that is not stochastic. We find

$$P(\alpha_2) = \begin{pmatrix} -7.9 & 7.9 & 1.0 \\ 6.1 & -6.1 & 1.0 \\ 0.6 & 0.4 & 0.0 \end{pmatrix}.$$

Its eigenvalues are 1.0, −1.0, and −14.0. The left-hand eigenvector corresponding to the unit eigenvalue for all three matrices is

$$\pi = (.22333,\ .27667,\ .50000)$$

—the stationary probability vector of the original stochastic matrix.

This theorem has important consequences for the convergence of certain iterative methods used to find stationary distributions of discrete- and continuous-time Markov chains. As we shall see later, the rate of convergence of these methods depends upon the separation of the largest (in modulus) and second largest eigenvalues. This theorem may be used to induce a larger separation, as can be seen in Example 10.5 in moving from $P$ to $P(\alpha_1)$, with a resulting faster convergence rate.

### *10.1.3 The Effect of Discretization*

Let us now find the values of $\Delta t > 0$ for which the matrix $P = Q\Delta t + I$ is stochastic.

**Example 10.3** Consider a two-state Markov chain whose infinitesimal generator is

$$Q = \begin{pmatrix} -q_1 & q_1 \\ q_2 & -q_2 \end{pmatrix},$$

with $q_1, q_2 \geq 0$. The transition probability matrix is then

$$P = Q\Delta t + I = \begin{pmatrix} 1 - q_1\Delta t & q_1\Delta t \\ q_2\Delta t & 1 - q_2\Delta t \end{pmatrix}.$$

Obviously the row sums are equal to 1. To ensure that $0 \leq q_1\Delta t \leq 1$ and $0 \leq q_2\Delta t \leq 1$, we require that $0 \leq \Delta t \leq q_1^{-1}$ and $0 \leq \Delta t \leq q_2^{-1}$. Let us assume, without loss of generality, that $q_1 \geq q_2$. Then $0 \leq \Delta t \leq q_1^{-1}$ satisfies both conditions. To ensure that $0 \leq 1 - q_1\Delta t \leq 1$ and $0 \leq 1 - q_2\Delta t \leq 1$ again requires that $\Delta t \leq q_1^{-1}$. Consequently the maximum value that we can assign to $\Delta t$ subject to the condition that $P$ be stochastic is $\Delta t = 1/\max_i |q_i|$.

Similar results hold for a general stochastic matrix $P = Q\Delta t + I$ to be stochastic, given that $Q$ is an infinitesimal generator. As before, for any value of $\Delta t$, the row sums of $P$ are unity, since by definition the row sums of $Q$ are zero. Therefore we must concern ourselves with the values of $\Delta t$ that guarantee the elements of $P$ lie in the interval [0,1]. Let $q$ be the size of the largest off-diagonal element:

$$q = \max_{i,j\ i \neq j}(q_{ij}) \quad \text{and} \quad q_{ij} \geq 0 \quad \text{for all } i, j.$$

Then $0 \leq p_{ij} \leq 1$ holds if $0 \leq q_{ij}\Delta t \leq 1$, which is true if $\Delta t \leq q^{-1}$. Now consider a diagonal element $p_{ii} = q_{ii}\Delta t + 1$. We have

$$0 \leq q_{ii}\Delta t + 1 \leq 1$$

or

$$-1 \leq q_{ii}\Delta t \leq 0.$$

The right-hand inequality holds for all $\Delta t \geq 0$, since $q_{ii}$ is negative. The left-hand inequality $q_{ii}\Delta t \geq -1$ is true if $\Delta t \leq -q_{ii}^{-1}$, i.e., if $\Delta t \leq |q_{ii}|^{-1}$. It follows then that, if $0 \leq \Delta t \leq (\max_i |q_{ii}|)^{-1}$, the matrix $P$ is stochastic. (Since the diagonal elements of $Q$ equal the negated sum of the off-diagonal elements in a row, we have $\max_i |q_{ii}| \geq \max_{i \neq j}(q_{ij})$.) Thus, $\Delta t$ must be less than or equal to the reciprocal of the absolute value of the largest diagonal element of $Q$.

The choice of a suitable value for $\Delta t$ plays a crucial role in some iterative methods for determining the stationary probability vector from Equation (10.3). As we mentioned above, the rate of convergence is intimately related to the magnitude of the eigenvalues of $P$. As a general rule, the closer the magnitudes of the subdominant eigenvalues are to 1, the slower the convergence rate. We would therefore like to maximize the distance between the largest eigenvalue, $\lambda_1 = 1$, and the subdominant eigenvalue (the eigenvalue that in modulus is closest to 1). Notice that, as $\Delta t \to 0$, the eigenvalues of $P$ *all* tend to unity. This would suggest that we choose $\Delta t$ to be as large as possible, subject only to the constraint that $P$ be a stochastic matrix. However, choosing $\Delta t = (\max_i |q_{ii}|)^{-1}$ does not necessarily guarantee the maximum separation of dominant and subdominant eigenvalues. It is simply a good heuristic.

**Example 10.4** Consider the $(2 \times 2)$ case as an example. The eigenvalues of $P$ are the roots of the characteristic equation $|P - \lambda I| = 0$, i.e.,

$$\begin{vmatrix} 1 - q_1 \Delta t - \lambda & q_1 \Delta t \\ q_2 \Delta t & 1 - q_2 \Delta t - \lambda \end{vmatrix} = 0.$$

These roots are $\lambda_1 = 1$ and $\lambda_2 = 1 - \Delta t(q_1 + q_2)$. As $\Delta t \to 0$, $\lambda_2 \to \lambda_1 = 1$. Also notice that the left-hand eigenvector corresponding to the unit eigenvalue $\lambda_1$ is independent of the choice of $\Delta t$. We have

$$(q_2/(q_1 + q_2) \;\; q_1/(q_1 + q_2)) \begin{pmatrix} 1 - q_1 \Delta t & q_1 \Delta t \\ q_2 \Delta t & 1 - q_2 \Delta t \end{pmatrix} = (q_2/(q_1 + q_2) \;\; q_1/(q_1 + q_2)).$$

This eigenvector is, of course, the stationary probability vector of the Markov chain and as such must be independent of $\Delta t$. The parameter $\Delta t$ can only affect the speed at which matrix iterative methods converge to this vector. As we mentioned above, it is often advantageous to choose $\Delta t$ to be as large as possible, subject only to the constraint that the matrix $P$ be a stochastic matrix. Intuitively, we may think that by choosing a large value of $\Delta t$ we are marching more quickly toward the stationary distribution. With small values we essentially take only small steps, and therefore it takes longer to arrive at our destination, the stationary distribution.

## 10.2 Direct Methods for Stationary Distributions

### *10.2.1 Iterative versus Direct Solution Methods*

There are two basic types of solution method in the field of numerical analysis: solution methods that are *iterative* and solution methods that are *direct*. Iterative methods begin with an initial approximation to the solution vector and proceed to modify this approximation in such a way that, at each step or iteration, it becomes closer and closer to the true solution. Eventually, it becomes equal to the true solution. At least that is what we hope. If no initial approximation is known, then a guess is made or an arbitrary initial vector is chosen instead. Sometimes iterative methods fail to converge to the solution. On the other hand, a direct method attempts to go straight to the final solution. A certain number of well defined steps must be taken, at the end of which the solution has been computed. However, all is not that rosy, for sometimes the number of steps that must be taken is prohibitively large and the buildup of rounding error can, in certain cases, be substantial.

Iterative methods of one type or another are by far the most commonly used methods for obtaining the stationary probability vector from either the stochastic transition probability matrix or from the infinitesimal generator. There are several important reasons for this choice. First, an examination of the standard iterative methods shows that the only operation in which the matrices are involved, is their multiplication with one or more vectors—an operation which leaves the transition matrices unaltered. Thus compact storage schemes, which minimize the amount of memory required to store the matrix and which in addition are well suited to matrix multiplication, may be conveniently implemented. Since the matrices involved are usually large and very sparse, the savings made by such schemes can be considerable. With direct equation solving methods, the elimination of one nonzero element of the matrix during the reduction phase often results in the creation of several nonzero elements in positions which previously contained zero. This is called *fill-in* and not only does it make the organization of a compact storage scheme more difficult, since provision must be made for the deletion and the insertion of elements, but in addition, the amount of fill-in can often be so extensive that available memory can be exhausted. A successful direct method must incorporate a means of overcoming these difficulties.

Iterative methods have other advantages. Use may be made of good initial approximations to the solution vector and this is especially beneficial when a series of related experiments is being conducted. In such circumstances the parameters of one experiment often differ only slightly from those of the previous; many will remain unchanged. Consequently, it is to be expected that the solution to the new experiment will be close to that of the previous and it is advantageous to use the previous result as the new initial approximation. If indeed there is little change, we should expect to compute the new result in relatively few iterations.

Also, an iterative process may be halted once a prespecified tolerance criterion has been satisfied, and this may be relatively lax. For example, it may be wasteful to compute the solution of a mathematical model correct to full machine precision when the model itself contains errors of the order of 5–10%. In contrast, a direct method must continue until the final specified operation has been carried out. And lastly, with iterative methods, the matrix is never altered and hence the buildup of rounding error is, to all intents and purposes, nonexistent.

For these reasons, iterative methods have traditionally been preferred to direct methods. However, iterative methods have a major disadvantage in that often they require a very long time to converge to the desired solution. Direct methods have the advantage that an upper bound on the time required to obtain the solution may be determined before the calculation is initiated. More important, for certain classes of problems, direct methods can result in a much more accurate answer being obtained in *less time*. Since iterative method will in general require less memory than direct methods, these latter can only be recommended if they obtain the solution in less time. Unfortunately, it is often difficult to predict when a direct solver will be more efficient than an iterative solver.

In this section we consider direct methods for computing the stationary distribution of Markov chains while in the next section we consider basic iterative methods. Some methods, such as *preconditioned projection methods* may be thought of as a combination of the direct and iterative approaches, but their study is beyond the scope of this text. Readers wishing further information on projection methods should consult one of the standard texts, such as [47, 50].

### *10.2.2 Gaussian Elimination and LU Factorizations*

Direct equation solving methods for obtaining the stationary distribution of Markov chains are applied to the system of equations

$$\pi Q = 0, \ \ \pi \geq 0, \ \ \pi e = 1, \tag{10.6}$$

i.e., a homogeneous system of $n$ linear equations in which the $n$ unknowns $\pi_i, \ i = 1, 2, \ldots n$ are the components of the stationary distribution vector, $\pi$. The vector $e$ is a column vector whose elements are all equal to 1. The system of equations (10.6) has a solution other than the trivial solution ($\pi_i = 0$, for all $i$) if and only if the determinant of the coefficient matrix is zero, i.e., if and only if the coefficient matrix is singular. Since the determinant of a matrix is equal to the product of its eigenvalues and since $Q$ possesses a zero eigenvalue, the singularity of $Q$ and hence the existence of a non-trivial solution, follows. The standard direct approaches for solving systems of linear equations are based on the method of Gaussian elimination (GE) and related $LU$ factorizations. Gaussian elimination is composed of two phases, a reduction phase during which the coefficient matrix is brought to upper triangular form, and a backsubstitution phase which generates the solution from the reduced coefficient matrix. The first (reduction) phase is the computationally expensive part of the algorithm, having an $O(n^3)$ operation count when the matrix is full. The backsubstitution phase has order $O(n^2)$. A detailed discussion of Gaussian elimination and $LU$ factorizations can be found in Appendix B. Readers unfamiliar with the implementation of these methods should consult this appendix before proceeding.

### Gaussian Elimination for Markov Chains

Consider the Gaussian elimination approach for computing the stationary distribution of a finite, irreducible continuous-time Markov chain. In this case the system of equations, $\pi Q = 0$, is homogeneous, the coefficient matrix is singular and there exists a unique solution vector $\pi$. We proceed by means of an example.

**Example 10.5** Suppose we seek to solve

$$(\pi_1, \pi_2, \pi_3, \pi_4) \begin{pmatrix} -4.0 & 1.0 & 2.0 & 1.0 \\ 4.0 & -9.0 & 2.0 & 3.0 \\ 0.0 & 1.0 & -3.0 & 2.0 \\ 0.0 & 0.0 & 5.0 & -5.0 \end{pmatrix} = (0, 0, 0, 0).$$

Although not strictly necessary, we begin by transposing both sides of this equation, thereby putting it into the standard textbook form, $Ax = b$, for implementing Gaussian elimination. We get

$$Ax = \begin{pmatrix} -4.0 & 4.0 & 0.0 & 0.0 \\ 1.0 & -9.0 & 1.0 & 0.0 \\ 2.0 & 2.0 & -3.0 & 5.0 \\ 1.0 & 3.0 & 2.0 & -5.0 \end{pmatrix} \begin{pmatrix} \pi_1 \\ \pi_2 \\ \pi_3 \\ \pi_4 \end{pmatrix} = \begin{pmatrix} 0 \\ 0 \\ 0 \\ 0 \end{pmatrix} = b.$$

We now proceed to perform the reduction phase of Gaussian elimination. The first step is to use the first equation to eliminate the nonzero elements in column 1 that lie below the pivot element, $a_{11} = -4$. This is accomplished by adding multiples of row 1 into rows 2 through $n = 4$. Once this has been done, the next step is to eliminate nonzero elements in column 2 that lie below the pivot element in row 2 (the newly modified $a'_{22} = -8$) by adding multiples of row 2 into rows 3 through $n = 4$. And so the reduction phase of Gaussian elimination continues until all nonzero elements below the diagonal have been eliminated—at which point the reduction phase of the algorithm has been completed. Following the notation introduced in Appendix B, we obtain the *reduction phase:*

$$\begin{array}{r|} \text{Multipliers} \\ 0.25 \\ 0.50 \\ 0.25 \end{array} \begin{pmatrix} -4.0 & 4.0 & 0.0 & 0.0 \\ \hline 1.0 & -9.0 & 1.0 & 0.0 \\ 2.0 & 2.0 & -3.0 & 5.0 \\ 1.0 & 3.0 & 2.0 & -5.0 \end{pmatrix} \Longrightarrow \begin{pmatrix} -4.0 & \vline & 4.0 & 0.0 & 0.0 \\ \hline 0.0 & \vline & -8.0 & 1.0 & 0.0 \\ 0.0 & \vline & 4.0 & -3.0 & 5.0 \\ 0.0 & \vline & 4.0 & 2.0 & -5.0 \end{pmatrix} \begin{pmatrix} \pi_1 \\ \pi_2 \\ \pi_3 \\ \pi_4 \end{pmatrix} = \begin{pmatrix} 0 \\ 0 \\ 0 \\ 0 \end{pmatrix},$$

$$
\begin{array}{l} \text{Multipliers} \\ \\ 0.50 \\ 0.50 \end{array}
\left|
\left(\begin{array}{r|rrr}
-4.0 & 4.0 & 0.0 & 0.0 \\
0.0 & -8.0 & 1.0 & 0.0 \\ \hline
0.0 & 4.0 & -3.0 & 5.0 \\
0.0 & 4.0 & 2.0 & -5.0
\end{array}\right)
\right.
\Longrightarrow
\left(\begin{array}{rr|rr}
-4.0 & 4.0 & 0.0 & 0.0 \\
0.0 & -8.0 & 1.0 & 0.0 \\ \hline
0.0 & 0.0 & -2.5 & 5.0 \\
0.0 & 0.0 & 2.5 & -5.0
\end{array}\right)
\begin{pmatrix} \pi_1 \\ \pi_2 \\ \pi_3 \\ \pi_4 \end{pmatrix}
=
\begin{pmatrix} 0 \\ 0 \\ 0 \\ 0 \end{pmatrix},
$$

$$
\begin{array}{r} \text{Multipliers} \\ \\ \\ 1.0 \end{array}
\left|
\left(\begin{array}{rr|rr}
-4.0 & 4.0 & 0.0 & 0.0 \\
0.0 & -8.0 & 1.0 & 0.0 \\
0.0 & 0.0 & -2.5 & 5.0 \\ \hline
0.0 & 0.0 & 2.5 & -5.0
\end{array}\right)
\right.
\Longrightarrow
\left(\begin{array}{rrr|r}
-4.0 & 4.0 & 0.0 & 0.0 \\
0.0 & -8.0 & 1.0 & 0.0 \\
0.0 & 0.0 & -2.5 & 5.0 \\ \hline
0.0 & 0.0 & 0.0 & 0.0
\end{array}\right)
\begin{pmatrix} \pi_1 \\ \pi_2 \\ \pi_3 \\ \pi_4 \end{pmatrix}
=
\begin{pmatrix} 0 \\ 0 \\ 0 \\ 0 \end{pmatrix}.
$$

At the end of these three steps, the coefficient matrix has the following upper triangular form;

$$
\begin{pmatrix}
-4.0 & 4.0 & 0.0 & 0.0 \\
0.0 & -8.0 & 1.0 & 0.0 \\
0.0 & 0.0 & -2.5 & 5.0 \\
0.0 & 0.0 & 0.0 & 0.0
\end{pmatrix}
\begin{pmatrix} \pi_1 \\ \pi_2 \\ \pi_3 \\ \pi_4 \end{pmatrix}
=
\begin{pmatrix} 0 \\ 0 \\ 0 \\ 0 \end{pmatrix}.
$$

Observe that, since the pivotal elements are the largest in each column, the multipliers do not exceed 1. Explicit pivoting which is used in general systems of linear equation to ensure that multipliers do not exceed 1, is generally not needed for solving Markov chain problems, since the elements along the diagonal are the largest in each column and this property is maintained during the reduction phase.

We make one additional remark concerning the reduction process. In some cases it can be useful to keep the multipliers for future reference. Since elements below the diagonal are set to zero during the reduction, these array positions provide a suitable location into which the multipliers may be stored. In this case, the coefficient matrix ends up as

$$
\begin{pmatrix}
-4.0 & 4.0 & 0.0 & 0.0 \\
0.25 & -8.0 & 1.0 & 0.0 \\
0.50 & 0.5 & -2.5 & 5.0 \\
0.25 & 0.5 & 1.0 & 0.0
\end{pmatrix}
$$

but only the part on and above the diagonal participates in the backsubstitution phase.

We now turn to the backsubstitution phase wherein $\pi_4$ is found from the last equation of the reduced system, $\pi_3$ found from the second last equation and so on until $\pi_1$ is obtained from the first equation. However, the last row contains all zeros (we ignore all multipliers that may have been stored during the reduction step) and does not allow us to compute the value of $\pi_4$. This should not be unexpected, because the coefficient matrix is singular and we can only compute the solution to a multiplicative constant. One possibility is to assign the value 1 to $\pi_4$ and to compute the value of the other $\pi_i$ in terms of this value. Doing so, we obtain

$$
\begin{array}{lrcl}
\text{Equation 4:} & & & \pi_4 = 1, \\
\text{Equation 3:} & -2.5\,\pi_3 + 5 \times 1 = 0 & \Longrightarrow & \pi_3 = 2, \\
\text{Equation 2:} & -8\,\pi_2 + 1 \times 2 = 0 & \Longrightarrow & \pi_2 = 0.25, \\
\text{Equation 1:} & -4\,\pi_1 + 4 \times 0.25 = 0 & \Longrightarrow & \pi_1 = 0.25.
\end{array}
$$

Therefore our computed solution is $(1/4,\ 1/4,\ 2,\ 1)$. However, this is not the stationary distribution vector just yet, since the sum of its elements does not add to 1. The stationary distribution vector is obtained after normalization, i.e., once we divide each element of this vector by the sum of all its components. Since all the components are positive, this is the same as dividing each element by the 1-norm of the vector. In our example, we have

$$
\|\pi\|_1 = |\pi_1| + |\pi_2| + |\pi_3| + |\pi_4| = 3.5,
$$

and the stationary distribution vector, the normalized computed solution, is therefore

$$\pi = \frac{2}{7}(1/4,\ 1/4,\ 2,\ 1).$$

Another way to look on this situation is to observe that the system of equations $\pi Q = 0$ does not tell the whole story. We also know that $\pi e = 1$. The $n$ equations of $\pi Q = 0$ provide only $n-1$ *linearly independent* equations, but together with $\pi e = 1$, we have a complete basis set. For example, it is possible to replace the last equation of the original system with $\pi e = 1$ which eliminates the need for any further normalization. In this case the coefficient matrix becomes nonsingular, the right-hand side becomes nonzero and a unique solution, the stationary probability distribution, is computed. Of course, it is not necessary to replace the last equation of the system by this normalization equation. Indeed, any equation could be replaced. However, this is generally undesirable, for it will entail more numerical computation. For example, if the first equation is replaced, the first row of the coefficient matrix will contain all ones and the right-hand side will be $e_1 = (1, 0, \ldots, 0)^T$. The first consequence of this is that during the reduction phase, the entire sequence of elementary row operations must be performed on the right-hand side vector, $e_1$, whereas if the last equation is replaced, the right-hand side is unaffected by the elementary row operations. The second and more damaging consequence is that substantial fill-in, the situation in which elements of the coefficient matrix that were previously zero become nonzero, will occur since a multiple of the first row, which contains all ones, will be added into higher numbered rows and a cascading effect will undoubtedly occur in all subsequent reduction steps. An equally viable alternative to replacing the last equation with $\pi e = 1$ is to set the last component $\pi_n$ to one, perform the back substitution and then to normalize the solution thus computed, just as we did in Example 10.5.

To summarize, Gaussian elimination for solving the balance equations arising from finite, irreducible, continuous-time Markov chains, consists of the following three-step algorithm, where we have taken $A = Q^T$ (i.e., the element $a_{ij}$ in the algorithm below is the rate of transition from state $j$ into state $i$).

ALGORITHM 10.1: GAUSSIAN ELIMINATION FOR CONTINUOUS-TIME MARKOV CHAINS

1. The reduction step:

   For $i = 1, 2, \ldots, n-1$ :
   $a_{ji} = -a_{ji}/a_{ii}$ for all $j > i$ % multiplier for row $j$
   $a_{jk} = a_{jk} + a_{ji}a_{ik}$ for all $j, k > i$ % reduce using row $i$

2. The backsubstitution step:

   $x_n = 1$ % set last component equal to 1

   For $i = n-1, n-2, \ldots, 1$ :

   $$x_i = -\left[\sum_{j=i+1}^{n} a_{ij}x_j\right] / a_{ii}$$ % backsubstitute to get $x_i$

3. The final normalization step:

   norm $= \sum_{j=1}^{n} x_j$ % sum components

   For $i = 1, 2, \ldots, n$ :

   $\pi_i = x_i/$ norm % component $i$ of stationary probability vector

An alternative to the straightforward application of Gaussian elimination to Markov chains as described above, is the *scaled Gaussian elimination* algorithm. In this approach, each equation of the system $Q^T\pi^T = 0$ is scaled so that the element on each diagonal is equal to $-1$. This is accomplished by dividing each element of a row by the absolute value of its diagonal element, which is just the same as dividing both left and right sides of each equation by the same value.

**Example 10.6** Observe that the two sets of equations

$$\begin{pmatrix} -4.0 & 4.0 & 0.0 & 0.0 \\ 1.0 & -9.0 & 1.0 & 0.0 \\ 2.0 & 2.0 & -3.0 & 5.0 \\ 1.0 & 3.0 & 2.0 & -5.0 \end{pmatrix} \begin{pmatrix} \pi_1 \\ \pi_2 \\ \pi_3 \\ \pi_4 \end{pmatrix} = \begin{pmatrix} 0 \\ 0 \\ 0 \\ 0 \end{pmatrix}$$

and

$$\begin{pmatrix} -1 & 1 & 0 & 0 \\ 1/9 & -1 & 1/9 & 0 \\ 2/3 & 2/3 & -1 & 5/3 \\ 1/5 & 3/5 & 2/5 & -1 \end{pmatrix} \begin{pmatrix} \pi_1 \\ \pi_2 \\ \pi_3 \\ \pi_4 \end{pmatrix} = \begin{pmatrix} 0 \\ 0 \\ 0 \\ 0 \end{pmatrix}$$

are identical from an equation solving point of view.

This has the effect of simplifying the backsubstitution phase. It does not significantly decrease the the number of numerical operations that must be performed, but rather facilitates programming the Gaussian elimination method for Markov chains. The previous three-step algorithm now becomes:

ALGORITHM 10.2: SCALED GAUSSIAN ELIMINATION FOR CONTINUOUS-TIME MARKOV CHAINS

1. The reduction step:

   For $i = 1, 2, \ldots, n-1$:

   $a_{ik} = -a_{ik}/a_{ii}$ for all $k > i$ % scale row $i$

   $a_{jk} = a_{jk} + a_{ji}a_{ik}$ for all $j, k > i$ % reduce using row $i$

2. The backsubstitution step:

   $x_n = 1$ % set last component equal to1

   For $i = n-1, n-2, \ldots, 1$:

   $x_i = \sum_{j=i+1}^{n} a_{ij}x_j$ % backsubstitute to get $x_i$

3. The final normalization step:

   $\text{norm} = \sum_{j=1}^{n} x_j$ % sum components

   For $i = 1, 2, \ldots, n$:

   $\pi_i = x_i/\text{ norm}$ % component $i$of stationary probability vector

Observe that during the reduction step no attempt is made actually to set the diagonal elements to $-1$: it suffices to realize that this must be the case. Nor is any element below the diagonal set to zero, for exactly the same reason. If the multipliers are to be kept, then they may overwrite these subdiagonal positions. At the end of the reduction step, the elements *strictly above* the diagonal of $A$ contain that portion of the upper triangular matrix $U$ needed for the backsubstitution step.

**Example 10.7** Beginning with

$$A = \begin{pmatrix} -4.0 & 4.0 & 0.0 & 0.0 \\ 1.0 & -9.0 & 1.0 & 0.0 \\ 2.0 & 2.0 & -3.0 & 5.0 \\ 1.0 & 3.0 & 2.0 & -5.0 \end{pmatrix},$$

the matrices obtained for each different value of $i = 1, 2, 3$ during the reduction phase are

$$A_1 = \begin{pmatrix} -4.0 & 1.0 & 0.0 & 0.0 \\ 1.0 & -8.0 & 1.0 & 0.0 \\ 2.0 & 4.0 & -3.0 & 5.0 \\ 1.0 & 4.0 & 2.0 & -5.0 \end{pmatrix}, \quad A_2 = \begin{pmatrix} -4.0 & 1.0 & 0.0 & 0.0 \\ 1.0 & -8.0 & 0.125 & 0.0 \\ 2.0 & 4.0 & -2.5 & 5.0 \\ 1.0 & 4.0 & 2.5 & -5.0 \end{pmatrix},$$

$$A_3 = \begin{pmatrix} -4.0 & 1.0 & 0.0 & 0.0 \\ 1.0 & -8.0 & 0.125 & 0.0 \\ 2.0 & 4.0 & -2.5 & 2.0 \\ 1.0 & 4.0 & 2.5 & 0.0 \end{pmatrix}.$$

In this last matrix, only the elements above the diagonal contribute to the backsubstitution step. Given that the diagonal elements are taken equal to $-1$, this means that the appropriate upper triangular matrix is

$$U = \begin{pmatrix} -1.0 & 1.0 & 0.0 & 0.0 \\ 0.0 & -1.0 & 0.125 & 0.0 \\ 0.0 & 0.0 & -1.0 & 2.0 \\ 0.0 & 0.0 & 0.0 & 0.0 \end{pmatrix}.$$

The backsubstitution step, beginning with $x_4 = 1$, successively gives $x_3 = 2 \times 1 = 2$, $x_2 = 0.125 \times 2 = 0.25$ and $x_1 = 1.0 \times 0.25 = 0.25$ which when normalized gives exactly the same result as before.

### *LU* Factorizations for Markov Chains

When the coefficient matrix in a system of linear equations $Ax = b$ can be written as the product of a lower triangular matrix $L$ and an upper triangular matrix $U$, then

$$Ax = LUx = b$$

and the solution can be found by first solving (by forward substitution) $Lz = b$ for an intermediate vector $z$ and then solving (by backward substitution) $Ux = z$ for the solution $x$. The upper triangular matrix obtained by the reduction phase of Gaussian elimination provides an upper triangular matrix $U$ to go along with a lower triangular matrix $L$ whose diagonal elements are all equal to 1 and whose subdiagonal elements are the multipliers with a minus sign in front of them. In the Markov chain context, the system of equations is homogeneous and the coefficient matrix is singular,

$$Q^T x = (LU)x = 0.$$

If we now set $Ux = z$ and attempt to solve $Lz = 0$, we find that, since $L$ is nonsingular (it is a triangular matrix whose diagonal elements are all equal to 1) we must have $z = 0$. This means that we may proceed directly to the back substitution on $Ux = z = 0$ with $u_{nn} = 0$. It is evident that we may assign any nonzero value to $x_n$, say $x_n = \eta$, and then determine, by simple backsubstitution, the remaining elements of the vector $x$ in terms of $\eta$. We have $x_i = c_i\eta$ for some constants $c_i$, $i = 1, 2, ..., n$, and $c_n = 1$. Thus the solution obtained depends on the value of $\eta$.

There still remains one equation that the elements of a probability vector must satisfy, namely that the sum of the probabilities must be 1. Normalizing the solution obtained from solving $Ux = 0$ yields the desired unique stationary probability vector $\pi$.

**Example 10.8** With the matrix $Q^T$ of Example 10.5, given by

$$Q^T = \begin{pmatrix} -4.0 & 4.0 & 0.0 & 0.0 \\ 1.0 & -9.0 & 1.0 & 0.0 \\ 2.0 & 2.0 & -3.0 & 5.0 \\ 1.0 & 3.0 & 2.0 & -5.0 \end{pmatrix},$$

we find the following $LU$ decomposition:

$$L = \begin{pmatrix} 1.00 & 0.0 & 0.0 & 0.0 \\ -0.25 & 1.0 & 0.0 & 0.0 \\ -0.50 & -0.5 & 1.0 & 0.0 \\ -0.25 & -0.5 & -1.0 & 1.0 \end{pmatrix}, \quad U = \begin{pmatrix} -4.0 & 4.0 & 0.0 & 0.0 \\ 0.0 & -8.0 & 1.0 & 0.0 \\ 0.0 & 0.0 & -2.5 & 5.0 \\ 0.0 & 0.0 & 0.0 & 0.0 \end{pmatrix},$$

whose product $LU$ is equal to $Q^T$. The upper triangular matrix $U$ is just the reduced matrix given by Gaussian elimination while the lower triangular matrix $L$ is the identity matrix with the negated multipliers placed below the diagonal. Forward substitution on $Lz = 0$ successively gives $z_1 = 0$, $z_2 = 0$, $z_3 = 0$, and $z_4 = 0$. Since we know in advance that this must be the case, there is no real need to carry out these operations, but instead we may go straight to the backsubstitution stage to compute $x$ from $Ux = z = 0$. The sequence of operations is identical to that of the backsubstitution phase of Example 10.5.

The above discussion concerns *finite, irreducible* Markov chains. If the Markov chain is decomposable into $k$ irreducible closed communicating classes, these $k$ irreducible components may be solved independently as separate irreducible Markov chains. If Gaussian elimination is applied to the entire transition rate matrix of a Markov chain with $k > 1$ separable components, the reduction phase will lead to a situation in which there are $k$ rows whose elements are all equal to zero. This is because such a matrix possesses $k$ eigenvalues all equal to zero; only $n - k$ of the equations are linearly independent. Although our algorithms can be adjusted to take this situation into account, it is much easier to treat each of the $k$ components separately.

An issue of concern in the implementation of direct methods is that of the data structure used to hold the coefficient matrix. Frequently the matrices generated from Markov models are too large to permit regular two-dimensional arrays to be used to store them in computer memory. Since these matrices are usually very sparse, it is economical, and indeed necessary, to use some sort of packing scheme whereby only the nonzero elements and their positions in the matrix are stored. The most suitable candidates for solution by direct methods are Markov chains whose transition matrix is small, of the order of a few hundred, or when it is *banded*, i.e., the only nonzero elements of the coefficient matrix are not too far from the diagonal. In this latter case, it means that an ordering can be imposed on the states such that no single step transition from state $i$ will take it to states numbered greater than $i + \delta$ or less than $i - \delta$. All fill-in will occur within a distance $\delta$ of the diagonal, and the amount of computation per step is proportional to $\delta^2$.

Direct methods are generally not recommended when the transition matrix is large and *not* banded, due to the amount of fill-in that can quickly overwhelm available storage capacity. When the coefficient matrix is generated row by row the following approach allows matrices of the order of several thousand to be handled by a direct method. The first row is generated and stored in a compacted form, i.e., only the nonzero element and its position in the row is stored. Immediately after the second row has been obtained, it is possible to eliminate the element in position (2,1) by adding a multiple of the first row to it. Row 2 may now be compacted and stored. This process may be continued so that when the i-th row of the coefficient matrix is generated, rows 1 through

$(i-1)$ have been derived, reduced to upper triangular form, compacted and stored. The first $(i-1)$ rows may therefore be used to eliminate all nonzero elements in row $i$ from column positions $(i, 1)$ through $(i, i-1)$, thus putting it into the desired triangular form. Note that since this reduction is performed on $Q^T$, it is the columns of the infinitesimal generator that are required to be generated one at a time and not its rows.

This method has a distinct advantage in that once a row has been generated in this fashion, no more fill-in will occur into this row. It is suggested that a separate storage area be reserved to hold temporarily a single unreduced row. The reduction is performed in this storage area. Once completed, the reduced row may be compacted into any convenient form and appended to the rows which have already been reduced. In this way no storage space is wasted holding subdiagonal elements which, due to elimination, have become zero, nor in reserving space for the inclusion of additional elements. The storage scheme should be chosen bearing in mind the fact that these rows will be used in the reduction of further rows and also later in the algorithm during the back-substitution phase.

This approach cannot be used for solving general systems of linear equations because it inhibits a pivoting strategy from being implemented. It is valid when solving irreducible Markov chains since pivoting is not required in order for the reduction phase to be performed in a stable manner.

### The Grassmann–Taksar–Heyman Advantage

It is appropriate at this point to mention a version of Gaussian elimination that has attributes that appear to make it even more stable than the usual version. This procedure is commonly referred to as the GTH (Grassmann–Taksar–Heyman) algorithm. In GTH the diagonal elements are obtained by summing off-diagonal elements rather than performing a subtraction: it is known that subtractions can sometimes lead to loss of significance in numerical computations. These subtractions occur in forming the diagonal elements during the reduction process. Happily, it turns out that at the end of each reduction step, the unreduced portion of the matrix is the transpose of a transition rate matrix in its own right. This has a probabilistic interpretation based on the restriction of the Markov chain to a reduced set of states and it is in this context that the algorithm is generally developed. It also means that the diagonal elements may be formed by adding off-diagonal elements and placing a minus sign in front of this sum instead of performing a single subtraction. When in addition, the concept of scaling is also introduced into the GTH algorithm the need to actually form the diagonal elements disappears (all are taken to be equal to $-1$). In this case, the scale factor is obtained by taking the reciprocal of the sum of off-diagonal elements.

**Example 10.9** Let us return to Example 10.7 and the matrices obtained at each step of Gaussian elimination with scaling.

$$\left(\begin{array}{|cccc} \mathbf{-4.0} & 4.0 & 0.0 & 0.0 \\ \mathbf{1.0} & -9.0 & 1.0 & 0.0 \\ \mathbf{2.0} & 2.0 & -3.0 & 5.0 \\ \mathbf{1.0} & 3.0 & 2.0 & -5.0 \end{array}\right) \longrightarrow \left(\begin{array}{c|ccc} -4.0 & 1.0 & 0.0 & 0.0 \\ \hline 1.0 & \mathbf{-8.0} & 1.0 & 0.0 \\ 2.0 & \mathbf{4.0} & -3.0 & 5.0 \\ 1.0 & \mathbf{4.0} & 2.0 & -5.0 \end{array}\right)$$

$$\longrightarrow \left(\begin{array}{cc|cc} -4.0 & 1.0 & 0.0 & 0.0 \\ 1.0 & -8.0 & 0.125 & 0.0 \\ \hline 2.0 & 4.0 & \mathbf{-2.5} & 5.0 \\ 1.0 & 4.0 & \mathbf{2.5} & -5.0 \end{array}\right) \longrightarrow \left(\begin{array}{ccc|c} -4.0 & 1.0 & 0.0 & 0.0 \\ 1.0 & -8.0 & 0.125 & 0.0 \\ 2.0 & 4.0 & -2.5 & 2.0 \\ \hline 1.0 & 4.0 & 2.5 & \mathbf{0.0} \end{array}\right).$$

Observe that the submatrix contained in each lower right-hand block is the transpose of a transition rate matrix with one less state than its predecessor. Consider the first reduction step, whereby elements in positions $a_{21}$, $a_{31}$, and $a_{41}$ are to be eliminated. Off-diagonal elements in row 1 are scaled by dividing each by the sum $a_{21} + a_{31} + a_{41} = 1 + 2 + 1$, although for this first row, it is just as easy to use the absolute value of the diagonal element. After the first row is scaled, it must be added into the second row to eliminate the $a_{21}$ element. While this should result in the subtraction $-9+1$ into position $a_{22}$, we choose instead to ignore this particular operation and leave the $a_{22}$ element equal to $-9$. Throughout the entire reduction process, the diagonal elements are left unaltered. Once the elements $a_{21}$, $a_{31}$ and $a_{41}$ have been eliminated, we are ready to start the second reduction step, which begins by forming the next scale factor as $a_{32}+a_{42}=4+4$, the sum of elements below $a_{22}$. The process continues in this fashion, first computing the scale factor by summing below the current diagonal element, then scaling all elements to the right of the current diagonal element and and finally by carrying out the reduction step. These steps are shown below.

$$\left(\begin{array}{|cccc}\hline \mathbf{-4.0} & 4.0 & 0.0 & 0.0\\ \mathbf{1.0} & -9.0 & 1.0 & 0.0\\ \mathbf{2.0} & 2.0 & -3.0 & 5.0\\ \mathbf{1.0} & 3.0 & 2.0 & -5.0\end{array}\right) \begin{array}{c}\text{scale}\\ \longrightarrow\end{array} \left(\begin{array}{|cccc}\mathbf{-4.0} & 1.0 & 0.0 & 0.0\\ \hline \mathbf{1.0} & -9.0 & 1.0 & 0.0\\ \mathbf{2.0} & 2.0 & -3.0 & 5.0\\ \mathbf{1.0} & 3.0 & 2.0 & -5.0\end{array}\right)$$

$$\begin{array}{c}\text{reduce}\\ \longrightarrow\end{array} \left(\begin{array}{c|ccc}-4.0 & 1.0 & 0.0 & 0.0\\ \hline 1.0 & \mathbf{-9.0} & 1.0 & 0.0\\ 2.0 & \mathbf{4.0} & -3.0 & 5.0\\ 1.0 & \mathbf{4.0} & 2.0 & -5.0\end{array}\right) \begin{array}{c}\text{scale}\\ \longrightarrow\end{array} \left(\begin{array}{c|ccc}-4.0 & 1.0 & 0.0 & 0.0\\ 1.0 & \mathbf{-9.0} & 0.125 & 0.0\\ \hline 2.0 & \mathbf{4.0} & -3.0 & 5.0\\ 1.0 & \mathbf{4.0} & 2.0 & -5.0\end{array}\right)$$

$$\begin{array}{c}\text{reduce}\\ \longrightarrow\end{array} \left(\begin{array}{cc|cc}-4.0 & 1.0 & 0.0 & 0.0\\ 1.0 & -9.0 & 0.125 & 0.0\\ \hline 2.0 & 4.0 & \mathbf{-3.0} & 5.0\\ 1.0 & 4.0 & \mathbf{2.5} & -5.0\end{array}\right) \begin{array}{c}\text{scale}\\ \longrightarrow\end{array} \left(\begin{array}{cc|cc}-4.0 & 1.0 & 0.0 & 0.0\\ 1.0 & -9.0 & 0.125 & 0.0\\ 2.0 & 4.0 & -3.0 & 2.0\\ \hline 1.0 & 4.0 & 2.5 & \mathbf{-5.0}\end{array}\right).$$

ALGORITHM 10.3: GTH FOR CONTINUOUS-TIME MARKOV CHAINS WITH $A=Q^T$

1. The reduction step:

   For $i=1,2,\ldots,n-1$:
   $a_{ik}=a_{ik}/\sum_{j=i+1}^{n}a_{ji}$ for all $k>i$ % scale row $i$
   $a_{jk}=a_{jk}+a_{ji}a_{ik}$ for all $j,k>i,\ k\neq j$ % reduce using row $i$

2. The backsubstitution step:

   $x_n=1$ % set last component equal to1
   For $i=n-1,n-2,\ldots,1$:
   $x_i=\sum_{j=i+1}^{n}a_{ij}x_j$ % backsubstitute to get $x_i$

3. The final normalization step:

   $\text{norm}=\sum_{j=1}^{n}x_j$ %sum components

   For $i=1,2,\ldots,n$:

   $\pi_i=x_i/\text{norm}$ % component $i$ of stationary probability vector

Comparing this algorithm with the scaled Gaussian elimination algorithm, we see that only the first step has changed, and within that step only in the computation of the scale factor. The GTH implementation requires more numerical operations than the standard implementation but this may be offset by a gain in precision when the matrix $Q$ is ill conditioned. The extra additions are not very costly when compared with the overall cost of the elimination procedure, which leads to the conclusion that the GTH advantage should be exploited where possible in elimination procedures.

If the transition rate matrix is stored in a two-dimensional or band storage structure, access is easily available to both the rows and columns of $Q$, and there is no difficulty in implementing GTH. Unfortunately, the application of GTH is not quite so simple when the coefficient matrix is stored in certain compacted representations.

### *Matlab code for Gaussian Elimination*

```
function [pi] = GE(Q)

A = Q';
n = size(A);
for i=1:n-1
    for j=i+1:n
        A(j,i) = -A(j,i)/A(i,i);
    end
    for j=i+1:n
        for k=i+1:n
            A(j,k) = A(j,k) + A(j,i)*A(i,k);
        end
    end
end

x(n) = 1;
for i=n-1:-1:1
    for j=i+1:n
        x(i) = x(i) + A(i,j)*x(j);
    end
    x(i) = -x(i)/A(i,i);
end

pi = x/norm(x,1);
```

### *Matlab code for Scaled Gaussian Elimination*

```
function [pi] = ScaledGE(Q)

A = Q';
n = size(A);
for i=1:n-1
   for k=i+1:n
      A(i,k) = -A(i,k)/A(i,i);
   end
   for j=i+1:n
      for k=i+1:n
         A(j,k) = A(j,k) + A(j,i)*A(i,k);
      end
   end
end

x(n) = 1;
for i=n-1:-1:1
```

```
        for j=i+1:n
            x(i) = x(i) + A(i,j)*x(j);
        end
    end

    pi = x/norm(x,1);
```

*Matlab code for GTH*

```
function [pi] = GTH(Q)

A = Q';
n = size(A);
for i=1:n-1
    scale = sum(A(i+1:n,i));
    for k=i+1:n
        A(i,k) = A(i,k)/scale;
    end
    for j=i+1:n
        for k=i+1:n
            A(j,k) = A(j,k) + A(j,i)*A(i,k);
        end
    end
end

x(n) = 1;
for i=n-1:-1:1
    for j=i+1:n
        x(i) = x(i) + A(i,j)*x(j);
    end
end

pi = x/norm(x,1);
```

## 10.3 Basic Iterative Methods for Stationary Distributions

Iterative methods for solving systems of equations begin with some approximation to, or guess at, the solution, and successively apply numerical operations designed to make this approximation approach the true solution. The coefficient matrix is not altered during the execution of the algorithm which makes iterative methods well suited to compacted storage schemes. An item of constant concern with iterative methods is their rate of convergence, the speed at which the initial approximation approaches the solution.

### *10.3.1 The Power Method*

Perhaps the approach that first comes to mind when we need to find the stationary distribution of an finite, ergodic, discrete-time Markov chain, is to let the chain evolve over time, step by step, until it reaches its stationary distribution. Once the probability vector no longer changes as the process evolves from some step $n$ to step $n+1$, that vector can be taken as the stationary probability vector, since at that point we have $zP = z$.

**Example 10.10** Consider a discrete-time Markov chain whose matrix of transition probabilities is

$$P = \begin{pmatrix} .0 & .8 & .2 \\ .0 & .1 & .9 \\ .6 & .0 & .4 \end{pmatrix}. \tag{10.7}$$

If the system starts in state 1, the initial probability vector is given by

$$\pi^{(0)} = (1,\ 0,\ 0).$$

Immediately after the first transition, the system will be either in state 2, with probability .8, or in state 3, with probability .2. The vector $\pi^{(1)}$, which denotes the probability distribution after one transition (or one step) is thus

$$\pi^{(1)} = (0,\ .8,\ .2).$$

Notice that this result may be obtained by forming the product $\pi^{(0)}P$.

The probability of being in state 1 after two time steps is obtained by summing (over all $i$) the probability of being in state $i$ after one step, given by $\pi_i^{(1)}$, multiplied by the probability of making a transition from state $i$ to state 1. We have

$$\sum_{i=1}^{3} \pi_i^{(1)} p_{i1} = \pi_1^{(1)} \times .0 + \pi_2^{(1)} \times .0 + \pi_3^{(1)} \times .6 = .12.$$

Likewise, the system will be in state 2 after two steps with probability .08 (= .0×.8+.8×.1+.2×.0), and in state 3 with probability .8 (= .0 × .2 + .8 × .9 + .2 × .4). Thus, given that the system begins in state 1, we have the following probability distribution after two steps:

$$\pi^{(2)} = (.12,\ .08,\ .8).$$

Notice once again that $\pi^{(2)}$ may be obtained by forming the product $\pi^{(1)}P$:

$$\pi^{(2)} = (.12,\ .08,\ .8) = (0.0,\ 0.8,\ 0.2) \begin{pmatrix} .0 & .8 & .2 \\ .0 & .1 & .9 \\ .6 & .0 & .4 \end{pmatrix} = \pi^{(1)}P.$$

We may continue in this fashion, computing the probability distribution after each transition step. For any integer $k$, the state of the system after $k$ transitions is obtained by multiplying the probability vector obtained after $(k-1)$ transitions by $P$. Thus

$$\pi^{(k)} = \pi^{(k-1)}P = \pi^{(k-2)}P^2 = \cdots = \pi^{(0)}P^k.$$

At step $k = 25$, we find the probability distribution to be

$$\pi = (.2813,\ .2500,\ .4688),$$

and thereafter, correct to four decimal places,

$$(.2813,\ .2500,\ .4688) \begin{pmatrix} .0 & .8 & .2 \\ .0 & .1 & .9 \\ .6 & .0 & .4 \end{pmatrix} = (.2813,\ .2500,\ .4688),$$

which we may now take to be the stationary distribution (correct to four decimal places).

When the Markov chain is finite, aperiodic, and irreducible (as in Example 10.10), the vectors $\pi^{(k)}$ converge to the stationary probability vector $\pi$ regardless of the choice of initial vector. We have

$$\lim_{k\to\infty} \pi^{(k)} = \pi.$$

This method of determining the stationary probability vector is referred to as the *power method* or *power iteration*. The power method is well known in the context of determining the right-hand eigenvector corresponding to a dominant eigenvalue of a matrix, $A$, and it is in this context that we shall examine its convergence properties. However, recall that the stationary distribution of a Markov chain is obtained from the *left-hand* eigenvector, so that in a Markov chain context, the matrix $A$ must be replaced with $P^T$, the transpose of the transition probability matrix. Let $A$ be a square matrix of order $n$. The power method is described by the iterative procedure

$$z^{(k+1)} = \frac{1}{\xi_k} A z^{(k)}, \tag{10.8}$$

where $\xi_k$ is a normalizing factor, typically $\xi_k = \|Az^{(k)}\|_\infty$, and $z^{(0)}$ is an arbitrary starting vector. Although this formulation of the power method incorporates a normalization at *each* iteration, whereby each element of the newly formed iterate is divided by $\xi_k$, this is not strictly necessary. Normalization may be performed less frequently.

To examine the rate of convergence of the power method, let $A$ have eigensolution

$$Ax_i = \lambda_i x_i, \qquad i = 1, 2, \ldots, n,$$

and suppose that

$$|\lambda_1| > |\lambda_2| \geq |\lambda_3| \geq \cdots \geq |\lambda_n|.$$

Let us further assume that the initial vector may be written as a linear combination of the eigenvectors of $A$, i.e.,

$$z^{(0)} = \sum_{i=1}^{n} \alpha_i x_i.$$

The rate of convergence of the power method may then be determined from the relationship

$$z^{(k)} = A^k z^{(0)} = A^k \sum_{i=1}^{n} \alpha_i x_i = \sum_{i=1}^{n} \alpha_i A^k x_i = \sum_{i=1}^{n} \alpha_i \lambda_i^k x_i = \lambda_1^k \left\{ \alpha_1 x_1 + \sum_{i=2}^{n} \alpha_i \left(\frac{\lambda_i}{\lambda_1}\right)^k x_i \right\}. \tag{10.9}$$

It may be observed that the process converges to the dominant eigenvector $x_1$. The rate of convergence depends on the ratios $|\lambda_i|/|\lambda_1|$ for $i = 2, 3, \ldots, n$. The smaller these ratios, the quicker the summation on the right-hand side tends to zero. It is, in particular, the magnitude of the subdominant eigenvalue, $\lambda_2$, that determines the convergence rate. The power method will not perform satisfactorily when $|\lambda_2| \approx |\lambda_1|$. Obviously major difficulties arise when $|\lambda_2| = |\lambda_1|$.

**Example 10.11** Returning to the $3 \times 3$ example given by Equation (10.7), the eigenvalues of $P$ are $\lambda_1 = 1$ and $\lambda_{2,3} = -.25 \pm .5979i$. Thus $|\lambda_2| \approx .65$. Notice that $.65^{10} \approx .01$, $.65^{25} \approx 2 \times 10^{-5}$, and $.65^{100} \approx 2 \times 10^{-19}$. Table 10.1 presents the probability distribution of the states of this example at specific steps, for each of three different starting configurations. After 25 iterations, no further changes are observed in the first four digits for any of the starting configurations. The table shows that approximately two decimal places of accuracy have been obtained after 10 iterations and four places after 25 iterations, which coincides with convergence guidelines obtained from the subdominant eigenvalue. Furthermore, after 100 iterations it is found that the solution is accurate to full machine precision, again as suggested by the value of $.65^{100}$.

We hasten to point out that the magnitude of $|\lambda_2|^k$ does not *guarantee* a certain number of decimal places of accuracy in the solution, as might be construed from the preceding example. As a general rule, the number of decimal places accuracy is determined from a *relative* error norm; a relative error norm of $10^{-j}$ yielding approximately $j$ decimal places of accuracy. However, some flexibility in the Markov chain context may be appropriate since both matrix and vectors have unit 1-norms.

Table 10.1. Convergence in power method.

| Step | Initial state | | | Initial state | | | Initial state | | |
|---|---|---|---|---|---|---|---|---|---|
| | 1.0 | .0 | .0 | .0 | 1.0 | .0 | .0 | .0 | 1.0 |
| 1 | .0000 | .8000 | .2000 | .0000 | .1000 | .9000 | .6000 | .0000 | .4000 |
| 2 | .1200 | .0800 | .8000 | .5400 | .0100 | .4500 | .2400 | .4800 | .2800 |
| 3 | .4800 | .1040 | .4160 | .2700 | .4330 | .2970 | .1680 | .2400 | .5920 |
| 4 | .2496 | .3944 | .3560 | .1782 | .2593 | .5626 | .3552 | .1584 | .4864 |
| ⋮ | ⋮ | ⋮ | ⋮ | ⋮ | ⋮ | ⋮ | ⋮ | ⋮ | ⋮ |
| 10 | .2860 | .2555 | .4584 | .2731 | .2573 | .4696 | .2827 | .2428 | .4745 |
| ⋮ | ⋮ | ⋮ | ⋮ | ⋮ | ⋮ | ⋮ | ⋮ | ⋮ | ⋮ |
| 25 | .2813 | .2500 | .4688 | .2813 | .2500 | .4688 | .2813 | .2500 | .4688 |

Although omitted from Equation (10.9), in the general formulation of the power method, it is usually necessary to normalize successive iterates, since otherwise the term $\lambda_1^k$ may cause successive approximations to become too large (if $\lambda_1 > 1$) or too small (if $\lambda_1 < 1$) and may result in overflow or underflow. Additionally, this normalization is required to provide a standardized vector with which to implement convergence testing. However, in Markov chain problems, the coefficient matrix has 1 as a dominant eigenvalue ($\lambda_1 = 1$) and the requirement for periodic normalization of iterates in the power method disappears. Indeed, if the initial starting approximation is a probability vector, all successive approximations will also be probability vectors. As we noted earlier, when the power method is applied in the Markov chain context, it is the left-hand eigenvector corresponding to a unit eigenvalue that is required and so the matrix to which the method is applied is $P^T$ and the above iteration, Equation (10.8), takes the form

$$z^{(k+1)} = P^T z^{(k)}. \tag{10.10}$$

It is known that the unit eigenvalue of a stochastic matrix is a dominant eigenvalue and that if the matrix is irreducible, there are no other unit eigenvalues. When the matrix is periodic, however, there exist other eigenvalues on the unit circle, which are different from 1 but whose modulus is equal to 1. A straightforward application of the power method in this case will fail. This situation may be circumvented by a slight modification that leaves the unit eigenvalue and its corresponding eigenvector unchanged. The matrix $P$ is usually obtained from the infinitesimal generator by means of the relationship

$$P = (Q\Delta t + I),$$

where $\Delta t \leq 1/\max_i |q_{ii}|$. If $\Delta t$ is chosen so that $\Delta t < 1/\max_i |q_{ii}|$, the resulting stochastic matrix has diagonal elements $p_{ii} > 0$ and therefore cannot be periodic. Under these conditions (irreducible and aperiodic), the power method can be guaranteed to converge. Its rate of convergence is governed by the ratio $|\lambda_2|/|\lambda_1|$, i.e., by $|\lambda_2|$.

Unfortunately, the difference between theoretical conditions for the convergence of an iterative method and its observed behavior in practical situations can be quite drastic. What in theory will converge may take such a large number of iterations that for all practical purposes the method should be considered unworkable. This occurs in the power method when the modulus of the subdominant eigenvalue, $|\lambda_2|$, is close to unity. For example, stochastic matrices that are nearly completely decomposable (NCD) arise frequently in modeling physical and mathematical systems; such matrices have subdominant eigenvalues that are necessarily close to 1. In these cases the power method will converge extremely slowly.

Given the fact that it often takes many iterations to achieve convergence, it may be thought that a more economical approach is to repeatedly square the matrix $P$. Let $k = 2^m$ for some integer $m$. Using the basic iterative formula $\pi^{(k)} = \pi^{(k-1)}P$ requires $k$ iterations to obtain $\pi^{(k)}$; each iteration includes a matrix-vector product. Repeatedly squaring the matrix requires only $m$ matrix products to determine $P^{2^m}$, from which $\pi^{(k)} = \pi^{(0)}P^k$ is quickly computed. It may be further speculated that, since a matrix-vector product requires $n^2$ multiplications and a matrix-matrix product requires $n^3$, that the squaring approach is to be recommended when $mn^3 < 2^m n^2$, i.e., when $nm < 2^m$. Unfortunately, this analysis completely omits the fact that the matrix $P$ is usually large and sparse. Thus, a matrix-vector product requires only $n_z$ multiplications, where $n_z$ is the number of nonzero elements in $P$. The matrix-squaring operation will increase the number of nonzero elements in the matrix (in fact, for an irreducible matrix, convergence will not be attained before all the elements have become nonzero), thereby increasing not only the number of multiplications needed but also the amount of memory needed. It is perhaps memory requirements more than time constraints that limit the applicability of matrix powering.

### 10.3.2 The Iterative Methods of Jacobi and Gauss–Seidel

Iterative solution methods are frequently obtained from the specification of a problem as an equation of the form $f(x) = 0$. The function $f(x)$ may be a linear function, a nonlinear function, or even a system of linear equations, in which case $f(x) = Ax - b$. An iterative method is derived from $f(x) = 0$ by writing it in the form $x = g(x)$ and then constructing the iterative process

$$x^{(k+1)} = g(x^{(k)})$$

with some initial approximation $x^{(0)}$. In other words, the new iterate is obtained by inserting the value at the previous iterate into the right-hand side. The standard and well-known iterative methods for the solution of systems of linear equations are the methods of Jacobi, Gauss–Seidel, and successive overrelaxation (SOR). These methods derive from a nonhomogeneous system of linear equations

$$Ax = b, \quad \text{or, equivalently,} \quad Ax - b = 0,$$

an iterative formula of the form

$$x^{(k+1)} = Hx^{(k)} + c, \qquad k = 0, 1, \ldots. \tag{10.11}$$

This is accomplished by splitting the coefficient matrix $A$. Given a splitting

$$A = M - N$$

with nonsingular $M$, we have

$$(M - N)x = b$$

or

$$Mx = Nx + b,$$

which leads to the iterative procedure

$$x^{(k+1)} = M^{-1}Nx^{(k)} + M^{-1}b = Hx^{(k)} + c, \qquad k = 0, 1, \ldots.$$

The matrix $H = M^{-1}N$ is called the *iteration* matrix and it is the eigenvalues of this matrix that determine the rate of convergence of the iterative method. The methods of Jacobi and Gauss–Seidel differ in their choice $M$ and $N$. We begin with the method of Jacobi.

Consider a nonhomogeneous system of linear equations, $Ax = b$ in which $A \in \Re^{(4\times4)}$ is nonsingular and $b \neq 0$. Writing this in full, we have

$$\begin{aligned} a_{11}x_1 + a_{12}x_2 + a_{13}x_3 + a_{14}x_4 &= b_1,\\ a_{21}x_1 + a_{22}x_2 + a_{23}x_3 + a_{24}x_4 &= b_2,\\ a_{31}x_1 + a_{32}x_2 + a_{33}x_3 + a_{34}x_4 &= b_3,\\ a_{41}x_1 + a_{42}x_2 + a_{43}x_3 + a_{44}x_4 &= b_4. \end{aligned}$$

Bringing all terms with off-diagonal components $a_{ij},\ i \neq j$ to the right-hand side, we obtain

$$\begin{aligned} a_{11}x_1 &= \phantom{-a_{21}x_1} -a_{12}x_2 - a_{13}x_3 - a_{14}x_4 + b_1,\\ a_{22}x_2 &= -a_{21}x_1 \phantom{- a_{32}x_2} - a_{23}x_3 - a_{24}x_4 + b_2,\\ a_{33}x_3 &= -a_{31}x_1 - a_{32}x_2 \phantom{- a_{43}x_3} - a_{34}x_4 + b_3,\\ a_{44}x_4 &= -a_{41}x_1 - a_{42}x_2 - a_{43}x_3 \phantom{- a_{34}x_4} + b_4. \end{aligned}$$

We are now ready to convert this into an iterative procedure. Taking components of $x$ on the right-hand side to be the old values (the values computed at iteration $k$) allows us to assign new values as follows:

$$\begin{aligned} a_{11}x_1^{(k+1)} &= \phantom{-a_{21}x_1^{(k)}} -a_{12}x_2^{(k)} - a_{13}x_3^{(k)} - a_{14}x_4^{(k)} + b_1,\\ a_{22}x_2^{(k+1)} &= -a_{21}x_1^{(k)} \phantom{- a_{32}x_2^{(k)}} - a_{23}x_3^{(k)} - a_{24}x_4^{(k)} + b_2,\\ a_{33}x_3^{(k+1)} &= -a_{31}x_1^{(k)} - a_{32}x_2^{(k)} \phantom{- a_{43}x_3^{(k)}} - a_{34}x_4^{(k)} + b_3,\\ a_{44}x_4^{(k+1)} &= -a_{41}x_1^{(k)} - a_{42}x_2^{(k)} - a_{43}x_3^{(k)} \phantom{- a_{34}x_4^{(k)}} + b_4. \end{aligned} \tag{10.12}$$

This is the *Jacobi* iterative method. In matrix form, $A$ is *split* as $A = D - L - U$ where[2]

- $D$ is a diagonal matrix,
- $L$ is a strictly lower triangular matrix,
- $U$ is a strictly upper triangular matrix,

and so the method of Jacobi becomes equivalent to

$$Dx^{(k+1)} = (L + U)x^{(k)} + b$$

or

$$x^{(k+1)} = D^{-1}(L + U)x^{(k)} + D^{-1}b.$$

Notice that the diagonal matrix $D$ must be nonsingular for this method to be applicable. Thus the method of *Jacobi* corresponds to the splitting $M = D$ and $N = (L + U)$. Its iteration matrix is given by

$$H_J = D^{-1}(L + U).$$

In the Markov chain context, the system of equations whose solution we seek is

$$\pi Q = 0, \quad \text{or, equivalently,} \quad Q^T\pi^T = 0.$$

For notational convenience, set $x = \pi^T$ and let $Q^T = D - (L + U)$. The matrix $D$ is nonsingular, since for all $j$, $d_{jj} \neq 0$ and so $D^{-1}$ exists. Once the $k^{\text{th}}$ approximation $x^{(k)}$ has been formed, the next approximation is obtained by solving the system of equations

$$Dx^{(k+1)} = (L + U)x^{(k)}$$

[2] The matrices $L$ and $U$ should not be confused with the $LU$ factors obtained from direct methods such as Gaussian elimination.

or

$$x^{(k+1)} = D^{-1}(L+U)x^{(k)}.$$

In scalar form,

$$x_i^{(k+1)} = \frac{1}{d_{ii}} \left\{ \sum_{j \neq i} \left( l_{ij} + u_{ij} \right) x_j^{(k)} \right\}, \qquad i = 1, 2, \ldots, n. \tag{10.13}$$

**Example 10.12** Consider a four-state Markov chain with stochastic transition probability matrix

$$P = \begin{pmatrix} .5 & .5 & 0 & 0 \\ 0 & .5 & .5 & 0 \\ 0 & 0 & .5 & .5 \\ .125 & .125 & .25 & .5 \end{pmatrix}.$$

Since we are given $P$ rather than $Q$, we need to write $\pi P = \pi$ as $\pi(P-I) = 0$ and take $Q = P - I$:

$$Q = \begin{pmatrix} -.5 & .5 & 0 & 0 \\ 0 & -.5 & .5 & 0 \\ 0 & 0 & -.5 & .5 \\ .125 & .125 & .25 & -.5 \end{pmatrix}.$$

Transposing this, we obtain the system of equations

$$\begin{pmatrix} -.5 & 0 & 0 & .125 \\ .5 & -.5 & 0 & .125 \\ 0 & .5 & -.5 & .250 \\ 0 & 0 & .5 & -.500 \end{pmatrix} \begin{pmatrix} \pi_1 \\ \pi_2 \\ \pi_3 \\ \pi_4 \end{pmatrix} = \begin{pmatrix} 0 \\ 0 \\ 0 \\ 0 \end{pmatrix}.$$

Writing this in full, we have

$$\begin{aligned} -.5\pi_1 + 0\pi_2 + 0\pi_3 + .125\pi_4 &= 0, \\ .5\pi_1 - .5\pi_2 + 0\pi_3 + .125\pi_4 &= 0, \\ 0\pi_1 + .5\pi_2 - .5\pi_3 + .25\pi_4 &= 0, \\ 0\pi_1 + 0\pi_2 + .5\pi_3 - .5\pi_4 &= 0, \end{aligned}$$

or

$$\begin{aligned} -.5\pi_1 &= & & & - .125\pi_4, \\ -.5\pi_2 &= -.5\pi_1 & & & - .125\pi_4, \\ -.5\pi_3 &= & -.5\pi_2 & & - .25\pi_4, \\ -.5\pi_4 &= & & -.5\pi_3. \end{aligned} \tag{10.14}$$

From this we can write the iterative version,

$$\begin{aligned} -.5\pi_1^{(k+1)} &= & & & - .125\pi_4^{(k)}, \\ -.5\pi_2^{(k+1)} &= -.5\pi_1^{(k)} & & & - .125\pi_4^{(k)}, \\ -.5\pi_3^{(k+1)} &= & -.5\pi_2^{(k)} & & - .25\pi_4^{(k)}, \\ -.5\pi_4^{(k+1)} &= & & -.5\pi_3^{(k)}, \end{aligned}$$

which leads to

$$\pi_1^{(k+1)} = .25\pi_4^{(k)}, \quad \pi_2^{(k+1)} = \pi_1^{(k)} + .25\pi_4^{(k)}, \quad \pi_3^{(k+1)} = \pi_2^{(k)} + .5\pi_4^{(k)}, \quad \pi_4^{(k+1)} = \pi_3^{(k)}.$$

We may now begin the iterative process. Starting with

$$\pi^{(0)} = (.5,\ .25,\ .125,\ .125),$$

we obtain

$$\begin{aligned}
\pi_1^{(1)} &= & & .25\pi_4^{(0)} = .25 \times .125 = .03125,\\
\pi_2^{(1)} &= \pi_1^{(0)} & & + .25\pi_4^{(0)} = .5 + .25 \times .125 = .53125,\\
\pi_3^{(1)} &= & \pi_2^{(0)} & + .5\pi_4^{(0)} = .25 + .5 \times .125 = .31250,\\
\pi_4^{(1)} &= & \pi_3^{(0)} & = .12500.
\end{aligned}$$

In this particular example, the sum of the components of $\pi^{(1)}$ is equal to 1, so a further normalization is not necessary. In fact, since at any iteration $k+1$,

$$\sum_{i=1}^{4} \pi_i^{(k+1)} = .25\pi_4^{(k)} + \pi_1^{(k)} + .25\pi_4^{(k)} + \pi_2^{(k)} + .50\pi_4^{(k)} + \pi_3^{(k)} = \sum_{i=1}^{4} \pi_i^{(k)} = 1,$$

the sum of the components of all the approximations to the stationary distribution will always be equal to 1, provided the initial approximation has components that sum to 1. Continuing with the iterative Jacobi procedure, we obtain the following sequence of approximations:

$$\begin{aligned}
\pi^{(0)} &= (.50000,\ .25000,\ .12500,\ .12500),\\
\pi^{(1)} &= (.03125,\ .53125,\ .31250,\ .12500),\\
\pi^{(2)} &= (.03125,\ .06250,\ .59375,\ .31250),\\
\pi^{(3)} &= (.078125,\ .109375,\ .21875,\ .59375).\\
&\ \ \vdots
\end{aligned}$$

Notice that substitution of $Q^T = D - (L+U)$ into $Q^T x = 0$ gives $(L+U)x = Dx$, and since $D$ is nonsingular, this yields the eigenvalue equation

$$D^{-1}(L+U)x = x, \tag{10.15}$$

in which $x$ is seen to be the right-hand eigenvector corresponding to a unit eigenvalue of the matrix $D^{-1}(L+U)$. This matrix will immediately be recognized as the iteration matrix for the method of Jacobi, $H_J$. That $H_J$ has a unit eigenvalue is obvious from Equation (10.15). Furthermore, from the zero-column-sum property of $Q^T$, we have

$$d_{jj} = \sum_{i=1,\ i\neq j}^{n} (l_{ij} + u_{ij}), \qquad j = 1, 2, \ldots,$$

with $l_{ij},\ u_{ij} \le 0$ for all $i,\ j,\ i \neq j$, and it follows directly from the theorem of Gerschgorin that no eigenvalue of $H_J$ can have modulus greater than unity. This theorem states that the eigenvalues of any square matrix $A$ of order $n$ lie in the union of the $n$ circular disks with centers $c_i = a_{ii}$ and radii $r_i = \sum_{j=1, j\neq i}^{n} |a_{ij}|$. The stationary probability vector $\pi$ is therefore the eigenvector corresponding to a dominant eigenvalue of $H_J$, and the method of Jacobi is identical to the power method applied to the iteration matrix $H_J$.

**Example 10.13** Returning to the previous example, the Jacobi iteration matrix is given by

$$H_J = \begin{pmatrix} -.5 & 0 & 0 & 0\\ 0 & -.5 & 0 & 0\\ 0 & 0 & -.5 & 0\\ 0 & 0 & 0 & -.5 \end{pmatrix}^{-1} \begin{pmatrix} 0 & 0 & 0 & -.125\\ -.5 & 0 & 0 & -.125\\ 0 & -.5 & 0 & -.250\\ 0 & 0 & -.5 & 0 \end{pmatrix} = \begin{pmatrix} 0 & 0 & 0 & .25\\ 1.0 & 0 & 0 & .25\\ 0 & 1.0 & 0 & .50\\ 0 & 0 & 1.0 & 0 \end{pmatrix}.$$

The four eigenvalues of this matrix are

$$\lambda_1 = 1.0, \quad \lambda_2 = -.7718, \quad \lambda_{3,4} = -0.1141 \pm 0.5576i.$$

The modulus of $\lambda_2$ is equal to 0.7718. Observe that $0.7718^{50} = .00000237$, which suggests about five to six places of accuracy after 50 iterations.

We now move on to the method of *Gauss–Seidel*. Usually the computations specified by Equation (10.13) in the method of Jacobi are carried out sequentially; the components of the vector $x^{(k+1)}$ are obtained one after the other as $x_1^{(k+1)}, x_2^{(k+1)}, \ldots, x_n^{(k+1)}$. When evaluating $x_i^{(k+1)}$, only components of the previous iteration $x^{(k)}$ are used, even though elements from the current iteration $x_j^{(k+1)}$ for $j < i$ are available and are (we hope) more accurate. The Gauss–Seidel method makes use of these most recently available component approximations. This may be accomplished by simply overwriting elements as soon as a new approximation is determined.

Referring back to Equation (10.12) and rewriting it using the most recent values, we obtain

$$\begin{aligned}
a_{11}x_1^{(k+1)} &= \qquad\qquad\qquad\; - a_{12}x_2^{(k)} - a_{13}x_3^{(k)} - a_{14}x_4^{(k)} + b_1,\\
a_{22}x_2^{(k+1)} &= -a_{21}x_1^{(k+1)} \qquad\qquad\; - a_{23}x_3^{(k)} - a_{24}x_4^{(k)} + b_2,\\
a_{33}x_3^{(k+1)} &= -a_{31}x_1^{(k+1)} - a_{32}x_2^{(k+1)} \qquad\quad - a_{34}x_4^{(k)} + b_3,\\
a_{44}x_4^{(k+1)} &= -a_{41}x_1^{(k+1)} - a_{42}x_2^{(k+1)} - a_{43}x_3^{(k+1)} \qquad + b_4.
\end{aligned}$$

Observe that in the second equation, the value of the newly computed first component, $x_1$, is used, i.e., we use $x_1^{(k+1)}$ rather than $x_1^{(k)}$. Similarly, in the third equation we use the new values of $x_1$ and $x_2$, and finally in the last equation we use the new values of all components other than the last. With $n$ linear equations in $n$ unknowns, the $i^{\text{th}}$ equation is written as

$$a_{ii}x_i^{(k+1)} = \left(b_i - \sum_{j=1}^{i-1} a_{ij}x_j^{(k+1)} - \sum_{j=i+1}^{n} a_{ij}x_j^{(k)}\right), \quad i = 1, 2, \ldots, n. \tag{10.16}$$

Rearranging these equations so that all new values appear on the left-hand side, we find

$$\begin{aligned}
a_{11}x_1^{(k+1)} &= - a_{12}x_2^{(k)} - a_{13}x_3^{(k)} - a_{14}x_4^{(k)} + b_1,\\
a_{21}x_1^{(k+1)} + a_{22}x_2^{(k+1)} &= - a_{23}x_3^{(k)} - a_{24}x_4^{(k)} + b_2,\\
a_{31}x_1^{(k+1)} + a_{32}x_2^{(k+1)} + a_{33}x_3^{(k+1)} &= - a_{34}x_4^{(k)} + b_3,\\
a_{41}x_1^{(k+1)} + a_{42}x_2^{(k+1)} + a_{43}x_3^{(k+1)} + a_{44}x_4^{(k+1)} &= b_4.
\end{aligned} \tag{10.17}$$

Using the same $D - L - U$ splitting as for Jacobi, the Gauss–Seidel iterative method is equivalent to

$$(D - L)x^{(k+1)} = Ux^{(k)} + b. \tag{10.18}$$

This is just the matrix representation of the system of equations (10.17). It may be written as

$$x^{(k+1)} = D^{-1}(Lx^{(k+1)} + Ux^{(k)} + b)$$

or

$$x^{(k+1)} = (D - L)^{-1}Ux^{(k)} + (D - L)^{-1}b. \tag{10.19}$$

Thus the iteration matrix for the method of Gauss–Seidel is given by

$$H_{GS} = (D - L)^{-1}U.$$

This iterative method corresponds to the splitting $M = (D - L)$ and $N = U$ and is applicable only when the matrix $D - L$ is nonsingular. Gauss–Seidel usually, but not always, converges faster than Jacobi.

For homogeneous systems of equations such as those which arise in Markov chains, the right-hand side is zero and Equation (10.19) becomes

$$x^{(k+1)} = (D - L)^{-1} U x^{(k)}, \quad \text{i.e.,} \quad x^{(k+1)} = H_{GS} x^{(k)}.$$

Furthermore, since the diagonal elements of $D$ are all nonzero, the inverse, $(D - L)^{-1}$, exists. The stationary probability vector $\pi = x^T$ obviously satisfies $H_{GS} x = x$, which shows that $x$ is the right-hand eigenvector corresponding to a unit eigenvalue of $H_{GS}$. As a consequence of the Stein–Rosenberg theorem ([53], p. 70) and the fact that the corresponding Jacobi iteration matrix $H_J$ possesses a dominant unit eigenvalue, the unit eigenvalue of the matrix $H_{GS}$ is a dominant eigenvalue. The method of Gauss–Seidel is therefore identical to the power method applied to $H_{GS}$.

**Example 10.14** We now apply Gauss–Seidel to the problem previously solved by the method of Jacobi:

$$\begin{aligned}
-.5\pi_1 &= & & & &- .125\pi_4,\\
-.5\pi_2 &= -.5\pi_1 & & & &- .125\pi_4,\\
-.5\pi_3 &= & -.5\pi_2 & & &- .25\pi_4,\\
-.5\pi_4 &= & & -.5\pi_3. & &
\end{aligned}$$

In the Gauss–Seidel iterative scheme, this becomes

$$\begin{aligned}
-.5\pi_1^{(k+1)} &= & & & &- .125\pi_4^{(k)},\\
-.5\pi_2^{(k+1)} &= -.5\pi_1^{(k+1)} & & & &- .125\pi_4^{(k)},\\
-.5\pi_3^{(k+1)} &= & -.5\pi_2^{(k+1)} & & &- .25\pi_4^{(k)},\\
-.5\pi_4^{(k+1)} &= & & -.5\pi_3^{(k+1)}, & &
\end{aligned}$$

$$\pi_1^{(k+1)} = .25\pi_4^{(k)}, \quad \pi_2^{(k+1)} = \pi_1^{(k+1)} + .25\pi_4^{(k)}, \quad \pi_3^{(k+1)} = \pi_2^{(k+1)} + .5\pi_4^{(k)}, \quad \pi_4^{(k+1)} = \pi_3^{(k+1)}.$$

We may now begin the iterative process. Starting with

$$\pi^{(0)} = (.5,\ .25,\ .125,\ .125),$$

we obtain

$$\begin{aligned}
\pi_1^{(1)} &= & & .25\pi_4^{(0)} &&= .25 \times .125 = .03125,\\
\pi_2^{(1)} &= \pi_1^{(1)} & & + .25\pi_4^{(0)} &&= .03125 + .25 \times .125 = .06250,\\
\pi_3^{(1)} &= & \pi_2^{(1)} & + .5\pi_4^{(0)} &&= .06250 + .5 \times .125 = .12500,\\
\pi_4^{(1)} &= & \pi_3^{(1)} & &&= .12500.
\end{aligned}$$

Notice that the sum of elements in $\pi^{(1)}$ does not add up to 1, and so it becomes necessary to normalize. We have

$$\|\pi^{(1)}\|_1 = 0.34375,$$

so dividing each element by 0.34375, we obtain

$$\pi^{(1)} = (0.090909,\ 0.181818,\ 0.363636,\ .363636) = \frac{1}{11}(1,\ 2,\ 4,\ 4).$$

By continuing this procedure, the following sequence of approximations is computed:

$$\pi^{(1)} = (0.090909,\ 0.181818,\ 0.363636,\ .363636),$$
$$\pi^{(2)} = (0.090909,\ 0.181818,\ 0.363636,\ .363636),$$
$$\pi^{(3)} = (0.090909,\ 0.181818,\ 0.363636,\ .363636),$$

and so on. For this particular example, Gauss–Seidel converges in only one iteration! To see why this is so, we need to examine the iteration matrix

$$H_{GS} = (D - L)^{-1}U. \tag{10.20}$$

In the example considered above, we have

$$D = \begin{pmatrix} -.5 & 0 & 0 & 0 \\ 0 & -.5 & 0 & 0 \\ 0 & 0 & -.5 & 0 \\ 0 & 0 & 0 & -.5 \end{pmatrix}, \quad L = \begin{pmatrix} 0 & 0 & 0 & 0 \\ -.5 & 0 & 0 & 0 \\ 0 & -.5 & 0 & 0 \\ 0 & 0 & -.5 & 0 \end{pmatrix}, \quad U = \begin{pmatrix} 0 & 0 & 0 & -.125 \\ 0 & 0 & 0 & -.125 \\ 0 & 0 & 0 & -.25 \\ 0 & 0 & 0 & 0 \end{pmatrix},$$

and

$$H_{GS} = (D - L)^{-1}U = \begin{pmatrix} 0 & 0 & 0 & .25 \\ 0 & 0 & 0 & .50 \\ 0 & 0 & 0 & 1.0 \\ 0 & 0 & 0 & 1.0 \end{pmatrix}.$$

Since $U$ has nonzeros only in the last column, it must be the case that $H_{GS}$ has nonzeros only in the last column and hence the only nonzero eigenvalue of $H_{GS}$ must be the last diagonal element. More generally, the eigenvalues of upper or lower triangular matrices are equal to the diagonal elements of the matrix. Since the rate of convergence of the power method when applied to Markov chain problems depends on the magnitude of the subdominant eigenvalue (here equal to 0), convergence must occur immediately after the first iteration which is indeed what we observe in this example.

As indicated in Equation (10.13), the method of Gauss–Seidel corresponds to computing the $i^{\text{th}}$ component of the current approximation from $i = 1$ through $n$, i.e., from top to bottom. To denote specificially the direction of solution, this is sometimes referred to as *forward* Gauss–Seidel. A *backward* Gauss–Seidel iteration takes the form

$$(D - U)x^{(k+1)} = Lx^{(k)}, \quad k = 0, 1, \ldots,$$

and corresponds to computing the components from bottom to top. Forward and backward iterations in a Jacobi setting are meaningless, since in Jacobi only components of the previous iteration are used in the updating procedure. As a general rule of thumb, a forward iterative method is usually recommended when the preponderance of the elemental mass is to be found below the diagonal, for in this case the iterative method essentially works with the inverse of the lower triangular portion of the matrix, $(D - L)^{-1}$, and, intuitively, the closer this is to the inverse of the entire matrix, the faster the convergence. Ideally, in a general context, a splitting should be such that $M$ is chosen as close to $Q^T$ as possible, subject only to the constraint that $M^{-1}$ be easy to find. On the other hand, a backward iterative scheme works with the inverse of the upper triangular portion, $(D - U)^{-1}$, and is generally recommended when most of the nonzero mass lies above the diagonal. However, some examples that run counter to this "intuition" are known to exist.

### *10.3.3 The Method of Successive Overrelaxation*

In many ways, the *successive overrelaxation method* (SOR) resembles the Gauss–Seidel method. When applied to $Ax = b$, a linear system of $n$ equations in $n$ unknowns, the $i^{\text{th}}$ component of the

$(k+1)^{\text{th}}$ iteration is obtained from

$$a_{ii}x_i^{(k+1)} = a_{ii}(1-\omega)x_i^{(k)} + \omega\left(b_i - \sum_{j=1}^{i-1} a_{ij}x_j^{(k+1)} - \sum_{j=i+1}^{n} a_{ij}x_j^{(k)}\right), \quad i = 1, 2, \ldots, n.$$

Observe that the expression within the large parentheses on the right-hand side completely constitutes the method of Gauss–Seidel as defined by Equation (10.16) and that SOR reduces to Gauss–Seidel when $\omega$ is set equal to 1. A *backward* SOR relaxation may also be written. For $\omega > 1$, the process is said to be one of *overrelaxation*; for $\omega < 1$ it is said to be *underrelaxation*. It may be shown that the SOR method converges only if $0 < \omega < 2$. This is a necessary, but not sufficient, condition for convergence.

The choice of an optimal, or even a reasonable, value for $\omega$ has been the subject of much study, especially for problems arising in the numerical solution of partial differential equations. Some results have been obtained for certain classes of matrices but unfortunately, little is known at present about the optimal choice of $\omega$ for arbitrary nonsymmetric linear systems. If a series of related experiments is to be conducted, it may well be worthwhile to carry out some numerical experiments to try to determine a suitable value; some sort of adaptive procedure might be incorporated into the algorithm. For example, it is possible to begin iterating with a value of $\omega = 1$ and, after some iterations have been carried out, to estimate the rate of convergence from the computed approximations. The value of $\omega$ may now be augmented to 1.1, say, and after some further iterations a new estimate of the rate of convergence computed. If this is better than before, $\omega$ should again be augmented, to 1.2, say, and the same procedure used. If the rate of convergence is not as good, the value of $\omega$ should be diminished.

When applied to the homogeneous system $Q^T x = (D-L-U)x = 0$, the SOR method becomes

$$x_i^{(k+1)} = (1-\omega)x_i^{(k)} + \omega\left\{\frac{1}{d_{ii}}\left(\sum_{j=1}^{i-1} l_{ij}x_j^{(k+1)} + \sum_{j=i+1}^{n} u_{ij}x_j^{(k)}\right)\right\}, \quad i = 1, 2, \ldots, n,$$

or in matrix form

$$x^{(k+1)} = (1-\omega)x^{(k)} + \omega\left\{D^{-1}\left(Lx^{(k+1)} + Ux^{(k)}\right)\right\}. \tag{10.21}$$

Rearranging, we find

$$(D-\omega L)x^{(k+1)} = [(1-\omega)D + \omega U]x^{(k)}$$

or

$$x^{(k+1)} = (D-\omega L)^{-1}[(1-\omega)D + \omega U]x^{(k)}, \tag{10.22}$$

and thus the iteration matrix for the SOR method is

$$H_\omega = (D-\omega L)^{-1}[(1-\omega)D + \omega U].$$

It corresponds to the splitting $M = \omega^{-1}[D-\omega L]$ and $N = \omega^{-1}[(1-\omega)D + \omega U]$.

From Equation (10.22), it is evident that the stationary probability vector is the eigenvector corresponding to a unit eigenvalue of the SOR iteration matrix. However, with the SOR method, it is not necessarily true that this unit eigenvalue is the dominant eigenvalue, because the eigenvalues depend on the choice of the relaxation parameter $\omega$. It is possible that $H_\omega$ has eigenvalues which are strictly greater than 1. When the unit eigenvalue is the dominant eigenvalue, then the SOR method is identical to the power method applied to $H_\omega$. The value of $\omega$ that maximizes the difference between this unit eigenvalue and the subdominant eigenvalue of $H_\omega$ is the optimal choice for the relaxation parameter, and the convergence rate achieved with this value of $\omega$ can be a considerable improvement over that of Gauss–Seidel.

To summarize to this point, we have now seen that the power method may be used to obtain $\pi$ from one of four sources: $P^T$, $H_J$, $H_{GS}$, and $H_\omega$. The eigenvalues (with the exception of the unit eigenvalue) will not be the same from one matrix to the next, and sometimes a considerable difference in the number of iterations required to obtain convergence may be observed. Since the computational effort to perform an iteration step is the same in all four cases, it is desirable to apply the power method to the matrix that yields convergence in the smallest number of iterations, i.e., to the matrix whose subdominant eigenvalues are, in modulus, furthest from unity.

The following Matlab code performs a fixed number of iterations of the basic Jacobi, forward Gauss–Seidel, or SOR method on the system of equation $Ax = b$. It accepts an input matrix $A$ (which may be set equal to an infinitesimal generator $Q^T$), an initial approximation $x_0$, a right-hand side vector $b$ (which may be set to zero—in which case, a normalization must also be performed), and the number of iterations to be carried out, *itmax*. It returns the computed solution, *soln*, and a vector containing the residual computed at each iteration, *resid*. It is not designed to be a "production" code, but rather simply a way to generate some results in order to get some intuition into the performance characteristics of these methods.

### *Matlab code for Jacobi/Gauss–Seidel/SOR*

```
function [soln,resid] = gs(A,x0,b,itmax)
% Performs ''itmax'' iterations of Jacobi/Gauss-Seidel/SOR on Ax = b

  [n,n] = size(A); L = zeros(n,n); U = L;  D = diag(diag(A));
  for i = 1:n,
      for j = 1:n,
          if i<j, U(i,j) = -A(i,j); end
          if i>j, L(i,j) = -A(i,j); end
      end
  end
  M = inv(D-L);  B = M*U;          % B is GS iteration matrix
 %M = inv(D);     B = M*(L+U); % Use this for Jacobi
 %w = 1.1; b = w*b; M = inv(D-w*L); B = M*((1-w)*D + w*U) % Use this for SOR

  for iter = 1:itmax,
      soln = B*x0+M*b;
      if norm(b,2) == 0 soln = soln/norm(soln,1); end % Normalize when b=0.
      resid(iter) = norm(A*soln-b,2);
      x0 = soln;
  end
  resid = resid';
  if norm(b,2) == 0 soln = soln/norm(soln,1); end  % Normalize when b = 0.
```

### *10.3.4 Data Structures for Large Sparse Matrices*

We focus next on some algorithmic details that must be taken into account when implementing iterative methods for solving large-scale Markov chains. When the transition matrix is larger than several hundred it becomes impractical to keep it in a two-dimensional array, the format in which we are used to seeing matrices. In most cases the transition matrix is sparse, since each state generally can reach only a small number of states in a single step, and hence there are only a few nonzero elements in each row. In this section we shall consider approaches for storing this matrix efficiently, by taking advantage of its sparsity. One of the major advantages that iterative methods have over direct methods is that no modification of the elements of the transition matrix occurs during the execution of the algorithm. Thus, the matrix may be stored once and for all in some convenient compact form without the need to provide mechanisms to handle insertions (due to zero elements becoming nonzero) and deletions (due to the elimination of nonzero elements). As a constraint, the storage scheme used should not hinder the numerical operations that must be conducted on the matrix. The basic numerical operation performed by the iterative methods we consider, in fact, the only numerical operation performed on the matrix, is its pre- and postmultiplication by a vector, i.e., $z = Ax$ and $z = A^T x$.

One simple approach is to use a real (double-precision) one-dimensional array *aa* to store the nonzero elements of the matrix and two integer arrays *ia* and *ja* to indicate, respectively, the row and column positions of these elements. Thus, if the nonzero element $a_{ij}$ is stored in position $k$ of $aa$, i.e., $aa(k) = a_{ij}$, we have $ia(k) = i$ and $ja(k) = j$.

**Example 10.15** The $(4 \times 4)$ matrix $A$ given by

$$A = \begin{pmatrix} -2.1 & 0.0 & 1.7 & 0.4 \\ 0.8 & -0.8 & 0.0 & 0.0 \\ 0.2 & 1.5 & -1.7 & 0.0 \\ 0.0 & 0.3 & 0.2 & -0.5 \end{pmatrix}$$

may be stored as

| $aa$ : | −2.1 | 1.7 | 0.4 | −0.8 | 0.8 | −1.7 | 0.2 | 1.5 | −0.5 | 0.3 | 0.2 |
|---|---|---|---|---|---|---|---|---|---|---|---|
| $ia$ : | 1 | 1 | 1 | 2 | 2 | 3 | 3 | 3 | 4 | 4 | 4 |
| $ja$ : | 1 | 3 | 4 | 2 | 1 | 3 | 1 | 2 | 4 | 2 | 3 |

or as

| $aa$ : | −2.1 | 0.8 | 0.2 | −0.8 | 1.5 | 0.3 | 1.7 | −1.7 | 0.2 | 0.4 | −0.5 |
|---|---|---|---|---|---|---|---|---|---|---|---|
| $ia$ : | 1 | 2 | 3 | 2 | 3 | 4 | 1 | 3 | 4 | 1 | 4 |
| $ja$ : | 1 | 1 | 1 | 2 | 2 | 2 | 3 | 3 | 3 | 4 | 4 |

or yet again as

| $aa$ : | −2.1 | −0.8 | −1.7 | −0.5 | 1.7 | 0.8 | 0.4 | 0.2 | 0.3 | 1.5 | 0.2 |
|---|---|---|---|---|---|---|---|---|---|---|---|
| $ia$ : | 1 | 2 | 3 | 4 | 1 | 2 | 1 | 4 | 4 | 3 | 3 |
| $ja$ : | 1 | 2 | 3 | 4 | 3 | 1 | 4 | 3 | 2 | 2 | 1 |

and so on. In the first case, the matrix is stored by rows, by columns in the second case, and in a random fashion in the third (although diagonal elements are given first).

Irrespective of the order in which the elements of the matrix $A$ are entered into *aa*, the following algorithm, in which $n_z$ denotes the number of nonzero elements in $A$, computes the product $z = Ax$.

ALGORITHM 10.4: SPARSE MATRIX-VECTOR MULTIPLICATION I

1. Set $z(i) = 0$ for all $i$.
2. For $next = 1$ to $n_z$ do
   - Set $nrow = ia(next)$.
   - Set $ncol = ja(next)$.
   - Compute $z(nrow) = z(nrow) + aa(next) \times x(ncol)$.

To perform the product $z = A^T x$ it suffices simply to interchange the arrays *ia* and *ja*. This algorithm is based on the fact that when multiplying a matrix by a vector, each element of the matrix is used only once: element $a_{ij}$ is multiplied with $x_j$ and constitutes one term of the inner product $\sum_{j=1}^{n} a_{ij} x_j = z_i$, the $i^{\text{th}}$ component of the result $z$. It follows that the elements in the array *aa* can be treated consecutively from first to last, at the end of which the matrix-vector product will have been formed.

A more efficient storage scheme can be implemented if a partial ordering is imposed on the positions of the nonzero elements in the array $aa$. Consider the case when the nonzero elements of the matrix are stored by rows; elements of row $i$ precede those of row $i+1$ but elements within a row may or may not be in order. This is frequently the case with Markov chains, since it is usual to generate all the states that may be reached in a single step from a given state $i$ before generating the states that can be reached from the next state, $i+1$. Hence the matrix is generated row by row. When the nonzero elements are stored by rows in this fashion, it is possible to dispense with the integer array *ia* and to replace it with a smaller array. The most commonly used compact storage scheme uses the elements of *ia* as pointers into the arrays $aa$ and *ja*. The $k^{\text{th}}$ element of *ia* denotes the position in $aa$ and *ja* at which the first element of row $k$ is stored. Thus, we always have $ia(1) = 1$. Additionally, it is usual to store the first empty position of $aa$ and *ja* in position $(n+1)$ of *ia*. Most often this means that $ia(n+1) = n_z + 1$. The number of nonzero elements in row $i$ is then given by $ia(i+1) - ia(i)$. This makes it easy to immediately go to any row of the matrix, even though it is stored in a compact form—the $ia(i+1) - ia(i)$ nonzero elements of row $i$ begin at $aa[ia(i)]$. This *row-wise* packing scheme is sometimes referred to as the *Harwell-Boeing format*.

**Example 10.16** Consider, once again, the same $4 \times 4$ matrix:

$$A = \begin{pmatrix} -2.1 & 0.0 & 1.7 & 0.4 \\ 0.8 & -0.8 & 0.0 & 0.0 \\ 0.2 & 1.5 & -1.7 & 0.0 \\ 0.0 & 0.3 & 0.2 & -0.5 \end{pmatrix}.$$

In this row-wise packing scheme, $A$ may be stored as

$$\begin{array}{rrrrrrrrrrrr} aa: & -2.1 & 1.7 & 0.4 & -0.8 & 0.8 & -1.7 & 0.2 & 1.5 & -0.5 & 0.3 & 0.2 \\ ja: & 1 & 3 & 4 & 2 & 1 & 3 & 1 & 2 & 4 & 2 & 3 \\ ia: & 1 & 4 & 6 & 9 & 12 & & & & & & \end{array}$$

It is not necessary for the elements in any row to be in order; it suffices that all the nonzero elements of row $i$ come before those of row $i+1$ and after those of row $i-1$. Using this storage scheme, the matrix-vector product $z = Ax$ may be computed by

ALGORITHM 10.5: SPARSE MATRIX-VECTOR MULTIPLICATION II

1. For $i = 1$ to $n$ do
   - Set $sum = 0$.
   - Set $initial = ia(i)$.
   - Set $last = ia(i+1) - 1$.
   - For $j = initial$ to $last$ do
     - Compute $sum = sum + aa(j) \times x(ja(j))$.
   - Set $z(i) = sum$.

In our discussion of the SOR algorithm, it may have appeared to have been numerically more complex than the simple power method or Gauss–Seidel and that incorporation of a sparse storage data structure would be more challenging. However, this is not the case. The SOR method requires only a matrix-vector multiplication per iteration, the same as the power method, or for that matter the Jacobi or Gauss–Seidel method. When programming SOR, we use the formula

$$x_i^{(k+1)} = (1-\omega)x_i^{(k)} + \frac{\omega}{a_{ii}}\left(b_i - \sum_{j=1}^{i-1} a_{ij}x_j^{(k+1)} - \sum_{j=i+1}^{n} a_{ij}x_j^{(k)}\right).$$

By scaling the matrix so that $a_{ii} = 1$ for all $i$ and setting $b_i = 0$, for all $i$, this reduces to

$$x_i^{(k+1)} = (1-\omega)x_i^{(k)} - \omega\left(\sum_{j=1}^{i-1} a_{ij}x_j^{(k+1)} + \sum_{j=i+1}^{n} a_{ij}x_j^{(k)}\right)$$

$$= x_i^{(k)} - \omega\left(\sum_{j=1}^{i-1} a_{ij}x_j^{(k+1)} + a_{ii}x_i^{(k)} + \sum_{j=i+1}^{n} a_{ij}x_j^{(k)}\right).$$

At iteration $k$, the program may be written (assuming $A$ is stored in the row-wise compact form just described) simply as

ALGORITHM 10.6: SPARSE SOR

1. For $i = 1$ to $n$ do
   - Set $sum = 0$.
   - Set $initial = ia(i)$.
   - Compute $last = ia(i+1) - 1$.
   - For $j = initial$ to $last$ do
     - Compute $sum = sum + aa(j) \times x(ja(j))$.
   - Compute $x(i) = x(i) - \omega \times sum$.

Observe that the only difference between this and the straightforward matrix-vector multiply algorithm, Algorithm 10.5, occurs in the very last line. In the SOR algorithm, the elements of $x$ are determined sequentially and hence it is possible to overwrite them with their new values as soon as they have been computed. The computation of element $x(i+1)$ does not begin until the new value of $x(i)$ has already been computed.

When using the SOR algorithm to obtain the stationary distribution of a Markov chain, the matrix $A$ above must be replaced by $Q^T$, the transpose of the infinitesimal generator. This may pose a problem if $Q$ is generated by rows, as is often the case. If it is not possible to generate $Q$ by columns (which means that for each state we need to find the states that can access this state in one step), then it becomes necessary to transpose the matrix and to do so without expanding it into full two-dimensional format. If sufficient storage is available to store the compacted matrix *and* its compacted transpose, then the operation of transposition can be effected in $O(n_z)$ operations, where $n_z$ is the number of nonzero elements stored. If space is not available for a second compacted copy, then the transposition may be carried out in place in $O(n_z \log n_z)$ operations, using a standard sorting procedure. The moral is obviously to try to store the matrix $Q$ by columns. Unfortunately, in many Markov chain applications, it is much more convenient to determine all destination states that occur from a given source state (row-wise generation) than to determine all source states which lead to a given destination state (column-wise generation).

### *10.3.5 Initial Approximations, Normalization, and Convergence*

During the iterative process, a suitably chosen initial approximation is successively modified until it converges to the solution. This poses three problems:

- What should we choose as the initial vector?
- What happens to this vector at each iteration?
- How do we know when convergence has occurred?

These three questions are answered in this section. When choosing an initial starting vector for an iterative method, it is tempting to choose something simple, such as a vector whose components are all zero except for one or two entries. If such a choice is made, care must be taken to ensure that the initial vector is not deficient in some component of the basis of the solution vector, otherwise a vector containing all zeros may arise, and the process will never converge.

**Example 10.17** Consider the $(2 \times 2)$ transition rate matrix

$$Q^T = \begin{pmatrix} -\lambda & \mu \\ \lambda & -\mu \end{pmatrix} = (D - L - U).$$

Applying the Gauss–Seidel method with $x^{(0)} = (1, 0)^T$ yields $x^{(k)} = (0,\ 0)^T$ for all $k \geq 1$, since

$$x^{(1)} = (D - L)^{-1} U x^{(0)} = (D - L)^{-1} \begin{pmatrix} 0 & -\mu \\ 0 & 0 \end{pmatrix} \begin{pmatrix} 1 \\ 0 \end{pmatrix} = (D - L)^{-1} \begin{pmatrix} 0 \\ 0 \end{pmatrix} = \begin{pmatrix} 0 \\ 0 \end{pmatrix},$$

and all successive vectors $x^{(k)}$, $k = 1, 2, \ldots$, are identically equal to zero.

If some approximation to the solution is known, then it should be used. If this is not possible, the elements of the initial iterate may be assigned random numbers uniformly distributed between 0 and 1 and then normalized to produce a probability vector. A simpler approach is to set each element equal to $1/n$, where $n$ is the number of states.

We now turn to the second question. At each iteration the current approximation must be multiplied by the iteration matrix. In methods other than the power method, the result of this multiplication may not be a probability vector, i.e., the sum of the components of the computed vector may not be one, even when the initial approximation is chosen to be a probability vector. This vector, however, may be made into a probability vector by normalization, by dividing each component by the sum of the components. In large-scale Markov chains, this can be a costly operation, since the sum of all $n$ components must be computed and then followed by $n$ divisions—indeed it can almost double the complexity per iteration. In most cases, it is unnecessary to constrain successive approximations to be probability vectors. Normalization is only required prior to conducting a test for convergence, since we need to compare standardized vectors, and to prevent problems of overflow and underflow of the elements of the successive approximations.

*Underflow.* While this is not fatal, it often gives rise to an undesirable error message. This may be avoided by periodically checking the magnitude of the elements and by setting those that are less than a certain threshold (e.g., $10^{-25}$) to zero. Note that the underflow problem can arise even when the approximations are normalized at each iteration, since some elements of the solution vector, although strictly positive, may be extremely small. The concern when normalization is not performed is that *all* of the elements will become smaller with each iteration until they are all set to zero. This problem can be avoided completely by a periodic scan of the approximation to ensure that at least one element exceeds a certain minimum threshold and to initiate a normalization when this test fails.

*Overflow.* With a reasonable starting vector, overflow is unlikely to occur, since the eigenvalues of the iteration matrices should not exceed one. All doubt can be eliminated by keeping a check on the magnitude of the largest element and normalizing the iterate if this element exceeds a certain maximum threshold, say $10^{10}$.

The final question to be answered concerns knowing when convergence has occurred. This generally comes under the heading of *convergence testing and numerical accuracy*. The number of iterations $k$ needed to satisfy a tolerance criterion $\epsilon$ may be obtained approximately from the relationship

$$\rho^k = \epsilon, \quad \text{i.e., } k = \frac{\log \epsilon}{\log \rho},$$

where $\rho$ is the spectral radius of the iteration matrix. In Markov chain problems, the magnitude of the subdominant eigenvalue is used in place of $\rho$. Thus, when we wish to have six decimal places of accuracy, we set $\epsilon = 10^{-6}$ and find that the number of iterations needed for different spectral radii is as follows:

| $\rho$ | .1 | .5 | .6 | .7 | .8 | .9 | .95 | .99 | .995 | .999 |
|---|---|---|---|---|---|---|---|---|---|---|
| $k$ | 6 | 20 | 27 | 39 | 62 | 131 | 269 | 1,375 | 2,756 | 13,809 |

Since the size of the subdominant eigenvalue is seldom known in advance, the usual method of testing for convergence is to examine some norm of the difference of successive iterates. When this difference becomes less than a certain prespecified tolerance, the iterative procedure is stopped. This is a satisfactory approach when the procedure converges relatively rapidly and the magnitude of the largest element of the vector is of order unity. However, when the procedure is converging slowly, it is possible that the difference in successive iterates is smaller than the tolerance specified, even though the vector may be far from the solution.

**Example 10.18** Consider the infinitesimal generator

$$Q = \begin{pmatrix} -.6 & 0 & .6 & 0 \\ .0002 & -.7 & 0 & .6998 \\ .1999 & .0001 & -.2 & 0 \\ 0 & .5 & 0 & -.5 \end{pmatrix}.$$

Using the Gauss–Seidel method with initial approximation $\pi^{(0)} = (.25, .25, .25, .25)$, we find, after 199 and 200 iterations,

$$\pi^{(199)} = (0.112758,\ 0.228774,\ 0.338275,\ 0.3201922),$$
$$\pi^{(200)} = (0.112774,\ 0.228748,\ 0.338322,\ 0.3201560),$$

which appears to give about four decimal places of accuracy. If the tolerance criterion had been set to $\epsilon = 0.001$, the iterative procedure would have stopped prior to iteration 200. However, the true solution is

$$\pi = (0.131589,\ 0.197384,\ 0.394768,\ 0.276259)$$

and the Gauss–Seidel method will eventually converge onto this solution. In this example, the subdominant eigenvalue of the Gauss–Seidel iteration matrix $H_{GS}$ is 0.9992 and over 6,000 iterations are needed to achieve an accuracy of just $\epsilon = .01$.

This problem may be overcome by testing, not successive iterates $\|\pi^{(k)} - \pi^{(k-1)}\| < \epsilon$, but rather iterates spaced further apart, e.g.,

$$\|\pi^{(k)} - \pi^{(k-m)}\| < \epsilon.$$

Ideally, $m$ should be determined as a function of the convergence rate. A simple but less desirable alternative is to allow $m$ to assume different values during the iteration procedure. For example, when the iteration number $k$ is

$$\begin{array}{rl} k < 100 & \text{let} m = 5, \\ 100 \le k < 500 & \text{let } m = 10, \\ 500 \le k < 1{,}000 & \text{let } m = 20, \\ k \ge 1,000 & \text{let } m = 50. \end{array}$$

A second problem arises when the approximations converge to a vector in which the elements are all small. Suppose the problem has 100,000 states, and all are approximately equally probable. Then each element of the solution is approximately equal to $10^{-5}$. If the tolerance criterion is set at $10^{-3}$ and the initial vector is chosen such that $\pi_i^{(0)} = 1/n$ for all $i$, then the process will probably "converge" after one iteration! This same problem can arise in a more subtle context if the approximations are not normalized before convergence testing *and* the choice of initial vector results in the iterative method converging to a vector in which all components are small. This may happen even though some elements of the solution may be large relative to others. For example, it happens in the $(2 \times 2)$ Markov chain of Example 10.17 when the initial approximation is chosen as $(\xi,\ 10^{-6})$ for any value of $\xi$. (Note that the opposite effect will occur if the approximations converge to a vector with all elements relatively large, e.g., if the initial vector is $(\xi,\ 10^{6})$ in Example 10.17.)

A solution to this latter aspect of the problem (when vectors are not normalized) is, of course, to normalize the iterates before testing for convergence. If this normalization is such that it produces a probability vector, the original problem (all of the components may be small) still remains. A better choice of normalization in this instance is $\|\pi^{(k)}\|_\infty = 1$ (i.e., normalize so that the largest element of $\pi^{(k)}$ is equal to 1—only the final normalization needs to produce a probability vector). A better solution, however, and the one that is recommended, is to use a relative measure, e.g.,

$$\max_i \left( \frac{|\pi_i^{(k)} - \pi_i^{(k-m)}|}{|\pi_i^{(k)}|} \right) < \epsilon.$$

This effectively removes the exponent from consideration in the convergence test and hence gives a better estimate of the precision that has been achieved.

Another criterion that has been used for convergence testing is to check the size of the residuals (the magnitude of $\|\pi^{(k)} Q\|$), which should be small. Residuals will work fine in many and perhaps even most modeling problems. Unfortunately, a small residual does not always imply that the error in the solution vector is also small. In ill-conditioned systems the residual may be very small indeed, yet the computed solution may be hopelessly inaccurate. A small residual is a necessary condition for the error in the solution vector to be small—but it is not a sufficient condition. The most suitable approach is to check the residual after the relative convergence test indicates that convergence has been achieved. In fact, it is best to envisage a battery of convergence tests, all of which must be satisfied before the approximation is accepted as being sufficiently accurate.

We now turn to the frequency with which the convergence test should be administered. Often it is performed during each iteration. This may be wasteful, especially when the matrix is very large and the iterative method is converging slowly. Sometimes it is possible to estimate the rate of convergence and to determine from this rate the approximate numbers of iterations that must be performed. It is now possible to proceed "full steam ahead" and carry out this number of iterations without testing for convergence or normalizing. For rapidly converging problems this may result in more iterations being performed than is strictly necessary, thereby achieving a more accurate result than the user actually needs. When the matrix is large and an iteration costly, this may be undesirable. One possibility is to carry out only a proportion (say 80–90%) of the estimated number of iterations before beginning to implement the battery of convergence tests.

An alternative approach is to implement a relatively inexpensive convergence test (e.g., using the relative difference in the first nonzero component of successive approximations) at each iteration. When this simple test is satisfied, more rigorous convergence tests may be initiated.

## 10.4 Block Iterative Methods

The iterative methods we have examined so far are sometimes referred to as *point* iterative methods in order to distinguish them from their *block* counterparts. Block iterative methods are generalizations of point iterative methods and can be particularly beneficial in Markov chain problems in which the state space can be meaningfully partitioned into subsets. In general such block iterative methods require more computation per iteration, but this is offset by a faster rate of convergence. Let us partition the defining homogeneous system of equations $\pi Q = 0$ as

$$(\pi_1, \pi_2, \ldots, \pi_N) \begin{pmatrix} Q_{11} & Q_{12} & \cdots & Q_{1N} \\ Q_{21} & Q_{22} & \cdots & Q_{2N} \\ \vdots & \vdots & \ddots & \vdots \\ Q_{N1} & Q_{N2} & \cdots & Q_{NN} \end{pmatrix} = 0.$$

We now introduce the block splitting

$$Q^T = D_N - (L_N + U_N),$$

where $D_N$ is a block diagonal matrix and $L_N$ and $U_N$ are, respectively, strictly lower and upper triangular block matrices. We have

$$D_N = \begin{pmatrix} D_{11} & 0 & \cdots & 0 \\ 0 & D_{22} & \cdots & 0 \\ \vdots & \vdots & \ddots & \vdots \\ 0 & 0 & \cdots & D_{NN} \end{pmatrix},$$

$$L_N = \begin{pmatrix} 0 & 0 & \cdots & 0 \\ L_{21} & 0 & \cdots & 0 \\ \vdots & \vdots & \ddots & \vdots \\ L_{N1} & L_{N2} & \cdots & 0 \end{pmatrix}, \quad U_N = \begin{pmatrix} 0 & U_{12} & \cdots & U_{1N} \\ 0 & 0 & \cdots & U_{2N} \\ \vdots & \vdots & \ddots & \vdots \\ 0 & 0 & \cdots & 0 \end{pmatrix}.$$

In analogy with Equation (10.18), the block Gauss–Seidel method is given by

$$(D_N - L_N)x^{(k+1)} = U_N x^{(k)}$$

and corresponds to the splitting $M = (D_N - L_N)$; $N = U_N$. The $i^{\text{th}}$ block equation is given by

$$D_{ii}x_i^{(k+1)} = \left( \sum_{j=1}^{i-1} L_{ij}x_j^{(k+1)} + \sum_{j=i+1}^{N} U_{ij}x_j^{(k)} \right) \quad i = 1, 2, \ldots, N, \tag{10.23}$$

where the subvectors $x_i$ are partitioned conformally with $D_{ii}$, $i = 1, 2, \ldots, N$. This implies that at each iteration we must now solve $N$ systems of linear equations

$$D_{ii}x_i^{(k+1)} = z_i, \quad i = 1, 2, \ldots, N, \tag{10.24}$$

where

$$z_i = \left( \sum_{j=1}^{i-1} L_{ij}x_j^{(k+1)} + \sum_{j=i+1}^{N} U_{ij}x_j^{(k)} \right), \quad i = 1, 2, \ldots, N.$$

The right-hand side $z_i$ can always be computed before the $i^{\text{th}}$ system has to be solved. If our Markov chain is irreducible, the $N$ systems of equations (10.24) are nonhomogeneous and have nonsingular coefficient matrices. We may use either direct or iterative methods to solve them. Naturally, there is no requirement to use the same method to solve all the diagonal blocks. Instead, it is possible to tailor methods to particular block structures. If a direct method is used, then an $LU$ decomposition of block $D_{ii}$ may be formed once and for all before beginning the iteration, so that solving $D_{ii}x_i^{(k+1)} = z_i$, $i = 1, \ldots, N$, in each global iteration simplifies to a forward and backward substitution. The nonzero structure of the blocks may be such that this is a particularly efficient approach. For example, if the diagonal blocks are themselves diagonal matrices, or if they are upper or lower triangular matrices or even tridiagonal matrices, then it is very easy to obtain their $LU$ decomposition, and a block iterative method becomes very attractive.

If the diagonal blocks are large and do not possess a suitable nonzero structure, it may be appropriate to use matrix iterative methods (such as point Gauss–Seidel) to solve these block equations—in which case we can have multiple inner (or local) iterative methods (one per block thus analyzed) within an outer (or global) iteration. A number of tricks may be used to speed up this process. First, the solution computed using any block $D_{ii}$ at iteration $k$ should be used as the initial approximation to the solution using this same block at iteration $k+1$. Second, it is hardly worthwhile computing a highly accurate solution in early (outer) iterations. We should require only a small number of digits of accuracy until the global process begins to converge. One convenient way to achieve this is to carry out only a fixed, small number of iterations for each inner solution. Initially, this will not give much accuracy, but when combined with the first suggestion, the accuracy achieved will increase from one outer iteration to the next.

Intuitively, it is expected that for a given transition rate matrix $Q$, the larger the block sizes (and thus the smaller the number of blocks), the fewer the number of (outer) iterations needed for convergence. This has been shown to be true under fairly general assumptions on the coefficient matrix for general systems of equations (see [53]). In the special case of only one block, the method degenerates to a standard direct method and we compute the solution in a single "iteration." The reduction in the number of iterations that usually accompanies larger blocks is offset to a certain degree by an increase in the number of operations that must be performed at each iteration. However, in some important cases it may be shown that there is no increase. For example, when the matrix is block tridiagonal (as in quasi-birth-death processes) and the diagonal blocks are also tridiagonal, it may be shown that the computational effort per iteration is the same for both point and block iterative methods. In this case the reduction in the number of iterations makes the block methods very efficient indeed.

In a similar vein to the block Gauss–Seidel method, we may also define a *block* Jacobi method

$$D_{ii}x_i^{(k+1)} = \left( \sum_{j=1}^{i-1} L_{ij}x_j^{(k)} + \sum_{j=i+1}^{N} U_{ij}x_j^{(k)} \right), \quad i = 1, 2, \ldots, N,$$

and a *block* SOR method

$$x_i^{(k+1)} = (1-\omega)x_i^{(k)} + \omega \left\{ D_{ii}^{-1} \left( \sum_{j=1}^{i-1} L_{ij}x_j^{(k+1)} + \sum_{j=i+1}^{N} U_{ij}x_j^{(k)} \right) \right\}, \quad i = 1, 2, \ldots, N.$$

**Example 10.19** We apply the block Gauss–Seidel method to find the stationary distribution of the continuous-time Markov chain with infinitesimal generator given by

$$Q = \begin{pmatrix} -4.0 & 2.0 & 1.0 & 0.5 & 0.5 \\ 0.0 & -3.0 & 3.0 & 0.0 & 0.0 \\ 0.0 & 0.0 & -1.0 & 0.0 & 1.0 \\ 1.0 & 0.0 & 0.0 & -5.0 & 4.0 \\ 1.0 & 0.0 & 0.0 & 1.0 & -2.0 \end{pmatrix}.$$

We put the first three states in the first subset and the remaining two states into a second subsest. Transposing $Q$ and writing out the system of equations, we have

$$Q^T x = \begin{pmatrix} D_{11} & -U_{12} \\ -L_{21} & D_{22} \end{pmatrix} \begin{pmatrix} x_1 \\ x_2 \end{pmatrix} = \left(\begin{array}{ccc|cc} -4.0 & 0.0 & 0.0 & 1.0 & 1.0 \\ 2.0 & -3.0 & 0.0 & 0.0 & 0.0 \\ 1.0 & 3.0 & -1.0 & 0.0 & 0.0 \\ \hline 0.5 & 0.0 & 0.0 & -5.0 & 1.0 \\ 0.5 & 0.0 & 1.0 & 4.0 & -2.0 \end{array}\right) \begin{pmatrix} \pi_1 \\ \pi_2 \\ \pi_3 \\ \hline \pi_4 \\ \pi_5 \end{pmatrix} = \begin{pmatrix} 0 \\ 0 \\ 0 \\ \hline 0 \\ 0 \end{pmatrix}.$$

Equation (10.23) becomes

$$D_{ii}x_i^{(k+1)} = \left( \sum_{j=1}^{i-1} L_{ij}x_j^{(k+1)} + \sum_{j=i+1}^{2} U_{ij}x_j^{(k)} \right), \quad i = 1, 2,$$

and leads to the two block equations

$$i = 1: \quad D_{11}x_1^{(k+1)} = \left( \sum_{j=1}^{0} L_{1j}x_j^{(k+1)} + \sum_{j=2}^{2} U_{1j}x_j^{(k)} \right) = U_{12}x_2^{(k)},$$

$$i = 2: \quad D_{22}x_2^{(k+1)} = \left( \sum_{j=1}^{1} L_{2j}x_j^{(k+1)} + \sum_{j=3}^{2} U_{2j}x_j^{(k)} \right) = L_{21}x_1^{(k+1)}.$$

Writing these block equations in full, they become

$$\begin{pmatrix} -4.0 & 0.0 & 0.0 \\ 2.0 & -3.0 & 0.0 \\ 1.0 & 3.0 & -1.0 \end{pmatrix} \begin{pmatrix} \pi_1^{(k+1)} \\ \pi_2^{(k+1)} \\ \pi_3^{(k+1)} \end{pmatrix} = - \begin{pmatrix} 1.0 & 1.0 \\ 0.0 & 0.0 \\ 0.0 & 0.0 \end{pmatrix} \begin{pmatrix} \pi_4^{(k)} \\ \pi_5^{(k)} \end{pmatrix}$$

and

$$\begin{pmatrix} -5.0 & 1.0 \\ 4.0 & -2.0 \end{pmatrix} \begin{pmatrix} \pi_4^{(k+1)} \\ \pi_5^{(k+1)} \end{pmatrix} = - \begin{pmatrix} 0.5 & 0.0 & 0.0 \\ 0.5 & 0.0 & 1.0 \end{pmatrix} \begin{pmatrix} \pi_1^{(k+1)} \\ \pi_2^{(k+1)} \\ \pi_3^{(k+1)} \end{pmatrix}.$$

We shall solve these block equations using $LU$ decompositions. Since $D_{11}$ is lower triangular, the first subsystem may be solved by *forward substitution* alone:

$$D_{11} = \begin{pmatrix} -4.0 & 0.0 & 0.0 \\ 2.0 & -3.0 & 0.0 \\ 1.0 & 3.0 & -1.0 \end{pmatrix} = L \times I.$$

Forming an $LU$ decomposition of the second subsystem, we have

$$D_{22} = \begin{pmatrix} -5.0 & 1.0 \\ 4.0 & -2.0 \end{pmatrix} = \begin{pmatrix} 1.0 & 0.0 \\ -0.8 & 1.0 \end{pmatrix} \begin{pmatrix} -5.0 & 1.0 \\ 0.0 & -1.2 \end{pmatrix} = L \times U.$$

Taking the initial distribution to be

$$\pi^{(0)} = (0.2,\ 0.2,\ 0.2,\ 0.2,\ 0.2)^T$$

and substituting the first block equation, we find

$$\begin{pmatrix} -4.0 & 0.0 & 0.0 \\ 2.0 & -3.0 & 0.0 \\ 1.0 & 3.0 & -1.0 \end{pmatrix} \begin{pmatrix} \pi_1^{(1)} \\ \pi_2^{(1)} \\ \pi_3^{(1)} \end{pmatrix} = - \begin{pmatrix} 1.0 & 1.0 \\ 0.0 & 0.0 \\ 0.0 & 0.0 \end{pmatrix} \begin{pmatrix} 0.2 \\ 0.2 \end{pmatrix} = - \begin{pmatrix} 0.4 \\ 0.0 \\ 0.0 \end{pmatrix}.$$

Forward substitution successively gives

$$\pi_1^{(1)} = 0.1000, \quad \pi_2^{(1)} = 0.0667, \quad \text{and} \quad \pi_3^{(1)} = 0.3000.$$

The second block system now becomes

$$\begin{pmatrix} 1.0 & 0.0 \\ -0.8 & 1.0 \end{pmatrix} \begin{pmatrix} -5.0 & 1.0 \\ 0.0 & -1.2 \end{pmatrix} \begin{pmatrix} \pi_4^{(1)} \\ \pi_5^{(1)} \end{pmatrix} = - \begin{pmatrix} 0.5 & 0.0 & 0.0 \\ 0.5 & 0.0 & 1.0 \end{pmatrix} \begin{pmatrix} 0.1000 \\ 0.0667 \\ 0.3000 \end{pmatrix} = - \begin{pmatrix} 0.0500 \\ 0.3500 \end{pmatrix}.$$

This is in the standard form $LUx = b$ from which the solution is computed by first obtaining $z$ from $Lz = b$ and then $x$ from $Ux = z$. From

$$\begin{pmatrix} 1.0 & 0.0 \\ -0.8 & 1.0 \end{pmatrix} \begin{pmatrix} z_1 \\ z_2 \end{pmatrix} = - \begin{pmatrix} 0.0500 \\ 0.3500 \end{pmatrix},$$

we obtain

$$z_1 = -0.0500 \quad \text{and} \quad z_2 = -0.3900.$$

Now, solving

$$\begin{pmatrix} -5.0 & 1.0 \\ 0.0 & -1.2 \end{pmatrix} \begin{pmatrix} \pi_4^{(1)} \\ \pi_5^{(1)} \end{pmatrix} = \begin{pmatrix} -0.0500 \\ -0.3900 \end{pmatrix},$$

we find

$$\pi_4^{(1)} = 0.0750 \quad \text{and} \quad \pi_5^{(1)} = 0.3250.$$

We now have

$$\pi^{(1)} = (0.1000,\ 0.0667,\ 0.3000,\ 0.0750,\ 0.3250)$$

which, when normalized so that the elements sum to 1, gives

$$\pi^{(1)} = (0.1154,\ 0.0769,\ 0.3462,\ 0.0865,\ 0.3750).$$

The next iteration may now be initiated. Interested readers who perform some additional iterations will see that, in this example, all subsequent iterations yield exactly the same result. Indeed, for this specific example, the block Gauss–Seidel method requires only a single iteration to obtain the solution to full machine precision. A examination of the matrix $U_N$ reveals that this matrix has only two nonzero elements in positions $(1, 4)$ and $(1, 5)$ and they are both equal. It must follow that the iteration matrix can have only two nonzero columns, the fourth and fifth, and they must be identical. It may be concluded that the iteration matrix has an eigenvalue equal to 1 and four equal to 0, which explains the fact that convergence is achieved in a single iteration.

The following code is a Matlab implementation of the *block Gauss–Seidel method*. The program accepts a stochastic matrix $P$; a partitioning vector $ni$, whose $i^{\text{th}}$ component stores the length of the $i^{\text{th}}$ block; and two integers: *itmax1*, which denotes the number of outer iterations to perform, and *itmax2*, which denotes the number of iterations to use to solve each of the blocks *if* Gauss–Seidel is used to solve these blocks. The program returns the solution vector $\pi$ and a vector of residuals. It calls the Gauss–Seidel program given previously.

### *Matlab code for Block Gauss–Seidel*

```
function [x,res] = bgs(P,ni,itmax1,itmax2)
[n,n] = size(P); [na,nb] = size(ni);

%          BLOCK Gauss-Seidel FOR P^T x = x

bl(1) = 1;                                    % Get beginning and end
for k = 1:nb, bl(k+1) = bl(k)+ni(k); end      % points of each block
x = ones(n,1)/n;                              % Initial approximation

%%%%%%%%%%%%%%%%%%%%%%%%%% BEGIN OUTER LOOP %%%%%%%%%%%%%%%%%%%%%%%%%%
for iter = 1:itmax1,
    for m = 1:nb,                             % All diagonal blocks
```

```
            A = P(bl(m):bl(m+1)-1,bl(m):bl(m+1)-1)';  % Get A_mm
            b = -P(1:n,bl(m):bl(m+1)-1)'*x+A*x(bl(m):bl(m+1)-1);  % RHS
            z = inv(A-eye(ni(m)))*b;                     % Solve for z

%           ***  To solve the blocks using Gauss--Seidel      ***
%           ***  instead of a direct method, substitute      ***
%           ***  the next two lines for the previous one.    ***
%**         x0 = x(bl(m):bl(m+1)-1);                    % Get starting vector
%**         [z,r] = gs(A-eye(ni(m)),x0,b,itmax2);       % Solve for z

            x(bl(m):bl(m+1)-1) = z;                      % Update x
        end
        res(iter) = norm((P'-eye(n))*x,2);               % Compute residual
end
x = x/norm(x,1); res = res';
```

## 10.5 Decomposition and Aggregation Methods

A decompositional approach to solving Markov chains is intuitively very attractive since it appeals to the principle of divide and conquer: if the model is too large or complex to analyze in toto, it is divided into subsystems, each of which is analyzed separately, and a global solution is then constructed from the partial solutions. Ideally the problem is broken into subproblems that can be solved independently and the global solution is obtained by "pasting" together the subproblem solutions. Although it is rare to find Markov chains that can be divided into independent subchains, it is not unusual to have Markov chains in which this condition almost holds. An important class of problems that frequently arise in Markov modeling are those in which the state space may be partitioned into disjoint subsets with strong interactions among the states of a subset but with weak interactions among the subsets themselves. Such problems are sometimes referred to as *nearly completely decomposable (NCD), nearly uncoupled,* or *nearly separable.* It is apparent that the assumption that the subsystems are independent and can therefore be solved separately does not hold. Consequently an error arises. This error will be small if the assumption is approximately true. An irreducible NCD stochastic matrix $P$ may be written as

$$P = \begin{pmatrix} P_{11} & P_{12} & \dots & P_{1N} \\ P_{21} & P_{22} & \dots & P_{2N} \\ \vdots & \vdots & \ddots & \vdots \\ P_{N1} & P_{N2} & \dots & P_{NN} \end{pmatrix},$$

where

$$||P_{ii}|| = O(1), \qquad i = 1, 2, \dots, N,$$

and

$$||P_{ij}|| = O(\epsilon), \qquad i \neq j.$$

In addition to its unit eigenvalue, such a matrix possesss $N - 1$ eigenvalues extremely close to 1. None of the point iterative methods discussed previously is effective in handling this situation. On the other hand, block and decompositional methods can be very effective.

We begin by examining what happens when the off-diagonal blocks are all zero, i.e.,

$$(\pi_1, \pi_2, \ldots, \pi_N) \begin{pmatrix} P_{11} & 0 & \ldots & 0 & 0 \\ 0 & P_{22} & \ldots & 0 & 0 \\ \vdots & \vdots & \ddots & \vdots & \vdots \\ 0 & 0 & \ldots & P_{N-1N-1} & 0 \\ 0 & 0 & \ldots & 0 & P_{NN} \end{pmatrix} = (\pi_1, \pi_2, \ldots, \pi_N).$$

In this case, each $P_{ii}$ is a stochastic matrix. The Markov chain is reducible into $N$ distinct irreducible classes and a stationary distribution can be found for each class. Each $\pi_i$ can be found directly from

$$\pi_i P_{ii} = \pi_i, \quad i = 1, 2, \ldots, N.$$

This allows us to envisage the following approximate solution procedure when the Markov chain is NCD rather than completely decomposable:

(a) Solve the diagonal blocks as if they are independent.
The solution obtained for each block should provide an approximation to the probability of being in the different states of that block, conditioned on being in that block.
(b) Estimate the probability of being in each block.
This will allow us to remove the condition in part (a).
(c) Combine (a) and (b) into a global approximate solution.

We now examine each of these three steps in more detail.

**Step (a): Solve blocks as if independent**

The initial step in approximating the solution of $\pi P = \pi$ when $P_{ij} \neq 0$ is to assume that the system is completely decomposable and to compute the stationary probability distribution for each component. A first problem that arises is that the $P_{ii}$ are not stochastic but rather strictly substochastic. A simple way around this problem is to simply ignore it; i.e., to work directly with the substochastic matrices $P_{ii}$ themselves. In other words, we may use the normalized eigenvector corresponding to the Perron root (the eigenvalue closest to 1) of block $P_{ii}$ as the probability vector whose elements denote the probabilities of being in the different states of the block, conditioned on being in this block.

**Example 10.20** To help, we shall illustrate the procedure using the $8 \times 8$ Courtois matrix, an examination of which reveals that it is nearly completely decomposable (NCD) into a block of order 3, a second of order 2, and a third of order 3:

$$P = \left(\begin{array}{ccc|cc|ccc} .85 & .0 & .149 & .0009 & .0 & .00005 & .0 & .00005 \\ .1 & .65 & .249 & .0 & .0009 & .00005 & .0 & .00005 \\ .1 & .8 & .0996 & .0003 & .0 & .0 & .0001 & .0 \\ \hline .0 & .0004 & .0 & .7 & .2995 & .0 & .0001 & .0 \\ .0005 & .0 & .0004 & .399 & .6 & .0001 & .0 & .0 \\ \hline .0 & .00005 & .0 & .0 & .00005 & .6 & .2499 & .15 \\ .00003 & .0 & .00003 & .00004 & .0 & .1 & .8 & .0999 \\ .0 & .00005 & .0 & .0 & .00005 & .1999 & .25 & .55 \end{array}\right).$$

For this matrix, we have the following blocks, Perron roots, and corresponding left-hand eigenvectors:

$$P_{11} = \begin{pmatrix} .85 & .0 & .149 \\ .1 & .65 & .249 \\ .1 & .8 & .0996 \end{pmatrix}, \quad \lambda_{1_1} = .99911, \quad u_1 = (.40143,\ .41672,\ .18185),$$

$$P_{22} = \begin{pmatrix} .7 & .2995 \\ .399 & .6 \end{pmatrix}, \quad \lambda_{2_1} = .99929, \quad u_2 = (.57140,\ .42860),$$

$$P_{33} = \begin{pmatrix} .6 & .2499 & .15 \\ .1 & .8 & .0999 \\ .1999 & .25 & .55 \end{pmatrix}, \quad \lambda_{3_1} = .9999, \quad u_3 = (.24074,\ .55563,\ .20364).$$

To summarize, in part (a), the left-hand eigenvector $u_i$ of length $n_i$ corresponding to the eigenvalue closest to 1, $\lambda_{i_1}$, in each block $i$, $1 \le i \le N$, is computed. In other words, we solve the $N$ eigenvalue problems

$$u_i P_{ii} = \lambda_{i_1} u_i, \qquad u_i e = 1, \quad i = 1, 2, \ldots, N.$$

**Step (b): Estimate block probabilities**

The second problem is that, once we have computed the stationary probability vector for each block, simply concatenating them together will not give a probability vector. The elements of each subvector sum to 1. We still need to weight each subvector by the probability of being in its subblock of states. The probability distributions computed from the $P_{ii}$ are conditional probabilities in the sense that they express the probability of being in a given state of subset $i$, $i = 1, 2, \ldots, N$, conditioned on the fact that the Markov chain is in one of the states of that subset. We need to remove that condition. To determine the probability of being in a given block of states we need to construct a matrix whose element $ij$ gives the probability of a transition from block $i$ to block $j$. This is an $(N \times N)$ stochastic matrix that characterizes the interactions among blocks. To construct this matrix, which is called the *coupling matrix*, we need to shrink each block $P_{ij}$ of $P$ down to a single element. The stationary distribution of the coupling matrix provides the block probabilities, the weights needed to form the global approximation.

In terms of the running example, we need to find weights $\xi_1$, $\xi_2$, and $\xi_3$ such that

$$(\xi_1 u_1,\ \xi_2 u_2,\ \xi_3 u_3)$$

is an approximate solution to $\pi$. Here $\xi_i$ is the proportion of time we spend in block $i$. We need to shrink our original $(8\times 8)$ stochastic matrix down to a $(3\times 3)$ stochastic matrix. This is accomplished by first replacing each row of each block by the sum of its elements. The sum of the elements of row $k$ of block $ij$ gives the probability of leaving state $k$ of block $i$ and entering into (one of the states of) block $j$. It no longer matters which particular state of block $j$ is this destination state. Mathematically, the operation performed for each block is $P_{ij}e$.

**Example 10.21** Summing across the block rows of the Courtois matrix gives

$$\left(\begin{array}{c|c|c} .999 & .0009 & .0001 \\ .999 & .0009 & .0001 \\ .9996 & .0003 & .0001 \\ \hline .0004 & .9995 & .0001 \\ .0009 & .999 & .0001 \\ \hline .00005 & .00005 & .9999 \\ .00006 & .00004 & .9999 \\ .00005 & .00005 & .9999 \end{array}\right).$$

The next step is to use these results to find the probability of leaving (any state of) block $i$ to enter (any state of) block $j$. This means that we must reduce each column subvector, $P_{ij}e$, to a scalar. As we have just noted, the $k^{\text{th}}$ element of $P_{ij}e$ is the probability of leaving state $k$ of block $i$ and entering into block $j$. To determine the total probability of leaving (any state of) block $i$ to enter into (any state of) block $j$ we need to sum the elements of this vector after each element has been weighed by the probability of being in that state (given that the Markov chain is in one of the states of that block). These weighing factors may be obtained from the elements of the stationary probability vector. They are the components of $\pi_i/||\pi_i||_1$. The $ij^{\text{th}}$ element of the reduced $(N \times N)$ matrix is therefore given by

$$(C)_{ij} = \frac{\pi_i}{||\pi_i||_1} P_{ij} e = \phi_i P_{ij} e,$$

where $\phi_i = \pi_i/||\pi_i||_1$. If $P$ is an irreducible stochastic matrix, then $C$ also is irreducible and stochastic. Let $\xi$ denote its left eigenvector, i.e., $\xi C = \xi$ and $\xi e = 1$. The $i^{\text{th}}$ component of $\xi$ is the stationary probability of being in (one of the states of) block $i$. It is easy to show that

$$\xi = (||\pi_1||_1,\ ||\pi_2||_1,\ \ldots,\ ||\pi_N||_1).$$

Of course, the vector $\pi$ is not yet known, so that it is not possible to compute the weights $||\pi_i||_1$. However, they may be approximated by using the probability vector computed from each of the individual $P_{ii}$, i.e., by setting $\phi_i = u_i$. Consequently, the weights $\xi_i$ can be estimated and an approximate solution to the stationary probability vector $\pi$ obtained. Any of the previously used methods to find stationary probability vectors may be used.

**Example 10.22** Performing these operations, we obtain the following coupling matrix for the Courtois example:

$$(.40143, .41672, .18185,\ |\ .57140, .42860,\ |\ .24074, .55563, .20364)\begin{pmatrix} .999 & .0009 & .0001 \\ .999 & .0009 & .0001 \\ .9996 & .0003 & .0001 \\ \hline .0004 & .9995 & .0001 \\ .0009 & .999 & .0001 \\ \hline .00005 & .00005 & .9999 \\ .00006 & .00004 & .9999 \\ .00005 & .00005 & .9999 \end{pmatrix}$$

$$= \begin{pmatrix} .99911 & .00079 & .00010 \\ \hline .00061 & .99929 & .00010 \\ \hline .00006 & .00004 & .99990 \end{pmatrix} = C.$$

Its eigenvalues are 1.0, .9998, and .9985 and its stationary probability vector is

$$\xi = (.22252,\ .27748,\ .50000).$$

**Step (c): Compute global approximation**
We are now in a position to form the final approximation to the stationary probability vector. It is given by the approximation

$$\pi \approx (\xi_1 u_1,\ \xi_2 u_2,\ \ldots,\ \xi_N u_N),$$

where the $u_i$ are approximations to $\pi_i/\|\pi_i\|_1$.

**Example 10.23** In the running example we obtain the approximate solution

$$\pi^* = (.08932,\ .09273,\ .04046,\ .15855,\ .11893,\ .12037,\ .27781,\ .10182),$$

which may be compared to the exact solution

$$\pi = (.08928,\ .09276,\ .04049,\ .15853,\ .11894,\ .12039,\ .27780,\ .10182).$$

To summarize, an approximation to the stationary probability vector of an NCD Markov chain may be obtained by first solving each of the blocks separately, then forming and solving the coupling matrix, and finally constructing the approximate solution from these pieces. It is implicitly understood that the states have been ordered so that the transition probability matrix has the required NCD block structure. Algorithmically, the entire procedure may be written as

ALGORITHM 10.7: NCD DECOMPOSITION APPROXIMATION

1. Solve the individual blocks: $u_i P_{ii} = \lambda_{i_1} u_i, \quad u_i e = 1$ for $i = 1, 2, \ldots, N$.
2. (a) Form the coupling matrix: $(C)_{ij} = u_i P_{ij} e$.
   (b) Solve the coupling problem: $\xi = \xi C, \quad \xi e = 1$.
3. Construct the approximate solution: $\pi^* = (\xi_1 u_1, \ \xi_2 u_2, \ \ldots, \ \xi_N u_N)$.

The question now arises as to whether we can incorporate this approximation back into the decomposition algorithm to get an even better approximation. Notice that the $u_i$ are used to form the coupling matrix. If we replace them with the new approximations $\xi_i u_i$, we will obtain exactly the same solution to the coupling matrix as before. Hence, there will be no change in the computed approximation. However, it was found that applying a power step to the approximation before plugging it back into the decomposition method had a very salutory effect. Later this power step was replaced by a block Gauss–Seidel step and became known as a disaggregation step; forming and solving the matrix $C$ being the aggregation step. The entire procedure is referred to as *iterative aggregation/disaggregation (IAD)*.

The algorithm is presented below. The iteration number is indicated by a superscript in parentheses on the appropriate variable names. The initial vector is designated as $\pi^{(0)}$. This may be the approximation obtained from the simple decomposition approach, or it may be just a random selection. Observe that many of the steps have corresponding steps in the decomposition algorithm. For instance, the formation of the coupling matrix (Step 3) is identical in both. Step 4(a) has its counterpart in the formation of the computed approximation in the decomposition algorithm, while Step 4(b) corresponds to solving the different blocks, except that in the IAD algorithm, a block Gauss–Seidel step is used.

ALGORITHM 10.8: ITERATIVE AGGREGATION/DISAGGREGATION

1. Let $\pi^{(0)} = (\pi_1^{(0)}, \pi_2^{(0)}, \ldots, \pi_N^{(0)})$ be a given initial approximation to the solution $\pi$, and set $m = 1$.
2. Compute $\phi^{(m-1)} = (\phi_1^{(m-1)}, \phi_2^{(m-1)}, \ldots, \phi_N^{(m-1)})$, where

$$\phi_i^{(m-1)} = \frac{\pi_i^{(m-1)}}{||\pi_i^{(m-1)}||_1}, \qquad i = 1, \ 2, \ \ldots, \ N.$$

3. (a) Form the coupling matrix: $C_{ij}^{(m-1)} = \phi_i^{(m-1)} P_{ij} e$ for $i, j = 1, 2, \ldots, N$.
   (b) Solve the coupling problem: $\xi^{(m-1)} C^{(m-1)} = \xi^{(m-1)}, \quad \xi^{(m-1)} e = 1$.
4. (a) Construct the row vector

$$z^{(m)} = (\xi_1^{(m-1)} \phi_1^{(m-1)}, \ \xi_2^{(m-1)} \phi_2^{(m-1)}, \ \ldots, \ \xi_N^{(m-1)} \phi_N^{(m-1)}).$$

   (b) Solve the following $N$ systems of equations to find $\pi^{(m)}$:

$$\pi_k^{(m)} = \pi_k^{(m)} P_{kk} + \sum_{j>k} z_j^{(m)} P_{jk} + \sum_{j<k} \pi_j^{(m)} P_{jk}, \quad k = 1, 2, \ \ldots, \ N. \tag{10.25}$$

5. Normalize and conduct a test for convergence.
   If satisfactory, then stop and take $\pi^{(m)}$ to be the required solution vector.
   Otherwise set $m = m + 1$ and go to step 2.

In these methods, it is important that the matrix has the block structure needed by the algorithms and it may be necessary to reorder the states to get this property. Only after reordering the states can we guarantee that the resulting transition matrix will have a property that directly reflects the structural characteristics of the NCD system. If the partitioning provided to the algorithm does not match the decomposability characteristics of the matrix, the convergence behavior may be much less satisfactory. A reordering of the states can be accomplished by treating the Markov chain as a directed graph in which edges with small weights (probabilities) are removed. A graph algorithm must then be used to find the connected components of $\hat{P} + \hat{P}^T$, where $\hat{P}$ is the modified transition probability matrix. The complexity of the algorithm is $O(|V| + |E|)$, where $|V|$ is the number of vertices and $|E|$ is the number of edges in the graph. Details are provided in [14].

**Example 10.24** When applied to the Courtois matrix, we find that both the iterative aggregation and disaggregation (IAD) method and the block Gauss–Seidel (BGS) method are very effective. The table below shows that convergence is achieved to full machine precision in only four iterations with the IAD method and nine iterations with BGS. In both cases, the diagonal block equations are solved using $LU$ decomposition.

IAD and BGS residuals for the Courtois NCD matrix

| Iteration | IAD residual | BGS residual |
|---|---|---|
| | $1.0e-05\times$ | $1.0e-05\times$ |
| 1 | 0.93581293961421 | 0.94805408435419 |
| 2 | 0.00052482104506 | 0.01093707688215 |
| 3 | 0.00000000280606 | 0.00046904081241 |
| 4 | 0.00000000000498 | 0.00002012500900 |
| 5 | 0.00000000000412 | 0.00000086349742 |
| 6 | 0.00000000000351 | 0.00000003705098 |
| 7 | 0.00000000000397 | 0.00000000158929 |
| 8 | 0.00000000000529 | 0.00000000006641 |
| 9 | 0.00000000000408 | 0.00000000000596 |
| 10 | 0.00000000000379 | 0.00000000000395 |

We now turn our attention to some implementation details. The critical points are Steps 3 and 4(b). In Step 3, it is more efficient to compute $P_{ij}e$ only once for each block and to store it somewhere for use in all future iterations. This is only possible if sufficient memory is available; otherwise it is necessary to compute it each time it is needed. To obtain the vector $\xi$ in Step 3(b), any of the methods discussed in the previous section may be used, since the vector $\xi$ is simply the stationary probability vector of an irreducible stochastic matrix $C$.

In Step 4(b), each of the $N$ systems of equations in (10.25) can be written as $Bx = r$ where $B = (I - P_{kk})^T$ and

$$r^T = \sum_{j>k} z_j P_{jk} + \sum_{j<k} \pi_j P_{jk}, \quad k = 1, 2, \ldots, N.$$

In all cases, $P_{kk}$ is a strictly substochastic matrix so that $B$ is nonsingular. The vector $r$ will have small norm if the system is NCD. If a direct method is used, the $LU$ decomposition of $(I - P_{kk})$, $k = 1, 2, \ldots, N$, need only be performed once, since this remains unchanged from one iteration to the next. If an iterative method is used we have an iteration algorithm within an iteration algorithm. In this case it is advantageous to perform only a small number of iterations, (e.g., 8–12 of the Gauss–Seidel method) each time a solution of $(I - P_{kk})^T x = r$ is needed but to use the final approximation at one step as the initial approximation the next time the solution of that same subsystem is needed.

**Example 10.25** Returning again to the Courtois matrix, the table below shows the number of iterations needed to achieve full machine precision when the diagonal block equations in the IAD method are solved using the Gauss–Seidel iterative method. It can be seen that now an additional iteration is needed.

| IAD: Gauss–Seidel for Block solutions | |
|---|---|
| Iteration | Residual: $\hat{\pi}(I - P)$ |
| | $1.0e-03\times$ |
| 1 | 0.14117911369086 |
| 2 | 0.00016634452597 |
| 3 | 0.00000017031189 |
| 4 | 0.00000000015278 |
| 5 | 0.00000000000014 |
| 6 | 0.00000000000007 |
| 7 | 0.00000000000006 |
| 8 | 0.00000000000003 |
| 9 | 0.00000000000003 |
| 10 | 0.00000000000006 |

Also, if we check the convergence of the inner Gauss–Seidel method for the diagonal blocks—as shown in the table below for the first diagonal block of size $(3 \times 3)$—it can be seen that the iterations stagnate after just a few steps. After about six iterations during each global iteration, progress slows to a crawl. It is for this reason that only a small number of iterations should be used when iterative methods are used to solve the (inner) block equations.

| | Global iteration | | | |
|---|---|---|---|---|
| Inner Iteration | 1 | 2 | 3 | 4 |
| 1 | 0.0131607445 | 0.000009106488 | 0.00000002197727 | 0.00000000002345 |
| 2 | 0.0032775892 | 0.000002280232 | 0.00000000554827 | 0.00000000000593 |
| 3 | 0.0008932908 | 0.000000605958 | 0.00000000142318 | 0.00000000000151 |
| 4 | 0.0002001278 | 0.000000136332 | 0.00000000034441 | 0.00000000000037 |
| 5 | 0.0001468896 | 0.000000077107 | 0.00000000010961 | 0.00000000000011 |
| 6 | 0.0001124823 | 0.000000051518 | 0.00000000003470 | 0.00000000000003 |
| 7 | 0.0001178683 | 0.000000055123 | 0.00000000003872 | 0.00000000000002 |
| 8 | 0.0001156634 | 0.000000053697 | 0.00000000003543 | 0.00000000000002 |
| 9 | 0.0001155802 | 0.000000053752 | 0.00000000003596 | 0.00000000000002 |
| 10 | 0.0001149744 | 0.000000053446 | 0.00000000003562 | 0.00000000000002 |
| 11 | 0.0001145044 | 0.000000053234 | 0.00000000003552 | 0.00000000000002 |
| 12 | 0.0001140028 | 0.000000052999 | 0.00000000003535 | 0.00000000000002 |
| 13 | 0.0001135119 | 0.000000052772 | 0.00000000003520 | 0.00000000000002 |
| 14 | 0.0001130210 | 0.000000052543 | 0.00000000003505 | 0.00000000000002 |
| 15 | 0.0001125327 | 0.000000052316 | 0.00000000003490 | 0.00000000000002 |
| 16 | 0.0001120464 | 0.000000052090 | 0.00000000003475 | 0.00000000000002 |

The following Matlab code may be used to experiment with IAD methods. It mirrors the block Gauss–Seidel method in its parameters. It was used to produce the results provided above.

### *Matlab code for Iterative Aggregation/Disaggregation*

```
function [soln,res] = kms(P,ni,itmax1,itmax2)
[n,n] = size(P); [na,nb] = size(ni);

%%%%%%%% ITERATIVE AGGREGATION/DISAGGREGATION FOR pi*P = pi %%%%%%%%%%%%%%%

bl(1) = 1;                                          % Get beginning and end
for k = 1:nb, bl(k+1) = bl(k)+ni(k); end            % points of each block

E = zeros(n,nb);                                    % Form (n x nb) matrix E
next = 0;                                      % (This is needed in forming
for i = 1:nb,                                       %  the coupling matrix: A )
    for k = 1:ni(i); next = next+1; E(next,i) = 1; end
end
Pije = P*E;                 % Compute constant part of coupling matrix
Phi = zeros(nb,n);          % Phi, used in forming the coupling matrix,
for m = 1:nb,               % keeps normalized parts of approximation
    for j = 1:ni(m), Phi(m,bl(m)+j-1) = 1/ni(m); end
end

A = Phi*Pije;                                       % Form the coupling matrix A
AA = (A-eye(nb))';     en = [zeros(nb-1,1);1];
xi = inv([AA(1:nb-1,1:nb);ones(1,nb)])*en;     % Solve the coupling matrix
z = Phi'*xi;                                        % Initial approximation


%%%%%%%%%%%%%%%%%%%%%%%%%%%  BEGIN OUTER LOOP  %%%%%%%%%%%%%%%%%%%%%%%%%%%%%%%


for iter = 1:itmax1,

    for m = 1:nb,                 % Solve all diag. blocks; Pmm y = b
        Pmm = P(bl(m):bl(m+1)-1,bl(m):bl(m+1)-1);
                                       % Get coefficient block, Pmm
        b = z(bl(m):bl(m+1)-1)'*Pmm-z'*P(1:n,bl(m):bl(m+1)-1);      % RHS

        y = inv(Pmm'-eye(ni(m)))*b';  % Substitute this line for the 2
                      % lines below it, to solve Pmm by iterative method.
        %x0 = z(bl(m):bl(m+1)-1);        % Get new starting vector
        %[y,resid] = gs(Pmm'-eye(ni(m)),x0,b',itmax2);
                                           % y is soln for block Pmm
        for j = 1:ni(m), z(bl(m)+j-1) = y(j); end % Update solution vector
        y = y/norm(y,1);                          % Normalize y

        for j = 1:ni(m),
            Phi(m,bl(m)+j-1) = y(j);              % Update Phi
        end
    end
    pi = z;
```

```
    res(iter) = norm((P'-eye(n))*pi,2);              % Compute residual
    A = Phi*Pije; AA = (A-eye(nb))';                 % Form the coupling matrix A
    xi  = inv([AA(1:nb-1,1:nb);ones(1,nb)])*en;      % Solve the coupling matrix
    z   = Phi'*xi;                                   % Compute new approximation
end

soln = pi; res = res';
```

## 10.6 The Matrix Geometric/Analytic Methods for Structured Markov Chains

We now turn to the numerical solution of Markov chains whose transition matrices have a special block structure—a block structure that arises frequently when modeling queueing systems—and examine an approach pioneered by Neuts [39, 40]. In the simplest case, these matrices are infinite block tridiagonal matrices in which the three diagonal blocks repeat after some initial period. To capture this nonzero structure, we write such a matrix as

$$\begin{pmatrix} B_{00} & B_{01} & 0 & 0 & 0 & 0 & \cdots \\ B_{10} & A_1 & A_2 & 0 & 0 & 0 & \cdots \\ 0 & A_0 & A_1 & A_2 & 0 & 0 & \cdots \\ 0 & 0 & A_0 & A_1 & A_2 & 0 & \cdots \\ & & & \ddots & \ddots & \ddots & \\ & & & & & & \\ \vdots & \vdots & \vdots & \vdots & \vdots & \vdots & \vdots \end{pmatrix}, \tag{10.26}$$

in which submatrices $A_0,\ A_1$, and $A_2$ are square and have the same dimension; the matrix $B_{00}$ is also square and need not have the same size as $A_1$, and the dimensions of $B_{01}$ and $B_{10}$ are defined to be in accordance with the dimensions of $B_{00}$ and $A_1$. A transition matrix having this structure arises when each state of the Markov chain can be represented as a pair $\{(\eta, k),\ \eta \geq 0,\ 1 \leq k \leq K\}$ and the states ordered, first according to increasing value of the parameter $\eta$ and, for states with the same $\eta$ value, by increasing value of $k$. This has the effect of grouping the states into "levels" according to their $\eta$ value. The block tridiagonal effect is achieved when transitions are permitted only between states of the same level (diagonal blocks), to states in the next highest level (superdiagonal blocks), and to states in the adjacent lower level (subdiagonal blocks). The repetitive nature of the blocks themselves arises if, after boundary conditions are taken into consideration (which gives the initial blocks $B_{00},\ B_{01}$, and $B_{10}$) the transition rates/probabilities are identical from level to level. A Markov chain whose transition matrix has this block tridiagonal structure is said to belong to the class of *quasi-birth-death* (QBD) processes.

**Example 10.26** Consider the Markov chain whose state transition diagram is shown in Figure 10.1.

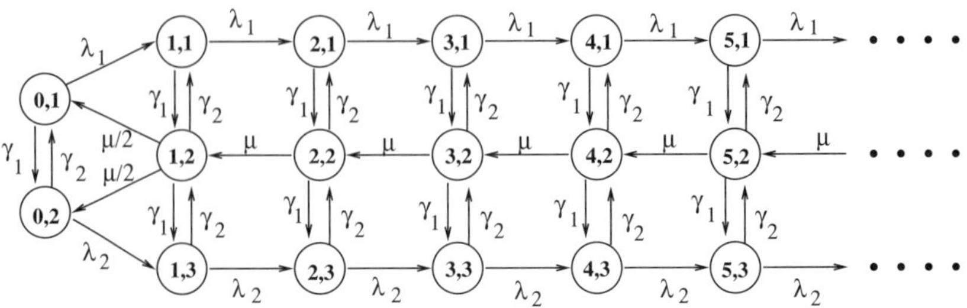


Figure 10.1. State transition diagram for an *M/M/*1-type process.

Its transition rate matrix is

$$Q = \left(\begin{array}{cc|ccc|ccc|ccc|ccc|c|c}
* & \gamma_1 & \lambda_1 & & & & & & & & & & & & & \\
\gamma_2 & * & & & \lambda_2 & & & & & & & & & & & \\
\hline
 & & * & \gamma_1 & & \lambda_1 & & & & & & & & & & \\
\mu/2 & \mu/2 & \gamma_2 & * & \gamma_1 & & & & & & & & & & & \\
 & & & \gamma_2 & * & & & \lambda_2 & & & & & & & & \\
\hline
 & & & & & * & \gamma_1 & & \lambda_1 & & & & & & & \\
 & & & \mu & & \gamma_2 & * & \gamma_1 & & & & & & & & \\
 & & & & & & \gamma_2 & * & & & \lambda_2 & & & & & \\
\hline
 & & & & & & & & * & \gamma_1 & & \lambda_1 & & & & \\
 & & & & & & \mu & & \gamma_2 & * & \gamma_1 & & & & & \\
 & & & & & & & & & \gamma_2 & * & & & \lambda_2 & & \\
\hline
 & & & & & & & & & & & * & \gamma_1 & & \lambda_1 & \\
 & & & & & & & & & \mu & & \gamma_2 & * & \gamma_1 & & \\
 & & & & & & & & & & & & \gamma_2 & * & \lambda_2 & \\
\hline
 & & & & & & & & & & & & \ddots & & \ddots & \ddots
\end{array}\right)$$

and has the typical block tridiagonal structure which makes it an ideal candidate for solution by the numerical techniques described in this section. Its diagonal elements, marked by asterisks, are such that the sum across each row is zero. We have the following block matrices:

$$A_0 = \begin{pmatrix} 0 & 0 & 0 \\ 0 & \mu & 0 \\ 0 & 0 & 0 \end{pmatrix}, \quad A_1 = \begin{pmatrix} -(\gamma_1 + \lambda_1) & \gamma_1 & 0 \\ \gamma_2 & -(\mu + \gamma_1 + \gamma_2) & \gamma_1 \\ 0 & \gamma_2 & -(\gamma_2 + \lambda_2) \end{pmatrix}, \quad A_2 = \begin{pmatrix} \lambda_1 & 0 & 0 \\ 0 & 0 & 0 \\ 0 & 0 & \lambda_2 \end{pmatrix},$$

and

$$B_{00} = \begin{pmatrix} -(\gamma_1 + \lambda_1) & \gamma_1 \\ \gamma_2 & -(\gamma_2 + \lambda_2) \end{pmatrix}, \quad B_{01} = \begin{pmatrix} \lambda_1 & 0 & 0 \\ 0 & 0 & \lambda_2 \end{pmatrix}, \quad B_{10} = \begin{pmatrix} 0 & 0 \\ \mu/2 & \mu/2 \\ 0 & 0 \end{pmatrix}.$$

The most common extensions of this simplest case are matrices which are block upper Hessenberg (referred to as *M/G/*1 type and solved using the matrix analytic approach) and those that are block lower Hessenberg (referred to as *G/M/*1 type, solved using the matrix geometric approach). Both are treated in this text. Results are also available for the case in which the parameters $\eta$ and $k$ are both finite, which results in a finite Markov chain, as well as for the case in which transitions may be dependent on the level $\eta$. These last extensions are beyond the scope of this text and the interested reader is invited to consult the references.

### 10.6.1 The Quasi-Birth-Death Case

Quasi-birth-death processes can be conveniently and efficiently solved using the *matrix geometric method*. We begin by considering the case when the blocks of a QBD process are reduced to single elements. Consider, for example, the following infinite infinitesimal generator:

$$Q = \begin{pmatrix} -\lambda & \lambda & & & & \\ \mu & -(\lambda+\mu) & \lambda & & & \\ & \mu & -(\lambda+\mu) & \lambda & & \\ & & \mu & -(\lambda+\mu) & \lambda & \\ & & & \ddots & \ddots & \ddots \end{pmatrix}.$$

This may be associated with a random walk problem in which the probability of moving from any state $k$ to state $k+1$ is $\lambda/(\lambda+\mu)$ and the probability of moving from state $k$ to state $k-1$ is $\mu/(\lambda+\mu)$. In a later chapter, we shall associate this infinitesimal generator with the *M/M/*1 queue. From $\pi Q = 0$, we may write $-\lambda\pi_0 + \mu\pi_1 = 0$, which gives $\pi_1 = (\lambda/\mu)\pi_0$, and in general

$$\lambda\pi_{i-1} - (\lambda+\mu)\pi_i + \mu\pi_{i+1} = 0.$$

We proceed by induction to show that $\pi_{i+1} = (\lambda/\mu)\pi_i$ for $i = 1, 2, \ldots$. We have already established the basis clause $\pi_1 = (\lambda/\mu)\pi_0$. From the inductive hypothesis, we have $\pi_i = (\lambda/\mu)\pi_{i-1}$ and hence

$$\pi_{i+1} = \left(\frac{\lambda+\mu}{\mu}\right)\pi_i - \left(\frac{\lambda}{\mu}\right)\pi_{i-1} = \left(\frac{\lambda}{\mu}\right)\pi_i,$$

which is the desired result. It now follows that

$$\pi_i = \left(\frac{\lambda}{\mu}\right)^i \pi_0 = \rho^i\pi_0,$$

where $\rho = \lambda/\mu$. Thus, once $\pi_0$ is known, the remaining values, $\pi_i$, $i = 1, 2, \ldots$, may be determined recursively. A similar result exists when $Q$ is a QBD process: In this case, the parameter $\rho$ becomes a square matrix $R$ of order $K$ and the components $\pi_i$ of the stationary distribution become subvectors of length $K$.

Let $Q$ be the infinitesimal generator of a QBD process. Then

$$Q = \begin{pmatrix} B_{00} & B_{01} & 0 & 0 & 0 & 0 & \cdots \\ B_{10} & A_1 & A_2 & 0 & 0 & 0 & \cdots \\ 0 & A_0 & A_1 & A_2 & 0 & 0 & \cdots \\ 0 & 0 & A_0 & A_1 & A_2 & 0 & \cdots \\ & & & \ddots & \ddots & \ddots & \\ \vdots & \vdots & \vdots & \vdots & \vdots & \vdots & \vdots \end{pmatrix},$$

and the stationary distribution is obtained from $\pi Q = 0$. Let $\pi$ be partitioned conformally with $Q$, i.e.,

$$\pi = (\pi_0, \pi_1, \pi_2, \ldots),$$

where

$$\pi_i = (\pi(i,1), \pi(i,2), \ldots \pi(i,K))$$

for $i = 0, 1, \ldots$, and $\pi(i,k)$ is the probability of finding the system in state $(i,k)$ at steady state. This gives the following equations:

$$\begin{aligned} \pi_0 B_{00} + \pi_1 B_{10} &= 0, \\ \pi_0 B_{01} + \pi_1 A_1 + \pi_2 A_0 &= 0, \\ \pi_1 A_2 + \pi_2 A_1 + \pi_3 A_0 &= 0, \\ &\vdots \\ \pi_{i-1} A_2 + \pi_i A_1 + \pi_{i+1} A_0 &= 0, \quad i = 2, 3, \ldots. \end{aligned}$$

In analogy with the point situation, it may be shown that there exists a constant matrix $R$ such that

$$\pi_i = \pi_{i-1}R \quad \text{for } i = 2, 3, \ldots. \tag{10.27}$$

The subvectors $\pi_i$ are said to be *geometrically* related to each other since

$$\pi_i = \pi_1 R^{i-1} \quad \text{for } i = 2, 3, \ldots. \tag{10.28}$$

If the subvectors $\pi_0$ and $\pi_1$ and the *rate matrix* $R$ can be found then the remaining subvectors of the stationary distribution may be formed using Equation (10.27). Returning to

$$\pi_{i-1}A_2 + \pi_i A_1 + \pi_{i+1}A_0 = 0$$

and substituting from Equation (10.28), we obtain, for $i = 2, 3, \ldots,$

$$\pi_1 R^{i-2} A_2 + \pi_1 R^{i-1} A_1 + \pi_1 R^i A_0 = 0,$$

i.e.,

$$\pi_1 R^{i-2}\left(A_2 + RA_1 + R^2 A_0\right) = 0.$$

It is now apparent that $R$ can be computed from

$$\left(A_2 + RA_1 + R^2 A_0\right) = 0. \tag{10.29}$$

The simplest way to accomplish this is by successive substitution. From Equation (10.29) and using the fact that $A_1$ must be nonsingular, we have

$$A_2 A_1^{-1} + R + R^2 A_0 A_1^{-1} = 0,$$

i.e.,

$$R = -A_2 A_1^{-1} - R^2 A_0 A_1^{-1} = -V - R^2 W,$$

where $V = A_2 A_1^{-1}$ and $W = A_0 A_1^{-1}$. This leads to the successive substitution procedure proposed by Neuts, namely,

$$R_{(0)} = 0, \quad R_{(k+1)} = -V - R_{(k)}^2 W, \quad k = 0, 1, 2, \ldots. \tag{10.30}$$

Neuts has shown that the sequence of matrices $R_{(k)}$, $k = 0, 1, 2, \ldots,$ is nondecreasing and converges to the rate matrix $R$. The process is halted once successive differences are less than a specified tolerance criterion. Unfortunately, this simple approach has the disadvantage of frequently requiring many iterations before a sufficiently accurate matrix $R$ is obtained. On the other hand, a *logarithmic reduction algorithm* developed by Latouche and Ramaswami [27] has extremely fast quadratic convergence (the number of decimal places doubles at each iteration). The development of this algorithm is beyond the scope of this text: it is presented in pseudocode and without additional comment, at the end of this section. There are also situations in which the rate matrix $R$ can be computed explicitly, without the need to conduct any iterations at all.

The only remaining problem is the derivation of $\pi_0$ and $\pi_1$. The first two equations of $\pi Q = 0$ are

$$\pi_0 B_{00} + \pi_1 B_{10} = 0,$$
$$\pi_0 B_{01} + \pi_1 A_1 + \pi_2 A_0 = 0.$$

Replacing $\pi_2$ with $\pi_1 R$ and writing these equations in matrix form, we obtain

$$(\pi_0, \pi_1)\begin{pmatrix} B_{00} & B_{01} \\ B_{10} & A_1 + RA_0 \end{pmatrix} = (0, 0). \tag{10.31}$$

Given the rate matrix $R$ and blocks $B_{00}$, $B_{01}$, $B_{10}$, $A_1$, and $A_0$, this system may be solved to obtain $\pi_0$ and $\pi_1$. Since this is a homogeneous system of equations, the computed solution needs to be normalized so that the components of $\pi$ sum to 1. In other words, we insist that $\pi e = 1$. Thus

$$\begin{aligned} 1 = \pi e &= \pi_0 e + \pi_1 e + \sum_{i=2}^{\infty} \pi_i e \\ &= \pi_0 e + \pi_1 e + \sum_{i=2}^{\infty} \pi_1 R^{i-1} e \\ &= \pi_0 e + \sum_{i=1}^{\infty} \pi_1 R^{i-1} e = \pi_0 e + \sum_{i=0}^{\infty} \pi_1 R^{i} e. \end{aligned}$$

This implies the condition

$$\pi_0 e + \pi_1 \left( \sum_{i=0}^{\infty} R^i \right) e = 1.$$

The eigenvalues of $R$ lie *inside* the unit circle, which means that $(I - R)$ is nonsingular and hence

$$\left( \sum_{i=0}^{\infty} R^i \right) = (I - R)^{-1}. \tag{10.32}$$

This enables us to complete the normalization of the vectors $\pi_0$ and $\pi_1$ by computing

$$\alpha = \pi_0 e + \pi_1 (I - R)^{-1} e$$

and dividing the computed subvectors $\pi_0$ and $\pi_1$ by $\alpha$.

In the simpler case, when $B_{00}$ is the same size as the $A$ blocks and when $B_{01} = A_2$, then $\pi_i = \pi_0 R^i$ for $i = 1, 2, \ldots$. In this case the system of equations (10.31) can be replaced with the simpler system, $\pi_0(B_{00} + RB_{10}) = 0$, from which $\pi$ can be computed and then normalized so that $\pi_0 (I - R)^{-1} e = 1$.

In a previous discussion about a random walk problem, it was observed that the Markov chain is positive recurrent only if the probability of moving to higher-numbered states is strictly less than the probability of moving to lower-numbered states. A similar condition exists for a QBD process to be ergodic, namely, that the *drift* to higher-numbered levels must be strictly less than the *drift* to lower levels. Let the stationary distribution of the infinitesimal generator $A = A_0 + A_1 + A_2$ be denoted by $\pi_A$. For a QBD process to be ergodic, the following condition must hold:

$$\pi_A A_2 e \ < \ \pi_A A_0 e. \tag{10.33}$$

Recall that the elements of $A_2$ move the process up to a higher-numbered level while those of $A_0$ move it down a level. Indeed, it is from this condition that Neuts shows that the spectral radius of $R$ is strictly less than 1 and consequently that the matrix $I - R$ is nonsingular.

When the Markov chain under consideration is a discrete-time Markov chain characterized by a stochastic transition probability matrix rather than the continuous-time version discussed in this section, then an almost identical analysis can be performed: it suffices to replace $-A_1^{-1}$ with $(I - A_1)^{-1}$. But be careful: in the first case, $A_1$ is the repeating diagonal block in a transition rate matrix while in the second it is the repeating diagonal block in a stochastic matrix. The formulation of the equations is the same, just the values in the blocks change according to whether the global matrix is an infinitesimal generator or a stochastic matrix. Before proceeding to an example, we summarize the steps that must be undertaken when solving a QBD process by the matrix geometric method:

1. Ensure that the matrix has the requisite block structure.
2. Use Equation (10.33) to ensure that the Markov chain is ergodic.

3. Use Equation (10.30) to compute the matrix $R$.
4. Solve the system of equations (10.31) for $\pi_0$ and $\pi_1$.
5. Compute the normalizing constant $\alpha$ and normalize $\pi_0$ and $\pi_1$.
6. Use Equation (10.27) to compute the remaining components of the stationary distribution vector.

**Example 10.27** We shall apply the matrix geometric method to the Markov chain of Example 10.26 using the following values of the parameters:

$$\lambda_1 = 1, \ \lambda_2 = .5, \ \mu = 4, \ \gamma_1 = 5, \ \gamma_2 = 3.$$

The infinitesimal generator is then given by

$$Q = \left(\begin{array}{cc|ccc|ccc|ccc|c}
-6 & 5.0 & 1 & & & & & & & & & \\
3 & -3.5 & & & .5 & & & & & & & \\ \hline
 & & -6 & 5 & & 1 & & & & & & \\
2 & 2.0 & 3 & -12 & 5.0 & & & & & & & \\
 & & & 3 & -3.5 & & & .5 & & & & \\ \hline
 & & & & & -6 & 5 & & 1 & & & \\
 & & & 4 & & 3 & -12 & 5.0 & & & & \\
 & & & & & & 3 & -3.5 & & & .5 & \\ \hline
 & & & & & & \ddots & & & \ddots & & \ddots
\end{array}\right).$$

1. The matrix obviously has the correct QBD structure.
2. We check that the system is stable by verifying Equation (10.33). The infinitesimal generator matrix

$$A = A_0 + A_1 + A_2 = \begin{pmatrix} -5 & 5 & 0 \\ 3 & -8 & 5 \\ 0 & 3 & -3 \end{pmatrix}$$

has stationary probability vector

$$\pi_A = (.1837, \ .3061, \ .5102)$$

and

$$.4388 = \pi_A A_2 e \ < \ \pi_A A_0 e = 1.2245.$$

3. We now initiate the iterative procedure to compute the rate matrix $R$. The inverse of $A_1$ is

$$A_1^{-1} = \begin{pmatrix} -.2466 & -.1598 & -.2283 \\ -.0959 & -.1918 & -.2740 \\ -.0822 & -.1644 & -.5205 \end{pmatrix},$$

which allows us to compute

$$V = A_2 A_1^{-1} = \begin{pmatrix} -.2466 & -.1598 & -.2283 \\ 0 & 0 & 0 \\ -.0411 & -.0822 & -.2603 \end{pmatrix} \quad \text{and}$$

$$W = A_0 A_1^{-1} = \begin{pmatrix} 0 & 0 & 0 \\ -.3836 & -.7671 & -1.0959 \\ 0 & 0 & 0 \end{pmatrix}.$$

Equation (10.30) becomes

$$R_{(k+1)} = \begin{pmatrix} .2466 & .1598 & .2283 \\ 0 & 0 & 0 \\ .0411 & .0822 & .2603 \end{pmatrix} + R_{(k)}^2 \begin{pmatrix} 0 & 0 & 0 \\ .3836 & .7671 & 1.0959 \\ 0 & 0 & 0 \end{pmatrix},$$

and iterating successively, beginning with $R_{(0)} = 0$, we find

$$R_{(1)} = \begin{pmatrix} .2466 & .1598 & .2283 \\ 0 & 0 & 0 \\ .0411 & .0822 & .2603 \end{pmatrix}, \quad R_{(2)} = \begin{pmatrix} .2689 & .2044 & .2921 \\ 0 & 0 & 0 \\ .0518 & .1036 & .2909 \end{pmatrix},$$

$$R_{(3)} = \begin{pmatrix} .2793 & .2252 & .3217 \\ 0 & 0 & 0 \\ .0567 & .1134 & .3049 \end{pmatrix}, \quad \cdots .$$

Observe that the elements are nondecreasing, as predicted by Neuts. After 48 iterations, successive differences are smaller than $10^{-12}$, at which point

$$R_{(48)} = \begin{pmatrix} .2917 & .2500 & .3571 \\ 0 & 0 & 0 \\ .0625 & .1250 & .3214 \end{pmatrix}.$$

4. Proceeding to the boundary conditions,

$$(\pi_0, \pi_1) \begin{pmatrix} B_{00} & B_{01} \\ B_{10} & A_1 + RA_0 \end{pmatrix} = (\pi_0, \pi_1) \left(\begin{array}{cc|ccc} -6 & 5.0 & 1 & 0 & 0 \\ 3 & -3.5 & 0 & 0 & .5 \\ \hline 0 & 0 & -6 & 6.0 & 0 \\ 2 & 2.0 & 3 & -12.0 & 5.0 \\ 0 & 0 & 0 & 3.5 & -3.5 \end{array}\right) = (0, 0).$$

We can solve this by replacing the last equation with the equation $\pi_{0_1} = 1$, i.e., setting the first component of the subvector $\pi_0$ to 1. The system of equations becomes

$$(\pi_0, \pi_1) \left(\begin{array}{cc|ccc} -6 & 5.0 & 1 & 0 & 1 \\ 3 & -3.5 & 0 & 0 & 0 \\ \hline 0 & 0 & -6 & 6.0 & 0 \\ 2 & 2.0 & 3 & -12.0 & 0 \\ 0 & 0 & 0 & 3.5 & 0 \end{array}\right) = (0, 0 \mid 0, 0, 1)$$

with solution

$$(\pi_0, \pi_1) = (1.0,\ 1.6923,\ \mid .3974,\ .4615,\ .9011).$$

5. The normalization constant is

$$\begin{aligned} \alpha &= \pi_0 e + \pi_1 (I - R)^{-1} e \\ &= (1.0,\ 1.6923)e + (.3974,\ .4615,\ .9011) \begin{pmatrix} 1.4805 & .4675 & .7792 \\ 0 & 1 & 0 \\ .1364 & .2273 & .15455 \end{pmatrix} e \\ &= 2.6923 + 3.2657 = 5.9580, \end{aligned}$$

which allows us to compute

$$\pi_0/\alpha = (.1678,\ .2840), \quad \text{and} \quad \pi_1/\alpha = (.0667,\ .0775,\ .1512).$$

6. Successive subcomponents of the stationary distribution are now computed from $\pi_k = \pi_{k-1}R$. For example,

$$\pi_2 = \pi_1 R = (.0667,\ .0775,\ .1512)\begin{pmatrix} .2917 & .2500 & .3571 \\ 0 & 0 & 0 \\ .0625 & .1250 & .3214 \end{pmatrix} = (.0289,\ .0356,\ .0724),$$

$$\pi_3 = \pi_2 R = (.0289,\ .0356,\ .0724)\begin{pmatrix} .2917 & .2500 & .3571 \\ 0 & 0 & 0 \\ .0625 & .1250 & .3214 \end{pmatrix} = (.0130,\ .0356,\ .0336),$$

and so on.

Steps 2 through 6 of the basic matrix geometric method for QBD processes are embedded in the following Matlab code in which $n$ is the order of the blocks $A_i$ and $m$ is the order of $B_{00}$:

### *Matlab code for QBD processes*

```
function[R,pi0,pi1] = mgm(n,A0,A1,A2,m,B00,B01,B10)

%              Matrix Geometric Method

%%%%%%%%%%%%%  Form Neuts' R matrix   %%%%%%%%%%%%%%%
     V = A2 * inv(A1);   W = A0 * inv(A1);
     R = zeros(n,n);   Rbis = -V - R*R * W;
     iter = 1;
     while (norm(R-Rbis,1) > 1.0e-10)
        R = Rbis;   Rbis = -V - R*R * W;
        iter = iter+1;
     end
     R = Rbis;

%%%%%%%  Form and solve boundary equations    %%%%%%%
     C = [B00,B01;B10,A1+R*A0];
     z = zeros(n+m,1); z(1) = 1;
     Cn = [C(1:n+m,1:n+m-1),z];
     rhs = zeros(1,n+m); rhs(n+m) = 1;
     pi = rhs*inv(Cn);
     pi0 = pi(1:m);    pi1 = pi(m+1:m+n);

%%%%%%%%%%  Normalize computed solution   %%%%%%%%%%
     alpha = norm(pi0,1)+ norm(pi1*inv(eye(n)-R),1);
     pi0 = pi0/alpha;  pi1 = pi1/alpha;

%%%%%%%%%%  Compute successive subvectors   %%%%%%%%%
     pi2 = pi1*R;  pi3 = pi2*R;   % and so on.
```

ALGORITHM 10.10: LOGARITHMIC REDUCTION—QUADRATIC CONVERGENCE FOR QBD PROCESSES

1. *Initialize:* $i = 0$;

| | | | |
|---|---|---|---|
| | $C_0 = (I - A_1)^{-1}A_0$; | $C_2 = (I - A_1)^{-1}A_2$; | if stochastic, or |
| %%% | $C_0 = -A_1^{-1}A_0$; | $C_2 = -A_1^{-1}A_2$; | if infinitesimal generator. |
| | $T = C_0$; | $S = C_2$; | |

2. *While* $\|e - Se\|_\infty < \epsilon$:
   - $i = i + 1;$
   - $A_1^* = C_0C_2 + C_2C_0; \quad A_0^* = C_0^2; \quad A_2^* = C_2^2;$
   - $C_0 = (I - A_1^*)^{-1}A_0^*; \quad C_2 = (I - A_1^*)^{-1}A_2^*;$
   - $S = S + TC_2; \qquad T = TC_0;$
3. *Termination*
   $G = S; \quad U = A_1 + A_0G; \quad R = A_0(I - U)^{-1}.$

### *10.6.2 Block Lower Hessenberg Markov Chains*

In a lower Hessenberg matrix $A$, all elements $a_{ij}$ must be zero for values of $j > i + 1$. In other words, if moving from top right to bottom left, we designate the three diagonals of a tridiagonal matrix as the superdiagonal, the diagonal, and the subdiagonal, then a lower Hessenberg matrix can have nonzero elements only on and below the superdiagonal. For example, the following matrix is a $6 \times 6$ lower Hessenberg matrix:

$$A = \begin{pmatrix} a_{00} & a_{01} & 0 & 0 & 0 & 0 \\ a_{10} & a_{11} & a_{12} & 0 & 0 & 0 \\ a_{20} & a_{21} & a_{22} & a_{23} & 0 & 0 \\ a_{30} & a_{31} & a_{32} & a_{33} & a_{34} & 0 \\ a_{40} & a_{41} & a_{42} & a_{43} & a_{44} & a_{45} \\ a_{50} & a_{51} & a_{52} & a_{53} & a_{54} & a_{55} \end{pmatrix}.$$

Block lower Hessenberg matrices, found in *G/M/*1-type stochastic processes, are just the block counterparts of lower Hessenberg matrices. In a similar manner, we can define block *upper* Hessenberg matrices, which are found in *M/G/*1-type stochastic processes, as block matrices whose only nonzero blocks are on or above the diagonal blocks and along the subdiagonal block. One is essentially the block transpose of the other. This leads us to caution the reader about the notation used to designate the individual blocks. With block lower Hessenberg matrices, the nonzero blocks are numbered from right to left as $A_0, A_1, A_2, \ldots$, while in block upper Hessenberg matrices they are numbered in the reverse order, from left to right. As we have just seen, we choose to denote the blocks in QBD processes (which are simultaneously both upper and lower block Hessenberg) from left to right, i.e., using the block upper Hesssenberg notation. Our interest in this section is with Markov chains whose infinitesimal generators $Q$ have the following repetitive block lower Hessenberg structure:

$$Q = \begin{pmatrix} B_{00} & B_{01} & 0 & 0 & 0 & 0 & 0 & \cdots \\ B_{10} & B_{11} & A_0 & 0 & 0 & 0 & 0 & \cdots \\ B_{20} & B_{21} & A_1 & A_0 & 0 & 0 & 0 & \cdots \\ B_{30} & B_{31} & A_2 & A_1 & A_0 & 0 & 0 & \cdots \\ B_{40} & B_{41} & A_3 & A_2 & A_1 & A_0 & 0 & \cdots \\ \vdots & & \ddots & \ddots & \ddots & \ddots & \ddots & \cdots \\ \vdots & \vdots & \vdots & \vdots & \vdots & \vdots & \vdots & \cdots \end{pmatrix}.$$

Markov chains with this structure occur when the states are grouped into levels, similar to those for QBD processes, but now transitions are no longer confined to interlevel and to adjacent neighboring levels; transitions are also permitted from any level to any *lower* level. In this particular case, we perform the analysis using two boundary columns ($B_{i0}$ and $B_{i1}, i = 0, 1, 2, \ldots$). Certain applications give rise to more than two boundary columns, which may necessitate a restructuring of the matrix. This is considered at the end of this section.

As always, our objective is to compute the stationary probability vector $\pi$ from the system of equations $\pi Q = 0$. As for QBD processes, let $\pi$ be partitioned conformally with $Q$, i.e. $\pi = (\pi_0, \pi_1 \pi_2, \ldots)$ where $\pi_i = (\pi(i,1)\pi(i,2), \ldots \pi(i,K))$ for $i = 0, 1, \ldots$, and $\pi(i,k)$ is the probability of finding the Markov chain in state $(i,k)$ at statistical equilibrium. Neuts has shown that there exists a matrix geometric solution to this problem and that it mirrors that of a QBD process, specifically that there exists a positive matrix $R$ such that

$$\pi_i = \pi_{i-1} R \quad \text{for } i = 2, 3, \ldots,$$

i.e., that

$$\pi_i = \pi_1 R^{i-1} \quad \text{for } i = 2, 3, \ldots.$$

Observe that from $\pi Q = 0$ we have

$$\sum_{k=0}^{\infty} \pi_{k+j} A_k = 0, \qquad j = 1, 2, \ldots,$$

and, in particular,

$$\pi_1 A_0 + \pi_2 A_1 + \sum_{k=2}^{\infty} \pi_{k+1} A_k = 0.$$

Substituting $\pi_i = \pi_1 R^{i-1}$

$$\pi_1 A_0 + \pi_1 R A_1 + \sum_{k=2}^{\infty} \pi_1 R^k A_k = 0$$

or

$$\pi_1 \left( A_0 + R A_1 + \sum_{k=2}^{\infty} R^k A_k \right) = 0$$

provides the following relation from which the matrix $R$ may be computed:

$$A_0 + R A_1 + \sum_{k=2}^{\infty} R^k A_k = 0. \tag{10.34}$$

Notice that Equation (10.34) reduces to Equation (10.29) when $A_k = 0$ for $k > 2$. Rearranging Equation (10.34), we find

$$R = -A_0 A_1^{-1} - \sum_{k=2}^{\infty} R^k A_k A_1^{-1},$$

which leads to the iterative procedure

$$R_{(0)} = 0, \quad R_{(l+1)} = -A_0 A_1^{-1} - \sum_{k=2}^{\infty} R_{(l)}^k A_k A_1^{-1}, \quad l = 0, 1, 2, \ldots,$$

which is, as Neuts has shown, nondecreasing and converges to the matrix $R$. In many cases, the structure of the infinitesimal generator is such that the blocks $A_i$ are zero for relatively small values of $i$, which limits the computational effort needed in each iteration. As before, the number of iterations needed for convergence is frequently large, and now the extremely efficient logarithmic reduction algorithm is no longer applicable—it is designed for QBD processes only. However, more efficient but also more complex algorithms have been developed and may be found in the current literature [3, 36, 43, 44].

We now turn to the derivation of the initial subvectors $\pi_0$ and $\pi_1$. From the first equation of $\pi Q = 0$, we have

$$\sum_{i=0}^{\infty} \pi_i B_{i0} = 0$$

and we may write

$$\pi_0 B_{00} + \sum_{i=1}^{\infty} \pi_i B_{i0} = \pi_0 B_{00} + \sum_{i=1}^{\infty} \pi_1 R^{i-1} B_{i0} = \pi_0 B_{00} + \pi_1 \left( \sum_{i=1}^{\infty} R^{i-1} B_{i0} \right) = 0, \qquad (10.35)$$

while from the second equation of $\pi Q = 0$,

$$\pi_0 B_{01} + \sum_{i=1}^{\infty} \pi_i B_{i1} = 0, \quad \text{i.e.,} \quad \pi_0 B_{01} + \pi_1 \sum_{i=1}^{\infty} R^{i-1} B_{i1} = 0. \qquad (10.36)$$

Putting Equations (10.35) and (10.36) together in matrix form, we see that we can compute $\pi_0$ and $\pi_1$ from

$$(\pi_0, \pi_1) \begin{pmatrix} B_{00} & B_{01} \\ \sum_{i=1}^{\infty} R^{i-1} B_{i0} & \sum_{i=1}^{\infty} R^{i-1} B_{i1} \end{pmatrix} = (0, 0).$$

The computed values of $\pi_0$ and $\pi_1$ must now be normalized by dividing them by

$$\alpha = \pi_0 e + \pi_1 \left( \sum_{i=1}^{\infty} R^{i-1} \right) e = \pi_0 e + \pi_1 (I - R)^{-1} e.$$

**Example 10.28** We shall apply the matrix geometric method to the Markov chain of Example 10.27 now modified so that it incorporates additional transitions ($\xi_1 = .25$ and $\xi_2 = .75$) to lower non-neighboring states, Figure 10.2.

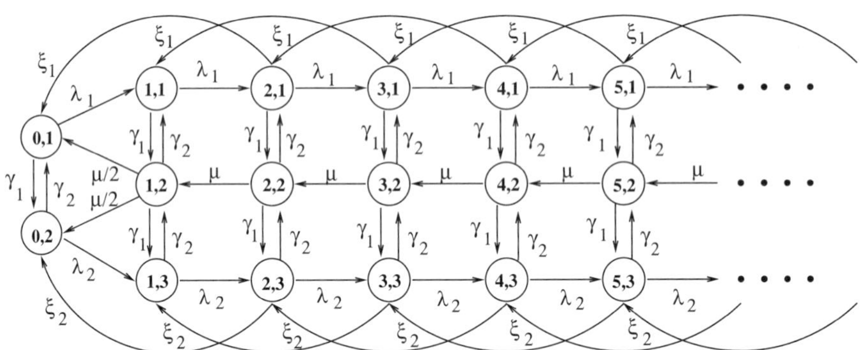


Figure 10.2. State transition diagram for a *G/M/*1-type process.

$$Q = \left( \begin{array}{cc|ccc|ccc|ccc|cc|c}
-6 & 5.0 & 1 & & & & & & & & & & & \\
3 & -3.5 & & & .5 & & & & & & & & & \\
\hline
 & & -6 & 5 & & 1 & & & & & & & & \\
2 & 2 & 3 & -12 & 5.0 & & & & & & & & & \\
 & & & 3 & -3.5 & & & .5 & & & & & & \\
\hline
.25 & & & & & -6.25 & 5 & & 1 & & & & & \\
 & & & 4 & & 3.00 & -12 & 5.00 & & & & & & \\
 & .75 & & & & & 3 & -4.25 & & & .5 & & & \\
\hline
 & & .25 & & & & & & -6.25 & 5 & & 1 & & \\
 & & & & & & 4 & & 3.00 & -12 & 5.00 & & & \\
 & & & & .75 & & & & & 3 & -4.25 & & .5 & \\
\hline
 & & & & & \ddots & & & \ddots & & & \ddots & & \ddots
\end{array} \right).$$

The computation of the matrix $R$ proceeds as previously: The inverse of $A_1$ is now

$$A_1^{-1} = \begin{pmatrix} -.2233 & -.1318 & -.1550 \\ -.0791 & -.1647 & -.1938 \\ -.0558 & -.1163 & -.3721 \end{pmatrix},$$

which allows us to compute

$$A_0 A_1^{-1} = \begin{pmatrix} -.2233 & -.1318 & -.1550 \\ 0 & 0 & 0 \\ -.0279 & -.0581 & -.1860 \end{pmatrix}, \quad A_2 A_1^{-1} = \begin{pmatrix} 0 & 0 & 0 \\ -.3163 & -.6589 & -.7752 \\ 0 & 0 & 0 \end{pmatrix},$$

$$A_3 A_1^{-1} = \begin{pmatrix} -.0558 & -.0329 & -.0388 \\ 0 & 0 & 0 \\ -.0419 & -.0872 & -.2791 \end{pmatrix}.$$

The iterative process is

$$R_{(k+1)} = \begin{pmatrix} .2233 & .1318 & .1550 \\ 0 & 0 & 0 \\ .0279 & .0581 & .1860 \end{pmatrix} + R_{(k)}^2 \begin{pmatrix} 0 & 0 & 0 \\ .3163 & .6589 & .7752 \\ 0 & 0 & 0 \end{pmatrix} + R_{(k)}^3 \begin{pmatrix} .0558 & .0329 & .0388 \\ 0 & 0 & 0 \\ .0419 & .0872 & .2791 \end{pmatrix},$$

and iterating successively, beginning with $R_{(0)} = 0$, we find

$$R_{(1)} = \begin{pmatrix} .2233 & .1318 & .1550 \\ 0 & 0 & 0 \\ .0279 & .0581 & .1860 \end{pmatrix}, \quad R_{(2)} = \begin{pmatrix} .2370 & .1593 & .1910 \\ 0 & 0 & 0 \\ .0331 & .0686 & .1999 \end{pmatrix},$$

$$R_{(3)} = \begin{pmatrix} .2415 & .1684 & .2031 \\ 0 & 0 & 0 \\ .0347 & .0719 & .2043 \end{pmatrix}, \quad \ldots.$$

After 27 iterations, successive differences are smaller than $10^{-12}$, at which point

$$R_{(27)} = \begin{pmatrix} .2440 & .1734 & .2100 \\ 0 & 0 & 0 \\ .0356 & .0736 & .1669 \end{pmatrix}.$$

The boundary conditions are now

$$(\pi_0, \pi_1) \begin{pmatrix} B_{00} & B_{01} \\ B_{10} + RB_{20} & B_{11} + RB_{21} + R^2 B_{31} \end{pmatrix} = (0,\ 0),$$

i.e.,

$$= (\pi_0, \pi_1) \left( \begin{array}{cc|ccc} -6.0 & 5.0 & 1 & 0 & 0 \\ 3.0 & -3.5 & 0 & 0 & .5 \\ \hline .0610 & .1575 & -5.9832 & 5.6938 & .0710 \\ 2.0000 & 2.000 & 3.000 & -12.0000 & 5.0000 \\ .0089 & .1555 & .0040 & 3.2945 & -3.4624 \end{array} \right) = (0,\ 0).$$

As before, we solve this by replacing the last equation with the equation $\pi_{0_1} = 1$. The system of equation becomes

$$(\pi_0, \pi_1) \left( \begin{array}{cc|ccc} -6.0 & 5.0 & 1 & 0 & 1 \\ 3.0 & -3.5 & 0 & 0 & 0 \\ \hline .0610 & .1575 & -5.9832 & 5.6938 & 0 \\ 2.0000 & 2.000 & 3.000 & -12.0000 & 0 \\ .0089 & .1555 & .0040 & 3.2945 & 0 \end{array} \right) = (0, 0 \mid 0, 0, 1)$$

with solution

$$(\pi_0, \pi_1) = (1.0,\ 1.7169 \mid .3730,\ .4095,\ .8470).$$

The normalization constant is

$$\begin{aligned}\alpha &= \pi_0 e + \pi_1 (I - R)^{-1} e \\ &= (1.0,\ 1.7169)e + (.3730,\ .4095,\ .8470) \begin{pmatrix} 1.3395 & .2584 & .3546 \\ 0 & 1 & 0 \\ .0600 & .1044 & 1.2764 \end{pmatrix} e \\ &= 2.7169 + 2.3582 = 5.0751,\end{aligned}$$

which allows us to compute

$$\pi_0/\alpha = (.1970,\ .3383) \quad \text{and} \quad \pi_1/\alpha = (.0735,\ .0807,\ .1669).$$

Successive subcomponents of the stationary distribution are now computed from $\pi_k = \pi_{k-1} R$. For example,

$$\pi_2 = \pi_1 R = (.0735,\ .0807,\ .1669) \begin{pmatrix} .2440 & .1734 & .2100 \\ 0 & 0 & 0 \\ .0356 & .0736 & .1669 \end{pmatrix} = (.0239,\ .0250,\ .0499)$$

and

$$\pi_3 = \pi_2 R = (.0239,\ .0250,\ .0499) \begin{pmatrix} .2440 & .1734 & .2100 \\ 0 & 0 & 0 \\ .0356 & .0736 & .1669 \end{pmatrix} = (.0076,\ .0078,\ .0135),$$

and so on.

Some simplifications occur when the initial $B$ blocks have the same dimensions as the $A$ blocks and when the infinitesimal generator can be written in the following commonly occurring form:

$$Q = \begin{pmatrix} B_{00} & A_0 & 0 & 0 & 0 & 0 & 0 & \cdots \\ B_{10} & A_1 & A_0 & 0 & 0 & 0 & 0 & \cdots \\ B_{20} & A_2 & A_1 & A_0 & 0 & 0 & 0 & \cdots \\ B_{30} & A_3 & A_2 & A_1 & A_0 & 0 & 0 & \cdots \\ B_{40} & A_4 & A_3 & A_2 & A_1 & A_0 & 0 & \cdots \\ \vdots & & \ddots & \ddots & \ddots & \ddots & \ddots & \cdots \\ \vdots & \vdots & \vdots & \vdots & \vdots & \vdots & \vdots & \cdots \end{pmatrix}.$$

In this case,

$$\pi_i = \pi_0 R^i \quad \text{for } i = 1, 2, \ldots.$$

Furthermore, $\sum_{i=0}^{\infty} R^i B_{i0}$ is an infinitesimal generator and the subvector $\pi_0$ is the stationary probability vector of $\sum_{i=0}^{\infty} R^i B_{i0}$ normalized so that $\pi_0(I - R)^{-1} e = 1$.

In some applications, such as queueing systems with bulk arrivals, more than two boundary columns can occur. Consider, for example the generator matrix

$$Q = \begin{pmatrix} B_{00} & B_{01} & B_{02} & A_0 & & & & & & & & \\ B_{10} & B_{11} & B_{12} & A_1 & A_0 & & & & & & & \\ B_{20} & B_{21} & B_{22} & A_2 & A_1 & A_0 & & & & & & \\ B_{30} & B_{31} & B_{32} & A_3 & A_2 & A_1 & A_0 & & & & & \\ B_{40} & B_{41} & B_{42} & A_4 & A_3 & A_2 & A_1 & A_0 & & & & \\ B_{50} & B_{51} & B_{52} & A_5 & A_4 & A_3 & A_2 & A_1 & A_0 & & & \\ B_{60} & B_{61} & B_{62} & A_6 & A_5 & A_4 & A_3 & A_2 & A_1 & A_0 & & \\ B_{70} & B_{71} & B_{72} & A_7 & A_6 & A_5 & A_4 & A_3 & A_2 & A_1 & A_0 & \\ B_{80} & B_{81} & B_{82} & A_8 & A_7 & A_6 & A_5 & A_4 & A_3 & A_2 & A_1 & A_0 \\ \vdots & \vdots & \vdots & & \ddots & \ddots & \ddots & \ddots & \ddots & \ddots & \ddots & \ddots & \ddots \end{pmatrix}.$$

At present, this matrix is *not* block lower Hessenberg. However, if it is restructured into the form

$$\overline{Q} = \begin{pmatrix} \overline{B_{00}} & \overline{A_0} & 0 & 0 & \cdots \\ \overline{B_{10}} & \overline{A_1} & \overline{A_0} & 0 & \cdots \\ \overline{B_{20}} & \overline{A_2} & \overline{A_1} & \overline{A_0} & \\ \vdots & & \ddots & \ddots & \ddots \end{pmatrix}$$

$$= \left(\begin{array}{ccc|ccc|ccc|ccc|c} B_{00} & B_{01} & B_{02} & A_0 & & & & & & & & & \\ B_{10} & B_{11} & B_{12} & A_1 & A_0 & & & & & & & & \\ B_{20} & B_{21} & B_{22} & A_2 & A_1 & A_0 & & & & & & & \\ \hline B_{30} & B_{31} & B_{32} & A_3 & A_2 & A_1 & A_0 & & & & & & \\ B_{40} & B_{41} & B_{42} & A_4 & A_3 & A_2 & A_1 & A_0 & & & & & \\ B_{50} & B_{51} & B_{52} & A_5 & A_4 & A_3 & A_2 & A_1 & A_0 & & & & \\ \hline B_{60} & B_{61} & B_{62} & A_6 & A_5 & A_4 & A_3 & A_2 & A_1 & A_0 & & & \\ B_{70} & B_{71} & B_{72} & A_7 & A_6 & A_5 & A_4 & A_3 & A_2 & A_1 & A_0 & & \\ B_{80} & B_{81} & B_{82} & A_8 & A_7 & A_6 & A_5 & A_4 & A_3 & A_2 & A_1 & A_0 & \\ \hline \vdots & \vdots & \vdots & \vdots & \vdots & \vdots & & \ddots & \ddots & \ddots & \ddots & \ddots & \ddots \end{array}\right)$$

with

$$\overline{A_0} = \begin{pmatrix} A_0 & & \\ A_1 & A_0 & \\ A_2 & A_1 & A_0 \end{pmatrix}, \quad \overline{A_1} = \begin{pmatrix} A_3 & A_2 & A_1 \\ A_4 & A_3 & A_2 \\ A_5 & A_4 & A_3 \end{pmatrix}, \quad \overline{B_{00}} = \begin{pmatrix} B_{00} & B_{01} & B_{02} \\ B_{10} & B_{11} & B_{12} \\ B_{20} & B_{21} & B_{22} \end{pmatrix}, \quad \ldots$$

then the matrix $\overline{Q}$ does have the desired block lower Hessenberg form and hence the techniques described in this section may be successfully applied to it.

In the case of discrete-time Markov chains, as opposed to the continuous-time case just outlined, it suffices to replace $-A_1^{-1}$ with $(I - A_1)^{-1}$, as we described for QBD processes.

### *10.6.3 Block Upper Hessenberg Markov Chains*

We now move to block upper Hessenberg Markov chains, also called *M/G/*1-type processes and solved using the *matrix analytic method.* In the past two sections concerning QBD and *G/M/*1-type processes, we posed the problem in terms of continuous-time Markov chains, and mentioned that discrete-time Markov chains can be treated if the matrix inverse $A_1^{-1}$ is replaced with the inverse

$(I - A_1)^{-1}$ where it is understood that $A_1$ is taken from a stochastic matrix. This time we shall consider the discrete-time case. Specifically, we consider the case when the stochastic transition probability matrix is irreducible and has the structure

$$P = \begin{pmatrix} B_{00} & B_{01} & B_{02} & B_{03} & \cdots & B_{0j} & \cdots \\ B_{10} & A_1 & A_2 & A_3 & \cdots & A_j & \cdots \\ 0 & A_0 & A_1 & A_2 & \cdots & A_{j-1} & \cdots \\ 0 & 0 & A_0 & A_1 & \cdots & A_{j-2} & \cdots \\ 0 & 0 & 0 & A_0 & \cdots & A_{j-3} & \cdots \\ \vdots & \vdots & \vdots & \ddots & \ddots & \ddots & \vdots \end{pmatrix},$$

in which all submatrices $A_j$, $j = 0, 1, 2, \ldots$ are square and of order $K$, and $B_{00}$ is square but not necessarily of order $K$. Be aware that whereas previously the block $A_0$ was the rightmost nonzero block of our global matrix, this time it is the leftmost nonzero block. Notice that the matrix $A = \sum_{i=0}^{\infty} A_i$ is a stochastic matrix. We shall further assume that $A$ is irreducible, the commonly observed case in practice. Instances in which $A$ is not irreducible are treated by Neuts.

$$\pi_A A = \pi_A \quad \text{and} \quad \pi_A e = 1.$$

The Markov chain $P$ is known to be positive recurrent if the following condition holds:

$$\pi_A \left( \sum_{i=1}^{\infty} i A_i \, e \right) \equiv \pi_A \, b \; < 1. \tag{10.37}$$

Our objective is the computation of the stationary probability vector $\pi$ from the system of equations $\pi P = \pi$. As before, we partition $\pi$ conformally with $P$, i.e.,

$$\pi = (\pi_0, \pi_1, \pi_2, \ldots),$$

where

$$\pi_i = (\pi(i, 1), \pi(i, 2), \ldots, \pi(i, K))$$

for $i = 0, 1, \ldots$ and $\pi(i, k)$ is the probability of finding the system in state $(i, k)$ at statistical equilibrium. The analysis of *M/G/*1-type processes is more complicated than that of QBD or *G/M/*1-type processes because the subvectors $\pi_i$ no longer have a matrix geometric relationship with one another.

The key to solving upper block Hessenberg structured Markov chains is the computation of a certain matrix $G$ which is stochastic if the Markov chain is recurrent, which we assume to be the case. This is the same matrix $G$ that appears in the logarithmic reduction algorithm and has an important probabilistic interpretation. The element $G_{ij}$ of this matrix is the conditional probability that starting in state $i$ of any level $n \geq 2$, the process enters level $n - 1$ for the first time by arriving at state $j$ of that level. This matrix satisfies the fixed point equation

$$G = \sum_{i=0}^{\infty} A_i G^i$$

and is indeed the minimal non-negative solution of

$$X = \sum_{i=0}^{\infty} A_i X^i.$$

It can be found by means of the iteration

$$G_{(0)} = 0, \quad G_{(k+1)} = \sum_{i=0}^{\infty} A_i G_{(k)}^i, \quad k = 0, 1, \ldots.$$

Once the matrix $G$ has been computed, then successive components of $\pi$ can be obtained from a relationship, called Ramaswami's formula, which we now develop. We follow the algebraic approach of Bini and Meini [3, 36], rather than the original probabilistic approach of Ramaswami. We begin by writing the system of equations $\pi P = \pi$ as $\pi(I - P) = 0$, i.e.,

$$(\pi_0,\ \pi_1,\ \ldots,\ \pi_j,\ \ldots)\left(\begin{array}{c|cccccc} I - B_{00} & -B_{01} & -B_{02} & -B_{03} & \cdots & -B_{0j} & \cdots \\ \hline -B_{10} & I - A_1 & -A_2 & -A_3 & \cdots & -A_j & \cdots \\ 0 & -A_0 & I - A_1 & -A_2 & \cdots & -A_{j-1} & \cdots \\ 0 & 0 & -A_0 & I - A_1 & \cdots & -A_{j-2} & \cdots \\ 0 & 0 & 0 & -A_0 & \cdots & -A_{j-3} & \cdots \\ \vdots & \vdots & \vdots & \vdots & \vdots & \vdots & \vdots \end{array}\right)$$

$$= (0,\ 0,\ \ldots,\ 0,\ \ldots). \tag{10.38}$$

The submatrix in the lower right is block Toeplitz. Bini and Meini have shown that there exists a decomposition of this Toeplitz matrix into a block upper triangular matrix $U$ and block lower triangular matrix $L$ with

$$U = \begin{pmatrix} A_1^* & A_2^* & A_3^* & A_4^* & \cdots \\ 0 & A_1^* & A_2^* & A_3^* & \cdots \\ 0 & 0 & A_1^* & A_2^* & \cdots \\ 0 & 0 & 0 & A_1^* & \cdots \\ \vdots & \vdots & \vdots & & \ddots \end{pmatrix} \quad \text{and} \quad L = \begin{pmatrix} I & 0 & 0 & 0 & \cdots \\ -G & I & 0 & 0 & \cdots \\ 0 & -G & I & 0 & \cdots \\ 0 & 0 & -G & I & \cdots \\ \vdots & \vdots & & \ddots & \ddots \end{pmatrix}.$$

We denote the nonzero blocks of $U$ as $A_i^*$ rather than $U_i$ since we shall see later that these blocks are formed using the $A_i$ blocks of $P$. Once the matrix $G$ has been formed then $L$ is known. Observe that the inverse of $L$ can be written in terms of the powers of $G$. For example,

$$\begin{pmatrix} I & 0 & 0 & 0 & \cdots \\ -G & I & 0 & 0 & \cdots \\ 0 & -G & I & 0 & \cdots \\ 0 & 0 & -G & I & \cdots \\ \vdots & \vdots & & \ddots & \ddots \end{pmatrix}\begin{pmatrix} I & 0 & 0 & 0 & \cdots \\ G & I & 0 & 0 & \cdots \\ G^2 & G & I & 0 & \cdots \\ G^3 & G^2 & G & I & \cdots \\ \vdots & \vdots & & \ddots & \ddots \end{pmatrix} = \begin{pmatrix} 1 & 0 & 0 & 0 & \cdots \\ 0 & 1 & 0 & 0 & \cdots \\ 0 & 0 & 1 & 0 & \cdots \\ 0 & 0 & 0 & 1 & \cdots \\ \vdots & \vdots & \vdots & & \ddots \end{pmatrix}.$$

From Equation (10.38), we have

$$(\pi_0,\ \pi_1,\ \ldots,\ \pi_j,\ \ldots)\left(\begin{array}{c|c} I - B_{00} & -B_{01}\ \ -B_{02}\ \ -B_{03}\ \cdots\ -B_{0j}\ \cdots \\ \hline -B_{10} & \\ 0 & \\ 0 & UL \\ 0 & \\ \vdots & \end{array}\right) = (0,\ 0,\ \ldots,\ 0,\ \ldots),$$

which allows us to write

$$\pi_0(-B_{01},\ -B_{02},\ \ldots) + (\pi_1,\ \pi_2,\ \ldots)UL = 0$$

or

$$\pi_0(B_{01},\ B_{02},\ \ldots)L^{-1} = (\pi_1,\ \pi_2,\ \ldots)U,$$

i.e.,

$$\pi_0\,(B_{01},\ B_{02},\ \dots)\begin{pmatrix} I & 0 & 0 & 0 & \cdots \\ G & I & 0 & 0 & \cdots \\ G^2 & G & I & 0 & \cdots \\ G^3 & G^2 & G & I & \cdots \\ \vdots & \vdots & & \ddots & \ddots \end{pmatrix} = (\pi_1,\ \pi_2,\ \dots)\,U.$$

Forming the product of $(B_{01},\ B_{02},\ \dots)$ and $L^{-1}$ leads to the important result

$$\pi_0\left(B_{01}^*,\ B_{02}^*,\ \dots\right) = (\pi_1,\ \pi_2,\ \dots)\,U, \tag{10.39}$$

where

$$B_{01}^* = B_{01} + B_{02}G + B_{03}G^2 + \cdots = \sum_{k=1}^{\infty} B_{0k}G^{k-1},$$

$$B_{02}^* = B_{02} + B_{03}G + B_{04}G^2 + \cdots = \sum_{k=2}^{\infty} B_{0k}G^{k-2},$$

$$\vdots$$

$$B_{0i}^* = B_{0i} + B_{0,i+1}G + B_{0,i+2}G^2 + \cdots = \sum_{k=i}^{\infty} B_{0k}G^{k-i}.$$

The system of equations (10.39) will allow us to compute the successive components of the vector $\pi$ once the initial component $\pi_0$ and the matrix $U$ are known. To see this, write Equation (10.39) as

$$\pi_0\left(B_{01}^*,\ B_{02}^*,\ \cdots\right) = (\pi_1,\ \pi_2,\ \cdots)\begin{pmatrix} A_1^* & A_2^* & A_3^* & A_4^* & \cdots \\ 0 & A_1^* & A_2^* & A_3^* & \cdots \\ 0 & 0 & A_1^* & A_2^* & \cdots \\ 0 & 0 & 0 & A_1^* & \cdots \\ \vdots & \vdots & \vdots & & \ddots \end{pmatrix}$$

and observe that

$$\begin{aligned} \pi_0 B_{01}^* &= \pi_1 A_1^* &\implies \pi_1 &= \pi_0 B_{01}^* A_1^{*-1}, \\ \pi_0 B_{02}^* &= \pi_1 A_2^* + \pi_2 A_1^* &\implies \pi_2 &= \pi_0 B_{02}^* A_1^{*-1} - \pi_1 A_2^* A_1^{*-1}, \\ \pi_0 B_{03}^* &= \pi_1 A_3^* + \pi_2 A_2^* + \pi_3 A_1^* &\implies \pi_3 &= \pi_0 B_{03}^* A_1^{*-1} - \pi_1 A_3^* A_1^{*-1} - \pi_2 A_2^* A_1^{*-1}, \\ &\vdots \end{aligned}$$

In general, we find

$$\begin{aligned} \pi_i &= \left(\pi_0 B_{0i}^* - \pi_1 A_i^* - \pi_2 A_{i-1}^* - \cdots - \pi_{i-1} A_2^*\right) A_1^{*-1} \\ &= \left(\pi_0 B_{0i}^* - \sum_{k=1}^{i-1} \pi_k A_{i-k+1}^*\right) A_1^{*-1}, \quad i = 1, 2, \dots. \end{aligned}$$

To compute the first subvector $\pi_0$ we return to

$$\pi_0\left(B_{01}^*,\ B_{02}^*,\ \dots\right) = (\pi_1,\ \pi_2,\ \dots)\,U$$

and write it as

$$(\pi_0, \pi_1, \ldots, \pi_j, \ldots)\left(\begin{array}{c|cccccc} I - B_{00} & -B^*_{01} & -B^*_{02} & -B^*_{03} & \cdots & -B^*_{0j} & \cdots \\ \hline -B_{10} & A^*_1 & A^*_2 & A^*_3 & \cdots & A^*_j & \cdots \\ 0 & 0 & A^*_1 & A^*_2 & \cdots & A^*_{j-1} & \cdots \\ 0 & 0 & 0 & A^*_1 & \cdots & A^*_{j-2} & \cdots \\ & & & & & \vdots & \\ 0 & 0 & 0 & 0 & \cdots & & \cdots \\ \vdots & \vdots & \vdots & \vdots & \vdots & \vdots & \vdots \end{array}\right) = (0, 0, \ldots, 0, \ldots).$$

From the first two equations, we have

$$\pi_0 (I - B_{00}) - \pi_1 B_{10} = 0$$

and

$$-\pi_0 B^*_{01} + \pi_1 A^*_1 = 0.$$

This latter gives

$$\pi_1 = \pi_0 B^*_{01} A^{*-1}_1$$

which when substituted into the first gives

$$\pi_0 (I - B_{00}) - \pi_0 B^*_{01} A^{*-1}_1 B_{10} = 0$$

or

$$\pi_0 \left(I - B_{00} - B^*_{01} A^{*-1}_1 B_{10}\right) = 0,$$

from which we may now compute $\pi_0$, but correct only to a multiplicative constant. It must be normalized so that $\sum_{i=0}^{\infty} \pi_i = 1$ which may be accomplished by enforcing the condition

$$\pi_0 e + \pi_0 \left(\sum_{i=1}^{\infty} B^*_{0i}\right)\left(\sum_{i=1}^{\infty} A^*_i\right)^{-1} e = 1. \tag{10.40}$$

We now turn our attention to the computation of the matrix $U$. Since

$$UL = \begin{pmatrix} I - A_1 & -A_2 & -A_3 & \cdots & -A_j & \cdots \\ -A_0 & I - A_1 & -A_2 & \cdots & -A_{j-1} & \cdots \\ 0 & -A_0 & I - A_1 & \cdots & -A_{j-2} & \cdots \\ 0 & 0 & -A_0 & \cdots & -A_{j-3} & \cdots \\ \vdots & \vdots & \vdots & \vdots & \vdots & \vdots \end{pmatrix},$$

we have

$$\begin{pmatrix} A^*_1 & A^*_2 & A^*_3 & A^*_4 & \cdots \\ 0 & A^*_1 & A^*_2 & A^*_3 & \cdots \\ 0 & 0 & A^*_1 & A^*_2 & \cdots \\ 0 & 0 & 0 & A^*_1 & \cdots \\ \vdots & \vdots & \vdots & & \ddots \end{pmatrix} = \begin{pmatrix} I - A_1 & -A_2 & -A_3 & \cdots & -A_j & \cdots \\ -A_0 & I - A_1 & -A_2 & \cdots & -A_{j-1} & \cdots \\ 0 & -A_0 & I - A_1 & \cdots & -A_{j-2} & \cdots \\ 0 & 0 & -A_0 & \cdots & -A_{j-3} & \cdots \\ \vdots & \vdots & \vdots & \vdots & \vdots & \vdots \end{pmatrix}\begin{pmatrix} I & 0 & 0 & 0 & \cdots \\ G & I & 0 & 0 & \cdots \\ G^2 & G & I & 0 & \cdots \\ G^3 & G^2 & G & I & \cdots \\ \vdots & \vdots & & \ddots & \ddots \end{pmatrix},$$

and it is now apparent that

$$A_1^* = I - A_1 - A_2G - A_3G^2 - A_4G^3 - \cdots = I - \sum_{k=1}^{\infty} A_k G^{k-1},$$

$$A_2^* = -A_2 - A_3G - A_4G^2 - A_5G^3 - \cdots = -\sum_{k=2}^{\infty} A_k G^{k-2},$$

$$A_3^* = -A_3 - A_4G - A_5G^2 - A_6G^3 - \cdots = -\sum_{k=3}^{\infty} A_k G^{k-3},$$

$$\vdots$$

$$A_i^* = -A_i - A_{i+1}G - A_{i+2}G^2 - A_{i+3}G^3 - \cdots = -\sum_{k=i}^{\infty} A_k G^{k-i}, \quad i \geq 2.$$

We now have all the results we need. The basic algorithm is

- Construct the matrix $G$.
- Obtain $\pi_0$ by solving the system of equations $\pi_0 \left(I - B_{00} - B_{01}^* A_1^{*-1} B_{10}\right) = 0$, subject to the normalizing condition, Equation (10.40).
- Compute $\pi_1$ from $\pi_1 = \pi_0 B_{01}^* A_1^{*-1}$.
- Find all other required $\pi_i$ from $\pi_i = \left(\pi_0 B_{0i}^* - \sum_{k=1}^{i-1} \pi_k A_{i-k+1}^*\right) A_1^{*-1}$,

where

$$B_{0i}^* = \sum_{k=i}^{\infty} B_{0k} G^{k-i}, \quad i \geq 1; \quad A_1^* = I - \sum_{k=1}^{\infty} A_k G^{k-1} \quad \text{and} \quad A_i^* = -\sum_{k=i}^{\infty} A_k G^{k-i}, \quad i \geq 2.$$

This obviously gives rise to a number of computational questions. The first is the actual computation of the matrix $G$. We mentioned previously that it can be obtained from its defining equation by means of the iterative procedure

$$G_{(0)} = 0, \quad G_{(k+1)} = \sum_{i=0}^{\infty} A_i G_{(k)}^i, \quad k = 0, 1, \ldots.$$

However, this is rather slow. Neuts proposed a variant that converges faster, namely,

$$G_{(0)} = 0; \quad G_{(k+1)} = (I - A_1)^{-1} \left(A_0 + \sum_{i=2}^{\infty} A_i G_{(k)}^i\right), \quad k = 0, 1, \ldots$$

Among fixed point iterations such as these, the following suggested by Bini and Meini [3, 36], has the fastest convergence:

$$G_{(0)} = 0, \quad G_{(k+1)} = \left(I - \sum_{i=1}^{\infty} A_i G_{(k)}^{i-1}\right)^{-1} A_0, \quad k = 0, 1, \ldots.$$

Nevertheless, fixed point iterations can be very slow in certain instances. More advanced techniques based on cyclic reduction have been developed and converge much faster.

The second major problem is the computation of the infinite summations that appear in the formulae. Frequently the structure of the matrix is such that $A_k$ and $B_k$ are zero for relatively small (single-digit integer) values of $k$, so that forming these summations is not too onerous. In all cases, the fact that $\sum_{k=0}^{\infty} A_k$ and $\sum_{k=0}^{\infty} B_k$ are stochastic implies that the matrices $A_k$ and $B_k$ are negligibly small for large values of $k$ and can be set to zero once $k$ exceeds some threshold $k_M$. In this case, we take $\sum_{k=0}^{k_M} A_k$ and $\sum_{k=0}^{k_M} B_k$ to be stochastic. More precisely, a nonnegative matrix $P$ is said to be *numerically stochastic* if $Pe < \mu e$ where $\mu$ is the precision of the computer. When $k_M$ is not small, finite summations of the type $\sum_{k=i}^{k_M} A_k G^{k-i}$ should be evaluated using Horner's rule. For example, if $i = 1$ and $k_M = 5$:

$$A_1^* = \sum_{k=1}^{5} A_k G^{k-1} = A_1 + A_2 G + A_3 G^2 + A_4 G^3 + A_5 G^4$$

should be evaluated from the innermost parentheses outward as

$$A_1^* = A_1 + (A_2 + (A_3 + (A_4 + A_5 G)G)G)G.$$

**Example 10.29** We shall apply the matrix analytic method to the Markov chain of Example 10.27, now modified so that it incorporates additional transitions ($\zeta_1 = 1/48$ and $\zeta_2 = 1/16$) to higher-numbered non-neighboring states. The state transition diagram is shown in Figure 10.3.

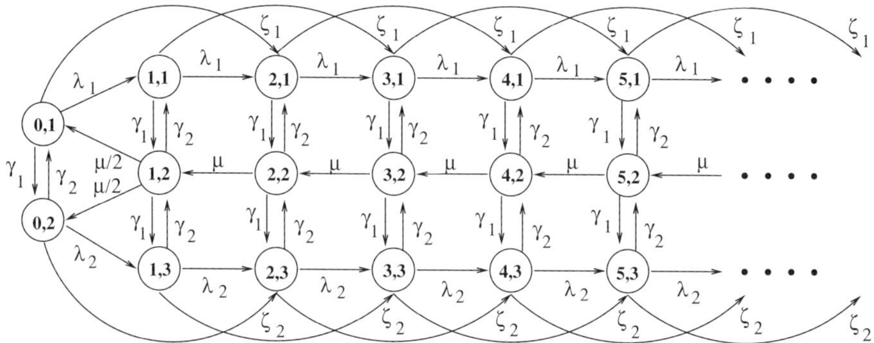


Figure 10.3. State transition diagram for an *M/G/*1-type process.

$$P = \left(\begin{array}{cc|ccc|ccc|ccc|c}
23/48 & 5/12 & 1/12 & & & 1/48 & & & & & & \\
1/4 & 31/48 & & & 1/24 & & & 1/16 & & & & \\
\hline
 & & 23/48 & 5/12 & & 1/12 & & & 1/48 & & & \\
1/3 & 1/3 & 1/4 & & 1/12 & & & & & & & \\
 & & & 1/4 & 31/48 & & & 1/24 & & & 1/16 & \\
\hline
 & & & & & 23/48 & 5/12 & & 1/12 & & & \\
 & & & 2/3 & & 1/4 & & 1/12 & & & & \\
 & & & & & & 1/4 & 31/48 & & & 1/24 & \\
\hline
 & & & & & & & & 23/48 & 5/12 & & \\
 & & & & & & 2/3 & & 1/4 & & 1/12 & \\
 & & & & & & & & & 1/4 & 31/48 & \\
\hline
 & & & & & & \ddots & & & \ddots & & \ddots
\end{array}\right).$$

We have the following block matrices:

$$A_0 = \begin{pmatrix} 0 & 0 & 0 \\ 0 & 2/3 & 0 \\ 0 & 0 & 0 \end{pmatrix}, \quad A_1 = \begin{pmatrix} 23/48 & 5/12 & 0 \\ 1/4 & 0 & 1/12 \\ 0 & 1/4 & 31/48 \end{pmatrix}, \quad A_2 = \begin{pmatrix} 1/12 & 0 & 0 \\ 0 & 0 & 0 \\ 0 & 0 & 1/24 \end{pmatrix},$$

$$A_3 = \begin{pmatrix} 1/48 & 0 & 0 \\ 0 & 0 & 0 \\ 0 & 0 & 1/16 \end{pmatrix}, \quad B_{00} = \begin{pmatrix} 23/48 & 5/12 \\ 1/4 & 31/48 \end{pmatrix}, \quad B_{01} = \begin{pmatrix} 1/12 & 0 & 0 \\ 0 & 0 & 1/24 \end{pmatrix},$$

$$B_{02} = \begin{pmatrix} 1/48 & 0 & 0 \\ 0 & 0 & 1/16 \end{pmatrix}, \quad \text{and} \quad B_{10} = \begin{pmatrix} 0 & 0 \\ 1/3 & 1/3 \\ 0 & 0 \end{pmatrix}.$$

First, using Equation (10.37), we verify that the Markov chain with transition probability matrix $P$ is positive recurrent,

$$A = A_0 + A_1 + A_2 + A_3 = \begin{pmatrix} .583333 & .416667 & 0 \\ .250000 & .666667 & .083333 \\ 0 & .250000 & .750000 \end{pmatrix},$$

from which we can compute

$$\pi_A = (.310345,\ .517241,\ .172414).$$

Also

$$b = (A_1 + 2A_2 + 3A_3)e = \begin{pmatrix} .708333 & .416667 & 0 \\ .250000 & 0 & .083333 \\ 0 & .250000 & .916667 \end{pmatrix} \begin{pmatrix} 1 \\ 1 \\ 1 \end{pmatrix} = \begin{pmatrix} 1.125000 \\ 0.333333 \\ 1.166667 \end{pmatrix}.$$

Since

$$\pi_A\, b = (.310345,\ .517241,\ .172414) \begin{pmatrix} 1.125000 \\ 0.333333 \\ 1.166667 \end{pmatrix} = .722701 < 1$$

the Markov chain is positive recurrent.

Let us consider the computation of the matrix $G$. As we noted previously, the $ij$ element of $G$ is the conditional probability that starting in state $i$ of any level $n \geq 2$, the process enters level $n-1$ for the first time by arriving at state $j$ of that level. What this means for this particular example is that the elements in column 2 of $G$ must all be equal to 1 and all other elements must be zero, since the only transitions from any level $n$ to level $n-1$ are from and to the second element. Nevertheless, it is interesting to see how each of the three different fixed point formula given for computing $G$ actually perform on this example. The fact that we know the answer in advance will help in this comparison. We take the initial value $G_{(0)}$ to be zero.

Formula 1: $G_{(k+1)} = \sum_{i=0}^{\infty} A_i G_{(k)}^i, \quad k = 0, 1, \ldots,$

$$G_{(k+1)} = A_0 + A_1 G_{(k)} + A_2 G_{(k)}^2 + A_3 G_{(k)}^3.$$

After ten iterations, the computed matrix is

$$G_{(10)} = \begin{pmatrix} 0 & .867394 & 0 \\ 0 & .937152 & 0 \\ 0 & .766886 & 0 \end{pmatrix}.$$

Formula 2: $G_{(k+1)} = (I - A_1)^{-1} \left( A_0 + \sum_{i=2}^{\infty} A_i G_{(k)}^i \right), \quad k = 0, 1, \ldots,$

$$G_{(k+1)} = (I - A_1)^{-1} \left( A_0 + A_2 G_{(k)}^2 + A_3 G_{(k)}^3 \right).$$

In this case, after ten iterations, the computed matrix is

$$G_{(10)} = \begin{pmatrix} 0 & .999844 & 0 \\ 0 & .999934 & 0 \\ 0 & .999677 & 0 \end{pmatrix}.$$

Formula 3: $G_{(k+1)} = \left( I - \sum_{i=1}^{\infty} A_i G_{(k)}^{i-1} \right)^{-1} A_0, \quad k = 0, 1, \ldots,$

$$G_{(k+1)} = \left( I - A_1 - A_2 G_{(k)} - A_3 G_{(k)}^2 \right)^{-1} A_0.$$

This is the fastest of the three and after ten iterations we have

$$G_{(10)} = \begin{pmatrix} 0 & .999954 & 0 \\ 0 & .999979 & 0 \\ 0 & .999889 & 0 \end{pmatrix}.$$

We now continue with this example and shall use the exact value of $G$ in the remaining parts of the algorithm. In preparation, we compute the following quantities, using the fact that $A_k = 0$ for $k > 3$ and $B_{0k} = 0$ for $k > 2$:

$$A_1^* = I - \sum_{k=1}^{\infty} A_k G^{k-1} = I - A_1 - A_2 G - A_3 G^2 = \begin{pmatrix} .520833 & -.520833 & 0 \\ -.250000 & 1 & -.083333 \\ 0 & -.354167 & .354167 \end{pmatrix},$$

$$A_2^* = -\sum_{k=2}^{\infty} A_k G^{k-2} = -(A_2 + A_3 G) = \begin{pmatrix} -.083333 & -.020833 & 0 \\ 0 & 0 & 0 \\ 0 & -.062500 & -.041667 \end{pmatrix},$$

$$A_3^* = -\sum_{k=3}^{\infty} A_k G^{k-3} = -A_3 = \begin{pmatrix} -.020833 & 0 & 0 \\ 0 & 0 & 0 \\ 0 & 0 & -.062500 \end{pmatrix},$$

$$B_{01}^* = \sum_{k=1}^{\infty} B_{0k} G^{k-1} = B_{01} + B_{02} G = \begin{pmatrix} .083333 & .020833 & 0 \\ 0 & .062500 & .041667 \end{pmatrix},$$

$$B_{02}^* = \sum_{k=2}^{\infty} B_{0k} G^{k-2} = B_{02} = \begin{pmatrix} .020833 & 0 & 0 \\ 0 & 0 & .062500 \end{pmatrix},$$

$$A_1^{*-1} = \begin{pmatrix} 2.640 & 1.50 & .352941 \\ .720 & 1.50 & .352941 \\ .720 & 1.50 & 3.176470 \end{pmatrix}.$$

We may now compute the initial subvector $\pi_0$ from

$$0 = \pi_0 \left( I - B_{00} - B_{01}^* A_1^{*-1} B_{10} \right) = \pi_0 \begin{pmatrix} .468750 & -.468750 \\ -.302083 & .302083 \end{pmatrix},$$

from which we find

$$\pi_0 = (.541701, \ .840571).$$

We now normalize this so that

$$\pi_0 e + \pi_0 \left( \sum_{i=1}^{\infty} B_{0i}^* \right) \left( \sum_{i=1}^{\infty} A_i^* \right)^{-1} e = 1,$$

i.e.,

$$\pi_0 e + \pi_0 \left( B_{01}^* + B_{02}^* \right) \left( A_1^* + A_2^* + A_3^* \right)^{-1} e = 1.$$

Evaluating,

$$\left(B_{01}^* + B_{02}^*\right)\left(A_1^* + A_2^* + A_3^*\right)^{-1} = \begin{pmatrix} .104167 & .020833 & 0 \\ 0 & .062500 & .104167 \end{pmatrix}$$

$$\times \begin{pmatrix} .416667 & -.541667 & 0 \\ -.250000 & 1 & -.083333 \\ 0 & -.416667 & .250000 \end{pmatrix}^{-1} = \begin{pmatrix} .424870 & .291451 & .097150 \\ .264249 & .440415 & .563472 \end{pmatrix}.$$

Thus

$$(.541701,\ .840571)\begin{pmatrix} 1 \\ 1 \end{pmatrix} + (.541701,\ .840571)\begin{pmatrix} .424870 & .291451 & .097150 \\ .264249 & .440415 & .563472 \end{pmatrix}\begin{pmatrix} 1 \\ 1 \\ 1 \end{pmatrix} = 2.888888,$$

which leads to

$$\pi_0 = (.541701,\ .840571)/2.888888 = (.187512,\ .290967)$$

as the initial subvector.
We can now find $\pi_1$ from the relationship $\pi_1 = \pi_0 B_{01}^* A_1^{*-1}$. This gives

$$\pi_1 = (.187512,\ .290967)\begin{pmatrix} .083333 & .020833 & 0 \\ 0 & .062500 & .041667 \end{pmatrix}\begin{pmatrix} 2.640 & 1.50 & .352941 \\ .720 & 1.50 & .352941 \\ .720 & 1.50 & 3.176470 \end{pmatrix}$$

$$= (.065888,\ .074762,\ .0518225).$$

Finally, all needed remaining subcomponents of $\pi$ can be found from

$$\pi_i = \left(\pi_0 B_{0i}^* - \sum_{k=1}^{i-1} \pi_k A_{i-k+1}^*\right) A_1^{*-1}.$$

For example, we have

$$\begin{aligned}
\pi_2 &= \left(\pi_0 B_{02}^* - \pi_1 A_2^*\right) A_1^{*-1} \\
&= (.042777,\ .051530,\ .069569), \\
\pi_3 &= \left(\pi_0 B_{03}^* - \pi_1 A_3^* - \pi_2 A_2^*\right) A_1^{*-1} = \left(-\pi_1 A_3^* - \pi_2 A_2^*\right) A_1^{*-1} \\
&= (.0212261,\ .024471,\ .023088), \\
\pi_4 &= \left(\pi_0 B_{04}^* - \pi_1 A_4^* - \pi_2 A_3^* - \pi_3 A_2^*\right) A_1^{*-1} = \left(-\pi_2 A_3^* - \pi_3 A_2^*\right) A_1^{*-1} \\
&= (.012203,\ .014783,\ .018471), \\
&\ \vdots
\end{aligned}$$

The probability that the Markov chain is in any level $i$ is given by $\|\pi_i\|_1$. The probabilities of this Markov chain being in the first five levels are given as

$$\|\pi_0\|_1 = .478479,\ \|\pi_1\|_1 = .192473,\ \|\pi_2\|_1 = .163876,\ \|\pi_3\|_1 = .068785,\ \|\pi_4\|_1 = .045457.$$

The sum of these five probabilities is 0.949070.

## 10.7 Transient Distributions

So far, our concern has been with the computation of the stationary distribution $\pi$ of discrete- and continuous-time Markov chains. We now turn our attention to the computation of state probability distributions at an arbitrary point of time. In the case of a discrete-time Markov chain, this means finding the distribution at some arbitrary time step $n$. This distribution is denoted $\pi^{(n)}$, a row vector whose $i^{\text{th}}$ component is the probability that the Markov chain is in state $i$ at time step $n$. As we saw previously, it satisfies the relationship

$$\pi^{(n)} = \pi^{(n-1)} P(n-1) = \pi^{(0)} P(0) P(1) \cdots P(n-1),$$

where $P(i)$ is the transition probability matrix at step $i$. For a homogeneous discrete-time Markov chain, this reduces to

$$\pi^{(n)} = \pi^{(n-1)} P = \pi^{(0)} P^n,$$

where now $P(0) = P(1) = \cdots = P$. In this section, we shall consider only homogeneous Markov chains. For a continuous-time Markov chain with infinitesimal generator $Q$, we seek the distribution at any time $t$. Such a distribution is denoted $\pi(t)$, a row vector whose component $\pi_i(t)$ is the probability that the Markov chain is in state $i$ at time $t$. We have previously seen that this vector satisfies the relationship

$$\pi(t) = \pi(0) e^{Qt},$$

where $e^{Qt}$ is the matrix exponential defined by

$$e^{Qt} = \sum_{k=0}^{\infty} (Qt)^k / k!.$$

In both cases, what is usually required is seldom the probability distribution $\pi^{(n)}$ or $\pi(t)$ itself, but rather some linear combination of the components of these vectors, such as the probability that the Markov chain is in a single state $i$, ($\pi^{(n)} e_i$ or $\pi(t)\, e_i$, where $e_i$ is a column vector whose elements are all zero except the $i^{\text{th}}$, which is equal to 1) or the probability that the Markov chain is in a subset of states $E_i$ $\left(\sum_{i \in E_i} \pi^{(n)} e_i \text{ or } \sum_{i \in E_i} \pi(t) e_i\right)$ or yet again a weighted sum of the probabilities wherein the unit component of $e_i$ is replaced by an arbitrary scaler. Also, the evolution of this statistic from the initial time to the desired time step $n$ or time $t$ may be required rather than just its (single) value at the final time point.

The computation of transient distributions of discrete-time Markov chains rarely poses major problems. The procedure consists of repeatedly multiplying the probability distribution vector obtained at step $k-1$ with the stochastic transition probability matrix to obtain the probability distribution at step $k$, for $k = 1, 2, \ldots, n$. If $n$ is large and the number of states in the Markov chain is small (not exceeding several hundreds), then some savings in computation time can be obtained by successively squaring the transition probability matrix $j$ times, where $j$ is the largest integer such that $2^j \leq n$. This gives the matrix $P^{2^j}$ which can now be multiplied by $P$ (and powers of $P$) until the value $P^n$ is obtained. The distribution at time step $n$ is now found from $\pi^{(n)} = \pi^{(0)} P^n$. If the matrix $P$ is sparse, this sparsity is lost in the computation of $P^n$ so this approach is not appropriate for large sparse Markov chains. Also, if a time trajectory of a statistic of the distribution is needed, this approach may be less than satisfactory, because only distributions at computed values of $P^{2^j}$ will be available. One final point worth noting is that for large values of $n$, it may be beneficial to maintain a check on convergence to the stationary distribution, since this may occur, correct to some desired computational accuracy, prior to step $n$. Any additional vector–matrix multiplications after this point will not alter the distribution.

**Example 10.30** Let us consider a discrete-time Markov chain with transition probability matrix $P$ given by

$$P = \begin{pmatrix} .4 & 0 & .6 & 0 \\ .0002 & .3 & 0 & .6998 \\ .1999 & .0001 & .8 & 0 \\ 0 & .5 & 0 & .5 \end{pmatrix}.$$

Let us assume that this Markov chain starts in state 1, i.e., $\pi^{(0)} = (1, 0, 0, 0)$, and the value of being in the different states at some time step $n$ is given by the vector $a = (0, 4, 0, 10)^T$. In other words, the Markov chain in states 1 and 3 is worthless, but worth 4 (arbitrary units) in state 2 and 10 in state 4. The value at time step $n$ is $\pi^{(n)}a$ and we now wish to compute this statistic for various values of $n$.

For small values of $n = 1, 2, \ldots$, it suffices to compute $\pi^{(1)} = \pi^{(0)}P$, $\pi^{(2)} = \pi^{(1)}P$, $\pi^{(3)} = \pi^{(2)}P$, ..., and we find

$$\begin{aligned} \pi^{(1)} &= (.4,\ 0,\ .6,\ 0), \\ \pi^{(2)} &= (.27994,\ .00006,\ .7200,\ 0), \\ \pi^{(3)} &= (.255904,\ .00009,\ .743964,\ .000042), \\ \pi^{(4)} &= (.251080,\ .000122,\ .748714,\ .000084), \\ &\ \ \vdots, \end{aligned}$$

which gives the following values of the statistic $\pi^{(n)}a$:

$$0,\quad .00024,\quad .000780,\quad .001329,\quad \ldots.$$

These values can be plotted to show the evolution of the statistic over a number of time steps.

If the transient distribution is required at much greater values of $n$, then advantage should be taken of the small size of the matrix to compute successive powers of the matrix. For example, if the distribution is required at time step 1000, then $P^{1000}$ can be found from powering the matrix $P$. If, in addition, we are asked to plot the behavior of the statistic, until that time, we may wish to compute

$$\begin{aligned} \pi^{(100)} &= (.248093,\ .003099,\ .744507,\ .004251), \\ \pi^{(200)} &= (.246263,\ .006151,\ .739062,\ .008524), \\ \pi^{(300)} &= (.244461,\ .009156,\ .733652,\ .012731), \\ \pi^{(400)} &= (.242688,\ .012113,\ .728328,\ .016871), \\ &\ \ \vdots \\ \pi^{(1000)} &= (.232618,\ .028906,\ .698094,\ .040382), \end{aligned}$$

which gives the following values of the statistic $\pi^{(n)}a$:

$$.054900,\quad .109846,\quad .163928,\quad .217161,\quad \ldots,\quad .519446.$$

Finally, we point out that the distribution at time step $n = 1,000$ is a long way from the stationary distribution given by $\pi = (.131589,\ .197384,\ .394768,\ .276259)$, which has the statistic $\pi a = 3.552123$.

In the remainder of this section, we devote our attention to the computation of transient distributions of continuous-time Markov chains, i.e., the computation of $\pi(t)$ from

$$\pi(t) = \pi(0)e^{Qt},$$

where $Q$ is the infinitesimal generator of an irreducible continuous-time Markov chain. Depending on the numerical approach adopted, $\pi(t)$ may be computed by first forming $e^{Qt}$ and then premultiplying this with the initial probability vector $\pi(0)$. Such is the case with the matrix-scaling and -powering methods described in Section 10.7.1. They are most suitable when the transition rate matrix is small. The Matlab function *expm* falls into this category. In other cases, $\pi(t)$ may be computed directly without explicitly forming $e^{Qt}$. This is the approach taken by the uniformization method of Section 10.7.2 and the ordinary differential equation (ODE) solvers of Section 10.7.3. Both are suitable for large-scale Markov chains. Indeed, the uniformization method is a widely used and often extremely efficient approach that is applicable both to small transition rate matrices that may be either dense or sparse and to large sparse transition matrices. It is less suitable for stiff Markov chains. Methods for solving systems of ordinary differential equations have been actively researched by the numerical analysis community and provide an alternative to the uniformization method. In this section we suggest both single-step and multistep methods as possible ODE solution approaches.

### *10.7.1 Matrix Scaling and Powering Methods for Small State Spaces*

Moler and Van Loan [37] discuss nineteen *dubious* ways to compute the exponential of a relatively small-order matrix. A major problem in all these methods is that the accuracy of the approximations depends heavily on the norm of the matrix. Thus, when the norm of $Q$ is large or $t$ is large, attempting to compute $e^{Qt}$ directly is likely to yield unsatisfactory results. It becomes necessary to divide the interval $[0, t]$ into subintervals (called panels) $[0, t_0], [t_0, t_1], \ldots, [t_{m-1}, t_m = t]$ and to compute the transient solution at each time $t_j$, $j = 0, 1, \ldots, m$ using the solution at the beginning of the panel as the starting point. It often happens that this is exactly what is required by a user who can, as a result, see the evolution of certain system performance measures with time.

Matrix-scaling and -powering methods arise from a property that is unique to the exponential function, namely,

$$e^{Qt} = \left(e^{Qt/2}\right)^2. \tag{10.41}$$

The basic idea is to compute $e^{Qt_0}$ for some small value $t_0$ such that $t = 2^m t_0$ and subsequently to form $e^{Qt}$ by repeated application of the relation (10.41). It is in the computation of the initial $e^{Qt_0}$ that methods differ, and we shall return to this point momentarily.

Let $Q$ be the infinitesimal generator of an ergodic, continuous-time Markov chain, and let $\pi(0)$ be the probability distribution at time $t = 0$. We seek $\pi(t)$, the transient solution at time $t$. Let us introduce an integer $m$ and a time $t_0 \neq 0$ for which $t = 2^m t_0$. Then

$$\pi(t) = \pi(2^m t_0).$$

Writing $t_j = 2t_{j-1}$, we shall compute the matrices $e^{Qt_j}$ for $j = 0, 1, \ldots, m$ and consequently, by multiplication with $\pi(0)$, the transient solution at times $t_0, 2t_0, 2^2 t_0, \ldots, 2^m t_0 = t$. Note that $P(t_j) \equiv e^{Qt_j}$ is a stochastic matrix and that, from the Chapman–Kolmogorov equations,

$$P(t_j) = P(t_{j-1})P(t_{j-1}). \tag{10.42}$$

Thus, once $P(t_0)$ has been computed, each of the remaining $P(t_j)$ may be computed from Equation (10.42) by squaring the previous $P(t_{j-1})$. Thus matrix-powering methods, in the course of their computation, provide the transient solution at the intermediate times $t_0, 2t_0, 2^2 t_0, \ldots, 2^{m-1} t_0$. However, a disadvantage of matrix-powering methods, besides computational costs proportional to $n^3$ and memory requirements of $n^2$, is that repeated squaring may induce rounding error buildup, particularly in instances in which $m \gg 1$.

**Example 10.31** Suppose we need to find the transition distribution at some time $t = 10$, of a Markov chain with infinitesimal generator

$$Q = \begin{pmatrix} -.6 & 0 & .6 & 0 \\ .1 & -.9 & .1 & .7 \\ .4 & .3 & -.8 & .1 \\ 0 & .5 & 0 & -.5 \end{pmatrix}.$$

To use the matrix-scaling and -powering method, we need to find a $t_0$ such that $t_0$ is small and there exists an integer $m$ so that $t = 2^m t_0$. Setting $m = 5$ and solving for $t_0$ gives $t_0 = 10/32$ which we take to be sufficiently small for the purpose of this example. The reader may now wish to verify that

$$P(t_0) = e^{10Q/32} = \begin{pmatrix} .838640 & .007076 & .151351 & .002933 \\ .026528 & .769552 & .026528 & .177393 \\ .102080 & .074581 & .789369 & .033970 \\ .002075 & .126413 & .002075 & .869437 \end{pmatrix}$$

and that performing the operations $P(t_j) = P(t_{j-1})P(t_{j-1})$ for $j = 1, 2, \ldots, 5$ gives the result

$$P(t_5) = \begin{pmatrix} .175703 & .265518 & .175700 & .383082 \\ .136693 & .289526 & .136659 & .437156 \\ .161384 & .274324 & .161390 & .402903 \\ .129865 & .293702 & .129865 & .446568 \end{pmatrix} = e^{10Q}.$$

When premultiplied by an initial probability distribution, the distribution at time $t = 10$ can be found.

We now turn our attention to the computation of $e^{Qt_0}$. Since $t_0$ is small, methods based on approximations around zero are possible candidates. One good choice is that of rational Padé approximations around the origin. The $(p, q)$ Padé approximant to the matrix exponential $e^X$ is, by definition, the unique $(p, q)$ rational function $R_{pq}(X)$,

$$R_{pq}(X) \equiv \frac{N_{pq}(X)}{D_{pq}(X)},$$

which matches the Taylor series expansion of $e^X$ through terms to the power $p + q$. Its coefficients are determined by solving the algebraic equations

$$\sum_{j=0}^{\infty} \frac{X^j}{j!} - \frac{N_{pq}(X)}{D_{pq}(X)} = O\left(X^{p+q+1}\right),$$

which yields

$$N_{pq}(X) = \sum_{j=0}^{p} \frac{(p+q-j)!p!}{(p+q)!j!(p-j)!} X^j$$

and

$$D_{pq}(X) = \sum_{j=0}^{q} \frac{(p+q-j)!q!}{(p+q)!j!(q-j)!} (-X)^j.$$

For more detailed information on Padé approximants, the interested reader should consult the text by Baker [1]. A major disadvantage of Padé approximants is that they are accurate only near the origin and so should not be used when $\|X\|_2$ is large. However, since we shall be using them in the context of a matrix-scaling and -powering procedure, we may (and shall) choose $t_0$ so that $\|Qt_0\|_2$ is sufficiently small that the Padé approximant to $e^{Qt_0}$ may be obtained with acceptable accuracy, even for relatively low-degree approximants.

When $p = q$, we obtain the *diagonal Padé approximants*, and there are two main reasons why this choice is to be recommended. First, they are more stable [37]. In Markov chain problems, all the eigenvalues of $X = Qt$ are to be found in the left half plane. In this case the computed approximants $R_{pq}(X)$ for $p \neq q$ have larger rounding errors, because either $p > q$ and cancellation problems may arise, or $p < q$ and $D_{pq}(X)$ may be badly conditioned. Second, we obtain a higher-order method with the same amount of computation. To compute $R_{pq}(X)$ with $p < q$ requires about $qn^3$ flops and yields an approximant that has order $p + q$. To compute $R_{qq}(X)$ requires essentially the same number of flops but produces an approximant of order $2q > p+q$. Similar statements may be made when $p > q$.

For diagonal Padé approximants we find

$$R_{pp}(X) = \frac{N_{pp}(X)}{N_{pp}(-X)}, \tag{10.43}$$

where

$$N_{pp}(X) = \sum_{j=0}^{p} \frac{(2p-j)!p!}{(2p)!j!(p-j)!} X^j \equiv \sum_{j=0}^{p} c_j X^j.$$

The coefficients $c_j$ can be conveniently constructed by means of the recursion

$$c_0 = 1; \quad c_j = c_{j-1} \frac{p+1-j}{j(2p+1-j)}.$$

For actual implementation purposes, the following irreducible form offers considerable savings in computation time at the expense of additional memory locations:

$$R_{pp}(X) = \begin{cases} I + 2\dfrac{X\sum_{k=0}^{p/2-1} c_{2k+1}X^{2k}}{\sum_{k=0}^{p/2} c_{2k}X^{2k} - X\sum_{k=0}^{p/2-1} c_{2k+1}X^{2k}} & \text{if } p \text{ is even,} \\[2ex] -I - 2\dfrac{\sum_{k=0}^{(p-1)/2} c_{2k}X^{2k}}{X\sum_{k=0}^{(p-1)/2} c_{2k+1}X^{2k} - \sum_{k=0}^{(p-1)/2} c_{2k}X^{2k}} & \text{if } p \text{ is odd.} \end{cases} \tag{10.44}$$

Thus, for even values of $p$,

$$R_{pp}(X) = I + 2\frac{S_e}{T_e - S_e},$$

where

$$S_e = c_1X + c_3X^3 + \cdots + c_{p-1}X^{p-1} \quad \text{and} \quad T_e = c_0 + c_2X^2 + c_4X^4 + \cdots + c_pX^p,$$

while for odd values of $p$,

$$R_{pp}(X) = -\left(I + 2\frac{S_o}{T_o - S_o}\right),$$

where now

$$S_o = c_0 + c_2X^2 + c_4X^4 + \cdots + c_{p-1}X^{p-1} \quad \text{and} \quad T_o = c_1X + c_3X^3 + \cdots + c_pX^p.$$

These computations may be conveniently combined, and they cry out for a Horner-type evaluation procedure. Indeed, Horner evaluations of the numerator and the denominator in Equation (10.44) need only one-half the operations of a straightforward implementation of Equation (10.43).

The following four steps, adapted from that presented in [42], implements a Padé variant of the matrix-powering and -scaling approach for the computation of $e^X$. In this implementation, the integer $m$ is chosen as $m = \lfloor \log \|X\|_\infty / \log 2 \rfloor + 1$. To compute the transient solution at time $t$ of a Markov chain with generator $Q$ and initial state $\pi(t_0)$, it suffices to apply this algorithm with $X = Qt$ and then to form $\pi(t_0)R$, where $R$ is the approximation to $e^X$ computed by the algorithm.

1. *Find appropriate scaling factor:*
   - Compute $m = \max(0, \lfloor \log \|X\|_\infty / \log 2 \rfloor + 1)$.
2. *Compute coefficients and initialize:*
   - Set $c_0 = 1$.
   - For $j = 1, 2, \ldots, p$ do
     - Compute $c_j = c_{j-1} \times (p + 1 - j)/(j(2p + 1 - j))$.
   - Compute $X1 = 2^{-m}X;\ \ X2 = X1^2;\ \ T = c_p I;\ \ S = c_{p-1} I$.
3. *Application of Horner scheme:*
   - Set $odd = 1$.
   - For $j = p - 1, \ldots, 2, 1$ do
     - if $odd = 1$, then
       * Compute $T = T \times X2 + c_{j-1} I$;

       else
       * Compute $S = S \times X2 + c_{j-1} I$.
     - Set $odd = 1 - odd$.
   - If $odd = 0$, then
     - Compute $S = S \times X1;\ \ R = I + 2 \times (T - S)^{-1} \times S$;

     else
     - Compute $T = T \times X1;\ \ R = -(I + 2 \times (T - S)^{-1} \times S)$.
4. *Raise matrix to power $2^m$ by repeated squaring:*
   - For $j = 1$ to $m$ do
     - Compute $R = R \times R$.

This Padé approximation for $e^X$ requires a total of approximately $(p + m + \frac{4}{3})n^3$ multiplications. It may be implemented with three double-precision arrays, each of size $n^2$, in addition to the storage required for the matrix itself.

This leaves us with the choice of $p$. In the appendix of [37], a backward error analysis of the Padé approximation is presented, in which it is shown that if $\|X\|_2/2^m \le \frac{1}{2}$, then

$$\left[R_{pp}(2^{-m}X)\right]^{2^m} = e^{X+E},$$

where

$$\frac{\|E\|_2}{\|X\|_2} \le \left(\frac{1}{2}\right)^{2p-3} \frac{(p!)^2}{(2p)!(2p+1)!} \approx \begin{cases} 0.77 \times 10^{-12}\ (p = 5), \\ 0.34 \times 10^{-15}\ (p = 6), \\ 0.11 \times 10^{-18}\ (p = 7), \\ 0.27 \times 10^{-22}\ (p = 8). \end{cases} \tag{10.45}$$

This suggests that relatively low-degree Padé approximants are adequate. However, the above analysis does not take rounding error into account. This aspect has been examined by Ward [54], who proposes certain criteria for selecting appropriate values for some computers. Additionally, a discussion on "the degree of best rational approximation to the exponential function" is provided by Saff [48]. Finally, numerical experiments on Markov chains by Philippe and Sidje [42] find that even values of $p$ are better than odd values and that the value $p = 6$ is generally satisfactory.

### 10.7.2 The Uniformization Method for Large State Spaces

For large-scale Markov chains, methods currently used to obtain transient solutions are based either on readily available differential equation solvers such as Runge-Kutta methods or the Adams formulae and backward differentiation formulae (BDF) or on the method of uniformization (also called Jensen's method or the method of randomization). Most methods experience difficulty when both $\max_j |q_{jj}|$ (the largest exit rate from any state) and $t$ (the time at which the solution is required) are large, and there appears to be little to recommend a single method for all situations. In this section we discuss the *uniformization method*. This method has attracted much attention, is extremely simple to program and often outperforms other methods, particularly when the solution is needed at a single time point close to the origin. If the solution is required at many points, or if plots need to be drawn to show the evolution of certain performance measures, then a method based on one of the differential equation solvers may be preferable. An extensive discussion on all these methods may be found in [50].

The uniformization method revolves around a discrete-time Markov chain that is embedded in the continuous-time process. The transition probability matrix of this discrete-time chain is constructed as

$$P = Q\Delta t + I$$

with $\Delta t \leq 1/\max_i |q_{ii}|$. In this Markov chain all state transitions occur at a *uniform* rate equal to $1/\Delta t$—hence the name *uniformization*. Letting $\gamma = \max_i |q_{ii}|$, we may write

$$Q = \gamma(P - I)$$

and inserting this into the Kolmogorov forward differential equations we get

$$\pi(t) = \pi(0)e^{Qt} = \pi(0)e^{\gamma(P-I)t} = \pi(0)e^{-\gamma t}e^{\gamma Pt}.$$

Expanding $e^{\gamma Pt}$ in a Taylor series, we obtain

$$\pi(t) = \pi(0)e^{-\gamma t}\sum_{k=0}^{\infty}\frac{(\gamma t)^k P^k}{k!},$$

i.e.,

$$\pi(t) = \sum_{k=0}^{\infty}\pi(0)P^k\frac{(\gamma t)^k}{k!}e^{-\gamma t}. \tag{10.46}$$

Two observations may be made. First, the term $\pi(0)P^k$ may be recognized as the vector that provides the probability distribution after $k$ steps of the discrete-time Markov chain whose stochastic transition probability matrix is $P$ and with initial distribution $\pi(0)$. Second, the term $e^{-\gamma t}(\gamma t)^k/k!$ may be recognized from the Poisson process with rate $\gamma$ as the probability of $k$ events occurring in $[0, t)$. It is the probability that the discrete-time Markov chain makes $k$ transition steps in the interval $[0, t)$. These probabilities may be interpreted as weights that when multiplied with the distribution of the discrete-time Markov chain after $k$ steps and summed over all possible number of steps, (in effect, *unconditioning* the transient distribution which has been written in terms of power series of $P$) yields the transient distribution $\pi(t)$. The uniformization method computes the distribution $\pi(t)$ directly from Equation (10.46). Writing it in the form

$$\pi(t) = e^{-\gamma t}\sum_{k=0}^{\infty}\left(\pi(0)P^{k-1}\frac{(\gamma t)^{k-1}}{(k-1)!}\right)P\frac{\gamma t}{k} \tag{10.47}$$

exposes a convenient recursive formulation. Setting $y = \pi = \pi(0)$ and iterating sufficiently long with

$$y = y\left(P\frac{\gamma t}{k}\right), \qquad \pi = \pi + y,$$

allows the transient distribution to be computed as $\pi(t) = e^{-\gamma t}\,\pi$.

The curious reader may wonder why we bother with Equation (10.46) when $\pi(t)$ might be thought to be more easily computed directly from the Chapman-Kolmogorov differential equations by means of the formula

$$\pi(t) = \pi(0)\sum_{k=0}^{\infty}\frac{(Qt)^k}{k!}.$$

However, the matrix $Q$ contains both positive and negative elements and these may be greater than one which leads to a less stable algorithm than one based on the matrix $P$ which, being a stochastic matrix, has all positive elements lying in the range [0, 1].

Among the numerical advantages of the uniformization technique is the ease with which it can be translated into computer code and the control it gives over the truncation error. Let us first discuss the truncation error. In implementing the uniformization method, we need to truncate the infinite series in (10.46). Let

$$\pi^*(t) = \sum_{k=0}^{K}\pi(0)P^k e^{-\gamma t}(\gamma t)^k/k! \tag{10.48}$$

and let $\delta(t) = \pi(t) - \pi^*(t)$. For any consistent vector norm, $||\delta(t)||$ is the truncation error. It is not difficult to numerically bound this error. If we choose $K$ sufficiently large that

$$1 - \sum_{k=0}^{K} e^{-\gamma t}(\gamma t)^k/k! \le \epsilon,$$

or, equivalently, that

$$\sum_{k=0}^{K}\frac{(\gamma t)^k}{k!} \ge \frac{1-\epsilon}{e^{-\gamma t}} = (1-\epsilon)e^{\gamma t}, \tag{10.49}$$

where $\epsilon$ is some prespecified truncation criterion, then it follows that

$$\|\pi(t) - \pi^*(t)\|_\infty \le \epsilon.$$

To see this, observe that

$$\left\|\pi(t) - \pi^*(t)\right\|_\infty$$

$$= \left\|\sum_{k=0}^{\infty}\pi(0)P^k e^{-\gamma t}(\gamma t)^k/k! - \sum_{k=0}^{K}\pi(0)P^k e^{-\gamma t}(\gamma t)^k/k!\right\|_\infty$$

$$= \left\|\sum_{k=K+1}^{\infty}\pi(0)P^k e^{-\gamma t}(\gamma t)^k/k!\right\|_\infty \le \sum_{k=K+1}^{\infty} e^{-\gamma t}(\gamma t)^k/k!$$

$$= \sum_{k=0}^{\infty} e^{-\gamma t}(\gamma t)^k/k! - \sum_{k=0}^{K} e^{-\gamma t}(\gamma t)^k/k!$$

$$= 1 - \sum_{k=0}^{K} e^{-\gamma t}(\gamma t)^k / k! \;\; \leq \epsilon.$$

**Example 10.32** Consider a continuous-time Markov chain with infinitesimal generator

$$Q = \begin{pmatrix} -5 & 2 & 3 \\ 1 & -2 & 1 \\ 6 & 4 & -10 \end{pmatrix}.$$

Given $\pi(0) = (1, 0, 0)$ and $t = 1$, we wish to examine the behavior of Equation (10.46) as the number of terms in the summation increases. The uniformization process gives $\gamma = 10$ and

$$P = \begin{pmatrix} .5 & .2 & .3 \\ .1 & .8 & .1 \\ .6 & .4 & 0 \end{pmatrix}.$$

Suppose we require an accuracy of $\epsilon = 10^{-6}$. To find the value of $K$, the number of terms to be included in the summation, we proceed on a step-by-step basis, incrementing $k$ until

$$\sigma_K = \sum_{k=0}^{K} \frac{(\gamma t)^k}{k!} \;\geq\; (1-\epsilon)e^{\gamma t} = (1 - 10^{-6})e^{10} = 22,026.4438.$$

Observe that successive terms in the summation satisfy

$$\xi_{K+1} = \xi_K \frac{\gamma t}{K+1} \quad \text{with } \xi_0 = 1$$

and that

$$\sigma_{K+1} = \sigma_K + \xi_{K+1} \quad \text{with } \sigma_0 = 1,$$

and so, beginning with $K = 0$, and using this recursion, we successively compute

$$\sigma_0 = 1,\; \sigma_1 = 11,\; \sigma_2 = 61,\; \sigma_3 = 227.6667, \sigma_4 = 644.33331,\; \ldots,\; \sigma_{28} = 22,026.4490.$$

Thus $K = 28$ terms are needed in the summation. To obtain our approximation to the transient distribution at time $t = 1$, we need to compute

$$\pi(t) \approx \sum_{k=0}^{28} \pi(0)P^k e^{-\gamma t}\frac{(\gamma t)^k}{k!} \tag{10.50}$$

for some initial distribution, $\pi(0) = (1, 0, 0)$, say. Using the recursion relation of Equation (10.47), we find

$k = 0:$ $y = (1,\ 0,\ 0);$ $\pi = (1,\ 0,\ 0),$

$k = 1:$ $y = (5,\ 2,\ 3);$ $\pi = (6,\ 2,\ 3),$

$k = 2:$ $y = (22.5,\ 19,\ 8.5);$ $\pi = (28.5,\ 21,\ 11.5),$

$k = 3:$ $y = (60.8333,\ 77,\ 28.8333);$ $\pi = (89.3333,\ 98,\ 40.3333),$

$\vdots$

$k = 28:$ $y = (.009371,\ .018742,\ .004686);$ $\pi = (6,416.9883,\ 12,424.44968,\ 3,184.4922).$

In this example, the elements of $y$ increase until they reach $y = (793.8925,\ 1,566.1563,\ 395.6831)$ for $k = 10$ and then they begin to decrease. Multiplying the final $\pi$ vector by $e^{-10}$ produces the desired transient distribution:

$$\pi(1) = e^{-10}(6,416.9883,\ 12,424.44968,\ 3,184.4922) = (.291331,\ .564093,\ .144576).$$

In implementing the uniformization technique, we may code Equation (10.46) exactly as it appears, with the understanding that $\pi(0)P^k$ is computed iteratively, i.e., we do not construct the $k^{\text{th}}$ power of $P$ and premultiply it with $\pi(0)$, but instead form the sequence of vectors, $\psi(j+1) = \psi(j)P$, with $\psi(0) = \pi(0)$, so that $\pi(0)P^k$ is given by $\psi(k)$. Alternatively we may partition Equation (10.46) into time steps $0 = t_0, t_1, t_2, \ldots, t_m = t$ and write code to implement

$$\pi(t_{i+1}) = \sum_{k=0}^{\infty} \pi(t_i)P^k e^{-\gamma(t_{i+1}-t_i)}\gamma^k(t_{i+1}-t_i)^k/k!$$

recursively for $i = 0, 1, \ldots, m-1$. This second approach is the obvious way to perform the computation if the transient solution is required at various points $t_1, t_2, \ldots$ between the initial time $t_0$ and the final time $t$. It is computationally more expensive if the transient solution is required only at a single terminal point. However, it may prove useful when the numerical values of $\gamma$ and $t$ are such that the computer underflows when computing $e^{-\gamma t}$. Such instances can be detected a priori and appropriate action taken. For example, one may decide not to allow values of $\gamma t$ to exceed 100. When such a situation is detected, the time $t$ may be divided into $d = 1 + \lfloor \gamma t/100 \rfloor$ equal intervals and the transient solution computed at times $t/d,\ 2t/d,\ 3t/d,\ \ldots, t$. Care must be taken in implementing such a procedure since the error in the computation of the intermediate values $\pi(t_i)$ may propagate along into later values, $\pi(t_j),\ j > i$.

An alternative to dividing a large interval $\gamma t$ into more manageable pieces may be to omit from the summation in Equation (10.48), terms for which the value of $e^{-\gamma t}(\gamma t)^k/k!$ is so small that it could cause numerical difficulties. This may be accomplished by choosing a *left* truncation point, $l$, as the largest value for which

$$\sum_{k=0}^{l-1} e^{-\gamma t}\frac{(\gamma t)^k}{k!} \le \epsilon_l$$

for some lower limit $\epsilon_l$. The required transient distribution vector is now computed as

$$\pi^*(t) = \sum_{k=l}^{K} \psi(k)e^{-\gamma t}(\gamma t)^k/k!$$

with $\psi(k)$ computed recursively as before. The value of $l$ is easily computed using the same procedure that was used to compute the upper limit of the summation, $K$. The amount of work involved in summing from $k = l$ is basically just the same as that in summing from $k = 0$, since $\psi(k)$ from $k = 0$ to $k = K$ must be computed in all cases. The only reason in summing from $k = l$ is that of computational stability.

One other wrinkle may be added to the uniformization method when it is used to compute transient distributions at large values of $\gamma t$—the *stationary distribution* of the uniformized chain may be reached well before the last term in the summation (10.48). If this is the case, it means that, from the point at which steady state is reached, the values $\psi(k)$ no longer change. It is possible to monitor convergence of the uniformized chain and to determine the point at which it reaches steady state. Assume this occurs when $k$ has the value $k_s$. The transient distribution at time $t$ may then be computed more efficiently as

$$\pi^*(t) = \sum_{k=l}^{k_s} \psi(k)e^{-\gamma t}(\gamma t)^k/k! + \left(\sum_{k=k_s+1}^{K} e^{-\gamma t}(\gamma t)^k/k!\right)\psi(k_s).$$

The following two step implementation computes the transient solution $\pi(t)$ at time $t$ given the probability distribution $\pi(0)$ at time $t = 0$; $P$, the stochastic transition probability matrix of the discrete-time Markov chain; $\gamma$, the parameter of the Poisson process; and $\epsilon$, a tolerance criterion. It is designed for the case in which the number of states in the Markov chain is large. The only operation involving the matrix $P$ is its multiplication with a vector. This implementation does not incorporate a lower limit nor a test for convergence to the steady-state.

1. *Use Equation (10.49) to compute $K$, the number of terms in the summation:*
   - Set $K = 0$; $\xi = 1$; $\sigma = 1$; $\eta = (1-\epsilon)/e^{-\gamma t}$.
   - While $\sigma < \eta$ do
     - Compute $K = K + 1$; $\xi = \xi \times (\gamma t)/K$; $\sigma = \sigma + \xi$.
2. *Approximate $\pi(t)$ from Equation (10.48):*
   - Set $\pi = \pi(0)$; $y = \pi(0)$.
   - For $k = 1$ to $K$ do
     - Compute $y = yP \times (\gamma t)/k$; $\pi = \pi + y$.
   - Compute $\pi(t) = e^{-\gamma t}\pi$.

### *10.7.3 Ordinary Differential Equation Solvers*

The transient distribution we seek to compute, $\pi(t) = \pi(0)e^{Qt}$, is the solution of the Chapman–Kolmogorov differential equations

$$\frac{d\pi(t)}{dt} = \pi(t)Q$$

with initial conditions $\pi(t=0) = \pi(0)$. The solution of ordinary differential equations (ODEs) has been (and continues to be) a subject of extensive research, and there are numerous possibilities for applying ODE procedures to compute transient solutions of Markov chains. An immediate advantage of such an approach is that, unlike uniformization and matrix scaling and powering, numerical methods for the solution of ODEs are applicable to *nonhomogeneous* Markov chains, i.e., Markov chains whose infinitesimal generators are a function of time, $Q(t)$. However, since the subject of ODEs is so vast, and our objectives in this chapter rather modest, we shall be content to give only the briefest of introductions to this approach. Given a first-order differential equation $y' = f(t, y)$ and an initial condition $y(t_0) = y_0$, a solution is a differentiable function $y(t)$ such that

$$y(t_0) = y_0, \quad \frac{d}{dt}y(t) = f(t, y(t)). \tag{10.51}$$

In the context of Markov chains, the solution $y(t)$ is the row vector $\pi(t)$ and the function $f(t, y(t))$ is simply $\pi(t)Q$. We shall use the standard notation of Equation(10.51) for introducing general concepts, and the Markov chain notation when the occasion so warrants, such as in examples and Matlab code.

Numerical procedures to compute the solution of Equation (10.51) attempt to follow a unique solution curve from its value at an initially specified point to its value at some other prescribed point $\tau$. This usually involves a discretization procedure on the interval $[0, \tau]$ and the computation of approximations to the solution at the intermediate points. Given a discrete set of points (or *mesh*) $\{0 = t_0, t_1, t_2, \ldots, t_\eta = \tau\}$ in $[0, \tau]$, we denote the exact solution of the differential Equation (10.51) at time $t_i$ by $y(t_i)$. The *step size* or *panel width* at step $i$ is defined as $h_i = t_i - t_{i-1}$. A numerical method generates a sequence $\{y_1, y_2, \ldots, y_\eta\}$ such that $y_i$ is an approximation[3] to $y(t_i)$. In some instances the solution at time $\tau$ is all that is required, and in that case the computed solution is taken to be $y_\eta$. In other cases the value of $y(t)$ for all $t_0 \le t \le \tau$ is required, often in the form of a graph, and this is obtained by fitting a suitable curve to the values $y_0, y_1, y_2, \ldots, y_\eta$.

[3] The reader should be careful not to confuse $y_i$ with $y(t_i)$. We use $y_i$ for $i \neq 0$ to denote a *computed approximation* to the *exact value* $y(t_i)$. For $i = 0$, the initial condition gives $y_0 = y(t_0)$.

### *Single-Step Euler Method*

If the solution is continuous and differentiable, then in a small neighborhood of the point $(t_0, y_0)$ we can approximate the solution curve by its tangent $y'_0$ at $(t_0, y_0)$ and thereby move from $(t_0, y_0)$ to the next point $(t_1, y_1)$. Since this means that

$$y'_0 = \frac{y_1 - y_0}{t_1 - t_0},$$

we obtain the formula

$$y_1 = y_0 + h_1 y'_0 = y_0 + h_1 f(t_0, y_0), \tag{10.52}$$

where $h_1 = t_1 - t_0$. This process may be repeated from the computed approximation $y_1$ to obtain

$$y_2 = y_1 + h_2 f(t_1, y_1), \tag{10.53}$$

where $h_2 = t_2 - t_1$, and so on until the final destination is reached,

$$y_\eta = y_{\eta-1} + h_\eta f(t_{\eta-1}, y_{\eta-1}) \tag{10.54}$$

with $h_\eta = t_\eta - t_{\eta-1}$. This method is called the *forward Euler method (FEM)* or *explicit Euler method.* From Equations (10.52)–(10.54), the FEM method may be written as

$$y_{i+1} = y_i + h_{i+1} f(t_i, y_i), \quad i = 0, 1, \ldots, \eta - 1, \tag{10.55}$$

where $h_{i+1} = t_{i+1} - t_i$. In computing $y_{i+1}$, a method may incorporate the values of previously computed approximations $y_j$ for $j = 0, 1, \ldots, i$, or even previous approximations to $y(t_{i+1})$. A method that uses only $(t_i, y_i)$ to compute $y_{i+1}$ is said to be an *explicit single-step method.* It is said to be a *multistep method* if it uses approximations at several previous steps to compute its new approximation. A method is said to be *implicit* if computation of $y_{i+1}$ requires an approximation to $y(t_{i+1})$; otherwise, it is said to be *explicit.* An *implicit Euler method,* also called a *backward Euler method,* is defined by using the slope at the point $(t_{i+1}, y_{i+1})$. This gives

$$y_{i+1} = y_i + h_{i+1} f(t_{i+1}, y_{i+1}).$$

The *modified Euler method* incorporates the average of the slopes at both points under the assumption that this will provide a better average approximation of the slope over the entire panel $[t_i, t_{i+1}]$. The formula is given by

$$y_{i+1} = y_i + h_{i+1} \frac{f(t_i, y_i) + f(t_{i+1}, y_{i+1})}{2}. \tag{10.56}$$

This is also referred to as the *trapezoid rule.* A final formula in this category is the *implicit midpoint rule,* which involves the slope at the midpoint of the interval. The formula is given by

$$y_{i+1} = y_i + h_{i+1}\, f\left(t_i + \frac{h_i}{2}, \frac{y_i + y_{i+1}}{2}\right).$$

These are all single-step methods, since approximations preceding $(t_i, y_i)$ are not used in the generation of the next approximation $y_{i+1}$.

In Markov chain problems the explicit Euler method (Equation10.55) is given by

$$\pi_{(i+1)} = \pi_{(i)} + h_{i+1} \pi_{(i)} Q,$$

i.e.,

$$\pi_{(i+1)} = \pi_{(i)} (I + h_{i+1} Q). \tag{10.57}$$

Note that $\pi_{(i)}$ is the *state vector* of probabilities at time $t_i$. We use this notation, rather than $\pi_i$, so as not to confuse the $i^{\text{th}}$ component of the vector with the entire vector at time $t_i$. Thus, in the explicit

Euler method applied to Markov chains, moving from one time step to the next is accomplished by a scalar-matrix product and a vector-matrix product.

**Example 10.33** Consider a continuous-time Markov chain with infinitesimal generator

$$Q = \begin{pmatrix} -2 & 1 & 1 \\ 3 & -8 & 5 \\ 1 & 2 & -3 \end{pmatrix}.$$

Suppose we wish to compute the transient distribution at time $t = 1$ given that the Markov chain begins in state 1. For illustrative purposes, we shall use a constant step size equal to 0.1 and compute $\pi_{(i)},\ i = 1, 2, \ldots, 10$. From the FEM formula, we obtain

$$\pi(.1) \approx \pi_{(1)} = \pi_{(0)}(I + .1Q) = (1, 0, 0)\left[\begin{pmatrix} 1 & 0 & 0 \\ 0 & 1 & 0 \\ 0 & 0 & 1 \end{pmatrix} + .1 \times \begin{pmatrix} -2 & 1 & 1 \\ 3 & -8 & 5 \\ 1 & 2 & -3 \end{pmatrix}\right]$$

$$= (1, 0, 0)\begin{pmatrix} .8 & .1 & .1 \\ .3 & .2 & .5 \\ .1 & .2 & .7 \end{pmatrix} = (.8, .1, .1).$$

$$\pi(.2) \approx \pi_{(2)} = (.8, .1, .1)\begin{pmatrix} .8 & .1 & .1 \\ .3 & .2 & .5 \\ .1 & .2 & .7 \end{pmatrix} = (.68, .12, .20).$$

Continuing in this fashion we finally compute

$$\pi(1.0) \approx \pi_{(10)} = (.452220,\ .154053,\ .393726)\begin{pmatrix} .8 & .1 & .1 \\ .3 & .2 & .5 \\ .1 & .2 & .7 \end{pmatrix} = (.447365,\ .154778,\ .397857).$$

The correct answer, obtained by a different method, is

$$\pi(1.0) = (.457446,\ .153269,\ .389285),$$

and a measure of the error is given as $\|\pi(1.0) - \pi_{(10)}\|_2 = .013319$.

The modified Euler or trapezoid rule, Equation (10.56), when applied to Markov chains, becomes

$$\pi_{(i+1)} = \pi_{(i)} + \frac{h_{i+1}}{2}\left(\pi_{(i)}Q + \pi_{(i+1)}Q\right),$$

i.e.,

$$\pi_{(i+1)}\left(I - \frac{h_{i+1}}{2}Q\right) = \pi_{(i)}\left(I + \frac{h_{i+1}}{2}Q\right), \tag{10.58}$$

which requires, in addition to the operations needed by the explicit Euler method, the solution of a system of equations at each step (plus a scalar-matrix product). These additional computations per step are offset to a certain extent by the better accuracy achieved with the trapezoid rule. If the size of the Markov chain is small and the step size is kept constant, the matrix

$$\left(I + \frac{h_{i+1}}{2}Q\right)\left(I - \frac{h_{i+1}}{2}Q\right)^{-1}$$

may be computed at the outset before beginning the stepping process, so that the computation per step required by the modified Euler method becomes identical to that of the explicit Euler. If the Markov chain is large, then the inverse should not be formed explicitly. Instead an $LU$ decomposition should be computed and the inverse replaced with a backward and forward substitution process with $U$ and $L$ respectively. When the step size is not kept constant, each different value of $h$ used requires that a system of linear equations be solved. Depending on the size and sparsity pattern of $Q$, the work required by the trapezoid rule to compute the solution to a specified precision may or may not be less than that required by the explicit Euler. The trade-off is between an implicit method requiring more computation per step but fewer steps and an explicit method requiring more steps but less work per step!

**Example 10.34** Let us return to Example 10.33 and this time apply the trapezoid rule. Again, we shall use a step size of 0.1 and successively compute $\pi_{(i)}$, $i = 1, 2, \ldots, 10$. From the Trapezoid formula, we obtain

$$\pi_{(i+1)} = \pi_{(i)} \left(I + \frac{h_{i+1}}{2} Q\right) \left(I - \frac{h_{i+1}}{2} Q\right)^{-1} = \pi_{(i)} \begin{pmatrix} .90 & .05 & .05 \\ .15 & .60 & .25 \\ .05 & .10 & .85 \end{pmatrix} \begin{pmatrix} 1.10 & -.05 & -.05 \\ -.15 & 1.40 & -.25 \\ -.05 & -.10 & 1.15 \end{pmatrix}^{-1},$$

i.e.,

$$\pi_{(i+1)} = \pi_{(i)} \begin{pmatrix} .832370 & .072254 & .095376 \\ .213873 & .459538 & .326590 \\ .098266 & .130058 & .771676 \end{pmatrix}.$$

Beginning with $\pi_{(0)} = (1, 0, 0)$, this allows us to compute

$$\begin{aligned} \pi(.1) \approx \pi_{(1)} &= (.832370,\ .072254,\ .095376), \\ \pi(.2) \approx \pi_{(2)} &= (.717665,\ .105750,\ .176585), \\ \pi(.3) \approx \pi_{(3)} &= (.637332,\ .123417,\ .239251), \\ &\vdots \\ \pi(1.0) \approx \pi_{(10)} &= (.456848,\ .153361,\ .389791). \end{aligned}$$

The correct answer is still

$$\pi(1.0) = (.457446,\ .153269,\ .389285),$$

and a measure of the error is given as $\|\pi(1.0) - \pi_{(10)}\|_2 = .0007889$ which is considerably better than that obtained with the FEM method.

**Example 10.35** This time we examine the effect of decreasing the step size using the same infinitesimal generator, taking $\pi_{(0)} = (1, 0, 0)$ and the length of the interval of integration to be $\tau = 1$, as before.

$$Q = \begin{pmatrix} -2 & 1 & 1 \\ 3 & -8 & 5 \\ 1 & 2 & -3 \end{pmatrix}.$$

Table 10.2 shows the results obtained when Equations (10.57) and (10.58) are implemented in Matlab.[4] The first column gives the step size used throughout the range, the second is the 2-norm of the absolute error using the explicit Euler method, and the third column gives the 2-norm of the absolute error when the trapezoid method is used. Observe that the accuracy achieved with the explicit Euler method has a direct relationship with the step size. Observe also the much more accurate results obtained with the trapezoid method. Since equal step sizes were used in these examples, an essentially identical amount of computation is required by both. The trapezoid rule needs only an additional $(3 \times 3)$ matrix inversion and a matrix-matrix product.

[4] A listing of all the Matlab implementations of ODE methods used to generate results in this chapter is given at the end of this section.

Table 10.2. Euler and Trapezoid Method Results

| Step size | Explicit Euler $\|y(t_1) - y_1\|_2$ | Trapezoid $\|y(t_1) - y_1\|_2$ |
|---|---|---|
| $h = .1$ | $0.133193e{-01}$ | $0.788907e{-03}$ |
| $h = .01$ | $0.142577e{-02}$ | $0.787774e{-05}$ |
| $h = .001$ | $0.143269e{-03}$ | $0.787763e{-07}$ |
| $h = .0001$ | $0.143336e{-04}$ | $0.787519e{-09}$ |
| $h = .00001$ | $0.143343e{-05}$ | $0.719906e{-11}$ |

### *Taylor Series Methods*

We now turn our attention to *Taylor series methods*. The forward Euler method corresponds to taking the first two terms of the Taylor series expansion of $y(t)$ around the current approximation point. Expanding $y(t)$ as a power series in $h$, we have

$$y(t+h) = y(t) + hy'(t) + \frac{h^2}{2!}y''(t) + \cdots . \tag{10.59}$$

If we know $y(t)$ exactly, we may compute as many terms in the series as we like by repeated differentiation of $y' = f(t, y(t))$, the only constraint being the differentiability of the function $f(t, y(t))$. Different *Taylor series algorithms* are obtained by evaluating different numbers of terms in the series (10.59). For example, using the obvious notational correspondence with Equation (10.59), the Taylor's algorithm of order 2 may be written

$$y_{i+1} = y_i + hT_2(t_i, y_i),$$

where

$$T_2(t_i, y_i) = f(t_i, y_i) + \frac{h}{2}f'(t_i, y_i).$$

The order-1 Taylor series algorithm corresponds to the FEM method, since we have

$$y_{i+1} = y_i + hT_1(t_i, y_i) = y_i + hf(t_i, y_i).$$

Taylor series methods are not widely used because of the need to construct derivatives of the function $f(t, y(t))$. Their use is that they provide a mechanism to develop other methods which have an accuracy comparable to that of Taylor series of order $p$ but which do not require derivatives of $f(t, y(t))$. This leads us introduce the concept of *order of convergence*.

**Definition 10.7.1 (Order of convergence)** *A numerical method is said to be convergent to order p if the sequence of approximations $\{y_0, y_1, y_2, \ldots, y_\eta\}$ generated by the method (with constant step size $h = \tau/\eta$) satisfies*

$$\|y_i - y(t_i)\| = O(h^p), \quad i = 0, 1, \ldots, \eta.$$

**Example 10.36** Euler's method, described by Equation (10.55), is convergent to order 1; the error behaves like $Mh$, where $M$ is a constant that depends on the problem and $h$ is the maximum step size.

The order of convergence of a method concerns only the truncation error made in terminating the infinite series (10.59). In particular, it does not take into account any roundoff error that might occur. Indeed, as the truncation error tends to zero with decreasing $h$, the roundoff error has the opposite effect. The smaller the step size, the greater the number of steps that must be taken, and hence the greater the amount of computation that must be performed. With this greater amount of computation comes the inevitable increase in roundoff error. As $h$ becomes smaller, the total error (truncation error plus rounding error) decreases, but invariably there comes a point at which it begins to increase again. Thus, it is important that $h$ not be chosen too small.

### *Runge-Kutta Methods*

An extremely important and effective class of single step methods is the class of Runge–Kutta methods. A Runge–Kutta algorithm of order $p$ provides an accuracy comparable to a Taylor series algorithm of order $p$, but without the need to determine and evaluate the derivatives $f', f'', \ldots, f^{(p-1)}$, requiring instead the evaluation of $f(t, y)$ at selected points. The derivation of an order $p$ Runge–Kutta method is obtained from a comparison with the terms through $h^p$ in the Taylor series method for the first step, i.e., the computation of $y_1$ from the initial condition $(t_0, y_0)$. This is because the analysis assumes that the value around which the expansion is performed is known exactly. A Runge–Kutta method is said to be of order $p$ if the Taylor series for the exact solution $y(t_0 + h)$ and the Taylor series for the computed solution $y_1$ coincide up to and including the term $h^p$.

For example, a Runge–Kutta method of order 2 (RK2) is designed to give the same order of convergence as a Taylor series method of order 2, without the need for differentiation. It is written as

$$y_{i+1} = y_i + h(ak_1 + bk_2), \tag{10.60}$$

where $a$ and $b$ are constants,

$$\begin{aligned} k_1 &= f(t_i, y_i), \\ k_2 &= f(t_i + \alpha h, y_i + h\beta k_1), \end{aligned}$$

and $\alpha$ and $\beta$ are constants. The constants are chosen so that the Taylor series approach agrees with Equation (10.60) at $i = 0$, through the term in $h^2$. Finding these constants requires computing the derivatives of Equation (10.60) at $i = 0$, and then comparing coefficients with those of the first and second derivatives obtained from the exact Taylor series expansion. We leave this as an exercise. Equating the coefficients imposes the following restrictions on the constants $a, b, \alpha, \beta$:

$$a + b = 1, \quad \alpha b = \beta b = 1/2.$$

These are satisfied by the choice $a = b = \frac{1}{2}$ and $\alpha = \beta = 1$, so the resulting RK2 method is

$$y_{i+1} = y_i + \frac{h}{2} f(t_i, y_i) + \frac{h}{2} f(t_i + h, y_i + hf(t_i, y_i)).$$

Other choices for the constants are also possible.

The most widely used Runge–Kutta methods are of order 4. The standard RK4 method requires four function evaluations per step and is given by

$$y_{i+1} = y_i + \frac{h}{6}(k_1 + 2k_2 + 2k_3 + k_4), \tag{10.61}$$

where

$$\begin{aligned} k_1 &= f(t_i, y_i), \\ k_2 &= f(t_i + h/2, y_i + hk_1/2), \\ k_3 &= f(t_i + h/2, y_i + hk_2/2), \\ k_4 &= f(t_i + h, y_i + hk_3). \end{aligned}$$

Notice that each $k_i$ corresponds to a point in the $(t, y)$ plane at which the evaluation of $f$ is required. Each is referred to as a *stage*. Thus, the method described by Equation (10.61) is a fourth-order, four-stage, explicit Runge–Kutta method.

When the standard explicit fourth-order Runge–Kutta method given by Equation (10.61) is applied to the Chapman–Kolmogorov equations $\pi' = \pi Q$, the sequence of operations to be performed to move from $\pi_{(i)}$ to the next time step $\pi_{(i+1)}$ is as follows:

$$\pi_{(i+1)} = \pi_{(i)} + h(k_1 + 2k_2 + 2k_3 + k_4)/6,$$

where

$$\begin{aligned} k_1 &= \pi_{(i)} Q, \\ k_2 &= (\pi_{(i)} + hk_1/2)Q, \\ k_3 &= (\pi_{(i)} + hk_2/2)Q, \\ k_4 &= (\pi_{(i)} + hk_3)Q. \end{aligned}$$

**Example 10.37** We return to the continuous-time Markov chain of the previous examples and this time apply the RK4 method to obtain the same transient distribution. With $\pi_{(0)} = (1, 0, 0)$ and

$$Q = \begin{pmatrix} -2 & 1 & 1 \\ 3 & -8 & 5 \\ 1 & 2 & -3 \end{pmatrix},$$

we find, taking steps of size .1,

$$\begin{aligned} k_1 &= \pi_{(0)} Q = (-2,\ 1,\ 1), \\ k_2 &= (\pi_{(0)} + .05k_1)Q = (.9,\ .05,\ .05)Q = (-1.6,\ .6,\ 1), \\ k_3 &= (\pi_{(0)} + .05k_2)Q = (.92,\ .03,\ .05)Q = (-1.7,\ .78,\ .92), \\ k_4 &= (\pi_{(0)} + .1k_3)Q = (.830,\ .078,\ .092)Q = (-1.334,\ .390,\ .944), \end{aligned}$$

which allows us to compute

$$\pi(.1) \approx \pi_{(1)} = \pi_{(0)} + .1(k_1 + 2k_2 + 2k_3 + k_4)/6 = (.834433,\ .069167,\ .0964).$$

Continuing in this fashion we find

$$\begin{aligned} \pi(.2) \approx \pi_{(2)} &= (.720017,\ .103365,\ .176618), \\ \pi(.3) \approx \pi_{(3)} &= (.639530,\ .121971,\ .238499), \\ &\ \ \vdots \\ \pi(1.0) \approx \pi_{(10)} &= (.457455,\ .153267,\ .389278), \end{aligned}$$

and the final error is

$$\|\pi(1.0) - \pi_{(10)}\|_2 = .00001256.$$

An important characterization of initial-value ODEs has yet to be discussed—that of *stiffness*. Explicit methods have enormous difficulty solving stiff ODEs, indeed to such an extent that one definition of stiff equations is "problems for which explicit methods don't work" [20]! Many factors contribute to stiffness, including the eigenvalues of the Jacobian $\partial f/\partial y$ and the length of the interval of integration. The problems stem from the fact that the solutions of stiff systems of differential equations contain rapidly decaying transient terms.

**Example 10.38** As an example, consider the infinitesimal generator

$$Q = \begin{pmatrix} -1 & 1 \\ 100 & -100 \end{pmatrix}$$

with initial probability vector $\pi_{(0)} = (1, 0)$. $Q$ has the decomposition $Q = S\Lambda S^{-1}$, where

$$S = \begin{pmatrix} 1 & -.01 \\ 1 & 1 \end{pmatrix} \quad \text{and} \quad \Lambda = \begin{pmatrix} 0 & 0 \\ 0 & -101 \end{pmatrix}.$$

Since $\pi_{(t)} = \pi_{(0)}e^{Qt} = \pi_{(0)}Se^{\Lambda t}S^{-1}$, we find that

$$\pi_{(t)} = \frac{1}{1.01}\left(e^0 + .01e^{-101t},\ .01e^0 - .01e^{-101t}\right).$$

The exponents $e^{-101t}$ in this expression tend rapidly to zero, leaving the stationary probability distribution. In spite of this, however, when an explicit method is used, small step sizes must be used over the entire period of integration. It is not possible to increase the step size once the terms in $e^{-101t}$ are effectively zero.

The classical definition asserts that a system of ODEs is stiff when certain eigenvalues of the Jacobian matrix (with elements $\partial f_i/\partial y_i$) have large negative real parts when compared to others. Using this definition, Markov chain problems (where the Jacobian is given by $Q$, the infinitesimal generator) are stiff when $\max_k |\mathrm{Re}(\lambda_k)| \gg 0$. The Gerschgorin disk theorem is useful in bounding this quantity. From the special properties of infinitesimal generator matrices, this theorem implies that

$$\max_k |\mathrm{Re}(\lambda_k)| \leq 2 \max_j |q_{jj}|,$$

i.e., twice the largest total exit rate from any one state in the Markov chain.

### *Multistep BDF Methods*

The so-called *BDF (backward differentiation formulae)* methods [16] are a class of multistep methods that are widely used for stiff systems. Only implicit versions are in current use, since low-order explicit versions correspond to the explicit Euler method ($k = 1$) and the midpoint rule ($k = 2$), and higher-order explicit versions ($k \geq 3$) are not stable. Implicit versions are constructed by generating an interpolating polynomial $z(t)$ through the points $(t_j, y_j)$ for $j = i-k+1, \ldots, i+1$. However, now $y_{i+1}$ is determined so that $z(t)$ satisfies the differential equation at $t_{i+1}$, i.e., in such a way that $z'(t_{i+1}) = f(t_{i+1}, y_{i+1})$. It is usual to express the interpolating polynomial in terms of backward differences

$$\nabla^0 f_i = f_i, \quad \nabla^{j+1} f_i = \nabla^j f_i - \nabla^j f_{i-1}.$$

With this notation, the general formula for implicit BDF methods is

$$\sum_{j=1}^{k} \frac{1}{j}\nabla^j y_{i+1} = hf_{i+1}$$

and the first three rules are

$$\begin{aligned} k = 1:&\quad y_{i+1} - y_i = hf_{i+1} \qquad \text{(implicit Euler)},\\ k = 2:&\quad 3y_{i+1}/2 - 2y_i + y_{i-1}/2 = hf_{i+1},\\ k = 3:&\quad 11y_{i+1}/6 - 3y_i + 3y_{i-1}/2 - y_{i-2}/3 = hf_{i+1}. \end{aligned}$$

The BDF formulae are known to be stable for $k \le 6$ and to be unstable for other values of $k$. When applied to Markov chains, the implicit ($k = 2$) BDF formula is

$$3\pi_{(i+1)}/2 - 2\pi_{(i)} + \pi_{(i-1)}/2 = h\pi_{(i+1)}Q,$$

so that the system of equations to be solved at each step is

$$\pi_{(i+1)}(3I/2 - hQ) = 2\pi_{(i)} - \pi_{(i-1)}/2. \qquad (10.62)$$

The solutions at two prior points $\pi_{(i-1)}$ and $\pi_{(i)}$ are used in the computation of $\pi_{(i+1)}$. To get the procedure started, the initial starting point $\pi_{(0)}$ is used to generate $\pi_{(1)}$ using the trapezoid rule, and then both $\pi_{(0)}$ and $\pi_{(1)}$ are used to generate $\pi_{(2)}$, using Equation (10.62).

**Example 10.39** We shall apply the implicit ($k = 2$) BDF formula to the same continuous-time Markov chain that we have used in previous ODE examples. When applying the trapezoid rule in Example 10.34, we found $\pi_{(1)} = (.832370, .072254, .095376)$. We now have the two starting points $\pi_{(0)}$ and $\pi_{(1)}$ needed to launch the implicit BDF formula. We first compute

$$S = (1.5I - .1Q)^{-1} = \begin{pmatrix} 1.7 & -.1 & -.1 \\ -.3 & 2.3 & -.5 \\ -.1 & -.2 & 1.8 \end{pmatrix}^{-1} = \begin{pmatrix} .595870 & .029485 & .041298 \\ .087021 & .449853 & .129794 \\ .042773 & .051622 & .572271 \end{pmatrix}.$$

Had the Markov chains been large, we would have computed an $LU$ factorization and applied this decomposition to solve the appropriate system of equations instead of forming the inverse. We now proceed by computing

$$\pi(.2) \approx \pi_{(2)} = \left(2\pi_{(1)} - .5\pi_{(0)}\right) S = (.714768, .109213, .176019),$$

$$\pi(.3) \approx \pi_{(3)} = \left(2\pi_{(2)} - .5\pi_{(1)}\right) S = (.623707, .127611, .239682),$$

$$\vdots$$

$$\pi(1.0) \approx \pi_{(10)} = \left(2\pi_{(9)} - .5\pi_{(8)}\right) S = (.454655, .153688, .391657),$$

and the final error is

$$\|\pi(1.0) - \pi_{(10)}\|_2 = .003687.$$

### *Matlab Code for ODE Methods*

The following code segments may be used to experiment with ODE methods for obtaining transient distributions of relatively small Markov chains. In each case, the computed result is checked against the results obtained from the Matlab matrix exponential function, *expm*. The input to each function is the transition rate matrix $Q$, the initial vector $y_0 = \pi(0)$, the time $t$ at which the distribution is to be found and the step size, $h$. The output is an approximation to the transient distribution at time $t$, i.e., $\pi(t)$, and an estimate of the error, *err*. Two methods are included that are not discussed in the section itself, an implicit Runge-Kutta method and an implicit Adam's method, both of which may be used on stiff problems.

```
*************************************************************************
function [y,err] = euler(Q,y0,t,h);    %%%%%%%%%%%%%%%% Explicit Euler

[n,n] = size(Q);  I = eye(n);  eta = t/h;   y = y0;
yex = y0*expm(t*Q);                         %%%%% Solution using Matlab fcn

R = I + h*Q;
```

```
for i=1:eta,
    y = y*R;
end
err = norm(y-yex,2);

**********************************************************************
function [y,err] = ieuler(Q,y0,t,h);  %%%%%%%%%%%%%%%%  Implicit Euler

[n,n] = size(Q);  I = eye(n);  eta = t/h;  y = y0;
yex = y0*expm(t*Q);                          %%%%% Solution using Matlab fcn

R = inv(I-h*Q);
for i=1:eta,
    y = y*R;
end
err = norm(y-yex,2);

**********************************************************************
function [y,err] = trap(Q,y0,t,h);    %%%%%%%  Trapezoid/Modified Euler

[n,n] = size(Q);  I = eye(n);  eta = t/h;  y = y0;
yex = y0*expm(t*Q);                          %%%%% Solution using Matlab fcn

R1 = I + h*Q/2;  R2 = inv(I-h*Q/2);  R = R1*R2;
for i=1:eta,
    y = y*R;
end
err = norm(y-yex,2);

**********************************************************************
function [y,err] = rk4(Q,y0,t,h);   %  Explicit Runge-Kutta --- Order 4

[n,n] = size(Q);  I = eye(n);  eta = t/h;   y = y0;
yex = y0*expm(t*Q);                        % Solution using Matlab fcn

for i=1:eta,
    k1 = y*Q;
    k2 = (y + h*k1/2)*Q;
    k3 = (y + h*k2/2)*Q;
    k4 = (y + h*k3)*Q;
    y = y + h*(k1 + 2*k2 + 2*k3 + k4)/6;
end
err = norm(y-yex,2);

**********************************************************************
function [y,err] = irk(Q,y0,t,h);   %%%%%%%%%%%%  Implicit Runge-Kutta

[n,n] = size(Q);  I = eye(n);  eta = t/h;  y = y0;
yex = y0*expm(t*Q);                        %%%%%% Solution using Matlab fcn
```

```
R1 = I + h*Q/3;  R2 = inv(I-2*h*Q/3+ h*h*Q*Q/6);  R = R1*R2;
for i=1:eta,
    y = y*R;
end
err = norm(y-yex,2);

**********************************************************************


function [y,err] = adams(Q,y0,t,h);   %%%%%   Adams --- Implicit: k=2

[n,n] = size(Q);  I = eye(n);  eta = t/h;
yex = y0*expm(t*Q);                   %%%% Solution using Matlab fcn

R1 = I + h*Q/2;  R2 = inv(I-h*Q/2);
y1 = y0*R1*R2;                        %%%%% Trapezoid rule for step 1

S1 = (I+2*h*Q/3);  S2 =  h*Q/12;  S3 = inv(I-5*h*Q/12);
for i=2:eta,                          %%%%%%% Adams for steps 2 to end
    y = (y1*S1 - y0*S2)*S3;
    y0 = y1; y1 = y;
end
err = norm(y-yex,2);

**********************************************************************
function [y,err] = bdf(Q,y0,t,h);   %%%%%%%%%%%  BDF --- Implicit: k=2

[n,n] = size(Q);  I = eye(n);  eta = t/h;
yex = y0*expm(t*Q);                  %%%%% Solution using Matlab fcn

R1 = I + h*Q/2;  R2 = inv(I-h*Q/2);
y1 = y0*R1*R2;                       %%%%%%% Trapezoid rule for step 1

S1 = inv(3*I/2 - h*Q);
for i=2:eta,                         %%%%%%%%%%%% BDF for steps 2 to end
    y = (2*y1 - y0/2)*S1;
    y0 = y1; y1 = y;
end
err = norm(y-yex,2);

**********************************************************************
```

## 10.8 Exercises

**Exercise 10.1.1** State Gerschgorin's theorem and use it to prove that the eigenvalues of a stochastic matrix cannot exceed 1 in magnitude.

**Exercise 10.1.2** Show that if $\lambda$ is an eigenvalue of $P$ and $x$ its associated left-hand eigenvector, then $1 - \alpha(1 - \lambda)$ is an eigenvalue of $P(\alpha) = I - \alpha(I - P)$ and has the same left-hand eigenvector $x$, where

$\alpha \in \Re' \equiv (-\infty, \infty)\backslash\{0\}$ (i.e., the real line with zero deleted). Consider the matrix

$$P(\alpha) = \begin{pmatrix} -1.2 & 2.0 & 0.2 \\ 0.6 & -0.4 & 0.8 \\ 0.0 & 2.0 & -1.0 \end{pmatrix}$$

obtained from the previous procedure. What range of the parameter $\alpha$ results in the matrix $P$ being stochastic? Compute the stationary distribution of $P$.

**Exercise 10.1.3** Look up the Perron-Frobenius theorem from one of the standard texts. Use this theorem to show that the transition probability matrix of an *irreducible* Markov chain possesses a *simple unit* eigenvalue.

**Exercise 10.1.4** Part of the Perron-Frobenius theorem provides information concerning the eigenvalues of transition probability matrices derived from periodic Markov chains. State this part of the Perron-Frobenius theorem and use it to compute all the eigenvalues of the following two stochastic matrices given that the first has an eigenvalue equal to 0 and the second has an eigenvalue equal to $-0.4$:

$$P_1 = \begin{pmatrix} 0 & 0 & 0 & .5 & 0 & 0 & 0 & .5 \\ .5 & 0 & 0 & 0 & .5 & 0 & 0 & 0 \\ 0 & .5 & 0 & 0 & 0 & .5 & 0 & 0 \\ 0 & 0 & .5 & 0 & 0 & 0 & .5 & 0 \\ 0 & 0 & 0 & .5 & 0 & 0 & 0 & .5 \\ .5 & 0 & 0 & 0 & .5 & 0 & 0 & 0 \\ 0 & .5 & 0 & 0 & 0 & .5 & 0 & 0 \\ 0 & 0 & .5 & 0 & 0 & 0 & .5 & 0 \end{pmatrix}, \quad P_2 = \begin{pmatrix} 0 & 0 & 0 & .8 & 0 & 0 & 0 & .2 \\ .7 & 0 & 0 & 0 & .3 & 0 & 0 & 0 \\ 0 & .6 & 0 & 0 & 0 & .4 & 0 & 0 \\ 0 & 0 & .5 & 0 & 0 & 0 & .5 & 0 \\ 0 & 0 & 0 & .4 & 0 & 0 & 0 & .6 \\ .7 & 0 & 0 & 0 & .3 & 0 & 0 & 0 \\ 0 & .8 & 0 & 0 & 0 & .2 & 0 & 0 \\ 0 & 0 & .9 & 0 & 0 & 0 & .1 & 0 \end{pmatrix}.$$

**Exercise 10.2.1** The state transition diagram for a discrete-time Markov chain is shown in Figure 10.4. Use Gaussian elimination to find its stationary distribution.

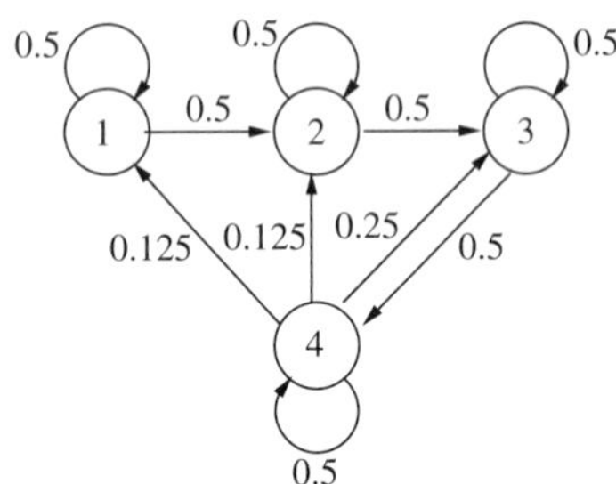


Figure 10.4. State transition diagram for Exercise 10.2.1.

**Exercise 10.2.2** The state transition diagram for a continuous-time Markov chain is shown in Figure 10.5. Use Gaussian elimination to find its stationary distribution.

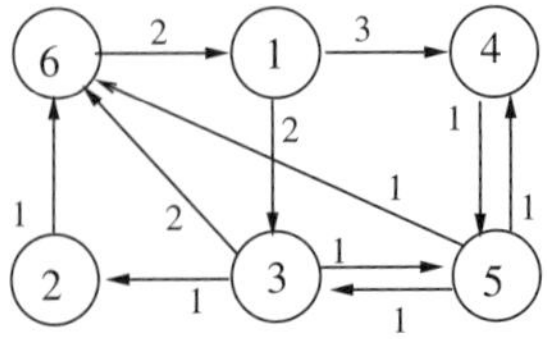


Figure 10.5. State transition diagram for Exercise 10.2.2.

**Exercise 10.2.3** What is the Doolittle $LU$ decomposition for the Markov chain of Exercise 10.2.2? Verify your answer by forming the product of $L$ and $U$ to obtain the infinitesimal generator matrix of this Markov chain.

**Exercise 10.2.4** Use Gaussian elimination to find the stationary distribution of the continuous-time Markov chain whose transition rate matrix is

$$Q = \begin{pmatrix} -1 & 1 & & & & \\ & -2 & 2 & & & \\ & & -3 & 3 & & \\ & & & -4 & 4 & \\ & & & & -5 & 5 \\ 6 & & & & & -6 \end{pmatrix}.$$

Compare the number of multiplications/divisions needed if the singularity of the coefficient matrix is handled by

(a) replacing the first equation by $\pi e = 1$;
(b) replacing the last equation by $\pi e = 1$;
(c) replacing the row of zeros obtained during the final reduction step by $\pi_6 = 1$ and using a final normalization.

**Exercise 10.2.5** Use Gaussian elimination to compute the stationary distribution of the discrete-time Markov chain whose transition probability matrix $P$ is

$$P = \begin{pmatrix} 0 & 0 & .6 & 0 & 0 & .4 \\ 0 & .3 & 0 & 0 & 0 & .7 \\ 0 & 0 & 0 & .4 & .6 & 0 \\ 0 & .5 & 0 & .5 & 0 & 0 \\ .6 & .4 & 0 & 0 & 0 & 0 \\ 0 & 0 & 0 & .8 & 0 & .2 \end{pmatrix}.$$

Is this stationary distribution a limiting distribution? Explain your answer. *Hint: Draw a picture.*

**Exercise 10.2.6** Use Gaussian elimination to compute the stationary distribution of the continuous-time Markov chain whose infinitesimal generator $Q$ is

$$Q = \begin{pmatrix} -1 & 0 & 1 & 0 & 0 & 0 & 0 \\ 0 & -5 & 0 & 3 & 0 & 2 & 0 \\ 0 & 0 & -3 & 0 & 3 & 0 & 0 \\ 0 & 0 & 0 & -4 & 0 & 4 & 0 \\ 0 & 0 & 0 & 0 & -5 & 0 & 5 \\ 0 & 4 & 0 & 0 & 0 & -4 & 0 \\ 2 & 0 & 2 & 0 & 2 & 0 & -6 \end{pmatrix}.$$

Is the stationary distribution you compute unique? Explain your answer. *Hint: Draw a picture.*

**Exercise 10.2.7** Use the Grassman-Taskar-Heyman (GTH) algorithm to compute the stationary distribution of the Markov chain whose transition rate matrix is

$$Q = \begin{pmatrix} -5 & 0 & 2 & 3 & 0 & 0 \\ 0 & -1 & 0 & 0 & 0 & 1 \\ 0 & 1 & -4 & 0 & 1 & 2 \\ 0 & 0 & 0 & -1 & 1 & 0 \\ 0 & 0 & 1 & 1 & -3 & 1 \\ 2 & 0 & 0 & 0 & 0 & -2 \end{pmatrix}.$$

**Exercise 10.3.1** The stochastic transition probability matrix of a three-state Markov chain is given by

$$P = \begin{pmatrix} 0.0 & 0.8 & 0.2 \\ 0.0 & 0.1 & 0.9 \\ 0.6 & 0.0 & 0.4 \end{pmatrix}.$$

Carry out three iterations of the power method and the method of Jacobi using $(0.5, 0.25, 0.25)$ as an initial approximation to the solution. Find the magnitude of the subdominant eigenvalue of the iteration matrix in both cases to see which will converge faster asymptotically. Approximately how many decimal places of accuracy would the subdominant eigenvalue suggest have been obtained after 100 iterations of both methods?

**Exercise 10.3.2** Compute three iterations of the Gauss–Seidel method on the three-state Markov chain of Exercise 10.3.1 using the same initial approximation. Construct the Gauss–Seidel iteration matrix and compute the number of iterations needed to achieve ten decimal places of accuracy.

**Exercise 10.3.3** The matrix $P$ below is the stochastic transition probability matrix of a five-state Markov chain. Carry out three iterations of the power method, using an initial vector approximation whose components are all equal:

$$P = \begin{pmatrix} 0.2 & 0.0 & 0.005 & 0.795 & 0.0 \\ 0.0 & 0.0 & 0.998 & 0.002 & 0.0 \\ 0.002 & 0.0 & 0.0 & 0.0 & 0.998 \\ 0.8 & 0.001 & 0.0 & 0.198 & 0.001 \\ 0.0 & 0.998 & 0.0 & 0.002 & 0.0 \end{pmatrix}.$$

**Exercise 10.3.4** Returning to the stochastic transition probability matrix of Exercise 10.3.3, and using the same initial starting approximation $x^{(0)}$ in which all components are equal, apply the standard (point) Gauss–Seidel method to obtain approximations $x^{(1)}$, $x^{(2)}$, and $x^{(3)}$.

**Exercise 10.3.5** Given the stochastic transition probability matrix

$$P = \begin{pmatrix} 0.2 & 0.0 & 0.005 & 0.795 & 0.0 \\ 0.0 & 0.0 & 0.998 & 0.002 & 0.0 \\ 0.002 & 0.0 & 0.0 & 0.0 & 0.998 \\ 0.8 & 0.001 & 0.0 & 0.198 & 0.001 \\ 0.0 & 0.998 & 0.0 & 0.002 & 0.0 \end{pmatrix},$$

provide a Harwell-Boeing compact form for the corresponding transition rate matrix $Q^T = P^T - I$, as it might be used, for example, in the method of Gauss–Seidel.

**Exercise 10.3.6** Consider the following stochastic transition probability matrix, provided in the Harwell-Boeing compact storage format.

$$\begin{array}{ccccccc} .4 & .6 & .4 & .4 & .2 & .2 & .8 \\ 3 & 1 & 2 & 1 & 3 & 3 & 2 \\ 1 & 3 & 6 & 8 & & & \end{array}$$

Use the Gauss–Seidel iterative method and an initial approximation $x^{(0)} = (0, 1, 0)^T$ to compute $x^{(1)}$ and $x^{(2)}$.

**Exercise 10.4.1** Consider the stochastic transition probability matrix of Exercise 10.3.3 which is in fact nearly completely decomposable. Permute this matrix to expose its NCD block structure and then perform two iterations of block Gauss–Seidel, beginning with an initial approximation in which all components are equal.

**Exercise 10.5.1** Apply the NCD approximation procedure of Section 10.5 to the following stochastic transition probability matrix:

$$P = \begin{pmatrix} 0.2 & 0.0 & 0.005 & 0.795 & 0.0 \\ 0.0 & 0.0 & 0.998 & 0.002 & 0.0 \\ 0.002 & 0.0 & 0.0 & 0.0 & 0.998 \\ 0.8 & 0.001 & 0.0 & 0.198 & 0.001 \\ 0.0 & 0.998 & 0.0 & 0.002 & 0.0 \end{pmatrix}.$$

**Exercise 10.5.2** Apply one complete step of the IAD method of Section 10.5 to the transition probability matrix of Exercise 10.5.1. Choose an equiprobable distribution as your starting approximation.

**Exercise 10.6.1** Prove Equation (10.32), i.e., for any matrix $R$ with spectral radius strictly less than 1 show that

$$\sum_{i=0}^{\infty} R^i = (I - R)^{-1}.$$

**Exercise 10.6.2** Implement the logarithmic reduction algorithm outlined in the text. How many iterations are required to obtain the rate matrix $R$ correct to ten decimal places, for a QBD process whose block matrices $A_0$, $A_1$, and $A_2$ are given below? Approximately how many iterations are required if the successive substitution algorithm is used?

$$A_0 = \begin{pmatrix} 4 & 0 & 0 & 0 \\ 0 & 4 & 0 & 0 \\ 0 & 0 & 4 & 0 \\ 0 & 0 & 0 & 4 \end{pmatrix}, \quad A_1 = \begin{pmatrix} -16.0 & 12 & 0 & 0 \\ 0 & -16.0 & 12 & 0 \\ 0 & 0 & -16.0 & 12 \\ 0 & 0 & 0 & -16.0 \end{pmatrix}, \quad A_2 = \begin{pmatrix} 0 & 0 & 0 & 0 \\ 0 & 0 & 0 & 0 \\ 0 & 0 & 0 & 0 \\ 12 & 0 & 0 & 0 \end{pmatrix}.$$

**Exercise 10.6.3** Consider a QBD process of the form

$$Q = \begin{pmatrix} B_{00} & A_2 & 0 & 0 & \cdots \\ A_0 & A_1 & A_2 & 0 & \cdots \\ 0 & A_0 & A_1 & A_2 & \\ \vdots & & \ddots & \ddots & \ddots \end{pmatrix},$$

where $A_0$, $A_1$, and $A_2$ are the same as in Exercise 10.6.2 and

$$B_{00} = \begin{pmatrix} -12 & 12 & 0 & 0 \\ 0 & -12 & 12 & 0 \\ 0 & 0 & -12 & 12 \\ 0 & 0 & 0 & -12 \end{pmatrix}.$$

Find the probability that this QBD process is in level 0 or 1.

**Exercise 10.6.4** Consider a discrete-time Markov chain whose stochastic matrix $P$ has the block structure

$$P = \begin{pmatrix} B_{00} & A_0 & & & \\ B_{10} & A_1 & A_0 & & \\ B_{20} & A_2 & A_1 & A_0 & \\ & A_3 & A_2 & A_1 & A_0 \\ \vdots & & \ddots & \ddots & \ddots & \ddots \end{pmatrix}$$

with

$$A_0 = \begin{pmatrix} 1/8 & 1/32 \\ 1/16 & 1/16 \end{pmatrix}, \quad A_1 = \begin{pmatrix} 1/16 & 1/32 \\ 1/8 & 0 \end{pmatrix}, \quad A_1 = \begin{pmatrix} 0 & 1/4 \\ 1/4 & 0 \end{pmatrix}, \quad A_0 = \begin{pmatrix} 1/4 & 1/4 \\ 0 & 1/2 \end{pmatrix}$$

$$B_{00} = \begin{pmatrix} 3/4 & 3/32 \\ 3/8 & 1/2 \end{pmatrix}, \quad B_{10} = \begin{pmatrix} 1/2 & 1/4 \\ 1/4 & 1/2 \end{pmatrix}, \quad B_{20} = \begin{pmatrix} 1/4 & 1/4 \\ 1/4 & 1/4 \end{pmatrix}.$$

Beginning with $R_{(0)} = 0$, compute $R_{(1)}$ and $R_{(2)}$ using the successive substitution algorithm. What is the magnitude of the difference in successive iterations after 10 steps? Finally, compute the probability of the Markov chain being in a state of level 1.

**Exercise 10.6.5** The transition rate matrix of a continuous-time Markov chain has the form

$$Q = \begin{pmatrix} B_{00} & B_{01} & & & & & & & \\ B_{10} & B_{11} & A_0 & & & & & & \\ B_{20} & 0 & A_1 & A_0 & & & & & \\ B_{30} & 0 & 0 & A_1 & A_0 & & & & \\ 0 & B_{41} & 0 & 0 & A_1 & A_0 & & & \\ 0 & 0 & A_4 & 0 & 0 & A_1 & A_0 & & \\ 0 & 0 & 0 & A_4 & 0 & 0 & A_1 & A_0 & \\ \vdots & \vdots & \vdots & & \ddots & \ddots & \ddots & \ddots & \ddots \end{pmatrix}$$

with

$$A_0 = \begin{pmatrix} 4 & 0 \\ 0 & 4 \end{pmatrix}, \quad A_1 = \begin{pmatrix} -12 & 0 \\ 0 & -10 \end{pmatrix}, \quad A_4 = \begin{pmatrix} 4 & 4 \\ 3 & 3 \end{pmatrix},$$

$$B_{00} = \begin{pmatrix} -8 & 6 \\ 2 & -4 \end{pmatrix}, \quad B_{01} = \begin{pmatrix} 2 & 0 \\ 0 & 2 \end{pmatrix}, \quad B_{10} = \begin{pmatrix} 1 & 3 \\ 2 & 0 \end{pmatrix}, \quad B_{11} = \begin{pmatrix} -8 & 0 \\ 0 & -6 \end{pmatrix},$$

$$B_{20} = \begin{pmatrix} 0 & 8 \\ 6 & 0 \end{pmatrix}, \quad B_{30} = \begin{pmatrix} 4 & 4 \\ 1 & 5 \end{pmatrix}, \quad B_{41} = \begin{pmatrix} 6 & 2 \\ 3 & 3 \end{pmatrix}.$$

Construct Neuts' rate matrix $R$ and the probability distribution of the states that constitute level 2.

**Exercise 10.6.6** Which of the following three sequences of $A$ blocks are taken from a positive-recurrent Markov chain of *M/G/*1 type? For those that are positive recurrent, compute the approximation to $G$ that is obtained after 20 iterations using one of the three successive substitution strategies. Multiply your computed value of $G$ on the right by the vector $e$ and estimate the accuracy obtained.

(a)

$$A_0 = \begin{pmatrix} 1/4 & 1/4 & 0 \\ 0 & 0 & 0 \\ 0 & 1/4 & 1/4 \end{pmatrix}, \quad A_1 = \begin{pmatrix} 0 & 1/4 & 0 \\ 0 & 0 & 1/2 \\ 1/4 & 0 & 0 \end{pmatrix}, \quad A_2 = \begin{pmatrix} 1/8 & 0 & 0 \\ 0 & 1/2 & 0 \\ 0 & 0 & 1/8 \end{pmatrix},$$

$$A_3 = \begin{pmatrix} 1/8 & 0 & 0 \\ 0 & 0 & 0 \\ 0 & 0 & 1/8 \end{pmatrix}.$$

(b)

$$A_0 = \begin{pmatrix} 1/4 & 1/4 & 0 \\ 1/4 & 0 & 1/4 \\ 0 & 1/4 & 1/4 \end{pmatrix}, \quad A_1 = \begin{pmatrix} 0 & 1/4 & 0 \\ 0 & 0 & 1/2 \\ 1/4 & 0 & 0 \end{pmatrix}, \quad A_2 = \begin{pmatrix} 3/16 & 0 & 0 \\ 0 & 0 & 0 \\ 0 & 0 & 3/16 \end{pmatrix},$$

$$A_3 = \begin{pmatrix} 1/16 & 0 & 0 \\ 0 & 0 & 0 \\ 0 & 0 & 1/16 \end{pmatrix}.$$

(c)

$$A_0 = \begin{pmatrix} 1/4 & 1/4 & 0 \\ 1/4 & 0 & 1/4 \\ 0 & 1/4 & 1/4 \end{pmatrix}, \quad A_1 = \begin{pmatrix} 0 & 1/4 & 0 \\ 0 & 0 & 1/4 \\ 1/4 & 0 & 0 \end{pmatrix}, \quad A_2 = \begin{pmatrix} 3/16 & 0 & 0 \\ 0 & 0 & 0 \\ 0 & 0 & 3/16 \end{pmatrix},$$

$$A_3 = \begin{pmatrix} 1/16 & 0 & 0 \\ 0 & 0 & 0 \\ 0 & 0 & 1/16 \end{pmatrix}, \quad A_4 = \begin{pmatrix} 0 & 0 & 0 \\ 1/8 & 0 & 1/8 \\ 0 & 0 & 0 \end{pmatrix}.$$

**Exercise 10.6.7** Consider a Markov chain whose infinitesimal generator is given by

$$Q = \left(\begin{array}{ccc|ccc|ccc|ccc|ccc}
-8 & 4 & 0 & 3 & 0 & 0 & 1 & 0 & 0 & & & & & & \\
0 & -4 & 4 & 0 & 0 & 0 & 0 & 0 & 0 & & & & & & \\
4 & 0 & -8 & 0 & 0 & 3 & 0 & 0 & 1 & & & & & & \\
\hline
5 & 5 & 0 & -18 & 4 & 0 & 3 & 0 & 0 & 1 & 0 & 0 & & & \\
5 & 0 & 5 & 0 & -14 & 4 & 0 & 0 & 0 & 0 & 0 & 0 & & & \\
0 & 5 & 5 & 4 & 0 & -18 & 0 & 0 & 3 & 0 & 0 & 1 & & & \\
\hline
 & & & 5 & 5 & 0 & -18 & 4 & 0 & 3 & 0 & 0 & 1 & 0 & 0 \\
 & & & 5 & 0 & 5 & 0 & -14 & 4 & 0 & 0 & 0 & 0 & 0 & 0 \\
 & & & 0 & 5 & 5 & 4 & 0 & -18 & 0 & 0 & 3 & 0 & 0 & 1 \\
\hline
 & & & & & & & \ddots & & & \ddots & & & \ddots & \ddots
\end{array}\right).$$

Verify that the states of this Markov chain are positive recurrent and find the probability that this Markov chain is in level 2.

**Exercise 10.7.1** Given that it starts in state 1, find the transient distribution at time steps $n = 10, 20$, and 30 of the discrete-time Markov chain whose transition probability matrix is given by

$$P = \begin{pmatrix} .4 & 0 & .6 & 0 \\ .1 & .1 & .1 & .7 \\ .4 & .3 & .2 & .1 \\ 0 & .5 & 0 & .5 \end{pmatrix}.$$

**Exercise 10.7.2** When the stationary distribution $\pi$ of a discrete-time Markov chain with stochastic transition probability matrix $P$ is required, we can find it from either $\pi P = \pi$ or $\pi Q = 0$, where $Q = P - I$. Can the same be said for computing transient distributions? Compare the transient distribution at time step $n = 10$ obtained in the previous question (using the same initial distribution $\pi^{(0)} = (1, 0, 0, 0)$) and the distribution obtained by forming $\pi^{(0)} e^{10Q}$.

**Exercise 10.7.3** Write Matlab code to implement the Padé algorithm for computing $e^{Qt}$. Check your algorithm by testing it on the infinitesimal generator

$$Q = \left(\begin{array}{ccc|cc|ccc}
-.15 & .0 & .149 & .0009 & .0 & .00005 & .0 & .00005 \\
.1 & -.35 & .249 & .0 & .0009 & .00005 & .0 & .00005 \\
.1 & .8 & -.9004 & .0003 & .0 & .0 & .0001 & .0 \\
\hline
.0 & .0004 & .0 & -.3 & .2995 & .0 & .0001 & .0 \\
.0005 & .0 & .0004 & .399 & -.4 & .0001 & .0 & .0 \\
\hline
.0 & .00005 & .0 & .0 & .00005 & -.4 & .2499 & .15 \\
.00003 & .0 & .00003 & .00004 & .0 & .1 & -.2 & .0999 \\
.0 & .00005 & .0 & .0 & .00005 & .1999 & .25 & -.45
\end{array}\right)$$

and comparing it with the answer obtained by the Matlab function expm. Choose a value of $t = 10$.

**Exercise 10.7.4** In the uniformization method, what value should be assigned to the integer $K$ in Equation (10.48) so that an accuracy of $10^{-6}$ is achieved when computing transient distributions for the Markov chains with infinitesimal generators:

(a)

$$Q = \begin{pmatrix} -4 & 2 & 1 & 1 \\ 1 & -4 & 2 & 1 \\ 1 & 1 & -4 & 2 \\ 2 & 1 & 1 & -4 \end{pmatrix} \quad \text{at time } t = 3.0;$$

(b)

$$Q = \begin{pmatrix} -1 & 1 & 0 & 0 \\ 1 & -2 & 1 & 0 \\ .5 & .5 & -1.5 & .5 \\ 0 & 0 & 1 & -1 \end{pmatrix} \quad \text{at time } t = 6.0.$$

**Exercise 10.7.5** Returning to the previous exercise, find the distribution at the specified times if each Markov chain begins in state 4.

**Exercise 10.7.6** The infinitesimal generator of a continuous-time Markov chain is

$$Q = \begin{pmatrix} -1 & 1 & 0 & 0 \\ 1 & -2 & 1 & 0 \\ .5 & .5 & -1.5 & .5 \\ 0 & 0 & 1 & -1 \end{pmatrix}.$$

If this Markov chain begins in state 4, what is the transient distribution at time $t = 6$ computed by (a) the explicit Euler method and (b) the trapezoid method? Use a step size of $h = .25$ in both methods.

**Exercise 10.7.7** In the text, a Runge-Kutta method of order 2 was described as having the form

$$y_{n+1} = y_n + ak_1 + bk_2$$

with

$$k_1 = hf(x_n, y_n), \quad k_2 = hf(x_n + \alpha h, y_n + \beta k_1).$$

Show that equating this with the Taylor series expansion of order 2 leads to the conditions

$$a + b = 1 \quad \text{and} \quad \alpha b = \beta b = 1/2.$$

**Exercise 10.7.8** Write Matlab code to implement the implicit ($k = 3$) BDF method. Compare the results obtained with this method and the implicit ($k = 2$) BDF method when applied to a continuous-time Markov chain with infinitesimal generator given by

$$Q = \begin{pmatrix} -0.800 & 0.0 & 0.005 & 0.795 & 0.0 \\ 0.0 & -1.000 & 0.998 & 0.002 & 0.0 \\ 0.002 & 0.0 & -1.000 & 0.0 & 0.998 \\ 0.800 & 0.001 & 0.0 & -0.802 & 0.001 \\ 0.0 & 0.998 & 0.0 & 0.002 & -1.000 \end{pmatrix}.$$

Use a step size of $h = .25$, initial starting distribution of (1,0,0,0,0) and compute the transient distribution at time $t = 12$.

# Part III

# QUEUEING MODELS

# Chapter 11

# *Elementary Queueing Theory*

## 11.1 Introduction and Basic Definitions

Queues form an annoying, yet indispensible, part of our everyday lives. They occur whenever there is competition for limited resources. For example, we line up in queues outside doctors' offices and supermarket checkout counters; airplanes queue along airport runways; and requests to satisfy operating system page faults queue at a computer hard drive. The entities that queue for service are called customers, or users, or jobs, depending on what is appropriate for the situation at hand. The limited resources are dispensed by servers and waiting lines of requests form in front of these servers. Customers who require service are said to "arrive" at the service facility and place service "demands" on the resource. At a doctor's office, the waiting line consists of the patients awaiting their turn to see the doctor; the doctor is the server who dispenses a limited resource, her time. Likewise, departing airplanes may have to wait in a line to use one of a small number of runways. The planes may be viewed as users and the runway viewed as a resource that is assigned by an air traffic controller. The resources are of finite capacity meaning that there is not an infinity of them nor can they work infinitely fast. Furthermore arrivals place demands upon the resource and these demands are unpredictable in their arrival times and unpredictable in their size. Taken together, limited resources and unpredictable demands imply a conflict for the use of the resource and hence queues of waiting customers.

Our ability to model and analyze systems of queues helps to minimize their inconveniences and maximize the use of the limited resources. An analysis may tell us something about the expected time that a resource will be in use, or the expected time that a customer must wait. This information may then be used to make decisions as to when and how to upgrade the system: for an overworked doctor to take on an associate or an airport to add a new runway, for example. We begin our discussion and analysis by considering a simple, single-server queueing system, such as that presented in Figure 11.1.

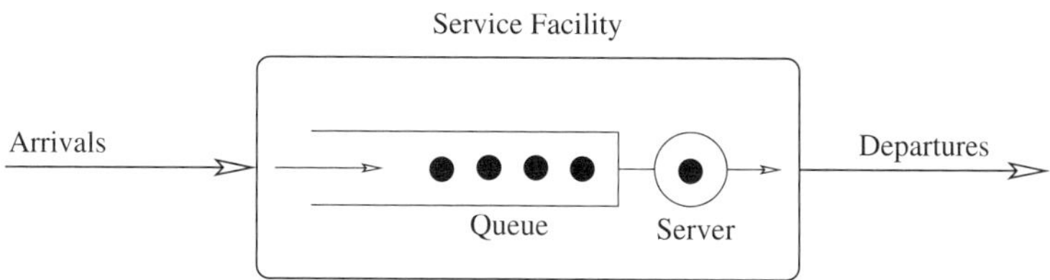


Figure 11.1. Single-server queue.

The resource and the dispenser of the resource and lumped together and referred to collectively as the "server." In some circumstances this is very natural, as when the server is a doctor and the resource that is dispensed is the doctor's time. On the other hand, for the airport scenario, the resource is a runway and the dispenser is an air traffic controller. But even here both runway and controller are collectively referred to as the server. The server(s) and the queue into which arriving customers are directed together constitute the *service facility*, or *queueing system*, or simply *system*. Also, somewhat confusingly, it frequently happens that both queue and server are collectively referred to as a *queue*, such as the *M/M/*1 queue or the $M/G/1$ queue; when multiple service facilities are combined into a network, the term *node* is commonly used instead of service facility. One after the other, customers arrive to the service facility, eventually receive service and then depart. We make the following assumptions about all the queueing systems we analyze:

- If the server is free, i.e., not already serving a customer, an arriving customer goes immediately into service. No time at all is spent in the queue.
- If the server is not free (busy), then the customer joins a queue of waiting customers and stays in the queue until entering service.
- When the server becomes free, a customer is chosen from the queue according to a scheduling policy and immediately enters into service. The time between the departure of one customer and the start of service of the next customer is zero.
- Customers remain in the facility until their service is completed and then they depart. Customers do not become impatient and leave before they receive service.

It is evident that much more detail needs to be provided before we can hope to analyze this queueing system. For example, we must specify the manner in which arrivals occur, how the next customer to be served is chosen from all those waiting, and so on. We address these topics in the following sections.

### *11.1.1 Arrivals and Service*

Congestion (i.e., queues) in a queueing system depend on the system irregularities (statistical fluctuations) and not just on average properties. In other words, the length of a queue depends on the complete probabilistic description of the arrival and service processes—the demands being placed on the resource—and of the capacity of the resource to service these demands. It would seem evident that, if the average service demands of arriving customers are greater than the system service capacity, the system will break down since unbounded queues will form. On the other hand, if the average arrival rate is less than the system service capacity, then, due to statistical fluctuations, we still get queues. Even when the average arrival and service rates are held constant, an increase in the variation of arrivals or service increases the congestion. Furthermore, as the average demand tends to the system service capacity, the effects of the fluctuations are magnified. These fluctuations are described in terms of probability distributions. Thus we use elementary probability theory to predict average waiting times, average queue length, distribution of queue length, etc., on the basis of

- the arrival pattern of customers to the resource,
- the service pattern of customers, and
- the scheduling algorithm, or the manner in which the next customer to be served is chosen.

In some queueing scenarios, the space available to hold customers waiting for service may be limited. When the queue has reached its maximum capacity, it is said to be "full"; customers who arrive to find the queue full are said to be "lost."

**The Arrival Process**
The customer arrival process may be described in two ways:

1. By characterizing the number of arrivals per unit time (the arrival rate);
2. By characterizing the time between successive arrivals (the interarrival time).

We use the variable $\lambda$ to denote the mean arrival rate. In this case, $1/\lambda$ denotes the mean time between arrivals. If the arrival pattern is not deterministic, the input process is a stochastic process, which means that we need its associated probability distribution. The probability distribution of the interarrival time of customers is denoted by $A(t)$ where

$$A(t) = \text{Prob}\{\text{time between arrivals } \leq t\}$$

and

$$\frac{1}{\lambda} = \int_0^\infty t \, dA(t),$$

where $dA(t)$ is the probability that the interarrival time is between $t$ and $t + dt$. Here we assume that these interarrival times are independent and identically distributed, which means that only $A(t)$ is of significance. If there are different types of customers, different customer classes is the usual connotation, then each class may have its own probability distribution function to describe its arrival process. The manner in which the arrival pattern changes in time may be important (e.g., the number of customers who arrive at a supermarket may be greater in late afternoon than in early morning.) An arrival pattern that does not change with time (i.e., the form and values of the parameters of $A(t)$ are time independent) is said to be a *homogeneous* arrival process. If it is invariant to shifts in the time origin, it is said to be a *stationary* arrival process.

**The Service Process**

Just like the arrival pattern, the service pattern may be described by a *rate*, the number of customers served per unit time, or by a *time*, the time required to serve a customer. The parameter $\mu$ is used to denote the mean service rate, and hence $1/\mu$ denotes the mean service time. We shall use $B(x)$ to denote the probability distribution of the demand placed on the system, i.e.,

$$B(x) = \text{Prob}\{\text{Service time } \leq x\}.$$

Thus

$$\frac{1}{\mu} = \int_0^\infty x \, dB(x),$$

where $dB(x)$ is the probability that the service time is between $x$ and $x + dx$.

Notice that the service time is equal to the length of time spent in service and does not include the time spent waiting in the queue. Furthermore service rates are conditioned on the fact that the system is not empty. If the system is empty, then the server must be idle. Although it is usual to associate the service time distribution with the server, the service time is actually the time that is requested or needed by the customer who is taken into service. Obviously, it does not make sense for a server to arbitrarily dispense service to customers without regard for their needs. The service may be *batch* or *single*. With batch service, several customers can be served simultaneously as is the case, for example, of customers who wait in line for taxis or buses. Also, the service rate may depend on the following factors.

- The number of customers present (called *state-dependent* or *load-dependent* service). For example, a server may speed up when the queue starts to become full, or slow down as it starts to empty.
- The time (called *time-dependent* or *nonhomogeneous* service). This is, for example, the case of a server that starts slowly in the morning and gradually speeds up during the day.

The total number of servers available at a queueing facility is denoted by $c$. Two possibilities arise when there is more than one server.

1. Each server has its own queue. For example, each supermarket checkout lane has its own queue. However, the effect of jockeying and lane changing may mean that a supermarket

checkout system may be more accurately modeled as a single queue in front of all checkout lanes.

2. There are fewer queues than servers. In most cases, there is a single queue for all servers. For example, a single queue usually forms in front of multiple bank tellers.

The servers may or may not be identical, i.e., $B(x)$ may be different for different servers. Also, given multiple classes of customers, the same server may give different service to different classes. The capacity (space) that a service facility has to hold waiting customers (called simply, the system capacity) is often taken to be infinite. When this is not the case, then the system is referred to as a finite queueing system.

**Poisson Arrivals and Exponential Service**

In stochastic modeling, numerous random variables are frequently modeled as exponentials. These include

- interarrival time (the time between two successive arrivals to a service center),
- service time (the time it takes for a service request to be satisfied),
- time to failure of a component, and
- time required to repair a component.

The assertion that the above distributions are exponential should not be taken as fact, but as an assumption. Experimental verification of this asumption should be sought before relying on the results of any analyses that use them. Recall that the cumulative distribution function for an exponential random variable, $X$, with parameter $\lambda > 0$, is given by

$$F(x) = \begin{cases} 1 - e^{-\lambda x}, & x \geq 0, \\ 0 & \text{otherwise,} \end{cases}$$

and its corresponding probability density function, obtained simply by taking the derivative of $F(x)$ with respect to $x$, is

$$f(x) = \begin{cases} \lambda e^{-\lambda x}, & x \geq 0, \\ 0 & \text{otherwise.} \end{cases}$$

Its mean, second moment, and variance are given, respectively, by

$$E[X] = \frac{1}{\lambda}, \quad E[X^2] = \frac{2}{\lambda^2}, \quad \text{and} \quad \text{Var}\,[X] = \frac{1}{\lambda^2}.$$

The reader may wish to review the material on the exponential distribution that was presented earlier in Section 7.2 of this text. Reasons for the use of the exponential distribution include

1. its memoryless (Markov) property and resulting analytical tractability;
2. its relationship to the Poisson process.

In Section 9.11 we introduced a counting process, $\{N(t),\ t \geq 0\}$, as a stochastic process which counts the number of events that occur by time $t$. Thus any counting process $N(t)$ is integer valued with the properties that $N(t) \geq 0$ and $N(t_1) \leq N(t_2)$ if $t_1$ is smaller than $t_2$. We defined a *renewal process* as a counting process for which the distributions of time between any two events (or renewals) are independent and identically distributed and we saw that when this distribution is exponentially distributed, the renewal process has a Poisson distribution and is in fact called a *Poisson process*. A counting process is said to have *independent increments* if the number of events that occur in nonoverlapping time intervals are independent and is further said to have *stationary increments* if the distribution of the number of events in any time interval depends only on the length of that interval. It turns out that the Poisson process is the only *continuous* renewal process that has independent and stationary increments.

**Definition 11.1.1** *A Poisson process* $\{N(t),\ t \geq 0\}$ *having rate* $\lambda \geq 0$ *is a counting process with independent and stationary increments, with* $N(0) = 0$*, and is such that the number of events that occur in any time interval of length* $t$ *has a Poisson distribution with mean* $\lambda t$*. This means that*

$$\text{Prob}\{N(t+s) - N(s) = n\} = e^{-\lambda t}\frac{(\lambda t)^n}{n!}, \qquad n = 0, 1, 2, \ldots.$$

Thus a Poisson process is a continuous-parameter, discrete-state process. We shall use it extensively to denote the number of arrivals to a queueing system by time $t$. Let $\lambda$ be the average rate at which customers arrive to a service facility, then $\lambda t$ customers arrive on average in an interval of length $t$. There are several other ways in which a Poisson process can be defined. We shall have use for two of these, which we now present.

**Definition 11.1.2** *Let the number of events that occur in* $(0, t]$ *be denoted by* $N(t)$*. When these events occur successively in time and are such that the times between successive events are independent and identically* exponentially *distributed, then the stochastic process* $\{N(t),\ t \geq 0\}$ *is a Poisson process with parameter (mean rate)* $\lambda$.

**Definition 11.1.3** *Let* $N(t)$ *be the number of events that occur in an interval* $(0, t]$*. When the following four conditions are true, then* $\{N(t),\ t \geq 0\}$ *is a Poisson process:*

1. $N(0) = 0$.
2. *Events that occur in nonoverlapping time intervals are mutually independent.*
3. *The number of events that occur in any interval depends only on the length of the interval and not on the past history of the system.*
4. *For sufficiently small* $h$ *and some positive constant* $\lambda$*, we have*

$$\begin{aligned}\text{Prob}\{\text{One event in } (t, t+h]\} &= \lambda h + o(h),\\ \text{Prob}\{\text{Zero events in } (t, t+h]\} &= 1 - \lambda h + o(h),\\ \text{Prob}\{\text{More than one event in } (t, t+h]\} &= o(h),\end{aligned}$$

*where* $o(h)$*—"little oh"—is any quantity that tends to zero faster than* $h$*; i.e.,*

$$\lim_{h \to 0} \frac{o(h)}{h} = 0.$$

Definitions 11.1.1, 11.1.2, and 11.1.3 are equivalent definitions of the Poisson process.

Let $p_n(t) = \text{Prob}\{N(t) = n\}$ be the probability that $n$ arrivals occur in the interval $(0, t]$. Then $p_0(t) = \text{Prob}\{\text{No arrivals in } (0, t]\}$. We shall now write an expression for $p_0(t + h)$, the probability that no arrivals occur in the interval $(0, t + h]$. We have, as is illustrated graphically in Figure 11.2,

$$\begin{aligned}p_0(t+h) &= \text{Prob}\{\text{No arrivals in } (0, t+h]\}\\ &= p_0(t)\text{Prob}\{\text{No arrivals in } (t, t+h]\} = p_0(t)[1 - \lambda h + o(h)],\end{aligned}$$

since the number of arrivals in $(0, t]$ is independent of the number of arrivals in $(t, t + h]$.

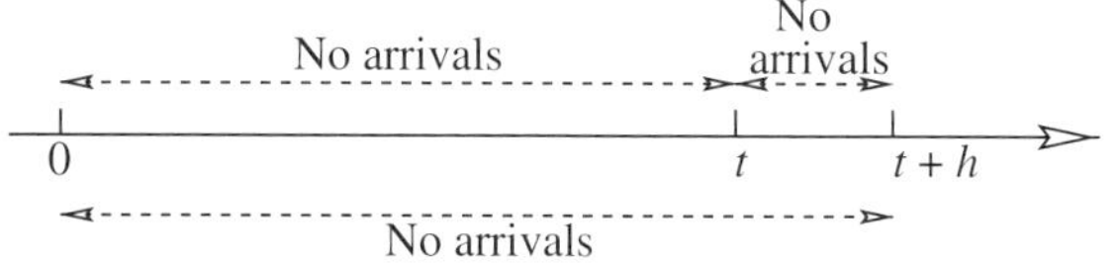


Figure 11.2. Zero arrivals.

From this we obtain

$$\frac{p_0(t+h) - p_0(t)}{h} = -\lambda p_0(t) + \frac{o(h)}{h} p_0(t)$$

and in the limit as $h \to 0$

$$\frac{dp_0(t)}{dt} = -\lambda p_0(t),$$

which implies that

$$p_0(t) = \alpha e^{-\lambda t}.$$

The initial condition $p_0(0) = 1$ implies that $\alpha = 1$ and hence

$$p_0(t) = e^{-\lambda t},$$

i.e., the probability that zero arrivals occur in the interval $(0, t]$ is equal to $e^{-\lambda t}$.

We now use a proof by induction to show that

$$p_n(t) = e^{-\lambda t} \frac{(\lambda t)^n}{n!}, \quad n = 0, 1, 2, \ldots. \tag{11.1}$$

We have just shown that the base clause holds, i.e., that $p_0(t) = e^{-\lambda t}$. Now consider $p_1(t+h)$, the probability of a single arrival between time 0 and time $t+h$. In this case, either an arrival occurred in the interval $(0, t]$ and no arrivals occurred in $(t, t+h]$ or no arrival occurred in the first interval $(0, t]$ but one did in the second, $(t, t+h]$. We have

$$\begin{aligned} p_1(t+h) &= p_1(t)[1 - \lambda h + o(h)] + p_0(t)[\lambda h + o(h)] \\ &= p_1(t) - \lambda h p_1(t) + \lambda h p_0(t) + o(h), \end{aligned}$$

and hence

$$\frac{p_1(t+h) - p_1(t)}{h} = -\lambda p_1(t) + \lambda p_0(t) + \frac{o(h)}{h},$$

$$\frac{dp_1(t)}{dt} = \lambda p_0(t) - \lambda p_1(t),$$

which yields the solution $p_1(t) = \lambda t e^{-\lambda t}$ as can be verified by direct substitution.

We now assume that Equation (11.1) holds for all integers through $n-1$ and thereby prove it must also hold for $n$. Using the law of total probability, we may write

$$p_n(t+h) = \sum_{k=0}^{n} \text{Prob}\{k \text{ arrivals in } (0, t]\} \text{ Prob}\{(n-k) \text{ arrivals in } (t, t+h]\} = \sum_{k=0}^{n} p_k(t) p_{n-k}(h).$$

Now applying the properties of the Poisson process given under Definition 11.1.3, we find

$$\begin{aligned} p_n(t+h) = \sum_{k=0}^{n} p_k(t) p_{n-k}(h) &= p_n(t) p_0(h) + p_{n-1}(t) p_1(h) + o(h) \\ &= p_n(t)(1 - \lambda h) + p_{n-1}(t)\lambda h + o(h). \end{aligned}$$

Hence

$$\frac{p_n(t+h) - p_n(h)}{h} = -\lambda p_n(t) + \lambda p_{n-1}(t) + \frac{o(h)}{h}.$$

Taking the limit as $h$ tends to zero yields

$$\frac{dp_n(t)}{dt} = -\lambda p_n(t) + \lambda p_{n-1}(t), \quad n > 0,$$

which is a set of differential-difference equations having the solution

$$p_n(t) = e^{-\lambda t}\frac{(\lambda t)^n}{n!}, \quad n = 0, 1, 2, \ldots,$$

which once again can be readily verified by direct substitution. Therefore, the number of arrivals, $N(t)$, in the interval $(0, t]$ has a Poisson distribution with parameter $\lambda t$.

To compute the mean number of arrivals in an interval of length $t$, we proceed as follows:

$$E[N(t)] = \sum_{k=1}^{\infty} k p_k(t) = \sum_{k=1}^{\infty} k\, e^{-\lambda t}\frac{(\lambda t)^k}{k!} = \lambda t \left(\sum_{k=1}^{\infty} \frac{(\lambda t)^{k-1}}{(k-1)!}\right) e^{-\lambda t}$$

$$= \lambda t \left(\sum_{k=0}^{\infty} \frac{(\lambda t)^k}{k!}\right) e^{-\lambda t} \quad = \lambda t.$$

It is now evident why $\lambda$ is referred to as the *rate* of the Poisson process, since the mean number of arrivals per unit time, $E[N(t)]/t$, is equal to $\lambda$. It may be shown, and indeed has already been shown in Part I of this text, that the variance of a Poisson process is equal to its expectation. Recall that in the usual notation, found in statistics and probability texts, the Poisson distribution is written as

$$f(k;\alpha) = e^{-\alpha}\frac{\alpha^k}{k!}, \quad k = 0, 1, 2, \ldots,$$

and has mean $\alpha$ and variance $\alpha$. In our case, we have $\alpha = \lambda t$.

**Example 11.1** The arrival of jobs to a supercomputing center follows a Poisson distribution with a mean interarrival time of 15 minutes. Thus the rate of arrivals is $\lambda = 4$ jobs per hour. We have the following:

- Prob{Time between arrivals $\le \tau$ hours} $= 1 - e^{4\tau}$.
- Prob{$k$ arrivals in $\tau$ hours} $= e^{-4\tau}(4\tau)^k/k!$.

**Poisson Processes Have Exponential Interarrival Times**

We now show that if the arrival process is a Poisson process, then the associated random variable defined as the time between successive arrivals (the interarrival time) has an exponential distribution.

**Example 11.2** Continuing with the previous example of arrivals to a supercomputing center at a rate of one every 15 minutes, suppose 45 minutes have passed without an arrival. Then the expected time until the next arrival is still just 15 minutes, a result of the memoryless property of the exponential distribution.

Since the arrival process is Poisson, we have

$$p_n(t) = e^{-\lambda t}\frac{(\lambda t)^n}{n!}, \quad n = 0, 1, 2, \ldots,$$

which describes the number of arrivals that have occurred by time $t$. Let $X$ be the random variable that denotes the time between successive events (arrivals). Its probability distribution function is given by

$$A(t) = \text{Prob}\{X \le t\} = 1 - \text{Prob}\{X > t\}.$$

But Prob$\{X > t\}$ = Prob{0 arrivals in $(0, t]$} $= p_0(t)$. Thus

$$A(t) = 1 - p_0(t) = 1 - e^{-\lambda t}, \quad t \ge 0,$$

with corresponding density function, obtained by differentiation,

$$a(x) = \lambda e^{-\lambda t}, \quad t \ge 0.$$

This defines the exponential distribution. Thus $X$ is exponentially distributed with mean $1/\lambda$ and consequently, for a Poisson arrival process, the time between arrivals is exponentially distributed.

Let $X_1$ be the random variable that represents the time until the first arrival. Then

$$\text{Prob}\{X_1 > t\} = \text{Prob}\{N(t) = 0\} = e^{-\lambda t}.$$

If $X_i,\ i > 1$, is the random variable that denotes the time between arrival $i-1$ and arrival $i$, then $X_i$ has exactly the same distribution as $X_1$. Suppose arrival $i-1$ occurs at some time $s$ and our concern is the time $t$ until the next arrival. We have

$$\begin{aligned}\text{Prob}\{X_i > t \mid A_{i-1} = s\} &= \text{Prob}\{0 \text{ arrivals in } (s, s+t] \mid A_{i-1} = s\}\\ &= \text{Prob}\{0 \text{ arrivals in } (s, s+t]\}\\ &= e^{-\lambda t},\end{aligned}$$

where the condition vanishes due to the independent nature of arrivals, while the number of arrivals in any interval depends only on the length of that interval. The time until the $n^{\text{th}}$ arrival may now be found as the sum $S_n = \sum_{i=1}^{n} X_i$ and was previously seen (Section 7.6.2) to have an Erlang-$n$ distribution:

$$F_{S_n}(t) = 1 - \sum_{i=0}^{n-1} e^{-\lambda t}\frac{(\lambda t)^i}{i!}, \qquad f_{S_n}(t) = \lambda e^{-\lambda t}\frac{(\lambda t)^{n-1}}{(n-1)!}, \qquad t \ge 0.$$

In Chapter 7 we used the Laplace transform to justify this result. We now use a more direct approach by arguing that arrival $n$ will occur before time $t$ if and only if $N(t) \ge n$. Thus

$$F_{S_n}(t) = \text{Prob}\{S_n \le t\} = \text{Prob}\{N(t) \ge n\} = \sum_{i=n}^{\infty} p_i(t) = \sum_{i=n}^{\infty} e^{-\lambda t}\frac{(\lambda t)^i}{i!} = 1 - \sum_{i=0}^{n-1} e^{-\lambda t}\frac{(\lambda t)^i}{i!}.$$

The corresponding density function is obtained by differentiation. We have

$$f_{S_n}(t) = \lambda e^{-\lambda t}\sum_{i=n}^{\infty}\frac{(\lambda t)^{i-1}}{(i-1)!} - \lambda e^{-\lambda t}\sum_{i=n}^{\infty}\frac{(\lambda t)^i}{i!} = \lambda e^{-\lambda t}\frac{(\lambda t)^{n-1}}{(n-1)!}.$$

Let us now examine the distribution of these arrival times. Suppose we are told that exactly one arrival occurred in the interval $[0, t]$. We would like to find the distribution of the time at which it actually arrived, i.e., we seek the distribution $\text{Prob}\{X_1 < s \mid N(t) = 1\}$ for $s \le t$. We have

$$\begin{aligned}\text{Prob}\{X_1 < s \mid N(t) = 1\} &= \frac{\text{Prob}\{X_1 < s,\ N(t) = 1\}}{\text{Prob}\{N(t) = 1\}}\\ &= \frac{\text{Prob}\{1 \text{ arrival in } [0, s)\} \times \text{Prob}\{0 \text{ arrivals in } [s, t)\}}{\text{Prob}\{N(t) = 1\}}\\ &= \frac{\lambda s e^{-\lambda s} e^{-\lambda(t-s)}}{\lambda t e^{-\lambda t}} = \frac{s}{t}.\end{aligned}$$

In other words, the distribution of the time at which the arrival occurs is uniformly distributed over the interval $[0, t]$. This result can be extended. If we are told that exactly two arrivals occur in $[0, t]$, then the two arrival instants are independently and uniformly distributed over $[0, t]$; with $n$ arrivals, all $n$ are independently and uniformly distributed over $[0, t]$. It is for this reason that arrivals which occur according to a Poisson process are sometimes referred to as *random* arrivals.

**Example 11.3** Customers arrive at a service center according to a Poisson process with rate $\lambda$. Suppose that three customers have been observed to arrived in the first hour. Let us find the probability that all three arrived in the first 15 minutes and the probability that at least one of the three arrived in the first 15 minutes. The key to answering this question lies in the fact that the arrivals are uniformly distributed over the one hour interval.

Since the arrival instants are uniformly distributed over the first hour, the probability of a customer arriving in the first 15 minutes is equal to $15/60 = 1/4$. Furthermore since the arrivals are independent, the probability that all three customers arrive in the first 15 minutes is $1/4^3 = 1/64$. With three customers and 1/4 hour intervals, we may visualized the situation as a unit cube partitioned into 64 smaller cubes each having dimension $1/4 \times 1/4 \times 1/4$. Only the small cube at the origin corresponds to all three customers arriving in the first 15 minutes.

To find the probability that at least one of the three arrived in the first 15 minutes, we simply need to sum the number of small cubes in sides that correspond to one of the three arriving in the first 15 minutes and dividing this total by 64, being sure not to count the same small cube more than once. This gives the probability to be $(16 + 12 + 9)/64 = 37/64$. Alternatively, the different possibilities are

- one customer arrives in the first fifteen minutes (probability 1/4) and the other two arrive at any time,
- one customer does not arrive in the first 15 minutes, one arrives in the first 15 minutes and the third arrives at any time (probability $3/4 \times 1/4$),
- two customers do not arrive in the first 15 minutes, but the third does (probability $3/4 \times 3/4 \times 1/4 = 9/64$).

The sum of these probabilities is equal to 37/64. Yet another possibility is to compute the probability that none of the three arrived in the first 15 minutes and subtract this from one. This gives

$$1 - \left(\frac{3}{4}\right)^3 = 1 - \frac{27}{64} = \frac{37}{64}.$$

Notice that in this example, the actual rate of the Poisson process does not figure in the computation.

**Superposition/Decomposition of Poisson Streams**

When two or more *independent* Poisson streams merge, the resulting stream is also a Poisson stream. It follows then that multiple arrival processes to a single service center can be merged to constitute a single arrival process whose interarrival times are exponentially distributed, so long as the interarrivals times of the individual arrival processes are exponentially distributed (although possibly having different parameters), as illustrated in Figure 11.3. If the $i^{\text{th}}$ stream, $i = 1, 2, \ldots, n$, has parameter $\lambda_i$, then the parameter for the merged stream (also called the *pooled* stream) is given by

$$\lambda = \sum_{i=1}^{n} \lambda_i.$$

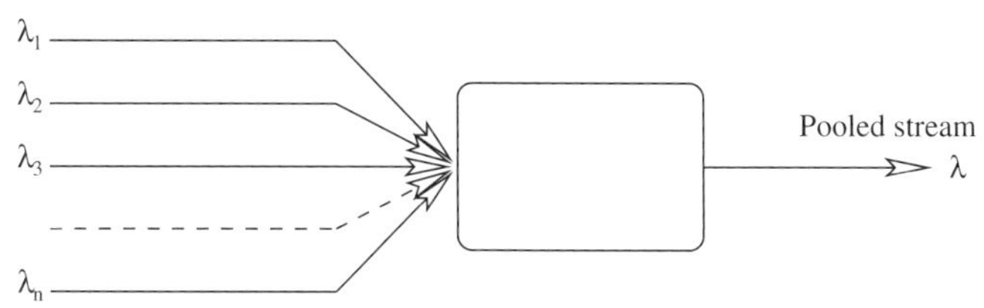

Figure 11.3. Superposition of Poisson streams.

The reverse operation is also possible. A single Poisson stream may be decomposed into multiple *independent* Poisson streams if the decomposition is such that customers are directed into substreams $i,\ i = 1, 2, \ldots, n$, with probability $p_i$ and $\sum_{i=1}^{n} p_i = 1$, as illustrated in Figure 11.4. The proof of these results is left to the exercises, where some helpful hints are provided.

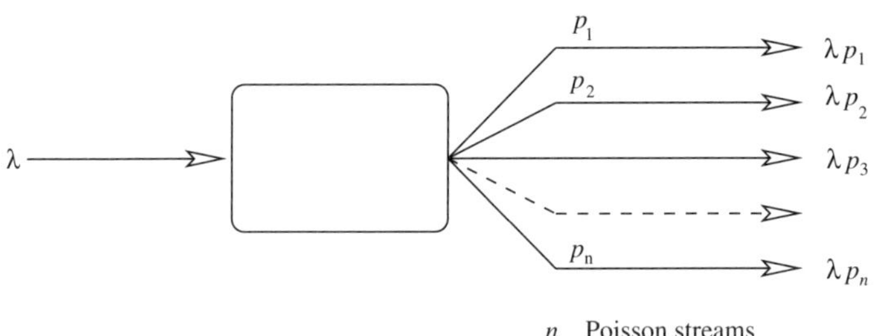


Figure 11.4. Decomposition of Poisson streams.

**Example 11.4** To stress the importance of the independence of the decomposed streams, consider the following. Customers enter a shoe store according to a Poisson process with a mean of 20 per hour. Four out of every five customers who enter are female. Thus, on average 16 female and 4 male customers enter the store each hour and we can consider the original Poisson stream to be decomposed into a Poisson stream of female customers at a rate of 16 per hour and a Poisson stream of male customers with a rate of 4 per hour. Suppose now that we are told that 32 female customers entered the store during an hour. The expected number of all customers who entered during that hour is equal to $32 + 4 = 36$. Because the streams are independent, the fact that considerably more female customers entered during that hour than should be expected has no influence on the expected number of male customers, which must remain equal to 4. The expected total number of customers who entered is thus equal to 36.

**PASTA: Poisson Arrivals See Time Averages**

An important property of the Poisson arrival process is that the distribution of customers seen by an arrival to a queueing facility is, stochastically, the same as the limiting distribution of customers at that facility. In other words, once the queueing system has reached steady state, each arrival from a Poisson process finds the system at equilibrium. If $p_n$ is the probability that the system contains $n$ customers at equilibrium and $a_n$ denotes the probability that an arriving customer finds $n$ customers already present, then PASTA says that $a_n = p_n$. This implies that the Poisson process sees the same distribution as a *random observer*, i.e., at equilibrium, Poisson arrivals take a random look at the system.

This result is a direct consequence of the memoryless property of the interarrival time distribution of customers to a queueing system fed by a Poisson process. In particular, it does not depend on the service time distribution. To prove the PASTA property, we proceed as follows. Let

$$\begin{aligned} N(t) &= \text{Actual number of customers in system at time } t, \\ p_n(t) &= \text{Prob}\{\text{System is in state } n \text{ at time } t\} = \text{Prob}\{N(t) = n\}, \\ a_n(t) &= \text{Prob}\{\text{Arrival at time } t \text{ finds system in state } n\}, \\ A(t, t+\delta t] &= \text{The event ``an arrival occurs in } (t, t+\delta t].\text{''} \end{aligned}$$

Then

$$\begin{aligned} a_n(t) &= \lim_{\delta t \to 0} \text{Prob}\{N(t) = n | A(t, t + \delta t]\} \\ &= \lim_{\delta t \to 0} \frac{\text{Prob}\{N(t) = n \text{ and } A(t, t + \delta t]\}}{\text{Prob}\{A(t, t + \delta t]\}} \\ &= \lim_{\delta t \to 0} \frac{\text{Prob}\{A(t, t + \delta t] | N(t) = n\}\text{Prob}\{N(t) = n\}}{\text{Prob}\{A(t, t + \delta t]\}} \\ &= \lim_{\delta t \to 0} \frac{\text{Prob}\{A(t, t + \delta t]\}\text{Prob}\{N(t) = n\}}{\text{Prob}\{A(t, t + \delta t]\}} \\ &= \text{Prob}\{N(t) = n\} = p_n(t). \end{aligned}$$

The crucial step in the above argument is

$$\text{Prob}\{A(t, t + \delta t] | N(t) = n\} = \text{Prob}\{A(t, t + \delta t]\}.$$

This results from the fact that, since interarrival times possess the memoryless property, $\text{Prob}\{A(t, t + \delta t]\}$ is independent of the past history of the arrival process and hence independent of the current state of the queueing system. With the Poisson arrival process having (constant) rate $\lambda$, the probability of having an arrival in $(t, t + \delta t]\}$ is equal to $\lambda \delta t + o(\delta t)$ which does not depend on $N(t)$.

The PASTA property does not hold for arrival processes other than Poisson. Consider, for example, a queueing system in which arrivals occur at fixed intervals of time, exactly one minute apart, and which require a fixed service length of 30 seconds. It follows that for exactly half the time, there is a single customer in service; the system is empty for the other half. We have $p_0 = p_1 = 0.5$ and $p_n = 0$ for $n \geq 2$. However, an arriving customer *always* finds the system empty: the previous customer arrived 60 seconds ago, and left after 30. In this case, $a_0 = 1$ and $a_n = 0$ for all $n \geq 1$.

### *11.1.2 Scheduling Disciplines*

We now turn our attention to the manner in which customers are selected from the queue and taken into service. This is called the scheduling policy, scheduling algorithm, or scheduling discipline. Unless otherwise stated, it is assumed that the time it takes to select a customer and to enter that customer into service, is zero: the two events, the departure of a customer and a waiting customer begins service, occur instantaneously. We now distinguish between policies that can interrupt the service of the customer in service (called preemptive policies) and those that cannot (Figure 11.5).

*Preemptive policies* may be used when there are different types of customers, having different service priorities: high priority customers are served before low priority ones. When a high priority customer arrives and finds a customer of lower priority being served, the service of the lower priority customer is *interrupted* and later resumed (called *preempt-resume*). The preempted customer is reinserted into the queue. This means that in a preemptive system several customers may be at various stages of completion at the same time. When a previously preempted customer returns to

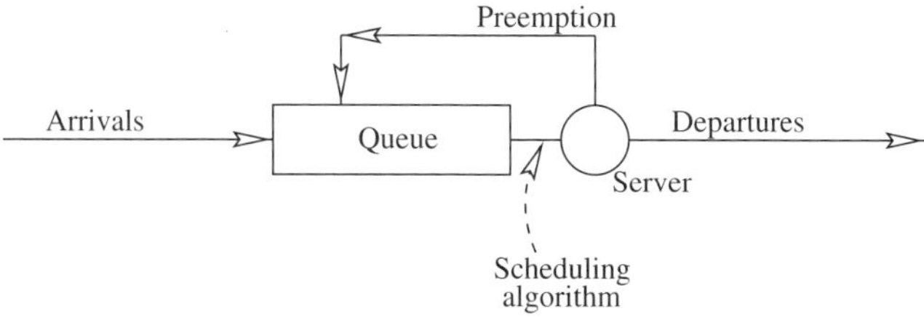


Figure 11.5. Scheduling customers for service.

service it may not be possible to continue from the point at which the preemption occurred. Some previously completed work may have been lost. Indeed in some cases it may even be necessary to begin the service all over again (called *preempt-restart*).

In a *nonpreemptive system* customers are served one at a time to completion. The server will commence the service of a low priority customer if no customer of higher priority is present. But, once the server has selected a customer, it is committed to serving that customer even if customers of higher priority arrive during its service.

Within the context of preemptive and nonpreemptive service, several scheduling disciplines exist. We briefly describe the most common.

- FCFS (first come, first served). Also called FIFO (first-in, first-out). This is a nonpreemptive policy that selects the customer at the head of the queue, the one that has waited the longest, as the next one to be served.
- LCFS (last come, first served). The opposite of FCFS. The last customer to arrive is the one next chosen to be served. While this may seem to be unfair, it is exactly the situation used when a stack data structure is implemented in a computer program.
- LCFS-PR (last come, first served, preempt resume). This is the situation when the preempt resume policy is applied to the LCFS scheduling discipline.
- SIRO (service in random order). The next customer to be served is chosen at random among all those waiting.
- RR (round robin). In this scheduling discipline, the duration of service provided each time a customer begins service is limited to a fixed amount (frequently called a *time slice*). If a customer completes service at any point during the time slice, it immediately departs from the service center and the customer at the head of the queue begins to be served. If, after the time slice has finished, a customer has still not completed its service requirement, that customer is reinserted back into the end of the queue to await its turn for a second, or indeed third, fourth, or more time slice.
- PS (processor sharing). This is the limiting case of RR in which the time slice goes to zero. It is used to approximate RR, because it is much more tractable analytically.
- IS (infinite server). The server has the ability to serve all customers simultaneously. It follows that an arriving customer goes immediately into service and there is no need for a queue.
- PRIO (priority scheduling). In this case the customer chosen from the queue is the one with the highest priority. FCFS is used to handle the case in which multiple customers of the same priority are waiting.

Combinations of the above scheduling disciplines are possible and some will be considered later. They all describe the order in which customers are taken from the queue and allowed into service.

### *11.1.3 Kendall's Notation*

Kendall's notation is used to characterize queueing systems. It is given by $A/B/C/X/Y/Z$, where

$A$ indicates the interarrival time distribution,
$B$ indicates the service time distribution,
$C$ indicates the number of servers,
$X$ indicates the system capacity,
$Y$ indicates the size of the customer population, and
$Z$ indicates the queue scheduling discipline.

Some possible distributions (for $A$ and/or $B$) are $M$ (for Markovian), $E$ (for Erlang), $C_k$ (for Coxian of order $k$), $G$ (for general), and $D$ for deterministic (or constant). For a general arrival process, the notation $GI$ is sometimes used to replace $G$ to indicate that, although the interarrival time distribution may be completely general, successive arrivals are independent of each other. The number of servers ($C$) is often taken to be 1. These first three parameters are always provided. Thus, for example, the *M/M/*1 queue means that the arrival process and service process are both Markovian (although we usually say that the arrival process is Poisson and the service time distribution is exponential) and there is a single server. This simplest of all queues is discussed in detail later. The letters that specify the system capacity, customer population and scheduling discipline may be omitted with the understanding that the default values for capacity and population size is infinite and the default for the scheduling discipline is FCFS. Thus the *M/M/*1 queue has an unlimited amount of space in which to hold waiting customers, an infinite population from which customers are drawn and applies a FCFS scheduling policy. We shall provide additional details on Kendall's notation as and when it is needed.

**Example 11.5** Consider as an example the case of travelers who arrive at a train station and purchase tickets at one of six ticket distribution machines. It is reasonable to assume that the time taken to purchase a ticket is constant whereas it may be observed that arrivals follow a Poisson process. Such a system could be represented as an *M/D/*6 queue. If the time needed to purchase a ticket is almost, but not quite constant, perhaps an $M/E_5/6$ queue may be more appropriate, since an Erlang-5 queue has some limited amount of variability associated with it. If no more than 50 travelers can squeeze into the ticket purchasing office, then an $M/E_5/6/50$ queue should be used. Other scenarios are readily envisaged.

### *11.1.4 Graphical Representations of Queues*

There exist a number of ways of graphing queueing systems and these can frequently provide insight into the behavior of the system. For a FCFS, single server queue, one possibility is shown in Figure 11.6. We let $C_n$ denote the $n^{\text{th}}$ customer to enter the queueing system. If the service discipline is not FCFS, then $C_n$ may not be the $n^{\text{th}}$ customer to depart. We let $\tau_n$ denote the time at which $C_n$ arrives in the system and $t_n$ the interarrival time between the arrival of $C_{n-1}$ and $C_n$ (i.e., $t_n = \tau_n - \tau_{n-1}$). We shall assume that the $t_n$ are drawn from a distribution such that $\text{Prob}\{t_n \le t\} = A(t)$, i.e., independent of $n$, where $A(t)$ is a completely arbitrary interarrival time distribution. Similarly, we shall define $x_n$ to be the service time for $C_n$ and $\text{Prob}\{x_n \le x\} = B(x)$, where $B(x)$ is a completely arbitrary (but independent) service time distribution. We are now in a position to graph $N(t)$, the number of customers in the system at time $t$. In Figure 11.6, $w_n$ represents the waiting time (in queue) for $C_n$, and $s_n$ represents the system time (queue plus service) for $C_n$. (Note that $s_n = w_n + x_n$).

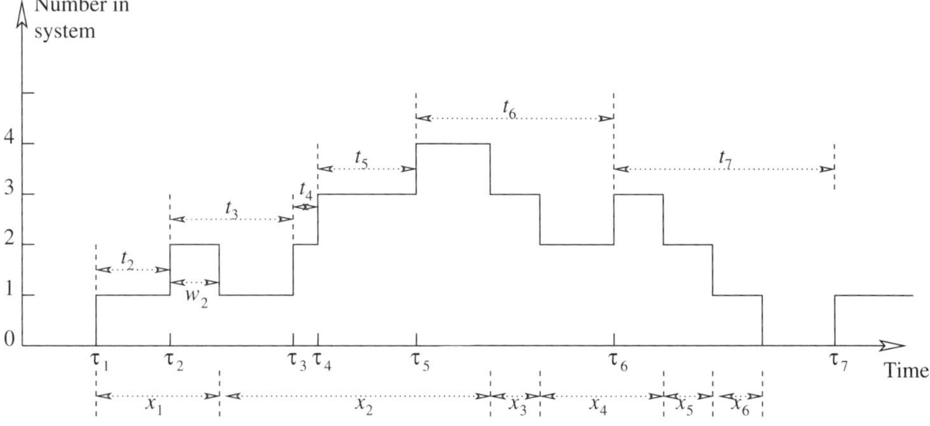


Figure 11.6. Graphical representation No. 1.

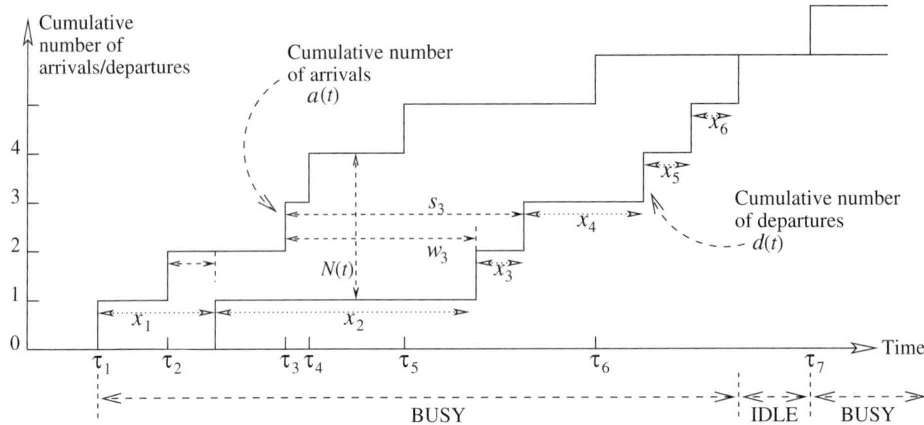


Figure 11.7. Graphical representation No. 2.

An alternative graphical representation is given in Figure 11.7. In this graph $N(t) = a(t) - d(t)$, where $a(t)$ denotes the number of arrivals in $[0, t)$ and $d(t)$ denotes the number of departures in this same time period.

There is also a double-time-axis graphical representation which is shown in Figure 11.8 (for the single-server, FCFS case). This particular variant is relatively easy to generalize to many servers and to disciplines other than FCFS.

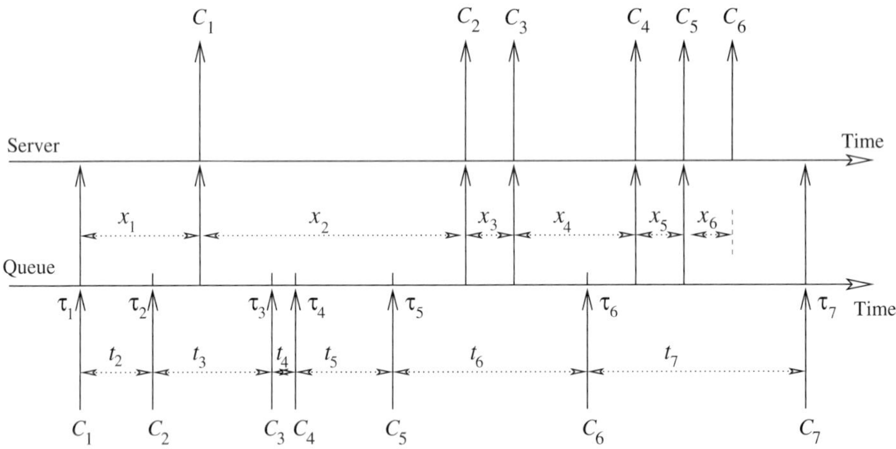


Figure 11.8. Double-time-axis notation.

### *11.1.5 Performance Measures—Measures of Effectiveness*

When we analyze a queueing system we do so for the purpose of obtaining the values of certain system properties. For example, we may wish to find

- the number of customers in the system;
- the waiting time for a customer;
- the length of a busy or idle period;
- the current work backload (in units of time).

These are called *measures of effectiveness*. They are all random variables and, whereas we might wish to know their complete probabilistic descriptions (i.e., their PDFs), most times we must be content to be able to derive just their first few moments (mean, variance, etc.). In addition, mathematical models of queueing systems provide quantitative answers to specific questions, such as: "By how much will the congestion be reduced if the variation of service time is reduced by 10%? 50%?" or "What happens if we replace two slow servers by a single fast server?", and so on. Also, a queueing analyst may wish to design an *optimal* system which implies the need to balance waiting time with idle time according to some inherent cost structure; e.g., to find the optimal number of servers, for example. We now consider the most commonly obtained measures of effectiveness.

**Number of customers**

We shall let $N$ be the random variable that describes the number of customers in the system at steady state. The probability that at steady state the number of customers present in the system is $n$ is denoted by $p_n$,

$$p_n = \text{Prob}\{N = n\},$$

and the average number in the system at steady state is

$$L = E[N] = \sum_{i=0}^{\infty} np_n.$$

Within the queueing system, customers may be present in the queue waiting for their turn to receive service or they may be receiving service. We shall let $N_q$ be the random variable that describes the number of customers waiting in the queue and we shall denote its mean by $L_q = E[N_q]$.

**System Time and Queueing Time**

The time that a customer spends in the system, from the instant of its arrival to the queue to the instant of its departure from the server, is called the *response time* or *sojourn time*. We shall denote the random variable that describes response time by $R$, and its mean value by $E[R]$.

The response time is composed of the time that the customer spends waiting in the queue, called the waiting time, plus the time the customer spends receiving service, called the service time. We shall let $W_q$ be the random variable that describes the time the customer spends waiting in the queue and its mean will be denoted by $E[W_q]$.

**System Utilization**

In a queueing system with a single server ($c = 1$), the utilization $U$ is defined as the fraction of time that the server is busy. If the rate at which customers arrive at, and are admitted into, a queueing facility is $\lambda$ and if $\mu$ is the rate at which these customers are served, then the utilization is equal to $\lambda/\mu$. Over a period of time $T$, this queueing system, in steady state, receives an average of $\lambda T$ customers, which are served in an average of $\lambda T/\mu$ seconds. In many queueing systems, the Greek letter $\rho$ is defined as $\rho = \lambda/\mu$ and consequently is identified with the utilization. However, $\lambda$ is generally defined as the arrival rate to the system and this may or may not be the rate at which customers actually enter the queueing facility. Thus it is not always the case that $\lambda/\mu$ correctly defines the utilization. Some customers may be refused admission (they are said to be "lost" customers) so that the effective arrival rate into the queueing facility is less than $\lambda$ and hence the utilization is less than $\rho = \lambda/\mu$. However, unless stated otherwise, we assume that all customers who arrive at a queueing facility are admitted.

In a $G/G/1$ queue where $p_0$ is the probability that the system is empty, it must follow that $U = 1 - p_0$. In a stable system (i.e., one in which the queue does not grow without bound), the server cannot be busy 100% of the time. This implies that we must have $\lambda/\mu < 1$ for the queueing system to be stable. Thus, in any time interval, the average number of customers that arrive must be strictly less than the average number of customers that the server can handle.

In the case of queueing systems with multiple servers ($c > 1$), the utilization is defined as the average fraction of servers that are active—which is just the rate at which work enters the system divided by the maximum rate (capacity) at which the system can perform this work, i.e., $U = \lambda/(c\mu)$. In multiserver systems, it is usual to define $\rho$ as $\rho = \lambda/(c\mu)$ with the same caveat as before concerning the identification of $\rho$ as the utilization.

**System Throughput**

The throughput of a queueing system is equal to its departure rate, i.e., the average number of customers that are processed per unit time. It is denoted by $X$. In a queueing system in which all customers that arrive are eventually served and leave the system, the throughput is equal to the arrival rate, $\lambda$. This is not the case in queueing systems with finite buffer, since arrivals may be lost before receiving service.

**Traffic Intensity**

We define the traffic intensity as the rate at which work enters the system, so it is therefore given as the product of the average arrival rate of customers and the mean service time, i.e., $\lambda\bar{x} = \lambda/\mu$, where $\bar{x} = 1/\mu$ and $\mu$ is the mean service rate. Notice that in single-server systems the traffic intensity is equal to the utilization. For multiple servers, the traffic intensity is equal to $cU$.

### *11.1.6 Little's Law*

Little's law is perhaps the most widely used formula in queueing theory. It is simple to state and intuitive, widely applicable, and depends only on weak assumptions about the properties of the queueing system. Little's law equates the number of customers in a system to the product of the effective arrival rate and the time spend in the system. We present only an intuitive justification. Consider the graph of Figure 11.7 once again, which represents the activity in a single-server, FCFS queueing system. Recall that $a(t)$ denotes the number of arrivals in $[0, t)$ and $d(t)$ denotes the number of departures in this same period. The number of customers in the system at any time $t$ is given by $a(t) - d(t)$. Let the total area between the two curves up to some point $t$ be denoted by $g(t)$. This represents the total time (the total number of seconds, for example) all customers have spent in the system during the interval $[0, t)$.

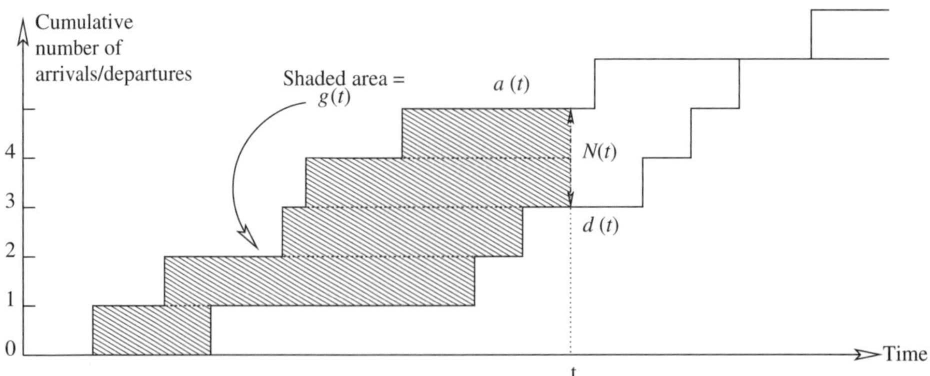


Figure 11.9. Graph No. 2 again.

From Figure 11.9, it is apparent that

$$\lambda_t \equiv \frac{a(t)}{t}$$

is the average arrival rate in the interval (e.g., number of customers arrivals per second);

$$R_t \equiv \frac{g(t)}{a(t)}$$

is the average system time (time spent queueing plus time spent in service) per customer up to time $t$, and

$$L_t \equiv \frac{g(t)}{t}$$

is the average number of customers in the system between 0 and $t$. It follows then that

$$L_t = \frac{g(t)}{t} = \frac{g(t)}{a(t)} \times \frac{a(t)}{t} = R_t \lambda_t.$$

Now let $t \to \infty$ and assume the following limits exist:

$$\lim_{t\to\infty} L_t = L = E[N], \quad \lim_{t\to\infty} \lambda_t = \lambda, \quad \lim_{t\to\infty} R_t = E[R].$$

Then

$$E[N] = \lambda E[R].$$

This says that the average number of customers in the system is equal to the average arrival rate of customer to the system multiplied by the average system time per customer, and is known as Little's law. Historically, Little's law has been written as

$$L = \lambda W$$

and in this usage it must be remembered that $W$ is defined as "response time," the time spent in the queue *and* at the server, and not just simply as the time spent "waiting" to be served. As before $L$ refers to the mean number of customers in the system and $\lambda$ the arrival rate.

Little's law can be applied when we relate $L$ to the number of customers *waiting* to receive service and $W$ to the time spent *waiting* for service. The same applies also to the servicing aspect itself. In other words, Little's law may be applied individually to the different parts of a queueing facility, namely the queue and the server. Given that $L_q$ is the average number of customers waiting in the queue, $L_s$ is the average number of customers receiving service, $W_q$ the average time spent in the queue, and $\bar{x}$ the average time spent receiving service, then, from Little's law,

$$L_q = \lambda W_q \qquad \text{and} \qquad L_s = \lambda \bar{x},$$

and we have

$$L = L_q + L_s = \lambda W_q + \lambda \bar{x} = \lambda(W_q + \bar{x}) = \lambda W.$$

Little's law may be applied even more generally than we have shown here. For example, it may be applied to separate parts of much larger queueing systems, such as subsystems in a queueing network. In such a case, $L$ should be defined with respect to the number of customers in a subsystem and $W$ with respect to the total time in that subsystem. Little's law may also refer to a specific class of customer in a queueing system, or to subgroups of customers, and so on. Its range of applicability is very wide indeed. Finally, we comment on the amazing fact that the proof of Little's law turns out to be independent of

- specific asumptions regarding the arrival distribution $A(t)$;
- specific asumptions regarding the service time distribution $B(t)$;
- the number of servers;
- the particular queueing discipline.

## 11.2 Birth-Death Processes: The *M/M/*1 Queue

Birth–death processes are continuous-time[1] Markov chains with a very special structure. If the states of the Markov chain are indexed by the integers $0, 1, 2, \ldots$, then transitions are permitted only from state $i > 0$ to its nearest neighbors, namely, states $i - 1$ and $i + 1$. As for state $i = 0$, on exiting this state, the Markov chain must enter state 1. Such processes are also called *skip-free* processes because to go from any state $i$ to any other state $j$, each intermediate state must be visited: no state between these two can be skipped. As we shall see in this chapter, birth-death processes arise in a variety of simple single-server queueing systems and the special structure makes their stationary distributions relatively easy to compute. An arrival to the queueing system results in one additional unit in the system, and is identified with a *birth*, whereas a departure removes a unit from the system as is referred to as a *death*. Observe that in this analogy, a birth can occur when the system is empty!

The simplest of all queueing systems is the *M/M/*1 queue. This is a single server queue with FCFS scheduling discipline, an arrival process that is Poisson and service time that is exponentially distributed. It is shown graphically in Figure 11.10 where $\lambda$ is the parameter of the Poisson arrival process and $\mu$ is the exponential service rate. The mean time between arrivals (which is exponentially distributed) is $1/\lambda$ and the mean service time is $1/\mu$.

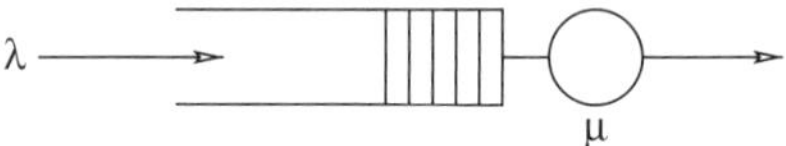


Figure 11.10. The *M/M/*1 queue.

### *11.2.1 Description and Steady-State Solution*

At any time $t$, the state of an *M/M/*1 queue is completely characterized by specifying the number of customers present. We shall use the integers $0, 1, 2, \ldots$ to represent these states: $n$ denotes the state in which there are $n$ customers in the system, including the one in service. We would like to be able to compute the *state probabilities*, i.e., the probability that the system is in any given state $n$ at any time $t$. We write these as

$$p_n(t) = \text{Prob}\{n \text{ in system at time } t\}.$$

This is often a difficult task, even for this simplest of queueing processes. Instead, at least for the moment, we shall look for the steady-state probabilities,

$$p_n = \lim_{t \to \infty} p_n(t).$$

If this limit exists, then the probability of finding the system in any particular state eventually becomes independent of the starting state, so that no matter when we query the system after it settles into steady state, the probability of finding $n$ customers present does not change. The steady state probabilities $p_n$ can be interpreted as the probability of finding $n$ customers in the system at an arbitrary point in time after the process has reached steady state. It is not true that all systems reach steady state, i.e., for some queueing systems it is possible that $\lim_{t\to\infty} p_n(t)$ may not yield a true probability distribution.

We first seek to compute the probability that the system is in state $n$ at time $t + \Delta t$, namely $p_n(t + \Delta t)$. It is evident that the system will be in the state $n$ at time $t + \Delta t$ if one of the following

[1] Birth-death processes in discrete time may be developed analogously to that of the continuous-time version considered in this chapter.

(mutually exclusive and collectively exhaustive) events occur:

1. The system is in state $n$ at time $t$ and no change occurs in $(t, t + \Delta t]$.
2. The system is in state $n - 1$ at time $t$ and an arrival occurs in $(t, t + \Delta t]$.
3. The system is in state $n + 1$ at time $t$ and a departure occurs in $(t, t + \Delta t]$.

For sufficiently small values of $\Delta t$, we have

$$\begin{aligned}
\text{Prob}\{1 \text{ arrival in } (t, t + \Delta t]\} &= \lambda \Delta t + o(\Delta t),\\
\text{Prob}\{1 \text{ departure in } (t, t + \Delta t]\} &= \mu \Delta t + o(\Delta t),\\
\text{Prob}\{0 \text{ arrivals in } (t, t + \Delta t]\} &= 1 - \lambda \Delta t + o(\Delta t),\\
\text{Prob}\{0 \text{ departures in } (t, t + \Delta t]\} &= 1 - \mu \Delta t + o(\Delta t).
\end{aligned}$$

These assumptions (which follow from the exponential nature of the arrivals and service) imply that the probability of multiple arrivals, of multiple departures, or of any near simultaneous combinations of births and deaths in some small interval of time $\Delta t$, is negligibly small, i.e., of order $o(\Delta t)$. It follows that the state transition probability diagram is as shown in Figure 11.11.

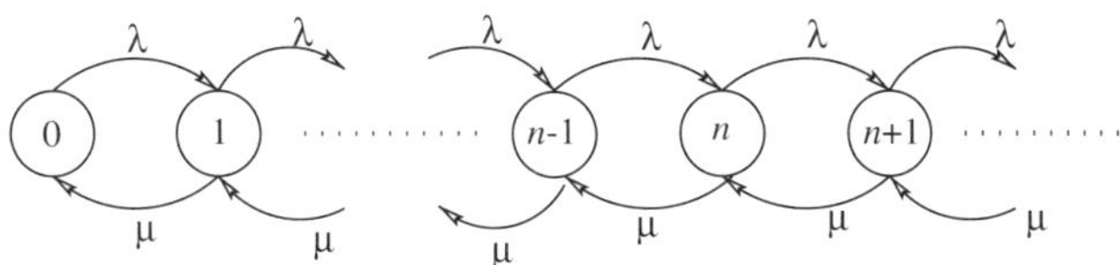


Figure 11.11. State transitions in the *M/M/*1 queue.

For $n \geq 1$, we may write

$$\begin{aligned}
p_n(t + \Delta t) = p_n(t)&[1 - \lambda \Delta t + o(\Delta t)][1 - \mu \Delta t + o(\Delta t)]\\
&+ p_{n-1}(t)[\lambda \Delta t + o(\Delta t)]\\
&+ p_{n+1}(t)[\mu \Delta t + o(\Delta t)].
\end{aligned}$$

When $n = 0$, we have

$$\begin{aligned}
p_0(t + \Delta t) = p_0(t)&[1 - \lambda \Delta t + o(\Delta t)]\\
&+ p_1(t)[\mu \Delta t + o(\Delta t)].
\end{aligned}$$

Expanding the right-hand side,

$$\begin{aligned}
p_n(t + \Delta t) &= p_n(t) - (\lambda + \mu)\Delta t p_n(t) + \lambda \Delta t p_{n-1}(t) + \mu \Delta t p_{n+1}(t) + o(\Delta t), \quad n \geq 1,\\
p_0(t + \Delta t) &= p_0(t) - \lambda \Delta t p_0(t) + \mu \Delta t p_1(t) + o(\Delta t).
\end{aligned}$$

Subtracting $p_n(t)$ from each side and dividing by $\Delta t$,

$$\begin{aligned}
\frac{p_n(t + \Delta t) - p_n(t)}{\Delta t} &= -(\lambda + \mu)p_n(t) + \lambda p_{n-1}(t) + \mu p_{n+1}(t) + \frac{o(\Delta t)}{\Delta t}, \quad n \geq 1,\\
\frac{p_0(t + \Delta t) - p_0(t)}{\Delta t} &= -\lambda p_0(t) + \mu p_1(t) + \frac{o(\Delta t)}{\Delta t}.
\end{aligned}$$

Taking the limit as $\Delta t \rightarrow 0$ yields

$$\frac{dp_n(t)}{dt} = -(\lambda + \mu)p_n(t) + \lambda p_{n-1}(t) + \mu p_{n+1}(t), \quad n \geq 1, \tag{11.2}$$

$$\frac{dp_0(t)}{dt} = -\lambda p_0(t) + \mu p_1(t), \tag{11.3}$$

i.e., a set of differential difference equations which represents the dynamics of the system. This turns out to be hard to solve for $p_n(t)$, but not for the stationary distribution. Let us assume that the steady state exists. Then

$$\frac{dp_n(t)}{dt} = 0$$

and we get

$$0 = -(\lambda + \mu)p_n + \mu p_{n+1} + \lambda p_{n-1}, \quad n \geq 1,$$
$$0 = -\lambda p_0 + \mu p_1 \quad \Longrightarrow \quad p_1 = \frac{\lambda}{\mu} p_0,$$

which is a set of second-order difference equations with constant coefficients. There exists a variety of ways of solving such equations. The direct approach is left as an exercise and some hints are provided. In a later section we shall use the generating function approach, but for the moment we proceed by using a recursive argument. Rearranging the set of difference equations, we obtain

$$p_{n+1} = \frac{\lambda + \mu}{\mu} p_n - \frac{\lambda}{\mu} p_{n-1}, \quad n \geq 1.$$

Substituting $n = 1$ yields

$$\begin{aligned} p_2 &= \frac{\lambda + \mu}{\mu}\left(\frac{\lambda}{\mu}\right) p_0 - \frac{\lambda}{\mu} p_0 \\ &= \frac{\lambda}{\mu}\left(\frac{\lambda + \mu}{\mu} - 1\right) p_0 \\ &= \frac{\lambda}{\mu}\left(\frac{\lambda}{\mu}\right) p_0 = \frac{\lambda^2}{\mu^2} p_0. \end{aligned}$$

Similarly, we may show that

$$p_3 = \left(\frac{\lambda}{\mu}\right)^3 p_0,$$

which leads us to conjecture that

$$p_n = \left(\frac{\lambda}{\mu}\right)^n p_0.$$

The proof is by induction.

The formula clearly holds for $n = 1$ and 2, as we have just shown.

Let us assume it to be true for $n - 1$ and $n$ and prove it true for $n + 1$. We have

$$\begin{aligned} p_{n+1} &= \frac{\lambda + \mu}{\mu}\left(\frac{\lambda}{\mu}\right)^n p_0 - \frac{\lambda}{\mu}\left(\frac{\lambda}{\mu}\right)^{n-1} p_0 \\ &= \left(\frac{\lambda}{\mu}\right)^n \left[\frac{\lambda + \mu}{\mu} - 1\right] p_0 \\ &= \left(\frac{\lambda}{\mu}\right)^{n+1} p_0. \end{aligned}$$

Now it only remains to determine $p_0$. We have

$$1 = \sum_{n=0}^{\infty} p_n = \sum_{n=0}^{\infty} \left(\frac{\lambda}{\mu}\right)^n p_0 = \sum_{n=0}^{\infty} \rho^n p_0$$

and therefore

$$p_0 = \frac{1}{\sum_{n=0}^{\infty} \rho^n},$$

where $\rho \equiv \lambda/\mu$, is the *utilization factor*, a measure of the average use of the service facility, i.e., it is the expected number of arrivals per mean service time.

Notice that $\sum_{n=0}^{\infty} \rho^n$ is the geometric series and converges if and only if $|\rho| < 1$. Therefore, the existence of a steady state solution demands that $\rho = \lambda/\mu < 1$; i.e., that $\lambda < \mu$. Note that if $\lambda > \mu$ the mean arrival rate is greater than the mean service rate and the server will get further and further behind. Thus the system size will keep increasingly without limit. It is not as intuitively obvious as to why this limitless increase occurs when $\lambda = \mu$. From the mathematics of the situation, we know that the geometric series does not converge when $\rho = 1$ and we have previously seen that the corresponding random walk problem is null-recurrent when $\lambda = \mu$.

For $\rho < 1$,

$$\sum_{n=0}^{\infty} \rho^n = \frac{1}{1-\rho}$$

and hence $p_0 = 1 - \rho$, or alternatively $\rho = 1 - p_0$. In summary, the steady-state solution for the *M/M/*1 queue is given by

$$p_n = \rho^n(1-\rho) \quad \text{for} \quad \rho = \lambda/\mu < 1,$$

which is the probability mass function of a modified geometric random variable. Observe that the equilibrium solution of the *M/M/*1 queue depends on the average arrival rate $\lambda$ and the average service rate $\mu$ only through their ratio $\rho$.

The probability that the queue contains at least $k$ customers has a particularly nice formula. We have

$$\begin{aligned}\text{Prob}\{n \geq k\} &= \sum_{i=k}^{\infty} p_i = (1-\rho)\sum_{i=k}^{\infty} \rho^i = (1-\rho)\left(\sum_{i=0}^{\infty} \rho^i - \sum_{i=0}^{k-1} \rho^i\right) \\ &= (1-\rho)\left(\frac{1}{1-\rho} - \frac{1-\rho^k}{1-\rho}\right) = \rho^k.\end{aligned}$$

**Example 11.6** The arrival pattern of cars to the local oil change center follows a Poisson distribution at a rate of four per hour. If the time to perform an oil change is exponentially distributed and requires on average of 12 minutes to carry out, what is the probability of finding more than 3 cars waiting for the single available mechanic to service their car?

This question requires us to compute $\sum_{i=4}^{\infty} p_i = 1 - p_0 - p_1 - p_2 - p_3$ from an *M/M/*1 queue with $\lambda = 4$ and $1/\mu = 12/60$ or $\mu = 5$. Thus $\rho = 4/5$ is strictly less than 1 and the system is stable. Also $p_0 = 1 - \rho = 1/5$ and

$$p_1 = \rho(1-\rho) = 4/25, \quad p_2 = \rho p_1 = 16/125, \quad p_3 = \rho p_2 = 64/625,$$

which allows us to compute the answer as

$$1 - \frac{1}{5} - \frac{4}{25} - \frac{16}{125} - \frac{64}{625} = .4096$$

which is exactly equal to $\rho^4$.

**Matrix Formulation of the *M/M/*1 Queue**

The *M/M/*1 queue with service rate $\mu$ and arrival rate $\lambda$ has the following infinite infinitesimal generator:

$$Q = \begin{pmatrix} -\lambda & \lambda & & & \\ \mu & -(\lambda+\mu) & \lambda & & \\ & \mu & -(\lambda+\mu) & \lambda & \\ & & \mu & -(\lambda+\mu) & \lambda \\ & & & \ddots & \ddots & \ddots \end{pmatrix}.$$

In the notation used for Markov chains, we have $\pi Q = 0$ (with $\pi_i = p_i$ for all $i$) and it is obvious that $-\lambda\pi_0 + \mu\pi_1 = 0$, i.e., that $\pi_1 = (\lambda/\mu)\pi_0$. In general, we have

$$\lambda\pi_{i-1} - (\lambda+\mu)\pi_i + \mu\pi_{i+1} = 0,$$

from which, by induction, we may derive

$$\pi_{i+1} = ((\lambda+\mu)/\mu)\pi_i - (\lambda/\mu)\pi_{i-1} = (\lambda/\mu)\pi_i.$$

Thus, once $\pi_0$ is known, the remaining values $\pi_i$, $i = 1, 2, \ldots$, may be determined recursively just as before. For the *M/M/*1 queue it has already been shown that the probability that the system is empty is given by $\pi_0 = (1 - \lambda/\mu)$.

Observe that the coefficient matrix is tridiagonal. and that once $p_0$ is known, the solution is just a forward elimination procedure. However, there is no computational advantage to be gained by setting up and solving the matrix equation, rather than using the previously developed recursive relations. We show this formulation at this time because it will become useful in other, more complex, cases when the matrix is Hessenberg, rather than tridiagonal.

### *11.2.2 Performance Measures*

We now turn our attention to computing various performance measures concerning the *M/M/*1 queue, such as mean number in system, mean queue length, and so on.

**Mean Number in System**

Let $N$ be the random variable that describes the number of customers in the system at steady state, and let $L = E[N]$. Then

$$L = \sum_{n=0}^{\infty} np_n = \sum_{n=0}^{\infty} n(1-\rho)\rho^n = (1-\rho)\sum_{n=0}^{\infty} n\rho^n = (1-\rho)\rho\sum_{n=0}^{\infty} n\rho^{n-1}. \tag{11.4}$$

If we assume that the system is stable, then $\rho < 1$ and

$$\sum_{n=0}^{\infty} n\rho^{n-1} = \frac{\partial}{\partial\rho}\left[\sum_{n=0}^{\infty}\rho^n\right] = \frac{\partial}{\partial\rho}\left[\frac{1}{1-\rho}\right] = \frac{1}{(1-\rho)^2}. \tag{11.5}$$

It now follows from Equation (11.4) that the mean number of customers in the *M/M/*1 queue is given by

$$L = (1-\rho)\frac{\rho}{(1-\rho)^2} = \frac{\rho}{1-\rho} = \frac{\lambda}{\mu-\lambda}.$$

**Variance of Number in System**

To compute the variance of the number of customers in an *M/M/*1 queue, we use the formula

$$\text{Var}\,[N] = E[N^2] - E[N]^2.$$

The second moment is computed as follows:

$$E[N^2] = \sum_{n=0}^{\infty} n^2 p_n = \sum_{n=0}^{\infty} n^2 (1-\rho)\rho^n = (1-\rho)\sum_{n=0}^{\infty} n^2\rho^n = (1-\rho)\rho\sum_{n=0}^{\infty} n^2\rho^{n-1}$$

$$= (1-\rho)\rho\frac{\partial}{\partial\rho}\left[\sum_{n=0}^{\infty} n\rho^n\right].$$

Substituting from Equation (11.5) we have

$$E[N^2] = (1-\rho)\rho\frac{\partial}{\partial\rho}\left[\frac{\rho}{(1-\rho)^2}\right] = (1-\rho)\rho\frac{(1-\rho)^2 + 2(1-\rho)\rho}{(1-\rho)^4} = \rho\frac{1+\rho}{(1-\rho)^2}.$$

Finally, the variance is computed as

$$\text{Var}\,[N] = \rho\frac{1+\rho}{(1-\rho)^2} - \left(\frac{\rho}{1-\rho}\right)^2 = \frac{\rho}{(1-\rho)^2}.$$

**Mean Queue Length**

Let $N_q$ be the random variable that describes the number of customers waiting in the queue at steady state, and let $L_q = E[N_q]$. Then

$$L_q = E[N_q] = 0 \times p_0 + \sum_{n=1}^{\infty}(n-1)p_n,$$

$$L_q = \sum_{n=1}^{\infty} np_n - \sum_{n=1}^{\infty} p_n = L - (1-p_0) = \frac{\rho}{1-\rho} - \rho = \frac{\rho^2}{1-\rho} = \rho L.$$

Thus,

$$L_q = \rho L = L - \rho.$$

**Average Response Time**

Let $R$ be the random variable that describes the response time of customers in the system. Little's law states that the mean number of customers in a queueing system in steady state is equal to the product of the arrival rate and the mean response time, i.e.,

$$E[N] = \lambda E[R] \quad (L = \lambda W).$$

Then

$$E[R] = \frac{1}{\lambda}E[N] = \frac{1}{\lambda}\frac{\rho}{1-\rho} = \frac{1/\mu}{1-\rho} = \frac{1}{\mu-\lambda}$$

which is the average service time divided by the probability that the server is idle. Thus the congestion in the system, and hence the delay, builds rapidly as the traffic intensity increases, (as $\rho \to 1$).

Notice that we have not specified any particular scheduling discipline in our computation of the mean response time and it may be (erroneously) thought that this result applies only to FCFS scheduling. It is manifestly true in the FCFS case, since an arriving customer must wait while those already in the queue, including the one in service, are served. With $k$ already present, and all $k$ having a mean service time of $1/\mu$ (including the one being served), the arriving customer will spend a total time equal to $\sum_{k=0}^{\infty} p_k(k+1)/\mu$ where the factor $(k+1)/\mu$ is the time needed to complete service for the $k$ customers already present plus the time needed to serve the entering customer. We evaluate this as follows:

$$\sum_{k=0}^{\infty} p_k(k+1)\frac{1}{\mu} = \sum_{k=0}^{\infty} \rho^k(1-\rho)(k+1)\frac{1}{\mu} = \frac{1-\rho}{\mu}\sum_{k=0}^{\infty} \rho^k(k+1) = \frac{1-\rho}{\mu}\sum_{k=0}^{\infty}\left(k\rho^k + \rho^k\right)$$

$$= \frac{1-\rho}{\mu}\left(\rho\frac{\partial}{\partial\rho}\sum_{k=0}^{\infty}\rho^k + \sum_{k=0}^{\infty}\rho^k\right) = \frac{1-\rho}{\mu}\left(\frac{\rho}{(1-\rho)^2} + \frac{1}{1-\rho}\right) = \frac{1}{\mu(1-\rho)}.$$

However, this same result applies to all scheduling disciplines that have the property that the server is never idle when there are customers present. This is true only for the mean value and not for higher moments. We shall consider the distribution of response time momentarily.

**Average Waiting Time $W_q$**
We have, from Little's law,

$$L_q = \lambda W_q$$

and so

$$W_q = \frac{\lambda}{\mu(\mu-\lambda)} = \frac{\rho}{\mu-\lambda}.$$

Similar comments regarding the scheduling discipline, made in our discussion of the mean response time, also apply here.

**Effective Queue Size for Nonempty Queue, $L'_q$**
In this case we ignore instants in which the queue is empty.

$$L'_q = E[N_q | N_q \neq 0] = \sum_{n=1}^{\infty}(n-1)p'_n = \sum_{n=2}^{\infty}(n-1)p'_n$$

where $p'_n = \text{Prob}\{n \text{ in system} \mid n \geq 2\}$. It therefore follows that

$$p'_n = \frac{\text{Prob}\{n \text{ in system and } n \geq 2\}}{\text{Prob}\{n \geq 2\}} = \frac{p_n}{\sum_{n=2}^{\infty} p_n}$$

$$= \frac{p_n}{1-p_0-p_1} = \frac{p_n}{1-(1-\rho)-(1-\rho)\rho} = \frac{p_n}{\rho^2}, \quad n \geq 2.$$

Notice that the probability distribution $p'_n$ is the probability distribution $p_n$ normalized when the cases $n = 0$ and $n = 1$ are omitted. It now follows that

$$L'_q = \sum_{n=2}^{\infty}(n-1)\frac{p_n}{\rho^2} = \frac{1}{\rho^2}\left[\sum_{n=2}^{\infty} np_n - \sum_{n=2}^{\infty} p_n\right] = \frac{1}{\rho^2}\left[(L-p_1)-(1-p_0-p_1)\right]$$

$$= \frac{1}{\rho^2}\left[\rho/(1-\rho)-(1-\rho)\rho - 1 + (1-\rho) + (1-\rho)\rho\right]$$

$$= \frac{1}{\rho^2}\left[\rho/(1-\rho)-\rho\right] = \frac{1}{\rho}\left[1/(1-\rho)-1\right] = \frac{1}{1-\rho}.$$

Thus

$$L'_q = \frac{1}{1-\rho} = \frac{\mu}{\mu-\lambda}.$$

Collecting these results together in one convenient location, we have

$$L = E[N] = \frac{\rho}{1-\rho}, \qquad L_q = E[N_q] = \frac{\rho^2}{1-\rho}, \qquad \text{Var}\,[N] = \frac{\rho}{(1-\rho)^2}, \tag{11.6}$$

$$W = E[R] = \left(\frac{1}{1-\rho}\right) E[S], \quad W_q = \left(\frac{\rho}{1-\rho}\right) E[S], \quad L'_q = \frac{1}{1-\rho}, \tag{11.7}$$

where $E[S] = 1/\mu$ is the mean service time.

**Example 11.7** Let us compute these performance measures for the oil change center of Example 11.6. Using the parameters $\lambda = 4$, $\mu = 5$, and $\rho = 4/5$, we have the following measures:

| | |
|---|---|
| Mean number in system: | $\rho/(1-\rho) = 4$ cars, |
| Variance of number in system: | $\rho/(1-\rho)^2 = 20$ cars, |
| Average response time: | $1/(\mu - \lambda) = 1$ hour, |
| Average time in queue prior to service: | $\rho/(\mu - \lambda) = .8$ hours (or 48 minutes), |
| Average size of queue when it is not empty: | $\mu/(\mu - \lambda) = 5$ cars. |

**Throughput, Utilization, and Traffic Intensity**

It is apparent from the definitions given in the previous section that the throughput of the *M/M*/1 queue is just equal to the arrival rate, i.e., $X = \lambda$, and that the utilization is equal to the traffic intensity and is given by $U = \lambda/\mu$.

**Example 11.8** Consider a cable modem which is used to transmit 8-bit characters and which has a capacity of 4 megabits per second (Mbps). Thus the maximum rate is 500,000 characters per second (cps). Given that traffic arrives at a rate of 450,000 cps, let us compute some standard performance measures, when this system is modelled as an *M/M*/1 queue.

In this example, we have $\lambda = 450{,}000$ cps and $\mu = 500{,}000$ cps. Thus the utilization (and hence the traffic intensity) of the cable modem is $\rho = \lambda/\mu = 0.9$. The mean number of characters in the system $L$, and the mean number of characters waiting to be transmitted $L_q$, are

$$L = \frac{\rho}{1-\rho} = \frac{0.9}{0.1} = 9 \quad \text{and} \quad L_q = \frac{\rho^2}{1-\rho} = \frac{0.9^2}{0.1} = 8.1;$$

the average transmission time per character (the response time) is equal to the expected number in the system divided by $\lambda$ (from Little's law) which gives

$$\frac{\rho}{1-\rho} \times \frac{1}{\lambda} = \frac{0.9}{0.1} \times \frac{1}{450{,}000} = .00002 \text{ seconds.}$$

The throughput is just equal to $\lambda = 450{,}000$ cps.

**Distribution of Time Spent in an *M/M*/1 Queue (Response Time)**

Most performance measures presented so far have been average values: the mean number of customers in the system, the mean time spent waiting prior to service, and so on. We may also wish to find the distributions of response time and queueing time, just as in $p_n, n = 0, 1, 2, \ldots,$ we have the distribution of customer population. Whereas queue length and $p_n$ are unaffected by the scheduling policy employed by a queueing system, it should be apparent that the distribution of time that a customer spends in the system is a function of this policy. Here our concern is with the *M/M*/1 queue and the FCFS scheduling discipline.

We first consider the total time (response time) a customer has to spend in the system, which includes the time spent in the queue waiting for service plus the time spent actually receiving service. Denote this random variable, the response time, by $R$, its PDF by $W_r(t)$, its density by $w_r(t)$, and its expected value by $E[R]$. We shall show that

$$w_r(t) = (\mu - \lambda)e^{-(\mu-\lambda)t}, \quad t > 0,$$

and

$$E[R] = \frac{1}{\mu - \lambda}.$$

We use the following two properties of an $M/M/1$ queue:

1. From the PASTA property of a Poisson stream, an arriving customer sees the steady-state distribution of the number of customers in the system. This arriving customer, with probability $p_n$, finds $n$ customers already waiting or in service.
2. From the memoryless property of the exponential distribution, if the new arrival finds a customer in service, the remaining service time of that customer is distributed exponentially with mean $1/\mu$, i.e., identical to the service requirements of all waiting customers.

The response time of an arriving customer who finds $n$ customers in the system is therefore the sum of $(n+1)$ exponentially distributed random variables, the $n$ already present plus the arriving customer itself. Such a sum has an $(n+1)$ stage Erlang density function. Then, if $B_k$ is the service time of customer $k$, we have

$$\text{Prob}\{R > t\} = \text{Prob}\left\{\sum_{k=1}^{n+1} B_k > t\right\}.$$

Now, conditioning on the number present when the arrival occurs and using the independence of arrivals and service, we obtain

$$\begin{aligned}\text{Prob}\{R > t\} &= \sum_{n=0}^{\infty}\left(\text{Prob}\left\{\sum_{k=1}^{n+1} B_k > t\right\}\right) p_n = \sum_{n=0}^{\infty}\left(e^{-\mu t}\sum_{k=0}^{n}\frac{(\mu t)^k}{k!}\right)(1-\rho)\rho^n \\ &= \sum_{k=0}^{n}\sum_{n=k}^{\infty}\left(e^{-\mu t}\frac{(\mu t)^k}{k!}\right)(1-\rho)\rho^n = \sum_{k=0}^{n}\left(e^{-\mu t}\frac{(\mu t)^k}{k!}\right)\sum_{n=k}^{\infty}(1-\rho)\rho^n \\ &= \sum_{k=0}^{n}\left(e^{-\mu t}\frac{(\mu t)^k}{k!}\right)\rho^k = e^{-\mu(1-\rho)t}, \quad t \ge 0.\end{aligned}$$

Hence the probability distribution function for the response time in an $M/M/1$ queue is

$$\text{Prob}\{R \le t\} = W_r(t) = 1 - e^{-(\mu-\lambda)t},$$

i.e., the exponential distribution with mean

$$E[R] = \frac{1}{\mu - \lambda} = \frac{1}{\mu(1-\rho)}.$$

Notice that it is not possible to write this performance measure just in terms of $\rho$. It depends on $\lambda$ and $\mu$ but not just through their ratio $\rho$. This means that it is possible to assign values to $\lambda$ and $\mu$ in such a way that the system can be almost saturated with large queues but still have a very short expected response time.

The probability density function for the response time can be immediately found as the density function of the exponential distribution with parameter $\mu(1-\rho)$ or evaluated directly as

$$w_r(t) = \sum_{n=0}^{\infty} p_n g_{n+1}(t) = (1-\rho)\mu e^{-\mu t}\sum_{n=0}^{\infty}\frac{(\rho\mu t)^n}{n!} = (\mu - \lambda)e^{-(\mu-\lambda)t},$$

where

$$g_{n+1}(t) = \frac{\mu(\mu t)^n e^{-\mu t}}{n!}$$

is the density function for the Erlang distribution with $(n+1)$ phases.

### Queueing Time Distribution in an *M/M/*1 Queue

The waiting time distribution is the distribution of time that an arriving customer must spend waiting in the queue before entering service. It is possible that a customer does not have to wait at all: an arriving customer can, with positive probability, find that the system is empty and enter immediately into service. The time spent waiting in this case is zero. In all other cases, an arrival finds the server busy and must wait. This means that the random variable, $T_q$, which defines the time spent waiting in the queue, is part discrete and part continuous. Let $W_q(t)$ be defined as the probability distribution function of $T_q$. It follows from the Poisson arrival property that

$$W_q(0) = \text{Prob}\{T_q \le 0\} = \text{Prob}\{T_q = 0\} = \text{Prob}\{\text{System is empty upon arrival}\} = p_0 = 1 - \rho.$$

We now need to find $W_q(t)$ when $T_q > 0$. If there are $n$ customers already in the system upon the arrival of a tagged customer at time $t = 0$, then in order for the tagged customer to go into service at some future time between 0 and $t$, all $n$ prior customers must have been served by time $t$. The service received by each of these $n$ customers is independent and exponentially distributed, and the sum of $n$ of them (the convolution of $n$ exponential random variables) is an Erlang-$n$ distribution. In this case the density function, which describes the time needed for the completion of service of these $n$ customers is given by

$$g_n(x) = \frac{\mu(\mu x)^{n-1}}{(n-1)!} e^{-\mu x}, \quad x \ge 0.$$

Since the input is Poisson, the probability that an arrival finds $n$ in the system is identical to the stationary distribution of system size (PASTA). Therefore

$$W_q(t) = \text{Prob}\{T_q \le t\}$$

$$= \sum_{n=1}^{\infty} \left[\text{Prob}\{n \text{ completions in } \le t \mid \text{arrival finds } n \text{ in system}\} \times p_n\right] + W_q(0)$$

$$= (1-\rho)\sum_{n=1}^{\infty} \rho^n \int_0^t \frac{\mu(\mu x)^{n-1}}{(n-1)!} e^{-\mu x} dx + (1-\rho)$$

$$= (1-\rho)\rho \int_0^t \mu e^{-\mu x} \sum_{n=1}^{\infty} \frac{(\mu x \rho)^{n-1}}{(n-1)!} dx + (1-\rho)$$

$$= \rho(1-\rho) \int_0^t \mu e^{-\mu(1-\rho)x} dx + (1-\rho)$$

$$= 1 - \rho e^{-\mu(1-\rho)t}, \quad t > 0.$$

Since this also holds when $t = 0$, we may write the distribution of waiting time in the queue as

$$W_q(t) = 1 - \rho e^{-\mu(1-\rho)t}, \quad t \ge 0.$$

The corresponding probability density function may be found by differentiation. We have

$$w_q(t) = \frac{d}{dt}\left(1 - \rho e^{-\mu(1-\rho)t}\right) = \rho\mu(1-\rho)e^{-\mu(1-\rho)t}.$$

Now we can get the expected waiting time:

$$\begin{aligned} W_q = E[T_q] &= \int_0^\infty t\rho\mu(1-\rho)e^{-\mu(1-\rho)t}dt \\ &= \int_0^\infty t\frac{\lambda}{\mu}(\mu-\lambda)e^{-(\mu-\lambda)t}dt \\ &= \frac{\lambda}{\mu}\int_0^\infty t(\mu-\lambda)e^{-(\mu-\lambda)t}dt \\ &= \frac{\lambda}{\mu}\frac{1}{\mu-\lambda}, \end{aligned}$$

since

$$\int_0^\infty t(\mu-\lambda)e^{-(\mu-\lambda)t}dt = \frac{1}{\mu-\lambda}$$

is the mean of the exponential distribution. Hence

$$W_q = \frac{\lambda}{\mu(\mu-\lambda)}.$$

### *11.2.3 Transient Behavior*

It is possible to derive equations that characterize the transient solution of the *M/M/*1 queue. However, here we shall be content to outline the steps needed to arrive at this solution, rather than go through the tedious procedure of deriving the solution. For those who seek more details, we recommend the analysis provided in Gross and Harris [19].

The transient solution of the *M/M/*1 queue is obtained by solving the set of differential-difference equations which we derived in Section 11.2.1. These equations are

$$\frac{dp_n(t)}{dt} = -(\lambda+\mu)p_n(t) + \lambda p_{n-1}(t) + \mu p_{n+1}(t), \quad n \ge 1,$$

$$\frac{dp_0(t)}{dt} = -\lambda p_0(t) + \mu p_1(t).$$

The solution is obtained by first defining the time-dependent transform

$$P(z,t) \equiv \sum_{n=0}^{\infty} p_n(t)z^n \quad \text{for complex } z,$$

multiplying the $n^{\text{th}}$ equation by $z^n$, and summing over all permissible $n$ to obtain a single differential equation for the $z$-transform of $p_n(t)$. This yields a linear, first-order partial differential equation for $P(z,t)$,

$$z\frac{\partial}{\partial t}P(z,t) = (1-z)[(\mu-\lambda z)P(z,t) - \mu p_0(t)].$$

This equation must once again be transformed, this time using the Laplace transform for $P(z,t)$, given by

$$P^*(z,s) \equiv \int_{0+}^\infty e^{-st}P(z,t)dt.$$

We obtain

$$P^*(z,s) = \frac{zP(z,0) - \mu(1-z)P_0^*(s)}{sz - (1-z)(\mu - \lambda z)}.$$

Thus we have transformed the set of differential-difference equations for $p_n(t)$ both on the discrete variable $n$ and on the continuous variable $t$. It is now necessary to turn to Rouché's theorem to determine the unknown function $P_0^*(s)$, by appealing to the analyticity of the transform. After much work, this leads to an explicit expression for the double transform which must then be inverted on both transform variables. The final solution for the transient solution of the *M/M/*1 queue is obtained as

$$p_n(t) = e^{-(\lambda+\mu)t}\left[\rho^{(n-i)/2}I_{n-i}(at) + \rho^{(n-i-1)/2}I_{n+i+1}(at) + (1-\rho)\rho^n \sum_{j=n+i+2}^{\infty} \rho^{-j/2}I_j(at)\right],$$

where $i$ is the initial system size (i.e., $p_n(0) = 1$ if $n = i$, and $p_n(0) = 0$ if $n \neq i$), $\rho = \lambda/\mu$, $a = 2\mu\rho^{1/2}$, and

$$I_n(x) \equiv \sum_{m=0}^{\infty} \frac{(x/2)^{n+2m}}{(n+m)!m!}$$

is the modified Bessel function of the first kind and of order $n$.

## 11.3 General Birth-Death Processes

A birth-death process may be viewed as a generalization of the *M/M/*1 queueing system. Perhaps more correctly, the *M/M/*1 queue is a special type of birth-death process. Whereas the birth and death rates in the *M/M/*1 queue ($\lambda$ and $\mu$, respectively) are the same irrespective of the number of customers in the system, a general birth-death process allows for different rates depending on the number of customers present. Arrivals to the system continue to be referred to as *births* and departures as *deaths* but now we introduce a birth rate $\lambda_n$, which is defined as the rate at which births occur when the population is of size $n$, and a death rate $\mu_n$, defined as the rate at which deaths occur when the population is of size $n$. Notice that, for all $n$, both $\lambda_n$ and $\mu_n$ are independent of $t$ which means that our concern is with continuous-time *homogeneous* Markov chains. On the other hand, the parameters $\lambda_n$ and $\mu_n$ can, and frequently do, depend on the currently occupied state of the system, namely, state $n$.

### *11.3.1 Derivation of the State Equations*

As for the *M/M/*1 queue, a state of the system at any time is characterized completely by specifying the size of the population at that time. Let $p_n(t)$ be the probability that the population is of size $n$ at time $t$. We assume that births and deaths are independent and that

$$\begin{aligned}
\text{Prob}\{\text{One birth in } (t, t+\Delta t] \mid n \text{ in population}\} &= \lambda_n \Delta t + o(\Delta t), \\
\text{Prob}\{\text{One death in } (t, t+\Delta t] \mid n \text{ in population}\} &= \mu_n \Delta t + o(\Delta t), \\
\text{Prob}\{\text{Zero births in } (t, t+\Delta t] \mid n \text{ in population}\} &= 1 - \lambda_n \Delta t + o(\Delta t), \\
\text{Prob}\{\text{Zero deaths in } (t, t+\Delta t] \mid n \text{ in population}\} &= 1 - \mu_n \Delta t + o(\Delta t).
\end{aligned}$$

These assumptions mean that the probability of two or more births, of two or more deaths, or of near simultaneous births and deaths in some small interval of time $\Delta t$ is negligibly small, i.e., of order $o(\Delta t)$.

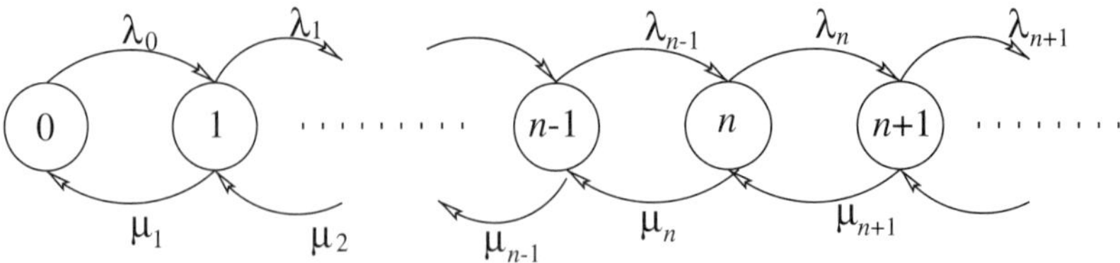

Figure 11.12. State transitions in a birth-death process.

The state transition diagram is shown in Figure 11.12. Notice that we make the assumption that it is impossible to have a death when the population is of size 0 (i.e., $\mu_0 = 0$) but that one can indeed have a birth when the population is zero (i.e., $\lambda_0 > 0$). We now need to develop the set of differential-difference equations that define the dynamics of this birth-death process, and to do so, we could proceed as for the *M/M/*1 queue, by deriving a relationship for $p_n(t + \Delta t)$ and taking the limit of $\left[p_n(t + \Delta t) - p_n(t)\right] / \Delta t$ as $\Delta t$ tends to zero. Instead, we shall use a short cut based on the fact that the rate at which probability "accumulates" in any state $n$ is the difference between the rates at which the system enters and leaves that state. The rate at which probability "flows" into state $n$ at time $t$ is

$$\lambda_{n-1} p_{n-1}(t) + \mu_{n+1} p_{n+1}(t),$$

while the rate at which it "flows" out is

$$(\lambda_n + \mu_n) p_n(t).$$

Subtracting these gives the *effective* probability flow rate into $n$, i.e.,

$$\frac{dp_n(t)}{dt} = \lambda_{n-1} p_{n-1}(t) + \mu_{n+1} p_{n+1}(t) - (\lambda_n + \mu_n) p_n(t), \quad n \geq 1, \tag{11.8}$$

and

$$\frac{dp_0(t)}{dt} = \mu_1 p_1(t) - \lambda_0 p_0(t),$$

which are forms of the Chapman–Kolmogorov forward equations. This shortcut is illustrated in Figure 11.13 where we have separated out state $n$ by surrounding it with a dashed line. The net rate of probability flow into $n$ is found by computing the flow across this boundary, using opposing signs for entering and leaving. Notice that if $\lambda_n = \lambda$ and $\mu_n = \mu$ for all $n$, we get exactly the same equation that we previously derived for the *M/M/*1 queue.

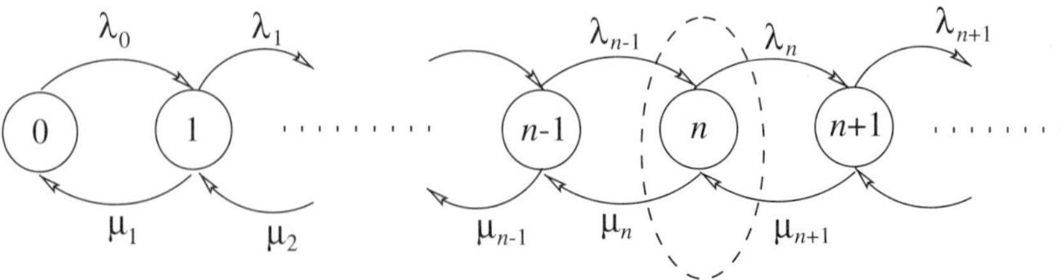

Figure 11.13. Transitions from/to state $n$.

**Example 11.9** A pure birth process.

In a pure birth process, there are no deaths, only births. The rates of transition are given by

$$\lambda_n = \lambda \quad \text{for all } n,$$

$$\mu_n = 0 \quad \text{for all } n.$$

The equations become

$$\frac{dp_n(t)}{dt} = -\lambda p_n(t) + \lambda p_{n-1}(t) \quad \text{for } n \geq 1,$$

$$\frac{dp_0(t)}{dt} = -\lambda p_0(t).$$

If initially $p_n(0) = 1$ if $n = 0$ and is equal to 0 if $n \geq 1$, which means that the system begins empty, then

$$p_0(t) = e^{-\lambda t},$$

since the solution is given by $p_0(t) = ce^{-\lambda t}$ where $c$, the constant of integration, must be equal to one ($p_0(0) = 1 = ce^0$). Furthermore, from

$$\frac{dp_1(t)}{dt} = -\lambda p_1(t) + \lambda e^{-\lambda t}$$

the solution is clearly

$$p_1(t) = \lambda t e^{-\lambda t}$$

since application of the product rule $(uv)' = udv + vdu$ with $u = \lambda t$ and $v = e^{-\lambda t}$ yields

$$\lambda t \left(-\lambda e^{-\lambda t}\right) + \lambda e^{-\lambda t} = -\lambda p_1(t) + \lambda e^{-\lambda t}.$$

Similarly, continuing by induction, we obtain

$$p_n(t) = e^{-\lambda t} \frac{(\lambda t)^n}{n!}, \qquad n \geq 0, \quad t \geq 0.$$

This is now seen as the formula for the Poisson distribution with mean $\lambda t$. In other words, the Poisson process is a pure birth process with constant birth rate $\lambda$.

### 11.3.2 Steady-State Solution

In the steady state, which we assume exists,

$$\frac{dp_n(t)}{dt} = 0,$$

and hence Equation (11.8) becomes

$$\begin{aligned} 0 &= \lambda_{n-1} p_{n-1} + \mu_{n+1} p_{n+1} - (\lambda_n + \mu_n) p_n, \qquad n \geq 1, \\ 0 &= -\lambda_0 p_0 + \mu_1 p_1 \quad \Longrightarrow \quad p_1 = \frac{\lambda_0}{\mu_1} p_0, \end{aligned} \tag{11.9}$$

where $p_n$ is defined as the limiting probability, $p_n = \lim_{t\to\infty} p_n(t)$, the probability that the system contains $n$ customers once it reaches steady state where all influence of the starting state has been erased. These equations are called the global balance equations. They may also be written as

$$p_{n+1} = \frac{\lambda_n + \mu_n}{\mu_{n+1}} p_n - \frac{\lambda_{n-1}}{\mu_{n+1}} p_{n-1}, \qquad n \geq 1,$$

$$p_1 = \frac{\lambda_0}{\mu_1} p_0.$$

We obtain a solution by iteration

$$p_2 = \frac{\lambda_1 + \mu_1}{\mu_2} p_1 - \frac{\lambda_0}{\mu_2} p_0 = \frac{\lambda_1 + \mu_1}{\mu_2} \left( \frac{\lambda_0}{\mu_1} p_0 \right) - \frac{\lambda_0}{\mu_2} p_0 = \frac{\lambda_1 \lambda_0}{\mu_2 \mu_1} p_0.$$

Similarly

$$p_3 = \frac{\lambda_2 \lambda_1 \lambda_0}{\mu_3 \mu_2 \mu_1} p_0.$$

This leads us to conjecture that

$$p_n = \frac{\lambda_{n-1} \lambda_{n-2} \dots \lambda_0}{\mu_n \mu_{n-1} \dots \mu_1} p_0 = p_0 \prod_{i=1}^{n} \frac{\lambda_{i-1}}{\mu_i}, \quad n \geq 1.$$

We prove this by induction. We assume it to be true for all $n \leq k$, and prove it true for $n = k + 1$. We have

$$\begin{aligned} p_{k+1} &= \frac{\lambda_k + \mu_k}{\mu_{k+1}} p_0 \prod_{i=1}^{k} \frac{\lambda_{i-1}}{\mu_i} - \frac{\lambda_{k-1}}{\mu_{k+1}} p_0 \prod_{i=1}^{k-1} \frac{\lambda_{i-1}}{\mu_i} \\ &= p_0 \frac{\lambda_k}{\mu_{k+1}} \prod_{i=1}^{k} \frac{\lambda_{i-1}}{\mu_i} + p_0 \frac{\mu_k}{\mu_{k+1}} \prod_{i=1}^{k} \frac{\lambda_{i-1}}{\mu_i} - p_0 \left[ \prod_{i=1}^{k} \frac{\lambda_{i-1}}{\mu_i} \right] \frac{\mu_k}{\mu_{k+1}} \\ &= p_0 \prod_{i=1}^{k+1} \frac{\lambda_{i-1}}{\mu_i}. \end{aligned}$$

We shall return to this "product-form" equation frequently in the remainder of this chapter. Indeed, it has been referred to as the principal equation in elementary queueing theory. Notice that by returning to Equation (11.9) and rearranging, we get

$$\begin{aligned} \lambda_n p_n - \mu_{n+1} p_{n+1} &= \lambda_{n-1} p_{n-1} - \mu_n p_n \\ &= \lambda_{n-2} p_{n-2} - \mu_{n-1} p_{n-1} \\ &= \cdots \\ &= \lambda_0 p_0 - \mu_1 p_1 = 0. \end{aligned}$$

Thus

$$\lambda_n p_n - \mu_{n+1} p_{n+1} = 0 \quad \text{or} \quad p_{n+1} = \frac{\lambda_n}{\mu_{n+1}} p_n, \quad \text{for } n \geq 0 \tag{11.10}$$

and

$$p_{n+1} = \frac{\lambda_n}{\mu_{n+1}} \frac{\lambda_{n-1}}{\mu_n} p_{n-1}, \quad n \geq 1$$

which eventually yields

$$p_{n+1} = \frac{\lambda_n}{\mu_{n+1}} \frac{\lambda_{n-1}}{\mu_n} \cdots \frac{\lambda_0}{\mu_1} p_0 = p_0 \prod_{i=0}^{n} \frac{\lambda_i}{\mu_{i+1}}, \quad n \geq 0$$

$$p_n = p_0 \prod_{i=0}^{n-1} \frac{\lambda_i}{\mu_{i+1}} = p_0 \prod_{i=1}^{n} \frac{\lambda_{i-1}}{\mu_i}, \quad n \geq 1.$$

Recall that the set of differential-difference equations that represent the dynamics of a general birth-death process can be obtained by examining the flow into and out of each individual state.

Consider what happens to that flow once the system reaches equilibrium. Since the system is in equilibrium, the rate into any state must be equal to the rate of flow out of that state. More generally, the rate of flow into and out of any closed set of states must possess this same property of conservation of probability flow. Let us choose special sets of states that allow us to generate Equation (11.10). Specifically, let us choose the following sequence of sets of states: the first set contains state 0, the second set contains states 0 and 1, the third contains state 0, 1 and 2, and so on, as is illustrated in Figure 11.14. Observe now that equating the flow of across the $n^{\text{th}}$ boundary gives us what we need (Equation11.10), namely

$$\lambda_{n-1} p_{n-1} = \mu_n p_n.$$

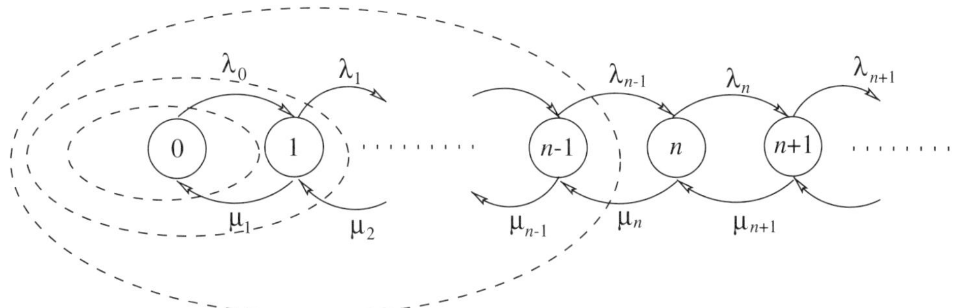


Figure 11.14. Transitions across groups of states.

This method of generating the equilibrium balance equations turns out to be useful in many different circumstances. The reader might wonder why we cluster the states in this fashion and observe the flow into and out of each cluster, rather than simply draw a vertical line to separate neighboring states and equate flow across this line. In this particular case, there is no difference, but in more complex situations, for example, when states are drawn as a two-dimensional grid, separating states with a single line becomes more onerous and error-prone.

We now turn our attention to the computation of $p_0$. We have

$$1 = \sum_{n\geq 0} p_n = p_0 + \sum_{n\geq 1} p_n = p_0 + \sum_{n\geq 1} p_0 \prod_{i=1}^{n} \frac{\lambda_{i-1}}{\mu_i} = p_0 \left[ 1 + \sum_{n\geq 1} \prod_{i=1}^{n} \frac{\lambda_{i-1}}{\mu_i} \right].$$

Setting $\rho_i = \lambda_{i-1}/\mu_i$, we obtain

$$p_0 = \frac{1}{1 + \sum_{n\geq 1} \prod_{i=1}^{n} \rho_i}.$$

To simplify the notation somewhat, let

$$\xi_k = \begin{cases} 1, & k = 0, \\ \prod_{i=1}^{k} \rho_i, & k > 0. \end{cases}$$

Then

$$p_n = p_0 \xi_n = \frac{\xi_n}{\sum_{k=0}^{\infty} \xi_k} \quad \text{for } n = 0, 1, 2, \ldots.$$

The limiting distribution $(p_0, p_1, p_2, \ldots)$ is now completely determined. The limiting probabilities are nonzero provided that $p_0 > 0$; otherwise $p_i = 0$ for all $i = 0, 1, 2, \ldots$ and there is no steady-state distribution. Thus the existence of a stationary distribution depends on whether or not there exists a finite probability that the queueing system empties itself. If it does then the steady-state distribution exists, otherwise it does not. To characterize the states of the system we need to introduce two quantities, namely,

$$\alpha = \sum_{k=0}^{\infty} \xi_k = 1 + \frac{\lambda_0}{\mu_1} + \frac{\lambda_0\lambda_1}{\mu_1\mu_2} + \frac{\lambda_0\lambda_1\lambda_2}{\mu_1\mu_2\mu_3} + \cdots \quad \text{and} \quad \beta = \sum_{k=0}^{\infty} \left( \frac{1}{\lambda_k \xi_k} \right). \tag{11.11}$$

The states of a general birth-death process are

- transient if and only if $\beta < \infty$,
- null recurrent if and only if $\alpha = \infty$ and $\beta = \infty$,
- ergodic if and only if $\alpha < \infty$ .

The last of these conditions is what we need for the existence of a steady-state distribution. However, from a practical point of view, it is usually easier to simply show that $p_0 > 0$. An alternative, and equally facile condition for ergodicity, is to show that there exists an integer $k_0$ and a constant $A < 1$ such that

$$\frac{\lambda_k}{\mu_{k+1}} \le A < 1 \quad \text{for all } k \ge k_0.$$

In this case, there is a point in the state space such that the rate of flow of arrivals into any state at or beyond this point is less than the rate of departures from that state. This effectively prevents the population from growing without bound.

**Example 11.10** For the *M/M/*1 queue, with $\lambda_n = \lambda$ and $\mu_n = \mu$ for all $n$ and with $\lambda/\mu = \rho < 1$, we find

$$\alpha = \sum_{k=0}^{\infty} \xi_k = 1 + \sum_{k=1}^{\infty} \prod_{i=1}^{k} \frac{\lambda_{i-1}}{\mu_i} = 1 + \sum_{k=1}^{\infty} \prod_{i=1}^{k} \frac{\lambda}{\mu} = 1 + \sum_{k=1}^{\infty} \rho^k = \sum_{k=0}^{\infty} \rho^k = \frac{1}{1-\rho} < \infty,$$

which shows that the *M/M/*1 queue is stable when $\rho < 1$. Notice also that

$$\beta = \sum_{k=0}^{\infty} \left( \frac{1}{\lambda_k \xi_k} \right) = \frac{1}{\lambda} \sum_{k=0}^{\infty} \frac{1}{\rho^k} = \infty$$

since $\lim_{k\to\infty} \rho^k = 0$.

### Matrix Formulation for Birth-Death Processes

As for the *M/M/*1 queue, we can write the infinite infinitesimal generator of a birth-death process with birth rates $\lambda_k$ and death rates $\mu_k$:

$$Q = \begin{pmatrix} -\lambda_0 & \lambda_0 & & & \\ \mu_1 & -(\lambda_1+\mu_1) & \lambda_1 & & \\ & \mu_2 & -(\lambda_2+\mu_2) & \lambda_2 & \\ & & \mu_3 & -(\lambda_3+\mu_3) & \lambda_3 \\ & & & \ddots & \ddots & \ddots \end{pmatrix}.$$

From $\pi Q = 0$ , it is obvious that $-\lambda_0\pi_0 + \mu_1\pi_1 = 0$, i.e., that $\pi_1 = (\lambda_0/\mu_1)\pi_0$. In general, we have

$$\lambda_{i-1}\pi_{i-1} - (\lambda_i + \mu_i)\pi_i + \mu_{i+1}\pi_{i+1} = 0,$$

i.e.,

$$\pi_{i+1} = \frac{\lambda_i + \mu_i}{\mu_{i+1}}\pi_i - \frac{\lambda_{i-1}}{\mu_{i+1}}\pi_{i-1},$$

and from which, by induction, we may derive

$$\pi_{k+1} = \pi_0 \prod_{i=1}^{k+1} \frac{\lambda_{i-1}}{\mu_i}.$$

Thus, once again, if $\pi_0$ is known, the remaining values $\pi_i$, $i = 1, 2, \ldots$, may be determined recursively.

## 11.4 Multiserver Systems

### *11.4.1 The M/M/c Queue*

Substitution of $\lambda_n = \lambda$ and $\mu_n = \mu$ for all $n$ turns the general birth-death system into the *M/M/*1 queue. In the sections that follow, we will analyze a number of important queueing systems, simply by assigning particular values to $\lambda_n$ and $\mu_n$ and solving the resulting birth-death equations. We begin with the case in which there are $c$ identical servers, the *M/M/c* queue. This is sometimes also referred to as a queue with parallel channels. Figure 11.15 illustrates this situation.

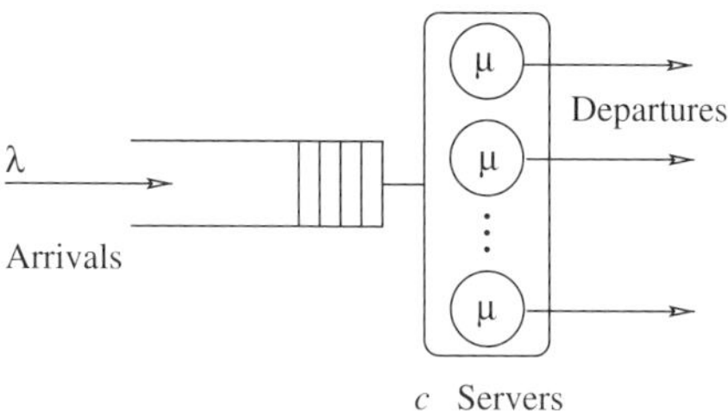


Figure 11.15. The *M/M/c* queue.

Customers arrive according to a Poisson process with rate $\lambda_n = \lambda$ for all $n$ and are served in first-come first-served order by any available server. Each of the $c$ servers provides independent and identically distributed exponential service at rate $\mu$. When the number of customers is greater than or equal to $c$, i.e., $n \geq c$, then all the servers are busy and the *effective* service rate, also called the mean system output rate (MSOR), is equal to $c\mu$. If the number of customers is less than $c$, then only $n$ out of the $c$ servers are busy and the MSOR is equal to $n\mu$. The term *load dependent* service center is used to designate a service center in which the customers departure rate is a function of the number of customers present. With $c$ identical exponential servers, each providing service at rate $\mu$, and a total of $n$ customers present, the load dependent service rate, written $\mu(n)$, is $\mu(n) = \min(n, c)\mu$. The state transition diagram is shown in Figure 11.16.

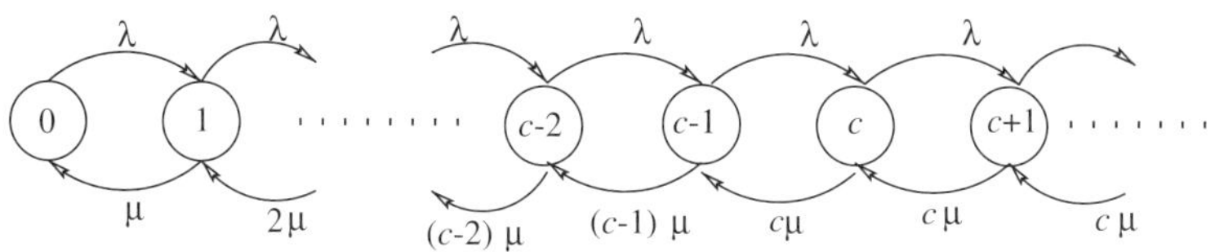


Figure 11.16. State transitions in the *M/M/c* queue.

It follows that the appropriate birth and death rates to define the *M/M/c* queue are given by

$$\lambda_n = \lambda \quad \text{for all } n,$$

$$\mu_n = \begin{cases} n\mu, & 1 \leq n \leq c, \\ c\mu, & n \geq c. \end{cases}$$

The original birth-death system gave rise to the equations

$$p_n = p_0 \prod_{i=1}^{n} \frac{\lambda_{i-1}}{\mu_i},$$

which in the *M*/*M*/*c* queue become

$$p_n = p_0 \prod_{i=1}^{n} \frac{\lambda}{i\mu} = p_0 \left(\frac{\lambda}{\mu}\right)^n \frac{1}{n!} \quad \text{if } 1 \le n \le c,$$

and

$$p_n = p_0 \prod_{i=1}^{c} \frac{\lambda}{i\mu} \prod_{i=c+1}^{n} \frac{\lambda}{c\mu} = p_0 \left(\frac{\lambda}{\mu}\right)^n \frac{1}{c!} \left(\frac{1}{c}\right)^{n-c} \quad \text{if } n \ge c.$$

We define $\rho = \lambda/(c\mu)$, and in order for the system to be stable, we must have $\rho < 1$. This implies that the mean arrival rate must be less than the mean maximum potential rate with which customers can be served. This expression for $\rho$ is consistent with our definition in terms of the expected fraction of busy servers, the utilization in an *M*/*M*/*c* queue, since the expected number of busy servers is equal to $c\rho = \lambda/\mu$.

Returning to the previous equation and substituting $c\rho$ for $\lambda/\mu$, we obtain

$$p_n = p_0 \frac{(c\rho)^n}{n!} \quad \text{for } n \le c,$$

$$p_n = p_0 \frac{(c\rho)^n}{c^{n-c} c!} = p_0 \frac{\rho^n c^c}{c!} \quad \text{for } n \ge c.$$

All that remains is to solve for $p_0$:

$$1 = \sum_{n=0}^{\infty} p_n = p_0 + \sum_{n=1}^{\infty} p_n = p_0 \left[1 + \sum_{n=1}^{c-1} \frac{(c\rho)^n}{n!} + \sum_{n=c}^{\infty} \frac{\rho^n c^c}{c!}\right],$$

i.e.,

$$p_0 = \left[1 + \sum_{n=1}^{c-1} \frac{(c\rho)^n}{n!} + \sum_{n=c}^{\infty} \frac{\rho^n c^c}{c!}\right]^{-1}.$$

Since

$$\sum_{n=c}^{\infty} \frac{\rho^n c^c}{c!} = \frac{1}{c!} \sum_{n=c}^{\infty} \rho^n c^c = \frac{(c\rho)^c}{c!} \sum_{n=c}^{\infty} \rho^{n-c} = \frac{(c\rho)^c}{c!} \frac{1}{1-\rho},$$

$$p_0 = \left[1 + \sum_{n=1}^{c-1} \frac{(c\rho)^n}{n!} + \frac{(c\rho)^c}{c!} \frac{1}{1-\rho}\right]^{-1}.$$

This gives us what we need. To summarize, the steady-state distribution of customers in an *M*/*M*/*c* queue is given by

$$p_n = p_0 \frac{(c\rho)^n}{n!} \quad \text{for } n \le c,$$

$$p_n = p_0 \frac{(c\rho)^n}{c^{n-c} c!} = p_0 \frac{\rho^n c^c}{c!} \quad \text{for } n \ge c,$$

where

$$p_0 = \left[ \sum_{n=0}^{c-1} \frac{(c\rho)^n}{n!} + \frac{(c\rho)^c}{c!} \frac{1}{1-\rho} \right]^{-1}.$$

**Performance Measures for the *M/M/c* Queue**

It is clearly easier to begin by computing the expected queue length $L_q$ rather than the mean number in the system, since the only $p_n$'s involved are those for which $n \geq c$. These are the only states for which customers must wait in the queue. We have

$$L_q = \sum_{n=c}^{\infty} (n-c) p_n$$

with

$$p_n = \frac{(\rho c)^n}{c^{n-c} c!} p_0 \quad \text{for } n \geq c,$$

which implies that

$$L_q = \sum_{n=c}^{\infty} \frac{n}{c^{n-c} c!} (\rho c)^n p_0 - \sum_{n=c}^{\infty} \frac{c}{c^{n-c} c!} (\rho c)^n p_0.$$

Let us consider each of the terms on the right-hand side separately. The first term is given by

$$\frac{p_0}{c!} \sum_{n=c}^{\infty} \frac{n(\rho c)^n}{c^{n-c}}.$$

Notice that, since

$$\frac{n\rho^n c^n}{c^{n-c}} = n\rho^n c^c = \frac{n\rho^{n-c-1} \rho^{c+1} c^{c+1}}{c} = \frac{(\rho c)^{c+1}}{c} n\rho^{n-c-1},$$

we have

$$\begin{aligned} \frac{p_0}{c!} \sum_{n=c}^{\infty} \frac{n(\rho c)^n}{c^{n-c}} &= \frac{p_0}{c!} \left[ \frac{(\rho c)^{c+1}}{c} \right] \sum_{n=c}^{\infty} \left[ (n-c)\rho^{n-c-1} + c\rho^{n-c-1} \right] \\ &= \frac{p_0}{c!} \left[ \frac{(\rho c)^{c+1}}{c} \right] \left\{ \sum_{n=c}^{\infty} (n-c)\rho^{n-c-1} + \sum_{n=c}^{\infty} c\rho^{n-c-1} \right\} \\ &= \frac{p_0}{c!} \left[ \frac{(\rho c)^{c+1}}{c} \right] \left\{ \frac{1}{(1-\rho)^2} + \frac{c}{\rho} \frac{1}{1-\rho} \right\}, \end{aligned}$$

using derivatives of the geometric series,

$$= \frac{p_0}{c!} \left[ \frac{(\rho c)^{c+1}}{c} \right] \left\{ \frac{1}{(1-\rho)^2} + \frac{c/\rho}{1-\rho} \right\}.$$

Now for the second term. We have

$$\begin{aligned} \frac{p_0}{c!} \sum_{n=c}^{\infty} \frac{c(\rho c)^n}{c^{n-c}} &= \frac{p_0}{c!} \sum_{n=c}^{\infty} c\rho^c \rho^{n-c} c^c = \frac{p_0}{c!} c(\rho c)^c \sum_{n=c}^{\infty} \rho^{n-c} \\ &= \frac{p_0}{c!} \frac{c(\rho c)^c}{1-\rho} = \frac{p_0}{c!} \left[ \frac{(\rho c)^{c+1}}{c} \right] \frac{c/\rho}{1-\rho}. \end{aligned}$$

Bringing these two terms together we obtain the mean number of customers waiting in the queue as

$$L_q = \frac{p_0}{c!}\left[\frac{(\rho c)^{c+1}}{c}\right]\left\{\frac{1}{(1-\rho)^2} + \frac{c/\rho}{1-\rho} - \frac{c/\rho}{1-\rho}\right\}$$

and thus

$$L_q = \frac{(\rho c)^{c+1}/c}{c!(1-\rho)^2}p_0$$

or, alternatively,

$$L_q = \frac{(\lambda/\mu)^c\lambda\mu}{(c-1)!(c\mu-\lambda)^2}p_0.$$

Having computed $L_q$, we are now in a position to find other performance measures, either by means of Little's law or from simple relationships among the measures. To find $W_q$, the mean time spent waiting prior to service; $W$, the mean time spent in the system, and $L$, the mean number of customers in the system, we proceed as follows:

1. Use $L_q = \lambda W_q$ to find $W_q$.
2. Use $W = W_q + 1/\mu$ to find $W$.
3. Use $L = \lambda W$ to find $L$.

We obtain

$$W_q = \left[\frac{(\lambda/\mu)^c\mu}{(c-1)!(c\mu-\lambda)^2}\right]p_0,$$

$$W = \left[\frac{(\lambda/\mu)^c\mu}{(c-1)!(c\mu-\lambda)^2}\right]p_0 + \frac{1}{\mu},$$

$$L = \left[\frac{(\lambda/\mu)^c\lambda\mu}{(c-1)!(c\mu-\lambda)^2}\right]p_0 + \frac{\lambda}{\mu}.$$

The probability that an arriving customer is forced to wait in the queue, which means that there is no server available, leads to the "Erlang-$C$ formula." It is the probability that all servers are busy and is given by

$$\begin{aligned}\text{Prob}\{\text{queueing}\} &= \sum_{n=c}^{\infty} p_n = p_0\sum_{n=c}^{\infty}\frac{c^c}{c!}\rho^n = p_0\frac{c^c}{c!}\left[\frac{\rho^c}{1-\rho}\right]\\ &= \frac{(c\rho)^c}{c!(1-\rho)}p_0 = \frac{(\lambda/\mu)^c\mu}{(c-1)!(c\mu-\lambda)}p_0.\end{aligned}$$

This then is the Erlang-$C$ formula. It is denoted by $C(c, \lambda/\mu)$. Observe that mean queue/system lengths and mean waiting times can all be written in terms of this formula. We have

$$L_q = \frac{(\rho c)^{c+1}/c}{c!(1-\rho)^2}p_0 = \frac{(\rho c)^c}{c!(1-\rho)}p_0 \times \frac{\rho}{(1-\rho)} = \frac{\rho}{(1-\rho)}C(c, \lambda/\mu) = \frac{\lambda}{c\mu-\lambda}C(c, \lambda/\mu).$$

$$\begin{aligned}W_q &= \frac{1}{\lambda}\frac{\rho C(c,\lambda/\mu)}{(1-\rho)} = \frac{C(c,\lambda/\mu)}{c\mu-\lambda},\\ W &= \frac{1}{\lambda}\frac{\rho C(c,\lambda/\mu)}{(1-\rho)} + \frac{1}{\mu} = \frac{C(c,\lambda/\mu)}{c\mu-\lambda} + \frac{1}{\mu},\\ L &= \frac{\rho C(c,\lambda/\mu)}{(1-\rho)} + c\rho = \frac{\lambda C(c,\lambda/\mu)}{c\mu-\lambda} + \frac{\lambda}{\mu}.\end{aligned}$$

**Example 11.11** Let us compare, on the basis of average response time, the performance of two identical servers each with its own separate queue, to the case when there is only a single queue to hold customers for both servers. The systems to compare are illustrated in Figure 11.17. We shall also check to see how these two possibilities compare to a single processor working twice as fast.

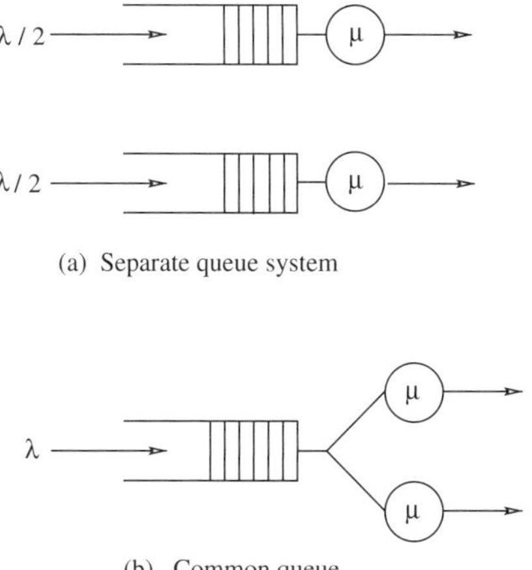


Figure 11.17. Two configurations.

In the first case, we have two independent *M/M/*1 queues, each with arrival rate $\lambda/2$ and service rate $\mu$. It follows that $\rho = (\lambda/2)/\mu = \lambda/(2\mu)$. The mean number in each *M/M/*1 queue is given by $\rho/(1-\rho)$ so that the mean number of customers in this first scenario is given as

$$L_1 = E[N_1] = 2 \times \frac{\rho}{1-\rho} = \frac{2\rho}{1-\rho}.$$

The average response time can now be found using Little's law. We have

$$E[R_1] = \frac{1}{\lambda} E[N_1] = \frac{1}{\lambda} \frac{2\rho}{(1-\rho)} = \frac{2}{2\mu - \lambda}.$$

Now consider the second scenario in which the system may be represented as an *M/M/*2 queue. To find $E[R_2]$, we first must find $E[N_2]$ $(= L_2)$. The mean number of customers in an *M/M/c* queue with arrival rate $\lambda$ and service rate $\mu$ per server is given by

$$E[N_2] = \frac{\lambda}{\mu} + \frac{(\lambda/\mu)^c \lambda\mu}{(c-1)!(c\mu - \lambda)^2} p_0 \quad \text{with} \quad \frac{\lambda}{c\mu} = \rho \quad \text{or} \quad \lambda/\mu = c\rho.$$

With $c = 2$, we obtain

$$\begin{aligned} L_2 = E[N_2] = \frac{\lambda}{\mu} + \frac{(\lambda/\mu)^2 \lambda\mu}{(2\mu - \lambda)^2} p_0 &= \frac{\lambda}{\mu} + \frac{(\lambda/\mu)^2 (\lambda/\mu)}{(1/\mu^2)(2\mu - \lambda)^2} p_0 \\ &= \frac{\lambda}{\mu} + \frac{(\lambda/\mu)^3}{(2 - \lambda/\mu)^2} p_0 \\ &= 2\rho + \frac{(2\rho)^3}{(2 - 2\rho)^2} p_0. \end{aligned}$$

The probability of the system being empty, $p_0$, is computed as

$$p_0 = \left[1 + \sum_{n=1}^{c-1} \frac{(c\rho)^n}{n!} + \frac{(c\rho)^c}{c!}\left(\frac{1}{1-\rho}\right)\right]^{-1}$$

$$= \left[1 + 2\rho + \frac{(2\rho)^2}{2!}\frac{1}{1-\rho}\right]^{-1} = \frac{1-\rho}{1+\rho}.$$

Thus,

$$L_2 = 2\rho + \frac{8\rho^3(1-\rho)}{4(1-\rho)^2(1+\rho)} = \frac{2\rho(1-\rho)(1+\rho)+2\rho^3}{(1-\rho)(1+\rho)} = \frac{2\rho}{1-\rho^2}.$$

Now, using Little's formula,

$$E[R_2] = \frac{1}{\lambda}E[N_2] = \frac{2\rho/\lambda}{1-\rho^2} = \frac{1/\mu}{1-\rho^2} = \frac{4\mu}{4\mu^2-\lambda^2}.$$

Finally, to compare these to a single superserver working twice as fast, we use the *M*/*M*/1 queue with arrival rate $\lambda$ and service rate $2\mu$. We obtain

$$L_3 = E[N_3] = \frac{\rho}{1-\rho} \quad \text{and} \quad E[R_3] = \frac{1/2\mu}{1-\lambda/2\mu} = \frac{1}{2\mu-\lambda}.$$

These results are summarized below.

$$E[N_1] = \frac{2\rho}{1-\rho} \quad \ge \quad E[N_2] = \frac{2\rho}{1-\rho}\cdot\frac{1}{1+\rho} \quad \ge \quad E[N_3] = \frac{\rho}{1-\rho},$$

$$E[R_1] = \frac{2}{2\mu-\lambda} \quad \ge \quad E[R_2] = \frac{1/\mu}{1-\rho^2} \quad \ge \quad E[R_3] = \frac{1}{2\mu-\lambda},$$

or, setting $\alpha = 2\rho/(1-\rho)$ and $\beta = 2/(2\mu-\lambda)$—notice that $\alpha = \lambda\beta$,

$$E[N_1] = \alpha \quad \ge \quad E[N_2] = \alpha\cdot\frac{1}{1+\rho} \quad \ge \quad E[N_3] = \alpha/2,$$

$$E[R_1] = \beta \quad \ge \quad E[R_2] = \beta\cdot\frac{1}{1+\rho} \quad \ge \quad E[R_3] = \beta/2.$$

These results show that the first scenario, that of two separate *M*/*M*/1 queues, is the worst of the three scenarios and indeed is twice as bad as the single super *M*/*M*/1 queue which works twice as fast. The second scenario, that of an *M*/*M*/2 queue, performs somewhere between the other two. Since $0 \le \rho \le 1$, we must have $1 \le 1+\rho \le 2$, meaning that the performance of the *M*/*M*/2 queue approaches that of the pair of *M*/*M*/1 queues as the traffic intensity decreases ($\rho \to 0$), and approaches that of the super *M*/*M*/1 queue when the traffic intensity increases,[2] i.e., as $\rho \to 1$.

Results similar to those in this illustrative example are readily obtained when the number of separate, independent *M*/*M*/1 queues is increased to $c$, the *M*/*M*/2 queue replaced by an *M*/*M*/*c* queue, and the single super *M*/*M*/1 queue works at rate $c\mu$. In particular, the response time of $c$ individual *M*/*M*/1 queues is exactly $c$ times as long as that of the super *M*/*M*/1 queue. These results are also intuitive. One should expect $c$ independent *M*/*M*/1 queues to perform worse than a single *M*/*M*/*c* queue, since, with the multiple queues, it is possible for some to be working and others to be idle. Furthermore, situations will evolve in which there are even waiting lines in front of some of the servers of the individual *M*/*M*/1 queues, while other servers are idle. As for the *M*/*M*/*c* queue

[2] Be aware, when interpreting these results, that $\alpha$ (and hence $\beta$) also depends on $\rho$.

and the super $M/M/1$ queue, both will perform identically as long as there are at least $c$ customers present, since in this case all $c$ servers will be busy. The super $M/M/1$ queue will outperform the $M/M/c$ queue when the number of customers is strictly less than $c$, since not all of the $c$ servers in the $M/M/c$ queue can be working while the super $M/M/1$ queue continues to operate at rate $c\mu$. Thus a single queue is better that a separate queue for each server. This minimizes variations in $W_q$ and $L_q$, and as a general rule, customers prefer less variation—a truism even when it means a longer overall wait. Also, one fast server is better than two servers fed by a single queue each working half as fast. This result, however, does not consider the question of reliability, of the possibility of some of the servers breaking down and becoming unavailable.

### 11.4.2 The M/M/∞ Queue

We now consider the case when $c$ tends to infinity. This gives the $M/M/\infty$ queue, also called the infinite server, responsive server or ample servers queue or queues with unlimited service. Examples are self-service-type situations. We have

$$\lambda_n = \lambda \quad \text{for all } n, \qquad \mu_n = n\mu \quad \text{for all } n,$$

which leads to

$$p_n = \frac{\lambda^n}{n\mu(n-1)\mu \cdots 2\mu 1\mu} p_0 = \frac{\lambda^n}{n!\mu^n} p_0.$$

Since

$$\sum_{n=0}^{\infty} p_n = 1,$$

we obtain

$$p_0 = \left[ \sum_{n=0}^{\infty} \frac{\lambda^n}{n!\mu^n} \right]^{-1} = e^{-\lambda/\mu},$$

and this is applicable for all finite $\lambda/\mu$. Therefore

$$p_n = \frac{(\lambda/\mu)^n e^{-\lambda/\mu}}{n!},$$

which is the Poisson distribution with parameter $\lambda/\mu$. Notice that $\lambda/\mu$ is not in any way restricted for the existence of a steady-state solution. This result also holds for the $M/G/\infty$ queue. The probability $p_n$ depends only on the mean of the service time distribution.

Since no customer will ever wait in an $M/M/\infty$ queue, the mean time spent in the system must be equal to the mean service time $1/\mu$, and so the mean number of customers present must be equal to $\lambda$ times this or $L = \lambda/\mu$. This is also easy to verify from first principles since

$$L = \sum_{n=1}^{\infty} np_n = e^{-\lambda/\mu} \sum_{n=1}^{\infty} \frac{(\lambda/\mu)^n}{(n-1)!} = e^{-\lambda/\mu} \frac{\lambda}{\mu} \sum_{n=1}^{\infty} \frac{(\lambda/\mu)^{n-1}}{(n-1)!} = \frac{\lambda}{\mu}.$$

Obviously $L_q = 0$ and $W_q = 0$, so that $W = 1/\mu$. It is also obvious that the response time distribution must be identical to the service time distribution. Since there is no waiting time before entering service, an $M/M/\infty$ queue is occasionally referred to as a *delay* (sometimes *pure delay*) station, the probability distribution of the delay being that of the service time.

## 11.5 Finite-Capacity Systems—The *M/M/1/K* Queue

We now turn our attention to the *M/M/1/K* queue, illustrated graphically in Figure 11.18. Customers arrive according to a Poisson process at rate $\lambda$ and receive service that is exponentially distributed with a mean service time of $1/\mu$ from a single server. The difference from the *M/M/1* queue is that at most $K$ customers are allowed into the system. A customer who arrives to find the system full simply disappears. That customer is not permitted to enter the system. There is no halt to the arrival process, which continues to churn out customers according to a Poisson process at rate $\lambda$, but only those who arrive to find strictly fewer than $K$ present are permitted to enter. In communication systems, rejected customers are called "lost" customers and the system is called a "loss" system. In the particular case of $K = 1$, in which a customer is permitted to enter only if the server is idle, the term "blocked calls cleared" is used. In other contexts, the *M/M/1/K* queue is referred to as a queue with truncation.

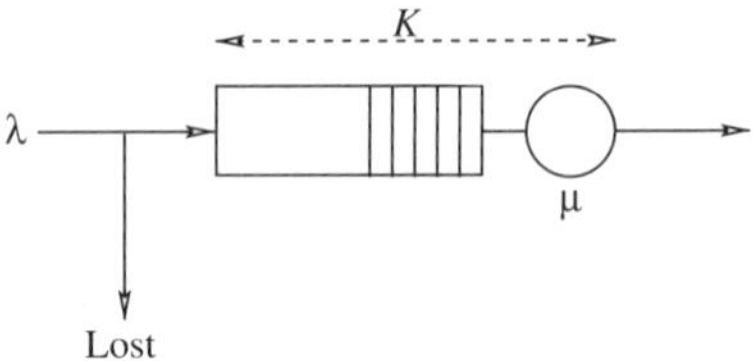


Figure 11.18. The *M/M/1/K* queue.

To analyze the *M/M/1/K* queue, we use the birth-death model and choose the parameters so that the Poisson arrivals stop as soon as $K$ customers are present. The parameters of the exponential service time distribution are unchanged. Thus

$$\lambda_n = \begin{cases} \lambda, & n < K, \\ 0, & n \geq K, \end{cases}$$

$$\mu_n = \mu \quad \text{for } n = 1, 2, \ldots, K.$$

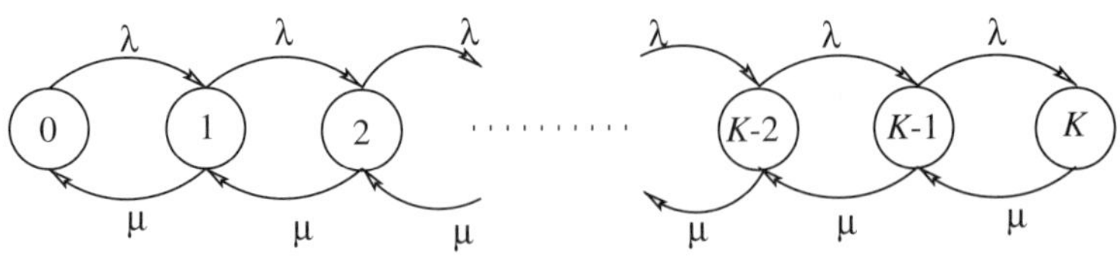


Figure 11.19. State transitions in *M/M/1/K* queue.

The state transition diagram is shown in Figure 11.19. The principal equation for birth-death processes is given by

$$p_n = p_0 \prod_{i=1}^{n} \frac{\lambda_{i-1}}{\mu_i}, \quad n \geq 1,$$

and this equation continues to hold in the case when the state space is truncated. Since this continuous-time Markov chain is finite and irreducible for all strictly positive $\lambda$ and $\mu$, it follows that all states are ergodic. We have

$$p_1 = \frac{\lambda}{\mu} p_0,$$

$$p_{n+1} = \frac{\lambda + \mu}{\mu} p_n - \frac{\lambda}{\mu} p_{n-1}, \quad 1 \le n < K - 1,$$

$$p_K = \frac{\lambda}{\mu} p_{K-1}.$$

Thus, for all possible values of $n$, and setting $\rho = \lambda/\mu$,

$$p_n = \left(\frac{\lambda}{\mu}\right)^n p_0 = \rho^n p_0, \quad 0 \le n \le K.$$

Notice, in particular, that this holds when $n = K$. We now need to compute $p_0$. We use

$$\sum_{n=0}^{K} p_n = 1$$

and thus

$$p_0 = \frac{1}{\sum_{n=0}^{K} \rho^n}.$$

The divisor is a finite geometric series and

$$\sum_{n=0}^{K} \rho^n = \frac{1 - \rho^{K+1}}{1 - \rho} \quad \text{if } \rho \ne 1$$

and is equal to $K + 1$ if $\rho = 1$. Hence

$$p_0 = \begin{cases} (1 - \rho)/(1 - \rho^{K+1}) & \text{if } \rho \ne 1 \\ 1/(K + 1) & \text{if } \rho = 1. \end{cases}$$

Therefore, for all $0 \le n \le K$,

$$p_n = \frac{(1 - \rho)\rho^n}{1 - \rho^{K+1}} \quad \text{if } \rho \ne 1,$$

and is equal to $1/(K + 1)$ if $\rho = 1$. Thus, the steady-state solution always exists, even for $\rho \ge 1$. The system is stable for all positive values of $\lambda$ and $\mu$. When $\lambda > \mu$, the number of customers in the system will increase, but it is bound from above by $K$. Also notice that $\rho$ no longer represents the utilization. We shall derive an expression for this in just a moment. Finally, notice what happens as $K \to \infty$ and $\rho < 1$. We have

$$\lim_{K \to \infty} \frac{1 - \rho}{1 - \rho^{K+1}} \rho^n = (1 - \rho)\rho^n,$$

which is the result previously obtained for the *M/M/1* queue.

**Example 11.12** Perhaps the simplest example we can present is the *M/M/1/1* queue. This gives rise to a two-state birth–death process, as illustrated in Figure 11.20.

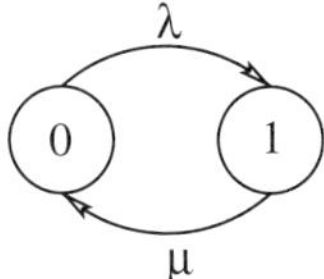


Figure 11.20. State transition diagram for the *M/M/1/1* queue.

Given the following results for the *M/M/1/K* queue:

$$p_n = \left(\frac{\lambda}{\mu}\right)^n p_0 \quad \text{for} \quad n \geq 1 \quad \text{and} \quad p_0 = \left[\sum_{n=0}^{K} \rho^n\right]^{-1},$$

and applying them to the *M/M/1/1* queue, we have

$$p_0 = \frac{1}{1+\rho} \quad \text{and} \quad p_1 = \frac{\rho}{1+\rho}.$$

The *M/M/1/1* example has application to the so-called machine breakdown problem, wherein a machine (or machine component) fails and is subsequently repaired. State 0 is used to represent the machine in its working condition while state 1 represents the machine undergoing repair. To qualify as a birth-death process, it is necessary that both failures and repairs occur according to exponential distributions with rates $\lambda$ and $\mu$, respectively. Thus the mean time to failure (MTTF) is equal to $1/\lambda$, the mean time to repair (MTTR) is equal to $1/\mu$, and the steady-state availability of the machine is just the steady-state probability of the system being in state 0. The availability is therefore given by

$$\frac{1}{1+\rho} = \frac{1}{1+\lambda/\mu} = \frac{1/\lambda}{1/\lambda + 1/\mu} = \frac{\text{MTTF}}{\text{MTTF} + \text{MTTR}}.$$

**Performance Measures for the *M/M/1/K* Queue**

We consider the case when $\rho \neq 1$, and let $L$ be the expected system size. Then

$$L = \sum_{n=0}^{K} n p_n \quad \text{and} \quad p_n = \rho^n p_0.$$

We have

$$\begin{aligned}
L = p_0\rho \sum_{n=0}^{K} n\rho^{n-1} &= p_0\rho \sum_{n=0}^{K} \frac{d}{d\rho}(\rho^n) = p_0\rho \frac{d}{d\rho}\left(\sum_{n=0}^{K} \rho^n\right) \\
&= p_0\rho \left(\frac{d}{d\rho}\left[\frac{1-\rho^{K+1}}{1-\rho}\right]\right) = p_0\rho \left(\frac{[1-\rho][-(K+1)\rho^K] + 1 - \rho^{K+1}}{(1-\rho)^2}\right) \\
&= p_0\rho \left(\frac{-(K+1)\rho^K + (K+1)\rho^{K+1} + 1 - \rho^{K+1}}{(1-\rho)^2}\right) \\
&= p_0\rho \left(\frac{1 - (K+1)\rho^K + K\rho^{K+1}}{(1-\rho)^2}\right).
\end{aligned}$$

Using

$$p_0 = \frac{1-\rho}{1-\rho^{K+1}}$$

we find

$$\begin{aligned}
L &= \frac{\rho\left[1 - (K+1)\rho^K + K\rho^{K+1}\right]}{(1-\rho)^2} \frac{1-\rho}{1-\rho^{K+1}} \\
&= \frac{\rho\left[1 - (K+1)\rho^K + K\rho^{K+1}\right]}{(1-\rho)(1-\rho^{K+1})}.
\end{aligned}$$

To compute $L_q$, we use

$$L_q = L - (1 - p_0) = L - \left(1 - \frac{1-\rho}{1-\rho^{K+1}}\right) = L - \frac{1 - \rho^{K+1} - 1 + \rho}{1-\rho^{K+1}} = L - \frac{\rho(1-\rho^K)}{1-\rho^{K+1}}.$$

To find $W$, we use Little's law, $(W = L/\lambda)$. However, what we need is the *effective* arrival rate, which we denote by $\lambda'$. This is the mean rate of customers actually entering the system and is given by

$$\lambda' = \lambda(1 - p_K)$$

since customers only enter when space is available. $W$ and $W_q$ may now be obtained from Little's law,

$$W = \frac{1}{\lambda'}L \quad \text{and} \quad W_q = \frac{1}{\lambda'}L_q.$$

Alternatively, we could also use

$$W_q = W - \frac{1}{\mu}.$$

Notice that in computing $L_q$ we did not use the formula $L_q = L - \rho$ as we did in the *M/M/*1 queue with $\rho = \lambda/\mu$, instead choosing to write this as $L_q = L - (1 - p_0)$. The reason should now be clear, as we require the effective arrival rate $\lambda'$ and not $\lambda$. Thus, we could have used $L_q = L - \lambda'/\mu$.

**Throughput and Utilization in the *M/M/1/K* Queue**

In a queueing situation in which customers may be lost, the throughput cannot be defined as being equal to the customer arrival rate—not all arriving customers actually enter the queue. The probability that an arriving customer is lost is equal to the probability that there are already $K$ customers in the system, i.e., $p_K$. The probability that the queue is *not* full, and hence the probability that an arriving customer is accepted into the queueing system is $1 - p_K$. Thus the throughput, $X$, is given by

$$X = \lambda(1 - p_K).$$

It is interesting to examine the throughput in terms of what actually leaves the *M/M/1/K* queue. So long as there are customers present, the server serves these customers at rate $\mu$. The probability that no customers are present is $p_0$, so the probability that the server is working is given by $1 - p_0$. It follows that the throughput must be given by $X = \mu(1 - p_0)$. Thus

$$X = \lambda(1 - p_K) = \mu(1 - p_0).$$

As we have just seen,

$$p_0 = \frac{(1-\rho)}{(1-\rho^{K+1})}$$

and

$$p_n = \frac{(1-\rho)\rho^n}{1-\rho^{K+1}}.$$

Therefore

$$\frac{1-p_0}{1-p_K} = \frac{1 - \left[(1-\rho)/(1-\rho^{K+1})\right]}{1 - \left[(1-\rho)\rho^K/(1-\rho^{K+1})\right]} = \frac{(1-\rho^{K+1}) - (1-\rho)}{(1-\rho^{K+1}) - (1-\rho)\rho^K} = \frac{\rho - \rho^{K+1}}{1-\rho^K} = \rho$$

and hence

$$\lambda(1 - p_K) = \mu(1 - p_0).$$

Observe that $(1 - p_K)$ is the probability that the queue is not full: it is the probability that new customers can enter the system and $\lambda(1-p_K)$ is the effective arrival rate, the rate at which customers enter the system. Similarly, $(1 - p_0)$ is the probability that the system is busy and so $\mu(1 - p_0)$ is the effective departure rate from the system. The equation $\lambda(1 - p_K) = \mu(1 - p_0)$ simply says that the effective rate into and out of the system are the same at steady state.

The utilization of the *M/M/1/K* queue is the probability that the server is busy. We have just seen that this is given by

$$U = 1 - p_0 = \frac{1}{\mu}X = \rho(1 - p_K).$$

Finally, observe that $X = \mu U$.

**Example 11.13** In a multiprogramming system, computer processes utilize a common CPU and an I/O device. After computing for an exponentially distributed amount of time with mean $1/\mu$, a process either joins the queue at the I/O device (with probability $r$) or exits the system (with probability $1 - r$). At the I/O device, a process spends an exponentially distributed amount of time with mean $1/\lambda$ and then joins the CPU queue. Successive CPU and I/O bursts are assumed to be independent of each other. The number of processes allowed in the system at any time is fixed at a constant degree of multiprogramming equal to $K$—the system is set up so that as soon as a process exits after a CPU burst, an identical process enters the CPU queue. Since the processes are assumed to be stochastically identical, this may be represented by feeding a departing process back into the CPU queue. This situation is shown graphically in Figure 11.21.

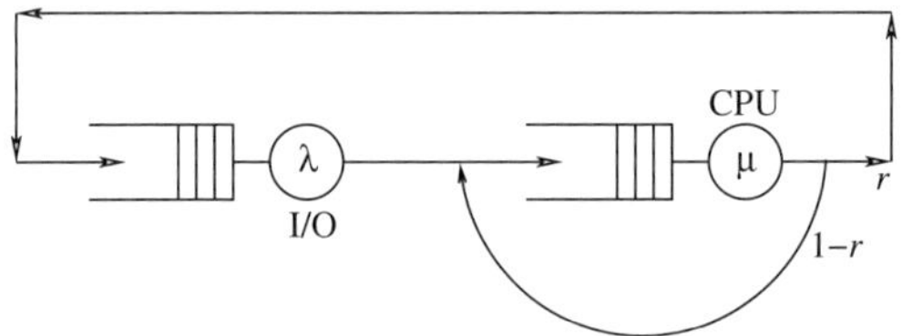


Figure 11.21. Model of a multiprogramming system.

Let the number of processes in the CPU denote the state of the system. It is not necessary to simultaneously give the number of processes in the CPU queue and at the I/O device to completely characterize a state of this system. If there are $n$ processes at the CPU, there must be $K - n$ at the I/O device. Thus we use the number of processes at the CPU to represent the state of the system. Figure 11.22 displays the state transition diagram. Notice that this is identical to the *M/M/1/K* queue with $\mu$ replaced by $r\mu$.

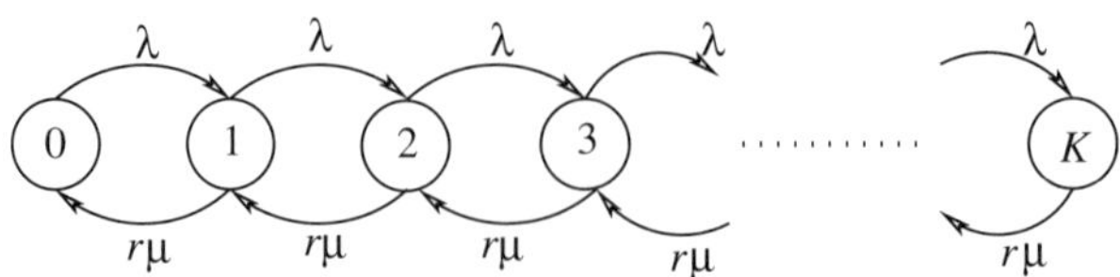


Figure 11.22. State transitions for a multiprogramming system.

Setting $\rho = \lambda/(r\mu)$, the steady-state probability is given by

$$p_i = (\rho)^i p_0,$$

with

$$p_0 = \left[\sum_{i=0}^{K} \rho^i\right]^{-1} = \begin{cases} (1-\rho)/(1-\rho^{K+1}), & \rho \neq 1, \\ 1/(K+1), & \rho = 1. \end{cases}$$

The CPU utilization, $U_{\text{CPU}} = 1 - p_0$, is given by

$$U_{\text{CPU}} = \frac{\rho - \rho^{K+1}}{1 - \rho^{K+1}} \quad \text{if } \rho \neq 1,$$

$$U_{\text{CPU}} = \frac{K}{K+1} \quad \text{if } \rho = 1.$$

To calculate the throughput, we need to be careful about using the previous definition given for the *M/M/1/K* queue, namely that $X = \mu U$. In the multiprogramming scenario, the throughput should relate to the processes that actually leave the system rather than those which depart from the CPU. Whenever the CPU is busy (given by $U_{\text{CPU}}$), it pushes processes through at a rate of $\mu$ per time unit but of these only a fraction (given by $1 - r$) actually depart from the system. Therefore the average system throughput is given by $(1 - r)\mu\, U_{\text{CPU}}$.

**Transient behavior of the *M/M/1/1* Queue**

It is relatively easy to compute the transient behavior of the *M/M/1/K* queue when only one customer is allowed into the system ($K = 1$). In this case the birth-death process has only two states, one in which no one is present in the system, and the second in which the server is busy serving a customer and no other customer is allowed to enter. The state transition diagram is shown in Figure 11.23.

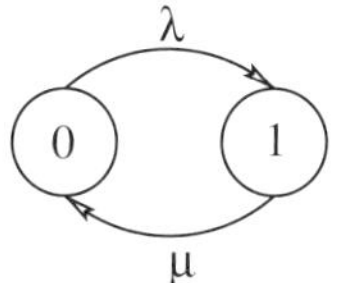


Figure 11.23. State transitions in the *M/M/1/1* queue.

We have

$$p_1(t) + p_0(t) = 1 \quad \text{at any time } t$$

and

$$\frac{dp_0(t)}{dt} = -\lambda p_0(t) + \mu p_1(t),$$

$$\frac{dp_1(t)}{dt} = -\mu p_1(t) + \lambda p_0(t).$$

Also,

$$p_0(t) = 1 - p_1(t) \quad \text{and} \quad p_1(t) = 1 - p_0(t)$$

and so

$$\frac{dp_1(t)}{dt} = -\mu p_1(t) + \lambda(1 - p_1(t)),$$

$$\frac{dp_1(t)}{dt} + (\lambda + \mu)p_1(t) = \lambda.$$

This is a first-order, linear, ordinary differential equation with constant coefficients. We begin by multiplying through with $e^{(\lambda+\mu)t}$ to obtain

$$\frac{dp_1(t)}{dt} e^{(\lambda+\mu)t} + (\lambda + \mu)e^{(\lambda+\mu)t} p_1(t) = \lambda e^{(\lambda+\mu)t}.$$

Observe that the left-hand side is the derivative of the product of $p_1(t)$ and $e^{(\lambda+\mu)t}$ and therefore

$$\frac{d}{dt}\left(e^{(\lambda+\mu)t} p_1(t)\right) = \lambda e^{(\lambda+\mu)t}.$$

Integrating, we obtain

$$e^{(\lambda+\mu)t}p_1(t) = \frac{\lambda}{\lambda+\mu}e^{(\lambda+\mu)t} + c$$

and so the general solution is given as

$$p_1(t) = \frac{\lambda}{\lambda+\mu} + ce^{-(\lambda+\mu)t}.$$

We shall assume the initial condition, $p_1(t) = p_1(0)$ at $t = 0$, which results in

$$p_1(0) = c + \frac{\lambda}{\lambda+\mu}$$

and thus

$$c = p_1(0) - \frac{\lambda}{\lambda+\mu}.$$

The complete solution is

$$\begin{aligned} p_1(t) &= \frac{\lambda}{\lambda+\mu} + \left(p_1(0) - \frac{\lambda}{\lambda+\mu}\right)e^{-(\lambda+\mu)t} \\ &= \frac{\lambda}{\lambda+\mu}\left(1 - e^{-(\lambda+\mu)t}\right) + p_1(0)e^{-(\lambda+\mu)t}, \end{aligned}$$

and

$$p_0(t) = 1 - p_1(t) = \frac{\mu}{\lambda+\mu}\left[1 - e^{-(\lambda+\mu)t}\right] + p_0(0)e^{-(\lambda+\mu)t}.$$

To obtain the steady-state solution from this, we let $t \to \infty$ and obtain

$$p_1 = \lim_{t\to\infty} p_1(t) = \frac{\lambda}{\lambda+\mu}$$

since

$$\lim_{t\to\infty} e^{-(\lambda+\mu)t} = 0.$$

Also, this must mean that

$$p_0 = \frac{\mu}{\lambda+\mu}.$$

## 11.6 Multiserver, Finite-Capacity Systems—The *M/M/c/K* Queue

We now proceed to a combination of two of the previous queueing systems, the *M/M/c* queue and the *M/M/1/K* queue. The resulting queue is denoted *M/M/c/K*, otherwise known as a queue with parallel channels and truncation, or the "*c*-server loss system." The state transition diagram for this system is shown in Figure 11.24.

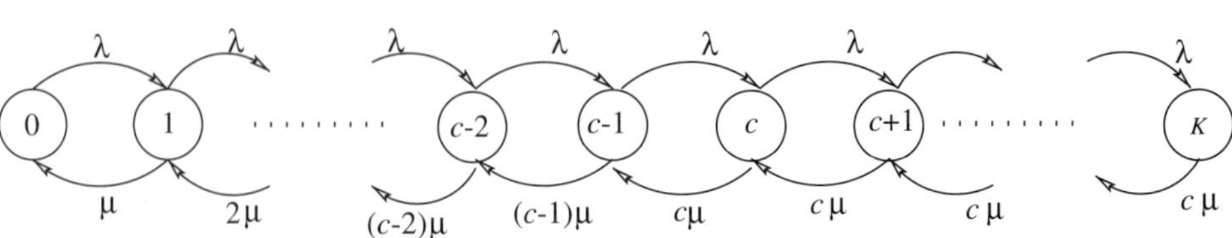


Figure 11.24. State transitions in an *M/M/c/K* queue.

As we mentioned earlier, the principal equation in elementary queueing theory, the steady-state solution of a general birth-death process, is given by

$$p_n = p_0 \prod_{i=1}^{n} \frac{\lambda_{i-1}}{\mu_i}$$

in which $p_0$ is computed from $\sum_{n=0}^{\infty} p_n = 1$. We have seen that results for a variety of simple queueing systems are obtained from this by substituting appropriate values for $\lambda_n$ and $\mu_n$. In particular, for the *M/M/c* queue, we substituted as follows

$$\lambda_n = \lambda \; \forall \, n, \qquad \mu_n = \begin{cases} n\mu, & 1 \le n \le c, \\ c\mu, & n \ge c, \end{cases}$$

while for the *M/M/1/K* queue we used

$$\lambda_k = \begin{cases} \lambda, & k < K, \\ 0, & k \ge K, \end{cases} \qquad \mu_k = \mu \quad \text{for all } k.$$

For the *M/M/c/K* queue, the values for the arrival and service rates are taken to be

$$\lambda_n = \begin{cases} \lambda, & 0 \le n < K, \\ 0, & n \ge K, \end{cases} \qquad \mu_n = \begin{cases} n\mu, & 1 \le n \le c, \\ c\mu, & c \le n \le K. \end{cases}$$

Returning to the fundamental equation,

$$p_n = p_0 \prod_{i=1}^{n} \frac{\lambda_{i-1}}{\mu_i},$$

and making the appropriate substitutions, we get

$$p_n = \frac{\lambda^n}{n\mu(n-1)\mu \cdots 2\mu 1\mu} p_0, \quad 0 \le n \le c,$$
$$p_n = \frac{\lambda^n}{c\mu c\mu \cdots c\mu c\mu(c-1)\mu(c-2)\mu \cdots 2\mu 1\mu} p_0, \quad c \le n \le K,$$

where the $c\mu$ in the denominator appears $n - c + 1$ times. Thus

$$p_n = \frac{1}{n!}\left(\frac{\lambda}{\mu}\right)^n p_0, \quad 0 \le n \le c,$$

$$p_n = \frac{1}{c^{n-c}c!}\left(\frac{\lambda}{\mu}\right)^n p_0, \quad c \le n \le K.$$

Using

$$\sum_{i=0}^{K} p_i = 1,$$

we find that

$$p_0 = \left[\sum_{n=0}^{c-1} \frac{1}{n!}\left(\frac{\lambda}{\mu}\right)^n + \sum_{n=c}^{K} \frac{1}{c^{n-c}c!}\left(\frac{\lambda}{\mu}\right)^n\right]^{-1}.$$

The interested reader may wish to check that the case $c = 1$ leads to the results previously obtained for the *M/M/1/K* queue, and that letting $K$ tend to infinity reproduces the *M/M/c* results.

To determine performance measures, we first compute

$$L_q = \sum_{n=c}^{K}(n-c)p_n = p_0\frac{(\lambda/\mu)^c\,\lambda/c\mu}{c!(1-\lambda/c\mu)^2}\left[1-\left(\frac{\lambda}{c\mu}\right)^{K-c+1} - \left(1-\frac{\lambda}{c\mu}\right)(K-c+1)\left(\frac{\lambda}{c\mu}\right)^{K-c}\right]$$

$$= p_0\frac{(c\rho)^c\rho}{c!(1-\rho)^2}\left[1-\rho^{K-c+1}-(1-\rho)(K-c+1)\rho^{K-c}\right],$$

where we have defined $\rho = \lambda/(c\mu)$. This allows us to obtain the average number of customers in the system (see also Exercise 11.6.1):

$$L = L_q + \left(c - \sum_{n=0}^{c-1}(c-n)p_n\right) = L_q + c - \sum_{n=0}^{c-1}(c-n)\frac{(\lambda/\mu)^n}{n!}p_0, \tag{11.12}$$

which is the number waiting in the queue plus the average number of busy servers. In the special case of $c = 1$, this reduces to

$$L = L_q + (1-p_0) = L_q + \frac{\lambda'}{\mu},$$

where $\lambda' = \lambda(1-p_K)$ is the effective arrival rate. The time spent in the system can now obtained from $W = L/\lambda'$. Finally, the mean time spent in the queue waiting for service to begin is found from $W_q = W - 1/\mu$ or $W_q = L_q/\lambda'$.

**Erlang's Loss Formula**

For the special case when $K = c$, i.e., the *M/M/c/c* queue, the so-called "blocked calls cleared with $c$ servers" system, these results lead to another well-known formula associated with the name Erlang. The probability that there are $n$ customers in the *M/M/c/c* queue is given as

$$p_n = \frac{(\lambda/\mu)^n/n!}{\sum_{i=0}^{c}(\lambda/\mu)^i/i!} = p_0\frac{(\lambda/\mu)^n}{n!}.$$

The formula for $p_c$ is called "Erlang's loss formula" and is the fraction of time that all $c$ servers are busy. It is written as $B(c, \lambda/\mu)$ and called "Erlang's $B$ formula":

$$B(c,\lambda/\mu) = \frac{(\lambda/\mu)^c/c!}{\sum_{i=0}^{c}(\lambda/\mu)^i/i!} = p_0\frac{(\lambda/\mu)^c}{c!}.$$

Notice that the probability that an arrival is lost is equal to the probability that all channels are busy. Erlang's loss formula is also valid for the *M/G/c/c* queue. In other words, the steady-state probabilities are a function only of the mean service time, and not of the complete underlying cumulative distribution function. An efficient recursive algorithm for computing $B(c, \lambda/\mu)$ is given by

$$B(0,\lambda/\mu) = 1, \quad B(c,\lambda/\mu) = \frac{(\lambda/\mu)B(c-1,\lambda/\mu)}{c+(\lambda/\mu)B(c-1,\lambda/\mu)}.$$

We leave its proof as an exercise. It is also possible to express Erlang's C formula in terms of $B(c, \lambda/\mu)$ and again, we leave this as an exercise.

## 11.7 Finite-Source Systems—The *M/M/c//M* Queue

The *M/M/c//M* queue is a $c$-server system with a finite customer population. No longer do we have a Poisson input process with an infinite user population. There is a total of $M$ customers and a customer is either in the system, or else is outside the system and is in some sense "arriving." When a customer is in the arriving condition, then the time it takes for that particular customer to arrive is a random variable with exponential distribution whose mean is $1/\lambda$. This is the mean time spent per customer in the "arriving" condition. All customers act independently of each other so that if there are $k$ customers in the "arriving" state, the total average arrival rate is $k\lambda$.

**Example 11.14** Consider the situation of $M$ students in a computer lab, each student in front of a computer terminal. All terminals are connected to a single ($c = 1$) central processor. Each student thinks for a period of time that is exponentially distributed at rate $\lambda$, then enters a command and waits for a reply: students alternate between a "thinking" state and an "idle" state. At a time when the processor is handling $k$ requests, there must be $M-k$ students thinking so that the rate of requests to the processor is equal to $(M - k)\lambda$. Observe that the arrival process itself may be modeled as an *M/M/*$\infty$ queue. When a student receives a reply from the processor, that student does not need to queue before beginning the thinking process. The response from the central processor adds one unit (one thinking student) to the *M/M/*$\infty$ queue.

To summarize then, the total average arrival rate to an *M/M/c//M* queue in state $n$ is $\lambda(M - n)$. Also, each of the $c$ servers works at rate $\mu$. It follows that the parameters to insert into the birth-death equations are given by

$$\lambda_n = \begin{cases} \lambda(M - n), & 0 \le n < M, \\ 0, & n \ge M, \end{cases}$$

$$\mu_n = \begin{cases} n\mu, & 0 \le n \le c, \\ c\mu, & n \ge c. \end{cases}$$

From the fundamental equation

$$p_n = p_0 \prod_{i=1}^{n} \frac{\lambda_{i-1}}{\mu_i},$$

we obtain

$$p_n = \frac{M!/(M-n)!}{n!}\left(\frac{\lambda}{\mu}\right)^n p_0 = \binom{M}{n}\left(\frac{\lambda}{\mu}\right)^n p_0 \quad \text{if} \quad 0 \le n \le c, \tag{11.13}$$

$$p_n = \frac{M!/(M-n)!}{c^{n-c}c!}\left(\frac{\lambda}{\mu}\right)^n p_0 = \binom{M}{n}\frac{n!}{c^{n-c}c!}\left(\frac{\lambda}{\mu}\right)^n p_0 \quad \text{if} \quad c \le n \le M, \tag{11.14}$$

where

$$\binom{M}{n} = \frac{M!}{n!(M-n)!}$$

is the binomial coefficient.

As usual, $p_0$ is found from the normalization equation and is given by

$$p_0 = \left[1 + \sum_{n=1}^{c-1}\binom{M}{n}\left(\frac{\lambda}{\mu}\right)^n + \sum_{n=c}^{M}\binom{M}{n}\frac{n!}{c^{n-c}c!}\left(\frac{\lambda}{\mu}\right)^n\right]^{-1}.$$

A special case occurs with $c = 1$, the single-server problem. In this case, the formulae are

$$p_n = p_0\left(\frac{\lambda}{\mu}\right)^n \frac{M!}{(M-n)!}, \quad 0 \le n \le M,$$

with $p_n = 0$ when $n > M$. Also, in this case ($c = 1$)

$$p_0 = \left[ \sum_{n=0}^{M} \left(\frac{\lambda}{\mu}\right)^n \frac{M!}{(M-n)!} \right]^{-1}.$$

In the general case, ($c \geq 1$), the average number of customers in the system, may be computed from $L = \sum_{n=0}^{M} np_n$ using Equations (11.13) and (11.14). We have

$$L = \sum_{n=0}^{M} np_n = \sum_{n=0}^{c-1} n \binom{M}{n} \left(\frac{\lambda}{\mu}\right)^n p_0 + \sum_{n=c}^{M} n \binom{M}{n} \frac{n!}{c^{n-c}c!} \left(\frac{\lambda}{\mu}\right)^n p_0$$

$$= p_0 \left[ \sum_{n=0}^{c-1} n \binom{M}{n} \left(\frac{\lambda}{\mu}\right)^n + \frac{1}{c!} \sum_{n=c}^{M} n \binom{M}{n} \frac{n!}{c^{n-c}} \left(\frac{\lambda}{\mu}\right)^n \right].$$

There is no neater expression for $L$. First $p_0$ must be calculated and then multiplied by the sum of two series. When $L$ has been computed, it then becomes possible to compute other quantities such as

$$L_q = L - \frac{\lambda'}{\mu}, \quad W = \frac{L}{\lambda'}, \quad W_q = \frac{L_q}{\lambda'},$$

where $\lambda'$ is equal to the mean arrival rate into the system. To compute this *effective* arrival rate, observe that, when $n$ customers are present, the arrival rate is $\lambda(M - n)$ and thus,

$$\lambda' = \sum_{n=0}^{M} \lambda(M-n)p_n = \lambda M \sum_{n=0}^{M} p_n - \lambda \sum_{n=0}^{M} np_n = \lambda M - \lambda L = \lambda(M - L).$$

Hence

$$L_q = L - \frac{\lambda(M-L)}{\mu}, \quad W = \frac{L}{\lambda(M-L)}, \quad W_q = \frac{L_q}{\lambda(M-L)},$$

**Example 11.15** The machine repairman problem.

In the classical version of this problem there are $M$ machines that are subject to breakdown and a single repairman to fix them. Each machine operates independently of the other machines and its time to failure is governed by an exponential distribution with parameter $\lambda$. Thus each machine fails at rate $\lambda$ per unit time. The single repair man requires $1/\mu$ times units to repair a failed machine and again the repair process is independent of the failure process and has an exponential distribution. If we allow $i$ to denote the case in which $i$ machines have failed, then the state transition diagram is as shown in Figure 11.25. Observe that the rate of transition from state $i$, in which $M - i$ machines are functioning, to state $i + 1$ is $(M - i)\lambda$. It is apparent that this system falls into the category of the *M/M/1//M* queue, the finite population queueing model.

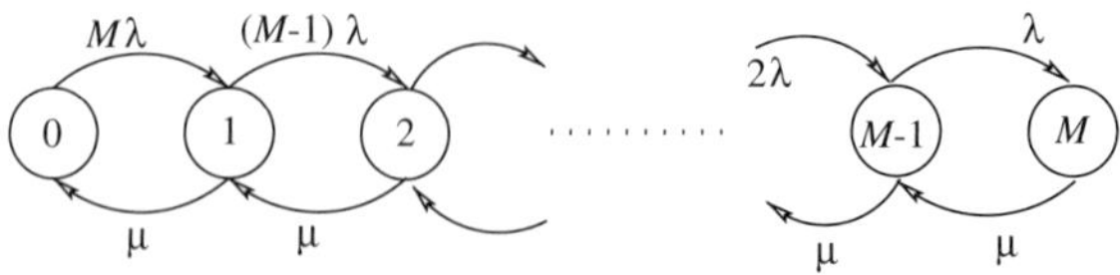


Figure 11.25. State transitions for machine repairman problem.

If there are $c$ repair men, rather than just one, then we obtain the *M/M/c//M* queue. Systems like this are also called *parallel redundant system*. The system is considered to be functional if there is at least one component working correctly. The objective is to compute the system availability, the probability that the system is available, i.e., the probability $1 - p_M$. For the *M/M/1//M* queue, the availability is given by

$$A = 1 - p_M = 1 - p_0 \left(\frac{\lambda}{\mu}\right)^M M! = 1 - \frac{(\lambda/\mu)^M M!}{\sum_{k=0}^{M} (\lambda/\mu)^k \frac{M!}{(M-k)!}}.$$

For example, if the mean time to repair a component is one-fifth the mean time to failure, then $\lambda/\mu = .2$ and we obtain the availability for a system with different numbers of components.

| $M$ | 1 | 2 | 3 | 4 | 5 | ... | 10 |
|---|---|---|---|---|---|---|---|
| Availability | .833333 | .945946 | .974576 | .984704 | .989061 | ... | .993168 |

Obviously, the availability of the system must increase as the number of components increases. In most practical situations the value of $\lambda/\mu$ is considerably smaller than $1/5$ and just a small number of components is sufficient to give a system availability very close to one.

## 11.8 State-Dependent Service

As a final example of the type of queueing system that lends itself to analysis through the general birth-death equations, we consider systems in which the mean service rate depends on the state of the system. We consider the case in which the server works at rate $\mu_1$ until there are $k$ customers in the system, at which point it changes to a different rate $\mu_2$. The mean service times now depend explicitly on the system state. We have

$$\lambda_n = \lambda \quad \text{for all } n,$$

$$\mu_n = \begin{cases} \mu_1, & 1 \le n < k, \\ \mu_2, & n \ge k. \end{cases}$$

The extension to three, four, or more different rates is straightforward. The balance equations become

$$p_n = \left(\frac{\lambda}{\mu_1}\right)^n p_0 \quad \text{if } 0 \le n < k,$$

$$p_n = \frac{\lambda^n}{\mu_1^{k-1}\mu_2^{n-k+1}} p_0 \quad \text{if } n \ge k,$$

and

$$p_0 = \left[\sum_{n=0}^{k-1} \left(\frac{\lambda}{\mu_1}\right)^n + \sum_{n=k}^{\infty} \frac{\lambda^n}{\mu_1^{k-1}\mu_2^{n-k+1}}\right]^{-1} = \left[\sum_{n=0}^{k-1} \left(\frac{\lambda}{\mu_1}\right)^n + \sum_{n=k}^{\infty} \frac{\lambda^{k-1}}{\mu_1^{k-1}} \frac{\lambda^{n-k+1}}{\mu_2^{n-k+1}}\right]^{-1}$$

$$= \left[\sum_{n=0}^{k-1} \left(\frac{\lambda}{\mu_1}\right)^n + \rho_1^{k-1}\rho_2 \sum_{n=k}^{\infty} \rho_2^{n-k}\right]^{-1}$$

where $\rho_1 = \lambda/\mu_1$ and $\rho_2 = \lambda/\mu_2 < 1$. This gives

$$p_0 = \left[\frac{1 - \rho_1^k}{1 - \rho_1} + \frac{\rho_2 \rho_1^{k-1}}{1 - \rho_2}\right]^{-1}.$$

Notice that if $\mu_1 = \mu_2$, we get the usual *M/M/1* queue.

To find the expected system size, we have

$$L = \sum_{n=0}^{\infty} np_n = p_0 \left[ \sum_{n=0}^{k-1} n\rho_1^n + \sum_{n=k}^{\infty} n\rho_1^{k-1}\rho_2^{n-k+1} \right]$$

$$= p_0 \left[ \sum_{n=0}^{k-1} n\rho_1^n + \rho_1^{k-1}\rho_2 \sum_{n=k}^{\infty} n\rho_2^{n-k} \right]$$

$$= p_0 \left[ \sum_{n=0}^{k-1} n\rho_1^n + \rho_1^{k-1}\rho_2 \left( \sum_{n=k}^{\infty} (n-k)\rho_2^{n-k} + \sum_{n=k}^{\infty} k\rho_2^{n-k} \right) \right]$$

$$= p_0 \left[ \sum_{n=0}^{k-1} n\rho_1^n + \rho_1^{k-1}\rho_2 \left( \sum_{n=0}^{\infty} n\rho_2^n + k \sum_{n=0}^{\infty} \rho_2^n \right) \right].$$

The first term within the brackets can be handled as follows:

$$\sum_{n=0}^{k-1} n\rho_1^n = \rho_1 \sum_{n=0}^{k-1} \frac{d}{d\rho_1}\rho_1^n = \rho_1 \frac{d}{d\rho_1}\left( \frac{1-\rho_1^k}{1-\rho_1} \right).$$

The remaining terms can be treated in a similar fashion and the entire equation simplifies to

$$L = p_0 \left[ \frac{\rho_1(1 + (k-1)\rho_1^k - k\rho_1^{k-1})}{(1-\rho_1)^2} + \frac{\rho_2\rho_1^{k-1}[k - (k-1)\rho_2]}{(1-\rho_2)^2} \right].$$

From this we can calculate

$$L_q = L - (1 - p_0), \quad W = L/\lambda, \quad \text{and} \quad W_q = L_q/\lambda.$$

Here we cannot use the relationship $W = W_q + 1/\mu$, since $\mu$ is not constant.

## 11.9 Exercises

**Exercise 11.1.1** Customers arrive at a service center according to a Poisson process with a mean interarrival time of 15 minutes.

(a) What is the probability that no arrivals occur in the first half hour?
(b) What is the expected time until the tenth arrival occurs?
(c) What is the probability of more than five arrivals occurring in any half-hour period?
(d) If two customers were observed to have arrived in the first hour, what is the probability that both arrived in the last 10 minutes of that hour?
(e) If two customers were observed to have arrived in the first hour, what is the probability that at least one arrived in the last 10 minutes of that hour?

**Exercise 11.1.2** Superposition of Poisson streams is Poisson: let $X_j,\ j = 1, 2, \ldots, n$ be $n$ independent Poisson processes with parameters $\lambda_1, \lambda_2, \ldots, \lambda_n$ respectively and let $X = \sum_{j=1}^{n} X_j$. By expanding the $z$-transform of $X$, $E[z^X]$, show that $X$ is a Poisson process. Recall that the $z$-transform of the Poisson process $X_j$ is given as $e^{\alpha_j(z-1)}$, where here we take $\alpha = \lambda_j t$.

**Exercise 11.1.3** Decomposition of a Poisson stream is Poisson: consider a Poisson process $X$ with parameter $\lambda$ that is decomposed into $k$ separate Poisson streams $X_j,\ j = 1, 2, \ldots, k$. The decomposition is such that a customer in the original stream is directed into sub-stream $X_j$ with probability $p_j$ and $\sum_{j=1}^{k} p_j = 1$.

Conditioning on $n$, the number of arrivals in the original stream up to some time $t$, show first that

$$\text{Prob}\{X_1 = n_1,\ X_2 = n_2,\ \ldots,\ X_n = n_k \mid X = n\} = \frac{n!}{n_1!n_2!\cdots n_k!} p_1^{n_1} p_2^{n_2} \cdots p_k^{n_k},$$

where $\sum_{j=1}^{k} n_j = n$. Now remove the condition to obtain the desired result, that

$$\text{Prob}\{X_1 = n_1,\ X_2 = n_2,\ \ldots,\ X_n = n_k\} = \prod_{j=1}^{k} \frac{(\lambda p_j t)^{n_j}}{n_j!} e^{-\lambda p_j t},$$

which shows that the $k$ decomposed streams are indeed Poisson processes with parameters $\lambda p_j$.

**Exercise 11.1.4** Let $(\tau_i, x_i)$ , for $i = 1, 2, \ldots, 12$ denote the arrival time and service requirement of the first 12 customers to arrive at a service center that contains two identical servers. Use the third graphical approach to draw this situation given the following arrivals and services:

$$(1, 10);\ (2, 6);\ (4, 1);\ (6, 6);\ (7, 2);\ (9, 4);\ (10, 6);\ (11, 3);\ (15, 2);\ (24, 2);\ (25, 5);\ (28, 4).$$

Indicate on your graph, the arrival and departure instant of each customer and the busy and idle periods for each server.

**Exercise 11.1.5** The single server at a service center appears to be busy for four minutes out of every five, on average. Also the mean service time has been observed to be equal to half a minute. If the time spent waiting for service to begin is equal to 2.5 minutes, what is the mean number of customers in the queueing system and the mean response time?

**Exercise 11.2.1** It is known that the solution of a set of second-order difference equations with constant coefficients such as

$$0 = -\lambda x_0 + \mu x_1,$$
$$0 = \lambda x_{n-1} - (\lambda + \mu) x_n + \mu x_{n+1}, \quad n \geq 1,$$

is given by

$$x_n = \alpha_1 x_1^n + \alpha_2 x_2^n, \quad n = 0, 1, 2, \ldots,$$

where $r_1$ and $r_2$ are the roots of the quadratic polynomial equation

$$\mu r^2 - (\lambda + \mu) r + \lambda = 0.$$

Use this information to compute the equilibrium solution of the *M*/*M*/1 queue.

**Exercise 11.2.2** The transmission rate of a communication node is 480,000 bits per second. If the average size of an arriving data packet is 120 8-bit characters and the average arrival rate is 18,000 packets per minute, find (a) the average number of packets in the node, (b) the average time spent in the node by a packet and (c) the probability that the number of packets in the node exceeds 5. Assume the arrival process to be Poisson and the service time to be exponentially distributed.

**Exercise 11.2.3** Each evening Bin Peng, a graduate teaching assistant, works the takeaway counter at Goodberry's frozen yogurt parlor. Arrivals to the counter appear to follow a Poisson distribution with mean of ten per hour. Each customer is served one at a time by Bin and the service time appears to follow an exponential distribution with a mean service time of four minutes.

(a) What is the probability of having a queue?
(b) What is the average queue length?
(c) What is the average time a customer spends in the system?
(d) How fast, on average, does Bin need to serve customers in order for the average total time a customer spends in Goodberry's to be less than 7.5 minutes?
(e) If Bin can spend his idle time grading papers and if he can grade, on average, 24 papers an hour, how many papers per hour he can average while working at Goodberry's?

**Exercise 11.2.4** Consider a barber's shop with a single barber who takes on average 15 minutes to cut a client's hair. Model this situation as an *M/M/*1 queue and find the largest number of incoming clients per hour that can be handled so that the average waiting time will be less than 12 minutes.

**Exercise 11.2.5** Customers arrive at Bunkey's car wash service at a rate of one every 20 minutes and the average time it takes for a car to proceed through their single wash station is 8 minutes. Answer the following questions under the assumption of Poisson arrivals and exponential service.

(a) What is the probability that an arriving customer will have to wait?
(b) What is the average number of cars waiting to begin their wash?
(c) What is the probability that there are more than five cars altogether?
(d) What is the probability that a customer will spend more than 12 minutes actually waiting before her car begins to be washed?
(e) Bunkey is looking to expand and he can justify the creation of a second wash station so that two cars can be washed simultaneously if the average time spent waiting prior to service exceeds 8 minutes. How much does the arrival rate have to increase in order for Bunkey to justify installing this second wash station?

**Exercise 11.2.6** Let $W_r(t) = \text{Prob}\{R \leq t\}$ be the response time distribution function in an *M/M/*1 queue. Derive an expression for $W_r(t)$ in terms of the probability of an arriving customer finding $n$ customers already present. Now use the PASTA property to show that

$$W_r(t) = 1 - e^{-(\mu-\lambda)t}.$$

**Exercise 11.3.1** The rate at which unemployed day workers arrive at a job employment center is a function of the number already present: they reason that the larger the number already present, the less likely they are to find a job that day. Let this function be such that when $n$ workers are already in the job queue, the arrival rate is given by $\alpha^n\lambda$ for some $\alpha < 1$. The system is to be modeled as a birth-death process with constant departure rate equal to $\mu$.

(a) Draw the state transition rate diagram for this system.
(b) Show that this system is stable.
(c) What are the equilibrium equations?
(d) Solve the equilibrium equations to obtain the steady-state probabilities.
(e) What is the traffic intensity $\rho$?

**Exercise 11.3.2** An *M/M/*1 queue with discouraged arrivals has parameters

$$\lambda_n = \frac{\lambda}{n+1} \quad \text{and} \quad \mu_n = \mu \quad \text{for all relevant } n.$$

(a) Does the PASTA property apply to this queue? Justify your answer.
(b) Use Equation (11.11) to find the conditions necessary for all states to be ergodic (and which guarantees the stability of this queue).
(c) Compute the stationary distribution under the assumption that the conditions of part (b) apply.
(d) What is the effective arrival rate needed for the application of Little's law?
(e) Find the mean number in the queueing system and the mean response time.

**Exercise 11.3.3** Consider a birth-death process with the following birth and death rates:

$$\lambda_n = \begin{cases} \lambda_1 + \lambda_2, & n < K, \\ \lambda_1, & n \geq K, \end{cases} \qquad \mu_n = \mu \quad \text{for all } n.$$

Such a process may be used to model a communication system consisting of a single channel that handles two different types of packets: voice packets that arrive at rate $\lambda_1$ and data packets that arrive at rate $\lambda_2$. Data packets are refused once the total number of packets reaches a preset limit $K$. The sizes of both voice and data packets are assumed to be exponentially distributed with mean equal to $1/\mu$.

Draw the state transition diagram for this birth-death process and then derive and solve the steady-state balance equations.

**Exercise 11.3.4** Consider a two-server queueing system in which one of the servers is faster than the other. Both servers provide exponential service, the first at rate $\mu_1$ and the second at rate $\mu_2$. We asume that $\mu_1 > \mu_2$. Let $\lambda$ be the arrival rate. Define a state of the system to be the pair $(n_1, n_2)$ where $n_1 \geq 0$ is the number of customers waiting in the queue plus 1 if the faster server is busy, and $n_2 \in \{0, 1\}$ is the number of customers being served by the slower server. For example, the state $(5, 1)$ means that both servers are busy and four customers are waiting for one of the servers to become free; the state $(0, 1)$ means that only the slow server is busy and that server is serving the only customer in the system; $(2, 0)$ is not a possible state. The scheduling discipine is FCFS and an arrival that occurs when the system is empty goes to the faster server.

(a) Draw the state transition rate diagram for this system.
(b) What are the equilibrium equations?
(c) Solve the equilibrium equations to obtain the steady-state probabilities.
(d) Compute the system utilization $\rho$.

*Hint: Observe that not all states have transitions only to linearly adjacent neighbors so that, strictly speaking, this is not a birth-death process. It can be handled by the birth-death solution procedure by paying special attention to the balance equations involving the initial states.*

**Exercise 11.3.5** For each of the following queueing systems, draw the state transition rate diagram, then write the equilibrium flow equations and compute the probablity that the system is empty.

(a) The $M/E_r/1/1$ queue.
(b) The $M/H_2/1/1$ queue with $\mu_1 = \mu$ and $\mu_2 = 2\mu$.

**Exercise 11.4.1** Prove that the stability condition for an $M/M/c$ queue whose Poisson arrival process has a mean interarrival time equal to $1/\lambda$ and whose $c$ servers provide independent and identically exponentially distributed service with rate $\mu$, is that $\rho = \lambda/c\mu < 1$.

**Exercise 11.4.2** Consider an $M/M/c$ queue with the following parameters: $\lambda = 8$, $\mu = 3$, and $c = 4$. What is the probability that an arriving customer finds all servers busy and is forced to wait? What is the average time spent waiting like this? and what is the mean number of customers in the system?

**Exercise 11.4.3** A crisis center is manned by a staff of trained volunteers who answer phone calls from people in distress. Experience has taught them that as Christmas approaches they need to be able to cater to a peak demand when phone calls arrive at a rate of six per hour. Each call requires approximately 20 minutes to calm and advise the distressed person. At present the center plans to have five volunteers on hand to handle this peak period. Analyze this situation as an $M/M/c$ queue and advise the center on whether the five volunteers are enough so that the average time a distressed person spends waiting to talk to a counselor is less than 15 seconds. What will this average time be if the center provides six volunteers?

**Exercise 11.4.4** The local photographic club is putting on an exhibition in a vast amphitheatre. Visitors arrive at a rate of two every five minutes and spend on average a total of 20 minutes looking over the photos. If arrivals follow a Poisson process and the time spent in the exhibition is exponentially distributed, what is the average number of visitors present at any one time? What is the probability of finding more than ten visitors in attendance?

**Exercise 11.5.1** A shoe-shine stand at the local airport has two chairs and a single attendant. A person arriving to find one chair free and the attendant busy with a customer seated on the other chair, waits his turn in the available chair. Potential customers finding both chairs occupied simply leave. Under the assumption that potential customers arrive randomly according to a Poisson process at rate ten per hour and that the time taken to polish shoes is exponentially distributed with mean 5 minutes, answer the following questions:

(a) Write down the balance equations and find the distributions of occupied chairs.
(b) Find the percentage time the attendant is busy and the mean number of customers served in an 8-hour day.
(c) What percentage of potential customers succeed in getting their shoes shined?

**Exercise 11.5.2** Suppose that in the multiprocessor system of Example 11.13, the I/O service rate $\lambda_k$ is a function of the number of requests, $k$, that it has to handle. In particular, assume that

$$1/\lambda_k = \frac{\tau}{k+1} + \frac{1}{\alpha} = \frac{\alpha\tau + k + 1}{\alpha(k+1)},$$

where $1/\alpha$ is the mean record transmission time, and $\tau$ is the mean rotation time, of the I/O device. Derive expressions for the CPU utilization and the average system throughput.

**Exercise 11.6.1** The mean number of customers actually waiting in an *M*/*M*/*c*/*K* queue is given by $L_q = \sum_{n=c}^{K}(n-c)p_n$. Work to incorporate $L = \sum_{n=0}^{K} np_n$, the mean number of customers in an *M*/*M*/*c*/*K* queueing center, into the right-hand side of this equation and hence deduce Equation (11.12).

**Exercise 11.6.2** In the airport shoe-shine scenario of Exercise 11.5.1, assume that there are now two attendants who work at the same rate.

(a) Write down the balance equations and find the distributions of occupied chairs.
(b) Find the mean number of customers served in an 8-hour day.
(c) What percentage of potential customers fail to get their shoes shined?

**Exercise 11.6.3** A bank manager observes that when all three of his tellers are busy and the line of customers waiting to reach a teller reaches six, further arriving customers give this line one glance and promptly leave, thinking of coming back some other time or even changing banks in frustration. Assuming that the arrival process of bank customers is such that a potential customer enters the bank every 90 seconds and the time to serve a customer averages five minutes, is the bank manager better off by hiring another teller, or by means of incentives, persuading his three tellers to work about 20% faster so that the time to serve each customer is reduced to four minutes? You may assume that the arrival process of customers is Poisson, that the service time distribution is exponential, and that only one of the two options is available to the manager. The main criterion for choosing one option over the other should be to minimize the probability of customers leaving because the line is too long. You should also consider waiting time as a secondary criterion.

**Exercise 11.6.4** Derive the following recurrence relation for Erlang's $B$ (loss) formula:

$$B(0, \lambda/\mu) = 1, \quad B(c, \lambda/\mu) = \frac{(\lambda/\mu)B(c-1, \lambda/\mu)}{c + (\lambda/\mu)B(c-1, \lambda/\mu)}.$$

**Exercise 11.6.5** What is wrong with the following analysis? Given

$$C(c, \lambda/\mu) = \frac{(\lambda/\mu)^c \mu}{(c-1)!(c\mu - \lambda)} p_0,$$

it follows that

$$C(c, \lambda/\mu) = \frac{(\lambda/\mu)^c}{(c-1)!} p_0 \frac{\mu}{c\mu - \lambda} = \frac{(\lambda/\mu)^c}{c!} p_0 \frac{c\mu}{c\mu - \lambda} = B(c, \lambda/\mu) \frac{c\mu}{c\mu - \lambda} = \frac{B(c, \lambda/\mu)}{1 - \lambda/(c\mu)}.$$

Show that the correct relationship between the Erlang $B$ and $C$ formulae is given by

$$C(c, \lambda/\mu) = \frac{B(c, \lambda/\mu)}{1 - \lambda/(c\mu)\left[1 - B(c, \lambda/\mu)\right]}.$$

**Exercise 11.7.1** Suppose a single repairman has been assigned the responsibility of maintaining three machines. For each machine, the probability distribution of running time before a breakdown is exponential with a mean of 9 hours. The repair time is also exponentially distributed with a mean of 2 hours.

(a) Calculate the steady state probability distribution and the expected number of machines that are not running.
(b) As a crude approximation, it could be assumed that the calling population is infinite so that the input process is Poisson with a mean arrival rate of 3 every 9 hours. Compare the result of part 1 of this question with those obtained from (i) an *M*/*M*/1 model, and (ii) an *M*/*M*/1/3 model.

**Exercise 11.7.2** A service station has one gasoline pump. Cars wanting gas arrive according to a Poisson process at a mean rate of 20/hour. However, if the pump is in use, these potential customers may balk (refuse to join the queue). To be more precise, if there are $n$ cars already at the service station, the probability that an arriving potential customer will balk is $n/4$ for $n = 1, 2, 3, 4$. The time required to serve a car has an exponential distribution with a mean service time of 3 minutes.

(a) Construct the transition rate diagram for this system.
(b) For all $k$, find the differential-difference equations for $p_k(t) = \text{Prob}\{k \text{ in system at time } t\}$.
(c) Determine the stationary probability distribution and hence find the average number of cars at the station.

# Chapter 12

# *Queues with Phase-Type Laws: Neuts' Matrix-Geometric Method*

In the single-server queues that we have considered so far, the only probability law used to model the interarrival or service time distributions, is the exponential distribution. This gave rise to queues that we collectively referred to as *birth-death* processes. In these systems, transitions from any state are to adjacent states only and the resulting structure of the transition matrix is tridiagonal. But sometimes the exponential distribution is simply not adequate. Phase-type distributions allow us to consider more general situations. Such distributions were discussed previously in Chapter 7 and were seen to include such common distributions as the Erlang and hypergeometric distributions. In this chapter we consider how phase-type distributions may be incorporated into single-server queues. Queueing system with phase-type arrival or service mechanisms give rise to transition matrices that are *block* tridiagonal and are referred to as *quasi*-birth-death (QBD) processes. Whereas a simple birth-death process gives rise to a tridiagonal matrix in which elements below the diagonal represent service completions (departures from the system) and elements above the diagonal describe customer arrivals, subdiagonal blocks in a QBD process represent a more complex departure process and superdiagonal blocks represent a more complex arrival process. In the past, the mathematical techniques used to solve such queueing systems depended upon the use of the $z$-transform. Nowadays, with the advent of high-speed computers and efficient algorithms, the matrix-geometric approach introduced by Neuts is more commonly employed. We begin by analyzing the $M/E_r/1$ queue using Neuts' method.

## 12.1 The Erlang-$r$ Service Model—The $M/E_r/1$ Queue

Consider a single server system for which the arrival process is Poisson with rate $\lambda$ and the service time, having an Erlang-$r$ distribution, is represented as a sequence of $r$ exponential services each with rate $r\mu$, as shown in Figure 12.1.

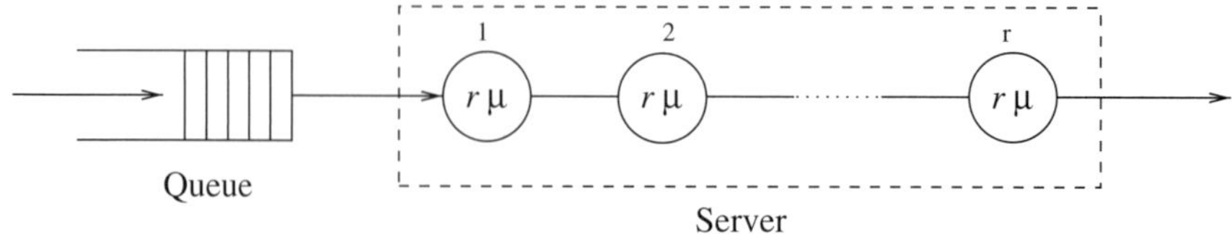


Figure 12.1. The $M/E_r/1$ queue.

The reader should be careful not to associate these $r$ phases with $r$ distinct servers. There cannot be more than one customer receiving service at any time—the single server provides each customer taken into service with $r$ consecutive service phases and then ejects that customer from the system.

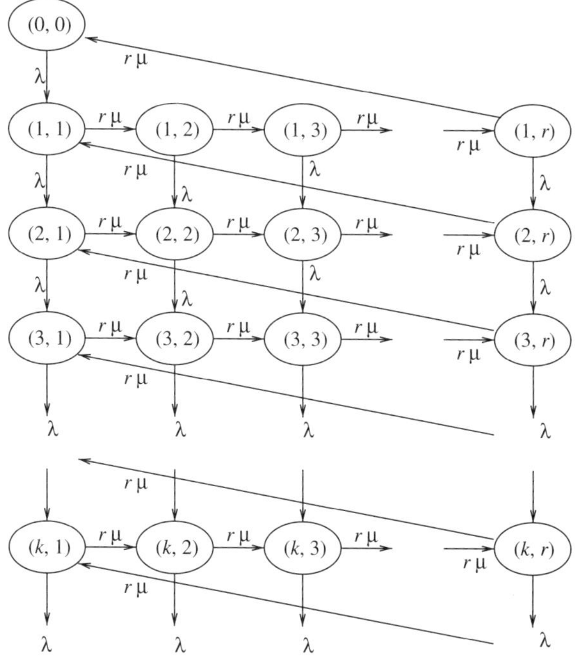


Figure 12.2. State transition diagram for the $M/E_r/1$ queue.

The density functions of the arrival and service processes are given by

$$a(t) = \lambda e^{-\lambda t}, \quad t \geq 0,$$

$$b(x) = \frac{r\mu(r\mu x)^{r-1}e^{-r\mu x}}{(r-1)!}, \quad x \geq 0.$$

Our first task is to formulate a state descriptor for this queueing system. In an $M/E_r/1$ queue, we must record both the number of customers present and the current phase of service of the customer in service (if any). Because of the exponential nature of the distribution of both the times between arrivals and the service times at each of the $r$ phases, a knowledge of the number of customers in the system and the current phase of service is sufficient to capture all the relevant past history of this system. It follows that a state of the system can be completely described by the pair $(k, i)$, where $k (k \geq 0)$ is the number of customers in the system, including the one in service, and $i (1 \leq i \leq r)$ denotes the current phase of service. If $k = 0$, then the value of $i$ is irrelevant. If $k > 0$, then $r - i + 1$ denotes the number of phases of service yet to be completed by the customer in service. The future evolution of the system is a function only of the current state $(k, i)$ and consequently a Markovian analysis may be carried out.

The state transition diagram is shown in Figure 12.2 where the states are arranged into levels according to the number of customers present. The states that have exactly $k$ customers are said to constitute level $k$. From the structure of this state transition diagram, it is apparent that the transition rate matrix has the typical block-tridiagonal (or QBD—quasi-birth-death) form:

$$Q = \begin{pmatrix} B_{00} & B_{01} & 0 & 0 & 0 & \cdots \\ B_{10} & A_1 & A_2 & 0 & 0 & \cdots \\ 0 & A_0 & A_1 & A_2 & 0 & \cdots \\ 0 & 0 & A_0 & A_1 & A_2 & \cdots \\ 0 & 0 & 0 & A_0 & A_1 & \cdots \\ \vdots & \vdots & \vdots & \vdots & \vdots & \ddots \end{pmatrix}$$

in which the matrices $A_i$, $i \geq 0$ are square and of order $r$. The matrices $A_0$ represent service completions at rate $r\mu$ from the last service phase at some level $k > 0$ to the first service phase at level $k-1$, i.e., transitions from state $(k, r)$ to state $(k-1, 1)$. Thus the only nonzero element in $A_0$ is the $r1$ element $A_0(r, 1)$ which has the value $r\mu$. The matrices $A_2$ represent arrivals at rate $\lambda$ which can occur during service at any phase $i$ at any level $k > 0$, i.e., transitions from state $(k, i)$ to state $(k+1, i)$. It follows that the only nonzero elements in this matrix are the diagonal elements which are all equal to $\lambda$. The superdiagonal elements of the matrices $A_1$ represent phase completion at rate $r\mu$ in service phase $i < r$ at level $k > 0$, i.e., transitions from state $(k, i)$ to state $(k, i+1)$. The diagonal elements are set equal to the negated sum of the off-diagonal elements of $Q$. All other elements in $A_1$ are equal to zero. Thus,

$$A_0 = \begin{pmatrix} 0 & 0 & 0 & 0 & \cdots & 0 \\ 0 & 0 & 0 & 0 & \cdots & 0 \\ 0 & 0 & 0 & 0 & \cdots & 0 \\ 0 & 0 & 0 & 0 & \cdots & 0 \\ \vdots & \vdots & \vdots & \vdots & \ddots & \vdots \\ r\mu & 0 & 0 & 0 & \cdots & 0 \end{pmatrix}, \quad A_2 = \lambda I, \quad \text{and}$$

$$A_1 = \begin{pmatrix} -\lambda - r\mu & r\mu & 0 & 0 & \cdots & 0 \\ 0 & -\lambda - r\mu & r\mu & 0 & \cdots & 0 \\ 0 & 0 & -\lambda - r\mu & r\mu & \cdots & 0 \\ \vdots & \vdots & \vdots & \ddots & \ddots & \vdots \\ 0 & 0 & 0 & 0 & \vdots & r\mu \\ 0 & 0 & 0 & 0 & \cdots & -\lambda - r\mu \end{pmatrix}.$$

The matrix $B_{01}$ is a $1 \times r$ row matrix all of whose elements are zero except for the first which is equal to $\lambda$. This is the rate of transition from state $(0, 0)$ to state $(1, 1)$ and corresponds to an arrival to an empty system. The matrix $B_{10}$ is a $r \times 1$ column matrix all of whose elements are zero except for the last which is equal to $r\mu$. This is the rate of transition from state $(1, r)$ to state $(0, 0)$ and corresponds to the complete service termination of the only customer in the system. The matrix $B_{00}$ is a $1 \times 1$ matrix whose nonzero element is $-\lambda$ and reflects the fact that the sum across the first row must be zero.

**Example 12.1** Throughout this section we shall illustrate the matrix-geometric approach by means of an $M/E_r/1$ queue with parameters $\lambda = 1$, $\mu = 1.5$, and $r = 3$. In this case, the infinitesimal generator $Q$ and its submatrices are as follows:

$$Q = \left(\begin{array}{c|ccc|ccc|ccc|c} -1 & 1 & 0 & 0 & 0 & 0 & 0 & 0 & 0 & 0 & \cdots \\ \hline 0 & -5.5 & 4.5 & 0 & 1 & 0 & 0 & 0 & 0 & 0 & \cdots \\ 0 & 0 & -5.5 & 4.5 & 0 & 1 & 0 & 0 & 0 & 0 & \cdots \\ 4.5 & 0 & 0 & -5.5 & 0 & 0 & 1 & 0 & 0 & 0 & \cdots \\ \hline 0 & 0 & 0 & 0 & -5.5 & 4.5 & 0 & 1 & 0 & 0 & \cdots \\ 0 & 0 & 0 & 0 & 0 & -5.5 & 4.5 & 0 & 1 & 0 & \cdots \\ 0 & 4.5 & 0 & 0 & 0 & 0 & -5.5 & 0 & 0 & 1 & \cdots \\ \hline 0 & 0 & 0 & 0 & 0 & 0 & 0 & -5.5 & 4.5 & 0 & \cdots \\ 0 & 0 & 0 & 0 & 0 & 0 & 0 & 0 & -5.5 & 4.5 & \cdots \\ 0 & 0 & 0 & 0 & 4.5 & 0 & 0 & 0 & 0 & -5.5 & \cdots \\ \hline \vdots & \vdots & \vdots & \vdots & \vdots & \vdots & \vdots & \vdots & \vdots & \vdots & \ddots \end{array}\right),$$

$$A_0 = \begin{pmatrix} 0 & 0 & 0 \\ 0 & 0 & 0 \\ 4.5 & 0 & 0 \end{pmatrix}, \quad A_1 = \begin{pmatrix} -5.5 & 4.5 & 0 \\ 0 & -5.5 & 4.5 \\ 0 & 0 & -5.5 \end{pmatrix}, \quad A_2 = \begin{pmatrix} 1 & 0 & 0 \\ 0 & 1 & 0 \\ 0 & 0 & 1 \end{pmatrix}.$$

$$B_{00} = -1, \quad B_{01} = (1, 0, 0), \quad B_{10} = \begin{pmatrix} 0 \\ 0 \\ 4.5 \end{pmatrix}.$$

Our quest is to compute the stationary probability vector $\pi$ from the system of homogeneous linear equations $\pi Q = 0$ using the matrix geometric method. This method is described in Chapter 10 in the context of solving Markov chains with structured transition rate matrices. The $M/E_r/1$ queue, and indeed all the queues of this chapter fall into this category. The reader may wish to review the matrix geometric solution procedure before proceeding. We begin by writing the stationary probability vector $\pi$ as

$$\pi = (\pi_0, \pi_1, \pi_2, \ldots, \pi_k, \ldots),$$

where $\pi_0$ is vector of length 1 (whose value is equal to the probability of an empty system) and $\pi_k,\ k = 1, 2, \ldots$ is a row vector of length $r$. Its $i^{\text{th}}$ component gives the probability of being in phase $i$ when there are $k$ customers in the system. It is shown in Chapter 10 that successive subvectors of $\pi$ satisfy the relationship $\pi_{i+1} = \pi_i R$ for $i = 1, 2, \ldots,$ where $R$ is the so-called Neuts' rate matrix. Thus the first step in applying the matrix-geometric approach is the computation of this matrix $R$. One possibility for a block tridiagonal matrix $Q$ is to use the iterative scheme, called *successive substitution*:

$$R_{l+1} = -(V + R_l^2 W), \quad l = 0, 1, 2, \ldots,$$

where $V = A_2 A_1^{-1}$ and $W = A_0 A_1^{-1}$ and using $R_0 = 0$ to initiate the procedure. As is shown by Neuts, the sequence $R_l$ is monotone increasing and converges to $R$. Due to the nonzero structure of $A_1$, nonzero elements along the diagonal and superdiagonal and zero elements everywhere else, its inverse is explicitly available. The $ij$ element of the inverse of any matrix of the form

$$M = \begin{pmatrix} d & a & 0 & 0 & \cdots & 0 \\ 0 & d & a & 0 & \cdots & 0 \\ 0 & 0 & d & a & \cdots & 0 \\ \vdots & \vdots & \vdots & \ddots & \ddots & \vdots \\ 0 & 0 & 0 & 0 & \vdots & a \\ 0 & 0 & 0 & 0 & \cdots & d \end{pmatrix}$$

is given as

$$\left(M^{-1}\right)_{ij} = (-1)^{j-i} \frac{1}{d} \left(\frac{a}{d}\right)^{j-i} \quad \text{if } i \le j \le r, \tag{12.1}$$

and is equal to zero otherwise. For the $M/E_r/1$ queue, we have $d = -(\lambda + r\mu)$ and $a = r\mu$. This facilitates the computation of the matrices $V$ and $W$ used in the iterative scheme. Since $W = A_0 A_1^{-1}$ and $A_0$ has a single nonzero element equal to $r\mu$ in position $(r, 1)$ it follows that $W$ has only one nonzero row, the last with elements given by $r\mu$ times the first row of $A_1^{-1}$. Therefore

$$W_{ri} = -\left(\frac{r\mu}{\lambda + r\mu}\right)^i \quad \text{for } 1 \le i \le r$$

and $W_{ki} = 0$ for $1 \le k < r$ and $1 \le i \le r$. Also, since $A_2 = \lambda I$, the computation of $V = A_2 A_1^{-1}$ is also easy to find. It suffices to multiply each element of $A_1^{-1}$ by $\lambda$. With the matrices $V$ and $W$ in hand, the iterative procedure can be initiated and Neuts' $R$ matrix computed.

**Example 12.2** Continuing our example of the $M/E_3/1$ queue with parameters $\lambda = 1$ and $\mu = 1.5$, we compute $A_1^{-1}$ using Equation (12.1) and obtain

$$A_1^{-1} = \begin{pmatrix} -2/11 & -18/121 & -162/1331 \\ 0 & -2/11 & -18/121 \\ 0 & 0 & -2/11 \end{pmatrix}$$

and hence

$$V = \begin{pmatrix} -2/11 & -18/121 & -162/1331 \\ 0 & -2/11 & -18/121 \\ 0 & 0 & -2/11 \end{pmatrix} \quad \text{and} \quad W = \begin{pmatrix} 0 & 0 & 0 \\ 0 & 0 & 0 \\ -9/11 & -81/121 & -729/1331 \end{pmatrix}.$$

We may now begin the iterative scheme

$$R_{l+1} = -V - R_l^2 W$$

with $R_0 = 0$. We obtain successively,

$$R_1 = \begin{pmatrix} 2/11 & 18/121 & 162/1331 \\ 0 & 2/11 & 18/121 \\ 0 & 0 & 2/11 \end{pmatrix}, \quad R_2 = \begin{pmatrix} 0.236136 & 0.193202 & 0.158075 \\ 0.044259 & 0.218030 & 0.178388 \\ 0.027047 & 0.022130 & 0.199924 \end{pmatrix}, \quad \text{etc.}$$

Continuing in this fashion, we obtain

$$R_{50} = \begin{pmatrix} 0.331961 & 0.271605 & 0.222222 \\ 0.109739 & 0.271605 & 0.222222 \\ 0.060357 & 0.049383 & 0.222222 \end{pmatrix} = R.$$

The second step in using the matrix-geometric approach is the computation of initial vectors so that successive subvectors $\pi_{i+1}$ can be computed from $\pi_{i+1} = \pi_i R$. This requires us to find $\pi_0$ and $\pi_1$. From $\pi Q = 0$, i.e., from

$$(\pi_0, \pi_1, \pi_2, \ldots, \pi_i, \ldots) \begin{pmatrix} B_{00} & B_{01} & 0 & 0 & 0 & \cdots \\ B_{10} & A_1 & A_2 & 0 & 0 & \cdots \\ 0 & A_0 & A_1 & A_2 & 0 & \cdots \\ 0 & 0 & A_0 & A_1 & A_2 & \cdots \\ 0 & 0 & 0 & A_0 & A_1 & \cdots \\ \vdots & \vdots & \vdots & \vdots & \vdots & \ddots \end{pmatrix} = (0, 0, 0, \ldots, 0, \ldots),$$

we have

$$\pi_0 B_{00} + \pi_1 B_{10} = 0,$$
$$\pi_0 B_{01} + \pi_1 A_1 + \pi_2 A_0 = 0.$$

Writing $\pi_2$ as $\pi_1 R$, we obtain

$$(\pi_0, \pi_1) \begin{pmatrix} B_{00} & B_{01} \\ B_{10} & A_1 + RA_0 \end{pmatrix} = (0, 0). \tag{12.2}$$

Observe in this system of equation that $\pi_0$ is a scalar quantity and $\pi_1$ is a row vector of length $r$. This system of equations can be solved using standard techniques from numerical linear algebra. Since it is unlikely that $r$ will be large, Gaussian elimination can be recommended. One final point must be taken into account: there is no unique solution to the system of Equation (12.2) and so the computed $\pi$ must be normalized so that the sum of its components is equal to one. Letting $e$ denote a column vector of length $r$ whose components are all equal to 1, then

$$1 = \pi_0 + \sum_{k=1}^{\infty} \pi_k e = \pi_0 + \sum_{k=0}^{\infty} \pi_1 R^k e = \pi_0 + \pi_1 (I - R)^{-1} e$$

provides the means by which $\pi_0$ and $\pi_1$ may be uniquely determined. Once $\pi_1$ has been computed, successive subvectors of the stationary probability distribution can be computed from

$$\pi_{k+1} = \pi_k R.$$

**Example 12.3** Returning to the $M/E_3/1$ example, we now seek to compute $\pi_0$ and $\pi_1$. Observe that $4.5 \times 0.222222 = 1$ and so

$$RA_0 = \begin{pmatrix} 1 & 0 & 0 \\ 1 & 0 & 0 \\ 1 & 0 & 0 \end{pmatrix} \quad \text{and} \quad A_1 + RA_0 = \begin{pmatrix} -4.5 & 4.5 & 0 \\ 1 & -5.5 & 4.5 \\ 1 & 0 & -5.5 \end{pmatrix}.$$

This allows us to compute $\pi_0$ and $\pi_1$ from

$$(\pi_0, \pi_1) \left( \begin{array}{c|ccc} -1 & 1 & 0 & 0 \\ \hline 0 & -4.5 & 4.5 & 0 \\ 0 & 1 & -5.5 & 4.5 \\ 4.5 & 1 & 0 & -5.5 \end{array} \right) = (0, 0).$$

The coefficient matrix has rank 3, so arbitrarily setting $\pi_0 = 1$ we may convert this system to the following nonhomogeneous system of equations with nonsingular coefficient matrix:

$$(\pi_0, \pi_{1_1}, \pi_{1_2}, \pi_{1_3}) \left( \begin{array}{c|ccc} -1 & 1 & 0 & 1 \\ \hline 0 & -4.5 & 4.5 & 0 \\ 0 & 1 & -5.5 & 0 \\ 4.5 & 1 & 0 & 0 \end{array} \right) = (0, 0, 0, 1).$$

Its solution is readily computed and we have

$$(\pi_0, \pi_{1_1}, \pi_{1_2}, \pi_{1_3}) = (1, 0.331962, 0.271605, 0.222222).$$

This solution needs to be normalized so that

$$\pi_0 + \pi_1 (I - R)^{-1} e = 1.$$

Substituting, we obtain

$$1 + (0.331962, 0.271605, 0.222222) \begin{pmatrix} 1.666666 & 0.666666 & 0.666666 \\ 0.296296 & 1.518518 & 0.518518 \\ 0.148148 & 0.148148 & 1.370370 \end{pmatrix} \begin{pmatrix} 1 \\ 1 \\ 1 \end{pmatrix} = 3.$$

Thus, the normalized solution is given as

$$\begin{aligned} (\pi_0, \pi_{1_1}, \pi_{1_2}, \pi_{1_3}) &= (1/3, 0.331962/3, 0.271605/3, 0.222222/3) \\ &= (1/3,\ 0.110654,\ 0.090535,\ 0.0740741). \end{aligned}$$

Additional probabilities may now be computed from $\pi_{k+1} = \pi_k R$. For example, we have

$$\begin{aligned} \pi_2 = \pi_1 R &= (0.051139, 0.058302, 0.061170), \\ \pi_3 = \pi_2 R &= (0.027067, 0.032745, 0.037913), \\ \pi_4 = \pi_3 R &= (0.014867, 0.018117, 0.021717), \\ \pi_5 = \pi_4 R &= (0.008234, 0.010031, 0.012156), \end{aligned}$$

and so on. The probability of having 0, 1, 2, ... customers is found by adding the components of these subvectors. We have

$$p_0 = 1/3, \quad p_1 = 0.275263, \quad p_2 = 0.170610, \quad p_3 = 0.097725, \; \ldots .$$

We shall defer questions concerning stability and performance measures until towards the end of this chapter, when we shall provide results that are applicable to all types of $Ph/Ph/1$ queues.

## 12.2 The Erlang-*r* Arrival Model—The $E_r/M/1$ Queue

We now move on to the case of a single server queue for which the arrival process is Erlang-$r$ and the service time is exponential with rate $\mu$ as shown in Figure 12.3.

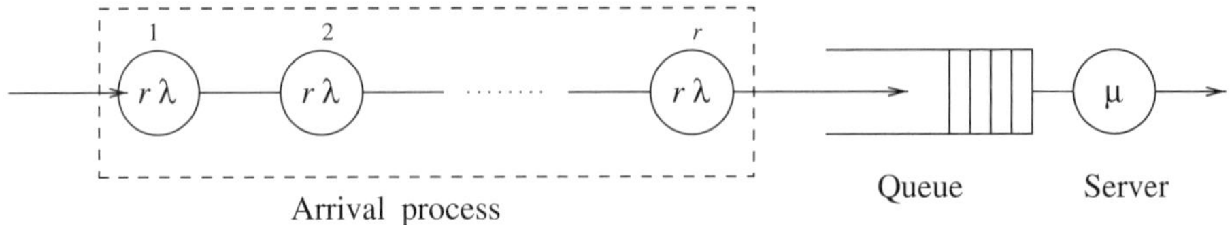


Figure 12.3. The $E_r/M/1$ queue.

The density functions of the arrival and service processes, respectively, are given by

$$a(t) = \frac{r\lambda(r\lambda t)^{r-1}e^{-r\lambda t}}{(r-1)!}, \quad t \geq 0,$$

$$b(x) = \mu e^{-\mu x}, \quad x \geq 0.$$

Before actually appearing in the queue proper, an arriving customer must pass through $r$ exponential phases each with parameter $r\lambda$. When a customer completes the arrival process another immediately begins. In other words, it is assumed that there is an infinite pool of available customers waiting to enter the arrival mechanism, and since only one can be "arriving" at any given instant of time, a new customer cannot enter the left-most exponential phase until all arrival phases have been completed. The instant that one customer completes the arrival process, a second customer simultaneously begins the process.

The state descriptor for the $E_r/M/1$ queue must specify the number of customers in the system, $k$, and also, the phase in which the arriving customer is to be found, $i$. In this way, the state diagram can be set up on a two-dimensional grid $(k, i)$, arranged into levels according to the number of customers present, in a manner similar to that undertaken for the $M/E_r/1$ queue. The state transition diagram for the $E_r/M/1$ queue is shown in Figure 12.4.

The transition rate matrix has the typical block tridiagonal structure found in quasi-birth-death processes:

$$Q = \begin{pmatrix} B_{00} & A_2 & 0 & 0 & 0 & 0 & \cdots \\ A_0 & A_1 & A_2 & 0 & 0 & 0 & \cdots \\ 0 & A_0 & A_1 & A_2 & 0 & 0 & \cdots \\ 0 & 0 & A_0 & A_1 & A_2 & 0 & \cdots \\ \vdots & \vdots & & \ddots & \ddots & \ddots & \vdots \end{pmatrix},$$

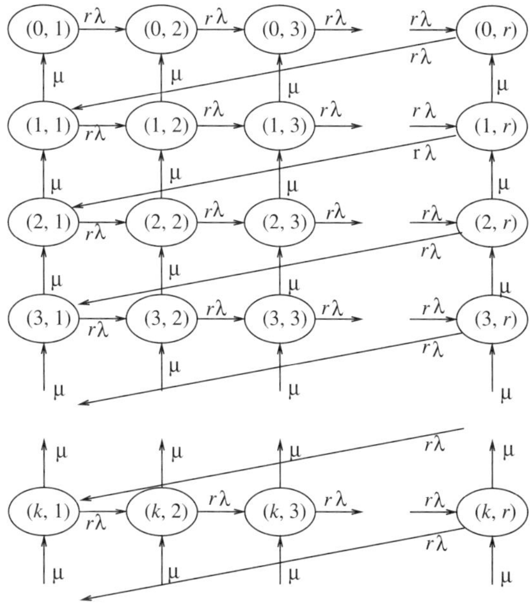


Figure 12.4. State transition diagram for the $E_r/M/1$ queue.

where all the subblocks are square and of order $r$. Specificially, we have

$$A_0 = \mu I, \quad A_1 = \begin{pmatrix} -\mu - r\lambda & r\lambda & 0 & 0 & \cdots & 0 \\ 0 & -\mu - r\lambda & r\lambda & 0 & \cdots & 0 \\ 0 & 0 & -\mu - r\lambda & r\lambda & \cdots & 0 \\ \vdots & \vdots & & \ddots & \ddots & \vdots \\ 0 & 0 & 0 & 0 & \vdots & r\lambda \\ 0 & 0 & 0 & 0 & \cdots & -\mu - r\lambda \end{pmatrix},$$

$$\text{and} \quad A_2 = \begin{pmatrix} 0 & 0 & 0 & 0 & \cdots & 0 \\ 0 & 0 & 0 & 0 & \cdots & 0 \\ 0 & 0 & 0 & 0 & \cdots & 0 \\ 0 & 0 & 0 & 0 & \cdots & 0 \\ \vdots & \vdots & \vdots & \vdots & \ddots & \vdots \\ r\lambda & 0 & 0 & 0 & \cdots & 0 \end{pmatrix}.$$

The matrices $A_0$ represent service completions at rate $\mu$ from a state at some level $k > 0$ and some number $i - 1$ of completed arrival phases to the state with the same number of completed arrival phases, but with one fewer customer, i.e., transitions from state $(k, i)$ to state $(k - 1, i)$. It follows that the only nonzero elements in this matrix are the diagonal elements which are all equal to $\mu$. The matrices $A_2$ represent an actual arrival to the system. If the arrival process is in its last phase, then at rate $r\lambda$, this final phase terminates and the number of customers actually present increases by 1. At this same instant, the arrival process begins again from phase 1. Thus the only nonzero element in $A_2$ is the $r1$ element $A_2(r, 1)$ which has the value $r\lambda$ and represents transitions from state $(k, r)$ to state $(k + 1, 1)$. The superdiagonal elements of the matrices $A_1$ represent the completion of one arrival phase $i < r$, at rate $r\lambda$, and the initiation of the next one, $i + 1$, i.e., transitions from state $(k, i)$ to state $(k, i + 1)$. The diagonal elements are set equal to the negated sum of the off-diagonal elements of $Q$. All other elements in $A_1$ are equal to zero. The matrix $B_{00}$ differs from $A_1$ only in that its diagonal elements are all equal to $-r\lambda$, whereas those of $A_1$ are each equal to $-\mu - r\lambda$. Like the diagonal elements of $A_1$, the diagonal elements of $B_{00}$ ensure that the sum of the elements across each row of $Q$ is equal to zero.

In comparing these matrices with the corresponding ones for the $M/E_r/1$ queue, the reader will notice that the structure of $A_0$ in the $E_r/M/1$ queue is identical to that of $A_2$ in the $M/E_r/1$ queue and that the structure of $A_2$ in the $E_r/M/1$ queue is identical to that of $A_0$ in the $M/E_r/1$ queue. Furthermore, as far as the actual nonzero values in these matrices are concerned, the $\lambda$'s in one are replaced by $\mu$'s in the other and vice versa.

**Example 12.4** We consider an $E_r/M/1$ queue with parameters $\lambda = 1.0$, $\mu = 1.5$ and $r = 3$. In this case, the infinitesimal generator and its submatrices are

$$Q = \left(\begin{array}{ccc|ccc|ccc|c} -3 & 3 & 0 & 0 & 0 & 0 & 0 & 0 & 0 & \cdots \\ 0 & -3 & 3 & 0 & 0 & 0 & 0 & 0 & 0 & \cdots \\ 0 & 0 & -3 & 3 & 0 & 0 & 0 & 0 & 0 & \cdots \\ \hline 1.5 & 0 & 0 & -4.5 & 3 & 0 & 0 & 0 & 0 & \cdots \\ 0 & 1.5 & 0 & 0 & -4.5 & 3 & 0 & 0 & 0 & \cdots \\ 0 & 0 & 1.5 & 0 & 0 & -4.5 & 3 & 0 & 0 & \cdots \\ \hline 0 & 0 & 0 & 1.5 & 0 & 0 & -4.5 & 3 & 0 & \cdots \\ 0 & 0 & 0 & 0 & 1.5 & 0 & 0 & -4.5 & 3 & \cdots \\ 0 & 0 & 0 & 0 & 0 & 1.5 & 0 & 0 & -4.5 & \cdots \\ \hline \vdots & \vdots & \vdots & \vdots & \vdots & \vdots & \vdots & \vdots & \vdots & \ddots \end{array}\right),$$

$$A_0 = \begin{pmatrix} 1.5 & 0 & 0 \\ 0 & 1.5 & 0 \\ 0 & 0 & 1.5 \end{pmatrix}, \quad A_1 = \begin{pmatrix} -4.5 & 3 & 0 \\ 0 & -4.5 & 3 \\ 0 & 0 & -4.5 \end{pmatrix}, \quad A_2 = \begin{pmatrix} 0 & 0 & 0 \\ 0 & 0 & 0 \\ 3 & 0 & 0 \end{pmatrix},$$

$$B_{00} = \begin{pmatrix} -3 & 3 & 0 \\ 0 & -3 & 3 \\ 0 & 0 & -3 \end{pmatrix}.$$

Once again, our quest is the computation of the stationary probability vector $\pi$, with $\pi Q = 0$, by means of the matrix geometric approach. This stationary probability vector $\pi$ may be written as

$$\pi = (\pi_0, \pi_1, \pi_2, \ldots, \pi_k, \ldots)$$

where each $\pi_k$, $k = 0, 1, 2, \ldots$ is a row vector of length $r$ whose $i^{\text{th}}$ component gives the probability of the arrival process having completed exactly $i - 1$ phases when there are $k$ customers present in the system and waiting for or receiving service. As before, the successive subvectors of $\pi$ satisfy the relationship $\pi_{i+1} = \pi_i R$ for $i = 1, 2, \ldots$, where $R$ is obtained from the iterative scheme

$$R_{l+1} = -(V + R_l^2 W), \quad l = 0, 1, 2, \ldots,$$

with $V = A_2 A_1^{-1}$ and $W = A_0 A_1^{-1}$ and using $R_0 = 0$ to initiate the procedure. The inverse of $A_1$, is an upper-trangular matrix which may be computed in exactly the same manner used in the $M/E_r/1$ queue. The nonzero $ij$ elements of $A_1^{-1}$ for the $E_r/M/1$ queue are given by

$$(A_1^{-1})_{ij} = (-1)^{j-i} \frac{1}{d} \left(\frac{a}{d}\right)^{j-i} \quad \text{for } i \le j \le r,$$

where $d = -(\mu + r\lambda)$ and $a = r\lambda$. Since $V = A_2 A_1^{-1}$ and $A_2$ has a single nonzero element equal to $r\lambda$ in position $(r, 1)$ it follows that $V$ has only one nonzero row, the last, with elements given by

$r\lambda$ times the first row of $A_1^{-1}$. Therefore

$$V_{ri} = -\left(\frac{r\lambda}{\mu + r\lambda}\right)^i \quad \text{for } 1 \le i \le r$$

and $V_{ki} = 0$ for $1 \le k < r$ and $1 \le i \le r$. Also, since $A_0 = \mu I$, the computation of $W = A_0 A_1^{-1}$ is easy to find. It suffices to multiply each element of $A_1^{-1}$ by $\mu$. With the matrices $V$ and $W$ in hand, the iterative procedure can be initiated and $R$ computed via the iterative procedure.

**Example 12.5** We shall continue with the $E_3/M/1$ example and compute the matrix $R$. First we need to compute $A_1^{-1}$. Using (12.1), we find

$$A_1^{-1} = \begin{pmatrix} -2/9 & -4/27 & -8/81 \\ 0 & -2/9 & -4/27 \\ 0 & 0 & -2/9 \end{pmatrix}$$

and hence

$$W = A_0 A_1^{-1} = \begin{pmatrix} -1/3 & -2/9 & -4/27 \\ 0 & -1/3 & -2/9 \\ 0 & 0 & -1/3 \end{pmatrix} \quad \text{and} \quad V = A_2 A_1^{-1} = \begin{pmatrix} 0 & 0 & 0 \\ 0 & 0 & 0 \\ -2/3 & -4/9 & -8/27 \end{pmatrix}.$$

Beginning with $R_0 = 0$, and iterating according to $R_{l+1} = -V - R_l^2 W$, we find

$$R_1 = \begin{pmatrix} 0 & 0 & 0 \\ 0 & 0 & 0 \\ 2/3 & 4/9 & 8/27 \end{pmatrix}, \quad R_2 = \begin{pmatrix} 0 & 0 & 0 \\ 0 & 0 & 0 \\ 0.732510 & 0.532236 & 0.3840878 \end{pmatrix}, \ldots,$$

which eventually converges to

$$R_{50} = \begin{pmatrix} 0 & 0 & 0 \\ 0 & 0 & 0 \\ 0.810536 & 0.656968 & 0.532496 \end{pmatrix} = R.$$

Observe that the matrix $R$ has only one nonzero row. This is due to the fact that $A_2$ has only one nonzero row (which propagates into $V$) and since $R$ is initialized to be the zero matrix, it never has the possibility of introducing nonzero elements into rows corresponding to zero rows of $A_2$.

The boundary equations are different in the $E_r/M/1$ queue from those in the $M/E_r/1$ queue. There is only a single $B$ block, namely $B_{00}$, whereas in the previous model there were three. Furthermore, in the $E_r/M/1$ queue, the $B_{00}$ block has the same dimensions as all other blocks, $r \times r$. Because of the structure of these boundary equations, in particular the form of $B_{00}$, only a single subvector $\pi_0$ needs to be found before all other subvectors can be constructed. We have

$$\pi_{i+1} = \pi_i R = \pi_0 R^{i+1} \quad \text{for } i = 0, 1, 2, \ldots$$

From $\pi Q = 0$, i.e., from

$$(\pi_0, \pi_1, \pi_2, \ldots, \pi_i, \ldots) \begin{pmatrix} B_{00} & A_2 & 0 & 0 & 0 & \cdots \\ A_0 & A_1 & A_2 & 0 & 0 & \cdots \\ 0 & A_0 & A_1 & A_2 & 0 & \cdots \\ 0 & 0 & A_0 & A_1 & A_2 & \cdots \\ 0 & 0 & 0 & A_0 & A_1 & \cdots \\ \vdots & \vdots & \vdots & \vdots & \vdots & \ddots \end{pmatrix} = (0, 0, 0, \ldots, 0, \ldots),$$

we have

$$\pi_0 B_{00} + \pi_1 A_0 = \pi_0 B_{00} + \pi_0 R A_0 = \pi_0 (B_{00} + R A_0) = 0.$$

A solution to this system of equations may be obtained using Gaussian elimination; a unique solution is found by enforcing the constraint that the sum of all components in $\pi$ be equal to one. This constraint implies that

$$1 = \sum_{k=0}^{\infty} \pi_k e = \sum_{k=0}^{\infty} \pi_0 R^k e = \pi_0 (I - R)^{-1} e$$

and provides the means by which $\pi_0$ may be uniquely determined.

**Example 12.6** Finishing off the example, we are now in a position to compute the initial subvector $\pi_0$. To do so, we need to solve the system of equations $\pi_0(B_{00} + RA_0) = 0$. Since this is a homogeneous system of equations without a unique solution, we substitute the last equation with $\pi_{01} = 1$ to obtain a specific solution. We have

$$\pi_0(B_{00} + RA_0) = (\pi_{01}, \pi_{02}, \pi_{03}) \begin{pmatrix} -3 & 3 & 1 \\ 0 & -3 & 0 \\ 1.215803 & 0.98545 & 0 \end{pmatrix} = (0, 0, 1).$$

The solution is $\pi_0 = (1,\ 1.810536,\ 2.467504)$ which must now be normalized in such a way that $\pi_0(I - R)^{-1}e = 1$. Since

$$(1,\ 1.810536,\ 2.467504) \begin{pmatrix} 1 & 0 & 0 \\ 0 & 1 & 0 \\ -0.810536 & -0.656968 & 0.467504 \end{pmatrix}^{-1} \begin{pmatrix} 1 \\ 1 \\ 1 \end{pmatrix} = 15.834116,$$

we divide each component of $\pi_0$ by 15.834116 to get the correctly normalized answer:

$$\pi_0 = (0.063155,\ 0.114344,\ 0.155835).$$

The remaining subvectors of $\pi$ may now be found from $\pi_k = \pi_{k-1} R = \pi_0 R^k$. We have

$$\pi_1 = \pi_0 R = (0.126310,\ 0.102378,\ 0.082981),$$

$$\pi_2 = \pi_1 R = (0.067259,\ 0.054516,\ 0.044187),$$

$$\pi_3 = \pi_2 R = (0.035815,\ 0.029030,\ 0.023530),$$

$$\pi_4 = \pi_3 R = (0.019072,\ 0.015458,\ 0.012529),$$

etc.

The probability of having $0, 1, 2, \ldots$ customers is found by adding the components of these subvectors. We have

$$p_0 = 1/3, \quad p_1 = 0.311669, \quad p_2 = 0.165963, \quad p_3 = 0.088374, \ldots.$$

## 12.3 The $M/H_2/1$ and $H_2/M/1$ Queues

Single-server queues with either hyperexponential service or arrival distributions can be handled by the matrix geometric approach just as easily as was the case for Erlang distributions. In this

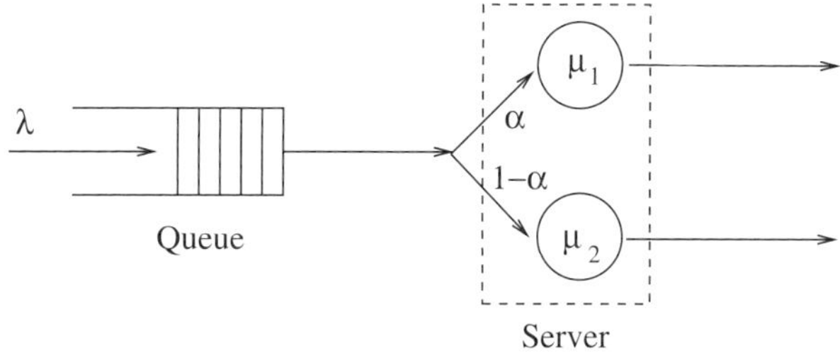


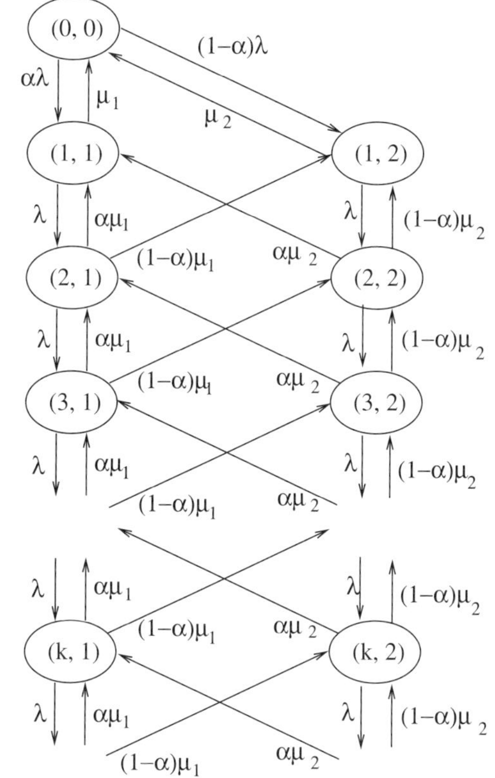


Figure 12.5. The $M/H_2/1$ queue (above) and its transition rate diagram.

section we shall consider only $H_2$ hyperexponential distributions, but the extension to higher orders is straightforward. Consider first the $M/H_2/1$ queue, as illustrated in Figure 12.5. Arrivals are generated according to a Poisson distribution at rate $\lambda$, while the service is represented by a two-phase hyperexponential distribution. With probability $\alpha$ a customer entering service receives service at rate $\mu_1$, while with probability $1 - \alpha$ this customer receives service at rate $\mu_2$. As for the $M/E_r/1$ queue, a state of this queueing system is given by the pair $(n, i)$ where $n$ is the total number present and $i$ is the current service phase.

The transition rate diagram for the $M/H_2/1$ queue is also shown in Figure 12.5. Increasing the number of phases from 2 to $r$ has the effect of incorporating additional columns into this figure. With $r$ possible service choices, a departure from any state $(n, i)$, $n > 1$ can take the system to any of the states $(n-1, i)$ at rate $\alpha_i \mu_i$ where $\alpha_i$ is the probability that an arriving customer enters service phase $i$, for $i = 1, 2, \ldots, r$. The transition rate matrix for the $M/H_2/1$ queue is given by

$$Q = \left(\begin{array}{c|cc|cc|cc|c}
-\lambda & \alpha\lambda & (1-\alpha)\lambda & 0 & 0 & 0 & 0 & \cdots \\
\hline
\mu_1 & -(\lambda+\mu_1) & 0 & \lambda & 0 & 0 & 0 & \cdots \\
\mu_2 & 0 & -(\lambda+\mu_2) & 0 & \lambda & 0 & 0 & \cdots \\
\hline
0 & \alpha\mu_1 & (1-\alpha)\mu_1 & -(\lambda+\mu_1) & 0 & \lambda & 0 & \\
0 & \alpha\mu_2 & (1-\alpha)\mu_2 & 0 & -(\lambda+\mu_2) & 0 & \lambda & \\
\hline
0 & 0 & 0 & \alpha\mu_1 & (1-\alpha)\mu_1 & -(\lambda+\mu_1) & 0 & \ddots \\
0 & 0 & 0 & \alpha\mu_2 & (1-\alpha)\mu_2 & 0 & -(\lambda+\mu_2) & \ddots \\
\hline
\vdots & \vdots & \vdots & & & \ddots & \ddots & \ddots
\end{array}\right)$$

and clearly has the block tridiagonal structure from which the blocks $A_0$, $A_1$, $A_2$, $B_{00}$, $B_{01}$, and $B_{10}$ can be identified. We have

$$A_0 = \begin{pmatrix} \alpha\mu_1 & (1-\alpha)\mu_1 \\ \alpha\mu_2 & (1-\alpha)\mu_2 \end{pmatrix}, \quad A_1 = \begin{pmatrix} -(\lambda+\mu_1) & 0 \\ 0 & -(\lambda+\mu_2) \end{pmatrix}, \quad A_2 = \begin{pmatrix} \lambda & 0 \\ 0 & \lambda \end{pmatrix},$$

$$B_{00} = (-\lambda), \quad B_{01} = (\alpha\lambda, (1-\alpha)\lambda), \quad B_{10} = \begin{pmatrix} \mu_1 \\ \mu_2 \end{pmatrix}.$$

The procedure is now identical to that carried out for the $M/E_r/1$ queue in that Neuts' $R$ matrix must be formed, the boundary equations set up and solved for $\pi_0$ and $\pi_1$ and from this all other components of the stationary probability vector computed.

Rather than working our way through an example at this point, we shall move on to the $H_2/M/1$ queue and postpone examples until after we formulate a Matlab program suitable for analyzing general QBD queueing systems. The $H_2/M/1$ queue is shown in Figure 12.6 and its transition rate diagram in Figure 12.7.

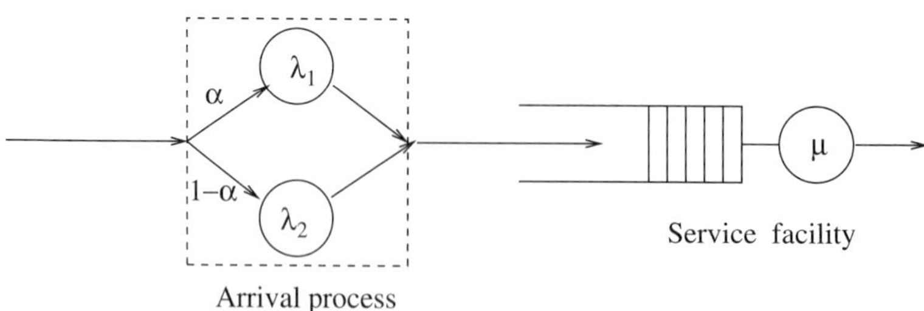


Figure 12.6. The $H_2/M/1$ queue.

The instant a customer enters the queue proper, a new customer immediately initiates its arrival process. The time between the moment at which this new customer initiates its arrival process and the time at which it actually arrives in the queue is exponentially distributed. With probability $\alpha$ this exponential distribution has rate $\lambda_1$, while with probability $1-\alpha$ it has rate $\lambda_2$. The transition rate

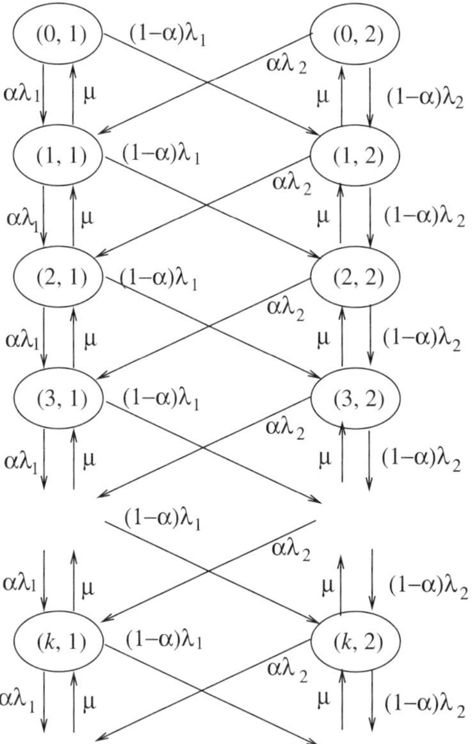


Figure 12.7. Transition diagram for the $H_2/M/1$ queue.

matrix for the $H_2/M/1$ queue is given by

$$Q=\left(\begin{array}{cc|cc|cc|cc|c} -\lambda_1 & 0 & \alpha\lambda_1 & (1-\alpha)\lambda_1 & 0 & 0 & 0 & 0 & \cdots \\ 0 & -\lambda_2 & \alpha\lambda_2 & (1-\alpha)\lambda_2 & 0 & 0 & 0 & 0 & \cdots \\ \hline \mu & 0 & -(\lambda_1+\mu) & 0 & \alpha\lambda_1 & (1-\alpha)\lambda_1 & 0 & 0 & \cdots \\ 0 & \mu & 0 & -(\lambda_2+\mu) & \alpha\lambda_2 & (1-\alpha)\lambda_2 & 0 & 0 & \cdots \\ \hline 0 & 0 & \mu & 0 & -(\lambda_1+\mu) & 0 & \alpha\lambda_1 & (1-\alpha)\lambda_1 & \\ 0 & 0 & 0 & \mu & 0 & -(\lambda_2+\mu) & \alpha\lambda_2 & (1-\alpha)\lambda_2 & \\ \hline 0 & 0 & 0 & 0 & \mu & 0 & -(\lambda_1+\mu) & 0 & \\ 0 & 0 & 0 & 0 & 0 & \mu & 0 & -(\lambda_2+\mu) & \\ \hline \vdots & \vdots & \vdots & \vdots & & & \ddots & \ddots & \ddots \end{array}\right)$$

and the block submatrices are

$$A_0=\begin{pmatrix}\mu & 0\\ 0 & \mu\end{pmatrix},\quad A_1=\begin{pmatrix}-(\lambda_1+\mu) & 0\\ 0 & -(\lambda_2+\mu)\end{pmatrix},\quad A_2=\begin{pmatrix}\alpha\lambda_1 & (1-\alpha)\lambda_1\\ \alpha\lambda_2 & (1-\alpha)\lambda_2\end{pmatrix},$$

$$B_{00}=\begin{pmatrix}-\lambda_1 & 0\\ 0 & -\lambda_2\end{pmatrix},\quad B_{01}=\begin{pmatrix}\alpha\lambda_1 & (1-\alpha)\lambda_1\\ \alpha\lambda_2 & (1-\alpha)\lambda_2\end{pmatrix}=A_2,\quad B_{10}=\begin{pmatrix}\mu & 0\\ 0 & \mu\end{pmatrix}=A_0.$$

The matrix-geometric approach which may now be initiated, is identical to that of the $E_r/M/1$ queue. Neuts' $R$ matrix is computed, the boundary equations solved and successive components of $\pi$ formed from the initial subvector $\pi_0$. Because of the repetitive nature of this analysis, it behooves us to write a simple Matlab program to solve systems of this type.

## 12.4 Automating the Analysis of Single-Server Phase-Type Queues

The procedure for solving phase-type queueing systems by means of the matrix-geometric approach has four steps, namely,

1. Construct the block submatrices.
2. Form Neuts' $R$ matrix.
3. Solve the boundary equations.
4. Generate successive components of the solution.

We shall write Matlab code for each of these four steps separately; a complete Matlab program is obtained by concatenating these four sections of code. In moving from one phase-type queueing system to another only the first of these sections should change. In other words we shall write the Matlab program so that steps 2, 3, and 4 will work for all single-server phase-type queues.

To avoid any confusion with the interpretation of the code, it is important to maintain a consistent notation throughout. Thus, in the code, we shall use the keyword `lambda` to represent the reciprocal of the expectation of an exponential interarrival time distribution and the keyword `mu` to represent the reciprocal of the expectation of an exponential service time distribution. When a distribution is represented by more than one exponential phase, we shall append indices $(1, 2, \ldots)$ to these keywords to represent the parameter of the exponential distribution at these phases.

**Step 1: The Block Submatrices**
In the examples treated so far, the submatrices $A_0$, $A_1$, and $A_2$ have all been of size $r \times r$, while the size of the boundary blocks, $B_{00}$, $B_{01}$, and $B_{10}$, were either of size $r \times r$, for the $E_r/M/1$ and $H_2/M/1$ or else of size $1 \times 1$, $1 \times r$, and $r \times 1$ respectively for the $M/E_r/1$ and $M/H_2/1$ queues. We shall introduce a parameter, $l$, to denote the size of the square block $B_{00}$. For example, the following Matlab codeforms the required block submatrices for the $M/H_2/1$ queue.

```
%%%%%  Construct submatrices for M/H_2/1 Queue  %%%%
     l = 1;   r = 2;   lambda = 1;
     alpha = 0.25;   mu1 = 1.5;   mu2 = 2.0;

     A0 = zeros(r,r);
     A0(1,1) = alpha * mu1; A0(1,2) = (1-alpha) * mu1;
     A0(2,1) = alpha * mu2; A0(2,2) = (1-alpha) * mu2;

     A1 = zeros(r,r);
     A1(1,1) = -lambda -mu1; A1(2,2) = -lambda -mu2;

     A2 = lambda * eye(r);

     B00 = -lambda * eye(l);
     B01 = [alpha*lambda, (1-alpha)*lambda];
     B10 = [mu1;mu2];
```

**Step 2: Neuts' $R$ Matrix**
The boundary blocks (i.e., the $B$ blocks) do not enter into the computation of Neut's $R$ matrix. We need to form $V = A_2 A_1^{-1}$ and $W = A_0 A_1^{-1}$ and then iterative successively with $R_{k+1} = -V - R_k^2 W$ with $R_0 = 0$ until $R_k$ converges to the matrix $R$. In the Matlab code given below we iterative until $\|R_{k+1} - R_k\|_1 \le 10^{-10}$.

```
%%%%%%%%%%%%%  Form Neuts' R matrix   %%%%%%%%%%%%%
     V = A2 * inv(A1);  W = A0 * inv(A1);

     R = zeros(r,r);  Rbis = -V - R*R * W;
     iter = 1;
     while (norm(R-Rbis,1) > 1.0e-10)
        R = Rbis;  Rbis = -V - R*R * W;
        iter = iter+1;
     end
     R = Rbis;
```

**Step 3: Solve the Boundary Equations**

In our analysis of the $E_r/M/1$ queue, we took the boundary equations to be

$$\pi_0(B_{00} + RA_0) = 0,$$

solved for $\pi_0$, and then used $\pi_k = \pi_0 R^k$, $k = 1, 2, \ldots$, to get the other block components of the stationary probability vector. For the $M/E_r/1$ queue we used

$$(\pi_0, \pi_1)\begin{pmatrix} B_{00} & B_{01} \\ B_{10} & A_1 + RA_0 \end{pmatrix} = 0,$$

from which we computed both $\pi_0$ and $\pi_1$. The other block components of the solution were obtained from $\pi_k = \pi_1 R^{k-1}$, $k = 2, 3, \ldots$. Since this latter approach will work in all cases, this is the version that we shall implement. To obtain a unique solution to this system of equations we replace one of the equations by an equation in which the first component of the subvector $\pi_0$ is set equal to 1. A final normalization so that

$$1 = \sum_{k=0}^{\infty} \pi_k e = \pi_0 e + \sum_{k=1}^{\infty} \pi_k e = \pi_0 e + \sum_{k=0}^{\infty} \pi_1 R^k e = \pi_0 e + \pi_1 (I - R)^{-1} e$$

allows us to correctly compute the subvectors $\pi_0$ and $\pi_1$. In the above equation, $e$ is a column vector of 1's whose size is determined by its context. The following Matlab code implements these concepts.

```
%%%%%%%%%%%%%  Solve boundary equations   %%%%%%%%%%%%
     N = [B00,B01;B10,A1+R*A0];  % Boundary equations

     N(1,r+l) = 1;               % First component equals 1
     for k=2:r+l
        N(k,r+l) = 0;
     end
     rhs = zeros(1,r+l); rhs(r+l)= 1;
     soln = rhs * inv(N);        % Un-normalized pi_0 and pi_1

     pi0 = zeros(1,l);  pi1 = zeros(1,r);
     for k=1:l                   % Extract pi_0 from soln
       pi0(k) = soln(k);
     end
```

```
    for k=1:r                      % Extract pi_1 from soln
      pi1(k) = soln(k+1);
    end
    e = ones(r,1);                 % Normalize solution
    sum = norm(pi0,1) +  pi1 * inv(eye(r)-R) * e;
    pi0 = pi0/sum;  pi1 = pi1/sum;
```

**Step 4: Generate Successive Components of the Solution**
The last step is to construct the different subblocks of the stationary probability vector. In addition to this, the Matlab code below also computes and prints the probability distribution of customers in the system (up to some given maximum number).

```
%%%% Generate successive components of solution  %%%%
    max = 10;          %  Maximum population requested
    pop = zeros(max+1,1);
    pop(1) =  norm(pi0,1);
    for k=1:max
      pi = pi1 * R^(k-1);   % Successive components of pi
      pop(k+1) = norm(pi,1);
    end
    pop                   %  Print population distribution
```

As mentioned previously, this code allows for all types of single-server phase-type queueing systems: it suffices to provide step 1 with the correct submatrices of the transition rate matrix. As a further example, the code for the $H_2/M/1$ queue is provided below. Other examples are provided in the subsections that follow.

```
%%%%%%%%%% Construct submatrices for H_2/M/1 queue  %%%%%%%%%%
     l = 2; r = 2;  lambda1 = 1;  lambda2 = 1.5;  mu = 2.0;  alpha = 0.25;

     A0 = mu * eye(r);

     A1 = zeros(r,r);
     A1(1,1) = -lambda1 - mu; A1(2,2) = -lambda2 - mu;

     A2 = [alpha*lambda1, (1-alpha)*lambda1;alpha*lambda2,
           (1-alpha)*lambda2];

     B00 =  [-lambda1,0;0,-lambda2];
     B01 = A2;
     B10 = A0;
```

## 12.5 The $H_2/E_3/1$ Queue and General *Ph/Ph/1* Queues

We analyze the $H_2/E_3/1$ queue in order to provide some insight into the analysis of more complex *Ph/Ph/1* queues. The arrival process of the $H_2/E_3/1$ queue is a two-phase hyperexponential distribution while the service process is an Erlang-3 distribution both of which have been considered in previous sections. This queueing system is shown graphically in Figure 12.8 where, for a proper

Erlang-3 distribution, we require $\mu_1 = \mu_2 = \mu_3 = 3\mu$. When these parameters do not all have the same value, then the distribution is more properly called a hypoexponential distribution.

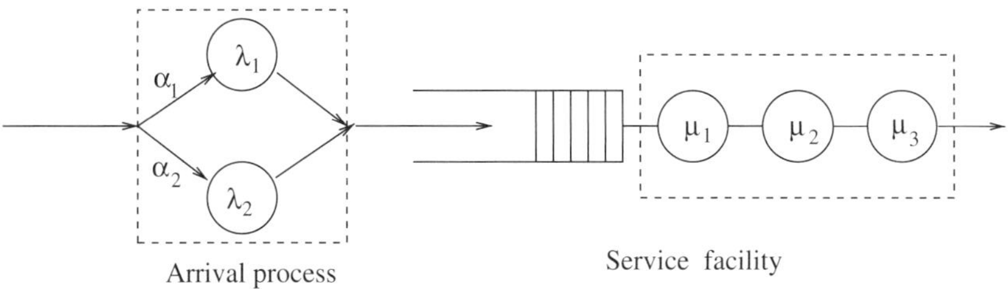


Figure 12.8. The $H_2/E_3/1$ queue.

To describe a state of this system we need three parameters, the first, $k$, to denote the number of customers actually present, the second, $a$, to describe the arrival phase of the "arriving" customer, and the last, $s$, to indicate the current phase of service. In drawing the state transition rate diagram, we shall present the states in rows (or levels) according to the number of customers present. Within each level $k$, states are ordered first according to the arrival phase of the hyperexponential distribution and second according to the phase of service. In other words, the states $(k, a, s)$ are ordered lexicographically. Since the state transition rate diagram can become rather complicated, we show the states and the transitions generated by the arrival process in one diagram, and in a separate diagram, we show the states and the transitions generated by the service process. Obviously the complete diagram is obtained by superimposing both sets of transitions onto the same state space. The first three levels of states and transitions generated by the service process are shown in Figure 12.9.

Transitions at rate $\mu_1$ move the system from states $(k, a, 1)$ to state $(k, a, 2)$; those at rate $\mu_2$ move the system from states $(k, a, 2)$ to state $(k, a, 3)$ while those at at rate $\mu_3$ move the system from states $(k, a, 3)$ to state $(k - 1, a, 1)$, for $k = 1, 2, \ldots$. It is only when the last service phase is

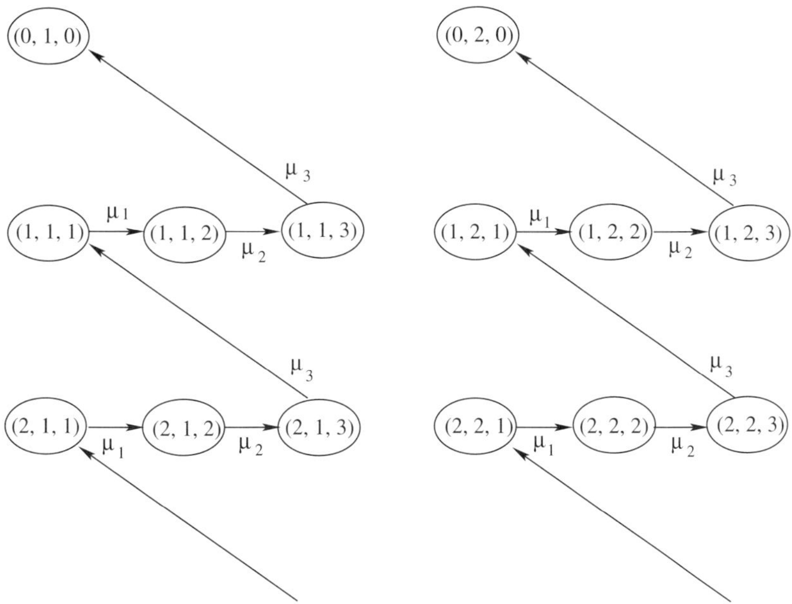


Figure 12.9. Transition diagram for the $H_2/E_3/1$ queue, part (a).

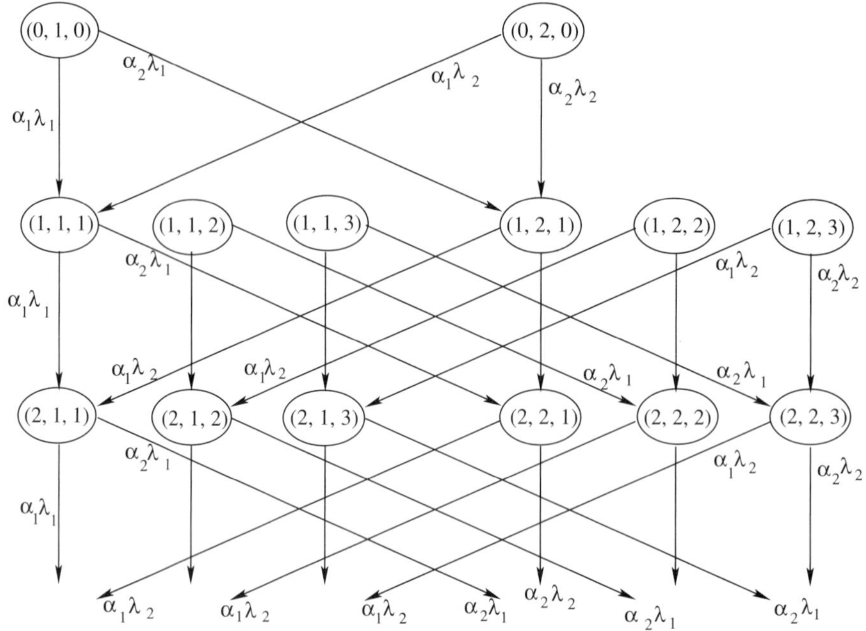


Figure 12.10. Transition diagram for the $H_2/E_3/1$ queue, part (b).

completed that a departure actually occurs. Transitions among the same set of states generated by the arrival process are illustrated in Figure 12.10.

From states $(k, 1, s)$ transitions occur at rate $\lambda_1$. With probability $\alpha_1$ the next customer to begin its arrival process enters phase 1, so that with rate $\alpha_1\lambda_1$, the system moves from state $(k, 1, s)$ to state $(k+1, 1, s)$; with probability $\alpha_2$ the customer beginning its arrival phase chooses phase 2 and so at rate $\alpha_2\lambda_1$ the system moves from state $(k, 1, s)$ to state $(k+1, 2, s)$. A similar set of transitions is obtained from states in which the customer arrives from phase 2. We obtain transitions at rate $\alpha_1\lambda_2$ from states $(k, 2, s)$ to states $(k+1, 1, s)$ and transitions at rate $\alpha_2\lambda_2$ from states $(k, 2, s)$ to $(k+1, 2, s)$. The first few rows of the complete matrix are given by

$$\left(\begin{array}{cc|cccccc|cccccc|c}
-\lambda_1 & 0 & \alpha_1\lambda_1 & 0 & 0 & \alpha_2\lambda_1 & 0 & 0 & 0 & 0 & 0 & 0 & 0 & 0 & \cdots \\
0 & -\lambda_2 & \alpha_1\lambda_2 & 0 & 0 & \alpha_2\lambda_2 & 0 & 0 & 0 & 0 & 0 & 0 & 0 & 0 & \cdots \\
\hline
0 & 0 & * & \mu_1 & 0 & 0 & 0 & 0 & \alpha_1\lambda_1 & 0 & 0 & \alpha_2\lambda_1 & 0 & 0 & \cdots \\
0 & 0 & 0 & * & \mu_2 & 0 & 0 & 0 & 0 & \alpha_1\lambda_1 & 0 & 0 & \alpha_2\lambda_1 & 0 & \cdots \\
\mu_3 & 0 & 0 & 0 & * & 0 & 0 & 0 & 0 & 0 & \alpha_1\lambda_1 & 0 & 0 & \alpha_2\lambda_1 & \cdots \\
0 & 0 & 0 & 0 & 0 & * & \mu_1 & 0 & \alpha_1\lambda_2 & 0 & 0 & \alpha_2\lambda_2 & 0 & 0 & \cdots \\
0 & 0 & 0 & 0 & 0 & 0 & * & \mu_2 & 0 & \alpha_1\lambda_2 & 0 & 0 & \alpha_2\lambda_2 & 0 & \cdots \\
0 & \mu_3 & 0 & 0 & 0 & 0 & 0 & * & 0 & 0 & \alpha_1\lambda_2 & 0 & 0 & \alpha_2\lambda_2 & \cdots \\
\hline
0 & 0 & 0 & 0 & 0 & 0 & 0 & 0 & * & \mu_1 & 0 & 0 & 0 & 0 & \cdots \\
0 & 0 & 0 & 0 & 0 & 0 & 0 & 0 & 0 & * & \mu_2 & 0 & 0 & 0 & \cdots \\
0 & 0 & \mu_3 & 0 & 0 & 0 & 0 & 0 & 0 & 0 & * & 0 & 0 & 0 & \cdots \\
0 & 0 & 0 & 0 & 0 & 0 & 0 & 0 & 0 & 0 & 0 & * & \mu_1 & 0 & \cdots \\
0 & 0 & 0 & 0 & 0 & 0 & 0 & 0 & 0 & 0 & 0 & 0 & * & \mu_2 & \cdots \\
0 & 0 & 0 & 0 & 0 & \mu_3 & 0 & 0 & 0 & 0 & 0 & 0 & 0 & * & \cdots \\
\hline
0 & 0 & 0 & 0 & 0 & 0 & 0 & 0 & 0 & 0 & 0 & 0 & 0 & 0 & \cdots \\
\vdots & \vdots & \vdots & \vdots & \vdots & \vdots & \vdots & \vdots & \vdots & \vdots & \vdots & \vdots & \vdots & \vdots & \vdots
\end{array}\right)$$

where the elements marked by an asterisk in each diagonal block are given by

$$\text{diag}\{-(\lambda_1+\mu_1),\ -(\lambda_1+\mu_2),\ -(\lambda_1+\mu_3),\ -(\lambda_2+\mu_1),\ -(\lambda_2+\mu_2),\ -(\lambda_2+\mu_3)\}$$

so that the sum across each row is equal to zero.

With diagrams of all possible transitions among states, it is possible to construct the block submatrices $A_0$, $A_1$, $A_2$, $B_{00}$, $B_{01}$, and $B_{10}$ and then apply the matrix-geometric approach. However, it is evident that this can become quite messy so we look for a better way to form these blocks. The phase-type distributions that we have considered so far, the Erlang, hyperexponential, and even Coxian distributions, all have very specific structures—the phases are laid out either in series or in parallel. However, as described in Section 7.6.6, it is possible to build much more general phase-type distributions. An arbitrary Markov chain with a single absorbing state and an initial probability distribution contains the essence of a phase-type distribution. The phase-type distribution is defined as the distribution of the time to absorption into the single absorbing state when the Markov chain is started with the given initial probability distribution.

**Example 12.7** For a three-stage hypoexponential distribution with parameters $\mu_1$, $\mu_2$, and $\mu_3$ this Markov chain is completely defined by the following transition rate matrix $S'$ (we shall use $S$ for service processes) and initial probability distribution $\sigma'$:

$$S' = \left(\begin{array}{ccc|c} -\mu_1 & \mu_1 & 0 & 0 \\ 0 & -\mu_2 & \mu_2 & 0 \\ 0 & 0 & -\mu_3 & \mu_3 \\ \hline 0 & 0 & 0 & 0 \end{array}\right) = \begin{pmatrix} S & S^0 \\ 0 & 0 \end{pmatrix}, \qquad \sigma' = (1 \quad 0 \quad 0 \,|\, 0) = (\sigma \quad 0).$$

For a two stage hyperexponential distribution with branching probabilities $\alpha_1$ and $\alpha_2$ $(= 1-\alpha_1)$ and exponential phases with rates $\lambda_1$ and $\lambda_2$, the Markov chain is completely defined by the transition rate matrix $T'$ (we use $T$ for arrival processes) and initial probability distribution $\xi'$ given as

$$T' = \left(\begin{array}{cc|c} -\lambda_1 & 0 & \lambda_1 \\ 0 & -\lambda_2 & \lambda_2 \\ \hline 0 & 0 & 0 \end{array}\right) = \begin{pmatrix} T & T^0 \\ 0 & 0 \end{pmatrix}, \qquad \xi' = (\alpha_1 \quad \alpha_2 \,|\, 0) = (\xi \quad 0).$$

We shall use the partitioned matrices, $S$, $S^0$, $\sigma$ and $T$, $T^0$ and $\xi$ to construct formulae for the specification of the block submatrices needed in the matrix-geometric approach. Indeed, for a *Ph/Ph/1* queue with $r_a$ phases in the description of the arrival process and $r_s$ phases in the description of the service process, the block submatrices required for the application of the matrix geometric approach are given by

$$A_0 = I_{r_a} \otimes (S^0 \cdot \sigma), \quad A_1 = T \otimes I_{r_s} + I_{r_a} \otimes S, \quad \text{and} \quad A_2 = (T^0 \cdot \xi) \otimes I_{r_s}$$

$$B_{00} = T, \quad B_{01} = (T^0 \cdot \xi) \otimes \sigma, \quad \text{and} \quad B_{10} = I_{r_a} \otimes S^0$$

where $I_n$ is the identity matrix of order $n$ and the symbol $\otimes$ denotes the Kronecker (or tensor) product. The Kronecker product of an $m \times n$ matrix $A$ with a matrix $B$ is given by

$$A \otimes B = \begin{pmatrix} a_{11}B & a_{12}B & a_{13}B & \cdots & a_{1n}B \\ a_{21}B & a_{22}B & a_{23}B & \cdots & a_{2n}B \\ a_{31}B & a_{32}B & a_{33}B & \cdots & a_{3n}B \\ \vdots & \vdots & \vdots & \vdots & \vdots \\ a_{m1}B & a_{m2}B & a_{m3}B & \cdots & a_{mn}B \end{pmatrix}.$$

For example, the Kronecker product of

$$A = \begin{pmatrix} a & b & c \\ d & e & f \end{pmatrix} \quad \text{and} \quad B = \begin{pmatrix} \alpha & \beta \\ \gamma & \delta \end{pmatrix}$$

is

$$A \otimes B = \begin{pmatrix} aB & bB & cB \\ dB & eB & fB \end{pmatrix} = \left(\begin{array}{cc|cc|cc} a\alpha & a\beta & b\alpha & b\beta & c\alpha & c\beta \\ a\gamma & a\delta & b\gamma & b\delta & c\gamma & c\delta \\ \hline d\alpha & d\beta & e\alpha & e\beta & f\alpha & f\beta \\ d\gamma & d\delta & e\gamma & e\delta & f\gamma & f\delta \end{array}\right).$$

With these definitions in hand, we may compute the block submatrices for the $H_2/E_3/1$ queue (with $r_a = 2$ and $r_s = 3$) as follows:

$$A_0 = I_2 \otimes (S^0 \cdot \sigma) = I_2 \otimes \begin{pmatrix} 0 \\ 0 \\ \mu_3 \end{pmatrix} (1 \quad 0 \quad 0) = I_2 \otimes \begin{pmatrix} 0 & 0 & 0 \\ 0 & 0 & 0 \\ \mu_3 & 0 & 0 \end{pmatrix} = \begin{pmatrix} 0 & 0 & 0 & 0 & 0 & 0 \\ 0 & 0 & 0 & 0 & 0 & 0 \\ \mu_3 & 0 & 0 & 0 & 0 & 0 \\ 0 & 0 & 0 & 0 & 0 & 0 \\ 0 & 0 & 0 & 0 & 0 & 0 \\ 0 & 0 & 0 & \mu_3 & 0 & 0 \end{pmatrix},$$

$$A_1 = T \otimes I_3 + I_2 \otimes S = \begin{pmatrix} -\lambda_1 & 0 \\ 0 & -\lambda_2 \end{pmatrix} \otimes I_3 + I_2 \otimes \begin{pmatrix} -\mu_1 & \mu_1 & 0 \\ 0 & -\mu_2 & \mu_2 \\ 0 & 0 & -\mu_3 \end{pmatrix}$$

$$= \left(\begin{array}{ccc|ccc} -\lambda_1 & 0 & 0 & 0 & 0 & 0 \\ 0 & -\lambda_1 & 0 & 0 & 0 & 0 \\ 0 & 0 & -\lambda_1 & 0 & 0 & 0 \\ \hline 0 & 0 & 0 & -\lambda_2 & 0 & 0 \\ 0 & 0 & 0 & 0 & -\lambda_2 & 0 \\ 0 & 0 & 0 & 0 & 0 & -\lambda_2 \end{array}\right) + \left(\begin{array}{ccc|ccc} -\mu_1 & \mu_1 & 0 & 0 & 0 & 0 \\ 0 & -\mu_2 & \mu_2 & 0 & 0 & 0 \\ 0 & 0 & -\mu_3 & 0 & 0 & 0 \\ \hline 0 & 0 & 0 & -\mu_1 & \mu_1 & 0 \\ 0 & 0 & 0 & 0 & -\mu_2 & \mu_2 \\ 0 & 0 & 0 & 0 & 0 & -\mu_3 \end{array}\right),$$

$$A_2 = (T^0 \cdot \xi) \otimes I_3 = \begin{pmatrix} \lambda_1 \\ \lambda_2 \end{pmatrix} (\alpha_1 \; \alpha_2) \otimes I_3 = \begin{pmatrix} \alpha_1\lambda_1 & \alpha_2\lambda_1 \\ \alpha_1\lambda_2 & \alpha_2\lambda_2 \end{pmatrix} \otimes I_3$$

$$= \left(\begin{array}{ccc|ccc} \alpha_1\lambda_1 & 0 & 0 & \alpha_2\lambda_1 & 0 & 0 \\ 0 & \alpha_1\lambda_1 & 0 & 0 & \alpha_2\lambda_1 & 0 \\ 0 & 0 & \alpha_1\lambda_1 & 0 & 0 & \alpha_2\lambda_1 \\ \hline \alpha_1\lambda_2 & 0 & 0 & \alpha_2\lambda_2 & 0 & 0 \\ 0 & \alpha_1\lambda_2 & 0 & 0 & \alpha_2\lambda_2 & 0 \\ 0 & 0 & \alpha_1\lambda_2 & 0 & 0 & \alpha_2\lambda_2 \end{array}\right),$$

$$B_{00} = \begin{pmatrix} -\lambda_1 & 0 \\ 0 & -\lambda_2 \end{pmatrix},$$

$$B_{01} = (T^0 \cdot \xi) \otimes \sigma = \begin{pmatrix} \lambda_1 \\ \lambda_2 \end{pmatrix} (\alpha_1 \quad \alpha_2) \otimes (1 \quad 0 \quad 0) = \begin{pmatrix} \alpha_1\lambda_1 & \alpha_2\lambda_1 \\ \alpha_1\lambda_2 & \alpha_2\lambda_2 \end{pmatrix} \otimes (1 \quad 0 \quad 0)$$

$$= \begin{pmatrix} \alpha_1\lambda_1 & 0 & 0 & \alpha_2\lambda_1 & 0 & 0 \\ \alpha_1\lambda_2 & 0 & 0 & \alpha_2\lambda_2 & 0 & 0 \end{pmatrix},$$

$$B_{10} = I_2 \otimes S^0 = I_2 \otimes \begin{pmatrix} 0 \\ 0 \\ \mu_3 \end{pmatrix} = \begin{pmatrix} 0 & 0 \\ 0 & 0 \\ \mu_3 & 0 \\ 0 & 0 \\ 0 & 0 \\ 0 & \mu_3 \end{pmatrix}.$$

It is now clear that these formulae give the same results as those obtained previously by constructing the matrix directly from the transition rate diagram. This technique is easily incorporated into Matlab code using the Matlab operator *kron(A,B)* to form the Kronecker product of $A$ and $B$. Furthermore, the size of the blocks may be obtained directly by means of the Matlab *size* operator: *size(A,2)* gives the number of columns in a matrix $A$. As an example, the following Matlab code constructs the block matrices for the $H_2/E_3/1$ queue. Observe that only the first eight lines of code (four for each of the arrival and service processes) are specific to this queue. The last eight are appropriate for all *Ph/Ph/1* queues.

```
%%%%%%%%%  Construct submatrices for H_2/E_3/1 Queue  %%%%%
%%%  Specific parameters for H_2/E_3/1 queue:

%%%    H_2 Arrival Process:
       alpha1 = 0.4; alpha2 = 0.6; lambda1 = 1; lambda2 = 2;
       T = [-lambda1, 0 ; 0, -lambda2];
       T0 = [lambda1;lambda2];
       xi = [alpha1, alpha2];

%%%    E_3 Service Process:
       mu1 = 4; mu2 = 8; mu3 = 8;
       S = [-mu1, mu1, 0; 0, -mu2, mu2; 0,0, -mu3];
       S0 = [0;0;mu3];
       sigma = [1,0,0];

%%% Block Submatrices for all types of queues:
       ra = size(T,2);  rs = size(S,2);
       A0 = kron(eye(ra), S0*sigma);
       A1 = kron(T, eye(rs)) + kron(eye(ra), S);
       A2 = kron(T0*xi, eye(rs));
       B00 = T;
       B01 = kron(T0*xi,sigma);
       B10 = kron(eye(ra),S0);
       l = size(B00,2);  r = size(A0,2);
```

The remaining sections of the Matlab code, those for generating Neuts' $R$ matrix, solving the boundary equations and for constructing successive solution components, remain unchanged and may be appended directly to the code just given to form a complete Matlab program that may be used to solve any *Ph/Ph/1* queue. Only one issue remains: that of determining the stability of the queue.

## 12.6 Stability Results for *Ph/Ph/1* Queues

The $M/M/1$ queue is stable—the number of customers does not increase indefinitely—when the mean arrival rate $\lambda$ is strictly less than the mean service rate $\mu$. If $E[A]$ is the mean interarrival time and $E[S]$ is the mean service time, then this stability condition can be written as

$$\frac{1}{E[A]} < \frac{1}{E[S]} \quad \text{or} \quad E[S] < E[A],$$

which means that the $M/M/1$ queue is stable when the average time it takes to serve a customer is strictly less than the average time between customer arrivals. A similar condition holds for all *Ph/Ph/1* queues: the mean interarrival time must be greater than the mean service time. This condition may be readily verified, especially when the phase-type distributions involved are standard distributions such as the Erlang and hyperexponential distributions used in the previous examples.

**Example 12.8** The expectation of a two-phase hyperexponential distribution used to model an arrival process is given by $E[A]=\alpha_1/\lambda_1 + \alpha_2/\lambda_2$, while the expectation of a three-phase hypoexponential distribution used to model a service process is $E[S] = 1/\mu_1 + 1/\mu_2 + 1/\mu_3$. Using the values of previous examples, ($\alpha_1 = 0.4,\ \alpha_2 = 0.6,\ \lambda_1 = 1,\ \lambda_2 = 2,\ \mu_1 = 4,\ \mu_2 = 8$, and $\mu_3 = 8$), we find

$$E[S] = \frac{1}{4} + \frac{1}{8} + \frac{1}{8} = 0.5 \ < \ \frac{0.4}{1} + \frac{0.6}{2} = 0.7 = E[A].$$

Alternatively, if we let $\lambda$ be the effective arrival rate into the $H_2/E_3/1$ and $\mu$ be the effective service rate, then $\lambda = 1/0.7 = 1.428571$ and $\mu = 1/0.5 = 2$ and the condition $\rho = \lambda/\mu < 1$ is satisfied.

For a general phase-type distribution, with an infinitesimal generator matrix given by

$$Z' = \begin{pmatrix} Z & Z_0 \\ 0 & 0 \end{pmatrix},$$

and an initial probability distribution $\zeta$, the expectation of the time to absorption (see Section 9.6.2) is given by

$$E[A] = \| - \zeta Z^{-1} \|_1.$$

**Example 12.9** Using the above formulation, the average interarrival time in the $H_2/E_3/1$ queue is obtained as

$$E[A] = \left\| -(\alpha_1, \alpha_2) \begin{pmatrix} -\lambda_1 & 0 \\ 0 & -\lambda_2 \end{pmatrix}^{-1} \right\|_1 = \left\| -(0.4,\ 0.6) \begin{pmatrix} -1 & 0 \\ 0 & -2 \end{pmatrix}^{-1} \right\|_1 = \|(0.4, 0.3)\|_1 = 0.7.$$

Similarly, the average service time may also be computed.

The same stability condition may be derived from the matrices $A_0$, $A_1$ and $A_2$ used in the matrix-geometric method. The matrix $A = A_0 + A_1 + A_2$ is an infinitesimal generator in its own right. In other words, its off-diagonal elements are nonnegative, its diagonal elements are nonpositive and the sum across all rows is equal to zero. As such it has a stationary probability vector which we shall call $\gamma$, i.e.,

$$\gamma A = \gamma(A_0 + A_1 + A_2) = 0.$$

The nonzero elements of the block subdiagonal matrices $A_0$ are responsible for moving the system down a level, i.e., from some level (number of customers) $l$ to the next lower level $l - 1$. This relates directly to service completions in a *Ph/Ph/1* queue. On the other hand, the nonzero elements of the superdiagonal blocks $A_2$ move the system up from some level $l$ to the next level $l + 1$: the number of customers in the system increases by one. For stability we require the effect of the matrices $A_2$ to be less than the effect of the matrices $A_0$: the system must have a tendency to drift to the point of zero customers rather than the opposite—similar to the condition that was found to be necessary for the states of a random walk problem to be positive recurrent. The matrix $A = A_0 + A_1 + A_2$ encapsulates these drift tendencies and the condition for stability can be written as

$$\|\gamma A_2\|_1 < \|\gamma A_0\|_1.$$

**Example 12.10** Substituting previously used values for the $H_2/E_3/1$ queue, we find

$$A = \begin{pmatrix} -4.6 & 4.0 & 0 & 0.6 & 0 & 0 \\ 0 & -8.6 & 8.0 & 0 & 0.6 & 0 \\ 8.0 & 0 & -8.6 & 0 & 0 & 0.6 \\ 0.8 & 0 & 0 & -4.8 & 4.0 & 0 \\ 0 & 0.8 & 0 & 0 & -8.8 & 8.0 \\ 0 & 0 & 0.8 & 8.0 & 0 & -8.8 \end{pmatrix}$$

and its stationary probability vector, obtained by solving $\gamma A = 0$ with $\|\gamma\|_1 = 1$, is

$$\gamma = (0.285714,\ 0.142857,\ 0.142857,\ 0.214286,\ 0.107143,\ 0.107143).$$

Computing $\|\gamma A_2\|_1$ and $\|\gamma A_0\|_1$, we find

$$\lambda = \|\gamma A_2\|_1 = \left\| \gamma \begin{pmatrix} 0.4 & 0 & 0 & 0.6 & 0 & 0 \\ 0 & 0.4 & 0 & 0 & 0.6 & 0 \\ 0 & 0 & 0.4 & 0 & 0 & 0.6 \\ 0.8 & 0 & 0 & 1.2 & 0 & 0 \\ 0 & 0.8 & 0 & 0 & 1.2 & 0 \\ 0 & 0 & 0.8 & 0 & 0 & 1.2 \end{pmatrix} \right\|_1$$

$$= \|(0.285714,\ 0.142857,\ 0.142857,\ 0.428571,\ 0.214286,\ 0.214286)\|_1 = 1.428571$$

and

$$\mu = \|\gamma A_0\|_1 = \left\| \gamma \begin{pmatrix} 0 & 0 & 0 & 0 & 0 & 0 \\ 0 & 0 & 0 & 0 & 0 & 0 \\ 8 & 0 & 0 & 0 & 0 & 0 \\ 0 & 0 & 0 & 0 & 0 & 0 \\ 0 & 0 & 0 & 0 & 0 & 0 \\ 0 & 0 & 0 & 8 & 0 & 0 \end{pmatrix} \right\|_1$$

$$= \|(1.142857,\ 0,\ 0,\ 0.857143,\ 0,\ 0)\|_1 = 2.0$$

which is the same as that obtained previously.

It is an interesting exercise to vary the parameters of the $H_2/E_3/1$ queue so that different values are obtained for $\rho$ and to observe the effect of this on the number of iterations needed for convergence to Neuts' $R$ matrix. Some results obtained by varying only the parameter $\lambda_1$, are shown in Table 12.1.

The numbers under the column heading SS are the numbers of iterations needed to converge when the *successive substitution* algorithm presented in the previous Matlab code segments is used. This table clearly shows that the closer $\rho$ is to 1, the slower the convergence. However, recall from Chapter 11 that there are alternative, more sophisticated, approaches to computing Neuts' $R$ matrix when the model is of the quasi-birth-death type, as in this case. In particular, the *logarithmic reduction* algorithm may be used to great effect. The number of iterations needed to obtain $R$ to the same precision using this method is given in the table under the column heading LR. Matlab code which implements the logarithmic-reduction algorithm is provided in the program at the end of this section.

Table 12.1. Effect of varying $\lambda_1$ on $\rho$ and convergence to $R$.

| $\lambda_1$ | $\rho$ | SS | LR |
|---|---|---|---|
| 0.1 | 0.1163 | 28 | 5 |
| 0.5 | 0.4545 | 50 | 6 |
| 1.0 | 0.7143 | 98 | 6 |
| 1.5 | 0.8824 | 237 | 8 |
| 1.6 | 0.9091 | 303 | 8 |
| 1.7 | 0.9341 | 412 | 8 |
| 1.8 | 0.9574 | 620 | 9 |
| 1.9 | 0.9794 | 1197 | 10 |
| 1.95 | 0.9898 | 2234 | 11 |
| 2.0 | 1.0 | $\infty$ | $\infty$ |

## 12.7 Performance Measures for *Ph/Ph/1* Queues

So far the only performance measures we have obtained for *Ph/Ph/1* queues are the stationary probabilities of the underlying Markov chains. It is of course, possible to get much useful information directly from these. For example, the probability that there are $k$ customers present in the queueing system, for $k \geq 1$, is obtained by adding the components of the $k^{\text{th}}$ subvector, i.e.,

$$p_k = \|\pi_k\|_1 = \|\pi_1 R^{k-1}\|_1.$$

In particular, the probability that the system is empty is given by $p_0 = \|\pi_0\|_1$ while the probability that the system is busy is $1 - p_0$. Of course, these results could also be obtained as $p_0 = 1 - \rho = 1 - \lambda/\mu$ where $\lambda$ and $\mu$ are computed from $\lambda = 1/E[A]$ and $\mu = 1/E[S]$ and

$$E[A] = \| -\xi T^{-1}\|_1 \quad \text{and} \quad E[S] = \| -\sigma S^{-1}\|_1.$$

The probability that there are $k$ or more customers present can also be obtained relatively easily. We have

$$\text{Prob}\{N \geq k\} = \sum_{j=k}^{\infty} \|\pi_j\|_1 = \left\| \pi_1 \sum_{j=k}^{\infty} R^{j-1} \right\|_1 = \left\| \pi_1 R^{k-1} \sum_{j=0}^{\infty} R^j \right\|_1 = \left\|\pi_1 R^{k-1}(I-R)^{-1}\right\|_1.$$

As in this last case, Neuts' $R$ matrix can be used to compute the mean number of customers in a *Ph/Ph/1* queueing system and this, together with Little's law and other relationships, allow us to compute the expected number of customers waiting for service and the expected response and waiting times. The average number of customers in a *Ph/Ph/1* queueing system is obtained as

$$E[N] = \sum_{k=1}^{\infty} k\,\|\pi_k\|_1 = \sum_{k=1}^{\infty} k\,\|\pi_1 R^{k-1}\|_1 = \left\| \pi_1 \sum_{k=1}^{\infty} \frac{d}{dR} R^k \right\|_1 = \left\| \pi_1 \frac{d}{dR} \left( \sum_{k=1}^{\infty} R^k \right) \right\|_1$$

$$= \left\| \pi_1 \frac{d}{dR} \left( (I-R)^{-1} - I \right) \right\|_1 = \left\|\pi_1 (I-R)^{-2}\right\|_1.$$

The mean number of customers waiting in the queue, $E[N_q]$; the average response time, $E[R]$ and the average time spent waiting in the queue, $E[W_q]$ may now be obtained from the standard formulae. We have

$$E[N_q] = E[N] - \lambda/\mu,$$

$$E[R] = E[N]/\lambda,$$

$$E[W_q] = E[N_q]/\lambda.$$

## 12.8 Matlab code for *Ph/Ph/1* Queues

A combination of the codes and procedures developed in this chapter results in the following Matlab program.

```
     Function ph_ph_1()
%%%%%%%%%%%%%%%%%%%%%%%%%%%%%%%%%%%%%%%%%%%%%%%%%%%%%%%%%%%%%%

%%%  Example 1:  M/E_4/1 Queue
%%%  Exponential arrival:
%    lambda = 4;
%    T = [-lambda]; T0=[lambda]; xi = [1];


%%%   Erlang-4 Service (use mu_i = r*mu per phase)
%     mu1 = 20; mu2 = 20; mu3 = 20; mu4 = 20;
%     S = [-mu1, mu1, 0,0; 0, -mu2, mu2,0; 0,0 -mu3,mu3;0,0,0, -mu4];
%     S0 = [0;0;0;mu4];
%     sigma = [1,0,0,0];


%%%   Example 2: H_2/Ph/1 queue:
%%%   H_2 Arrival Process:
      alpha1 = 0.4; alpha2 = 0.6; lambda1 = 1.9; lambda2 = 2;
      T = [-lambda1, 0 ; 0, -lambda2];
      T0 = [lambda1;lambda2];
      xi = [alpha1, alpha2];
%%%   Hypo-exponential-3 Service Process:
      mu1 = 4; mu2 = 8; mu3 = 8;
      S = [-mu1, mu1, 0; 0, -mu2, mu2; 0,0, -mu3];
      S0 = [0;0;mu3];
      sigma = [1,0,0];


%%%%%%%%%  Block Submatrices for all types of queues:  %%%%%%
      ra = size(T,2);  rs = size(S,2);
      A0 = kron(eye(ra), S0*sigma);
      A1 = kron(T, eye(rs)) + kron(eye(ra), S);
      A2 = kron(T0*xi, eye(rs));
      B00 = T;
      B01 = kron(T0*xi,sigma);
      B10 = kron(eye(ra),S0);
      l = size(B00,2);  r = size(A0,2);


%%%%%%%%%  Check stability  %%%%%%%%%%%%%%%%%%%%%%%%%%%%%%%%%%%

      meanLambda = 1/norm(-xi* inv(T),1);
      meanMu = 1/norm(-sigma * inv(S),1);
```

```
      rho = meanLambda/meanMu
%%%%%%%%%%%  Alternatively:   %%%%%%%%
      A = A0+A1+A2;
      for k=1:r
          A(k,r) = 1;
      end
      rhs = zeros(1,r); rhs(r)= 1;
      ss = rhs*inv(A);
      rho = norm(ss*A2,1)/norm(ss*A0,1);
%%%%%%%%%%%%%%%%%%%%%%%%%%%%%%%%%%%%%%%%

      if  rho >=0.999999
          error('Unstable System');
      else
          disp('Stable system')
      end


%%%%%%%%%%     Form Neuts' R matrix    %%%%%%%%%%%%%%%%%%%%%%%%%%%%
%%%%%%%%%%              by             %%%%%%%%%%%%%%%%%%%%%%%%%%%%
%%%%%%%%%%   Successive Substitution   %%%%%%%%%%%%%%%%%%%%%%%%%%%%

     V = A2 * inv(A1);  W = A0 * inv(A1);

     R = -V;      Rbis = -V - R*R * W;
     iter = 1;
     while (norm(R-Rbis,1)> 1.0e-10 & iter<100000)
        R = Rbis;  Rbis = -V - R*R * W;
        iter = iter+1;
     end
     iter
     R = Rbis;

%%%%%%%%%%%%%           or  by          %%%%%%%%%%%%%%%%%%%%%%%%%%%%
%%%%%%%%%%%%%    Logarithmic Reduction %%%%%%%%%%%%%%%%%%%%%%%%%%%%

%    Bz = -inv(A1)*A2;  Bt = -inv(A1)*A0;
%    T = Bz;  S = Bt;
%    iter = 1;
%    while (norm(ones(r,1)-S*ones(r,1) ,1)> 1.0e-10 & iter<100000)
%       D = Bz*Bt + Bt*Bz;
%       Bz = inv(eye(r)-D) *Bz*Bz;
%       Bt = inv(eye(r)-D) *Bt*Bt;
%       S = S + T*Bt;
%       T = T*Bz;
%       iter = iter+1;
%    end
%    iter
%    U = A1 + A2*S;
%    R = -A2 * inv(U)
```

```
%%%%%%%%%%%%%  Solve boundary equations  %%%%%%%%%%%%%%%%%%%%
     N = [B00,B01;B10,A1+R*A0];  % Set up boundary equations

     N(1,r+l) = 1;                % Set first component equal to 1
     for k=2:r+l
        N(k,r+l) = 0;
     end
     rhs = zeros(1,r+l); rhs(r+l)= 1;
     soln = rhs * inv(N);          % Un-normalized pi_0 and pi_1

     pi0 = zeros(1,l);  pi1 = zeros(1,r);
     for k=1:l
       pi0(k) = soln(k);          % Extract pi_0
     end
     for k=1:r
       pi1(k) = soln(k+l);        % Extract pi_1
     end
     e = ones(r,1);
     sum = norm(pi0,1) +  pi1 * inv(eye(r)-R) * e;    % Normalize solution
     pi0 = pi0/sum; pi1 = pi1/sum;

%%%%%%%%%%%%%%%%  Print results %%%%%%%%%%%%%%%%%%%%%%%%%%%%
     max = 10;     %  maximum population requested
     pop = zeros(max+1,1);
     pop(1) =  norm(pi0,1);
     for k=1:max
       pi = pi1 * R^(k-1);           % Get successive components of pi
       pop(k+1) = norm(pi,1);
     end
     pop

%%%%%%%%%%%%  Measures of Effectiveness  %%%%%%%%%%%%%%%%%%%%
     EN = norm(pi1*inv(eye(r)-R)^2,1)
%    ENq = EN-meanLambda/meanMu
%    ER =  EN/meanLambda
%    EWq = ENq/meanLambda

%%%%%%%%%%%%%%%%%%%%%%%%%%%%%%%%%%%%%%%%%%%%%%%%%%%%%%%%%%%%%%
```

## 12.9 Exercises

**Exercise 12.1.1** Use Equation (12.1) to form the inverse of the following matrices:

$$M_1 = \begin{pmatrix} 1 & 1 & 0 & 0 & 0 \\ 0 & 1 & 1 & 0 & 0 \\ 0 & 0 & 1 & 1 & 0 \\ 0 & 0 & 0 & 1 & 1 \\ 0 & 0 & 0 & 0 & 1 \end{pmatrix}, \quad M_2 = \begin{pmatrix} -5 & 4 & 0 & 0 & 0 \\ 0 & -5 & 4 & 0 & 0 \\ 0 & 0 & -5 & 4 & 0 \\ 0 & 0 & 0 & -5 & 4 \\ 0 & 0 & 0 & 0 & -5 \end{pmatrix}.$$

**Exercise 12.1.2** Prove that the $ij$ element of the inverse of any matrix of the form

$$M = \begin{pmatrix} d & a & 0 & 0 & \cdots & 0 \\ 0 & d & a & 0 & \cdots & 0 \\ 0 & 0 & d & a & \cdots & 0 \\ \vdots & \vdots & \vdots & \ddots & \ddots & \vdots \\ 0 & 0 & 0 & 0 & \vdots & a \\ 0 & 0 & 0 & 0 & \cdots & d \end{pmatrix}$$

is

$$M_{ij}^{-1} = (-1)^{j-i}\frac{1}{d}\left(\frac{a}{d}\right)^{j-i} \quad \text{for } i \le j \le r,$$

and is equal to zero otherwise.

**Exercises 12.1.3–12.3.2 should be answered from first principles, following the manner of Sections 12.1–12.3.**

**Exercise 12.1.3** Draw the state transition diagram of an $M/E_4/1$ queue with mean interarrival time of 1/4 and mean service time of 1/5. Form the submatrices $A_0$, $A_1$, $A_2$, $B_{00}$, $B_{01}$, and $B_{10}$ that are needed in solving this queueing system by the method of Neuts.

**Exercise 12.1.4** For the following $M/E_r/1$ queues, compute the matrices $V$ and $W$ to be used in the successive substitution iterative algorithm for computing Neuts' $R$ matrix. Then, beginning with $R_0 = 0$, compute $R_1$ and $R_2$.

(a) $M/E_2/1$ with $\lambda = 4.0$ and, at each service phase, $\mu_i = 10.0$.
(b) $M/E_4/1$ with $\lambda = 4.0$ and, at each service phase, $\mu_i = 20.0$.

**Exercise 12.1.5** It is known that Neuts' $R$ matrix for the $M/E_2/1$ queue with mean interarrival time $E[A] = 1/4$ and mean service time $E[S] = 1/5$ is given by

$$R = \begin{pmatrix} 0.5600 & 0.4000 \\ 0.1600 & 0.4000 \end{pmatrix},$$

while for the $M/E_4/1$ queue with the same expected interarrival and service times, it is given by

$$R = \begin{pmatrix} 0.3456 & 0.2880 & 0.2400 & 0.2000 \\ 0.1456 & 0.2880 & 0.2400 & 0.2000 \\ 0.1056 & 0.0880 & 0.2400 & 0.2000 \\ 0.0576 & 0.0480 & 0.0400 & 0.2000 \end{pmatrix}.$$

For each of these queues, compute the probability that

(a) the system is empty;
(b) the system contains exactly three customers;
(c) the system contains three or more customers.

**Exercise 12.2.1** Draw the state transition diagram of an $E_4/M/1$ queue with mean interarrival time of 1/3 and mean service time of 1/4. Form the submatrices $A_0$, $A_1$, $A_2$, $B_{00}$, $B_{01}$, and $B_{10}$ that are needed in solving this queueing system by the method of Neuts.

**Exercise 12.2.2** For the following $E_r/M/1$ queues, compute the matrices $V$ and $W$ to be used in the successive substitution iterative algorithm for computing Neuts' $R$ matrix. Then, beginning with $R_0 = 0$, compute $R_1$ and $R_2$.

(a) $E_2/M/1$ with mean interarrival time of 1/3 and mean service time of 1/4.
(b) $E_4/M/1$ with mean interarrival time of 1/3 and mean service time of 1/4.

**Exercise 12.2.3** It is known that Neuts' $R$ matrix for the $E_2/M/1$ queue with mean interarrival time of 1/3 and mean service time of 1/4 is given by

$$R = \begin{pmatrix} 0 & 0 \\ 0.8229 & 0.6771 \end{pmatrix},$$

while for the $E_4/M/1$ queue with mean interarrival time of 1/3 and mean service time of 1/4 it is given by

$$R = \begin{pmatrix} 0 & 0 & 0 & 0 \\ 0 & 0 & 0 & 0 \\ 0 & 0 & 0 & 0 \\ 0.8882 & 0.7889 & 0.7007 & 0.6223 \end{pmatrix}.$$

For each of these queues, compute the probability that

(a) The system is empty;
(b) The system contains exactly three customers;
(c) The system contains more than three customers.

**Exercise 12.3.1** Draw the state transition diagram and the transition rate matrix for each of the following queues. In both cases, generate the matrices $V$ and $W$ needed in the computation of Neuts' $R$ matrix, and carry out two iterations of the successive substitution method (beginning with $R_0 = 0$).

(a) The $M/H_3/1$ queue with $\lambda = 2$, $\alpha_1 = 0.5$, $\alpha_2 = 0.3$, $\alpha_3 = 0.2$, $\mu_1 = 4$, $\mu_2 = 2$, and $\mu_3 = 1$.
(b) The $H_3/M/1$ queue with $\alpha_1 = 0.6$, $\alpha_2 = 0.3$, $\alpha_3 = 0.1$, $\lambda_1 = 4$, $\lambda_2 = 3$, $\lambda_3 = 2$, and $\mu = 4$.

**Exercise 12.3.2** It is known that Neuts' $R$ matrix for the $M/H_3/1$ queue with $\lambda = 2$, $\alpha_1 = 0.5$, $\alpha_2 = 0.3$, $\alpha_3 = 0.2$, $\mu_1 = 4$, $\mu_2 = 2$, and $\mu_3 = 1$ is given by

$$R = \begin{pmatrix} 0.4343 & 0.0909 & 0.0808 \\ 0.1515 & 0.6364 & 0.1212 \\ 0.2020 & 0.1818 & 0.8283 \end{pmatrix},$$

while for the $H_3/M/1$ queue with $\alpha_1 = 0.6$, $\alpha_2 = 0.3$, $\alpha_3 = 0.1$, $\lambda_1 = 4$, $\lambda_2 = 3$, $\lambda_3 = 2$, and $\mu = 4$. It is given by

$$R = \begin{pmatrix} 0.5179 & 0.3302 & 0.1519 \\ 0.3884 & 0.2477 & 0.1139 \\ 0.2590 & 0.1651 & 0.0759 \end{pmatrix}.$$

For each of these queues, compute the probability that

(a) the system is empty;
(b) the system contains exactly three customers;
(c) the system contains more than three customers.

**Exercise 12.4.1** Develop the Matlab code necessary to construct the block submatrices for an $M/E_r/1$ queue and show that when $r$ is taken to be equal to 3 and this code is incorporated with the Matlab code of Section 12.4, that the resulting program produces the same results as those obtained in the text.

**Exercise 12.4.2** Develop the Matlab code necessary to construct the block submatrices for an $E_r/M/1$ queue and show that when $r$ is taken to be equal to 3 and this code is incorporated with the Matlab code of Section 12.4, that the resulting program produces the same results as those obtained in the text.

**Exercises 12.5.1–12.5.10 may be answered using the Matlab code in Section 12.8.**

**Exercise 12.5.1** Consider a queueing system in which the arrival process is Poisson at rate $\lambda = 2$ and for which the density function of the service process is given as

$$b(x) = 64xe^{-8x}, \quad x \geq 0.$$

Compute the probability of having three or more customers in the system.

**Exercise 12.5.2** Consider a queueing system in which the service process is exponential at rate $\mu = 1.2$ and for which the density function of the arrival process is given as

$$a(x) = \frac{3(3x)^2 e^{-3x}}{2}, \quad x \geq 0.$$

Compute the probability of having exactly three customers in the system.

**Exercise 12.5.3** Cars arriving at Bunkey's Car Wash follow a Poisson distribution with a mean interarrival time of 10 minutes. These cars are successively vacuumed, washed, and hand-dried, and the time to perform each of the three tasks is exponentially distributed with a mean of 3 minutes. How long should an arriving customer expect to wait before vacuuming begins on her car? Bunkey realizes that this is excessive and hires additional help so that the time to vacuum and hand-dry a car is reduced to 1 minute each. By how much does this reduce the customer's waiting time?

**Exercise 12.5.4** The distribution of the time it takes to perform an oil change at QuikLube has an expectation of 4 minutes and a variance of 4. If car arrivals follow a Poisson process with mean interarrival time of 5 minutes, find

(a) the probability of having more than 2 cars at QuikLube;
(b) the mean number of cars present;
(c) the mean time spent waiting prior to the commencement of service.

**Exercise 12.5.5** The mean interarrival time between patients to a doctor's office is 16 minutes and has a variance of 32. The time spent with the doctor is exponentially distributed with mean 12 minutes. What is the probability of finding at least two patients sitting in the waiting room? Also find the mean number in the doctor's office and the expected time spent in the waiting room.

**Exercise 12.5.6** What is the Markov chain and initial probability vector that represent a four-phase Coxian distribution with parameters $\mu_i, \alpha_i, i = 1, 2, 3, 4$.

**Exercise 12.5.7** Draw the phase-type distribution that corresponds to the following Markov chain and initial probability vector:

$$Z' = \left(\begin{array}{cccc|c} -\mu_1 & 0 & \alpha_1\mu_1 & (1-\alpha_1)\mu_1 & 0 \\ \mu_2 & -\mu_2 & 0 & 0 & 0 \\ 0 & \alpha_3\mu_3 & -\mu_3 & (1-\alpha_3)\mu_3 & 0 \\ 0 & \alpha_4\mu_4 & 0 & -\mu_4 & (1-\alpha_4)\mu_4 \\ \hline 0 & 0 & 0 & 0 & 0 \end{array}\right), \quad \zeta' = (0.6 \quad 0.4 \quad 0 \quad 0 \,|\, 0).$$

**Exercise 12.5.8** Consider a single-server queueing system in which the service process has a mean service time of $E[S] = 3$ and standard deviation $\sigma_S = 4$ while the arrival process has a mean interarrival time of $E[A] = 4$ and variance $\text{Var}[A] = 5$. Model the arrival and service processes by phase-type distributions and analyze the resulting $Ph/Ph/1$ queue to determine the probability of having exactly $i$ (for $i = 0, 1, 2, 3$, and 4) customers in the system.

**Exercise 12.5.9** Use the Matlab code developed in the text to investigate the effect of reducing the variance in the service time distribution of a single serve queue with a hyperexponential arrival process. In particular, assume that the arrival process has an $H_2$ distribution with parameters $\alpha_1 = 0.4$, $\lambda_1 = 1$, and $\lambda_2 = 4$ and that the service process has a mean service rate equal to 2 (mean service time equal to 1/2). Examine the effect of modeling the service process with an $E_r$ distribution for $r = 1, 2, 4$, and 8 and investigate how the expected number of customers in the system, $E[N]$, varies with $r$.

**Exercise 12.5.10** The same as the previous question, but this time reverse the arrival and service time laws. Assume that the service process has an $H_2$ distribution with parameters $\alpha_1 = 0.4$, $\mu_1 = 1.0$, and $\mu_2 = 8.0$ and that the arrival process has a mean interarrival time equal to 1/2 (mean arrival rate equal to 2). Examine the effect of modeling the arrival process with an $E_r$ distribution for $r = 1, 2, 4$, and 8 and investigate how the expected number of customers in the system, $E[N]$, varies with $r$.

# Chapter 13

# *The z-Transform Approach to Solving Markovian Queues*

Queueing systems that can be represented by a set of states in which the sojourn time is exponentially distributed are called Markovian queues. Examples include all of the queues discussed so far, from the $M/M/1$ queue and its extensions of Chapter 11 through the *Ph/Ph/1* type queues of Chapter 12. When the states can be arranged in a linear fashion and the transitions are to nearest neighbors only, the process is called a birth-death process; the transition rate matrix is tridiagonal and efficient solution approaches are readily available as was seen in Chapter 11. When states can be arranged into equally sized levels and transitions are permitted within a level and to neighboring levels only, the process is called a quasi-birth-death process; the transition rate matrix is block diagonal and the efficient matrix geometric method can be used as in Chapter 12. In both cases, a limited number of boundary states that do not meet the transition restriction to nearest neighbor states (or levels) can be handled. Prior to Neuts' development of the matrix geometric approach, the standard method for solving advanced Markovian models was the $z$-transform approach which we develop in this chapter, the only other alternative being the numerical solution methods which were discussed extensively in Part II of this text. Although the $z$-transform approach is sometimes considered to be an analytic as opposed to a numerical procedure, it is not entirely analytic, because, as we shall see in later sections, the roots of a polynomial equation frequently must be found numerically and used in various formulae.

## 13.1 The $z$-Transform

In Chapter 5, the probability mass function of a discrete random variable is shown to be uniquely determined by its $z$-transform (also called its probability generating function). Furthermore this $z$-transform provides a convenient means for computing the mean and higher moments of the distribution. Indeed, in Chapter 5 our only interest in the $z$-transform is in using it to compute expectations and higher moments. The property of the $z$-transform that is of interest to us in this chapter is that it provides an effective method for solving the sets of difference equations that arise in analyzing certain queueing models. The approach is to combine the difference equations into a single equation from which the $z$-transform can be extracted and moments subsequently computed. In some cases, it is possible to invert the transform to compute the actual probability distribution of customers in the system. We begin by recalling the definition of the $z$-transform.

**Definition:** *The z-transform of a sequence* $p_k,\ k = 0,\ 1,\ \ldots,$ *is defined as*

$$P(z) = \sum_{k=0}^{\infty} p_k z^k,$$

*where z is a complex variable and is such that* $P(z)$ *is analytic i.e.,* $\sum_{k=0}^{\infty} p_k z^k < \infty$.

If the sequence $p_k$ represents the probability distribution of a discrete random variable X, then a valid range for $z$ includes the unit disk, $|z| \le 1$. Observe that the $z$-transform has combined the sequence into a single number. Each term has been "tagged" by a different power of $z$, and this enables the reconstruction of the sequence from the transform. In some cases, the easiest way to invert the transform and retrieve the sequence is to expand the transform in powers of $z$ and to associate the coefficient of $z^k$ with $p_k$.

**Example 13.1** The $z$-transform of a sequence is given by

$$P(z) = \frac{1}{1 - \alpha z}$$

for some constant $\alpha$. Expanding this transform in terms of $z$, we find

$$\frac{1}{1-\alpha z} = \sum_{k=0}^{\infty} \alpha^k z^k = z^0 + \alpha z^1 + \alpha^2 z^2 + \alpha^3 z^3 + \cdots .$$

Identifying the coefficients of the powers of $z$ with successive terms in the sequence shows that the sequence is given by

$$1,\ \alpha,\ \alpha^2,\ \alpha^3,\ \ldots .$$

In this case, for the transform to be analytic, we require that $|z| < 1/\alpha$. The constant $\alpha$ itself may be greater than or less than 1.

In other cases, it may be more appropriate to compute individual elements of the sequence from the formula

$$p_k = \frac{1}{k!} \frac{d^k}{dz^k} P(z) \bigg|_{z=0} \quad \text{for } k = 0,\ 1,\ \ldots . \tag{13.1}$$

For example, we have

$$\frac{1}{0!} P(z) \bigg|_{z=0} = p_0 \times 1\,|_{z=0} + p_1 \times z^1\,|_{z=0} + p_2 \times z^2\,|_{z=0} + p_3 \times z^3\,|_{z=0} \cdots = p_0,$$

$$\frac{1}{1!} \frac{d}{dz} P(z) \bigg|_{z=0} = p_0 \times 0\,|_{z=0} + p_1 \times 1\,|_{z=0} + p_2 \times 2z\,|_{z=0} + p_3 \times 3z^2\,|_{z=0} \cdots = p_1,$$

$$\frac{1}{2!} \frac{d^2}{dz^2} P(z) \bigg|_{z=0} = \frac{1}{2} (p_1 \times 0\,|_{z=0} + p_2 \times 2\,|_{z=0} + p_3 \times 6z\,|_{z=0} \cdots ) = p_2,$$

and so on. The mean and higher moments of the probability distribution may be conveniently computed directly from the $z$-transform by evaluating these same derivatives at $z = 1$ rather than at $z = 0$. The mean is obtained as

$$E[X] = \frac{d}{dz} P(z) \bigg|_{z=1} .$$

Second and higher moments may be computed from correspondingly higher derivatives. The $k^{\text{th}}$ derivative of the $z$-transform evaluated at $z = 1$ gives the $k^{\text{th}}$ factorial moment:

$$\lim_{z \to 1} P^{(k)}(z) = E[X(X-1) \ldots .(X-k+1)].$$

Thus, for example,

$$\lim_{z \to 1} P''(z) = \sum_{k=0}^{\infty} k(k-1) p_k = E[X(X-1)] = E[X^2] - E[X]$$

and the variance may now be found as

$$\text{Var}[X] = E[X(X-1)] + E[X] - E^2[X].$$

**Example 13.2** Consider the probability sequence $p_0 = 0.5$, $p_1 = 0.25$, $p_2 = 0.25$ and $p_k = 0$ for all $k \geq 3$. Its $z$-transform is given by $P(z) = 0.5 \times z^0 + 0.25 \times z^1 + 0.25 \times z^2 = 0.25z^2 + 0.25z + 0.5$. Observe that the terms of the sequence are reconstructed as

$$\begin{aligned}
p_0 &= P(z)|_{z=0} = 0.25z^2 + 0.25z + 0.5\big|_{z=0} = 0.5,\\
p_1 &= \frac{d}{dz}P(z)\bigg|_{z=0} = 2 \times 0.25z + 0.25|_{z=0} = 0.25,\\
p_2 &= \frac{1}{2}\frac{d^2}{dz^2}P(z)\bigg|_{z=0} = \frac{1}{2} \times 2 \times 0.25\bigg|_{z=0} = 0.25,
\end{aligned}$$

and for higher values $k \geq 3$, the $k^{\text{th}}$ derivative of $P(z)$ is zero. The mean of this distribution is given by

$$E[X] = \frac{d}{dz}P(z)\,|_{z=1} = (2 \times 0.25z + 0.25)\,|_{z=1} = 0.75,$$

which may be verified by direct calculation of $0 \times p_0 + 1 \times p_1 + 2 \times p_2$.

**Example 13.3** Consider now the sequence $p_k = 1/2^{k+1}$ for all $k \geq 0$. This is the geometric sequence $\{1/2, 1/4, 1/8, \ldots\}$ whose sum is 1. Its $z$-transform is given by

$$P(z) = \sum_{k=0}^{\infty} p_k z^k = \sum_{k=0}^{\infty} \frac{z^k}{2^{k+1}} = \frac{1}{2}\sum_{k=0}^{\infty}\left(\frac{z}{2}\right)^k.$$

Since we assume $|z| < 1$, we may apply the well-known formula for the sum of an infinite geometric series to obtain

$$P(z) = \frac{1}{2}\left(\frac{1}{1 - z/2}\right) = \frac{1}{2 - z}.$$

Once again, we observe how taking successive derivatives with respect to $z$ and evaluating the results at $z = 0$ returns the individual terms of the sequence. We have

$$\begin{aligned}
p_0 &= P(z)|_{z=0} = \frac{1}{2-z}\bigg|_{z=0} = 0.5,\\
p_1 &= \frac{d}{dz}P(z)\bigg|_{z=0} = \frac{1}{(2-z)^2}\bigg|_{z=0} = 0.25,\\
p_2 &= \frac{1}{2}\frac{d^2}{dz^2}P(z)\bigg|_{z=0} = \frac{1}{2}\frac{d}{dz}\frac{1}{(2-z)^2}\bigg|_{z=0} = \left(\frac{1}{2}\right)\frac{2(2-z)}{(2-z)^4}\bigg|_{z=0} = 0.125, \qquad \text{etc.}
\end{aligned}$$

The mean of this distribution can be found by evaluating

$$E[X] = \frac{d}{dz}P(z)\bigg|_{z=1} = \frac{1}{(2-z)^2}\bigg|_{z=1} = 1.0.$$

This may be verified by computing the mean directly. We have

$$\begin{aligned}
E[X] &= 0 \times (1/2) + 1 \times (1/4) + 2 \times (1/8) + 3 \times (1/16) + \cdots\\
&= \frac{1}{2}[1 \times (1/2)^1 + 2 \times (1/2)^2 + 3 \times (1/2)^3 + \cdots] = \frac{1}{2}\sum_{i=1}^{\infty} i2^{-i} = \frac{1}{2} \times 2.
\end{aligned}$$

In Example 13.1, the $z$-transform of the sequence $\alpha^i$, $i = 0, 1, 2, \ldots$, for some constant $\alpha$ was shown to be

$$P(z) = \sum_{k=0}^{\infty} \alpha^k z^k = \frac{1}{1 - \alpha z}$$

with $|z| < 1/\alpha$ for the transform to be analytic. The sequence and its transform are said to form a pair and are written as

$$\alpha^k \Longleftrightarrow \frac{1}{1 - \alpha z}.$$

A similar reasoning shows that, for constant $A$,

$$A\alpha^k \Longleftrightarrow \frac{A}{1 - \alpha z}, \tag{13.2}$$

and for sequences $\alpha^k$ and $\beta^k$, $k = 0, 1, 2, \ldots$, and constants $A$ and $B$

$$A\alpha^k + B\beta^k \Longleftrightarrow \frac{A}{1 - \alpha z} + \frac{B}{1 - \beta z}.$$

In general, for sequences $p_k$ and $q_k$, $k = 0, 1, 2 \ldots$, with transforms $P(z)$ and $Q(z)$, respectively, and constants $A$ and $B$,

$$Ap_k + Bq_k \Longleftrightarrow AP(z) + BQ(z).$$

The double arrow symbol ($\Longleftrightarrow$) is used to associate a sequence with its transform. Thus

$$p_k \Longleftrightarrow P(z)$$

indicates that the $z$-transform of the sequence $p_k$, $k = 0, 1, \ldots$, is given by $P(z)$ and vice versa. Also, sequences are denoted by lower-case letters and their transforms by the corresponding upper-case letter. The value of $p_k$ is assumed to be zero for $k < 0$.

## 13.2 The Inversion Process

The derivative method described earlier for inverting $z$-transforms (using Equation (13.1)) is appropriate if only the first few terms in the sequence are needed. It is not convenient when we require the entire sequence. Fortunately, it is frequently possible to obtain the entire sequence just by inspection, using known relationships between the transform and its inverse such as the transform pair of Equation (13.2):

$$A\,\alpha^k \Longleftrightarrow \frac{A}{1 - \alpha z}.$$

The most common of these transform-inverse pairs are given in tabular form at the end of this section.

We now describe the approach to use when the inverse is not known or is not directly available from tables. The basic idea is to take the transform $P(z)$ and to write it as a sum of much simpler terms, each of which can be individually transformed using known transform pairs. If the transform can be written as the quotient of two polynomials in $z$, then one possibility is to obtain a partial fraction expansion of $P(z)$. Let us write the quotient as

$$P(z) = N(z)/D(z),$$

where the denominator polynomial has constant term (coefficient of $z^0$) equal to 1. If the constant term of the denominator polynomial is different from 1, then dividing top and bottom by the constant term achieves the desired effect. Consider first the case when the degree of the denominator $D(z)$ is strictly greater than the degree of the numerator, $N(z)$. Let the degree of the polynomial $D(z)$ be $r$. Then $D(z)$ possesses $r$ roots and if, for the moment, we assume all $r$ roots to be distinct, this polynomial may be written as

$$D(z) = (1 - z/z_1)(1 - z/z_2)\cdots(1 - z/z_r)$$

where $z_1, z_2, \ldots, z_r$ are the $r$ roots. With such a $D(z)$, we may obtain a partial fraction expansion of $P(z)$ as

$$\begin{aligned} P(z) = \frac{N(z)}{D(z)} &= \frac{N(z)}{(1 - z/z_1)(1 - z/z_2)\cdots(1 - z/z_r)} \\ &= \frac{A_1}{(1 - z/z_1)} + \frac{A_2}{(1 - z/z_2)} + \cdots + \frac{A_r}{(1 - z/z_r)}, \end{aligned} \tag{13.3}$$

where the constants $A_i$ are determined from

$$A_i = (1 - z/z_i)\frac{N(z)}{D(z)}\bigg|_{z=z_i}.$$

Observe that each of the terms of the summation in Equation (13.3) can be inverted very simply by substituting $\alpha = 1/z_i$ into the transform pair:

$$\alpha^k \Longleftrightarrow \frac{1}{1 - \alpha z}.$$

We have, for each term $i$ in the summation,

$$A_i\left(\frac{1}{z_i}\right)^k \Longleftrightarrow \frac{A_i}{1 - z/z_i}.$$

Now, repeatedly using the property $Ap_k + Bq_k \Longleftrightarrow AP(z) + BQ(z)$, we can show that the sequence corresponding to the transform $P(z)$ is as follows:

$$P(z) = \frac{A_1}{1 - z/z_1} + \frac{A_2}{1 - z/z_2} + \cdots + \frac{A_r}{1 - z/z_r} \Longleftrightarrow A_1\left(\frac{1}{z_1}\right)^k + A_2\left(\frac{1}{z_2}\right)^k + \cdots + A_r\left(\frac{1}{z_r}\right)^k.$$

**Example 13.4** The $z$-transform of a sequence is given by

$$P(z) = \frac{(1 - z)}{2z^2 - 7z + 3}.$$

We wish to determine the sequence. We must first confirm that the degree of the numerator polynomial (=1) is strictly less than the degree of the denominator polynomial (=2), which it is. Our next task is to factor the denominator to obtain

$$2z^2 - 7z + 3 = (2z - 1)(z - 3) = 3(1 - z/z_1)(1 - z/z_2)$$

with $z_1 = 1/2$ and $z_2 = 3$. This gives

$$P(z) = \frac{(1 - z)/3}{(1 - z/z_1)(1 - z/z_2)} = \frac{N(z)}{D(z)},$$

which we must now write in partial fraction form. We get

$$P(z) = \frac{A_1}{1 - z/z_1} + \frac{A_2}{1 - z/z_2}$$

with

$$A_1 = (1 - z/z_1)\frac{N(z)}{D(z)}\bigg|_{z=z_1} = \frac{(1-z)/3}{(1-z/z_2)}\bigg|_{z=z_1} = \frac{(1-z_1)/3}{(1-z_1/z_2)} = \frac{(1-1/2)/3}{(1-(1/2)/3)} = \frac{1/6}{5/6} = \frac{1}{5}$$

and

$$A_2 = (1 - z/z_2)\frac{N(z)}{D(z)}\bigg|_{z=z_2} = \frac{(1-z)/3}{(1-z/z_1)}\bigg|_{z=z_2} = \frac{(1-z_2)/3}{(1-z_2/z_1)} = \frac{(1-3)/3}{(1-3\times 2)} = \frac{-2/3}{-5} = \frac{2}{15}.$$

Notice that

$$P(z) = \frac{A_1}{1-z/z_1} + \frac{A_2}{1-z/z_2} = \frac{1/5}{1-z/(1/2)} + \frac{2/15}{1-z/3} = \frac{(1/5)\times(1-z/3) + (2/15)\times(1-2z)}{(1-2z)(1-z/3)}$$

$$= \frac{(1/5) + (2/15) - (z/15) - (4z/15)}{(1-2z)(1-z/3)} = \frac{(1/3)(1-z)}{(1-2z)(1-z/3)} = \frac{(1-z)}{2z^2 - 7z + 3},$$

as indeed it should be. We now have the transform $P(z)$ in a form that is easy to transpose. Using the transform pair of Equation (13.2) we have

$$P(z) = \frac{1/5}{1-2z} + \frac{2/15}{1-z/3} \Longleftrightarrow \frac{1}{5}2^k + \frac{2}{15}\left(\frac{1}{3}\right)^k.$$

The sequence for which we have been looking is therefore given by

$$p_k = \frac{1}{5}\left(2^k + \frac{2}{3}\left(\frac{1}{3}\right)^k\right) \quad \text{for } k = 0,\ 1, \ldots .$$

We now consider the case when the degree of the numerator polynomial $N(z)$ is greater than or equal to the degree of the denominator polynomial $D(z)$. To make use of the partial fraction expansion technique that we have just discussed, it becomes necessary to rewrite the transform $P(z)$. Ideally, it may be possible to factor out some power of the numerator so that we can write

$$P(z) = z^i\frac{N(z)}{D(z)}$$

with the degree of $N(z)$ strictly less than that of $D(z)$, and then use the transform property

$$p_{k-i} \Longleftrightarrow z^i P(z)$$

or perhaps to factor out a term like $(1 - z^i)$ and use the transform property

$$p_k - p_{k-i} \Longleftrightarrow (1 - z^i)P(z).$$

**Example 13.5** Suppose we have

$$P(z) = \frac{z^2(1-z)}{2z^2 - 7z + 3}.$$

We can write this as

$$P(z) = z^2\frac{(1-z)}{2z^2 - 7z + 3} = z^2\frac{N(z)}{D(z)}.$$

Now, using the previously computed sequence corresponding to $N(z)/D(z)$ and applying the appropriate transform property, we obtain

$$\frac{N(z)}{D(z)} \Longleftrightarrow \frac{1}{5}\left(2^k + \frac{2}{3}\left(\frac{1}{3}\right)^k\right)$$

and

$$P(z) = z^2\frac{N(z)}{D(z)} \Longleftrightarrow \frac{1}{5}\left(2^{k-2} + \frac{2}{3}\left(\frac{1}{3}\right)^{k-2}\right).$$

If it is not possible to factor the numerator polynomial, it is generally possible to divide the denominator into the numerator to obtain a polynomial in $z$ plus a remainder term. If, for example

$$P(z) = \frac{N(z)}{D(z)} \quad \text{with} \quad N(z) = D(z)M(z) + R(z)$$

then

$$P(z) = M(z) + \frac{R(z)}{D(z)}.$$

The individual terms in the polynomial $M(z)$ can be inverted separately, and generally without difficulty, while the degree of the numerator polynomial obtained as the remainder $R(z)$ will be strictly less than the degree of the denominator $D(z)$ and may now be expanded into partial fractions.

Obtaining the partial fraction expansion of a quotient of two polynomials is more complex when the denominator polynomial contains roots that are not simple. For example, the polynomial $z^2 - 2z + 1 = (z-1)^2$ has a root of multiplicity 2 at $z = 1$. We shall assume that the denominator polynomial $D(z)$ has $j$ distinct roots and that the $i^{\text{th}}$ root has multiplicity $m_i$. It follows that $D(z)$ is a polynomial of degree $\sum_{i=1}^{j} m_i$. We shall assume that the degree of the numerator polynomial is strictly less than this, for otherwise, we will need to adopt the procedures described in the paragraph above. We can now write

$$D(z) = \prod_{i=1}^{j}(1 - z/z_i)^{m_i},$$

where the $j$ distinct roots are given by $z_1, z_2, \ldots, z_j$. In this case, it may be shown that the partial fraction expansion of $P(z)$ is given by

$$\begin{aligned} P(z) = {} & \frac{A_{11}}{(1-z/z_1)^{m_1}} + \frac{A_{12}}{(1-z/z_1)^{m_1-1}} + \cdots + \frac{A_{1m_1}}{(1-z/z_1)} \\ & + \frac{A_{21}}{(1-z/z_2)^{m_2}} + \frac{A_{22}}{(1-z/z_2)^{m_2-1}} + \cdots + \frac{A_{2m_2}}{(1-z/z_2)} + \cdots \\ & + \frac{A_{j1}}{(1-z/z_j)^{m_j}} + \frac{A_{j2}}{(1-z/z_j)^{m_j-1}} + \cdots + \frac{A_{jm_j}}{(1-z/z_j)}. \end{aligned}$$

The constant terms are computed from

$$A_{ij} = \frac{1}{(j-1)!}(-z_i)^{j-1}\frac{d^{j-1}}{dz^{j-1}}\left[(1-z/z_i)^{m_i}\frac{N(z)}{D(z)}\right]\Bigg|_{z=z_i}.$$

**Example 13.6** In Example 13.4 we had

$$P(z) = \frac{(1-z)}{2z^2 - 7z + 3} = \frac{(1/3)(1-z)}{(1-2z)(1-z/3)} = \frac{(1/3)(1-z)}{(1-z/z_1)(1-z/z_2)}.$$

Let us modify this by supposing that

$$P(z) = \frac{N(z)}{D(z)} = \frac{(1/3)(1-z)}{(1-2z)(1-z/3)^2} = \frac{(1/3)(1-z)}{(1-z/z_1)(1-z/z_2)^2}.$$

In this case, the denominator is now a cubic polynomial with a simple root ($m_1 = 1$) at $z = 1/2$ and a double root ($m_2 = 2$) at $z = 3$. When we expand $P(z)$ as a partial fraction we will obtain three terms

$$P(z) = \frac{A_{11}}{(1 - z/z_1)} + \frac{A_{21}}{(1 - z/z_2)^2} + \frac{A_{22}}{(1 - z/z_2)}$$

with $z_1 = 1/2$ and $z_2 = 3$. The $A_{ij}$ are computed from the formula as follows:

$$A_{11} = (1 - z/z_1)\frac{N(z)}{D(z)}\bigg|_{z=z_1} = \frac{(1 - z)/3}{(1 - z/z_2)^2}\bigg|_{z=z_1} = \frac{(1 - z_1)/3}{(1 - z_1/z_2)^2} = \frac{(1 - 1/2)/3}{(1 - (1/2)/3)^2} = \frac{1/6}{25/36} = \frac{6}{25}$$

$$A_{21} = (1 - z/z_2)^2\frac{N(z)}{D(z)}\bigg|_{z=z_2} = \frac{(1 - z)/3}{(1 - z/z_1)}\bigg|_{z=z_2} = \frac{(1 - z_2)/3}{(1 - z_2/z_1)} = \frac{(1 - 3)/3}{(1 - 3 \times 2)} = \frac{-2/3}{-5} = \frac{2}{15}$$

$$A_{22} = \frac{1}{1!}(-z_2)^1\frac{d}{dz}\left[(1 - z/z_2)^2\frac{N(z)}{D(z)}\right]\bigg|_{z=z_2} = -z_2\frac{d}{dz}\left[\frac{(1 - z)/3}{(1 - z/z_1)}\right]\bigg|_{z=z_2} = -3\frac{d}{dz}\left[\frac{(1 - z)/3}{(1 - 2z)}\right]\bigg|_{z=z_2}$$

$$= -\frac{d}{dz}\left[\frac{(1 - z)}{(1 - 2z)}\right]\bigg|_{z=z_2} = -\frac{(1 - 2z)(-1) - (1 - z)(-2)}{(1 - 2z)^2}\bigg|_{z=z_2} = -\frac{1}{(1 - 2z)^2}\bigg|_{z=z_2} = -\frac{1}{25}.$$

We therefore have

$$P(z) = \frac{A_{11}}{(1 - z/z_1)} + \frac{A_{21}}{(1 - z/z_2)^2} + \frac{A_{22}}{(1 - z/z_2)}$$

$$= \frac{6/25}{(1 - 2z)} + \frac{2/15}{(1 - z/3)^2} - \frac{1/25}{(1 - z/3)}. \tag{13.4}$$

The reader should verify this result by checking that

$$\frac{6/25}{(1 - 2z)} + \frac{2/15}{(1 - z/3)^2} - \frac{1/25}{(1 - z/3)} = \frac{(1/3)(1 - z)}{(1 - 2z)(1 - z/3)^2} = P(z).$$

We are now in a position to invert $P(z)$,

$$P(z) = \frac{6/25}{(1 - 2z)} + \frac{2/15}{(1 - z/3)^2} - \frac{1/25}{(1 - z/3)},$$

and determine the sequence. The first and third terms fall into the category of transform pairs previously discussed. For the second term, we need to use the relation

$$\frac{1}{(1 - \alpha z)^2} \Longleftrightarrow (k + 1)\alpha^k.$$

Treating each term of Equation (13.4) separately, we obtain

$$\frac{6/25}{(1 - 2z)} \Longleftrightarrow \frac{6}{25}2^k,$$

$$\frac{2/15}{(1 - z/3)^2} \Longleftrightarrow \frac{2}{15}(k + 1)\frac{1}{3^k},$$

$$\frac{1/25}{(1 - z/3)} \Longleftrightarrow \frac{1}{25}\frac{1}{3^k},$$

which when taken together provide the final result

$$P(z) \Longleftrightarrow \frac{6 \times 2^k}{25} + \frac{2(k+1)}{15 \times 3^k} - \frac{1}{25 \times 3^k} = \frac{1}{75}\left[18 \times 2^k + (10k+7)3^{-k}\right].$$

Table 13.1 shows some of the most commonly used transform pairs. In this table, the symbol $\star$ implies convolution and $u_k$ is the unit function which has the value 1 if $k = 0$ and the value 0 otherwise.

Table 13.1. Some common transform pairs.

| Sequence | | Transform |
|---|---|---|
| $Ap_k + Bq_k$ | $\Longleftrightarrow$ | $AP(z) + BQ(z)$ |
| $p_k \star q_k$ | $\Longleftrightarrow$ | $P(z)Q(z)$ |
| $\alpha^k p_k$ | $\Longleftrightarrow$ | $P(\alpha z)$ |
| $p_{k+1}$ | $\Longleftrightarrow$ | $[P(z) - p_0]/z$ |
| $p_{k+i}$ | $\Longleftrightarrow$ | $\left[P(z) - \sum_{j=1}^{i} z^{j-1} p_{j-1}\right]/z^i$ |
| $p_{k-1}$ | $\Longleftrightarrow$ | $zP(z)$ |
| $p_{k-i}$ | $\Longleftrightarrow$ | $z^i P(z)$ |
| $p_k - p_{k-1}$ | $\Longleftrightarrow$ | $(1-z)P(z)$ |
| $p_k - p_{k-i}$ | $\Longleftrightarrow$ | $(1-z^i)P(z)$ |

| Sequence | | Transform |
|---|---|---|
| $u_k$ | $\Longleftrightarrow$ | $1$ |
| $u_{k-i}$ | $\Longleftrightarrow$ | $z^i$ |
| $\{1, \alpha, \alpha^2, \ldots\} = \alpha^k$ | $\Longleftrightarrow$ | $1/(1-\alpha z)$ |
| $\{0, \alpha, 2\alpha^2, 3\alpha^3, \ldots\} = k\alpha^k$ | $\Longleftrightarrow$ | $\alpha z/(1-\alpha z)^2$ |
| $\{0, 1, 2, 3, \ldots\} = k$ | $\Longleftrightarrow$ | $z/(1-z)^2$ |
| $\{1, 2, 3, \ldots\} = k+1$ | $\Longleftrightarrow$ | $1/(1-z)^2$ |
| $\frac{1}{m!}(k+m)(k+m-1)\ldots(k+1)\alpha^k$ | $\Longleftrightarrow$ | $1/(1-\alpha z)^{m+1}$ |

## 13.3 Solving Markovian Queues using $z$-Transforms

### *13.3.1 The $z$-Transform Procedure*

As we mentioned previously, the $z$-transform provides an effective method for solving sets of difference equations, and it is to the difference equations that arise in Markovian queueing systems that we now apply it. The general procedure is specified in the following steps.

1. By considering transitions into and out of an arbitrary state, generate the difference equations for the queueing system.
2. Multiply the $j^{\text{th}}$ equation by $z^j$ and sum over all applicable $j$.
3. Identify the $z$-transform $P(z)$ and isolate it on the left-hand side.
4. Eliminate all unknowns from the right-hand side. Once this step has been completed, various moments of customer occupancy may be found.
5. Invert the $z$-transform to obtain the stationary probability distribution of customers in the system.

In general, it is this last step that creates the most difficulties.

### *13.3.2 The M/M/1 Queue Solved using $z$-Transforms*

We illustrate the $z$-transform approach by an analysis of the $M/M/1$ queue.

**Step 1:** The following set of difference equations arises when solving for the stationary distribution of customers in the $M/M/1$ queue. The probability that the queueing system contains $n+1$ customers ($p_{n+1}$) is related to the probabilities that it contains $n$ customers ($p_n$) and $n-1$ customers ($p_{n-1}$) through the difference equations

$$p_{n+1} = \frac{\lambda+\mu}{\mu}p_n - \frac{\lambda}{\mu}p_{n-1} = (\rho+1)p_n - \rho p_{n-1}, \quad n \ge 1,$$

and

$$p_1 = \frac{\lambda}{\mu}p_0 = \rho p_0.$$

**Step 2:** To solve these equations using the $z$-transform, we must first multiply the $n^{\text{th}}$ equation by $z^n$ and sum over all applicable $n$. We obtain

$$p_{n+1}z^n = (\rho+1)p_n z^n - \rho p_{n-1}z^n, \quad n \ge 1,$$

$$z^{-1}p_{n+1}z^{n+1} = (\rho+1)p_n z^n - \rho z p_{n-1}z^{n-1}, \quad n \ge 1,$$

and hence

$$z^{-1}\sum_{n=1}^{\infty} p_{n+1}z^{n+1} = (\rho+1)\sum_{n=1}^{\infty} p_n z^n - \rho z\sum_{n=1}^{\infty} p_{n-1}z^{n-1}. \tag{13.5}$$

**Step 3:** We identify the $z$-transform $P(z)$ by writing Equation (13.5) as

$$z^{-1}\left\{\sum_{n=-1}^{\infty} p_{n+1}z^{n+1} - p_1 z - p_0\right\} = (\rho+1)\left\{\sum_{n=0}^{\infty} p_n z^n - p_0\right\} - \rho z\sum_{n=1}^{\infty} p_{n-1}z^{n-1},$$

and observing that

$$P(z) \equiv \sum_{n=-1}^{\infty} p_{n+1}z^{n+1} = \sum_{n=0}^{\infty} p_n z^n = \sum_{n=1}^{\infty} p_{n-1}z^{n-1}.$$

Therefore

$$z^{-1}\{P(z) - p_1 z - p_0\} = (\rho + 1)\{P(z) - p_0\} - \rho z P(z)$$

or

$$P(z) - p_1 z - p_0 = z(\rho + 1)\{P(z) - p_0\} - \rho z^2 P(z).$$

Bringing terms in $P(z)$ to the left-hand side,

$$P(z) - z(\rho + 1)P(z) + \rho z^2 P(z) = p_1 z + p_0 - z(\rho + 1)p_0,$$

we obtain

$$P(z) = \frac{p_1 z + p_0 - z(\rho + 1)p_0}{1 - z(\rho + 1) + \rho z^2}.$$

**Step 4:** Two unknowns, $p_0$ and $p_1$, need to be removed from the right-hand side. The second of these can be eliminated by using the fact that $p_1 = \rho p_0$. This gives

$$P(z) = \frac{\rho p_0 z + p_0 - z(\rho + 1)p_0}{1 - z\rho - z + \rho z^2} = \frac{\rho z + 1 - z\rho - z}{1 - z\rho - z + \rho z^2} p_0 = \frac{(1 - z)}{(1 - z)(1 - z\rho)} p_0.$$

We therefore conclude that

$$P(z) = \frac{p_0}{1 - z\rho}.$$

To eliminate the remaining unknown, $p_0$, we use the fact that the sum of the probabilities must be equal to 1. Substituting $z = 1$ gives $P(1) = \sum_{n=0}^{\infty} p_n = 1$. Thus $1 = p_0/(1 - \rho)$ and hence $p_0 = 1 - \rho$. It follows that the $z$-transform for the $M/M/1$ queue is given by

$$P(z) = \frac{1 - \rho}{1 - z\rho}.$$

**Step 5:** To invert this transform we can expand $P(z) = (1 - \rho)/(1 - \rho z)$ as a power series and pick off the terms. Observe that $1/(1 - z\rho)$ is equal to the sum of a geometric series:

$$\frac{1}{1 - z\rho} = 1 + z\rho + (z\rho)^2 + (z\rho)^3 + \cdots.$$

Therefore

$$P(z) = \sum_{k=0}^{\infty} (1 - \rho)\rho^k z^k$$

and so the coefficient of $z^k$ (which we previously specified to be $p_k$) is equal to $(1-\rho)\rho^k$. This gives the final solution as $p_k = (1 - \rho)\rho^k$ which is the same as we obtained before.

Notice that we could also have inverted the transform using known transform pairs. Identifying the right-hand side of the transform pair

$$A\alpha^n \Longleftrightarrow \frac{A}{1 - \alpha z}$$

with the transform

$$P(z) = \frac{1 - \rho}{1 - \rho z},$$

where $A$ has the value $A = 1 - \rho$, we directly obtain the sequence

$$p_k = (1 - \rho)\rho^k.$$

### 13.3.3 The M/M/1 Queue with Arrivals in Pairs

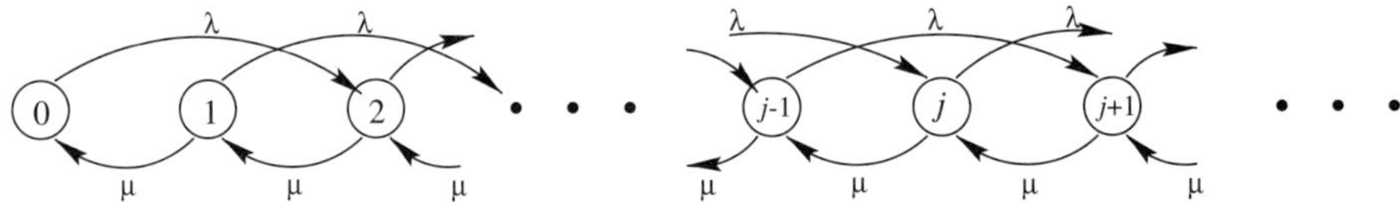


Figure 13.1. Arrivals in groups of size 2.

Consider an $M/M/1$ queue in which at each arrival instant exactly two customers arrive. The state transition rate diagram for this queueing system is shown in Figure 13.1 and its transition rate matrix is given by

$$Q = \left(\begin{array}{cc|cc|cc|c}
-\lambda & 0 & \lambda & 0 & & & \\
\mu & -(\lambda+\mu) & 0 & \lambda & & & \\
\hline
0 & \mu & -(\lambda+\mu) & 0 & \lambda & 0 & \ddots \\
0 & 0 & \mu & -(\lambda+\mu) & 0 & \lambda & \ddots \\
\hline
 & & 0 & \mu & -(\lambda+\mu) & 0 & \ddots \\
 & & 0 & 0 & \mu & -(\lambda+\mu) & \ddots \\
\hline
 & & & & & & \ddots
\end{array}\right).$$

Observe that $Q$ has a block tridiagonal (quasi-birth-death) structure and hence the matrix-geometric approach, discussed in the previous chapter, may be used to solve this system. Leaving this approach as Exercise 13.3.1, let us instead proceed to move through the various steps involved in solving this system by the method of $z$-transforms.

**Step 1:** By inspection, the balance equations are

$$\begin{aligned}
\lambda p_0 &= \mu p_1, && j = 0,\\
(\lambda+\mu)p_1 &= \mu p_2, && j = 1,\\
(\lambda+\mu)p_j &= \lambda p_{j-2} + \mu p_{j+1}, && j \ge 2.
\end{aligned}$$

**Step 2:** Multiply the $j^{\text{th}}$ equation by $z^j$ and sum over all $j \ge 0$:

$$\begin{aligned}
\lambda p_0 &= \mu p_1, && j = 0,\\
(\lambda+\mu)p_1 z &= \mu p_2 z, && j = 1\\
(\lambda+\mu)p_j z^j &= \lambda p_{j-2} z^j + \mu p_{j+1} z^j, && j \ge 2.
\end{aligned}$$

Thus

$$\lambda p_0 + (\lambda+\mu)\, p_1 z + (\lambda+\mu)\sum_{j=2}^{\infty} p_j z^j = \mu p_1 + \mu p_2 z + \lambda \sum_{j=2}^{\infty} p_{j-2} z^j + \mu \sum_{j=2}^{\infty} p_{j+1} z^j. \quad (13.6)$$

**Step 3:** Identify the $z$-transform $P(z) = \sum_{j=0}^{\infty} p_j z^j$ and bring it to the left-hand side: Simplifying Equation (13.6) and subsequently identifying $P(z)$, we obtain

$$\lambda \sum_{j=0}^{\infty} p_j z^j + \mu \sum_{j=1}^{\infty} p_j z^j = \lambda \sum_{j=2}^{\infty} p_{j-2} z^j + \mu \sum_{j=0}^{\infty} p_{j+1} z^j$$

$$\lambda P(z) + \mu[P(z) - p_0] = \lambda z^2 P(z) + \frac{\mu}{z}[P(z) - p_0]$$

$$P(z) = \frac{\mu p_0(1 - 1/z)}{\lambda(1 - z^2) + \mu(1 - 1/z)} = \frac{\mu p_0}{\mu - \lambda z(z + 1)}.$$

**Step 4:** Eliminate the unknown $p_0$ from right-hand side. Since the average arrival rate $\bar{\lambda}$ is equal to $2\lambda$, we have $\rho = \bar{\lambda}\bar{x} = 2\lambda/\mu$. Thus

$$P(z) = \frac{2p_0}{2 - \rho z(z + 1)}.$$

Also observe that $P(1) = 1 = 2p_0/(2 - 2\rho)$, i.e., $p_0 = 1 - \rho$, and hence

$$P(z) = \frac{2(1 - \rho)}{2 - \rho z(z + 1)}.$$

The mean number of customers in the system can be obtained at this point:

$$E[N] = \left.\frac{dP(z)}{dz}\right|_{z=1} = \left.\frac{2(1 - \rho)\rho(2z + 1)}{[2 - \rho z(z + 1)]^2}\right|_{z=1} = \frac{2(1 - \rho)\rho(3)}{(2 - 2\rho)^2} = \frac{3}{2}\frac{\rho}{1 - \rho}.$$

**Step 5:** Invert the transform. We have

$$P(z) = \frac{2(1 - \rho)}{2 - \rho z(z + 1)} = \frac{2(1 - \rho)/\rho}{2/\rho - z - z^2}.$$

Let the roots of $z^2 + z - 2/\rho$ be $z_1$ and $z_2$, i.e.,

$$z^2 + z - 2/\rho = (z - z_1)(z - z_2) \quad \text{with } z_{1,2} = -\frac{1}{2} \pm \frac{\sqrt{1 + 8/\rho}}{2}.$$

Observe that $z_1 z_2 = -2/\rho$. We have

$$\begin{aligned} P(z) &= \frac{2(1 - \rho)/\rho}{2/\rho - z - z^2} = \frac{(1 - \rho)(-2/\rho)}{z^2 + z - 2/\rho} = \frac{(1 - \rho)(-2/\rho)}{(z - z_1)(z - z_2)} \\ &= \frac{(1 - \rho)(z_1 z_2)}{(z - z_1)(z - z_2)} = \frac{(1 - \rho)}{(1 - z/z_1)(1 - z/z_2)}. \end{aligned}$$

Defining $N(z)$ and $D(z)$ from

$$P(z) = \frac{(1 - \rho)}{(1 - z/z_1)(1 - z/z_2)} = \frac{N(z)}{D(z)},$$

we proceed to form the partial fraction expansion of $P(z)$ as

$$P(z) = \frac{A_1}{1 - z/z_1} + \frac{A_2}{1 - z/z_2}$$

with

$$A_1 = (1 - z/z_1)\left.\frac{N(z)}{D(z)}\right|_{z=z_1} = \frac{(1 - \rho)}{(1 - z_1/z_2)} \quad \text{and} \quad A_2 = (1 - z/z_2)\left.\frac{N(z)}{D(z)}\right|_{z=z_2} = \frac{(1 - \rho)}{(1 - z_2/z_1)}.$$

This allows us to obtain the successive probabilities from the transform pair,

$$P(z) \Longleftrightarrow A_1\left(\frac{1}{z_1}\right)^k + A_2\left(\frac{1}{z_2}\right)^k.$$

Suppose, for example, that $\lambda = 4$ and $\mu = 15$. This gives $\rho = 8/15$ and the roots of the quadratic polynomial are $z_1 = 3/2$ and $z_2 = -5/2$. The values of $A_1$ and $A_2$ are

$$A_1 = \frac{(1-\rho)}{(1-z_1/z_2)} = \frac{7}{24} \quad \text{and} \quad A_2 = \frac{(1-\rho)}{(1-z_2/z_1)} = \frac{7}{40}$$

and successive values of $p_k = \text{Prob}\{N = k\}$ are found as

$$p_k = \frac{7}{24}\left(\frac{2}{3}\right)^k + \frac{7}{40}\left(-\frac{2}{5}\right)^k.$$

Using this formula, we see, for example, that

$$p_0 = 0.466667, \quad p_1 = 0.124444, \quad p_2 = 0.157630, \quad p_3 = 0.075220, \quad p_4 = 0.062093, \quad \text{etc.}$$

Substituting $k = 10$ directly into the formula gives $p_{10} = 0.005076$.

### 13.3.4 The $M/E_r/1$ Queue Solved using $z$-Transforms

It is possible to implement the $z$-transform approach to solve the $M/E_r/1$ queue using the two-dimensional state descriptor $(k, i)$ seen in the matrix geometric method of the previous chapter. This is the approach developed in the text by Gross and Harris [19]. It is also possible to combine the components $k$ and $i$ into a single quantity, namely, the number of service phases yet to be completed by all customers in the system and to apply the $z$-transform to the difference equations derived using this compact descriptor. This is the approach developed in the text by Kleinrock [24]. We choose to follow this second approach, since it is somewhat simpler. Indeed our application of the $z$-transform to the queueing systems of the next three sections closely mirrors that of Kleinrock.

When an $M/E_r/1$ queue contains $k$ customers and the customer in service is in phase $i$, the total number of service phases still to be completed is

$$j = (k-1)r + (r-i+1) = rk - i + 1.$$

Let $P_j$ denote the probability of having $j$ phases in the system at steady state:

$$P_j = \text{Prob}\{\text{Number of phases in system } = j\}.$$

It follows that, if we let $p_k$ denote the equilibrium probability of having $k$ customers in the system, i.e.,

$$p_k = \text{Prob}\{\text{Number of customers in system } = k\},$$

then

$$p_k = \sum_{j=(k-1)r+1}^{kr} P_j, \quad k = 1, 2, 3, \ldots,$$

and therefore the distribution of customers in the system at any time may be easily computed from a knowledge of the number of service phases remaining at that time. Using the one-dimensional state descriptor, the state transition diagram may be drawn as shown in Figure 13.2.

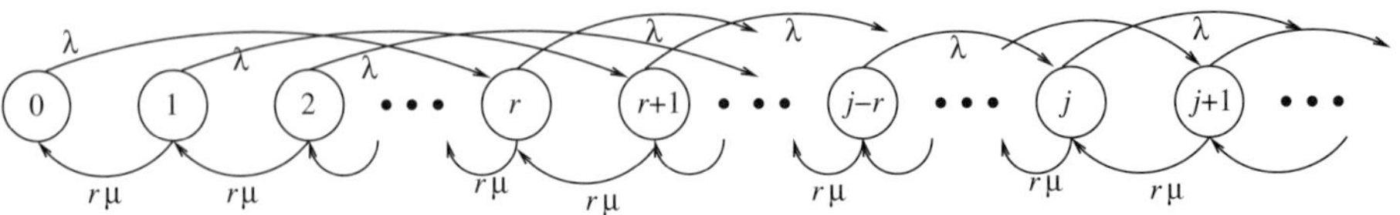


Figure 13.2. Alternate state transition diagram for the $M/E_r/1$ queue.

**Step 1:** We obtain the difference equations for this queueing system by considering transitions into and out of an arbitrary state $j \geq r$. This state is entered from state $j - r$ due to an arrival and from state $j + 1$ due to a departure. State $j$ is exited either due to an arrival, at rate $\lambda$, or due to a service completion, at rate $r\mu$. Special boundary conditions exist for states $0, 1, \ldots, r - 1$.

The difference equations are

$$\begin{aligned}
\lambda P_0 &= r\mu P_1, \\
(\lambda + r\mu) P_1 &= r\mu P_2, \\
(\lambda + r\mu) P_j &= r\mu P_{j+1}, \quad 1 \leq j \leq r - 1, \\
(\lambda + r\mu) P_j &= \lambda P_{j-r} + r\mu P_{j+1}, \quad j \geq r.
\end{aligned}$$

**Step 2:** We solve these difference equations using $z$-transforms, where the $z$-transform of the sequence $\{P_0, P_1, P_2, \ldots\}$ is defined as

$$P(z) \equiv \sum_{j=0}^{\infty} P_j z^j.$$

To do so, we multiply the $j^{\text{th}}$ equation by $z^j$ and sum over all $j \geq 1$. We obtain

$$\sum_{j=1}^{\infty} (\lambda + r\mu) P_j z^j = \sum_{j=r}^{\infty} \lambda P_{j-r} z^j + \sum_{j=1}^{\infty} r\mu P_{j+1} z^j.$$

**Step 3:** We now arrange to identify the $z$-transform. We have

$$(\lambda + r\mu) \left[ \sum_{j=0}^{\infty} P_j z^j - P_0 \right] = \lambda z^r \sum_{j=r}^{\infty} P_{j-r} z^{j-r} + \frac{r\mu}{z} \sum_{j=1}^{\infty} P_{j+1} z^{j+1},$$

$$(\lambda + r\mu) \left[ P(z) - P_0 \right] = \lambda z^r P(z) + \frac{r\mu}{z} \left[ P(z) - P_0 - P_1 z \right],$$

$$P(z) = \frac{P_0 \left[ \lambda + r\mu - (r\mu)/z \right] - r\mu P_1}{\lambda + r\mu - \lambda z^r - (r\mu)/z}.$$

**Step 4:** To eliminate the unknowns $P_0$ and $P_1$ from the right-hand side, we first make use of the equation $\lambda P_0 = r\mu P_1$, and obtain

$$P(z) = \frac{r\mu P_0 \left[ 1 - 1/z \right]}{\lambda + r\mu - \lambda z^r - (r\mu)/z} = \frac{r\mu P_0 \left[ 1 - z \right]}{r\mu + \lambda z^{r+1} - (\lambda + r\mu) z}.$$

We now use the fact that $P(1) = 1$ to evaluate $P_0$. To do so, we need L'Hôpital's rule which allows us to find the limit (if any) of a quotient of functions when both numerator and denominator approach zero. If

$$\lim_{x \to c} f(x) = \lim_{x \to c} g(x) = 0$$

and if

$$\lim_{x \to c} \left( \frac{f'(x)}{g'(x)} \right) = L$$

then

$$\lim_{x \to c} \left( \frac{f(x)}{g(x)} \right) = L.$$

Thus, using L'Hôpital's rule (since $P(1) = 0/0$),

$$1 = P(1) = \lim_{z\to 1} P(z) = \lim_{z\to 1}\left[\frac{r\mu P_0[1-z]}{r\mu + \lambda z^{r+1} - (\lambda + r\mu)z}\right]$$

$$= \lim_{z\to 1}\left[\frac{-r\mu P_0}{\lambda(r+1)z^r - (\lambda + r\mu)}\right] = \frac{-r\mu P_0}{\lambda(r+1) - (\lambda + r\mu)} = \frac{r\mu P_0}{r\mu - \lambda r} = \frac{P_0}{1 - (\lambda r)/(r\mu)},$$

which implies that $P_0 = 1 - \lambda/\mu = 1 - \rho$ (which might have been expected).

In this system, the arrival rate is $\lambda$ and the average service time is held fixed at $1/\mu$ independent of $r$. Therefore the utilization factor is given by

$$\rho = \lambda \bar{x} = \lambda/\mu.$$

It follows that

$$P(z) = \frac{r\mu(1-\rho)(1-z)}{r\mu + \lambda z^{r+1} - (\lambda + r\mu)z}.$$

Notice that substituting $r = 1$ in the above expression yields

$$P(z) = \frac{\mu(1-\rho)(1-z)}{\mu + \lambda z^2 - (\lambda + \mu)z} = \frac{\mu(1-\rho)(1-z)}{(\lambda z - \mu)(z-1)}$$

$$= \frac{\mu(1-\rho)}{(\mu - \lambda z)} = \frac{1-\rho}{1-\rho z},$$

which is the $z$-transform for the $M/M/1$ queue, namely,

$$P(z) = \frac{1-\rho}{1-\rho z} \iff (1-\rho)\rho^j,$$

the geometric distribution as expected.

**Step 5:** To obtain the distribution of the number of phases in the system, we need to invert this transform. The usual approach is to make a partial fraction expansion and to invert each term by inspection. But, to carry out the expansion, we need to identify the $r + 1$ zeros of the denominator polynomial,

$$\lambda z^{r+1} - (\lambda + r\mu)z + r\mu = 0.$$

Unity must be one of the roots, since $(1 - z)$ is a factor in the numerator. Dividing $(1 - z)$ into $D(z)$ gives

$$r\mu - \lambda(z + z^2 + \cdots + z^r).$$

The same result may be obtained by observing that

$$\begin{aligned}(1-z)&\left[r\mu - \lambda(z + z^2 + \cdots + z^r)\right]\\ &= r\mu - \lambda z - \lambda z^2 - \lambda z^3 - \cdots - \lambda z^r\\ &\quad -zr\mu + \lambda z^2 + \lambda z^3 + \cdots + \lambda z^r + \lambda z^{r+1}\\ &= r\mu - (\lambda + r\mu)z + \lambda z^{r+1}.\end{aligned}$$

We shall also factor out $r\mu$ so that both top and bottom of $P(z)$ can be divided by $r\mu(1 - z)$:

$$P(z) = \frac{(1-\rho)}{1 - \lambda(z + z^2 + \cdots + z^r)/r\mu}. \tag{13.7}$$

Denote the $r$ yet to be determined roots by $z_1, z_2, \ldots, z_r$. These are the roots of the denominator in Equation (13.7). Once we find them, we can write the denominator as

$$(1 - z/z_1)(1 - z/z_2)\cdots(1 - z/z_r),$$

and so the expression for the $z$-transform becomes

$$P(z) = \frac{1-\rho}{(1 - z/z_1)(1 - z/z_2)\cdots(1 - z/z_r)} = \frac{N(z)}{D(z)}.$$

In this form, the partial fraction expansion is easy to compute. It is given by

$$P(z) = \sum_{i=1}^{r} \frac{A_i}{(1 - z/z_i)},$$

where

$$A_i = (1 - z/z_i)P(z)\Big|_{z=z_i}.$$

We may now invert this transform by repeated use of the transform relationship

$$\frac{A_i}{(1 - z/z_i)} \iff A_i\left(\frac{1}{z_i}\right)^j.$$

It then follows that $P_j$, the probability of $j$ service phases, can be obtained as

$$P(z) = \sum_{i=1}^{r} \frac{A_i}{(1 - z/z_i)} \iff \sum_{i=1}^{r} A_i\left(\frac{1}{z_i}\right)^j = P_j.$$

The equilibrium distribution of customers in the $M/E_r/1$ queue may now be computed as

$$p_k = \sum_{j=(k-1)r+1}^{kr} P_j, \quad k = 1, 2, 3, \ldots.$$

The expected number of phases in the system $E[N]$ is obtained by evaluating $P'(z)|_{z=1}$. We have

$$P'(z)|_{z=1} = \frac{-[r\mu + \lambda z^{r+1} - (\lambda + r\mu)z] - (1-z)[-(\lambda + r\mu) + \lambda(r+1)z^r]}{[r\mu + \lambda z^{r+1} - (\lambda + r\mu)z]^2} r\mu(1-\rho).$$

To evaluate at $z = 1$, we must use L'Hôpital's rule (twice) to get

$$E[N] = P'(1) = \frac{(r+1)\rho}{2(1-\rho)},$$

the expected number of phases. The expected time spent in the queue waiting for service, $W_q$, is given by

$$W_q = \left(P'(z)|_{z=1}\right)\frac{1}{r\mu} = \frac{1}{r\mu}\frac{(r+1)\rho}{2(1-\rho)} = \frac{(r+1)\rho}{2r\mu(1-\rho)},$$

since, from PASTA, an arriving customer finds $E[N]$ service phases already present and all of these must be completed (at a mean service time of $1/r\mu$ each) before service can begin on the newly arriving customer.

We may now obtain $W$ from $W = W_q + 1/\mu$ and $L_q$ and $L$ from Little's law:

$$L_q = \lambda W_q = \frac{(r+1)\rho}{2r(1-\rho)}\frac{\lambda}{\mu} = \frac{r+1}{2r}\frac{\lambda^2}{\mu(\mu-\lambda)}$$

and $L = \lambda W$.

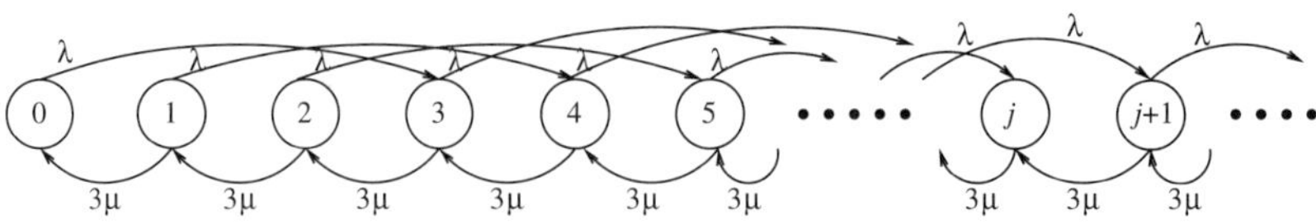


Figure 13.3. State transition diagram for the $M/E_3/1$ queue.

**Example 13.7** The Erlang-3 Service Model

Let us consider an $M/E_r/1$ queue with parameters $\lambda = 1.0$, $\mu = 1.5$, and $r = 3$. We would like to compute the probability of having $k$ customers in this system and to plot this information graphically. As before, let $k$ denote the number of customers and $j$ denote the number of service phases in the system. Also

$$p_k = \text{Prob}\{\text{Number of customers in system } = k\}$$

and

$$P_j = \text{Prob}\{\text{Number of phases in system } = j\}.$$

It follows that

$$p_k = \sum_{j=(k-1)r+1}^{kr} P_j = \sum_{j=3(k-1)+1}^{3k} P_j = P_{3k-2} + P_{3k-1} + P_{3k}, \quad k = 1, 2, 3, \ldots .$$

The balance equations, obtained by inspection from the state transition diagram (Figure 13.3) are given by

$$\begin{aligned}
\lambda P_0 &= 3\mu P_1, \\
(\lambda + 3\mu)P_1 &= 3\mu P_2, \\
(\lambda + 3\mu)P_2 &= 3\mu P_3, \\
(\lambda + 3\mu)P_j &= \lambda P_{j-3} + 3\mu P_{j+1}, \qquad j \geq 3.
\end{aligned}$$

To prepare to generate the $z$-transform, we multiply the $j^{\text{th}}$ equation by $z^j$ and obtain

$$\begin{aligned}
\lambda P_0 &= 3\mu P_1, \\
(\lambda + 3\mu)P_1 z &= 3\mu P_2 z, \\
(\lambda + 3\mu)P_2 z^2 &= 3\mu P_3 z^2, \\
(\lambda + 3\mu)P_j z^j &= \lambda P_{j-3} z^j + 3\mu P_{j+1} z^j, \qquad j \geq 3.
\end{aligned}$$

Summing both sides over all permissible values of $j$ gives

$$\lambda P_0 + (\lambda + 3\mu)\sum_{j=1}^{\infty} P_j z^j = \lambda \sum_{j=3}^{\infty} P_{j-3} z^j + 3\mu \sum_{j=0}^{\infty} P_{j+1} z^j,$$

$$(\lambda + 3\mu)\sum_{j=0}^{\infty} P_j z^j - 3\mu P_0 = \lambda z^3 \sum_{j=0}^{\infty} P_j z^j + 3\mu z^{-1}\left(\sum_{j=0}^{\infty} P_j z^j - P_0\right),$$

$$(\lambda + 3\mu)P(z) - 3\mu P_0 = \lambda z^3 P(z) + 3\mu z^{-1}(P(z) - P_0),$$

$$(\lambda + 3\mu - \lambda z^3 - 3\mu z^{-1})P(z) = 3\mu P_0 - 3\mu z^{-1} P_0.$$

It follows then that

$$P(z) = \frac{3\mu P_0(1-z^{-1})}{\lambda + 3\mu - \lambda z^3 - 3\mu z^{-1}} = \frac{3\mu P_0(1-z)}{\lambda z^4 - (\lambda+3\mu)z + 3\mu} = \frac{(9/2)P_0(1-z)}{z^4 - (1+9/2)z + 9/2}$$

$$= \frac{9P_0(1-z)}{2z^4 - 11z + 9} = \frac{9P_0(1-z)}{(z-1)(2z^3+2z^2+2z-9)} = \frac{9P_0}{9-2z(z^2+z+1)}.$$

From $P(1) = 1$, we obtain

$$9 - 2(1+1+1) = 9P_0 \implies P_0 = 1/3.$$

The roots of the denominator are

$$z_1 = 1.2169,$$
$$z_{2,\,3} = -1.1085 \pm 1.5714i.$$

Notice that $z_1 z_2 z_3 = 9/2$. These roots may be determined in a number of ways. It is possible to use pocket calculators, software such as Matlab, root finding algorithms, such as the secant method or Newton's method, and so on. Also, for cubics, a closed form exists. The cubic[1] $x^3 + 3ax^2 + 3bx + c = 0$ has three roots given by

$$x_1 = (s_1 + s_2) - a,$$
$$x_2 = -\left(\frac{s_1+s_2}{2} - a\right) + i\frac{\sqrt{3}}{2}(s_1 - s_2),$$
$$x_3 = -\left(\frac{s_1+s_2}{2} - a\right) - i\frac{\sqrt{3}}{2}(s_1 - s_2),$$

where

$$s_1 = \left[r + \sqrt{q^3+r^2}\right]^{1/3}, \quad s_2 = \left[r - \sqrt{q^3+r^2}\right]^{1/3}, \quad \text{and } q = b - a^2, \quad r = (3ab-c)/2 - a^3.$$

We may now obtain the partial fraction expansion as

$$P(z) = \frac{3}{9-2z(z^2+z+1)} = \frac{3}{-2(z-z_1)(z-z_2)(z-z_3)}$$

$$= \frac{3/(z_1z_2z_3)}{2[(1-z/z_1)(1-z/z_2)(1-z/z_3)]}$$

$$= \frac{1/3}{(1-z/z_1)(1-z/z_2)(1-z/z_3)} = \frac{A_1}{(1-z/z_1)} + \frac{A_2}{(1-z/z_2)} + \frac{A_3}{(1-z/z_3)},$$

where $A_1 = .15649$, $A_2 = .088421 - .0017774i$, $A_3 = .088421 + .0017774i$. The $A_i$ are obtained as indicated previously. For example,

$$A_1 = (1-z/z_1)P(z)\big|_{z=z_1} = \frac{1/3}{(1-z/z_2)(1-z/z_3)}\big|_{z=z_1} = 0.15649.$$

Therefore, using

$$P_k = A_1\left(\frac{1}{z_1}\right)^k + A_2\left(\frac{1}{z_2}\right)^k + A_3\left(\frac{1}{z_3}\right)^k,$$

[1] Note that all cubic equations may be written in this form.

we get

$$p_0 = P_0 = A_1 + A_2 + A_3 = .33333,$$

$$p_k = P_{3k-2} + P_{3k-1} + P_{3k} =$$

$$A_1 \left(\frac{1}{z_1}\right)^{3k-2} \left[1 + \left(\frac{1}{z_1}\right) + \left(\frac{1}{z_1}\right)^2\right]$$

$$+ A_2 \left(\frac{1}{z_2}\right)^{3k-2} \left[1 + \left(\frac{1}{z_2}\right) + \left(\frac{1}{z_2}\right)^2\right]$$

$$+ A_3 \left(\frac{1}{z_3}\right)^{3k-2} \left[1 + \left(\frac{1}{z_3}\right) + \left(\frac{1}{z_3}\right)^2\right].$$

Notice that we could have obtained these solutions more simply by just inserting the appropriate values into the general formulae that we have already developed. To follow this approach we first compute the roots $z_1, z_2, \ldots, z_r$ of the denominator polynomial

$$r\mu - \lambda(z + z^2 + \cdots + z^r).$$

With these roots we then compute the coefficients

$$A_i = (1 - z/z_i)P(z)\big|_{z=z_i}.$$

We can now find the probabilities for the number of service phases in the system as

$$P_j = (1-\rho)\sum_{i=1}^{r} \frac{A_i}{z_i^j}, \quad j = 1, 2, \ldots,$$

and finally we can compute (and plot) the required probabilities of the number of customers in the system from

$$p_k = \sum_{j=(k-1)r+1}^{kr} P_j, \quad k = 1, 2, \ldots.$$

The Matlab program given below carries out these operations. Notice that two of the roots of the denominator polynomial turn out to be a complex pair. However, the probabilities $P_j$ are always real, as the following analysis shows. First, both $z_3$ and $A_3$ are real. Let

$$z_{1,2} = a \pm ib \implies z_{1,2} = \xi e^{\pm i\theta},$$

where $\xi = \sqrt{a^2 + b^2}$ and $\theta = \tan^{-1}(b/a)$, and let

$$A_{1,2} = c \pm id \implies A_{1,2} = \gamma e^{\pm i\alpha}.$$

Since

$$\frac{A_1}{z_1^j} + \frac{A_2}{z_2^j} = \frac{\gamma e^{+i\alpha}}{(\xi e^{+i\theta})^j} + \frac{\gamma e^{-i\alpha}}{(\xi e^{-i\theta})^j} = \frac{\gamma}{\xi^j}\left(e^{i(\alpha-\theta j)} + e^{-i(\alpha-\theta j)}\right)$$

$$= \frac{2\gamma \cos(\alpha - \theta_j)}{\xi^j}$$

is real, it follows that

$$P_j = (1-\rho)\left[\left(\frac{A_1}{z_1^j} + \frac{A_2}{z_2^j}\right) + \frac{A_3}{z_3^j}\right]$$

is also real. The following Matlab code computes and plots (Figure 13.4) the distribution of customers in the $M/E_3/1$ queue.

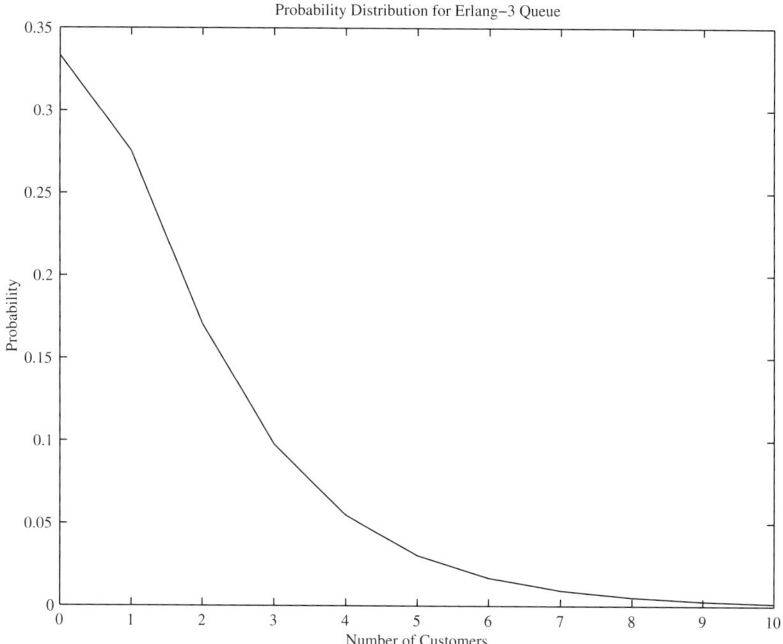


Figure 13.4. Probability distribution for Erlang service model as plotted by Matlab.

```
Function E3service()
poly=[-1 -1 -1 4.5];
z = roots(poly); rho = 1/1.5;
PO = 1 - rho;
A1 = PO/((1-z(1)/z(2))*(1-z(1)/z(3)));
A2 = PO/((1-z(2)/z(1))*(1-z(2)/z(3)));
A3 = PO/((1-z(3)/z(1))*(1-z(3)/z(2)));
for j=1:30
    P(j) = (A1*z(1)^(-j)+A2*z(2)^(-j)+A3*z(3)^(-j));
end
for k=1:10
    p(k) = sum(P(3*k-2:3*k));
end
k = 0:10;
Prob=[PO p]'
plot(k,Prob)
title('Probability Distribution for Erlang-3 Queue')
xlabel('Number of Customers')
ylabel('Probability')
```

The roots of the denominator polynomial and the coefficients of the partial fraction expansion, as computed by Matlab, are

$$z_1 = -1.1085 + 1.5714i, \qquad A1 = 0.0884 - 0.0018i,$$
$$z_2 = -1.1085 - 1.5714i, \qquad A2 = 0.0884 + 0.0018i,$$
$$z_3 = 1.2169, \qquad A3 = 0.1565.$$

The initial entries of the computed probability distribution are

$$p_0 = 0.33333333333333,$$
$$p_1 = 0.27526291723823,$$
$$p_2 = 0.17061034683687,$$
$$p_3 = 0.09772495709866,$$
$$p_4 = 0.05470074691842,$$
$$p_5 = 0.03042090224876,$$
$$p_6 = 0.01688928695838,$$
$$p_7 = 0.00937279128319,$$
$$p_8 = 0.00520098215039,$$
$$p_9 = 0.00288598001521.$$

### 13.3.5 The $E_r/M/1$ Queue Solved using z-Transforms

In applying the $z$-transform approach to the $E_r/M/1$ queue we shall again follow Kleinrock [24] and adopt a procedure similar to that used for the $M/E_r/1$ queue. We begin by converting the two-dimensional state description of the matrix-geometric approach to a one-dimensional one. In the two-dimensional description, a state $(k, i)$ indicates that $k$ customers are present in the system and the customer who is in the process of arriving is in arrival phase $i$. Each customer who has arrived (but not yet departed; there are $k$ of them) has completed all $r$ "arrival" phases. To this we shall add the number of arrival phases actually completed by an arriving customer, $(i - 1)$, and use the total number of arrival phases in the system as a one-dimensional state descriptor. Thus, when there are $k$ customers in the system and the arriving customer is in phase $i$, the total number of arrival phases is

$$j = rk + (i - 1).$$

This then is the state descriptor for the $E_r/M/1$ queue. The state transition diagram is shown in Figure 13.5 where it may be seen that a service completion results in the removal of $r$ arrival phases.

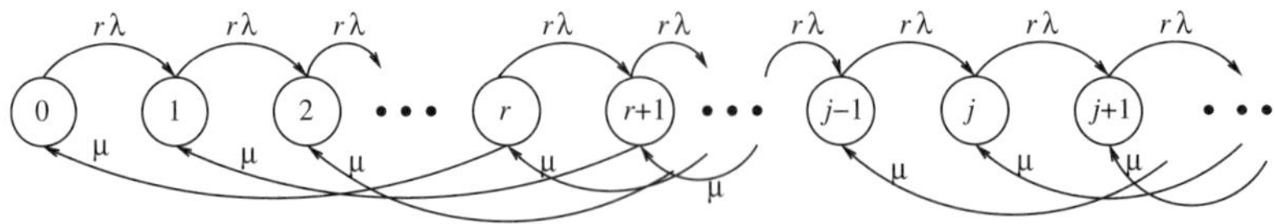


Figure 13.5. State transition diagram for the $E_r/M/1$ queue.

Let $P_j$ be the equilibrium probability of having $j$ arrival phases in the system,

$$P_j = \text{Prob}\{\text{Number of arrival phases in system } = j\},$$

and let $p_k$ be the equilibrium probability of having $k$ customers present,

$$p_k = \text{Prob}\{\text{Number of customers in system } = k\}.$$

Then

$$p_0 = \sum_{j=0}^{r-1} P_j, \quad p_1 = \sum_{j=r}^{2r-1} P_j, \quad \text{and in general,} \quad p_k = \sum_{j=rk}^{r(k+1)-1} P_j, \quad k = 0, 1, 2, \ldots .$$

The equilibrium equations can be found by inspection and are

$$\begin{aligned} r\lambda P_0 &= \mu P_r, \\ r\lambda P_j &= r\lambda P_{j-1} + \mu P_{j+r}, \quad 1 \le j \le r-1, \\ (r\lambda + \mu) P_j &= r\lambda P_{j-1} + \mu P_{j+r}, \quad j \ge r. \end{aligned}$$

Multiplying the $j^{\text{th}}$ equation by $z^j$ and adding over all equations, we obtain, on the left-hand side,

$$\begin{aligned} \sum_{j=0}^{\infty} r\lambda P_j z^j + \sum_{j=r}^{\infty} \mu P_j z^j &= r\lambda P(z) + \mu P(z) - \mu \sum_{j=0}^{r-1} P_j z^j \\ &= (r\lambda + \mu) P(z) - \mu \sum_{j=0}^{r-1} P_j z^j, \end{aligned}$$

and for the right-hand side,

$$\sum_{j=1}^{\infty} r\lambda P_{j-1} z^j + \sum_{j=0}^{\infty} \mu P_{j+r} z^j = r\lambda z \sum_{j=1}^{\infty} P_{j-1} z^{j-1} + \frac{\mu}{z^r} \sum_{j=0}^{\infty} P_{j+r} z^{j+r}$$

$$= r\lambda z P(z) + \frac{\mu}{z^r} \left[ \sum_{j=0}^{\infty} P_j z^j - \sum_{j=0}^{r-1} P_j z^j \right] = r\lambda z P(z) + \frac{\mu}{z^r} P(z) - \frac{\mu}{z^r} \sum_{j=0}^{r-1} P_j z^j.$$

Equating both sides gives

$$(r\lambda + \mu) P(z) - \mu \sum_{j=0}^{r-1} P_j z^j = r\lambda z P(z) + \frac{\mu}{z^r} P(z) - \frac{\mu}{z^r} \sum_{j=0}^{r-1} P_j z^j.$$

Bringing all terms in $P(z)$ to the right-hand side and all other terms to the left-hand side,

$$P(z) \left[ (\mu + r\lambda) - r\lambda z - \frac{\mu}{z^r} \right] = \mu \sum_{j=0}^{r-1} P_j z^j - \frac{\mu}{z^r} \sum_{j=0}^{r-1} P_j z^j,$$

and finally solving for $P(z)$

$$P(z) = \frac{(\mu/z^r) \sum_{j=0}^{r-1} P_j z^j - \mu \sum_{j=0}^{r-1} P_j z^j}{r\lambda z - (\mu + r\lambda) + (\mu/z^r)} = \frac{\mu \sum_{j=0}^{r-1} P_j z^j - \mu z^r \sum_{j=0}^{r-1} P_j z^j}{r\lambda z^{r+1} - (\mu + r\lambda) z^r + \mu},$$

i.e.,

$$P(z) = \frac{(1 - z^r) \sum_{j=0}^{r-1} P_j z^j}{r\rho z^{r+1} - (1 + r\rho) z^r + 1}, \tag{13.8}$$

where, as always, $\rho = \lambda \bar{x} = \lambda/\mu$.

The denominator polynomial, $r\rho z^{r+1} - (1 + r\rho) z^r + 1$, being of degree $r+1$, has $r+1$ zeros, of which one is unity. Of the remaining $r$ zeros, it can be shown from Rouché's theorem that exactly $r-1$ of them lie in the range $|z| < 1$ and the last, which we denote $z_0$, has modulus strictly greater than 1. We now turn our attention to the numerator $(1 - z^r) \sum_{j=0}^{r-1} P_j z^j$. Unlike the numerator of $P(z)$ in the $M/E_r/1$ queue, which contains only one unknown probability $P_0$, the numerator for the $E_r/M/1$ queue contains $r$ unknown probabilities $P_j$, $j = 0, 1, \ldots, r-1$. To overcome this difficulty, we use the fact that the $z$-transform of a probability distribution *must* be analytic in the range $|z| < 1$, i.e.,

$$P(z) = \sum_{j=0}^{\infty} P_j z^j < \infty \quad \text{for } |z| < 1.$$

Since we have just seen that the denominator has $r - 1$ zeros in this range, the numerator must also have zeros at exactly the same $r - 1$ points, for otherwise $P(z)$ blows up. Observe that the numerator consists of two factors. All of the zeros of the first factor, $(1 - z^r)$, have absolute value equal to unity. None of these can compensate for the $r$ roots of the denominator that lie in the range $|z| < 1$. The second factor, a summation, is a polynomial of degree $r - 1$ and therefore has $r - 1$ zeros. Therefore the "compensating" zeros in the numerator must come from its summation. This means that once the roots $z = 1$ and $z = z_0$ are factored out of the denominator, we can set the result equal to some constant $K$ times the summation in the numerator. This gives

$$\frac{r\rho z^{r+1} - (1 + r\rho)z^r + 1}{(1 - z)(1 - z/z_0)} = K \sum_{j=0}^{r-1} P_j z^j$$

and allows us to conclude that

$$P(z) = \frac{(1 - z^r)}{K(1 - z)(1 - z/z_0)}.$$

The constant $K$ can be found by using the fact that $P(1) = 1$:

$$1 = \lim_{z \to 1} P(z) = \lim_{z \to 1} \frac{1 - z^r}{K(1 - z)(1 - z/z_0)}.$$

To continue, we need to use L'Hôpital's rule to get

$$1 = \lim_{z \to 1} \frac{-rz^{r-1}}{K[-(1 - z/z_0) - (1/z_0)(1 - z)]} = \frac{-r}{-K[1 - 1/z_0]}$$

and hence

$$K = \frac{r}{(1 - 1/z_0)}$$

and

$$P(z) = \frac{(1 - z^r)(1 - 1/z_0)/r}{(1 - z)(1 - z/z_0)}.$$

This then is the generating function for the number of phases in the $E_r/M/1$ queue. Taking $r = 1$ should yield the same results as previously for the $M/M/1$ queue. Performing this substitution gives

$$P(z) = \frac{(1 - z)(1 - 1/z_0)}{(1 - z)(1 - z/z_0)} = \frac{(1 - 1/z_0)}{(1 - z/z_0)}.$$

To compute $z_0$ we need to find the second root of the quadratic

$$\rho z^2 - (1 + \rho)z + 1 = 0.$$

These are $z = 1$ and $z_0 = 1/\rho$ and thus

$$P(z) = \frac{1 - \rho}{1 - \rho z}$$

as before.

To invert the transform in the case when $r > 1$, we must first write it in partial fraction form. Since the degree of the numerator is greater than the degree of the denominator, we must write the transform as

$$P(z) = (1 - z^r)\left[\frac{(1 - 1/z_0)}{r(1 - z)(1 - z/z_0)}\right]$$

and now compute the partial fraction expansion of the part within the brackets. This gives

$$P(z) = (1 - z^r)\left[\frac{1/r}{1 - z} - \frac{1/(rz_0)}{1 - z/z_0}\right].$$

The factor $(1 - z^r)$ implies that if $f_j$ is the inverse transform of the terms within the brackets, the inverse transform of $P(z)$ itself is given by $f_j - f_{j-r}$, i.e., $P_j = f_j - f_{j-r}$. Using the transform relationship

$$A\alpha^n \iff \frac{A}{1 - \alpha z},$$

we can invert the terms within the brackets and get

$$\frac{1/r}{1 - z} - \frac{1/(rz_0)}{1 - z/z_0} \iff f_j = \frac{1}{r}1^j - \frac{1}{rz_0}\left(\frac{1}{z_0}\right)^j.$$

Thus

$$f_j = \frac{1}{r}(1 - z_0^{-j-1}), \quad j \geq 0, \quad \text{and} \quad f_j = 0, \quad j < 0.$$

It follows then that

$$P_j = \frac{1}{r}(1 - z_0^{-j-1}) - \frac{1}{r}(1 - z_0^{-j+r-1}) = \frac{1}{r} - \frac{1}{r}z_0^{-j-1} - \frac{1}{r} + \frac{1}{r}z_0^{-j+r-1},$$

i.e.,

$$P_j = \frac{1}{r}z_0^{r-j-1}(1 - z_0^{-r}) \quad \text{for } j \geq r.$$

Notice that, since $z_0$ is a root of $D(z) = r\rho z^{r+1} - (1 + r\rho)z^r + 1$,

$$r\rho z^{r+1} - (1 + r\rho)z^r + 1 \,\big|_{z=z_0} = 0,$$

$$r\rho z_0 - (1 + r\rho) + z_0^{-r} = 0,$$

which yields

$$r\rho(z_0 - 1) = 1 - z_0^{-r},$$

and therefore

$$P_j = \frac{1}{r}z_0^{r-j-1}(1 - z_0^{-r}) = \frac{1}{r}z_0^{r-j-1}r\rho(z_0 - 1) = \rho(z_0 - 1)z_0^{r-j-1}, \quad j \geq r.$$

Also, for $0 \leq j < r$, we have $f_{j-r} = 0$. To conclude then, we have

$$P_j = \begin{cases} \left(1 - z_0^{-j-1}\right)/r, & 0 \leq j < r, \\ \rho(z_0 - 1)z_0^{r-j-1}, & j \geq r. \end{cases}$$

To find the distribution of the number of customers present, we use our earlier relation between $p_k$ and $P_j$. We proceed first for the case when $k > 0$. For $k > 0$,

$$\begin{aligned} p_k = \sum_{j=rk}^{r(k+1)-1} P_j &= \rho \sum_{j=rk}^{r(k+1)-1} (z_0 - 1)z_0^{r-j-1} \\ &= \rho(z_0 - 1)z_0^{r-1} \sum_{j=rk}^{r(k+1)-1} \left(\frac{1}{z_0}\right)^j \\ &= \rho(z_0 - 1)z_0^{r-1} \left[\frac{(1/z_0)^{rk} - (1/z_0)^{r(k+1)}}{1 - 1/z_0}\right] \\ &= \rho(z_0 - 1)z_0^{r} \left[\frac{(1/z_0)^{rk} - (1/z_0)^{r(k+1)}}{z_0 - 1}\right] \\ &= \rho z_0^r \left[z_0^{-rk}(1 - z_0^{-r})\right], \end{aligned}$$

which yields the result:

$$p_k = \rho(z_0^r - 1)z_0^{-rk} \quad \text{for} \quad k > 0.$$

For a single server, we know that $p_0 = 1 - \rho$. We wish to verify that $\sum_{j=0}^{r-1} P_j$ gives this result.

$$\begin{aligned} p_0 = \sum_{j=0}^{r-1} P_j &= \sum_{j=0}^{r-1} \frac{1}{r}\left(1 - z_0^{-j-1}\right) \\ = 1 - \frac{1}{r}\sum_{j=0}^{r-1} z_0^{-j-1} &= 1 - \frac{1}{rz_0}\sum_{j=0}^{r-1}\left(\frac{1}{z_0}\right)^j \\ &= 1 - \frac{1}{rz_0}\left(\frac{1 - (1/z_0)^r}{1 - 1/z_0}\right), \end{aligned}$$

i.e.,

$$p_0 = 1 - \frac{1 - (1/z_0)^r}{r(z_0 - 1)}.$$

Thus our task is to show that

$$\rho = \frac{1 - (1/z_0)^r}{r(z_0 - 1)}.$$

This equation may be written as

$$r(z_0 - 1)\rho = 1 - (1/z_0)^r,$$

i.e.,

$$r\rho z_0^{r+1} - r\rho z_0^r = z_0^r - 1$$

or

$$r\rho z_0^{r+1} - (1 + r\rho)z_0^r + 1 = 0.$$

This equation must be true, since $z_0$ is a root of

$$r\rho z^{r+1} - (1 + r\rho)z^r + 1 = 0,$$

the denominator of $P(z)$.

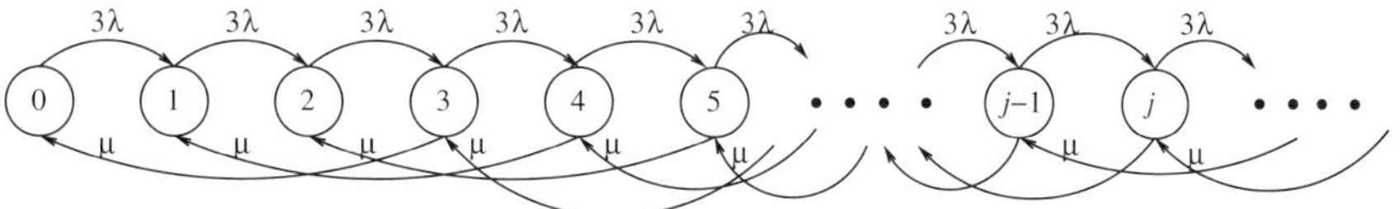


Figure 13.6. State transition diagram for the $E_3/M/1$ queue.

**Example 13.8** The Erlang-3 arrival model.

Let us consider an $E_r/M/1$ queue with parameters $\lambda = 1.0$, $\mu = 1.5$, and $r = 3$ (Figure 13.6). We would like to compute the probability of having $k$ customers in this system and to plot this information graphically. As before, let $k$ denote the number of customers and $j$ denote the number of arrival phases in the system. Also

$$p_k = \text{Prob}\{\text{Number of customers in system } = k\}$$

and

$$P_j = \text{Prob}\{\text{Number of phases in system } = j\}.$$

It follows that

$$p_k = \sum_{j=rk}^{r(k+1)-1} P_j = \sum_{j=3k}^{3(k+1)-1} P_j = P_{3k} + P_{3k+1} + P_{3k+2}, \quad k = 0, 1, 2, \ldots .$$

The balance equations, which are obtained by inspection from the state transition diagram, are given by

$$\begin{aligned} 3\lambda P_j &= \mu P_{j+3}, & j &= 0, \\ 3\lambda P_j &= 3\lambda P_{j-1} + \mu P_{j+3}, & j &= 1, 2, \\ (3\lambda + \mu) P_j &= 3\lambda P_{j-1} + \mu P_{j+3}, & j &\geq 3. \end{aligned}$$

To prepare to generate the $z$-transform, we multiply the $j^{\text{th}}$ equation by $z^j$ and obtain

$$\begin{aligned} 3\lambda P_0 &= \mu P_3, \\ 3\lambda P_1 z &= 3\lambda P_0 z + \mu P_4 z, \\ 3\lambda P_2 z^2 &= 3\lambda P_1 z^2 + \mu P_5 z^2, \\ (3\lambda + \mu) P_j z^j &= 3\lambda P_{j-1} z^j + \mu P_{j+3} z^j, \qquad j \geq 3. \end{aligned}$$

Summing both sides over all permissible values of $j$ gives

$$3\lambda \textstyle\sum_{j=0}^{\infty} P_j z^j + \mu \sum_{j=3}^{\infty} P_j z^j = 3\lambda \sum_{j=1}^{\infty} P_{j-1} z^j + \mu \sum_{j=0}^{\infty} P_{j+3} z^j,$$

i.e.,

$$\begin{aligned} 3\lambda P(z) + \mu P(z) - \mu \left(P_0 + P_1 z + P_2 z^2\right) &= 3\lambda z P(z) + \frac{\mu}{z^3} \sum_{j=0}^{\infty} P_{j+3} z^{j+3} \\ &= 3\lambda z P(z) + \frac{\mu}{z^3} P(z) - \frac{\mu}{z^3} \left(P_0 + P_1 z + P_2 z^2\right) \end{aligned}$$

$$P(z) \left(3\lambda + \mu - 3\lambda z - \frac{\mu}{z^3}\right) = \mu \left(P_0 + P_1 z + P_2 z^2\right) - \frac{\mu}{z^3} \left(P_0 + P_1 z + P_2 z^2\right),$$

and we find

$$P(z) = \frac{\left(\mu - \mu/z^3\right)\left(P_0 + P_1 z + P_2 z^2\right)}{3\lambda(1-z) + \mu(1 - 1/z^3)}$$

$$= \frac{(z^3 - 1)\left(P_0 + P_1 z + P_2 z^2\right)}{3\rho z^3(1-z) + (z^3 - 1)}$$

$$= \frac{(1 - z^3)\left(P_0 + P_1 z + P_2 z^2\right)}{3\rho z^4 - (1+3\rho)z^3 + 1},$$

which is identical to Equation (13.8) with $r = 3$. Since $(z-1)$ is a factor in both numerator and denominator, we may divide through by $(z-1)$ and obtain

$$P(z) = \frac{(1 + z + z^2)(P_0 + P_1 z + P_2 z^2)}{1 + z + z^2 - 3\rho z^3}.$$

From $P(1) = 1$, we obtain

$$1 + 1 + 1 - 3 \times 2/3 = (1 + 1 + 1)(P_0 + P_1 + P_2) \implies P_0 + P_1 + P_2 = 1/3.$$

Notice that

$$p_0 = P_0 + P_1 + P_2 = 1/3 = 1 - \lambda/\mu = 1 - \rho.$$

The roots of the denominator are

$$z_0 = 1.2338,$$

$$z_{1,2} = -.36688 \pm .52026i,$$

and only the first of these (the only root that is greater than 1) is of interest to us. We may then write the final result for the probability distribution of customers in the $E_3/M/1$ queue:

$$p_0 = 1 - \rho = 1/3,$$

$$p_k = \rho(z_0^3 - 1)z_0^{-3k} = \left(\frac{2}{3}\right)\left(\frac{1.2338^3 - 1}{1.2338^{3k}}\right), \quad k = 1, 2, \ldots .$$

The following Matlab code computes and plots (Figure 13.7) the distribution of customers in the $E_3/M/1$ queue.

```
Function E3arrival()
poly=[2 -3 0 0 1];
z = roots(poly); rho = 1/1.5;
p0 = 1 - rho;
z0 = z(1);
for k=1:10
    p(k) = (z0^3-1)/(1.5*(z0^(3*k)));
 end
 k = 0:10;
 Prob=[p0 p]'
 plot(k,Prob)
 title('Probability Distribution for Erlang-3 Arrival Model')
 xlabel('Number of Customers')
 ylabel('Probability')
```

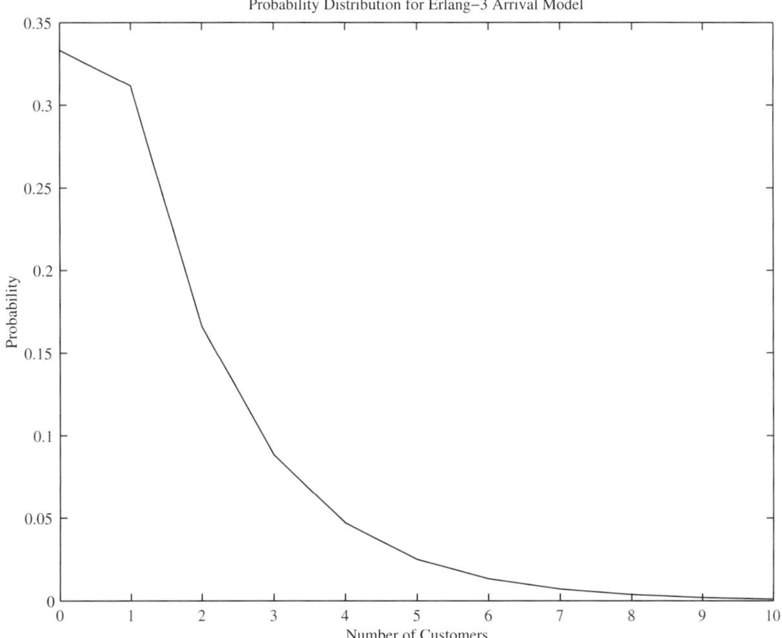


Figure 13.7. Probability distribution for Erlang arrival model as plotted by Matlab.

The roots of the denominator polynomial as computed by Matlab are

$$\begin{aligned} z_0 &= 1.2338, \\ z_1 &= 1.0000, \\ z_2 &= -0.3669 + 0.5203i, \\ z_3 &= -0.3669 - 0.5203i. \end{aligned}$$

The initial entries of the computed probability distribution are

$$\begin{aligned} p_0 &= 0.33333333333333, \\ p_1 &= 0.31166923803768, \\ p_2 &= 0.16596266712920, \\ p_3 &= 0.08837448011891, \\ p_4 &= 0.04705906979796, \\ p_5 &= 0.02505877315792, \\ p_6 &= 0.01334370005349, \\ p_7 &= 0.00710546881108, \\ p_8 &= 0.00378363473570, \\ p_9 &= 0.00201477090307. \end{aligned}$$

### *13.3.6 Bulk Queueing Systems*

There is a connection between queueing systems in which multiple customers may arrive simultaneously (as a bulk) and the $M/E_r/1$ queue. In the $M/E_r/1$ queue each customer must pass through

$r$ phases of service to complete its total service. When we analyzed this queue, we counted the number of service phases each customer contributed upon arrival and used the total number of phases present to define a state of the system. Now look at this another way. Consider each customer arrival to be in reality the arrival of $r$ customers, each of these $r$ customers requiring only a single phase of service. These two points of view define identical systems: the $M/E_r/1$ queue and the $M/M/1$ queue with bulk arrivals. In this case, the bulk size is fixed at $r$. Similarly, the $E_r/M/1$ queue may be viewed as a queue with bulk service! The single server takes $r$ customers from the queue and gives exponential service to the bulk. All $r$ leave at the same instant. If there are fewer than $r$ customers, the server waits until there are $r$ in the queue. In the $E_r/M/1$ queue, service completion is equivalent to losing $r$ service phases.

Let us now consider a more general case and permit bulks of arbitrary sizes to arrive at each (Poisson) arrival instant. We take the arrival rate of bulks to be $\lambda$ and the probability mass function of bulk size to be

$$g_i \equiv \text{Prob}\{\text{Bulk size is } i\}.$$

A state of the system is completely characterized by the number of customers present. Since we have not imposed any limit to the size of bulks, transitions out of any state $k$ can occur to any state $k+i,\ i = 1, 2, \ldots,$ at rate $\lambda g_i$ and so the net exit rate from state $k$ due to arrivals is

$$\lambda g_1 + \lambda g_2 + \cdots = \lambda \sum_{i=1}^{\infty} g_i = \lambda.$$

Also from any state $k > 0$ a transition can occur due to a service completion at rate $\mu$ and move the system to state $k-1$, and so the overall departure rate from state $k > 0$ is $\lambda + \mu$.

State $k$ may be entered from any state $k-i,\ i = 1, 2, \ldots, k$, at rate $\lambda g_i$, since the arrival of a bulk of size $i$ in state $k-i$ will move the system to state $k$. A transition is also possible from state $k+1$, due to a service completion at rate $\mu$, to state $k$. The equilibrium equations are therefore given by

$$(\lambda + \mu) p_k = \mu p_{k+1} + \sum_{i=0}^{k-1} p_i \lambda g_{k-i}, \quad k \geq 1,$$

with

$$\lambda p_0 = \mu p_1,$$

and we may solve them by the use of $z$-transforms,

$$(\lambda + \mu) \sum_{k=1}^{\infty} p_k z^k = \frac{\mu}{z} \sum_{k=1}^{\infty} p_{k+1} z^{k+1} + \sum_{k=1}^{\infty} \sum_{i=0}^{k-1} p_i \lambda g_{k-i} z^k. \tag{13.9}$$

Figure 13.8 shows that the double summation may be rewritten as

$$\sum_{k=1}^{\infty} \sum_{i=0}^{k-1} = \sum_{i=0}^{\infty} \sum_{k=i+1}^{\infty}.$$

All the terms to be included are at the intersection of the dashed and dotted lines in this figure, and it is immaterial whether these are traversed vertically, following the dotted lines as in the original order, or horizontally, following the dashed lines.

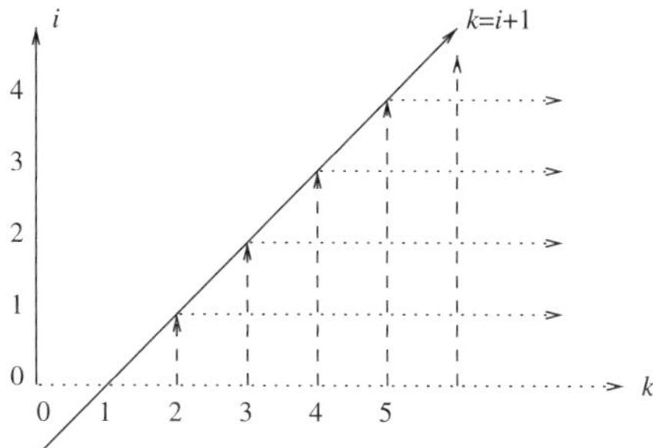


Figure 13.8. Double summations horizontally and vertically.

This yields

$$\sum_{k=1}^{\infty}\sum_{i=0}^{k-1} p_i \lambda g_{k-i} z^k = \lambda \sum_{i=0}^{\infty} p_i z^i \sum_{k=i+1}^{\infty} g_{k-i} z^{k-i}$$

$$= \lambda \sum_{i=0}^{\infty} p_i z^i \sum_{j=1}^{\infty} g_j z^j,$$

and Equation (13.9) can be written as

$$(\lambda + \mu)\sum_{k=1}^{\infty} p_k z^k = \frac{\mu}{z}\sum_{k=1}^{\infty} p_{k+1} z^{k+1} + \lambda \sum_{i=0}^{\infty} p_i z^i \sum_{j=1}^{\infty} g_j z^j.$$

Let $G(z)$ be the $z$-transform of the distribution of bulk size:

$$G(z) \equiv \sum_{k=1}^{\infty} g_k z^k.$$

Then

$$(\lambda + \mu)[P(z) - p_0] = \frac{\mu}{z}\left[P(z) - p_0 - p_1 z\right] + \lambda P(z)G(z).$$

Using $\lambda p_0 = \mu p_1$ and simplifying, we obtain

$$P(z) = \frac{\mu p_0(1-z)}{\mu(1-z) - \lambda z(1 - G(z))}.$$

To eliminate $p_0$, we would like to use the fact that $P(1) = 1$, but since unity is a root of both the numerator and the denominator, we must have recourse to L'Hôpital's rule:

$$1 = \lim_{z \to 1} \frac{-\mu p_0}{\lambda G(z) + \lambda z G'(z) - \lambda - \mu}$$

$$= \frac{-\mu p_0}{\lambda + \lambda G'(1) - \lambda - \mu} = \frac{\mu p_0}{\mu - \lambda G'(1)}.$$

Therefore

$$p_0 = \frac{\mu - \lambda G'(1)}{\mu} = 1 - \frac{\lambda G'(1)}{\mu}.$$

Observe that $\lambda G'(1)/\mu$ is just the utilization factor, $\rho$, for this queue, is the quotient of the average effective arrival rate and the average service rate ($G'(1)$ is the average bulk size). Hence $p_0 = 1 - \rho$.

This then yields the final result

$$P(z) = \frac{\mu(1-\rho)(1-z)}{\mu(1-z) - \lambda z(1-G(z))}.$$

Once the sequence $\{g_k\}$ is given, we can compute $G(z)$ and then possibly invert the transform $P(z)$. In any case, the mean and variance of the number of customers may be obtained from $P(z)$.

**Example 13.9** Let us consider the queueing system whose state transition rate diagram is shown in Figure 13.9. This may be associated with an $M/M/1$ queue with the difference that each arrival brings either one or two customers with equal probability. Let us find the $z$-transform of this system by identifying it with an $M/M/1$ queue with bulk arrivals.

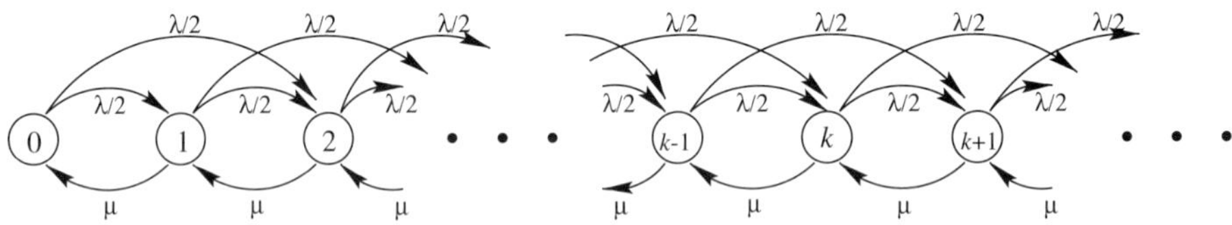


Figure 13.9. Transition diagram for Example 13.9.

To compute $P(z)$, we may use the expression just derived, namely,

$$P(z) = \frac{\mu(1-\rho)(1-z)}{\mu(1-z) - \lambda z[1-G(z)]}.$$

In this example, $g_1 = g_2 = 1/2$ and therefore

$$G(z) = \frac{1}{2}z + \frac{1}{2}z^2 = \frac{1}{2}z(1+z).$$

Also, $G'(z) = \frac{1}{2} + z$, and

$$\rho = \frac{\lambda G'(1)}{\mu} = \frac{3\lambda}{2\mu}.$$

Substituting $G(z)$ and $\rho$ into $P(z)$ above and simplifying, we get

$$P(z) = \frac{1-\rho}{1-\rho z(z+2)/3}.$$

The expected number of customers in the system is obtained as

$$E[N] = \left.\frac{dP(z)}{dz}\right|_{z=1} = (1-\rho)\left.\frac{2\rho(z+1)/3}{[1-\rho z(z+2)/3]^2}\right|_{z=1}$$

$$= \frac{4}{3}\frac{\rho}{1-\rho}.$$

## 13.4 Exercises

**Exercise 13.1.1** Compute the $z$-transform of the Poisson distribution. Use your answer to find the mean and variance of this distribution.

**Exercise 13.1.2** Find the $z$-transform of the probability sequence

$$p_0 = 3 - e, \quad p_1 = 0, \quad p_k = 1/k! \text{ for } k \geq 2.$$

**Exercise 13.1.3** Prove the following relations, taken from Table 13.1:

| Sequence | | Transform |
|---|---|---|
| $p_k \star q_k$ | $\Longleftrightarrow$ | $P(z)Q(z)$ |
| $p_{k+1}$ | $\Longleftrightarrow$ | $[P(z) - p_0]/z$ |
| $p_{k-1}$ | $\Longleftrightarrow$ | $zP(z)$ |
| $u_{k-i}$ | $\Longleftrightarrow$ | $z^i$ |
| $\{0,\ \alpha,\ 2\alpha^2,\ 3\alpha^3,\ \ldots\} \ = \ k\alpha^k$ | $\Longleftrightarrow$ | $\alpha z/(1-\alpha z)^2$ |
| $\{1,\ 2,\ 3,\ \ldots\} \ = \ k+1$ | $\Longleftrightarrow$ | $1/(1-z)^2$ |

**Exercise 13.1.4** Show that the $z$-transform of the sequence $p_k = -2u_k - 2u_{k-1} + 3$, where $u_k$ is the unit function, is given by

$$P(z) = \frac{1+2z^2}{1-z}.$$

**Exercise 13.1.5** Consider the infinite sequence whose $k^{\text{th}}$ term is given by $p_k = (1-\rho)\rho^k$, $k = 0, 1, \ldots,$ with $0 < \rho < 1$. Find the $z$-transform of this sequence and show that the $k^{\text{th}}$ term in the sequence is recovered when the $k^{\text{th}}$ derivative of the transform is formed and evaluated at $z = 0$.

**Exercise 13.1.6** The $z$-transform of a probability sequence is given by

$$P(z) = \frac{1+z^4}{10} + \frac{z+z^2+z^3}{5}.$$

Find the sequence

(a) by writing the transform as a power series and comparing coefficients;
(b) by differentiation and evaluation of the transform.

**Exercise 13.2.1** Find the sequence whose $z$-transform is given by

$$P(z) = \frac{1+3z}{2-z-z^2}.$$

**Exercise 13.2.2** Find the sequence whose $z$-transform is given by

$$P(z) = \frac{14-27z}{24-84z+36z^2}.$$

**Exercise 13.2.3** Find the sequence whose $z$-transform is given below. Observe that unity is one of the roots of the denominator.

$$P(z) = \frac{30-57z+9z^2}{40-42z+2z^3}.$$

**Exercise 13.2.4** Find the sequence whose $z$-transform is given by

$$P(z) = \frac{48z-18z^2}{40-10z-5z^2}.$$

**Exercise 13.2.5** Find the sequence whose $z$-transform is given by

$$P(z) = \frac{1+z^2}{1-2z}.$$

**Exercise 13.2.6** Find the sequence whose $z$-transform is given by

$$P(z) = \frac{2z-1/2}{(1-2z)^2}.$$

**Exercise 13.2.7** Find the sequence whose $z$-transform is given by

$$P(z) = \frac{-1+62z+49z^2-60z^3}{4+24z+4z^2-96z^3+64z^4}.$$

*Hint: $z_1 = 1$ is a double root of the denominator.*

**Exercise 13.3.1** For the queuing system with arrivals in pairs discussed in Section 13.3.3, what are the block matrices needed to apply the matrix-geometric method? Apply the Matlab code of Chapter 12 to find the probabilities $p_i$, $i = 0, 1, 2$, and 3.

**Exercise 13.3.2** Consider a single-server queueing system with bulk arrivals and for which the probability density of bulk sizes is

$$g_1 = .5,\ \ g_2 = .25,\ \ g_3 = 0,\ \ g_4 = .25,\ \ g_k = 0,\ k > 4.$$

(a) Derive the $z$-transform $P(z)$ of this system
- by identifying this system with an $M/M/1$ queue with bulk arrivals;
- by writing down the equilibrium probabilities and finding the $z$-transform from first principles.

(b) Use the $z$-transform to compute the mean number of customers present at steady state.

(c) Using the values $\lambda = 1$ and $\mu = 4$, invert the transform to find the distribution of customers in this queueing system.

# Chapter 14

# *The M/G/1 and G/M/1 Queues*

In this chapter we examine two important single-server queues, the *M/G/*1 queue and its dual, the *G/M/*1 queue. Our concern is with the stationary behavior of these queues, and it turns out that this can be obtained rather conveniently by constructing and then solving a Markov chain that is embedded at certain well-specified instants of time within the stochastic processes that define the systems. For the *M/G/*1 queue, we choose the instants at which customers depart and it turns out that the solution at these instants is also the stationary solution we seek. For the *G/M/*1 queue, we choose arrival instants which allows us to find the distribution of customers at arrival instants. Although this solution is not equal to the distribution as seen by an arbitrary observer of the system (which is what we seek), it does open the way for the computation of the stationary distribution. Unfortunately, the embedded Markov chain approach does not apply to the *G/G/*1 queue and more advanced methods, such as the use of the Lindsley integral equation, must be used. Since this is beyond the level of this text, we do not consider the *G/G/*1 queue further except to say that one possibility might be to model the general arrival and service time distributions by phase-type distributions and to apply the matrix geometric methods of Chapter 12.

## 14.1 Introduction to the *M/G/*1 Queue

The *M/G/*1 queue is a single-server queue, illustrated graphically in Figure 14.1.

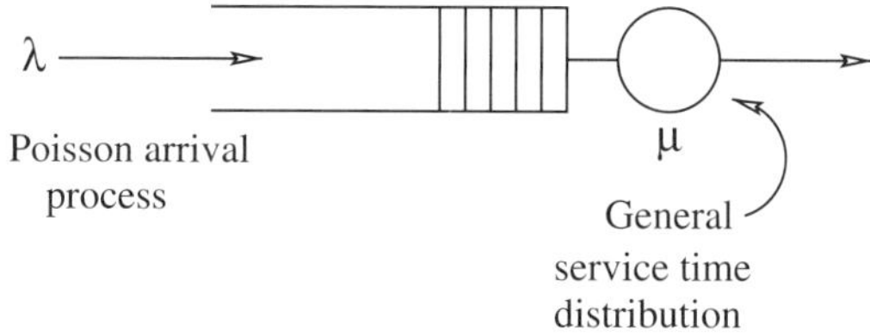


Figure 14.1. The *M/G/*1 queue.

The arrival process is Poisson with rate $\lambda$, i.e., its distribution function is

$$A(t) = 1 - e^{-\lambda t}, \quad t \geq 0.$$

The service times of customers are independent and identically distributed and obey an unspecified arbitrary or general distribution function. In particular, the remaining service time may no longer be independent of the service already received. As usual, the mean service rate is denoted by $\mu$. We denote the service time distribution function as

$$B(x) = \text{Prob}\{S \leq x\},$$

where $S$ is the random variable "service time"; its density function, denoted $b(x)$, is given by

$$b(x)dx = \text{Prob}\{x < S \leq x + dx\}.$$

We shall take the scheduling discipline to be FCFS although the results derived in this chapter frequently apply to many other nonpreemptive scheduling policies. As a special case of the *M/G/*1 queue, if we let $B(x)$ be the exponential distribution with parameter $\mu$, we obtain the *M/M/*1 queueing system. If the service times are assumed to be constant, then we have the *M/D/1* queueing system.

Recall that with the *M/M/*1 queue, all that is required to summarize its entire past history is a specification of the number of customers present, $N(t)$, and, in this case, the stochastic process $\{N(t), t \geq 0\}$ is a Markov process. For the *M/G/*1 queue, the stochastic process $\{N(t), t \geq 0\}$ is *not* a Markov process since, when $N(t) \geq 1$, a customer is in service and the time already spent by that customer in service must be taken into account—a result of the fact that the service process need not possess the memoryless property of the exponential distribution.

Let us examine this time dependence of service a little more closely. Consider the *conditional* probability $C(x)$ that the service finishes before $x + dx$ knowing that its duration is greater than $x$,

$$C(x) = \text{Prob}\{S \leq x + dx | S > x\} = \frac{\text{Prob}\{x < S \leq x + dx\}}{\text{Prob}\{S > x\}} = \frac{b(x)dx}{1 - B(x)}.$$

Generally $C(x)$ depends on $x$. However, if $B(x)$ is the exponential distribution, $B(x) = 1 - e^{-\mu x}$, $b(x) = \mu e^{-\mu x}$, and

$$C(x) = \frac{\mu e^{-\mu x} dx}{1 - 1 + e^{-\mu x}} = \mu dx,$$

which is independent of $x$. In this particular case, if we start to observe the service in progress at an arbitrary time $x$, the probability that the service completes on the interval $(x, x + dx]$ does not depend on $x$, the duration of service already received by the customer. However, for the *M/G/*1 queue, where $C(x)$ generally depends on $x$, the process $\{N(t), t \geq 0\}$ is not Markovian and if we start to observe the process at an arbitrary time $x$, the probability that the service completes on the interval $(x, x+dx]$, which implies a transition of the process $N(t)$, is not independent of $x$, the time already spent in service for the customer currently in service. In other words, the probability that a transition will occur depends on its past history, and therefore the process is non-Markovian. What this means is that, if at some time $t$ we want to summarize the complete relevent past history of an *M/G/*1 queue, we must specify both

1. $N(t)$, the number of customers present at time $t$, *and*
2. $S_0(t)$, the service time already spent by the customer in service at time $t$.

Notice that, while $N(t)$ is not Markovian, $[N(t), S_0(t)]$ *is* a Markov process, since it provides all the past history necessary for describing the future evolution of an *M/G/*1 queue. The component $S_0(t)$ is called a *supplementary variable* and the approach of using this state description to solve the *M/G/*1 queue is called the method of supplementary variables. We shall not dwell on this approach, which involves working with two components, the first discrete and the second continuous, but rather we shall seek an alternative solution based on a single discrete component: the embedded Markov chain approach.

## 14.2 Solution via an Embedded Markov Chain

As its name suggests, in the embedded Markov chain approach, we look for a Markov chain within (i.e., at certain time instants within) the stochastic process $[N(t), S_0(t)]$ and solve for the distribution of customers at these times. One convenient set of time instants is the set of *departure* instants. These are the times at which a customer is observed to terminate service and leave the queueing system. At precisely these instants, the customer entering service has received exactly zero seconds of service. At these instants we know both $N(t)$ and $S_0(t)$, the latter being equal to zero. It follows that this embedded Markov chain approach allows us to replace the two-dimensional state description $[N(t), S_0(t)]$ with a one-dimensional description $N_k$, where $N_k$ denotes the number of customers left behind by the $k^{\text{th}}$ departing customer. Of course, when we solve the embedded Markov chain, we obtain the distribution of customers at departure instants. Conveniently, however, for the *M/G/*1 queue, it turns out that the distribution of customers at departure instants is also identical to the distribution of customers at any time instant.

Let us now examine the distribution of time between customer departures. When a customer departs and leaves at least one customer behind, the time until the next departure is the same as the distribution of the service time. On the other hand, when a customer departs and leaves behind an empty system, the next departure will not occur until after a new customer arrives *and* has been served. In this case the distribution of time between departures is equal to the convolution of the exponential interarrival time distribution and the general service time distribution. Thus the system that we have just described is actually a *semi-Markov process*, since the time between state transitions (departure instants) obeys an arbitrary probability distribution whereas in a discrete-time Markov chain the time between state transitions must be *geometrically* distributed. As we shall see, we need only be concerned with the Markov chain embedded within this semi-Markov process and not with the semi-Markov process in its entirety.

At this point is is worthwhile digressing a little to discuss the concept of *regeneration points*, or *renewal instants*—see also Section 9.11 and the discussion on renewal processes. As the name suggests, a regeneration point is an instant in time at which the future behavior of a system depends only on its state at that instant. It is completely independent of the evolutionary path that led to its current state. The system is reborn. For example, each departure instant in an *M/G/*1 queue is a regeneration point. Let $\{X(t), t \geq 0\}$ be an arbitrary stochastic process and let $t_0$ be an arbitrary instant. Generally we do not have

$$\text{Prob}\{X(t)|X(t_0)\} = \text{Prob}\{X(t)|X(s), s \leq t_0\}.$$

If this relation is true for all $t > t_0$, then $t_0$ is a regeneration point. Notice that a stochastic process $\{X(t), t \geq 0\}$ is a Markov process if and only if every instant $t_0$ is a regeneration point. Furthermore, if there exists a sequence $\{t_k\}_{k=1,2,\ldots}$ of regeneration points, then $\{X(t_k), k = 1, 2, \ldots\} \equiv \{X_k, k = 1, 2, \ldots\}$ is a Markov chain. It is the Markov chain embedded in the stochastic process $\{X(t), t \geq 0\}$, since, from the definition of regeneration points,

$$\text{Prob}\{X_{k+l}|X_k\} = \text{Prob}\{X_{k+l}|X_k, X_{k-1}, \ldots\} \quad \text{for all } k.$$

Consider now the stochastic process $\{N(t), t \geq 0\}$ associated with the *M/G/*1 queue and the sequence $\{t_k\}_{k=1,2,\ldots}$ where $t_k$ is defined as the instant of the end of service for the $k^{\text{th}}$ customer. Let $N_k \equiv N(t_k)_{k=1,2,\ldots}$ be the number of customers in the system just after the departure of the $k^{\text{th}}$ customer to be served. Given $N_k$, the value of $N_{k+l}$ depends just on the number of arrivals between $t_k$ and $t_{k+l}$, since the number of departures is exactly equal to $l$. However, the arrival process in the *M/G/*1 queue is Poisson which means that the number of arrivals between $t_k$ and $t_{k+l}$ does not depend on the past, i.e., on $N_{k-j}$, $j = 1, 2, \ldots$. Therefore

$$\text{Prob}\{N_{k+l}|N_k\} = \text{Prob}\{N_{k+l}|N_k, N_{k-1}, \ldots\}$$

and so $\{N_k, \; k = 1, 2, \ldots\}$ is a Markov chain. It is a Markov chain *embedded* within the stochastic process $\{N(t), t \geq 0\}$.

It only remains to show that the solution obtained at departure instants is also the solution at any point of time, and we do so in two steps. First, we have previously defined $a_n$ to be the probability that an arriving customer finds $n$ customers already present. Now let $d_n$ be the probability that a departing customer leaves $n$ customers behind. Since the overall arrival rate is equal to $\lambda$, the rate of transition from state $n$ to state $n+1$—a transition occasioned by an arrival finding $n$ customers present—is given by $\lambda\, a_n$. Furthermore, at equilibrium, the departure rate is also equal to $\lambda$ and so the rate of transition from state $n + 1$ to state $n$—a transition occasioned by a departure leaving behind $n$ customers—is equal to $\lambda\, d_n$. At equilibrium, the rate of transition from state $n$ to $n + 1$ is equal to the rate of transition from state $n + 1$ to $n$ and thus it follows that $\lambda\, a_n = \lambda\, d_n$, i.e., that $a_n = d_n$. This result applies more generally: the limiting distribution of customers found by a new arrival in any system that changes states by increments of plus or minus one is the same as the limiting distribution of customers left behind by a departure, assuming that these limits actually exist. In the *M/G/*1 queue this condition holds since customers arrive and depart individually and the limits exist when the mean interarrival time is strictly greater than the mean service time ($\lambda/\mu < 1$). Therefore the solution at departure instants in an *M/G/*1 queue is also the solution at arrival instants: in other words, customers arriving to the system see the same customer population distribution as that seen by customers departing the system.

Second, due to the Poisson nature of arrivals to the *M/G/*1 queue, we know from the PASTA (Poisson arrivals see time averages) property that an arriving customer sees the same distribution as a random observer. Therefore, in the *M/G/*1 queue, the distribution of customers at arrival instants is also the same as the distribution of customers at any time instant. Taking these two facts together, we may conclude that the solution obtained at departure instants in an *M/G/*1 queue is also the solution at any arbitrary instant of time.

Let $A_k$ be the random variable that denotes the number of customers that arrive during the service time of the $k^{\text{th}}$ customer. We now derive a relationship for the number of customers left behind by the $(k + 1)^{\text{th}}$ customer in terms of the number left behind by its predecessor and $A_{k+1}$. First, if $N_k = i > 0$, we must have

$$N_{k+1} = N_k - 1 + A_{k+1},$$

since there are $N_k$ present in the system when the $(k + 1)^{\text{th}}$ customer enters service, an additional $A_{k+1}$ arrive while this customer is being served, and the number in the system is reduced by 1 when this customer finally exits. Similarly, if $N_k = 0$ we must have

$$N_{k+1} = A_{k+1}.$$

We may combine these into a single equation with the use of the function $\delta(N_k)$ defined as

$$\delta(N_k) = \begin{cases} 1, & N_k > 0, \\ 0, & N_k = 0. \end{cases}$$

Since

$$\begin{aligned} N_{k+1} &= N_k - 1 + A_{k+1}, \qquad & N_k > 0, \\ N_{k+1} &= A_{k+1}, & N_k = 0, \end{aligned}$$

we have

$$N_{k+1} = N_k - \delta(N_k) + A. \tag{14.1}$$

Recall that $N_k$ is the number of customers the $k^{\text{th}}$ customer leaves behind, and $A_k$ is the number of customers who arrive during the service time of the $k^{\text{th}}$ customer. Since the service times are independent and identically distributed and independent of the interarrival times, which are also independent and identically distributed, it follows that the random variables $A_k$ are independent and identically distributed. Thus $A_k$ must be independent of $k$, which allows us to write it simply as $A$ in Equation (14.1).

We shall now find the stochastic transition probability matrix for the embedded Markov chain $\{N_k,\ k = 1, 2, \ldots\}$. The $ij^{\text{th}}$ element of this matrix is given by

$$f_{ij}(k) = \text{Prob}\{N_{k+1} = j | N_k = i\},$$

which in words says that the probability the $(k+1)^{\text{th}}$ departing customer leaves behind $j$ customers, given that the $k^{\text{th}}$ departure leaves behind $i$, is $f_{ij}(k)$. As might be expected, and as we shall see momentarily, this is independent of $k$, and so we write the transition probability matrix simply as $F$. Let $p$ be the stationary distribution of this Markov chain:

$$pF = p.$$

The $j^{\text{th}}$ component of $p$ gives the stationary probability of state $j$, i.e., the probability that a departing customer leaves $j$ customers behind. It provides us with the probability distribution of customers at departure instants. The single-step transition probability matrix is given by

$$F = \begin{pmatrix} \alpha_0 & \alpha_1 & \alpha_2 & \alpha_3 & \alpha_4 & \cdots \\ \alpha_0 & \alpha_1 & \alpha_2 & \alpha_3 & \alpha_4 & \cdots \\ 0 & \alpha_0 & \alpha_1 & \alpha_2 & \alpha_3 & \cdots \\ 0 & 0 & \alpha_0 & \alpha_1 & \alpha_2 & \cdots \\ \vdots & \vdots & \ddots & \ddots & \ddots & \ddots \\ 0 & 0 & 0 & 0 & 0 & \end{pmatrix}, \tag{14.2}$$

where $\alpha_i$ is the probability that $i$ arrivals occur during an arbitrary service. Since a departure cannot remove more than one customer, all elements in the matrix $F$ that lie below the subdiagonal must be zero. These are the elements $f_{ij}(k)$ for which $i > j + 1$. Also, all elements that lie above the diagonal are strictly positive, since any number of customers can arrive during the service of the $k^{\text{th}}$ customer. Observe that it is possible to reach all states from any given state, since $\alpha_i > 0$ for all $i \geq 0$, and so the Markov chain is irreducible. If $i \leq j$, we can go from state $i$ to state $j$ in a single step; if $i > j$, we need $(i - j)$ steps to go from state $i$ to state $j$. Furthermore, since the diagonal elements are nonzero, the Markov chain is aperiodic.

Given that the random variables $A_k$ are independent and identically distributed, and assuming that the $k^{\text{th}}$ departing customer leaves at least one customer behind ($N_k = i > 0$), we have

$$\text{Prob}\{N_{k+1} = j | N_k = i\} \equiv f_{ij}(k) = \alpha_{j-i+1} \quad \text{for } j = i-1, i, i+1, i+2, \ldots.$$

In words, if there are $i > 0$ customers at the start and one departs, then we need $j - i + 1$ arrivals to end up with a total of $j$, since $i - 1 + (j - i + 1) = j$. On the other hand, in the case of the $k^{\text{th}}$ departure leaving behind an empty system ($N_k = i = 0$), we have

$$\text{Prob}\{N_{k+1} = j | N_k = 0\} \equiv f_{0j}(k) = \alpha_j \quad \text{for } j = 0, 1, 2, \ldots.$$

Thus the transition probabilities $f_{ij}(k)$ do not depend on $k$, which means that the Markov chain $\{N_k, k = 1, 2, \ldots\}$ is homogeneous.

**Example 14.1** Keeping in mind the fact that the indices $i$ and $j$ range from an initial value of 0 and not 1, we have the following examples: The $j^{\text{th}}$ component of row 0 is $f_{0j} = \alpha_j$. Thus the fourth component in row 0, corresponding to $j = 3$, is equal to $\alpha_3$ and gives the probability that the previous customer left behind an empty system ($i = 0$) and that during the service of the customer who arrived to this empty system, exactly $j$ (i.e., 3) customers arrived.

**Example 14.2** Now consider the fifth component in the third row of the matrix $F$ which is equal to $\alpha_3$. This element is the probability

$$f_{24} = \text{Prob}\{N_{k+1} = 4 | N_k = 2\} = \alpha_{4-2+1} = \alpha_3.$$

For customer $k + 1$ to leave behind four customers, given that the previous customer left behind only two, three new customers must arrive during the time that customer $k + 1$ is being served.

We now need to calculate the probabilities $\alpha_i$, $i = 0, 1, 2, \ldots$. From our definition of $S$ as the random variable "service time" and $b(x)$ the probability density function of this random variable, we have

$$\text{Prob}\{A = i \text{ and } x < S \leq x + dx\} = \text{Prob}\{A = i \mid x < S \leq x + dx\}\text{Prob}\{x < S \leq x + dx\}.$$

The first term in this product is Poisson and equal to $((\lambda x)^i/i!)e^{-\lambda x}$; the second is equal to $b(x)dx$. Therefore

$$\alpha_i = \text{Prob}\{A = i\} = \int_0^\infty \frac{(\lambda x)^i}{i!} e^{-\lambda x} b(x)dx. \tag{14.3}$$

To accommodate all types of service time distributions (i.e., discrete, continuous, and mixed), the Stieltjes integral form should be used and $b(x)dx$ replaced with $dB(x)$ in Equation (14.3). This then completely specifies the transition probability matrix $F$.

If the $\alpha_i$'s of Equation (14.3) can be conveniently computed, then the upper Hessenberg structure of $F$ facilitates the computation of the stationary probability vector of the *M/G/*1 queue, especially since we frequently know, a priori, the probability that the system is empty, i.e., $p_0 = 1 - \lambda/\mu = 1 - \rho$. Successively writing out the equations, we have

$$\begin{aligned} \alpha_0 p_0 + \alpha_0 p_1 &= p_0, \\ \alpha_1 p_0 + \alpha_1 p_1 + \alpha_0 p_2 &= p_1, \\ \alpha_2 p_0 + \alpha_2 p_1 + \alpha_1 p_2 + \alpha_0 p_3 &= p_2, \\ &\vdots \end{aligned}$$

Rearranging,

$$\begin{aligned} p_1 &= \frac{1-\alpha_0}{\alpha_0} p_0, \\ p_2 &= \frac{1-\alpha_1}{\alpha_0} p_1 - \frac{\alpha_1}{\alpha_0} p_0, \\ p_3 &= \frac{1-\alpha_1}{\alpha_0} p_2 - \frac{\alpha_2}{\alpha_0} p_1 - \frac{\alpha_2}{\alpha_0} p_0, \\ &\vdots \end{aligned}$$

and, in general,

$$p_j = \frac{1-\alpha_1}{\alpha_0} p_{j-1} - \frac{\alpha_2}{\alpha_0} p_{j-2} - \cdots - \frac{\alpha_{j-1}}{\alpha_0} p_1 - \frac{\alpha_{j-1}}{\alpha_0} p_0.$$

To compute any element $p_j$, we only need to know the values of previous elements $p_i$, $i < j$, and given that we know the value of $p_0 = 1 - \rho$, a recursive procedure may be set in motion to compute all required components of the stationary distribution vector. Furthermore, since successive elements are monotonic and concave, we will eventually reach a point at which component values are sufficiently small as to be negligible. However, from a computational standpoint, it is preferable to stop once the sum of successive elements is sufficiently close to one rather than stopping once a component is reached that is less than some specified small value.

**Example 14.3** Consider an *M/D/*1 queue for which $\lambda = 1/2$ and $\mu = 1$. The elements of the transition probability matrix can be found from

$$\alpha_i = \frac{0.5^i}{i!} e^{-0.5}.$$

This yields the following values:

$$\alpha_0 = 0.606531, \quad \alpha_1 = 0.303265, \quad \alpha_2 = 0.075816, \quad \alpha_3 = 0.012636,$$

$$\alpha_4 = 0.001580, \quad \alpha_5 = 0.000158, \quad \alpha_6 = 0.000013, \quad \alpha_7 = 0.000001.$$

Using the recursive procedure and beginning with $p_0 = 0.5$, we obtain the following:

$$p_1 = 0.324361: \quad \sum_{i=0}^{1} p_i = 0.824361,$$

$$p_2 = 0.122600: \quad \sum_{i=0}^{2} p_i = 0.946961,$$

$$p_3 = 0.037788: \quad \sum_{i=0}^{3} p_i = 0.984749,$$

$$p_4 = 0.010909: \quad \sum_{i=0}^{4} p_i = 0.995658,$$

$$p_5 = 0.003107: \quad \sum_{i=0}^{5} p_i = 0.998764,$$

$$p_6 = 0.000884: \quad \sum_{i=0}^{6} p_i = 0.999648,$$

$$\vdots \qquad \vdots$$

When the elements $\alpha_i$ of the transition probability matrix $F$ can be computed, but the initial value to begin the recursion, $p_0 = 1 - \rho$, is not known, it is still possible to use the same solution procedure. It suffices to assign $p_0$ an arbitrary value, say $p_0 = 1$, and to compute all other components from this value. In this case the process must be continued until a $p_{k+1}$ (or better still a sum of consecutive elements $\sum_{i=k+1}^{i=k+l} p_i$—the larger the value of $l$ the better) is found which is so small that it can be taken to be zero. At this point, the computed values of $p_0$ through $p_k$ must be normalized by dividing each by the sum $\sum_{i=0}^{i=k} p_i$.

## 14.3 Performance Measures for the *M/G/*1 Queue

### *14.3.1 The Pollaczek-Khintchine Mean Value Formula*

Let us return to Equation (14.1), i.e., $N_{k+1} = N_k - \delta(N_k) + A$, and use it to obtain the mean value for $N_k$. We assume the existence of a steady state and write

$$\lim_{k\to\infty} E[N_{k+1}] = \lim_{k\to\infty} E[N_k] = L.$$

Whereas it might be more appropriate to include a superscript $(D)$ on the quantity $L$ so as to denote departure instants, we now know that the solution at departure instants is also the solution at any point in time, and so we write it simply as $L$. We begin by taking the expected value of Equation (14.1) as $k \to \infty$ and obtain

$$L = L - \lim_{k\to\infty} E[\delta(N_k)] + \lim_{k\to\infty} E[A].$$

This implies that

$$E[A] = \lim_{k\to\infty} E[\delta(N_k)] = \lim_{k\to\infty} \sum_{i=0}^{\infty} \delta(N_k)\text{Prob}\{N_k = i\}$$

$$= \lim_{k\to\infty} \sum_{i=1}^{\infty} \text{Prob}\{N_k = i\} = \lim_{k\to\infty} \text{Prob}\{N_k > 0\}$$

$$= \lim_{k\to\infty} \text{Prob}\{\text{server is busy}\}.$$

The final equality holds since the *M/G/*1 queue has but a single server. Therefore

$$E[A] = \lim_{k\to\infty} \text{Prob}\{\text{server is busy}\} = \rho,$$

i.e., the average number of arrivals in a service period is equal to $\rho$.

While this is interesting, it is not what we are looking for—it does not give us the mean number of customers in the *M/G/*1 queue. To proceed further, we square both sides of Equation (14.1) and obtain

$$\begin{aligned} N_{k+1}^2 &= N_k^2 + \delta(N_k)^2 + A^2 - 2N_k\delta(N_k) - 2\delta(N_k)A + 2N_kA \\ &= N_k^2 + \delta(N_k) + A^2 - 2N_k - 2\delta(N_k)A + 2N_kA. \end{aligned}$$

Now, taking the expectation of each side, we find

$$E[N_{k+1}^2] = E[N_k^2] + E[\delta(N_k)] + E[A^2] - 2E[N_k] - 2E[A\delta(N_k)] + 2E[AN_k].$$

Taking the limit as $k \to \infty$ and letting $N = \lim_{k\to\infty} N_k$ gives

$$\begin{aligned} 0 &= E[\delta(N)] + E[A^2] - 2E[N] - 2E[A\delta(N)] + 2E[AN] \\ &= \rho + E[A^2] - 2E[N] - 2E[A]E[\delta(N)] + 2E[A]E[N] \\ &= \rho + E[A^2] - 2E[N] - 2\rho^2 + 2\rho E[N], \end{aligned} \tag{14.4}$$

where we have made use of the relationships $E[A] = E[\delta(N)] = \rho$ and the independence of arrivals and the number of customers at departure instants. Rearranging Equation (14.4), we obtain

$$E[N](2 - 2\rho) = \rho + E[A^2] - 2\rho^2$$

and hence

$$L = E[N] = \frac{\rho - 2\rho^2 + E[A^2]}{2(1-\rho)}.$$

It now only remains to find $E[A^2]$. A number of approaches may be used to show that $E[A^2] = \rho + \lambda^2 E[S^2]$ where $E[S^2] = \sigma_s^2 + E[S]^2$ is the second moment of the service time distribution and $\sigma_s^2$ is the variance of the service time. Exercise 14.3.1 at the end of this chapter guides the reader through one approach. With this result we finally obtain

$$L = \frac{\rho - 2\rho^2 + \rho + \lambda^2 E[S^2]}{2(1-\rho)} = \frac{2\rho(1-\rho) + \lambda^2 E[S^2]}{2(1-\rho)} = \rho + \frac{\lambda^2 E[S^2]}{2(1-\rho)}. \tag{14.5}$$

This result is frequently given in terms of the squared coefficient of variation:

$$\begin{aligned} L &= \rho + \frac{\lambda^2 E[S^2]}{2(1-\rho)} = \rho + \frac{\lambda^2\left(\sigma_s^2 + 1/\mu^2\right)}{2(1-\rho)} = \rho + \rho^2\frac{(C_s^2+1)}{2(1-\rho)} \\ &= \rho + \left(\frac{1+C_s^2}{2}\right)\frac{\rho^2}{1-\rho}, \end{aligned}$$

where $C_s^2 = \mu^2\sigma_s^2$ is the squared coefficient of variation. Equation (14.5) is called the *Pollaczek-Khintchine mean value formula*: it gives the average number of customers in the *M/G/*1 queue. Notice that this average depends only upon the first two moments of the service time distribution and that the mean number in the system increases *linearly* with the squared coefficient of variation, $C_s^2$. The expected response time (total time spent in the system) may now be computed using Little's formula, $L = \lambda W$. Thus

$$W = \frac{1}{\mu} + \frac{\lambda E[S^2]}{2(1-\rho)} = \frac{1}{\mu} + \frac{\lambda[(1/\mu)^2 + \sigma_s^2]}{2(1-\lambda/\mu)}.$$

Given $W$ and $L$, we may also obtain $W_q$ and $L_q$ from $W = W_q + 1/\mu$ and $L_q = L - \rho$, respectively. We have

$$W_q = \frac{\lambda[(1/\mu)^2 + \sigma_s^2]}{2(1-\lambda/\mu)} = \frac{\lambda E[S^2]}{2(1-\rho)} \quad \text{and} \quad L_q = \frac{\lambda^2 E[S^2]}{2(1-\rho)}.$$

The above equations for $W$, $W_q$, and $L_q$ are also referred to as *Pollaczek-Khintchine mean value formulae*. Comparing these results with those obtained in the *M/M/*1 context reveals an interesting relationship. The Pollaczek-Khintchine formulae we have just derived may be written as

$$L = \rho + \left(\frac{1+C_s^2}{2}\right)\frac{\rho^2}{1-\rho}, \qquad L_q = \left(\frac{1+C_s^2}{2}\right)\frac{\rho^2}{1-\rho},$$

$$W = \left(1 + \left(\frac{1+C_s^2}{2}\right)\frac{\rho}{1-\rho}\right)E[S], \qquad W_q = \left(\frac{1+C_s^2}{2}\right)\frac{\rho}{1-\rho}E[S],$$

where $E[S] = 1/\mu$ is the mean service time. Comparing the formulae in this form with the corresponding *M/M/*1 formulae given in Equations (11.6) and (11.7), we see that the only difference is the extra factor $(1 + C_s^2)/2$.

We shall see one more Pollaczek-Khintchine mean value formula, one that concerns the residual service time. However, before venturing there, we shall derive the Pollaczek-Khintchine transform equations for the distributions of customers, response time, and queueing time in an *M/G/*1 queue.

**Example 14.4** Consider the application of the Pollaczek-Khintchine mean value formula to the *M/M/*1 and *M/D/*1 queues. For the *M/M/*1 queue, we have

$$L = \rho + \rho^2\frac{(1+1)}{2(1-\rho)} = \frac{\rho}{1-\rho},$$

while for the *M/D/*1 queue, we have

$$L = \rho + \rho^2\frac{1}{2(1-\rho)} = \frac{\rho}{1-\rho} - \frac{\rho^2}{2(1-\rho)},$$

which shows that, on average, the *M/D/*1 queue has $\rho^2/2(1-\rho)$ fewer customers than the *M/M/*1 queue. As we have already noted in the past, congestion increases with the variance of service time.

**Example 14.5** Let us now consider the $M/H_2/1$ queue of Figure 14.2. The arrival rate is $\lambda = 2$; service at phase 1 is exponential with parameter $\mu_1 = 2$ and exponential at phase 2 with parameter $\mu_2 = 3$. Phase 1 is chosen with probability $\alpha = 0.25$, while phase 2 is chosen with probability $1 - \alpha = 0.75$.

The probability density function of the service time is given by

$$b(x) = \frac{1}{4}(2)e^{-2x} + \frac{3}{4}(3)e^{-3x}, \quad x \geq 0.$$

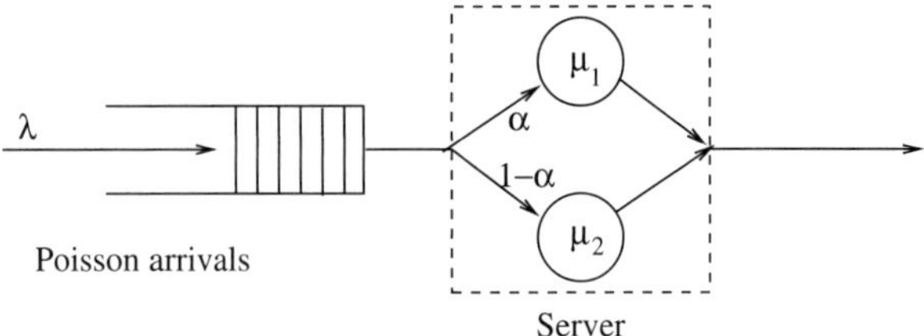


Figure 14.2. The $M/H_2/1$ queue.

The mean service time $\bar{x} = 1/\mu$ is given by

$$\bar{x} = \frac{1/4}{2} + \frac{3/4}{3} = \frac{3}{8},$$

and the variance $\sigma_s^2$ is equal to $29/192$ since

$$\sigma_s^2 = \overline{x^2} - \bar{x}^2 = 2\left(\frac{1/4}{4} + \frac{3/4}{9}\right) - \frac{9}{64} = \frac{29}{192}.$$

(Recall that the second moment of an $r$-phase hyperexponential distribution is $\overline{x^2} = 2\sum_{i=1}^{r} \alpha_i/\mu_i^2$.) So,

$$\mu^2\sigma_s^2 = C_s^2 = \frac{64}{9} \times \frac{29}{192} = \frac{29}{27}.$$

Therefore,

$$L = \rho + \frac{\rho^2(1 + 29/27)}{2(1-\rho)} = \frac{\rho}{1-\rho} + \frac{(2/27)\rho^2}{2(1-\rho)} = 3.083333.$$

### *14.3.2 The Pollaczek-Khintchine Transform Equations*

**Distribution of Customers in System**

To derive a relationship for the distribution of the number of customers in the *M/G/*1 queue, we need to return to the state equations

$$(p_0, p_1, p_2, \ldots) = (p_0, p_1, p_2, \ldots)\begin{pmatrix} \alpha_0 & \alpha_1 & \alpha_2 & \alpha_3 & \alpha_4 & \cdots \\ \alpha_0 & \alpha_1 & \alpha_2 & \alpha_3 & \alpha_4 & \cdots \\ 0 & \alpha_0 & \alpha_1 & \alpha_2 & \alpha_3 & \cdots \\ 0 & 0 & \alpha_0 & \alpha_1 & \alpha_2 & \cdots \\ \vdots & \vdots & \ddots & \ddots & \ddots & \ddots \\ 0 & 0 & 0 & 0 & 0 & \end{pmatrix},$$

where $p_j$ is the limiting probability of being in state $j$. The $j^{\text{th}}$ equation is given by

$$p_j = p_0\alpha_j + \sum_{i=1}^{j+1} p_i\alpha_{j-i+1}, \quad j = 0, 1, 2, \ldots.$$

Multiplying each term of this equation by $z^j$ and summing over all applicable $j$, we find

$$\sum_{j=0}^{\infty} p_j z^j = \sum_{j=0}^{\infty} p_0\alpha_j z^j + \sum_{j=0}^{\infty}\sum_{i=1}^{j+1} p_i\alpha_{j-i+1} z^j. \tag{14.6}$$

Figure 14.3 shows that the double summation $\sum_{j=0}^{\infty}\sum_{i=1}^{j+1}$ may be rewritten as $\sum_{i=1}^{\infty}\sum_{j=i-1}^{\infty}$. All the terms to be included are at the intersection of the dashed and dotted lines in this figure, and

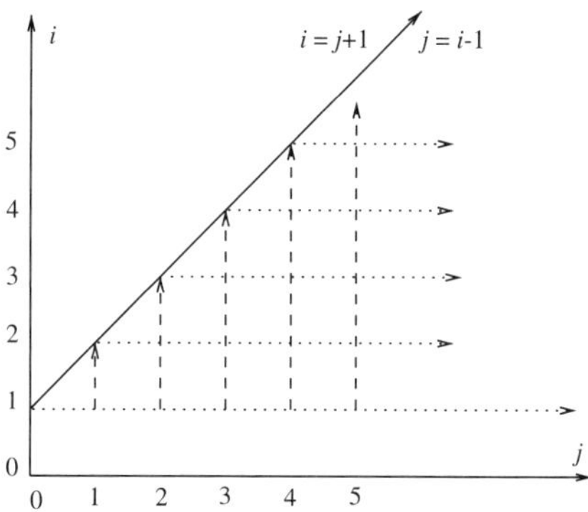


Figure 14.3. Double summations horizontally and vertically.

it is immaterial whether these are traversed vertically, following the dotted lines as in the original order, or horizontally, following the dashed lines.

Replacing the double summation $\sum_{j=0}^{\infty}\sum_{i=1}^{j+1}$ with $\sum_{i=1}^{\infty}\sum_{j=i-1}^{\infty}$ in Equation (14.6) and identifying the generating function $P(z) \equiv \sum_{j=0}^{\infty} p_j z^j$, we obtain

$$\begin{aligned} P(z) &= p_0 \sum_{j=0}^{\infty} \alpha_j z^j + \sum_{i=1}^{\infty} \sum_{j=i-1}^{\infty} p_i \alpha_{j-i+1} z^j \\ &= p_0 \sum_{j=0}^{\infty} \alpha_j z^j + \sum_{i=1}^{\infty} \sum_{k=0}^{\infty} p_i \alpha_k z^{k+i-1} \\ &= p_0 \sum_{j=0}^{\infty} \alpha_j z^j + \frac{1}{z} \left[ \sum_{i=1}^{\infty} p_i z^i \sum_{k=0}^{\infty} \alpha_k z^k \right]. \end{aligned}$$

Let the $z$-transform of the number of arrivals that occur during a service period be denoted by $G_A(z)$, i.e.,

$$G_A(z) \equiv \sum_{j=0}^{\infty} \alpha_j z^j.$$

Then

$$P(z) = p_0 G_A(z) + \frac{1}{z}[P(z) - p_0] G_A(z)$$

or

$$P(z) = \frac{(z-1) p_0 G_A(z)}{z - G_A(z)}.$$

This equation contains two unknowns, $p_0$ and $G_A(z)$. We determine $p_0$ first. Since

$$G_A(1) = \sum_{j=0}^{\infty} \alpha_j z^j \bigg|_{z=1} = \sum_{j=0}^{\infty} \alpha_j = 1,$$

we find that $\lim_{z\to 1} P(z)$ has the indeterminate form 0/0, forcing us to apply L'Hôpital's rule. We obtain

$$1 = \lim_{z\to 1} P(z) = \lim_{z\to 1} \left[ p_0 \frac{(z-1)G'_A(z) + G_A(z)}{1 - G'_A(z)} \right] = p_0 \frac{1}{1 - G'_A(1)}.$$

We now show that $G'_A(1) = \lambda/\mu$. We have

$$\frac{d}{dz}G_A(1) = \sum_{j=0}^{\infty} j\alpha_j z^{j-1}\Bigg|_{z=1} = \sum_{j=0}^{\infty} j\alpha_j.$$

Since $\alpha_j$ is the probability of $j$ arrivals during an arbitrary service, it follows that $\sum_{j=0}^{\infty} j\alpha_j$ is just equal to the expected number of customers that arrive during a service which in turn is just equal to $\lambda/\mu = \rho$. Therefore

$$1 = \frac{p_0}{1-\rho},$$

i.e., $p_0 = 1-\rho$, and hence

$$P(z) = \frac{(1-\rho)(z-1)G_A(z)}{z - G_A(z)}. \tag{14.7}$$

We now turn our attention to the second unknown, namely, $G_A(z)$. From Equation (14.3) we have

$$\alpha_i = \int_0^\infty \frac{(\lambda x)^i}{i!} e^{-\lambda x} b(x)dx,$$

which means that

$$\begin{aligned} G_A(z) = \sum_{j=0}^{\infty} \alpha_j z^j &= \sum_{j=0}^{\infty} \int_0^\infty \frac{(\lambda x)^j}{j!} z^j e^{-\lambda x} b(x)dx \\ &= \int_0^\infty e^{-\lambda x} \sum_{j=0}^{\infty} \frac{(\lambda x z)^j}{j!} b(x)dx \\ &= \int_0^\infty e^{-\lambda x(1-z)} b(x)dx = B^*[\lambda(1-z)], \end{aligned}$$

where $B^*[\lambda(1-z)]$ is the Laplace transform of the service time distribution evaluated at $s = \lambda(1-z)$. (Recall that the Laplace transform of a function $f(t)$ is given by $F^*(s) = \int_{0^-}^\infty f(t)e^{-st}dt$.) Substituting this into Equation (14.7), we get the Pollaczek-Khintchine transform equation No. 1 (for the distribution of the number of customers in the system):

$$P(z) = \frac{(1-\rho)(z-1)B^*[\lambda(1-z)]}{z - B^*[\lambda(1-z)]}.$$

The average number of customers in the system at steady state may be found by taking the derivative with respect to $z$ and then taking the limit as $z \to 1$. The reader may wish to verify that the result derived in this manner is the same as that given by Equation (14.5).

**Example 14.6** The *M/M/*1 queue.
We have

$$b(x) = \mu e^{-\mu x} \quad \text{for } x > 0 \quad \text{and} \quad B^*(s) = \frac{\mu}{s+\mu}.$$

Therefore

$$B^*[\lambda(1-z)] = \frac{\mu}{\lambda(1-z)+\mu}$$

and hence

$$P(z) = \frac{\mu}{\lambda(1-z)+\mu}\left[\frac{(1-\rho)(1-z)}{\mu/(\lambda(1-z)+\mu) - z}\right] = \frac{1-\rho}{1-\rho z}.$$

We can now take derivatives and evaluate at $z = 1$. However, in this case, it is easy to invert the transform, which yields

$$p_k = (1-\rho)\rho^k \quad \text{for } k = 0, 1, 2, \ldots.$$

**Example 14.7** Let us consider an $M/H_2/1$ queue with the following parameters: $\lambda = 1$, $\alpha = 0.25$, $\mu_1 = 7/2$, and $\mu_2 = 9/2$. These values, which might appear random, actually allow us to obtain the inverse of the Pollaczek-Khintchine transform equation quite easily. We have

$$b(x) = \frac{1}{4}(7/2)e^{-7x/2} + \frac{3}{4}(9/2)e^{-9x/2}, \quad x \geq 0,$$

$$B^*(s) = \left(\frac{1}{4}\right)\frac{7/2}{s+7/2} + \left(\frac{3}{4}\right)\frac{9/2}{s+9/2} = \frac{17s+63}{4(s+9/2)(s+7/2)}.$$

Since $s = \lambda(1-z) = 1-z$,

$$B^*(1-z) = \frac{63+17(1-z)}{4(11/2-z)(9/2-z)}$$

and hence

$$\begin{aligned} P(z) &= \frac{(1-\rho)(1-z)[63+17(1-z)]}{63+17(1-z)-4(11/2-z)(9/2-z)z} \\ &= \frac{(1-\rho)(1-z)[63+17(1-z)]}{80-116z+40z^2-4z^3} \\ &= \frac{(1-\rho)(1-z)[80-17z]}{4(1-z)(4-z)(5-z)} = \frac{(1-\rho)[1-(17/80)z)]}{(1-z/5)(1-z/4)}. \end{aligned}$$

Factorizing the cubic in the denominator on the second line is facilitated by the fact that we know that $z = 1$ must be one of the roots. Expanding into partial fractions we obtain

$$P(z) = (1-\rho)\left[\frac{1/4}{1-z/5} + \frac{3/4}{1-z/4}\right],$$

which may be inverted by inspection to yield

$$p_k = (1-\rho)\left[\frac{1}{4}\left(\frac{1}{5}\right)^k + \frac{3}{4}\left(\frac{1}{4}\right)^k\right] \quad \text{for } k = 0, 1, 2, \ldots.$$

**Response Time Distributions**

Let $R$ be the random variable that describes the response time of a customer and let $w(x)$ be its probability density function. Then the probability that $i$ arrivals occur during the response time of a customer is equal to

$$p_i = \int_0^\infty \frac{(\lambda x)^i}{i!} e^{-\lambda x} w(x)dx.$$

The derivation of this equation is identical to that used to derive Equation (14.3). Multiplying $p_i$ by $z^i$ and summing over all $i$, we obtain

$$P(z) = \sum_{i=0}^{\infty} p_i z^i = \sum_{i=0}^{\infty} \int_0^\infty \frac{(\lambda x z)^i}{i!} e^{-\lambda x} w(x)dx = \int_0^\infty e^{-\lambda x(1-z)} w(x)dx = W^*[\lambda(1-z)], \quad (14.8)$$

where $W^*$ is the Laplace transform of customer response time evaluated at $s = \lambda(1-z)$.

It is interesting to derive Equation (14.8) in an alternative fashion, by examining the number of arrivals that occur during two different time periods, the first period being that of an arbitrary service time and the second that of an arbitrary response time. We have just seen that

$$G_A(z) \equiv \sum_{j=0}^{\infty} \alpha_j z^j = B^*[\lambda(1-z)],$$

where $B^*[\lambda(1-z)]$ is the Laplace transform of the service time distribution evaluated at $s = \lambda(1-z)$ and $\alpha_j = \text{Prob}\{A = j\}$ is the probability of $j$ arrivals during an arbitrary service. In other words, $G_A(z)$ is the $z$-transform of the distribution of the number of customer arrivals in a particular interval, where the arrival process is Poisson at rate $\lambda$ customers per second—the particular interval is a service interval with distribution $B(x)$ and Laplace transform $B^*(s)$. Observe that this is a relationship between two random variables, the first which counts the number of Poisson events that occur during a certain period of time and the second which characterizes this period of time. If we now change the distribution of the time period but leave the Poisson nature of the occurrence of the events unchanged, the *form* of this relationship must remain unaltered.

With this in mind, let us change the time period to "time spent in the system," i.e., response time, instead of "service time." In other words, our concern is now with the number of arrivals during the response time of a customer instead of the number of arrivals during the service time of a customer. This means that in the above formula, we need to replace $\alpha_j$ (the probability of $j$ arrivals during an arbitrary service), with $p_j$, (the probability of $j$ arrivals during an arbitrary sojourn time of a customer in the *M/G/*1 queue). Remember that $p_j$ has previously been defined as the stationary probability that the system contains $j$ customers, but since this is the distribution seen by a customer arriving to an *M/G/*1 queue which is turn must be equal to the number left behind by a departing customer, it must also be the number of arrivals that occur during the time that a customer spends in the system—*assuming a FCFS discipline*. Thus we may immediately write

$$P(z) = W^*[\lambda(1-z)],$$

where $W^*$ denotes the Laplace transform of customer response time, and hence we are back to Equation (14.8).

Recalling our earlier expression for $P(z)$ given by the Pollaczek-Khintchine transform equation No. 1,

$$P(z) = \frac{(1-\rho)(z-1)B^*[\lambda(1-z)]}{z - B^*[\lambda(1-z)]},$$

and applying Equation (14.8), we conclude that

$$W^*[\lambda(1-z)] = \frac{(1-\rho)(z-1)B^*[\lambda(1-z)]}{z - B^*[\lambda(1-z)]}.$$

Letting $s = \lambda - \lambda z$ gives $z = 1 - s/\lambda$ and hence

$$W^*(s) = B^*(s)\frac{s(1-\rho)}{s - \lambda + \lambda B^*(s)}.$$

This is called the Pollaczek-Khintchine transform equation number No. 2. It provides the Laplace transform of the distribution of response time (total time spent waiting for and receiving service) in an *M/G/*1 queue with service provided in first-come first-served order.

**Example 14.8** The *M/M/*1 queue.

$$W^*(s) = B^*(s)\frac{s(1-\rho)}{s-\lambda+\lambda B^*(s)}$$
$$= \frac{\mu}{s+\mu}\left[\frac{s(1-\rho)}{s-\lambda+\lambda\mu/(s+\mu)}\right] = \frac{\mu(1-\rho)}{s+\mu(1-\rho)}.$$

This can be inverted to obtain the probability distribution and density functions of total time spent in the *M/M/*1 queue, distributions that have already been described in Section 11.2.2.

**Queueing Time Distributions**

The probability distribution function of the time spent waiting in the queue before beginning service may be computed from the previous results. We have

$$R = T_q + S,$$

where $R$ is the random variable that characterizes the total time (response time) in the system, $T_q$ the random variable for time spent in the queue and $S$ the random variable that describes service time. From the convolution property of transforms, we have

$$W^*(s) = W_q^*(s)B^*(s),$$

i.e.,

$$B^*(s)\frac{s(1-\rho)}{s-\lambda+\lambda B^*(s)} = W_q^*(s)B^*(s)$$

and hence

$$W_q^*(s) = \frac{s(1-\rho)}{s-\lambda+\lambda B^*(s)}.$$

This is known as the Pollaczek-Khintchine transform equation No. 3: it gives the Laplace transform of the distribution of time spent waiting in the queue.

## 14.4 The *M/G/*1 Residual Time: Remaining Service Time

When a customer arrives at an *M/G/*1 queue and finds at least one customer already present, at that arrival instant a customer is in the process of being served. In this section our concern is with the time that remains until the completion of that service, the so-called *residual (service) time*. In the more general context of an arbitrary stochastic process, the terms *residual lifetime* and *forward recurrence time* are also employed. The time that has elapsed from the moment service began until the current time is called the *backward recurrence time*. Since we have previously used $R$ to denote the random variable "response time," we shall let $\mathcal{R}$ (i.e., calligraphic $\mathcal{R}$ as opposed to italic $R$) be the random variable that describes the residual service time and we shall denote its probability density function by $f_{\mathcal{R}}(x)$. If the system is empty, then $\mathcal{R} = 0$. As always, we shall let $S$ be the random variable "service time" and denote its density function by $b(x)$. In an *M/M/*1 queue, we know that the remaining service time is distributed exactly like the service time itself and in this case $f_{\mathcal{R}}(x) = b(x) = \mu e^{-\mu x},\ x > 0$.

The mean residual service time is easily found from the Pollaczek-Khintchine mean value formulae in the context of a first-come, first-served scheduling policy. Given that we have an expression for the expected time an arriving customer must wait until its service begins, $W_q$, and another expression for the expected number of customers waiting in the queue, $L_q$, the mean residual service time is obtained by subtracting $1/\mu$ times the second from the first. Since

$$W_q = \frac{\lambda E[S^2]}{2(1-\rho)} \quad \text{and} \quad L_q = \frac{\lambda^2 E[S^2]}{2(1-\rho)},$$

it follows that

$$E[\mathcal{R}] = W_q - \frac{1}{\mu}L_q = \frac{\lambda E[S^2]}{2(1-\rho)} - \frac{1}{\mu}\frac{\lambda^2 E[S^2]}{2(1-\rho)} = \frac{\lambda E[S^2]}{2(1-\rho)}(1-\rho) = \frac{\lambda E[S^2]}{2}.$$

The expression $E[\mathcal{R}] = \lambda E[S^2]/2$ is another Pollaczek-Khintchine mean value formula. It provides the expected residual service time as seen by an arriving customer. However, since the arrival process is Poisson, this is also the expected residual service time as seen by a random observer, i.e., the expected residual service at any time. Although derived in the context of a first-come first-served policy, it can be shown that this result is also applicable to any work conserving scheduling algorithm, i.e., when the server continues working as long as there are customers present, and customers wait until their service is completed. Finally, observe the interesting relationship between $E[\mathcal{R}]$ and $W_q$, i.e., that $E[\mathcal{R}] = (1-\rho)W_q$, where $(1-\rho)$ is the probability that the server is idle.

It is a useful exercise to derive the result $E[\mathcal{R}] = \lambda E[S^2]/2$ from first principles. Let $S_i = 1/\mu_i$ be the mean service time required by customer $i$. Figure 14.4 displays a graph of $\mathcal{R}(t)$ and shows the residual service requirement at any time $t$. Immediately prior to the initiation of service for a customer, this must be zero. The moment the server begins to serve a customer, the residual service time must be equal to the total service requirement of that customer, i.e., $S_i$ for customer $i$. As time passes, the server reduces this service requirement at the rate of one unit per unit time; hence the slope of $-1$ from the moment service begins until the service requirement of the customer has been completely satisfied, at which time the remaining service time is equal to zero. If at this instant another customer is waiting in the queue, the residual service time jumps an amount equal to the service required by that customer. Otherwise, the server becomes idle and remains so until a customer arrives to the queueing system.

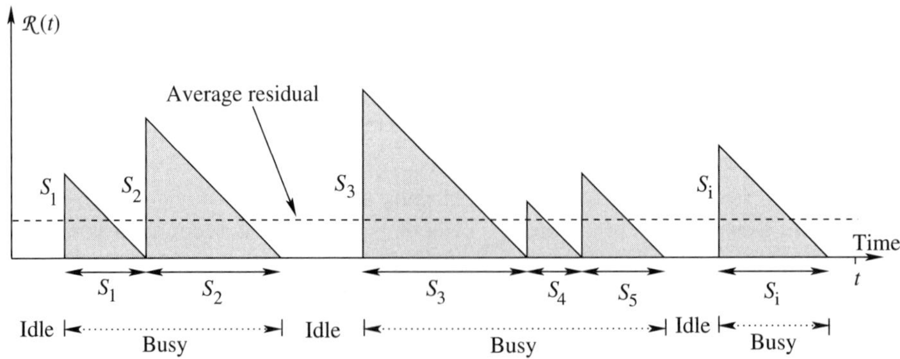


Figure 14.4. $\mathcal{R}(t)$: Residual service time in an M/G/1 queue.

Let us assume that at time $t = 0$ the system is empty and let us choose a time $t$ at which the system is once again empty. If $\lambda < \mu$ we are guaranteed that such times occur infinitely often. Let $M(t)$ be the number of customers served by time $t$. The average residual service time in $[0, t]$ is the area under the curve $\mathcal{R}(t)$ divided by $t$ and since the area of each right angle triangle with base and height equal to $S_i$ is $S_i^2/2$, we find

$$\frac{1}{t}\int_0^t \mathcal{R}(\tau)d\tau = \frac{1}{t}\sum_{i=1}^{M(t)} \frac{S_i^2}{2} = \frac{1}{2} \times \frac{M(t)}{t} \times \frac{\sum_{i=1}^{M(t)} S_i^2}{M(t)}.$$

Taking the limit as $t \to \infty$ gives the desired result:

$$E[\mathcal{R}] = \lim_{t\to\infty} \frac{1}{t}\int_0^t \mathcal{R}(\tau)d\tau = \frac{1}{2}\lim_{t\to\infty}\frac{M(t)}{t}\lim_{t\to\infty}\frac{\sum_{i=1}^{M(t)} S_i^2}{M(t)} = \frac{1}{2}\lambda E[S^2],$$

where

$$\lambda = \lim_{t\to\infty}\frac{M(t)}{t} \quad \text{and} \quad E[S^2] = \lim_{t\to\infty}\frac{\sum_{i=1}^{M(t)} S_i^2}{M(t)}.$$

These limits exist since our system is ergodic. Notice that we set $\lambda$ equal to the mean output rate, which is appropriate since at equilibrium the arrival rate is equal to the departure rate. We now turn our attention to the derivation of the probability distribution of the residual service time, and we do so with the understanding that it is conditioned on the server being busy. We shall denote this random variable $\mathcal{R}_b$. Let $X$ be the random variable that denotes the service time of the customer in service when an arrival occurs and let $f_X(x)$ be its density function. Observe that $X$ and $S$ describe different random variables. Since the service time is a random variable having a general distribution function, the service received by some customers will be long while that received by other customers will be short. Everything else being equal, it is apparent that an arrival is more likely to occur during a large service time than in a small service interval. Therefore the probability that $X$ is of duration $x$ should be proportional to the length $x$. The other factor that plays into this scenario is the frequency of occurrence of service intervals having length $x$, namely, $b(x)dx$: both length of service and frequency of occurrence must be taken into account. Thus

$$f_X(x)dx = \text{Prob}\{x \le X \le x + dx\} = \alpha\, x\, b(x)dx,$$

where the role of $\alpha$ is to ensure that this is a proper density function, i.e., that

$$\int_0^\infty \alpha\, x\, b(x)dx = 1.$$

Since $E[S] = \int_0^\infty x\, b(x)dx$, it follows that $\alpha = 1/E[S]$ and hence

$$f_X(x) = \frac{x\, b(x)}{E[S]}.$$

So much for the distribution of service periods at an arrival instant. Let us now consider the placement of an arrival within these service periods. Since arrivals are Poisson, hence random, an arrival is uniformly distributed over the service interval $(0, x)$. This means that the probability that the remaining service time is less than or equal to $t$, $0 \le t \le x$, given that the arrival occurs in a service period of length $x$, is equal to $t/x$, i.e., $\text{Prob}\{\mathcal{R}_b \le t \mid X = x\} = t/x$. It now follows that

$$\text{Prob}\{t \le \mathcal{R}_b \le t + dt \mid X = x\} = \frac{dt}{x}, \quad t \le x.$$

Removing the condition, by integrating over all possible $x$, allows us to obtain probability distribution function of the residual service time conditioned on the server being busy. We have

$$\text{Prob}\{t \le \mathcal{R}_b \le t + dt\} = f_{\mathcal{R}_b}(t)dt = \int_{x=t}^\infty \frac{dt}{x} f_X(x)dx = \int_{x=t}^\infty \frac{b(x)}{E[S]}dx\, dt = \frac{1 - B(t)}{E[S]}dt$$

and hence

$$f_{\mathcal{R}_b}(t) = \frac{1 - B(t)}{E[S]}.$$

The mean residual time is found from

$$E[\mathcal{R}_b] = \int_0^\infty t f_{\mathcal{R}_b}(t)dt = \frac{1}{E[S]}\int_0^\infty t\,(1 - B(t))dt.$$

Taking $u = [1 - B(t)]$, $du = -b(t)\,dt$, $dv = t\,dt$, $v = t^2/2$, and integrating by parts ($\int u\,dv = uv - \int v\,du$), we find

$$E[\mathcal{R}_b] = \frac{1}{E[S]}\left[ [1 - B(t)]\frac{t^2}{2}\Big|_0^\infty + \int_0^\infty \frac{t^2}{2} b(t)dt \right]$$

$$= \frac{1}{2E[S]}\int_0^\infty t^2\, b(t)dt = \frac{E[S^2]}{2E[S]} = \frac{\mu E[S^2]}{2}.$$

Higher moments are obtained analogously. We have

$$E[\mathcal{R}_b^{k-1}] = \frac{\mu E[S^k]}{k}, \quad k = 2, 3, \ldots.$$

Finally, notice that

$$E[\mathcal{R}] = \rho E[\mathcal{R}_b],$$

where $\rho = 1 - p_0$ is the probability that the server is busy.

Exactly the same argument we have just made for the forward recurrence time applies also to the backward recurrence time, and in particular the mean backward recurrence time must also be equal to $E[S^2]/(2E[S])$. Therefore the sum of the mean forward and backward recurrence times is not equal to the mean service time! This is generally referred to as the *paradox of residual life*.

**Example 14.9** Observe that, when the service time is exponentially distributed, the remaining service time is the same as the service time itself. We have

$$f_{\mathcal{R}_b}(t) = \frac{1 - B(x)}{E[S]} = \frac{1 - (1 - e^{-\mu t})}{1/\mu} = \mu e^{-\mu t}.$$

The mean forward and backward recurrence times are both equal to $1/\mu$ and hence their sum is equal to twice the mean service time.

**Example 14.10** Given that the first and second moments of a random variable having an Erlang-$r$ distribution ($r$ exponential phases each with parameter $r\mu$), are $E[S] = 1/\mu$ and $E[S^2] = r(r+1)/(r\mu)^2$, respectively, the expected residual time in an $M/E_r/1$ queue is

$$E[\mathcal{R}_b] = \frac{(1 + 1/r)/\mu^2}{2/\mu} = \frac{(1 + 1/r)}{2\mu}.$$

As $r \to \infty$, $E[\mathcal{R}_b] \to 1/(2\mu) = E[S]/2$, which, as may be verified independently, is the expected residual service time in an *M/G/*1 queue when the service process is deterministic. With deterministic service times, $E[S^2] = E[S]^2$ and the sum of the mean forward and backward recurrence times is equal to the mean service time.

## 14.5 The *M/G/*1 Busy Period

A stable queueing system always alternates between periods in which there are no customers present (called *idle periods*) and periods in which customers are present and being served by the server (called *busy periods*). More precisely, a busy period is the length of time between the instant a customer arrives to an empty system (thereby terminating an idle period) and the moment at which the system empties (thereby initiating the next idle period). This is illustrated in Figure 14.5.

In this figure the horizontal line is the time axis and, at any time $t$, the corresponding point on the solid line represents the amount of work that remains for the server to accomplish before it becomes idle. This line has a jump when a new customer arrives (represented by an upward pointing arrow), bringing with it the amount of work that must be performed to serve this customer. This is

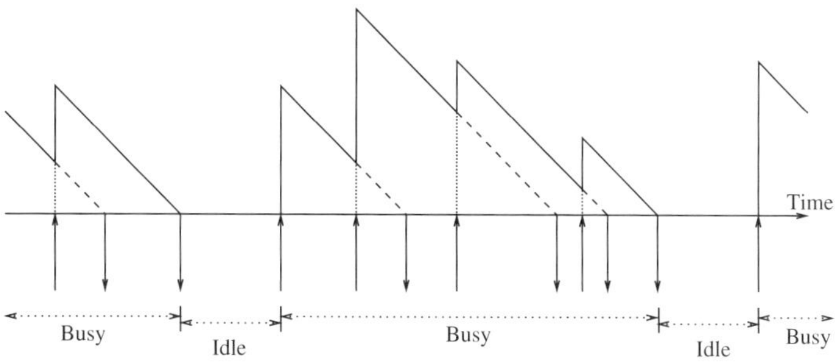


Figure 14.5. Busy and idle periods in an *M/G/*1 queue.

added to the amount of work already in the system (represented by a dotted vertical line). The solid line decreases at a rate of one unit per unit time until it reaches zero and the server become idle once again. The dashed line leading to a downward pointing arrow represents the continuation and eventually the completion of the service and the subsequent departure of a customer when the scheduling algorithm is FCFS.

An idle period begins at the end of a busy period and terminates with an arrival. Since arrivals in an *M/G/*1 queue have a (memoryless) Poisson distribution, the time until the idle period terminates must also have this same distribution. Thus the distribution of idle periods is given by $1 - e^{-\lambda t}$. Finding the distribution of the busy period is somewhat more difficult. We begin by finding its expectation. Let $Y$ be the random variable that describes the length of a busy period and let

$$G(y) \equiv \text{Prob}\{Y \leq y\}$$

be the busy period distribution. We have already seen that the probability of having zero customers (idle period) in an *M/G/*1 queue is $1 - \rho$. Notice also that the expected length of time between the start of two successive busy periods is $1/\lambda + E[Y]$ and, since $1/\lambda$ is the length of the idle period, the probability of the system being idle is the ratio of these. Combining these results we have

$$1 - \rho = \frac{1/\lambda}{1/\lambda + E[Y]}.$$

Rearranging, we find

$$\lambda E[Y] = \frac{1}{1-\rho} - 1 = \frac{\rho}{1-\rho}$$

or

$$E[Y] = \frac{1/\mu}{1-\rho} = \frac{1}{\mu - \lambda},$$

and so the average length of a busy period depends only on the mean values $\lambda$ and $\mu$. Observe, furthermore, that the average length of a busy period in an *M/G/*1 queue is exactly equal to the average amount of time a customer spend in an *M/M/*1 queue!

In Figure 14.5, the implied scheduling of customers for service is FCFS. While a particular scheduling algorithm can make a significant difference to customer waiting times, it does not have any effect on the distribution of the busy period *if* the scheduling discipline is a work-conserving discipline. A work-conserving discipline is one in which the server continues to remain busy while there are customers present and customers cannot exit the system until they have been completely served. The calculation of the busy period distribution is made much easier by choosing a scheduling procedure that is very different from FCFS. Let the first customer who ends an idle period be called $c_0$. This customer on arrival is immediately taken into service and served until it has finished. Assume that during the service time of this customer, $n$ other customers arrive. Let these be called

$c_1, c_2, \ldots, c_n$. Once customer $c_0$ exits the system, the service of customer $c_1$ begins and continues uninterrupted until it too exits the system. At this point the scheduling algorithm changes. During the service time of $c_1$ other customers arrive and these customers are to be served before service begins on customer $c_2$. Indeed, the service of customer $c_2$ does not begin until the only remaining customers in the system are $c_2, c_3, \ldots, c_n$. The process continues in this fashion: the $n$ customers who arrive during the service time of the first customer are served in order, but the service of any one of them does not begin until all later arrivals (customers who arrive after the departure of $c_0$) have been served. In this manner, the time to serve $c_1$ and all the customers who arrive during $c_1$'s service has the same distribution as the time to serve $c_2$ and the customers who arrive during $c_2$'s service, and so on. Furthermore, this common distribution for all $n$ arrivals must be distributed exactly the same as the busy period we seek! Essentially, the beginning of service of any of the first $n$ customers, say $c_i$, may be viewed as the beginning of an "included" or "sub-" busy period that terminates when the last customer who arrives during the service of $c_i$ leaves the system. The distribution of the "included" busy period begun by the initiation of service of any of the customers $c_i$, $i = 1, 2, \ldots, n$, must be identical for all $n$ of them, and must also be the same as the distribution of the busy period initiated by the arrival of $c_0$.

With this understanding, we may now proceed to find the busy period distribution itself. Let

$$G^*(s) \equiv \int_0^\infty e^{-sy} dG(y) = E[e^{-sY}]$$

be the Laplace transform for the probability density function associated with $Y$, the random variable that describes the length of a busy period. We shall show that

$$G^*(s) = B^*[s + \lambda - \lambda G^*(s)],$$

where $B^*[s+\lambda-\lambda G^*(s)]$ is the previously defined Laplace transform of the service time distribution, this time evaluated at the point $s+\lambda-\lambda G^*(s)$. We have just seen that the duration of the busy period initiated by the arrival of customer $c_0$ is the sum of the random variable that describes the service time for $c_0$ and the $n$ random variables that describe the $n$ "included" busy periods associated with customers $c_1$ through $c_n$. Let $Y_i$ be the random variable which characterizes the "included" busy period associated with $c_i$, $i = 1, 2, \ldots, n$. Conditioning $Y$ on the number of arrivals, $A = n$, that occur while $c_0$ is being served and on $S = x$, the length of service provided to $c_0$, we have

$$\begin{aligned} E[e^{-sY} \mid A = n, \; S = x] = E[e^{-s(x+Y_1+Y_2+\cdots+Y_n)}] &= E[e^{-sx} e^{-sY_1} e^{-sY_2} \cdots e^{-sY_n}] \\ &= E[e^{-sx}]E[e^{-sY_1}]E[e^{-sY_2}] \cdots E[e^{-sY_n}] \\ &= e^{-sx} E[e^{-sY_1}]E[e^{-sY_2}] \cdots E[e^{-sY_n}]. \end{aligned}$$

The last equality arises from the fact that $x$ is a given constant, while the equality preceding the last is a result of the independence of the "included" busy periods. Since the "included" busy periods are identically distributed and equal to that of the overall busy period, we have

$$E[e^{-sY_i}] = E[e^{-sY}] = G^*(s) \quad \text{for } i = 1, 2, \ldots, n,$$

and hence

$$E[e^{-sY} \mid n, x] = e^{-sx}[G^*(s)]^n.$$

We now remove the two conditions, first on $n$, the number of arrivals during the service of $c_0$, and then on $x$, the service time of $c_0$. Since the arrival process is Poisson with rate $\lambda$, the distribution of

$n$ must be Poisson with mean value $\lambda x$. We obtain

$$E[e^{-sY} \mid x] = \sum_{k=0}^{\infty} E[e^{-sY} \mid n, x] \text{ Prob}\{n = k\}$$
$$= \sum_{k=0}^{\infty} e^{-sx}[G^*(s)]^k e^{-\lambda x} \frac{(\lambda x)^k}{k!}$$
$$= \sum_{k=0}^{\infty} e^{-sx} e^{-\lambda x} \frac{[\lambda G^*(s) x]^k}{k!}$$
$$= e^{-x(s+\lambda-\lambda G^*(s))}.$$

Now, removing the condition on $x$ by integrating with respect to the service time distribution function $B(x)$, we find

$$G^*(s) = E[e^{-sY}] = \int_0^{\infty} e^{-x(s+\lambda-\lambda G^*(s))} dB(x),$$

i.e.,

$$G^*(s) = B^*[s + \lambda - \lambda G^*(s)]. \tag{14.9}$$

This is a functional equation which must be solved numerically for $G^*(s)$ at any given value of $s$. For example, it may be shown that the iterative scheme

$$G^*_{j+1}(s) = B^*[s + \lambda - \lambda G^*_j(s)]$$

converges to $G^*(s)$ when $\rho < 1$ and the initial value $G^*_0(s)$ is chosen to lie in the interval [0, 1].

We can use Equation (14.9) to calculate moments of the busy period. We already formed the expected busy period length using a different approach and saw that

$$E[Y] = \frac{1}{\mu - \lambda}.$$

The derivation of this result by means of differentiation is left as an exercise, as is the formation of the second moment and the variance, given respectively by

$$E[Y^2] = \frac{E[S^2]}{(1-\rho)^3}$$

and

$$\text{Var}[Y] = \frac{\text{Var}[S] + \rho E[S]^2}{(1-\rho)^3},$$

where we recall that $S$ is the random variable "service time."

**Example 14.11** In general, it is not possible to invert Equation (14.9). One instance in which it is possible is the *M/M/*1 queue. The transform for the service time distribution in the *M/M/*1 queue is given by $B^*(s) = \mu/(s + \mu)$. Replacing $s$ with $s + \lambda - \lambda G^*(s)$ and identifying the result with Equation (14.9) results in

$$G^*(s) = \frac{\mu}{[s + \lambda - \lambda G^*(s)] + \mu}.$$

This gives the quadratic equation

$$\lambda[G^*(s)]^2 - (\mu + \lambda + s)G^*(s) + \mu = 0,$$

which may be solved for $G^*(s)$. Making the appropriate choice between the two roots ($|G^*(s)| \leq 1$), the result obtained is

$$G^*(s) = \frac{(\mu + \lambda + s) - \sqrt{(\mu + \lambda + s)^2 - 4\lambda\mu}}{2\lambda}.$$

This can be inverted and the probability density function of the busy period obtained as

$$g(y) = \frac{1}{y\sqrt{\rho}} e^{-(\lambda+\mu)y} I_1(2y\sqrt{\lambda\mu}),$$

where $I_1$ is the modified Bessel function of the first kind of order 1.

**Distribution of Customers Served in a Busy Period**

A similar analysis may be used to derive a functional equation for the transform of the number of customers served during a busy period. Since arrivals are Poisson and the lengths of "included" busy periods are independent and identically distributed and equal to the distribution of the length of the entire busy period, the number of customers served during each 'included" busy period is independent and identically distributed and equal to the distribution of the number served in a complete busy period. If $N_i$ is the random variable that denotes the number of customers served during the $i^{\text{th}}$, $i = 1, 2, \ldots, n$, "included" busy period, then $N_b$, the number served during the entire busy period, is given by

$$N_b = 1 + N_1 + N_2 + \cdots + N_n.$$

Let $h_k = \text{Prob}\{N_b = k\}$ and

$$H(z) = E[z^{N_b}] = \sum_{k=0}^{\infty} h_k z^k.$$

Conditioning on the number of arrivals during the service period of customer $c_0$, we have

$$E[z^{N_b} \mid A = n] = E[z^{1+N_1+N_2+\cdots+N_n}] = z \prod_{k=1}^{n} E[z^{N_k}] = zH(z)^n,$$

so that when we now remove the condition, we obtain

$$\begin{aligned} H(z) = E[z^{N_b}] &= \sum_{n=0}^{\infty} E[z^{N_b} \mid A = n]\text{Prob}\{A = n\} \\ &= \sum_{n=0}^{\infty} zH(z)^n \int_0^{\infty} \frac{(\lambda x)^n}{n!} e^{-\lambda x} b(x)dx \\ &= z \int_0^{\infty} e^{-[\lambda - \lambda H(z)]x}\, b(x)dx \\ &= zB^*[\lambda - \lambda H(z)]. \end{aligned} \qquad (14.10)$$

The expectation, second moment, and variance of the number served in a busy period may be computed from Equation (14.10). They are given respectively by

$$\text{Mean:} \qquad \frac{1}{1-\rho},$$

$$\text{Second moment:} \quad \frac{2\rho(1-\rho)+\lambda^2\overline{x^2}}{(1-\rho)^3}+\frac{1}{1-\rho},$$

$$\text{Variance:} \qquad \frac{\rho(1-\rho)+\lambda^2\overline{x^2}}{(1-\rho)^3}.$$

## 14.6 Priority Scheduling

In a queueing system in which customers are distinguished by class, it is usual to assign priorities according to the perceived importance of the customer. The most important class of customers is assigned priority 1; classes of customers of lesser importance are assigned priorities 2, 3, .... When the system contains customers of different classes, those with priority $j$ are served before those of priority $j+1$, $j = 1, 2, \ldots$. Customers within each class are served in first-come, first-served order. The question remains as to how a customer in service should be treated when a higher-priority customer arrives. This gives rise to two scheduling policies.

The first policy is called *preemptive priority* and in this case, a lower-priority customer in service is ejected from service the moment a higher-priority customer arrives. The interrupted customer is allowed back into service once the queue contains no customer having a higher priority. The interruption of service may mean that all of the progress made toward satisfying the ejected customer's service requirement has been lost, so that it becomes necessary to start this service from the beginning once again. This is called *preempt-restart*. Happily, in many cases, the work completed on the ejected customer up to the point of interruption is not lost so that when that customer is again taken into service, the service process can continue from where it left off. This is called *preempt-resume*.

The second policy is nonpreemptive. The service of a low-priority customer will begin when there are no higher-priority customers present. Now, however, once service has been initiated on a low-priority customer, the server is obligated to serve this customer to completion, even if one or more higher-priority customers arrive during this service.

### *14.6.1 M/M/1: Priority Queue with Two Customer Classes*

To ease us into the analysis of such queues, we shall start with the $M/M/1$ queue with just two customer classes that operate under the preemptive priority policy. Customers of class 1 arrive according to a Poisson process with rate $\lambda_1$; those of class 2 arrive according to a second (independent) Poisson process having rate $\lambda_2$. Customers of both classes receive the same exponentially distributed service at rate $\mu$. We shall let $\rho_1 = \lambda_1/\mu$, $\rho_2 = \lambda_2/\mu$, and assume that $\rho = \rho_1 + \rho_2 < 1$. Given that the service is exponentially distributed, it matters little whether we designate preempt-resume or preempt-restart—thanks to the memoryless property of the exponential distribution. Furthermore, since we have conveniently chosen the service time of all customers to be the same, the total number of customers present, $N$, is independent of the scheduling policy chosen. It then follows from standard $M/M/1$ results that

$$E[N] = \frac{\rho}{1-\rho}.$$

From the perspective of a class 1 customer, class 2 customers do not exist, since service to customers of class 2 is immediately interrupted upon the arrival of a class 1 customer. To a class 1 customer, the system behaves exactly like an *M/M/*1 queue with arrival rate $\lambda_1$ and service rate $\mu$. The mean number of class 1 customers present and the mean response time for such a customer are given, respectively, by

$$E[N_1] = \frac{\rho_1}{1-\rho_1} \quad \text{and} \quad E[R_1] = \frac{1}{\mu(1-\rho_1)}.$$

We may now compute the mean number of class 2 customers present from

$$E[N_2] = E[N] - E[N_1] = \frac{\rho}{1-\rho} - \frac{\rho_1}{1-\rho_1} = \frac{\rho_1+\rho_2}{1-\rho_1-\rho_2} - \frac{\rho_1}{1-\rho_1} = \frac{\rho_2}{(1-\rho_1)(1-\rho_1-\rho_2)}.$$

The mean response time of class 2 customers follows from the application of Little's law. We have

$$E[R_2] = \frac{1/\mu}{(1-\rho_1)(1-\rho_1-\rho_2)}.$$

Let us now move on to the nonpreemptive scheduling policy. This time an arriving class 1 customer finding a class 2 customer in service is forced to wait until that class 2 customer finishes its service. From PASTA, we know that an arriving class 1 customer will find, on average, $E[N_1]$ class 1 customers already present, each of which requires $1/\mu$ time units to complete its service. The arriving customer also has a mean service time of $1/\mu$, and if the arriving customer finds a class 2 customer in service, a further $1/\mu$ time units must be added into the total time the arriving class 1 customer spends in the system. The probability of an arriving customer finding a class 2 customer in service is equal to $\rho_2$—recall that the probability the system has at least one customer is $\rho = \rho_1 + \rho_2$. Summing these three time periods together, we compute the mean response time for a class 1 customer as

$$E[R_1] = \frac{E[N_1]}{\mu} + \frac{1}{\mu} + \frac{\rho_2}{\mu}.$$

Now, using Little's law, we have $E[N_1] = \lambda_1 E[R_1]$, which when substituted into the previous equation gives

$$E[R_1] = \frac{\lambda_1 E[R_1]}{\mu} + \frac{1}{\mu} + \frac{\rho_2}{\mu}.$$

Solving for $E[R_1]$ yields

$$E[R_1] = \frac{(1+\rho_2)/\mu}{1-\rho_1}$$

and we now find the mean number of class 1 customers present as

$$E[N_1] = \frac{(1+\rho_2)\rho_1}{1-\rho_1}.$$

As before, the mean number of class 2 customers can be found from

$$\begin{aligned} E[N_2] = E[N] - E[N_1] &= \frac{\rho_1+\rho_2}{1-\rho_1-\rho_2} - \frac{(1+\rho_2)\rho_1}{1-\rho_1} \\ &= \frac{\rho_2 - \rho_1\rho_2 + \rho_1^2\rho_2 + \rho_1\rho_2^2}{(1-\rho_1)(1-\rho_1-\rho_2)} \\ &= \frac{\rho_2[1-\rho_1(1-\rho_1-\rho_2)]}{(1-\rho_1)(1-\rho_1-\rho_2)}, \end{aligned}$$

and finally, from Little's law, the mean response time for class 2 customers is

$$E[R_2] = \frac{E[N_2]}{\lambda_2} = \frac{[1-\rho_1(1-\rho_1-\rho_2)]/\mu}{(1-\rho_1)(1-\rho_1-\rho_2)}.$$

**Example 14.12** Consider an *M/M/*1 queue in which class 1 customers arrive at rate $\lambda_1 = .3$ and class 2 customers at rate $\lambda_2 = .5$ per unit time. Let the mean of the exponential service time be 1 time unit, i.e., $\mu = 1$. Treating both classes identically results in a standard *M/M/*1 queue with $\lambda = .8$ and $\mu = 1$. The mean number of customers and mean response time are given by

$$E[N] = \frac{.8}{1-.8} = 4, \quad E[R] = \frac{1}{1-.8} = 5.$$

With the priority policies we have

$$\rho_1 = .3 \text{ and } \rho_2 = .5.$$

When the preemptive priority policy is applied, we find

$$E[N_1] = \frac{.3}{.7} = .428571, \quad E[N_2] = \frac{.5}{.7\times .2} = 3.571429,$$

$$E[R_1] = \frac{1}{.7} = 1.428571, \quad E[R_2] = \frac{1}{.7\times .2} = 7.142857.$$

When the nonpreemptive priority policy is applied, we have

$$E[N_1] = \frac{1.5\times .3}{.7} = .642857, \quad E[N_2] = \frac{.5(1-.3\times .2)}{.7\times .2} = 3.357143,$$

$$E[R_1] = \frac{.642857}{.3} = 2.142857, \quad E[R_2] = \frac{3.357143}{.5} = 6.714286.$$

Thus the mean response time for class 2 customers increases from 5.0 with the FCFS policy to 6.71 with nonpreemptive priority to 7.14 with preemptive priority. Observe that the mean number of customers present in all three cases is $E[N] = E[N_1] + E[N_2] = 4$. Furthermore, something similar is apparent with the waiting times: the mean waiting time in the FCFS case is equal to a weighted sum of the waiting times in the two priority cases, the weights being none other than $\rho_1$ and $\rho_2$. With the first-come, first-served policy we have

$$\rho W^q = \rho\frac{\rho}{\mu-\lambda} = .8\times\frac{.8}{1-.8} = 3.2,$$

while in both priority systems we have

$$\rho_1 W_1^q + \rho_2 W_2^q = .3\times(E[R_1]-1) + .5\times(E[R_2]-1) = 3.2.$$

We shall see later that there exist a conservation law that explains this phenomenon. However, in the case of preempt-resume, it applies only when the service time is exponentially distributed, as in the present example. It applies to all distributions when the policy is nonpreemptive. More details are provided in a later section.

### *14.6.2 M/G/1: Nonpreemptive Priority Scheduling*

Let us now consider the general case of $J \geq 2$ different classes of customer in which the different classes can have different service requirements. We assume that the arrival process of class $j$ customers, $j = 1, 2, \ldots, J$, is Poisson with parameter $\lambda_j$ and that the service time distribution of this class of customers is general, having a probability density function denoted by $b_j(x)$, $x \geq 0$, and expectation $\overline{x_j} = 1/\mu_j$. We shall let $\rho_j = \lambda_j/\mu_j$ and assume that $\rho = \sum_{j=1}^{J}\rho_j < 1$. We shall let $L_j$ and $L_j^q$ be the mean number of class $j$ customers in the system and waiting in the queue respectively; we shall let $E[R_j]$ and $W_j^q$ be the mean response time and mean time spent waiting respectively; and we shall let $E[\mathcal{R}_j]$ be the expected residual service time of a class $j$ customer.

We consider first the non-preemptive priority scheduling case. Here the time an arriving class $j$ customer, to whom we refer to as the "tagged" customer, spends waiting in the queue is the sum of the following three time periods.

1. The residual service time of the customer in service.
2. The sum of the service time of all customers of classes 1 through $j$ that are already present the moment the tagged customer arrives.
3. The sum of the service time of all higher-priority customers who arrive during the time the tagged customer spends waiting in the queue.

The response time of the tagged customer is found by adding its service requirement to this total. Two features allow us to treat the first of these three periods relatively simply. First, an arriving customer, whatever the class, must permit the customer in service to finish its service: the probability that the customer in service is of class $j$ is given by $\rho_j$ (observe that $\rho = \rho_1 + \rho_2 + \cdots + \rho_J$ is the probability that the server is busy). Second, we know that the expected remaining service time of any customer in service as seen by an arriving customer whose arrival process is Poisson is $E[\mathcal{R}_j]$ if the customer in service is of class $j$. In Section 14.4 we saw how to compute these residuals. Thus the expected residual service time as experienced by the tagged customer is

$$E[\mathcal{R}] = \sum_{i=1}^{J} \rho_i E[\mathcal{R}_i].$$

The second time period is found by using the PASTA property: the tagged customer finds the queueing system at steady state, and hence observes the stationary distribution of all classes of customer already present. Given that, at equilibrium, the mean number of class $i$ customers waiting in the queue is $L_i^q$, the mean duration to serve all customers of equal or higher priority found by the arriving tagged customer is

$$\sum_{i=1}^{j} L_i^q \overline{x_i}.$$

With just these two time periods, we already have sufficient information to compute results for the highest-priority customers. The mean length of time such customers spend waiting in the queue is

$$W_1^q = L_1^q \overline{x_1} + \sum_{i=1}^{J} \rho_i E[\mathcal{R}_i].$$

Now, applying Little's law, $L_1^q = \lambda_1 W_1^q$,

$$W_1^q = \lambda_1 W_1^q \overline{x_1} + \sum_{i=1}^{J} \rho_i E[\mathcal{R}_i] = \rho_1 W_1^q + \sum_{i=1}^{J} \rho_i E[\mathcal{R}_i],$$

which leads to

$$W_1^q = \frac{\sum_{i=1}^{J} \rho_i E[\mathcal{R}_i]}{1 - \rho_1}. \tag{14.11}$$

From this result, the response time of class 1 customers can be computed and then, using Little's law, the mean number in the system and the mean number waiting in the queue. Later on,

Equation (14.11) will serve as the basis clause of a recurrence relation involving customers of lower priority classes.

For customers of class 2 or greater, we need to compute the third time period, the time spent waiting for the service completion of all higher-priority customers who arrive while the tagged customer is waiting. Given that we have defined $W_j^q$ to be the total time spent waiting by a class $j$ customer, the time spent serving higher-priority customers who arrive during this wait is $\sum_{i=1}^{j-1} \lambda_i W_j^q \overline{x_i} = W_j^q \sum_{i=1}^{j-1} \rho_i$. Thus the total time spent waiting by a class $j$ customer is

$$W_j^q = \sum_{i=1}^{J} \rho_i E[\mathcal{R}_i] + \sum_{i=1}^{j} L_i^q \overline{x_i} + W_j^q \sum_{i=1}^{j-1} \rho_i.$$

Using Little's law to replace $L_i^q$ with $\lambda_i W_i^q$, we find

$$\begin{aligned} W_j^q \left(1 - \sum_{i=1}^{j-1} \rho_i\right) &= \sum_{i=1}^{J} \rho_i E[\mathcal{R}_i] + \sum_{i=1}^{j} \lambda_i W_i^q \overline{x_i} \\ &= \sum_{i=1}^{J} \rho_i E[\mathcal{R}_i] + \sum_{i=1}^{j-1} \rho_i W_i^q + \rho_j W_j^q, \end{aligned} \tag{14.12}$$

which leads to

$$W_j^q \left(1 - \sum_{i=1}^{j} \rho_i\right) = \sum_{i=1}^{J} \rho_i E[\mathcal{R}_i] + \sum_{i=1}^{j-1} \rho_i W_i^q.$$

Comparing the right-hand side of this equation with the right-hand side of Equation (14.12), we see that

$$W_j^q \left(1 - \sum_{i=1}^{j} \rho_i\right) = W_{j-1}^q \left(1 - \sum_{i=1}^{j-2} \rho_i\right),$$

and multiplying both sides with $(1 - \sum_{i=1}^{j-1} \rho_i)$ yields the convenient recursive relationship

$$W_j^q \left(1 - \sum_{i=1}^{j} \rho_i\right)\left(1 - \sum_{i=1}^{j-1} \rho_i\right) = W_{j-1}^q \left(1 - \sum_{i=1}^{j-1} \rho_i\right)\left(1 - \sum_{i=1}^{j-2} \rho_i\right).$$

It was with this relationship in mind that we derived Equation (14.11) for the highest-priority customers. Repeated application of this recurrence leads to

$$W_j^q \left(1 - \sum_{i=1}^{j} \rho_i\right)\left(1 - \sum_{i=1}^{j-1} \rho_i\right) = W_1^q (1 - \rho_1),$$

so that using Equation (14.11) we finally obtain

$$W_j^q = \frac{\sum_{i=1}^{J} \rho_i E[\mathcal{R}_i]}{\left(1 - \sum_{i=1}^{j} \rho_i\right)\left(1 - \sum_{i=1}^{j-1} \rho_i\right)}, \quad j = 1, 2, \ldots, J. \tag{14.13}$$

The mean response time of class $j$ customers can now be found by adding $\overline{x_j}$ to this equation, and then Little's law can be used to determine the mean number of class $k$ customers in the system and waiting in the queue.

### 14.6.3 M/G/1: Preempt-Resume Priority Scheduling

Our final objective in this section is to derive the corresponding results when the scheduling policy is such that a low-priority customer in service is interrupted to allow an arriving customer of a higher priority to begin service immediately. The interrupted customer is later scheduled to continue its service from the point at which it was interrupted. With this policy, customers of class $j+1, j+2, \ldots, J$ do not affect the progress of class $j$ customers; customers with lower priorities are essentially invisible to higher-priority customers. In light of this we may set $\lambda_k = 0$ for $k = j+1, j+2, \ldots, J$ when analyzing the performance of class $j$ customers. There are two common ways of determining $W_j^q$, the time that a class $j$ customer spends waiting in the queue. We refer to them as approaches $A$ and $B$, respectively. The first is to compute the time $T_1^A$ it takes to serve all customers of equal or higher priority that are present at the moment the tagged customer arrives and then to add to this $T_2^A$, the time that is spent serving all higher-priority customers that arrive during the total time that the tagged customer spends in the system. The second approach is to compute $T_1^B$, the time spent waiting until the tagged class $j$ customer is taken into service for the first time and to add to this $T_2^B$, the time taken by all higher-priority customers who arrive after the tagged customer first enters service. We begin with approach $A$.

Let $T_1^A$ be the average time it takes to serve all customers of equal or higher-priority that are present at the moment a class $j$ customer arrives. This time must be identical to the time that an arriving customer must wait prior to being served in a standard (no priority) $M/G/1$ queue. The amount of work present at the moment of arrival must all be finished and it matters little in which order, i.e., which (work-preserving) scheduling policy is used. The total amount of work is independent of the order in which customers are served. $T_1^A$ is equal to the sum of the residual service time of the customer in service and the time required to serve all waiting customers: i.e.,

$$T_1^A = \sum_{i=1}^{j} \rho_i E[\mathcal{R}_i] + \sum_{i=1}^{j} \overline{x_i} L_i^q,$$

where $E[\mathcal{R}_i]$ is the residual service time of a class $i$ customer as seen by a Poisson arrival, $\overline{x_i}$ is the mean service time of a class $i$ customer, and $L_i^q$ is the average number of class $i$ customers found waiting in the queue at equilibrium. We continue the analogy with the (standard) $M/G/1$ queue. The number of class $i$ customers who arrive during a period of length $T_1^A$ is $\lambda_i T_1^A$. Since the number of customers present in the queue at a departure instant is equal to the number present at an arrival instant, we must have $L_i^q = \lambda_i T_1^A$. Thus

$$T_1^A = \sum_{i=1}^{j} \rho_i E[\mathcal{R}_i] + \sum_{i=1}^{j} \rho_i T_1^A,$$

which leads to

$$T_1^A = \frac{\sum_{i=1}^{j} \rho_i E[\mathcal{R}_i]}{1 - \sum_{i=1}^{j} \rho_i}.$$

Now we need to find $T_2^A$, the time taken by higher-priority arrivals during the time that the tagged customer is in the system, i.e., during the mean response time of customer $j$, $E[R_j]$. The number of class $i$ arrivals during this period is $\lambda_i E[R_j]$. Therefore the time to serve these customer is

$$T_2^A = \sum_{i=1}^{j-1} \rho_i E[R_j] = (W_j^q + 1/\mu_j) \sum_{i=1}^{j-1} \rho_i.$$

The total waiting time for a class $j$ customer is then equal to

$$W_j^q = T_1^A + T_2^A = \frac{\sum_{i=1}^{j} \rho_i E[\mathcal{R}_i]}{1 - \sum_{i=1}^{j} \rho_i} + (W_j^q + 1/\mu_j)\sum_{i=1}^{j-1} \rho_i.$$

Solving for $W_j^q$ gives

$$W_j^q = \frac{\sum_{i=1}^{j} \rho_i E[\mathcal{R}_i]}{\left(1 - \sum_{i=1}^{j} \rho_i\right)\left(1 - \sum_{i=1}^{j-1} \rho_i\right)} + \frac{1/\mu_j \sum_{i=1}^{j-1} \rho_i}{\left(1 - \sum_{i=1}^{j-1} \rho_i\right)}. \tag{14.14}$$

The expected total time spent in the system, the mean response or sojourn time of a class $j$ customer, can now be found as

$$E[R_j] = W_j^q + \frac{1}{\mu_j} = \frac{\sum_{i=1}^{j} \rho_i E[\mathcal{R}_i]}{\left(1 - \sum_{i=1}^{j} \rho_i\right)\left(1 - \sum_{i=1}^{j-1} \rho_i\right)} + \frac{1/\mu_j}{\left(1 - \sum_{i=1}^{j-1} \rho_i\right)}.$$

It is also possible to solve for the mean response time directly from

$$E[R_j] = T_1^A + T_2^A + \frac{1}{\mu_j} = \frac{\sum_{i=1}^{j} \rho_i E[\mathcal{R}_i]}{1 - \sum_{i=1}^{j} \rho_i} + E[R_j]\sum_{i=1}^{j-1} \rho_i + \frac{1}{\mu_j}$$

so that, solving for $E[R_j]$, being sure not to confuse italic $R$ with calligraphic $\mathcal{R}$, gives

$$E[R_j] = \frac{1}{1 - \sum_{i=1}^{j-1} \rho_i}\left(\frac{\sum_{i=1}^{j} \rho_i E[\mathcal{R}_i]}{1 - \sum_{i=1}^{j} \rho_i} + \frac{1}{\mu_j}\right)$$

as before. The mean number of class $j$ customers present in the system, $L_j$, and the number waiting in the queue, $L_j^q$, can now be obtained from Little's law.

We now consider approach $B$, the second common way of analyzing the performance characteristics of a class $j$ customer when the scheduling policy is preempt-resume. As we indicated earlier, this is to compute the sum of the time spent waiting until a class $j$ customer enters service for the first time, $T_1^B$, and the remaining time, $T_2^B$, that is spent in service and in interrupted periods caused by higher priority customers who arrive after the customer first starts its service.

Under a non-preemptive policy, the first time a customer enters service is also the only time it does so, whereas in a preempt-resume policy, a customer may enter service, be interrupted, enter service again, and so on. We now relate the time that a customer of class $j$ spends waiting, prior to entering service for the first time, in both policies. The reader may have noticed that the first term on the right hand side of Equation (14.14) is almost identical to the right-hand side of Equation (14.13), which gives the time spent prior to entering service (for the first time) when the priority policy is non-preemptive. The only difference is that the summation is over all classes 1 through $J$ in the nonpreemptive case while it is just over classes 1 through $j$ in the preempt-resume policy. Under both policies, all equal- or higher-priority customers that are present upon an arrival must first be served, as must all higher priority customers who arrive while the class $j$ customer is waiting. In addition, in the nonpreemptive case, and only in the nonpreemptive case, any lower-priority customer that happens to be in service when a class $j$ customer arrives, must be permitted to complete its service. This explains the difference in both formulae: the difference in the upper limits of the summation arises since a class $j$ customer under the preempt-resume policy ignores lower-priority customers so that $\lambda_i = 0$ for $i = j+1, j+2, \ldots, J$ and hence $\rho_i = 0$ for $i = j+1, j+2, \ldots, J$ and can be omitted from the summation. Thus, under a preempt-resume scheduling discipline, the time spent waiting by a class $j$ customer prior to entering service for the first time, is given by

$$T_1^B = \frac{\sum_{i=1}^{j} \rho_i E[\mathcal{R}_i]}{\left(1 - \sum_{i=1}^{j} \rho_i\right)\left(1 - \sum_{i=1}^{j-1} \rho_i\right)}.$$

It must then be the case that the second term of Equation (14.14) is equal to $T_2^B$, i.e.,

$$T_2^B = \frac{1/\mu_j \sum_{i=1}^{j-1} \rho_i}{1 - \sum_{i=1}^{j-1} \rho_i}.$$

This may be shown independently as follows. Observe that $T_2^B + 1/\mu$ is the time that elapses from the moment the tagged customer enters service for the first time until it leaves the queueing system. During this time, $\lambda_i \left(T_2^B + 1/\mu\right)$ class $i$ customers arrive and each has a service requirement equal to $\overline{x_i}$. Therefore

$$T_2^B = \sum_{i=1}^{j-1} \lambda_i \overline{x_i} \left(T_2^B + 1/\mu\right) = \left(T_2^B + 1/\mu\right) \sum_{i=1}^{j-1} \rho_i.$$

Solving for $T_2^B$ gives the desired result.

**Example 14.13** Consider a queueing system which caters to three different classes of customers whose arrival processes are all Poisson. The most important customers require $\overline{x_1} = 1$ time unit of service and have a mean interarrival period of $1/\lambda_1 = 4$ time units. The corresponding values for classes 2 and 3 are $\overline{x_2} = 5$, $1/\lambda_2 = 20$, and $\overline{x_3} = 20$, $1/\lambda_3 = 50$, respectively. Thus $\rho_1 = 1/4$, $\rho_2 = 5/20$, $\rho_3 = 20/50$, and $\rho = \rho_1 + \rho_2 + \rho_3 = .9 < 1$. To facilitate the computation of the residual service times, we shall assume that all service time distributions are deterministic. Thus $\mathcal{R}_1 = .5$, $\mathcal{R}_2 = 2.5$, and $\mathcal{R}_3 = 10.0$.

With the nonpreemptive priority policy, the times spent waiting in the queue by a customer of each of the three classes are as follows:

$$W_1^q = \frac{\rho_1\mathcal{R}_1 + \rho_2\mathcal{R}_2 + \rho_3\mathcal{R}_3}{(1-\rho_1)} = \frac{4.75}{.75} = 6.3333,$$

$$W_2^q = \frac{\rho_1\mathcal{R}_1 + \rho_2\mathcal{R}_2 + \rho_3\mathcal{R}_3}{(1-\rho_1-\rho_2)(1-\rho_1)} = \frac{4.75}{.50 \times .75} = 12.6667,$$

$$W_3^q = \frac{\rho_1\mathcal{R}_1 + \rho_2\mathcal{R}_2 + \rho_3\mathcal{R}_3}{(1-\rho_1-\rho_2-\rho_3)(1-\rho_1-\rho_2)} = \frac{4.75}{.10 \times .50} = 95.0.$$

With the preempt-resume policy, the corresponding waiting times are

$$W_1^q = \frac{\rho_1\mathcal{R}_1}{(1-\rho_1)} = \frac{.125}{.75} = 0.16667,$$

$$W_2^q = \frac{\rho_1\mathcal{R}_1 + \rho_2\mathcal{R}_2}{(1-\rho_1-\rho_2)(1-\rho_1)} + \frac{\rho_1/\mu_2}{1-\rho_1} = \frac{.75}{.50 \times .75} + \frac{1.25}{.75} = 3.6667,$$

$$W_3^q = \frac{\rho_1\mathcal{R}_1 + \rho_2\mathcal{R}_2 + \rho_3\mathcal{R}_3}{(1-\rho_1-\rho_2-\rho_3)(1-\rho_1-\rho_2)} + \frac{(\rho_1+\rho_2)/\mu_3}{(1-\rho_1-\rho_2)} = \frac{4.75}{.10 \times .50} + \frac{10}{.5} = 115.0.$$

### *14.6.4 A Conservation Law and SPTF Scheduling*

It is true what they say—there is no free lunch, and this applies to queueing systems as much as to everything else. When some classes of customers are privileged and have short waiting times, it is at the expense of other customers who pay for this by having longer waiting times. Under certain conditions, it may be shown that a weighted sum of the mean time spent waiting by all customer classes is constant, so that it becomes possible to quantify the penalty paid by low-priority customers. When the queueing system is work preserving (i.e., the server remains busy as long as there are customers present and customers do not leave before having been served), when the service times are independent of the scheduling policy, and when the only preemptive policy permitted is preempt-resume (in which case all service time distributions are exponential with the same parameter for all customer classes), then

$$\sum_{j=1}^{J} \rho_j W_j^q = C,$$

where $C$ is a constant. To justify this result, we proceed as follows. We consider an $M/G/1$ queue with a nonpreemptive scheduling algorithm. Let $U$ be the average amount of unfinished work in the system. This is equal to the expected work yet to be performed on the customer in service plus the sum, over all customer classes $j,\ j = 1, 2, \ldots, J$, of the mean number of class $j$ customers present times the mean service time each class $j$ customer requires. Thus

$$U = \sum_{j=1}^{J} \rho_j E[\mathcal{R}_j] + \sum_{j=1}^{J} L_j^q E[S_j] = E[\mathcal{R}] + \sum_{j=1}^{J} \rho_j W_j^q, \tag{14.15}$$

where, as usual, $E[\mathcal{R}]$ is the expected remaining service time of the customer in service, $L_j^q$ is the mean number of class $j$ customers waiting in the queue and $W_j^q$ is the mean time spent by a class $j$ customer prior to entering service. Observe that this quantity $U$ is independent of the scheduling algorithm and thus has the same value for all scheduling algorithms. No matter the order in which the customers are served, the remaining work is reduced by the single server at a rate of one unit of work per unit time. Furthermore, in a nonpreemptive system, $E[\mathcal{R}]$ is independent of the scheduling algorithm—once a customer is taken into service, that customer is served to completion regardless of the scheduling policy. Now since the right-hand side of Equation (14.15) is independent of the scheduling policy, as is its first term, $E[\mathcal{R}]$, it must be the case that $\sum_{j=1}^{J} \rho_j W_j^q$ is the same for all nonpreemptive scheduling algorithms and hence

$$\sum_{j=1}^{J} \rho_j W_j^q = C.$$

The astute reader will have noticed that it is the required independence of $E[\mathcal{R}]$ that limits preempt-resume policies to those in which the service times are exponentially distributed with the same parameter for all classes.

We now relate this to the mean waiting time in a standard $M/G/1$ queue. The unfinished work in a FCFS $M/G/1$ queue is given by

$$U_{\text{FCFS}} = E[\mathcal{R}] + L^q E[S] = E[\mathcal{R}] + \rho W^q.$$

In this case the constant $C$ that is independent of the scheduling algorithm is easy to find. We have

$$\rho W^q = \rho \frac{\lambda E[S^2]}{2(1-\rho)} = \frac{\rho}{1-\rho} \frac{\lambda E[S^2]}{2} = \frac{\rho}{1-\rho} E[\mathcal{R}]$$

and hence

$$\sum_{j=1}^{J} \rho_j W_j^q = \frac{\rho}{1-\rho} E[\mathcal{R}].$$

This is called the Kleinrock conservation law.

**Example 14.14** Consider a nonpreemptive priority *M/G/1* queue with three classes of customer. The first two moments of the highest priority customers are given by $E[S_1] = 5$ and $E[S_1^2] = 28$, respectively. The corresponding values for classes 2 and 3 are $E[S_2] = 4$; $E[S_2^2] = 24$ and $E[S_3] = 22$; $E[S_3^2] = 1{,}184$, respectively. Arrival rates for the three classes are such that $\rho_1 = 1/3$, $\rho_2 = 11/30$, and $\rho_3 = 11/60$, respectively, and so $\rho = \rho_1 + \rho_2 + \rho_3 = 53/60$. The mean residual service times are

$$E[\mathcal{R}_1] = \frac{E[S_1^2]}{2E[S_1]} = \frac{28}{10} = 2.8, \qquad E[\mathcal{R}_2] = \frac{E[S_2^2]}{2E[S_2]} = \frac{24}{8} = 3,$$

$$E[\mathcal{R}_3] = \frac{E[S_3^2]}{2E[S_3]} = \frac{1184}{44} = 26.909091.$$

This allows us to compute the mean time spent waiting by customers of each class. Given that $\sum_{j=1}^{3} \rho_j E[\mathcal{R}_j] = 6.966667$,

$$W_1^q = \frac{\sum_{j=1}^{3} \rho_j E[\mathcal{R}_j]}{1-\rho_1} = \frac{6.966667}{2/3} = 10.45,$$

$$W_2^q = \frac{\sum_{j=1}^{3} \rho_j E[\mathcal{R}_j]}{(1-\rho_1-\rho_2)(1-\rho_1)} = \frac{6.966667}{3/10 \times 2/3} = 34.833333,$$

$$W_3^q = \frac{\sum_{j=1}^{3} \rho_j E[\mathcal{R}_j]}{(1-\rho_1-\rho_2-\rho_3)(1-\rho_1-\rho_2)} = \frac{6.966667}{7/60 \times 3/10} = 199.047619,$$

and hence

$$\sum_{j=1}^{3} \rho_i W_i^q = \frac{1}{3} \times 10.45 + \frac{11}{30} \times 34.833333 + \frac{11}{60} \times 199.047619 = 52.747619.$$

This is the constant that is independent of the scheduling policy. In the same *M/G/1* queue without priority, we find $E[\mathcal{R}] = 6.966667$ so that

$$\frac{\rho}{1-\rho} E[\mathcal{R}] = \frac{53/60}{7/60} \times 6.966667 = 52.747619,$$

as expected.

We now turn to a scheduling algorithm known as the *shortest processing time first* (SPTF), also called the *shortest job next* (SJN), policy. As its name implies, among all waiting customers, the next to be brought into service by an SPFT algorithm is the customer whose service requirement is least. Once a customer is taken into service, that customer continues uninterrupted until its service has been completed. Viewed in this light, the SPTF policy is a nonpreemptive scheduling policy in which customers with the shortest service time get priority over those with longer service requirements. The length of service can be given in integer units (discrete) but here we assume that any service duration is possible (continuous). We consider an *M/G/1* queue whose Poisson arrival process has rate $\lambda$ and which operates the SPTF policy. As always, the random variable "service time" is denoted by $S$ and its probability density function is denoted by $b_S(x)$. Thus

$$\text{Prob}\{x \le S \le x + dx\} = b_S(x)dx.$$

A customer for which $x \le S \le x + dx$, where $dx \to 0$, is said to have priority $x$. The set of all customers whose service time is $x$ constitutes class $x$. Such customers have a fixed service requirement equal to $x$ and so their residual service time is

$$E[\mathcal{R}_x] = x/2.$$

The arrival rate of customers of class-$x$, denoted $\lambda_x$, is given by

$$\lambda_x = \lambda\, \text{Prob}\{x \le S \le x + dx\} = \lambda\, b_S(x)dx.$$

Since the service time of these customers is equal to $x$, the utilization of the system generated by class-$x$ customers is

$$\rho_x = \lambda_x x = \lambda x b_S(x)dx,$$

and so the total utilization is

$$\rho = \lambda \int_0^\infty x b_S(x)dx = \lambda E[S].$$

We now proceed by analogy with the nonpreemptive priority system having $J$ classes of customers as discussed in Section 14.6.2. We make use of two correspondences. First, we previously found the mean residual time to be

$$E[\mathcal{R}] = \sum_{j=1}^{J} \rho_j E[\mathcal{R}_j].$$

Similarly, with the SPTF policy, using $\rho_x = \lambda x b_S(x)dx$ and $E[\mathcal{R}_x] = x/2$ and integrating rather than summing, we have

$$\int_0^\infty \lambda x b_S(x)\frac{x}{2}dx = \frac{\lambda}{2}\int_0^\infty x^2 b_S(x)dx = \frac{\lambda E[S^2]}{2} = \rho\frac{E[S^2]}{2E[S]}.$$

Second, in Section 14.6.2, we formed the expected waiting time of a class $j$ customer as

$$W_j^q = \frac{\sum_{i=1}^{J} \rho_i E[\mathcal{R}_i]}{\left(1 - \sum_{i=1}^{j} \rho_i\right)\left(1 - \sum_{i=1}^{j-1} \rho_i\right)}, \quad j = 1, 2, \ldots, J.$$

In the continuous SPTF case, we very naturally replace $\sum_{i=1}^{j} \rho_i$ with $\int_0^x \lambda t b_S(t)dt$. We also replace $\sum_{i=1}^{j-1} \rho_i$ with the exact same quantity, $\int_0^x \lambda t b_S(t)dt$. This is because the difference between a class-$x$ customer and a class-$(x+\delta)$ customer tends to zero as $\delta$ tends to 0. Thus we obtain the mean waiting time of a class-$x$ customer to be

$$W_x^q = \frac{\lambda E[S^2]/2}{\left(1 - \lambda\int_0^x t b_S(t)dt\right)^2}.$$

The expected time spent waiting in the queue by an arbitrary customer is obtained as

$$W^q = \int_0^\infty W_x^q b_S(x)dx = \int_0^\infty \frac{\lambda E[S^2] b_S(x)dx}{2\left(1 - \lambda\int_0^x t b_S(t)dt\right)^2} = \frac{\lambda E[S^2]}{2}\int_0^\infty \frac{b_S(x)dx}{\left(1 - \lambda\int_0^x t b_S(t)dt\right)^2}.$$

**Example 14.15** Let us show that the SPTF policy obeys the same conservation law as other nonpreemptive policies. We have

$$\rho_x = \lambda x b_S(x)dx \quad \text{and} \quad W_x^q = \frac{\lambda E[S^2]/2}{\left(1 - \lambda\int_0^x t b_S(t)dt\right)^2}$$

and we wish to show that

$$\int_0^\infty \rho_x W_x^q dx = \frac{\rho}{1-\rho} E[\mathcal{R}].$$

Therefore

$$\begin{aligned}
\int_0^\infty \rho_x W_x^q dx &= \frac{\lambda E[S^2]}{2} \int_0^\infty \frac{\lambda x b_S(x)dx}{\left(1 - \int_0^x \lambda t b_S(t)dt\right)^2} \\
&= \frac{\lambda E[S^2]}{2} \left. \frac{1}{\left(1 - \int_0^x \lambda t b_S(t)dt\right)} \right|_0^\infty \\
&= \frac{\lambda E[S^2]}{2} \left( \frac{1}{1 - \lambda E[S]} - 1 \right) = \frac{\rho}{1-\rho} E[\mathcal{R}],
\end{aligned}$$

as required.

## 14.7 The *M/G/1/K* Queue

As for the *M/G/1* queue, arrivals to an *M/G/1/K* queue follow a Poisson process with rate $\lambda$ and the service time distribution is general, its probability density function denoted by $b(x)$. However, this time a maximum of $K$ customers can be in the system at any one time: $K-1$ awaiting service plus a single customer receiving service. Arrivals to an *M/G/1/K* queue that occur when there are already $K$ customers present are lost. Consider now the embedded Markov chain obtained at departure instants in an *M/G/1/K* queue. This Markov chain can have only $K$ states, since a customer leaving an *M/G/1/K* queue can leave only $0, 1, 2, \ldots, K-1$ customers behind. In particular, a departing customer cannot leave $K$ customers behind since this would imply that $K+1$ customers were present prior to its departure. Furthermore, the first $K-1$ columns of the transition probability matrix $F^{(K)}$ of this embedded Markov chain must be identical to those of the transition probability matrix $F$ in an *M/G/1* queue: when the *M/G/1/K* queue contains $0 < k < K$ customers, the next departure can leave behind $k-1$ customers with probability $\alpha_0$, $k$ with probability $\alpha_1$, $k+1$ with probability $\alpha_2$, and so on, up to the penultimate value; a departure can leave behind $K-2$ customers with probability $\alpha_{K-1-k}$, where, as before, $\alpha_i$ is the probability of exactly $i$ arrivals into the system during the service period of an arbitrary customer. This, we saw, is equal to

$$\alpha_i = \int_0^\infty \frac{(\lambda x)^i}{i!} e^{-\lambda x} b(x) dx.$$

Only the elements in the last column differ. The largest number any departure can leave behind is $K-1$. When the system contains $k$ customers, the next departure can leave $K-1$ customers behind only when $K-k$ or more customers arrive during a service (all but the first $K-k$ of which are denied access). This happens with probability $1 - \alpha_0 - \alpha_1 - \cdots - \alpha_{K-1-k}$. Alternatively, since the matrix is stochastic each row sum is equal to 1, i.e., $\sum_{k=0}^{K-1} f_{ik}^{(K)} = 1$ for $i = 0, 1, \ldots, K-1$ and so the elements in the last column are given by

$$f_{i,K-1}^{(K)} = 1 - \sum_{k=0}^{K-2} f_{ik}^{(K)}, \quad i = 0, 1, \ldots, K-1.$$

Thus the $K \times K$ matrix of transition probabilities is

$$F^{(K)} = \begin{pmatrix} \alpha_0 & \alpha_1 & \alpha_2 & \cdots & \alpha_{K-2} & 1 - \sum_{k=0}^{K-2} \alpha_k \\ \alpha_0 & \alpha_1 & \alpha_2 & \cdots & \alpha_{K-2} & 1 - \sum_{k=0}^{K-2} \alpha_k \\ 0 & \alpha_0 & \alpha_1 & \cdots & \alpha_{K-3} & 1 - \sum_{k=0}^{K-3} \alpha_k \\ 0 & 0 & \alpha_0 & \cdots & \alpha_{K-4} & 1 - \sum_{k=0}^{K-4} \alpha_k \\ \vdots & \vdots & \vdots & \ddots & \vdots & \vdots \\ 0 & 0 & 0 & \cdots & \alpha_0 & 1 - \alpha_0 \end{pmatrix}.$$

Given the upper Hessenberg structure of this matrix, it is now possible to solve it and obtain the probability distribution of customers at departure instants, using the same numerical procedure (recursive solution) adopted for the *M*/*G*/1 queue. This time however, we do not know, a priori, as we did for the *M*/*G*/1 queue, the first component of the solution vector. Instead, the first component can be assigned an arbitrary value and a later normalization used to construct the correct probability distribution.

Letting $\pi_i^{(K)}$ be the probability that a departing customer in an *M*/*G*/1/*K* queue leaves behind $i$ customers, the system to be solved is

$$\pi^{(K)} \left( F^{(K)} - I \right) = 0$$

where the first $K - 1$ equations are identical to the corresponding first $K - 1$ equations in an *M*/*G*/1 queue. Given this homogeneous system of equations in which the coefficient matrix is singular, we replace one equation (the $K$th) by the normalization equation and the solution obtained is proportional to a vector whose $K$ elements are the first $K$ elements in the solution of the *M*/*G*/1 queue, i.e., we have

$$\pi_i^{(K)} = \frac{p_i^{(\infty)}}{\sum_{k=0}^{K-1} p_k^{(\infty)}}, \quad 0 \le i < K,$$

where $1/\sum_{k=0}^{K-1} p_k^{(\infty)}$ is the constant of proportionality and where we now explicitly write $p_i^{(\infty)}$ to denote the probability of a random observer finding $i$ customers present in an *M*/*G*/1 queue.

In this way, we can compute the distribution of customers in an *M*/*G*/1/*K* queue at departure instants. However, what we most often require is the distribution of customers at any instant and not just at departure instants. Since the PASTA property holds in an *M*/*G*/1/*K* queue, the distribution of customers at arrival instants is the same as the distribution of customers as seen by a random observer. But, unlike the *M*/*G*/1 queue, arrivals and departures in an *M*/*G*/1/*K* queue do *not* see the same distribution. This is most clearly demonstrated by the fact that a customer arriving to an *M*/*G*/1/*K* queue will, with nonzero probability, find that the system is full ($K$ customers present) and is therefore lost, whereas the probability that a departure in an *M*/*G*/1/*K* queue will leave behind $K$ customers must be equal to zero. However, the departure probabilities $\pi_i^{(K)}$ are the same as those seen by an arriving customer *given that that customer is admitted into the system.*

We shall let $p_i^{(K)}$ be the stationary probability of finding $i$ customers in an *M*/*G*/1/*K* queue. In particular, the probability that an *M*/*G*/1/*K* queue contains $K$ customers is $p_K^{(K)}$. This is the probability that an arriving customer is refused admission. Alternatively, $1 - p_K^{(K)}$ is the probability that an arriving customer is admitted. It follows that the probability of an arriving customer finding $i < K$ customers in an *M*/*G*/1/*K* queue is

$$p_i^{(K)} = \text{Prob}\{N = i \mid \text{arriving customer is admitted}\} = \pi_i^{(K)} \left(1 - p_K^{(K)}\right), \quad 0 \le i < K,$$

and, in particular,

$$p_0^{(K)} = \pi_0^{(K)}\left(1 - p_K^{(K)}\right).$$

This leads us to seek an expression for $p_K^{(K)}$. At equilibrium, it is known that the flow into and out of the queueing system must be equal. The flow into the system, the effective arrival rate, is given by $\lambda(1-p_K^{(K)})$, since arrivals occur at rate $\lambda$ as long as the system is not full. On the other hand, the flow out of the system is given by $\mu(1-p_0^{(K)})$ since customers continue to be served at rate $\mu$ while there is at least one present. This gives us our equation for $p_K^{(K)}$:

$$\lambda\left(1 - p_K^{(K)}\right) = \mu\left(1 - p_0^{(K)}\right) = \mu\left(1 - \pi_0^{(K)}\left[1 - p_K^{(K)}\right]\right).$$

Solving for $p_K^{(K)}$ we find

$$p_K^{(K)} = \frac{\rho + \pi_0^{(K)} - 1}{\rho + \pi_0^{(K)}}$$

or

$$1 - p_K^{(K)} = \frac{1}{\rho + \pi_0^{(K)}},$$

where, as usual, $\rho = \lambda/\mu$. It now follows that

$$p_i^{(K)} = \frac{\pi_i^{(K)}}{\rho + \pi_0^{(K)}}, \quad 0 \le i < K. \tag{14.16}$$

Although we know how to compute the departure probabilities $\pi_i^{(K)}$ for $i = 0, 1, \ldots, K-1$, it is more usual to present the results for the distribution of customers in an *M/G/*1*/K* queue in terms of the probability distributions in an *M/G/*1 queue. Returning to the equation

$$\pi_i^{(K)} = \frac{p_i^{(\infty)}}{\sum_{k=0}^{K-1} p_k^{(\infty)}}, \quad 0 \le i < K,$$

and writing

$$\sigma_K = \sum_{k=K}^{\infty} p_k^{(\infty)} = 1 - \sum_{k=0}^{K-1} p_k^{(\infty)},$$

we have

$$\pi_i^{(K)} = \frac{p_i^{(\infty)}}{1-\sigma_K}, \quad 0 \le i < K \quad \text{with} \quad \pi_0^{(K)} = \frac{p_0^{(\infty)}}{1-\sigma_K} = \frac{1-\rho}{1-\sigma_K}, \quad 0 \le i < K.$$

Substituting into Equation (14.16), we obtain

$$p_i^{(K)} = \frac{p_i^{(\infty)}/(1-\sigma_K)}{\rho + (1-\rho)/(1-\sigma_K)} = \frac{p_i^{(\infty)}}{1-\rho\sigma_K}, \quad 0 \le i < K.$$

Observe that, when $\sigma_K$ is small, the probability distribution in an *M/G/*1*/K* queue is approximately equal to that in the corresponding *M/G/*1 queue. Finally, for $p_K^{(K)}$ we obtain

$$p_K^{(K)} = \frac{\rho + (1-\rho)/(1-\sigma_K) - 1}{\rho + (1-\rho)/(1-\sigma_K)} = \frac{(1-\rho)\sigma_K}{1-\rho\sigma_K}.$$

**Example 14.16** Consider an *M/D/1/4* queue for which $\lambda = 1/2$ and $\mu = 1$. When we examined the corresponding *M/D/1* queue (Example 14.3) having these parameters, we saw that

$$\alpha_i = \frac{0.5^i}{i!}e^{-0.5},$$

which gave the following values:

$$\alpha_0 = 0.606531, \quad \alpha_1 = 0.303265, \quad \alpha_2 = 0.075816, \quad \alpha_3 = 0.012636.$$

Therefore the $(4 \times 4)$ transition probability matrix for the embedded Markov chain at departure instants is

$$F^{(4)} = \begin{pmatrix} \alpha_0 & \alpha_1 & \alpha_2 & 1-\sum_{k=0}^{2}\alpha_k \\ \alpha_0 & \alpha_1 & \alpha_2 & 1-\sum_{k=0}^{2}\alpha_k \\ 0 & \alpha_0 & \alpha_1 & 1-\sum_{k=0}^{1}\alpha_k \\ 0 & 0 & \alpha_0 & 1-\alpha_0 \end{pmatrix} = \begin{pmatrix} 0.606531 & 0.303265 & 0.075816 & 0.014388 \\ 0.606531 & 0.303265 & 0.075816 & 0.014388 \\ 0 & 0.606531 & 0.303265 & 0.090204 \\ 0 & 0 & 0.606531 & 0.393469 \end{pmatrix}.$$

We may now solve the system of equations, $\pi^{(4)}(F^{(4)} - I) = 0$ using the recursive procedure discussed previously, beginning with $\pi_1^{(4)} = 1$, and then renormalizing once all four components have been found. We shall leave this as an exercise. Alternatively, we may replace the last equation with the normalizing equation directly and solve using Matlab. The system of equations then becomes

$$\pi^{(4)} \begin{pmatrix} -0.393469 & 0.303265 & 0.075816 & 1.0 \\ 0.606531 & -0.696735 & 0.075816 & 1.0 \\ 0 & 0.606531 & -0.696735 & 1.0 \\ 0 & 0 & 0.606531 & 1.0 \end{pmatrix} = (0,\ 0,\ 0,\ 1)$$

and the solution obtained by Matlab is

$$\pi^{(4)} = (0.507744,\ 0.329384,\ 0.124499,\ 0.038373).$$

This is the probability distribution as seen by departing customers. Given that $\rho = \lambda/\mu = 1/2$, we may now compute $p_4^{(4)}$ from

$$p_4^{(4)} = \frac{\rho + \pi_0^{(4)} - 1}{\rho + \pi_0^{(4)}} = \frac{1/2 + 0.507744 - 1.0}{1/2 + 0.507744} = 0.007684,$$

and the rest of the probability distribution as seen by a random observer from

$$p_i^{(4)} = \frac{\pi_i^{(4)}}{\rho + \pi_0^{(4)}} = \frac{\pi_i^{(4)}}{0.5 + 0.507744}, \quad 0 \le i < 4.$$

This results in the following probability distribution of the number of customer in the *M/D/1/4* queue:

$$p^{(4)} = (0.503842,\ 0.326853,\ 0.123542,\ 0.038078,\ 0.007684).$$

Let us now compute these same results using the probability distribution obtained for the *M/D/1* queue of Example 14.3. In this example, we found the probability distribution to be

$$p^{(\infty)} = (0.5,\ 0.324361,\ 0.122600,\ 0.037788,\ 0.010909,\ 0.003107,\ \ldots).$$

Given that

$$\sigma_4 = 1 - \sum_{k=0}^{3} p_k^{(\infty)} = 0.015251,$$

and using the results

$$p_i^{(4)} = \frac{p_i^{(\infty)}}{1-\rho\sigma_4}, \quad i = 0, 1, 2, 3; \quad p_4^{(4)} = \frac{(1-\rho)\sigma_4}{1-\rho\sigma_4},$$

we obtain

$$p^{(4)} = \left(\frac{1}{1-0.015251/2}[0.5,\ 0.324361,\ 0.122600,\ 0.037788,\ \ 0.015251/2]\right),$$

which when evaluated gives the same result as before.

## 14.8 The *G/M/*1 Queue

Similarly to the *M/G/*1 queue, the *G/M/*1 queue is a single-server queue, but this time the distributions of arrivals and service times are reversed: the service process has an exponential distribution with mean service time $1/\mu$, i.e.,

$$B(x) = 1 - e^{-\mu x}, \quad x \geq 0,$$

while the arrival process is general with mean interarrival time equal to $1/\lambda$. Customers arrive individually and their interarrival times are independent and identically distributed. As a result, the notation *GI/M/*1 is sometimes used to stress this independence of arrivals. We shall denote the arrival distribution by $A(t)$ and its probability density function by $a(t)$.

To represent this system by a Markov chain, it is necessary to keep track of the time that passes between arrivals, since the distribution of interarrival times does not in general possess the memoryless property of the exponential. As was the case for the *M/G/*1 queue, a two-component state descriptor may be used; the first to indicate the number of customers present and the second to indicate the elapsed time since the previous arrival. In this way, the *G/M/*1 queue can be solved using the method of supplementary variables. It is also possible to define a Markov chain embedded within the *G/M/*1 queue, and this is the approach that we shall follow here. The embedded time instants are precisely the instants of customer arrivals, since the elapsed interarrival time at these moments is known—it is exactly equal to zero. This allows us to form a transition probability matrix and to compute the distribution of customers as seen by an arriving customer. Unfortunately, the PASTA property no longer holds (we do not have Poisson arrivals) and so we cannot conclude that the distribution as seen by an arrival is the same as that seen by a random observer: indeed they are not the same.

We shall now construct the transition probability matrix of the Markov chain embedded at arrival instants and write its solution. Let $M_k$ be the number of customers present in a *G/M/*1 queue just prior to the $k^{\text{th}}$ arrival. Let $B_{k+1}$ be the number of service completions that occur between the arrival of the $k^{\text{th}}$ customer and that of the $(k+1)^{\text{th}}$ customer. It follows that

$$M_{k+1} = M_k + 1 - B_{k+1}.$$

The $ij$ element of the transition probability matrix $F$, namely,

$$f_{ij} = \text{Prob}\{M_{k+1} = j \mid M_k = i\},$$

is equal to the probability that $i + 1 - j$ customers are served during an arbitrary interarrival time. This must be equal to zero when $i < j - 1$ (the second of two consecutive arrivals cannot find more than one additional customer to the number that the first finds). Also, an arrival can find any number of customers from a minimum of zero to a maximum of one more than its predecessor finds. In other words, the transition probability matrix has a lower Hessenberg structure.

The probability that $i$ customers complete their service during the period between the $k^{\text{th}}$ and $(k+1)^{\text{th}}$ arrivals is given by

$$\beta_i = \text{Prob}\{B_{k+1} = i\} = \int_0^\infty e^{-\mu t}\frac{(\mu t)^i}{i!}dA(t),$$

and, given the independence and identical distribution of interarrival times, the transition probability matrix is given by

$$F = \begin{pmatrix} 1-\beta_0 & \beta_0 & 0 & 0 & \cdots \\ 1-\sum_{i=0}^{1}\beta_i & \beta_1 & \beta_0 & 0 & \cdots \\ 1-\sum_{i=0}^{2}\beta_i & \beta_2 & \beta_1 & \beta_0 & \cdots \\ \vdots & \vdots & \vdots & \vdots & \ddots \end{pmatrix}.$$

It is apparent that this matrix is irreducible and aperiodic when $\beta_0 > 0$ and $\beta_0 + \beta_1 < 1$. It turns out that the solution to this Markov chain is geometric in form. In other words, there exists a $\xi$, $0 < \xi < 1$, such that

$$\pi_i = C\xi^i, \quad i \geq 0,$$

where $\pi_i$ is now the stationary probability of an arrival finding $i$ customers already present. To show that this is the case, we shall replace $\pi_i$ with $C\xi^i$ in the system of equations, $\pi = \pi F$, and find the restrictions that this imposes on $\xi$ so that $\pi_i = C\xi^i$ is indeed the solution we seek. Extracting the $k^{\text{th}}$ equation from this system of equations, we find

$$\pi_k = \sum_{i=0}^{\infty}\pi_i f_{ik} = \sum_{i=k-1}^{\infty}\pi_i f_{ik} = \sum_{i=k-1}^{\infty}\pi_i\beta_{i+1-k} \quad \text{for } k \geq 1.$$

We do not need to be concerned with the case $k = 0$, since in considering $\pi(F - I) = 0$, the first equation is a linear combination of all the others. Substituting $\pi_k = C\xi^k$ gives

$$C\xi^k = \sum_{i=k-1}^{\infty} C\xi^i\beta_{i+1-k} \quad \text{for } k \geq 1.$$

Now, when we divide through by $C\xi^{k-1}$, we obtain

$$\xi = \sum_{i=k-1}^{\infty}\xi^{i-k+1}\beta_{i+1-k} = \sum_{j=0}^{\infty}\xi^j\beta_j = G_B(\xi),$$

where $G_B(z)$ is the $z$-transform of the number of service completions that occur during an interarrival period. This then is the condition that we must impose on $\xi$ in order for the solution to be given by $\pi_i = C\xi^i$: namely, that $\xi$ be a root of the equation $z = G_B(z)$. One root of this system is obviously $\xi = 1$; however, we need a root that is strictly less than 1. When the steady state exists, the strict convexity of $G_B(z)$ implies that there is exactly one root of $z = G_B(z)$ that lies strictly between zero and one: this root is the value $\xi$ for which $C\xi^i$ is the probability of an arrival finding $i$ customers already present. Furthermore,

$$\xi = \sum_{j=0}^{\infty} \xi^j \beta_j = \sum_{j=0}^{\infty} \xi^j \int_0^{\infty} \frac{(\mu t)^j}{j!} e^{-\mu t} dA(t) = \int_0^{\infty} e^{-(\mu-\mu\xi)t} dA(t) = F_A^*(\mu - \mu\xi).$$

Observe that the right-hand side is the Laplace transform of the probability density function of interarrival time, evaluated at the point $\mu - \mu\xi$. The solution $\xi$ to this functional equation may be obtained by successively iterating with

$$\xi^{(j+1)} = F_A^*(\mu - \mu\xi^{(j)}) \tag{14.17}$$

and taking the initial approximation, $\xi^{(0)}$, to lie strictly between zero and one. As for the constant $C$, it may be determined from the normalization equation. We have

$$1 = \sum_{i=0}^{\infty} C\xi^i = C\frac{1}{1-\xi}$$

i.e., $C = (1 - \xi)$. This leads us to conclude that

$$\pi_i = (1 - \xi)\xi^i.$$

It is impossible not to recognize the similarity between this formula and the formula for the number of customers in an *M/M/*1 queue, namely, $p_i = (1 - \rho)\rho^i$, $i \geq 0$, where $\rho = \lambda/\mu$. It follows by analogy with the *M/M/*1 queue that performance measures, such as the mean number of customers present, can be obtained by replacing $\rho$ with $\xi$ in the corresponding formulae for the *M/M/*1 queue. Thus, for example, whereas $1 - \rho$ is the probability that no customers are present in an *M/M/*1 queue, $1 - \xi$ is the probability that an arrival in a *G/M/*1 queue finds it empty; the mean number in an *M/M/*1 queue is $\rho/(1 - \rho)$ and the variance is $\rho/(1 - \rho)^2$, while the mean and variance of the number of customers seen by an arrival in a *G/M/*1 queue are $\xi/(1-\xi)$ and $\xi/(1-\xi)^2$, respectively, and so on. It is important to remember, however, that $\pi$ is the probability distribution as seen by an arrival to a *G/M/*1 queue and that this in not equal to the stationary distribution of this queue. If, indeed, the two are the same, it necessarily follows that $G = M$.

Finally, results for the stationary distribution of customers in a *G/M/*1 queue—the equilibrium distribution as seen by a random observer, rather that that seen by an arriving customer—are readily available from the elements of the vector $\pi$. First, although $\pi_0 = 1 - \xi$ is that probability that an arrival finds the system empty, the stationary probability of the system being empty is actually $p_0 = 1 - \rho$. Furthermore Cohen [10] has shown that the stationary probability of a *G/M/*1 queue having $k > 0$ customers is given by

$$p_k = \rho(1 - \xi)\xi^{k-1} = \rho\pi_{k-1} \quad \text{for } k > 0.$$

Thus once the variable $\xi$ has been computed from Equation (14.17), the stationary distribution is quickly recovered.

**Example 14.17** Let us show that $\xi = \rho$ when the arrival process in a *G/M/*1 queue is Poisson, i.e., when $G = M$. To show this, we need to solve the functional equation $\xi = F_A^*(\mu - \mu\xi)$ when $a(t) = \lambda e^{-\lambda t}$. The Laplace transform is now

$$F_A^*(s) = \int_0^{\infty} e^{-st}\lambda e^{-\lambda t} dt = -\frac{\lambda}{s+\lambda} e^{-(s+\lambda)t}\Big|_0^{\infty} = \frac{\lambda}{s+\lambda},$$

so that the functional equation becomes

$$\xi = \frac{\lambda}{\mu(1 - \xi) + \lambda}$$

or

$$(\xi - 1)(\mu\xi - \lambda) = 0,$$

with the two solutions $\xi = 1$ and $\xi = \lambda/\mu = \rho$. Only the latter solution satisfies the requirement that $0 < \xi < 1$ and this is exactly what we wanted to show.

**Example 14.18** The *D/M/1* Queue.

In this case, the interarrival time is constant and equal to $1/\lambda$. The probability distribution function has the value 0 for $t < 1/\lambda$ and has the value 1 for $t \geq 1/\lambda$. The density function is a Dirac impulse at the point $t = 1/\lambda$ and its Laplace transform is known to be $F_A^*(s) = e^{-s/\lambda}$. Thus the functional equation we need to solve for $\xi$ is

$$\xi = e^{-(\mu-\mu\xi)/\lambda} = e^{-(1-\xi)/\rho}.$$

To proceed any further it is necessary to give a numeric value to $\rho$ and solve using an iterative procedure. Let us take $\rho = 3/4$ and begin the iterative process with $\xi^{(0)} = 0.5$. Successive iterations of

$$\xi^{(j+1)} = \exp\left(\frac{\xi^{(j)} - 1}{0.75}\right)$$

give

$$0.5,\ 0.513417,\ 0..522685,\ 0.529183,\ 0.533788, 0.537076,\ \ldots,$$

which eventually converges to $\xi = 0.545605$. The mean number in this system at arrival epochs is given by

$$E[N_A] = \frac{\xi}{1-\xi} = 1.200729,$$

while the probability that an arrival to this system finds it empty is

$$\pi_0 = 1 - \xi = 0.454395.$$

**Example 14.19** Consider a *G/M/1* queue in which the exponential service distribution has mean service time equal to 1/2 and in which the arrival process has a hypoexponential distribution function, represented as a passage through two exponential phases, the first with parameter $\lambda_1 = 2$ and the second with parameter $\lambda_2 = 4$. The Laplace transform (see Section 7.6.3) of this distribution is given by

$$F_A^*(s) = \left(\frac{\lambda_1}{s+\lambda_1}\right)\left(\frac{\lambda_2}{s+\lambda_2}\right).$$

Its expectation is $1/\lambda_1 + 1/\lambda_2 = 1/2 + 1/4 = 3/4$, which allows us to compute $\rho = (4/3)/2 = 2/3$. Substituting the values for $\lambda_1$ and $\lambda_2$ we find

$$F_A^*(s) = \frac{8}{(s+2)(s+4)}$$

so that

$$F_A^*(\mu - \mu\xi) = \frac{8}{(2 - 2\xi + 2)(2 - 2\xi + 4)} = \frac{2}{(\xi - 2)(\xi - 3)},$$

and the fixed point equation $\xi = F_A^*(\mu - \mu\xi)$ now becomes

$$\xi^3 - 5\xi^2 + 6\xi - 2 = 0.$$

Since this is a cubic equation, its roots may be computed directly without having to resort to an iterative procedure. Furthermore, since we know that $\xi = 1$ is a root we can reduce this cubic to a quadratic, thereby making the computation of the remaining roots easy. Dividing the cubic equation by $\xi - 1$, we obtain the quadratic

$$\xi^2 - 4\xi + 2 = 0,$$

which has the two roots $2 \pm \sqrt{2}$. We need the root that is strictly less than 1, i.e., the root $\xi = 2 - \sqrt{2} = 0.585786$. This then allows us to find any and all performance measures for this system. For example, the probabilities that it contains zero, one, or two customers are given, respectively, by

$$p_0 = 1 - \rho = 1/3, \quad p_1 = \rho(1-\xi)\xi^0 = 0.27614, \quad p_2 = \rho(1-\xi)\xi^1 = 0.161760;$$

the probability that an arrival finds the system empty is $1 - \xi = 0.414214$; the mean number of customers seen by an arrival is $\xi/(1-\xi) = 1.414214$ and so on.

### Waiting Time Distributions in a *G/M/*1 Queue

Whereas the distribution of customers at arrival epochs is generally not that which is sought (the distribution of customers as seen by a random observer being the more usual), it is exactly that which is needed to compute the distribution of the time spent waiting in a *G/M/*1 queue before beginning service. When the scheduling policy is first come, first served, an arriving customer must wait until all the customers found on arrival are served before this arriving customer can begin its service. If there are $n$ customers already present, then an arriving customer must wait through $n$ services, all independent and exponentially distributed with mean service time $1/\mu$. An arriving customer that with probability $\pi_n = (1-\xi)\xi^n$ finds $n > 0$ customers already present, experiences a waiting time that has an Erlang-$n$ distribution. There is also a finite probability $\pi_0 = 1 - \xi$ that the arriving customer does not have to wait at all. The probability distribution function of the random variable $T_q$ that represents the time spent waiting for service will therefore have a jump equal to $1 - \xi$ at the point $t = 0$. For $t > 0$, we may write

$$\begin{aligned} W_q(t) = \text{Prob}\{T_q \le t\} &= \sum_{n=1}^{\infty} \int_0^t \frac{\mu(\mu x)^{n-1}}{(n-1)!} e^{-\mu x} dx \, (1-\xi)\xi^n \; + \; (1-\xi) \\ &= \xi(1-\xi) \int_0^t \mu e^{-\mu x(1-\xi)} dx \; + (1-\xi) \\ &= 1 - \xi e^{-\mu(1-\xi)t}, \quad t > 0. \end{aligned}$$

This formula also gives the correct result $1 - \xi$ when $t = 0$ and so we may write

$$W_q(t) = 1 - \xi e^{-\mu(1-\xi)t}, \quad t \ge 0, \tag{14.18}$$

and thus the time spent queueing in a *G/M/*1 queue is exponentially distributed with a jump of $(1-\xi)$ at $t = 0$. Observe that we have just witnessed the (well-known) fact that a geometric sum of exponentially distributed random variables is itself exponentially distributed. The probability density function of queueing time may be found by differentiating Equation (14.18). We have

$$w_q(t) = \frac{d}{dt}\left(1 - \xi e^{-\mu(1-\xi)t}\right) = \xi\mu(1-\xi)e^{-\mu(1-\xi)t}.$$

This allows us to compute the mean queueing time as

$$W_q = E[T_q] = \frac{\xi}{\mu(1-\xi)}. \tag{14.19}$$

Of course, all these results could have been obtained directly from the corresponding results for the *M/M/*1 queue by simply replacing $\rho$ with $\xi$. For example, we saw that the mean waiting time in the *M/M/*1 queue is given by

$$W_q = \frac{\lambda}{\mu(\mu - \lambda)}.$$

Since $\lambda/\mu = \rho$, replacing $\lambda$ with $\mu\xi$ gives Equation (14.19).

For the sake of completeness, we provide also the probability distribution and density functions for the response time (total system time) in a *G/M/*1 queue. By analogy with the corresponding results in an *M/M/*1 queue, we find

$$W(t) = 1 - e^{-\mu(1-\xi)t}, \quad t \geq 0,$$

and

$$w(t) = \mu(1 - \xi)e^{-\mu(1-\xi)t}, \quad t \geq 0.$$

**Example 14.20** Returning to Example 14.19, we can write the distribution function of this *G/M/*1 queue as

$$W_q(t) = 1 - \left(2 - \sqrt{2}\right) e^{-2(1-2+\sqrt{2})t} = 1 - \left(2 - \sqrt{2}\right) e^{-2(\sqrt{2}-1)t}, \quad t \geq 0,$$

which has the value $\sqrt{2} - 1$ when $t = 0$. The mean time spent waiting in this system is

$$W_q = \frac{\xi}{\mu(1 - \xi)} = \frac{2 - \sqrt{2}}{2(\sqrt{2} - 1)} = 0.707107.$$

## 14.9 The *G/M/1/K* Queue

When we examined the *M/G/*1*/K* queue we saw that the matrix of transition probabilities is of size $K \times K$, since the number of customers that a departing customer leaves behind has to lie between zero and $K - 1$ inclusive. In particular, a departing customer could not leave $K$ customers behind. In the *G/M/*1*/K* queue, the transition probability matrix is of size $(K + 1) \times (K + 1)$ since an arrival can find any number of customers present between zero and $K$ inclusive. Of course, when an arrival finds $K$ customers present, that customer is not admitted to the system, but is lost. It follows then that the matrix of transition probabilities for the embedded Markov chain is given by

$$F^{(K)} = \begin{pmatrix} 1 - \beta_0 & \beta_0 & 0 & 0 & \cdots & 0 \\ 1 - \sum_{i=0}^{1} \beta_i & \beta_1 & \beta_0 & 0 & \cdots & 0 \\ 1 - \sum_{i=0}^{2} \beta_i & \beta_2 & \beta_1 & \beta_0 & \cdots & 0 \\ \vdots & \vdots & \vdots & \vdots & \ddots & \vdots \\ 1 - \sum_{i=0}^{K-1} \beta_i & \beta_{K-1} & \beta_{K-2} & \beta_{K-3} & \cdots & \beta_0 \\ 1 - \sum_{i=0}^{K-1} \beta_i & \beta_{K-1} & \beta_{K-2} & \beta_{K-3} & \cdots & \beta_0 \end{pmatrix}.$$

For example, if an arrival finds two customers already present (row 3 of the matrix), the next arrival can find three customers (with probability $\beta_0$ there were no service completions between the two arrivals), or two customers (with probability $\beta_1$ there was a single service completion between the two arrivals), or one or no customers present. The last two rows are identical because whether an arrival finds $K - 1$ or $K$ customers present, the next arrival can find any number between $K$ and 0 according to the probabilities $\beta_0$ through $1 - \sum_{i=0}^{K-1} \beta_i$.

The probabilities $\pi_i$, $i = 0, 1, \ldots, K$, which denote the probabilities of an arrival finding $i$ customers, can be obtained from the system of equations $\pi(F^{(K)} - I) = 0$ by means of a reverse recurrence procedure. The last equation is

$$\pi_K = \beta_0 \pi_{K-1} + \beta_0 \pi_K$$

which we write as

$$\beta_0 \pi_{K-1} + (\beta_0 - 1)\pi_K = 0.$$

If we assign a value to the last component of the solution, such as $\pi_K = 1$, then we can obtain $\pi_{K-1}$ as

$$\pi_{K-1} = \frac{1 - \beta_0}{\beta_0}.$$

With the value thus computed for $\pi_{K-1}$, together with the chosen value of $\pi_K$, we can now use the second to last equation to obtain a value for $\pi_{K-2}$. This equation, namely,

$$\beta_0 \pi_{K-2} + (\beta_1 - 1)\pi_{K-1} + \beta_1 \pi_K = 0,$$

when solved for $\pi_{K-2}$ gives

$$\pi_{K-2} = \frac{(1 - \beta_1)\pi_{K-1}}{\beta_0} - \frac{\beta_1}{\beta_0}.$$

The next equation is

$$\beta_0 \pi_{K-3} + (\beta_1 - 1)\pi_{K-2} + \beta_2 \pi_{K-1} + \beta_2 \pi_K = 0,$$

which when solved for $\pi_{K-3}$ gives

$$\pi_{K-3} = \frac{(1 - \beta_1)\pi_{K-2}}{\beta_0} - \frac{\beta_2 \pi_{K-1}}{\beta_0} - \frac{\beta_2}{\beta_0}.$$

In this manner, a value can be computed for all $\pi_i$, $i = 0, 1, \ldots, K$. At this point a final normalization to force the sum of all $K + 1$ components to be equal to 1 will yield the correct stationary probability distribution of customers at arrival epochs in a *G/M/1/K* queue.

**Example 14.21** Consider a *D/M/1/5* queue in which customers arrive at a rate of one per unit time and for which the mean of the exponential service time is equal to 3/4. We first compute the elements of the transition probability matrix from the relation

$$\beta_i = \frac{\mu^i}{i!} e^{-\mu} = \frac{(4/3)^i}{i!} e^{-4/3} = \frac{(4/3)^i}{i!} 0.263597.$$

This gives the following values for $\beta_i$, $i = 0, 1, \ldots, 5$:

$$\beta_0 = 0.263597, \quad \beta_1 = 0.351463, \quad \beta_2 = 0.234309,$$
$$\beta_3 = 0.104137, \quad \beta_4 = 0.034712, \quad \beta_5 = 0.009257.$$

The transition probability matrix is therefore given by

$$F^{(K)} = \begin{pmatrix} 1-\beta_0 & \beta_0 & 0 & 0 & 0 & 0 \\ 1-\sum_{i=0}^{1}\beta_i & \beta_1 & \beta_0 & 0 & 0 & 0 \\ 1-\sum_{i=0}^{2}\beta_i & \beta_2 & \beta_1 & \beta_0 & 0 & 0 \\ 1-\sum_{i=0}^{3}\beta_i & \beta_3 & \beta_2 & \beta_1 & \beta_0 & 0 \\ 1-\sum_{i=0}^{K-1}\beta_i & \beta_4 & \beta_3 & \beta_2 & \beta_1 & \beta_0 \\ 1-\sum_{i=0}^{K-1}\beta_i & \beta_4 & \beta_3 & \beta_2 & \beta_1 & \beta_0 \end{pmatrix}$$

$$= \begin{pmatrix} 0.736403 & 0.263597 & 0.0 & 0.0 & 0.0 & 0.0 \\ 0.384940 & 0.351463 & 0.263597 & 0.0 & 0.0 & 0.0 \\ 0.150631 & 0.234309 & 0.351463 & 0.263597 & 0.0 & 0.0 \\ 0.046494 & 0.104137 & 0.234309 & 0.351463 & 0.263597 & 0.0 \\ 0.011782 & 0.034712 & 0.104137 & 0.234309 & 0.351463 & 0.263597 \\ 0.011782 & 0.034712 & 0.104137 & 0.234309 & 0.351463 & 0.263597 \end{pmatrix}.$$

We now set $\pi_5 = 1.0$ and apply the reverse recurrence procedure to the system of equations $\pi(F^{(K)} - I) = 0$. This gives

$$\pi_4 = \frac{1-\beta_0}{\beta_0} = 2.793668, \qquad \pi_3 = \frac{(1-\beta_1)\pi_4}{\beta_0} - \frac{\beta_1}{\beta_0} = 5.540024,$$

$$\pi_2 = \frac{(1-\beta_1)\pi_3}{\beta_0} - \frac{\beta_2\pi_4}{\beta_0} - \frac{\beta_2}{\beta_0} = 10.258164,$$

and so on. Continuing the process yields

$$\pi_1 = 18.815317 \quad \text{and finally} \quad \pi_0 = 34.485376.$$

The sum of all five elements is given by

$$\|\pi\|_1 = 72.892549,$$

so that, dividing through by this sum, we obtain the final answer

$$\pi = (0.473099,\ 0.258124,\ 0.14073,\ 0.076003,\ 0.038326,\ 0.013719).$$

It is important to notice that the elements become excessively larger and care must be taken to preserve numerical accuracy as the size of the matrix grows. This may be accomplished by periodically renormalizing the partial solution obtained to that point.

## 14.10 Exercises

**Exercise 14.2.1** Show that the following two conditions are both necessary and sufficient for the transition probability matrix of the *M/G/*1 queue, i.e., the matrix $F$ of Equation (14.2), to be irreducible.

(a) $\alpha_0 > 0$,
(b) $\alpha_0 + \alpha_1 < 1$.

**Exercise 14.2.2** Consider an *M/D/*1 queue for which $\lambda = 0.8$ and $\mu = 2.0$. Find

(a) the probability that the system contains exactly four customers;
(b) the probability that it contains more than four customers.

**Exercise 14.2.3** A company has a very large number of computers which fail at a rate of two per 8-hour day and are sent to the company's repair shop. It has been experimentally verified that the occurrence of these failures satisfies a Poisson distribution. It has also been observed that the failures fall into three categories: the first which happens 60% of the time requires exactly one hour to fix; the second which occurs 30% of the time requires 3 hours to fix. The third type of failure requires 5 hours to repair. Find $\alpha_i$, the probability of $i$ arrivals to the repair shop during an average repair time and the probability that the repair shop has more than three computers waiting for or undergoing service.

**Exercise 14.2.4** A university professor observes that the arrival process of students to her office follows a Poisson distribution. Students' questions are answered individually in the professor's office. This professor has calculated that the number of students who arrive during the time she is answering the questions of a single student has the following distribution:

$$\begin{aligned}
\text{Prob}\{0 \text{ students arrive}\} &= 0.45,\\
\text{Prob}\{1 \text{ student arrives}\} &= 0.40,\\
\text{Prob}\{2 \text{ students arrive}\} &= 0.10,\\
\text{Prob}\{3 \text{ students arrive}\} &= 0.05.
\end{aligned}$$

Using the fact that the traffic intensity $\rho$ is equal to the expected number of arrivals during one service, find the probability of having exactly three students waiting outside the professor's office.

**Exercise 14.3.1** It was shown that the probability of $k$ arrivals to an *M/G/*1 queue during the service time of an arbitrary customer is gven by

$$\alpha_k = \text{Prob}\{A = k\} = \int_0^\infty \frac{(\lambda x)^k}{k!} e^{-\lambda x} b(x)dx.$$

We also know that the $j^{\text{th}}$ derivative of the Laplace transform of the probability density function of a random variable evaluated at zero is equal to the $j^{\text{th}}$ moment of the distribution multiplied by $(-1)^j$.

(a) Form the $z$-transform for the random variable $A$, denote it by $V(z)$, and show that $V(z) = B^*(\lambda - \lambda z)$, where $B^*$ is the Laplace transform of the probability density function of service time.
(b) Compute the first derivative of $V(z)$ with respect to $z$, evaluate it at $z = 1$ and show that $V'(1) = \rho$.
(c) Show that the second derivative of $V(z)$ with respect to $z$, evaluated at $z = 1$, is equal to $\lambda^2\overline{x^2}$.
(d) Finally, using the fact that successive derivatives of $V(z)$ evaluated at $z = 1$ give factorial moments, show that $E[A^2] = \rho + \lambda^2 E[S^2] = \rho + \lambda^2\sigma_s^2 + \rho^2$.

**Exercise 14.3.2** Use the Pollaczek-Khintchine mean value formula to find the mean number of customers in an *M/E$_r$/*1 queue with arrival rate $\lambda$ and mean service time equal to $1/\mu$.

**Exercise 14.3.3** Consider an *M/H$_2$/*1 queue. Customers arrive according to a Poisson distribution with parameter $\lambda$. These customers receive service at the hyperexponential server at rate $\lambda$ with probability 0.25 or at rate $2\lambda$ with probability 0.75.

(a) Write down the density function for the hyperexponential server.
(b) Compute the mean, variance, and squared coefficient of variation of the service time.
(c) Use the Pollaczek-Khintchine mean value formula to compute the mean number of customers in this system.
(d) Compare this result with the results obtained from an *M/M/*1 queue and an *M/D/*1 queue whose mean service times are the same as that of the hyperexponential server.

**Exercise 14.3.4** Six out of every ten patrons of a hairdressing salon are female and and the average time they require to have their hair done is 50 minutes. Male customers require only 25 minutes on average. If on average a new customer arrives every hour, what is the expected number of customers in the salon and how long does the average patron spend there? How many minutes on average does a customer wait for his/her turn with the hairdresser? Assume that the arrival process is poisson.

**Exercise 14.3.5** A manufacturing process that deposits items onto a production line sometimes fails and produces less than perfect samples. These imperfect items are detected during their progress through the factory and are automatically transferred to a repair station where they are repaired and returned to the production line. It has been observed that, on average, 32 faulty items are detected and repaired each day and that the mean repair time is 8 minutes with a variance of 25. Furthermore, it has been estimated that each hour an item spends in the repair facility costs the company \$240 (per item in the facility). Management has recently become aware of the existence of a greatly improved repair machine, which can repair items in an average of two minutes with variance also equal to two. However, this machine costs \$400,000 to purchase and has an additional operating cost of \$150,000 per year. Should the company purchase this new machine if it wishes to recoup its cost in two years? Assume that arrivals to the repair facility have a Poisson distribution and that the factory operates for 8 hours per day, 250 days per year.

**Exercise 14.3.6** Arrivals to a service center follow a Poisson process with mean rate $\lambda = 1/4$. The probability density function of the service mechanism is given by

$$b(x) = 2/x^3 \quad \text{if } x \geq 1,$$

and is zero otherwise. Show that $\rho = 1/2$, which means that the queueing system is stable, but that in spite of this, the expected number in the system is infinite. (This awkward situation arises because the stability of a queueing system is defined in terms of the first moments of the arrival and service processes, while in the $M/G/1$ queue, increased variance in service time leads to increased queue lengths: the variance of the service process in this example is infinite!)

**Exercise 14.3.7** Consider an $M/H_2/1$ queue in which the arrival rate is given by $\lambda = 1$. The first of the two server phases is chosen with probability $\alpha = 1/4$ and has a mean service time given by $1/\mu_1 = 1$; the second service phase is selected with probability $1 - \alpha$ and the mean service time at this phase is given by $1/\mu_2 = .5$. Find the Pollaczek-Khintchine transform for the distribution of customers in this queue, invert this transform and then compute the probability that the system contains at least two customers.

**Exercise 14.4.1** Derive the following version of the Pollaczek-Khintchine mean value formula:

$$E[R] = \frac{\rho}{(1-\rho)} E[\mathcal{R}] + E[S].$$

**Exercise 14.4.2** Find the expected residual service time, conditioned on the server being busy, when the service time distribution is

(a) uniformly distributed between 2 and 8 minutes;
(b) distributed according to a two-phase Erlang distribution in which the mean time spent at each phase is 2 minutes;
(c) distributed according to a two-phase hyperexponential distribution in which the mean service time at phase 1 is equal to 40 minutes and that of phase 2 is equal to 16 minutes. A customer entering service chooses phase 1 with probability $\alpha = .25$.

**Exercise 14.4.3** Customers of three different classes arrive at an $M/G/1$ queue according to a Poisson process with mean interarrival time equal to 6 minutes. On average, eight out of every twenty customers belong to class 1; eleven out of every twenty belong to class 2 and the remainder belong to class 3. The service requirements of each class are those of the previous question—the service time distribution of customers of class 1 is uniformly distributed on [2, 8]; those of class 2 have an Erlang-2 distribution and those of class 3 are hyperexponentially distributed and the parameters of these latter two distributions are given in the previous question. Find the mean residual service time.

**Exercise 14.5.1** Use the transform equation

$$G^*(s) = B^*[s + \lambda - \lambda G^*(s)]$$

to compute the expectation, second moment and variance of the busy period distribution in an $M/G/1$ queue.

**Exercise 14.5.2** Consider an *M/G/1* queue in which the arrival rate is given by $\lambda = 0.75$ and in which the mean service time is equal to $1/\mu = 1$. Find the mean and variance of the busy period when the service time distribution is (a) exponential, (b) Erlang-2, (c) Erlang-4, and (d) constant.

**Exercise 14.5.3** In the *M/G/1* queue of Exercise 14.3.4, find

(a) the mean and variance of the busy period and
(b) the mean and variance of the number of customers served during the busy period.

**Exercise 14.6.1** Customers of two different types arrive at an *M/M/1* queue. Customers of class 1 arrive at a rate of fifteen per hour while those of class 2 arrive at a rate of six per hour. The mean service requirement of all customers is 2 minutes. Find the mean response time when the customers are handled in first-come, first-served order. How does this performance measure change when class 1 customers are given (a) nonpreemptive priority and (b) preempt-resume priority, over class 2?

**Exercise 14.6.2** This time the *M/M/1* queue of the previous exercise is changed so that class 2 customers require 4 minutes of service. All other parameters remain unchanged. Find both the mean response time and the mean number of customers present under all three possible scheduling policies: FCFS, preempt-resume priority, and nonpreemptive priority. Summarize your results in a table.

**Exercise 14.6.3** Travelers arrive individually at the single check-in counter operated by a local airline company. Coach-class travelers arrive at a rate of 15 per hour and each requires exactly 2 minutes to get boarding tickets and luggage checked in. First-class customers arrive at a rate of four per hour and since they need be pampered, require a fixed time of 5 minutes to check-in. Lastly, at a rate of two per hour, late-arriving customers, about to miss their flights, require an exponentially distributed amount of time with a mean check-in time of 1 minute. The arrival process for all types of customers is Poisson.

(a) The airline company, in the interest of fairness to all, at first operated on a first-come, first-served basis. Determine the mean time spent by a traveler at the check-in facility while this policy was enforced.
(b) As the airline company began to lose money, it decided to implement a priority policy whereby first-class customers have nonpreemptive priority over coach travelers and late-arriving customers have preemptive priority over both first- and coach-class customers. Find the mean time spent at the check-in facility by an arbitrary traveler under this new system.

**Exercise 14.6.4** Computer programs (jobs) be to executed arrive at a computing unit at a rate of 24 every hour. On average, two of these jobs are compute intensive, requiring $E[S_3] = 10$ minutes of compute time and having a standard deviation equal to 6.0; 12 of the jobs are regular jobs with $E[S_2] = 2$ minutes and a standard deviation equal to 3.0; the remaining jobs are short jobs with $E[S_1] = .5$ and having a standard deviation equal to 1. What is the average response time for a typical job? If short jobs are given the highest priority and long jobs the lowest priority, find the response time for all three job classes under both the preempt-resume policy and the nonpreemptive policy.

**Exercise 14.6.5** Printed circuit boards arrive at a component insertion facility according to a Poisson distribution at a rate of one every 6 seconds. The number of components to be inserted onto a board has a geometric distribution with parameter $p = 0.5$. The time to insert each component is a constant $\tau$ seconds.

(a) Compute the mean, second moment, and variance of the time required to insert the components onto an arbitrary printed circuit board.
(b) Assuming that the printed circuit boards are serviced in first-come, first-served order, find the mean time spent by an arbitrary circuit board at the insertion facility.
(c) Now assume that the servicing of printed circuit boards requiring the insertion of only one component (called type-1 circuit boards) are given nonpreemptive priority over the servicing of all other circuit boards (called type-2 circuit boards). Compute the mean response time of type-1 and type-2 circuit boards in this case.
(d) Given that $\tau = 2$, find the mean response time of an arbitrary printed circuit board under the nonpreemptive priority scheduling of part (c) and compute the relative improvement over the FCFS order.

**Exercise 14.6.6** Consider an *M/G/*1 queue operating under the SPTF scheduling policy. The Poisson arrival process has parameter $\lambda = 1/10$ and the service time is uniformly distributed over [2, 8]. Write an expression for the mean time spent waiting to enter service for a customer whose service time is $\tau$, with $2 \leq \tau \leq 8$. Compute the expected waiting time for $\tau = 2, 4, 6$, and 8. Compare this with the mean waiting time when the FCFS scheduling is used instead of SPTF.

**Exercise 14.6.7** Consider an *M/M/*1 queue with arrival rate $\lambda$, service rate $\mu$ and which operates under the SPTF scheduling algorithm. Derive an expression for the mean waiting time in this queueing system.

**Exercise 14.7.1** Use the recursive approach to compute the distribution of customers at departure instants in the *M/D/*1/4 queue of Example 14.16.

**Exercise 14.7.2** Assume that the repair shop of the company described in Exercise 14.2.3 can hold at most four failed computers. Broken computers that arrive when the workshop is full are sent to an external facility (which turns out to be very expensive). Find the probability that a broken computer is shipped to an external facility.

**Exercise 14.7.3** Assume in the scenario of Exercise 14.2.4 that a student who arrives and finds five students already present becomes discouraged and leaves. Compute the probability of this happening.

**Exercise 14.8.1** Consider an $E_2$*/M/*1 queue in which each exponential phase of the arrival process has mean rate $\lambda_i = 2.0$, $i = 1, 2$, and the mean service rate is $\mu = 3.0$. Compute the probability of having at least three customers in the system.

**Exercise 14.8.2** Generalize the previous question by showing that the root $\xi$ for an $E_2$*/M/*1 queue whose exponential arrival phases each have rate $2\lambda$ and for which the exponential service has mean service time $1/\mu$ is given by

$$\xi = \frac{1 + 4\rho - \sqrt{1 + 8\rho}}{2}.$$

Prove that this value of $\xi$ satisfies $0 < \xi < 1$ whenever $\rho < 1$.

**Exercise 14.8.3** At the outbreak of a serious infectious disease in a remote part of the globe, volunteer doctors and nurses work in pairs (doctor plus nurse) to provide some form of relief. They arrange their service, one volunteer pair per affected village, in such a way that the population of the village first must pass some preliminary tests with the nurse before being seen by the doctor. These preliminary tests, of which there are two, may be assumed to be exponentially distributed with means of 1/4 and 1/6 hours, respectively. The time spent with the doctor may be assumed to be exponentially distributed with mean service time equal to 1/9 hours.

Explain how the system at a single village may be modeled as a *G/M/*1 queue and compute the mean waiting time a villager spends between leaving the nurse and being called by the doctor.

**Exercise 14.8.4** In the scenario of the previous question, suppose that the doctor can assign some of his work to the nurse, so that the nurse must now perform three tasks (assumed exponential) with mean durations of 4, 6, and 2 minutes, and the time now needed with the doctor is reduced to 8 minutes. What is the time now spent by patients between nurse and doctor?

**Exercise 14.8.5** Arrivals to a $H_2$*/M/*1 queue choose the first exponential phase which has rate $\lambda_1 = 2$ with probability $\alpha_1 = 0.75$; the rate at the second exponential phase is $\lambda_2 = 1$. The service time is exponentially distributed with rate $\mu = 2$. Compute the probability that this queueing system contains exactly three customers, and find the mean response time.

**Exercise 14.8.6** Consider the situation in which the arrival process in a *G/M/*1 queue is given in terms of a discrete probability distribution containing $k$ points. Write down the functional equation that must be solved to obtain the parameter $\xi$.

Consider now the specific case in which the distribution of the arrival process in a *G/M/1* queue is given in the following table.

| $t_i$ | $a(t)$ | $A(t)$ |
|---|---|---|
| 1 | 0.2 | 0.2 |
| 2 | 0.3 | 0.5 |
| 3 | 0.4 | 0.9 |
| 4 | 0.1 | 1.0 |

If the exponential service time distribution has a mean of $1/\mu = 2.0$, what is the mean number of customers in this system as seen by an arrival, and what is the mean response time?

# Chapter 15

# *Queueing Networks*

## 15.1 Introduction

### *15.1.1 Basic Definitions*

So far, we have considered queueing systems in which each customer arrives at a service center, receives a single service operation from this center, and then departs, never to return. This is sometimes called a "single-node" system. A "multiple-node" system is one in which a customer requires service at more than one node. Such a system may be viewed as a *network of nodes*, in which each node is a service center having storage room for queues to form and perhaps with multiple servers to handle customer requests. Throughout this chapter we shall use the words "node" and "service center" interchangeably. Customers enter the system by arriving at one of the service centers, queue for and eventually receive service at this center, and upon departure either proceed to some other service center in the network to receive additional service, or else leave the network completely.

It is not difficult to envisage examples of such networks. When patients arrive at a doctor's office, they often need to fill out documentation, for insurance purposes or to update their medical records, then it's off to the nurse's station for various measurements like weight, blood pressure, and so on. The next stop is generally to queue (i.e., wait patiently) for one of the doctors to arrive and begin the consultation and examination. Perhaps it may be necessary to have some X-rays taken, an ultrasound may be called for, and so on. After these procedures have been completed, it may be necessary to talk with the doctor once again. The final center through which the patient must pass is always the billing office. At each of these different points, the patient may have to wait while other patients are being treated. Thus the patient arrives, receives service/treatment at a number of different points, and finally leaves to return to work or to go home.

A number of new considerations arise when one considers networks of queues. For instance, the topological structure of the network is important since it describes the permissible transitions between service centers. Thus it would be inappropriate for a patient to leave the nurse's station and go directly for an X-ray before consulting with the doctor. Also, the path taken by individual customers must somehow be described. Some patients may leave the nurse's station to see Dr. Sawbones while others may leave the same nurse's station to see Dr. Doolittle. Some of Dr. Sawbones' patients may need to go to the X-ray lab, others to have an ultrasound, while others may be sent directly to the billing office.

In queueing networks we are faced with a situation in which customers departing from one service center may mix with customers leaving a second service center and the combined flow may be destined to enter a third service center. Thus there is *interaction* among the queues in the network and this ultimately may have a complicating effect on the arrival process at downstream service centers. Customer arrival times can become correlated with customer service times once customers proceed past their point of entry into the network and this correlation can make the mathematical analysis of these queues difficult.

Another difficulty is that of the actual customer service requirements. In the example of the doctors' office, it is reasonable to assume that the service required at one service center is relatively independent of the service required at other service centers. However, for other queueing models this may not be the case. Consider a queueing network representing a data network of transmission lines and routers. Messages pass through the network from one point to another, and may need to transit through several routers on their journey. Messages are queued at the routers while waiting for the transmission line to become free. If we assume that the delay across the transmission line depends on the size of the message, and that the size of the message does not change, then the service requirement of a given message must be the same at every transmission line in the message's journey across the data network.

**Example 15.1** Consider the simple case of two transmission lines in tandem, both having equal capacity (service rate). This is shown graphically in Figure 15.1.

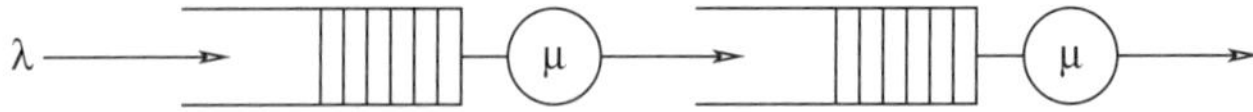


Figure 15.1. Representation of two transmission lines in tandem.

Assume that the arrival process of packets to the first node (transmission point) is Poisson with rate $\lambda$ packets/second, and assume furthermore that all packets have *equal* length with transmission time equal to $1/\mu$. This means that the first node is essentially an $M/D/1$ queue and the average packet delay may be obtained from the Pollaczek-Khintchine formula. Now consider what happens at the second node. Here the interarrival time *cannot* be less than $1/\mu$ since this is the time that it takes a package to move through the first node. As long as the first node is not empty, packets will arrive regularly every $1/\mu$ seconds. If the first node is empty, a packet will not arrive at the second node until one arrives at the first node and then spends $1/\mu$ seconds in transmission. Thus there is a strong correlation among interarrival times and packet lengths. But this is not all. Since the packet transmission times are the same at both nodes, a packet that arrives at the second node will take exactly the same amount of time that the next packet needs at the first node so that the first packet will exit the second node at the same instant that the second packet leaves the first node. Thus there is *no waiting at the second node.*

The fact that all packets have equal length obviously complicated matters. Even when the packet lengths are exponentially and independently distributed and are independent of arrival times at the first queue, this difficulty remains. It stems from the fact that the service time requirements of a packet do not change as it moves from the first node to the second. The first node may be modeled as an $M/M/1$, but not the second since the interarrival times at the second node are strongly correlated with the packet lengths. The interarrival time at the second node must be greater than or equal to the transmission time at the first node. Furthermore, consider what happens to long packets. Typically they will have shorter waiting times at the second node because the length of time they require to exit the first node gives packets at the second node more time to empty out.

Queueing networks, even the simplest, in which each customer has identical service requirements at multiple service centers, such as the packets in the previous example, are difficult to solve. We will not consider this possibility further. Instead we shall assume that the service requirements of customers change as they move through the network, just as was illustrated in the example of the doctors' office.

### *15.1.2 The Departure Process—Burke's Theorem*

Our concern is with the stochastic processes that describe the flow of customers through the network. If we know the various properties of the variables that are associated with the service centers/nodes, the number of servers, their service time distribution, the scheduling discipline and so on, then the only item missing is the stochastic process that describes the arrivals. This is linked to the departure processes from nodes that feed the particular service center as well as the flow of customers directly from the exterior into the center. For example, in the case of a tandem queueing system where customers departing from node $i$ immediately enter node $i+1$, we see that the interdeparture times from the former generate the interarrival times of the latter. For this reason, we now examine departure processes.

Consider an $M/G/1$ system and let $A^*(s)$, $B^*(s)$, and $D^*(s)$ denote the Laplace transform of the PDFs describing the interarrival times, the service times, and the interdeparture times, respectively. Let us calculate $D^*(s)$. When a customer departs from the server, either another customer is available in the queue and ready to be taken into service immediately, or the queue is empty. In the first case, the time until the next customer departs from the server will be distributed exactly as a service time and we have

$$D^*(s)|_{\text{server nonempty}} = B^*(s).$$

On the other hand, if the first customer leaves an empty system behind, we must wait for the sum of two intervals: (a) the time until the next customer arrives and (b) the service time of this next customer. Since these two intervals are independently distributed, the transform of the sum PDF is the product of the transforms of the individual PDFs, and we have

$$D^*(s)|_{\text{server empty}} = A^*(s)B^*(s) = \frac{\lambda}{s+\lambda}B^*(s).$$

In our discussion of the $M/G/1$ queue, we saw that the probability of a departure leaving behind an empty system is the same as the stationary probability of an empty system, i.e., $1-\rho$. Thus we may write the unconditional transform for the interdeparture PDF as

$$D^*(s) = \rho B^*(s) + (1-\rho)\frac{\lambda}{s+\lambda}B^*(s).$$

In the particular case of exponential service times (the $M/M/1$ system), the Laplace transform is given by

$$B^*(s) = \frac{\mu}{s+\mu},$$

where $\mu$ is the service rate. Substituting this into the previous expression for $D^*(s)$ yields

$$\begin{aligned} D^*(s) &= \frac{\lambda}{\mu}\frac{\mu}{s+\mu} + \left(1-\frac{\lambda}{\mu}\right)\frac{\lambda}{s+\lambda}\frac{\mu}{s+\mu} \\ &= \frac{\lambda}{s+\mu} + \frac{\lambda}{s+\lambda}\frac{\mu}{s+\mu} - \frac{\lambda}{\mu}\frac{\lambda}{s+\lambda}\frac{\mu}{s+\mu} \\ &= \frac{\lambda(s+\lambda)}{(s+\mu)(s+\lambda)} + \frac{\lambda\mu-\lambda^2}{(s+\lambda)(s+\mu)} \\ &= \frac{\lambda(s+\mu)}{(s+\mu)(s+\lambda)} = \frac{\lambda}{s+\lambda}, \end{aligned}$$

and so the interdeparture time distribution is given by

$$D(t) = 1 - e^{-\lambda t}, \quad t \geq 0.$$

Thus we must conclude that the interdeparture times are exponentially distributed with the same parameter as the interarrival times. This result can be written more generally as Burke's theorem [6] which we now give without proof.

**Theorem 15.1.1 (Burke)** *Consider an M/M/1, M/M/c, or M/M/∞ system with arrival rate λ. Suppose that the system is in steady state. Then the following hold true:*

1. *The departure process is Poisson with rate λ.*
2. *At each time t, the number of customers in the system is independent of the sequence of departure times prior to t.*

We shall have need of the second part of Burke's theorem later. It says that we cannot make any statements about the number of customers at a node after having observed a sequence of departures. For example, a spate of closely spaced departures does not imply that the last of these leaves behind a large number of customers, nor does the end of an observation period having few departures imply that these departures leave behind a largely empty system. Burke's theorem simply does not allow us to draw these conclusions.

As a final note, it has been shown that the above systems, (i.e., $M/M/1$, $M/M/c$, or $M/M/\infty$) are the only such FCFS queues with this property. In other words, if the departure process of an $M/G/1$ or $G/M/1$ system is Poisson, then $G = M$.

### *15.1.3 Two M/M/1 Queues in Tandem*

Consider a queueing network with Poisson arrivals and two nodes in tandem as illustrated in Figure 15.2. As we mentioned a moment ago, we now assume that the service times of a customer at the first and second nodes are mutually independent as well as independent of the arrival process. We now show that, as a result of this assumption and Burke's theorem, the distribution of customers in the two nodes is the same as if they were two isolated independent $M/M/1$ queues.

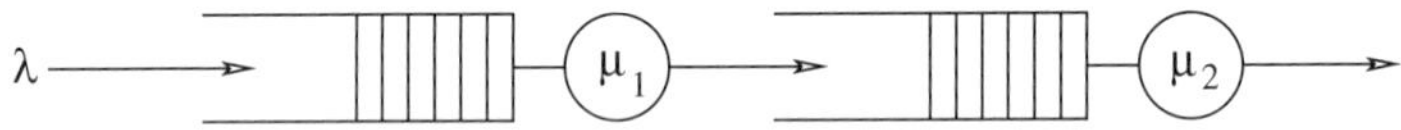


Figure 15.2. Two $M/M/1$ queues in tandem.

Let the rate of the Poisson arrival process be $\lambda$ and let the mean service times at servers 1 and 2 be $1/\mu_1$ and $1/\mu_2$, respectively. Let $\rho_1 = \lambda/\mu_1$ and $\rho_2 = \lambda/\mu_2$ be the corresponding utilization factors, and assume that $\rho_1 < 1$ and $\rho_2 < 1$. Notice that node 1 is an $M/M/1$ queue, and so by part 1 of Burke's theorem, its departure process (and hence the arrival process to node 2) is Poisson. By assumption, the service time at node 2 is independent of node 1. Therefore node 2 behaves like an $M/M/1$ queue, and can be analyzed independently of node 1. Thus, from our earlier results on $M/M/1$ queues

$$\text{Prob}\{n \text{ at node } 1\} = \rho_1^n(1-\rho_1),$$
$$\text{Prob}\{m \text{ at node } 2\} = \rho_2^m(1-\rho_2).$$

From part 2 of Burke's theorem, it follows that the number of customers present in node 1 is independent of the sequence of earlier arrivals at node 2. Consequently, it must also be independent of the number of customers present in node 2. This implies that

$$\begin{aligned}\text{Prob}\{n \text{ at node 1 AND } m \text{ at node } 2\} &= \text{Prob}\{n \text{ at node } 1\} \times \text{Prob}\{m \text{ at node } 2\}\\ &= \rho_1^n(1-\rho_1)\rho_2^m(1-\rho_2).\end{aligned}$$

Thus, under steady state conditions, the number of customers at node 1 and node 2 at any given time are independent and the joint distribution of customers in the tandem queueing system has a *product-form* solution.

## 15.2 Open Queueing Networks

### *15.2.1 Feedforward Networks*

It is apparent that Burke's theorem allows us to string together, one after the other in tandem fashion, any number of multiple-server nodes, each server providing service that is exponentially distributed, and the decomposition just witnessed continues to apply. Each node may be analyzed in isolation and the joint distribution obtained as the product of these individual solutions. Furthermore, more general arrangements of nodes are possible and this product-form solution continues to be valid. The input to each node in the network will be Poisson if the network is a *feedforward* network such as the one illustrated in Figure 15.3. Indeed, since the aggregation of mutually independent Poisson processes is a Poisson process, the input to nodes that are fed by the combined output of other nodes is Poisson. Likewise, since the decomposition of a Poisson process yields multiple Poisson processes, a single node can provide Poisson input to multiple downstream nodes.

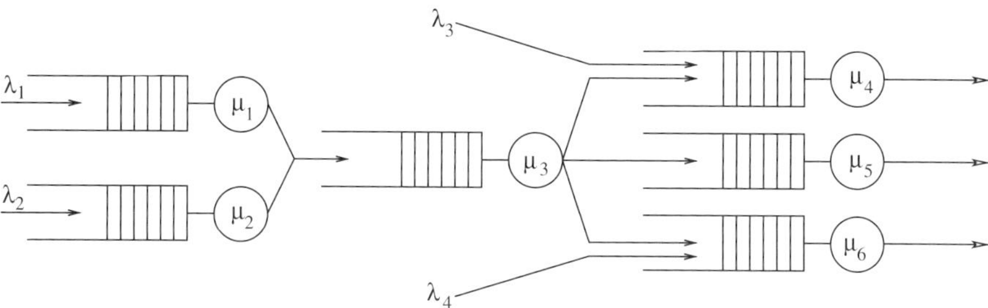


Figure 15.3. A feedforward queueing network.

Feedforward networks do not permit feedback paths since this may destroy the Poisson nature of the feedback stream. To see this, consider the simple single-server queue with feedback illustrated in Figure 15.4. If $\mu$ is much larger than $\lambda$, then, relatively speaking, customer arrivals from the exterior occur rather infrequently. These external arrivals are served rather quickly, but then, with rather high probability, the customer is rerouted back to the queue. The overall effect from the server's point of view is that a burst of arrivals is triggered each time an external customer enters the node. Such as arrival process is clearly not Poisson.

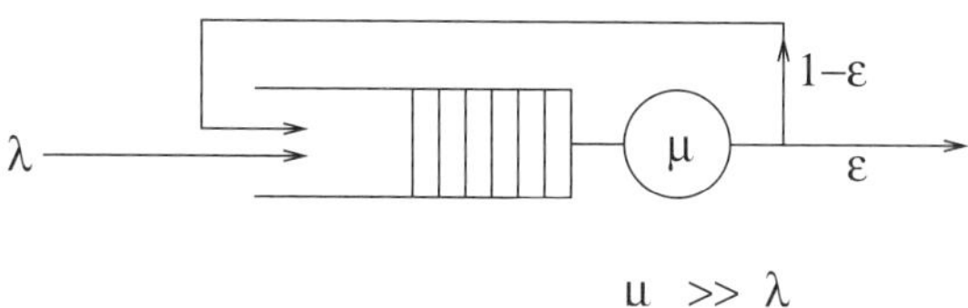


Figure 15.4. A single queue with feedback.

### *15.2.2 Jackson Networks*

Burke's theorem allows us to analyze any feedforward queueing network in which all the servers are exponential. The work of Jackson [22, 23] shows that, even in the presence of feedback loops, the individual nodes behave *as if* they were fed by Poisson arrivals, when in fact they are not. We now consider *Jackson networks*. Consider a network consisting of $M$ nodes. The network is *open*, i.e., there is at least one node to which customers arrive from an external source and there is at least one node from which customers depart from the system. The system is a Jackson network if the following are true for $i, j = 1, 2, \ldots, M$:

- Node $i$ consists of an infinite FCFS queue and $c_i$ exponential servers, each with parameter $\mu_i$.
- External arrivals to node $i$ are Poisson with rate $\gamma_i$.
- After completing service at node $i$, a customer will proceed to node $j$ with probability $r_{ij}$ independent of past history (notice that this permits the case where $r_{ii} \geq 0$) or will depart from the system, never to return again, with probability $1 - \sum_{j=1}^{M} r_{ij}$.

The service centers in this network may be *load dependent*, meaning that the rate at which customers depart from service can be a function of the number of customers present in the node. For example, when node $i$ contains $c_i$ identical exponential servers, each providing service at rate $\mu_i$, and there are $k$ customers present in this node, then the load dependent service rate, written $\mu_i(k)$, is given by $\mu_i(k) = \min(k, c_i)\mu_i$. The routing matrix $R = [r_{ij}]$ determines the permissible transitions between service centers and it is from this matrix that we determine the total average arrival rate of customers to each center. We shall denote the total average arrival rate of customers to service center $i$ by $\lambda_i$. This total rate is given by the sum of the (Poisson) arrivals from outside the system plus arrivals (not necessarily Poisson) from all internal service centers. Thus we obtain the following set of traffic equations:

$$\lambda_i = \gamma_i + \sum_{j=1}^{M} \lambda_j r_{ji}, \quad i = 1, 2, \ldots, M. \tag{15.1}$$

In steady state we also have a traffic equation for the network as a whole:

$$\sum_{i=1}^{M} \gamma_i = \gamma = \sum_{i=1}^{M} \lambda_i \left(1 - \sum_{j=1}^{M} r_{ij}\right).$$

This simply states that, at equilibrium, the total rate at which customers arrive into the queueing network from the outside, $\gamma$, must be equal to the rate at which they leave the network. Equation (15.1) constitutes a nonhomogeneous system of linear equations (nonzero right-hand side) with non-singular coefficient matrix. Its unique solution may be found using standard methods such as Gaussian elimination and the $i^{\text{th}}$ component of this solution is the effective arrival rate into node $i$.

A Jackson network may be viewed as a continuous-time Markov chain with state descriptor vector

$$k = (k_1, k_2, \ldots, k_M),$$

in which $k_i$ denotes the number of customers at service center $i$. At a given state $k$, the possible successor states correspond to a single customer arrival from the exterior, a departure from the network or the movement of a customer from one node to another. The transition from state $k$ to state

$$k(i^+) = (k_1, \ldots, k_{i-1}, k_i + 1, k_{i+1}, \ldots, k_M)$$

corresponds to an external arrival to node $i$ and has transition rate $\gamma_i$. The transition rate from state $k$ to state

$$k(i^-) = (k_1, \ldots, k_{i-1}, k_i - 1, k_{i+1}, \ldots, k_M)$$

corresponds to a departure from center $i$ to the exterior and has transition rate $\mu_i(k_i)(1 - \sum_{j=1}^{M} r_{ij})$. Finally the transition from state $k$ to state

$$k(i^+, j^-) = (k_1, \ldots, k_{i-1}, k_i + 1, k_{i+1}, \ldots, k_{j-1}, k_j - 1, k_{j+1}, \ldots, k_M)$$

corresponds to a customer moving from node $j$ to node $i$ (possibly with $i = j$) and has transition rate $\mu_j(k_j)r_{ji}$. Let

$$P(k) = P(k_1, k_2, \ldots, k_M)$$

denote the stationary distribution of the Markov chain. In other words, $P(k)$ is the equilibrium probability associated with state $k$. We denote the marginal distribution of finding $k_i$ customers in node $i$ by $P_i(k_i)$. Jackson's theorem, which we now state without proof, provides the steady-state joint distribution for the states of the network.

**Theorem 15.2.1 (Jackson)** *For a Jackson network with effective arrival rate $\lambda_i$ to node $i$, and assuming that $\lambda_i < c_i\mu_i$ for all $i$, the following are true in steady state:*

1. *Node $i$ behaves stochastically as if it were subjected to Poisson arrivals with rate $\lambda_i$.*
2. *The number of customers at any node is independent of the number of customers at every other node.*

In other words, Jackson's theorem shows that the joint distribution for all nodes factors into the product of the marginal distributions, that is

$$P(k) = P_1(k_1)P_2(k_2)\cdots P_M(k_M), \tag{15.2}$$

where $P_i(k_i)$ is given as the solution to the classical $M/M/1$-type queue studied earlier. It is an example of a *product-form solution*. In the special case when all the service centers contain a single exponential server ($c_i = 1$ for all $i$), the above expression becomes

$$P(k) = \prod_{i=1}^{M}(1-\rho_i)\rho_i^{k_i} \quad \text{where} \quad \rho_i = \frac{\lambda_i}{\mu_i}.$$

The obvious stability condition for a Jackson network is that $\rho_i = \lambda_i/(c_i\mu_i) < 1$ for $i = 1, 2, \ldots, M$. To solve a Jackson network, we therefore first need to solve the traffic equations to obtain the effective arrival rate at each node. Each node is then solved in isolation and the global solution formed from the product of the individual solutions.

**Example 15.2** Consider the open queueing network illustrated in Figure 15.5. Node 1 consists of two identical exponential servers that operate at rate $\mu_1 = 20$. Customers departing from this node go next to node 2 with probability 0.4 or to node 3 with probability 0.6. Nodes 2 and 3 each have a single exponential server with rates $\mu_2 = 13.75$ and $\mu_3 = 34$, respectively. On exiting either of these nodes, a customer goes to node 4 with probability 1. The last node, node 4 is a pure delay station at which customers spend an exponentially distributed amount of time with mean equal to 1/7 time units. All arrivals from the exterior are Poisson and occur at rate $\gamma_1 = 6$ to node 1 and at rates $\gamma_2 = 3$ and $\gamma_3 = 5$ to nodes 2 and 3 respectively. Departures to the exterior are from node 4 only. Customers departing from this node either depart the network with probability 0.5 or go to node 1 with probability 0.5. Let us compute $P(0, 0, 0, 4)$, the probability that there are 4 customers in the network and that all four are at node 4. Arrivals from the exterior are given by

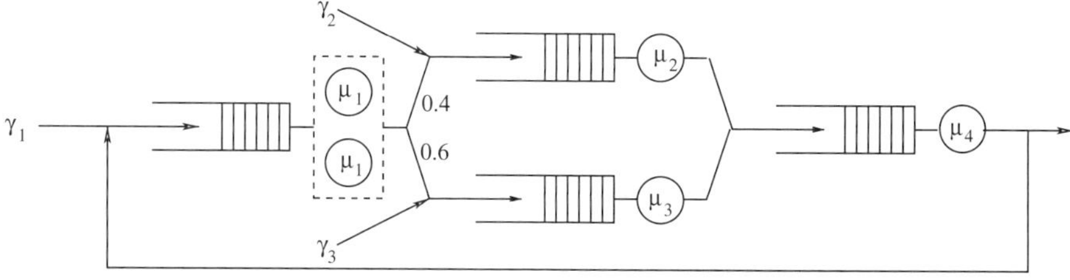


Figure 15.5. The open queueing network of Example 15.2.

$\gamma_1 = 6,\ \gamma_2 = 3,\ \gamma_3 = 5$, and $\gamma_4 = 0$. The routing probability matrix is

$$R = \begin{pmatrix} 0 & .4 & .6 & 0 \\ 0 & 0 & 0 & 1 \\ 0 & 0 & 0 & 1 \\ .5 & 0 & 0 & 0 \end{pmatrix}.$$

We now solve the traffic equations

$$\lambda_i = \gamma_i + \sum_{j=1}^{4} \lambda_j r_{ji}, \quad i = 1, 2, 3, 4.$$

Writing these in full, we have

$$\begin{aligned} \lambda_1 &= \gamma_1 + \lambda_1 r_{11} + \lambda_2 r_{21} + \lambda_3 r_{31} + \lambda_4 r_{41} = 6 + .5\lambda_4, \\ \lambda_2 &= \gamma_2 + \lambda_1 r_{12} + \lambda_2 r_{22} + \lambda_3 r_{32} + \lambda_4 r_{42} = 3 + .4\lambda_1, \\ \lambda_3 &= \gamma_3 + \lambda_1 r_{13} + \lambda_2 r_{23} + \lambda_3 r_{33} + \lambda_4 r_{43} = 5 + .6\lambda_1, \\ \lambda_4 &= \gamma_4 + \lambda_1 r_{14} + \lambda_2 r_{24} + \lambda_3 r_{34} + \lambda_4 r_{44} = \lambda_2 + \lambda_3. \end{aligned}$$

Solving this linear system, we find its unique solution to be $\lambda_1 = 20$, $\lambda_2 = 11$, $\lambda_3 = 17$, and $\lambda_4 = 28$. We use the notation

$$\rho_1 = \lambda_1/(2\mu_1) \quad \text{and} \quad \rho_i = \lambda_i/\mu_i, \quad i = 2, 3, 4,$$

which gives $\rho_1 = 20/(2 \times 20) = .5$, $\rho_2 = 11/13.75 = .8$, $\rho_3 = 17/34 = .5$, and $\rho_4 = 28/7 = 4$. We wish to find the probabilities $P(0, 0, 0, 4) = P_1(0)P_2(0)P_3(0)P_4(4)$ where we treat node 1 as an $M/M/2$ queue, nodes 2 and 3 as $M/M/1$ queues and node 4 as an $M/M/\infty$ queue. The probability of zero customers in an $M/M/2$ queue is obtained using the results of Chapter 12. We find

$$P_1(0) = \left[1 + 2\rho_1 + \frac{(2\rho_1)^2}{2(1-\rho_1)}\right]^{-1} = \left[1 + 1 + \frac{1}{2(1-.5)}\right]^{-1} = \frac{1}{3}.$$

Since nodes 2 and 3 are $M/M/1$ queues, we have

$$P_2(0) = 1 - \rho_2 = .2 \quad \text{and} \quad P_3(0) = 1 - \rho_3 = .5.$$

Finally, the probability of finding four customers in an $M/M/\infty$ queue is

$$P_4(4) = \frac{\rho_4^n e^{-\rho_4}}{4!} = \frac{4^4 e^{-4}}{24} = 10.6667 \times .018312 = .1954.$$

Hence we may conclude that

$$P(0, 0, 0, 4) = P_1(0)P_2(0)P_3(0)P_4(4) = .3333 \times .2 \times .5 \times .1954 = .006512.$$

Other state probabilities may be computed in a similar manner. For example, we find

$$P(0, 1, 1, 1) = P_1(0)P_2(1)P_3(1)P_4(1) = .3333 \times .16 \times .25 \times .07326 = .0009768.$$

Although arrivals at each service center in a Jackson network are not generally Poisson, customers arriving at a service center "see" the same distribution of customers in the service center as a random observer in steady state. This implies that the response time distribution at each node $i$ is exponential with parameter $\mu_i - \lambda_i$, just like in the isolated $M/M/1$ system. This does not follow from Jackson's theorem since this theorem considers steady state in continuous time. Our concern here is with the steady state at arrival instants. We have the following theorem.

**Theorem 15.2.2 (Random observer property)** *In a Jackson queueing network, let $P(k)$ be the stationary probability of state $k$. Then the probability that the network is in state $k$ immediately prior to an arrival to any service center is also $P(k)$.*

### 15.2.3 *Performance Measures for Jackson Networks*

In computing performance measure for Jackson networks, we shall assume that all nodes in the network are $M/M/1$ queues, that the service rate at node $i$ is fixed and equal to $\mu_i$, and that the utilization factor is $\rho_i = \lambda_i/\mu_i < 1$. The results in the more general case of multiple-server nodes are readily obtained by extension.

**Mean Number of Customers**
The mean number of customers in node $i$, $L_i$, is that of an isolated $M/M/1$ queue with arrival rate $\lambda_i$. It is given by

$$L_i = \frac{\rho_i}{1-\rho_i}, \quad i = 1, 2, \ldots, M.$$

It follows that the total mean number of customers in the network is given by

$$L = L_1 + L_2 + \cdots + L_M = \sum_{i=1}^{M} \frac{\rho_i}{1-\rho_i}.$$

**Example 15.3** Let us return to Example 15.2 and compute $L_i = E[N_i]$, $i = 1, 2, 3, 4$, the mean number of customers at each of the four service centers. The mean number of customers present at node 1, an $M/M/2$ queue, is obtained from the standard formula and using the previously computed value of $P_1(0) = 1/3$ we find

$$E[N_1] = 2\rho_1 + \frac{(2\rho_1)^3}{(2-2\rho_1)^2} P_1(0) = 1 + \frac{1}{1} P_1(0) = \frac{4}{3}.$$

Nodes 2 and 3 are each $M/M/1$ queues, so

$$E[N_2] = \frac{\rho_2}{1-\rho_2} = \frac{11/13.75}{2.75/13.75} = \frac{.8}{.2} = 4,$$

$$E[N_3] = \frac{\rho_3}{1-\rho_3} = \frac{17/34}{17/34} = 1.$$

Node 4 is an $M/M/\infty$ queue which allows us to compute

$$E[N_4] = \rho_4 = \frac{28}{7} = 4.$$

The mean number of customers in the network is therefore given by $L = E[N_1] + E[N_2] + E[N_3] + E[N_4] = 10.3333$.

**Mean Time Spent in Network**
The average waiting times are just as easy to derive. The average time that a customer spends in the network may be found from Little's law. It is given by

$$W = \frac{L}{\gamma},$$

where $\gamma = \sum_{i=1}^{M} \gamma_i$ is the total average arrival rate. The average time that a customer spends at node $i$, *during each visit* to node $i$, is given by

$$W_i = \frac{L_i}{\lambda_i} = \frac{1}{\mu_i(1-\rho_i)}.$$

Notice that $W \neq W_1 + W_2 + \cdots + W_M$.

The average time that a customer spends queueing prior to service on a visit to node $i$ is

$$W_i^Q = W_i - \frac{1}{\mu_i} = \frac{\rho_i}{\mu_i(1-\rho_i)}.$$

**Throughput**

The throughput of a single node is the rate at which customers leave that node. The throughput of node $i$ is denoted by $X_i$ and is given by

$$X_i = \sum_{k=1}^{\infty} P_i(k)\mu_i(k),$$

where $\mu_i(k)$ is the load-dependent service rate. At a service center in a Jackson network, and indeed in many other queueing networks, the throughput, at equilibrium, is equal to the arrival rate. However, if a node contains a finite buffer which causes customers to be lost if full, then not all customers will be served, and its throughput will be less than its arrival rate.

The throughput of the entire open queueing network is the rate at which customers leave the network. If customers are not lost to the network, nor combined nor split in any way, then at steady state, the throughput of the network is equal to the rate at which customers enter the network.

## 15.3 Closed Queueing Networks

### *15.3.1 Definitions*

When we visualize a system of queues and servers, we typically think of customers who arrive from the exterior, spend some time moving among the various service centers and eventually depart altogether. However, in a closed queueing network, a fixed number of customers forever cycles among the service centers, never to depart. Although this may seem rather strange, it does have important applications. For example, it has been used in modeling a multiprogramming computer system into which only a fixed number of processes can be handled at any one time. If the processes are identical from a stochastic point of view, then the departure of one is immediately countered by the arrival of a new, stochastically identical, process. This new arrival is modeled by routing the exiting process to the entry point of the new process. Similar situations in other contexts are readily envisaged. Closed Jackson queueing networks are more frequently called *Gordon and Newell* [17, 18] networks, after the names of the pioneers of closed queueing networks.

Consider a closed single-class queueing network consisting of $M$ nodes. Let $n_i$ be the number of customers at node $i$. Then the total customer population is given by

$$N = n_1 + n_2 + \cdots + n_M.$$

A state of the network is defined as follows:

$$n = (n_1, n_2, \ldots, n_M), \quad n_i \geq 0, \quad \sum_{i=1}^{M} n_i = N.$$

The total number of states is given by the binomial coefficient

$$\binom{N+M-1}{M-1}.$$

This is the number of ways in which $N$ customers can be placed among the $M$ nodes. The queueing discipline at each node is FCFS and the service time distributions are exponential. Let $\mu_i$ be the mean service rate at node $i$. When the service rate is load dependent, we shall denote the service rate by $\mu_i(k)$.

Let $p_{ij}$ be the fraction of departures from node $i$ that go next to node $j$ and let $\lambda_i$ be the overall arrival rate to node $i$. Since the rate of departure is equal to the rate of arrival, when the network is in steady state, we may write

$$\lambda_i = \sum_{j=1}^{M} \lambda_j p_{ji}, \quad i = 1, 2, \ldots, M,$$

i.e.,

$$\lambda = \lambda P \quad \text{or} \quad (P^T - I)\lambda^T = 0. \tag{15.3}$$

The linear equations defined by (15.3) are called the traffic equations. Unlike Equation (15.1), which relates to the situation found in open queueing networks, this system of equations does not have a unique solution. Observe that $P$ is a stochastic matrix and we find ourselves in the situation discussed when solving Markov chains numerically: the $\lambda_i$ can be computed only to a multiplicative constant.

We now introduce $v_i$, the *visit ratio* at node $i$. This gives the mean number of visits to node $i$ *relative* to a specified node, which in our case will always be node 1. Thus $v_1 = 1$ and $v_i$ is the mean number of visits to node $i$ between two consecutive visits to node 1. Notice that $v_i$ may be greater (or smaller) than 1. If $X_i(N)$ is the throughput of node $i$ in a closed network with $N$ customers, then we must have $v_i = X_i(N)/X_1(N)$. At steady state, the rate of arrival to a node is equal to the rate of departure (read throughput) so that we may write $v_i = \lambda_i/\lambda_1$ which means that the $v_i$ *uniquely* satisfy the system of equations

$$v_1 = 1 \quad \text{and} \quad v_i = \sum_{j=1}^{M} v_j p_{ji}, \quad i = 1, 2, \ldots, M.$$

Viewed another way, the visit ratios are just the arrival rates normalized so that $\lambda_1 = 1$.

The major result for this closed queueing network is that the equilibrium distribution is given by the product form

$$P(n) = P(n_1, n_2, \ldots, n_M) = \frac{1}{G(N)} \prod_{i=1}^{M} f_i(n_i), \tag{15.4}$$

where

$$f_i(n_i) = \frac{v_i^{n_i}}{\prod_{k=1}^{n_i} \mu_i(k)} \quad \text{if node } i \text{ is load dependent,} \tag{15.5}$$

$$f_i(n_i) = \left(\frac{v_i}{\mu_i}\right)^{n_i} = Y_i^{n_i} \quad \text{if node } i \text{ is load independent.} \tag{15.6}$$

The constant $G(N)$ is called the *normalization* constant. It ensures that the sum of the probabilities is equal to one. It is apparent from Equation (15.4) that once $G(N)$ is known the stationary distribution of any state can readily be computed using only $G(N)$ and easily computable network parameters.

To characterize the normalization constant we proceed as follows: Let the set of all possible states be denoted by $S(N, M)$. Then

$$S(N, M) = \left\{ (n_1, n_2, \ldots, n_M) \,\middle|\, \sum_{i=1}^{M} n_i = N,\ n_i \geq 0,\ i = 1, 2, \ldots, M \right\}$$

and we must have

$$1 = \sum_{S(N,M)} P(n) = \frac{1}{G(N)} \sum_{S(N,M)} \prod_{i=1}^{M} f_i(n_i),$$

which implies that

$$G(N) = \sum_{S(N,M)} \left[ \prod_{i=1}^{M} f_i(n_i) \right].$$

Notice that, although $G(N)$ is written as a function of $N$ only, it is also a function of $M$, $p_{ij}$, $v_i$, and so on.

### *15.3.2 Computation of the Normalization Constant: Buzen's Algorithm*

We shall use the convolution algorithm, also referred to as Buzen's algorithm [7], to evaluate the normalization constant. To proceed, we define the auxiliary function

$$g_m(n) = \sum_{S(n,m)} \prod_{i=1}^{m} f_i(n_i).$$

Observe that $G(N) = g_M(N)$. Notice also that

$$g_M(n) = G(n), \quad n = 0, 1, 2, \ldots, N,$$

which is the normalizing constant for the network with only $n$ customers. We now derive a recursive procedure to compute $g_m(n)$. We have

$$g_m(n) = \sum_{S(n,m)} \prod_{i=1}^{m} f_i(n_i) = \sum_{k=0}^{n} \left[ \sum_{\substack{S(n,m)\\ n_m=k}} \prod_{i=1}^{m} f_i(n_i) \right]$$

$$= \sum_{k=0}^{n} f_m(k) \left[ \sum_{S(n-k,m-1)} \prod_{i=1}^{m-1} f_i(n_i) \right].$$

This gives the recursive formula

$$g_m(n) = \sum_{k=0}^{n} f_m(k) g_{m-1}(n-k). \tag{15.7}$$

Some simplifications are possible when node $i$ is load independent. In this case, and using

$$f_m(k) = Y_m^k = Y_m f_m(k-1),$$

we find

$$g_m(n) = \sum_{k=0}^{n} Y_m^k g_{m-1}(n-k) = Y_m^0 g_{m-1}(n) + \sum_{k=1}^{n} Y_m f_m(k-1) g_{m-1}(n-k)$$

$$= g_{m-1}(n) + Y_m \sum_{l=0}^{n-1} f_m(l) g_{m-1}(n-1-l),$$

which gives the result

$$g_m(n) = g_{m-1}(n) + Y_m g_m(n-1). \tag{15.8}$$

Initial conditions for these recursions are obtained from

$$g_1(n) = \sum_{S(n,1)} \prod_{i=1}^{1} f_i(n_i) = \sum_{\{(n)\}} f_1(n) = f_1(n), \quad n = 0, 1, 2, \ldots, N, \tag{15.9}$$

and

$$g_m(0) = 1, \quad m = 1, 2, \ldots, M.$$

This latter is obtained from

$$g_m(0) = \sum_{S(0,m)} f_1(n_1) f_2(n_2) \cdots f_m(n_m)$$

which, for $n = (0, 0, \ldots, 0)$, is just

$$g_m(0) = f_1(0) f_2(0) \cdots f_m(0).$$

For example, in the load-independent case, this becomes

$$g_m(0) = Y_1^0 \, Y_2^0 \, \cdots \, Y_m^0 = 1.$$

Equations (15.7) and (15.8) are to be used in computing the normalization constant.

**The Convolution Algorithm: Implementation Details**

The only value we need from the convolution algorithm is $G(N)$. However, if we look closely at Equations (15.7) and (15.8), we see that we need to keep a certain number of values during the execution of the algorithm. For convenience we reproduce these equations side by side. We have, for the load-independent and load-dependent cases, respectively,

$$\text{for } m = 1, 2, \ldots, M:$$

$$g_m(n) = g_{m-1}(n) + Y_m g_m(n-1), \quad n = 0, 1, 2, \ldots, N, \quad \text{LI case}, \quad (15.7)$$

$$g_m(n) = \sum_{k=0}^{n} f_m(k) g_{m-1}(n-k), \quad n = 0, 1, 2, \ldots, N, \quad \text{LD case}. \quad (15.8)$$

Notice that the recursion proceeds along two directions, $n$ and $m$, with $n$ assuming values from 0 through $N$ and $m$ assuming value from 1 through $M$. The first of these corresponds to increasing numbers of customers in the network, and the second to a progression through the nodes of the network. All the values of $g_m(n)$ may be visualized as the elements in a two-dimensional table having $N+1$ rows and $M$ columns, as illustrated in Figure 15.6.

| n \ m | 1 | 2 | ••• | m−1 | m | ••• | M |
|---|---|---|---|---|---|---|---|
| 0 | 1 | 1 | ••• | 1 | 1 | ••• | 1 |
| 1 | $f_1(1)$ | | | | | | |
| 2 | $f_1(2)$ | | | | | | |
| ⋮ | ⋮ | | | | $g_m(n-1)$ | | |
| n | $f_1(n)$ | | | $g_{m-1}(n)$ | $g_m(n)$ | | |
| ⋮ | ⋮ | | | | | | $G(N-1)$ |
| N | $f_1(N)$ | | | | | | $G(N)$ |

Figure 15.6. Tabular layout for computation of normalization constant.

The elements may be filled in one after the other, column by column until the table is completed and the required element, $G_M(N)$, is the final value in the last column. The first row and column are obtained from initial conditions: the first row contains all ones (since $g_m(0) = 1,\ m = 1, 2, \ldots, M)$ and the first column is obtained as $g_1(n) = f_1(n),\ n = 0, 1, \ldots, N$. The implementation proceeds by considering the nodes of the network one after the other. The actual order in which they are taken is not an immediate concern. As $m$ takes on values from 1 through $M$, in other words, as we proceed through all the nodes in the network, we use one or the other of the Equations (15.7) and (15.8), depending on whether the current value of $m$ corresponds to a load independent or a load dependent node.

A small disadvantage of this visualization offered by Figure 15.6 is that it may lead one to think that a two-dimensional array is required to implement Buzen's algorithm. This is not the case. As we now show, a single column, a vector of length $N+1$, suffices. Since each of these equations involves $g_{m-1}$ as well as $g_m$, it is apparent that at least some of the values computed for each preceding node will be needed in the evaluation of the $g$ values for the current node. Let us first consider the case when node $m$ is a load-independent node.

We assume that we have already processed nodes 1 through $m - 1$ and in particular, we have all the values $g_{m-1}(n)$ for $n = 0, 1, \ldots, N$ and that these values are in consecutive locations in a one-dimensional array. Equation (15.7) is the appropriate equation in the load-independent case:

$$g_m(n) = g_{m-1}(n) + Y_m g_m(n-1), \qquad n = 0, 1, 2, \ldots, N.$$

The first element is the same for all values of $m$: from the initial conditions we have, for all $m$, $g_m(0) = 1$. As successive elements are computed, i.e., as the value of $n$ in $g_m(n)$ increases from 0 to 1 to 2 and so on, the newly computed value of $g_m(n)$ overwrites the previous value of $g_{m-1}(n)$ in the array. This is illustrated graphically in Figure 15.7. The figure shows the situation immediately before and immediately after the element $g_m(n)$ is computed from Equation (15.7). In this figure, the current contents of the array are contained within the bold lines. Overwritten values of $g_{m-1}(n)$ are

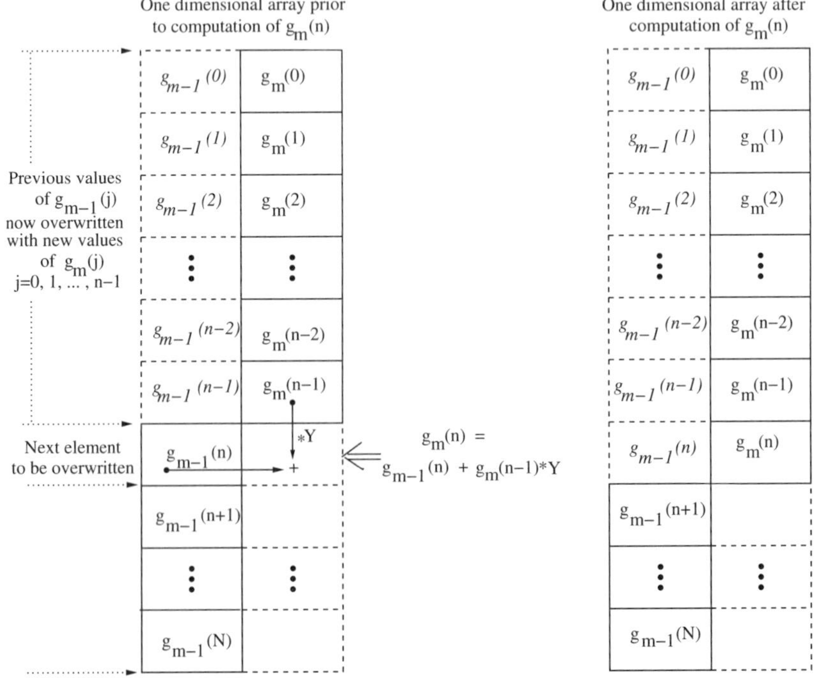


Figure 15.7. Array status for a load-independent node.

displayed within dashed lines on the upper left and yet to be computed values of $g_m(n)$ are displayed within dashed lines on the lower right. It is apparent that a single one-dimensional array is all that is needed in this case.

Let us now consider the case when node $m$ is a load-dependent node, and as before, we assume that all nodes prior to node $m$, whether load dependent or load independent, have already been handled. This time, Equation (15.8) is the appropriate equation to use:

$$g_m(n) = \sum_{k=0}^{n} f_m(k) g_{m-1}(n-k), \qquad n = 0, 1, 2, \ldots, N.$$

Observe that, whereas in the load-independent case, we require only a single value from the previous node, $g_{m-1}(n)$, this time we need $n+1$ values, namely, $g_{m-1}(j)$ for $j = n, \; n-1, \; \ldots, \; 0$. It is instructive to write Equation (15.8) as shown below. Notice that we begin with $g_m(1)$, since we know that $g_m(0) = 1$ for all $m$:

$$\begin{aligned}
g_m(1) &= & & & f_m(0)g_{m-1}(1) + f_m(1)g_{m-1}(0),\\
g_m(2) &= & & f_m(0)g_{m-1}(2) + f_m(1)g_{m-1}(1) + f_m(2)g_{m-1}(0),\\
g_m(3) &= f_m(0)g_{m-1}(3) + f_m(1)g_{m-1}(2) + f_m(2)g_{m-1}(1) + f_m(3)g_{m-1}(0),\\
&\;\vdots\\
g_m(N-1) &= & & f_m(0)g_{m-1}(N-1) + \cdots + f_m(N-1)g_{m-1}(0),\\
g_m(N) &= & f_m(0)g_{m-1}(N) + f_m(1)g_{m-1}(N-1) + \cdots + f_m(N)g_{m-1}(0).
\end{aligned}$$

This shows that $g_{m-1}(1)$ is needed in computing all $g_m(n)$ for $n = 1, 2, \ldots, N$ and cannot be overwritten until all these values have been computed. Likewise, $g_{m-1}(2)$ is needed in computing all $g_m(n)$ for $n = 2, 3, \ldots, N$ and cannot be overwritten before then. On the other hand, $g_{m-1}(N)$ is used only once, in computing $g_m(N)$. Once $g_m(N)$ has been computed it may be placed into the

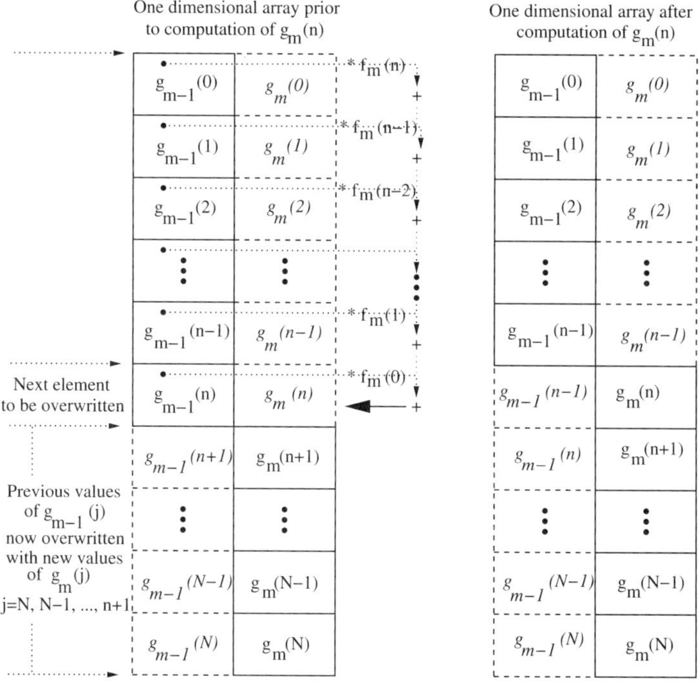


Figure 15.8. Array status for a load-dependent node.

array location previously occupied by $g_{m-1}(N)$. Also, $g_{m-1}(N-1)$ is used only twice, in computing $g_m(N-1)$ and $g_m(N)$. Its position in the array may be overwritten when it is no longer needed, i.e., once both $g_m(N)$ and $g_m(N-1)$ have been computed. In this way, we may proceed from the last element of the array to the first, overwriting elements $g_{m-1}(j)$ with $g_m(j)$ from $j = N, N-1, \ldots, 1$. This is illustrated in Figure 15.8.

There is one instance in which this more complicated approach for handing load dependent nodes can be avoided—by designating the load dependent node as node number 1. Then the column (column number 1) is formed from the initial conditions rather than from the convolution formula. Of course, if there is more than one load dependent node, only one of them can be treated in this fashion.

One final point concerning the convolution algorithm is worth noting. The coefficients $f_m(n)$ may be computed from Equations (15.5) and (15.6) and incorporated directly into the scheme just described for updating the one-dimensional array. In particular, it is not necessary to generate the $f_m(n)$ in advance nor to store them in a two-dimensional array. It suffices that the values of $v_i$ and $\mu_i(k)$ be available. The complete algorithm, written in pseudocode, is now presented. A Java implementation is presented at the end of this chapter.

```
Buzen's Algorithm

    int N;                 \\ Number of customers
    int M;                 \\ Number of nodes
    boolean LI[1:M];       \\ True if node m is load independent; otherwise false
    double G[0:N];         \\ Array to hold normalizing constants

    G[0] = 1.0;                       \\ Handle first node first
    if (LI[1]) then {                       \\ Node 1 is Load Independent
       double Y = lambda[1]/mu[1];
       for (int n=1; n<=N; n++) {
          G[n] = G[n-1]*Y;
       }
    }
    else {                                  \\ Node 1 is Load Dependent
       double Y = lambda[1];
       for (int n=1; n<=N; n++) {
          G[n] = G[n-1]*Y/mu[1,n]
       }
    }

    for (int m=2, m<=M; m++) {        \\ Handle remaining nodes
       if (LI[m]) then {                    \\ Node m is Load Dependent
          double Y = lambda[m]/mu[m];             \\ Compute coefficient
          for (int n=1; n<=N; n++) {
             G[n] = G[n] + G[n-1]*Y;              \\ Compute auxiliary function
          }
       }
       else {                               \\ Node m is Load Dependent
          f[0] = 1;
          for (int n=1; n<=N; n++) {
             f[n] = f[n-1]*lambda[m]/mu[m,n];     \\ Compute coefficient
```

```
        }
        for (int n=N; n>=1; n--) {
           double sum = G[n];
           for (int k=1; k<=n; k++) {
              sum = sum + f[k]*G[n-k];          \\ Compute auxiliary function
           }
        G[n] = sum;
      }
   }
```

**Example 15.4** As an example we shall apply Buzen's algorithm to compute the normalization constant for the central server model shown in Figure 15.9. In this model, the central server (node 1 in the diagram) possesses two exponential servers each having rate $\mu = 1.0$. The three nodes that feed the central server (nodes 2, 3, and 4 in the diagram) each possess a single server providing exponential service at rates 0.5, 1.0, and 2.0 respectively. On exiting the central server, customers go to node $i = 2, 3, 4$ with probabilities 0.5, 0.3, and 0.2, respectively. We shall assume that $N = 3$ customers circulate in this network.

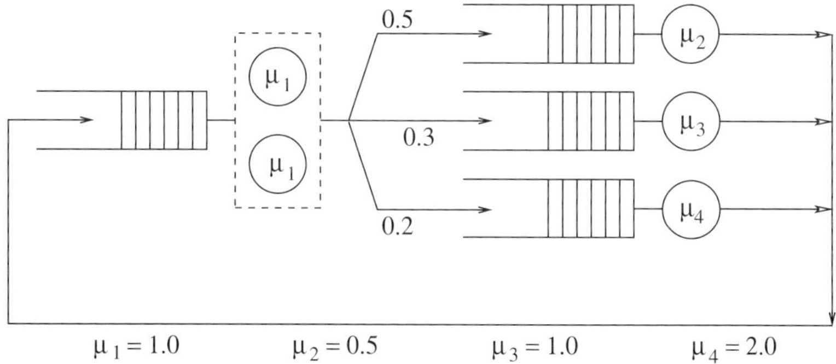


Figure 15.9. Central server queueing model.

The first step is to construct the routing matrix and then solve the traffic equations to obtain the $v$'s. From the diagram, we see that the routing matrix is given by

$$R = \begin{pmatrix} 0 & .5 & .3 & .2 \\ 1 & 0 & 0 & 0 \\ 1 & 0 & 0 & 0 \\ 1 & 0 & 0 & 0 \end{pmatrix}.$$

The traffic equations are given by

$$(v_1,\ v_2,\ v_3,\ v_4,) = (v_1,\ v_2,\ v_3,\ v_4,) \begin{pmatrix} 0 & .5 & .3 & .2 \\ 1 & 0 & 0 & 0 \\ 1 & 0 & 0 & 0 \\ 1 & 0 & 0 & 0 \end{pmatrix},$$

which may be written more conveniently as

$$\begin{aligned} v_1 &= v_2 + v_3 + v_4, \\ v_2 &= .5v_1, \\ v_3 &= .3v_1, \\ v_4 &= .2v_1. \end{aligned}$$

By setting $v_1 = 1$, these equations yield

$$v_1 = 1, \quad v_2 = .5, \quad v_3 = .3, \quad v_4 = .2.$$

We saw that in the program segment given above that the values of the $f_m(n)$ may be incorporated directly into the algorithm. However, in this example, we shall compute them explicitly in order to show as clearly as possible, how the normalization constant is formed. Node 1 is a load-dependent node and hence the values of $f_m(n)$ are computed from Equation (15.5). Rewriting this equation again, we have

$$f_i(n_i) = \frac{v_i^{n_i}}{\prod_{k=1}^{n_i} \mu_i(k)}$$

with $i$ taking the value 1 (node 1) and $n_1$ taking all possible values from 0 through 3 (all possible number of customers present at node 1). This gives

$$\begin{aligned}
f_1(0) &= v_1^0 / \prod_{k=1}^{0} \mu_1(k) = 1, \\
f_1(1) &= v_1^1/\mu_1(1) = 1/1 = 1, \\
f_1(2) &= v_1^2/[\mu_1(1)\mu_1(2)] = 1/(1 \times 2) = 0.5, \\
f_1(3) &= v_1^3/[\mu_1(1)\mu_1(2)\mu_1(3)] = 1/(1 \times 2 \times 2) = 0.25.
\end{aligned}$$

Nodes 2, 3, and 4 are all load-independent nodes and so we use Equation (15.6):

$$f_i(n_i) = \left(\frac{v_i}{\mu_i}\right)^{n_i} = (v_i s_i)^{n_i}.$$

We obtain

$$\begin{aligned}
f_2(0) &= 1, & f_3(0) &= 1, & f_4(0) &= 1, \\
f_2(1) = v_2/\mu_2 &= .5/.5 = 1, & f_3(1) = v_3/\mu_3 &= .3/1 = .3, & f_4(1) = v_4/\mu_4 &= .2/2 = .1, \\
f_2(2) = (v_2/\mu_2)^2 &= 1, & f_3(2) = (v_3/\mu_3)^2 &= (.3)^2 = .09, & f_4(2) = (v_4/\mu_4)^2 &= (.1)^2 = .01, \\
f_2(3) = (v_2/\mu_2)^3 &= 1, & f_3(3) = (v_3/\mu_3)^3 &= (.3)^3 = .027, & f_4(3) = (v_4/\mu_4)^3 &= (.1)^3 = .001.
\end{aligned}$$

We now move on to the auxilary functions and the computation of $G(N)$ itself. We will compute the elements of the array $G$ node by node. All nodes other than the first are load independent and hence successive updates for $m = 2$, 3, and 4 are obtained from

$$g_m(n) = g_{m-1}(n) + Y_m g_m(n-1),$$

with

$$Y_2 = .5/.5 = 1, \quad Y_3 = .3/1 = .3, \quad Y_4 = .2/2 = .1.$$

The values for node 1 are obtained from the initial conditions, Equation (15.9) having designated this single load-dependent node as node number 1, precisely to avoid the use of the convolution formula. We have

$$g_1(n) = f_1(n) \quad \text{for } n = 0, 1, 2, \ldots, N$$

and so the initial values in $G$ are

$$g_1[0] = 1,$$
$$g_1[1] = 1,$$
$$g_1[2] = 0.5,$$
$$g_1[3] = 0.25.$$

Given these initial values, we successively obtain

$$\begin{array}{lll} g_2[0] = 1, & g_3[0] = 1, & g_4[0] = 1, \\ g_2[1] = 1 + 1 \times 1 = 2, & g_3[1] = 2 + .3 \times 1 = 2.3, & g_4[1] = 2.3 + .1 \times 1 = 2.4, \\ g_2[2] = .5 + 1 \times 2 = 2.5, & g_3[2] = 2.5 + .3 \times 2.3 = 3.19, & g_4[2] = 3.19 + .1 \times 2.4 = 3.43, \\ g_2[3] = .25 + 1 \times 2.5 = 2.75, & g_3[3] = 2.75 + .3 \times 3.19 = 3.707, & g_4[3] = 3.707 + .1 \times 3.43 = 4.05. \end{array}$$

We are now in a position to compute the stationary distribution of any state of this queueing network. For example, the probability of having all three customers at the central server is given by

$$\begin{aligned} P(n) = P(3, 0, 0, 0) &= \frac{1}{G(N)} \prod_{i=1}^{4} f_i(n_i) \\ &= \frac{1}{4.05} f_1(3) f_2(0) f_3(0) f_4(0) = \frac{0.25}{4.05} = 0.0617. \end{aligned}$$

### *15.3.3 Performance Measures*

We now proceed to develop a number of important measures of effectiveness for this type of queueing network. We shall compute the marginal queue length distributions, the throughput and utilization of a node and the mean queue lengths.

**Marginal Queue Length Distributions**

We seek the probability of finding $n$ customers at node $i$, which we denote as $p_i(n, N)$, for $n = 0, 1, 2, \ldots, N$. We have

$$p_i(n, N) = \sum_{\substack{S(N,M) \\ n_i = n}} P(n_1, \ldots, n_{i-1}, n, n_{i+1}, \ldots, n_M) \quad \text{for } n = 0, 1, 2, \ldots, N.$$

This gives

$$p_i(n, N) = \frac{1}{G(N)} \sum_{\substack{S(N,M) \\ n_i = n}} \prod_{j=1}^{M} f_j(n_j) = \frac{f_i(n)}{G(N)} \sum_{\substack{S(N,M) \\ n_i = n}} \prod_{j=1, j \neq i}^{M} f_j(n_j).$$

To evaluate this, we use another auxilary function defined as

$$g_m^i(n) = \sum_{\substack{S(N,m) \\ n_i = N-n}} \prod_{j=1, j \neq i}^{m} f_j(n_j).$$

This is the normalization constant of a new network with node $i$ removed and only $n$ customers. Thus

$$p_i(n, N) = \frac{f_i(n)}{G(N)} g_M^i(N - n) \quad \text{for } n = 0, 1, 2, \ldots, N.$$

Notice that

$$g_M^M(n) = g_{M-1}(n) \quad \text{for } n = 0, 1, 2, \ldots, N,$$

since

$$g_M^M(n) = \sum_{\substack{S(N,M) \\ n_M=N-n}} \prod_{j=1, j\neq M}^{M} f_j(n_j) = \sum_{S(n,M-1)} \prod_{j=1}^{M-1} f_j(n_j) = g_{M-1}(n).$$

We now need an algorithm to compute this new auxilary function. Since the sum of the marginal probabilities must be equal to 1, we have

$$1 = \sum_{k=0}^{N} p_i(k, N) = \sum_{k=0}^{N} \frac{f_i(k)}{G(N)} g_M^i(N-k)$$

and therefore

$$G(N) = \sum_{k=0}^{N} f_i(k) g_M^i(N-k).$$

We may therefore write the recursive formula

$$g_M^i(N) = G(N) - \sum_{k=1}^{N} f_i(k) g_M^i(N-k). \tag{15.10}$$

Observe that this formula also holds when $N$ is replaced by $n$ for any permissible value of $n$. The values of the auxilary function may be computed iteratively using the initial conditions

$$g_M^i(0) = G(0) = 1, \quad f_i(0) = 1, \quad f_i(k) = \frac{v_i}{\mu_i(k)} f_i(k-1).$$

Some simplifications are possible in the load-independent case. We have

$$\begin{aligned} g_M^i(n) &= G(n) - \sum_{k=1}^{n} f_i(k) g_M^i(n-k) \\ &= G(n) - \sum_{k=0}^{n-1} f_i(k+1) g_M^i(n-1-k) \\ &= G(n) - Y_i \sum_{k=0}^{n-1} f_i(k) g_M^i(n-1-k) \\ &= G(n) - Y_i G(n-1) \end{aligned}$$

with $G(j) = 0$ when $j < 0$. To summarize, for a load-independent node we have

$$p_i(n, N) = \frac{Y_i^n}{G(N)} \left[ G(N-n) - Y_i G(N-1-n) \right], \tag{15.11}$$

with $G(j) = 0$ when $j < 0$, while for a load-dependent node we have

$$p_i(n, N) = \frac{f_i(n)}{G(N)} g_M^i(N-n). \tag{15.12}$$

**Example 15.5** Continuation of Example 15.4. We now compute the marginals for the central server node. To do so we first need to compute $g_M^1(n)$ for $n = 0, 1, 2, 3$. From Equation (15.10), with

$M = 4$, $i = 1$, and $N$ replaced with $n$ which takes values from 0 through 3, we obtain

$$\begin{aligned}
g_4^1(0) &= G(0) = 1, \\
g_4^1(1) &= G(1) - f_1(1)g_4^1(0) \\
&= 2.4 - 1 \times 1 = 1.4, \\
g_4^1(2) &= G(2) - f_1(1)g_4^1(1) - f_1(2)g_4^1(0) \\
&= 3.43 - 1 \times 1.4 - .5 \times 1 = 1.53, \\
g_4^1(3) &= G(3) - f_1(1)g_4^1(2) - f_1(2)g_4^1(1) - f_1(3)g_4^1(0) \\
&= 4.05 - 1 \times 1.53 - .5 \times 1.4 - .25 \times 1 = 1.57.
\end{aligned}$$

We are now ready to compute the distribution of the number of customers at the central server. Setting $M = 4$ and $i = 1$ in Equation (15.12) we obtain

$$\begin{aligned}
p_1(0) &= \frac{f_1(0)}{G(3)} g_4^1(3) = \frac{1}{4.05} 1.57 = .3877, \\
p_1(1) &= \frac{f_1(1)}{G(3)} g_4^1(2) = \frac{1}{4.05} 1.53 \; = .3778, \\
p_1(2) &= \frac{f_1(2)}{G(3)} g_4^1(1) = \frac{.5}{4.05} 1.4 \; = .1728, \\
p_1(3) &= \frac{f_1(3)}{G(3)} g_4^1(0) = \frac{.25}{4.05} 1.0 \; = .0617.
\end{aligned}$$

**Throughput of a Node**

The throughput of a node is defined as the rate at which customers leave that node. Let $X_i(N)$ denote the throughput of node $i$ in a network with $N$ customers. Then

$$\begin{aligned}
X_i(N) &= \sum_{n=1}^{N} p_i(n, N)\mu_i(n) = \sum_{n=1}^{N} \frac{f_i(n)}{G(N)} g_M^i(N-n)\mu_i(n) \\
&= \sum_{n=1}^{N} \frac{v_i}{\mu_i(n)} \frac{f_i(n-1)}{G(N)} g_M^i(N-n)\mu_i(n) \\
&= \frac{v_i}{G(N)} \sum_{n=1}^{N} f_i(n-1) g_M^i(N-n) = \frac{v_i}{G(N)} \sum_{n=0}^{N-1} f_i(n) g_M^i(N-1-n) \\
&= v_i \frac{G(N-1)}{G(N)}.
\end{aligned}$$

In a closed queueing network, the overall throughput of the network is defined as the throughput of the node for which the visit ratio is equal to 1. Thus

$$X(N) = X_1(N) = \frac{X_i(N)}{v_i}.$$

It then follows that

$$X(N) = \frac{G(N-1)}{G(N)}. \tag{15.13}$$

**Example 15.6** Continuing with Example 15.4, let us compute the throughput of the three load-independent nodes. We have

$$X_2 = .5\frac{3.43}{4.05} = .4235,$$

$$X_3 = .3\frac{3.43}{4.05} = .2541,$$

$$X_4 = .2\frac{3.43}{4.05} = .1694.$$

The overall throughput of the network is

$$X = X_1 = \frac{X_i}{v_i} = \frac{3.43}{4.05} = 0.85.$$

**Utilization of a Node**

Let $U_i(N)$ denote the utilization of node $i$ in a network with $N$ customers. Then

$$U_i(N) = \sum_{n=1}^{N} p_i(n, N) = 1 - p_i(0, N)$$

$$= 1 - \frac{g_M^i(N)}{G(N)}.$$

There is a simpler formula available for a load-independent node. Buzen has shown that in this case

$$U_i(N) = \frac{1}{\mu_i} X_i(N)$$

which implies that, for a load-independent node,

$$U_i(N) = Y_i \frac{G(N-1)}{G(N)}.$$

**Example 15.7** Continuation of Example 15.4. The utilizations of the three load-independent nodes are given by

$$U_2 = \frac{X_2}{\mu_2} = \frac{.4235}{.5} = .8470,$$

$$U_3 = \frac{X_3}{\mu_3} = \frac{.2541}{1} = .2541,$$

$$U_4 = \frac{X_4}{\mu_4} = \frac{.1694}{2} = .0847.$$

**Mean Number of Customers at a Node**

The mean number of customers at node $i$ in a network with $N$ customers is given by

$$\bar{n}_i(N) = \sum_{n=1}^{N} n p_i(n, N).$$

This is most easily evaluated for the case of load-independent nodes, where it may be shown that the mean can be computed from the simple recursive formula

$$\bar{n}_i(N) = U_i(N)\,[1 + \bar{n}_i(N-1)] \tag{15.14}$$

with $\bar{n}_i(0) = 0$. We prove this in two steps. First we show that

$$\bar{n}_i(N) = \sum_{n=1}^{N} Y_i^n \frac{G(N-n)}{G(N)} \tag{15.15}$$

and then, from this equation, we obtain the recursive formula, Equation (15.14). To prove Equation (15.15) we first define

$$Q_i(n) = \sum_{\substack{S(N,M) \\ n_i \geq n}} P(n_1, n_2, \ldots, n_M).$$

Observe that the quantity we require may be written in terms of $Q_i(n)$, since

$$\bar{n}_i(N) = \sum_{n=1}^{N} Q_i(n).$$

From the definition of $Q_i(n)$ we have

$$\begin{aligned}
Q_i(n) &= \sum_{\substack{S(N,M) \\ n_i \geq n}} \frac{1}{G(N)} \prod_{j=1}^{M} f_j(n_j) \\
&= \frac{f_i(n)}{G(N)} \sum_{S(N-n,M)} \prod_{j=1}^{M} f_j(n_j) \\
&= \frac{Y_i^n}{G(N)} G(N-n),
\end{aligned}$$

and Equation (15.15) immediately follows. We now derive the recursive equation. We have

$$\begin{aligned}
\bar{n}_i(N) = \sum_{n=1}^{N} Q_i(n) &= \sum_{n=1}^{N} Y_i^n \frac{G(N-n)}{G(N)} \\
&= Y_i \frac{G(N-1)}{G(N)} + \sum_{n=2}^{N} Y_i^n \frac{G(N-n)}{G(N)} \\
&= U_i(N) + \frac{Y_i}{G(N)} \sum_{n=1}^{N-1} Y_i^n G(N-1-n) \\
&= U_i(N) + U_i(N) \sum_{n=1}^{N-1} Y_i^n \frac{G(N-1-n)}{G(N-1)} \\
&= U_i(N)\,[1 + \bar{n}_i(N-1)]\,,
\end{aligned}$$

which is the result we seek.

The computation is messier for load-dependent servers. It must be evaluted from its definition. However, it can be sped up by performing it in conjunction with the evaluation of the normalizing constant, if we allow it to be the final (i.e., $M^{\text{th}}$) node. We have

$$\bar{n}_M(N) = \sum_{n=1}^{N} n p_M(n, N) = \sum_{n=1}^{N} n \frac{f_M(n)}{G(N)} g_M^M(N-n)$$
$$= \frac{1}{G(N)} \sum_{n=1}^{N} n f_M(n) g_{M-1}(N-n).$$

Notice the similarity between this summation and the last step in computing $G(N)$. Recall that for $G(N)$ we have

$$G(N) = \sum_{n=0}^{N} f_M(n) g_{M-1}(N-n).$$

**Example 15.8** Continuation of Example 15.4. We now find the mean number of customers at the three load independent nodes in the central server model. We shall use Equation (15.15), which we may now write as

$$\bar{n}_i = Y_i \frac{G(2)}{G(3)} + Y_i^2 \frac{G(1)}{G(3)} + Y_i^3 \frac{G(0)}{G(3)}.$$

This gives

$$\bar{n}_2 = \left(\frac{.5}{.5}\right) \times \frac{3.43}{4.05} + \left(\frac{.5}{.5}\right)^2 \times \frac{2.4}{4.05} + \left(\frac{.5}{.5}\right)^3 \times \frac{1}{4.05} = \frac{6.83}{4.05} = 1.6864,$$

$$\bar{n}_3 = \left(\frac{.3}{1}\right) \times \frac{3.43}{4.05} + \left(\frac{.3}{1}\right)^2 \times \frac{2.4}{4.05} + \left(\frac{.3}{1}\right)^3 \times \frac{1}{4.05} = \frac{1.2721}{4.05} = 0.3141,$$

$$\bar{n}_4 = \left(\frac{.2}{2}\right) \times \frac{3.43}{4.05} + \left(\frac{.2}{2}\right)^2 \times \frac{2.4}{4.05} + \left(\frac{.2}{2}\right)^3 \times \frac{1}{4.05} = \frac{0.3681}{4.05} = 0.0909.$$

## 15.4 Mean Value Analysis for Closed Queueing Networks

The convolution algorithm provides a means of computing the complete set of marginal probabilities of each node in a product-form queueing network. The disadvantage of this method is that it requires forming the normalization constant and this can be numerically unstable, as well as time consuming. Furthermore, in some circumstances, the complete set of marginal probabilities is not needed; all that is required in many cases are mean values, such as the mean number of customers at each node. The *mean value analysis (MVA)* approach allows us to compute mean values without having to first compute a normalization constant. This approach is based on the *arrival theorem* which was proved independently by Lavenberg and Reiser [28] and by Sevcik and Mitrani [49]. We state this theorem without proof.

**Theorem 15.4.1 (Arrival theorem)** *In a closed queueing network, the stationary distribution observed by a customer arriving at a node of the network (having left one node and about to enter another) is equal to the stationary distribution of customers in the same network with one fewer customer.*

In other words, at the moment of arrival to a node, a customer in a closed queueing network containing $k$ customers sees the same stationary distribution of customers over the nodes of the network that an external observer sees when the network contains only $k - 1$ customers. This theorem hints at a recursion based on the number of customers in the network. Beginning with an analysis of the network with a single customer, we shall seek to successively add customers one at a time until we have the results for the network with the required number of customers, $N$. The arrival theorem by itself is insufficient to allow us to do this. We also need to use Little's law, applied to the overall network as well as to the individual nodes of the network. Thus, the MVA approach is based on

1. the arrival theorem—to compute the average response time at each node;
2. Little's law applied to the entire network—to compute the overall throughput of the network and from it, the throughput of the individual nodes;
3. Little's law applied to each node—to compute the mean number of customers at each node.

We begin by considering a closed queueing network in which each of the $M$ service nodes contains a single FCFS exponential server. The service rate at node $i$ is given by $\mu_i$. We recall the following notation for the network at steady state and containing $k$ customers:

- $L_i(k) = E[N_i(k)]$ is the mean number of customers at node $i$, for $i = 1, 2, \ldots, M$.
- $R_i(k)$ is the response time at node $i$, for $i = 1, 2, \ldots, M$. It is equal to the time spent queueing plus the time being served.
- $X_i(k)$ is the throughput at node $i$, for $i = 1, 2, \ldots, M$.
- $v_i$ is the visit ratio for node $i$, for $i = 1, 2, \ldots, M$. It gives the number of times node $i$ is visited between two successive visits to node 1.

Consider now what the arrival theorem tells us at the instant a customer, departing from some node in the network, arrives at node $i$. This customer observes a certain number of other customers already present. The arrival theorem tells us how many. The mean number already present at the moment of arrival is equal to the mean number at this node when only $k - 1$ customers circulate in the network. This mean number is given by $L_i(k - 1)$. Thus the arriving customer finds $L_i(k - 1)$ customers already present, all of which will be served before the arriving customer, a result of the FCFS scheduling policy. The arriving customer will wait in the queue until all are served, (i.e., $L_i(k - 1)/\mu_i$) and will then spend a time period equal to $1/\mu_i$ being served itself. The average response time (per visit to node $i$) for the arriving customer is equal to the sum of these, i.e.,

$$R_i(k) = \frac{1}{\mu_i}(L_i(k-1) + 1), \quad i = 1, 2, \ldots, M. \tag{15.16}$$

Although it is not intuitively obvious, this result which was derived for exponential servers and FCFS scheduling, is also applicable to PS nodes and to LCFS-PR nodes and in these cases, the service time distributions need not be exponential. For IS nodes, no customer waits for service and in this case the average response time is just equal to the average service time:

$$R_i(k) = \frac{1}{\mu_i}, \quad i = 1, 2, \ldots, M.$$

We now apply Little's law to the entire network. The mean number of customers in the network is simply the fixed number of customers that circulate in the network, i.e., $k$. By Little's law this must be equal to the product of the arrival rate and the mean response time. Since the network is in steady state, the arrival rate is equal to the overall throughput of the network i.e., the throughput of node 1 which is $X_1(k) = X(k)$. The mean response time is equal to $\sum_{i=1}^{M} v_i R_i(k)$. Expressing the throughput in terms of the two other quantities, we have

$$X(k) = \frac{k}{\sum_{i=1}^{M} v_i R_i(k)}. \tag{15.17}$$

Observe that the throughput of the individual nodes may be computed from this overall throughput. We have

$$X_i(k) = v_i X(k), \quad i = 1, 2, \ldots, M.$$

Finally, we need to apply Little's law to the individual nodes of the network. Equating the arrival rate to the throughput, $X_i(k)$, at node $i$, we have

$$L_i(k) = X_i(k)R_i(k) = X(k)v_i R_i(k), \quad i = 1, 2, \ldots, M. \tag{15.18}$$

With these new values of $L_i(k)$, we may return to equation (15.16) and begin all over again, with $k$ incremented by one. Thus, the MVA algorithm is obtained by recursively applying Equations (15.16), (15.17), and (15.18) beginning with the boundary conditions

$$L_i(0) = 0, \quad i = 1, 2, \ldots, M.$$

In this manner we may compute the mean response time, the mean number of customers, as well as the throughput and utilization at each node of the network. What we cannot compute are the marginal distributions at each node. For this we need the normalization constant $G(N)$. However, using the results obtained previously during our discussion of the convolution algorithm, we saw that the throughput could be written as

$$X(k) = \frac{G(k-1)}{G(k)}, \quad \text{i.e.,} \quad G(k) = \frac{G(k-1)}{X(k)}.$$

This may be incorporated into the second step of the MVA algorithm and the normalization constant computed recursively using the initial condition $G(0) = 1$. Once the normalization constant has been obtained, the marginals follow directly from Equation (15.11). We illustrate these concepts by means of the following example.

**Example 15.9** Consider the closed queueing network of the central server type illustrated in Figure 15.10. Node 1 (the central server) is an infinite server with service rate $\mu_1 = 2.0$. Customers leaving the central server go to node 2, 3, or 4 with probability 0.2, 0.3, and 0.5, respectively. These nodes have a single exponential server which provide service at rates $\mu_2 = 1.0$, $\mu_3 = 2.0$, and $\mu_4 = 4.0$ respectively. Customers leaving nodes 2, 3, and 4 return to the central server.

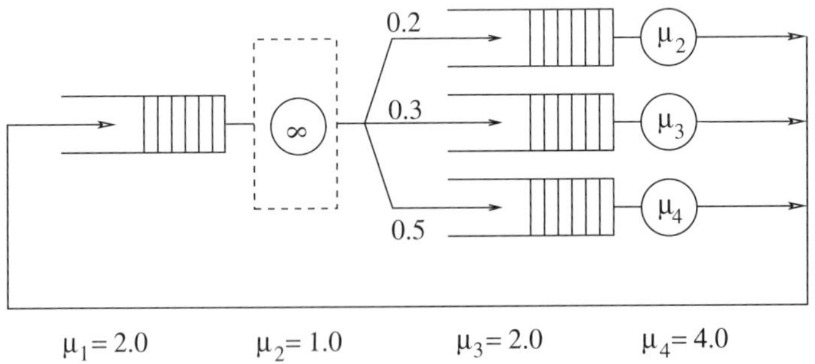


Figure 15.10. The central server model of Example 15.9.

We begin the recursive process using Equation (15.16) with $k = 1$ and obtain

$$R_1(1) = \mu_1^{-1} = 0.5,$$
$$R_2(1) = \mu_2^{-1}(L_2(0) + 1) = \mu_2^{-1}(0 + 1) = 1.0,$$
$$R_3(1) = \mu_3^{-1}(L_3(0) + 1) = \mu_3^{-1}(0 + 1) = 0.5,$$
$$R_4(1) = \mu_4^{-1}(L_4(0) + 1) = \mu_4^{-1}(0 + 1) = 0.25.$$

We next use Equation (15.17) to compute the overall throughput. We have

$$X(1) = \frac{1}{\sum_{i=1}^{4} v_i R_i(1)} = \frac{1}{1.0 \times 0.5 + 0.2 \times 1.0 + 0.3 \times 0.5 + 0.5 \times 0.25} = \frac{1}{0.975} = 1.0256.$$

At this point we may now compute $G(1)$. We have

$$G(1) = \frac{G(0)}{X(1)} = \frac{1}{1.0256} = 0.975.$$

The final calculation for this first round is to use Equation (15.18) to determine the mean customer population at each node. We have

$$L_1(1) = X(1)v_1 R_1(1) = 1.0256 \times 1.0 \times 0.5 = 0.5128,$$

$$L_2(1) = X(1)v_2 R_2(1) = 1.0256 \times 0.2 \times 1.0 = 0.2051,$$

$$L_3(1) = X(1)v_3 R_3(1) = 1.0256 \times 0.3 \times 0.5 = 0.1539,$$

$$L_4(1) = X(1)v_4 R_4(1) = 1.0256 \times 0.5 \times 0.25 = 0.1282.$$

We are now ready to begin the second round. Returning to Equation (15.16), this time using $k = 2$, we obtain

$$R_1(2) = \mu_1^{-1} = 0.5,$$

$$R_2(2) = \mu_2^{-1}(L_2(1) + 1) = \mu_2^{-1}(0.2051 + 1) = 1.2051,$$

$$R_3(2) = \mu_3^{-1}(L_3(1) + 1) = \mu_3^{-1}(0.1539 + 1) = 0.5770,$$

$$R_4(2) = \mu_4^{-1}(L_4(1) + 1) = \mu_4^{-1}(0.1282 + 1) = 0.2821.$$

Again we use Equation (15.17) to compute the overall throughput. We have

$$\begin{aligned} X(2) &= \frac{2}{\sum_{i=1}^{4} v_i R_i(2)} = \frac{2}{1.0 \times 0.5 + 0.2 \times 1.2051 + 0.3 \times 0.5770 + 0.5 \times 0.2821} \\ &= \frac{2}{1.0552} = 1.8954. \end{aligned}$$

At this point we may now compute $G(2)$. We have

$$G(2) = \frac{G(1)}{X(2)} = \frac{0.975}{1.8954} = 0.5144.$$

The final calculation for this second round is to use Equation (15.18) to determine the mean population size at each node. We have

$$L_1(2) = X(2)v_1 R_1(2) = 1.8954 \times 1.0 \times 0.5000 = 0.9477,$$

$$L_2(2) = X(2)v_2 R_2(2) = 1.8954 \times 0.2 \times 1.2051 = 0.4568,$$

$$L_3(2) = X(2)v_3 R_3(2) = 1.8954 \times 0.3 \times 0.5770 = 0.3281,$$

$$L_4(2) = X(2)v_4 R_4(2) = 1.8954 \times 0.5 \times 0.2821 = 0.2674.$$

The process continues in this fashion until the network contains the desired number of customers. Tables 15.1 and 15.2 show the results obtained after one more step, $k = 3$ and with $k = 10$.

It is interesting to observe that the size of the normalization constant becomes smaller with large population sizes. Indeed, if we continue the experiments, we find that $G(20) = 1.3462\text{E–}12$ and $G(50) = 1.4631\text{E–}33$. As we mentioned previously, the computation of the normalization constant can be computationally unstable.

Table 15.1. Performance measures for $k = 3$.

| Service center | 1 | 2 | 3 | 4 |
|---|---|---|---|---|
| Mean number of customers $L_i(3)$ | 1.3055 | 0.7608 | 0.52013 | 0.4136 |
| Mean response time $R_i(3)$ | 0.5000 | 1.4569 | 0.6640 | 0.3168 |
| Throughput $X_i(3)$ | 2.6110 | 0.5222 | 0.7833 | 1.3055 |
| Normalization constant $G(3)$ | 0.1970052083333333 | | | |

Table 15.2. Performance measures for $k = 10$.

| Service center | 1 | 2 | 3 | 4 |
|---|---|---|---|---|
| Mean number of customers $L_i(10)$ | 2.3347 | 4.4554 | 1.9305 | 1.2794 |
| Mean response time $R_i(10)$ | 0.5000 | 4.7709 | 1.3782 | 0.5480 |
| Throughput $X_i(10)$ | 4.6693 | 0.9339 | 1.4008 | 2.3347 |
| Normalization constant $G(10)$ | 1.0820657639390957E-5 | | | |

Finally, to compute the marginal distribution of customers at node 2, for example, we use the formula for a load-independent node namely,

$$p_i(n, N) = \frac{Y_i^n}{G(N)}\,[G(N-n) - Y_i G(N-1-n)]$$

with $Y_2 = .2/1 = .2$ and $G(0) = 1$, $G(1) = .975$, $G(2) = .5144$ and $G(3) = .1970$ to obtain

$$p_2(0, 3) = \frac{.2^0}{G(3)}\,[G(3) - .2G(2)] = .4778,$$

$$p_2(1, 3) = \frac{.2^1}{G(3)}\,[G(2) - .2G(1)] = .3243,$$

$$p_2(2, 3) = \frac{.2^2}{G(3)}\,[G(1) - .2G(0)] = .1574,$$

$$p_2(3, 3) = \frac{.2^3}{G(3)}\,[G(0)] = .0406.$$

**MVA and Load-Dependent Service Centers**

There remains one case that we have not yet discussed, the case of load-dependent nodes. This is the case that arises when a node contains $c > 1$ identical servers, for example. In our discussions of the convolution algorithm, we saw that the analysis was somewhat more complex in this case, and the same is true for the MVA algorithm. The difficulty arises in computing the mean response time in load-dependent nodes. We shall let $\mu_i \alpha_i(k)$ be the service rate when $k$ customers are present. A suitable choice of $\alpha_i(k)$ allows us to handle all possible cases. For example, setting $\alpha_i(k) = \min(c_i, k)$, is what is needed for a node with $c_i$ identical servers. Setting $\alpha_i(k) = \beta_i(k)/\mu_i$ sets the service rate to $\beta_i(k)$.

In the load-independent case, it was sufficient to know the mean number of customers present at a node to compute the mean response time, since the server works at the same rate no matter how many are present. In a load dependent node, this is no longer the case: the rate of service is a function of the number of customers present. Now it is necessary to know the probability distribution of customers. We shall let $p_i(j, k)$ denote the equilibrium probability of having $j$ customer present at node $i$ when a total of $k$ customers circulates in the network. These are the marginal probabilities of node $i$ in a network of $k$ customers.

Consider what happens when a customer arrives at load-dependent server $i$ and finds $j-1$, $j=1,2,\ldots,k$, customers already present. From the arrival theorem, we know that the probability of this happening is given by $p_i(j-1,k-1)$. Once the customer has arrived, the node will contain $j$ customers all of which must be served. With $j$ customers, the mean service time is $1/\mu_i\alpha_i(j)$. It follows then that the response time at node $i$ is given by

$$R_i(k)=\sum_{j=1}^{k}\frac{j}{\mu_i\alpha_i(j)}p_i(j-1,k-1).$$

This equation replaces Equation (15.16) when the node is load dependent. Of course this entails a knowledge of the marginal distribution of customers at a load-dependent node and it is to this that we now turn our attention. It is fortunate that these too may be computed from the marginal distribution of customers in a network with one fewer customer. From the results obtained using the convolution algorithm, Equation (15.12), we saw that the marginal distribution of customers at node $i$ is given by

$$p_i(j,k)=f_i(j)\frac{g_M^i(k-j)}{G(k)}.$$

Since

$$f_i(j)=\frac{v_i}{\mu_i\alpha_i(j)}f_i(j-1)$$

and writing $g_M^i(k-j)=g_M^i(k-1-j+1)=g_M^i((k-1)-(j-1))$, we have

$$\begin{aligned}p_i(j,k)&=\frac{v_i}{\mu_i\alpha_i(j)}f_i(j-1)\frac{g_M^i((k-1)-(j-1))}{G(k)}\\&=\frac{v_i}{\mu_i\alpha_i(j)}\frac{G(k-1)}{G(k)}f_i(j-1)\frac{g_M^i((k-1)-(j-1))}{G(k-1)}\\&=\frac{X_i(k)}{\mu_i\alpha_i(j)}p_i(j-1,k-1).\end{aligned}\tag{15.19}$$

Thus, given the throughput of a load-dependent node and its marginal probability distribution when the network has $k-1$ customers, we can compute the marginals when the network has $k$ customers. The algorithm is bootstrapped from the initial condition $p_i(0,0)=1$, which simply says that there cannot be any customers at service center $i$ if there are none anywhere in the network. For any value $k>0$, Equation (15.19) allows us to compute the probabilities $p_i(j,k)$ for values of $j=1,2,\ldots,k$. This still leaves the problem of computing $p_i(0,k)$. However, since the sum of the marginal probabilities at any node must be 1, i.e., $\sum_{j=0}^{k}p_i(j,k)=1$, we must have

$$p_i(0,k)=1-\sum_{j=1}^{k}p_i(j,k).$$

Unfortunately, this subtraction can lead to a loss of significance and is a major difficulty in implementing a robust version of the MVA algorithm. Observe that it is not necessary to use a full two-dimensional array for the computation of these probabilities. A single vector of length $N+1$ is sufficient, provided the computation is conducted from the last value to the first. This is illustrated below, where the double up-arrows indicate the positions into which the new probability will be placed. Notice that if the insertions are made from right to left, i.e., $p_i(j,k)$ is computed and inserted before $p_i(j-1, k)$, then the values needed in the computation of any $p_i(j, k)$ are not overwritten until they are no longer needed.

$$
\begin{array}{llll}
[\ p(0,0) & & & \\
\Uparrow & \Uparrow & & \\
[\ p(0,1) & p(1,1) & & \\
\Uparrow & \Uparrow & \Uparrow & \\
[\ p(0,2) & p(1,2) & p(2,2) & \\
\Uparrow & \Uparrow & \Uparrow & \Uparrow \\
[\ p(0,3) & p(1,3) & p(2,3) & p(3,3)
\end{array}
$$

We now present an example to illustrate these concepts.

**Example 15.10** We shall modify the previous example to include a load dependent server. The modified queueing network is shown in Figure 15.11.

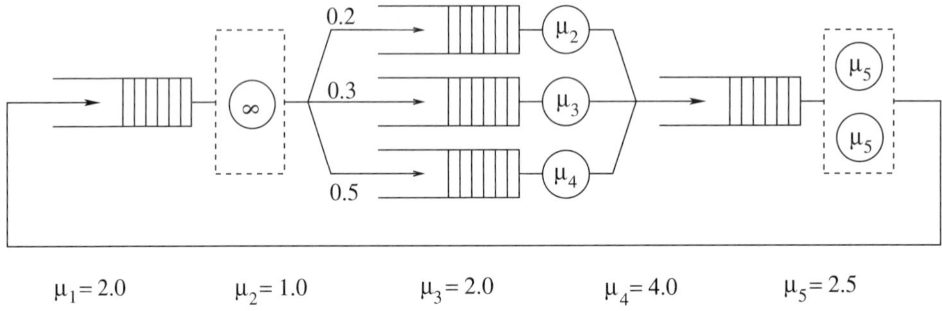


Figure 15.11. The modified queueing network of Example 15.10.

In this example, rather than returning to service center 1, customers departing from service centers 2, 3, and 4 proceed on to the new node. This node, denoted service center 5, contains $m_5 = 2$ identical exponential servers with parameter $\mu_5 = 2.5$. Thus the load dependent service rate is given by $\mu(j) = \mu_5\alpha_5(j)$ where $\alpha_5(1) = 1$ and $\alpha_5(j) = 2$ for $j > 1$. Customers leaving this service center return to service center 1. As previously, service center 1 is an infinite server with service rate $\mu_1 = 2.0$. Customers leaving this node go to nodes 2, 3, or 4 with probabilities 0.2, 0.3, and 0.5 respectively. These nodes have a single exponential server which provide service at rates $\mu_2 = 1.0$, $\mu_3 = 2.0$, and $\mu_4 = 4.0$ respectively.

We begin the recursive process with $k = 1$. From Equation (15.16), we have

$$R_1(1) = \mu_1^{-1} = 0.5,$$

$$R_2(1) = \mu_2^{-1}(L_2(0) + 1) = \mu_2^{-1}(0 + 1) = 1.0,$$

$$R_3(1) = \mu_3^{-1}(L_3(0) + 1) = \mu_3^{-1}(0 + 1) = 0.5,$$

$$R_4(1) = \mu_4^{-1}(L_4(0) + 1) = \mu_4^{-1}(0 + 1) = 0.25,$$

$$R_5(1) = (\alpha_5(1)\mu_5)^{-1} p_5(0, 0) = \mu_5^{-1} = 0.4,$$

where we have used the initial condition $p_5(0, 0) = 1$. We next use Equation (15.17) to compute the overall throughput. We have

$$X(1) = \frac{1}{\sum_{i=1}^{5} v_i R_i(1)} = \frac{1}{1.0 \times 0.5 + 0.2 \times 1.0 + 0.3 \times 0.5 + 0.5 \times 0.25 + 1.0 \times 0.4} = 0.7273.$$

At this point we may now compute $G(1)$. We have

$$G(1) = \frac{G(0)}{X(1)} = \frac{1}{.7273} = 1.375.$$

The next step is to use Equation (15.18) to determine the mean number of customers at each node. We have

$$L_1(1) = X(1)v_1 R_1(1) = 0.7273 \times 1.0 \times 0.5 = 0.3636,$$

$$L_2(1) = X(1)v_2 R_2(1) = 0.7273 \times 0.2 \times 1.0 = 0.1455,$$

$$L_3(1) = X(1)v_3 R_3(1) = 0.7273 \times 0.3 \times 0.5 = 0.1091,$$

$$L_4(1) = X(1)v_4 R_4(1) = 0.7273 \times 0.5 \times 0.25 = 0.0909,$$

$$L_5(1) = X(1)v_5 R_5(1) = 0.7273 \times 1.0 \times 0.4 = 0.2909.$$

Before finishing with $k = 1$, we must first compute the marginals for node 5. In particular we need $p_5(0, 1)$ and $p_5(1, 1)$. We first compute $p_5(1, 1)$ from Equation (15.19). We have

$$p_5(1, 1) = \frac{X_5(1)}{\mu_5 \alpha_5(1)} p_5(0, 0) = \frac{0.7273}{2.5} = 0.2909.$$

and, from $p_5(0, 1) = 1 - p_5(1, 1)$, we have $p_5(0, 1) = 0.7091$.

We are now ready to begin the second round, this time with $k = 2$. We first compute the mean response time for all five nodes:

$$\begin{aligned}
R_1(2) &= \mu_1^{-1} = 0.5,\\
R_2(2) &= \mu_2^{-1}(L_2(1) + 1) = \mu_2^{-1}(0.1455 + 1) = 1.1455,\\
R_3(2) &= \mu_3^{-1}(L_3(1) + 1) = \mu_3^{-1}(0.1091 + 1) = 0.5546,\\
R_4(2) &= \mu_4^{-1}(L_4(1) + 1) = \mu_4^{-1}(0.0909 + 1) = 0.2727,\\
R_5(2) &= \frac{1}{\alpha_5(1)\mu_5} p_5(0, 1) + \frac{2}{\alpha_5(2)\mu_5} p_5(1, 1)\\
&= \frac{1}{1 \times 2.5} \times 0.7091 + \frac{2}{2 \times 2.5} \times 0.2909 = 0.4.
\end{aligned}$$

Again we use Equation (15.17) to compute the overall throughput. We have

$$\begin{aligned}
X(2) = \frac{2}{\sum_{i=1}^{5} v_i R_i(2)} &= \frac{2}{1.0 \times 0.5 + 0.2 \times 1.1455 + 0.3 \times 0.5546 + 0.5 \times 0.2727 + 1.0 \times 0.4}\\
&= 1.3968.
\end{aligned}$$

At this point we may now compute $G(2)$. We have

$$G(2) = \frac{G(1)}{X(2)} = \frac{1.375}{1.3968} = 0.9844.$$

The final calculation for this second round is to use Equation (15.18) to determine the mean population sizes. We have

$$L_1(2) = X(2)v_1 R_1(2) = 1.3968 \times 1.0 \times 0.5000 = 0.6984,$$

$$L_2(2) = X(2)v_2 R_2(2) = 1.3968 \times 0.2 \times 1.1455 = 0.3200,$$

$$L_3(2) = X(2)v_3 R_3(2) = 1.3968 \times 0.3 \times 0.5546 = 0.2324,$$

$$L_4(2) = X(2)v_4 R_4(2) = 1.3968 \times 0.5 \times 0.2727 = 0.1905,$$

$$L_5(2) = X(2)v_5 R_5(2) = 1.3968 \times 1.0 \times 0.4000 = 0.5587.$$

The marginals at node 5 may now be computed. We have, from Equation (15.19),

$$p_5(1,2) = \frac{X_5(2)}{\alpha_5(1)\mu_5} p_5(0,1) = \frac{1.3968}{1 \times 2.5} \times 0.7091 = 0.3962,$$

$$p_5(2,2) = \frac{X_5(2)}{\alpha_5(2)\mu_5} p_5(1,1) = \frac{1.3968}{2 \times 2.5} \times 0.2909 = 0.0813,$$

and hence $p_5(0,2) = 1.0 - p_5(1,2) - p_5(2,2) = 0.5225$.

We shall do one more step, now augmenting $k$ to 3. Again the first quantities to compute, from Equation (15.16), are the mean response times. We obtain

$$R_1(3) = \mu_1^{-1} = 0.5,$$

$$R_2(3) = \mu_2^{-1}(L_2(2) + 1) = \mu_2^{-1}(0.3200 + 1) = 1.3200,$$

$$R_3(3) = \mu_3^{-1}(L_3(2) + 1) = \mu_3^{-1}(0.2324 + 1) = 0.6162,$$

$$R_4(3) = \mu_4^{-1}(L_4(2) + 1) = \mu_4^{-1}(0.1905 + 1) = 0.2976,$$

$$R_5(3) = \frac{1}{\alpha_5(1)\mu_5} p_5(0,2) + \frac{2}{\alpha_5(2)\mu_5} p_5(1,2) + \frac{3}{\alpha_5(3)\mu_5} p_5(2,2)$$

$$= \frac{1}{1 \times 2.5} \times 0.5225 + \frac{2}{2 \times 2.5} \times 0.3962 + \frac{3}{2 \times 2.5} \times 0.0813 = 0.4163.$$

Again we use Equation (15.17) to compute the overall throughput. We have

$$X(3) = \frac{3}{\sum_{i=1}^{5} v_i R_i(3)} = \frac{3}{1.0 \times 0.5 + 0.2 \times 1.3200 + 0.3 \times 0.6162 + 0.5 \times 0.2976 + 1.0 \times 0.4163}$$

$$= 1.9816.$$

The normalization constant is obtained as

$$G(3) = \frac{G(2)}{X(3)} = \frac{0.9844}{1.9816} = 0.4968.$$

The final calculation for this third round is to use Equation (15.18) to determine the means:

$$L_1(3) = X(3)v_1 R_1(3) = 1.9816 \times 1.0 \times 0.5000 = 0.9908,$$

$$L_2(3) = X(3)v_2 R_2(3) = 1.9816 \times 0.2 \times 1.3200 = 0.5231,$$

$$L_3(3) = X(3)v_3 R_3(3) = 1.9816 \times 0.3 \times 0.6162 = 0.3663,$$

$$L_4(3) = X(3)v_4 R_4(3) = 1.9816 \times 0.5 \times 0.2976 = 0.2949,$$

$$L_5(3) = X(3)v_5 R_5(3) = 1.9816 \times 1.0 \times 0.4163 = 0.8249.$$

The marginals at node 5 may now be computed. We have, from Equation (15.19),

$$p_5(1,3) = \frac{X_5(3)}{\alpha_5(1)\mu_5} p_5(0,2) = \frac{1.9816}{1 \times 2.5} \times 0.5225 = 0.4142,$$

$$p_5(2,3) = \frac{X_5(3)}{\alpha_5(2)\mu_5} p_5(1,2) = \frac{1.9816}{2 \times 2.5} \times 0.3962 = 0.1570,$$

$$p_5(3,3) = \frac{X_5(3)}{\alpha_5(3)\mu_5} p_5(2,2) = \frac{1.9816}{2 \times 2.5} \times 0.0813 = 0.0322,$$

and hence $p_5(0,3) = 1.0 - p_5(1,3) - p_5(2,3) - p_5(3,3) = 0.3966$.

And so the process continues in this fashion until the network contains the desired number of customers. Table 15.3 shows the results obtained once we arrive at $k = 10$. The marginal distribution of customers at node 5, when the network contains $k = 10$ customers is shown in Table 15.4.

Table 15.3. Performance measures for $k = 10$.

| Service center | 1 | 2 | 3 | 4 | 5 |
|---|---|---|---|---|---|
| Mean number of customers $L_i(10)$ | 2.0122 | 2.6327 | 1.3338 | 0.9391 | 3.0823 |
| Mean response time $R_i(10)$ | 0.5000 | 3.2710 | 1.1047 | 0.4667 | 0.7659 |
| Throughput $X_i(10)$ | 4.0243 | 0.8049 | 1.2073 | 2.0122 | 4.0244 |
| Normalization constant $G(10)$ | | 1.0724123796354306E-4 | | | |

Table 15.4. Marginal distribution at node 5 with $k = 10$.

| $n$ | 0 | 1 | 2 | 3 | 4 | 5 | 6 | 7 | 8 | 9 | 10 |
|---|---|---|---|---|---|---|---|---|---|---|---|
| $p_5(n)$ | 0.1009 | 0.1885 | 0.1719 | 0.1519 | 0.1283 | 0.1018 | 0.0739 | 0.0470 | 0.0246 | 0.0093 | 0.0019 |

## 15.5 The Flow-Equivalent Server Method

As a final possibility for solving closed queueing networks, we briefly review the so-called "flow-equivalent server" method. This method is based on Norton's theorem from electric circuit theory and was shown to be applicable to queueing networks. In its simplest form, one node in the network is selected for analysis and the remainder of the network replaced by a single server whose flow is "equivalent" to that of the part of the network it replaces. This gives rise to a reduced system consisting of the selected node and the *flow-equivalent* server. Norton's theorem may now be applied to show that the behavior of the selected node is identical in both the reduced network and in the original network. The flow-equivalent server has load-dependent service rates, $\mu(k)$, $k = 1, 2, \ldots, N$ where $N$ is the number of customers who circulate in the network. These rates are determined by short-circuiting the selected node in the network (simply set its mean service time to zero) and computing the throughput across that short-circuited path. This throughput is computed for each possible nonzero number of customers in the network, $k = 1, 2, \ldots, N$ and provides the values for $\mu(k)$, $k = 1, 2, \ldots, N$. The visit ratio of both nodes in the reduced network is taken to be equal to the visit ratio of the selected node in the original network. The normalization constant, $G(N)$, of the reduced network is also the normalization constant of the original network and now Equation (15.4) can be used to find the stationary probability of any state of the queueing network. These concepts are best illustrated by means of an example.

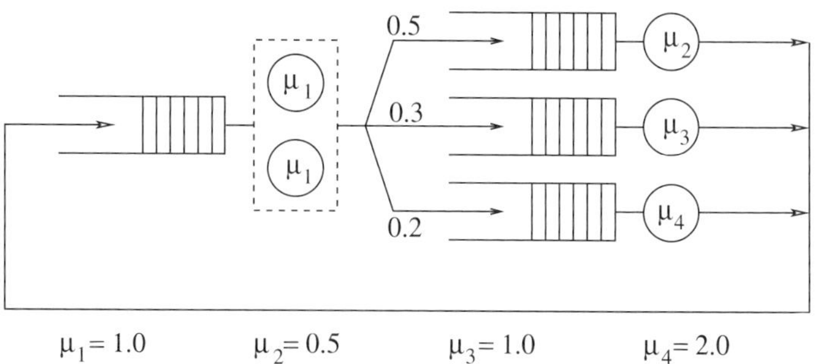


Figure 15.12. A central server queueing model.

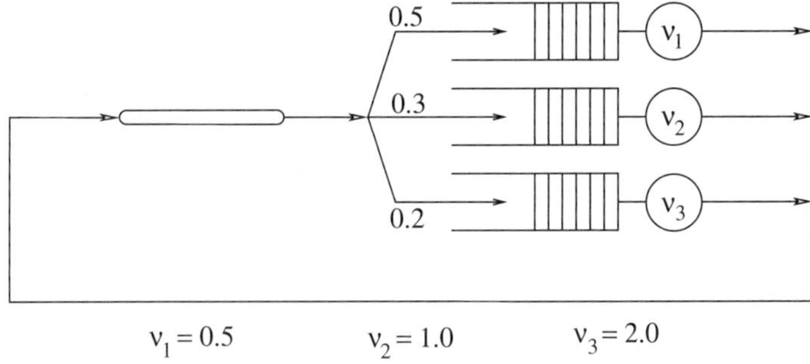


Figure 15.13. The short-circuited model.

**Example 15.11** We shall use the model of Example 15.4 which is illustrated (again) in Figure 15.12. In this model, the central server (node 1 in the diagram) possesses two exponential servers each having rate $\mu = 1.0$. The three nodes that feed the central server (nodes 2, 3, and 4 in the diagram) each possess a single server providing exponential service at rates 0.5, 1.0, and 2.0 respectively. On exiting the central server, customers go to node $i = 2, 3, 4$ with probabilities 0.5, 0.3, and 0.2, respectively. We assume that $N = 3$ customers circulate in this network.

We choose to compute the marginal distribution at the central server using the flow-equivalent method so our first step is to short-circuit this node and compute the throughput across its path for $k = 1, 2, 3$ customers. The short-circuited network is shown in Figure 15.13.

Either the convolution algorithm or MVA approach is possible and here we go with the Buzen's convolution algorithm. The visit ratios are given by $\gamma_1 = .5$, $\gamma_2 = .3$, and $\gamma_3 = .2$ and we have $\nu_1 = .5$, $\nu_2 = 1$, and $\nu_3 = 2$. Since the nodes are all load independent, we use the formula

$$f_i(n_i) = \left(\frac{\gamma_i}{\nu_i}\right)^{n_i},$$

which allows us to compute

$$f_1(n_1) = 1^{n_1}, \quad f_2(n_2) = .3^{n_2}, \quad f_3(n_3) = .1^{n_3}.$$

Progressing through the computation of the normalization constant by Buzen's algorithm, we find the successive columns of $G(n)$ to be

| | | |
|---|---|---|
| 1 | 1 | 1 |
| 1 | 1.3 | 1.4 |
| 1 | 1.39 | 1.53 |
| 1 | 1.417 | 1.57 |

$$G(0) = 1, \quad G(1) = 1.4, \quad G(2) = 1.53, \quad G(3) = G(N) = 1.57,$$

so that we can compute the throughput with one, two, and three customers in the network as

$$X(1) = G(0)/G(1) = .7143, \quad X(2) = G(1)/G(2) = .9150, \quad X(3) = G(2)/G(3) = .9745.$$

These constitute the load-dependent rates at the flow-equivalent server. Note carefully that had the visit ratio to the central server in the original network been different from 1, then these rates would have to be multiplied by that visit ratio.

We are now ready to analyze the two-node reduced model. This is shown in Figure 15.14. In this model the visit ratios are both equal to 1 and the service rates are

$$\mu_1(1) = 1, \quad \mu_1(2) = 2, \quad \mu_1(3) = 2, \quad \mu_2(1) = .7143, \quad \mu_2(2) = .9150, \quad \mu_2(3) = .9745.$$

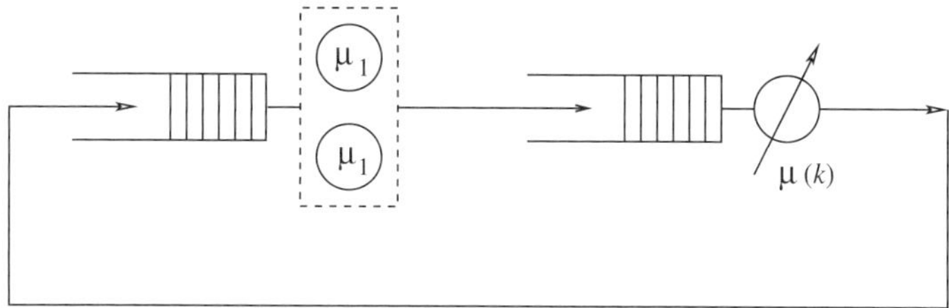


Figure 15.14. The reduced model.

We now form the quantities $f_i(n_i)$ for $i = 1, 2$ and $n_i = 0, 1, 2, 3$. Both nodes being load dependent, we use

$$f_i(n_i) = \frac{v_i^{n_i}}{\prod_{k=1}^{n_i} \mu_i(k)}.$$

This gives

$$f_1(0) = 1,\; f_1(1) = \frac{v_1}{\mu_1(1)} = 1,\; f_1(2) = \frac{v_1^2}{\mu_1(1)\mu_1(2)} = 1/2,\; f_1(3) = \frac{v_1^3}{\mu_1(1)\mu_1(2)\mu_1(3)} = 1/4,$$

$$f_2(0) = 1,\; f_2(1) = \frac{v_2}{\mu_2(1)} = 1.4,\; f_2(2) = \frac{v_2^2}{\mu_2(1)\mu_2(2)} = 1.53,\; f_2(3) = \frac{v_2^3}{\mu_2(1)\mu_2(2)\mu_2(3)} = 1.57.$$

We now begin the computation of the normalizing constant by means of

$$g_m(n) = \sum_{k=0}^{n} f_m(k) g_{m-1}(n-k), \quad \text{load-dependent case,}$$

and we obtain the following table.

| | $m = 1$ | $m = 2$ |
|---|---|---|
| $n = 0$ | 1 | 1 |
| $n = 1$ | $f_1(1) = 1$ | $1 \times 1 + 1.4 \times 1 = 2.4$ |
| $n = 2$ | $f_1(2) = .5$ | $1 \times .5 + 1.4 \times 1 + 1.53 \times 1 = 3.43$ |
| $n = 3$ | $f_1(3) = .25$ | $1 \times .25 + 1.4 \times .5 + 1.53 \times 1 + 1.57 \times 1 = 4.05$ |

Thus we have

$$G(0) = 1, \quad G(1) = 2.4, \quad G(2) = 3.43, \quad G(N) = G(3) = 4.05.$$

Observe that the computed value of $G(N)$ is the same as that obtained for the original network in Example 15.4. Our goal is to compute the marginal distribution at the central server which means that we must first compute $g_M^1(n)$ for $n = 0, 1, 2, 3$ from

$$g_M^1(n) = G(n) - \sum_{k=1}^{n} f_1(k) g_M^1(n-k).$$

This gives

$$g_M^1(0) = G(0) = 1,$$

$$g_M^1(1) = G(1) - f_1(1)g_M^1(0) = 2.4 - 1 \times 1 = 1.4,$$

$$g_M^1(2) = G(2) - f_1(1)g_M^1(1) - f_1(2)g_M^1(0) = 3.43 - 1 \times 1.4 - .5 \times 1 = 1.53,$$

$$g_M^1(3) = G(3) - f_1(1)g_M^1(2) - f_1(2)g_M^1(1) - f_1(3)g_M^1(0) = 4.05 - 1 \times 1.53 - .5 \times 1.4 - .25 \times 1 = 1.57.$$

It follows that the marginals at the central server are

$$p_i(n, N) = \frac{f_i(n)}{G(N)} g^i_M(N-n) \quad \text{for } n = 0, 1, 2, 3,$$

which gives

$$p_1(0, N) = \frac{f_1(0)}{G(N)} g^1_M(3) = \frac{1}{4.05} \times 1.57 = .3877,$$

$$p_1(1, N) = \frac{f_1(1)}{G(N)} g^1_M(2) = \frac{1}{4.05} \times 1.53 = .3778,$$

$$p_1(2, N) = \frac{f_1(2)}{G(N)} g^1_M(1) = \frac{.5}{4.05} \times 1.4 = .1728,$$

$$p_1(3, N) = \frac{f_1(3)}{G(N)} g^1_M(0) = \frac{.25}{4.05} \times 1 = .0617,$$

and is identical to the distribution obtained in Example 15.4.

In the example we just studied, a single node was selected for analysis and the rest of the network was replaced by one flow-equivalent server. This allowed the chosen node to be analyzed using a reduced two-node tandem network. However, the flow-equivalent method has much wider applicability: it can be used when a subset of nodes is chosen for analysis. In this case all the nodes in the selected set are short circuited and the throughput within the remainder of the network computed for all $k = 1, 2, \ldots, N$. These throughputs constitute the service rates for a flow equivalent server which replaces the nonselected nodes in the network. An analysis of this reduced network, which need no longer be a tandem network, provides all required performance measures concerning the selected nodes.

The flow equivalent method does not have any advantages either in computation time or in memory requirements over the convolution algorithm or MVA when it is used to analyze a closed queueing network with a given fixed set of parameters. The real advantages of this method become apparent in two cases:

1. When the parameters of one node (or a selected subset of nodes) are to be varied while keeping the parameters of the remaining nodes of the network fixed. Rather than analyzing the entire network for each different set of parameters of the chosen node(s), the original network can be replaced with a smaller reduced network in which nodes with nonvarying parameters are replaced by a single flow equivalent server. Once the service rates of the flow equivalent server have been computed, all parameter-varying experiments can be conducted on the reduced network thereby resulting in saving in computation time. The performance measures for the selected node(s) are the same in the original network and in the reduced network.
2. The flow-equivalent server method provides the basis for a large number of approximation methods. One such approach is Marie's method which, in some respects, is similar to an iterated form of the flow-equivalent server approach.

## 15.6 Multiclass Queueing Networks and the BCMP Theorem

The convolution and MVA algorithms discussed previously in the context of closed single-class queueing networks form the basis of numerous approaches to solving more complex models such as queueing networks with multiple classes of customer, load-dependent arrival rates and routing probabilities, and, in some cases, having general service time distributions. As its name suggests, a multiclass queueing network allows customers to belong to different classes which may have different routing probabilities among the nodes of the network. Furthermore, at the same service facility, customers belonging to one class may be provided service at a different rate from customers of a different class. When some of the classes are *open*, meaning that customers of that class enter the network from the exterior and eventually depart to the exterior, and others are *closed*, the term *mixed* queueing network is used. In some networks customers may change from one class to another as they move from one node to another. All these possibilities come together in the theorem of Baskett, Chandy, Muntz, and Palacios, the so-called BCMP theorem, which holds that a network with one or more of these properties has a product form solution: the stationary distribution of any state can be written as the product of functions $f_i(y_i)$ which depend on the number of customers of each class and on the type of service facility to which node $i$ belongs. Such networks are often referred to generically as "product-form networks."

### *15.6.1 Product-Form Queueing Networks*

In previous sections of this chapter we saw that the stationary probability of the states of single-class open and closed networks could be written as a product of the solutions of the individual nodes in the network. Thus it is not necessary to solve the global balance equations directly—the computationally easier approach is to solve for the solution of each of the individual nodes and to present the global solution as the (possibly normalized) product of the separate solutions. At that time we did not ask ourselves what it is about these particular networks that makes this approach work. This is a topic that we now address, because it will enable us to ascertain whether other, more complex, queueing networks can be handled in a similar manner. Whereas it may be feasible to generate and solve the global balance equations for relatively simple models having few nodes and a limited number of customers in the closed case, this becomes computational impossible with complex models, having many nodes, large customer populations belonging to different classes, and more advanced scheduling algorithms and general service time distributions. To proceed, we must first introduce the concept of local balance and contrast it with global balance.

**Global Balance Equations**

We assume that the Markov chain that underlies the queueing network is finite and ergodic. This Markov chain has a unique stationary distribution $\pi$ whose element $\pi_i$ is the probability of finding the Markov chain in state $i$ at equilibrium. Its infinitesimal generator, or transition rate matrix, is denoted by $Q$. Element $q_{ij}$, $i \neq j$ of this matrix gives the rate at which the Markov chain at equilibrium moves from state $i$ into state $j$ and $q_{ii} = -\sum_{j \neq i} q_{ij}$. We have previously shown that $\pi Q = 0$, or more explicitly, for the $i$th equation, that $\sum_{all\ j} \pi_j q_{ji} = 0$. Hence

$$-\pi_i q_{ii} = \sum_{j \neq i} \pi_j q_{ji}$$

or

$$\pi_i \sum_{j \neq i} q_{ij} = \sum_{j \neq i} \pi_j q_{ji}.$$

Whereas the left-hand side of this equation give the flow out of state $i$, the right-hand side is the flow into state $i$. Thus the flow into and out of state $i$ are balanced. Since this is true for all states $i$, the set of equations represented by $\pi Q = 0$ are called the *global balance equations*. The business of solving a queueing network is the business of computing the stationary probability vector $\pi$. One possibility is to compute this vector directly by solving the global balance equations. However, in all but the simplest of queueing networks, this is computationally too expensive. In seeking a more efficient approach, we are led to the topic of local balance.

### Local Balance Equations

Whereas each global balance equation equates the total flow into a state to the total flow out of that state, local balance equations identify different balanced subflows into and out of the same state. Since they are balanced, each subflow into a state is equal to the corresponding subflow out of the state. Furthermore, when summed, the subflows revert to the global balanced equation from which they were extracted. The concept of local balance was introduced by Chandy in 1972 [8].

**Definition 15.6.1 (Local balance)** *A node of a queueing network is said to have local balance when the flow out of a state due to the completion of service and subsequent departure of a customer from the node is equal to the flow into the state caused by the arrival of a customer to that same node.*

In multiclass networks, the class of the customer must be taken into account. Thus a node is said to have *local balance* when the flow out of a state due to the service completion of a customer of class $r$ at the node is equal to the flow into that state due to the arrival of a class $r$ customer to that node.

**Example 15.12** Consider the simple central server model shown in Figure 15.15 in which $p_1 + p_2 = 1$. Assume that only two customers circulate in this network. Then there are six states, namely, $(2, 0, 0)$, $(1, 1, 0)$, $(1, 0, 1)$, $(0, 2, 0)$, $(0, 1, 1)$, and $(0, 0, 2)$, where the state $(n_1, n_2, n_3)$ has $n_1$ customers at the central server and $n_2$ and $n_3$ customers at nodes 2 and 3, respectively.

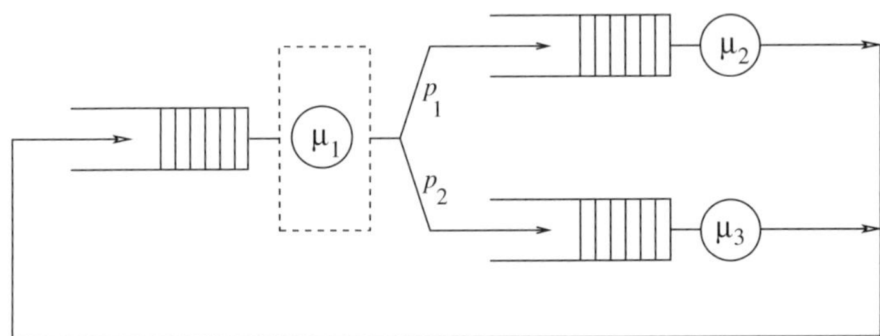


Figure 15.15. A central server queueing model.

The global balance equations are

$$\pi Q = (\pi_1, \pi_2, \pi_3, \pi_4, \pi_5, \pi_6)\begin{pmatrix} -\mu_1 & p_1\mu_1 & p_2\mu_1 & 0 & 0 & 0 \\ \mu_2 & -(\mu_1+\mu_2) & 0 & p_1\mu_1 & p_2\mu_1 & 0 \\ \mu_3 & 0 & -(\mu_1+\mu_3) & 0 & p_1\mu_1 & p_2\mu_1 \\ 0 & \mu_2 & 0 & -\mu_2 & 0 & 0 \\ 0 & \mu_3 & \mu_2 & 0 & -(\mu_2+\mu_3) & 0 \\ 0 & 0 & \mu_3 & 0 & 0 & -\mu_3 \end{pmatrix} = 0.$$

Consider the second global balance equation:

$$(\mu_1 + \mu_2)\pi_2 = p_1\mu_1\pi_1 + \mu_2\pi_4 + \mu_3\pi_5.$$

This separates into the following two local balance equations:

$$\mu_2\pi_2 = p_1\mu_1\pi_1,$$
$$\mu_1\pi_2 = \mu_2\pi_4 + \mu_3\pi_5.$$

In order to clarify particular states, we set $\pi_1 \equiv \pi_{(2,0,0)}$, $\pi_2 \equiv \pi_{(1,1,0)}$, $\ldots$, $\pi_6 \equiv \pi_{(0,0,2)}$. The first local balance equation is now written as

$$\mu_2\pi_{(1,1,0)} = p_1\mu_1\pi_{(2,0,0)},$$

and it is apparent that this equates the flow rate out of state $(1, 1, 0)$ due to a departure (at rate $\mu_2$) from node 2 with the flow into state $(1, 1, 0)$ due to an arrival (at rate $p_1\mu_1$) into node 2 from node 1.

The second of the two local balance equations, now written as

$$\mu_1\pi_{(1,1,0)} = \mu_2\pi_{(0,2,0)} + \mu_3\pi_{(0,1,1)},$$

shows that the flow out of state $(1, 1, 0)$ due to a departure from node 1 is equal to the flow into state $(1, 1, 0)$ due to an arrival at node 1. It may also be shown that global balance equations 3 and 5 can each be separated into two local balance equations, while the remaining three global balance equations are themselves local balance equations. It is a useful exercise to draw the transition rate diagram for the states of this queueing network and to identify on it, for each global balance equation, the corresponding local balance equations.

It has been shown that the following four types of queueing node possess local balance.

*Type 1:* A FCFS node having $c$ servers, each of which provides exponentially distributed service that is the same for all customer classes. The service rate can depend on the total number of customers present in the node.

*Type 2:* A node which operates in processor-sharing (PS) mode. If $n_i$ customers are present at PS node $i$, all $n_i$ are served simultaneously with each receiving $1/n_i$ of the service capacity.

*Type 3:* An infinite-server (IS) node, in which each customer has its own dedicated server.

*Type 4:* A last-come, first-served preempt-resume (LCFS-PR) node.

In all four types of node, the service rate can be a function of the total number of customers present in the node. Also, for nodes of type 2–4, the service time distribution can be general subject only to having a rational Laplace transform and furthermore, customers of different classes can have different general service time distributions. The rate of service at these nodes (types 2–4) can either be a function of the total number of customers present in the node, or alternatively the rate of service for a given customer class can depend on the number of customers of that class present in the node. We note in passing that the Coxian distribution has a rational Laplace transform and that any distribution can be modeled arbitrarily closely using a Coxian distribution.

Not all queueing systems have global balance equations that can be separated into a consistent set of local balance equations. Such is the case when a network has one or more FCFS nodes at which customers of different classes have different service time distributions. When a solution to the set of local balance equations can be found, the queueing system itself is said to have local balance and it follows that the solution of the local balance equations must also be the unique solution of the global balance equations—after all, each global balance equation is the sum of a subset of local balance equations. Thus we may think of local balance as being a sufficient, but not necessary, condition for global balance. Furthermore, it has also been shown that when all the nodes in a network have local balance, then the network itself has local balance and a solution to the local balance equations exists.

Although the local balance equations are less complex than the global balance equations, there are generally more of them and this might have been the end of the story but for a result due to Muntz [38]. In 1973, Muntz proved that, when a network possesses local balance, the global balance equations have a product-form solution: the stationary probability of the states of the network can be written as the product of factors which describe the state of the different nodes. We have already developed two product-form solutions, namely, Equation (15.2) in our analysis of open Jackson networks in Section 15.2 and Equation (15.4) when analyzing closed Gordon and Newell networks in Section 15.3, and seen the substantial benefits that derive from product form. Hence the importance of local balance. In a network in which all the nodes have local balance, the stationary distribution of the network can be determined from an analysis of each of the nodes in isolation. It is not necessary, and indeed computationally very inefficient, if not impossible, to generate and solve all of the global balance equations or even local balance equations in their totality.

### 15.6.2 The BCMP Theorem for Open, Closed, and Mixed Queueing Networks

The research of Chandy and Muntz, together with some related work of Baskett and Palacios merged into what is now called the theorem of Baskett, Chandy, Muntz, and Palacios, the BCMP theorem [2]. This is arguably the most widely referenced theorem in queueing networks, its only possible competitor being Little's theorem. A BCMP network has the following characteristics:

1. The network consists of $M$ nodes, each node belonging to one of types 1–4, as described previously.
2. Customers belong to one of $R$ classes. The network is said to be open if all customer classes are open, closed if all customer classes are closed, and mixed if both open classes and closed classes coexist in the network. The total number of customers in the network at any time is denoted by $n$, the total number at node $i$ is $n_i$ and the total number of class $r$ customers at node $i$ is $n_{ir}$. Thus
$$n = \sum_{i=1}^{M} n_i = \sum_{i=1}^{M}\sum_{r=1}^{R} n_{ir}.$$
3. If class switching is not permitted, the number of customers in each closed class must be constant.
   If class switching is permitted, then customer classes must be partitioned into (*routing*) *chains*. Customers of classes $r$ and $s$ belong to the same chain if it is possible for a customer of class $r$ to become a customer of class $s$. In this manner the number of customers in a closed chain is constant. A chain consists of all (node, class) pairs $(i, r)$ and $(j, s)$ such that a class $r$ customer leaving node $i$ can (eventually) enter node $j$ as a customer of class $s$. We shall let $C$ be the number of chains in the network and $C_o$ the number of open chains. If $C_o = 0$, the network is closed; if $C_o = C$, the network is open, and if $0 < C_o < C$, the network is mixed. The total number of customers in the network that belong to chain $c$ is denoted by $n_c$.
4. Open class customers arrive from the exterior according to a Poisson process. The rate of arrival can depend on the number of customers present in the network in one of two ways.
   (a) There is a single Poisson arrival process whose rate $\lambda(n)$ depends on the total number of customers present in the network at the moment of arrival. An arrival goes to node $i$ as a class $r$ customer with probability $p_{0,ir}$, where $\sum_{i=1}^{M}\sum_{r=1}^{R} p_{0,ir} = 1$. It is understood that $p_{0,ir} = 0$ if $r$ is a closed class.
   (b) There is a separate Poisson arrival process for each open chain. The rate of arrival $\lambda_c(n_c)$ for open chain $c$ depends on the number of customers of that chain currently present in the network. An arrival goes to node $i$ of the chain with probability $p_{0,ir}$, where $\sum_{i=1}^{M}\sum_{r=1}^{R} p_{0,ir} = 1$ and $p_{0,ir} = 0$ if $r$ does not belong to open chain $c$.

   A combination of (a) and (b) is not possible.
5. Each customer class has its own routing matrix, which describes how customers of that class transition among the nodes of the network. On exiting node $i$, a class-$r$ customer enters node $j$ as a class-$s$ customer with probability $p_{ir,js}$. With probability $p_{ir,0} = 1 - \sum_{j=1}^{M}\sum_{s=1}^{R} p_{ir,js}$, a class-$r$ customer exits the network from node $i$.

The concept of class change deserves some elaboration. It is only during the transition from one node to another that this transformation can take place and not while the customer is already present at a node. Each customer class is assigned to one and only one chain in such a way that if the routing probability matrix of a class-$r$ customer permits this customer to become a class-$s$ customer, or vice versa, then classes $r$ and $s$ are combined into the same chain. This means that a customer belonging to a class of one chain can never become a customer of a class belonging to any other chain. Since the number of customers in each closed chain is fixed, it is apparent that a customer from an open class cannot change to become a customer of a closed class and vice versa.

The conditions of the BCMP theorem call for the chains to be ergodic. The discrete-time Markov chain represented by the transition probability matrix whose states are the (node, class) pairs $(i, r)$ with $i = 0, 1, \ldots, M$ ($i = 0$ represents the exterior from which customers belonging to open classes emanate and to which they eventually proceed) and $r = 1, 2, \ldots, R$, is decomposable into $C$ chains, each of which is an ergodic Markov chain in its own right. Hence, if it is possible for a customer of closed class $r$ to become a customer of closed class $s$, the reverse must also be possible. If this were not the case, then eventually, there would remain no customers of class $r$ in the network; closed class-$r$ customers would have no effect whatever on the stationary distribution of the network and could be eliminated from the beginning. Since all open class customers eventually exit to the exterior, the same problem does not arise, although it may be useful to distinguish a "different" exterior for different open chains.

Corresponding to each chain $C_k \in C$ of the network, is a set of traffic equations which must be solved. These are given by

$$v_{ir} = p_{0,ir} + \sum_{\text{all } (j,s)\in C_k} v_{js} p_{js,ir} \quad \text{for all } (i, r) \in C_k. \tag{15.20}$$

The quantities $v_{ir}$ are visit ratios similar to those introduced in Section 15.3 in the context of single-class closed queueing networks. Equation (15.20) has a unique solution if the chain $C_k$ is open, i.e., if there is at least one pair $(i, r) \in C_k$ for which $p_{0,ir} > 0$. If the chain $C_k$ is closed, then $p_{0,ir} = 0$ for all $(i, r) \in C_k$ and the solution of Equation (15.20) can be found only to a multiplicative constant.

Having thus characterized a BCMP network, we are ready to define a state of its underlying Markov chain. Since the network contains $M$ nodes, a state $S$ is a vector quantity whose $i^{\text{th}}$ component $S_i$ is the state of node $i$:

$$S = (S_1,\ S_2,\ \ldots,\ S_M).$$

In turn, the state of each node is also a vector quantity which describes the number, class, state of service and any other relevant information concerning the customers at the node. This information varies depending on the type of node.

- For type-1 nodes, which have FCFS scheduling and identical exponential service time distribution for all classes of customer, it is necessary to specify the class of the customer at each position in the node. If node $i$ is a type-1 node and there are $n_i$ customers present, then $S_i$ is a vector of length $n_i$:

  $$S_i = (r_{i1},\ r_{i2},\ \ldots,\ r_{in_i}),$$

  where $r_{ik}$ is the class of the customer in position $k$, $k = 1, 2, \ldots, n_i$. If $n_i > 0$, the customer at position 1 is the customer currently being served.
- For nodes of type 2 or 3, it suffices to specify the number of class-$r$ customers, $r = 1, 2, \ldots, R$, that are in each of the possible service phases. Let $u_{ir}$ be the number of service phases in the Coxian representation of the service time distribution of class $r$ customers at node $i$. So, if node $i$ is a type-2 or -3 node, then $S_i$ is a vector of length $R$:

  $$S_i = (s_{i1},\ s_{i2},\ \ldots,\ s_{iR}),$$

  whose $r^{\text{th}}$ component $s_{ir}$ is a subvector of length equal to $u_{ir}$ given by

  $$s_{ir} = (\sigma_{1r},\ \sigma_{2r},\ \ldots,\ \sigma_{u_{ir}r}),$$

  and $\sigma_{kr}$ is the number of class $r$ customers in phase $k$.

It is also possible to define the state of a type-2 or -3 node in a manner similar to that used for type-1 nodes, with the understanding that, although arranged into order of arrival, this order is of no consequence, since all customers at type-2 and -3 nodes are served concurrently. Each component $r_{ik}$ of $S_i = (r_{i1},\ r_{i2},\ \ldots,\ r_{in_i})$ now becomes the pair $(r_{ik}, \eta_k)$ whose components provide the class and phase of service of the customer in position $k$. The first approach gives much shorter node descriptors when the network has many customers and the Coxian service time distributions have only a small number of phases.

- Nodes of type 4 have both an ordering component (LCFS-PR) and Coxian service time distributions. The representation of the node is similar to that of type-1 nodes, with the additional requirement that along with the class of the customer at position $k$ in the node, the current phase of service must also be provided. Thus, for nodes of type 4,

$$S_i = (s_{i1},\ s_{i2},\ \ldots,\ s_{in_i}),$$

where $n_i$ is the number of customers at node $i$ and $s_{ik} = (r_{ik}, \eta_k)$ is a vector of length 2 whose first component $r_{ik}$ gives the class of the customer in position $i$ in the node and whose second component $\eta_k$ gives the current phase of service of this customer.

The global balance equations, which relate the flow into and out of each state $S = (S_1, S_2, \ldots, S_M)$ can now be formed and it may be shown that the stationary probability of these states has a product form. As we have seen, these state descriptors contain much more information than is usually required, information such as the position of each customer in FCFS and LCFS-PR queues and the phase of service of all customers other than those in type-1 nodes. Most often, it suffices to know only the number of customers of each class at each of the nodes. The BCMP authors provide both. The product form involving the complete state description is provided, but then it is shown that by summing over groups of states, a more compact aggregated product form can be obtained. An aggregated state $S = (y_1, y_2, \ldots, y_M)$ is a vector of length $M$ whose $i^{\text{th}}$ component, $y_i = (n_{i1}, n_{i2}, \ldots, n_{iR})$, is a subvector of length $R$. The $r^{\text{th}}$ component of this subvector $n_{ir}$ is the number of class-$r$ customers at node $i$. The product form is

$$\text{Prob}\{S = (y_1, y_2, \ldots, y_M)\} = \frac{1}{G} d(S) \prod_{i=1}^{M} f_i(y_i),$$

where $G$ is a normalization constant which insures that the sum of the probabilities is equal to 1; $d(S)$ is a function of the arrival processes, and the functions $f_i(y_i)$ differ according to the type of node. The condition for stability depends only on the open chains: if the utilization at each node generated by open class customers is less than one, then the network is stable, irrespective of the number of closed class customers present. When the Poisson arrival processes are load independent, the utilization of node $i$ caused by open class customers is

$$\rho_i = \sum_{r=1}^{R} \lambda_r \frac{v_{ir}}{\mu_i} \quad \text{(single server, FCFS node)}; \quad \rho_i = \sum_{r=1}^{R} \lambda_r \frac{v_{ir}}{\mu_{ir}} \quad \text{(PS, IS, LCFS-PR nodes)}.$$

Let us consider the terms $d(S)$ and $f_i(y_i)$ in more detail. If the network has no open chains then $d(S) = 1$. Otherwise, $d(S)$ depends on which of the two arrival processes is in force. If there is a single Poisson arrival process whose rate depends on $n$, the total number of customers of all classes present, open and closed, then

$$d(S) = \prod_{k=0}^{n-1} \lambda(k),$$

where $\lambda(k)$ is the arrival rate when the network contains $k$ customers. If there are $C_o$ independent Poisson streams, one for each open chain, and the arrival process is of the second kind, then

$$d(S) = \prod_{c=1}^{C_o} \prod_{k=0}^{n_c - 1} \lambda_c(k),$$

where $n_c$ is the number of customers of chain $c$ present in the network and $\lambda_c(k)$ is the rate of arrival to chain $c$ when this chain contains $k$ customers.

The terms $f_i(y_i)$ are defined below for nodes of each type. In these formulae, $n_{ir}$ is the number of class-$r$ customers at node $i$ and $n_i = \sum_{r=1}^{R} n_{ir}$ is the total number of customers at node $i$, $1/\mu_{ir}$ is the mean service time of a class-$r$ customer at node $i$, and $v_{ir}$ is the visit ratio for a class-$r$ customer at node $i$, obtained by solving Equation (15.20):

$$\text{For nodes of type 1:} \quad f_i(y_i) = n_i! \left( \prod_{r=1}^{R} \frac{1}{n_{ir}!} v_{ir}^{n_{ir}} \right) \left( \frac{1}{\mu_i} \right)^{n_i}.$$

$$\text{For nodes of types 2 and 4:} \quad f_i(y_i) = n_i! \prod_{r=1}^{R} \frac{1}{n_{ir}!} \left( \frac{v_{ir}}{\mu_{ir}} \right)^{n_{ir}}.$$

$$\text{For nodes of type 3:} \quad f_i(y_i) = \prod_{r=1}^{R} \frac{1}{n_{ir}!} \left( \frac{v_{ir}}{\mu_{ir}} \right)^{n_{ir}}.$$

Observe that the definition of $f_i(y_i)$ presented for nodes 2 and 3 is also valid for nodes of type 1, since in nodes of type 1, $1/\mu_{ir} = 1/\mu_i$ for all $r = 1, 2, \ldots, R$, and $n_i = \sum_{r=1}^{R} n_{ir}$. When the service rates depend on the total number of customers at a node, the factors in the product form become

$$\text{For nodes of type 1:} \quad f_i(y_i) = n_i! \left( \prod_{r=1}^{R} \frac{1}{n_{ir}!} v_{ir}^{n_{ir}} \right) \left( \prod_{j=1}^{n_i} \frac{1}{\mu_i(j)} \right),$$

$$\text{For nodes of type 2 and 4:} \quad f_i(y_i) = n_i! \prod_{r=1}^{R} \left( \frac{1}{n_{ir}!} v_{ir}^{n_{ir}} \prod_{j=1}^{n_i} \frac{1}{\mu_{ir}(j)} \right),$$

$$\text{For nodes of type 3:} \quad f_i(y_i) = \prod_{r=1}^{R} \left( \frac{1}{n_{ir}!} v_{ir}^{n_{ir}} \prod_{j=1}^{n_i} \frac{1}{\mu_{ir}(j)} \right),$$

where $1/\mu_{ir}(j)$ is the mean service time of a class-$r$ customer at node $i$ when the customer population at the node is $j$. The BCMP theorem also treats the case when the service time of a class $r$ customer depends on the number of customers of class-$r$ present at node $i$, but we do not consider that possibility here. As noted explicitly by the authors of the BCMP theorem, these are surprising results. The only moment of the service time distributions to appear in these formulae is the first—the service time distributions for the different classes might as well have been exponential! Notice, nevertheless, that the class of a customer still plays a role, through the different mean service times $1/\mu_{ir}$.

There remains one problem, the computation of $G$, the constant of normalization. For closed multiclass networks, the convolution algorithm of Buzen developed in Section 15.3 can be extended by the introduction of vector quantities to replace the scalars used in the single-class environment. The modification of the auxiliary functions is also straightforward. For mixed queueing networks, it is usual to combine all open classes into a singe open chain and to compute the utilization at each node due to open class customers only. The normalization constant is then computed in the absence of open class customers. The open class utilizations can now be used to scale the factors of the product form. The MVA algorithm can also be modified to solve open, closed, and mixed multiclass networks. Indeed there is a large number of possible computational algorithms currently available, each having different advantages and disadvantages. Some of these include LBANC (local balance algorithm for normalizing constant) developed by Chandy and Sauer [9], the tree convolution algorithm of Lam and Lien [26], and RECAL (recursion by chain) by Conway and Georganas [11]. A later book by the last two authors [12] places many of these computational algorithms onto a common framework.

As might be expected, the addition of different classes of customer into a network substantially increases the computational complexity and memory requirements of both the convolution algorithm and the MVA algorithm. For example, the complexity of Buzen's algorithm when applied to a closed queueing network with $N$ customers and $M$ nodes each with a single load-independent exponential server is $O(MN)$; when the service is load dependent, this number becomes $O(MN^2)$; when there are $R$ customer classes and $N_r$ customers of class $r = 1, 2, \ldots, R$, with no class switching permitted and only load-independent service, the computational complexity has order $O(MR\prod_{r=1}^{R}(N_r + 1))$. The MVA algorithm has similar complexity. It will quickly be recognized that the amount of computation needed for large networks, particularly those with many different classes, can become huge, sometimes to such an extent that this approach must be abandoned and others sought.

Approximate methods are used in product form networks when the computation complexity is so large that it effectively prevents the application of either the convolution algorithm or the MVA method. Most approximate methods in these cases are based on the MVA algorithm and the results obtained are frequently more than adequate in terms of the accuracy achieved. Numerical problems associated with underflow and overflow in the computation of the normalization constant can be avoided. Approximations are also used when the queueing network does not have a product form solution. These commonly occur in the presence of a first-come, first-served scheduling policy with nonexponential service time distributions. One possibility is to replace the nonexponential servers with exponential servers having the same mean. Although this can work reasonably well in some cases, and in closed networks better than open networks, it is not always satisfactory. The number and type of approximation methods currently available is extensive and warrants a text all to itself. The book by Bolch et al. [4] comes as close as any to achieving this goal.

## 15.7 Java Code

### *The Convolution Algorithm*

```
import java.util.*;

class  buzen
{
   public static void main (String args[])
   {
                 // Example 1 from Notes
   int N = 3;                // Number of customers
   int M = 4;                // Number of nodes
   boolean [] LI = {false, true, true, true};       // True if node m
                                       // is load independent; otherwise false
   double [] lambda = {1.0, 0.5, 0.3, 0.2};
```

```
   double [][] mu = { {0.0,1.0,2.0,2.0},{0.5,0.5,0.5,0.5},{1.0,1.0,1.0,1.0},
                      {2.0,2.0,2.0,2.0} };
/*
              // Example 2 from Notes
   int N = 2;               // Number of customers
   int M = 4;               // Number of nodes
   boolean [] LI = {false, true, true, false};      // True if node m
                                    // is load independent; otherwise false
   double [] lambda = {1.0, 0.4, 0.6, 1.0};
   double [][] mu = { {0.0,1.0,2.0}, {0.8,0.8,0.8}, {0.4,0.4,0.4},
                      {0.0,0.5,1.0} };
*/
   double [] G = new double [N+1];    // Array to hold normalizing constants
   double [] f = new double [N+1];

   System.out.println("Node 0");
   G[0] = 1.0;
   if (LI[0]) {
      double Y = lambda[0]/mu[0][0];
      for (int n=1; n<=N; n++) {
         G[n] = G[n-1]*Y;
      }
      for (int n=0; n<=N; n++) {
         System.out.println("   G-LI = " + G[n]);
      }
   }
   else {
      double Y = lambda[0];
      for (int n=1; n<=N; n++) {
         G[n] = G[n-1]*Y/mu[0][n];
      }
      for (int n=0; n<=N; n++) {
         System.out.println("   G-LD = " + G[n]);
      }
   }

   for (int m=1; m<M; m++) {
   System.out.println("Node "+ m);
      if (LI[m]) {
         double Y = lambda[m]/mu[m][m];
         for (int n=1; n<=N; n++) {
            G[n] = G[n] + G[n-1]*Y;
         }
         for (int n=0; n<=N; n++) {
            System.out.println("   G-LI = " + G[n]);
         }
      }
      else {
         f[0] = 1;
         for (int n=1; n<=N; n++) {
```

```
            f[n] = f[n-1]*lambda[m]/mu[m][n];
            System.out.println("f(n) = " + f[n]);
         }
         for (int n=N; n>0; n--) {
            double sum = G[n];
            for (int k=1; k<=n; k++) {
               sum = sum + f[k]*G[n-k];
            }
            G[n] = sum;
         }
         G[0] = 1;
         for (int n=0; n<=N; n++) {
            System.out.println("   G-LD = " + G[n]);
         }
      }
   }
   }
}
```

### *The MVA Algorithm*

```
import java.util.*;
import java.text.*;

class mva
{
   public static void main (String args[]) {

/*
              ***** First part defines the queueing network  *****
          ***** This is the only part that is should be altered   *****
*/

//          Example from Notes  ============>
      int M = 5;                             // Number of nodes
      int N = 10;                            // Population size
      int [] qType = {0,1,2,2,2,3};          // Node types (1, 2 or 3)
      int [] mnum = {0,10,1,1,1,2};          // Number of servers per node
      double [] mu = {0,2,1,2,4,2.5};        // Service rates for each node
      double [] V  = {0,1.0,0.2,0.3,0.5,1.0}; // Visit ratios for each node
/*
//          Example from Bolch et al.  ============>
      int M = 4;
      int N = 3;
      int [] qType = {0,3,2,2,1};
      int [] mnum =  {0,2,1,1,3};
      double [] mu = { 0, 2, 1.66666666666, 1.25, 1};
      double [] V =  {0, 1.0, 0.5, 0.5, 1.0};
*/
```

```
/*
                  ***** The MVA algorithm  *****
         ***** There is no need to alter beyond this point  ****
*/

    double G = 1;                              // Normalization constant
    double [] L = new double [M+1];            // Mean number of customers
    double [] R = new double [M+1];            // Mean response times
    double [] X = new double [M+1];            // Throughputs

    int numLD = 0;
    for (int i=1; i<=M; i++) {
        if (qType[i] == 3) {numLD++;}          // Determine number of LD nodes
    }
    double [][] pp = new double [N+1][numLD+1];
    for (int m=0; m<=numLD; m++) {
       pp[0][m] = 1;                           // Marginals for LD nodes
    }




//              ***** Begin iterating over customer population  ***
    for (int k=1; k<=N; k++){

//     ***  MVA Step 1:  Compute mean response times  ***
       int LDcount = 0;                              // Indicator for LD nodes
       for (int m=1; m <= M; m++) {
          if (qType[m] == 1) { R[m] = 1/mu[m]; }  // Infinite server node
          if (qType[m] == 2) { R[m] = (L[m]+1)/mu[m]; }
                                                     // FCFS exponential, etc.
          if (qType[m] == 3) {                 // Load dependent node
             R[m] = 0;                         //   ... considered to be a
             for (int j=1; j <=k; j++) {       //   multiserver expon. node
             // In this implementation, alpha_i(j) is equal to min(m_i, j)
                 double alpha = j; if (mnum[m] < j){alpha = mnum[m];}
                 R[m] = R[m] + j/(mu[m] * alpha) * pp[j-1][LDcount];
             }
             LDcount++;
          }
       }

//     ***  MVA Step 2:   Compute throughputs
       double bottom = 0;
       for (int m=1; m<=M; m++){
           bottom = bottom + V[m]*R[m];
       }
       X[1] = k/bottom;                              // Overall throughput
       for (int m=2; m<=M; m++) {
           X[m] = V[m]*X[1];                         // Throughput of node m
```

```
          }
//                    Update normalization constant
          G = G/X[1];

//        *** MVA Step 3:   Compute mean number of customers   ***
          for (int m=1; m<= M; m++) {
             L[m] = X[m]*R[m];
          }

//                  Compute marginals for all load dependent nodes
          LDcount = 0;                                  //  Indicator for LD nodes
          for (int m=1; m<=M; m++) {
              if (qType[m] == 3) {
                 double p0 = 0;
                 for (int j=k; j >=1; j--) {
              // In this implementation, alpha_i(j) is equal to min(m_i, j)
                     double alpha = j; if (mnum[m] < j){alpha = mnum[m];}
                     pp[j][LDcount] = X[1]/(mu[m] * alpha) * pp[j-1][LDcount];
                     p0 = p0 + pp[j][LDcount];
                 }
                 pp[0][LDcount] = 1-p0;
                 LDcount++;
              }
          }
       }

//                 Output results
       NumberFormat fmat = NumberFormat.getNumberInstance();
       fmat.setMaximumFractionDigits(6);
       fmat.setMinimumFractionDigits(6);
       System.out.println("      " );
       System.out.println("                 R[m]        X[m]          L[m] " );
       for (int m=1; m <=M; m++) {
          System.out.println(" m = " + m + ":    " + fmat.format(R[m]) + " "
                      + fmat.format(X[m]) + "   " + fmat.format(L[m]));
       }
       System.out.println(" " );
       double sum = 0;
       for (int m=1; m <=M; m++) {
           sum = sum + L[m];
       }
       System.out.println("Sum check: Population size = "+ fmat.format(sum));
       System.out.println("Normalization constant: G = " + G);
       System.out.println("      " );
    }
}
```

## 15.8 Exercises

**Exercise 15.2.1** Consider the most basic of feedback networks, a single node with feedback as shown in Figure 15.16. This could be used, for example, to model a data transmission channel with probability $r$ that a message is transmitted unsuccessfully and requires retransmission. Find the distribution and mean number of customers in this network. Assume a Poisson arrival process and exponential service time distribution.

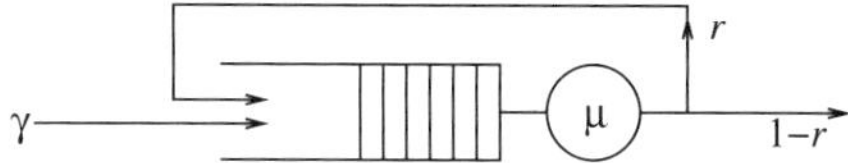


Figure 15.16. Single queue with feedback.

**Exercise 15.2.2** Consider an open queueing network of the central server type. Customers arrive at the central server according to a Poisson process with rate 0.1. The central server behaves stochastically like an $M/M/2$ queue in which each of the servers operates at rate 2.0. After receiving service at the central server, customers either leave the network (with probability 0.2) or else proceed to one of three other nodes with probabilities 0.3, 0.3, and 0.2 respectively. Service at each of the three service centers is exponentially distributed at rate 0.5, 1.0, and 2.0 respectively. After receiving service at one of these three nodes, customers return to the central server. Compute the probability that the central server is idle and the mean number of customers in the network.

**Exercise 15.2.3** Consider the queueing model shown in Figure 15.17. This may be used to model a computer CPU connected to an I/O device as indicated on the figure. Processes enter the system according to a Poisson process with rate $\gamma$, and use the CPU for an exponentially distributed amount of time with mean $1/\mu_1$. Upon exiting the CPU, with probability $r$, a process uses the I/O device for a time that is exponentially distributed with mean $1/\mu_2$, or with probability $1 - r$, it exits the system. Upon exit from the I/O device, a process again joins the CPU queue. Assume that all service times, including successive service times of the same process at the CPU or the I/O device are independent. Find the mean number of customers at the CPU and at the I/O device. What is the average time a process spends in the system? What is the utilization at the CPU and at the I/O device?

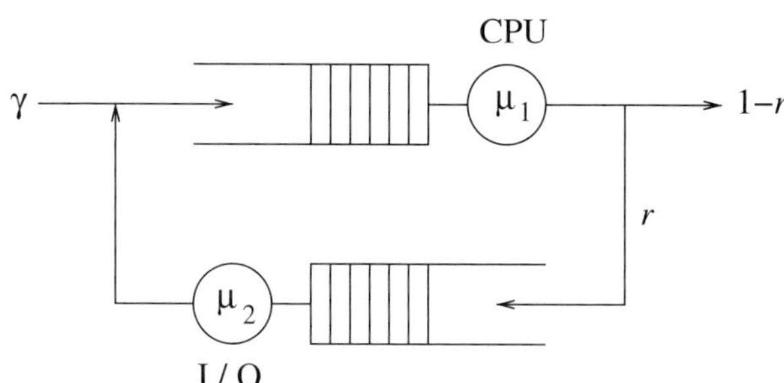


Figure 15.17. Queueing network representing a CPU and I/O device.

**Exercise 15.2.4** This question relates the queueing network with feedback of Exercise 15.2.3 to the two node tandem queueing network *without* feedback shown in Figure 15.18. In this tandem queueing network, the first node represents the CPU of the previous example and the second represents the I/O device. The arrival process is Poisson with rate $\gamma$. Each arriving process requires the CPU only once, and holds it for an average of $S_1$ seconds. On exiting the CPU, each process must go to the I/O device and spends an average of $S_2$ seconds there after which it departs never to return.

The following values for $S_1$ and $S_2$ are given:

$$S_1 = \frac{1}{\mu_1(1-r)}, \quad S_2 = \frac{r}{\mu_2(1-r)}.$$

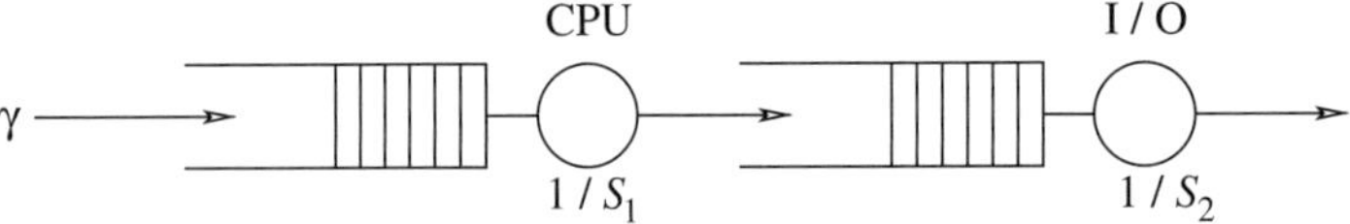


Figure 15.18. The related tandem queue.

(a) Show that the utilization factor in this tandem model is the same as the utilization factor in the feedback model of Exercise 15.2.3.
(b) Show that the response time (average time a process spends in the network) is the same in both models.
(c) Show that the probability distribution of customers is the same in both networks.
(d) Generate a counter-example to show that the waiting time distribution is *not* the same in both networks. *Hint: Choose* $\mu_1 \gg \mu_2$.
(e) How should $S_1$ and $S_2$ be interpreted in terms of CPU and I/O requirements?

**Exercise 15.3.1** Consider a closed queueing network of three nodes. Each node contains a single exponential server and the scheduling discipline is FCFS. The routing probability matrix is

$$\begin{pmatrix} 0 & 0.7 & 0.3 \\ 1 & 0 & 0 \\ 1 & 0 & 0 \end{pmatrix}.$$

The exponential service time distributions have a mean service time of 2.0 at node 1, 1.0 at node 2, and 0.5 at node 3. The number of customers that circulate in the network is equal to 3. Use the convolution algorithm to determine the marginal distribution of customers at node 1.

**Exercise 15.3.2** This time answer the previous question using a Markovian analysis. In other words, compute the infinitesimal generator, and solve it using a numerical method. Be sure your answers to this question and the previous are the same.

**Exercise 15.3.3** Consider a closed queueing network with three nodes each containing a single exponential server with rates $\mu_1 = 1$, $\mu_2 = 2$, and $\mu_3 = 4$ respectively. A total of three customers circulates in this network and scheduling at all nodes is first-come, first-served. The routing matrix is given by

$$R = \begin{pmatrix} .2 & .3 & .5 \\ .2 & .3 & .5 \\ .2 & .3 & .5 \end{pmatrix}.$$

Draw this queueing network and then use the convolution approach to find the throughput of the network and the distribution of customers at node 1.

**Exercise 15.3.4** Consider a closed queueing network with two service centers in tandem between which three customers circulate. The first service center contains two identical exponential servers which provide service at rate $\mu_1 = 1$ while the second service center is a pure delay node (infinite server) where customers spend an exponentially distributed amount of time with mean .5 time units. Use the convolution algorithm to compute the margin distribution of customers at the first service center.

**Exercise 15.3.5** Consider a closed queueing network in which only two identical customers circulate. The network is composed of four nodes; the first node contains two exponential servers each of which provides service at rate $\mu_1 = 1$; node 2 is an infinite-server node, in which each server provides exponentially distributed service at rate $\mu_2 = 0.5$; nodes 3 and 4 both contain a single exponential server with rates $\mu_3 = 0.8$ and $\mu_4 = 0.4$, respectively. On exiting from node 1, all customers go to node 2; on exiting node 2 customers will next go to node 3 with probability 0.4 or to node 4 with probability 0.6. On exiting either node 3 or 4, customers return to node 1. Use Buzen's algorithm to find the marginal distribution at nodes 1 and 2 and the mean number of customers at each of the four nodes.

**Exercise 15.4.1** Three service centers operate in tandem with the third feeding back into the first. The first node is a pure delay with mean equal to 1 time unit. The other two nodes are single server nodes whose service rates are given respectively by $\mu_2 = 2$ and $\mu_3 = 4$. A total of four customers circulates in this network . Use the MVA algorithm to compute the mean number of customers at each node and the throughput of the network.

**Exercise 15.4.2** Return to Exercise 15.3.3, but this time answer the question using the MVA algorithm.

**Exercise 15.4.3** Return to Exercise 15.3.4, but this time answer the question using the MVA algorithm.

**Exercise 15.6.1** Write down each of the six global balance equations belonging to the queueing network of Example 15.12. Separate each into one or more local balance equations and for each explain why they are balanced.

**Exercise 15.6.2** Consider a closed multiclass central server queueing model with two classes of customer as shown in Figure 15.19. The central server consists of a single exponential server that operates in LCFS-PR mode. The mean service time for class $i$ customers is $1/\mu_i$, $i = 1, 2$. On exiting the central server, class 1 customers go to the first of the two auxiliary nodes while class 2 customers go to the second. On leaving the auxiliary nodes, customers return to the central server. Both auxiliary nodes contain a single exponential server operating in FCFS order with rates as indicated on the figure. To keep the model simple, assume that only one customer of each class circulates in the model.

(a) Write down the global balance equations for this network.
(b) Separate the global balance equations into local balance equations and explain the rationale for each local balance equation.
(c) Now assume that the scheduling policy at the central server is changed to FCFS, while still maintaining different service rates for each class of customer. What are the global balance equations in this case?
(d) Which of the local balance equations obtained in part (b) are modified by the FCFS policy and are the new versions local balance equations? Explain why.

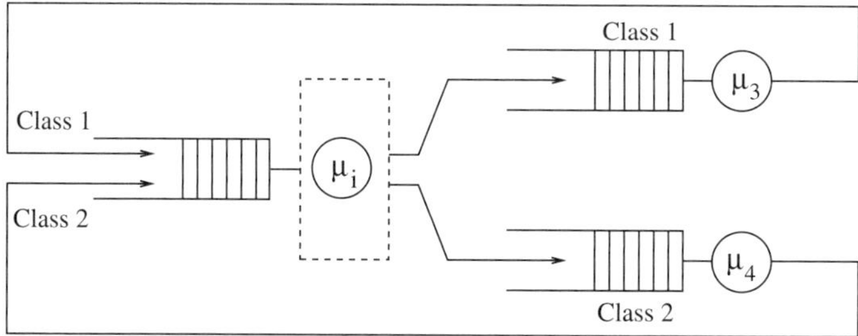


Figure 15.19. A central server multiclass queueing model.

# Part IV

# SIMULATION

# Chapter 16

# *Some Probabilistic and Deterministic Applications of Random Numbers*

In the very first chapter of this book, we introduced a certain number of probability axioms and used them to develop a theory by which different probabilistic scenarios could be analyzed mathematically. Behind all of this was the implied assumption that if a probability experiment could be conducted a very large number of times (in fact, an infinite number of times) the same results could be obtained without the theory. This we referred to as the frequency approach to probability. A finite, but sufficiently large number of times would give an excellent approximation, certainly good enough for most applications.

Unfortunately, conducting a probability experiment, even as simple as tossing a coin or choosing a card, for a large number of times can be very time consuming. A computer, however, could do such tasks extremely rapidly, if we could only decide what it means for a computer to toss a coin, or choose a card! This is where the concept of *random numbers* becomes useful. If we can use the computer to generate a sequence of real numbers that are randomly distributed in the interval $[0, 1)$, then we could multiply each number in the sequence by 2 and associate the result, if it is strictly less than 1, with the event that the experiment gives a tail. If the result is greater than or equal to 1, associate it with the event that a head is obtained. With a fair coin, the probability of getting a tail on a single toss, is 1/2. With a random sequence of reals in the interval $[0, 2)$ we should also expect to find a number less than 1 with probability 1/2. Similarly, if our concern is with choosing a card from a well shuffled regular deck of 52 cards, then a computer generated sequence of reals in the interval $[0, 1)$ could be multiplied by 52, and each resulting number in an interval $[i, i + 1)$, for $i = 0, 1, \ldots, 51$, could be associated with the appearance of a particular card. But what exactly does it mean to generate a *random sequence of reals* in any interval? We shall return to this question later in the chapter, but for the moment, intuitively, it means that if we partition the interval into a set of collectively exhaustive and mutually exclusive equal subintervals, the next number in the sequence should be as likely to fall into any one of these subintervals as another.

The fact that the numbers are generated at random does not mean that they cannot be applied to deterministic scenarios. Indeed, one of the first uses of randomly generated numbers was the evaluation of definite integrals. Towards the end of this chapter, we shall provide some examples of this for illustrative purposes.

## 16.1 Simulating Basic Probability Scenarios

For the moment, we shall assume that we have access to a randomly generated sequence of real numbers that lie in the interval $[0, 1)$. From this sequence we shall generate a sequence of integers, randomly distributed in the interval $[0, T - 1]$ by multiplying each real $r$ by $T$ and choosing the nearest integer that is less than or equal to the result. In other words, we compute $\lfloor rT \rfloor$. We shall use this sequence to answer probability questions that, in previous chapters, we solved mathematically by means of the probability axioms.

Consider the case when all the elementary events of a sample space are equiprobable. Let us assume that the sample space contains $T$ elementary events, $e_i$, $i = 1, 2, \ldots, T$. Our concern is to determine the probability associated with a certain event $E$, defined as a proper subset of these elementary events. As we proceed through the simulation, we shall generate random reals in the interval $[0, 1)$, multiply each by $T$ and take the floor of the result. If the product of the real number and $T$ lies in the interval $[i - 1, i)$, then the random integer obtained is $i - 1$ and it is associated with the occurrence of elementary event $e_i$. We may now test if the elementary event $e_i$ is in the event subset $E$, i.e., if the event $E$ occurred or not. We shall keep a count of the number of trials conducted, $n$, and the number of these, $n_E$, in which the desired event $E$ occurred. As $n$ becomes large, the ratio of these, i.e., the ratio $n_E/n$, tends to the probability of the event $E$. The algorithm is as follows:

```
Initialize event count: n_E = 0
For a total of "n" trials {
    * Perform the probability experiment
    * If event E occurs, increment n_E
}
Estimated probability of event E = n_E/n
```

The statement, "`* Perform the probability experiment`", means that a random integer between 0 and $T - 1$ is chosen to represent the outcome of the trial. The event $E$ occurs if this integer represents an elementary event in the subset that defines $E$. This is perhaps best illustrated by a number of examples. We shall use these examples to show how a simulation run mimics a trial of the probability experiment with particular emphasis on the number of random numbers that each trial requires and the large number of trials that are generally needed to obtain even modest precision.

**Probability Scenario 1**

Let us use simulation to determine the probability of getting a total of three spots or less when two fair dice are thrown simultaneously. In this case, there are 36 equiprobable elementary events in the sample space, denoted $(i, j)$ for $i, j = 1, 2, 3, 4, 5, 6$. We could begin by indexing the 36 elementary events from 0 through 35 as follows

| Index | 0 | 1 | 2 | 3 | 4 | 5 | 6 | ··· | 33 | 34 | 35 |
|---|---|---|---|---|---|---|---|---|---|---|---|
| E. Event | (1,1) | (2,1) | (1,2) | (3,1) | (2,2) | (1,3) | (4,1) | ··· | (6,5) | (5,6) | (6,6) |

The event "three spots or fewer" corresponds to the subset that contains the first three elementary events. We initialize the counters $n$ and $n_E$ to zero and begin generating random numbers in the interval $[0, 1)$. Each random number is multiplying by 36 and the counter $n$ is incremented. For each result that is strictly less than 3, the counter $n_E$ is also incremented, since a result of 0, 1, or 2 corresponds to the only three elementary events whose sum is 3 or less. Table 16.1 displays the results obtained using the Java random number generator, and appears to be converging to 0.08333, i.e., one twelfth, the correct answer.

Table 16.1. Probability of getting three or fewer spots when throwing two dice.

| $n$ | 50 | 100 | 1,000 | 10,000 | 50,000 | 100,000 | 500,000 | 1,000,000 |
|---|---|---|---|---|---|---|---|---|
| $n_E$ | 4 | 7 | 84 | 850 | 4,178 | 8,533 | 41,648 | 83,317 |
| $n_E/n$ | 0.0800 | 0.0700 | 0.0840 | 0.0850 | 0.08356 | 0.08533 | 0.083296 | 0.083317 |

These results were obtained using the Java code presented below. This code will be used in the remaining examples in this chapter, but we shall only provide the portion that lies between the `*** Begin Simulation ***` and `*** End Simulation ***` comments, since the initial six statements and the last statement are identical in all the remaining examples. Only the part that is unique to the simulation experiment will be given in future examples. A single run of the program actually performs eight simulations with an increasing number of trials ranging from 50 to 1,000,000. This enables us to observe that, generally speaking, accuracy increases with increased number of trials.

```
import java.util.Random;
class scenario {
  public static void main (String args[]) {
    int max [] = {50,100,1000,10000,50000,100000,500000,1000000};
    Random generator = new Random();
    for (int i=0; i<8; i++) {

// *** Begin Simulation ***************************
       int n_E = 0;                               // Initialize success count
       for (int n=1; n<max[i]; n++) {
                                        // Generate max[i] random integers ...
          int r_e = generator.nextInt(36);        //          between 0 and 35
          if (r_e < 3){n_E++;}                    // Increment success count
       }
       double prob = (double)n_E/max[i];          // Approximate probability
// *** End Simulation *****************************

       System.out.println("n_E = " + n_E + "  max = " + max[i] +
                               "  n_E/max[i] = " + prob);
    }
  }
}
```

In this example, we knew that there were 36 elementary events and that only three of them lead to success, so we effectively knew the answer was 3/36 before we even started. Had we not known the number of elementary events nor the number of them that lead to success, we would have had to perform the simulation in a different fashion, by more closely following the way in which the trials themselves are actually carried out. We would have had to simulate tossing one die by generating a random integer between 1 and 6, then tossing a second die by generating a second integer between 1 and 6, and then adding the two and checking to see if the result were less than or equal to three. The body of the simulation program would then have been as follows:

```
// *** Begin Simulation ***************************
       int n_E = 0;                                 // Initialize success count
       for (int n=0; n<max[i]; n++) {               // Perform max[i] trials ...
          int r_e1 = 1 + generator.nextInt(6);      // Die 1 (result 1,2,...,6)
          int r_e2 = 1 + generator.nextInt(6);      // Die 2 (result 1,2,...,6)
          if (r_e1+re2 <= 3){n_E++;}                // Increment success count
       }
       double prob = (double)n_E/max[i];            // Approximate probability
// *** End Simulation *****************************
```

This is, computationally, more expensive than the previous implementation, because each trial requires the generation of not one, but two random numbers. However, it is not unusual to need several random numbers in creating a trial, as we shall see in the next few examples.

**Probability Scenario 2**

In this example we shall use simulation to determine the probability of getting exactly three heads in five tosses of a fair coin. Each time we generate a random number $r$ in the interval $[0, 1)$ we shall multiply it by 2 and take the floor of the result, $\lfloor 2r \rfloor$. When this gives 0, we shall associate it with the event that a tail occurs; when it is 1, that a head occurs. In five consecutive tosses, we need to know when exactly 3 heads occur; in other words, when the total of five consecutive values of $\lfloor 2r \rfloor$ is exactly equal to 3. In this example the body of the Java code may be written as

```
// *** Begin Simulation ***************************
       int n_E = 0;                            // Initialize success count
       for (int n=0; n<max[i]; n++) {          // Perform max[i] trials
          int heads = 0;
          for (int j=1; j<=5; j++) {           // Trial consists of 5 tosses
            int r_e = generator.nextInt(2);    // 1 if heads; 0 if tails
            heads = heads+r_e;
          }
         if (heads == 3){n_E++;}               // Count successes
       }
       double prob = (double)n_E/max[i];       // Approximate probability
// *** End Simulation *****************************
```

Notice that we may replace the integer 3 in line 9 with any integer between 0 and 5 inclusive to determine the probability of getting that number of heads in five tosses of a coin. We may also replace the integer 5 in line 5 to allow for more or fewer tosses. In this case, the number of heads sought must obviously lie between 0 and the total number of tosses. One execution of this program led to the results in Table 16.2.

Table 16.2. Probability of getting exactly three heads in five tosses.

| $n$ | 50 | 100 | 1,000 | 10,000 | 50,000 | 100,000 | 500,000 | 1,000,000 |
|---|---|---|---|---|---|---|---|---|
| $n_E$ | 18 | 28 | 321 | 3,165 | 15,631 | 31,310 | 155,989 | 312,257 |
| $n_E/n$ | 0.3600 | 0.2800 | 0.3210 | 0.3165 | 0.31262 | 0.3131 | 0.311978 | 0.312257 |

It appears that the answer is converging to the correct result, 0.3125. Each trial in this simulation requires the generation of five random numbers, one for each toss of the coin. This is the "natural" way to simulate this problem. An alternative approach requiring less random numbers is to generate a single random integer between 0 and 31 and use it to represent all five tosses. This random integer may be converted to its 5-bit binary equivalent in which the number of 1's represents the number of heads thrown. If the number of 1's in this representation is exactly equal to 3, then the event "throw exactly three heads" has occurred and the success count should be incremented.

**Probability Scenario 3**

In this example, we wish to find the probability that in a class of 23 students, at least two students have the same birthday. To solve this problem using simulation, for each value of $n$ we shall randomly choose 23 numbers between 1 and 365 and increment $n_E$ only if these twenty-three numbers contains at least one repetition.

```
// *** Begin Simulation ***************************
       int n_E = 0;                               // Initialize success count
       for (int n=0; n<max[i]; n++) {             // Perform max[i] trials
          int [] birthday = new int [23];         // Array to store dates
          boolean no_match = true;
          int j = 0;
          while (no_match && j<23) {              // Generate up to 23 dates
              birthday[j] = generator.nextInt(365);   // New date
              for (int k=0; k<j; k++) {               // Compare with others
                if (birthday[j] == birthday[k]){no_match = false;} // Match?
              }
              j++;
          }
          if (no_match == false){n_E++;}           // Increment success count
       }
       double prob = (double)n_E/max[i];          // Approximate probability
// *** End Simulation ****************************
```

One execution of this program led to the results in Table 16.3.

Table 16.3. Probability of at least two people among 23 having the same birthday.

| $n$ | 50 | 100 | 1,000 | 10,000 | 50,000 | 100,000 | 500,000 | 1,000,000 |
|---|---|---|---|---|---|---|---|---|
| $n_E$ | 25 | 47 | 523 | 5,056 | 25,386 | 50,678 | 253,529 | 507,666 |
| $n_E/n$ | 0.5000 | 0.4700 | 0.5230 | 0.5056 | 0.50772 | 0.50678 | 0.507058 | 0.507666 |

Simulating a trial in this example requires the generation of up to 23 random numbers (in the case when no matches are found) and approximately up to $23^2/2$ comparisons (again, when no matches are found). It is evident that simulations requiring a large number of simulated trials can become computationally expensive.

**Probability Scenario 4**

The first of two boxes contains $b_1$ blue balls and $r_1$ red balls; the second contains $b_2$ blue balls and $r_2$ red balls. One ball is randomly chosen from the first box and put into the second. When this has been accomplished, a ball is chosen at random from the second box and put into the first. A ball is now chosen from the first box. What is the probability that it is blue?

```
// *** Begin Simulation ***************************
   int n_E = 0;                          // Initialize success count
   for (int n=0; n<max[i]; n++) {        // Perform max[i] trials
     int b1 = 3; int r1 = 7; int b2 = 4; int r2 = 3;  // Select colors

     int r_e = generator.nextInt(b1+r1);
     if (r_e < b1){b1--; b2++;}  // Blue ball moves from box 1 to box 2, or
        else {r1--; r2++;}       // Red  ball moves from box 1 to box 2
```

```
        r_e = generator.nextInt(b2+r2);
        if (r_e < b2){b2--; b1++;}  // Blue ball moves from box 2 to box 1, or
           else {r2--; r1++;}       // Red  ball moves from box 2 to box 1

        r_e = generator.nextInt(b1+r1);  // Choose a ball from box 1
        if (r_e < b1){n_E++;}            // Success if this ball is blue
     }
     double prob = (double)n_E/max[i];  // Approximate probability
  // *** End Simulation ****************************
```

We obtain the results in Table 16.4. The exact answer from Chapter 1 is equal to 0.32375. This simulation mirrors exactly the probability experiment of moving balls from bucket to bucket.

Table 16.4. Probability that the last ball chosen is blue.

| $n$ | 50 | 100 | 1,000 | 10,000 | 50,000 | 100,000 | 500,000 | 1,000,000 |
|---|---|---|---|---|---|---|---|---|
| $n_E$ | 18 | 34 | 320 | 3,263 | 16,345 | 32,662 | 161,896 | 323,807 |
| $n_E/n$ | 0.3600 | 0.3400 | 0.3200 | 0.3263 | 0.3269 | 0.32662 | 0.323792 | 0.323807 |

## 16.2 Simulating Conditional Probabilities, Means, and Variances

### Conditional Probabilities

Consider now the simulation of *conditional* probabilities. The basic algorithm needs to be modified. We must first test for the condition, and only when the condition is met do we test for the desired event. The answer is given by the ratio of the number of times the desired event is met to the number of times that the condition is satisfied. The basic algorithm to determine the probability of the event $A$ given the event $B$, i.e., Prob$\{A|B\}$, may be written as follows:

```
Initialize event counts for A and B:  n_A = n_B = 0;
For a total of "n" trials {
    * Perform the probability experiment
    * If event B occurs, then {
        ~ increment n_B;
        ~ if event A occurs, then { increment n_A }
      }
}
probability of A|B = n_A/n_B
```

### Probability Scenario 5

The following example involves conditional probabilities. A box contains two white balls, two red balls, and a black ball. Two balls are chosen without replacement from the box. What is the probability that the second ball is white given that the first ball chosen is white. Although it is pretty obvious that the answer is 0.25, we shall use simulation to illustrate how it is used with conditional probabilities. Here, event $B$ occurs when the first ball chosen is white, and event $A$ occurs when the second ball chosen is white.

```
// *** Begin Simulation ***************************
        int n_A = 0;  int n_B = 0;            // Initialize event counts
        for (int n=0; n<max[i]; n++) {        // Perform max[i] trials
           int w = 2; int r = 2; int b = 1;   // Select colors
```

```
        int r_e = generator.nextInt(w+r+b);     // Choose a ball
        if (r_e < w){                           // If it is white, then
           n_B++;                               // Event B has occurred, so
           r_e = generator.nextInt(w+r+b-1);    // Choose another ball
           if (r_e < w-1){n_A++;}               // Success, if this is white
        }
     }
     double prob = (double)n_A/n_B;             // Approximate probability
// *** End Simulation ****************************
```

We obtained the results in Table 16.5 on one execution of the simulation program.

Table 16.5. Probability that the second ball is white given that the first is white.

| $n$ | 50 | 100 | 1,000 | 10,000 | 50,000 | 100,000 | 500,000 | 1,000,000 |
|---|---|---|---|---|---|---|---|---|
| $n_A$ | 1 | 13 | 94 | 1,028 | 4,924 | 9,921 | 50,173 | 99,803 |
| $n_B$ | 20 | 37 | 403 | 3,972 | 19,969 | 39,923 | 199,752 | 399,812 |
| $n_A/n_B$ | 0.0500 | 0.35135 | 0.23325 | 0.25881 | 0.24658 | 0.25026 | 0.25118 | 0.24962 |

**Means and Variances by Simulation**

When we use simulation to determine the probabilities of certain events, we simulate the experiment, count the number of times that a successful outcome occurs and take the ratio of this number to the total number of trails generated. When using simulation to compute means and higher moments of a random variable $X$, we need to adopt a somewhat different strategy. In these cases, we simulate the probability experiment a large number of times, take note of the value that the random variable assumes at each trial and, to form the mean, take the average of these values. Higher moments are computed in a similar fashion. This is best illustrated by means of the following example, in which we compute the mean $E[X]$ and the variance as $\text{Var}[X] = E[X^2] - (E[X])^2$ of the random variable $X$.

**Probability Scenario 6**

Balls are drawn from an urn containing $w$ white balls and $b$ black balls until a white ball appears. The simulation program given below computes an estimate of the mean and the variance of the number of balls drawn, assuming that each ball is replaced after being drawn. In this program, the mean number incorporates a count for when the white ball is chosen. As mentioned when this example was first discussed in Chapter 1, the mean number of balls drawn *before* the white ball is drawn is one less than this number. The variance is unaffected. The following table shows the results of a succession of runs of this simulation program, each using the same number of trials, 10,000. The number of white balls is chosen to be $w = 2$ and the number of black balls to be $b = 8$. As expected, the results (Table 16.6) are not too far off the exact answers of $1 + b/w = 5.0$ and $b(w + b)/w^2 = 20.0$ respectively.

Table 16.6. Mean and variance of number of balls chosen.

| $E[X]$ | 5.0061 | 5.0065 | 4.9604 | 5.0235 | 4.9948 | 5.0081 | 4.9938 | 5.0390 |
|---|---|---|---|---|---|---|---|---|
| $\text{Var}[X]$ | 19.8725 | 20.24981 | 19.9034 | 20.0370 | 19.3104 | 20.2276 | 20.2590 | 20.7259 |

```
import java.util.Random;
class mean {
   public static void main (String args[]) {
      Random generator = new Random();
      int w = 2; int b = 8;                          // Select colors
      int max = 10000;
      double average = 0; double variance = 0;
      for (int n = 1; n<=max; n++) {                 // Perform "max" trails
         int n_d = 0;                                // Initialize drawn count
         boolean not_white = true;

         do {                                        // Do while unsuccessful
            int r_e = generator.nextInt(w+b);        // Pick a ball, and ...
            n_d++;                                   //    Increment total drawn
            if (r_e < w){                            // Is it white? ...
                not_white = false;                   //    Yes, so stop
            }
         }
         while (not_white);                          // This trial drew n_d balls

         average = average + n_d;                    // Building up average
         variance = variance + n_d*n_d;              // Building up variance
      }
      average = average/max;                          // Compute average
      variance = variance/max - average*average;      // Compute variance

      System.out.println("mean = " + average);        // Print results
      System.out.println("variance= " + variance);
   }
}
```

Given that the above six probability scenarios were all previously solved mathematically, the reader may wonder whether simulation serves a useful purpose or not. The answer must be a resounding yes, for there are many problems that are very hard, or even impossible, to solve mathematically, but yet are easily amenable to the simulation approach. Consider the problem of rolling three dice until six spots appear on all three of them. We might like to know the probability that this will take at least 30 throws. While it is not difficult to imagine how to write a simulation to approximate this probability, solving this problem mathematically is very hard indeed. When mathematical solutions can be found, then they should be used, for often their formulation provides more information about the underlying probability experiment itself. Simulation simply gives us a number. Additionally, complex simulation programs require time to program, may be hard to debug and may need extensive computational resources. However, in the absence of mathematical solutions, they can be invaluable.

## 16.3 The Computation of Definite Integrals

Random numbers are not limited to the evaluation of probability scenarios, but have a much wider field of application. However, since this takes us somewhat outside of intended scope of this text, we shall be content to briefly describe only one, albeit a rather important, application, namely the computation of definite integrals. Definite integration is usually described in terms of computing the area under a curve between two limits. Let $f(x)$ be a bounded function on $[a, b]$, i.e., there exists a constant $c$ such that $f(x) \le c$ for all $x \in [a, b]$. Then $I = \int_a^b f(x)dx$ is the area beneath the curve $y = f(x)$ between the limits $x = a$ and $x = b$. This area is contained in a bounding rectangle whose base has length equal to $b - a$ and with height equal to $c$, as illustrated in Figure 16.1. The area we seek to compute, namely $I$, the shaded portion of the figure, constitutes a certain percentage of this rectangle. If we were to generate random points within this rectangle, then the proportion of these points to fall into the shaded area would be the same as the percentage of the area $c(b - a)$ occupied by $I$. This gives us a means to evaluate the integral using random numbers.

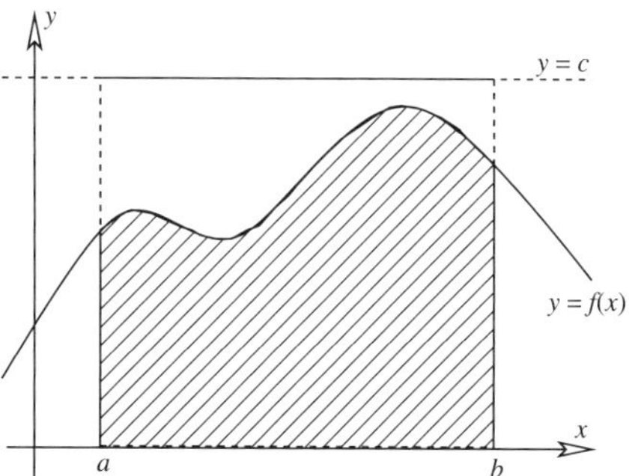


Figure 16.1. Illustration of definite integration.

If $r_1$ and $r_2$ are uniformly distributed on $[0, 1]$, then $x_1 = a+(b-a)r_1$ is uniformly distributed on $[a, b]$; $y_1 = cr_2$ is uniformly distributed on $[0, c]$ and the point $(x_1, y_1)$ is uniformly distributed over the rectangle with base length $b - a$ and height $c$. For any point generated like this, that point lies in the shaded area if $y_1 \le f(x_1)$. The percentage of the rectangle occupied by the shaded portion can be approximated by generating many points $(x_i, y_i)$ randomly and uniformly distributed over the rectangle and forming the ratio of the number that satisfy the constraint $y_i \le f(x_i)$ to the total number of points generated. The value of the integral is then obtained by multiplying this percentage by $c(b - a)$, the area of the rectangle. This approach is called the "accept–reject" method. A point $(x_i, y_i)$ for which $y_i \le f(x_i)$ is "accepted"—it lies in the shaded area; otherwise it is rejected. The proportion of the rectangle occupied by the area under $y = f(x)$ is given by the ratio of the number of accepted points to the total number of points generated. Later we shall see that this approach can be modified to provide a means of generating random numbers that are not uniformly distributed but which instead satisfy some arbitrary density function.

While this method of evaluating a definite integral is simple to understand, it is not very effective, requiring two uniformly distributed random numbers for each point in the rectangle and a very large number of points for reasonable accuracy. What we really need is the *average* value of $f(x)$ on the interval $[a, b]$ which could then be multiplied by $b - a$ to provide the value of $I$. To proceed in this direction, let us first consider the case when the limits of integration are $a = 0$ and $b = 1$ and hence $I = \int_0^1 f(x)dx$. If $r$ is a uniformly distributed random number in the interval $[0, 1]$ , then $f(r)$ is a point at random on the curve $y = f(x)$ within the region of interest. Generating many such points allows us to approximate the expected value of $f(x)$ for $x \in [0, 1]$. In particular, if $r$ is uniformly distributed on $[0, 1]$, then

$$I = E[f(r)],$$

the expected value of the function on $[0, 1]$ times the length of the interval, 1. Given a sequence $r_1, r_2, \ldots$ of uniformly distributed random numbers on $[0, 1]$, the corresponding random variables $f(r_1), f(r_2), \ldots$ are independent and identically distributed with mean value equal to $I$, the integral we seek. From the strong law of large numbers, we have, with probability 1,

$$\lim_{n\to\infty} \sum_{i=1}^{n} \frac{f(r_i)}{n} = E[f(r)] = I.$$

This approach to evaluating definite integrals is generally termed "Monte Carlo" integration.

**Example 16.1** As a simple example, let us compute $I = \int_0^1 e^x dx$ which we know to be equal to $e - 1 = 1.718282$. The following Java program follows the guidelines of the previous examples and estimates the value of $I$ using different numbers of uniformly distributed random numbers.

```
import java.util.Random;
import static java.lang.Math.*;

class integrate {
   public static void main (String args[]) {
      int max [] = {50,100,1000,10000,50000,100000,500000,1000000};
      Random generator = new Random();
      for (int i=0; i<8; i++) {

// ***    Begin Simulation ***************************
          double s = 0;
          for (int n=1; n<max[i]; n++) {          // Generate max[i] runs.

               double r = generator.nextDouble();
               s = s + exp(r);                    // f(x) = exp(x)
          }
          double Ivalue = s/(double)max[i];       // Approximate integral
// ***    End Simulation ***************************

          System.out.println("n = " + max[i] + "   Ivalue = " + Ivalue);
      }
   }
}
```

A single run of this program produced the results in Table 16.7.

Table 16.7. Monte Carlo evaluations of the definite integral $\int_0^1 e^x dx$.

| $n$ | 50 | 100 | 1,000 | 10,000 | 50,000 | 100,000 | 500,000 | 1,000,000 |
|---|---|---|---|---|---|---|---|---|
| Ivalue | 1.832951 | 1.704603 | 1.725699 | 1.717236 | 1.718372 | 1.718015 | 1.718362 | 1.718751 |

We now examine the case when the limits of integration are $a$ and $b$ rather than 0 and 1. Observe that if $y$ is uniformly distributed on $[0, 1]$ then $x = a + (b - a)y$ is uniformly distributed on $[a, b]$. This suggests a change of variable. Letting

$$y = \frac{x - a}{b - a}$$

gives

$$dy = \frac{dx}{b - a}$$

and thus

$$\int_a^b f(x)dx = \int_0^1 f(a+[b-a]y)(b-a)dy = (b-a)\int_0^1 f(a+[b-a]y)\,dy = (b-a)\int_0^1 g(y)dy,$$

where $g(y) = f(a + [b-a]y)$. This means that we can compute $I = \int_a^b f(x)dx$ by generating uniformly distributed random numbers $y \in [0, 1]$, estimating the average value of $g(y)$ on $[0, 1]$ and multiplying the result by $b - a$.

**Example 16.2** To use this approach to obtain approximations to $\int_1^3 e^x\,dx$ using the previous Java program, it suffices to modify the program as follows.

```
// ***   Begin Simulation **************************
        double s = 0;  double a = 1;  double b = 3;
        for (int n=1; n<max[i]; n++) {              // Generate max[i] runs.

             double r = generator.nextDouble();
             double y = a+(b-a)*r;
             s = s + exp(y);                        // g(y) = exp(y)
        }
        double Ivalue = (b-a)*s/(double)max[i];     // Approximate integral
// ***   End Simulation **************************
```

At this point we shall leave our discussion of the use of random numbers to approximate definite integrals. It is not difficult to extend these simple concepts to the case when one or both limits are infinite or to the case of multivariate integration. Indeed it is in this last area that the Monte Carlo method excels.

Our goal in this chapter has been to show some of the uses to which uniformly distributed random numbers can be put and to provide a feeling for some of the issues involved, such as the number of random numbers needed to mimic an event and the large number of trials usually needed to obtain a reasonably accurate result. This has set the stage for an examination of some important topics. In the next chapter we shall concern ourselves with mechanisms for generating sequences of uniformly distributed random numbers and procedures for testing if the numbers we generate are actually "random" and "uniformly distributed." Following that, in Chapter 18, we shall consider procedures for generating numbers that are not *uniformly* distributed over some interval, but rather are distributed according to some other distribution function, such as exponential or normal. In Chapter 19 we investigate how simulation programs should be structured and examine some of the problems involved in developing computer simulations of complex systems. A major issue in simulation concerns the accuracy of the results obtained by a simulation. Once we have written a simulation program and used it to solve a physical or hypothetical problem, how can we estimate the accuracy of the computed solution? This is the topic of the final chapter of this book.

## 16.4 Exercises

**Exercise 16.1.1** Write code similar to that in Section 16.1 to simulate each of the three scenarios described below. Also, for each scenario, compute the exact result mathematically and compare your simulation answers with the exact answer.

(a) A prisoner in a Kafkaesque prison is put in the following situation. A regular deck of 52 cards is placed in front of him. He must choose cards one at a time to determine their color. Once chosen, the card is replaced in the deck and the deck is shuffled. If the prisoner happens to select three consecutive red cards, he is executed. If he happens to select six cards before three consecutive red cards appear he is granted freedom. What is the probability that the prisoner is executed?

(b) Three cards are placed in a box; one is white on both sides, one is black on both sides, and the third is white on one side and black on the other. One card is chosen at random from the box and placed on a table. The (uppermost) face that shows is white. What is the probability that the hidden face is black?

(c) A factory has three machines that manufacture widgets. The percentages of a total day's production manufactured by the machines are 10%, 35%, and 55% respectively. Furthermore, it is known that 5%, 3%, and 1% of the outputs of the respective three machines are defective. What is the probability that a randomly selected widget at the end of the day's production runs will be defective?

**Exercise 16.1.2** Consider the problem of rolling three dice until six spots appear on all three of them. Simulate this scenario to estimate the probability that this will take at least 30 throws.

**Exercise 16.1.3** Consider the gambler's ruin problem in which a gambler begins with \$50, wins \$10 on each play with probability $p = 0.45$, or loses \$10 with probability $q = 0.55$. The gambler will quit once he doubles his money or has nothing left of his original \$50. Write a simulation program to compute the expected number of times the gambler has

(a) \$90 before ending up broke;
(b) \$90 before doubling his money;
(c) \$10 before doubling his money;
(d) \$10 before going broke.

**Exercise 16.1.4** By enclosing a circle of unit radius and centered at the origin inside a square box whose sides are of length 2, compute an approximation to the number $\pi$ using a sequence of uniformly distributed random numbers.

**Exercise 16.1.5** Evaluate the following integrals using the Monte Carlo method.

(a)

$$I_a = \int_0^2 \frac{x\,dx}{\sqrt{2x^2+1}}.$$

(b)

$$I_b = \int_{\pi/6}^{\pi/2} \cos^2\theta \sin\theta \, d\theta.$$

(c)

$$I_c = \int_0^{\ln 2} \frac{e^{2x}\,dx}{\sqrt[4]{e^x+1}}.$$

# Chapter 17

# *Uniformly Distributed "Random" Numbers*

In the previous chapter we assumed that we had access to a sequence of random numbers in the half-open interval $[0, 1)$. But now the question arises: "What is a random number?" Is 101 a random number? Is 33? We would be wrong to answer in the affirmative to either of these questions. A number by itself cannot be random. It is only when we look at a *sequence* of numbers that we may begin to ask meaningful questions about randomness. It (randomness) is not something that we can easily explain, but we believe that we know it when we see it! If we were to generate a sequence of $T$ random integers that lie the range $[0, 9]$, then we would expect 0 to occur $T/10$ times, and the same for the other nine integers. We would not accept a sequence of $T$ zeros as being random and yet the probability of getting $T$ zeros is just the same as the probability of getting any other specific sequence, including the one we end up with. It would appear that we wish to obtain random sequences that are not entirely random!

There is not as yet any satisfactory definition of randomness, although a number of authors have made valiant efforts. In particular, Knuth [25] devotes considerable attention to the topic and we highly recommend reading his work to flesh out the limited amount of space that we can devote to it in this chapter. We shall be content with the definition that we gave earlier, that, by random, we mean that if the interval were partitioned into equally sized subintervals, the next random number to be generated would be equally likely to fall into any of the subintervals. Random numbers having this property are said to be *uniformly distributed* random numbers.

Before the advent of computers, random numbers were generated by purely mechanical means. Random digits produced in this fashion are often said to be *truly* random, to distinguish them from computer-generated *pseudorandom* numbers. Urns containing numbered balls, much like those seen today in lotteries, roulette machines, and other specifically engineered devices were used to produce random integers which were arranged into tables and printed. Typically, books that contained tables of logarithms and the like, also included several pages of random numbers arranged into groups of five. A user began somewhere in the middle of the table and used the numbers from that point onward. Beginning in 1927 and continuing through 1955, a number of different sources produced tables containing from 1,600 through 250,000 random digits. Then, in 1955, the RAND corporation published a book containing one million random digits along with material on the generation process and the use of these tables. These were produced using electronically generated noise to increment electronic counters.

Once computers became more widely used, programs were written to produce random numbers. One of the earliest was the so-called *middle square* method von Neumann. This approach was very simple. The method begins with an initial number called the *seed*: the next number in the sequence is found by taking the middle part of the square of the current number. For example, an $n$-digit number is squared to produce a $2n$-digit number and some middle $n$-digits chosen as the next number in the "random" sequence. However, numbers generated like this are most certainly *not* random. Indeed, the sequence is completely deterministic. However, the idea is that, if examined in its entirety, the sequence of numbers *looks as if* it has the properties that we like to see in sequences of random numbers. Additionally, numbers generated in this fashion have the highly desirable property that they can be reproduced, which can be extremely helpful in many circumstances. Unfortunately, and despite its simplicity, the middle square method is not a good method, for the sequence repeats after a relatively short cycle, or else degenerates into a string of zeros. With all computer generated sequences of random numbers, the sequence will begin to repeat after a certain point. The length of the sequence before this point is called its cycle length. Current computer generated methods usually have extremely long cycles, as we shall see later. This is not a property that the middle square method shared.

The development of efficient and effective random number generators, generally abbreviated to RNG, is an ongoing hot topic of research and has application in many important branches of science, technology and entertainment, such as cryptology and gambling, to name just two. Substantial progress has been made with both true RGNs (TRNGs) and pseudo RNGs (PRGNs). The Distributed Systems Group at Trinity College, Dublin generates truly random sequences by means of atmospheric noise and makes them available to the general public for free from their web site *www.Random.org*. Since these numbers are genuinely random, they are not reproducible so a user wishing to repeat an experiment, to locate a problem for example, must store the random numbers as they are generated, to enable the exact same simulation experiment to be repeated at a later time. This need to store random number is a major disadvantage of TRNGs and explains why computer-generated sequences are often preferred. In situations when it is unnecessary to rerun an experiment, such as in gambling, TRNGs are often preferred. Not only are truly random numbers used, but the inability to reproduce the sequence is a deterrent to cheating.

Our interest in this chapter lies not with true RNGs but with pseudo-RNGs. Pseudorandom number generators are invariably provided as an integral function in programming languages. However, a user should be careful because the PRNG used in some programming languages leaves a lot to be desired. The paper by L'Ecuyer [30] is particularly enlightening in this regard. Before we leave this general discussion on RNGs, we mention one interesting possibility based on successive digits in the number $\pi$. Quartically and even quintically converging methods (in which the number of correct digits of $\pi$ is quintupled at each iteration) are now available and hundreds of billions of digits have been produced. There does not appear to be a finite cycle length, and mathematicians tell us that none exists. Additionally, statistical tests indicate that the successive digits in $\pi$ satisfy stringent requirements for randomness.

## 17.1 Linear Recurrence Methods

Our goal is to produce a sequence of random numbers in which each number is independently drawn from the continuous uniform distribution on the interval [0, 1], i.e., the distribution with density function

$$f_X(x) = \begin{cases} 1, & 0 \le x \le 1, \\ 0 & \text{otherwise.} \end{cases}$$

Thus each and every number in the sequence has mean equal to $1/2$ and variance equal to $1/12$. This means that if we partition the interval $[0,\ 1]$ into $n$ panels, a sequence of length $N$ will have an average of $N/n$ random numbers in each panel and the panel into which the next random number is placed is independent of the placement of all previous random numbers. This then is the goal. Unfortunately, computer generated sequences of random numbers, by virtue of a deterministic algorithm which generates them, cannot truly meet this goal. Consequently, the onus is on the user of such software products to ensure that the generated sequence is sufficiently close to the goal to satisfy his or her requirements, and in particular to verify that the numbers are uniformly distributed and that there is no apparent autocorrelation among them. The properties which all good random number generators should possess include the following.

1. To the greatest extent possible, the random numbers that they generate should be independently and identically distributed on [0, 1].
2. The cycle length must be long, since complex simulations require extremely large quantities of random numbers. For this same reason, the algorithm must also be fast.
3. The numbers should be replicable, so that identical runs of the simulation can be repeated to detect anomalies.
4. The software should be portable, so that the generator can be run and tested on a variety of computer systems and computer languages.
5. A "jump-ahead" feature, by which the random number at any distance $\eta$ from a given random number can be easily computed, is useful when the software is run in parallel.

The most commonly used methods for generating sequences of uniformly distributed random numbers today are extensions and variants of the linear congruent method. The *linear congruent method (LCM)*, first proposed by Lehmer [31], begins with an initial seed $z_0$ and generates random integers according to the recurrence formula

$$z_i = (az_{i-1} + c) \bmod m. \tag{17.1}$$

The parameter $a$ is called the *multiplier*, $c$ the *increment*, and $m$ the *modulus*. The initial value $z_0$ is called the *seed*. The fact that the arithmetic is carried out modulo $m$ means that Equation (17.1) will produce at most $m$ distinct integers. Also the process is guaranteed to be periodic: each time the recurrence formula produces a number that is equal to the seed, one cycle has been completed and the next cycle, which will be identical to the last, begins.

**Example 17.1** A poor choice of the parameters $a,\ c,\ z_0$, and $m$ results in a very poor sequence. For example, with $a = 1,\ c = 2,\ z_0 = 3$, and $m = 10$, we obtain the sequence

$$5, 7, 9, 1, 3, 5, 7, 9, 1, 3, 5, 7, 9, \ldots .$$

Not only do none of the even integers appear, but the odd integers are in increasing order.

When a cycle contains all possible integers from 0 through $m - 1$, it is said to have a *full period*. Such a cycle is a permutation of the integers $0, 1, \ldots, m - 1$, the particular permutation generated being a function of the multiplier $a$ and the increment $c$. Reals in the interval [0, 1) are generated as the successive numbers $u_i = z_i/m$. Observe that not all reals in the interval [0, 1) can be generated. Only real numbers equal to $i/m,\ i = 0, 1, \ldots, m - 1$ can possibly appear. Furthermore, it is blatantly apparent that these numbers are not "random" since each is determined mathematically from the previous; yet we expect to end up with a sequence which "appears" random. To achieve this appearance of randomness, the parameters $m,\ c$, and $a$ should be chosen to satisfy the following conditions:

- $a \geq 0, c \geq 0$,
- $m > \max(z_0,\ a,\ c)$,
- $c$ and $m$ are relatively prime, i.e., they have no common divisor,
- $a - 1$ is a multiple of $p$—for every prime $p$ that divides $m$, and
- $a - 1$ is a multiple of 4—if $m$ is a multiple of 4.

These conditions alone are not sufficient to produce satisfactory sequences of random numbers. The last three are meant to guarantee that a full period is achieved and, for this property, these three conditions are both necessary and sufficient. In general, $m$ should be chosen as large as possible, since the period can never be longer than this. A good choice is the largest integer that can be represented on the computer being used. The parameters $a$ and $c$ should be chosen to satisfy the above conditions. An example from the literature which satisfies these as well as some additional conditions and which appears to produce acceptable sequences is $m = 2^{31}$, $a = 314159265$, the first nine digits of $\pi$, and $c = 453806245$. In some references, the number $a$ appears as $a = 314159269$.

When $c$ is chosen to be zero, the algorithm reduces to

$$z_i = az_{i-1} \bmod m \tag{17.2}$$

and is called the *multiplicative congruent method (MCM)*. It involves fewer numeric operations and is consequently somewhat faster than the LCM. Although it is not possible to get a full period with this multiplicative alternative, periods of up to $m-1$ can be achieved. For a 32-bit machine, choosing $m$ as $m = 2^{31} - 1 = 2147483647$, which just happens to be a prime number, and $a = 7^5 = 16807$, gives a period that is equal to $m - 1$.

**Example 17.2** Given $a = 16807$, $m = 2^{31} - 1 = 2147483647$, and initial seed $z_0 = 230455$, we obtain the following:

$$\begin{aligned}
&z_i = (az_{i-1}) \bmod m && u_i = z_i/2^{31} \\
&z_1 = 1725773538 && u_1 = z_1/2^{31} = 0.80362592730671 \\
&z_2 = 1161716784 && u_2 = z_2/2^{31} = 0.54096653312445 \\
&z_3 = 52670164 && u_3 = z_3/2^{31} = 0.02452645637095 \\
&z_4 = 464183784 && u_4 = z_4/2^{31} = 0.21615241840482 \\
&z_5 = 1876251784 && u_5 = z_5/2^{31} = 0.87369782105088.
\end{aligned}$$

Notice that the division is with $m + 1$, a power of 2, rather than with $m$. This makes the division more efficient and has negligible consequences on the quality of the generator.

The MCM method with these specific values of $a$ and $m$ was used extensively until rather recently. It was the generator used in the widely distributed "Arena" software simulation package and apparently proved to be satisfactory over a period of many years. However, some deficiencies were found with this generator [30] and more sophisticated approaches are now used in Arena and elsewhere.

A consensus is building that even cycles of length $m = 2^{31}$ are insufficient for current critical applications and this has lead to the development of methods that are collectively referred to as *multiple recursive generators (MRGs)*. The basic multiplicative congruent method may be extended to incorporate not just the previous random integer into the formula, but a linear combination of several of the preceding random integers. This approach forms the basis of the most widely used RNGs today. For example, the sequence generated from

$$z_i = (a_1 z_{i-1} + a_2 z_{i-2}) \bmod m$$

incorporates the two previous random integers. In this case, two initial seeds $z_0$ and $z_1$ are required but this can be accomplished by running the standard multiplicative method with one initial seed to generate a second. A variant of this, called the *additive congruent method*, sets both coefficients equal to 1 and uses the previous random integer and one that is back a distance $\eta$. This yields

$$z_i = (z_{i-1} + z_{i-\eta}) \quad \bmod m,$$

and has the advantage that no multiplication is required. Since the performance characteristics of this generator are less well known than that of the MCM generator, Knuth recommends a careful validation of the sequence generated. Today it is common to use more than just the two previous integers to produce the next in the sequence. In general, the $i^{\text{th}}$ value is obtained from the preceding

$k$ values by means of the formula

$$z_i = \left(\sum_{j=1}^{k} a_j z_{i-j}\right) \bmod m.$$

This is referred to as a modulus $m$, order $k$, MRG. It requires $k$ initial seeds, $z_0, z_1, \ldots z_{k-1}$, all of which may be found from the basic multiplicative congruential method and a single starting seed. Uniformly distributed random numbers in the interval [0, 1) are obtained as $u_i = z_i/m$. With an appropriate selection of the coefficients $a_i$, cycles of up to $m^k - 1$ can be obtained [25]. However the number of different reals in the interval [0, 1) is not increased — this procedure's chief purpose is to increase the cycle length without sacrificing either efficiency or effectiveness.

**Example 17.3** Given initial seeds $s_0 = 3,\ s_1 = 5,\ s_2 = 7$ and parameters $a_1 = 7,\ a_2 = 1,\ a_3 = 3$, and $m = 11$:

- $z_i = (a_1 z_{i-1}) \bmod m$ has period equal to 10,
- $z_i = (a_1 z_{i-2} + a_2 z_{i-1}) \bmod m$ has period equal to 60, and
- $z_i = (a_1 z_{i-3} + a_2 z_{i-2} + a_3 z_{i-1}) \bmod m$ has period equal to 1330.

In all three cases, only ten distinct integers are produced, namely, 1–10 inclusive. The periods however, increase from 10 to 60 to 1330; the first and last of these are the maximum possible with $m = 11$ and order $k = 1$ and $k = 3$.

The RANMAR generator at Cern falls into this category. It requires 103 initial seeds which are usually set by a single integer. The period is approximately equal to $10^{43}$. Even larger periods can be obtained by combining the results of multiple MRGs that execute in parallel. One example is L'Ecuyer's MGR32k3a generator [30] which combines two MRGs of order 3. Let these two MRGs be as follows:

$$z_{1,i} = (a_{1,1} z_{1,i-1} + a_{1,2} z_{1,i-2} + a_{1,3} z_{1,i-3}) \bmod m_1,$$
$$z_{2,i} = (a_{2,1} z_{2,i-1} + a_{2,2} z_{2,i-2} + a_{2,3} z_{2,i-3}) \bmod m_2.$$

Then the next random integer in the sequence is chosen to be

$$x_i = (z_{1,i} - z_{2,i}) \bmod m_1.$$

L'Ecuyer recommends the following parameter values:

$$a_{1,1} = 0, \quad a_{1,2} = 1403580, \quad a_{1,3} = -810728, \quad m_1 = 2^{32} - 209,$$

$$a_{2,1} = 527612, \quad a_{2,2} = 0, \quad a_{2,3} = -1370589, \quad m_2 = 2^{32} - 22853,$$

and shows that the resulting period is close to $2^{191} \approx 3 \times 10^{57}$. But even the length this period pales besides that of the colorfully named *Mersenne twister* generator of Matsumoto and Nishimura [33] which has a period that is equal to $2^{19937} - 1$. Unlike the MGR32k3a generator, which uses the values $m_1 = 2^{32} - 209$ and $m_2 = 2^{32} - 22853$, the Mersenne twister is one of a class of generators that are based on recurrences modulo 2. Such generators produce uniformly distributed random numbers in the interval [0, 1) in a bitwise fashion.

We conclude this section with the following two observations:

1. When it is possible to jump directly from a given random integer $z_i$ to $z_{i+\eta}$, for some arbitrary integer $\eta$, a property that well-designed generators should possess, then multiple sequences can be obtained from the same recurrence formula. This is particularly appropriate for implementing in parallel.

2. The generally stated objective of a computer program that produces sequences of random numbers is to output numbers that are uniformly distributed on the interval [0, 1] and to do so efficiently. However the occurrence of either of the end points 0 and 1 could lead to difficulties in certain applications such as in the generation of random numbers that have a distribution other than uniform. Furthermore, if any $z_i$ takes the value 0 in Equation (17.2), all succeeding random integers will also be zero. Thus the output of a random number generator should always lie in the interval (0, 1). This may be accomplished by using $z_i + 1$ instead of $z_i$ and dividing the result by $m + 1$ rather than $m$. Unless noted otherwise, in what follows, we shall use the term "uniformly distributed random number" to designate a random number uniformly distributed in (0, 1). Having said this, for the purpose of analysis, the interval is usually taken to be [0, 1).

## 17.2 Validating Sequences of Random Numbers

There are two approaches to testing random number generators. The first, the empirical approach, involves running the generator many times and applying a series of statistical tests to evaluate the "randomness" of the sequences generated. This is the approach that we develop in this section and it can be applied to sequences of numbers obtained from both true and pseudorandom number generators. The second approach is a mathematical analysis of the formulae upon which the generator is based. This approach relies heavily on an understanding of mathematical number theory. Obviously, such an approach can only be applied to computer generated random numbers. Readers interested in this approach should consult Knuth [25].

Empirical approaches, of which there exists a large number, fall into one of two categories depending on their objective. The first category consists of tests whose purpose is to verify that the numbers in a given sequence are uniformly distributed over the interval [0, 1). We shall consider two tests in this category, the *chi-square "goodness-of-fit" test* and the *Kolmogorov-Smirnov test*. The null hypothesis $H_0$ for these tests is that the sequence of random numbers is indeed uniformly distributed over the unit interval. The reader is cautioned that a failure to reject the null hypothesis means only that no evidence of nonuniformity was observed and not that the generator is guaranteed to produce sequences that satisfy the uniform distribution criterion. The second category contains tests designed to ensure that the numbers in the sequence are independent, that there is no correlation among them. Here a large number of tests is at our disposition and we shall review a number of different *run tests*, a *gap test*, and the *poker test*. The null hypothesis for these tests is that the numbers satisfy our independence criterion, again with the same caveat as before. Each test requires the specification of a significance level, which we denote by $\alpha$ and which is usually of the order of 0.01 or 0.05. This is the probability of rejecting the null hypothesis given that the null hypothesis is in fact true. We have

$$\alpha = \text{Prob}\{H_0 \text{ is rejected} \mid H_0 \text{ is true}\}.$$

Thus if we have a random number generator that is known to have good uniformity properties and we perform a statistical test with a significance level of $\alpha = 0.05$ on 20 sequences of random numbers output by this generator, we should expect that, purely by chance, the null hypothesis is rejected once. When the statistical test is run on 100 generated sequences, we should expect the null hypothesis to be rejected $100\alpha = 5$ times. The fact that the generator fails the test on average 1 out of every 20 times does not mean that it is defective and should be discarded but only that the element of chance must be taken into account.

### *17.2.1 The Chi-Square "Goodness-of-Fit" Test*

The chi-square goodness-of-fit test allows us to compare a sample distribution against a theoretical one. In our case, we would like to check just how uniformly a particular sequence of random numbers is distributed. To proceed, we partition an interval containing $n$ pseudorandom numbers into $k$ equal subintervals and compare the count of random numbers in each subinterval to the theoretical count of $n/k$. To get meaningful results, $k$ should typically be larger than 10 and $n$ should be larger than $10 \times k$. The word "binning" is sometimes used to describe the placement of observations into different categories, or bins. What we seek to do is to compare the distribution of one (experimental) random variable, the one found by the experiment that consists of generating the $n$ pseudorandom numbers, with that of a second (theoretical) random variable, the one that consists of exactly $n/k$ numbers per subinterval. A very specific random variable has been constructed to allow us to perform this comparison. It is called the *chi-square*, $\chi^2$, random variable and can be used to measure the degree of similarity of two probability mass functions. If its value is smaller than a critical value, we may conclude that the two random variables used to form $\chi^2$ have identical distributions.

If $n_i$ denotes the number of random numbers found in the $i^{\text{th}}$ interval and $\bar{n}_i$ is the number we should theoretically expect to find, then the statistic

$$\chi^2 = \sum_{i=1}^{k} \frac{(n_i - \bar{n}_i)^2}{\bar{n}_i} \tag{17.3}$$

defines the random variable $\chi^2$. In our case, with a uniform distribution, $\bar{n}_i = n/k$ for all $i$. Values of the cumulative distribution function of a chi-square random variable are usually obtained from tables. These tables give the value of $\chi^2$ under different circumstances. Each row corresponds to a different number of degrees of freedom. This is the number of ways in which the two distributions may vary from each other. When associated with discrete densities, the number of degrees of freedom is equal to the number of subintervals, $k$, minus the number of differences in the density parameters. With the uniform density there can only be one, the mean value, so for the problem at hand, Equation (17.3) has $k - 1$ degrees of freedom and the $(k-1)^{\text{th}}$ row in the table of $\chi^2$ values must be used.

Each column in the table corresponds to a different probability, $\alpha$, typically .99, .95, .90, .50, .10, .05, and .01. The hypothesis that the random numbers are uniformly distributed, i.e., the *null* hypothesis, must be accepted if the computed value of $\chi^2$ is less than the value $\chi^2_\alpha$ obtained from the table. We have $\text{Prob}\{\chi^2 \le \chi^2_\alpha\} = 1 - \alpha$ which simply means that in many tests, we should expect the computed value of $\chi^2$ to exceed $\chi^2_\alpha$ about $100\alpha$ percent of the time. A significant variation from this percentage, either above *or below*, should be viewed with concern.

**Example 17.4** This may be confusing, but an example should help make the situation more clear. The Java program presented below generates 1,000 random numbers in the range 0–9 and then applies the chi-square test to the result. One particular run of this program produced the following:

$$\text{Count[]} = 94 \;\; 105 \;\; 90 \;\; 90 \;\; 108 \;\; 98 \;\; 100 \;\; 111 \;\; 101 \;\; 103$$

which results in the value $\chi^2 = 4.6$. Let us now ask what this means in terms of the "randomness" of the numbers generated. Since we choose $k = 10$, we need to look at the row that corresponds to nine degrees of freedom (DF). This row is given as

| DF | 0.995 | 0.990 | 0.975 | 0.950 | 0.900 | 0.500 | 0.100 | 0.050 | 0.025 | 0.010 | 0.005 |
|---|---|---|---|---|---|---|---|---|---|---|---|
| 8 | | | | | | | | | | | |
| 9 | 1.73 | 2.09 | 2.70 | 3.33 | 4.17 | 8.34 | 14.68 | 16.92 | 19.02 | 21.67 | 23.59 |
| 10 | | | | | | | | | | | |

The acceptance region for the null hypothesis, with nine degrees of freedom and a significance level of $\alpha = 0.05$, is $\chi^2 \le 16.92$ and our computed value of

$$\chi^2 = \frac{1}{100} \sum_{i=1}^{10} (n_i - 100)^2 = 4.6$$

which is clearly within this region. This means that the hypothesis that the random numbers are uniformly distributed should not be rejected. To simplify, it means that with probability 0.95, the numbers *are* uniformly distributed. As mention just above, these numbers were obtained with a single run of the Java program. In other runs, different values of $\chi^2$ were obtained. Consecutive values of 11.86, 2.94, 8.84, 9.44, 10.72 were all obtained prior to obtaining the above value of 4.6. In all these cases, $\chi^2$ is less than 16.92 indicating that the random numbers *are* uniformly distributed.

```
import java.util.Random;
class chi2 {
   public static void main (String args[]) {
      Random generator = new Random();
      int n = 1000;  int k = 10; double chiSquare = 0;
      int [] count = new int[k];

         for (int i=0; i<n; i++) {
                                 // Generate and partition 1000 random numbers
            int subinterval  = generator.nextInt(k);
            count[subinterval]++;
         }
         for (int i=0; i<k; i++) {        // Form chi-square statistic
           chiSquare = chiSquare + (count[i]-n/k) * (count[i]-n/k);
         }
         chiSquare = chiSquare/(n/k);

         for (int i=0; i<k; i++) {        // Print results
              System.out.println("Count[] = " + count[i]);
         }
         System.out.println("chi-Square = " + chiSquare);
   }
}
```

It is easy to see how two-, three-, and higher-dimensional versions of this test may be developed. The chi-square goodness-of-fit test in more than one dimension is called a *serial test*. Serial tests consider the numbers in the test, not one at a time, but rather in groups of two, three, or more. Suppose we have generated a sequence of random integers between 0 and 9. Partitioning these random numbers into consecutive pairs may be viewed as forming points on a two dimensional plane bounded by the corners (0,0), (9,0), (0,9), and (9,9). If the numbers are independent and identically, uniformly distributed then we should expect the points to be evenly distributed across the 100 unit squares. A two-dimensional chi-square test can be used to test this hypothesis. Similarly, if the sequence of random integers is partitioned into groups of three, each group of three may be viewed as a point in three-dimensional space, and we would expect them to be equally distributed among the 1,000 unit cubes. In general, forming groups of $d \ge 4$ integers, may be viewed as generating vectors of dimension $d$ in a hyperspace of dimension $d$. A $d$-dimensional chi-square test may be used to test the hypothesis that these vectors are uniformly distributed in this hyperspace. A disadvantage of high-dimension serial tests is that they require large numbers of random numbers, in order that each hypercube will collect some minimum number of points.

### *17.2.2 The Kolmogorov-Smirnov Test*

To set the stage for the Kolmogorov-Smirnov test, consider a sequence of $N$ uniformly distributed random numbers $z_1, z_2, \ldots, z_N$ in the interval $[0, 1)$. These numbers may be thought of as the values assumed by a discrete random variable $Z$, and the probability that this random variable assumes any particular number is taken to be $1/N$. If these $N$ numbers are now arranged in non-decreasing order, the probability distribution function of the random variable will have the typical staircase shape with steps occurring at each of the points $z_1, z_2, \ldots, z_N$ and height equal to $1/N$.

**Example 17.5** Suppose $N = 5$ and the uniformly distributed random numbers happen to be .70, .27, .16, .88 and .56. Arranging these into increasing order gives .16, .27, .56, .70, and .88 and the probability mass function of our supposed random variable is

| $z_i$ | .16 | .27 | .56 | .70 | .88 |
|---|---|---|---|---|---|
| $p_Z(z_i)$ | .2 | .2 | .2 | .2 | .2 |
| $\text{Prob}\{Z \leq z_i\}$ | .2 | .4 | .6 | .8 | 1.0 |

The probability distribution function is shown on the left in Figure 17.1. The distribution function on the right arises when ten rather than five random numbers are selected.

We shall refer to the staircase function, built around a sequence of $N$ uniformly distributed random numbers $z_1, z_2, \ldots, z_N$ as $S_N(x)$ and it is easy to see that $S_N(x)$ can be defined by

$$S_N(x) = \frac{\text{number of } z_1,\ z_2,\ \ldots,\ z_N \text{ that are } \leq x}{N}.$$

Figure 17.1 also shows the line $y = x$, and in the interval $[0, 1]$ this represents the continuous uniform distribution:

$$F_X(x) = x, \quad 0 \leq x \leq 1.$$

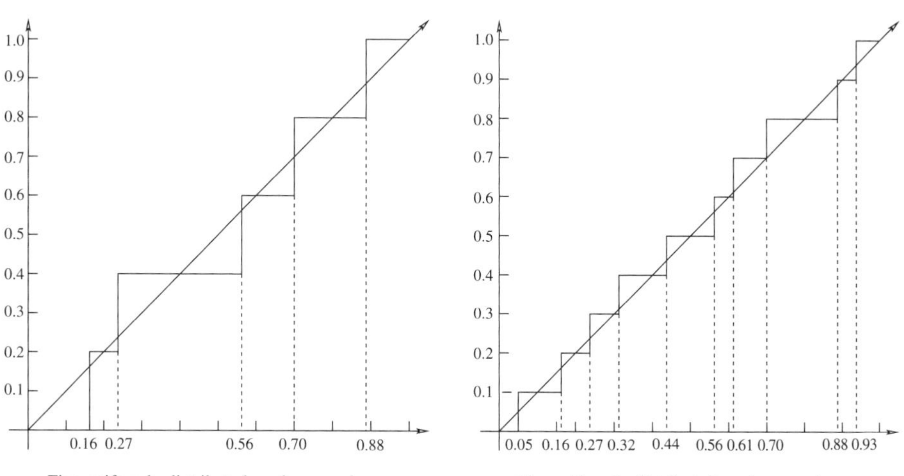


Figure 17.1. Illustration of the function $S_N(x)$ for two sets of random numbers.

As $N$ becomes large, we should expect $S_N(x)$ to approach $F_X(x)$, a characteristic we see in Figure 17.1 when $N$ is increased from 5 to 10. This is where the Kolmogorov-Smirnov test plays a role, for this test enables us to compare the continuous probability distribution function $F_X(x)$ to the probability distribution function obtained from the sample of $N$ random numbers. The Kolmogorov-Smirnov test is based on the statistic

$$D = \max |F_X(x) - S_N(x)|$$

over the range of $x$. We hasten to point out that this test is not only applicable when it is desirable to compare experimental observations to the continuous uniform distribution. Indeed, the Kolmogorov-Smirnov test is frequently used to compare customer arrivals with a Poisson distribution, population samples with the normal distribution, reliability statistics with the Weibull distribution, and so on.

As was the case for the chi-square random variable, the distribution of $D$ is available from statistical tables wherein the rows correspond to the value of $N$ and the columns correspond to typical significance level values such as .20, .10, .05 and .01. Since the maximum value of $|F_X(x) - S_N(x)|$ must occur at the points at which $S_N(x)$ increases in value, it is not necessary to evaluate $|F_X(x) - S_N(x)|$ at all possible values of $x \in [0, 1)$ but only at the step points $z_i$, $i = 1, 2, \ldots, N$. In summary, to apply the Kolmogorov-Smirnov test:

1. Generate a set of $N$ random numbers $z_1, z_2, \ldots, z_n$ in the interval $[0, 1)$ and arrange them into nondecreasing order.
2. Compute the quantities
$$D^+ = \max_{1 \le i \le N} \left( \frac{i}{N} - z_i \right), \quad D^- = \max_{1 \le i \le N} \left( z_i - \frac{i-1}{N} \right), \quad D = \max\left(D^+, D^-\right).$$
3. Consult tables of Kolmogorov-Smirnov critical values and choose the value $D_\alpha$ in row $N$ having the desired significance level $\alpha$.
4. If $D \le D_\alpha$ accept the hypothesis that the numbers are uniformly distributed.

Although both the chi-square test and the Kolmogorov-Smirnov test may be used when the value of $N$ is large ($N \ge 50$, for example), the Kolmogorov-Smirnov test is considered to be the more reliable of the two. When $N$ is small, only the Kolmogorov-Smirnov test should be used.

**Example 17.6** Let us run the Kolmogorov-Smirnov test on the first of the two data sets used to generate Figure 17.1. With $N = 5$, we have

| | | | | | |
|---|---|---|---|---|---|
| $z_i$ | .16 | .27 | .56 | .70 | .88 |
| $i/N$ | .20 | .40 | .60 | .80 | 1.0 |
| $i/N - z_i$ | .04 | .13 | .04 | .10 | .12 |
| $z_i - (i-1)/N$ | .16 | .07 | .16 | .10 | .08 |

and thus $D^+ = .13$, $D^- = .16$ and $D = .16$. Table 17.1 gives the critical values needed to complete the Kolmogorov-Smirnov test. For a 0.05 significance level, we see that $D_{.05} = .563$ and, since this is greater that $D = .16$, we accept the null hypothesis.

### *17.2.3 "Run" Tests*

The chi-square and Kolmogorov-Smirnov tests as implemented in the previous section may be used to test the distribution of randomly generated numbers against a given distribution and in particular, to test whether they appear to be uniform or not. However, they do not provide us with any information concerning the independence of the generated numbers. A number of *run* tests are available that, while not proving independence, allow us to be reasonably confident that the numbers

Table 17.1. Table of critical Kolmogorov-Smirnov values for some values of $N$.

| $N$ | $\alpha = .20$ | $\alpha = .10$ | $\alpha = .05$ | $\alpha = .01$ |
|---|---|---|---|---|
| ⋮ | | | | |
| 5 | .447 | .510 | .563 | .669 |
| ⋮ | | | | |
| 10 | .323 | .369 | .409 | .490 |
| ⋮ | | | | |
| 12 | .296 | .338 | .375 | .449 |
| ⋮ | | | | |
| 20 | .232 | .265 | .294 | .352 |
| ⋮ | | | | |
| $N > 35$ | $1.07/\sqrt{N}$ | $1.22/\sqrt{N}$ | $1.36/\sqrt{N}$ | $1.63/\sqrt{N}$ |

in the sequence appear to have this property. A *run-up* in a sequence of numbers is a subsequence in which each successive number is greater than the previous. A *run-down* is a subsequence with the opposite property. While this is easy to state, the devil is in the details. Currently, there is little consensus as to what constitutes the length of a subsequence. For some, the number on which a run-up ends is also the number at which a run-down must begin (for example, the number 8 in the subsequence 2, 8, 7, 6, 9), while for others the number that follows the last in a run is always discarded. Furthermore, for some, there must be at least two consecutive numbers to constitute a run-up or run-down—for them, a number in isolation cannot be a run. A run-up of length 1 begins with a number whose predecessor and successor are greater than it, and whose successor's successor is less than the successor, such as in the sequence 9, 2, 8, 7, where the numbers 2 and 8 constituting a run-up of length 1. For others, a run-up of length 1 can be achieved by a single number such as the number 7 in the subsequence 2, 8, 7, 6, 9, whereas 2, 8 and 6, 9 both constitute run-ups of length 2. The easy case is when the numbers involved are actually binary digits, which arises frequently, for here the length of a run of 1's is the number of ones bound before and after with a zero and the length of a run of 0's is the number of zeros bound between two ones: in essence, there is no run-up or run-down but only runs. We shall adopt the convention that when the tests involve the sum total of runs (both run-ups and run-downs), then a run must consist of at least two numbers. When only run-ups or run-downs are being considered we shall follow Knuth [25] and define the length of a run to be the number of numbers it contains.

**Example 17.7** Consider the sequence 3, 5, 9, 2, 8, 7, 6, 9, 1, 3, 5, 2, 2, 6, 7, 1, 5. The diagram below shows three run-ups of lengths 1 and three of length 2 (these are drawn above the numbers) and four run-downs of length 1 and one of length 2 (drawn below the numbers). This is the scheme we shall follow when comparing the total number of runs (run-ups plus run-downs) to the total number we should expect to find if the arrangement of the numbers in the sequence were in fact independent.

```
|-------|   |---|       |---|   |-------|       |-------|   |---|
3   5   9   2   8   7   6   9   1   3   5   2   2   6   7   1   5
        |---|   |-------|   |---|       |---|       |---|
```

On the other hand, when we wish to compare the number of run-ups of length 1, the number of run-ups of length 2, the number of run-ups of length 3, etc., i.e., the distribution of run-ups, to the number of run-ups of different lengths that should probabilistically occur if the numbers are indeed independent, we apply Knuth's numbering scheme and for the particular sequence given above, we find there are two run-ups of length 1, three of length 2 and three of length 3 as illustrated below.

```
|-------|  |---|  |  |---|  |-------|  |  |-------|  |---|
3   5   9  2   8  7  6   9  1   3   5  2  2   6   7  1   5
```

With these specifications in place we are now ready to describe a number of run tests.

**Run Test Number 1: The Total Number of Runs**

The sequence of ten random numbers

$$.05,\ .23,\ .27,\ .31,\ .36,\ .55,\ .56,\ .67\ .82,\ .93$$

has only one run, a run-up of length 9, while the sequence

$$.05,\ .27,\ .23,\ .36,\ .31,\ .56,\ .55,\ .82\ .67,\ .93$$

has nine runs, five run-ups and four run-downs. These represent the minimum and maximum number of runs possible in a sequence of ten numbers. If the random numbers in the sequence were independent, it is unlikely that either of these two arrangements would occur; more correctly, the probability of their occurrence would be small. With $N$ random numbers, the minimum number of runs must be 1 and the maximum must be $N - 1$ and the expected number would lie somewhere between these two. Let $X$ be the random variable that represents the total number of runs in a sequence of $N$ uniformly distributed and independent numbers. The expectation and variance of $X$ are known to be

$$E[X] = \frac{2N-1}{3} \quad \text{and} \quad \text{Var}[X] = \frac{16N-29}{90}.$$

When $N$ is not small, in excess of 20 or 30, for example, the probability distribution of $X$ may be approximated by a normal distribution having this mean and variance, the distribution $N(E[X], \text{Var}[X])$. Thus it becomes possible to test how well the number of runs $x$ found in a generated sequence of random numbers adheres to the theoretical estimate. The test statistic to use is

$$z = \frac{x - E[X]}{\sigma_X} = \frac{x - (2N-1)/3}{\sqrt{(16N-29)/90}}.$$

If we seek a significance level of $\alpha$, then the null hypothesis that the random numbers are independent should not be rejected if $-z_{\alpha/2} \le z \le z_{\alpha/2}$. For a typical significance level, such as $\alpha = 0.05$ for example, we need to determine the value of $z_{\alpha/2} = z_{.025}$. Standard normal tables provide the values of

$$\Phi_Z(z) = \frac{1}{\sqrt{2\pi}} \int_{-\infty}^{z} e^{-t^2/2} dt = \text{Prob}\{Z \le z\}.$$

Since $\text{Prob}\{Z > z\} = 1 - \text{Prob}\{Z \le z\}$, we check the table to obtain the value that corresponds to $1 - .025 = .975$ and find the critical value to be 1.96. Thus, for the 5% significance level, the condition for accepting the null hypothesis is

$$-z_{\alpha/2} \le z \le z_{\alpha/2} \quad \Longrightarrow \quad -z_{.025} \le z \le z_{.025} \quad \Longrightarrow \quad -1.96 \le z \le 1.96.$$

This means that 95% of the normal curve lies between the limits $-1.96$ and $+1.96$, often called the $2\sigma$ level (convention rounding 1.96 to 2) or two standard deviations.

**Example 17.8** A sequence of 50 random numbers is found to have a total of 30 runs. We wish to determine if, with a significance level of $\alpha = .05$, the null hypothesis of independence should be rejected. Given that $N = 50$ and $x = 30$, we have

$$z = \frac{30 - (100 - 1)/3}{\sqrt{(800 - 29)/90}} = \frac{-3}{2.9269} = -1.0250.$$

We now need to determine the value of $z_{\alpha/2} = z_{.025}$. Since we have just found this to be 1.96 and since $-1.96 \leq -1.0250 \leq 1.96$, the null hypothesis should not be rejected.

Suppose that instead of finding 30 runs, a total of 40 runs is found in the sequence of 50 random numbers. Should the hypothesis be rejected at the $\alpha = .10$ significance level?
This time we find

$$z = \frac{40 - (100 - 1)/3}{\sqrt{(800 - 29)/90}} = \frac{7}{2.9269} = 2.3916.$$

From standard normal tables, we seek the value that corresponds to $1 - .05 = .95$ and this time the critical value is 1.645. Since $2.3916 > 1.645$, the hypothesis should be rejected.

**Run Test Number 2: Runs Above or Below the Mean**

A somewhat similar test involves subsequences that occur above or below the mean. Consider the following sequence of 20 two-digit random numbers in the interval $[0, 1)$:

.05, .15, .23, .18, .06 .27, .31, .36, .32, .45, .55, .56, .51, .67, .62, .83, .96, .90, .70, .77.

That these numbers satisfy all the previous tests for uniformity and independence is left to the exercises. However, a generator that produces such a sequence should not be used: The first ten numbers are all less than the mean while the last ten are all greater than the mean. This is not what we should expect if the numbers are truly independent. Our second run test helps to detect such an anomaly. This time a run is defined as a subsequence of consecutive numbers that are less than, or greater than, the mean. The total number of such subsequences is computed and compared to a theoretical estimate. Let $X$ be the random variable that represents the sum of the total number of runs-below and runs-above the mean in a sequence of $N$ uniformly distributed and independent numbers. Also let $n_1$ be the number of random numbers in the entire sequence that are greater than the mean, and $n_2$ the number less than the mean. Numbers that are equal to the mean are ignored. In this case the expectation and variance of $X$ are known to be

$$E[X] = \frac{2n_1n_2}{n_1 + n_2} + 1 \quad \text{and} \quad \text{Var}[X] = \frac{2n_1n_2(2n_1n_2 - n_1 - n_2)}{(n_1 + n_2)^2(n_1 + n_2 - 1)}. \tag{17.4}$$

As long as one of $n_1$ or $n_2$ is sufficiently large (greater than 20, for example), then $X$ is approximately normally distributed and, as was the case for *run test number 1*, the test statistic we need to use is

$$z = \frac{(x \pm 0.5) - E[X]}{\sqrt{\text{Var}[X]}},$$

where $E[X]$ and $\text{Var}[X]$ are given in Equation (17.4) and the $\pm 0.5$ is a continuity correction; when $x > E[X]$ we use $-0.5$; when $x < E[X]$ we use $+0.5$. The continuity correction may be omitted when $n_1$ and $n_2$ are large.

**Example 17.9** The following 40 numbers were obtained from a single run of a Java program using the built-in Java random number generator:

.11, .15, .62, .94, .11, .92, .57, .02, .99, .70, .35, .67, .35, .36, .09, .51, .21, .12, .41, .00,

.06, .84, .94, .06, .60, .34, .51, .16, .81, .43, .59, .51, .15, .70, .77, .96, .96, .09, .13, .36.

In this sequence 19 numbers are greater than the mean ($n_1 = 19$) and 21 are smaller than the mean ($n_2 = 21$). The total number of runs is $x = 23$ which is illustrated below, where numbers less than the mean are represented by a minus sign and those greater than the mean, by a plus sign.

$$- \; - \; + \; + \; - \; + \; + \; - \; + \; + \; - \; + \; - \; - \; - \; + \; - \; - \; - \; -$$

$$- \; + \; + \; - \; + \; - \; + \; - \; + \; - \; + \; + \; - \; + \; + \; + \; + \; - \; - \; -$$

Let us now check to see if these numbers satisfy *run test number 2*. We have

$$E[X] = \frac{2 \times 19 \times 21}{40} + 1 = 20.95, \quad \text{Var}[X] = \frac{2 \times 19 \times 21(2 \times 19 \times 21 - 40)}{40^2 \times 39} = 9.6937,$$

which allows us to compute

$$z = \frac{(23.0 - 0.5) - 20.95}{\sqrt{9.6937}} = \frac{1.55}{3.1135} = .4978.$$

Since this is smaller than 1.96 (the critical value of the normal distribution for a significance level of $\alpha = 0.05$), the hypothesis that the random numbers are independent should not be rejected by the outcome of this test.

**Run Test Number 3: The Distribution of Run-ups (or Run-downs)**

In this test, a generated sequence of random numbers is traversed once and the number of run-ups (or run-downs) of length 1, 2, 3, 4, 5, and greater than or equal to 6, are counted. Since our concern is with either run-ups or run-downs, we count an isolated element as a run of length 1. Thus, as we mentioned previously, the sequence 3, 5, 9, 2, 8, 7, 6, 9, 1, 3, 5, 2, 2, 6, 7, 1, 5 has two run-ups of length 1, three of length 2, and three of length 3, as illustrated below.

```
|-------|  |---|  |  |---|  |-------|  |  |-------|  |---|
3   5   9  2   8  7  6   9  1   3   5  2  2   6   7  1   5
```

From the counts of the number of run-ups of different lengths, a random variable is constructed whose distribution tends to that of the $\chi^2$ distribution with six degrees of freedom. If we let $r_i$, $i = 1, 2, \ldots, 6$, denote the number of run-ups of length $i$, then for large values of $n$, i.e., $n \geq 4,000$, the random variable

$$X = \frac{1}{n} \sum_{j=1}^{6} \sum_{k=1}^{6} (r_j - n \times b_j)(r_k - n \times b_k) a_{jk}$$

tends to the $\chi^2$ distribution. The elements of $A$ and $B$ have been computed by Knuth and are as follows:

$$A = \begin{pmatrix} 4529 & 9045 & 13568 & 18091 & 22615 & 27892 \\ 9045 & 18097 & 27139 & 36187 & 45234 & 55789 \\ 13568 & 27139 & 40721 & 54281 & 67852 & 83685 \\ 18091 & 36187 & 54281 & 72414 & 90470 & 111580 \\ 22615 & 45234 & 67852 & 90470 & 113262 & 139476 \\ 27892 & 55789 & 83685 & 111580 & 139476 & 172860 \end{pmatrix},$$

$$B = \begin{pmatrix} 1/6 & 5/24 & 11/120 & 19/720 & 29/5040 & 1/840 \end{pmatrix}.$$

It is also possible to develop a similar test where the run lengths are taken to be the lengths of runs above (or below) the mean. However, we caution the reader that this test should be used only when the population from which the random numbers are drawn is large.

**Example 17.10** For the sample 17 random number given above with run-ups of $r_1 = 2$, $r_2 = 3$, $r_3 = 3$ and $r_4 = r_5 = r_6 = 0$, we obtain a value of $X = 2.324$. By examining the row corresponding to 6 degrees of freedom in the table of $\chi^2$ values, we may conclude, with confidence level of 0.75, that the sequence is independent.

The Java program below generates $n = 4,000$ random integers in the interval [0, 9], counts the number of run-ups of length 1 through 5 and greater than or equal to 6 and then computes the $\chi^2$ value. Removing the comment marks from the two statements that are commented out will run the 17 integer example described above.

```
import java.util.Random;
class run_up {
   public static void main (String args[]) {
      Random generator = new Random();
      int n = 4000;
      // n = 17;    int [] array = {3,5,9,2,8,7,6,9,1,3,5,2,2,6,7,1,5};
                                                                // Example

      int [] r = new int [6];
      int count = -1;
      int last = -999;
      for (int i=0; i<n; i++) {
          int next = generator.nextInt(100);      // Next random number is ...
          // next = array[i];        // Example
          if (next > last) {count++; last = next; }
                                                 // ... larger than previous
          else { if (count > 5) {count = 5;}     // ... smaller than previous
                 r[count]++;
                 count = 0; last = next;}
      }
      r[count]++;                                      // Print run-up counts:
      System.out.println("r = " + r[0] + "    " + r[1] + "    " + r[2] +
                       "    " +r[3] + "    " + r[4] + "    " + r[5]);

                                                    // A & B Parameter values
      double [][] A = { { 4529,  9045, 13568,  18091,  22615,  27892},
                        { 9045, 18097, 27139,  36187,  45234,  55789},
                        {13568, 27139, 40721,  54281,  67852,  83685},
                        {18091, 36187, 54281,  72414,  90470, 111580},
                        {22615, 45234, 67852,  90470, 113262, 139476},
                        {27892, 55789, 83685, 111580, 139476, 172860} };
      double [] B = {1.0/6, 5.0/24, 11.0/120, 19.0/720, 29.0/5040, 1.0/840};

      double X = 0;                                   // Compute chi-square
      for (int j=0; j<6; j++) {
          for (int k=0; k<6; k++) {
              X = X + (r[j] - n*B[j]) * (r[k] - n* B[k])* A[j][k];
          }
      }
      X = X/n;
      System.out.println("X = " + X);                 // Print chi-square
   }
}
```

### 17.2.4 The "Gap" Test

This test calculates the distance between the repetition of numbers in a given sequence and makes comparisons with what should be expected theoretically if the numbers are indeed independent. The Kolmogorov-Smirnov test is used to assess the statistical significance of the distance between repetitions. If the numbers are independent, then the length of the gaps should be geometrically distributed. When the total number of different numbers that can possibly occur is $\gamma$, the probability that a gap has length $k$ is

$$\text{Prob}\{\text{gap has length } k\} = \left(\frac{\gamma-1}{\gamma}\right)^k \left(\frac{1}{\gamma}\right),$$

since each individual number has probability $1/\gamma$ of occurring. The probability of a number occurring, other than the number being tested, is $(\gamma-1)/\gamma$.

**Example 17.11** In the following sequence of 35 single-digit integers, the integer 5 occurs in positions 2, 9, 17, 28, and 33:

| | | | | | | |
|---|---|---|---|---|---|---|
| 9 | 5 | 4 | 1 | 6 | 8 | 3 |
| 4 | 5 | 8 | 9 | 0 | 0 | 2 |
| 7 | 2 | 5 | 3 | 1 | 7 | 1 |
| 6 | 9 | 3 | 4 | 7 | 8 | 5 |
| 0 | 1 | 6 | 8 | 5 | 9 | 2 |

This gives four gaps, of length 6 ($= 9-2-1$), 7 ($= 17-9-1$), 10 ($= 28-17-1$), and 4 ($= 33-28-1$) where the length of a gap is the number of digits between the two repeated digits. With $\gamma = 10$, the probabilities of gaps of these lengths are

$$\text{Prob}\{L=6\} = 0.9^6 \times 0.1 = .05314, \quad \text{Prob}\{L=7\} = 0.9^7 \times 0.1 = .04783,$$

$$\text{Prob}\{L=10\} = 0.9^{10} \times 0.1 = .03487, \quad \text{Prob}\{L=4\} = 0.9^4 \times 0.1 = .06561.$$

When performing the gap test, it is not sufficient to check the gaps between the occurrences of one specific number, or even two or three. The distribution of the gaps for all possible numbers must be found. When the sequence to which the gap test is applied contains real numbers rather than integers, it is appropriate to establish intervals, such as $[0, .1)$, $[.1, .2)$, $\ldots$, $[.9, 1)$, and to tabulate the gaps between a real number falling into one interval and the next real number that falls into the same interval.

**Example 17.12** Continuing with the example of the 35 integers, the following table shows, for each integer, the length of each gap. Thus, for example, the integer 0 has two gaps, one of length 0 and the other of length 15 and, as we mentioned before, the integer 5 has 4 gaps of lengths 4, 6, 7, and 10.

| Integer | 0 | 1 | 2 | 3 | 4 | 5 | 6 | 7 | 8 | 9 |
|---|---|---|---|---|---|---|---|---|---|---|
| Gap size | 0 | 1 | 1 | 5 | 6 | 4 | 8 | 4 | 2 | 9 |
| | 15 | 8 | 18 | 10 | 16 | 6 | 16 | 5 | 3 | 10 |
| | | 14 | | | | 7 | | | 16 | 11 |
| | | | | | | 10 | | | | |

Observe that the total number of gaps is $35 - 10 = 25$. A moment's reflection should convince the reader that the number of gaps is always equal to $N - \gamma$, where $N$ is the number of numbers in the sequence and $\gamma$ is the number of different values possible. Using these numbers, the computed gap length distribution and the relative frequency of occurrence, for this set of 35 integers, is

| Gap length | 0–1 | 2–3 | 4–5 | 6–7 | 8–9 | 10–11 | 12–13 | 14–15 | 16–17 | 18–19 |
|---|---|---|---|---|---|---|---|---|---|---|
| Occurs | 3 | 2 | 4 | 3 | 3 | 4 | 0 | 2 | 3 | 1 |
| Relative Frequency | .12 | .08 | .16 | .12 | .12 | .16 | .00 | .08 | .12 | .04 |

This table illustrates the fact that the gap lengths are usually combined into groups of size 2, 3, 4, or 5.

The theoretical distribution, the distribution which arises if the numbers are truly independent and against which the computed distribution must be compared, via the Kolmogorov-Smirnov test, is as follows:

$$F_X(x) = \text{Prob}\{X \le x\} = \frac{1}{\gamma} \times \sum_{k=0}^{x} \left(\frac{\gamma - 1}{\gamma}\right)^k = 1 - \left(\frac{\gamma - 1}{\gamma}\right)^{x+1}.$$

**Example 17.13** Continuing with the example of 35 integers, the values of $x$ at which we evaluate $F_X(x) = \text{Prob}\{X \le x\}$ are $x = 1, 3, 5, 7, 9, 11, 13, 15, 17, 19$.

| Gap length | 0–1 | 2–3 | 4–5 | 6–7 | 8–9 | 10–11 | 12–13 | 14–15 | 16–17 | 18–19 |
|---|---|---|---|---|---|---|---|---|---|---|
| Relative Frequency | .12 | .08 | .16 | .12 | .12 | .16 | .00 | .08 | .12 | .04 |
| $S_N(x)$ | .12 | .20 | .36 | .48 | .60 | .76 | .76 | .84 | .96 | 1.0 |
| $F_X(x)$ | .1900 | .3439 | .4686 | .5695 | .6513 | .7176 | .7712 | .8417 | .8499 | .8784 |
| $\lvert F_X(x) - S_N(x)\rvert$ | .0700 | .1439 | .1086 | .0895 | .0513 | .0424 | .0112 | .0017 | .1101 | .1216 |

Thus $\max_x |F_X(x) - S_N(x)| = .1439$. The critical value $D_\alpha$ to use in the Kolmogorov-Smirnov test is found from tables as $D_{0.05} = .409$, and, since .1439 is less than this, the null hypothesis should not be rejected on the basis of this gap test.

### 17.2.5 The "Poker" Test

This test identifies sequences of random numbers with "hands" in a poker game. When the random numbers belong to a certain range, such as the integers 0 through 9, or have been "binned" into a fixed number of categories, then it becomes possible to take them in groups and test for the occurrence of different combinations. In the poker analogy, the numbers are taken in groups of five (a hand) and checked to see if the hand contains all different numbers (or belong to five different bins), a pair (two the same and the other three different from each other and different from the pair) and so on. If we let $A$, $B$, $C$, $D$, and $E$ be five different possible numbers, or the designation of five different bins, then examples of the possible poker hands are

| | |
|---|---|
| $ABCDE$ | All different, |
| $AACDE$ | One pair, |
| $AACCE$ | Two pair, |
| $AAADE$ | Three of a kind, |
| $AAABB$ | Full house, |
| $AAAAB$ | Four of a kind. |

If the numbers are uniformly and independently distributed, then it is possible to compute the probability of each of these hands and with this information it becomes possible to gauge the independence property of the random number generator. The chi-square test is used to determine if the random sequence conforms to the expected distribution. Even though it is not possible to get five of a kind in poker, it is possible to handle this situation ($AAAAA$) in the poker test. Also, the number of cards in a hand need not be equal to five, as the next example now shows.

**Example 17.14** Consider a sequence in which each element (hand) is a three-digit random number. This allows for the possibility of only three poker hands, with theoretical probabilities given by

$$\begin{aligned}\text{Prob}\{3 \text{ different digits}\} &= 1 \times .9 \times .8 = .72,\\ \text{Prob}\{3 \text{ of a kind}\} &= 1 \times .1 \times .1 = .01,\\ \text{Prob}\{1 \text{ pair}\} &= 1 - .72 - .01 = .27.\end{aligned}$$

Suppose a random number generator produces 1000 such numbers and it is found that 700 of then have three digits all different, 15 of then have all three digits the same and the rest, 285, have exactly one pair. Forming the chi-square statistic, we find

$$\chi^2 = \frac{(720-700)^2}{720} + \frac{(10-15)^2}{10} + \frac{(270-285)^2}{270} = .5556 + 2.5000 + .8333 = 3.8889.$$

Consulting tables of chi-square critical values, and using line 2 corresponding to two degrees of freedom (one less than the number of intervals) we find $\chi^2_{.05} = 5.99$. Since $\chi^2 < \chi^2_{.05}$, the hypothesis that the numbers are independent cannot be rejected.

In practice, the poker test is frequently replaced with a simpler test, but one that is easier to work with and whose performance is almost as good as the regular poker test. In this new version we simply count the number of different values in the hand. Thus the two categories, "two pairs" and "three of a kind" are combined into the single category "three different values," while both "full house" and "four of a kind" are combined into the single category, "two different values."

$r = 5$ all 5 cards are different,
$r = 4$ one pair,
$r = 3$ two pairs or three of a kind,
$r = 2$ full house or four of a kind,
$r = 1$ five of a kind.

The probabilities of these different possibilities when the digits are independent are:

| Number of different values, $r$ | 5 | 4 | 3 | 2 | 1 |
|---|---|---|---|---|---|
| Probability | .3024 | .5040 | .1800 | .0135 | .0001 |

For example, the probability of getting five different values is $1 \times .9 \times .8 \times .7 \times .6 = .3024$, while the probability of getting all five the same is $1 \times .1 \times .1 \times .1 \times .1 = .0001$. It can be seen that the probability of getting only one or two different values is quite small, .0001 and .0135, respectively. Before applying the chi-square test it is usual to lump low-probability categories together. In fact, this lumping of low probabilities is recommended whenever the chi-square test is used.

The formula for computing these probabilities in a more general context is given in terms of Sterling numbers of the second kind, $S(k, r)$, as

$$\text{Prob}\{r \text{ different values}\} = \frac{d(d-1)(d-2)\cdots(d-r+1)}{d^k} S(k, r),$$

where $d$ is the number of different "cards" ($d = 10$ in our example, the digits $0, 1, \ldots, 9$), $k$ is the number of cards in a hand ($k = 5$ in a regular poker hand) and $r$ is the number of different values in a hand ($r = 2$ for "four of a kind" or "full house"). The Sterling numbers can be computed by means of the recursive formula

$$S(k, r) = S(k-1, r-1) + r \times S(k-1, r).$$

This leads to the following convenient "Sterling's Triangle":

| | $r=1$ | $r=2$ | $r=3$ | $r=4$ | $r=5$ | $r=6$ |
|---|---|---|---|---|---|---|
| $k=1$ | 1 | | | | | |
| $k=2$ | 1 | 1 | | | | |
| $k=3$ | 1 | 3 | 1 | | | |
| $k=4$ | 1 | 7 | 6 | 1 | | |
| $k=5$ | 1 | 15 | 25 | 10 | 1 | |
| $k=6$ | 1 | 31 | 90 | 65 | 15 | 1 |

**Example 17.15** A sequence of 1,000 octal numbers generated at random shows the following distribution of different values found in consecutive groups of four:

| $r$ | 4 | 3 | 2 | 1 |
|---|---|---|---|---|
| Number | 99 | 120 | 28 | 3 |

Thus, for example, out of the 250 groups of four, exactly 99 have all different octal values. Given that $d = 8$ and $k = 4$ , the probabilities for obtaining $r = 4, 3, 2, 1$ different octal values, under the assumption that numbers in the sequence are independent of each other, are computed as

$$\text{Prob}\{4 \text{ different values}\} = \frac{8 \times 7 \times 6 \times 5}{8^4} \times 1 = .410156,$$

$$\text{Prob}\{3 \text{ different values}\} = \frac{8 \times 7 \times 6}{8^4} \times 6 = .492188,$$

$$\text{Prob}\{2 \text{ different values}\} = \frac{8 \times 7}{8^4} \times 7 = .095703,$$

$$\text{Prob}\{1 \text{ different value}\} = \frac{8}{8^4} \times 1 = .001953.$$

The expected frequency of occurrence is obtained by multiplying these probabilities by 250. Since the last case in which all four octal numbers are the same ($r = 1$) has a very small probability, we shall lump it in with the preceding case ($r = 2$) and we apply the chi-square test to the data

| $r$ | 4 | 3 | $\leq 2$ |
|---|---|---|---|
| Observed | 99 | 120 | 31 |
| Expected | 102.54 | 123.05 | 24.41 |

The appropriate chi-square statistic to use is

$$\frac{3.54^2}{102.54} + \frac{3.05^2}{123.05} + \frac{6.59^2}{24.41} = 1.9769.$$

With two degrees of freedom and a significance value of $\alpha = .05$, the critical chi-square value is $\chi^2_\alpha = 5.99$, and since 1.9769 is less than this we cannot reject the null hypothesis that the 1,000 numbers are independent.

### *17.2.6 Statistical Test Suites*

The statistical tests for uniformity and independence of the numbers produced by random number generators presented in this chapter constitute only a small portion of the tests that are possible and which have been incorporated into various software packages. Two test suites of renown are the NIST (National Institute of Standards and Technology) statistical test suite, which contains a collection of 16 tests and is available at `http://csrc.nist.gov/rng/index.html` and the DIEHARD suite developed by George Marsaglia at Florida State University (`http://stat.fsu.edu/pub/diehard/` which contains 12 different tests (more if we count the three binary rank test and the different overlapping tests separately). Many of these tests have colorful names such as "the squeeze test," "the birthday spacings test," "the parking lot test," and so on. Descriptions of all these tests along with software implementations can be found on the appropriate web page.

## 17.3 Exercises

**Exercise 17.1.1** A linear congruential random number generator has a multiplier of five, an increment of 3, and a modulus of 16.

(a) What is the period of this generator?
(b) Using $z_0 = 3$, generate the complete cycle of random integers and their associated random numbers in the interval [0, 1].

**Exercise 17.1.2** Given that a LCM has multiplier equal to 13, increment equal to 5, and the modulus equal to 64, does this LCM satisfy the conditions for a full period? Using these parameter values, compute $z_{453}$, given that $z_0 = 7$.

**Exercise 17.1.3** Consider the following two MRGs:

$$z_{1,i} = (a_{1,1}z_{1,i-1} + a_{1,2}z_{1,i-2} + a_{1,3}z_{1,i-3}) \bmod m_1,$$

$$z_{2,i} = (a_{2,1}z_{2,i-1} + a_{2,2}z_{2,i-2} + a_{2,3}z_{2,i-3}) \bmod m_2,$$

which are to be combined to produce the next random integer in the sequence as

$$x_i = (z_{1,i} - z_{2,i}) \bmod m_1.$$

Given the following starting values:

$$z_{1,1} = 45, \ z_{1,2} = 100, \ z_{1,3} = 57, \quad z_{2,1} = 17, \ z_{2,2} = 1, \ z_{2,3} = 24$$

and parameters

$$a_{1,1} = 12, \quad a_{1,2} = 0, \quad a_{1,3} = 89, \quad m_1 = 2^7 - 1 = 127,$$

$$a_{2,1} = 0, \quad a_{2,2} = 14, \quad a_{2,3} = 28, \quad m_2 = 2^5 - 1 = 31,$$

compute the first six random integers.

**Exercise 17.1.4** The following are well-known results on modulo arithmetic: If $m$ is a positive integer and $a_1, a_2, \ldots, a_k$ are integers, then

$$(a_1 + a_2 + \cdots + a_k) \bmod m = [(a_1 \bmod m) + (a_2 \bmod m) + \cdots + (a_k \bmod m)] \bmod m,$$

$$(a_1 a_2 \ldots a_k) \bmod m = [(a_1 \bmod m)(a_2 \bmod m) \cdots (a_k \bmod m)] \bmod m.$$

With these results, it is easy to see that the "jump ahead by $\eta$" computation from some random integer $z_i$ in an MCM generator with multiplier $a$ and modulus $m$ can be accomplished by forming

$$z_{i+\eta} = (a^\eta z_i) \mod m,$$

since

$$z_{i+1} = (az_i) \mod m,$$
$$z_{i+2} = (az_{i+1}) \mod m = a[(az_i) \mod m] \mod m = (a^2 z_i) \mod m,$$

and so on, where we have used the fact that $a \mod m = a$, since $a < m$. Derive a formula for the "jump ahead by $\eta$" procedure when the recurrence relation is given by

$$z_i = (a_1 z_{i-1} + a_2 z_{i-2}) \bmod m.$$

*Hint: Think in terms of matrices.* Generalize this to the case when the recurrence relation is

$$z_i = \left( \sum_{j=1}^{k} a_j z_{i-j} \right) \bmod m.$$

**Exercise 17.2.1** Consider the following sequence of 20 two-digit random numbers in the interval [0, 1):

.05, .15, .23, .18, .06, .27, .31, .36, .32, .45, .55, .56, .51, .67, .62, .83, .96, .90, .70, .77.

Do these numbers satisfy, at the $\alpha = .05$ significance level, the chi-square goodness of fit test and the Kolmogorov-Smirnov test for uniformly distributed random numbers?

**Exercise 17.2.2** A sequence of 500 random numbers in the interval [0, 1) is analyzed and partitioned into ten equal subintervals (or bins). The following distribution is found:

Count[] = 45 54 39 6251 36 65 40 47 56

Apply the chi-square goodness-of-fit test to this data and determine if the sequence from which it was drawn can be assumed to be uniformly distributed, based on the results of this test.

**Exercise 17.2.3** The following five numbers were produced by a random number generator: .11, .98, .66, .76, .75. Use the Kolmogorov-Smirnov test to determine whether this generator is likely to be satisfactory.

**Exercise 17.2.4** Dr. Harry Perros is examining data that he has just collected concerning arrivals and which he wishes to incorporate into his queueing model. He would be very happy if the data showed that the arrival pattern is Poisson. Here is the data he has collected on the number of customers that were observed to arrive in a one-hour period

| Number of Customers | 0–1 | 2–3 | 4–5 | 6–7 | 8–9 | 10–11 | 12–13 | 14–15 |
|---|---|---|---|---|---|---|---|---|
| Observed Frequency | 18 | 24 | 40 | 68 | 37 | 16 | 10 | 8 |

Show that the average arrival rate is $\lambda = 6.4005$. and use the Kolmogorov-Smirnov test to see if Dr. Perros can accept the hypothesis that his arrival process is Poisson. You may take $\lambda = 6.4$ which will enable you to use tables of the cumulative Poisson distribution function.

**Exercise 17.2.5** Apply each of the three different run tests to the following set of 60 two-digit random numbers:

.10, .23, .69, .72, .66, .57, .01, .61, .62, .08, .17, .72, .87, .91, .12, .20, .57, .71, .04, .17,

.36, .08, .13, .34, .53, .49, .35, .99, .95, .12, .08, .29, .55, .05, .13, .22, .36, .81, .02, .21,

.70, .26, .21, .55, .44, .62, .72, .11, .69, .94, .45, .29, .14, .18, .82, .03, .26, .46, .06, .67.

**Exercise 17.2.6** Let the interval [0, 1) be partitioned into five equal subintervals, [0, .2), [.2, .4), [.4, .6), [.6, .8) and [.8, 1)] and label these intervals 0–4. Then each of the 60 random numbers in Exercise 17.2.5 falls into exactly one of these subintervals and the sequence can now be written as 0, 1, 3, 3, 3, 2, .... Apply the gap test to this sequence.

**Exercise 17.2.7** Apply the modified poker test to the following sequence of 96 five-digit numbers to determine if the digits in the sequence of 96 $\times$ 5 numbers are independent.

| | | | | | | | |
|---|---|---|---|---|---|---|---|
| 21888 | 32305 | 11069 | 96023 | 55298 | 34930 | 93277 | 84330 |
| 79985 | 75668 | 61162 | 68190 | 77828 | 71216 | 46311 | 55800 |
| 09750 | 10581 | 30819 | 83187 | 42839 | 98288 | 57491 | 09273 |
| 55645 | 95543 | 31450 | 56349 | 75385 | 96466 | 63466 | 09912 |
| 10236 | 97266 | 57016 | 16208 | 17728 | 79112 | 20577 | 10417 |
| 36081 | 33453 | 49359 | 99512 | 08295 | 50513 | 22368 | 10221 |
| 70262 | 15544 | 62721 | 16994 | 45291 | 41882 | 03264 | 60667 |
| 36964 | 96614 | 87652 | 49247 | 83383 | 56255 | 37500 | 69181 |
| 83673 | 75866 | 24760 | 86738 | 00851 | 69676 | 38679 | 62215 |
| 16544 | 28211 | 67438 | 39437 | 69344 | 27255 | 68909 | 92754 |
| 91468 | 20987 | 61089 | 57695 | 64151 | 10170 | 38985 | 20017 |
| 77179 | 06903 | 44204 | 60726 | 15736 | 59590 | 92961 | 07884. |

# Chapter 18

# *Nonuniformly Distributed "Random" Numbers*

A fundamental concept in simulation is the notion of an *event*. An event may be defined as any action which causes the system undergoing study to change state, and the occurrence and effect of these events on the system must be examined and analyzed. The distribution of the occurrence of events is most likely not uniformly distributed so it becomes necessary to have access to sequences of random numbers that are perhaps exponentially distributed or normally distributed, or have some other appropriate distribution, rather than the uniform distribution. This is accomplished by generating a sequence of uniformly distributed random numbers and then transforming this sequence to a sequence having the required distribution. In this chapter we shall examine the principal approaches to performing this transformation. These approaches are the *inverse transformation* method, a "*mimicry*" approach, the *accept-reject* method, and *compositional* methods, including methods that partition the area under the density function, such as the *rectangle-wedge-tail* method and the *ziggurat* method. In some cases, a specific probability distribution is best suited to only one transformation method, while other distributions may be handled equally conveniently by more than one method. Random numbers that satisfy a particular distribution function are sometimes referred to as *random variates*.

## 18.1 The Inverse Transformation Method

The first method we discuss is that of *inversion*, also called the *inverse transformation method*, which may be used only when the distribution function can be analytically inverted. Suppose we wish to generate random numbers that are distributed according to some cumulative distribution function $F(x)$. The basic idea is to generate a sequence of uniformly distributed random numbers and to derive from this sequence a different sequence of random numbers that has the distribution $F(x)$. Observe that $F(x)$ is a probability, the probability that the random variable $X$ has a value less than or equal to $x$, i.e.,

$$F(x) \equiv F_X(x) = \text{Prob}\{X \leq x\},$$

and hence has a value that lies between 0 and 1, just like our uniformly distributed random numbers. If the distribution function $F(x)$ is a strictly increasing function, then there exists an *inverse* function, denoted by $F^{-1}$, that corresponds to $F$. In other words, if $x = F^{-1}(y)$, then $y = F(x)$.

We saw in a much earlier chapter that, given a random variable $X$, a new random variable $Y$ that is some function of $X$ may be created, such as $Y = X^2 + 2X$, $Y = e^X$, and so on. In the inverse transformation method we let the defining function of $Y$ be none other than the probability distribution function of $X$ itself. We now show that the distribution function of $Y$, defined in this manner, is the uniform distribution in the interval [0, 1]. We have

$$F_Y(y) = \text{Prob}\{Y \leq y\} = \text{Prob}\{F_X(x) \leq y\}.$$

If we assume that $F_X(x)$ can be inverted, then we have

$$\text{Prob}\{F_X(x) \leq y\} = \text{Prob}\{X \leq F_X^{-1}(y)\} \quad \text{for} \quad 0 \leq y \leq 1.$$

In other words

$$F_Y(y) = \text{Prob}\{X \le F_X^{-1}(y)\} \quad \text{for} \quad 0 \le y \le 1.$$

But $\text{Prob}\{X \le x\} = F_X(x)$, and replacing the $x$ with $F_X^{-1}(y)$, we obtain

$$F_Y(y) = F_X(F_X^{-1}(y)) = y \quad \text{for} \quad 0 \le y \le 1.$$

In other words, $Y$ is uniformly distributed on the interval [0, 1].

To summarize therefore, to generate random numbers according to some distribution function $F$ that is invertible, it suffices to generate random numbers that are uniformly distributed and to apply the inverse function $F^{-1}$ to these numbers. This is illustrated graphically in Figure 18.1.

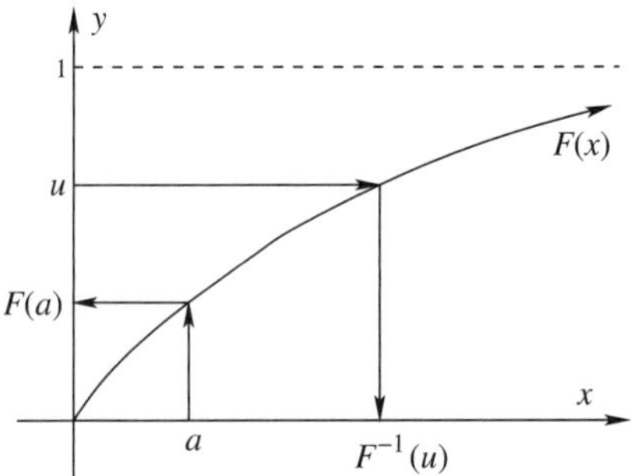


Figure 18.1. Illustration of an invertible function.

Given a uniformly distributed random number $u$, a line parallel to the $x$-axis is drawn until it reaches the curve $y = F(x)$ at which point a vertical line parallel to the $y$-axis is drawn. The point at which this line intersects the $x$-axis, $F^{-1}(u)$, is taken as a random number that satisfies the distribution $F$.

Not all distribution functions are invertible. For example, no discrete distribution function can be inverted, since such a function is a *stepwise* function rather than one that is strictly increasing.

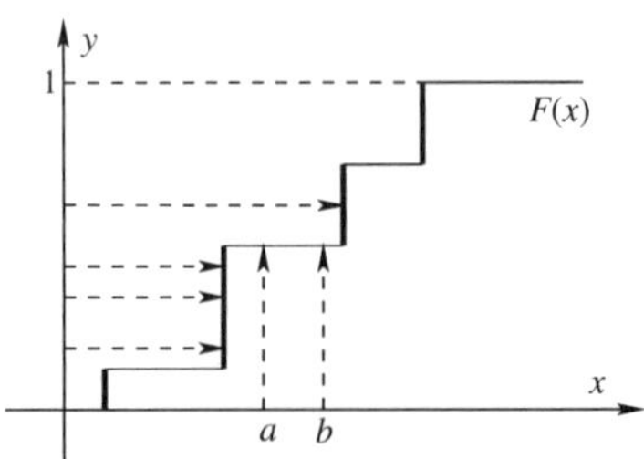


Figure 18.2. Illustration of step function.

In such cases, it is still possible to use the inversion approach. As before, a uniformly distributed random number is generated and a line parallel to the $x$-axis drawn through this number. This line intersects a *riser* within the step function. The $x$-value at this riser is taken as the value of the discrete random variable. Thus, a number that falls in the interval $[F(x_{i-1}), F(x_i))$ is associated with the event $X = x_i$. This is illustrated in Figure 18.2. Put another way, this approach partitions the interval [0, 1] into subintervals, and the length of each subinterval corresponds to the probability of one of the events that can occur. Since the sum of the probabilities of all the events is exactly equal to one, each subinterval corresponds to one, and only one, event. This is exactly the procedure we used in Chapter 16 when simulating simple probability experiments.

We now consider a number of probability distributions that can be handled by this method. Since we have formulated the analysis in terms of $y = F(x)$, we shall, in this section, refer to the sequences of uniformly distributed random numbers as $y_i$ rather than $u_i$.

### 18.1.1 The Continuous Uniform Distribution

The probability density and cumulative distribution functions for continuous random variables that are uniformly distribution on the interval $[a, b]$ are respectively given by

$$f(x) = \begin{cases} 1/(b-a), & x \in [a, b], \\ 0 & \text{otherwise}, \end{cases} \quad \text{and} \quad F(x) = \int_a^x \frac{1}{b-a}\,dt = \frac{x-a}{b-a}.$$

Setting $y = (x-a)/(b-a)$, we obtain the inverse by writing $x$ in terms of $y$. We have

$$x = a + (b-a)y.$$

Thus a sequence of uniformly distributed random numbers ($y$) in the interval $[0, 1]$ is converted to a sequence of random numbers ($x$) that is uniformly distributed in the interval $[a, b]$.

**Example 18.1** A sequence of uniformly distributed random numbers[1] that begins

$$y_i = .11,\ .15,\ .62,\ .94,\ .11,\ .92,\ .57,\ .02,\ .99,\ .20,\ \ldots$$

converts to the following sequence of random numbers uniformly distributed in the range (2, 6):

$$x_i = 2.44,\ 2.60,\ 4.48,\ 5.76,\ 2.44,\ 5.68,\ 4.28,\ 2.08,\ 5.96,\ 2.80,\ \ldots,$$

where each $x_i$ is obtained by multiplying $y_i$ by 4 and adding 2.

### 18.1.2 "Wedge-Shaped" Density Functions

The hypotenuse of a right angle triangle of height $h$ and width $w$ with right angle placed at the origin can, under certain constraints on $h$ and $w$, be taken as the density function of a random variable $X$ defined on $[0, w]$ as shown in Figure 18.3. The area of this right-angle triangle is $hw/2$. For the hypotenuse to represent a density function this area must be equal to 1 and thus we must have $hw/2 = 1$ or $h = 2/w$. The density function is given by

$$f_X(x) = \begin{cases} (1 - x/w)h, & 0 \le x \le w, \\ 0 & \text{otherwise}. \end{cases}$$

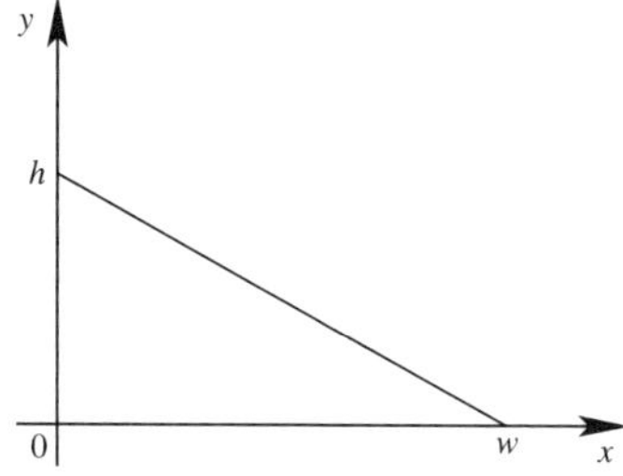


Figure 18.3. A "wedge-shaped" density function on $[0, w]$.

[1] Recall that, with respect to the generation of random numbers, we use the term *uniformly distributed random numbers* to mean random numbers that are uniformly distributed in (0, 1).

The cumulative distribution function is obtained by integrating $f_X(t)$ between 0 and $x$. We have, for $0 \le x \le w$,

$$\int_0^x f(t)dt = h\int_0^x \left(1-\frac{t}{w}\right)dt = h\left(t-\frac{t^2}{2w}\right)\Big|_0^x = hx - \frac{hx^2}{2w} = hx\left(1-\frac{x}{2w}\right) = \frac{2x}{w}\left(1-\frac{x}{2w}\right).$$

Hence the cumulative distribution function is

$$F_X(x) = \begin{cases} 0, & x < 0, \\ \dfrac{2x}{w}\left(1-\dfrac{x}{2w}\right), & 0 \le x \le w, \\ 1, & x > w. \end{cases}$$

We now invert this function in order to be able to generate random numbers that obey a wedge density function. Let

$$y = \frac{2x}{w}\left(1-\frac{x}{2w}\right).$$

This gives rise to the quadratic equation

$$x^2 - 2wx + w^2 y = 0$$

with roots

$$x = \frac{2w \pm \sqrt{4w^2 - 4w^2 y}}{2} = w \pm w\sqrt{1-y} = w(1 \pm \sqrt{1-y}).$$

We require the root $x = w(1-\sqrt{1-y})$, since our concern is with $x \in [0, w]$. This is the inverse transformation we need.

**Example 18.2** Let us consider the specific wedge density function having $h = 1$. It follows that $w = 2$ and, from the following sequence of uniformly distributed random numbers $y_i$:

$$y_i = .11, .15, .62, .94, .11, .92, .57, .02, .99, .70, .35, .67, .35, .36, .09, .51, .21, .12, .41, .01,$$

each $x_i$ is obtained as $x_i = 2(1-\sqrt{1-y_i})$. Thus the following sequence of random numbers satisfy the wedge density:

0.1132, 0.1561, 0.7671, 1.5101, 0.1132, 1.4343, 0.6885, 0.0201, 1.8000, 0.9046,

0.3875, 0.8511, 0.3875, 0.4000, 0.0921, 0.6000, 0.2224, 0.1238, 0.4638, 0.0100.

For example, $2(1-\sqrt{1-.11}) = .1132$, $2(1-\sqrt{1-.15}) = .1561$, and so on.

Naturally, it is possible to define a wedge density similar to the one given above, but having the right angle at some distance $a$ from the origin. It is also possible for the right angle to be on the right, rather than on the left as was the case discussed above. These possibilities are left to the exercises.

### *18.1.3 "Triangular" Density Functions*

Related to wedge-shaped density functions are the triangle density functions, one of which is illustrated in Figure 18.4. Given that the base is of length $(b-a)$, the height $h$ must be equal to $2/(b-a)$ for the function to be a genuine density function. The point $c$, the $x$-value at which the triangle peaks, is called the "mode."

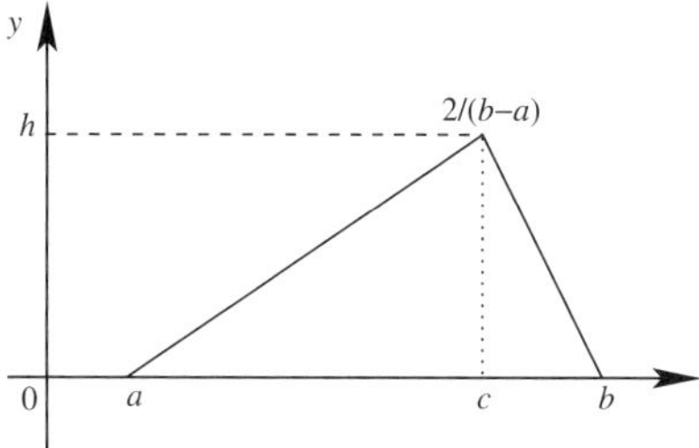


Figure 18.4. A "triangle" density function on $[a, b]$.

The reader may wish to verify that the density function is given by

$$f_X(x) = \begin{cases} \dfrac{2(x-a)}{(b-a)(c-a)}, & a \le x \le c, \\[2ex] \dfrac{2(b-x)}{(b-a)(b-c)}, & c \le x \le b, \end{cases}$$

and by integrating $f_X(x)$, the cumulative distribution function is found to be

$$F_X(x) = \begin{cases} 0, & x \le a, \\[1ex] \dfrac{(x-a)^2}{(b-a)(c-a)}, & a \le x \le c, \\[2ex] 1 - \dfrac{(b-x)^2}{(b-a)(b-c)}, & c \le x \le b, \\[2ex] 1, & x \ge b. \end{cases}$$

This distribution has mean and standard deviation

$$E[X] = \frac{a+b+c}{3}, \qquad \sigma = \frac{\sqrt{(a-b)^2 + (a-c)^2 + (b-c)^2}}{6}.$$

To generate random numbers having this distribution, it is necessary to invert the cumulative density function. This must be done in two parts, the first part corresponding to the base $(a, c)$ and the second corresponding to the base $(c, b)$. For $a \le x \le c$ we set

$$y = \frac{(x-a)^2}{(b-a)(c-a)},$$

which inverts to give

$$x = a + \sqrt{(b-a)(c-a)y}, \qquad 0 \le y \le \frac{c-a}{b-a}.$$

Notice that we need to pay attention to the range of values of $y$. When $x = a$, this implies that $y = 0$, while when $x = c$, we find that $y = (c-a)/(b-a)$.

In a similar manner, for $c \le x \le b$ we set

$$y = 1 - \frac{(b-x)^2}{(b-a)(b-c)},$$

which inverts to give

$$x = b - \sqrt{(b-a)(b-c)(1-y)}, \qquad \frac{c-a}{b-a} \le y \le 1.$$

**Example 18.3** Let us generate some random numbers according to a triangular density function with parameters $a = 2$, $c = 6$, and $b = 7$ and using the sequence of uniformly distributed random numbers

$$y_i = .11, .15, .62, .94, .11, .92, .57, .02, .99, .70, .35, .67, .35, .36, .09, .51, .21, .12, .41, .01.$$

When $y_i \le (c-a)/(b-a) = 4/5$ we use the formula

$$x_i = a + \sqrt{(b-a)(c-a)y} = 2 + \sqrt{5 \times 4 \times y_i} = 2(1 + \sqrt{5y_i}),$$

and when $y_i \ge 4/5$, we use

$$x_i = b - \sqrt{(b-a)(b-c)(1-y_i)} = 7 - \sqrt{5 \times 1 \times (1-y_i)} = 7 - \sqrt{5(1-y_i)}.$$

Observe that both give the same value when $y_i = 4/5$, namely, $x_i = 6$. The above sequence of uniformly distributed random numbers gives rise to the following sequence of random numbers that obey the designated triangular density function:

3.4832, 3.7321, 5.5214, 6.4523, 3.4832, 6.3675, 5.3764, 2.6325, 6.7764, 5.7417,

4.6458, 5.6606, 4.6458, 4.6833, 3.3416, 5.1937, 4.0494, 3.5492, 4.8636, 2.4472.

For example, $x_1 = 2(1 + \sqrt{.55}) = 3.4832$ and $x_4 = 7 - \sqrt{5(1-.94)} = 6.4523$.

### *18.1.4 The Exponential Distribution*

The probability density and cumulative distribution functions for an exponentially distributed random variable are as follows:

$$f(x) = \begin{cases} \lambda e^{-\lambda x}, & x \ge 0; \lambda > 0, \\ 0 & \text{otherwise}, \end{cases} \quad \text{and} \quad F(x) = \int_0^x f(t)dt = 1 - e^{-\lambda x}.$$

Since $F(x) = 1 - e^{-\lambda x}$, we obtain the inverse, $x$, as follows:

$$\begin{aligned} y &= 1 - e^{-\lambda x}, \\ e^{-\lambda x} &= 1 - y, \\ -\lambda x &= \ln(1-y), \\ x &= -\frac{1}{\lambda}\ln(1-y). \end{aligned} \tag{18.1}$$

Generating random numbers that are uniformly distributed on (0, 1) and substituting them into the right-hand side of Equation (18.1) results in a sequence of random numbers that are exponentially distributed with parameter $\lambda$. This formula may be simplified somewhat, by observing that if $y$ is uniformly distributed on (0, 1), then so also is $1 - y$ and we may write

$$x = -\frac{1}{\lambda}\ln(y), \tag{18.2}$$

thereby saving one arithmetic operation. However, some sources prefer to use the original Equation (18.1), arguing that the saving is minimal and preferring, on a purely aesthetic basis, that a large value of $y$ give a large value of $x$, and vice versa. Equation (18.2) has reverse monotonicity, meaning that large values of $y$ give small values of $x$, and vice versa.

**Example 18.4** A sequence of uniformly distributed random numbers in (0, 1) that begins

$$y_i = .11, .15, .62, .94, .11, .92, .57, .02, .99, .20, \ldots$$

converts to the following sequence of exponentially distributed random numbers with mean value 0.25 (parameter $\lambda = 4$):

$$x_i = .5518,\ .4743,\ .1195,\ .0155,\ .5518,\ .0208,\ .1405,\ .9780,\ .0025,\ .4024,\ \ldots.$$

Each $x_i$ is obtained as $x_i = -\ln(y_i)/4$.

A drawback with the inverse function method for generating random numbers that are exponentially distributed is the computational cost associated with the logarithmic function.

### 18.1.5 The Bernoulli Distribution

The Bernoulli probability mass function is given by

$$\text{Prob}\{X = 0\} = q,$$
$$\text{Prob}\{X = 1\} = p,$$

where $p + q = 1$. Its corresponding cumulative distribution function is given as

$$F(x) = \begin{cases} 0, & x < 0, \\ q, & 0 \le x < 1, \\ 1, & x \ge 1, \end{cases}$$

and hence has the typical step function graph. If the uniformly distributed random number that is generated lies in the interval $[0, q)$, the random variable is considered to have the value 0; if the random number lies in the interval $[q, 1)$ it is taken to have the value 1.

**Example 18.5** Suppose a biased coin turns up heads eight times for every ten times tossed. A sequence of uniformly distributed random numbers in (0, 1) that begins

$$y_i = .11,\ .15,\ .62,\ .94,\ .11,\ .92,\ .57,\ .02,\ .99,\ .20,\ \ldots$$

produces the following sequence of tosses:

$$x_i = H,\ H,\ H,\ T,\ H,\ T,\ H,\ H,\ T,\ H,\ \ldots.$$

A head $(H)$ is obtained every time $y_i$ is strictly less than 0.8; otherwise, a tail is obtained.

### 18.1.6 An Arbitrary Discrete Distribution

Consider a more general case of a random variable having a discrete probability mass function and probability distribution function. Suppose, for example, these are given as

$$p(x_i) = \text{Prob}\{X = x_i\} \quad \text{and} \quad F(x) = \text{Prob}\{X \le x\}.$$

We now generate uniformly distributed random numbers in the interval [0, 1). As we mentioned before, a number that falls in the interval $[F(x_{i-1}), F(x_i))$ is associated with the event $X = x_i$.

**Example 18.6** Suppose we have

| $x_i$ | 1 | 2 | 3 | 4 | 5 | 6 | 7 | 8 |
|---|---|---|---|---|---|---|---|---|
| $p(x_i)$ | 0.05 | 0.10 | 0.25 | 0.15 | 0.10 | 0.20 | 0.10 | 0.05 |

and

| $x_i$ | 1 | 2 | 3 | 4 | 5 | 6 | 7 | 8 |
|---|---|---|---|---|---|---|---|---|
| $F(x_i)$ | 0.05 | 0.15 | 0.40 | 0.55 | 0.65 | 0.85 | 0.95 | 1.00 |

If the uniformly distributed number that we generate is given by $y$, then

$$\begin{array}{rl} \text{if } y < 0.05, & \text{then } X = 1, \\ \text{if } 0.05 \le y < 0.15, & \text{then } X = 2, \\ \text{if } 0.15 \le y < 0.40, & \text{then } X = 3, \\ \text{if } 0.40 \le y < 0.55, & \text{then } X = 4, \\ \text{if } 0.55 \le y < 0.65, & \text{then } X = 5, \\ \text{if } 0.65 \le y < 0.85, & \text{then } X = 6, \\ \text{if } 0.85 \le y < 0.95, & \text{then } X = 7, \\ \text{if } 0.95 \le y < 1.00, & \text{then } X = 8. \end{array}$$

Thus, given the following sequence of uniformly distributed random numbers in (0, 1):

$$y_i = .11,\ .15,\ .62,\ .94,\ .11,\ .92,\ .57,\ .02,\ .99,\ .20,\ \ldots,$$

the random variable $X$ assumes the values

$$x_i = 2,\ 3,\ 5,\ 7,\ 2,\ 7,\ 5,\ 1,\ 8,\ 3,\ \ldots.$$

## 18.2 Discrete Random Variates by Mimicry

A number of important discrete random variables arise in modeling probabilistic scenarios and so it becomes possible to generate random numbers having the distribution of these random variables by mimicking (or simulating) the probability experiment. Binomial, geometric, and Poisson random variables fall into this category and the "mimicry" approach for generating random numbers having these distributions is very convenient. As we shall see, multiple uniformly distributed random numbers $u_i \in (0, 1)$ are required to produce a single random number having the sought-for distribution. It is important to remember that each $u_i$ can be used only once. For example, if three values are needed to produce a single random number having the required distribution, then $u_1$, $u_2$ and $u_3$ should be used to generate the first; $u_4$, $u_5$ and $u_6$ to generate the second and so on. In particular, it is wrong to use $u_1$, $u_2$, and $u_3$ for the first and then use $u_2$, $u_3$, and $u_4$ for the second.

### *18.2.1 The Binomial Distribution*

The probability mass function of a binomial random variable $X$ is given by

$$p_k = \text{Prob}\{X = k) = p_X(k) = \begin{cases} \binom{n}{k} p^k q^{n-k}, & 0 \le k \le n, \\ 0 & \text{otherwise.} \end{cases}$$

The corresponding cumulative distribution function is denoted by $B(t; n, p)$ and is given as

$$B(t; n, p) = F_X(t) = \sum_{i=0}^{\lfloor t \rfloor} \binom{n}{i} p^i (1-p)^{n-i}.$$

We previously identified this distribution with the number of successes obtained in a sequence of Bernoulli trials. We simply simulate this situation to generate random numbers that are distributed according to the binomial law. We generate a sequence of $n$ uniformly distributed random numbers in (0, 1) and count the number of times that we obtain a number less than $p$.

```
int k = 0;
for (int i=0; i< n; i++) {
   u = random();
   if ( u < p ) { k++; }
return k;
```

This approach can be very costly when $n$ is large. We shall return to this distribution in the next section as it will provide an example for the *accept-reject method* of generating nonuniform random numbers.

**Example 18.7** A simulation study incorporates a binomial random variable with parameters $n = 5$ and $p = .25$. The following sequence of uniformly distributed random numbers:

$u_i = .11,\ .15,\ .62,\ .94,\ .11,\ .92,\ .57,\ .02,\ .99,\ .70,\ .35,\ .67,\ .35, .36, .09, .51, .21, .12, .41, .01$

$.06,\ .84,\ .94,\ .06,\ .60,\ .34,\ .51,\ .16,\ .81,\ .43,\ .59,\ .51,\ .15,\ .70,\ .77,\ .96,\ .96,\ .09,\ .13,\ .36$

produces the following sequence of binomial random numbers:

$$k = 3,\ 1,\ 1,\ 3,\ 2,\ 1,\ 1,\ 2.$$

For example, the first group of five numbers, $u_1$–$u_5$, contains three that are strictly less than 0.25 and hence $k = 3$.

### *18.2.2 The Geometric Distribution*

The probability mass function of a random variable that has a geometric distribution is given by

$$p_X(k) = q^{k-1}p = p(1-p)^{k-1}, \quad k = 1, 2, \ldots .$$

This corresponds to a sequence of $k - 1$ failures (with probability $q = 1 - p$) followed by a single success (with probability $p$, $0 < p < 1$). Its image is the set of integers greater than or equal to one. The corresponding cumulative distribution function is given as

$$F_X(k) = \text{Prob}\{X \le k\} = \sum_{i=1}^{k} p(1-p)^{i-1} = 1 - (1-p)^k = 1 - q^k \quad \text{for } k = 1, 2, \ldots .$$

Thus we obtain geometrically distributed random numbers by generating uniformly distributed random numbers between 0 and 1 and counting the number of times the random number lies in the interval $[0, q)$ before we obtain one that does not satisfy this property. To this total, we add 1 for the "successful" outcome.

```
int k = 1;
u = random();
while ( u < q ) {  k++;   u = random();}
return k;
```

This algorithm is most efficient if the *while* loop is executed only a small number of times, i.e., if the probability of success is high ($q$ is small). Otherwise, inverting the cumulative distribution function may be more efficient. In this case, $k$ is obtained from

$$k = \left\lfloor \frac{\ln(y)}{\gamma} \right\rfloor + 1,$$

where $\gamma = \ln q$. Observe that $\gamma$ need only be computed once, and not for every random number generated. The proof of the inverse formula is left to the exercises.

**Example 18.8** Let us repeat the previous example, this time using the geometric distribution with parameter $q = 1 - p = .45$. The sequence of uniformly distributed random numbers, is

$.11,\ .15,\ .62,\ .94,\ .11,\ .92,\ .57,\ .02,\ .99,\ .70,\ .35,\ .67,\ .35,\ .36,\ .09,\ .51,\ .21,\ .12,\ .41,\ .01$

$.06,\ .84,\ .94,\ .06,\ .60,\ .34,\ .51,\ .16,\ .81,\ .43,\ .59,\ .51,\ .15,\ .70,\ .77,\ .96,\ .96,\ .09,\ .13,\ .46.$

The following geometric random numbers are generated:

$$k = 3,\ 1,\ 2,\ 1,\ 2,\ 1,\ 2,\ 4,\ 6,\ 1,\ 2,\ 2,\ 2,\ 2,\ 1,\ 2,\ 1,\ 1,\ 1,\ 3.$$

For example, since $u_1 = .11 < .45$ and $u_2 = .15 < .45$ but $u_3 = .62 \geq .45$, we find $k = 2 + 1 = 3$ to start the sequence of geometric random numbers.

### *18.2.3 The Poisson Distribution*

The probability mass function of a Poisson random variable $N$ with mean value $\alpha > 0$, usually denoted as $f(k, \alpha)$, is given by

$$\text{Prob}\{N = k\} = f(k, \alpha) = e^{-\alpha}\frac{\alpha^k}{k!} \quad \text{for } k = 0, 1, 2, \ldots .$$

The cumulative distribution function is given by

$$F(x) = e^{-\alpha}\sum_{k=0}^{j}\frac{\alpha^k}{k!} \quad \text{for} \quad j \leq x < j + 1.$$

We previously associated this distribution with the number of arrivals in a given period of time, or more generally, with the number of successes during a fixed time period. We also showed that, with a Poisson distribution, the time between arrivals is exponentially distributed. It is in this sense that we consider the mimicry approach to generating random numbers according to a Poisson process.

We interpret $N$ as the number of arrivals from a Poisson process in one time unit, and the times between successive arrivals, which we denote $A_1,\ A_2,\ \ldots$, are exponentially distributed. Therefore

$$N = k \quad \text{if and only if} \quad \sum_{i=1}^{k} A_i \leq 1 < \sum_{i=1}^{k+1} A_i.$$

In words, it is possible to have $N = k$ in one time unit only if at least $k$ arrivals occur *at or before* the end of the time unit and the next arrival, the $(k + 1)^{\text{th}}$, occurs *after* the end of the time unit, and vice versa. The technique is therefore to generate exponentially distributed random numbers with parameter $\alpha$, add them until they exceed 1 and then back of by one arrival. This gives

$$\sum_{i=1}^{k} -\frac{1}{\alpha}\ln(u_i) \leq 1 < \sum_{i=1}^{k+1} -\frac{1}{\alpha}\ln(u_i).$$

Applying the summation law for logarithms and multiplying through by $-\alpha$ we obtain

$$\ln\left(\prod_{i=1}^{k} u_i\right) \geq -\alpha > \ln\left(\prod_{i=1}^{k+1} u_i\right),$$

i.e.,

$$\prod_{i=1}^{k} u_i \geq e^{-\alpha} > \prod_{i=1}^{k+1} u_i.$$

Thus the procedure reduces to generating uniformly distributed random numbers in $(0, 1)$ and multiplying them together until the product is strictly less than $e^{-\alpha}$. The value of $k$ is one less than the number of products formed. Notice that when $N = k$, a total of $k + 1$ uniformly distributed random numbers will be needed. Given that the mean of the Poisson distribution is $\alpha$, it follows that, on average $\alpha + 1$ uniformly distributed random numbers will be needed to generate one Poisson distributed random number. This can be quite expensive when $\alpha$ is large.

**Example 18.9** We return to the previous example one more time, this time using the Poisson distribution with parameter $\alpha = 2.5$. The sequence of uniformly distributed random numbers, is

.11, .15, .62, .94, .11, .92, .57, .02, .99, .70, .35, .67, .35, .36, .09, .51, .21, .12, .41, .01

.06, .84, .94, .06, .60, .34, .51, .16, .81, .43, .59, .51, .15, .70, .77, .96, .96, .09, .13, .46.

We first compute $e^{-2.5} = .0821$. We now begin multiplying numbers in the sequence until the result falls below .0821 and then subtract one from the number of terms that formed the product. We find that the following Poisson random numbers are generated:

$$k = 1,\ 2,\ 2,\ 4,\ 1,\ 2,\ 1,\ 1,\ 1,\ 3,\ 4,\ 4,\ 1.$$

Thus, for example $u_1 > .0821$ but $u_1 \times u_2 < .0821$; $u_3 \times u_4 > .0821$ but $u_3 \times u_4 \times u_5 < .0821$, and so on.

A useful alternative to this approach when $\alpha$ is large is now presented. Random numbers having the Poisson distribution can be found with only a single uniformly distributed random number using the inverse transformation method, but it is first necessary to generate and store the Poisson probabilities

$$\begin{aligned} p_0 &= \text{Prob}\{N = 0\} = e^{-\alpha}, \\ p_1 &= \text{Prob}\{N = 1\} = \alpha e^{-\alpha}, \\ p_2 &= \text{Prob}\{N = 2\} = 1/2\alpha^2 e^{-\alpha}, \\ &\vdots \end{aligned}$$

The interval $[0, 1)$ must now be partitioned into consecutive subintervals of length $p_0, p_1, \ldots$. The interval into which each uniformly distributed random number falls gives the value of the random variable. Notice that the Poisson probabilities are easily computed by means of the recursion:

$$p_0 = e^{-\alpha}, \quad p_k = \frac{\alpha p_{k-1}}{k} \quad \text{for } k \geq 1.$$

Another possibility when $\alpha$ is large is to use the normal approximation. For large $\alpha$, $(N - \alpha)/\sqrt{\alpha}$ is approximately standard normal distributed. This means that procedures for generating random numbers that are normally distributed can be used to produce random numbers that are approximately Poisson distributed. If $Z$ is a random number having standard normal distribution, then

$$N = \lceil \alpha + \sqrt{\alpha} Z - .5 \rceil$$

has a Poisson distribution. The .5 in the formula is designed to force a "round-up" operation. The generation of random numbers that are normally distributed is considered in some detail in a later section.

## 18.3 The Accept-Reject Method

It is sometimes possible to determine the probability density function $f(x)$ for some random variable, but computation of the cumulative distribution function $F(x)$ (or its inverse) is difficult or impossible. An example is the *normal* random variable. In such instances, the *accept-reject* method, or more simply, the *rejection method*, can be rather effective. Let us assume that the density function is defined on a finite interval, or support, $[a, b]$ and further assume that there exists a constant $c < \infty$ which bounds $f(x)$ on this interval, i.e., $f(x) \leq c$ for all $a \leq x \leq b$. Usually we choose $c = \max f(x)$ in $[a, b]$. We now generate *two* uniformly distributed random numbers, $u_1$ and $u_2$ and set $x_1 = a + (b - a)u_1$. The number $x_1$ is taken as a random number that obeys the density function $f(x)$ whenever it satisfies the condition

$$cu_2 \le f(x_1). \tag{18.3}$$

If it does not satisfy this condition, it is rejected. To put it another way, we generate a random point $(u_1, u_2)$, uniformly distributed over a unit square, and convert it to a point $(a + (b - a)u_1, cu_2)$ uniformly distributed over a box of height $c$ and base $(b - a)$. If this point falls below the density curve, then $x_1 = a + (b - a)u_1$ is taken as a random number that satisfies the density, otherwise both $u_1$ and $u_2$ are rejected. The rejection method is illustrated in Figure 18.5 in which $x_1$ is accepted and $x_3$ is rejected.

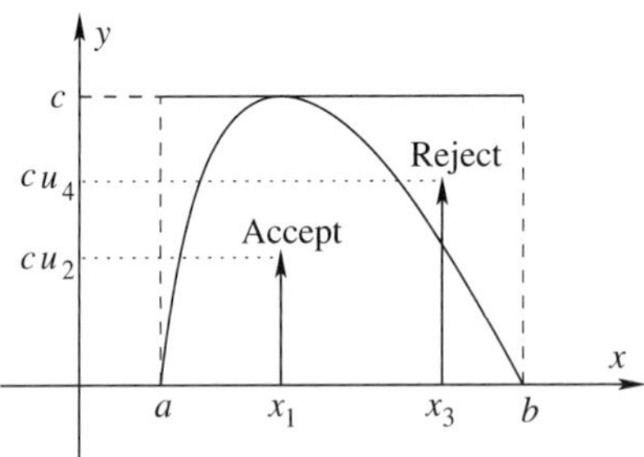


Figure 18.5. Illustration of the accept-reject method.

Observe that, unlike the method of inversion, the accept-reject approach requires the evaluation of the density function for different values of its argument, and this may be costly. The validity of the approach may be established by considering conditional probabilities. Let $X$ and $Y$ be two random variables, the first uniformly distributed on $[a, b]$ and the second uniformly distributed on $[0, c]$. Then

$$\begin{aligned}\text{Prob}\{x \le X \le x + dx \mid Y \le f_X(X)\} &= \frac{\text{Prob}\{x \le X \le x + dx,\ Y \le f_X(x)\}}{\text{Prob}\{Y \le f_X(X)\}} \\ &= \left(\frac{dx}{b-a}\right)\left(\frac{f_X(x)}{c}\right)\left(\frac{1}{c(b-a)}\right)^{-1} = f_X(x)dx.\end{aligned}$$

On the second line of this equation, the first term in parentheses arises because $X$ is uniformly distributed on $[a, b]$, the second in parentheses is because $Y$ is uniformly distributed on $[0, c]$ and the third because both $X$ and $Y$ are independent and the random pair is uniformly distributed over a rectangle of area $c(b - a)$. Finally, we note that the accept-reject method may be applied when the density function has nonzero values outside the range $[a, b]$, but in this case, only an approximation is obtained.

**Example 18.10** Consider the density function

$$f_X(x) = \begin{cases} -3x^2 + 4x, & 0 \le x \le 1, \\ 0 & \text{otherwise.} \end{cases}$$

The maximum value of this density function in $[0, 1]$ occurs at $x = 2/3$ with maximum value equal to $4/3$. This provides us with the value of $c$. The accept-reject method requires us to generate a uniformly distributed number $x_1$ in the range $[a, b]$, here equal to $[0, 1]$, and a second uniformly distributed number $u_2$ so that $cu_2$ lies in the range $[0, c]$, here equal to $[0, 4/3]$. The number $x_1$ is accepted if $cu_2 \le f(x_1) = -3x_1^2 + 4x_1$. Suppose the sequence of uniformly distributed random numbers generated is

$$u_i = .11,\ .15,\ .62,\ .94,\ .11,\ .92,\ .57,\ .02,\ .99,\ .70,\ .35,\ .67,\ .35,\ .36,\ .09,\ .51,\ .21,\ .12,\ .41,\ .01.$$

Then the following values of $x_i$, $cu_{i+1}$, $f(x_i)$ are obtained and the decision to accept or reject $x_1$ as having probability density $f(x)$ subsequently taken.

| $u_i$ | $u_{i+1}$ | $x_i$ | $cu_{i+1}$ | $f(x_i)$ | Acc/Rej |
|---|---|---|---|---|---|
| .11 | .15 | .11 | 0.20 | 0.40 | Acc |
| .62 | .94 | .62 | 1.25 | 1.33 | Acc |
| .11 | .92 | .11 | 1.23 | 0.40 | Rej |
| .57 | .02 | .57 | 0.03 | 1.31 | Acc |
| .99 | .70 | .99 | 0.93 | 1.02 | Acc |
| .35 | .67 | .35 | 0.89 | 1.03 | Acc |
| .35 | .36 | .35 | 0.48 | 1.03 | Acc |
| .09 | .51 | .09 | 0.68 | 0.34 | Rej |
| .21 | .12 | .21 | 0.16 | 0.71 | Acc |
| .41 | .01 | .41 | 0.01 | 1.14 | Acc |

The sequence of random numbers satisfying this density function is

$$.11,\ .62,\ .57,\ .99,\ .35,\ .35,\ .21,\ .41.$$

As well as requiring the evaluation of the function $f(x)$, which may be expensive, the accept-reject method requires a minimum two uniformly distributed random numbers per accepted random number with density $f(x)$. Each time a rejection occurs, two uniformly distributed random numbers have been generated but no random number satisfying the required density has been obtained. It follows then that the rejection method works best when the area under the density curve $f(x)$ occupies most of the area in the rectangle of length $b - a$ and height $c$ since in this case, an accept becomes much more common than a reject. In light of this, it is possible to replace the rectangle with a different object, but one which still bounds $f(x)$ on $[a, b]$. Let $g(x)$ be a probability density function on $[a, b]$ such that

$$\frac{f(x)}{g(x)} \leq c \text{ for all } a \leq x \leq b$$

for some $c \geq 1$. The constant $c$ is usually found by finding the maximum value of $f(x)/g(x)$ for $x \in [a, b]$ and the function $cg(x)$ is said to *majorize* $f(x)$. The following algorithm allows us to generate random numbers that satisfy the density function $f(x)$:

- Generate a random number $x_1$ having density function $g(x)$.
- Generate a uniformly distributed random number $u_2$ on $(0, 1)$.
- If $cu_2 \leq f(x_1)/g(x_1)$, accept $x_1$; otherwise reject.

This algorithm is most efficient when the number of rejections is low and the probability of success is high. Observe that success occurs when the test $u_2 \leq f(x_1)/cg(x_1)$ is satisfied: hence we accept $x_1$ with probability $f(x_1)/cg(x_1)$. Each test has, independently, a probability of acceptance equal to $1/c$ and so the number of tests (or iterations) for acceptance is geometrically distributed with parameter $1/c$ and mean value $c$. Thus $c$ must be greater than 1 and the closer it is to 1, the fewer the number of rejections.

When we bound $f(x)$ by the uniform density function in $[a, b]$, namely, $g(x) = 1/(b - a)$, then the value of $c$ is taken as the maximum value of $(b - a)f(x)$ in the interval $[a, b]$. In this case, the test condition is

$$cu_2 = u_2(b-a)\max_{a\le x\le b} f(x) \le f(x_1)/g(x_1) = (b-a)f(x_1)$$

or

$$u_2 \max_{a\le x\le b} f(x) \le f(x_1),$$

which returns us to Equation (18.3).

The advantage of choosing a majorizing function other than a bounding constant, is that we expect it to more closely bound $f(x)$ with a resulting reduction in the number of rejections. However, it has the additional cost of requiring the generation of a random number that satisfies the density function $g(x)$. While this is easy when $g(x)$ is uniform on $[a, b]$, it is more costly with other density functions. It is therefore evident that $g(x)$ should be chosen keeping in mind that random numbers having this distribution will need to be generated. Distributions for which the inverse transformation approach leads to simple formulae are frequently used. These include wedge-shaped density functions, the triangle density function, and the exponential density function.

### *18.3.1 The Lognormal Distribution*

If $Y$ is a normally distributed random variable, then the random variable, $X = e^Y$ is said to have a lognormal distribution. Lognormal random variables are frequently used to model service times. The probability density function of a lognormal random variable is given by

$$f_X(x) = \frac{1}{\sigma x\sqrt{2\pi}} \exp\left(-\frac{(\ln(x)-\mu)^2}{2\sigma^2}\right), \quad x > 0,$$

and is zero elsewhere. It should be noted that the preferred approach for generating random numbers that have a lognormal distribution, is to generate a normally distributed random number ($Y$) and to apply the defining transformation $X = e^Y$. We include the lognormal distribution here just to illustrate the application of the accept-reject method.

**Example 18.11** We shall take $\mu = 0$ and $\sigma = 1$ as parameters for the lognormal density and we shall use the exponential function $\exp(-0.8x)$, $x \ge 0$, as the bounding function. These are shown in Figure 18.6. Suppose our interest lies only in the lognormal density constrained to the interval $[0, 6]$. Since the area under the lognormal density between 0 and 6 is given by

$$\frac{1}{2} + \frac{1}{2}\text{erf}\left(\frac{\ln(6)}{\sqrt{2}}\right) = 0.9634,$$

we set $f(x) = f_{\text{LN}}/0.9634$ so that $f(x)$ is a true density function on the interval $[0, 6]$. We must also be sure that our bounding function is a true density function on $[0, 6]$. The area under $\exp(-0.8x)$ between 0 and 6 is equal to 1.2397 and so we choose the bounding density function to be

$$g(x) = \frac{e^{-0.8x}}{1.2397} \quad \text{for} \quad 0 \le x \le 6,$$

and zero otherwise. Now that we have both $f(x)$ and $g(x)$, we need to find $c$, the maximum value of $f(x)/g(x)$ on $[0, 6]$. This maximum occurs at $x = 6$ and is equal to 2.0875. Thus we shall set $c = 2.1$.

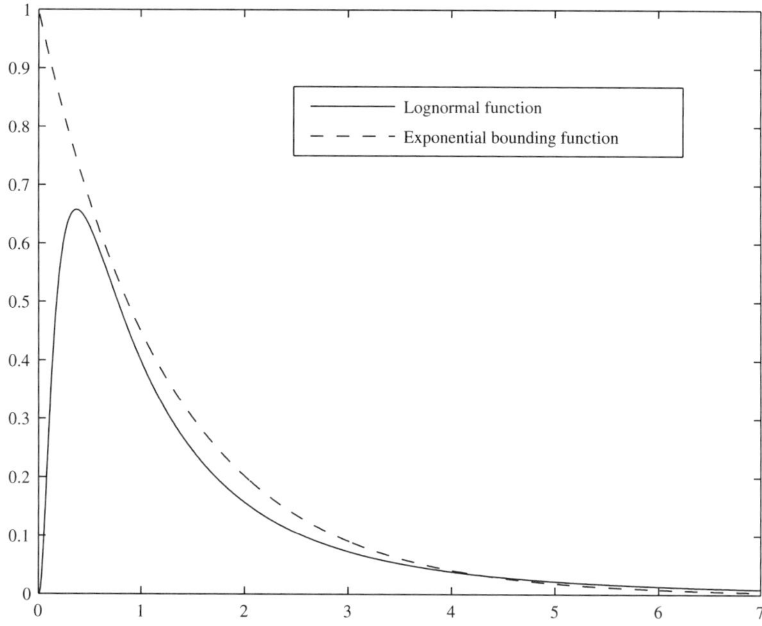


Figure 18.6. The lognormal density and its exponential bounding function.

With a sequence of uniformly distributed random numbers $u_1$, $u_2$, ... already available, the steps required to generate random numbers with this lognormal distribution are

1. Use $u_i$ to generate $x_i$ having density $g(x) = \exp(0.8x)/1.2397$ on $[0, 6]$. This is accomplished by using the inverse transformation for exponential densities, and gives
$$x_i = -\frac{\ln(1 - .9918u_i)}{0.8}.$$
The reader may wish to verify that a density function of the form $\exp(-\lambda x)/\gamma$, $\lambda > 0$, gives rise to the inversion formula $x = -\ln(1 - \lambda\gamma u)/\lambda$.
2. Accept $x_i$ if $cu_{i+1} \leq f(x_i)/g(x_i)$.

Given the following sequence of uniformly distributed random numbers:

$$u_i = .11, .15, .62, .94, .11, .92, .57, .02, .99, .70, .35, .67, .35, .36, .09, .51, .21, .12, .41, .01,$$

we obtain the following results:

| $u_i$ | $u_{i+1}$ | $x_i$ | $cu_{i+1}$ | $f(x_i)/g(x_i)$ | Acc/Rej |
|---|---|---|---|---|---|
| .11 | .15 | 0.1444 | 0.315 | 0.6135 | Acc |
| .62 | .94 | 1.1929 | 1.974 | 1.1003 | Rej |
| .11 | .92 | 0.1444 | 1.932 | 0.6135 | Rej |
| .57 | .02 | 1.0414 | 0.042 | 1.1331 | Acc |
| .99 | .70 | 5.0136 | 1.470 | 1.5410 | Acc |
| .35 | .67 | 0.5330 | 1.407 | 1.2103 | Rej |
| .35 | .36 | 0.5330 | 0.756 | 1.2103 | Acc |
| .09 | .51 | 0.1169 | 1.071 | 0.4817 | Rej |
| .21 | .12 | 0.2919 | 0.252 | 1.0408 | Acc |
| .41 | .01 | 0.6524 | 0.021 | 1.2105 | Acc |

The sequence of random numbers satisfying this density function is

$$0.1444, \ 1.0414, \ 5.0136, \ 0.5330, \ 0.2919, \ 0.6524.$$

## 18.4 The Composition Method

In some instances, random variables can be written in terms of other, simpler, random variables and this provides a mechanism to generate random numbers having complex distributions using their more convenient underlying distributions. This is the case of the Erlang-$r$ and hyperexponential distributions which consist of combinations of exponential distributions and which are considered below. An identical approach can be used to generate random numbers that satisfy any type of phase-type distribution. Furthermore, a more generic version of the composition method can be developed and applied to many random variables whose density functions do not decompose naturally into any particular structure. We refer to this extension as a density partitioning approach. It includes the *rectangle-wedge-tail* method and the *ziggurat* method.

### 18.4.1 The Erlang-$r$ Distribution

The probability density and cumulative distribution functions for a random variable that is distributed according to an Erlang-$r$ law with mean $r/\xi$ and variance $r/\xi^2$ are as follows:

$$b(x) = \frac{\xi^r x^{r-1} e^{-\xi x}}{(r-1)!}, \quad x \geq 0,$$

$$B(x) = 1 - \sum_{k=0}^{r-1} \frac{(\xi x)^k}{k!} e^{-\xi x}, \quad x \geq 0.$$

Since this distribution function is not readily inverted, it is more convenient to consider the Erlang-$r$ distribution as a sequence of exponential phases in tandem as represented in Figure 18.7 and discussed in previous chapters.

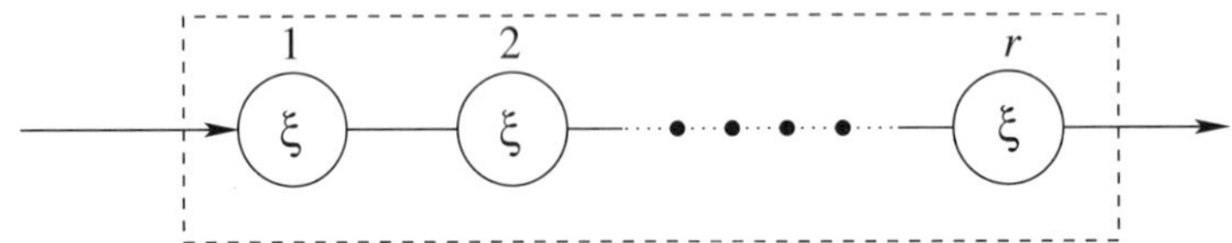


Figure 18.7. Erlang distribution: $r$ exponential service phases in tandem.

To generate random numbers that satisfy an Erlang-$r$ distribution, the straightforward approach would then be to generate $r$ random numbers according to an exponential distribution and to add these together. However, this would work out to be too expensive, requiring the computation of $r$ natural logarithms. Instead we generate $r$ uniformly distributed random numbers, $u_i, \ i = 1, \ldots, r$, set $x_i = -\ln(u_i)/\xi$ and, using the summation rule for logarithms, obtain a single Erlang-$r$ distributed random number as

$$\sum_{i=1}^{r} x_i = -\frac{1}{\xi} \sum_{i=1}^{r} \ln(u_i) = -\frac{1}{\xi} \ln\left(\prod_{i=1}^{r} u_i\right),$$

and a sequence $z_k$ of such numbers as

$$z_k = \sum_{i=r(k-1)+1}^{rk} x_i = -\frac{1}{\xi} \ln\left(\prod_{i=r(k-1)+1}^{rk} u_i\right), \quad k = 1, 2, \ldots .$$

Given the already large computation costs usually associated with simulation, computational saving devices such as this use of the summation rule for logarithms are extremely important.

**Example 18.12** We wish to use the following sequence of 40 uniformly distributed random numbers in the unit interval to generate a sequence of random numbers that have an Erlang-4 distribution with mean 2.0.

$$u_i = [.11, .15, .62, .94], [.11, .92, .57, .02], [.99, .70, .35, .67], [.35, .36, .09, .51], [.21, .12, .41, .01], [.06, .84, .94, .06], [.60, .34, .51, .16], [.81, .43, .59, .51], [.15, .70, .77, .96], [.96, .09, .13, .36].$$

Given that $r = 4$ and $r/\xi = 2$, this implies that $\xi = 2$ and successive random numbers from this Erlang-4 distribution are formed by multiplying four consecutive $u_i$ values together, taking the natural logarithm of the result and then dividing by $-2$. This gives

$$2.3222,\ 3.3824,\ .9085,\ 2.5764,\ 4.5888,\ 2.9315,\ 2.0478,\ 1.1278,\ 1.2780,\ 2.7553, \dots,$$

where

$$\frac{-\ln(.11 \times .15 \times .62 \times .94)}{2} = 2.3222, \quad \frac{-\ln(.11 \times .92 \times .57 \times .02)}{2} = 3.3824, \quad \dots.$$

### 18.4.2 The Hyperexponential Distribution

The probability density and cumulative distribution functions for the hyperexponential distribution are given by

$$b(x) = \sum_{i=1}^{r} \alpha_i \mu_i e^{-\mu_i x}, \quad x \geq 0; \quad B(x) = \sum_{i=1}^{r} \alpha_i (1 - e^{-\mu_i x}), \quad x \geq 0.$$

The first two moments are given, respectively, by

$$E[X] = \sum_{i=1}^{r} \frac{\alpha_i}{\mu_i} \quad \text{and} \quad E[X^2] = 2 \sum_{i=1}^{r} \frac{\alpha_i}{\mu_i^2}.$$

As is the case for the Erlang distribution function, a hyperexponential distribution function is not easy to invert and we resort to our earlier construction of the hyperexponential distribution as a number of exponential phases in parallel as illustrated in Figure 18.8. Exponential phase $i$, with rate $\mu_i$, is chosen with probability $\alpha_i$. The branching probabilities are defined so that $\sum_{i=1}^{r} \alpha_r = 1$.

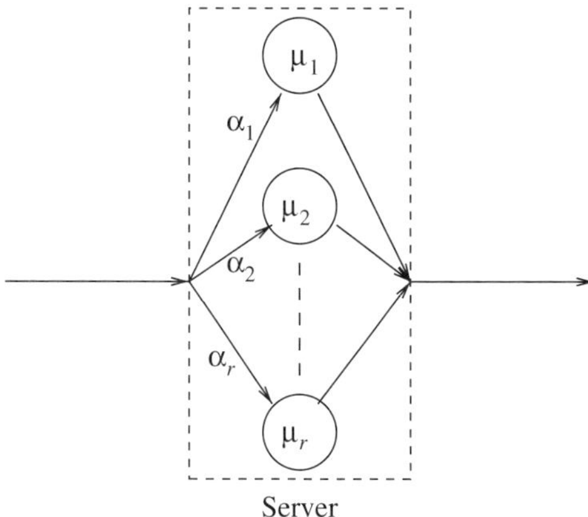


Figure 18.8. Hyperexponential distribution: $r$ exponential phases in parallel.

A convenient way to avoid having to invert the distribution function is to first generate a uniformly distributed random number to select a particular exponential phase $i$, followed by the generation of a random number that satisfies an exponential distribution with rate $\mu_i$. This allows us to generate hyperexponentially distributed random numbers rather easily.

**Example 18.13** Suppose we wish to model a scenario in which 20% of customers at a service center require 15 minutes of service while the other 80% need, on average only 10 minutes. This allows us to set $\alpha_1 = .2$, $\alpha_2 = .8$, $\mu_1 = 4$, and $\mu_2 = 6$, where we specify mean service time $(= 1/\mu)$ in units of hours. Let us use the following sequence of 20 uniformly distributed random numbers in the unit interval to generate a sequence of random numbers which obey this hyperexponential distribution:

$u_i = .11, .15, .62, .94, .11, .92, .57, .02, .99, .70, .35, .67, .35, .36, .09, .51, .21, .12, .41, .01.$

We take these random numbers in groups of two. If the first number in the pair is strictly less than 0.2, then we will use the second to generate an exponentially distributed random number having mean 1/4; if the first number in the pair is greater than or equal to 0.2, then we shall use the second to generate an exponentially distributed random number with mean 1/6. For example, the first number is .11, which means that we use the second, .15, to generate an exponentially distributed random number with mean 1/4, (or rate $\lambda = 4$). This gives $-\ln(.15)/4 = .4743$ hours, or 28.45 minutes. Continuing like this, we obtain the following sequence of hyperexponentially distributed random numbers:

$$.4743, .0103, .0208, .6520, .0594, .0667, .1703, .1683, .3534, .7675.$$

### *18.4.3 Partitioning of the Density Function*

We now turn to a more general version of the composition approach. Observe that if $f_1(x), f_2(x), \ldots, f_n(x)$ are $n$ density functions and $p_1, p_2, \ldots, p_n$ are $n$ probabilities such that $\sum_{i=1}^n p_i = 1$, then

$$f(x) = p_1 f_1(x) + p_2 f_2(x) + \cdots + p_n f_n(x) \tag{18.4}$$

is also a probability density function. It may be thought of as composed of a linear combination of the $n$ densities $f_i(n)$; hence the characterization of the approach we now describe as a compositional approach. Suppose that we require random numbers that are distributed according to some density function $f(x)$. Let the interval over which the density function is defined be partitioned into $n$, not necessarily equal, subintervals, $[x_0, x_1], [x_1, x_2], \ldots, [x_{n-1}, x_n]$. The $i^{\text{th}}$ subinterval, $[x_{i-1}, x_i]$, $i = 1, 2, \ldots, n$, gives rise to a panel whose base is $[x_{i-1}, x_i]$ and whose ceiling is formed by the density curve between $f(x_{i-1})$ and $f(x_i)$. The area of each panel is necessarily positive and less than 1 and so can be taken to be a probability. Let $p_i$ be the area of panel $i$. Furthermore, the sum of these $n$ areas must be equal to 1, since the entire area under the density curve is equal to 1. Thus it follows that these $p_i$, $i = 1, 2, \ldots, n$, can be taken to constitute the $n$ coefficients of the composition in Equation (18.4). The $n$ density functions $f_i(x)$ are defined by the original density function $f(x)$ restricted to the panel $[x_{i-1}, x_i]$.

So far, so good, but as yet nothing has been gained. However, a method for generating random numbers that *approximately* satisfy the distribution $f(x)$ is to replace the $i^{\text{th}}$ panel with a rectangle having the same base but with height given by $h_i = [f(x_{i-1}) + f(x_i)]/2$, the midpoint of the density function in this subinterval. This allows us to compute the area of panel $i$ as $\hat{p}_i = (x_i - x_{i-1})h_i$ and the density function $f_i(x)$ in this panel is replaced with a simpler uniform density function. Of course, the sum $\sum_{i=1}^n \hat{p}_i$ need no longer be equal to one, but an appropriate normalization can take care of this. This approach is illustrated in Figure 18.9.

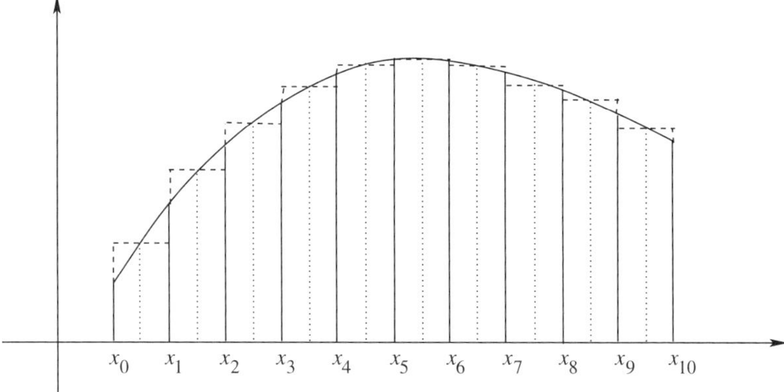


Figure 18.9. A density function $f(x)$ approximated by ten uniform densities.

The procedure for producing random numbers that approximate the density $f(x)$ is to first choose a particular panel according to its area. Specifically, choose panel $k$ with probability $\hat{p}_k$. Next generate a number $u$ that is uniformly distributed on (0, 1). Then $x_{k-1} + u(x_k - x_{k-1})$ satisfies, approximately, the density $f(x)$. What this means in terms of Equation (18.4) is that we first choose which density function $f_i(x)$ to use and then generate a random number which satisfies this density. The accuracy of the approximation depends, of course, on the particular density function and on the number of panels chosen, the greater the number of panels, the better the accuracy. However, with an excessive number of panels, time is wasted in deciding which one to choose (the selection based on $\hat{p}_k$), particularly, if the selection is done in a linear fashion instead of a binary search. This has led to the development of better approximations, which we now consider.

Let us once again consider the area under the density curve and this time divide it into rectangles that are contained completely within the area and wedgelike pieces that sit on top of the rectangles so that the entire area is covered by these $n$ rectangles (areas 1 through $n$) and $n$ wedgelike pieces (areas $n + 1$ through $2n$), as illustrated in Figure 18.10. We call the pieces that sit on top of the rectangles "wedgelike" rather than "wedges," reserving this latter for bounding functions in the accept-reject method.

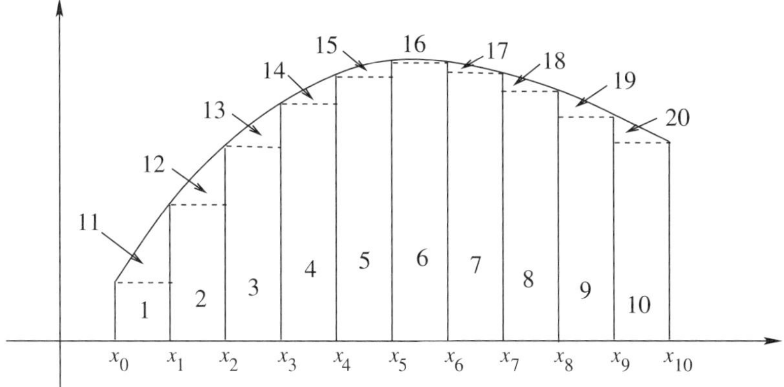


Figure 18.10. Area under $f(x)$ partitioned into ten rectangles and ten wedgelike pieces.

When the distribution has a tail, then this tail will constitute yet another wedgelike piece. It is expected that most of the area will be covered by the rectangles and only a small portion covered by the wedgelike pieces. This is important because generating random numbers that obey a rectangular density is very inexpensive when compared to the effort needed to generate numbers that obey the wedgelike densities. Each rectangle and wedgelike piece is assigned a probability equal to its area. In an ideal situation the sum of the rectangular areas should be close to 1 and the sum of the wedgelike areas close to zero. Thus the probability that when a component density function $f_i(x)$ is chosen it is likely that it will be a rectangular density. In the few cases when the chosen density is wedgelike, then a random number having this density must be generated using, for example, the accept-reject method. The accept-reject method is to be recommended since most of the areas above the rectangles are almost right-angle triangles themselves and can be bound by a close-matching wedge which makes the probability of a reject very small. The algorithm is very similar to the previous. Let the area under the density curve be partitioned into $n$ rectangles and $m$ wedge-like pieces and let $p_i$, $i = 1, 2, \ldots, n+m$, be the area of the $i^{\text{th}}$ partition.

- Choose partition $i$, $1 \le i \le n+m$, with probability $p_i$.
- If $i \le n$, generate $u$, uniformly distributed on $(0, 1)$, and take $x_{i-1} + u(x_i - x_{i-1})$ as the next random number that obeys $f(x)$.
- If $i > n$, use the accept-reject method to generate $y$ which obeys $f_i(x)$ on $[x_{i-1-n}, x_{i-n}]$.

When a wedgelike partition is selected, care must be taken to move this area down on to the $x$-axis and scale it so that it is a true density function, since, as we have seen, the two functions $f(x)$ and $g(x)$ used in the accept-reject method are density functions. Of course, all of this is taken care of before the generation of the random numbers begins. The relevant information is computed once at initiation, stored in tables and accessed as needed during the running of the generation algorithm.

**Example 18.14** Let us generate random numbers having a Cauchy distribution using this approach. The probability density function of a random variable having this distribution is

$$f_X(x) = \frac{1}{\pi(1+x^2)}, \qquad -\infty < x < \infty.$$

Since this is obviously symmetric, we shall generate random variates in the upper right quadrant and randomly (with probability .5) assign a $-$ or a $+$. The density function of $|X|$, shown in Figure 18.11, is

$$f(x) \equiv f_{|X|}(x) = \frac{2}{\pi(1+x^2)}, \qquad 0 \le x < \infty.$$

We shall partition the area under this density curve into seven pieces, three rectangles with bases $[0, 1]$, $[1, 2]$ and $[2, 3]$, three wedges that sit on top of these rectangles and a tail section, from coordinate $x = 3$ onward. A more rigorous approach might involve many more than seven partitions—here our only purpose is to present an example. The heights of the three rectangles are $2/[\pi(1+k^2)]$ for $k = 1, 2, 3$, the point at which their right sides intersect with the density curve. Hence the areas (and the associated probabilities) of these three rectangles are

$$p_1 = \frac{2}{\pi(1+1)} = \frac{1}{\pi} = .318310, \quad p_2 = \frac{2}{\pi(1+4)} = \frac{2}{5\pi} = .127324, \quad \text{and}$$

$$p_3 = \frac{2}{\pi(1+9)} = \frac{2}{10\pi} = .063662.$$

We now calculate the area of the three wedges. This is made easy since it is known that

$$\int \frac{dx}{1+x^2} = \arctan(x).$$

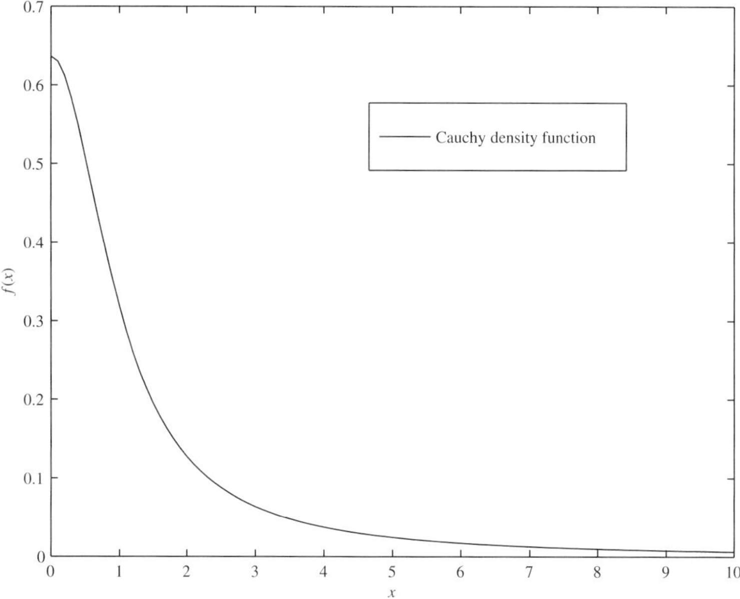


Figure 18.11. The density function of $|X|$ where $X$ is a Cauchy random variable.

The area under the density curve between 0 and any point $x$ is $2\arctan(x)/\pi$. It follows that the areas of the three wedges are

$$p_4 = \frac{2}{\pi}\arctan(1) - p_1 = 0.5 - \frac{1}{\pi} = .181690,$$

$$p_5 = \frac{2}{\pi}[\arctan(2) - \arctan(1)] - p_2 = 0.204828 - \frac{2}{5\pi} = .077509,$$

$$p_6 = \frac{2}{\pi}[\arctan(3) - \arctan(2)] - p_3 = 0.090334 - \frac{1}{5\pi} = .026672.$$

The area (and probability) of the tail is

$$p_7 = 1 - \frac{2}{\pi}\arctan(3) = 0.204833.$$

The reader may wish to verify that the sum of these areas is indeed equal to 1. The cumulative probabilities obtained from the $p_i,\ i = 1, 2, \ldots, 7$, are

$$0.318310, \quad 0.445634, \quad 0.509296, \quad 0.690986, \quad 0.768495, \quad 0.795167, \quad 1.0.$$

A uniformly distributed random number is used to select one of the seven partitions according to the particular interval into which it falls. If the random number $u_i$ falls into partition $k,\ k = 1, 2, 3$, we take $u_i + (k - 1)$ as a Cauchy random number. If $u_i$ falls into partition 4, we could use the accept-reject approach with a straight line through $2/\pi$ as the boundary. If it falls into partition 5 or 6, we use a wedge as the boundary. Finally, if it falls into the last partition, we use accept-reject taking $1/x^2$ as the bounding density function. This density function is the reciprocal of a random variable that is uniformly distributed on (0, 1), which simplifies the process. In fact this could also be used as the bounding function for wedgelike sections 5 and 6.

Suppose we wish to generate Cauchy random numbers from the following uniformly distributed random numbers:

$$u_i = .4627,\ .8926,\ .0145,\ .1352,\ .7246,\ .4502,\ .8819,\ .5654.$$

Since $u_1 \in [.445634, .509296)$, the third interval, we take $k = 3$ and compute $2 + .4627(3 - 2) = 2.4627$. To this we must add a plus sign (since $u_2 = .8926 > .5$) and take $y_1 = +2.4627$ as the first Cauchy random number.

Next we have $u_3 = .0145$ which places us in the first rectangle, $k = 1$ and we compute $0 + .0145(1 - 0)$ to which we append a minus sign (since $u_4 = .1352 < .5$). This gives our second Cauchy random number, $y_2 = -.0145$.

Since $u_5 = .7246$ falls in the interval $[.690986, .768495)$, which gives $k = 5$, we now have the opportunity to see how a wedgelike area is handled and the accept reject method used to generate the next Cauchy random number. This wedgelike area is situated on top of the rectangle based on the interval $[1, 2]$. It must be moved down to the $x$-axis and then scaled to give a true density function. i.e., we need to derive the density function $f_5(x)$ from the density function $f(x)$. Since it must be a true density, the area under $f_5(x)$ in the interval $[1, 2]$ must be equal to 1. Let $\alpha$ be the area under $f(x)$ in the interval $[1, 2]$. Then

$$\alpha = \frac{2}{\pi}(\arctan(2) - \arctan(1)) = .204833.$$

Hence the density function $f_5(x)$ is

$$f_5(x) = \frac{2}{\alpha\pi(1 + x^2)}, \qquad 1 \le x \le 2,$$

and is zero otherwise. This function is concave on $[1, 2]$ and is conveniently bound by the straight line that passes through the endpoints $f_5(1) = 1/(\alpha\pi)$ and $f_5(2) = 2/(5\alpha\pi)$. The equation of this straight line is

$$y = \frac{(8 - 3x)}{5\alpha\pi}.$$

Integrating this straight line between $x = 1$ and 2 gives

$$\frac{1}{5\alpha\pi}\int_1^2 (8 - 3x)dx = \frac{7}{10\alpha\pi},$$

and hence the density function used in the acceptance test, $g(x)$, is

$$g(x) = \frac{(8 - 3x)}{5\alpha\pi} \times \frac{10\alpha\pi}{7} = \frac{2}{7}(8 - 3x), \qquad 1 \le x \le 2,$$

and is zero otherwise. These functions are plotted in Figure 18.12.

Since we want $c$ to satisfy

$$cg(x) = \frac{(8 - 3x)}{5\alpha\pi},$$

this means that

$$c\frac{2(8 - 3x)}{7} = \frac{(8 - 3x)}{5\alpha\pi} \Longrightarrow c = \frac{7}{10\alpha\pi} = 1.0878.$$

Observe that $c$ is very close to 1, which means that we should expect the number of rejections to be relatively small. Also notice that

$$\frac{f_5(x)}{cg(x)} = \frac{10}{(1 + x^2)(8 - 3x)}$$

and hence

$$\frac{f_5(1)}{cg(1)} = \frac{f_5(2)}{cg(2)} = 1,$$

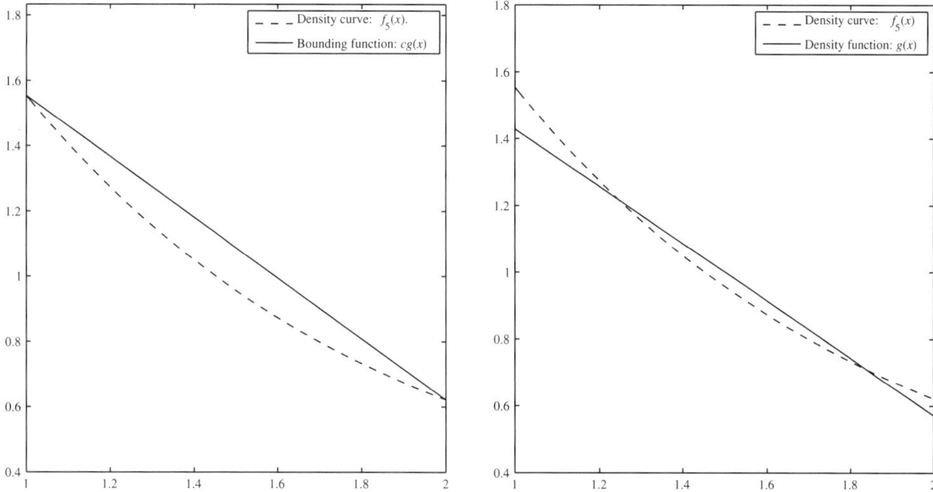


Figure 18.12. Left: $f_5(x)$ and straight line $y = (8 - 3x)/(5\alpha\pi)$. Right: $f_5(x)$ and the wedge density function $g(x)$.

which should be expected since $f_5(x)$ and $cg(x)$ are equal at the end points. Also, the maximum difference between $f_5(x)$ and $cg(x)$ occurs at $x = 1.453617$ and

$$\frac{f_5(1.453617)}{cg(1.453617)} = .882715.$$

At this point we have gathered together all the information we need to implement the accept-reject method, namely, the density function for the random variates $f(x)$, the bounding function $g(x)$, and the constant $c$. Recall the three steps needed to implement the accept-reject method.

1. Use $u_1$, uniformly distributed on (0, 1), to generate $x_1$ having density function $g(x)$.
2. Generate a uniformly distributed random number $u_2$ on (0, 1).
3. If $cu_2 \leq f(x_1)/g(x_1)$, accept $x_1$; otherwise reject.

1. We shall use the method of inversion to generate a random deviate having density function $g(x) = 2(8 - 3x)/7$. To do so, we first form the cumulative distribution function

$$G(x) = \int_1^x g(t)dt = \frac{1}{7}\int_1^x 2(8 - 3t)dt = \frac{1}{7}(-3x^2 + 16x - 13).$$

Setting $u = (-3x^2 + 16x - 13)/7$ and inverting, we obtain

$$x = \frac{16 \pm \sqrt{16^2 - 12(13 + 7u)}}{6} = \frac{16 \pm \sqrt{100 - 84u}}{6}. \tag{18.5}$$

With our next uniformly distributed random number $u_6 = .4502$, we find $x_1 = 2.6667 \pm 1.3143$. We choose the root that lies in [1, 2], namely, $x_1 = 2.6667 - 1.3143 = 1.3524$. This then completes step 1.

2. We use the next number in our sequence, $u_7 = .8819$.

3. Finally, we conduct the test. Given that

$$\frac{f_5(1.3524)}{cg(1.3524)} = .8965$$

and since $u_7 = .8819 < .8965$, we accept $y_3 = 1.3524$ as a Cauchy random number. We choose a positive result since $u_8 = .5654 > .5$.

While it might appear that the generation of a random number corresponding to a wedge-shaped partition is rather complex, remember that most of the work will have been done well before the first random number is generated. For each wedge-shaped partition, the function $g(x)$ and its inversion, as well as the constant $c$, are computed in advance.

## 18.5 Normally Distributed Random Numbers

The density function of a normally distributed random variable with mean value $\mu$ and variance $\sigma^2$ is

$$f(x) = \frac{1}{\sigma\sqrt{2\pi}} e^{-(x-\mu)^2/2\sigma^2} \quad \text{for } -\infty < x < \infty. \tag{18.6}$$

This normal density function has the familiar bell-shaped curve. Because of its widespread applicability, one should not be surprised to learn that procedures for the generation of random numbers that obey this distribution have received much study and that there exist many different approaches that may be used. We examine four approaches in this section. However, perhaps the best method for generating standard normal variates is the ziggurat method which is described in Section 18.6. As a general rule, the different methods generate random numbers that are standard normal, $N(0, 1)$, i.e., with $\mu = 0$ and $\sigma = 1$, and then convert these to random numbers that satisfy $N(\mu, \sigma^2)$. This is easily accomplished, since if $Z$ has a standard normal distribution $N(0, 1)$, the random variable $\mu + \sigma Z$ has the distribution $N(\mu, \sigma^2)$.

### *18.5.1 Normal Variates Via the Central Limit Theorem*

The central limit theorem states that a sum of $n$ independent and identically distributed random variables, $X_1, X_2, \ldots, X_n$ with mean value $\mu$ and variance $\sigma^2$ is approximately normally distributed with mean $n\mu$ and variance $n\sigma^2$, and the greater the value of $n$, the better the approximation. Thus the sum of $n$ uniformly distributed random numbers in (0, 1) approximates a normal distribution with mean $n/2$ and variance $n/12$, since a uniformly distributed random number in (0, 1) has mean value 1/2 and variance 1/12. In particular, when $n = 12$, the random variable $X = \sum_{i=1}^{12} X_i - 6$ is $N(0, 1)$. The algorithm therefore is to generate sequences of tewlve uniformly distributed random numbers in (0, 1), add them together and subtract the integer 6, to produce a single random variable that is standard normal. This is perhaps the simplest of all methods for generating normally distributed random numbers, but is also one of the slowest requiring the generation of twelve uniformly distributed random numbers for each normally distributed random number.

**Example 18.15** The following sequence of 36 uniformly distributed random numbers produces three standard normal random variates, obtained by adding them in groups of twelve and subtracting 6:

.11, .15, .62, .94, .11, .92, .57, .02, .99, .70, .35, .67,

.35, .36, .09, .51, .21, .12, .41, .01, .06, .84, .94, .06,

.60, .34, .51, .16, .81, .43, .59, .51, .15, .70, .77, .96.

The standard normal variates are 0.15, −2.04, and 0.53.

### *18.5.2 Normal Variates via Accept-Reject and Exponential Bounding Function*

The accept-reject method works best when the bounding function is close to the density function for then the probability of a reject is low. Since at least part of the standard normal density function looks similar to an exponential curve, we shall use a (negative) exponential function as the majorizing function. Also, since the standard normal distribution is symmetric about the origin, we shall produce normally distributed random values in the right plane only and then randomly (with probability 1/2) assign a + or − to this value. This means that we must be concerned with the absolute value of $Z$ for which the probability density function is

$$f(x) \equiv f_{|Z|}(x) = 2 \times \frac{1}{\sqrt{2\pi}} e^{-x^2/2} = \sqrt{2/\pi}\, e^{-x^2/2}, \quad 0 < x < \infty.$$

The exponential density function we use is given by

$$g(x) = e^{-x}, \quad 0 < x < \infty,$$

and we compute $c$ as the maximum value of $f(x)/g(x)$ for $0 < x < \infty$, i.e.,

$$c = \max_{0<x<\infty} f(x)/g(x) = \max_{0<x<\infty} \frac{\sqrt{2/\pi}\, e^{-x^2/2}}{e^{-x}} = \max_{0<x<\infty} \sqrt{2/\pi}\, e^{x-x^2/2}.$$

This maximum must occur when $x - x^2/2$ is maximized, which happens when $x = 1$. Thus

$$c = \sqrt{2/\pi}\, e^{1/2} = \sqrt{2e/\pi} = 1.3155.$$

This then allows us to establish the acceptance test, $cu_2 \le f(x_1)/g(x_1)$, as

$$\sqrt{2e/\pi}\, u_2 \le \sqrt{2/\pi}\, e^{x_1 - x_1^2/2} \quad \text{or} \quad u_2 \le e^{x_1 - x_1^2/2 - 1/2} = e^{-(x_1-1)^2/2},$$

where $x_1$ is exponentially distributed with mean value equal to 1. The complete algorithm is given as follows:

- Generate a uniformly distributed random number, $u_1$, in (0, 1) and set $x_1 = -\ln u_1$.
- Generate a uniformly distributed number $u_2$ on (0, 1).
- If $u_2 \le e^{-(x_1-1)^2/2}$ then
  - — generate a random sign, attach it to $x_1$ and accept $x_1$,
  - — otherwise reject.

**Example 18.16** Let us generate some standard normal random numbers from the following sequence of uniformly distributed random numbers. For the sign, we choose − when $0 \le u < .5$ and + otherwise.

$$u_i = .11,\ .15,\ .62,\ .94,\ .11,\ .92,\ .57,\ .02,\ .99,\ .70,\ .35,\ .67,\ .35,\ .36,\ .09,\ .51,\ .21,\ .12,\ .41,\ .01,\ .55.$$

| $u_i$ | $x_i = -\ln u_i$ | $u_{i+1}$ | $\exp(-(x_i-1)^2/2$ | Acc/Rej | $u_{i+2}$ | ± |
|---|---|---|---|---|---|---|
| .11 | 2.2073 | .15 | .4825 | Acc | .62 | + |
| .94 | 0.0619 | .11 | .6440 | Acc | .92 | + |
| .57 | 0.5621 | .02 | .9086 | Acc | .99 | + |
| .70 | 0.3567 | .35 | .8131 | Acc | .67 | + |
| .35 | 1.0498 | .36 | .9988 | Acc | .09 | − |
| .51 | 0.6733 | .21 | .9480 | Acc | .12 | − |
| .41 | 0.8916 | .01 | .9941 | Acc | .55 | + |

The sequence of random numbers satisfying this density function is

$$+2.2073,\ +0.0619,\ +0.5621,\ +0.3567,\ -1.0498,\ -0.6733,\ +0.8916.$$

### 18.5.3 Normal Variates via Polar Coordinates

As the previous example shows, each standard normal random number produced by the accept-reject method requires the generation of three uniformly distributed random numbers, the generation of an exponentially distributed random number, two function evaluations ($f(x)$ and $g(x)$) and a test. Sometimes, no standard normal random variate is even obtained. The polar method, which we now present, tries to avoid this cost. This method is based on the concept that a pair of uniformly distributed random numbers can represent a point in the plane, a point that can be transformed into polar coordinates. Box and Muller [5] show that if $u_1$ and $u_2$ are two uniformly distributed random numbers, then

$$x_1 = \cos(2\pi u_1)\sqrt{-2\ln u_2} \quad \text{and} \quad x_2 = \sin(2\pi u_1)\sqrt{-2\ln u_2}$$

are standard normal, $N(0, 1)$, random numbers. These may be transferred to the distribution $N(\mu, \sigma^2)$ by setting

$$x_1 = \sigma x_1 + \mu \quad \text{and} \quad x_2 = \sigma x_2 + \mu.$$

Notice that this method takes two uniformly distributed random numbers and produces *two* normally distributed random numbers. Unfortunately, however, it has the disadvantage of involving square roots, natural logarithms, and sine and cosine functions, all of which are, relatively speaking, computationally expensive to compute. This situation can be much improved by moving from these rectangular coordinates to polar coordinates. In this case, two numbers $(2u_1 - 1)$ and $(2u_2 - 1)$ that are uniformly distributed on the interval $(-1, +1)$ are generated. If both these numbers lie within the unit circle, i.e., if $r = (2u_1 - 1)^2 + (2u_2 - 1)^2 < 1$, then

$$y_1 = (2u_1 - 1)\rho \quad \text{and} \quad y_2 = (2u_2 - 1)\rho$$

both obey a standard normal distribution, where $\rho = \sqrt{-2\ln r/r}$. Since the area of the square is 4 and that of the circle is $\pi$, it should be expected that a reject will occur approximately 21% of the time.

**Example 18.17** Let us use the polar method to generate a sequence of normally distributed random numbers from the following sequence of uniformly distributed random numbers in $(-1, 1)$:

$$2u_i - 1 = .0597,\ .9119,\ -.7527,\ -.3070,\ .9723,\ .3205,\ -.7669,\ .2644,\ .8208,\ .9690,\ -.1384,\ -.2326.$$

For each pair $u_i$ and $u_{i+1}$, we form $r = (2u_i - 1)^2 + (2u_{i+1} - 1)^2$, and if this is less than 1, we take $y_i = (2u_i - 1)\rho$ and $y_{i+1} = (2u_{i+1} - 1)\rho$ to be standard normal variates, where $\rho = \sqrt{-2\ln r/r}$. This gives the following table of results:

| $2u_i - 1$ | $2u_{i+1} - 1$ | $r$ | $\rho$ | $y_i$ | $y_{i+1}$ |
|---|---|---|---|---|---|
| .0597 | .9119 | 0.8351 | 0.6569 | 0.0392 | 0.5990 |
| −.7527 | −.3070 | 0.6608 | 1.1198 | −0.8429 | −0.3438 |
| .9723 | .3205 | 1.0481 | | | |
| −.7669 | .2644 | 0.6580 | 1.1278 | −0.8649 | 0.2982 |
| .8208 | .9690 | 1.6127 | | | |
| −.1384 | −.2326 | 0.0733 | 8.4474 | −1.1691 | −1.9649 |

The sequence of normally distributed random numbers is

$$.0392,\ .5990,\ -.8429,\ -.3438,\ -.8649,\ .2982,\ -1.1691,\ -1.9649.$$

### *18.5.4 Normal Variates via Partitioning of the Density Function*

We now turn to the application of the partitioning approach to the generation of normally distributed random numbers. The method is due to Marsaglia [34] and we follow the approach presented by Knuth [25]. We choose $n = 15$ intervals of length .2 which gives rise to 15 rectangles, 15 wedgelike pieces, and a tail as shown in Figure 18.13. Thus the name, the *rectangle-wedge-tail* method.

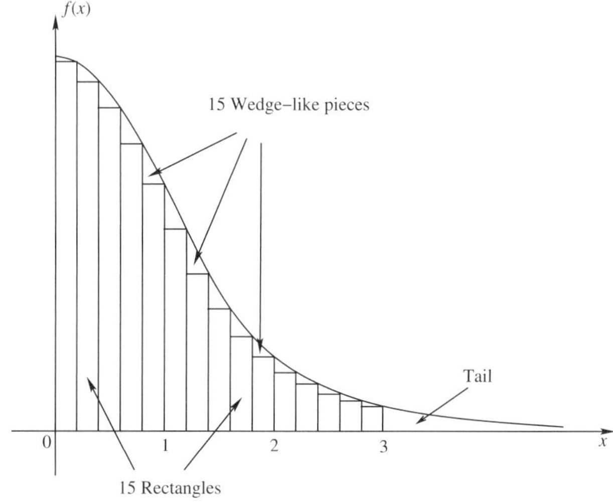


Figure 18.13. Area under the normal curve partitioned into rectangles, wedgelike pieces, and a tail.

The base of the $k^{\text{th}}$ rectangle, $k = 1, 2, \ldots, 15$, is $[.2(k-1), .2k]$; its height is $f(.2k)$. Given that $f(x) = \sqrt{2/\pi}e^{-x^2/2}$, this height is equal to $\sqrt{2/\pi}e^{-k^2/50}$. It follows that the area of the $k^{\text{th}}$ rectangle is

$$p_k = .2\sqrt{2/\pi}e^{-k^2/50} = \sqrt{\frac{2}{25\pi}}e^{-k^2/50}, \qquad k = 1, 2, \ldots, 15.$$

Since $\sum_{k=1}^{15} p_k = .9183$, the total remaining area, the sum of the areas of the remaining 16 pieces, is equal to .0817. The area of each can be approximated by replacing each with a wedge, followed by a normalization to insure that the sum of all 16 is equal to .0817. In this way, all 31 probabilities $p_k$ can be obtained.

Furthermore, since $\sum_{k=1}^{15} p_k = .9183$, random numbers that satisfy $f(x)$ are generated from a uniform distribution almost 92% of the time. They are obtained from the simple formula

$$x_i = u_i/5 + \xi,$$

where $\xi$ is an offset which is equal to $(k-1)/5$ with probability $p_k$. For the other 8% of the time either one of the wedgelike areas or the tail area is chosen and the accept-reject approach must be used to generate a random number having this density.

## 18.6 The Ziggurat Method

Marsaglia has improved upon this even further. His *ziggurat* method [35] results in a normally distributed random number being generated from a uniform distribution 99% of the time: a reject is expected only once in every 100 tries. Unlike the basic approach described above, the rectangles in the ziggurat method are arranged horizontally rather than vertically. The lowest section consists of a rectangle plus the tail of the distribution. The rectangular portion of each section that lies within the density function has been called the core. When the number of segments is large, it is to be expected that the core occupies almost all the rectangle which means that normally distributed random numbers will be produced with almost every try. The defining characteristic of the ziggurat method is that all the rectangles, (and the lowest section of rectangle plus tail) all have exactly the same area. This means that during the running of the generation procedure, the choice of a particular rectangle is immediate. This is illustrate graphically in Figure 18.14, which is not drawn to scale. The 99% acceptance rate is obtained with 255 rectangles plus the lowest section of rectangle and tail.

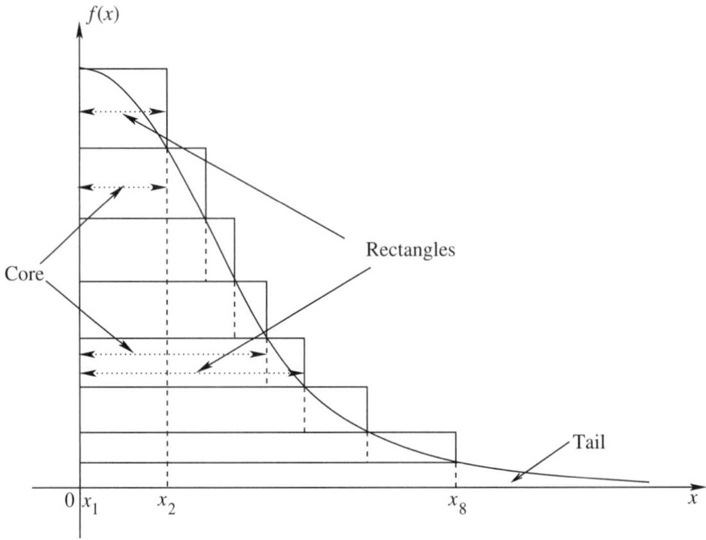


Figure 18.14. Ziggurat ($n = 8$): seven rectangles plus base section of rectangle and tail.

Marsaglia's ziggurat method applies to density functions other than standard normal and in particular appears to be a very effective method for generating exponentially distributed random numbers. However, here our concern is with the generation of standard normal variates where the density function is

$$f(x) = \sqrt{\frac{1}{2\pi}} e^{-x^2/2}, \qquad -\infty < x < \infty.$$

The term $\sqrt{1/2\pi}$ is usually ignored since it is just a normalizing constant—the ziggurat method works with $f(x) = \exp(-x^2/2)$. Let $x_k,\ k = 2, 3, \ldots, n$ be the coordinate of the right side of rectangle $k - 1$, where rectangle 1 is the topmost rectangle; $x_1 = 0$ is the left boundary of all rectangles. Let us postpone momentarily how these $x_i$ can be found and assume that they are already available. Let $\alpha_k = x_k/x_{k+1},\ k = 1, 2, \ldots, n-1$ and $\alpha_n = x_n f(x_n)/V$ where $V$ is the area of the base section (rectangle plus tail). Thus $\alpha_1 = 0$ and $\alpha_k,\ k = 2, 3, \ldots, n$ is the proportion of rectangle $k$ occupied by its core. The generation procedure consists of the following steps.

- Generate a uniformly distributed random number $u_1$ and select rectangle $k = \lceil nu_1 \rceil$.
- Generate a second uniformly distributed random number $u_2$:
  - if $u_2 \le \alpha_k$, take $(2u_2 - 1)\,x_k$ as a standard normal random number,
  - otherwise apply the accept-reject method to the wedge (or tail) in rectangle $k$.

We now return to the technical problems of computing $x_i$, $i = 2, 3, \ldots, n$. If the last intercept $x_n$ (the right coordinate of rectangle $n-1$ and the point at which rectangle and tail meet in the last section) is known, then is possible to compute the common area as

$$V = x_n f(x_n) + \int_{x_n}^{\infty} f(x)\,dx.$$

Furthermore, it now becomes possible to compute the right coordinate $x_k$ of each of the other rectangles using the fact that all rectangles have exactly the same area. The area of rectangle $k$ (and the area of all rectangles and the area of the bottom section) is given by (base times height)

$$V = x_{k+1} \times \left[f(x_k) - f(x_{k+1})\right], \quad k = 1, 2, \ldots, n-1,$$

so that we may compute $x_k$ as

$$x_k = f^{-1}\left(f(x_{k+1}) + \frac{V}{x_{k+1}}\right).$$

Thus once we know $V$ and $x_n$ we can find all of the points $x_i$, $i = n-1, \ldots, 2$. The problem, of course, lies in the computation of $V$ and $x_n$. Consider the function

$$V(r) = rf(r) + \int_{r}^{\infty} f(x)\,dx.$$

This is the area of a rectangle of height $f(r)$ and length $r$, plus the area of the tail of the distribution from the point $r$. At the point $r = x_n$, $V(r) = V$, the area we seek to determine. Let us now introduce a second function,

$$z(r) = x_2(r)[f(0) - f(x_2(r))] - V(r) = x_2(r)[1 - f(x_2(r))] - V(r), \tag{18.7}$$

using the fact that since $f(x) = \exp(-x^2/2)$, $f(x_1) = f(0) = 1$. In this equation $x_k(r)$ is the coordinate of the right edge of rectangle $k-1$ when $r$ is the point at which rectangle and tail meet in the lowest segment. Observe that $x_2(r)[1 - f(x_2(r))]$ is the area of the top rectangle and when $r$ is such that $x_2(r)[1 - f(x_2(r))] = V(r)$ we have found the value of $r$ we seek, the value of $x_n$. The problem then is to find a root of Equation (18.7): for a given value of $r$, the function $z(r)$ may be computed as

**Algorithm: Given $n$, $f(x)$, and $r$, find $z(r)$:**

$x_n = r$;
$V = rf(r) + \int_r^\infty f(x)\,dx$
for $k = n-1,\ -1,\ 2$
  $x_k = f^{-1}(f(x_{k+1}) + V/x_{k+1})$
$z = V - x_2(1 - f(x_2))$

The procedure then is to use a standard root finding procedure to solve $z(r) = 0$ using the algorithm just given to evaluate $z(r)$ as and when needed. The right coordinate of rectangle $n-1$, namely, $x_n$, is given as the solution of $z(r) = 0$. The area of each rectangle can now be found from

$$V = x_n f(x_n) + \int_{x_n}^{\infty} f(x)\,dx,$$

and all other values of $x_k$ found from

$$x_k = f^{-1}\left(f(x_{k+1}) + \frac{V}{x_{k+1}}\right), \quad k = n-1, \ldots, 2.$$

Naturally, these values are generated only once and stored in tables. Finding the proper root of Equation (18.7) can sometimes be quite challenging and for this reason Marsaglia provides the following values of $x_n$ and $V$:

| $n$ | $x_n$ | $V$ |
|---|---|---|
| 128 | 3.442619855899 | .0099125630353 |
| 256 | 3.6541528853610088 | .0049286732339 |

Software implementations of this method incorporate these values directly into the code and generate the tables from them. This is done once only, at initialization. Since the area of each rectangle is approximately equal to .0049287 when $n = 256$, an accept will occur 99.33% of the time. See Exercise 18.6.1.

To terminate this section, let us return once more to the exponential distribution. We saw previously that random numbers having an exponential distribution can be found quite conveniently using the natural logarithm function, the only drawback being the expense of computing a natural logarithm. It turns out that the ziggurat method can be conveniently applied to the exponential distribution and with 255 rectangles and a base section of rectangle and tail, it achieves an efficiency of 98.9%. Thus, approximately 99 times out of 100, an exponentially distributed random number can be obtained as $u\, x_k$ where $k$ is the selected rectangle. When $f(x) = e^{-x}$, Marsaglia provides the following values for $x_n$ and $V$:

| $n$ | $x_n$ | $V$ |
|---|---|---|
| 256 | 7.69711747013104972 | .0039496598225815571993 |

## 18.7 Exercises

**Exercise 18.1.1** The cumulative distribution function of a random variable $X$ is given by

$$F_X(x) = \begin{cases} 0, & x < 0, \\ x^3/8, & 0 \le x \le 2, \\ 1, & x > 2. \end{cases}$$

Use the inverse transformation method to obtain random variates with this distribution from the following sequence of uniformly distributed random numbers

$$y_i = .11,\ .15,\ .62,\ .94,\ .11,\ .92,\ .57,\ .02,\ .99,\ .70,\ .35,\ .67,\ .35,\ .36,\ .09,\ .51,\ .21,\ .12,\ .41,\ .01.$$

**Exercise 18.1.2** Consider a right-angle triangle of height $h$ and width $w$ with right angle placed at a distance $a + w$ from the origin. Under certain constraints on $h$ and $w$, the hypotenuse of this triangle can be taken as a density function of a random variable $X$ defined on $[a, a + w]$. Show that this density function is given by

$$f_X(x) = \begin{cases} h(x - a)/w, & a \le x \le a + w, \\ 0 & \text{otherwise}, \end{cases}$$

and specify the conditions that must be imposed on $h$ and $w$. Find the corresponding cumulative distribution function. The inverse transformation method is to be used to generate random numbers that obey the density function $f_X(x)$. Find the inverse function that is to be used and apply it to the following sequence of uniformly distributed random numbers in the specific case when $a = 2$ and $h = .5$:

$$y_i = .11,\ .15,\ .62,\ .94,\ .11,\ .92,\ .57,\ .02,\ .99,\ .70,\ .35,\ .67,\ .35,\ .36,\ .09,\ .51,\ .21,\ .12,\ .41,\ .01.$$

**Exercise 18.1.3** Let $X$ be a random variable with mean value $E[X] = 11/3$ and whose density function is triangular on the basis $(a, b) = (2, 6)$. From the following sequence of uniformly distributed random numbers generate a corresponding sequence of random numbers having this triangular density:

.06, .84, .94, .06, .60, .34, .51, .16, .81, .43, .59, .51, .15, .70, .77, .96, .96, .09, .13, .36.

**Exercise 18.1.4** The three-parameter Weibull density function is given as

$$f_X(x) = \frac{\beta}{\eta}\left(\frac{x-\xi}{\eta}\right)^{\beta-1} e^{-[(x-\xi)/\eta]^\beta} \quad \text{for } x \geq 0,$$

and is equal to zero otherwise. Derive the inverse function needed to generate random numbers having this distribution in the special case when the location parameter $\xi$ is equal to zero.

**Exercise 18.1.5** Prove that the inverse function used to generate sequences of geometrically distributed random numbers is given by

$$k = \left\lfloor \frac{\ln(y)}{\alpha} \right\rfloor + 1.$$

**Exercise 18.2.1** Use the following sequence of uniformly distributed random numbers to obtain random numbers that obey the probability mass function

$$p_X(k) = \begin{cases} \binom{8}{k} .15^k (.85)^{8-k}, & 0 \leq k \leq 8, \\ 0 & \text{otherwise.} \end{cases}$$

.68, .12, .09, .05, .94, .13, .25, .60, .46, .03, .82, .40, .34, .98, .81, .78,

.70, .43, .87, .04, .69, .14, .62, .57, .64, .96, .98, .87, .06, .86, .02, .94,

.64, .56, .84, .08, .68, .24, .47, .80, .94, .58, .64, .53, .17, .61, .16, .68.

**Exercise 18.2.2** Consider a machine that at the start of each day is either in working condition or has broken down. If it is working, then during the day, it will with probability $p = .05$ break down. Let $X$ be the random variable that denotes the number of consecutive days that the machine is found to be in working condition. Use the following sequence of uniformly distributed random numbers to generate a sequence of random numbers that simulate $X$.

.64, .36, .98, .87, .06, .86, .02, .94, .70, .43, .87, .04, .69, .14, .62, .57,

.94, .58, .64, .53, .17, .61, .16, .68, .64, .56, .84, .08, .68, .24, .47, .80,

.46, .03, .82, .40, .34, .98, .81, .78, .68, .12, .09, .05, .94, .13, .25, .96.

**Exercise 18.2.3** A bank opens its doors at 8:00 am each day and customers begin to arrive according to a Poisson process with parameter $\lambda = 15$ customers per hour. Given the following sequence of uniformly distributed random numbers

.73, .36, .45, .02, .94.

Simulate the time of arrival of the first five customers.

**Exercise 18.2.4** Returning to the bank of the previous exercise, and given the following sequence of uniformly distributed random numbers:

.73, .36, .45, .02, .94, .98, .87, .06, .60, .14, .62, .57, .43, .87, .04, .69,

.15, .16, .68, .58, .64, .53, .17, .61, .54, .24, .47, .80, .56, .84, .08, .68,

.25, .34, .98, .81, .78, .03, .82, .40, .88, .12, .25, .96, .09, .05, .94, .13, generate a sequence of Poisson distributed random numbers, with parameter $\lambda = 15$, by forming products of these uniformly distributed random numbers.

**Exercise 18.2.5** The same as the last exercise, except this time use the method of inversion and the following sequence of uniformly distributed random numbers:

$$.87,\ .06,\ .72,\ .25,\ .96,\ .09,\ .56,\ .35.$$

**Exercise 18.3.1** The density function of a $\beta(i, j)$ random variable is

$$f_X(x) = \begin{cases} \beta(i, j)x^{i-1}(1-x)^{j-1}, & 0 < x < 1, \\ 0 & \text{otherwise,} \end{cases}$$

where the $\beta$ function $\beta(i, j)$ is defined as

$$\beta(i, j) = \frac{(i+j-1)!}{(i-1)!(j-1)!}.$$

This exercise concerns the accept-reject method for generating random numbers having a $\beta(2, 2)$ distribution. Compute the maximum value of the density function in the interval $(0, 1)$ and use a straight line through this point as the bounding function. Use the following sequence of uniformly distributed random numbers to generate random numbers with a $\beta(2, 2)$ distribution:

$$u_i = .11,\ .15,\ .62,\ .94,\ .11,\ .92,\ .57,\ .02,\ .99,\ .70,\ .35,\ .67,\ .35,\ .36,\ .09,\ .51,\ .21,\ .12,\ .41,\ .01.$$

**Exercise 18.3.2** The same as Exercise 18.3.1 except this time use the $\beta(2, 4)$ distribution instead of the $\beta(2, 2)$ distribution.

**Exercise 18.4.1** Use the following sequence of uniformly distributed random numbers:

$$u_i = .06,\ .84,\ .94,\ .06,\ .60,\ .34,\ .51,\ .16,\ .81,\ .43,\ .59,\ .51,\ .15,\ .70,\ .77,\ .96,\ .96,\ .09,\ .13$$

to generate random numbers that satisfy the Coxian distribution shown in Figure 18.15. The values of the parameters of the exponential phases are $\mu_1 = 1/4$, $\mu_2 = 2$, and $\mu_3 = 5$.

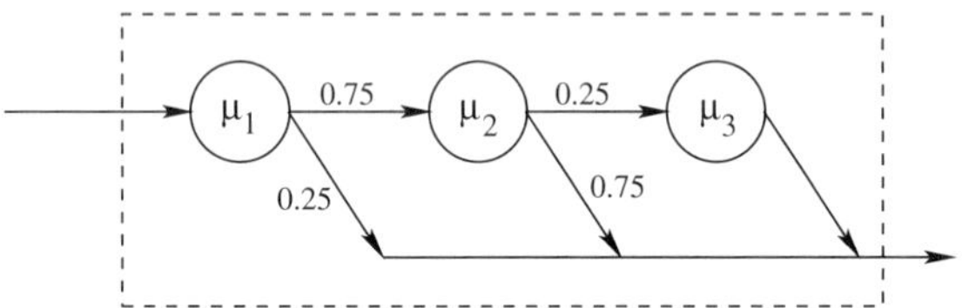


Figure 18.15. Coxian-3 distribution.

**Exercise 18.4.2** A function $g(x) = mx + d$ is a straight line between $x = a$ and $x = b$ and is zero elsewhere. Find the values of $m$ and $d$ in terms of $a$, $b$, $g(a)$ and $g(b)$ and derive a condition for this function to be a density function. (Such density functions are frequently used as bounding functions in the accept-reject method.) Assuming this condition is true and $g(x)$ is indeed a density function, find the corresponding cumulative distribution function $G(x)$, in terms of $m$, $d$ and $a$ and from it, the inverse relationship which converts a random number $u$ which is uniformly distributed on $(0, 1)$, to a random variate that satisfies the density function $g(x)$ on $(a, b)$.

**Exercise 18.4.3** Generate at least one Cauchy random number, using the same partitioning as in Example 18.14, from the following uniformly distributed random numbers:

$$u_i = .7824,\ .8723,\ .6473,\ .4375.$$

**Exercise 18.4.4** Like the previous exercise, this one also asks you to generate at least one Cauchy random number, using the same partitioning as in Example 18.14, from the following uniformly distributed random numbers:

$$u_i = .5542, .8723, .6390, .2943, .9171, .0107.$$

This time $u_1$ leads to the wedgelike section on top of the first rectangle. This area is to be bounded by a horizontal line.

**Exercise 18.4.5** One final exercise on the generation of Cauchy random variables. Generate at least one Cauchy random number, using the same partitioning as in Example 18.14, from the following uniformly distributed random numbers:

$$u_i = .8572, .9825, .3846, .4924.$$

This time $u_1$ leads to the final tail section and you should use the function $1/x^2$ as the bounding function.

**Exercise 18.5.1** Use the polar method to generate a sequence of normally distributed random numbers from the following sequence of uniformly distributed random numbers:

$$u_i = .0597, .9119, .7527, .3070, .9723, .3205, .7669, .2644, .8208, .9690, .1384, .2326.$$

**Exercise 18.6.1** Show that the efficiency of the ziggurat method in generating standard normal random numbers, using $n = 256$ rectangles/sections, is 99.33%. Show that its efficiency in generating exponentially distributed random numbers, using the same number of rectangles/section, is 98.9%.

**Exercise 18.6.2** This question concerns the application of a simple $n = 4$ ziggurat method to the generation of random numbers that have the exponential distribution, $f(x) = e^{-x}$. Given $r = x_n = 2.25991014168348$:

(a) Find the area of each of the three rectangles and the rectangle/tail base section.
(b) Find the coordinates $x_i$, $i = 3, 2, 1$.
(c) Compute the value of $z(r)$.
(d) Compute the area of the topmost rectangle and verify that it is the same as in part (a).
(e) Given the following sequence of uniformly distributed random numbers, use the $n = 4$ ziggurat method to generate a sequence of exponentially distributed random variables:

$$u_i = .5081, .2716, .4338, .0325, .3419, .6440, .9621, .1204, .0653, .4907.$$

# Chapter 19

# *Implementing Discrete-Event Simulations*

## 19.1 The Structure of a Simulation Model

In this chapter, our concern is the computer implementation of *discrete-event simulations*. Although the system being simulated exists in continuous time, changes in the simulation model take place only upon the occurrence of well-defined events which are discrete (hence the name, discrete-event simulation) and which take place at discrete instants of time. This is not the only type of simulation that is possible. The simulation of continuous processes, such as weather patterns, chemical reactions, and the like, are frequently simulated and modeled as systems of partial differential equations. Simulations that use sequences of random numbers and which are independent of time, such as the use of random numbers to evaluate definite integrals, are called *Monte Carlo simulations*. We begin by discussing in a general context, the key features involved in the development of a discrete event simulation model and then illustrate these features by means of a number of examples.

### The State Descriptor

One of the most important aspects of simulation is that of model construction. For large complex systems, it is seldom possible to model the system in its entirety, but rather, critical parts of the system are analyzed independently. In some cases this is all that is required, but in others it may be necessary to incorporate critical subsystems into a hierarchical model. The first task of the modeler is to determine which questions the simulation study is supposed to answer. This will help in an appropriate choice for representing the system. The representation of the system is called the *state descriptor vector* or simply, the *system state*. It is a function of time. In very simple models, it may be just a single integer. For example, if a system such as an auto-repair garage has a single repair bay, all that may be needed to represent its state is the number of cars waiting or undergoing repair. However, most commonly, the state descriptor vector consists of a number of items called components and each component has its own specific set of *attributes*. In fact, each component can be viewed as a discrete random variable which assumes the values of its attributes.

### Events and Their Management

Once all the components of the system state and their different attributes have been defined, the next task of the modeler is to define all possible events and specify their effects on the system state. Naturally, the specification of the state descriptor and the definition of the events will likely occur pretty much simultaneously in the mind of the modeler. We separate them here to provide an outline of the steps needed to build a simulation model. The occurrence (or firing) of an event can cause the components of the state descriptor to change values. A particular event may alter the value of only one component or of many components. The occurrence of an event may also leave the system state unchanged. Furthermore, the firing of an event can, in some simulation models, alter the scheduled firing times of other events.

**Example 19.1** In a simple *M/M*/1 queue, there are only two events:

- a new customer arrives at the queue;
- a customer completes service and departs.

A single integer which specifies the number of customers present suffices to represent the state of the system. An arrival increments the system state by 1; a service completion decrements it by one. In a single-server feedback queue, in which customers are returned to the queue with probability $\alpha$, a service completion will, with probability $\alpha$, leave the system state unchanged.

Events succeed each other in time and it is important in the simulation to schedule these correctly. Thus it is usual to keep an *event list* which holds the time for the next occurrence of each different possible event. Both unordered and ordered list-type data structures can be used to hold these times.

- If an unordered list data structure is used, each event is assigned its own memory location (either a location within an array or its own data record) which contains, at a minimum, the time of the next occurrence of that event. With an unordered data structure, it is necessary to search through the entire list to find the event that is next to fire. Once the event occurs, the time until the next occurrence of this same event is generated and inserted into its assigned position in the data structure.
- If an ordered list data structure is used, the event at the head of the list is the next to fire. An event which fires must be removed from the head of the list. The time of the next occurrence of this event is then generated and inserted into the list at the appropriate position—the position that maintains the ordered property of the list.

List data structures may be array-based or linked. Both singly linked and doubly linked variants are possible. Array-based lists are more memory efficient since they do not require pointer variables. In a linked list with $n$ elements, insertion and deletion are $O(1)$ while obtaining any element $i$ (called a *lookup*) is $O(n)$, whereas in an array-based list, insertion and deletion are $O(n)$ and lookup is $O(1)$. Searching for a key value is $O(n)$ in both cases, but in an array based list, ordered according to key value (here the times of the next occurrence of each event), a binary search algorithm can be used which reduces the cost to $\log_2(n)$. Indeed, when the key values are approximately uniformly distributed, a *dictionary* search having complexity $\log_2 \log_2 n$ can be used. A variant of binary search applied to a linked list is called a *skip list* but requires additional link pointers.

There are other data structures and algorithms available for maintaining event lists. Stacks and queues are restricted types of lists and can be either array based or linked. They do not have all the functionality of a general list data structure, but they can be implemented more simply which makes them more efficient. A *heap* data structure is essentially an array-based, complete, binary tree whose smallest element is at the root of the tree and hence imminently accessible. Insertion into and deletion from a heap are $\log_2(n)$. However, finding a specific key value is $O(n)$, since a heap is not unique. We shall not pursue the issue of appropriate data structures for implementing an event list any further. Someone wishing to program their own simulation should be sufficiently adept at handling the data structures typically found in an undergraduate computer science course as to be capable of doing this correctly. Otherwise, someone using a simulation language will have built-in data structures already available.

### Simulation Time

The only remaining question is the duration $T$ of the simulation experiment. This is not the time required to execute the simulation program, (sometimes called the *wall-clock time*) but rather the simulated time, the time over which the model is exercised (sometimes called the *internal clock time*). It can be either a fixed period of time or the time until a pre-specified number of one or more events has occurred. A variable $t$ is used to indicate the passage of time within the simulation. Each execution of a simulation from the point at which it is initiated, $t = 0$, until it is halted at $t = T$, is called a *simulation run* and defines a sample path of the simulation model. The evolution over time $t \in [0, T]$ of the state descriptor vector, which we described above as a vector of random variables, is a sample path. A number of performance statistics must to be collected during a simulation run. These might include the number of times that a particular event occurred, the average duration between occurrences of the same event, or indeed any other statistic that is deemed necessary. In order to provide for this, the simulation program must contain various variables, usually counters, which are updated when events occur. The values of these variables at the end of a simulation run are used to evaluate performance statistics.

The internal clock can be advanced in one of two ways: either at fixed increments of time, called *synchronous simulation*, or upon the firing of events, called *asynchronous simulation*, since the system state changes only at these firing instants. The first is generally easier to program but important decisions need to be made concerning the length of the fixed time increment. If this is chosen too long, then multiple events may occur in each interval which complicates event handling; if it is chosen too small, then there will be many intervals in which no events at all occur and computer time is wasted checking this situation. The usual approach to advancing the clock in discrete event simulation is the second, even though more complicated data structures are usually required.

**Program Structure**

Once a state descriptor vector and the set of events has been defined, there remains the problem of incorporating them into a computer program. The most straightforward way, and perhaps also the most efficient way, to program a discrete-event simulation is to generate a module (subroutine/method) for each different type of event, and a main routine whose responsibility is the sequencing of event occurrences and calling the appropriate event at the appropriate time. The event modules should include actions that

- update the internal clock;
- change the components of the state descriptor vector in accordance with the effect of the event on the system;
- increment a counter of the number of times this event has occurred;
- generate the time of the next occurrence of this event;
- update other variables that might be required for the computation of performance measures.

The main routine is responsible for

- initializing all variables for a new simulation run;
- selecting the next event to occur;
- passing control to the event module and waiting until control is returned;
- determining when the current simulation run finishes;
- computing performance statistics for the simulation run.

## 19.2 Some Common Simulation Examples

### *19.2.1 Simulating the M/M/1 Queue and Some Extensions*

We now consider a number of examples in more detail, beginning with the simple $M/M/1$ queue. This queue was treated analytically in Part III of this text, and although it is rather straightforward to simulate (and indeed to treat analytically), it is adequate for illustrating all the most important aspects of designing and implementing a simulation. To recap, the $M/M/1$ queue is a single server queue with first-come first-served scheduling of customers. The arrival process is Poisson with parameter $\lambda$, i.e., interarrival times have an exponential distribution with mean $1/\lambda$; the service time is also exponentially distributed and the mean service time is $1/\mu$.

The different variables that we shall employ are now specified:

- A single integer $n$ is sufficient to completely characterize a state of the system and thus is taken as the state descriptor vector.
- There are only two events: the arrival of a new customer to the system and the departure of a customer who has just completed service.
- Integer variables $n_a$ and $n_d$ count the number of arrivals and departures, respectively.
- We shall run the simulation until a total of $N$ customers have been served and then compute the mean interarrival time, the mean service time and the mean time spent waiting in the queue.
- The variable $t$ will be used to denote the internal clock time. The variables $t_a$ and $t_d$ denote the scheduled times of the next arrival and the next departure.
- We use $t_\lambda$ and $t_\mu$ to denote generated interarrival times and service times, respectively. During the simulation run, $tot_\lambda$ and $tot_\mu$ are running totals of arrival times and service times.
- Since we require the mean time spent in the system, we keep the arrival time of each customer. Element $i$ of array $T$ will hold the arrival time of customer $i$, for $i \leq N$. Subtracting $T[i]$ from the time customer $i$ departs gives customer $i$'s total system time (the variable *response*). The average queueing time (variable *wait*) is found by subtracting the sum of all service times from the sum of all response times and dividing by $N$.
- At simulation initialization, we shall set $t = 0$ and $t_d = \infty$ where $\infty$ is taken to be a number so large that there is little possibility of $t$ ever reaching this value during the simulation. The initialization will also generate the time of the first arrival and initialize $t_a$ to this value.

The following actions must be performed by the program modules that handle events:

1. An arrival:
   - Advance the internal clock to the time of arrival.
   - An additional customer has arrived so modify system state $n$ appropriately ($n = n + 1$).
   - Increment $n_a$, the number of arrivals so far and, if $n_a \leq N$, set $T[n_a] = t$.
   - Generate $t_\lambda$, the time until the next arrival; compute the next arrival instant $t_a = t + t_\lambda$ and add $t_\lambda$ into $tot_\lambda$.
   - If $n = 1$, signifying that the server had been idle prior to this arrival, generate $t_\mu$, the service time for this customer, compute its departure time $t_d = t + t_\mu$ and add $t_\mu$ into $tot_\mu$.
2. A departure:
   - Advance the internal clock to time of departure.
   - A customer has left so modify system state $n$ appropriately ($n = n - 1$).
   - Increment $n_d$, the number of departures so far. Compute the time this departing customer spent in the system and add into *response*.
   - If $n = 0$, set $t_d = \infty$. Otherwise generate $t_\mu$, the service time of the next customer; compute the next departure instant $t_d = t + t_\mu$ and add $t_\mu$ into $tot_\mu$.

All that remains is to specify the main module. This module must

- Perform all initialization functions as described above.
- If the number of departures is less than $N$:
  – initiate an arrival event if $t_a < t_d$: otherwise initiate a departure event ($t_a \geq t_d$).
- Generate terminal statistics and print report.

These features have been incorporated into the following Java program. Because of the simplicity of the model, we have not defined separate modules to handle arrival and departure events, but simply incorporated their actions into the body of the program. The statements in the program closely follow the description just given and eliminates the need for extensive comments. The model parameters are taken to be $N = 100$, $\lambda = 1.0$ and $\mu = 1.25$.

```
import java.util.Random;
import static java.lang.Math.*;
class mm1 {
   public static void main (String args[]) {

//     *** Variable definitions and initializations ***

       Random generator = new Random();   double u;
       int N = 100;  double lambda = 1.0; double mu = 1.25;
         double infty = 999*N*mu;
       int n = 0;  int n_a = 0;  int n_d = 0;
       double t = 0;  double t_a;  double t_d = infty;
       double t_lambda = 0; double t_mu = 0; double tot_lambda = 0;
        double tot_mu = 0;
       double[] TA = new double [N+1]; double response = 0; double wait = 0;
       u = generator.nextDouble(); t_a = -log(u)/lambda;   tot_lambda = t_a;

//     *** Begin  simulation proper ***

       while (n_d < N) {

          if (t_a < t_d) {                         // Arrival event
             t = t_a;    n++;
             n_a++;  if(n_a <= N) {TA[n_a] = t; }
             u = generator.nextDouble(); t_lambda = -log(u)/lambda;
             t_a  = t + t_lambda;  tot_lambda += t_lambda;
             if (n==1) {                          // Arrival to an empty system
                u = generator.nextDouble(); t_mu = -log(u)/mu;
                t_d = t+t_mu;   tot_mu += t_mu;
             }
          }
          else {                                  // Departure event
             t = t_d;   n--;   n_d++;
             response += t-TA[n_d];
             if (n==0) {
                t_d = infty;
             }
             else {
                u = generator.nextDouble(); t_mu = -log(u)/mu;
                t_d = t+t_mu;  tot_mu += t_mu;
             }
          }
       }

//     *** Generate statistics and print report ***
```

```
        wait = (response-tot_mu)/N;
        System.out.println("  ");
        System.out.println("Mean interarrival time: " + tot_lambda/n_a);
        System.out.println("Mean queueing time: " + wait);
        System.out.println("Mean service time: " + tot_mu/N);
    }
}
```

Sample runs of this program with different values of $N$ give the following results:

| $N$ | Mean interarrival time | Mean queueing time | Mean service time |
|---|---|---|---|
| $10^2$ | 0.8486 | 3.6909 | 0.8665 |
| $10^4$ | 0.9915 | 3.0980 | 0.7915 |
| $10^6$ | 0.9995 | 3.2153 | 0.8001 |
| Exact | 1.0 | 3.2 | 0.8 |

## Observations

1. There are many ways to write a program to simulate an $M/M/1$ queue and this particular program is only one of them. This Java program is not designed to be a robust production code, but should be used for illustration purposes only.
2. Since there are only two events, we choose not to use an event list. The next event to occur is found using a single *if* statement.
3. The simulation run ends once $N = 100$ customers have been served and we store the arrival times of these customers in an array $T$ of length $N + 1$, since Java indexes arrays from 0. Location zero is not used and the arrival time of customer $i$, $i = 1, 2, \ldots, 100$ is stored in location $i$.
4. Once $N$ customers have arrived, we could have set $t_a$, the time of the next arrival, to infinity and saved on some computation. In this case, care must be taken to correctly compute the mean interarrival time. This simulation program permits arrivals to continue to occur up until the time the $N^{\text{th}}$ customer departs.
5. Other implementations may seek to compute the lengths of busy and idle periods.

## Some Extensions to the *M/M/*1 Queue

1. Non-exponential interarrival and service time distributions.
   This can be handled quite easily. The previous code can be used with the exponential random number generator replaced by a random number generator for whatever type of distribution is required.
2. Limited waiting room ($M/M/1/K$).
   In this case, some customers will not be able to enter the queue and a counter is needed to keep track of the number of lost customers. Only the arrival event module needs to be modified. After the count of lost customers is incremented and the internal clock updated, the only remaining task is to generate the time of the next arrival, and update the sum of interarrival times.
3. Multiple servers ($M/M/c$).
   Both arrival and departure event modules need to be modified. If, upon an arrival, the number in the system is less than $c$, that arriving customer can go immediately into service—the condition $n == 1$ should be replaced with $n <= c$. Within the departure event module, the mechanism for computing the queueing time must be modified, since departures need not

be in the same order as arrivals. One possibility is to use an additional array $TS$ whose $i^{\text{th}}$ elements keeps the time at which customer $i$ enters service. Then $TS[i] - TA[i]$ is the time that customer $i$ waits in the queue.

4. The $M/M/1$ queue with probability $\alpha$ of feedback.
   Only the departure event module will need to be changed. It is necessary to generate a uniformly distributed random number $u \in (0, 1)$ to determine whether a departure is fedback or not. If $u < \alpha$ then the system state must be left unaltered. Everything else remains unchanged.
5. Different customer classes with nonpreemptive scheduling.
   In this case, the system state must be replaced by a vector of length $K$ where $K$ is the maximum number of possible classes. At each arrival instant, the arrival module must first generate a uniformly distributed random number $u \in (0, 1)$ and use it to select the class of the next arrival. When this has been accomplished, it must then generate the time of the next arrival for a customer of that class and increment the system state according. If upon arrival the system is empty, a service time specific to the class of the arriving customer is generated and the arriving customer goes immediately into service.
   At each departure instant, the departure event module must select the next customer to enter service (highest priority first), modify the appropriate component of the state descriptor vector and generate the service time of the next customer of the same class, if such a customer is present. Since departures are not necessarily in the same order as arrivals, the comments made with respect to the $M/M/c$ queue and the array $TS$ also apply here.

### *19.2.2 Simulating Closed Networks of Queues*

We consider a network consisting of $M$ service centers, each of which contains a single queue that feeds a single server. Service times are exponentially distributed but may be different at different centers. A fixed number of customers, $N$, circulates among these centers: after leaving service center $i$, a customer next proceeds to service center $j$ with probability $r_{ij}$ for $i, j = 1, 2, \ldots, M$ and $\sum_{j=1}^{M} r_{ij} = 1$ for $i = 1, 2, \ldots, M$. Customers do not arrive from the exterior, nor do they leave the network.

The different variables we use in simulating this network are as follows:

- The variable $t$ denotes the internal clock time.
- We shall run the simulation until the first departure from any service center after time $T$. Our objective is to find the bottleneck service center, the service center with the most customers.
- The system state is represented by a vector of length $M$ whose $i^{\text{th}}$ component holds the number of customers at service center $i$. We shall use the array $SS$ (for system state) to hold this vector.
- Only departure events need be considered, and there is one for each service center in the network. An array $EL$ (for event list) keeps the scheduled time of next departure from each service center. If service center $i$ is empty, then $EL[i] = \infty$.
- The service center from which a departure occurs is designated the "source" center; the service center to which a customer arrives is designated the "destination" center.
- The parameter of the exponential service time distribution at service center $i$ is stored in position $i$ of array $mu$.
- At initialization, we set $t = 0$ and partition the $N$ customers among the $M$ service centers. For each $i$ such that $SS[i] > 0$, we generate an exponentially distributed random number having parameter $mu\,[i]$ and store it in $EL[i]$. All other positions in $EL[i]$ are set equal to infinity.

The main simulation module contains all variable definitions and initializations. It is responsible for computing final statistics and printing the report at the end of the simulation. While the internal clock time $t$ does not exceed the simulation terminal time $T$, the main simulation module repeatedly searches through the event list to find the firing time of the next event and the index, "source," of the service center from which it occurs. It then passes control to the event module that handles departures from service center "source." The following actions must be performed by each event-handling module.

1. Advance the internal clock to the time of this event.
2. Generate a uniformly distributed random number and use it to determine the destination, "dest," of the transition from this, the source service center.
3. Move one customer from this service center to the destination service center.
4. Generate the time of the next transition from the source service center.
5. If the destination service center now contains only one customer, signifying that the arrival occurred to an empty station, generate the next departure time from this "destination" service center.

The Java program given below incorporates these features. It closely follow the description just given so extensive comments are not included. The model parameters are taken to be $M = 4$, $N = 10$ and $mu = (3, 1, .75, 2)$. The simulation begins at time $t = 0$ and ends at time $T = 100$.

```
import java.util.Random;
import static java.lang.Math.*;

class qn {
  public static void main (String args[]) {

//      *** Variable definitions and initializations ***

        int M=4;    int N=10;    double t=0;   double T=100;
        int [] SS = new int [M];    double [] EL = new double [M];
        double [] mu = {3, 1, .75, 2};
        double [][] RR = {{.2,.4,.4,1}, {.5,1,1,1}, {0,.4,.4,1}, {0,0,1,1}};
        double infty = 99999999.99;

        Random generator = new Random();    double u;
        SS[0] = N;
        for (int i=1; i<M; i++) { SS[i] = 0;}
        for (int i=0; i<M; i++) {
          EL[i] = infty;
          if ( SS[i] > 0 ) {
            u = generator.nextDouble(); EL[i] = -log(u)/mu[i];
          }
        }

//     *** Begin  simulation proper ***

       while (t <=  T) {

         int source = 0;    double small = EL[0];               // Get source
         for (int i=1; i<M; i++) {
           if (small > EL[i]) { small = EL[i]; source = i;}
         }
```

```
            u = generator.nextDouble();                    // Get destination
            int dest = 0;   while (u > RR[source][dest]) dest++;

            t = EL[source];  SS[source]--;  SS[dest]++;         // Next state
            EL[source] = infty;
            if (SS[source] > 0) {
                u = generator.nextDouble();  EL[source] = t-log(u)/mu[source];
            }
            if (SS[dest] == 1) {
                u = generator.nextDouble();  EL[dest] = t-log(u)/mu[dest];
            }
        }

//      *** Generate statistics and print report ***

        System.out.println("State at time T:");
        System.out.println(SS[0] + "  " + SS[1] + "  " + SS[2] + "  " + SS[3]);
    }
}
```

### Observations

1. In this program, the $M$ service centers are indexed from 0 through $M-1$.
2. To facilitate the choice of the destination service center, each row of the two-dimensional routing matrix contains cumulative probabilities. Thus $RR[i][j] = \sum_{k=0}^{j} r_{ik}$ where $r_{ik}$ is the probability that on leaving service center $i$ the destination is service center $k$. In this example, the routing probability matrix $R$ and its stored cumulative variant, $RR$, are

$$R = \begin{pmatrix} .2 & .2 & 0 & .6 \\ .5 & .5 & 0 & 0 \\ 0 & .4 & 0 & .6 \\ 0 & 0 & 1 & 0 \end{pmatrix}, \qquad RR = \begin{pmatrix} .2 & .4 & .4 & 1 \\ .5 & 1 & 1 & 1 \\ 0 & .4 & .4 & 1 \\ 0 & 0 & 1 & 1 \end{pmatrix}.$$

3. A linear search is used to find the next event. Extensions to this basic model can result in this list becoming much larger, in which case some other approach might need to be used.
4. Since all the service centers in this queueing network are identical, except for the rate of service they provide, it suffices to have only one event module to cater for departures from all $M$ service centers. The appropriate service rate can be picked up from the array *mu*. Also, since this is very straightforward, it is incorporated directly into the body of the program.
5. The objective of the simulation is to find the bottleneck service center. Several runs of this program produced $(0, 1, 9, 0)$, $(0, 3, 7, 0)$, $(0, 1, 8, 1)$ as the final state, from which it is apparent that, for this particular model, the third service center (with index 2) is the bottleneck.

### Some Extensions to the Basic Model

1. While some of the extensions identified previously for the *M/M/*1 queue can be applied in a straightforward manner to the individual service centers of a queueing network—exponentially distributed service times can be immediately replaced with more general distributions—others can induce considerable complexity. When limitations are placed on

the number of customers a service center can hold, blocking occurs. In a closed queueing network, customers cannot be lost, so those who are refused admission to a service center must be accommodated elsewhere.

2. Allowing for multiple customer classes, some of which have customers who arrive from the exterior and eventually depart from the network while others have customers who forever cycle among the service centers of the network, adds another order of magnitude to the complexity of the simulation. The length of the system state descriptor can become large, as can the number of possible events, requiring data structures and algorithms that are much more sophisticated than a linear search on an event list maintained as a short array—the approach adopted here.
3. To compute the marginal distribution of customers at any service center it is first necessary to run the simulation until the effect of the initial placement of customers in the network has faded. This can be accomplished by running the simulation until the system state appears to fluctuate around a certain set of values, instead of appearing to move in some particular direction. The reader should try running the Java program to observe this phenomenon. Once this point has been passed, then it is possible to begin to total the periods of time that the specified service center has $0, 1, \ldots, N$ customers. Dividing each by the total time minus the initialization period gives the required marginal distributions.

### *19.2.3 The Machine Repairman Problem*

In this scenario, a factory possesses a total of $M$ identical machines that are subject to breakdown. A certain minimum number of working machines $N < M$ is required for the factory to function correctly. When less than $N$ of the machines are in working order, the factory is forced to stop production. Machines in working order in excess of the $N$ needed for the factory to keep functioning are kept as spares. When a machine breaks down, it is immediately replaced by a spare, assuming one is available, and the broken machine is set aside for repair. We shall assume that machines break down independently and we shall let $B(t)$ be the (general) probability distribution function of the time from the moment a machine is placed in service until it breaks down. A single repair man is available and he repairs broken machines one at a time. The distribution of repair time has a (general) probability distribution function $R(t)$. Once repaired, a machine immediately becomes available as a spare. Spare machines are not used and so do not break down.
In simulating this system, we shall use the following variables:

- The total number of machines is $M$ and a minimum of $N$ of them are needed for the factory to continue to function. The variable $t$ will be used to track internal clock time.
- The number of machines that are in working order at time $t$ is denoted by $n$ while the number that are broken is denoted by $b$. Since at any time $t$, $n + b = M$, only one if these variables is actually needed, but we shall use both to facilitate clarity.
- Either $n$ or $b$ can be used as the system state descriptor.
- There are two different types of event—a machine breaks down, or a machine is repaired and returns to the spare category.
- Since the distribution of machine breakdown times need not be exponentially distributed, it is necessary to keep the next breakdown time for each of the $N$ working machines.
These times are stored in positions 0 through $N - 1$ of the event list, $EL$.
- The time until the next repair completion, which we take to be $\infty$ if $b = 0$ (the number of broken machines is zero) is stored in position $N$ of the event list.
- We shall use this simulation to compute the length of time until the number of working machines first falls below $N$ and the factory is forced to halt production.

As always, the main simulation module contains all variable definitions and initializations and is responsible for printing the report at the end of the simulation. So long as the number of working machines is greater than or equal to $N$, the main module searches the event list to determine the time and type of the next event and passes control to the appropriate event handing module. If the next event is a breakdown, the *index* of the machine that failed must also be obtained.

The following actions must be performed by the module that handles repair events.

1. Advance the internal clock to the time of this event.
2. Increment the number of working machines and decrement the number of broken machines.
3. If there is another machine to be repaired, compute the time at which the repair will be finished, and insert this time into the event list at position $N$. If the repair queue is empty, the time until the next repair completion is set to $\infty$.

The following actions must be performed by the module that handles breakdown events.

1. Advance the internal clock to the time of this event.
2. Decrement the number of working machines and increment the number of broken machines.
3. If there is a spare machine available ($n > N$), put it into service and generate a random number having distribution $B(t)$ with which to determine the time at which this new machine will fail. This time should be inserted into the event list at position *index*.
4. If $b = 1$, signifying that prior to this breakdown, the repairman was idle, generate a random number having distribution $R(t)$ and use it to determine the time at which the repair will be completed.

The following Java program incorporates these features. It closely follow the description just given so extensive comments are not included. The model parameters are taken to be $M = 10$ and $N = 6$. Both repair time and breakdown times are taken to be normally distributed: mean repair times are 15 time units with standard deviation $\sigma = 1$ while times between breakdown have mean 600 time units and standard deviation $\sigma = 100$. As was the case for the previous examples, and for exactly the same reason, the event modules are incorporated directly into the main program routine.

```
import java.util.Random;
import static java.lang.Math.*;
class mr {
   public static void main (String args[]) {

//       *** Variable definitions and initializations ***

         int M=10;    int N=6;    double t=0;   int n = M;   int b = 0;
         double [] EL = new double [N+1];
         double infty = 99999999.99;

         Random generator = new Random();   double u1;  double u2;
         double sigma1 = 100; double mu1 = 600;  double sigma2 = 1;
          double mu2 = 15;
         for (int i=0; i<N; i=i+2) {
            u1 = generator.nextDouble();    u2 = generator.nextDouble();
            double x1 = cos(2*PI*u1) * sqrt(-2 * log(u2));
            double x2 = sin(2*PI*u1) * sqrt(-2 * log(u2));
            EL[i] = sigma1*x1 + mu1; EL[i+1] = sigma1*x2 + mu1;
         }
         EL[N] = infty;
```

```
//      *** Begin  simulation proper ***

        while (n >= N) {

           int index= 0;     double small = EL[0];
           for (int i=1; i<=N; i++) {                        // Get next event
              if (small > EL[i]) { small = EL[i]; index = i;}
           }

           t = EL[index];
           if (index == N)  {               // Next event is a repair completion
              n++;  b--;
              EL[N] = infty;
              if (b > 0) {            // Get time until next repair completion
                 u1 = generator.nextDouble();   u2 = generator.nextDouble();
                 double x1 = cos(2*PI*u1) * sqrt(-2 * log(u2));
                 EL[N] = t + sigma2*x1 + mu2;
              }
           }
           else {                              // Next event is a breakdown
               n--;   b++;
               if ( n > N ) {         // Get time until the new machine fails
                 u1 = generator.nextDouble();   u2 = generator.nextDouble();
                 double x1 = cos(2*PI*u1) * sqrt(-2 * log(u2));
                 EL[index] = t + sigma1*x1 + mu1;
                }
               if (b==1) {                // Repair queue was empty
                 u1 = generator.nextDouble();   u2 = generator.nextDouble();
                 double x1 = cos(2*PI*u1) * sqrt(-2 * log(u2));
                 EL[N] = t + sigma2*x1 + mu2;
                }
            }
        }
//      *** Generate statistics and print report ***

        System.out.println("System halts at time " + t);
    }
}
```

### Observations

1. The polar method is used to compute normally distributed random numbers. This method uses two uniformly distributed random numbers $u_1$ and $u_1$ to generate two normally distributed random numbers $x_1$ and $x_2$. Both are used in the initialization phase. However in the event handling modules, when normally distributed random numbers are required one at a time, only $x_1$ is used. Some computation time can be saved by computing $x_2$ and using it the next time a normally distributed random number having the same distribution is needed.
2. The program does not contain a test to ensure that the generated normally distributed random numbers are positive. The particular normal distributions used in the example make the occurrence of negative numbers unlikely. This may not be so for other normal distributions.

3. The event list is not stored according to shortest firing time and a linear search is used to obtain the next event. This is appropriate when the number of machines is not large, less than 20, for example. When the number of machines is of the order of hundreds or greater, then a different data structure may be more appropriate.

**Some Extensions to the Basic Model**

1. The most common extension to the basic machine repairman problem is the incorporation of additional repairmen. In this case, the state descriptor needs to include the number of busy (or idle) repairmen and the event list must be augmented to include a slot for each repairman (unless repair times are exponentially distributed). If a machine breaks down while at least one repairman is idle, repairs on the machine can begin immediately.
2. This simulation halts once the number of working machines falls below the minimum required. In other scenarios, the percentage of unproductive time might be what is needed. In this case, the simulation continues until some predetermined ending time $T$ and the sum of the lengths of time for which $n < N$ is computed. The ratio of this to $T$ can be used to compute the percentage down-time during the simulation run.
3. In a factory, there are possibly different types of machine and some limited number of each is needed in order for production to continue. Thus spares of each of the different types are kept on standby. Furthermore, machines of different type may have different breakdown and repair time distributions and repairmen may be specialized to handle just one or two different types of machine. The objective of simulating such a system may be to determine the optimal number of spares of each machine type to keep on standby given certain pricing and cost constraints.

### *19.2.4 Simulating an Inventory Problem*

We consider a shop which sells a certain product. Customers arrive at the shop according to a Poisson process with parameter $\lambda$ and purchase a certain quantity of the product. We assume that the product is available in unit measurements and the number of units of the product purchased by a customer is a discrete random variable having an arbitrary probability mass function. If the shop has less units available than the customer wants, the customer purchases all available units but does not seek to back-order the rest from the shop. Perhaps the customer goes elsewhere to fill the remainder of his order. In any case, from the shop's point of view, this is lost business.

The shopkeeper purchases the product from a supplier at a cost of $\$c$ per unit and sells it at a cost of $\$d$ per unit. The amount of the product on hand in the shop, the number of units of the product available for sale, is called the shop inventory. There is a certain cost associated with holding this product. For example, in a car dealership, the dealership owner may have obtained a bank loan to cover his costs and the longer each car remains unsold in his inventory, the more interest he pays to the bank. The cost of holding a unit of stock is taken to be a linear function of the time it is held. When the inventory falls below a certain threshold, more is ordered from the supplier. Specifically, when the number of units available for sale falls to $s$ or less, a sufficient quantity is ordered to bring the amount in stock up to $S$. This is called an $(s, S)$ ordering policy. The shop owner incurs a cost in ordering additional units of the product over and above the cost of the units purchased. This is typically a fixed cost to cover transportation. Furthermore, there is usually a delay between sending the order to the supplier and the units of the product arriving at the store. We assume this delay has a general probability distribution given by $G(t)$. Only one order at a time is given to the supplier and the shopkeeper waits until that order has been filled before sending an additional order. We seek to determine the amount of profit (or loss) the shopkeeper makes during a time period equal to $T$, given that he starts with an initial $S$ units of the product.

In simulating this inventory scenario, we use the following variables:

- The variable $t$ denotes internal clock time. The simulation runs until the completion of the first event after time $T$.
- The state of the system is defined by the pair $(n, m)$, where $n$ is the number of units of the stock in the inventory and $m$ is the number of units on order and awaiting delivery to the shop.
- There are only two possible events:
  1. a customer arrives and requests a certain number of units of the product;
  2. the arrival of previously ordered units from the supplier.
- The first position in the event list $EL[0]$ stores the time of the next customer arrival; $EL[1]$ holds the next arrival time of new supplies to the shop.
- The shopkeeper's cost of one unit of the product is \$$c$, which he sells at \$$d$.
- The variable $h$ is the cost of holding one unit of the product for one time unit. At any time $t$, the variable $tot_H$ keeps the total cost of holding inventory up to time $t$.
- The variable $tot_S$ holds the total of supply costs up to time $t$: $tot_S$ includes the cost of the product to the shopkeeper plus the delivery cost, \$$z$ per delivery.
- A variable called $tot_R$ is used to keep the total revenue earned by the shopkeeper.

The main simulation module contains all variable definitions and initializations and is responsible for printing the report at the end of the simulation. While $t < T$, it checks the event list to determine the time of the next event and passes control to either the customer arrival module or the order arrival module; it then waits for control to be returned from the event handling module at which time it launches the next event.

The following actions must be performed by the module that handles customer arrival events.

1. Update total inventory holding costs: $tot_H = tot_H + n(EL[0] - t)h$.
2. Advance the internal clock to the time of this event: $t = EL[0]$.
3. Generate $x$, the number of units of the product required by the customer, from a general probability mass function.
4. Determine the number of units actually sold to the customer: $r = \min(n, x)$.
5. Update the inventory level: $n = n - r$.
6. Update shopkeeper's revenue: $tot_R = tot_R + r * d$.
7. If $n < s$ and $m = 0$, order $m = S - n$ units of the product, generate the time at which the order will arrive (general probability distribution $G(t)$) and store it in $EL[1]$.
8. Generate time of next customer arrival (exponential interarrival times) and store it in $EL[0]$.

The following actions must be performed by the module that handles inventory resupply events.

1. Update total inventory holding costs: $tot_H = tot_H + n(EL[1] - t)h$.
2. Advance the internal clock to the time of this event: $t = EL[1]$.
3. Update shopkeeper's total costs: $tot_S = tot_S + m \times c + z$.
4. Update state descriptor: $n = n + m$ and $m = 0$.
5. Set $EL[1] = \infty$ since there is no longer an outstanding order.

The following Java program incorporates these features. It closely follow the description just given so extensive comments are not included. The model parameters are taken to be $s = 4$ and $S = 10$. The cost and selling price of one unit of the product are given by $c = 3$ and $d = 5$ respectively. The inventory holding cost is taken to be $h = .5$ per unit of product per time unit and the fixed cost of transportation from the supplier to the shop is $z = 4$. The shopkeeper incurs an initial cost of $cS + z$, the price of the initial inventory. The probability $p_k$ that an arriving customer requests $k$ units of the product is

$$p_1 = .25, \quad p_2 = .55, \quad p_3 = .20, \quad \text{and} \quad p_k = 0 \text{ otherwise.}$$

We take the probability distribution of both customer arrival times and order arrival times to be Poisson, the former with parameter $\lambda = 2$ and the latter with parameter $\gamma = .4$.

```
import java.util.Random;
import static java.lang.Math.*;

class inv {
   public static void main (String args[]) {

//       *** Variable definitions and initializations ***

         double t=0;  double T = 100;   int s = 4;  int S = 10;
         int n = S;   int m = 0;        double [] EL = new double [2];
         double lambda = 2;  double gamma = .4;   double infty = 99999999.99;
         double c = 3;  double d = 5;  double h=.5;  double z = 4;
         double totH = 0;  double totS = S*c+z;  double totR = 0;

         Random generator = new Random();  double u = generator.nextDouble();
         EL[0] = -log(u)/lambda;    EL[1] = infty;

//       *** Begin  simulation proper ***

        while (t < T) {

           if (EL[0] < EL[1]) {          // Next event is a customer arrival.
              totH = totH+n*(EL[0]-t)*h;      t = EL[0];

                 u = generator.nextDouble();    int x = 1;
                 if (u >=.25 && u < .75) x = 2;
                 if (u >= .75) x = 3;

              int r = x;  if (r > n) r = n;
              n = n-r;    totR = totR + r*d;
              if (n<s && m==0) {
                  m = S-n;
                  u = generator.nextDouble();    EL[1] = t-log(u)/gamma;
              }
              u = generator.nextDouble();        EL[0] = t-log(u)/lambda;
           }
           else {                          // Next event is an order arrival.
              totH = totH+n*(EL[1]-t)*h;
              t = EL[1];
              totS = totS + m*c+z;
              n = n+m;  m=0;  EL[1] = infty;
            }
        }
```

```
//      *** Generate statistics and print report ***

        double profit = totR-totS-totH;  double meanProfit = profit/t;
        System.out.println(" " );
        System.out.println("System halts at time " + t);
        System.out.println("Profit/Loss: " + profit);
        System.out.println("Profit/Loss per unit time: " + meanProfit);
    }
}
```

One run of this program produced the following output:

```
System halts at time 100.53463369921157
Profit/Loss: 183.03161503538936
Profit/Loss per unit time: 1.8205827017085434
```

### Some Extensions to the Basic Model

1. The basic program provided above computes only the profit/loss over a fixed period of time. Statements could be inserted directly into the program to compute other quantities, such as the loss of revenue due to an insufficient supply of the product.
2. Often the purpose of running simulations of inventory systems is to determine optimal values for the parameters $s$ and $S$ of the ordering policy, given that the costs are all known.
3. In some inventory models, the product is only viable for a limited period of time and must be discarded at the expiration date. This is the case of many groceries in a grocery store, or daily newspapers at a newsagent's store. Sometimes the product has a residual value once its expiration date is reached. Unsold newspapers can perhaps be sold to a paper recycling facility for a small fraction of their face value.

## 19.3 Programming Projects

**Exercise 19.1.1** Modify the $M/M/1$ Java simulation program by replacing the exponential distributions with Erlang-4 distributions having the same means. Generate output for a total of six simulation runs. Compare your results with the exact answer, obtained by running the code for the $Ph/Ph/1$ queue provided in Part III of this text.

**Exercise 19.1.2** Extend the $M/M/1$ Java simulation program so that it simulates

(a) an $M/M/1/K$ queue with $K = 10$. Run your simulation six times and estimate the probability that an arriving customer is lost.
(b) an $M/M/c$ queue with $c = 2$. Run your simulation six times and estimate the probability that both servers are simultaneously busy.

**Exercise 19.1.3** Modify the Java simulation program *qn.java* so that it computes the marginal distribution at the first service center. Then use your program to simulate the following central server queueing network and determine the marginal distribution of customers at the first service center. Compare your output to the exact results obtained using the algorithms of Chapter 16 of this text.

The model consists of three service centers, each of which contains a single exponential server and the scheduling discipline is FCFS. The routing probability matrix is given as

$$\begin{pmatrix} 0 & 0.7 & 0.3 \\ 1 & 0 & 0 \\ 1 & 0 & 0 \end{pmatrix}.$$

The exponential service time distributions have a mean service time of 2.0 at server 1, 1.0 at server 2, and 0.5 at server 3. The number of customers that circulate in the network is equal to 3 and all service centers have sufficient space that blocking is not a problem.

**Exercise 19.1.4** Extend the Java simulation program *qn.java* to two classes of customer, the first of which arrives from, and departs to, the exterior, while the second cycles permanently among the service centers. Each class has its own routing matrix and service rates, and customers of class 1 have nonpreemptive priority over those of class 2.

**Exercise 19.1.5** Modify the machine repairman simulation code to incorporate two repairmen. Run the same experiment as in the text to determine by how much the additional repairman extends the time until the entire system halts.

**Exercise 19.1.6** Write a simulation program to determine the time to failure in the following multicomponent system. There are $M_1$ machines of the first type and $M_2$ of the second type. To function, the factory needs at least $N_1$ machines of the first type and $N_2$ of the second type. The probability distribution of failure times and repair times are all normally distributed. Machines of type 1 fail according to an $N(600, 10^4)$ distribution and are repaired according to an $N(15, 1)$ distribution. The corresponding distributions for machines of type 2 are $N(1200, 16^4)$ and $N(50, 4)$, respectively.

**Exercise 19.1.7** Modify the inventory model code so that it computes the number of units of product that the shopkeeper missed out on selling because his inventory was empty. Run your program ten times and compute the average of these values. Now modify the values of $s$ and $S$ and repeat the experiment. In your experiments, which values of $s$ and $S$ perform best?

# Chapter 20

# *Simulation Measurements and Accuracy*

So far we have seen how having access to a sequence of uniformly distributed random numbers allows us to obtain approximate answers to different probabilistic scenarios. We have investigated how to generate such sequences of random numbers and how to convert a sequence of uniformly distributed random numbers into distributions, both discrete and continuous, that are not uniform. We have seen how to implement simulation experiments and studied a number of simulation problems. In this final chapter we ask some questions on the nature of simulation itself.

When simulating a system, whether a simple probability experiment or a complex system simulation, each execution or run of the experiment produces a single result such as a head or a tail, the maximum number of customers in a waiting room, the length of time that patients wait to see a doctor, the number of messages lost in a communication system, the amount of unsold stock in an inventory, and so on. Each result (we shall also use the word output) need not be a single real number, but could also be the values that a set of parameters have acquired during the simulation. Nevertheless, each execution of the simulation produces just one output, be it scalar or vector, and the value of that result is a function of the random numbers used in the simulation. Running the simulation again will almost certainly produce a different result. This raises a number of questions that must be asked, including

- how many times do we perform the simulation?
- what do we choose as the final "answer" ?
- how do we measure the accuracy of the results obtained?

Basic to these questions is the concept of *sampling* to which we now turn, since it is useful to think of the output of a simulation run as a *sample* from the infinite sample space of all possible outputs.

## 20.1 Sampling

To motivate our discussion on sampling, consider the following hypothetical situation concerning a problem faced by the Cary, North Carolina, town council. The council is about to pass new laws concerning smoking in public places, but they have decided to first conduct a survey to determine the number of Cary residents who smoke. Cary is a town of approximately 100,000 inhabitants. The company charged with conducting the survey has a number of important decisions to make, including just how many of the town's population they need to contact and interview, and how these particular individuals should be chosen. They also need to infer the number of Cary citizens who smoke based on the information they collect. The set of individuals chosen to be interviewed is called the *sample set* and its size is called the *sample size*. The 100,000 inhabitants of Cary form the *population* for the sampling experiment.

Suppose the company decides that a sample size of 50 is adequate for sampling purposes. Leaving aside for the moment, a discussion of how large the sample size should be, an important question to pose is that of how to choose these 50 individuals. It would appear natural to choose them completely at random, so that each citizen had an equal chance of being selected. If this is indeed the case, the sample is said to be *unbiased*. On the other hand, *sampling bias* arises when the sample set does not represent the population. Suppose, for instance, that a travel agency needs information concerning the choice of holiday destinations for the residents of Cary, and establishes a polling booth at the local airport. The results would lead to inaccurate answers, because only airline travelers would be interviewed and no information would be collected concerning those who take motoring vacations, or vacations in the beautiful North Carolina mountains and beaches. Furthermore, no matter how large the sample size, this sample bias would not be removed.

There is another kind of error that can arise in choosing a sample set. It may just happen that an unrepresentative selection could occur by accident. For example suppose an unusually large number of pregnant mothers were chosen and interviewed as to their smoking habits. It is possible that many of them who previously smoked may have stopped smoking during the pregnancy and the number of smokers might be underreported. The same thing might occur if the survey is conducted during the first week of the year, when many might have made a New Year's resolution to quit. Or perhaps, just by accident, the survey is sent only to non-smokers. This is called the problem of *sampling variability* and is more likely to occur when the sample size is small. Unlike sample bias, sample variability can be reduced by choosing a larger sample size.

To make the sampling process more concrete, let us shrink the population of Cary from 100,000 to ten and let us assume that of these ten citizens, three smoke. Thus we would like our sampling experiment to return the value 0.3 as the probability that a given citizen is a smoker. Let us choose a sample size of 2 and let us also assume that the sampling is unbiased. Choosing a sample of two citizens from the ten can yield two, one, or zero smokers. Since the sample is assumed to be unbiased, the probability that both citizens are smokers is the product $3/10 \times 2/9 = 6/90$, the probability that neither are smokers is given as $7/10 \times 6/9 = 42/90$, whereas the probability that only one of them smokes is given as $3/10 \times 7/9 + 7/10 \times 3/9 = 42/90$. Observe that the individuals chosen in the sample can be modeled as random variables. Let $X_i$, $i = 1, 2, \ldots, n$, be the random variable that describes the $i^{\text{th}}$ citizen where $X_i = 1$ if that citizen is a smoker and $X_i = 0$ otherwise. Since they are random variables, $X_1$ and $X_2$ have distributions; indeed, they have exactly the same distribution, which, since the sample is unbiased, is also the distribution of the entire population, namely, $X_1 = X_2 = 1$ with probability 0.3 and $X_1 = X_2 = 0$ with probability 0.7. They must also have the same mean value as the population mean. In other words, if the population mean is $\mu$ (in our example, $\mu = 0.3$), then $E[X_1] = E[X_2] = \mu$. The same is true for the variance, i.e., $\text{Var}[X_i] = \sigma^2$, where $\sigma^2$ is the variance of the entire population, and indeed it is true for all higher moments. However, $X_1$ and $X_2$ are *not* independent. For example, if $X_1 = 1$ then $\text{Prob}\{X_2 = 1\} = 2/9$ whereas if $X_1 = 0$, then $\text{Prob}\{X_2 = 1\} = 3/9$. This is a result of the fact that the sampling is conducted *without* replacement. If the sampling were done *with* replacement, then $X_1$ and $X_2$ would be independent.

### *20.1.1 Point Estimators*

The mean value for a sample set of size $n$, called the *sample mean*, is defined as

$$\bar{X} = \frac{1}{n}\sum_{i=1}^{n} X_i.$$

Its expectation is exactly equal to the overall population mean, since, using the fact that $E[X_i] = \mu$ for $i = 1, 2, \ldots, n$, we have

$$E[\bar{X}] = \frac{1}{n}\sum_{i=1}^{n} E[X_i] = \frac{1}{n}\sum_{i=1}^{n} \mu = \mu.$$

In this case, the estimator is said to be *unbiased*. In general, an estimator of a parameter is said to be unbiased when its expectation is equal to the parameter. It is said to be *asymptotically unbiased*, a weaker property, when its expectation tends to the parameter in the limit as $n$ goes to infinity.

The sample mean is a random variable, since it is constituted from the random variables $X_i$, $i = 1, 2, \ldots, n$. It is used as an *estimator* of $\mu$. Recall that we do *not* know the value of $\mu$, but yet we wish to determine just how good $\bar{X}$ is as an estimator of $\mu$. More generally, any function of the observations $X_i$, $i = 1, 2, \ldots, n$, is called a *statistic* and since the $X_i$ are random variables, so are statistics defined as functions of them. The sample mean is a case in point. A *point estimator* is a statistic that is used to approximate some parameter of the distribution from which the $X_i$ are chosen. Often the only point estimators of interest are the (sample) mean and (sample) variance but others may be defined. There is no reason why we should not develop estimators, not only for the mean, and variance, but for any of the moments of the distribution function of the population.

We have just seen that the expectation of $\bar{X}$ is equal to the population mean. We now wish to examine the variance of this estimator. If the random variables $X_1, X_2, \ldots, X_n$ are *independent* then, using the facts that for any constants $\alpha$ and $\beta$,

$$\text{Var}[\alpha X + \beta] = \alpha^2 \text{Var}[X]$$

and for independent random variables $X_1, X_2, \ldots, X_n$,

$$\text{Var}[X_1 + X_2 + \cdots + X_n] = \text{Var}[X_1] + \text{Var}[X_2] + \cdots + \text{Var}[X_n],$$

we obtain

$$\begin{aligned}
\text{Var}\,[\bar{X}] &= \text{Var}\left[\frac{X_1 + X_2 + \cdots + X_n}{n}\right] \\
&= \frac{1}{n^2}\text{Var}[X_1 + X_2 + \cdots + X_n] \\
&= \frac{1}{n^2}(\text{Var}[X_1] + \text{Var}[X_1] + \cdots + \text{Var}[X_1]) \\
&= \frac{1}{n^2}(n\sigma^2) = \frac{\sigma^2}{n}.
\end{aligned}$$

In other words, the variance of $\bar{X}$ is equal to the population variance divided by $n$, the sample size. An estimator whose variance goes to zero with increased $n$ is said to be a *consistent* estimator. Thus the sample mean is an unbiased, consistent estimator. Since the variance of a random variable $\bar{X}$ having mean value $\mu$ is defined as $\text{Var}[\bar{X}] = E[(\bar{X} - \mu)^2]$, we have

$$E[(\bar{X} - \mu)^2] = \frac{\sigma^2}{n} \quad \text{or} \quad n\,E[(\bar{X} - \mu)^2] = \sigma^2, \tag{20.1}$$

which means that the larger the value of $n$, the closer $\bar{X}$ is to $\mu$, at least probabilistically. More correctly, we say that the sample mean is a good estimator of $\mu$ when the standard deviation, $\sigma/\sqrt{n}$, is small. We have seen, however, that the $X_i$ will not be independent if sampling is done *without* replacement. When the population size is much larger than the sample size, then there is little difference in sampling *with* and *without* replacement. This is the case in the example of the town of Cary. There is very little difference in sampling with or without replacement when the sample size is 200 and the population size is 100,000. In general, to be able to use the independence property that permits variances to be added, it is usual to assume that the sampling method used is sampling with replacement, even though this may not be strictly true. On the other hand, the fact that the expectation of the sample mean is equal to the population mean in unbiased sampling does *not* depend on the independence property.

Returning to Equation (20.1), i.e., $E[(\bar{X} - \mu)^2] = \sigma^2/n$, it is important to note that the accuracy of $\bar{X}$ as an estimator of the population mean $\mu$ increases as the sample size grows. Furthermore, this accuracy does *not* depend on the size of the actual population. Thus, if the analysis reveals that a sample size of 200 is sufficient to accurately estimate the percentage of smokers in Cary, a town with about 100,000 inhabitants, this same number is sufficient when applied to any other large city such as New York, London, or Paris with many millions of inhabitants. This may seem counterintuitive at first; it may be supposed that in some way or other, the *ratio* of sample size to population size should play some role in the accuracy of the estimator. However, this is not true. As is clearly shown in the formula, only the size of the sample space is important. Of course, this assumes that the sampling is conducted in an unbiased fashion.

The probability that the difference $|\bar{X} - \mu|$ is greater than some error bound $\epsilon$ may be obtained by using Chebychev's inequality. This inequality is generally given in the form

$$\text{Prob}\{|X - E[X]| \geq t\} \leq \frac{\sigma_X^2}{t^2}.$$

For the case of the sample mean, we make the substitutions $X = \bar{X}$, $E[X] = \mu$, $\sigma_X^2 = \sigma^2/n$. We also replace $t$ with $\epsilon$ and obtain

$$\text{Prob}\{|\bar{X} - \mu| \geq \epsilon\} \leq \frac{\text{Var}[\bar{X}]}{\epsilon^2} = \frac{\sigma^2}{n\epsilon^2},$$

which goes to zero as $n$ tends to infinity. This formula gives us a means of asking certain questions concerning the actual accuracy of $\bar{X}$. Unfortunately, the Chebychev inequality gives rather course bounds: much better are bounds obtained when $n$ is sufficiently large that it allows us to apply the *central limit theorem*. If the population distribution is not normal, then the distribution of $\bar{X}$ will not be normal either. However, the central limit theorem allows us to state that, if the underlying distribution has mean $\mu$ and variance $\sigma^2$ but is not necessarily normally distributed, then, as $n$ becomes large, $\bar{X}$ approaches the normal distribution with mean $\mu$ and variance $\sigma^2/n$. Put another way, when $n$ is larger than 30, for example, the distribution of the random variable $(\bar{X} - \mu)/(\sigma/\sqrt{n})$ is approximately $N(0, 1)$ and

$$\text{Prob}\{|\bar{X} - \mu| \geq \epsilon\} \approx \text{Prob}\left\{|Z| > \frac{\epsilon\sqrt{n}}{\sigma}\right\} = 2\left[1 - \Phi\left(\frac{\epsilon\sqrt{n}}{\sigma}\right)\right],$$

where $\Phi(x)$ is the cumulative standard normal distribution function. This typically gives a much tighter bound than the Chebychev bound. For example, with $\epsilon\sqrt{n}/\sigma = 1.645$ we find

$$\text{Prob}\{|\bar{X} - \mu| \geq \epsilon\} \approx \text{Prob}\{|Z| > 1.645\} = 2\,[1 - \Phi\,(1.645)] = 0.1$$

since $\Phi(1.645) = .95$, whereas with the Chebychev bound we find the much weaker result

$$\text{Prob}\{|\bar{X} - \mu| \geq \epsilon\} \leq \frac{\sigma^2}{n\epsilon^2} = \frac{1}{1.645^2} = .3695.$$

**Example 20.1** The duration of telephone calls taken by the Town of Cary is normally distributed with mean $\mu = 180$ seconds and standard deviation $\sigma = 30$ seconds. Ten calls are chosen at random and sampled. Since the duration of a call is normally distributed, the ten random variables $X_1, X_2, \ldots, X_{10}$ are also normally distributed, and from the linearity property of the normal distribution,

$$\bar{X} = \frac{1}{n}(X_1 + X_2 + \cdots + X_{10})$$

is normally distributed as well. Indeed, $\bar{X}$ has a $N(180, 90)$ distribution, since its variance is given by $\sigma^2/n$. The probability that $\bar{X}$ is in error by more than 15 seconds may be obtained by observing that the standard deviation of $\bar{X}$ is $\sqrt{30^2/10} = 9.4868$. Then

$$\begin{aligned}
\text{Prob}\{|\bar{X} - \mu| > 15\} &= 1 - \text{Prob}\{|\bar{X} - \mu| \leq 15\} \\
&= 1 - \text{Prob}\{-15 \leq \bar{X} - \mu \leq 15\} \\
&= 1 - \text{Prob}\left\{-\frac{15}{9.4868} \leq \frac{\bar{X} - \mu}{\sigma} \leq \frac{15}{9.4868}\right\} \\
&= 1 - \text{Prob}\{-1.5811 \leq Z \leq 1.5811\} \\
&= 1 - 2\,\text{Prob}\{Z \leq 1.5811\} \\
&= 1 - 2 \times 0.4429 = .1142,
\end{aligned}$$

where we have read off the value $\text{Prob}\{Z \leq 1.5811\} = 0.4429$ from tables of the cumulative normal distribution function. There is therefore an 11% chance that $\bar{X}$ misses the true mean by more than 15 seconds when only 10 samples are chosen. The error rises to 29.38% that $\bar{X}$ misses the true mean by more than 10 seconds with only 10 samples. The reader may wish to verify that, when 20 samples are chosen, we obtain

$$\text{Prob}\{|\bar{X} - \mu| > 15\} = 0.0250$$

and

$$\text{Prob}\{|\bar{X} - \mu| > 10\} = 0.1362.$$

The astute reader will have observed that we made the implicit assumption that we know the variance of the distribution of the population as a whole, but not its mean. We assumed that we knew this variance and used it to judge the accuracy of the sample mean, $\bar{X}$, as an estimator of $\mu$, the mean of the population distribution. This might seem strange given that our purpose is to derive an estimator of the population mean and it might seem odd that we know its variance but not its mean. We now take up this question concerning our knowledge (or lack thereof) of the variance of the source population and develop a point estimator for it. One possibility is to use the function $S^2$ defined as

$$S^2 = \frac{1}{n}\sum_{i=1}^{n}(X_i - \bar{X})^2.$$

Although this may at first appear to be a *natural* way to define an estimator of the variance, it turns out that it is a *biased* estimator. To see this, we use the fact that the expected value of a sum is equal to the sum of the expected values and proceed as follows:

$$\begin{aligned}
E[S^2] &= E\left[\frac{1}{n}\sum_{i=1}^{n}(X_i - \bar{X})^2\right] \\
&= \frac{1}{n}\sum_{i=1}^{n} E[(X_i - \bar{X})^2] = E[(X_1 - \bar{X})^2] \qquad (20.2) \\
&= E[X_1^2 - 2X_1\bar{X} + \bar{X}^2] = E[X_1^2] - 2E[X_1\bar{X}] + E[\bar{X}^2]. \qquad (20.3)
\end{aligned}$$

Equation (20.2) holds since

$$E[(X_1 - \bar{X})^2] = E[(X_2 - \bar{X})^2] = \cdots = E[(X_n - \bar{X})^2].$$

We now consider the terms on the right-hand side of Equation (20.3) one at a time. For the first term we use a property of variances, namely, $\text{Var}[X] = E[X^2] - E[X]^2$, to obtain

$$E[X_1^2] = \sigma^2 + \mu^2.$$

For the second term, we have

$$E[X_1\bar{X}] = E\left[X_1\frac{1}{n}\sum_{i=1}^{n} X_i\right] = \frac{1}{n}E\left[X_1^2 + \sum_{i=2}^{n} X_1X_i\right] = \frac{1}{n}\left(E[X_1^2] + \sum_{i=2}^{n} E[X_1X_i]\right)$$
$$= \frac{(\sigma^2+\mu^2)+(n-1)\mu^2}{n}.$$

To simplify the third term, we once again apply the same property of variances used to simplify the first term and obtain

$$E[\bar{X}^2] = \frac{\sigma^2}{n} + \mu^2.$$

Taking these three together, the $\mu^2$'s cancel out and we are finally left with

$$E[S^2] = \sigma^2 - \frac{\sigma^2}{n} = \frac{n-1}{n}\sigma^2.$$

For $S^2$ to be an *unbiased* estimator we need to have $E[S^2] = \sigma^2$, so this particular estimator of the variance turns out to be biased. It slightly underestimates the value of $\sigma^2$. However, a simple adjustment is sufficient to correct this bias. If we define $S^2$ as $S^2 = \sum_{i=1}^{n}(X_i - \bar{X})^2/(n-1)$ and not as $S^2 = \sum_{i=1}^{n}(X_i - \bar{X})^2/n$, then $S^2$ is an unbiased estimator of the variance. To show this we proceed as follows:

$$\begin{aligned}
E[S^2] &= \frac{1}{n-1}E\left[\sum_{i=1}^{n}(X_i - \bar{X})^2\right] \\
&= \frac{1}{n-1}E\left[\sum_{i=1}^{n}X_i^2 - 2\sum_{i=1}^{n}X_i\bar{X} + \sum_{i=1}^{n}\bar{X}^2\right] \\
&= \frac{1}{n-1}E\left[\sum_{i=1}^{n}X_i^2 - 2\bar{X}\sum_{i=1}^{n}X_i + n\bar{X}^2\right] \\
&= \frac{1}{n-1}E\left[\sum_{i=1}^{n}X_i^2 - 2\bar{X}\left(n\bar{X}\right) + n\bar{X}^2\right] \\
&= \frac{1}{n-1}E\left[\sum_{i=1}^{n}X_i^2 - n\bar{X}^2\right] \\
&= \frac{1}{n-1}\sum_{i=1}^{n}E[X_i^2] - \frac{n}{n-1}E[\bar{X}^2].
\end{aligned}$$

Substituting $E[X_i^2] = \sigma^2 + \mu^2$ and $E[\bar{X}^2] = \sigma^2/n + \mu^2$, obtained previously, this becomes

$$E[S^2] = \frac{n}{n-1}\left(\sigma^2+\mu^2\right) - \frac{n}{n-1}\left(\frac{\sigma^2}{n}+\mu^2\right) = \frac{n\sigma^2}{n-1} - \frac{\sigma^2}{n-1} = \sigma^2.$$

Hence this second formulation of $S^2$ is an unbiased estimator of the variance. However, an estimator may be biased and still serve a useful purpose, particularly if the bias is not great. This is the case of the original estimator for $S^2$, which is as close to being unbiased as $(n-1)/n$ is to 1. We shall see in the next section that this biased estimator provides a particularly simple formulation for confidence intervals when the samples $X_i$, $i = 1, 2, \ldots, n$, assume only the values 0 or 1, as in the case of the smoking survey example for the town of Cary.

Before leaving this section we briefly give two recursive formulae for computing the sample mean $\bar{X}$ and the sample variance $S^2$. For a given value of $n$, these quantities can be computed directly and efficiently from their respective formulae,

$$\bar{X} = \frac{1}{n}\sum_{i=1}^{n} X_i \quad \text{and} \quad S^2 = \frac{1}{n-1}\sum_{i=1}^{n}(X_i - \bar{X})^2.$$

However, it can happen that the first choice of $n$ is not sufficiently large, that the current set of samples does not provide sufficient accuracy and so it becomes necessary to generate enough additional samples to meet the accuracy requirements. The following recursive approach (see Ross [46]) is an efficient way to carry this out. Let

$$\bar{X}_n = \frac{1}{n}\sum_{i=1}^{n} X_i \quad \text{and} \quad S_n^2 = \frac{1}{n-1}\sum_{i=1}^{n}(X_i - \bar{X})^2$$

with initial conditions $\bar{X}_0 = 0$ and $S_0^2 = 0$. Then

$$\bar{X}_{n+1} = \bar{X}_n + \frac{X_{n+1} - \bar{X}_n}{n+1} \tag{20.4}$$

and

$$S_{n+1}^2 = (1 - 1/n)S_n^2 + (n+1)\left(\bar{X}_{n+1} - \bar{X}_n\right)^2. \tag{20.5}$$

Thus it is possible to incorporate additional data points as necessary. Since we have seen that $\bar{X}$ is a good estimator of $\mu$ when the standard deviation $\sigma/\sqrt{n}$ is small, the following approach can be used:

- Choose a value of $n$ (where $n$ is at least 30) and generate $n$ samples $X_1, X_2, \ldots, X_n$.
- Compute the sample mean $\bar{X}$ and the sample variance $S^2$.
- If $S/\sqrt{n}$ is sufficiently small, then stop.
  Otherwise generate additional samples and incorporate into the recursive formulae until the desired accuracy is attained.

For example, if $S/\sqrt{n}$, the standard deviation of our estimator $\bar{X}$, satisfies $S/\sqrt{n} < \delta$ then, with a 95% confidence level, we can assert that $\bar{X}$ does not differ from $\mu$ by more than $1.96\,\delta$. Thus it becomes possible to obtain a rough estimate of the number of samples needed so that the standard deviation of the estimator satisfies a given criterion. If it is estimated that a large number of additional samples is needed, then it will likely be more efficient to use the original formulas rather than the recursive ones.

**Example 20.2** Consider the following sample of $n = 16$ numbers:

11.6, 13.5, 9.1, 13.9, 12.1, 9.6, 10.7, 9.5, 12.7, 10.8, 8.1, 11.4, 12.4, 11.2, 9.8, 6.1.

The sample mean is $\bar{X} = 10.7812$ and the sample variance is $S^2 = 62.3244/15 = 4.1550$. The standard deviation of $\bar{X} = S/\sqrt{n} = 2.0384/4 = .5096$. At the 95% confidence we can can assert that the sample mean of 10.7812 does not differ from the true mean by more than $1.96 \times .5096 = .9988$.

To estimate the number of samples needed for the standard deviation of the estimator to be less than .25, we need the value of $n$ such that $S/\sqrt{n} < .25$. Given that $S = 2.0384$, the value of $n$ must be greater than 66.4793, i.e., 67.

### 20.1.2 Interval Estimators/Confidence Intervals

Confidence intervals are the usual means by which the accuracy of computer simulations is gauged. The estimators we have discussed so far generally go under the name of *point estimators* whereas confidence intervals are more general *interval estimators*. For example, the sample mean, $\bar{X}$, is a point estimator that provides us with a single point, an approximation to the population mean. Similarly, the sample variance, $S^2$, is a point estimator of the population variance. A point estimator simply lets us know that the estimated value of some parameter $\xi$ is $\alpha$. Specifically it does not give us any information about how much, if any, $\alpha$ differs from the true value, $\xi$. A case in point is taking the sample mean, $\alpha = \bar{X}$, as an estimator of the population mean, $\xi = \mu$. Interval estimators more generally allow us to find the probability, $\rho$, that the true value $\xi$ lies in some interval $(\alpha-\epsilon_1,\ \alpha+\epsilon_2)$. We have

$$\text{Prob}\{\alpha - \epsilon_1 < \xi < \alpha + \epsilon_2\} = \rho. \tag{20.6}$$

We call the interval $(\alpha - \epsilon_1, \alpha + \epsilon_2)$ a $100 \times \rho$ percent confidence interval for the parameter $\xi$, and $\rho$ is called the *confidence coefficient*. Equation (20.6) states that the probability that the true value of the parameter $\xi$ lies between $\alpha - \epsilon_1$ and $\alpha + \epsilon_2$ is $\rho$, but it does not *guarantee* that this interval even contains $\xi$. When $\alpha$ is an estimator based on some sample set, then $\alpha$ is a random variable. It follows that the *interval* $(\alpha - \epsilon_1, \alpha + \epsilon_2)$ is a *random interval* of length $\epsilon_1 + \epsilon_2$. This interval may, or may not, contain the true value $\xi$ of the parameter. The probability $\rho$ gives the *likelihood* that the interval does in fact contain $\xi$. If for example $\rho = 0.95$, then there is a 95% chance that the interval contains $\xi$. Put another way, if 100 sample sets are chosen under the same circumstances (same sample size, same method of selection, etc.) and 100 intervals constructed, one from each of the $\alpha$ values obtained from each sample set, then, of these 100 random intervals, one should expect the true value of the parameter $\xi$ to lie in 95 of them. One should also expect that five of the intervals do *not* contain $\xi$.

Let us now look at how we may derive confidence intervals for the sample mean. Confidence intervals for other statistics follow the same procedure. We return to Chebychev's inequality

$$\text{Prob}\{|\bar{X} - \mu| \geq \epsilon\} \leq \frac{\sigma^2/n}{\epsilon^2}$$

and write it as

$$\text{Prob}\{-\epsilon < \bar{X} - \mu < \epsilon\} = \text{Prob}\{|\bar{X} - \mu| < \epsilon\} \geq 1 - \frac{\sigma^2}{n\epsilon^2}.$$

Observe that we may make this probability arbitrarily close to 1 by choosing a sufficiently large value for the sample size $n$. Consider the situation in which the population from which the sample set is drawn is *normally* distributed with mean $\mu$ and variance $\sigma^2$, i.e., with distribution $N(\mu, \sigma^2)$. The objective is to compute confidence intervals for the sample mean, under the assumption that the true mean $\mu$ is unknown but the true variance $\sigma^2$ is known. Furthermore, it has previously been shown that the sample mean is normally distributed, having distribution $N(\mu, \sigma^2/n)$, and that the random variable $Z = (\bar{X}-\mu)/(\sigma/\sqrt{n})$ is distributed according to $N(0, 1)$. We may now use standard tables for the normal distribution to compute a value $z > 0$ such that $\text{Prob}\{|Z| < z\} = \rho$ and relate this back to $\bar{X}$. We have

$$\text{Prob}\{|Z| < z\} = \text{Prob}\left\{\frac{|\bar{X} - \mu|}{\sigma/\sqrt{n}} < z\right\} = \text{Prob}\left\{|\bar{X} - \mu| < \frac{z\sigma}{\sqrt{n}}\right\} = \rho,$$

i.e.,

$$\text{Prob}\left\{\mu - \frac{z\sigma}{\sqrt{n}} < \bar{X} < \mu + \frac{z\sigma}{\sqrt{n}}\right\} = \rho. \tag{20.7}$$

For convenience, let us define the complement of this probability to be $\zeta$, i.e., $\zeta = 1 - \rho$. Then the probability that the true mean lies outside these bounds is $\zeta$ and if we assume that the distribution is symmetric, as is the case of the normal distribution, then we may write

$$\text{Prob}\{Z < -z\} = \text{Prob}\{Z > z\} = \frac{\zeta}{2}.$$

The point $z$ for which $\text{Prob}\{Z > z\} = \zeta/2$ may now be read from readily available tables and the confidence interval constructed. This particular value of $z$ is denoted by $z_{\zeta/2}$ and the confidence interval is

$$\left(\bar{X} - \frac{z_{\zeta/2}\sigma}{\sqrt{n}},\ \bar{X} + \frac{z_{\zeta/2}\sigma}{\sqrt{n}}\right). \tag{20.8}$$

This is said to be a $100(1-\zeta)\%$ confidence interval for the population mean $\mu$. Some of the more commonly used values for $1-\zeta$ and the corresponding values of $z_{\zeta/2}$ are as follows.

| $1-\zeta$ | 0.80 | 0.90 | 0.95 | 0.98 | 0.99 | 0.998 |
|---|---|---|---|---|---|---|
| $z_{\zeta/2}$ | 1.282 | 1.645 | 1.960 | 2.326 | 2.576 | 3.090 |

**Example 20.3** Consider a set of six random samples chosen independently from a normal distribution having mean $\mu = 5.0$ and variance $\sigma^2 = 0.3$. Suppose the values of $X_i$, $i = 1, 2, \ldots, 6$, the chosen sample, are

4.789, 5.021, 4.856, 4.691, 5.255, 4.703.

The sample mean is given by

$$\bar{X} = \frac{1}{6}\sum_{i=1}^{6} X_i = 4.8858.$$

We now need to consult the table to estimate the value of $z_{\zeta/2}$. If we require a confidence level of 95%, then from the table we obtain $z_{\zeta/2} = 1.96$ and from it we determine the confidence interval to be

$$\left(\bar{X} - \frac{1.96\sigma}{\sqrt{n}},\ \bar{X} + \frac{1.96\sigma}{\sqrt{n}}\right) = \left(4.8858 - \frac{1.96 \times \sqrt{0.3}}{2.4495},\ 4.8858 + \frac{1.96 \times \sqrt{0.3}}{2.4495}\right)$$
$$= (4.4475,\ 5.3241),$$

which contains the true mean $\mu = 5.0$.

From Equation (20.7), it is apparent that $z\sigma/\sqrt{n}$, called the *half-width* of the confidence interval, is a measure of the accuracy of the sample mean. Given that $z$ and $\sigma$ remain (approximately) unchanged with $n$, then the accuracy of the estimate is proportional to the inverse of the square root of $n$—to double the accuracy requires a fourfold increase in the sample size! On the other hand, reducing the variance $\sigma$ by 50% has the same effect as quadrupling the sample size. Thus the popularity of variance reduction techniques, considered later in this chapter.

When the variance of the population is not known, then the unbiased estimator $S^2 = \sum_{i=1}^{n}(X_i - \bar{X})^2/(n-1)$ may be used in place of $\sigma^2$. However, it is not a good idea to simply replace $\sigma$ with $S$ in the above formula, derived for the normal distribution, unless the value of $n$ is large, at least 24 or preferably 30 or higher, because it gives inaccurate results—when $n$ is large, the effect of the central limit theorem comes to the rescue. For small values of $n$, all is not lost however, because the random variable

$$T = \frac{\bar{X} - \mu}{S/\sqrt{n}}$$

has a *Student T-distribution* with $n - 1$ degrees of freedom and tables are available from which to compute the requisite probabilities. This may be used for small values of $n$, although it is usually recommended that for meaningful results, $n$ should be at least 10. For large values of $n$, this distribution approaches the normal distribution.

**Example 20.4** Returning to the previous example, we compute the sample variance as

$$S^2 = \frac{1}{5}\sum_{i=1}^{6}(X_i - 4.8858)^2 = 0.1190.$$

Had we used this in place of 0.3 above, then $1.96S/\sqrt{n} = 0.2760$, and we get the interval $(4.6098,\ 5.1618)$ which also contains the true mean $\mu = 5$. Using the Student $T$ distribution with five degrees of freedom,

| $1-\zeta$ | 0.80 | 0.90 | 0.95 | 0.98 | 0.99 | 0.998 |
|---|---|---|---|---|---|---|
| $t_{\zeta/2}$ | 1.476 | 2.015 | 2.571 | 3.365 | 4.032 | 5.893 |

we find the value $t_{\zeta/2} = 2.571$ and hence from $2.571S/\sqrt{n} = 0.3621$, we obtain the interval $(4.5237,\ 5.2479)$ which contains the true mean $\mu = 5$.

We previously mentioned that biased estimators can sometimes be of value. This is true for the slightly biased estimator of the sample variance when each sample $X_i,\ i = 1, 2, \ldots, n$, is either 0 or 1. In this case $X_i^2 = X_i,\ i = 1, 2, \ldots, n$. Indeed $X_i$ raised to any power is 1 if $X_i = 1$ and is equal to 0 if $X_i = 0$ and it follows that the biased estimator simplifies to $S^2 = \bar{X}(1-\bar{X}) = \mu(1-\mu)$. We have

$$\begin{aligned} S^2 &= \frac{1}{n}\sum_{i=1}^{n}(X_i - \bar{X})^2 = \frac{1}{n}\sum_{i=1}^{n}\left(X_i^2 - 2X_i\bar{X} + \bar{X}^2\right) \\ &= \frac{1}{n}\left(\sum_{i=1}^{n}X_i^2 - 2\sum_{i=1}^{n}X_i\bar{X} + \sum_{i=1}^{n}\bar{X}^2\right) \\ &= \frac{1}{n}\sum_{i=1}^{n}X_i^2 - 2\bar{X}\frac{1}{n}\sum_{i=1}^{n}X_i + \bar{X}^2 \\ &= \frac{1}{n}\sum_{i=1}^{n}X_i - 2\bar{X}^2 + \bar{X}^2 = \bar{X} - \bar{X}^2 = \mu(1-\mu). \end{aligned}$$

Equation (20.8) then simplifies to become

$$\left(\bar{X} - z_{\zeta/2}\sqrt{\frac{\bar{X}(1-\bar{X})}{n}},\quad \bar{X} + z_{\zeta/2}\sqrt{\frac{\bar{X}(1-\bar{X})}{n}}\right).$$

**Example 20.5** In its survey of 100 residents, the Town of Cary finds that 15 of its citizens smoke. Let us compute a 90% and a 95% confidence interval for the number of smokers in the town.

We compute $\bar{X} = 0.15$ and $\sqrt{\bar{X}(1-\bar{X})/n} = 0.03571$. For a 90% confidence interval, we use the value $z_{\zeta/2} = 1.645$ and compute the interval to be

$$(0.15 - 0.05874,\ 0.15 + 0.05874) = (0.09126,\ 0.2087).$$

For a 95% confidence interval, we use the value $z_{\zeta/2} = 1.960$ and compute the interval to be

$$(0.15 - 0.06999,\ 0.15 + 0.06999) = (0.0800,\ 0.2200).$$

Since these are rather large intervals, the town decides to use a greater sample. Let us now compute the confidence intervals obtained when 500 citizens are sampled and 80 declare themselves to be smokers.

In this case, we compute $\bar{X} = 8/50 = 0.16$ and $\sqrt{0.16 \times 0.84/500} = 0.01952$. The 90% and 95% confidence intervals are given respectively by $0.16 \pm 1.645 \times 0.01952$ and $0.16 \pm 1.960 \times 0.01952$, i.e.,

$$(0.1279,\ 0.1921) \quad \text{and} \quad (0.1217,\ 0.1983).$$

## 20.2 Simulation and the Independence Criteria

The relationship between sampling and simulation is immediate and evident. In sampling we easily imagine a pollster conducting telephone interviews; each member of the population contacted constitutes one sample. In simulation, random numbers provide the means by which samples are generated. For example, in using simulation to estimate the probability of getting exactly three heads in five tosses of a fair coin, each and every sample is obtained from a set of five random boolean numbers. If the sample satisfies the criterion (the sum of the five booleans is 3) a counter is incremented and the estimated probability of getting three heads in five tosses is taken to be the ratio of the number of successes (the value of the counter) to the total number of tests conducted. In sampling terminology, the estimated probability is the sample mean, i.e., $\bar{X} = \sum_{i=1}^{n} X_i/n$, where $X_i$ is the random variable that has the value one if the sum of the five random boolean numbers is three, and is zero otherwise, and $n$ is the total number of sets of five random numbers generated.

This view of simulation allows us to apply the concepts of estimators and confidence intervals just discussed in the sampling context to simulation, and thereby gain some idea of the accuracy of the results obtained from the simulation. In view of the important role that the independence of the individual samples $X_i$, $i = 1, 2, \ldots, n$, plays in determining confidence intervals, it is imperative to consider the independence of the random variables used in our simulations. For elementary simulations such as the simple probability experiment just described, the independence of the samples $X_i$ derives from the fact that the random numbers generated are independent. In each group of five random numbers, all are independent of each other *and* are independent of all other random numbers generated either before or after these five.

Independence of the samples becomes more of a problem in complex simulations and it is useful to consider this question of independence in somewhat more detail as it has a considerable effect on determining the accuracy of simulation experiments. Consider the following example for motivational purposes. The patients at a medical facility are becoming increasingly irritated at the lengths of time they spend waiting to see a doctor, to the point that some are threatening to move elsewhere. The facility manager decides to carry out some simulations to assess the situation in a hurry: to order an actual survey would take many weeks, possibly months and the fear is that the patients may have gone elsewhere by then. Since patients log in their time of arrival and doctors log the time at which they see the patients, past records may be analyzed to determine the parameters needed for the simulation. It is decided that the random variable $X_i$ will represent the waiting time of the $i^{\text{th}}$ patient to arrive in the office. The manager realizes that her confidence intervals will be meaningless unless the $X_i$ are independent—but alas, this is not the case! If patient $i$ spends a long time waiting, because an emergency caused the doctor to get behind, then it is very likely that patient $i+1$ will also have to wait a long time. The random variables $X_i$ and $X_{i+1}$ are correlated. There is yet another effect that must also be considered. When the medical facility first opens in the morning, the facility is empty and the early scheduled patients wait little, if at all. Only after about an hour or so does the office settle into its regular pattern of behavior, and it is this pattern that the office manager wishes to simulate. If these problems can be solved then, with some certainty, the office manager can assert that the situation will improve—if additional doctors are incorporated into the practice, or if fewer new patients are accepted, and so on. The system can be accurately simulated and the office manager can have confidence in the results produced by the simulation analysis.

Intimately related to this question of independence is the duration and initial state of the simulation experiment. Two different types of simulation must be distinguished. The first, called a *transient* or *terminating* simulation, is one in which the simulated period is well defined and relatively short: consequently the state in which the system is started plays an important part in its evolution. The second, called a *steady state* or *nonterminating* simulation, is the opposite. There is no fixed beginning instant, other than the beginning must occur only after the initial conditions have faded, nor is there an ending instant. We provide a number of examples of the two types.

**Example 20.6** A bank opens its doors at 8:00 a.m. At that time, all the tellers and managers are available and ready to assist customers. The bank closes at 5:00 p.m. and any customers present at that time are served and depart. A simulation study might wish to find the average time a customer waits in a queue before reaching a teller/manager. This is an example of the first type. There is a well defined beginning and end, and the initial conditions play a role.

**Example 20.7** In modeling the survivability of an electronic unit under adverse conditions, the various components that constitute the unit are taken to be new (initial condition) and the simulation continues until the unit fails. Although the ending time is not known in advance, there is a well specified terminating event, whose occurrence triggers the end of the experiment. This also is an example of transient type simulation.

**Example 20.8** Nonterminating simulations occur in the analysis of processes that are taken to be permanent or ongoing, such as the emergency room at a hospital, the flow of data over a communication network, the operation of the computer system at a military installation, and so on. In these instances, there is no specific moment at which a start or end can be identified. This can pose a problem for simulations, because all simulations must begin from some initial point.

Transient simulations are fairly straightforward. Independent observations (or samples) can only be obtained by running the simulation many times, using exactly the same initial conditions each time and terminating at the same time instant or stopping event. This is called the method of *independent replications*. Since the duration of the simulation should be relatively short, it is possible to conduct many replications. Different suites of random numbers must be used in each replication so care must be taken to properly seed random number generators (since running with the same seed would just produce the same estimation). Let $N$ be the number of simulation runs conducted and assume that on each of these $N$ runs, $n$ samples are obtained. Denote the $k^{\text{th}}$ such sample obtained during the $i^{\text{th}}$ simulation run as $X_{ik}$. On simulation run $i$, $i = 1, 2, \ldots, N$, a point estimator $\bar{X}_i$ may be obtained from $X_{ik}$, $k = 1, 2, \ldots, n$. This means that each point estimator $\bar{X}_i$ will be formed from samples that are possibly correlated. However the point estimators $\bar{X}_i$ are independent, because the simulation runs from which they are derived are independent. Thus a final point estimator, formed from these as

$$\bar{\bar{X}} = \frac{1}{N}\sum_{i=1}^{N} \bar{X}_i \quad \text{where} \quad \bar{X}_i = \frac{1}{n}\sum_{k=1}^{n} X_{ik},$$

may be constructed. Furthermore, since the $\bar{X}_i$ are independent, they may also be used to form the sample variance and confidence intervals as described in the previous sections.

Steady state simulations are generally more computationally intensive than transient simulations and therein lies their problem. Simulation experiments of complex systems must be initiated with the system in some particular state. During the course of the simulation the system changes state until it reaches some sort of steady state or equilibrium situation and it is often the behavior of the system at equilibrium that is of interest to the modelers. This means that the simulation results obtained up to this equilibrium point should be discarded. Determining when this so-called *transient* phase has ended is generally not easy to compute. One possibility is to conduct the simulation experiment for such a long time that the simulated values obtained during this initial period are completely swamped by later values. However, this can be rather expensive. A better alternative is to monitor the values until they appear to settle into some sort of regular pattern. The values obtained up to this point are then omitted from all future analysis. For example, if samples $X_i$, $i = 1, 2, \ldots, n$, are generated and the effects of the initial state are apparent in the first $k$ of them, then the sample mean should be taken to be $\bar{X} = \sum_{k+1}^{n} X_i/(n-k)$. In all cases, if some information is available concerning the equilibrium state of the system, that information should guide the choice of initial conditions. For example, if a heavily loaded system is to be simulated, it makes more sense to initialize the system with a large load, rather than a light (or empty) load since in this case, the simulation will reach its equilibrium point more quickly—there is frequently a tendency to initialize queueing systems with empty queues and wait for them to fill up!

A second more serious problem with steady-state simulations is the (very) long time over which they must be run. If we can assume that the determination of the end of the transient period can be detected, then the problem of the independence of the individual $X_i$ during the simulation of the stationary phase arises. One possibility is to use the method of independent replication which we saw was suitable for transient simulations. However, it has a major drawback. Each simulation run requires the elimination of the transient portion and this can be expensive, particularly if the transient period is long. This elimination of the transient phase must be done for each of the $N$ simulation runs before the $N$ values of $\bar{X}_i$ are obtained. To offset this somewhat, when a certain fixed total number of samples is to be generated, it is generally recommended to keep the number of simulation runs ($N$) small, of the order of 10 or so, and the number of samples in each run ($n$) large, rather than the other way around.

The second approach we consider is referred to as the method of *batch means*. The basic idea is to perform just one simulation, remove the transient phase just for this one run and divide the remainder of the simulation into a set of $N$ partial simulations or *batches*. If the objective of the simulation relates to some measure of customer statistics, where the concept of customer is taken in the broadest possible sense, then the length of each batch should be determined by the length of time needed until a fixed number of customers, the same fixed number for all batches, have been processed. Otherwise the length (timed duration) of each batch should be exactly the same. The logic behind the idea of batching is that although the $X_i$ are correlated, this correlation is most strong with values of $X_j$ for which $j$ is close to $i$, and becomes weaker the further that $j$ moves away from $i$. Thus $X_{i-2}$, $X_{i-1}$, $X_{i+1}$ and $X_{i+2}$ may be strongly correlated with $X_i$, but the correlation between $X_i$ and $X_{i+100}$ is likely to be negligible. Each of these batch simulations is analyzed as if it were a complete simulation, minus the transient period. Estimators derived from the batches will be approximately independent, so long as the batch lengths are sufficiently long. If $L$ is the number of samples that are needed so that the correlation between $X_i$ and $X_{i+L}$ is negligible, then it is recommended that the length of a batch should be at least $5L$. In this case, one can be relatively certain that estimators obtained from non-adjacent batches will be independent. As for adjacent batches, only samples at the end of the first and at the beginning of the second will be correlated. But since the batch lengths are long, estimators constructed for the first batch will mostly be from samples that are uncorrelated from the majority of samples in the second so even estimators constructed from samples of adjacent batches will be approximately independent. The procedure to follow then resembles that developed for the method of independent replications. A sample mean is constructed for each batch. These are taken to be independent and from them, an overall sample mean, overall sample variance and confidence intervals may be constructed. This approach has the obvious drawback that the partial simulations will not be independent, although they are assumed to be so. Despite this drawback, the method of batch means is the approach usually adopted.

It is possible to monitor the degree of correlation among the $X_i$ and to take action whenever this is deemed to be excessive. As seen in Chapter 5, the *covariance* between two random variables $X_1$ with mean $\mu_1$ and variance $\sigma_1^2$ and $X_2$ with mean $\mu_2$ and variance $\sigma_2^2$ is defined as

$$\text{Cov}(X_1, X_2) = E[(X_1 - \mu_1)(X_2 - \mu_2)] = E[X_1 X_2] - \mu_1 \mu_2.$$

Since the covariance can take any value between $-\infty$ and $+\infty$, it is frequently replaced with the correlation, defined as

$$\rho = \text{Corr}(X_1, X_2) = \frac{\text{Cov}(X_1, X_2)}{\sigma_1 \sigma_2}$$

and it follows that $-1 \leq \text{Corr}(X_1, X_2) \leq 1$. A sequence of possibly dependent random variables $X_1, X_2, \ldots$ all having the same distribution (mean $\mu$ and variance $\sigma^2$), which is the case of interest to us in simulation, is referred to as a *time series*. In a time series, the covariance and correlation become the autocovariance and autocorrelation. In particular, $\text{Cov}(X_i, X_{i+k})$ and $\text{Corr}(X_i, X_{i+k})$ are called the *lag-k* autocovariance and *lag-k* autocorrelation. When these depends only on $k$ and not on $i$, the time series is said to be *covariance stationary* and in this case we write

$$\gamma_k = \text{Cov}(X_i, X_{i+k}), \qquad \rho_k = \text{Corr}(X_i, X_{i+k}).$$

When $\rho_k = 0$, the random variables are not autocorrelated; the autocorrelation increases as $\rho_k$ moves from zero. When $\rho_k > 0$, the time series is positively autocorrelated: as a general rule, large observations are followed by large observations and small one by small ones. When $\rho_k < 0$ the opposite effect occurs: large observations tend to be followed by small observations and vice versa.

It is recommended that a test be performed to check the size of the lag-1 autocorrelation of batch means. If $N$ is the number of batches, $\bar{X}_i$ the sample mean computed in batch $i$ and $\bar{X}$ the overall sample mean, then the lag-1 autocorrelation may be estimated as

$$\tilde{\rho}_1 = \frac{\sum_{i=1}^{N-1}(\bar{X}_i - \bar{X})(\bar{X}_{i+1} - \bar{X})}{\sum_{i=1}^{N}(\bar{X}_i - \bar{X})^2}.$$

If this is small, the problem of autocorrelation can be dismissed. On the other hand, in the presence of autocorrelation, confidence intervals computed from the usual sample variance will likely be in error. However, it is possible to generate a more accurate estimate of the sample variance, and hence a more meaningful confidence interval, using the autocorrelation functions. After a fair amount of algebra, it may be shown that the sample variance in the presence of autocorrelation is given by

$$\text{Var}(\bar{X}) = \frac{1}{n^2}\sum_{i=1}^{n}\sum_{j=1}^{n}\text{Cov}(X_i, X_j) = \frac{\gamma_0}{n} + \frac{2}{n^2}\left[\sum_{k=1}^{n-1}(n-k)\gamma_k\right], \tag{20.9}$$

where $\gamma_0 = \sigma^2$. The $\gamma_k$ can be estimated from the generated samples. When the length of the batches is long, only a small number of $\gamma_k$ need actually be formed. In this case, the summation in Equation (20.9) contains only a few terms, say $m$, and the approximation becomes

$$\text{Var}(\bar{X}) \approx \frac{\tilde{\gamma}_0}{n} + \frac{2}{n}\left[\sum_{k=1}^{m}\tilde{\gamma}_k\right]$$

where $\tilde{\gamma}_k$ are the approximations to $\gamma_k$ obtained from using the samples. This variance may now be used with the Student $T$-distribution to compute confidence intervals.

The problem of eliminating the transient portion of a steady-state simulation run is non-existent in the case of a *regenerative process*. As the name implies, a regenerative process is one in which the system "regenerates" itself from time to time, essentially beginning again from scratch. The lengths of time between consecutive regeneration instants are called *regeneration cycles* and random variables defined in one regeneration cycle have the highly desirable property of being independent from their counterparts in other cycles as well as being distributed identically to them. The evolution of the system depends only on the state occupied at a regeneration instant and must be the same for all regeneration instants.

**Example 20.9** In a $G/G/1$ queueing system, arrival instants which initiate a busy period are regeneration instants. These are the instants in which an arrival finds the system empty. In a $G/M/1$ queue, in which service times are exponentially distributed, each arrival instant is a regeneration instant while in an $M/G/1$ queue, each departure instant is a regeneration instant.

In the simulation context, each regeneration cycle constitutes one batch and the method of batch means becomes the *regenerative method*: successive batches coincide with consecutive regeneration cycles and there is no need to first remove a supposed transient period. Statistics gathered from one batch will be independent of, and identical to, those gathered from any other batch. Thus the regenerative method removes two problems: how to handle the transient portion of a steady-state simulation and how to ensure the independence of samples. Unfortunately, the regenerative method also has a problem. What might be taken as a point estimator, turns out to be biased and resulting confidence intervals might be seriously in error. If $L_i$ is the duration of the $i$th regeneration cycle and $Y_i$ a statistic gathered during this cycle, then the point estimator $\bar{Y}/\bar{L}$ is biased. Since the mean of a ratio is not equal to the ratio of the means

$$E\left[\frac{\bar{Y}}{\bar{L}}\right] \neq \frac{E[\bar{Y}]}{E[\bar{L}]} = \frac{E[Y_i]}{E[L_i]}.$$

Some procedures have been devised to combat this problem, such as the *Jackknife* method [52]. However, we shall defer from exploring this further since opportunities for the application of the regenerative approach in simulation are limited. Even systems that possess regeneration properties frequently suffer from the drawback that the regeneration cycles, although finite, are frequently long, too long to be considered as batches.

## 20.3 Variance Reduction Methods

Simulations require the generation of a great many uniformly distributed random numbers and, from them, random numbers satisfying different probability distributions. One cost-effective way to reduce this number is to reduce the variance of parameter estimators, for, as we have already seen, halving the variance results in a fourfold decrease in the sample size needed to satisfy a given confidence interval. A number of possibilities arise and in this section, we consider just two of them, namely antithetic variables and control variables. Other possibilities include stratified sampling, importance sampling and conditioning on expectations. Readers seeking additional information would do well to consult the text by Law [29] and/or the text by Ross [46]. The amount of extra computation involved in implementing a variance reduction procedure is usually minimal, but the decrease in variance can seldom be gauged in advance, unfortunately.

### *20.3.1 Antithetic Variables*

The sample mean $\bar{X}$ is computed from a set of $n$ observations $X_1, X_2, \ldots, X_n$, as

$$\bar{X} = \frac{1}{n}\sum_{i=1}^{n} X_i.$$

All $n$ random variables $X_i$ are identically distributed with mean $\mu$ and variance $\sigma^2$. For any two $X_1$ and $X_2$ (we choose 1 and 2 for ease of notation)

$$E\left[\frac{X_1 + X_2}{2}\right] = \frac{E[X_1] + E[X_2]}{2} = E[X]$$

and

$$\mathrm{Var}\left[\frac{X_1 + X_2}{2}\right] = \frac{1}{4}\left[\mathrm{Var}[X_1] + \mathrm{Var}[X_2] + 2\,\mathrm{Cov}(X_1, X_2)\right]$$

$$= \frac{\mathrm{Var}[X] + \mathrm{Cov}(X_1, X_2)}{2},$$

since $\mathrm{Var}[X_1] = \mathrm{Var}[X_2] = \mathrm{Var}[X]$. Here $\mathrm{Cov}(X_1, X_2)$ is the covariance and is equal to $E[X_1 X_2] - \mu^2$ when $X_1$ and $X_2$ have the same mean value $\mu$. Since the covariance can be any positive or negative real number, it would be advantageous if $X_1$ and $X_2$ were negatively correlated, rather than being independent, since in this case the variance is reduced. The method of antithetic variables is an attempt to introduce negative correlation into the samples. One simple approach is to manipulate the sequence of uniformly distributed random numbers. Let us make the dependence of $X_1$ on the sequence of random numbers explicit by writing $X_1(u_1, u_2, \ldots, u_k)$ thereby indicating that $X_1$ is a function of $k$ uniformly distributed random numbers $u_1, u_2, \ldots, u_k$. The same is true for $X_2$ which we now write as $X_2(v_1, v_2, \ldots, v_k)$ where $v_1, v_2, \ldots, v_k$ is a different sequence of $k$ uniformly distributed random numbers. The role played by $u_i$, for $i = 1, 2, \ldots, k$, in generating $X_1$ is identical to the role played by $v_i$ in generating $X_2$. If we choose $v_i = 1 - u_i$, then since $u_i$ is a uniformly distributed random number, so too is $v_i = 1 - u_i$ but now both $u_i$ and $v_i$ are negatively correlated. The hope, and expectation, is that $X_1$ is now negatively correlated with $X_2$. The benefits are twofold: a possible reduction in the variance which entails a reduction in the number of samples needed and an additional reduction in the number of uniformly distributed random numbers needed, since we use the sequence $1 - u_1, 1 - u_2, \ldots, 1 - u_k$ instead of generating a new sequence of $k$ random numbers.

The concept extends in a straightforward manner to the case of $n$ observations. Let us assume that $n$ is even and equal to $2m$ and let $U_j = \{u_{j1}, u_{j2}, \ldots, u_{jk}\},\ j = 1, 2, \ldots, m$, be $m$ independent sets each consisting of $k$ uniformly distributed random numbers. Each of these $m$ sets is used to generate a sample: let $U_j$ be the set that generates $X_j$, which we write as $X_j(U_j)$. Let $V_j,\ j = 1, 2, \ldots, m$ be another $m$ independent sets of $k$ uniformly distributed random numbers but having the property that $U_j$ and $V_j,\ j = 1, 2, \ldots, m$ are dependent (even though the elements of the sequence $\{U_j,\ j = 1, 2, \ldots, m\}$ are independent of each other, as are the elements of the sequence $\{V_j,\ j = 1, 2, \ldots, m\}$). These latter $m$ sequences are used to generate the remaining $m$ samples $X_{m+j}(V_j)$, for $j = 1, 2, \ldots, m$. Let

$$X^{\dagger} = \frac{1}{m}\sum_{j=1}^{m} X_j(U_j) \quad \text{and} \quad X^{\ddagger} = \frac{1}{m}\sum_{j=1}^{m} X_{m+j}(V_j).$$

Then

$$\bar{X} = \frac{X^{\dagger} + X^{\ddagger}}{2}$$

is an unbiased estimator for the sample mean and the sample variance is

$$\mathrm{Var}[\bar{X}] = \frac{1}{4}\left[\mathrm{Var}[X^{\dagger}] + \mathrm{Var}[X^{\ddagger}] + 2\,\mathrm{Cov}(X^{\dagger}, X^{\ddagger})\right]$$

As before, the objective is to choose the sequences $U_j$ and $V_j,\ j = 1, 2, \ldots, m$ so that $X^{\dagger}$ and $X^{\ddagger}$ are negatively correlated. Random variables that accomplish this are said to be *antithetic variables*.

One possibility is to assign the element $v_{ji}$ of set $V_j$, the value $1 - u_{ji}$ with the hope that replacing small values in $U$ with large values in $V$ and vice versa leads to the desired negative correlation in $X^\dagger$ and $X^\ddagger$.

**Example 20.10** Congestion in queueing systems increases as the arrival rate increases and the service rate decreases. On the other hand, congestion decreases when the arrival rate decreases and the service rate increases. Thus we should expect the antithetic variables approach to work well when random numbers that give rise to samples with large congestion values are combined with random numbers that generate samples of low congestion. When a low-valued sample is paired with a high-valued sample, both are negatively correlated and their average will be closer to their common mean, i.e., there will be less variance.

**Example 20.11** An alternative approach to replacing a uniformly distributed random number $u$ used in generating one sample with $1 - u$ in an opposing sample that can be effective in queueing situations is as follows. The sequence of random numbers used to generate arrival rates in one set of samples can be used to generate service rates in an opposing set. This works in general since arrival rates and service rates work in opposition to each other. Large arrival rates (and short service rates) mean large numbers of customers present; large service rates (and short arrival rates) means few customers present.

### *20.3.2 Control Variables*

The idea of control variables is to obtain a second set of identically distributed samples, $Y_i$, $i = 1, 2, \ldots, n$ such that $Y_i$ is correlated with $X_i$ and for which $E[Y_i] = \nu$, is known. The correlation may be either positive or negative and the more strongly correlated, the better. Leaving aside for the moment the problem of how $\nu$ is known, the value of $\bar{Y} = \sum_{i=1}^{n} Y_i/n$ can be used to modify $\bar{X}$ depending upon whether $\bar{Y}$ is greater than or less than $\nu$. Suppose that $\bar{Y}$ and $\bar{X}$ are positively correlated, so that a large value of $\bar{Y}$ implies that $\bar{X}$ is large and a small value of $\bar{Y}$ implies a small value of $\bar{X}$. Then, if during the course of a simulation, it is found that $\bar{Y} > \nu$, which we can check since we know $\nu$, then we should also expect that $\bar{X}$ will be greater than its mean $\mu$, since $\bar{X}$ is positively correlated with $\bar{Y}$. In this case $\bar{X}$ should be adjusted downwards. Similarly, if $\bar{Y}$ is lower than its mean, we can assume that $\bar{X}$ is also less than its mean value and should be adjusted upwards. The opposite effects take place when $\bar{X}$ and $\bar{Y}$ are negatively correlated. If $\bar{Y}$ is greater than its mean, we should suspect that $\bar{X}$ will be less than its mean and should be adjusted upwards and vice versa. Thus $\bar{Y}$ controls the value of the $\bar{X}$ and hence the name: the *method of control variables*.

**Example 20.12** In a queueing system, we should expect that longer-than-average service times ($\bar{Y} > \nu$) lead to longer-than-average waiting times ($\bar{X} > \mu$), while shorter-than-average service times ($\bar{Y} < \nu$) should lead to shorter-than-average waiting times ($\bar{X} < \mu$). Service times and waiting times can be taken to be positively correlated. On the other hand, shorter-than-average interarrival times are expected to generate longer-than-average waiting times, and vice versa, so interarrival times and waiting times are negatively related. Control variables of this nature, chosen as differing quantities of the system being modeled, are called *concomitant* variables.

We now tackle the question of the adjustment of $\bar{X}$. If $\bar{X}$ is an unbiased estimator of $X$, then so too is $\bar{X}_c$ where, for any value of $c$

$$\bar{X}_c = \bar{X} - c(\bar{Y} - \nu). \tag{20.10}$$

$\bar{X}_c$ is an unbiased estimator since

$$E[\bar{X}_c] = E[\bar{X}] - c(E[\bar{Y}] - E[\nu]) = E[\bar{X}] - c(\nu - \nu) = E[\bar{X}] = \mu.$$

The variance of this quantity (see Section 5.2) is given by

$$\text{Var}(\bar{X}_c) = \text{Var}\left[\bar{X} - c(\bar{Y} - \nu)\right] = \text{Var}\left[\bar{X} - c\bar{Y}\right] = \text{Var}\left[\bar{X}\right] + c^2\text{Var}\left[\bar{Y}\right] - 2c\,\text{Cov}(\bar{X}, \bar{Y}), \quad (20.11)$$

and naturally we wish to choose $c$ to minimize this variance. Observe that the variance of $\bar{X}_c$ is less than the variance of $\bar{X}$ if and only if

$$2c\,\text{Cov}(\bar{X}, \bar{Y}) > c^2\text{Var}\left[\bar{Y}\right].$$

Considering the right-hand side of Equation (20.11) as a quadratic equation in $c$, the value of $c$ that minimizes the variance is given by

$$c_{\text{opt}} = \frac{\text{Cov}(\bar{X}, \bar{Y})}{\text{Var}\left[\bar{Y}\right]}.$$

Thus the minimum value of the variance of the controlled estimator is

$$\begin{aligned}
\text{Var}\left[\bar{X}_c\right] &= \text{Var}\left[\bar{X}\right] + \left(\frac{\text{Cov}(\bar{X}, \bar{Y})}{\text{Var}\left[\bar{Y}\right]}\right)^2 \text{Var}\left[\bar{Y}\right] - 2\left(\frac{\text{Cov}(\bar{X}, \bar{Y})}{\text{Var}\left[\bar{Y}\right]}\right)\text{Cov}(\bar{X}, \bar{Y}) \\
&= \text{Var}\left[\bar{X}\right] - \frac{\left[\text{Cov}(\bar{X}, \bar{Y})\right]^2}{\text{Var}\left[\bar{Y}\right]}.
\end{aligned}$$

Dividing both sides by $\text{Var}(\bar{X})$ gives

$$\frac{\text{Var}\left[\bar{X}_c\right]}{\text{Var}[\bar{X}]} = 1 - [\text{Corr}(\bar{X}, \bar{Y})]^2,$$

where $\text{Corr}(\bar{X}, \bar{Y})$ is the correlation between $\bar{X}$ and $\bar{Y}$, which shows that if $\bar{X}$ and $\bar{Y}$ are at all correlated, no matter how insignificantly, the variance of $\bar{X}_c$ will be less that that of $\bar{X}$. Furthermore, with greater correlation comes greater variance reduction. The only fly in the ointment now appears: to compute the optimum value of $c$ we need the values of $\text{Var}[\bar{Y}]$ and $\text{Cov}(\bar{X}, \bar{Y})$. One possibility is to estimate these quantities from the samples $X_i$ and $Y_i$, $i = 1, 2, \ldots, n$, obtained during the running of the simulation. Thus approximations to the covariance and variance are taken as

$$\widehat{\text{Cov}}(\bar{X}, \bar{Y}) = \frac{\sum_{i=1}^n (X_i - \bar{X})(Y_i - \bar{Y})}{n-1} \quad \text{and} \quad \widehat{\text{Var}}[\bar{Y}] = \frac{\sum_{i=1}^n (Y_i - \bar{Y})^2}{n-1},$$

and they allow us to compute an approximation to the optimal $c$:

$$\hat{c}_{\text{opt}} = \frac{\sum_{i=1}^n (X_i - \bar{X})(Y_i - \bar{Y})}{\sum_{i=1}^n (Y_i - \bar{Y})^2}.$$

When substituted into Equation (20.10), we should obtain a more accurate estimator, $X_c$. The variance of the controlled estimator is taken as

$$S_c^2 \equiv S_{\bar{X}_c}^2 = \frac{S_X^2 + S_Y^2 - 2S_{X,Y}}{n},$$

where $S_X^2 = \sum_{i=1}^n (X_i - \bar{X})^2/(n-1)$, $S_Y^2 = \widehat{\text{Var}}[\bar{Y}]$, and $S_{X,Y} = \widehat{\text{Cov}}(\bar{X}, \bar{Y})$, which allows confidence intervals to be generated.

Finally we return to the question of the expected value $E[Y_i] = \nu$ which we assumed was known right from the start. If this is not the case and the value of $\nu$, or a close approximation to it, is not available, then some other method must be used to find it, perhaps even an independent simulation study itself. When an extended sequence of complex simulations is to be performed, it may well be appropriate to take the time and effort of running separate simulations to compute this mean value before beginning the extended sequence of simulations proper.

**Example 20.13** Consider the simulation of an $M/M/1$ queue in which the mean inter-arrival time is $1/\lambda = 5$ minutes and the mean service time is $1/\mu = 4$ minutes. A total of ten simulation runs are conducted. Each begins with the system empty and continues until the instant the 200th customer enters service. The objective is to obtain information concerning the average amount of time spent by customers waiting in the queue ($X_i$). The mean service time ($Y_i$) is also observed to act as a control variable. The following table of results is obtained, one row for each of the ten simulation runs.

| $i$ | $X_i$ | $Y_i$ |
|---|---|---|
| 1 | 18.79 | 4.15 |
| 2 | 17.97 | 3.82 |
| 3 | 18.54 | 4.09 |
| 4 | 9.58 | 3.53 |
| 5 | 14.51 | 4.01 |
| 6 | 11.61 | 3.90 |
| 7 | 9.87 | 3.78 |
| 8 | 16.66 | 3.94 |
| 9 | 27.09 | 5.24 |
| 10 | 22.92 | 4.97 |

We find that $\bar{X} = 16.7540$ and $\bar{Y} = 4.1430$. We also know that $E[Y_i] = \nu = 4$ and from results on the $M/M/1$ queue we have $E[\bar{X}_i] = (\lambda/\mu)/(\mu - \lambda) = 16.0$. Thus both $\bar{X}$ and $\bar{Y}$ overestimate their respective means which supports our contention that they are positively correlated. Let us now compute the different variances and covariances. We have

$$S_X^2 = \frac{1}{9}\sum_{i=1}^{10}(X_i - \bar{X})^2 = 31.5586, \quad S_Y^2 = \frac{1}{9}\sum_{i=1}^{10}(Y_i - \bar{Y})^2 = 0.2911,$$

$$S_{X,Y} = \frac{1}{9}\sum_{i=1}^{10}(X_i - \bar{X})(Y_i - \bar{Y}) = 2.7143.$$

The correlation between $X$ and $Y$ is found as

$$\text{Corr}(X, Y) = \frac{\text{Cov}(X, Y)}{\sigma_X \sigma_Y} = \frac{2.7143}{\sqrt{31.5586}\sqrt{.2911}} = .8955.$$

The optimum value for $c$ is taken to be

$$\hat{c}_{\text{opt}} = \frac{S_{X,Y}}{S_Y^2} = 9.3243,$$

and our final approximation is

$$\bar{X}_c = \bar{X} - \hat{C}_{\text{opt}}(\bar{Y} - \nu) = 16.7540 - 9.3243(4.1430 - 4) = 15.4207,$$

which is closer to the true mean, 16. Observe that the variance of the controlled estimator is given by

$$S_c^2 = \frac{S_X^2 + S_Y^2 - 2S_{X,Y}}{n} = \frac{31.5586 + 0.2911 - 2 \times 2.7143}{10} = 2.6421.$$

## 20.4 Exercises

**Exercise 20.1.1** Given the following 20 samples, compute the sample mean and sample variance. These samples were obtained as exponentially distributed random numbers with mean value $1/\lambda = 4.0$.

4.8622, 5.9821, 7.8550, 1.0610, 4.2413, 3.6297, 2.6534, 1.0022, 3.0925, 5.9598,

4.2795, 6.2796, 1.0365, 1.2278, 0.1223, 2.5259, 3.5029, 3.2223, 0.8769, 1.2340.

**Exercise 20.1.2** Given the following 30 samples, compute the sample mean and sample variance. These samples were obtained as $N(15, 4^2)$ normally distributed random numbers.

17.3950, 9.3055, 15.2807, 12.7739, 13.8892, 16.6974, 14.6249, 16.8984, 7.0533, 17.8103,

18.4537, 14.8095, 11.3635, 14.2145, 19.0939, 19.8625, 19.0527, 19.7766, 8.8417, 15.0815,

16.0902, 12.0137, 13.7863, 21.4924, 17.6435, 18.8750, 14.5378, 11.5988, 11.0676, 16.9576.

**Exercise 20.1.3** The management of a bank observes the duration of service provided to 20 customers. What is the probability that the mean duration time observed by the management (the sample mean) is in error by more than 30 seconds if the actual duration of customer service is $\mu = 5$ minutes and the variance is $\sigma^2 = 60$ minutes? What is the probability that the sample mean is in error by more than 60 seconds? Repeat your analysis for the case when the management observes 50 customers rather than 20.

**Exercise 20.1.4** Given the set of samples in Exercise 20.1.2, estimate the number of additional samples that are needed if we wish the standard deviation of the estimator $\bar{X}$ to be (a) less than .25; (b) less than .1.

**Exercise 20.1.5** Compute a 90% confidence interval for the sample set of Exercise 20.1.1, first in the case when the true variance $\sigma^2 = 16$ is used and in the second case, when the estimated sample variance $S^2$ is used.

**Exercise 20.1.6** Compute a 90% and a 95% confidence interval for the sample set of Exercise 20.1.2 using the sample variance as an estimator for the true variance.

**Exercise 20.2.1** Run the queueing network simulation, the Java program *qn.java* of the previous chapter and determine approximately the value of $t$ at which the initialization portion is reached. Does this particular program need to be modified to remove this transient portion of a simulation run in order to correctly determine the bottleneck station? Justify your answer.

**Exercise 20.2.2** In the $M/M/1$ Java simulation program *mm1.java*, insert statements to output statistics at each departure instant and use them to determine approximately when the transient period ends. Now modify the program to remove this transient section from the final results. What is the mean waiting time obtained by your modified program when the simulation ends after $N = 10{,}000$ departures?

**Exercise 20.2.3** This exercise relates to the previous one, but now you are asked to use the method of batch means. Use six batches each containing 500 customer departures and construct the mean waiting time as the average of the six batches. Is your answer more accurate than that of the previous question?

**Exercise 20.2.4** The following sample means were obtained from the method of batch means applied to a simulation experiment. Estimate the lag-1 autocorrelation of this data. Are the sample means correlated?

3.0602, 3.7349, 3.9410, 2.6324, 3.2106, 2.1325.

**Exercise 20.3.1** A total of 10 simulation runs are conducted on a queueing model. During run $i$, $i = 1, 2, \ldots, 10$, the computed mean time spend in the queue and mean service time, in seconds, are given below. A simplified model indicates that the actual mean service time is 4 seconds. Use this information and the method of control variables to obtain a better estimate of the mean waiting time.

| Number | Queue | Service |
|---|---|---|
| 1 | 17.20 | 3.97 |
| 2 | 23.89 | 4.10 |
| 3 | 14.88 | 3.98 |
| 4 | 14.89 | 4.01 |
| 5 | 20.25 | 4.00 |
| 6 | 17.55 | 4.01 |
| 7 | 14.91 | 4.02 |
| 8 | 16.47 | 3.98 |
| 9 | 13.88 | 4.03 |
| 10 | 17.08 | 4.04 |

(Just for curiosity's sake, these data were taken from a simulation of a system in which the exact queueing time is equal to 16 seconds.)

**Exercise 20.3.2** The following data concerning mean interarrival times, mean queueing times, and mean service times have been collected from ten simulations of a queueing system. It is known that the exact arrival rate is $\lambda = 0.9$ and the exact service rate is $\mu = 1.0$.

| Number | Arrival | Queue | Service |
|---|---|---|---|
| 1 | 1.0895 | 4.5633 | 0.9830 |
| 2 | 1.0743 | 7.4785 | 1.0115 |
| 3 | 1.1367 | 10.9648 | 1.0231 |
| 4 | 1.0912 | 10.1538 | 0.9747 |
| 5 | 1.0804 | 14.7951 | 1.0011 |
| 6 | 1.0983 | 13.0862 | 0.9448 |
| 7 | 1.1431 | 5.5254 | 0.9382 |
| 8 | 1.1189 | 3.7098 | 0.9466 |
| 9 | 1.1699 | 3.3439 | 0.9775 |
| 10 | 1.1149 | 3.5893 | 0.9349 |

(a) Compute the sample mean and sample variance of inter-arrival times, queueing times, and service times.
(b) Compute the covariance of arrival data and queueing data and the covariance of service data and queueing data.
(c) For the given data, show that queueing time and service time are positively correlated and that arrival data and queueing data are negatively correlated.
(d) Use the method of control variables to generate a better estimate of the sample mean queueing time using the exact arrival rate.
(e) Same as part (d), but this time use the exact service rate.

# Appendix A: The Greek Alphabet

| | | |
|---|---|---|
| Alpha | $A$ | $\alpha$ |
| Beta | $B$ | $\beta$ |
| Gamma | $\Gamma$ | $\gamma$ |
| Delta | $\Delta$ | $\delta$ |
| Epsilon | $E$ | $\epsilon$ |
| Zeta | $Z$ | $\zeta$ |
| Eta | $H$ | $\eta$ |
| Theta | $\Theta$ | $\theta$ |
| Iota | $I$ | $\iota$ |
| Kappa | $K$ | $\kappa$ |
| Lambda | $\Lambda$ | $\lambda$ |
| Mu | $M$ | $\mu$ |
| Nu | $N$ | $\nu$ |
| Xi | $\Xi$ | $\xi$ |
| Omicron | $O$ | $o$ |
| Pi | $\Pi$ | $\pi$ |
| Rho | $P$ | $\rho$ |
| Sigma | $\Sigma$ | $\sigma$ |
| Tau | $T$ | $\tau$ |
| Upsilon | $\Upsilon$ | $\upsilon$ |
| Phi | $\Phi$ | $\phi$ |
| Chi | $X$ | $\chi$ |
| Psi | $\Psi$ | $\psi$ |
| Omega | $\Omega$ | $\omega$ |

# Appendix B: Elements of Linear Algebra

In this appendix we briefly present some of the fundamentals of linear algebra. Throughout the body of this text, we freely introduced vectors and matrices when needed and implicitly assumed familiarity with the manner in which they can be manipulated. The matrices used were all square matrices with elements taken from the field of reals and representing transition probabilities or transition rates among the states of a Markov chain. The vectors were probability vectors, i.e., row vectors whose elements lie between 0 and 1 inclusive, and whose sum is 1. In the more general context of linear algebra, unless otherwise noted, vectors are taken to be column vectors, and matrices need not be square nor real. We shall adopt the convention of describing probability vectors as row vectors and all other vectors as column vectors. This will allow us to move freely within the linear algebra world in which all vectors are column vectors and the world of applied probability in which probability vectors are always row vectors. Vectors and matrices whose elements are from the set of reals are said to be *real* vectors and matrices respectively. The set of all real matrices having $m$ rows and $n$ columns is denoted by $\Re^{m \times n}$.

## B.1 Vectors and Matrices

As a general rule, upper-case latin characters will denote matrices, and lower-case latin characters will denote (column) vectors. We shall let $A$ be a matrix having $m$ rows and $n$ columns. The element on row $i$ and column $j$ of $A$ is denoted by $a_{ij}$. Similarly, let $B$ have $k$ rows and $l$ columns and $ij$ element given by $b_{ij}$. Many of the operations below are defined on matrices. However, they may also be applied to vectors, since a vector of length $m$ is an $m \times 1$ matrix and a row vector of length $n$ is a $1 \times n$ matrix.

A square matrix $A$ with the property that $a_{ij} = 0$ for $i \neq j$ is called a *diagonal matrix*; nonzero elements can only appear along the diagonal. If the elements of $A$ are such that $a_{ij} = 0$ for $i < j$, then $A$ is a lower triangular matrix; if $a_{ij} = 0$ for $i > j$, then it is upper triangular. A diagonal matrix is simultaneously upper and lower triangular. A diagonal matrix whose nonzero elements are all equal to 1 is called the identity matrix and is denoted by $I$. The matrix whose elements are all equal to zero is called the zero matrix and is denoted by 0.

The *transpose* of a matrix $A$ is obtained by interchanging rows and columns. The $i$th row of $A$ becomes the $i$th column of $A^T$. For example,

$$A = \begin{pmatrix} -8.0 & 5.0 & 3.0 \\ 2.0 & -4.0 & 2.0 \end{pmatrix}, \quad A^T = \begin{pmatrix} -8.0 & 2.0 \\ 5.0 & -4.0 \\ 3.0 & 2.0 \end{pmatrix}.$$

A matrix is said to be *symmetric* if it is equal to its transpose, i.e., if $A = A^T$. The following matrix is symmetric:

$$A = \begin{pmatrix} -8.0 & 5.0 & 3.0 \\ 5.0 & -9.0 & 4.0 \\ 3.0 & 4.0 & -7.0 \end{pmatrix}.$$

A matrix is said to be *skew-symmetric* if $A = -A^T$.

## B.2 Arithmetic on Matrices

Let $A \in \Re^{m \times n}$ and $B \in \Re^{k \times l}$ be two real matrices.

**Matrix Addition**

Two matrices may be added only if they have the same dimensions. If $m = k$ and $n = l$, then the $ij^{\text{th}}$ element of the sum $C = A + B$ is given by

$$c_{ij} = a_{ij} + b_{ij}.$$

**Vector Inner Product**

When each element of a row vector $x$ of length $n$ is multiplied by the corresponding element of a column vector $y$ of the same length and these $n$ products added, the result is a single number, called the inner product of the two vectors. The product of an $1 \times n$ matrix and a $n \times 1$ matrix is a $1 \times 1$ matrix, i.e., a scalar. We have

$$\zeta = \sum_{i=1}^{n} x_i y_i.$$

**Vector Outer Product**

When an $m \times 1$ column vector is multiplied by a $1 \times n$ row vector, the result is an $m \times n$ matrix, called the outer product of $a$ and $b$. Its $ij^{\text{th}}$ element is given by

$$Z_{ij} = x_i y_j.$$

**Matrix Multiplication**

To form the product $C = A \times B$, it is necessary that the number of columns of $A$ be equal to the number of rows of $B$, i.e., that $n = k$. In this case, the $ij$ element of the product is

$$c_{ij} = \sum_{\xi=1}^{n} a_{i\xi} b_{\xi j}.$$

Notice that each element of $C$ is formed by multiplying a row of $A$ with a column of $B$, i.e., as the inner product of a row of $A$ with a column of $B$.

**(Post)multiplication by a Vector $x$**

To form the product $y = Ax$, the vector $x$ must have the same number of elements as the number of columns of $A$. We have

$$y_i = \sum_{\xi=1}^{n} a_{i\xi} x_\xi.$$

**(Pre)multiplication by a Row Vector $u$**

To form the product $v = uA$, the row vector $u$ must have the same number of elements as the number of rows of $A$. The $j^{\text{th}}$ element of the resulting row vector is

$$v_j = \sum_{\xi=1}^{n} u_\xi a_{\xi j}.$$

**Multiplication by a Scalar $\rho$**

Each element of the matrix is multiplied by the scalar. The $ij^{\text{th}}$ element of $C = \rho A$ is given by

$$c_{ij} = \rho a_{ij}.$$

Matrix multiplication is associative and distributive, but not commutative. For example,

$$(AB)C = A(BC) \quad \text{and} \quad A(B + C) = AB + AC;$$

but, in most cases,

$$AB \neq BA.$$

A notable exception to the commutative law occurs when one of the matrices is the identity matrix. We have $AI = IA = A$.

**Matrix Inverse**

If, corresponding to a square matrix $A$, there exists a matrix $B$ such that $AB = I = BA$, then $B$ is said to be the *inverse* of $A$. It is usually written as $A^{-1}$, and we have

$$AA^{-1} = I = A^{-1}A.$$

Notice that, from the matrix size compatibility requirements of matrix multiplication, both $A$ and $A^{-1}$ must be square and have the same size. A square matrix that does not possess an inverse, is said to be *singular*.

## B.3 Vector and Matrix Norms

To characterize the magnitude of vectors and matrices we use vector and matrix norms. A norm utilizes the magnitude of some or all of the elements of the vector or matrix to obtain a single real number to represent the magnitude of the vector or matrix. For example, in two or three dimensional space (vectors $x$ in $\Re^2$ or $\Re^3$), the length of a line is computed as $(x_1^2 + x_2^2)^{1/2}$ or $(x_1^2 + x_2^2 + x_3^2)^{1/2}$. This is easily generalized to $n$ dimensional vectors and defines a norm, called the *Euclidean norm* or 2-norm.

This is not the only vector norm. More generally, a vector norm is any function $f$ from $\Re^n$ into $\Re$ with the following properties:

- $f(x) \geq 0$ for $x \in \Re^n$ and $f(x) = 0$ if and only if $x = 0$,
- $f(x + y) \leq f(x) + f(y), \quad x, y, \in \Re^n$,
- $f(\alpha x) = |\alpha| f(x), \quad \alpha \in \Re$ and $x \in \Re^n$.

Norms are denoted by a pair of straight lines on either side of the vector or matrix. For example, for the Euclidean norm we have

$$\|x\|_2 = \sqrt{\sum_{i=1}^{n} x_i^2}.$$

If the columns of a matrix $A$ are piled on top of each other, then the matrix becomes a vector and a matrix norm can be defined accordingly. A matrix norm would therefore satisfy the same three properties. We have

- $\|A\| \geq 0$ for $A \in \Re^{m\times n}$ and $\|A\| = 0$ if and only if $A = 0$,
- $\|A + B\| \leq \|A\| + \|B\|, \quad A, B \in \Re^{m\times n}$,
- $\|\alpha A\| = |\alpha| \, \|A\|, \quad \alpha \in \Re$ and $A \in \Re^{m\times n}$.

However, this is not the usual way in which matrix norms are defined. Given a vector norm $\|\cdot\|$, a corresponding matrix norm is defined as

$$\|A\| = \sup\left\{\|Au\| \, : \; u \in \Re^n, \; \|u\| = 1\right\}.$$

Such a matrix norm is said to be *subordinate* to its vector norm. It is not difficult to show that matrix norms defined in this fashion satisfy the three requirements listed above. In addition, subordinate matrix norms satisfy a submultiplicative property, namely,

- $\|AB\| \leq \|A\| \|B\|$ for $A \in \Re^{m\times n}$ and $B \in \Re^{n\times l}$.

It is precisely because they satisfy this property that subordinate norms are usually preferred to other matrix norms. Such norms are sometimes called *natural* matrix norms.

Some common vector norms ($x \in \Re^n$) and their corresponding subordinate matrix norms ($A \in \Re^{n \times n}$) are the following.

### The 1-Norm

The vector 1-norm is simply the sum of the absolute values of the elements of the vector. The matrix 1-norm is the maximum of the sums of all the absolute values in each column of the matrix (called the maximum column sum). We have

$$\|x\|_1 = \sum_{i=1}^{n} |x_i|, \qquad \|A\|_1 = \max_j \left( \sum_{i=1}^{n} |a_{ij}| \right).$$

### The $\infty$-Norm

The vector infinity norm is the maximum of the absolute values of the elements in the vector. The matrix infinity norm, in the style given above, is the maximum row sum. We have

$$\|x\|_\infty = \max_i |x_i|, \qquad \|A\|_\infty = \max_i \left( \sum_{j=1}^{n} |a_{ij}| \right).$$

It might have been thought that the *Frobenius* matrix norm, defined as

$$\|A\|_F = \sqrt{\sum_{i=1}^{n} \sum_{j=1}^{n} |a_{ij}|^2},$$

would be the matrix norm subordinate to the vector 2-norm, but it is not. Indeed, the Frobenius norm is not even a subordinate norm. Despite this, it has many useful properties, but that is outside the scope of this text.

## B.4 Vector Spaces

The set of all real vectors of length $n$ is said to form a *vector space* of length $n$. It is denoted by $\Re^n$.

### Linear Combinations of Vectors

Let $u_1, u_2, \ldots, u_k \in \Re^n$ be $k$ real vectors and let $\alpha_1, \alpha_2, \ldots, \alpha_k$ be $k$ scalars. Then

$$u = \alpha_1 u_1 + \alpha_2 u_2 + \cdots + \alpha_k u_k$$

is said to be a *linear combination* of the vectors $u_1, u_2, \ldots, u_k$. For example, in $\Re^3$, the vector $v$

$$v = \begin{pmatrix} 5 \\ 3 \\ 0 \end{pmatrix} = 2 \begin{pmatrix} 1 \\ 0 \\ 0 \end{pmatrix} + 3 \begin{pmatrix} 1 \\ 1 \\ 0 \end{pmatrix} = 2u_1 + 3u_2$$

is a linear combination of the vectors $u_1$ and $u_2$.

### Vector Subspaces

Let $u_1, u_2, \ldots, u_k \in \Re^n$ be $k$ real vectors. Then the set of all linear combinations of these $k$ vectors forms a *vector subspace*. Denote this subspace by $U$. Then $U \subset \Re^n$. For example, the vector subspace $U$ which consists of all linear combinations of $u_1$ and $u_2$, where

$$u_1 = \begin{pmatrix} 1 \\ 0 \\ 0 \end{pmatrix} \quad \text{and} \quad u_2 = \begin{pmatrix} 1 \\ 1 \\ 0 \end{pmatrix},$$

contains all vectors of length 3 whose third component is equal to zero.

Observe that the zero vector 0, the vector whose elements are all equal to zero, is an element of every subspace. It is obtained when the scaler coefficients of the vectors are all equal to zero. Vector spaces and subspaces are *closed* under the operations of addition and scalar multiplication.

**Linearly Independent Vectors**

A set of vectors, $u_1, u_2, \ldots, u_k$ is said to be *linearly independent* if all nontrivial linear combinations are nonzero, i.e.,

$$\alpha_1 u_1 + \alpha_2 u_2 + \cdots + \alpha_k u_k = 0 \quad \text{if and only if} \quad \alpha_1 = \alpha_2 = \cdots = \alpha_k = 0. \tag{B.1}$$

Otherwise the set of vectors is said to be *linearly dependent*. In this case, any one of the vectors may be written as a linear combination of the others. For example, with a set of linearly dependent vectors we may write

$$-\alpha_1 u_1 = \alpha_2 u_2 + \cdots + \alpha_k u_k$$

and hence

$$u_1 = \beta_2 u_2 + \beta_3 u_3 + \cdots + \beta_k u_k$$

with $\beta_i = -\alpha_i/\alpha_1$ for $i = 2, 3, \ldots, k$. For example, the vectors $u_1$ and $u_2$ given above are linearly independent, but the set of three vectors

$$u_1 = \begin{pmatrix} 1 \\ 0 \\ 0 \end{pmatrix}, \quad u_2 = \begin{pmatrix} 1 \\ 1 \\ 0 \end{pmatrix}, \quad \text{and} \quad u_3 = \begin{pmatrix} 0 \\ 1 \\ 0 \end{pmatrix}$$

are not, since any one may be written as a linear combination of the others. We have

$$u_1 = u_2 - u_3, \quad u_2 = u_1 + u_3 \quad \text{and} \quad u_3 = u_2 - u_1.$$

**The Span of a Subspace**

Consider any vector subspace $W$ and a set of vectors $w_1, w_2, \ldots, w_l$ in that subspace. If every vector $w \in W$ can be written as a linear combination of this set of vectors, i.e., if there exists coefficients $\gamma_1, \gamma_2 \ldots, \gamma_l$ for which

$$w = \gamma_1 w_1 + \gamma_2 w_2 + \cdots + \gamma_l w_l,$$

then the set is said to *span* the subspace $W$. As an example, the three vectors

$$u_1 = \begin{pmatrix} 1 \\ 0 \\ 0 \end{pmatrix}, \quad u_2 = \begin{pmatrix} 1 \\ 1 \\ 0 \end{pmatrix}, \quad \text{and} \quad u_3 = \begin{pmatrix} 0 \\ 1 \\ 0 \end{pmatrix} \tag{B.2}$$

span the subspace $U$.

**Basis Vectors**

If a set of vectors span a vector space and are linearly independent, then this set of vectors is said to form a *basis* for the space. The two properties

1. the vectors span the vector space, and
2. the vectors are linearly independent

are both required. Thus, the three vectors given in Equation (B.2) span $U$ but they do not constitute a basis for $U$, since they are not linearly independent. However, any two of these three are linearly independent and do span $U$. Thus, $u_1$ and $u_2$ constitute a basis for $U$ as do $u_1$ and $u_3$, and $u_2$ and $u_3$.

A particularly useful set of basis vectors for the space $\Re^n$ is the set of vectors $e_1, e_2, \ldots, e_n$, all of length $n$, given by

$$e_1 = \begin{pmatrix} 1 \\ 0 \\ 0 \\ \vdots \\ 0 \end{pmatrix}, \quad e_2 = \begin{pmatrix} 0 \\ 1 \\ 0 \\ \vdots \\ 0 \end{pmatrix}, \quad e_3 = \begin{pmatrix} 0 \\ 0 \\ 1 \\ \vdots \\ 0 \end{pmatrix}, \quad \ldots, \quad e_n = \begin{pmatrix} 0 \\ 0 \\ 0 \\ \vdots \\ 1 \end{pmatrix}.$$

It is easy to see how any vector in $\Re^n$ may be written as a linear combination of this basis set.

Every vector in a vector space can be written *uniquely* in terms of basis vectors. In other words, if the vectors $z_1, z_2, \ldots, z_k$ constitute a basis for a subspace that contains a vector $y$, and if

$$y = \alpha_1 z_1 + \alpha_2 z_2 + \cdots + \alpha_k z_k$$

and

$$y = \beta_1 z_1 + \beta_2 z_2 + \cdots + \beta_k z_k,$$

then $\alpha_i = \beta_i$ for $i = 1, 2, \ldots, k$.

The number of vectors in a basis set (which must be the same for all basis) is called the *dimension* of the vector space. Any set of vectors that constitutes a span for a vector space may be downsized (by eliminating some of its members until its size is equal to the dimension of the subspace) to yield a basis set. Similarly, any set of linearly independent vectors may be augmented to produce a basis set.

### The Rank of a Matrix

The $n$ columns in an $\Re^{m \times n}$ matrix $A$ constitute a subspace of $\Re^m$ which is called the *column space* of $A$, often referred to as the *range* of $A$. The dimension of the range of a matrix is called its *rank*. Similarly, the $m$ rows of $A$ constitute a subspace of $\Re^n$, called the *row space* of $A$. For each of these two spaces we may define spans and basis sets. It turns out that the dimension of the column space of a matrix is also equal to the dimension of its row space, from which it follows that any square matrix with linearly independent rows must also have linearly independent columns.

## B.5 Determinants

The *determinant* of a matrix $A \in \Re^{n \times n}$ is a single real number. It is defined as follows. If $A \in \Re^{2 \times 2}$ is given by

$$A = \begin{pmatrix} a & b \\ c & d \end{pmatrix},$$

then the determinant of $A$, written as $\det(A)$ or $|A|$ (we shall use both interchangeably) is the real number

$$\det(A) = ad - bc.$$

The determinant of a matrix $A \in \Re^{n \times n}$ that is larger than $2 \times 2$, is written in terms of the determinants of smaller submatrices of $A$. The matrix obtained from $A$ by removing its $i^{\text{th}}$ row and $j^{\text{th}}$ column is written as $M_{ij}$ and is called the *ij-Minor* of $A$. The determinant of this minor, together with a specific sign, is called a *cofactor* of $A$, and is denoted by $A_{ij}$. More specifically, the $ij$ cofactor of $A$ is given by

$$A_{ij} = (-1)^{i+j} M_{ij}.$$

We may now define the determinant of a matrix $A \in \Re^{n\times n}$. It is given by

$$\det(A) = a_{i1}A_{i1} + a_{i2}A_{i2} + \cdots + a_{in}A_{in}. \qquad \text{(B.3)}$$

Consider the following matrix as an example:

$$A = \begin{pmatrix} -8.0 & 5.0 & 3.0 \\ 2.0 & -4.0 & 2.0 \\ 1.0 & 6.0 & -7.0 \end{pmatrix}.$$

Some of its minors are

$$M_{11} = \begin{pmatrix} -4.0 & 2.0 \\ 6.0 & -7.0 \end{pmatrix}, \quad M_{22} = \begin{pmatrix} -8.0 & 3.0 \\ 1.0 & -7.0 \end{pmatrix}, \quad \text{and} \quad M_{32} = \begin{pmatrix} -8.0 & 3.0 \\ 2.0 & 2.0 \end{pmatrix}.$$

The cofactors corresponding to these minors are

$$A_{11} = (-1)^2 \begin{vmatrix} -4.0 & 2.0 \\ 6.0 & -7.0 \end{vmatrix} = 28 - 12 = 16,$$

$$A_{22} = (-1)^4 \begin{vmatrix} -8.0 & 3.0 \\ 1.0 & -7.0 \end{vmatrix} = 56 - 3 = 53, \quad \text{and}$$

$$A_{32} = (-1)^5 \begin{vmatrix} -8.0 & 3.0 \\ 2.0 & 2.0 \end{vmatrix} = -(16 - 6) = -22.$$

To compute the determinant of $A$, we need to expand across any row (or column). Equation (B.3) defines the expansion in terms of row $i$. For example, we have

$$\det(A) = a_{11}(-1)^{1+1}\det(M_{11}) + a_{12}(-1)^{1+2}\det(M_{12}) + a_{13}(-1)^{1+3}\det(M_{13})$$

$$= -8\begin{vmatrix} -4.0 & 2.0 \\ 6.0 & -7.0 \end{vmatrix} - 5\begin{vmatrix} 2.0 & 2.0 \\ 1.0 & -7.0 \end{vmatrix} + 3\begin{vmatrix} 2.0 & -4.0 \\ 1.0 & 6.0 \end{vmatrix}$$

$$= -8 \times 16 - 5 \times (-16) + 3 \times 16 = 0.$$

Thus, in this particular example, the determinant of the matrix is zero. The reader should observe that the matrix in this example is an infinitesimal generator matrix, the transition rate matrix of a 3-state continuous-time Markov chain. It is always the case that determinants of infinitesimal generator matrices are zero. The reader may wish to verify that expanding $A$ along any of the other two rows, or indeed along any of the columns, produces the same result. The reader may also wish to verify that the matrix $P$ produced from $A$ by discretization, i.e., by setting $P = I + 0.125A$, has a determinant that is nonzero.

The determinant of a matrix has many many properties, most of which will not be needed in this text. Indeed, the only reason that determinants are introduced at all is due to the important role they play in defining eigenvalues. This is discussed in Appendix B.7. Some properties of determinants are given below.

1. The determinant of a matrix changes sign when any two rows are interchanged.
2. Adding a multiple of any row into any other row leaves the determinant unchanged.
3. The determinant of a matrix having two identical rows is zero.
4. The determinant of a matrix whose rows are linearly dependent is zero.

5. The determinant of a matrix whose rows are linearly independent is nonzero.
6. A matrix is singular if and only if its determinant is zero.
7. The determinant of a triangular matrix is equal to the product of its diagonal elements.
8. $\det(A^T) = \det(A)$, and for two compatible matrices $A$ and $B$, $\det(AB) = \det(A)\det(B)$.

## B.6 Systems of Linear Equations

Generally our first introduction to systems of linear equations occurs when we are given a problem of the type,

```
1 apple,  1 banana and a cantaloupe cost $6.00, while
1 apple, 2 bananas and three cantaloupe cost $14.00 and
1 apple, 3 bananas and four cantaloupe cost $19.0.
How much does  a cantaloupe cost?
```

We quickly realize that we may manipulate these three equations with their three unknown quantities, $a$, the price of an apple, $b$, the price of a banana, and $c$, the price of a cantaloupe,

$$\begin{aligned} a + b + c &= \$6.0, \\ a + 2b + 3c &= \$14.0, \\ a + 3b + 4c &= \$19.0, \end{aligned}$$

to obtain a simpler system consisting of only two equations in two unknowns. Subtracting the first equation from the second and from the third gives the simpler system

$$\begin{aligned} b + 2c &= 8.0, \\ 2b + 3c &= 13.0. \end{aligned}$$

Now, on subtracting twice the first of these from the second, it becomes obvious that the price of a cantaloupe is \$3. Furthermore, it is now easy to determine the values of the other unknowns. Since $b + 2c = 8.0$ and $c = 3$, it follows that $b = 2$ and since $a + b + c = 6$, using $b = 2$ and $c = 3$ we must have $a = 1.0$.

The three equations

$$\begin{aligned} a + b + c &= 6.0, \\ a + 2b + 3c &= 14.0, \\ a + 3b + 4c &= 19.0 \end{aligned}$$

may be written in matrix form as

$$\begin{pmatrix} 1 & 1 & 1 \\ 1 & 2 & 3 \\ 1 & 3 & 4 \end{pmatrix} \begin{pmatrix} a \\ b \\ c \end{pmatrix} = \begin{pmatrix} 6.0 \\ 14.0 \\ 19.0 \end{pmatrix}.$$

If we denote the $3 \times 3$ matrix by $A$, the vector containing the unknowns by $x$, and the right-hand side vector by $b$, we have

$$Ax = b. \tag{B.4}$$

The matrix $A$ is called the *coefficient matrix*. The operations applied to this example reduced it to

$$\begin{pmatrix} 1 & 1 & 1 \\ 0 & 1 & 2 \\ 0 & 0 & -1 \end{pmatrix} \begin{pmatrix} a \\ b \\ c \end{pmatrix} = \begin{pmatrix} 6.0 \\ 8.0 \\ -3.0 \end{pmatrix}.$$

The coefficient matrix in the resulting system of equations is *upper* triangular and the three unknowns are then obtained by *backward substitution*.

The reduction of a coefficient matrix to triangular form is achieved by means of *elementary row operations*. Any two rows may be interchanged and a multiple of any row may be added into any other row. A matrix derived from another by means of a sequence of elementary row operations is completely equivalent to the original matrix; equivalent in the sense that none of the information content is lost. The same elementary row operations must also be applied to the elements of the right-hand side vector. The matrix and right-hand side vector taken together is called the *augmented* system.

In our example, the matrix $A$ has an inverse. It is given by

$$A^{-1} = \begin{pmatrix} 1.0 & 1.0 & -1.0 \\ 1.0 & -3.0 & 2.0 \\ -1.0 & 2.0 & -1.0 \end{pmatrix}.$$

Multiplying both sides of Equation (B.4) on the left by $A^{-1}$ yields

$$A^{-1}Ax = A^{-1}b \quad \text{and hence } x = A^{-1}b.$$

For the particular example considered above, we have

$$A^{-1}b = \begin{pmatrix} 1.0 & 1.0 & -1.0 \\ 1.0 & -3.0 & 2.0 \\ -1.0 & 2.0 & -1.0 \end{pmatrix} \begin{pmatrix} 6.0 \\ 14.0 \\ 19.0 \end{pmatrix} = \begin{pmatrix} 1.0 \\ 2.0 \\ 3.0 \end{pmatrix}.$$

Not all systems of linear equations have a solution. If the second equation in the example is replaced by $2a + 2b + 2c = 12$, then no solution can be found. In this case the second equation is redundant because it is just equal to twice the first. It does not provide us with any information that we cannot obtain from the first equation. If we now proceed as we did before, to add a multiple of the first into the second with the goal of eliminating one of the unknowns, the result is to eliminate *all* of the unknowns from the second equation. Each element in the second row of the matrix becomes zero. In this case, the rows of the matrix are not linearly independent, its determinant is zero and the matrix is singular.

It is, however, possible to derive some information from the first and third equations, since these two *are* linearly independent. Given only that

$$\begin{pmatrix} 1 & 1 & 1 \\ 1 & 3 & 4 \end{pmatrix} \begin{pmatrix} a \\ b \\ c \end{pmatrix} = \begin{pmatrix} 6.0 \\ 19.0 \end{pmatrix},$$

we may assert that *if* the cost of an apple is \$1.00 (essentially we are adding in a third equation, $1a = 1.00$) then the cost of a banana is \$2.0 and the cost of a cantaloupe is \$3.0, or again, *if* the cost of an apple is \$5.00, then the cost of a banana is \$10.0 and the cost of a cantaloupe is \$15.0. In reality, all we can do is to state the relative values of the three unknowns: a banana costs twice as much as an apple while a cantaloupe costs three times as much. There is no *unique* solution to this system of equations: any nonzero scalar multiple of a solution is also a solution. In a system of linear equations with $n$ unknowns and only $r < n$ linearly independent equations, $n - r$ of the unknowns can be assigned random values and the remaining unknowns obtained in terms of these assigned values. In the example described above, $n - r = 1$.

One other case is important for us, the case when the right-hand side vector in Equation (B.4) is zero. It is from such a system of linear equation that we compute the stationary distribution of a continuous-time Markov chain. There are two possibilities:

1. If the coefficient matrix $A$ is nonsingular, then it must follow that the only solution to $Ax = 0$ is $x = 0$. If $A$ is nonsingular, its determinant is nonzero and its $n$ columns form a set of $n$ linearly independent vectors. Let these columns be denoted by $c_1, c_2, \ldots, c_n$ and let the $n$ unknowns of the vector $x$ be denoted by $\xi_1, \xi_2, \ldots, \xi_n$. Then $Ax$ represents a linear

combination of the rows of $A$. We have

$$Ax = \xi_1 c_1 + \xi_2 c_2 + \cdots + \xi_n c_n.$$

However, from Equation (B.1), we know that $n$ vectors are linearly independent if and only if all nontrivial linear combinations are nonzero. In other words,

$$Ax = \xi_1 c_1 + \xi_2 c_2 + \cdots + \xi_n c_n = 0$$

if and only if $\xi_1 = \xi_2 = \cdots = \xi_n = 0$. Thus, for nonsingular $A$, the only solution to $Ax = 0$ is $x = 0$.

2. On the other hand, if $A$ is singular, then the number of linearly independent rows of $A$ is strictly less that $n$. Let the number of linearly independent rows be given by $r$. In this case, random values may be assigned to $n - r$ of the unknowns and the remaining $r$ unknowns computed as a function of these values. There is no unique solution.

To summarize, the system of equations $Ax = 0$ has a nontrivial solution if and only if $A$ is singular.

### *B.6.1 Gaussian Elimination and LU Decompositions*

#### Gaussian Elimination

Gaussian elimination (GE) is the standard algorithm for solving systems of linear equations, and due to its importance we now proceed to a detailed description of this method applied to the system $Ax = b$ in which $A$ is nonsingular and $b \neq 0$. Writing this system in full, we have

$$\begin{aligned}
a_{11}x_1 + a_{12}x_2 + a_{13}x_3 + \cdots + a_{1n}x_n &= b_1,\\
a_{21}x_1 + a_{22}x_2 + a_{23}x_3 + \cdots + a_{2n}x_n &= b_2,\\
a_{31}x_1 + a_{32}x_2 + a_{33}x_3 + \cdots + a_{3n}x_n &= b_3,\\
\vdots \qquad\qquad & \ \vdots \ \vdots\\
a_{n1}x_1 + a_{n2}x_2 + a_{n3}x_3 + \cdots + a_{nn}x_n &= b_n,
\end{aligned}$$

i.e., a system of $n$ linear equations in $n$ unknowns. The first step in Gaussian elimination is to use one of these equations to eliminate one of the unknowns in the other $n - 1$ equations. This is accomplished by adding a multiple of one row into the other rows; the particular multiple is chosen to zero out the coefficient of the unknown to be eliminated. The particular equation chosen is called the *pivotal equation* and the diagonal element in this equation is called the *pivot*. The equations that are modified are said to be *reduced*. The operations of multiplying one row by a scalar and adding or subtracting two rows are *elementary operations*; they leave the system of equations invariant. For instance, using the first equation to eliminate $x_1$ from each of the other equations gives the reduced system

$$\begin{aligned}
a_{11}x_1 + a_{12}x_2 + a_{13}x_3 + \cdots + a_{1n}x_n &= b_1,\\
0 + a'_{22}x_2 + a'_{23}x_3 + \cdots + a'_{2n}x_n &= b'_2,\\
0 + a'_{32}x_2 + a'_{33}x_3 + \cdots + a'_{3n}x_n &= b'_3,\\
\vdots \qquad\qquad & \ \vdots \ \vdots\\
0 + a'_{n2}x_2 + a'_{n3}x_3 + \cdots + a'_{nn}x_n &= b'_n.
\end{aligned} \tag{B.5}$$

The first coefficient in the second row, $a_{21}$, becomes zero when we multiply the first row by $-a_{21}/a_{11}$ and then add the first row into the second row. The other coefficients in the second row, and the right-hand side, are affected by this operation: $a_{2j}$ becomes $a'_{2j} = a_{2j} - a_{1j}/a_{11}$ for $j = 2, 3, \ldots, n$ and $b_2$ becomes $b'_2 = b_2 - b_1/a_{11}$. Similar effects occur in the others rows; row 1 must be multiplied by $-a_{31}/a_{11}$ and added into row 3 to eliminate $a_{31}$, and in general, to eliminate the coefficient

$a_{i1}$ in column $i$, row 1 must be multiplied by $-a_{i1}/a_{11}$ and added into row $i$. When this has been completed for all rows $i = 2, 3, \ldots, n$, the first step of Gaussian elimination is completed and the system of equations is that of Equation (B.5).

When the first step of Gaussian elimination is finished, one equation involves all $n$ unknowns while the other $n-1$ equations involve only $n-1$ unknowns. These $n-1$ equations may be treated independently of the first. They constitute a system of $n-1$ linear equations in $n-1$ unknowns, which we may now solve using Gaussian elimination. We use one of them to eliminate an unknown in the $n-2$ others. If we choose the first of the $n-1$ equations as the next pivotal equation and use it to eliminate $x_2$ from the other $n-2$ equations, we obtain

$$\begin{aligned} a_{11}x_1 + a_{12}x_2 + a_{13}x_3 + \cdots + a_{1n}x_n &= b_1, \\ 0 \;\; + a'_{22}x_2 + a'_{23}x_3 + \cdots + a'_{2n}x_n &= b'_2, \\ 0 \;\; + \;\; 0 \;\; + a''_{33}x_3 + \cdots + a''_{3n}x_n &= b''_3, \\ \vdots \qquad\qquad & \;\; \vdots \\ 0 \;\; + \;\; 0 \;\; + a''_{n3}x_3 + \cdots + a''_{nn}x_n &= b''_n. \end{aligned} \tag{B.6}$$

We now have one equation in $n$ unknowns, one equation in $n-1$ unknowns, and $n-2$ equations in $n-2$ unknowns. If we continue this procedure, we would hope to end up with one equation in one unknown, two equations in two unknowns, and so on up to $n$ equations in $n$ unknowns:

$$\begin{aligned} a_{11}x_1 + a_{12}x_2 + a_{13}x_3 + \quad \cdots \quad + a_{1n}x_n &= b_1, \\ 0 \;\; + a'_{22}x_2 + a'_{23}x_3 + \quad \cdots \quad + a'_{2n}x_n &= b'_2, \\ 0 \;\; + \;\; 0 \;\; + a''_{33}x_3 + \quad \cdots \quad + a''_{3n}x_n &= b''_3, \\ &\;\; \vdots \\ \bar{a}_{n-1,n-1}x_{n-1} + \bar{a}_{n-1,n}x_n &= \bar{b}_{n-1}, \\ \bar{a}_{nn}x_n &= \bar{b}_n. \end{aligned}$$

At this stage the so-called *elimination phase* or *reduction phase* of Gaussian elimination is finished and we proceed to the *backsubstitution phase* during which the values of the unknowns are computed. Given one equation in one unknown (the last equation in the reduced system), the value of that unknown is easily computed. With $\bar{a}_{nn}x_n = \bar{b}_n$, we compute $x_n = \bar{b}_n/\bar{a}_{nn}$. Since we now know $x_n$, the value of $x_{n-1}$ can be found from the penultimate equation in the reduced system:

$$\bar{a}_{n-1,n-1}x_{n-1} + \bar{a}_{n-1,n}x_n = \bar{b}_{n-1}.$$

We have

$$x_{n-1} = \frac{\bar{b}_{n-1} - \bar{a}_{n-1,n}x_n}{\bar{a}_{n-1,n-1}}.$$

The process of backsubstitution continues in this fashion until all the unknowns have been evaluated. The computationally intensive part of Gaussian elimination is generally the reduction phase. The number of numerical operations for reduction is of the order $n^3/3$, whereas the complexity of the backsubstitution phase is only $n^2$. At this point it is worthwhile doing an example in full.

**Example B.1** Consider the following system of four linear equation in four unknowns:

$$\begin{aligned} -4.0x_1 + 4.0x_2 + 0.0x_3 + 0.0x_4 &= 0.0, \\ 1.0x_1 - 9.0x_2 + 1.0x_3 + 0.0x_4 &= 0.0, \\ 2.0x_1 + 2.0x_2 - 3.0x_3 + 5.0x_4 &= 0.0, \\ 0.0x_1 + 2.0x_2 + 0.0x_3 + 0.0x_n &= 2.0. \end{aligned}$$

In the more convenient matrix representation this becomes

$$\begin{pmatrix} -4.0 & 4.0 & 0.0 & 0.0 \\ 1.0 & -9.0 & 1.0 & 0.0 \\ 2.0 & 2.0 & -3.0 & 5.0 \\ 0.0 & 2.0 & 0.0 & 0.0 \end{pmatrix} \begin{pmatrix} x_1 \\ x_2 \\ x_3 \\ x_4 \end{pmatrix} = \begin{pmatrix} 0.0 \\ 0.0 \\ 0.0 \\ 2.0 \end{pmatrix}.$$

We are now ready to begin the reduction phase. This will consist of three steps. In the first we shall eliminate the unknown $x_1$ from rows 2, 3, and 4; in the second we shall eliminate $x_2$ from rows 3 and 4; and in the last, we shall eliminate $x_3$ from row 4. These are shown below at the beginning of each step. To the left we indicate the multiplier, the value by which the pivot row must be multiplied prior to being added into the row to be reduced. Notice that, in this example, the right-hand side does not change. This is simply because the only nonzero element is the last one, so adding multiples of lower-indexed elements to higher-indexed elements will not change the right-hand side. In other cases, the right-hand side may need to be modified. In many texts, the right-hand side is appended to the matrix $A$ to give an $n \times (n+1)$ array, called the *augmented matrix*. All the elementary operations are carried out on the augmented matrix.

**Reduction Phase**

$$\begin{array}{r|} \text{Multipliers} \\ 0.25 \\ 0.50 \\ 0.00 \end{array} \begin{pmatrix} -4.0 & 4.0 & 0.0 & 0.0 \\ \hline 1.0 & -9.0 & 1.0 & 0.0 \\ 2.0 & 2.0 & -3.0 & 5.0 \\ 0.0 & 2.0 & 0.0 & 0.0 \end{pmatrix} \Longrightarrow \left(\begin{array}{c|ccc} -4.0 & 4.0 & 0.0 & 0.0 \\ \hline 0.0 & -8.0 & 1.0 & 0.0 \\ 0.0 & 4.0 & -3.0 & 5.0 \\ 0.0 & 2.0 & 0.0 & 0.0 \end{array}\right) \begin{pmatrix} x_1 \\ x_2 \\ x_3 \\ x_4 \end{pmatrix} = \begin{pmatrix} 0.0 \\ 0.0 \\ 0.0 \\ 2.0 \end{pmatrix},$$

$$\begin{array}{r|} \text{Multipliers} \\ \\ 0.50 \\ 0.25 \end{array} \left(\begin{array}{c|ccc} -4.0 & 4.0 & 0.0 & 0.0 \\ 0.0 & -8.0 & 1.0 & 0.0 \\ \hline 0.0 & 4.0 & -3.0 & 5.0 \\ 0.0 & 2.0 & 0.0 & 0.0 \end{array}\right) \Longrightarrow \left(\begin{array}{cc|cc} -4.0 & 4.0 & 0.00 & 0.0 \\ 0.0 & -8.0 & 1.00 & 0.0 \\ \hline 0.0 & 0.0 & -2.50 & 5.0 \\ 0.0 & 0.0 & 0.25 & 0.0 \end{array}\right) \begin{pmatrix} x_1 \\ x_2 \\ x_3 \\ x_4 \end{pmatrix} = \begin{pmatrix} 0.0 \\ 0.0 \\ 0.0 \\ 2.0 \end{pmatrix},$$

$$\begin{array}{r|} \text{Multipliers} \\ \\ \\ 0.10 \end{array} \left(\begin{array}{cc|cc} -4.0 & 4.0 & 0.00 & 0.0 \\ 0.0 & -8.0 & 1.00 & 0.0 \\ 0.0 & 0.0 & -2.50 & 5.0 \\ \hline 0.0 & 0.0 & 0.25 & 0.0 \end{array}\right) \Longrightarrow \left(\begin{array}{ccc|c} -4.0 & 4.0 & 0.0 & 0.0 \\ 0.0 & -8.0 & 1.0 & 0.0 \\ 0.0 & 0.0 & -2.5 & 5.0 \\ 0.0 & 0.0 & 0.0 & 0.5 \end{array}\right) \begin{pmatrix} x_1 \\ x_2 \\ x_3 \\ x_4 \end{pmatrix} = \begin{pmatrix} 0.0 \\ 0.0 \\ 0.0 \\ 2.0 \end{pmatrix}.$$

At the end of these three steps, the matrix has the following upper triangular form and we are now ready to begin the backsubstitution phase:

$$\begin{pmatrix} -4.0 & 4.0 & 0.0 & 0.0 \\ 0.0 & -8.0 & 1.0 & 0.0 \\ 0.0 & 0.0 & -2.5 & 5.0 \\ 0.0 & 0.0 & 0.0 & 0.5 \end{pmatrix} \begin{pmatrix} x_1 \\ x_2 \\ x_3 \\ x_4 \end{pmatrix} = \begin{pmatrix} 0.0 \\ 0.0 \\ 0.0 \\ 2.0 \end{pmatrix}.$$

**Backsubstitution Phase**

In the backsubstitution phase, we compute the components of the solution in reverse order, using the $i^{\text{th}}$ equation to compute the value of $x_i$. The values of previously computed components, $x_j,\ j = i+1, i+2, \ldots, n$, are substituted into the $i^{\text{th}}$ equation, so the only remaining unknown is $x_i$ itself. We have the following:

$$\begin{array}{lrcl} \text{Equation 4}: & 0.5\,x_4 = 2 & \Longrightarrow & x_4 = 4, \\ \text{Equation 3}: & -2.5\,x_3 + 5 \times 4 = 0 & \Longrightarrow & x_3 = 8, \\ \text{Equation 2}: & -8\,x_2 + 1 \times 8 = 0 & \Longrightarrow & x_2 = 1, \\ \text{Equation 1}: & -4\,x_1 + 4 \times 1 = 0 & \Longrightarrow & x_1 = 1. \end{array}$$

Therefore the complete solution is $x^T = (1,\ 1,\ 8,\ 4)$.

Observe in this example that the multipliers do not exceed 1.0. This is a desirable property for it lessens the likelihood of rounding error build-up. In this example the multipliers are less than 1 because at each step, the pivot element is larger than the elements to be reduced. Implementations of Gaussian elimination usually incorporate a *pivoting* strategy. Most often the chosen strategy is *partial pivoting* whereby at step $i$ of the algorithm, the row with the largest element in magnitude in the unreduced part of column $i$ is chosen as the pivotal equation. A more costly version is *complete pivoting* whereby the largest element in magnitude in the entire unreduced portion of the matrix is chosen to be the pivot. Explicit pivoting is generally not needed for solving Markov chain problems, since the elements along the diagonal are already the largest in magnitude in each column and this property is maintained during the reduction phase.

### *LU* Decompositions

There exists an interesting relationship between the multipliers and the upper triangular matrix that results from Gaussian elimination. Let $U$ be this upper triangular matrix and define a lower triangular matrix $L$ whose diagonal elements are all equal to 1 and whose subdiagonal elements are the negated multipliers.

**Example B.2** Returning to the previous example, we find

$$L = \begin{pmatrix} 1.00 & 0.00 & 0.0 & 0.0 \\ -0.25 & 1.00 & 0.0 & 0.0 \\ -0.50 & -0.50 & 1.0 & 0.0 \\ 0.00 & -0.25 & -0.1 & 1.0 \end{pmatrix}, \quad U = \begin{pmatrix} -4.0 & 4.0 & 0.0 & 0.0 \\ 0.0 & -8.0 & 1.0 & 0.0 \\ 0.0 & 0.0 & -2.5 & 5.0 \\ 0.0 & 0.0 & 0.0 & 0.5 \end{pmatrix},$$

and the product of these two triangular matrices is the original matrix $A$, i.e., $LU = A$:

$$\begin{pmatrix} 1.00 & 0.00 & 0.0 & 0.0 \\ -0.25 & 1.00 & 0.0 & 0.0 \\ -0.50 & -0.50 & 1.0 & 0.0 \\ 0.00 & -0.25 & -0.1 & 1.0 \end{pmatrix} \begin{pmatrix} -4.0 & 4.0 & 0.0 & 0.0 \\ 0.0 & -8.0 & 1.0 & 0.0 \\ 0.0 & 0.0 & -2.5 & 5.0 \\ 0.0 & 0.0 & 0.0 & 0.5 \end{pmatrix} = \begin{pmatrix} -4.0 & 4.0 & 0.0 & 0.0 \\ 1.0 & -9.0 & 1.0 & 0.0 \\ 2.0 & 2.0 & -3.0 & 5.0 \\ 0.0 & 2.0 & 0.0 & 0.0 \end{pmatrix}.$$

When a matrix $A$ can be written as the product of a lower triangular matrix $L$ and an upper triangular matrix $U$, these two triangular matrices are said to constitute an *LU decomposition* or *LU factorization* of $A$. If $A$ is nonsingular, then the factors $L$ and $U$ must both be nonsingular, as is the case in Example B.2. Given a system of equations $Ax = b$ and an $LU$ decomposition of $A$, the solution $x$ may be computed by means of a forward substitution step on $L$ followed by a backward substitution step on $U$. For example, suppose we are required to solve $Ax = b$ with $\det(A) \neq 0$ and $b \neq 0$ and suppose further that the decomposition $A = LU$ is available. We have

$$Ax = LUx = b.$$

Let $z = Ux$. Then the solution $x$ is obtained in two easy steps:

$$\text{Forward substitution: Solve } Lz = b \text{ for } z;$$
$$\text{Backward substitution: Solve } Ux = z \text{ for } x.$$

In other words, the vector $z$ may be obtained by forward substitution on $Lz = b$, since both $L$ and $b$ are known quantities. Subsequently, the solution $x$ may be obtained from $Ux = z$ by backward substitution since by this time both $U$ and $z$ are known quantities. When we implement Gaussian elimination, the first of these two operations is carried out automatically. The solution $z$ that results from the forward substitution step is just equal to the right hand side after the elementary row operations are performed.

**Example B.3** Solving $Lz = b$ (by forward substitution) for $z$:

$$\begin{pmatrix} 1.00 & 0.00 & 0.0 & 0.0 \\ -0.25 & 1.00 & 0.0 & 0.0 \\ -0.50 & -0.50 & 1.0 & 0.0 \\ 0.00 & -0.25 & -0.1 & 1.0 \end{pmatrix} \begin{pmatrix} z_1 \\ z_2 \\ z_3 \\ z_4 \end{pmatrix} = \begin{pmatrix} 0 \\ 0 \\ 0 \\ 2 \end{pmatrix}$$

successively gives $z_1 = 0, \ z_2 = 0, \ z_3 = 0$, and $z_4 = 2$.

Now solving $Ux = z$ (by backward substitution) for $x$:

$$\begin{pmatrix} -4.0 & 4.0 & 0.0 & 0.0 \\ 0.0 & -8.0 & 1.0 & 0.0 \\ 0.0 & 0.0 & -2.5 & 5.0 \\ 0.0 & 0.0 & 0.0 & 0.5 \end{pmatrix} \begin{pmatrix} x_1 \\ x_2 \\ x_3 \\ x_4 \end{pmatrix} = \begin{pmatrix} 0 \\ 0 \\ 0 \\ 2 \end{pmatrix}$$

successively gives $x_4 = 4, \ x_3 = 8, \ x_2 = 1$, and $x_1 = 1$.

An $LU$ decomposition, when it exists, is not unique. The particular one that arises out of Gaussian elimination, as shown above, is called the *Doolittle decomposition*, and is characterized by ones along the diagonal of $L$. Computing an $LU$ decomposition involves determining the $n(n+1)/2$ elements in the lower triangular part of $L$ and the $n(n+1)/2$ elements in the upper triangular part of $U$. To obtain these $n(n+1)$ elements we have only a total of $n^2$ equations; the inner product of the $i$th row of $L$ with the $j$th column of $U$ must be equal to $a_{ij}$. With $n^2$ equations and $n(n+1)$ unknowns, we have $n$ more unknowns than equations which means that we may assign values to $n$ of the unknowns and solve the $n^2$ equations to determine the remaining $n^2$ unknowns. It makes sense to do this assignment in such a way as to make the computation of the remaining $n^2$ unknowns as easy as possible. In the Doolitle decomposition, the diagonal elements of $L$ are set to 1. Another often used approach is the *Crout decomposition*, which assigns the value of 1 to each of the diagonal elements of $U$ and solves for the remaining unknowns in $L$ and $U$. As we have just seen, an $LU$ decomposition can be constructed during the course of Gaussian elimination. However, when pivoting is not required, an alternative approach allows the triangular factors to be formed via inner vector products which can be accumulated in double precision and which yields a more stable algorithm.

## B.7 Eigenvalues and Eigenvectors

Again in this section our concern is with real square matrices. Let $A \in \Re^{n\times n}$ and $0 \neq x \in \Re^n$ and consider the result obtained when the product $Ax$ is formed. If the result obtained is once again the vector $x$, i.e., if $Ax = x$, then $x$ is said to be a *right-hand eigenvector* of $A$. If instead of $x$, a scalar multiple of $x$ is obtained, i.e., if $Ax = \lambda x$ for some scalar $\lambda$, then $x$ is still a right-hand eigenvector of $A$. In this case $\lambda$ is said to be an *eigenvalue* of $A$ and $x$ is its corresponding right-hand eigenvector. In the case $Ax = x$, the vector $x$ is an eigenvector corresponding to a unit ($\lambda = 1$) eigenvalue. The eigenvalue/vector problem is then to find scalars $\lambda$ and vectors $x \neq 0$ which satisfy the equation

$$Ax = \lambda x. \tag{B.7}$$

This may be written as

$$(A - \lambda I)x = 0,$$

from which it is clear that a nontrivial $x$ exists if and only if the coefficient matrix $A - \lambda I$ is singular. In other words, the determinant, $\det(A)$, must be zero. This provides us with a means of obtaining the eigenvalues. They are precisely the values for which

$$|A - \lambda I| = 0. \tag{B.8}$$

Equation (B.8) is called the *characteristic equation*. If we write this out in more detail, we have

$$\begin{vmatrix} a_{11} - \lambda & a_{12} & a_{13} & \cdots & a_{1n} \\ a_{21} & a_{22} - \lambda & a_{23} & \cdots & a_{2n} \\ \vdots & \vdots & \vdots & \ddots & \vdots \\ a_{n1} & a_{n2} & a_{n3} & \cdots & a_{nn} - \lambda \end{vmatrix} = 0,$$

and it becomes apparent that this represents a polynomial of degree $n$ in $\lambda$, the roots of which are the eigenvalues of $A$. Consider the $2 \times 2$ case. We have

$$\begin{vmatrix} a_{11} - \lambda & a_{12} \\ a_{21} & a_{22} - \lambda \end{vmatrix} = (a_{11} - \lambda)(a_{22} - \lambda) - a_{12}a_{21} = 0,$$

which is the quadratic polynomial

$$\lambda^2 - (a_{11} + a_{22})\lambda + (a_{11}a_{22} - a_{12}a_{21}) = 0.$$

It has two and only two roots, which means that a $2 \times 2$ matrix has exactly two eigenvalues. They are denoted by $\lambda_1$ and $\lambda_2$. These two roots (eigenvalues) need not be distinct (for example, the identity matrix of order 2 has two eigenvalues both equal to 1) and even though the matrix has only real elements, the eigenvalues can be complex, forming a complex pair. The characteristic equation of a matrix of size $3 \times 3$ is a cubic polynomial with exactly three roots, and thus a $3 \times 3$ matrix has exactly three eigenvalues. In general, an $n \times n$ matrix has exactly $n$ eigenvalues, denoted by $\lambda_1, \lambda_2, \ldots, \lambda_n$. Finding the eigenvalues of a matrix thus reduces to finding the roots of a polynomial equation, although often this is much easier said than done.

**Example B.4** As an example, let us find the eigenvalues of the matrix

$$A = \begin{pmatrix} 8 & 6 \\ -3 & -1 \end{pmatrix}.$$

Expanding the characteristic equation, we find

$$\begin{vmatrix} (8 - \lambda) & 6 \\ -3 & (-1 - \lambda) \end{vmatrix} = (8 - \lambda)(-1 - \lambda) + 18 = \lambda^2 - 7\lambda + 10 = (\lambda - 5)(\lambda - 2) = 0.$$

The two eigenvalues are therefore given by $\lambda_1 = 5$ and $\lambda_2 = 2$.

Some important results concerning eigenvalues are the following:

- The eigenvalues of $A$ and $A^T$ are identical.
  Observe that since $\det(A) = \det(A^T)$, it follows that $\det(A - \lambda I) = \det(A^T - \lambda I^T) = \det(A^T - \lambda I)$ and hence the result follows.
- The sum of the diagonal elements of a matrix, $A$, is called its *trace* and is denoted by $\text{tr}(A)$. The sum of the eigenvalues of $A$ is equal to the trace of $A$, i.e.,

$$\sum_{i=1}^{n} \lambda_i = \text{tr}(A).$$

- The product of the eigenvalues of a matrix $A$ is equal to the determinant of $A$, i.e.,

$$\prod_{i=1}^{n} \lambda_i = \det(A).$$

The *spectrum* of a matrix is the set of its eigenvalues and the largest of the magnitudes of the eigenvalues is called the *spectral radius* of the matrix and is denoted by $\rho$. We have

$$\rho(A) = \max_s |\lambda_s(A)|,$$

where $\lambda_s(A)$ denotes the $s^{\text{th}}$ eigenvalue of $A$. The spectral radius of any matrix cannot exceed the value of any of its *natural* matrix norms, $\|\cdot\|$. We have

$$\rho(A) \leq ||A||.$$

We are now able to give the matrix norm that is subordinate to the vector 2-norm. It is given by

$$\|A\| = \sqrt{\rho(A^T A)},$$

i.e., the square root of the spectral radius of $A^T A$. Finally, an important theorem for the localization of the eigenvalues of a matrix is *Gerschgorin's theorem.*

**Theorem B.7.1** (*Gerschgorin*) *Every eigenvalue of A lies in at least one of the n circles with center $a_{ii}$ and radius $r_i = \sum_{k=1,k\neq i}^{n} |a_{ik}|$.*

Eigenvectors are associated with eigenvalues and they must have at least one nonzero element. To find the eigenvector corresponding to one of the eigenvalues of a matrix $A$, we return to the defining equation

$$(A - \lambda I)x = 0,$$

insert the value of $\lambda$, and solve for $x$. Observe that this system of equations is homogeneous (its right-hand side is zero) and its coefficient matrix is singular, since the value of $\lambda$ is one of the roots of $\det(A - \lambda I)$. Therefore a unique solution does not exist; if $x$ is an eigenvector corresponding to an eigenvalue $\lambda$, then $\alpha x$, for $\alpha \neq 0$, is also an eigenvector corresponding to the eigenvalue $\lambda$. This is easily seen by multiplying each side of Equation (B.7) by $\alpha$. From

$$Ax = \lambda x,$$

which shows $x$ to be an eigenvector corresponding to eigenvalue $\lambda$, we may write

$$\alpha(Ax) = \alpha(\lambda x) \quad \text{or} \quad A(\alpha x) = \lambda(\alpha x) \quad \text{for } \alpha \neq 0,$$

and hence

$$Ay = \lambda y \quad \text{where } y = \alpha x,$$

which shows that $y$ too is an eigenvector corresponding to eigenvalue $\lambda$.

Using the same example as before, let us compute the eigenvector corresponding to the eigenvalue $\lambda_1 = 5$. Substituting $\lambda = 5$ into Equation (B.7), we find

$$\begin{pmatrix} 8-5 & 6 \\ -3 & -1-5 \end{pmatrix} \begin{pmatrix} x_1 \\ x_2 \end{pmatrix} = \begin{pmatrix} 0 \\ 0 \end{pmatrix},$$

i.e.,

$$\begin{pmatrix} 3 & 6 \\ -3 & -6 \end{pmatrix} \begin{pmatrix} x_1 \\ x_2 \end{pmatrix} = \begin{pmatrix} 0 \\ 0 \end{pmatrix}.$$

As expected, this gives us only a single equation to work with. We have $x_1 + 2x_2 = 0$, and so, taking $x_1 = 2$, we find $x_2 = -1$. We can verify that this is indeed an eigenvector by substituting it back. We have

$$\begin{pmatrix} 8 & 6 \\ -3 & -1 \end{pmatrix} \begin{pmatrix} 2 \\ -1 \end{pmatrix} = 5 \begin{pmatrix} 2 \\ -1 \end{pmatrix}.$$

The reader may wish to verify that $(-1, +1)^T$ is an eigenvector corresponding to the eigenvalue $\lambda_2 = 2$. Observe also, that the two eigenvectors are linearly independent. It turns out that eigenvectors corresponding to distinct eigenvalues must be linearly independent.

Although an $n \times n$ matrix always has $n$ eigenvalues, it may have fewer than $n$ eigenvectors. This will only happen when some of the eigenvalues are identical. For example, the matrix

$$A = \begin{pmatrix} 1 & 0 \\ 0 & 1 \end{pmatrix}$$

has eigenvalues $\lambda_1 = \lambda_2 = 1$ and two distinct eigenvectors. We have

$$\begin{pmatrix} 1 & 0 \\ 0 & 1 \end{pmatrix} \begin{pmatrix} 1 \\ 0 \end{pmatrix} = 1 \begin{pmatrix} 1 \\ 0 \end{pmatrix} \quad \text{and} \quad \begin{pmatrix} 1 & 0 \\ 0 & 1 \end{pmatrix} \begin{pmatrix} 0 \\ 1 \end{pmatrix} = 1 \begin{pmatrix} 0 \\ 1 \end{pmatrix}.$$

These eigenvectors are linearly independent, even though the eigenvalues are identical. On the other hand, the matrix

$$A = \begin{pmatrix} 1 & 1 \\ 0 & 1 \end{pmatrix}$$

also has two eigenvalues, both equal to 1, but only a single eigenvector. We have

$$\begin{pmatrix} 1 & 1 \\ 0 & 1 \end{pmatrix} \begin{pmatrix} 1 \\ 0 \end{pmatrix} = 1 \begin{pmatrix} 1 \\ 0 \end{pmatrix} \quad \text{but} \quad \begin{pmatrix} 1 & 1 \\ 0 & 1 \end{pmatrix} \begin{pmatrix} 0 \\ 1 \end{pmatrix} \neq 1 \begin{pmatrix} 0 \\ 1 \end{pmatrix}.$$

No vector, other than a scalar multiple of $(1, 0)^T$, can be found to be an eigenvector of this matrix. This matrix is said to be *defective*. When the eigenvalues are all distinct, there exists a complete set of corresponding eigenvectors, and furthermore, these eigenvectors form a basis for $\Re^n$. Even when the eigenvalues are not distinct, there may exist $n$ linearly independent eigenvectors. Having distinct eigenvalues is a sufficient, but not necessary condition for the existence of $n$ linearly independent eigenvectors.

Now consider a matrix $A \in \Re^{n \times n}$ whose $n$ eigenvalues are $\lambda_1, \lambda_2, \ldots, \lambda_n$ and with corresponding, linearly independent, eigenvectors, $x_1, x_2, \ldots, x_n$. Let $X$ be the $n \times n$ matrix whose $n$ columns are these $n$ eigenvectors, i.e.,

$$X = (x_1,\ x_2,\ \ldots,\ x_n).$$

Notice that $X$ is not necessarily a real matrix. Some of its columns may be complex, corresponding to complex eigenvalues of $A$. Since the columns of $X$ are linearly independent, $\det(X) \neq 0$ and the inverse, $X^{-1}$, of $X$ exists. Let $\Lambda$ be the diagonal matrix whose elements are the eigenvalues, $\lambda_1, \lambda_2, \ldots, \lambda_n$ of $A$. Then, since $Ax_i = \lambda_i x_i$ for $i = 1, 2, \ldots n$, we have

$$AX = X\Lambda.$$

Notice that the correct effect is obtained only if $\Lambda$ multiplies $X$ on the right. We may rewrite this equation as

$$X^{-1}AX = \Lambda$$

which shows that $A$ may be diagonalized. When a matrix is multiplied on the left by $X^{-1}$ and on the right by $X$, it is said to have undergone a *similarity transformation*. Advanced methods for determining the eigenvalues of a matrix use sequences of similarity transformations to force the matrix to approach diagonal, or tridiagonal form.

The preceding discussions centered on eigenvalues and *right-hand* eigenvectors. The same discussions could have taken place using *left-hand* eigenvectors instead. A left-hand eigenvector corresponding to an eigenvalue $\lambda$ of $A$ is a vector $y \in \Re^n$ such that

$$y^T A = \lambda y^T.$$

It follows that *left-hand* eigenvectors of $A$ are the *right-hand* eigenvectors of $A^T$ since

$$y^T A = \lambda y^T \quad \text{implies that} \quad A^T y = \lambda y.$$

We saw previously that the eigenvalues of $A$ and $A^T$ are identical. However, this is not true for the eigenvectors. Left-hand and right-hand eigenvectors corresponding to the same eigenvalue are generally different. An exception is when the matrix is symmetric.

## B.8 Eigenproperties of Decomposable, Nearly Decomposable, and Cyclic Stochastic Matrices

### *B.8.1 Normal Form*

A nonnegative matrix is one in which the elements are all greater than or equal to zero. A square nonnegative matrix $A$ is said to be *decomposable* if it can be brought by a symmetric permutation of its rows and columns to the form

$$A = \begin{pmatrix} U & 0 \\ W & V \end{pmatrix}, \tag{B.9}$$

where $U$ and $V$ are square, nonzero matrices and $W$ is, in general, rectangular and nonzero. We may relate this concept of decomposability with that of reducibility in the context of Markov chains; in fact, the two terms are often used interchangeably. If a Markov chain is reducible, then there is an ordering of the state space such that the transition probability matrix has the form given in (B.9). If the matrices $U$ and $V$ are of order $n_1$ and $n_2$ respectively ($n_1 + n_2 = n$, where $n$ is the order of $A$), then the states $s_i$, $i = 1, 2, \ldots, n$ of the Markov chain may be decomposed into two nonintersecting sets

$$B_1 = \{s_1, s_2, \ldots, s_{n_1}\}$$

and

$$B_2 = \{s_{n_1+1}, \ldots, s_n\}.$$

Let us assume that $W \neq 0$. The nonzero structure of the matrix $A$ in Equation (B.9) shows that transitions are possible from states of $B_2$ to $B_1$ but not the converse, so that if the system is at any time in one of the states $s_1, \ldots, s_{n_1}$, it will never leave this set. If the system is initially in one of the states of $B_2$, it is only a question of time until it eventually finishes in $B_1$. The set $B_1$ is called *isolated* or *essential*, and $B_2$ is called *transient* or *nonessential*. Since the system eventually reduces to $B_1$, it is called *reducible* and the matrix of transition probabilities is said to be *decomposable* [45]. If $W = 0$, then $A$ is completely decomposable, and in this case both sets, $B_1$ and $B_2$, are essential.

It is possible for the matrix $U$ itself to be decomposable, in which case the set of states in $B_1$ may be reduced to a new set whose order must be less than $n_1$. This process of decomposition may be continued until $A$ is reduced to the form, called the *normal* form of a decomposable nonnegative matrix, given by

$$A = \begin{pmatrix} A_{11} & 0 & 0 & \cdots & 0 & 0 & \cdots & 0 \\ 0 & A_{22} & 0 & \cdots & 0 & 0 & \cdots & 0 \\ \vdots & \vdots & \vdots & \ddots & \vdots & \vdots & \ddots & \vdots \\ 0 & 0 & 0 & \cdots & A_{kk} & 0 & \cdots & 0 \\ A_{k+1,1} & A_{k+1,2} & \cdots & \cdots & A_{k+1,k} & A_{k+1,k+1} & \cdots & 0 \\ \vdots & \vdots & \vdots & \ddots & \vdots & \vdots & \ddots & \vdots \\ A_{m,1} & A_{m,2} & \cdots & \cdots & A_{m,k} & A_{m,k+1} & \cdots & A_{m,m} \end{pmatrix}. \tag{B.10}$$

The submatrices $A_{ii}, i = 1, \ldots, m$, are square, nonzero, and nondecomposable. All submatrices to the right of the diagonal blocks are zero, as are those to the left of $A_{ii}$ for $i = 1, 2, \ldots, k$. As for the remaining blocks, for each value of $i \in [k+1, m]$ there is at least one value of $j \in [1, i-1]$ for which $A_{ij} \neq 0$. The diagonal submatrices are as follows:

- $A_{ii}, \quad i = 1, \ldots, k$, isolated and nondecomposable,
- $A_{ii}, \quad i = k+1, \ldots, m$, transient, and again, nondecomposable.

If $B_i$ is the set of states represented by $A_{ii}$, then states of $B_i$ for $i = 1, \ldots, k$ have the property that once the system is in any state of $B_i$, it cannot move out of that set. For sets $B_i, i = k+1, \ldots, m$, any transition out of these sets is to one possessing a lower index only, so that the system eventually finishes in one of the isolated sets $B_1, \ldots, B_k$. If all of the off-diagonal blocks are zero, $(k = m)$, then $A$ is said to be a *completely decomposable* nonnegative matrix.

## *B.8.2 Eigenvalues of Decomposable Stochastic Matrices*

Let $P$ be a decomposable stochastic matrix of order $n$ written in normal decomposable form as

$$P = \begin{pmatrix} P_{11} & 0 & 0 & \cdots & 0 & 0 & \cdots & 0 \\ 0 & P_{22} & 0 & \cdots & 0 & 0 & \cdots & 0 \\ \vdots & \vdots & \vdots & \ddots & \vdots & \vdots & \ddots & \vdots \\ 0 & 0 & 0 & \cdots & P_{kk} & 0 & \cdots & 0 \\ P_{k+1,1} & P_{k+1,2} & \cdots & \cdots & P_{k+1,k} & P_{k+1,k+1} & \cdots & 0 \\ \vdots & \vdots & \vdots & \ddots & \vdots & \vdots & \ddots & \vdots \\ P_{m,1} & P_{m,2} & \cdots & \cdots & P_{m,k} & P_{m,k+1} & \cdots & P_{m,m} \end{pmatrix}. \tag{B.11}$$

The submatrices $P_{ii}, i = 1, \ldots, k$, are all nondecomposable stochastic matrices and thus each possesses one and only one unit eigenvalue. The matrices $P_{ii}, i = k+1, \ldots, m$, are said to be *substochastic* since the sum of elements on at least one row is strictly less than unity. The Perron-Frobenius theorem allows us to state that the spectral radius of any matrix that is substochastic is strictly less than one. Thus, the largest eigenvalue of each of the matrices $P_{ii}, i = k+1, \ldots, m$, of (B.12) is strictly less than unity in modulus.

Since the set of eigenvalues of a block (upper or lower) triangular matrix (such as $P$ in Equation (B.11)) is the union of the sets of eigenvalues of the individual blocks, a decomposable stochastic matrix of the form (B.11) has exactly as many unit eigenvalues as it has isolated subsets; i.e., $k$.

**Example B.5** Consider the Markov chain defined diagrammatically by Figure B.1.

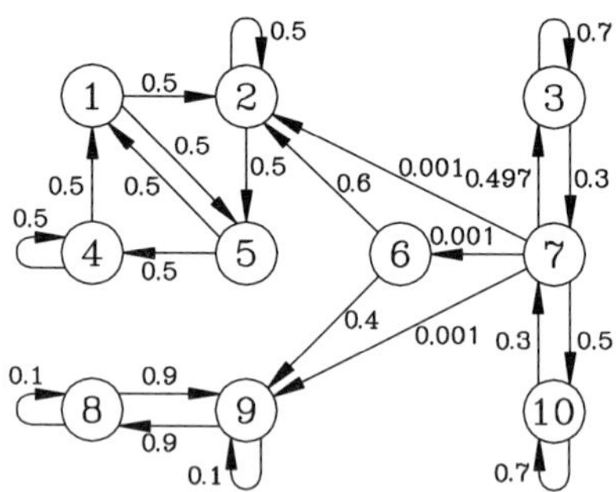


Figure B.1. A Markov chain.

The corresponding stochastic matrix is given by

$$P = \begin{pmatrix} & 0.5 & & & 0.5 & & & & & \\ & 0.5 & & & 0.5 & & & & & \\ & & 0.7 & & & & 0.3 & & & \\ 0.5 & & & 0.5 & & & & & & \\ 0.5 & & & 0.5 & & & & & & \\ & 0.6 & & & & & & & 0.4 & \\ & 0.001 & 0.497 & & & 0.001 & & & 0.001 & 0.5 \\ & & & & & & & 0.1 & 0.9 & \\ & & & & & & & 0.9 & 0.1 & \\ & & & & & & 0.3 & & & 0.7 \end{pmatrix},$$

which has no clearly defined form. However, if the rows and columns are permuted according to

| New state no. | 1 | 2 | 3 | 4 | 5 | 6 | 7 | 8 | 9 | 10 |
|---|---|---|---|---|---|---|---|---|---|---|
| Old state no. | 1 | 2 | 4 | 5 | 8 | 9 | 6 | 3 | 7 | 10 |

the matrix acquires the normal decomposable form (B.11) shown below. Clearly it may be seen that once the system is in set $B_1$ consisting of reordered states $\{1, 2, 3, 4\}$, or in $B_2$ with reordered states $\{5, 6\}$, it will never leave it. If it is initially in any other set, e.g., $B_3 = \{7\}$ or $B_4 = \{8, 9, 10\}$, it will in the long run leave it and move into $B_1$ or $B_2$. The set of eigenvalues of this matrix is

$$\{1.0, 1.0, 0.9993, -0.8000, 0.7000, -0.2993, 0.0000, 0.0000, 0.0000, 0.0\},$$

containing two unit eigenvalues and with the other eigenvalues having modulus strictly less than unity. It possesses one eigenvalue that is identically zero and three that are less than $10^{-5}$ but nonzero.

$$
P_{\text{permuted}} = \left( \begin{array}{ccccccccccccc}
 & 0.5 & & 0.5 & \vdots & & & \vdots & & \vdots & & & \\
 & 0.5 & & 0.5 & \vdots & & & \vdots & & \vdots & & & \\
0.5 & & 0.5 & & \vdots & & & \vdots & & \vdots & & & \\
0.5 & & 0.5 & & \vdots & & & \vdots & & \vdots & & & \\
\cdots & \cdots & \cdots & \cdots & \vdots & \cdots & \cdots & \vdots & \cdots & \vdots & \cdots & \cdots & \cdots \\
 & & & & \vdots & 0.1 & 0.9 & \vdots & & \vdots & & & \\
 & & & & \vdots & 0.9 & 0.1 & \vdots & & \vdots & & & \\
\cdots & \cdots & \cdots & \cdots & \vdots & \cdots & \cdots & \vdots & \cdots & \vdots & \cdots & \cdots & \cdots \\
 & 0.6 & & & \vdots & & 0.4 & \vdots & 0.0 & \vdots & & & \\
\cdots & \cdots & \cdots & \cdots & \vdots & \cdots & \cdots & \vdots & \cdots & \vdots & \cdots & \cdots & \cdots \\
 & & & & \vdots & & & \vdots & & \vdots & 0.7 & 0.3 & \\
 & 0.001 & & & \vdots & & 0.001 & \vdots & 0.001 & \vdots & 0.497 & & 0.5 \\
 & & & & \vdots & & & \vdots & & \vdots & & 0.3 & 0.7
\end{array} \right).
$$

### *B.8.3 Eigenvectors of Decomposable Stochastic Matrices*

Let the stochastic matrix $P$ be decomposable into the form (B.11), and let $y_l$, $l = 1, \ldots, k$, be a set of right-hand eigenvectors corresponding to the $k$ unit eigenvalues. Then by definition

$$
P y_l = y_l, \qquad l = 1, 2, \ldots, k.
$$

Let the $y_l$ be partitioned into $m$ disjoint parts corresponding to the $m$ different sets of essential and transient states, i.e.,

$$
y_l = \left( y_l^1,\ y_l^2,\ \ldots,\ y_l^i,\ \ldots,\ y_l^m \right)^T .
$$

If $n_i$ is the number of states in the $i$th set, then $y_l^i$ consists of elements $1+\sum_{j=1}^{i-1} n_j$ through $\sum_{j=1}^{i} n_j$ of the eigenvector $y_l$. We have

$$
\begin{pmatrix}
P_{11} & 0 & 0 & \ldots & 0 & 0 & \ldots & 0 \\
0 & P_{22} & 0 & \ldots & 0 & 0 & \ldots & 0 \\
\vdots & \vdots & \vdots & \ddots & \vdots & \vdots & \ddots & \vdots \\
0 & 0 & 0 & \ldots & P_{kk} & 0 & \ldots & 0 \\
P_{k+1,1} & P_{k+1,2} & 0 & \ldots & P_{k+1,k} & P_{k+1,k+1} & \ldots & 0 \\
\vdots & \vdots & \vdots & \ddots & \vdots & \vdots & \ddots & \vdots \\
P_{m,1} & P_{m,2} & 0 & \ldots & P_{m,k} & P_{m,k+1} & \ldots & P_{m,m}
\end{pmatrix}
\begin{pmatrix} y_l^1 \\ y_l^2 \\ \vdots \\ y_l^k \\ y_l^{k+1} \\ \vdots \\ y_l^m \end{pmatrix}
=
\begin{pmatrix} y_l^1 \\ y_l^2 \\ \vdots \\ y_l^k \\ y_l^{k+1} \\ \vdots \\ y_l^m \end{pmatrix},
$$

and therefore $P_{ii} y_l^i = y_l^i$ for $i = 1, 2, \ldots, k$, which implies that $y_l^i$ is a right eigenvector corresponding to a unit eigenvalue of $P_{ii}$. But for values of $i = 1, 2, \ldots, k$, $P_{ii}$ is a nondecomposable stochastic matrix and has a unique unit eigenvalue. The subvector $y_l^i$ is therefore the right eigenvector corresponding to the unique unit eigenvalue of $P_{ii}$, and as such all of its elements must be equal. But this is true for all $i = 1, 2, \ldots, k$. This allows us to state the following theorem:

**Theorem B.8.1** *In any right-hand eigenvector corresponding to a unit eigenvalue of the matrix $P$, the elements corresponding to states of the same essential class have identical values.*

The above reasoning applies only to the states belonging to essential classes. When the states of the Markov chain have not been ordered, it might be thought that the values of components belonging to transient states might make the analysis more difficult. However, it is possible to determine which components correspond to transient states. These are the only states that have zero components in *all* of the *left-hand* eigenvectors corresponding to the $k$ unit eigenvalues. These states must have a steady-state probability of zero. Consequently these states may be detected and eliminated from the analysis of the right-hand eigenvectors that is given above.

Notice finally that it is possible to choose linear combinations of the right-hand eigenvectors to construct a new set of eigenvectors $y_i'$, $i = 1, 2, \ldots, k$, in which the components of eigenvector $i$ corresponding to states of essential set $i$ are all equal to 1, while those corresponding to states belonging to other essential subsets are zero.

**Example B.6** Referring back to the previous example, the eigenvectors corresponding to the three dominant eigenvalues are printed below.
Eigenvalues:

$$\lambda_1 = 1.0, \qquad \lambda_2 = 1.0, \qquad \lambda_3 = 0.9993.$$

Right-hand eigenvectors:

$$y_1 = \begin{pmatrix} -.4380 \\ -.4380 \\ -.4380 \\ -.4380 \\ 0.0000 \\ 0.0000 \\ -.2628 \\ -.2336 \\ -.2336 \\ -.2336 \end{pmatrix}, \quad y_2 = \begin{pmatrix} .4521 \\ .4521 \\ .4521 \\ .4521 \\ -.1156 \\ -.1156 \\ .2250 \\ .1872 \\ .1872 \\ .1872 \end{pmatrix}, \quad y_3 = \begin{pmatrix} 0.0 \\ 0.0 \\ 0.0 \\ 0.0 \\ 0.0 \\ 0.0 \\ 0.0 \\ 0.5778 \\ 0.5765 \\ 0.5778 \end{pmatrix}.$$

Left-hand eigenvectors:

$$x_1 = \begin{pmatrix} -0.5 \\ -0.5 \\ -0.5 \\ -0.5 \\ 0.0 \\ 0.0 \\ 0.0 \\ 0.0 \\ 0.0 \\ 0.0 \end{pmatrix}, \quad x_2 = \begin{pmatrix} 0.0 \\ 0.0 \\ 0.0 \\ 0.0 \\ 0.7071 \\ 0.7071 \\ 0.0 \\ 0.0 \\ 0.0 \\ 0.0 \end{pmatrix}, \quad x_3 = \begin{pmatrix} 0.1836 \\ 0.1828 \\ 0.1836 \\ 0.1833 \\ 0.3210 \\ 0.3207 \\ -0.0003 \\ -0.5271 \\ -0.3174 \\ -0.5302 \end{pmatrix}.$$

Since 0.9993 is not exactly equal to unity, we must consider only the right-hand eigenvectors $y_1$ and $y_2$. In $y_1$, the elements belonging to the essential set $B_2$ are zero. We may readily find a linear combination of $y_1$ and $y_2$ in which the elements of the essential set $B_1$ are zero. Depending upon the particular software used to generate the eigenvectors, different linear combinations of the vectors may be produced. However, it is obvious from the vectors above that regardless of the combination chosen, the result will have the property that the components related to any essential set are all identical. One particular linear combination of interest is $\alpha y_1 + \beta y_2$, with

$$\alpha = \frac{(1 - \frac{.4521}{-.1156})}{-.4380} \quad \text{and} \quad \beta = \frac{1}{-.1156},$$

which gives $\alpha y_1 + \beta y_2 = e$.

Notice that in this example, the elements of the right-hand eigenvectors corresponding to the transient class $B_4$ are also identical. With respect to the left-hand eigenvectors, notice that regardless of the linear combination of $x_1$ and $x_2$ used, the components belonging to the two transient classes will always be zero.

### B.8.4 Nearly Decomposable Stochastic Matrices

If the matrix $P$ given by (B.11) is now altered so that the off-diagonal submatrices that are zero become nonzero, then the matrix is no longer decomposable, and $P$ will have only one unit eigenvalue. However, if these off-diagonal elements are small compared to the nonzero elements of the diagonal submatrices, then the matrix is said to be *nearly decomposable* in the sense that there are only weak interactions among the diagonal blocks. In this case there are multiple eigenvalues close to unity.

**Example B.7** In the simplest of all possible cases of completely decomposable matrices, we have

$$P = \begin{pmatrix} 1.0 & 0.0 \\ 0.0 & 1.0 \end{pmatrix}.$$

The eigenvalues are, of course, both equal to 1.0. When off-diagonal elements $\epsilon_1$ and $\epsilon_2$ are introduced, we have

$$P = \begin{pmatrix} 1 - \epsilon_1 & \epsilon_1 \\ \epsilon_2 & 1 - \epsilon_2 \end{pmatrix},$$

and the eigenvalues are now given by $\lambda_1 = 1.0$ and $\lambda_2 = 1.0 - \epsilon_1 - \epsilon_2$. As $\epsilon_1, \epsilon_2 \rightarrow 0$, the system becomes completely decomposable, and $\lambda_2 \rightarrow 1.0$.

In a strictly nondecomposable system a subdominant eigenvalue close to 1.0 is often indicative of a nearly decomposable matrix. When $P$ has the form given by (B.11), the characteristic equation $|\lambda I - P| = 0$ is given by

$$P(\lambda) = P_{11}(\lambda)P_{22}(\lambda)\cdots P_{mm}(\lambda) = 0, \tag{B.12}$$

where $P_{ii}(\lambda)$ is the characteristic equation for the block $P_{ii}$. Equation (B.12) has a root $\lambda = 1.0$ of multiplicity $k$. Suppose a small element $\epsilon$ is subtracted from $P_{ii}$, one of the first $k$ blocks, i.e., $i = 1, \ldots, k$, and added into one of the off-diagonal blocks in such a way that the matrix remains stochastic. The block $P_{ii}$ is then strictly substochastic, and hence its largest eigenvalue is strictly less than 1. The modified matrix now has a root $\lambda = 1.0$ of multiplicity $(k - 1)$, a root $\lambda = 1 - O(\epsilon)$, and $(m - k)$ other roots of modulus $< 1.0$; i.e., the system is decomposable into $(k - 1)$ isolated sets. Since the eigenvalues of a matrix are continuous functions of the elements of the matrix, it follows that as $\epsilon \rightarrow 0$, the subdominant eigenvalue tends to unity and the system reverts to its original $k$ isolated sets. If we consider the system represented by the previous figure, it may be observed that the transient set $B_4 = \{8, 9, 10\}$ is almost isolated, and as expected, the eigenvalues include one very close to unity. However, the existence of eigenvalues close to 1.0 is *not* a sufficient condition for the matrix to be nearly decomposable.

Nearly completely decomposable (NCD) Markov chains arise frequently in modelling applications [13]. In these applications the state space may be partitioned into disjoint subsets with strong interactions among the states of a subset but with weak interactions among the subsets themselves. Efficient computational procedures exist to compute the stationary distributions of these systems and are discussed in Chapter 10 of this text.

### *B.8.5 Cyclic Stochastic Matrices*

In an irreducible discrete-time Markov chain, when the number of single-step transitions required on leaving any state to return to that same state (by any path) is a multiple of some integer $p > 1$, the Markov chain is said to be *periodic of period* $p$, or *cyclic of index* $p$. One of the fundamental properties of a cyclic stochastic matrix of index $p$ is that it is possible by a permutation of its rows and columns to transform it to the form, called the *normal cyclic form*

$$P = \begin{pmatrix} 0 & P_{12} & 0 & \cdots & 0 \\ 0 & 0 & P_{23} & \cdots & 0 \\ \vdots & \vdots & \vdots & \ddots & \vdots \\ 0 & 0 & 0 & \cdots & P_{p-1,p} \\ P_{p1} & 0 & 0 & \cdots & 0 \end{pmatrix}, \tag{B.13}$$

in which the diagonal submatrices $P_{ii}$ are square, zero, and of order $n_i$ and the only nonzero submatrices are $P_{12}, P_{23}, \ldots, P_{p1}$. This corresponds to a partitioning of the states of the system into $p$ distinct subsets and an ordering imposed on the subsets. The ordering is such that once the system is in a state of subset $i$, it must exit this subset in the next time transition and enter a state of subset $(i \bmod p) + 1$. The matrix $P$ is said to be *cyclic of index* $p$, or $p$-*cyclic*. When the states of the Markov chain are not periodic, (i.e., $p = 1$), then $P$ is said to be *primitive* or *aperiodic* or *acyclic*.

The matrix given in equation (B.13) possesses $p$ eigenvalues of unit modulus. Indeed, it follows as a direct consequence of the Perron–Frobenius theorem that a cyclic stochastic matrix of index $p$ possesses, among its eigenvalues, all the roots of $\lambda^p - 1 = 0$. Thus a cyclic stochastic matrix of

- index 2 possesses eigenvalues $1, -1$;
- index 3 possesses eigenvalues $1, -(1 \pm \sqrt{3}i)/2$;
- index 4 possesses eigenvalues $1, -1, i, -i$;
- etc.

Notice that a cyclic stochastic matrix of index $p$ has $p$ eigenvalues equally spaced around the unit circle. Consider what happens as $p$ becomes large. The number of eigenvalues spaced equally on the unit circle increases. It follows that a cyclic stochastic matrix will have eigenvalues, other than the unit eigenvalue, that will tend to unity as $p \to \infty$. Such a matrix is *not* decomposable or nearly decomposable. As pointed out previously, eigenvalues close to unity do not necessarily imply that a Markov chain is nearly decomposable. On the other hand, a nearly completely decomposable Markov chain *must* have subdominant eigenvalues close to 1.

# Bibliography


[1] A. Baker. *Essentials of Padé Approximants.* Academic Press, New York, 1975.

[2] F. Baskett, K. M. Chandy, R. R. Muntz, and F. G. Palacios. Open, Closed, and Mixed Networks of Queues with Different Classes of Customers. *Journal of the ACM*, Vol. 22, No. 2, pp. 248–260, 1975.

[3] D. Bini and B. Meini. On Cyclic Reduction Applied to a Class of Toeplitz Matrices Arising in Queueing Problems. Pages 21–38 in W. J. Stewart, editor, *Computations with Markov Chains.* Kluwer Academic, Boston, Mass. 1995.

[4] G. Bolch, S. Greiner, H. de Meer, and K. S. Trivedi. *Queueing Networks and Markov Chains.* Wiley Interscience, New York, 1998.

[5] G. E. P. Box and M. F. Muller. A Note of the Generation of Random Normal Deviates. *Annals of Mathematical Statistics,* Vol. 29, pp. 610–611, 1958.

[6] P. J. Burke. The Output of a Queueing System. *Operations Research*, Vol. 4, No. 6, pp. 699–704, 1956.

[7] J. P. Buzen. Computational Algorithms for Closed Queueing Networks with Exponential Servers. *Comm. of the ACM*, Vol. 16, No. 9, pp. 527–531, 1973.

[8] K. M. Chandy. The Analysis and Solution of General Queueing Networks. Pages 224–228 in *Proc. 6th Annual Princeton Conf. on Information Sciences and Systems*, Princeton University Press, Princeton, N.J., 1972.

[9] K. M. Chandy and C. H. Sauer. Computational Algorithms for Product Form Queueing Networks. *Communication of the ACM*, Vol. 23, No. 10, pp. 573–583, 1980.

[10] J. W. Cohen. *The Single Server Queue.* Wiley, New York, 1969.

[11] A. E. Conway and N. D. Georganas. A New Method for Computing the Normalization Constant for Multiple-Chains Queueing Networks. *INFOR*, Vol. 24, No. 3, pp. 184–196, 1986.

[12] A. E. Conway and N. D. Georganas. *Queueing Networks—Exact Computational Algorithms.* MIT Press, Cambridge, Mass., 1989.

[13] P-J. Courtois. *Decomposablity: Queueing and Computer Systems Applications.* Academic Press, New York, 1977.

[14] T. Dayar. Permuting Markov Chains to Nearly Completely Decomposable Form. Technical Report BU-CEIS-9808, Department of Computer Engineering and Information Science, Bilkent University, Ankara, Turkey, August 1998.

[15] W. Feller. *An Introduction to Probability Theory and its Applications*, Vol. I and II. John Wiley and Sons, New York, 1968.

[16] C. W. Gear. *Numerical Initial Value Problems in Ordinary Differential Equations*, Prentice-Hall, Englewood Cliffs, N.J., 1971.

[17] W. J. Gordon and G. F. Newell. Closed Queueing Networks with Exponential Servers. *Operations Research*, Vol. 15, No. 2, pp. 254–265, 1967.

[18] W. J. Gordon and G. F. Newell. Cyclic Queueing Systems with Restricted Length Queues. *Operations Research*, Vol. 15, No. 2, pp. 266–277, 1967.

[19] D. Gross and C. M. Harris. *Fundamentals of Queueing Theory*, John Wiley and Sons, New York, 1974.

[20] E. Hairer and G. Wanner. *Solving Ordinary Differential Equations II: Stiff and Differential-Algebraic Problems*, Series in Computational Mathematics. Springer-Verlag, Berlin, 1991.

[21] B. R. Haverkort. *Performance of Computer Communication Systems*, John Wiley and Sons, Chichester, U.K., 1998.

[22] J. R. Jackson. Networks of Waiting Lines. *Operations Research*, Vol. 5, No. 4, pp. 518–521, 1957.

[23] J. R. Jackson. Jobshop Like Queueing Systems. *Management Science*, Vol. 10, No. 1, pp. 131–142, 1963.

[24] L. Kleinrock. *Queueing Systems, Vol. 1: Theory*. John Wiley and Sons, New York, 1975.

[25] D. E. Knuth. *The Art of Computer Programming, Vol. 2: Seminumerical Algorithms*. Third edition, Addison-Wesley, Reading, Mass., 1998.

[26] S. S. Lam and Y. I. Lien. A Tree Convolution Algorithm for the Solution of Queueing Networks. *Comm. of the ACM*, Vol. 26, No. 3, pp. 203–215, 1983.

[27] G. Latouche and Y. Ramaswami. A Logarithmic Reduction Algorithm for Quasi Birth and Death Processes. *J. Appl. Probab.*, Vol. 30, pp. 650–674, 1994.

[28] S. S. Lavenberg and M. Resier. Stationary State Probabilities at Arrival Instants for Closed Queueing Networks with Multiple Types of Customers. *Journal of Applied Probability*, Vol. 17, No. 4, pp. 1048–1061, 1980.

[29] A. M. Law and W. D. Kelton. *Simulation Modeling and Analysis*. Third edition, McGraw-Hill, New York, 2000.

[30] P. L'Ecuyer. Software for Uniform Random Number Generation: Distinguishing the Good and the Bad. Pages 95–105 in *Proceedings of the 2001 Winter Simulation Conference*, IEEE Press, 2001.

[31] D. H. Lehmer. Mathematical Methods in Large Scale Computing Units, Pages 141–146 in *Proceedings of the Second Symposium on Large-Scale Digital Computing Machinery*, Harvard University Press, Cambridge, Mass., 1951.

[32] R. Marie. Calculating Equilibrium Probabilities for $\lambda(n)/C_k/1/N$ Queues. *ACM Sigmetrics Performance Evaluation Review*, Vol. 9, No. 2, pp. 117–125, 1980.

[33] M. Matsumoto and T. Nishimura. Mersenne Twister: A 623-Dimensionally Equidistributed Uniform Pseudo-Random Number Generator. *ACM Transactions on Modeling and Computer Simulation*, Vol. 8, No. 1, pp 3–30, 1998.

[34] F. Marsaglia and T. A. Bray. A Convenient Method for Generating Normal Variables. *SIAM Review*, Vol. 6, No. 3, pp. 260–264, 1964.

[35] G. Marsaglia and W. W. Tsang, The Ziggurat Method for Generating Random Variables. *Journal of Statistical Software*, Vol. 5, No. 8, pp. 17, 2000.

[36] B. Meini. New Convergence Results on Functional Techniques for the Numerical Solution of $M/G/1$ Type Markov Chains. *Numer. Math.*, Vol. 78, pp. 39–58, 1997.

[37] C. B. Moler and C. F. Van Loan. Nineteen Dubious Ways to Compute the Exponential of a Matrix. *SIAM Review*, Vol. 20, No. 4, pp. 801–836, October 1978.

[38] R. R. Muntz. Poisson Departure Process and Queueing Networks. Pages 435–440 in *Proceedings of the 7th Annual Princeton Conference on Information Sciences and Systems*, Princeton University Press, Princeton, N.J., 1973.

[39] M. F. Neuts. *Matrix Geometric Solutions in Stochastic Models—An Algorithmic Approach*. Johns Hopkins University Press, Baltimore, MD, 1981.

[40] M. F. Neuts. *Structured Stochastic Matrices of M/G/1 Type and Their Applications*. Marcel Dekker, New York, 1989.

[41] T. Osogami and M. Harchol-Balter. Necessary and Sufficient Conditions for Representing General Distributions by Coxians. Carnegie Mellon University Technical Report, CMU-CS-02-178, 2002.

[42] B. Philippe and B. Sidje. Transient Solution of Markov Processes by Krylov Subspaces, Technical Report, IRISA—Campus de Beaulieu, Rennes, France, 1993.

[43] V. Ramaswami. A Stable Recursion for the Steady State Vector in Markov chains of M/G/1 type. *Commun. Statist. Stochastic Models*, Vol. 4, pp. 183–188, 1988.

[44] V. Ramaswami and M. F. Neuts. Some explicit formulas and computational methods for infinite server queues with phase type arrivals. *J. Appl. Probab.*, Vol. 17, pp. 498–514, 1980.

[45] V. I. Romanovsky. *Discrete Markov Chains*, Wolters-Noordhoff, Groningen, The Netherlands, 1970.

[46] S. M. Ross. *Simulation*. Second edition, Academic, Press, New York, 1997.

[47] Y. Saad. *Iterative Methods for Sparse Linear Systems*. Second edition, SIAM Publications, Philadelphia, 2003.

[48] E. B. Saff. On the Degree of the Best Rational Approximation to the Exponential Function. *Journal of Approximation Theory*, Vol. 9, pp. 97–101, 1973.

[49] K. C. Sevcik and I. Mitrani. The Distribution of Queueing Network States at Input and Output Instants. *Journal of the ACM*, Vol. 28, No. 2, pp. 358–371, 1981.

[50] W. J. Stewart. *Introduction to the Numerical Solution of Markov Chains*. Princeton University Press, Princeton, N.J., 1994.

[51] R. J. Tarjan. Depth First Search and Linear Graph Algorithms. *SIAM Journal of Computing*, Vol. 1, No. 2, pp. 146–160, 1972.

[52] J. W. Tukey. Bias and Confidence in not quite Large Samples. *Annals of Mathematical Statistics*, Vol. 29, page 614, 1958.

[53] R. S. Varga. *Matrix Iterative Analysis*. Prentice-Hall, Englewood Cliffs, N.J., 1962.

[54] R. C. Ward. Numerical Computation of the Matrix Exponential with Accuracy Estimate, *SIAM Journal on Numerical Analysis*, Vol. 14, No. 4, pp. 600–610, 1977.

# Index